Intermediate Algebra

Intermediate Algebra

10th EDITION

Marvin L. Bittinger

Indiana University Purdue University Indianapolis

PEARSON

Addison
Wesley

Boston   San Francisco   New York
London   Toronto   Sydney   Tokyo   Singapore   Madrid
Mexico City   Munich   Paris   Cape Town   Hong Kong   Montreal

Publisher	Greg Tobin
Editor in Chief	Maureen O'Connor
Executive Editor	Jennifer Crum
Executive Project Manager	Kari Heen
Project Editor	Katie Nopper
Editorial Assistant	Alison Macdonald
Managing Editor	Ron Hampton
Cover Designer	Dennis Schaefer
Supplements Supervisor	Emily Portwood
Media Producers	Sharon Smith, Ceci Fleming
Software Development	Mary Durnwald, TestGen; Rebecca Williams, MathXL
Marketing Manager	Jay Jenkins
Marketing Coordinator	Tracy Rabinowitz
Senior Author Support/ Technology Specialist	Joseph K. Vetere
Senior Prepress Supervisor	Caroline Fell
Rights and Permissions	Dana Weightman, Ellen Keohane
Senior Manufacturing Buyer	Evelyn M. Beaton
Art and Design Services	Geri Davis/The Davis Group, Inc.
Editorial and Production Services	Martha K. Morong/Quadrata, Inc.
Composition	Beacon Publishing Services
Illustrations	William Melvin, Network Graphics
Cover Photo	© Michio Hoshino/Minden Pictures

About the cover: The aurora borealis, or northern lights, as photographed in the skies over Alaska.

Photo credits appear on p. 826

Library of Congress Cataloging-in-Publication Data

Bittinger, Marvin L.
    Intermediate algebra.—10th ed./Marvin L. Bittinger.
        p. cm.
    Includes indexes.
    ISBN 0-321-31908-7 (Student's Edition)
    (Hardback Student's Edition) ISBN 0-321-38825-9
    1. Algebra—Textbooks. I. Title.

QA154.3.B578 2005
512.9—dc22   2005045307

   2 3 4 5 6 7 8 9 10–VH–09 08 07

Contents

 REVIEW OF BASIC ALGEBRA

 SOLVING LINEAR EQUATIONS AND INEQUALITIES

5 RATIONAL EXPRESSIONS, EQUATIONS, AND FUNCTIONS

6 RADICAL EXPRESSIONS, EQUATIONS, AND FUNCTIONS

QUADRATIC EQUATIONS AND FUNCTIONS

EXPONENTIAL AND LOGARITHMIC FUNCTIONS

9 CONIC SECTIONS

A APPENDIXES

Index of Applications

Index of Study Tips

Preface

It is with great pride and excitement that we present to you the tenth edition of *Intermediate Algebra*. The text has evolved dramatically over the past 36 years through your comments, responses, and opinions. This feedback, combined with our overall objective of presenting the material in a clear and accurate manner, drives each revision. It is our hope that *Intermediate Algebra*, Tenth Edition, and the supporting supplements will help provide an improved teaching and learning experience by matching the needs of instructors and successfully preparing students for their future.

This text is the fourth in a series that includes the following:

Bittinger: *Basic Mathematics,* Tenth Edition

Bittinger: *Fundamental Mathematics,* Fourth Edition

Bittinger: *Introductory Algebra,* Tenth Edition

Bittinger: *Intermediate Algebra,* Tenth Edition

Bittinger/Beecher: *Introductory and Intermediate Algebra,* Third Edition

Building Understanding through an Interactive Approach

The pedagogy of this text is designed to provide an interactive learning experience between the student and the exposition, annotated examples, art, margin exercises, and exercise sets. This unique approach, which has been developed and refined over ten editions and is illustrated at right, provides students with a clear set of learning

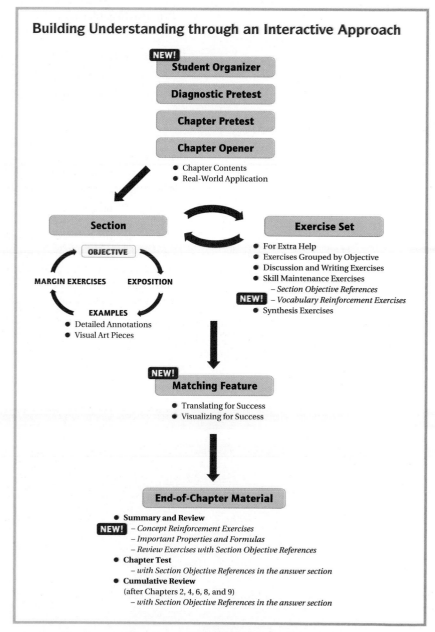

Building Understanding through an Interactive Approach

NEW!
Student Organizer

Diagnostic Pretest

Chapter Pretest

Chapter Opener
- Chapter Contents
- Real-World Application

Section
OBJECTIVE
MARGIN EXERCISES — EXPOSITION
EXAMPLES
- Detailed Annotations
- Visual Art Pieces

Exercise Set
- For Extra Help
- Exercises Grouped by Objective
- Discussion and Writing Exercises
- Skill Maintenance Exercises
  – Section Objective References
  NEW! – *Vocabulary Reinforcement Exercises*
- Synthesis Exercises

NEW!
Matching Feature
- Translating for Success
- Visualizing for Success

End-of-Chapter Material
- **Summary and Review**
  NEW! – *Concept Reinforcement Exercises*
  – *Important Properties and Formulas*
  – *Review Exercises with Section Objective References*
- **Chapter Test**
  – *with Section Objective References in the answer section*
- **Cumulative Review**
  (after Chapters 2, 4, 6, 8, and 9)
  – *with Section Objective References in the answer section*

objectives, involves them with the development of the material, and provides immediate and continual reinforcement and assessment through the margin exercises.

Let's Visit the Tenth Edition

The style, format, and approach of the ninth edition have been strengthened in this new edition in a number of ways. However, the accuracy that the Bittinger books are known for has not changed. This edition, as with all editions, has gone through an exhaustive checking process to ensure accuracy in the problem sets, mathematical art, and accompanying supplements. We know what a critical role the accuracy of a book plays in student learning and we value the reputation for accuracy that we have earned.

NEW! IN THE TENTH EDITION

Each revision gives us the opportunity to incorporate new elements and refine existing elements to provide a better experience for students and teachers alike. Below are four new features designed to help students succeed.

- Student Organizer
- Translating/Visualizing for Success matching exercises
- Vocabulary Reinforcement exercises
- Concept Reinforcement exercises

These features, along with the hallmark features of this book, are discussed in the pages that follow.

In addition, the tenth edition has been designed to be open and flexible, helping students focus their attention on details that are critical at this level through prominent headings, boxed definitions and rules, and clearly labeled objectives. *Chapter Pretests*, now located along with the Diagnostic Pretest in the *Printed Test Bank* and in MyMathLab, diagnose at the section and objective level and can be used to place students in a specific section, or objective, of the chapter, allowing them to concentrate on topics with which they have particular difficulty. Answers to these pretests are available in the *Printed Test Bank* and in MyMathLab.

NEW! STUDENT ORGANIZER

Along with the study tips found throughout the text, we have provided a pull-out card that will help students stay organized and increase their ability to be successful in this course. Students can use this card to keep track of important dates and useful contact information and to access information for technology. It can also be used to plan time for classes, study, commuting, work, family, and relaxation.

CHAPTER OPENERS

To engage students and prepare them for the upcoming chapter material, gateway chapter openers are designed with exceptional artwork that is tied to a motivating real-world application. (See pages 75, 157, and 243.)

OBJECTIVE BOXES

At the beginning of each section, a boxed list of objectives is keyed by letter not only to section subheadings, but also to the exercises in the Pretest (located in the *Printed Test Bank* and MyMathLab), the section exercise sets, and the Summary and Review exercises, as well as to the answers to the questions in the Chapter Tests and Cumulative Reviews. This correlation enables students to easily find appropriate review material if they need help with a particular exercise or skill at the objective level. (See pages 141, 174, and 326.)

ANNOTATED EXAMPLES

Detailed annotations and color highlights lead the student through the structured steps of the examples. The level of detail in these annotations is a significant reason for students' success with this book. (See pages 253, 338, and 429.)

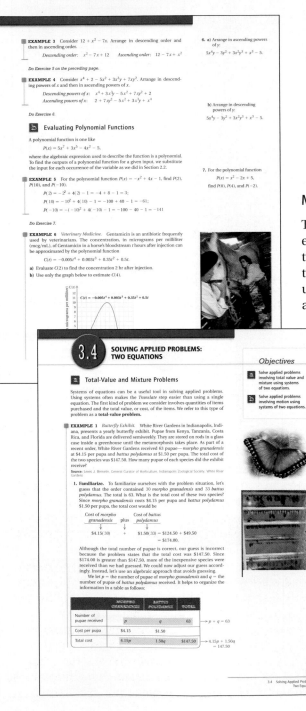

EXAMPLE 10 Add: $\dfrac{3x^2 + 3xy}{x^2 - y^2} + \dfrac{2 - 3x}{x - y}$.

We first find the LCD of the denominators. To do so, we first factor:

$$x^2 - y^2 = (x + y)(x - y) \qquad \text{LCD} = (x + y)(x - y).$$
$$x - y = x - y$$

We now multiply by 1 to get the LCD in the second expression. Then we add and simplify if possible.

$$\dfrac{3x^2 + 3xy}{(x + y)(x - y)} + \dfrac{2 - 3x}{x - y} \cdot \dfrac{x + y}{x + y} \quad \text{Multiplying by 1 to get the LCD}$$
$$= \dfrac{3x^2 + 3xy}{(x + y)(x - y)} + \dfrac{(2 - 3x)(x + y)}{(x - y)(x + y)}$$
$$= \dfrac{3x^2 + 3xy}{(x + y)(x - y)} + \dfrac{2x + 2y - 3x^2 - 3xy}{(x - y)(x + y)} \quad \text{Multiplying in the numerator}$$
$$= \dfrac{3x^2 + 3xy + 2x + 2y - 3x^2 - 3xy}{(x + y)(x - y)} \quad \text{Adding the numerators}$$
$$= \dfrac{2x + 2y}{(x + y)(x - y)} \quad \text{Combining like terms}$$
$$= \dfrac{2(x + y)}{(x + y)(x - y)} \quad \text{Factoring the numerator}$$
$$= \dfrac{2\cancel{(x + y)}}{\cancel{(x + y)}(x - y)} \quad \text{Removing a factor of 1: } \dfrac{x + y}{x + y} = 1$$
$$= \dfrac{2}{x - y}$$

Do Exercises 13 and 14.

EXAMPLE 11 Subtract: $\dfrac{2y + 1}{y^2 - 7y + 6} - \dfrac{y + 3}{y^2 - 5y - 6}$.

$$\dfrac{2y + 1}{y^2 - 7y + 6} - \dfrac{y + 3}{y^2 - 5y - 6}$$
$$= \dfrac{2y + 1}{(y - 6)(y - 1)} - \dfrac{y + 3}{(y - 6)(y + 1)} \quad \text{LCD} = (y - 6)(y - 1)(y + 1)$$
$$= \dfrac{2y + 1}{(y - 6)(y - 1)} \cdot \dfrac{y + 1}{y + 1} - \dfrac{y + 3}{(y - 6)(y + 1)} \cdot \dfrac{y - 1}{y - 1} \quad \text{Multiplying by 1 to get the LCD}$$
$$= \dfrac{(2y + 1)(y + 1) - (y + 3)(y - 1)}{(y - 6)(y - 1)(y + 1)} \quad \text{Subtracting the numerators}$$
$$= \dfrac{(2y^2 + 3y + 1) - (y^2 + 2y - 3)}{(y - 6)(y - 1)(y + 1)} \quad \text{Multiplying. Note the use of parentheses.}$$
$$= \dfrac{2y^2 + 3y + 1 - y^2 - 2y + 3}{(y - 6)(y - 1)(y + 1)}$$
$$= \dfrac{y^2 + y + 4}{(y - 6)(y - 1)(y + 1)} \quad \text{The numerator cannot be factored. The rational expression is simplified.}$$

We generally do not multiply out a numerator or a denominator if it has three or more factors (other than monomials). This will be helpful when we solve equations.

Do Exercises 15 and 16.

Add.

13. $\dfrac{3x}{7} + \dfrac{4y}{3x}$

14. $\dfrac{2xy - 2x^2}{x^2 - y^2} + \dfrac{2x + 3}{x + y}$

Subtract.

15. $\dfrac{a}{a + 3} - \dfrac{a - 4}{a}$

16. $\dfrac{4y - 5}{y^2 - 7y + 12} - \dfrac{y + 7}{y^2 + 2y - 15}$

Answers on page A-28

429

5.2 LCMs, LCDs, Addition, and Subtraction

EXAMPLE 3 Consider $12 + x^2 - 7x$. Arrange in descending order and then in ascending order.

Descending order: $x^2 - 7x + 12$ Ascending order: $12 - 7x + x^2$

Do Exercise 5 on the preceding page.

EXAMPLE 4 Consider $x^4 + 2 - 5x^2 + 3x^3y + 7xy^2$. Arrange in descending powers of x and then in ascending powers of x.

Descending powers of x: $x^4 + 3x^3y - 5x^2 + 7xy^2 + 2$
Ascending powers of x: $2 + 7xy^2 - 5x^2 + 3x^3y + x^4$

Do Exercise 6.

Evaluating Polynomial Functions

A polynomial function is one like

$$P(x) = 5x^7 + 3x^5 - 4x^2 - 5,$$

where the algebraic expression used to describe the function is a polynomial. To find the outputs of a polynomial function for a given input, we substitute the input for each occurrence of the variable as we did in Section 2.2.

EXAMPLE 5 For the polynomial function $P(x) = -x^2 + 4x - 1$, find $P(2)$, $P(10)$, and $P(-10)$.

$$P(2) = -2^2 + 4(2) - 1 = -4 + 8 - 1 = 3;$$
$$P(10) = -10^2 + 4(10) - 1 = -100 + 40 - 1 = -61;$$
$$P(-10) = -(-10)^2 + 4(-10) - 1 = -100 - 40 - 1 = -141$$

Do Exercise 7.

EXAMPLE 6 *Veterinary Medicine.* Gentamicin is an antibiotic frequently used by veterinarians. The concentration, in micrograms per milliliter (mcg/mL), of Gentamicin in a horse's bloodstream t hours after injection can be approximated by the polynomial function

$$C(t) = -0.005t^4 + 0.003t^3 + 0.35t^2 + 0.5t.$$

a) Evaluate $C(2)$ to find the concentration 2 hr after injection.
b) Use only the graph below to estimate $C(4)$.

$$C(t) = -0.005t^4 + 0.003t^3 + 0.35t^2 + 0.5t$$

6. a) Arrange in ascending powers of y:
$5x^4y - 3y^2 + 3x^2y^3 + x^3 - 5$.

b) Arrange in descending powers of y:
$5x^4y - 3y^2 + 3x^2y^3 + x^3 - 5$.

7. For the polynomial function
$P(x) = x^2 - 2x + 5$,
find $P(0)$, $P(4)$, and $P(-2)$.

3.4 SOLVING APPLIED PROBLEMS: TWO EQUATIONS

a) Total-Value and Mixture Problems

Systems of equations can be a useful tool in solving applied problems. Using systems often makes the *Translate* step easier than using a single equation. The first kind of problem we consider involves quantities of items purchased and the total value, or cost, of the items. We refer to this type of problem as a **total-value problem**.

EXAMPLE 1 *Butterfly Exhibit.* White River Gardens in Indianapolis, Indiana, presents a yearly butterfly exhibit. Pupae from Kenya, Tanzania, Costa Rica, and Florida are delivered semiweekly. They are stored on rods in a glass case inside a greenhouse until the metamorphosis takes place. As part of a recent order, White River Gardens received 63 pupae — *morpho granadensis* at $4.15 per pupa and *battus polydamus* at $1.50 per pupa. The total cost of the two species was $147.50. How many pupae of each species did the exhibit receive?
Source: Lewis J. Bemelin, General Curator of Horticulture, Indianapolis Zoological Society, White River Gardens

1. **Familiarize.** To familiarize ourselves with the problem situation, let's guess that the order contained 30 *morpho granadensis* and 33 *battus polydamus*. The total is 63. What is the total cost of these two species? Since *morpho granadensis* costs $4.15 per pupa and *battus polydamus* $1.50 per pupa, the total cost would be

Cost of *morpho granadensis* plus Cost of *battus polydamus*

$4.15(30) + $1.50(33) = $124.50 + $49.50
= $174.00.

Although the total number of pupae is correct, our guess is incorrect because the problem states that the total cost was $147.50. Since $174.00 is greater than $147.50, more of the inexpensive species were received than we had guessed. We could now adjust our guess accordingly. Instead, let's use an algebraic approach that avoids guessing.

We let p = the number of pupae of *morpho granadensis* and q = the number of pupae of *battus polydamus* received. It helps to organize the information in a table as follows:

	MORPHO GRANADENSIS	BATTUS POLYDAMUS	TOTAL	
Number of pupae received	p	q	63	→ $p + q = 63$
Cost per pupa	$4.15	$1.50		
Total cost	$4.15p$	$1.50q$	$147.50	→ $4.15p + 1.50q = 147.50$

269

3.4 Solving Applied Problems: Two Equations

Objectives

a) Solve applied problems involving total value and mixture using systems of two equations.

b) Solve applied problems involving motion using systems of two equations.

MARGIN EXERCISES

Throughout the text, students are directed to numerous margin exercises that provide immediate practice and reinforcement of the concepts covered in each section. Answers are provided at the back of the book so students can immediately self-assess their understanding of the skill or concept at hand. (See pages 160, 265, and 329.)

REAL-DATA APPLICATIONS

This text encourages students to see and interpret the mathematics that appears every day in the world around them. Throughout the writing process, an extensive and energetic search for real-data applications was conducted, and the result is a variety of examples and exercises that connect the mathematical content with the real world. A large number of the applications are new to this edition, and many are drawn from the fields of business and economics, life and physical sciences, social sciences, and areas of general interest such as sports and daily life. To further encourage students to understand the relevance of mathematics, many applications are enhanced by graphs and drawings similar to those found in today's newspapers and magazines, and feature source lines as well. (See pages 136, 293, and 600.)

NEW! TRANSLATING/VISUALIZING FOR SUCCESS

Translating for Success The goal of the matching exercises in this new feature is to practice step two, *Translate*, of the five-step problem-solving process. Students translate each of ten problems to an equation or an inequality and select the correct translation from fifteen given equations and inequalities. This feature appears once in each chapter that contains problems that can be solved using the five-step problem-solving process and reviews skills and concepts with problems from all preceding chapters. (See pages 122, 278, and 396.)

Visualizing for Success This new feature provides students with an opportunity to match an equation with its graph by focusing on characteristics of the equation or inequality and the corresponding attributes of the graph. This feature appears at least once in each chapter that contains graphing instruction and reviews graphing skills and concepts with exercises from all preceding chapters. (See pages 213, 378, and 633.)

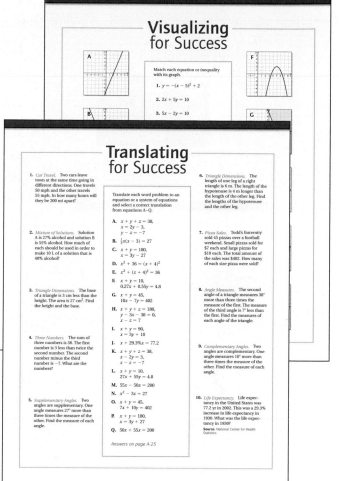

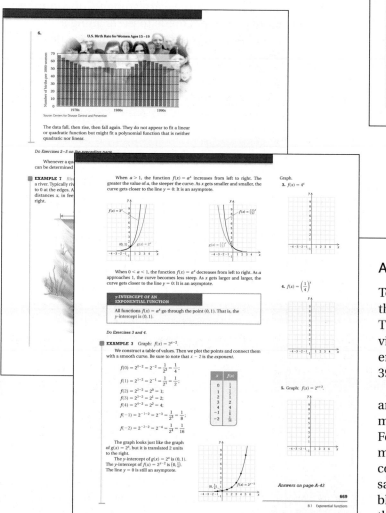

ART PROGRAM

Today's students are often visually oriented and their approach to a printed page is no exception. The art program is designed to improve the visualization of the mathematical concepts and to enhance the real-data applications. (See pages 269, 399, and 643.)

The use of color is carried out in a methodical and precise manner so that it conveys a consistent meaning, which enhances the readability of the text. For example, the use of both red and blue in mathematical art increases understanding of the concepts. When two lines are graphed using the same set of axes, one is usually red and the other blue. Note that equation labels are the same color as the corresponding line to aid in understanding.

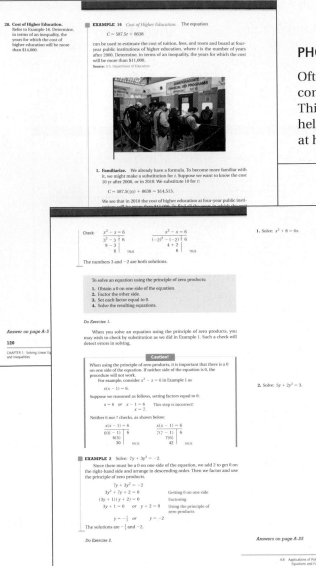

PHOTOGRAPHS

Often, an application becomes relevant to students when the connection to the real world is illustrated with a photograph. This text has numerous photographs throughout in order to help students see the relevance and visualize the application at hand. (See pages 120, 255, and 353.)

CAUTION BOXES

Found at relevant points throughout the text, boxes with the "Caution!" heading warn students of common misconceptions or errors made in performing a particular mathematics operation or skill. (See pages 176, 389, and 590.)

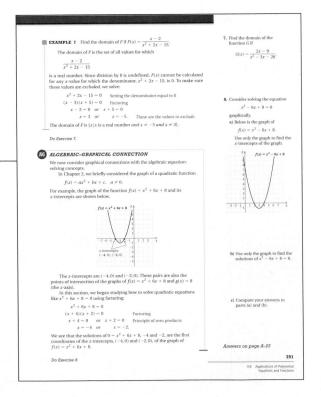

ALGEBRAIC–GRAPHICAL CONNECTIONS

To provide a visual understanding of algebra, algebraic–graphical connections are included in nearly every chapter. This feature gives the algebra more meaning by connecting it to a graphical interpretation. (See pages 249, 391, and 574.)

CALCULATOR CORNERS

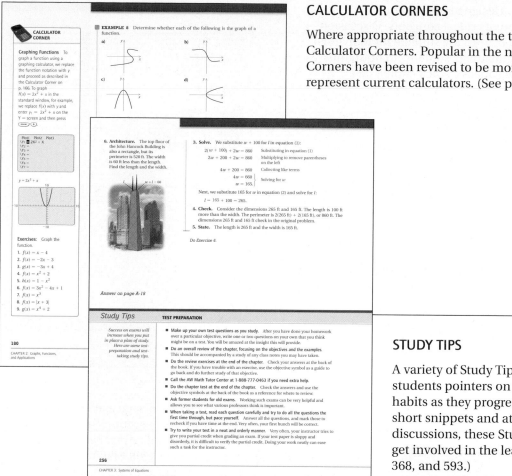

Where appropriate throughout the text, students will find optional Calculator Corners. Popular in the ninth edition, these Calculator Corners have been revised to be more accessible to students and to represent current calculators. (See pages 166, 392, and 640.)

STUDY TIPS

A variety of Study Tips throughout the text give students pointers on how to develop good study habits as they progress through the course. At times short snippets and at other times more lengthy discussions, these Study Tips encourage students to get involved in the learning process. (See pages 256, 368, and 593.)

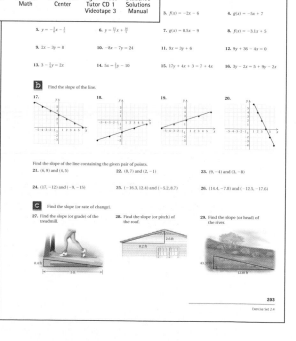

EXERCISE SETS

The exercise sets are a critical part of any math book. To give students ample opportunity to practice what they have learned, each section is followed by an extensive exercise set designed to reinforce the section concepts. In addition, students also have the opportunity to synthesize the objectives from the current section with those from preceding sections.

For Extra Help Many valuable study aids accompany this text. Located before each exercise set, "For Extra Help" references list appropriate video/CD, tutorial, and Web resources so that students can easily find related support materials.

Exercises Grouped by Objective Exercises in the section exercise sets are keyed by letter to the section objectives for easy review and remediation. This reinforces the objective-based structure of the book. (See pages 279, 353, and 647.)

Discussion and Writing Exercises Designed to help students develop a deeper comprehension of critical concepts, Discussion and Writing exercises (indicated by D_W) are suitable for individual or group work. These exercises encourage students to both think and write about key mathematical ideas in the chapter. (See pages 188, 319, and 696.)

Solve.

13. *Manufacturing Caps.* Martina's Custom Printing is planning on adding painter's caps to its product line. For the first year, the fixed costs for setting up production are $16,404. The variable costs for producing a dozen caps is $6.00. The revenue on each dozen caps is $18.00. Find the following.
a) The total cost $C(x)$ of producing x dozen caps
b) The total revenue $R(x)$ from the sale of x dozen caps
c) The total profit $P(x)$ from the production and sale of x dozen caps
d) The profit or loss from the production and sale of 3000 dozen caps; of 1000 dozen caps
e) The break-even point

14. *Sport Coat Production.* Sarducci's is planning a new line of sport coats. For the first year, the fixed costs for setting up production are $10,000. The variable costs for producing each coat are $20. The revenue from each coat is $100. Find the following.
a) The total cost $C(x)$ of producing x coats
b) The total revenue $R(x)$ from the sale of x coats
c) The total profit $P(x)$ from the production and sale of x coats
d) The profit or loss from the production and sale of 2000 coats; of 50 coats
e) The break-even point

b Find the equilibrium point for each of the following pairs of demand and supply functions.

15. $D(p) = 1000 - 10p;$ $S(p) = 230 + p$
16. $D(p) = 2000 - 60p;$ $S(p) = 460 + 94p$
17. $D(p) = 760 - 13p;$ $S(p) = 430 + 2p$
18. $D(p) = 800 - 43p;$ $S(p) = 210 + 16p$

19. $D(p) = 7500 - 25p;$ $S(p) = 6000 + 5p$
20. $D(p) = 8800 - 30p;$ $S(p) = 7000 + 15p$
21. $D(p) = 1600 - 53p;$ $S(p) = 320 + 75p$
22. $D(p) = 5500 - 40p;$ $S(p) = 1000 + 85p$

23. D_W Variable costs and fixed costs are often compared to the slope and the y-intercept, respectively, of an equation of a line. Explain why this analogy is valid.

24. D_W In this section, we examined supply and demand functions for coffee. Does it seem realistic to you for the graph of D to have a constant slope? Why or why not?

SKILL MAINTENANCE

Find the slope and the y-intercept. [2.4b]

25. $5y - 3x = 8$
26. $6x + 7y - 9 = 4$
27. $2y = 3.4x + 98$
28. $\frac{x}{3} + \frac{y}{4} = 1$

319
Exercise Set 3.8

SKILL MAINTENANCE

Solve. [1.1d]
49. $5(3x - 4) = -2(x + 5)$
50. $4(3x + 4) = 2 - x$
51. $2(x - 1) + 3(x - 2) - 4(x - 5) = 10$
52. $10x - 8(3x - 7) = 2(4x - 1)$
53. $5x + 7x = -144$
54. $0.5x - 2.34 + 2.4x = 7.8x - 9$

Given the function $f(x) = |2 - x|$, find each of the following function values. [2.2b]
55. $f(0)$
56. $f(-1)$
57. $f(1)$
58. $f(10)$
59. $f(-2)$
60. $f(2a)$
61. $f(-4)$
62. $f(1.8)$

SYNTHESIS

63. *Luggage Size.* Unless an additional fee is paid, most major airlines will not check any luggage for which the sum of the item's length, width, and height exceeds 62 in. The U.S. Postal Service will ship a package only if the sum of the package's length and girth (distance around its midsection) does not exceed 130 in. Video Promotions is ordering several 30-in. long cases that will be both mailed and checked as luggage. Using w and h for width and height (in inches), respectively, write and graph an inequality that represents all acceptable combinations of width and height.
Source: U.S. Postal Service

64. *Exercise Danger Zone.* It is dangerous to exercise when the weather is hot and humid. The solutions of the following system of inequalities give a "danger zone" for which it is dangerous to exercise intensely:
$$4H - 3F < 70,$$
$$F + H > 160,$$
$$2F + 3H > 390,$$
where F is the temperature, in degrees Fahrenheit, and H is the humidity.

Skill Maintenance Exercises Found in each exercise set, these exercises review concepts from other sections in the text to prepare students for their final examination. Section and objective codes appear next to each Skill Maintenance exercise for easy reference. (See pages 313, 370, and 745.)

---

23. D_W Consider Exercise 19. Suppose 92 points were scored with 50 baskets but no foul shots. Would there still be a solution? Why or why not?

24. D_W Consider Exercise 1. Suppose Jake collected $150. Could the problem still be solved? Why or why not?

SKILL MAINTENANCE

VOCABULARY REINFORCEMENT

In each of Exercises 25–32, fill in the blank with the correct term from the given list. Some of the choices may not be used.

25. The expression $x \leq q$ means x is _____ q. [1.4d]

26. When two sets have no elements in common, the intersection of the two sets is called the _____. [1.5a]

27. The graph of a(n) _____ equation is a line. [2.1c]

28. When the slope of a line is _____, the graph of the line slants down from left to right. [2.4b]

29. A(n) _____ system of equations has at least one solution. [3.1a]

30. Two lines are _____ if the product of their slopes is −1. [2.5d]

31. The _____ of the graph of $f(x) = mx + b$ is the point $(0, b)$. [2.4a]

32. When the slope of a line is zero, the graph of the line is _____. [2.4b]

parallel
perpendicular
union
empty set
consistent
inconsistent
linear
x-intercept
y-intercept
positive
zero
negative
vertical
horizontal
at least
at most

SYNTHESIS

33. Find the sum of the angle measures at the tips of the star in this figure.

34. *Sharing Raffle Tickets.* Hal gives Tom as many raffle tickets as Tom has and Gary has. In like manner, Tom then gives Hal and Gary as many tickets as each then has. Similarly, Gary gives Hal and Tom as many tickets as each then has. If each finally has 40 tickets, with how many tickets does Tom begin?

35. *Digits.* Find a three-digit positive integer such that the sum of all three digits is 14, the tens digit is 2 more than the ones digit, and if the digits are reversed, the number is unchanged.

36. *Ages.* Tammy's age is the sum of the ages of Carmen and Dennis. Carmen's age is 2 more than the sum of the ages of Dennis and Mark. Dennis's age is four times Mark's age. The sum of all four ages is 42. How old is Tammy?

297
Exercise Set 3.6

NEW! **Vocabulary Reinforcement Exercises**
This new feature checks and reviews students' understanding of the vocabulary introduced throughout the text. It appears once in every chapter, in the Skill Maintenance portion of an exercise set, and is intended to provide a continuing review of the terms that students must know in order to be able to communicate effectively in the language of mathematics. (See pages 151, 188, and 297.)

Synthesis Exercises In most exercise sets, Synthesis exercises help build critical-thinking skills by requiring students to synthesize or combine learning objectives from the current section as well as from preceding text sections. (See pages 322, 337, and 607.)

END-OF-CHAPTER MATERIAL

At the end of each chapter, students can practice all they have learned as well as tie the current chapter content to material covered in earlier chapters.

SUMMARY AND REVIEW

A three-part *Summary and Review* appears at the end of each chapter. The first part includes the Concept Reinforcement Exercises described below. The second part is a list of important properties and formulas, when applicable, and the third part provides an extensive set of review exercises. (See pages 152, 232, and 320.)

NEW! **Concept Reinforcement Exercises** Found in the Summary and Review of every chapter, these true/false exercises are designed to increase understanding of the concepts rather than merely assess students' skill at memorizing procedures. (See pages 320, 402, and 660.)

Important Properties and Formulas A list of the important properties and formulas discussed in the chapter is provided for students in an organized manner to help them prioritize topics learned and prepare for chapter tests. This list is only provided in those chapters in which new properties or formulas are presented. (See pages 402, 660, and 746.)

Review Exercises At the end of each chapter, students are provided with an extensive set of review exercises. Reference codes beside each exercise or direction line allow students to easily refer back to specific, objective-level content for remediation. (See pages 153, 233, and 320.)

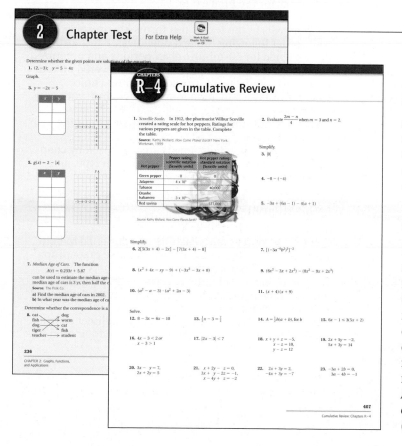

CHAPTER TEST

Following the Review Exercises, a sample Chapter Test allows students to review and test comprehension of chapter skills prior to taking an instructor's exam. Answers to all questions in the Chapter Test are given at the back of the book. Section and objective references for each question are included with the answers. (See pages 155, 236, and 323.)

CUMULATIVE REVIEW

Following Chapters 2, 4, 6, 8, and 9, students encounter a Cumulative Review. This exercise set reviews skills and concepts from all preceding chapters to help students recall previously learned material and prepare for a final exam. At the back of the book are answers to all Cumulative Review exercises, together with section and objective references, so that students know exactly what material to study if they miss a review exercise. Additional Cumulative Review Tests for every chapter are available in the *Printed Test Bank*. (See pages 240, 407, and 753.)

Ancillaries

The following ancillaries are available to help both instructors and students use this text more effectively.

Student Supplements

Student's Solutions Manual   (ISBN 0-321-30579-5)

- By Judith A. Penna, *Indiana University Purdue University Indianapolis*
- Contains completely worked-out solutions with step-by-step annotations for all the odd-numbered exercises in the text, with the exception of the discussion and writing exercises, as well as completely worked-out solutions to all the exercises in the Chapter Reviews, Chapter Tests, and Cumulative Reviews.

Collaborative Learning Activities Manual
(ISBN 0-321-30582-5)

- Features group activities tied to text sections and includes the focus, time estimate, suggested group size and materials, and background notes for each activity.
- Available as a stand-alone supplement sold in the bookstore, as a textbook bundle component for students, or as a classroom activity resource for instructors.

Videotapes   (ISBN 0-321-30576-0)

- To tie student learning to the pedagogy of the text, lectures are organized by objectives, which are indicated on the screen at the start of each new objective.
- Present a series of lectures correlated directly to the content of each section of the text.
- Feature an engaging team of instructors who present material in a format that stresses student interaction, often using examples and exercises from the text.

Digital Video Tutor   (ISBN 0-321-30580-9)

- Complete set of digitized videos (as described above) on CD-ROMs for student use at home or on campus.
- Ideal for distance learning or supplemental instruction.
- Are available with captioning on request. Contact your local Addison-Wesley representative for details.

NEW! Work It Out! Chapter Test Video on CD
(ISBN 0-321-42264-3)

- Presented by Judith A. Penna and Barbara Johnson
- Provides step-by-step solutions to every exercise in each Chapter Test from the text.
- Helps students prepare for chapter tests and synthesize content.

Math Study Skills for Students Video on CD
(ISBN 0-321-29745-8)

- Presented by author Marvin Bittinger
- Designed to help students make better use of their math study time and improve their retention of concepts and procedures taught in classes from basic mathematics through intermediate algebra.
- Through carefully crafted graphics and comprehensive on-camera explanation, focuses on study skills that are commonly overlooked.

Instructor Supplements

Annotated Instructor's Edition   (ISBN 0-321-30572-8)

- Includes answers to all exercises printed in blue on the same page as those exercises.

Instructor's Solutions Manual   (ISBN 0-321-30574-4)

- By Judith A. Penna, *Indiana University Purdue University Indianapolis*
- Contains brief solutions to the even-numbered exercises in the exercise sets, answers to all discussion and writing exercises, and completely worked-out solutions to all the exercises in the Chapter Reviews, Chapter Tests, and Cumulative Reviews.

Online Answer Book

- By Judith A. Penna, *Indiana University Purdue University Indianapolis*
- Available in electronic form from the instructor resource center. Contact your local Addison-Wesley representative for details.
- Contains answers to all the section exercises in the text.

Printed Test Bank   (ISBN 0-321-30570-1)

- By Mark Stevenson, *Oakland Community College, Auburn Hills*
- Contains one diagnostic test.
- Contains one pretest for each chapter.
- Provides 13 new test forms for every chapter and 8 new test forms for the final exam.
- For the chapter tests, 5 test forms are modeled after the chapter tests in the text, 3 test forms are organized by topic order following the text objectives, 3 test forms are designed for 50-minute class periods and organized so that each objective in the chapter is covered on one of the tests, and 2 test forms are multiple-choice. Chapter tests also include more challenging synthesis questions.
- Contains 2 cumulative tests per chapter beginning with Chapter 2.
- For the final exam, 3 test forms are organized by chapter, 3 test forms are organized by question type, and 2 test forms are multiple-choice.

NEW! Instructor and Adjunct Support Manual
(ISBN 0-321-30581-7)

- Includes Adjunct Support Manual material.
- Features resources and teaching tips designed to help both new and adjunct faculty with course preparation and classroom management.
- Resources include chapter reviews, extra practice sheets, conversion guide, video index, audio index, and transparency masters.
- Also available electronically so course/adjunct coordinators can customize material specific to their schools.

Student Supplements

Audio Recordings

- By Bill Saler
- Lead students through the material in each section of the text, explaining solution steps to examples, pointing out common errors, and focusing on margin exercises and solutions.
- Audio files are available to download in MP3 format. Contact your local Addison-Wesley representative for details.

Addison-Wesley Math Tutor Center

www.aw-bc.com/tutorcenter

- The Addison-Wesley Math Tutor Center is staffed by qualified mathematics instructors who provide students with tutoring on examples and odd-numbered exercises from the textbook. Tutoring is available via toll-free telephone, toll-free fax, e-mail, or the Internet. White Board technology allows tutors and students to actually see problems worked while they "talk" in real time over the Internet during tutoring sessions.

MathXL® Tutorials on CD   (ISBN 0-321-30577-9)

- Provides algorithmically generated practice exercises that correlate at the objective level to the content of the text.
- Includes an example and a guided solution to accompany every exercise and video clips for selected exercises.
- Recognizes student errors and provides feedback; generates printed summaries of students' progress.

Instructor Supplements

TestGen with Quizmaster   (ISBN 0-321-30573-6)

- Enables instructors to build, edit, print, and administer tests.
- Features a computerized bank of questions developed to cover all text objectives.
- Algorithmically based content allows instructors to create multiple but equivalent versions of the same question or test with a click of a button.
- Instructors can also modify test-bank questions or add new questions by using the built-in question editor, which allows users to create graphs, input graphics, and insert math notation, variable numbers, or text.
- Tests can be printed or administered online via the Internet or another network. Quizmaster allows students to take tests on a local area network.
- Available on a dual-platform Windows/Macintosh CD-ROM.

MathXL®   www.mathxl.com

MathXL is a powerful online homework, tutorial, and assessment system that accompanies Addison-Wesley textbooks in mathematics or statistics. With MathXL, instructors can create, edit, and assign online homework and tests using algorithmically generated exercises correlated at the objective level to the textbook. They can also create and assign their own online exercises and import TestGen tests for added flexibility. All student work is tracked in MathXL's online gradebook. Students can take chapter tests in MathXL and receive personalized study plans based on their test results. The study plan diagnoses weaknesses and links students directly to tutorial exercises for the objectives they need to study and retest. Students can also access supplemental animations and video clips directly from selected exercises. MathXL is available to qualified adopters. For more information, visit our Web site at www.mathxl.com or contact your Addison-Wesley sales representative.

MyMathLab   www.mymathlab.com

MyMathLab is a series of text-specific, easily customizable online courses for Addison-Wesley textbooks in mathematics and statistics. Powered by CourseCompass™ (Pearson Education's online teaching and learning environment) and MathXL® (our online homework, tutorial, and assessment system), MyMathLab gives instructors the tools they need to deliver all or a portion of their course online, whether students are in a lab setting or working from home. MyMathLab provides a rich and flexible set of course materials, featuring free-response exercises that are algorithmically generated for unlimited practice and mastery. Students can also use online tools, such as video lectures, animations, and a multimedia textbook, to independently improve their understanding and performance. Instructors can use MyMathLab's homework and test managers to select and assign online exercises correlated directly to the textbook, and they can also create and assign their own online exercises and import TestGen tests for added flexibility. MyMathLab's online gradebook—designed specifically for mathematics and statistics—automatically tracks students' homework and test results and gives the instructor control over how to calculate final grades. Instructors can also add offline (paper-and-pencil) grades to the gradebook. MyMathLab is available to qualified adopters. For more information, visit our Web site at www.mymathlab.com or contact your Addison-Wesley sales representative.

InterAct Math® Tutorial Web site www.interactmath.com

Get practice and tutorial help online! This interactive tutorial Web site provides algorithmically generated practice exercises that correlate directly to the exercises in the textbook. Students can retry an exercise as many times as they like with new values each time for unlimited practice and mastery. Every exercise is accompanied by an interactive guided solution that provides helpful feedback for incorrect answers, and students can also view a worked-out sample problem that steps them through an exercise similar to the one they're working on.

ADDISON-WESLEY MATH ADJUNCT SUPPORT CENTER

The Addison-Wesley Math Adjunct Support Center is staffed by qualified mathematics instructors with over 50 years of combined experience at both the community college and university level. Assistance is provided for faculty in the following areas:

- Suggested syllabus consultation
- Tips on using materials packaged with your book
- Book-specific content assistance
- Teaching suggestions including advice on classroom strategies

For more information, visit www.aw-bc.com/tutorcenter/math-adjunct.html

Acknowledgments and Reviewers

Many of you helped to shape the tenth edition by reviewing and spending time with us on your campuses. Our deepest appreciation to all of you and in particular to the following:

Ebrahim Ahmadizadeh, *Northampton Community College*

Anne Evans, *University of Massachusetts Boston*

Deborah A. Okonieski, *University of Akron*

Marilyn Platt, *Gaston College*

Jane Seeger, *Nash Community College*

Kenneth Takvorian, *Mount Wachusett Community College*

Joyce Wellington, *Southeastern Community College*

We wish to express our heartfelt appreciation to a number of people who have contributed in special ways to the development of this textbook. Our editor, Jennifer Crum, and marketing manager, Jay Jenkins, encouraged our vision and provided marketing insight. Kari Heen, the project manager, deserves special recognition for overseeing every phase of the project and keeping it moving. The unwavering support of the Developmental Math group, including Katie Nopper, project editor, Alison Macdonald, editorial assistant, Ron Hampton, managing editor, Dennis Schaefer, cover designer, and Sharon Smith and Ceci Fleming, media producers, and the endless hours of hard work by Martha Morong and Geri Davis have led to products of which we are immensely proud.

We also want to thank Judy Penna for writing the *Student's* and *Instructor's Solutions Manuals* and for her strong leadership in the preparation of the printed supplements, videotapes, and MyMathLab. Other strong support has come from Mark Stevenson for the *Printed Test Bank,* Bill Saler for the audio recordings, and Barbara Johnson, Jeremy Pletcher, and Michelle Lanosga for their accuracy checking of the manuscript. We also wish to recognize those who wrote scripts, presented lessons on camera, and checked the accuracy of the videotapes.

To the Student

As your author, I would like to welcome you to this study of *Intermediate Algebra.* Whatever your past experiences, I encourage you to look at this mathematics course as a fresh start. Approach this course with a positive attitude about mathematics. Mathematics is a base for life, for many majors, for personal finances, for most careers, or just for pleasure.

You are the most important factor in the success of your learning. In earlier experiences, you may have allowed yourself to sit back and let the instructor "pour in" the learning, with little or no follow-up on your part. But now you must take a more assertive and proactive stance. This may be the first adjustment you make in college. As soon as possible after class, you should thoroughly read the textbook and the supplements and do all you can to learn on your own. In other words, rid yourself of former habits and take responsibility for your own learning. Then, with all the help you have around you, your hard work will lead to success.

One of the most important suggestions I can make is to allow yourself enough *time* to learn. You can have the best book, the best instructor, and the best supplements, but if you do not give yourself time to learn, how can they be of benefit? Many other helpful suggestions are presented in the Study Tips that you will find throughout the book. You may want to read through all the Study Tips before you begin the text. An Index of Study Tips can be found at the front of the book.

M.L.B.

Bittinger Student Organizer

Study Tips

Throughout this text, you will find a feature called *Study Tips*. We discuss these in the Preface of this text. They are intended to help improve your math study skills. An index of all the *Study Tips* can be found at the front of the book.

For Extra Help

MyMathLab   MathXL

Student's Solutions Manual

Digital Video Tutor CD Videotape

Math Tutor Center

InterAct Math

Additional Resources

Basic Math Review Card (ISBN 0-321-39476-3)

Algebra Review Card (ISBN 0-321-39473-9)

Math for Allied Health Reference Card (ISBN 0-321-39474-7)

Graphing Calculator Reference Card (ISBN 0-321-39475-5)

Math Study Skills for Students Video on CD (ISBN 0-321-29745-8)

Go to www.aw-bc.com/math for more information.

On the first day of class, complete this chart and the weekly planner that follows on the reverse page.

Instructor Information:

Name _____

Office Hours and Location _____

_____

Phone Number _____

Fax Number _____

E-mail Address _____

Find the names of two students whom you could contact for class information or study questions:

1. Name _____

   Phone Number _____

   Fax Number _____

   E-mail Address _____

2. Name _____

   Phone Number _____

   Fax Number _____

   E-mail Address _____

Math Lab on Campus:

Location _____

Hours _____

Phone Number _____

Tutoring:

Campus Location _____

Hours _____

 To order the Addison-Wesley Math Tutor Center, call 1-888-777-0463.

(*See the Preface for important information concerning this tutoring.*)

Important Supplements: (*See the Preface for a complete list of available supplements.*)

Supplements recommended by the instructor _____

_____

Online Log-in Information (*include access code, password, Web address, etc.*)

_____

_____

Bittinger Student Organizer

Success is planned. On this page, plan a typical week. Consider time allotments for class, study, work, travel, family, and relaxation.

Important Dates

Midterm Exam

_____

Final Exam

_____

Holidays

_____

_____

_____

_____

Other

_____

_____

_____

_____

TIME	Sun.	Mon.	Tues.	Wed.	Thurs.	Fri.	Sat.
6:00 A.M.							
6:30							
7:00							
7:30							
8:00							
8:30							
9:00							
9:30							
10:00							
10:30							
11:00							
11:30							
12:00 P.M.							
12:30							
1:00							
1:30							
2:00							
2:30							
3:00							
3:30							
4:00							
4:30							
5:00							
5:30							
6:00							
6:30							
7:00							
7:30							
8:00							
8:30							
9:00							
9:30							
10:00							
10:30							
11:00							
11:30							
12:00 A.M.							

Review of Basic Algebra

R

Real-World Application

A gecko is an insect-eating lizard. Its feet will adhere to virtually any surface because they contain millions of miniscule hairs, or setae, that are 200 billionths of a meter wide. Write 200 billionths in scientific notation.

Source: The Proceedings of the National Academy of Sciences, Dr. Kellar Autumn and Wendy Hansen of Lewis and Clark College, Portland, Oregon.

This problem appears as Exercise 85 in Section R.7.

Objectives

a Use roster and set-builder notation to name sets, and distinguish among various kinds of real numbers.

b Determine which of two real numbers is greater and indicate which, using $<$ and $>$; given an inequality like $a < b$, write another inequality with the same meaning; and determine whether an inequality like $-2 \leq 3$ or $4 > 5$ is true.

c Graph inequalities on a number line.

d Find the absolute value of a real number.

Find the opposite of the number.

1. 9

2. -6

3. 0

To the student:

In the preface, at the front of the text, you will find a Student Organizer card. This pullout card will help you keep track of important dates and useful contact information. You can also use it to plan time for class, study, work, and relaxation. By managing your time wisely, you will provide yourself the best possible opportunity to be successful in this course.

Answers on page A-1

a Set Notation and the Set of Real Numbers

A **set** is a collection of objects. In mathematics, we usually consider sets of numbers. The set we consider most in algebra is **the set of real numbers.** There is a real number for every point on the real-number line. Some commonly used sets of numbers are **subsets** of, or sets contained within, the set of real numbers. We begin by examining some subsets of the set of real numbers.

The set containing the numbers -5, 0, and 3 can be named $\{-5, 0, 3\}$. This method of describing sets is known as the **roster method.** We use the roster method to describe three frequently used subsets of real numbers. Note that three dots are used to indicate that the pattern continues without end.

**NATURAL NUMBERS
(OR COUNTING NUMBERS)**

Natural numbers are those numbers used for counting: $\{1, 2, 3, \ldots\}$

WHOLE NUMBERS

Whole numbers are the set of natural numbers with 0 included: $\{0, 1, 2, 3, \ldots\}$

INTEGERS

Integers are the set of whole numbers and their opposites:

$$\{\ldots, -4, -3, -2, -1, 0, 1, 2, 3, 4, \ldots\}$$

The integers can be illustrated on the real-number line as follows.

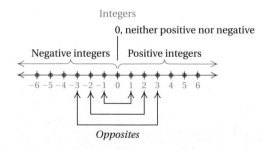

The set of integers extends infinitely to the left and to the right of 0. The **opposite** of a number is found by reflecting it across the number 0. Thus the opposite of 3 is -3. The opposite of -4 is 4. The opposite of 0 is 0. We read a symbol like -3 as either "the opposite of 3" or "negative 3."

The natural numbers are called **positive integers.** The opposites of the natural numbers (those to the left of 0) are called **negative integers.** Zero is neither positive nor negative.

Do Exercises 1–3 (in the margin at left).

Each point on the number line corresponds to a real number. In order to fill in the remaining numbers on our number line, we must describe two other subsets of the real numbers. And to do that, we need another type of set notation.

Set-builder notation is used to specify conditions under which a number is in a set. For example, the set of all odd natural numbers less than 9 can be described as follows:

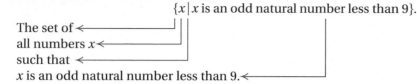

$$\{x \mid x \text{ is an odd natural number less than } 9\}.$$

The set of ←
all numbers x ←
such that ←
x is an odd natural number less than 9. ←

We can easily write another name for this set using roster notation, as follows:

$$\{1, 3, 5, 7\}.$$

EXAMPLE 1 Name the set consisting of the first six even whole numbers using both roster notation and set-builder notation.

Roster notation: $\{0, 2, 4, 6, 8, 10\}$

Set-builder notation: $\{x \mid x \text{ is one of the first six even whole numbers}\}$

Do Exercise 4.

The advantage of set-builder notation is that we can use it to describe very large sets that may be difficult to describe using roster notation. Such is the case when we try to name the set of **rational numbers.** Rational numbers can be named using fraction notation. The following are examples of rational numbers:

$$\frac{5}{8}, \quad \frac{12}{-7}, \quad \frac{-17}{15}, \quad -\frac{9}{7}, \quad \frac{39}{1}, \quad \frac{0}{6}.$$

We can now describe the set of rational numbers.

RATIONAL NUMBERS

A **rational number** can be expressed as an integer divided by a nonzero integer:

$$\left\{\frac{p}{q} \,\middle|\, p \text{ is an integer, } q \text{ is an integer, and } q \neq 0\right\}.$$

Rational numbers are numbers whose decimal representation either terminates or has a repeating block of digits.

For example, each of the following is a rational number:

$$\frac{5}{8} = \underbrace{0.625}_{} \quad \text{and} \quad \frac{6}{11} = \underbrace{0.545454\ldots}_{} = 0.\overline{54}.$$

Terminating   Repeating

The bar in $0.\overline{54}$ indicates the repeating block of digits in the decimal notation. Note that $\frac{39}{1} = 39$. Thus the set of rational numbers contains the integers.

Do Exercises 5 and 6.

4. Name the set consisting of the first seven odd whole numbers using both roster notation and set-builder notation.

Convert to decimal notation by long division and determine whether it is terminating or repeating.

5. $\dfrac{11}{16}$

6. $\dfrac{14}{13}$

Answers on page A-1

The real-number line has a point for every rational number.

$$-\frac{9}{2} \quad\quad -1.7 \quad\quad \frac{1}{10} \quad\quad 2\frac{7}{8} \quad 4.\overline{13}$$

$$\begin{array}{c|c|c|c|c|c|c|c|c|c|c}
\hline
-5 & -4 & -3 & -2 & -1 & 0 & 1 & 2 & 3 & 4 & 5
\end{array}$$

However, there are many points on the line for which there is no rational number. These points correspond to what are called **irrational numbers.**

$$-1.898898889\ldots \quad\quad \sqrt{13}$$
$$-\sqrt{10} \quad\quad\quad \sqrt{2} \quad \pi$$

$$\begin{array}{c|c|c|c|c|c|c|c|c|c|c}
\hline
-5 & -4 & -3 & -2 & -1 & 0 & 1 & 2 & 3 & 4 & 5
\end{array}$$

Numbers like π, $\sqrt{2}$, $-\sqrt{10}$, $\sqrt{13}$, and $-1.898898889\ldots$ are examples of irrational numbers. The decimal notation for an irrational number *neither* terminates *nor* repeats. Recall that decimal notation for rational numbers either terminates or has a repeating block of digits.

IRRATIONAL NUMBERS

Irrational numbers are numbers whose decimal representation neither terminates nor has a repeating block of digits. They cannot be represented as the quotient of two integers.

The irrational number $\sqrt{2}$ (read "the square root of 2") is the length of the diagonal of a square with sides of length 1. It is also the number that, when multiplied by itself, gives 2. No rational number can be multiplied by itself to get 2, although some approximations come close:

1.4 is an *approximation* of $\sqrt{2}$ because $(1.4)^2 = (1.4)(1.4) = 1.96$;

1.41 is a better approximation because $(1.41)^2 = (1.41)(1.41) = 1.9881$;

1.4142 is an even better approximation because $(1.4142)^2 = (1.4142)(1.4142) = 1.99996164$.

We say that 1.4142 is a rational approximation of $\sqrt{2}$ because

$$(1.4142)^2 = 1.99996164 \approx 2.$$

The symbol $\approx$ means "is approximately equal to." We can find rational approximations for square roots and other irrational numbers using a calculator.

4

The set of all rational numbers, combined with the set of all irrational numbers, gives us the set of **real numbers.**

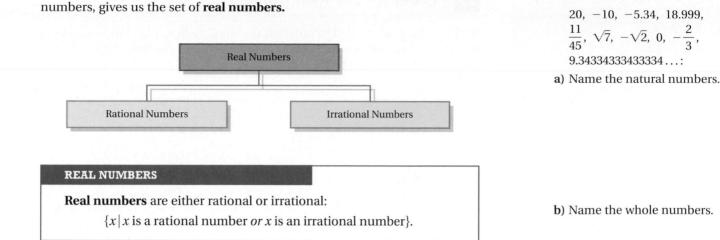

REAL NUMBERS

Real numbers are either rational or irrational:

$\{x \mid x \text{ is a rational number } or \ x \text{ is an irrational number}\}.$

Every point on the number line represents some real number and every real number is represented by some point on a number line.

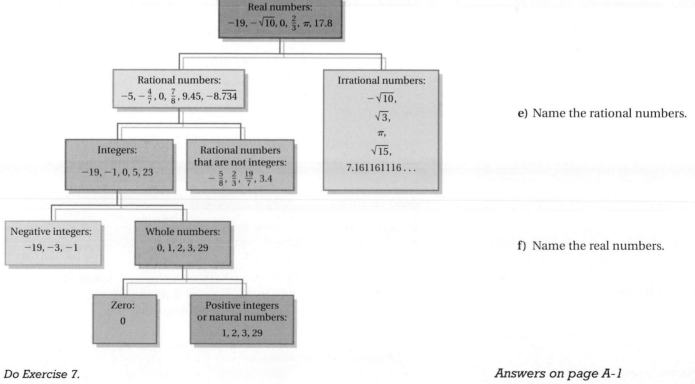

The following figure shows the relationships among various kinds of real numbers.

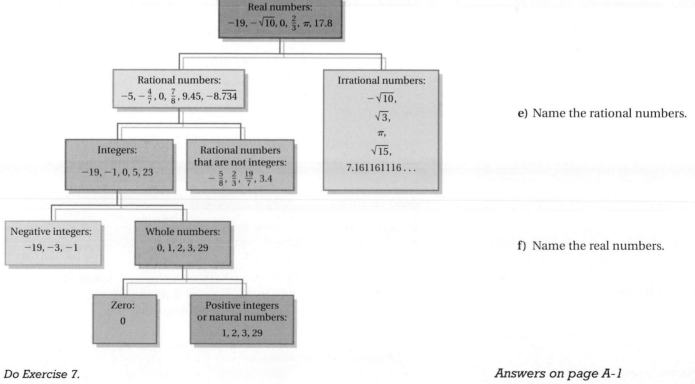

7. Given the numbers

$20, \ -10, \ -5.34, \ 18.999,$

$\dfrac{11}{45}, \ \sqrt{7}, \ -\sqrt{2}, \ 0, \ -\dfrac{2}{3},$

$9.34334333433334\ldots:$

a) Name the natural numbers.

b) Name the whole numbers.

c) Name the integers.

d) Name the irrational numbers.

e) Name the rational numbers.

f) Name the real numbers.

Insert < or > for ☐ to write a true sentence.

8. -5 ☐ -4

9. $-\dfrac{1}{4}$ ☐ $-\dfrac{1}{2}$

10. 87 ☐ 67

11. -9.8 ☐ $-4\dfrac{2}{3}$

12. 6.78 ☐ -6.77

13. $-\dfrac{4}{5}$ ☐ -0.86

14. $\dfrac{14}{29}$ ☐ $\dfrac{17}{32}$

15. $-\dfrac{12}{13}$ ☐ $-\dfrac{14}{15}$

16. 1.8 ☐ 1.08

Answers on page A-1

b   Order for the Real Numbers

Real numbers are named in order on the number line, with larger numbers named further to the right. For any two numbers on the line, the one to the left is less than the one to the right.

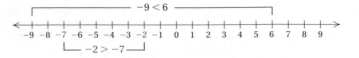

We use the symbol **<** to mean "**is less than.**" The sentence $-9 < 6$ means "-9 is less than 6." The symbol **>** means "**is greater than.**" The sentence $-2 > -7$ means "-2 is greater than -7." A handy mental device is to think of > or < as an arrowhead that points to the smaller number.

EXAMPLES   Use either < or > for ☐ to write a true sentence.

2. 4 ☐ 9 Since 4 is to the left of 9, 4 is less than 9, so $4 < 9$.

3. -8 ☐ 3 Since -8 is to the left of 3, we have $-8 < 3$.

4. 7 ☐ -12 Since 7 is to the right of -12, then $7 > -12$.

5. -21 ☐ -5 Since -21 is to the left of -5, we have $-21 < -5$.

6. -2.7 ☐ $-\dfrac{3}{2}$ Since $-\dfrac{3}{2} = -1.5$ and -2.7 is to the left of -1.5, we have $-2.7 < -\dfrac{3}{2}$.

7. $1\dfrac{1}{4}$ ☐ -2.7 The answer is $1\dfrac{1}{4} > -2.7$.

8. 4.79 ☐ 4.97 The answer is $4.79 < 4.97$.

9. -8.45 ☐ 1.32 The answer is $-8.45 < 1.32$.

10. $\dfrac{5}{8}$ ☐ $\dfrac{7}{11}$ We convert to decimal notation $\left(\dfrac{5}{8} = 0.625 \text{ and } \dfrac{7}{11} = 0.6363\ldots\right)$ and compare. Thus, $\dfrac{5}{8} < \dfrac{7}{11}$.

Do Exercises 8–16.

All positive real numbers are greater than zero and all negative real numbers are less than zero.

> If x is a positive real number, then $x > 0$.
> If x is a negative real number, then $x < 0$.

Note that $-8 < 5$ and $5 > -8$ are both true. These are **inequalities.** Every true inequality yields another true inequality if we interchange the numbers or variables and reverse the direction of the inequality sign.

> $a < b$ also has the meaning $b > a$.

EXAMPLES Write a different inequality with the same meaning.

11. $a < -5$ The inequality $-5 > a$ has the same meaning.

12. $-3 > -8$ The inequality $-8 < -3$ has the same meaning.

Do Exercises 17 and 18.

Expressions like $a \leq b$ and $b \geq a$ are also **inequalities.** We read **$a \leq b$** as "***a* is less than or equal to *b*.**" We read **$a \geq b$** as "***a* is greater than or equal to *b*.**" If a is nonnegative, then $a \geq 0$.

EXAMPLES Write true or false.

13. $-8 \leq 5.7$ True since $-8 < 5.7$ is true.

14. $-8 \leq -8$ True since $-8 = -8$ is true.

15. $-7 \geq 4\frac{1}{3}$ False since neither $-7 > 4\frac{1}{3}$ nor $-7 = 4\frac{1}{3}$ is true.

16. $-\frac{2}{3} \geq -\frac{5}{4}$ True since $-\frac{2}{3} = -0.666\ldots$ and $-\frac{5}{4} = -1.25$ and $-0.666\ldots \geq -1.25$.

Do Exercises 19–22.

C   Graphing Inequalities on a Number Line

Some replacements for the variable in an inequality make it true and some make it false. A replacement that makes it true is called a **solution.** The set of all solutions is called the **solution set.** A **graph** of an inequality is a drawing that represents its solution set.

EXAMPLE 17 Graph the inequality $x > -3$ on a number line.

The solutions consist of all real numbers greater than -3, so we shade all numbers greater than -3. Note that -3 is not a solution. We indicate this by using a parenthesis at -3.

$$\overset{\longleftarrow\;\mid\;\overset{(}{\mid}\;\mid\;\mid\;\mid\;\mid\;\mid\;\mid\;\longrightarrow}{\scriptstyle -4\;-3\;-2\;-1\;\;0\;\;1\;\;2\;\;3\;\;4}$$

The graph represents the solution set $\{x \mid x > -3\}$. Numbers in this set include -2.6, -1, 0, π, $\sqrt{2}$, $3\frac{7}{8}$, 5, and 123.

EXAMPLE 18 Graph the inequality $x \leq 2$ on a number line.

We make a drawing that represents the solution set $\{x \mid x \leq 2\}$. This time the graph consists of 2 as well as the numbers less than 2. We shade all numbers to the left of 2 and use a bracket at 2 to indicate that it is also a solution.

$$\overset{\longleftarrow\;\mid\;\mid\;\mid\;\mid\;\mid\;\overset{]}{\mid}\;\mid\;\mid\;\longrightarrow}{\scriptstyle -4\;-3\;-2\;-1\;\;0\;\;1\;\;2\;\;3\;\;4}$$

Do Exercises 23–28. (Exercises 25–28 are on the following page.)

Write a different inequality with the same meaning.

17. $x > 6$

18. $-4 < 7$

Write true or false.

19. $6 \geq -9.4$

20. $-18 \leq -18$

21. $-7.6 \leq -10\dfrac{4}{5}$

22. $-\dfrac{24}{27} \geq -\dfrac{25}{28}$

Graph the inequality.

23. $x > -1$

$$\overset{\longleftarrow\;\mid\;\mid\;\mid\;\mid\;\mid\;\mid\;\mid\;\mid\;\mid\;\mid\;\longrightarrow}{\scriptstyle -5\;-4\;-3\;-2\;-1\;\;0\;\;1\;\;2\;\;3\;\;4\;\;5}$$

24. $x \leq 5$

$$\overset{\longleftarrow\;\mid\;\mid\;\mid\;\mid\;\mid\;\mid\;\mid\;\mid\;\mid\;\mid\;\longrightarrow}{\scriptstyle -5\;-4\;-3\;-2\;-1\;\;0\;\;1\;\;2\;\;3\;\;4\;\;5}$$

Answers on page A-1

Match the inequality with one of the graphs shown below.

25. $x \le -\dfrac{5}{2}$ **26.** $x > 0$

27. $-4 > x$ **28.** $2 \le x$

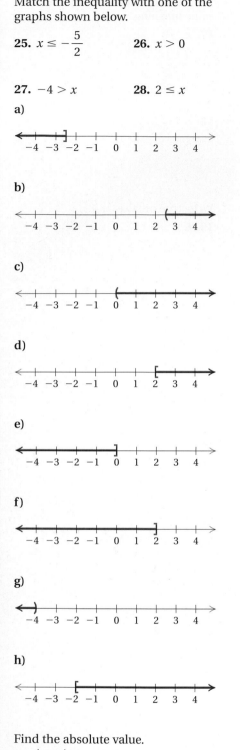

a)

b)

c)

d)

e)

f)

g)

h)

Find the absolute value.

29. $\left|-\dfrac{1}{4}\right|$ **30.** $|2|$

31. $\left|\dfrac{3}{2}\right|$ **32.** $|-2.3|$

d Absolute Value

From the number line, we see that numbers like -6 and 6 are the same distance from 0. Distance is always a nonnegative number. We call the distance of a number from zero on a number line the **absolute value** of the number.

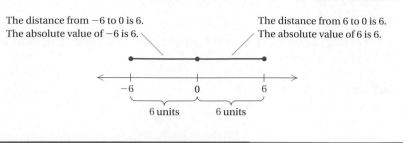

The distance from -6 to 0 is 6.
The absolute value of -6 is 6.

The distance from 6 to 0 is 6.
The absolute value of 6 is 6.

6 units   6 units

ABSOLUTE VALUE*

The **absolute value** of a number is its distance from zero on the number line. We use the symbol $|x|$ to represent the absolute value of a number x.

To find absolute value:

1. If a number is negative, make it positive.
2. If a number is positive or zero, leave it alone.

EXAMPLES   Find the absolute value.

19. $|-7|$ The distance of -7 from 0 is 7, so $|-7|$ is 7.

20. $|12|$ The distance of 12 from 0 is 12, so $|12|$ is 12.

21. $|0|$ The distance of 0 from 0 is 0, so $|0|$ is 0.

22. $\left|\dfrac{4}{5}\right| = \dfrac{4}{5}$

23. $|-3.86| = 3.86$

Do Exercises 29–32.

---

*A more formal definition of $|x|$ is given in Section R.2.

CALCULATOR CORNER

Graphing Calculators and Absolute Value   Although a calculator is *not* required for this textbook, the book contains a series of *optional* discussions on using a graphing calculator. The keystrokes for a TI-84 Plus graphing calculator will be shown throughout. For keystrokes for other models of calculators, consult the user's manual for your particular calculator.

Note that there are options above the keys as well as on them. To access the option written on a key, simply press the key. The options written in blue above the keys are accessed by first pressing the blue **2ND** key and then pressing the key corresponding to the desired option. The green options are accessed by first pressing the green **ALPHA** key.

To turn the calculator on, press the **ON** key at the bottom left-hand corner of the keypad. You should see a blinking rectangle, or cursor, on the screen. If you do not see the cursor, try adjusting the display contrast. To do this, first press **2ND** and then press and hold ⌃ to increase the contrast or ⌄ to decrease the contrast.

To turn the calculator off, press **2ND** **OFF**. (OFF is the second operation associated with the **ON** key.) The calculator will turn itself off automatically after about five minutes of no activity.

Press **MODE** to display the MODE settings. Initially, you should select the settings on the left side of the display.

```
NORMAL SCI ENG
FLOAT 0123456789
RADIAN DEGREE
FUNC PAR POL SEQ
CONNECTED DOT
SEQUENTIAL SIMUL
REAL a+bi re^θi
FULL HORIZ G-T
```

To change a setting on the MODE screen, use ⌄ or ⌃ to move the cursor to the line of that setting. Then use ▷ or ◁ to move the blinking cursor to the desired setting and press **ENTER** . Press **CLEAR** or **2ND** **QUIT** to leave the MODE screen. QUIT is the second operation associated with the **MODE** key.) Pressing **CLEAR** or **2ND** **QUIT** will take you to the home screen where computations are performed.

We can use a graphing calculator to find the absolute value of a number. The absolute-value operation is the first item in the Catalog on the TI-84 Plus graphing calculator. To find $|-7|$, as in Example 19, we first press **2ND** **CATALOG** **ENTER** to copy "abs(" to the home screen. (CATALOG is the second operation associated with the **0** numeric key.) Then we press **(-)** **7** **)** **ENTER** . The result is 7. To find $\left|\dfrac{4}{5}\right|$, as in Example 22, and to express the result as a fraction, we press **2ND** **CATALOG** **ENTER** **4** **÷** **5** **)** **MATH** **1** **ENTER** . The result is $\dfrac{4}{5}$. The keystrokes **MATH** **1** select the ▷FRAC option from the MATH menu, causing the result to be expressed in fraction form.

```
abs(-7)
                    7
abs(4/5)▶Frac
                  4/5
```

Exercises: Find the absolute value.

1. $|-3|$

2. $|14|$

3. $|0|$

4. $|5.67|$

5. $|-18.4|$

6. $|-0.8|$

7. $\left|-\dfrac{2}{3}\right|$

8. $\left|\dfrac{7}{9}\right|$

R.1

EXERCISE SET

For Extra Help

MathXL MyMathLab InterAct Math Math Tutor Center Digital Video Tutor CD 1 Videotape 1 Student's Solutions Manual

a   Given the numbers $-6, 0, 1, -\frac{1}{2}, -4, \frac{7}{9}, 12, -\frac{6}{5}, 3.45, 5\frac{1}{2}, \sqrt{3}, \sqrt{25}, -\frac{12}{3}, 0.131331333133331\ldots:$

1. Name the natural numbers.

2. Name the whole numbers.

3. Name the rational numbers.

4. Name the integers.

5. Name the real numbers.

6. Name the irrational numbers.

Given the numbers $-\sqrt{5}, -3.43, -11, 12, 0, \frac{11}{34}, -\frac{7}{13}, \pi, -3.565665666566665\ldots:$

7. Name the whole numbers.

8. Name the natural numbers.

9. Name the integers.

10. Name the rational numbers.

11. Name the irrational numbers.

12. Name the real numbers.

Use roster notation to name the set.

13. The set of all letters in the word "math"

14. The set of all letters in the word "solve"

15. The set of all positive integers less than 13

16. The set of all odd whole numbers less than 13

17. The set of all even natural numbers

18. The set of all negative integers greater than -4

Use set-builder notation to name the set.

19. $\{0, 1, 2, 3, 4, 5\}$

20. $\{4, 5, 6, 7, 8, 9, 10\}$

21. The set of all rational numbers

22. The set of all real numbers

23. The set of all real numbers greater than -3

24. The set of all real numbers less than or equal to 21

10

CHAPTER R: Review of Basic Algebra

Copyright © 2007 Pearson Education, Inc.

Use either < or > for ☐ to write a true sentence.

25. 13 ☐ 0

26. 18 ☐ 0

27. −8 ☐ 2

28. 7 ☐ −7

29. −8 ☐ 8

30. 0 ☐ −11

31. −8 ☐ −3

32. −6 ☐ −3

33. −2 ☐ −12

34. −7 ☐ −10

35. −9.9 ☐ −2.2

36. $-13\frac{1}{5}$ ☐ $\frac{11}{250}$

37. $37\frac{1}{5}$ ☐ $-1\frac{67}{100}$

38. −13.99 ☐ −8.45

39. $\frac{6}{13}$ ☐ $\frac{13}{25}$

40. $-\frac{14}{15}$ ☐ $-\frac{27}{53}$

Write a different inequality with the same meaning.

41. $-8 > x$

42. $x < 7$

43. $-12.7 \leq y$

44. $10\frac{2}{3} \geq t$

Write true or false.

45. $6 \leq -6$

46. $-7 \leq -7$

47. $5 \geq -8.4$

48. $-11 \geq -13\frac{1}{2}$

c Graph the inequality.

49. $x < -2$

50. $x < -1$

51. $x \leq -2$

52. $x \geq -1$

53. $x > -3.3$

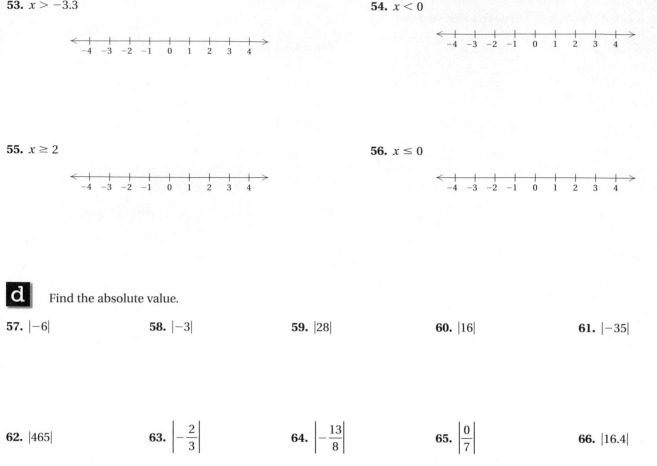

54. $x < 0$

55. $x \geq 2$

56. $x \leq 0$

d Find the absolute value.

57. $|-6|$ **58.** $|-3|$ **59.** $|28|$ **60.** $|16|$ **61.** $|-35|$

62. $|465|$ **63.** $\left|-\dfrac{2}{3}\right|$ **64.** $\left|-\dfrac{13}{8}\right|$ **65.** $\left|\dfrac{0}{7}\right|$ **66.** $|16.4|$

To the student and the instructor: The Discussion and Writing exercises, denoted by the symbol $\mathbf{D_W}$, are meant to be answered with one or more sentences. They can be discussed and answered collaboratively by the entire class or by small groups. Because of their open-ended nature, the answers to these exercises do not appear at the back of the book.

67. $\mathbf{D_W}$ Explain the difference between a rational number and an irrational number in as many ways as you can.

68. $\mathbf{D_W}$ List five examples of rational numbers that are not integers and explain why they are not.

SYNTHESIS

To the student and the instructor: The Synthesis exercises found at the end of every exercise set challenge students to combine concepts or skills studied in that section or in preceding parts of the text. Exercises marked with a [calculator symbol] symbol are meant to be solved using a calculator.

Use either $\leq$ or $\geq$ for $\square$ to write a true sentence.

69. $|-3| \square 5$ **70.** $|-5| \square |-2|$ **71.** $|4| \square |-7|$ **72.** $|-8| \square |8|$

73. List the following numbers in order from least to greatest.

$$\dfrac{1}{11}, \quad 1.1\%, \quad \dfrac{2}{7}, \quad 0.3\%, \quad 0.11, \quad \dfrac{1}{8}\%, \quad 0.009, \quad \dfrac{99}{1000}, \quad 0.286, \quad \dfrac{1}{8}, \quad 1\%, \quad \dfrac{9}{100}$$

R.2 OPERATIONS WITH
REAL NUMBERS

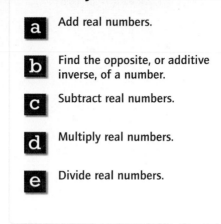

We now review addition, subtraction, multiplication, and division of real numbers.

a Addition

In this section, we review addition of real numbers. First, to gain an understanding, we add using a number line. Then we consider rules for addition.

ADDITION ON A NUMBER LINE

To do the addition $a + b$, we begin at 0. Next, we move to a and then move according to b.

- If b is positive, we move to the right.
- If b is negative, we move to the left.
- If b is 0, we stay at a.

Add using a number line.
1. $-5 + 9$

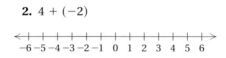

■ **EXAMPLES**

1. $6 + (-8) = -2$: We begin at 0 and move 6 units right since 6 is positive. Then we move 8 units left since -8 is negative. The answer is -2.

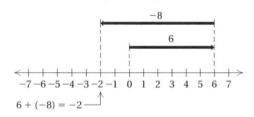

2. $4 + (-2)$

2. $-3 + 7 = 4$: We begin at 0 and move 3 units left since -3 is negative. Then we move 7 units right since 7 is positive. The answer is 4.

3. $3 + (-8)$

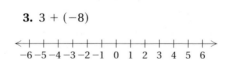

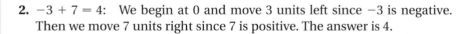

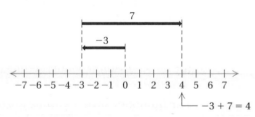

3. $-2 + (-5) = -7$: We begin at 0 and move 2 units left since -2 is negative. Then we move 5 units further left since -5 is negative. The answer is -7.

4. $-5 + 5$

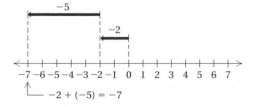

Answers on page A-1

Add.

5. $-7 + (-11)$

6. $-8.9 + (-9.7)$

7. $-\dfrac{6}{5} + \left(-\dfrac{23}{5}\right)$

8. $-\dfrac{3}{10} + \left(-\dfrac{2}{5}\right)$

9. $-7 + 7$

10. $-7.4 + 0$

11. $4 + (-7)$

12. $-7.8 + 4.5$

13. $\dfrac{3}{8} + \left(-\dfrac{5}{8}\right)$

14. $-\dfrac{3}{5} + \dfrac{7}{10}$

Do Exercises 1–4 on the preceding page.

You may have noticed some patterns in the preceding examples. These lead us to rules for adding without using the number line.

RULES FOR ADDITION OF REAL NUMBERS

1. *Positive numbers*: Add the numbers. The result is positive.
2. *Negative numbers*: Add absolute values. Make the answer negative.
3. *A positive and a negative number*: Subtract the smaller absolute value from the larger. Then:

   a) If the positive number has the greater absolute value, make the answer positive.
   b) If the negative number has the greater absolute value, make the answer negative.
   c) If the numbers have the same absolute value, the answer is 0.

4. *One number is zero*: The sum is the other number.

Rule 4 is known as the **Identity Property of 0.** It says that for any real number a, $a + 0 = a$.

EXAMPLES Add without using a number line.

4. $-13 + (-8) = -21$ Two negatives. Add the absolute values: $|-13| + |-8| = 13 + 8 = 21$. Make the answer *negative*: -21.

5. $-2.1 + 8.5 = 6.4$ One negative, one positive. Find the absolute values: $|-2.1| = 2.1$; $|8.5| = 8.5$. Subtract the smaller absolute value from the larger: $8.5 - 2.1 = 6.4$. The *positive* number, 8.5, has the larger absolute value, so the answer is *positive*, 6.4.

6. $-48 + 31 = -17$ One negative, one positive. Find the absolute values: $|-48| = 48$; $|31| = 31$. Subtract the smaller absolute value from the larger: $48 - 31 = 17$. The *negative* number, -48, has the larger absolute value, so the answer is *negative*, -17.

7. $2.6 + (-2.6) = 0$ The numbers have the same absolute value. The sum is 0.

8. $-\dfrac{5}{9} + 0 = -\dfrac{5}{9}$ One number is zero. The sum is $-\frac{5}{9}$.

9. $-\dfrac{3}{4} + \dfrac{9}{4} = \dfrac{6}{4} = \dfrac{3}{2}$

10. $-\dfrac{2}{3} + \dfrac{5}{8} = -\dfrac{16}{24} + \dfrac{15}{24} = -\dfrac{1}{24}$

Answers on page A-1

Do Exercises 5–14.

b Opposites, or Additive Inverses

Find the opposite, or additive inverse, of the number.

Suppose we add two numbers that are **opposites**, such as 4 and −4. The result is 0. When opposites are added, the result is always 0. Such numbers are also called **additive inverses.** Every real number has an opposite, or additive inverse.

15. −14

OPPOSITES, OR ADDITIVE INVERSES

Two numbers whose sum is 0 are called **opposites,** or **additive inverses,** of each other.

---

16. $\dfrac{2}{3}$

EXAMPLES   Find the opposite, or additive inverse, of the number.

11. 8.6   The opposite of 8.6 is −8.6 because 8.6 + (−8.6) = 0.

12. 0   The opposite of 0 is 0 because 0 + 0 = 0.

13. $-\dfrac{7}{9}$ The opposite of $-\dfrac{7}{9}$ is $\dfrac{7}{9}$ because $-\dfrac{7}{9} + \dfrac{7}{9} = 0.$

17. 0

To name the opposite, or additive inverse, we use the symbol −, and read the symbolism $-a$ as "the opposite of a" or "the additive inverse of a."

Do Exercises 15–17.

Caution!

A symbol such as −8 is usually read "negative 8." It could be read "the opposite of 8," because the opposite of 8 is −8. It could also be read "the additive inverse of 8," because the additive inverse of 8 is −8. When a variable is involved, as in a symbol like $-x$, it can be read "the opposite of x" or "the additive inverse of x" but *not* "negative x," because we do not know whether the symbol represents a positive number, a negative number, or 0. It is never correct to read −8 as "minus 8."

---

18. Evaluate $-a$ when $a = 9.$

19. Evaluate $-a$ when $a = -\dfrac{3}{5}.$

EXAMPLE 14   Evaluate $-x$ and $-(-x)$ when **(a)** $x = 23$ and **(b)** $x = -5.$

a) If $x = 23$, then $-x = -23 = -23.$ The opposite, or additive inverse, of 23 is −23.

If $x = 23$, then $-(-x) = -(-23) = 23.$ The opposite of the opposite of 23 is 23.

b) If $x = -5$, then $-x = -(-5) = 5.$

If $x = -5$, then $-(-x) = -(-(-5)) = -(5) = -5.$

20. Evaluate $-(-a)$ when $a = -5.9.$

Note in Example 14(b) that an extra set of parentheses is used to show that we are substituting the negative number −5 for x. Symbolism like $--x$ is not considered meaningful.

Do Exercises 18–21.

We can use the symbolism $-a$ for the opposite of a to restate the definition of opposite.

21. Evaluate $-(-a)$ when $a = \dfrac{2}{3}.$

THE SUM OF OPPOSITES

For any real number a, the **opposite,** or **additive inverse,** of a, which is $-a$, is such that

$$a + (-a) = (-a) + a = 0.$$

Answers on page A-1

22. Change the sign.

a) 11

b) −17

c) 0

d) x

e) −x

Subtract.

23. $8 - (-9)$

24. $-10 - 6$

25. $5 - 8$

26. $-23.7 - 5.9$

27. $-2 - (-5)$

28. $-\dfrac{11}{12} - \left(-\dfrac{23}{12}\right)$

29. $\dfrac{2}{3} - \left(-\dfrac{5}{6}\right)$

30. a) $17 - 23$

b) $-17 - 23$

c) $-17 - (-23)$

Answers on page A-1

SIGNS OF NUMBERS

A negative number is sometimes said to have a "negative sign." A positive number is said to have a "positive sign." When we replace a number with its opposite, or additive inverse, we can say that we have "changed its sign."

EXAMPLES   Change the sign. (Find the opposite, or additive inverse.)

15. -3 $-(-3) = 3$ **16.** $-\dfrac{3}{8}$ $-\left(-\dfrac{3}{8}\right) = \dfrac{3}{8}$

17. 0 $-0 = 0$ **18.** 14 $-(14) = -14$

Do Exercise 22.

We can now use the concept of opposite to give a more formal definition of absolute value.

ABSOLUTE VALUE

For any real number a, the **absolute value** of a, denoted $|a|$, is given by

$$|a| = \begin{cases} a, & \text{if } a \geq 0, \quad \text{For example, } |8| = 8 \text{ and } |0| = 0. \\ -a, & \text{if } a < 0. \quad \text{For example, } |-5| = -(-5) = 5. \end{cases}$$

(The absolute value of a is a if a is nonnegative. The absolute value of a is the opposite of a if a is negative.)

C   Subtraction

SUBTRACTION

The difference $a - b$ is the unique number c for which $a = b + c$. That is, $a - b = c$ if c is a number such that $a = b + c$.

For example, $3 - 5 = -2$ because $3 = 5 + (-2)$. That is, -2 is the number that when added to 5 gives 3. Although this illustrates the formal definition of subtraction, we generally use the following when we subtract.

SUBTRACTING BY ADDING THE OPPOSITE

For any real numbers a and b,

$$a - b = a + (-b).$$

(We can subtract by adding the opposite (additive inverse) of the number being subtracted.)

EXAMPLES   Subtract.

19. $3 - 5 = 3 + (-5) = -2$ Changing the sign of 5 and adding

20. $7 - (-3) = 7 + (3) = 10$ Changing the sign of -3 and adding

21. $-19.4 - 5.6 = -19.4 + (-5.6) = -25$

22. $-\dfrac{4}{3} - \left(-\dfrac{2}{5}\right) = -\dfrac{4}{3} + \dfrac{2}{5} = -\dfrac{20}{15} + \dfrac{6}{15} = -\dfrac{14}{15}$

Do Exercises 23–30 on the preceding page.

 Multiplication

We know how to multiply positive numbers. What happens when we multiply a positive number and a negative number?

Do Exercise 31.

> **THE PRODUCT OF A POSITIVE NUMBER AND A NEGATIVE NUMBER**
>
> To multiply a positive number and a negative number, multiply their absolute values. Then make the answer negative.

EXAMPLES Multiply.

23. $-3 \cdot 5 = -15$

24. $6 \cdot (-7) = -42$

25. $(-1.2)(4.5) = -5.4$

26. $3 \cdot \left(-\frac{1}{2}\right) = \frac{3}{1} \cdot \left(-\frac{1}{2}\right) = -\frac{3}{2}$

Note in Example 25 that the parentheses indicate multiplication.

Do Exercises 32–34.

What happens when we multiply two negative numbers?

Do Exercise 35.

> **THE PRODUCT OF TWO NEGATIVE NUMBERS**
>
> To multiply two negative numbers, multiply their absolute values. The answer is positive.

EXAMPLES Multiply.

27. $-3 \cdot (-5) = 15$

28. $-5.2(-10) = 52$

29. $(-8.8)(-3.5) = 30.8$

30. $\left(-\frac{3}{4}\right) \cdot \left(-\frac{5}{2}\right) = \frac{15}{8}$

Do Exercises 36–38.

 Division

> **DIVISION**
>
> The quotient $a \div b$, or $\frac{a}{b}$, where $b \neq 0$, is that unique real number c for which $a = b \cdot c$.

The definition of division parallels the one for subtraction. Using this definition and the rules for multiplying, we can see how to handle signs when dividing.

31. Look for a pattern and complete.

$4 \cdot 5 = 20$ $-2 \cdot 5 =$
$3 \cdot 5 = 15$ $-3 \cdot 5 =$
$2 \cdot 5 =$ $-4 \cdot 5 =$
$1 \cdot 5 =$ $-5 \cdot 5 =$
$0 \cdot 5 =$ $-6 \cdot 5 =$
$-1 \cdot 5 =$

Multiply.

32. $-4 \cdot 6$

33. $(3.5)(-8.1)$

34. $-\frac{4}{5} \cdot 10$

35. Look for a pattern and complete.

$4(-5) = -20$ $-1(-5) =$
$3(-5) = -15$ $-2(-5) =$
$2(-5) =$ $-3(-5) =$
$1(-5) =$ $-4(-5) =$
$0(-5) =$ $-5(-5) =$

Multiply.

36. $-8(-9)$

37. $\left(-\frac{4}{5}\right) \cdot \left(-\frac{2}{3}\right)$

38. $(-4.7)(-9.1)$

Answers on page A-1

Divide.

39. $\dfrac{-28}{-14}$

40. $125 \div (-5)$

41. $\dfrac{-75}{25}$

42. $-4.2 \div (-21)$

Divide, if possible.

43. $\dfrac{8}{0}$

44. $\dfrac{0}{9}$

45. $\dfrac{17}{2x - 2x}$

46. $\dfrac{3x - 3x}{x - x}$

Find the reciprocal of the number.

47. $\dfrac{3}{8}$

48. $-\dfrac{4}{5}$

49. 18

50. -4.3

51. 0.5

Answers on page A-1

EXAMPLES Divide.

31. $\dfrac{10}{-2} = -5$, because $-5 \cdot (-2) = 10$

32. $\dfrac{-32}{4} = -8$, because $-8 \cdot (4) = -32$

33. $\dfrac{-25}{-5} = 5$, because $5 \cdot (-5) = -25$

34. $\dfrac{40}{-4} = -10$ **35.** $-10 \div 5 = -2$

36. $\dfrac{-10}{-40} = \dfrac{1}{4}$, or 0.25   **37.** $\dfrac{-10}{-3} = \dfrac{10}{3}$, or $3.\overline{3}$

The rules for division and multiplication are the same.

> To multiply or divide two real numbers:
>
> 1. Multiply or divide the absolute values.
> 2. If the signs are the same, then the answer is positive.
> 3. If the signs are different, then the answer is negative.

Do Exercises 39–42.

EXCLUDING DIVISION BY ZERO

We cannot divide a nonzero number n by zero. By the definition of division, $n/0$ would be some number that when multiplied by 0 gives n. But when any number is multiplied by 0, the result is 0. The only possibility for n would be 0.

Consider $0/0$. We might say that it is 5 because $5 \cdot 0 = 0$. We might also say that it is -8 because $-8 \cdot 0 = 0$. In fact, $0/0$ could be any number at all. So, division by 0 does not make sense. Division by 0 is not defined and not possible.

EXAMPLES Divide, if possible.

38. $\dfrac{7}{0}$ Not defined: Division by 0.

39. $\dfrac{0}{7} = 0$ The quotient is 0 because $0 \cdot 7 = 0$.

40. $\dfrac{4}{x - x}$ Not defined: $x - x = 0$ for any x.

Do Exercises 43–46.

DIVISION AND RECIPROCALS

Two numbers whose product is 1 are called **reciprocals** (or **multiplicative inverses**) of each other.

> **PROPERTIES OF RECIPROCALS**
>
> Every nonzero real number a has a **reciprocal** (or **multiplicative inverse**) $1/a$. The reciprocal of a positive number is positive. The reciprocal of a negative number is negative.

EXAMPLES Find the reciprocal of the number.

41. $\dfrac{4}{5}$ The reciprocal is $\dfrac{5}{4}$, because $\dfrac{4}{5} \cdot \dfrac{5}{4} = 1$.

42. 8   The reciprocal is $\dfrac{1}{8}$, because $8 \cdot \dfrac{1}{8} = 1$.

43. $-\dfrac{2}{3}$ The reciprocal is $-\dfrac{3}{2}$, because $-\dfrac{2}{3} \cdot \left(-\dfrac{3}{2}\right) = 1$.

44. 0.25   The reciprocal is $\dfrac{1}{0.25}$ or 4, because $0.25 \cdot 4 = 1$.

Remember that a number and its reciprocal (multiplicative inverse) have the same sign. Do *not* change the sign when taking the reciprocal of a number. On the other hand, when finding an opposite (additive inverse), change the sign.

Do Exercises 47–52. (Exercises 47–51 are on the preceding page.)

We know that we can subtract by adding an opposite, or additive inverse. Similarly, we can divide by multiplying by a reciprocal.

RECIPROCALS AND DIVISION

For any real numbers a and b, $b \neq 0$,

$$a \div b = \frac{a}{b} = a \cdot \frac{1}{b}.$$

(To divide, we can multiply by the reciprocal of the divisor.)

We sometimes say that we "invert the divisor and multiply."

EXAMPLES Divide by multiplying by the reciprocal of the divisor.

45. $\dfrac{1}{4} \div \dfrac{3}{5} = \dfrac{1}{4} \cdot \dfrac{5}{3} = \dfrac{5}{12}$ "Inverting" the divisor, $\dfrac{3}{5}$, and multiplying

46. $\dfrac{2}{3} \div \left(-\dfrac{4}{9}\right) = \dfrac{2}{3} \cdot \left(-\dfrac{9}{4}\right) = -\dfrac{18}{12}$, or $-\dfrac{3}{2}$

47. $-\dfrac{5}{7} \div 3 = -\dfrac{5}{7} \cdot \dfrac{1}{3} = -\dfrac{5}{21}$

Do Exercises 53–56.

The following properties can be used to make sign changes.

SIGN CHANGES IN FRACTION NOTATION

For any numbers a and b, $b \neq 0$,

$$\frac{-a}{b} = \frac{a}{-b} = -\frac{a}{b} \quad \text{and} \quad \frac{-a}{-b} = \frac{a}{b}.$$

We can illustrate this property with $a = 4$ and $b = 9$:

$$\frac{-4}{9} = \frac{4}{-9} = -\frac{4}{9} \quad \text{and} \quad \frac{-4}{-9} = \frac{4}{9}.$$

52. Complete the following table.

NUMBER	OPPOSITE (Additive Inverse)	RECIPROCAL (Multiplicative Inverse)
$\dfrac{2}{3}$	$-\dfrac{2}{3}$	$\dfrac{3}{2}$
$\dfrac{4}{5}$		
$-\dfrac{3}{4}$		
0.25		
8		
-5		
0		

Divide by multiplying by the reciprocal of the divisor.

53. $-\dfrac{3}{4} \div \dfrac{7}{8}$

54. $-\dfrac{12}{5} \div \left(-\dfrac{7}{15}\right)$

55. $-\dfrac{3}{8} \div (-5)$

56. $\dfrac{4}{5} \div \left(-\dfrac{1}{10}\right)$

Answers on page A-1

Operations with Real Numbers   We can perform operations on the real numbers on a graphing calculator. Note that negative numbers are entered using the opposite key ⊙. This is different from the ⊝ key, which is used for the operation of subtraction. Consider the sum $-7.4 + (-3.9)$ shown in the window below. On paper, we use parentheses when we write this sum in order to separate the addition symbol and the "opposite of" symbol and thus make the expression more easily read. When we enter this calculation on a graphing calculator, however, the parentheses are not necessary. We can press ⊙ ⑦ · ④ ⊕ ⊙ ③ · ⑨ **ENTER**. The result is -11.3. Note that it is not incorrect to enter the parentheses. The result will be the same if this is done.

```
⁻7.4+⁻3.9
                    ⁻11.3
⁻7.4+(⁻3.9)
                    ⁻11.3
```

To find the difference $12 - (-6)$, we press ① ② ⊝ ⊙ ⑥ **ENTER**. The result is 18. We can also multiply and divide real numbers. To find $-4 \cdot (-8)$, we press ⊙ ④ ✕ ⊙ ⑧ **ENTER**, and to find $54 \div (-9)$, we press ⑤ ④ ÷ ⊙ ⑨ **ENTER** . Note in the window below that it is not necessary to use parentheses in any of these calculations, although expressions may be read more easily in the display when parentheses are included.

```
12-⁻6
                    18
⁻4∗⁻8
                    32
54/⁻9
                    ⁻6
```

Exercises: Use a calculator to perform the operation.

1. $-9 + 3$
2. $0.8 + (-1.6)$
3. $-8 + (-4)$
4. $-5.8 + (-4.6)$
5. $-9 - 3$
6. $0.8 - (-1.6)$

7. $-8 - (-4)$
8. $-5.8 - (-4.6)$
9. $-9 \cdot 6$
10. $0.8 \cdot (-1.6)$
11. $-8 \cdot (-4)$
12. $-5.8 \cdot (-4.6)$

13. $-9 \div 3$
14. $0.8 \div (-1.6)$
15. $-8 \div (-4)$
16. $-5.8 \div (-4.6)$

a   Add.

1. $-10 + (-18)$

2. $-13 + (-12)$

3. $7 + (-2)$

4. $7 + (-5)$

5. $-8 + (-8)$

6. $-6 + (-6)$

7. $7 + (-11)$

8. $9 + (-12)$

9. $-16 + 6$

10. $-21 + 11$

11. $-26 + 0$

12. $0 + (-32)$

13. $-8.4 + 9.6$

14. $-6.3 + 8.2$

15. $-2.62 + (-6.24)$

16. $-2.73 + (-8.46)$

17. $-\dfrac{5}{9} + \dfrac{2}{9}$

18. $-\dfrac{3}{7} + \dfrac{1}{7}$

19. $-\dfrac{11}{12} + \left(-\dfrac{5}{12}\right)$

20. $-\dfrac{3}{8} + \left(-\dfrac{7}{8}\right)$

21. $\dfrac{2}{5} + \left(-\dfrac{3}{10}\right)$

22. $-\dfrac{3}{4} + \dfrac{1}{8}$

23. $-\dfrac{2}{5} + \dfrac{3}{4}$

24. $-\dfrac{5}{6} + \left(-\dfrac{7}{8}\right)$

b   Evaluate $-a$ for each of the following.

25. $a = -4$

26. $a = -9$

27. $a = 3.7$

28. $a = 0$

Find the opposite (additive inverse).

29. 10

30. $-\dfrac{2}{3}$

31. 0

32. $-2x$

c   Subtract.

33. $3 - 7$

34. $8 - 13$

35. $-5 - 9$

36. $-6 - 14$

37. $23 - 23$

38. $23 - (-23)$

39. $-23 - 23$

40. $-23 - (-23)$

41. $-6 - (-11)$

42. $-7 - (-12)$

43. $10 - (-5)$

44. $28 - (-16)$

45. $15.8 - 27.4$

46. $17.2 - 34.9$

47. $-18.01 - 11.24$

48. $-19.04 - 15.76$

49. $-\dfrac{21}{4} - \left(-\dfrac{7}{4}\right)$

50. $-\dfrac{16}{5} - \left(-\dfrac{3}{5}\right)$

51. $-\dfrac{1}{3} - \left(-\dfrac{1}{12}\right)$

52. $-\dfrac{7}{8} - \left(-\dfrac{5}{2}\right)$

53. $-\dfrac{3}{4} - \dfrac{5}{6}$

54. $-\dfrac{2}{3} - \dfrac{4}{5}$

55. $\dfrac{1}{3} - \dfrac{4}{5}$

56. $-\dfrac{4}{7} - \left(-\dfrac{5}{9}\right)$

d Multiply.

57. $3(-7)$

58. $5(-8)$

59. $-2 \cdot 4$

60. $-5 \cdot 9$

61. $-8(-3)$

62. $-5(-7)$

63. $-7 \cdot 16$

64. $-8 \cdot 19$

65. $-6(-5.7)$

66. $-7(-6.1)$

67. $-\dfrac{3}{5} \cdot \dfrac{4}{7}$

68. $-\dfrac{5}{4} \cdot \dfrac{11}{3}$

69. $-3\left(-\dfrac{2}{3}\right)$

70. $-5\left(-\dfrac{3}{5}\right)$

71. $-3(-4)(5)$

72. $-6(-8)(9)$

73. $(-4.2)(-6.3)$

74. $(-7.4)(-9.6)$

75. $-\dfrac{9}{11} \cdot \left(-\dfrac{11}{9}\right)$

76. $-\dfrac{13}{7} \cdot \left(-\dfrac{5}{2}\right)$

77. $-\dfrac{2}{3} \cdot \left(-\dfrac{2}{3}\right) \cdot \left(-\dfrac{2}{3}\right)$

78. $-\dfrac{4}{5} \cdot \left(-\dfrac{4}{5}\right) \cdot \left(-\dfrac{4}{5}\right)$

CHAPTER R: Review of Basic Algebra

e   Divide, if possible.

79. $\dfrac{-8}{4}$

80. $\dfrac{-16}{2}$

81. $\dfrac{56}{-8}$

82. $\dfrac{63}{-7}$

83. $-77 \div (-11)$

84. $-48 \div (-6)$

85. $\dfrac{-5.4}{-18}$

86. $\dfrac{-8.4}{-12}$

87. $\dfrac{5}{0}$

88. $\dfrac{92}{0}$

89. $\dfrac{0}{32}$

90. $\dfrac{0}{17}$

91. $\dfrac{9}{y-y}$

92. $\dfrac{2x-2x}{2x-2x}$

Find the reciprocal of the number.

93. $\dfrac{3}{4}$

94. $\dfrac{9}{10}$

95. $-\dfrac{7}{8}$

96. $-\dfrac{5}{6}$

97. 25

98. -65

99. 0.2

100. 0.8

101. $-\dfrac{a}{b}$

102. $\dfrac{1}{8x}$

Divide.

103. $\dfrac{2}{7} \div \left(-\dfrac{11}{3}\right)$

104. $\dfrac{3}{5} \div \left(-\dfrac{6}{7}\right)$

105. $-\dfrac{10}{3} \div \left(-\dfrac{2}{15}\right)$

106. $-\dfrac{12}{5} \div \left(-\dfrac{3}{10}\right)$

107. $18.6 \div (-3.1)$

108. $39.9 \div (-13.3)$

109. $(-75.5) \div (-15.1)$

110. $(-12.1) \div (-0.11)$

111. $-48 \div 0.4$

112. $520 \div (-0.13)$

113. $\dfrac{3}{4} \div \left(-\dfrac{2}{3}\right)$

114. $\dfrac{5}{8} \div \left(-\dfrac{1}{2}\right)$

115. $-\dfrac{5}{4} \div \left(-\dfrac{3}{4}\right)$

116. $-\dfrac{5}{9} \div \left(-\dfrac{5}{6}\right)$

117. $-\dfrac{2}{3} \div \left(-\dfrac{4}{9}\right)$

118. $-\dfrac{3}{5} \div \left(-\dfrac{5}{8}\right)$

119. $-\dfrac{3}{8} \div \left(-\dfrac{8}{3}\right)$

120. $-\dfrac{5}{8} \div \left(-\dfrac{5}{6}\right)$

121. $-6.6 \div 3.3$

122. $-44.1 \div (-6.3)$

123. $\dfrac{-12}{-13}$

124. $\dfrac{-1.9}{20}$

125. $\dfrac{48.6}{-30}$

126. $\dfrac{-17.8}{3.2}$

127. $\dfrac{-9}{17-17}$

128. $\dfrac{-8}{-6+6}$

129. Complete the following table.

NUMBER	OPPOSITE (Additive Inverse)	RECIPROCAL (Multiplicative Inverse)
$\dfrac{2}{3}$		
$-\dfrac{5}{4}$		
0		Does not exist
1		
-4.5		
$x, x \neq 0$		

130. Complete the following table.

NUMBER	OPPOSITE (Additive Inverse)	RECIPROCAL (Multiplicative Inverse)
$-\dfrac{3}{8}$		
$\dfrac{7}{10}$		
-1		
0		Does not exist
-6.4		
$a, a \neq 0$		

131. $\mathbf{D_W}$ Explain in your own words why a positive number divided by a negative number is negative.

132. $\mathbf{D_W}$ Explain in your own words why $\dfrac{7}{0}$ is not defined.

SKILL MAINTENANCE

This heading indicates that the exercises that follow are *Skill Maintenance exercises,* which review any skill previously studied in the text. You can expect such exercises in every exercise set. Answers to *all* skill maintenance exercises are found at the back of the book. If you miss an exercise, restudy the objective shown in red.

Given the numbers $\sqrt{3}$, -12.47, -13, 26, π, 0, $-\dfrac{23}{32}$, $\dfrac{7}{11}$, $4.57557555755557\ldots$: [R.1a]

133. Name the whole numbers.

134. Name the natural numbers.

135. Name the integers.

136. Name the irrational numbers.

137. Name the rational numbers.

138. Name the real numbers.

Use either $<$ or $>$ for $\square$ to write a true sentence. [R.1b]

139. $-7 \; \square \; 8$

140. $5 \; \square \; \dfrac{3}{8}$

141. $-45.6 \; \square \; -23.8$

142. $123 \; \square \; -10$

SYNTHESIS

143. The reciprocal of an electric resistance is called *conductance.* When two resistors are connected in parallel, the conductance is the sum of the conductances,

$$\dfrac{1}{r_1} + \dfrac{1}{r_2}.$$

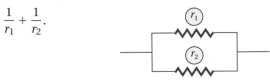

Find the conductance of two resistors of 12 ohms and 6 ohms when connected in parallel.

144. What number can be added to 11.7 to obtain $-7\dfrac{3}{4}$?

145. What number can be multiplied by -0.02 to obtain -625?

R.3

EXPONENTIAL NOTATION AND ORDER OF OPERATIONS

 a Rewrite expressions with whole-number exponents, and evaluate exponential expressions.

b Rewrite expressions with or without negative integers as exponents.

c Simplify expressions using the rules for order of operations.

a Exponential Notation

Exponential notation is a shorthand device. For $3 \cdot 3 \cdot 3 \cdot 3$, we write 3^4. In the **exponential notation** 3^4, the number 3 is called the **base** and the number 4 is called the **exponent.**

EXPONENTIAL NOTATION

Exponential notation a^n, where n is an integer greater than 1, means

$$\underbrace{a \cdot a \cdot a \cdots a \cdot a.}_{n \text{ factors}}$$

We read "a^n" as "a to the nth power," or simply "a to the nth."
We can read "a^2" as "a-squared" and "a^3" as "a-cubed."

Caution!

a^n does *not* mean to multiply n times a. For example, 3^2 means $3 \cdot 3$, or 9, not $3 \cdot 2$, or 6.

EXAMPLES Write exponential notation.

1. $7 \cdot 7 \cdot 7 = 7^3$

2. $xxxxx = x^5$

3. $\dfrac{2}{3} \cdot \dfrac{2}{3} \cdot \dfrac{2}{3} \cdot \dfrac{2}{3} = \left(\dfrac{2}{3}\right)^4$

Do Exercises 1–3.

EXAMPLES Evaluate.

4. $9^2 = 9 \cdot 9 = 81$

5. $\left(\dfrac{1}{2}\right)^3 = \dfrac{1}{2} \cdot \dfrac{1}{2} \cdot \dfrac{1}{2} = \dfrac{1}{8}$

6. $\left(\dfrac{7}{8}\right)^2 = \dfrac{7}{8} \cdot \dfrac{7}{8} = \dfrac{49}{64}$

7. $(0.1)^4 = (0.1)(0.1)(0.1)(0.1) = 0.0001$

8. $(-5)^3 = (-5)(-5)(-5) = -125$

9. $-(5^3) = -(5 \cdot 5 \cdot 5) = -125$

10. $-(10)^4 = -(10 \cdot 10 \cdot 10 \cdot 10) = -10,000$

11. $(-10)^4 = (-10)(-10)(-10)(-10) = 10,000$

Note that $-(10)^4 \neq (-10)^4$, as shown in Examples 10 and 11. In $-(10)^4$, the sign is *outside* the parentheses; in $(-10)^4$, the sign is *inside* the parentheses.

Do Exercises 4–10.

Write exponential notation.

1. $8 \cdot 8 \cdot 8 \cdot 8$

2. *mmmmmm*

3. $\dfrac{7}{8} \cdot \dfrac{7}{8} \cdot \dfrac{7}{8}$

Evaluate.

4. 3^4

5. $\left(\dfrac{1}{4}\right)^2$

6. $(-10)^6$

7. $(0.2)^3$

8. $(5.8)^4$

9. -4^4

10. $(-3)^4$

Answers on page A-2

Exponential Notation   We use the ⌃ key to evaluate exponential notation on a graphing calculator. To find $(-3)^4$, for example, we press ((−) 3) ⌃ 4 ENTER. The result is 81. Note that parentheses are necessary when we raise a negative number to an even power. If the parentheses were not used in entering $(-3)^4$, we would find -3^4, or the opposite of 3^4, which is -81.

To find $\left(\dfrac{5}{8}\right)^{-3}$ and express the result in fraction notation, we press (5 ÷ 8) ⌃ (−) 3 MATH 1 ENTER. Note that the parentheses are also necessary in this calculation. If they were not used, we would be calculating $5 \div 8^{-3}$, or $5 \div \dfrac{1}{8^3}$, which is 2560.

```
(−3)^4
                    81
(5/8)^−3▶Frac
              512/125
```

```
5.6²
                  31.36
5.6^2
                  31.36
```

The calculator has a special x^2 key that can be used to raise a number to the second power. To find 5.6^2, shown on the right above, for example, we press 5 . 6 x^2 ENTER. We could also use the ⌃ key to do this calculation, pressing 5 . 6 ⌃ 2 ENTER.

Exercises: Evaluate.

1. 5^4

2. 9^7

3. $(-24)^2$

4. 6.413^2

5. $(-1.6)^5$

6. 4.23^3

7. $\left(-\dfrac{3}{4}\right)^6$

8. $\left(-\dfrac{5}{3}\right)^5$

When an exponent is an integer greater than 1, it tells how many times the base occurs as a factor. What happens when the exponent is 1 or 0? We cannot have the base occurring as a factor 1 time or 0 times because there are no products. Look for a pattern below. Think of dividing by 10 on the right.

On this side, the exponents decrease by 1 at each step.

$$10^4 = 10 \cdot 10 \cdot 10 \cdot 10 = 10{,}000$$
$$10^3 = 10 \cdot 10 \cdot 10 = 1000$$
$$10^2 = 10 \cdot 10 = 100$$
$$10^1 = ?$$
$$10^0 = ?$$

On this side, we divide by 10 at each step.

In order for the pattern to continue, 10^1 would have to be 10 and 10^0 would have to be 1. We will *agree* that exponents of 1 and 0 have that meaning.

EXPONENTS OF 0 AND 1

For any number a, we agree that a^1 means a.

For any nonzero number a, we agree that a^0 means 1.

Rewrite without exponents.

11. 8^1

EXAMPLES Rewrite without an exponent.

12. $4^1 = 4$

13. $(-97)^1 = -97$

14. $6^0 = 1$

15. $(-37.4)^0 = 1$

12. $(-31)^1$

Let's consider a justification for not defining 0^0. By examining the pattern $3^0 = 1$, $2^0 = 1$, and $1^0 = 1$, we might think that 0^0 should be 1. However, by examining the pattern $0^3 = 0$, $0^2 = 0$, and $0^1 = 0$, we might think that 0^0 should be 0. To avoid this confusion, mathematicians agree *not* to define 0^0.

Do Exercises 11–15.

b Negative Integers as Exponents

How shall we define negative integers as exponents? Look for a pattern below. Again, think of dividing by 10 on the right.

13. 3^0

On this side, the exponents decrease by 1 at each step. $\Big\downarrow$ | $\begin{aligned} 10^2 &= 100 \\ 10^1 &= 10 \\ 10^0 &= 1 \\ 10^{-1} &= ? \\ 10^{-2} &= ? \end{aligned}$ | On this side, we divide by 10 at each step. $\Big\downarrow$

In order for the pattern to continue, 10^{-1} would have to be $\frac{1}{10}$ and 10^{-2} would have to be $\frac{1}{100}$. This leads to the following agreement.

14. $(-7)^0$

NEGATIVE EXPONENTS

For any real number a that is nonzero and any integer n,

$$a^{-n} = \frac{1}{a^n}.$$

EXAMPLES Rewrite using a positive exponent. Evaluate, if possible.

15. y^0, where $y \neq 0$

16. $y^{-5} = \dfrac{1}{y^5}$

17. $\dfrac{1}{t^{-4}} = t^4$

18. $(-2)^{-3} = \dfrac{1}{(-2)^3} = \dfrac{1}{(-2)(-2)(-2)} = \dfrac{1}{-8} = -\dfrac{1}{8}$

19. $\left(\dfrac{1}{2}\right)^{-3} = \dfrac{1}{\left(\frac{1}{2}\right)^3} = \dfrac{1}{\frac{1}{8}} = 1 \cdot \dfrac{8}{1} = 8$

20. $\left(\dfrac{2}{5}\right)^{-2} = \dfrac{1}{\left(\frac{2}{5}\right)^2} = \dfrac{1}{\frac{4}{25}} = 1 \cdot \dfrac{25}{4} = \dfrac{25}{4}$

Answers on page A-2

Rewrite using a positive exponent.

16. m^{-4}

17. $(-4)^{-3}$

18. $\dfrac{1}{x^{-3}}$

19. $\left(\dfrac{1}{5}\right)^{-3}$

20. $\left(\dfrac{3}{4}\right)^{-2}$

Rewrite using a negative exponent.

21. $\dfrac{1}{a^3}$

22. $\dfrac{1}{(-5)^4}$

The numbers a^n and a^{-n} are reciprocals because

$$a^n \cdot a^{-n} = a^n \cdot \frac{1}{a^n} = \frac{a^n}{a^n} = 1.$$

For example, y^3 and y^{-3} are reciprocals:

$$y^3 \cdot y^{-3} = y^3 \cdot \frac{1}{y^3} = \frac{y^3}{y^3} = 1.$$

Caution!

A negative exponent does *not* necessarily indicate that an answer is negative! For example, 3^{-2} means $1/3^2$, or $1/9$, not -9.

Do Exercises 16–20.

EXAMPLES Rewrite using a negative exponent.

21. $\dfrac{1}{x^2} = x^{-2}$

22. $\dfrac{1}{(-7)^4} = (-7)^{-4}$

Do Exercises 21 and 22.

C   Order of Operations

What does $8 + 2 \cdot 5^3$ mean? If we add 8 and 2 and multiply by 5^3, or 125, we get 1250. If we multiply 2 times 125 and add 8, we get 258. Both results cannot be correct. To avoid such difficulties, we make agreements about which operations should be done first.

RULES FOR ORDER OF OPERATIONS

1. Do all the calculations within grouping symbols, like parentheses, before operations outside.
2. Evaluate all exponential expressions.
3. Do all multiplications and divisions in order from left to right.
4. Do all additions and subtractions in order from left to right.

Most computers and calculators are programmed using these rules.

EXAMPLE 23 Simplify: $-43 \cdot 56 - 17$.

There are no parentheses or powers so we start with the third rule.

$$-43 \cdot 56 - 17 = -2408 - 17 \qquad \text{Carrying out all multiplications and divisions in order from left to right}$$

$$= -2425 \qquad \text{Carrying out all additions and subtractions in order from left to right}$$

Answers on page A-2

EXAMPLE 24 Simplify: $8 + 2 \cdot 5^3$.

$$8 + 2 \cdot 5^3 = 8 + 2 \cdot 125 \qquad \text{Evaluating the exponential expression}$$
$$= 8 + 250 \qquad \text{Doing the multiplication}$$
$$= 258 \qquad \text{Adding}$$

EXAMPLE 25 Simplify and compare: $(8 - 10)^2$ and $8^2 - 10^2$.

$$(8 - 10)^2 = (-2)^2 = 4;$$
$$8^2 - 10^2 = 64 - 100 = -36$$

We see that $(8 - 10)^2$ and $8^2 - 10^2$ are *not* the same.

EXAMPLE 26 Simplify: $3^4 + 62 \cdot 8 - 2(29 + 33 \cdot 4)$.

$$3^4 + 62 \cdot 8 - 2(29 + 33 \cdot 4)$$
$$= 3^4 + 62 \cdot 8 - 2(29 + 132) \qquad \text{Carrying out operations inside parentheses first; doing the multiplication}$$
$$= 3^4 + 62 \cdot 8 - 2(161) \qquad \text{Completing the addition inside parentheses}$$
$$= 81 + 62 \cdot 8 - 2(161) \qquad \text{Evaluating exponential expressions}$$
$$= 81 + 496 - 322 \qquad \text{Doing all multiplications}$$
$$= 577 - 322 \qquad \text{Doing all additions and subtractions in order from left to right}$$
$$= 255$$

Do Exercises 23–26.

When parentheses occur within parentheses, we can make them different shapes, such as [ ] (also called "brackets") and { } (usually called "braces"). All of these have the same meaning. When parentheses occur within parentheses, **computations in the *innermost* ones are to be done first.**

EXAMPLE 27 Simplify: $5 - \{6 - [3 - (7 + 3)]\}$.

$$5 - \{6 - [3 - (7 + 3)]\} = 5 - \{6 - [3 - 10]\} \qquad \text{Adding } 7 + 3$$
$$= 5 - \{6 - [-7]\} \qquad \text{Subtracting } 3 - 10$$
$$= 5 - 13 \qquad \text{Subtracting } 6 - (-7)$$
$$= -8$$

EXAMPLE 28 Simplify: $7 - [3(2 - 5) - 4(2 + 3)]$.

$$7 - [3(2 - 5) - 4(2 + 3)] = 7 - [3(-3) - 4(5)] \qquad \text{Doing the calculations in the innermost grouping symbols first}$$
$$= 7 - [-9 - 20]$$
$$= 7 - [-29]$$
$$= 36$$

Do Exercises 27 and 28.

Simplify.

23. $43 - 52 \cdot 80$

24. $3^5 \div 3^4 \cdot 3^2$

25. $62 \cdot 8 + 4^3 - (5^2 - 64 \div 4)$

26. Simplify and compare:

$(7 - 4)^2$ and $7^2 - 4^2$.

Simplify.

27. $6 - \{5 - [2 - (8 + 20)]\}$

28. $5 + \{6 - [2 + (5 - 2)]\}$

Answers on page A-2

Simplify.

29. $\dfrac{8 \cdot 7 - |6 - 8|}{5^2 + 6^3}$

In addition to the usual grouping symbols—parentheses, brackets, and braces—a fraction bar and absolute-value signs can act as grouping symbols.

EXAMPLE 29 Calculate: $\dfrac{12|7 - 9| + 8 \cdot 5}{3^2 + 2^3}$.

An equivalent expression with brackets as grouping symbols is

$$[12|7 - 9| + 8 \cdot 5] \div [3^2 + 2^3].$$

What this shows, in effect, is that we do the calculations in the numerator and in the denominator separately, and then divide the results:

30. $\dfrac{(8 - 3)^2 + (7 - 10)^2}{3^2 - 2^3}$

$$\dfrac{12|7 - 9| + 8 \cdot 5}{3^2 + 2^3} = \dfrac{12|-2| + 8 \cdot 5}{9 + 8}$$

$$= \dfrac{12(2) + 8 \cdot 5}{17}$$

Subtracting inside the absolute-value signs before taking the absolute value

$$= \dfrac{24 + 40}{17} = \dfrac{64}{17}.$$

Answers on page A-2

Do Exercises 29 and 30.

CALCULATOR CORNER

Order of Operations Computations are usually entered on a graphing calculator in the same way in which we would write them. To calculate $5 + 3 \cdot 4$, for example, we press ⑤ ➕ ③ ✖ ④ **ENTER**. The result is 17.

When an expression contains grouping symbols, we enter them using the ⦅ and ⦆ keys. To calculate $7(11 - 2) - 24$, we press ⑦ ⦅ ① ① ➖ ② ⦆ ➖ ② ④ **ENTER**. The result is 39.

Since a fraction bar acts as a grouping symbol, we must supply parentheses when entering some fraction expressions. To calculate $\dfrac{45 + 135}{2 - 17}$, for example, we think of rewriting it with grouping symbols as $(45 + 135) \div (2 - 17)$. We press ⦅ ④ ⑤ ➕ ① ③ ⑤ ⦆ ➗ ⦅ ② ➖ ① ⑦ ⦆ **ENTER**. The result is -12.

```
5+3*4
                    17
7(11-2)-24
                    39
(45+135)/(2-17)
                   -12
```

Exercises: Calculate.

1. $48 \div 2 \cdot 3 - 4 \cdot 4$

2. $48 - 8 \div 4 + 3 \cdot 5$

3. $48 \div (2 \cdot 3 - 4) \cdot 4$

4. $(3 + 6)^3 - 4(11 - 7)^2$

5. $41.4 - 7.6 \times 3.01 + 4(3^2 - 17.6)$

6. $5.6 - 4.7[87.2 - 2.1(60.3 - 59.4)]$

7. $\{(25 \cdot 30) \div [(2 \cdot 16) \div (4 \cdot 2)]\} + 15(45 \div 9)$

8. $\left(\dfrac{28}{89} - 42.8 \times 17.01\right)^3 \div \left(-\dfrac{678}{119} + \dfrac{23.2}{46.08}\right)^2$

9. $\dfrac{196 - 56}{7 + 28}$

10. $\dfrac{17^2 - 311}{16 - 7}$

11. $785 - \dfrac{285 - 5^4}{17 - 3 \cdot 51}$

12. $8^5 - 8^4 + 12^4 \div 12^2 - 28.7^3$

13. What result do you get if you ignore the parentheses when evaluating $(36 + 144) \div (3 - 8)$? How does the calculator do the calculation? What is the correct result using parentheses?

R.3 **EXERCISE SET** For Extra Help

MathXL | MyMathLab | InterAct Math | Math Tutor Center | Digital Video Tutor CD 1 Videotape 1 | Student's Solutions Manual

a   Write exponential notation.

1. $4 \cdot 4 \cdot 4 \cdot 4 \cdot 4$

2. $6 \cdot 6 \cdot 6$

3. $5 \cdot 5 \cdot 5 \cdot 5 \cdot 5 \cdot 5$

4. $x \cdot x \cdot x \cdot x$

5. mmm

6. $ttttt$

7. $\dfrac{7}{12} \cdot \dfrac{7}{12} \cdot \dfrac{7}{12} \cdot \dfrac{7}{12}$

8. $(3.8)(3.8)(3.8)(3.8)(3.8)$

9. $(123.7)(123.7)$

10. $\left(-\dfrac{4}{5}\right)\left(-\dfrac{4}{5}\right)\left(-\dfrac{4}{5}\right)$

Evaluate.

11. 2^7

12. 9^3

13. $(-2)^5$

14. $(-7)^2$

15. $\left(\dfrac{1}{3}\right)^4$

16. $(0.1)^6$

17. $(-4)^3$

18. $(-3)^4$

19. $(-5.6)^2$

20. $\left(\dfrac{2}{3}\right)^4$

21. 5^1

22. $\left(\sqrt{6}\right)^1$

23. 34^0

24. $\left(\dfrac{5}{2}\right)^1$

25. $\left(\sqrt{6}\right)^0$

26. $(-4)^0$

27. $\left(\dfrac{7}{8}\right)^1$

28. $(-87)^0$

b   Rewrite using a positive exponent. Evaluate, if possible.

29. $\left(\dfrac{1}{4}\right)^{-2}$

30. $\left(\dfrac{1}{5}\right)^{-3}$

31. $\left(\dfrac{2}{3}\right)^{-3}$

32. $\left(\dfrac{5}{2}\right)^{-4}$

33. y^{-5}

34. x^{-6}

35. $\dfrac{1}{a^{-2}}$

36. $\dfrac{1}{y^{-7}}$

37. $(-11)^{-1}$

38. $(-4)^{-3}$

Rewrite using a negative exponent.

39. $\dfrac{1}{3^4}$

40. $\dfrac{1}{9^2}$

41. $\dfrac{1}{b^3}$

42. $\dfrac{1}{n^5}$

43. $\dfrac{1}{(-16)^2}$

44. $\dfrac{1}{(-8)^6}$

C   Simplify.

45. $12 - 4(5 - 1)$

46. $6 - 4(8 - 5)$

47. $9[8 - 7(5 - 2)]$

48. $10[7 - 4(8 - 5)]$

49. $[5(8 - 6) + 12] - [24 - (8 - 4)]$

50. $[9(7 - 4) + 19] - [25 - (7 + 3)]$

51. $[64 \div (-4)] \div (-2)$

52. $[48 \div (-3)] \div \left(-\dfrac{1}{4}\right)$

53. $19(-22) + 60$

54. $30 \cdot 10 - 18 \cdot 25$

55. $(5 + 7)^2; \quad 5^2 + 7^2$

56. $(9 - 12)^2; \quad 9^2 - 12^2$

57. $2^3 + 2^4 - 20 \cdot 30$

58. $7 \cdot 8 - 3^2 - 2^3$

59. $5^3 + 36 \cdot 72 - (18 + 25 \cdot 4)$

60. $4^3 + 20 \cdot 10 + 7^2 - 23$

61. $(13 \cdot 2 - 8 \cdot 4)^2$

62. $(9 \cdot 8 + 3 \cdot 3)^2$

63. $4000 \cdot (1 + 0.12)^3$

64. $5000 \cdot (4 + 1.16)^2$

65. $(20 \cdot 4 + 13 \cdot 8)^2 - (39 \cdot 15)^3$

66. $(43 \cdot 6 - 14 \cdot 7)^3 + (33 \cdot 34)^2$

67. $18 - 2 \cdot 3 - 9$

68. $18 - (2 \cdot 3 - 9)$

69. $(18 - 2 \cdot 3) - 9$

70. $(18 - 2)(3 - 9)$

71. $[24 \div (-3)] \div \left(-\dfrac{1}{2}\right)$

72. $[(-32) \div (-2)] \div (-2)$

73. $15 \cdot (-24) + 50$

74. $30 \cdot 20 - 15 \cdot 24$

75. $4 \div (8 - 10)^2 + 1$

76. $16 \div (19 - 15)^2 - 7$

77. $6^3 + 25 \cdot 71 - (16 + 25 \cdot 4)$

78. $5^3 + 20 \cdot 40 + 8^2 - 29$

79. $5000 \cdot (1 + 0.16)^3$

80. $4000 \cdot (3 + 1.14)^2$

81. $4 \cdot 5 - 2 \cdot 6 + 4$

82. $8(7 - 3)/4$

83. $4 \cdot (6 + 8)/(4 + 3)$

84. $4^3/8$

85. $[2 \cdot (5 - 3)]^2$

86. $5^3 - 7^2$

87. $8(-7) + 6(-5)$

88. $10(-5) + 1(-1)$

89. $19 - 5(-3) + 3$

90. $14 - 2(-6) + 7$

91. $9 \div (-3) + 16 \div 8$

92. $-32 - 8 \div 4 - (-2)$

93. $7 + 10 - (-10 \div 2)$

94. $(3 - 8)^2$

95. $5^2 - 8^2$

96. $28 - 10^3$

97. $20 + 4^3 \div (-8)$

98. $2 \times 10^3 - 5000$

99. $-7(3^4) + 18$ **100.** $6[9 - (3 - 4)]$ **101.** $9[(8 - 11) - 13]$

102. $1000 \div (-100) \div 10$ **103.** $256 \div (-32) \div (-4)$ **104.** $\dfrac{20 - 6^2}{9^2 + 3^2}$

105. $\dfrac{5^2 - |4^3 - 8|}{9^2 - 2^2 - 1^5}$ **106.** $\dfrac{4|6 - 7| - 5 \cdot 4}{6 \cdot 7 - 8|4 - 1|}$

107. $\dfrac{30(8 - 3) - 4(10 - 3)}{10|2 - 6| - 2(5 + 2)}$ **108.** $\dfrac{5^3 - 3^2 + 12 \cdot 5}{-32 \div (-16) \div (-4)}$

109. $\mathbf{D_W}$ Explain the meaning of a negative exponent in as many ways as you can.

110. $\mathbf{D_W}$ Students sometimes use the memory device PEMDAS, or "Please Excuse My Dear Aunt Sally," to remember the rules for order of operations. Explain how this works.

SKILL MAINTENANCE

Find the absolute value.   [R.1d]

111. $\left| -\dfrac{9}{7} \right|$ **112.** $|2.3|$ **113.** $|0|$ **114.** $|-900|$

Compute.   [R.2a, c, d]

115. $23 - 56$ **116.** $-23 - 56$ **117.** $-23 - (-56)$ **118.** $-23 + (-56)$

119. $(-10)(2.3)$ **120.** $(-10)(-2.3)$ **121.** $10(-2.3)$ **122.** $\left(-\dfrac{2}{3} \right)\left(-\dfrac{15}{16} \right)$

SYNTHESIS

Simplify.

123. $(-2)^0 - (-2)^3 - (-2)^{-1} + (-2)^4 - (-2)^{-2}$ **124.** $2(6^1 \cdot 6^{-1} - 6^{-1} \cdot 6^0)$

125. Place parentheses in this statement to make it true: $9 \cdot 5 + 2 - 8 \cdot 3 + 1 = 22$.

The symbol 〰 means to use your calculator to work a particular exercise.

126. 〰 Find each of the following.

$12345679 \cdot 9 = ?$

$12345679 \cdot 18 = ?$

$12345679 \cdot 27 = ?$

Then look for a pattern and find $12345679 \cdot 36$ without the use of a calculator.

127. 〰 Find $(0.2)^{(-0.2)^{-1}}$.   **128.** 〰 Compare $(\pi)^{\sqrt{2}}$ and $(\sqrt{2})^{\pi}$.

129. Compare $(2 + 3)^{-1}$ and $2^{-1} + 3^{-1}$.

34

CHAPTER R: Review of Basic Algebra

PART 2 MANIPULATIONS
INTRODUCTION TO ALGEBRAIC EXPRESSIONS

a Translate a phrase to an algebraic expression.

b Evaluate an algebraic expression by substitution.

The study of algebra involves the use of equations to solve problems. Equations are constructed from algebraic expressions. The purpose of Part 2 of this chapter is to provide a review of the types of expressions encountered in algebra and ways in which we can manipulate them.

ALGEBRAIC EXPRESSIONS AND THEIR USE

In arithmetic, you worked with expressions such as

$$91 + 76, \quad 26 - 17, \quad 14 \cdot 35, \quad 7 \div 8, \quad \frac{7}{8}, \quad \text{and} \quad 5^2 - 3^2.$$

In algebra, we use these as well as expressions like

$$x + 76, \quad 26 - q, \quad 14 \cdot x, \quad d \div t, \quad \frac{d}{t}, \quad \text{and} \quad x^2 - y^2.$$

When a letter is used to stand for various numbers, it is called a **variable.** Let $t =$ the number of hours that a passenger jet has been flying. Then t is a variable, because t changes as the flight continues. If a letter represents one particular number, it is called a **constant.** Let $d =$ the number of hours in a day. Then d is a constant.

An **algebraic expression** consists of variables, numbers, and operation signs, such as $+, -, \cdot, \div$. All the expressions above are examples of algebraic expressions. When an equals sign, $=$, is placed between two expressions, an **equation** is formed.

We compare algebraic expressions with equations in the following table. Note that none of the expressions has equals signs ($=$).

ALGEBRAIC EXPRESSIONS	EQUATIONS
10	$t = 10$
$x - 5$	$x - 5 = 10$
$11x$	$x - 5 = 11x$
$\frac{1}{2}y^2 + 2y$	$\frac{1}{2}y^2 + 2y = 10 + y$

Do Exercise 1.

1. Which of the following are equations?

a) $3x + 7$

b) $-3x - 7 = 18$

c) $-3(x - 5) + 17$

d) $7 = t - 4$

Answers on page A-2

2. Refer to the graph. Translate to an equation and solve:

How many more cable-modem users were there in 2005 than in 2002?

Algebraic expressions and equations occur frequently in applications. For example, consider the following line graph that you might see in a newspaper or a magazine.

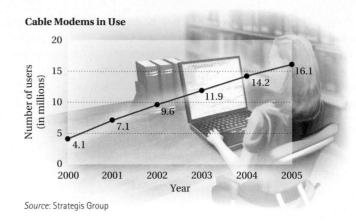

Cable Modems in Use

Source: Strategis Group

The use of high-speed cable modems to enhance the speed of Internet usage has increased the last few years, as the data in the graph indicate. Suppose we want to determine how many more cable-modem users there were in 2005 than in 2003. Using algebra, we can translate the problem to an equation. It might be done as follows.

Number of cable-modem users in 2003	plus	How many more users	is	Number of cable-modem users in 2005
↓	↓	↓	↓	↓
11.9	+	x	=	16.1

Note that we have an algebraic expression, $11.9 + x$, on the left. To find the number x, we can subtract 11.9 on both sides of the equation:

$$11.9 + x = 16.1$$
$$11.9 + x - 11.9 = 16.1 - 11.9 \qquad \text{Subtracting } 11.9$$
$$x = 4.2.$$

We now have the answer, 4.2 million.

Do Exercise 2.

 Translating to Algebraic Expressions

To translate problems to equations, we need to know that certain words correspond to certain symbols, as shown in the following table.

KEY WORDS

ADDITION	SUBTRACTION	MULTIPLICATION	DIVISION
add	subtract	multiply	divide
sum	difference	product	quotient
plus	minus	times	divided by
total	decreased by	twice	ratio
increased by	less than	of	per
more than			

Answer on page A-2

Expressions like rs represent products and can also be written as $r \cdot s$, $r \times s$, $(r)(s)$, or $r(s)$. The multipliers r and s are also called **factors**. A quotient $m \div 5$ can also be represented as $m/5$ or $\dfrac{m}{5}$.

3. Translate to an algebraic expression: Sixteen less than some number.

◼ **EXAMPLE 1** Translate to an algebraic expression: Eight less than some number.

We can use any variable we wish, such as x, y, t, m, n, and so on. Here we let t represent the number. If we knew the number to be 23, then the translation of "eight less than 23" would be $23 - 8$. If we knew the number to be 345, then the translation of "eight less than 345" would be $345 - 8$. Since we are using a variable for the number, the translation is

$\quad t - 8.$ *Caution!* $8 - t$ would be incorrect.

Do Exercise 3.

Translate to an algebraic expression.

4. Forty-seven more than some number

◼ **EXAMPLE 2** Translate to an algebraic expression: Twenty-two more than some number.

This time we let y represent the number. If we knew the number to be 47, then the translation would be $47 + 22$, or $22 + 47$. If we knew the number to be 17.95, then the translation would be $17.95 + 22$, or $22 + 17.95$. Since we are using a variable, the translation is

$\quad y + 22, \quad \text{or} \quad 22 + y.$

5. Sixteen minus some number

◼ **EXAMPLE 3** Translate to an algebraic expression: Five less than forty-three percent of the quotient of two numbers.

We let r and s represent the two numbers.

6. One-fourth of some number

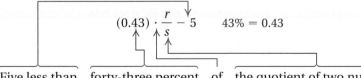

$$(0.43) \cdot \frac{r}{s} - 5 \qquad 43\% = 0.43$$

Five less than   forty-three percent   of   the quotient of two numbers

7. Six more than eight times some number

◼ **EXAMPLE 4** Translate each of the following to an algebraic expression.

PHRASE	ALGEBRAIC EXPRESSION
Five *more than* some number	$n + 5$, or $5 + n$
Half *of* a number	$\dfrac{1}{2}t$, or $\dfrac{t}{2}$
Five *more than* three *times* some number	$3p + 5$, or $5 + 3p$
The *difference* of two numbers	$x - y$
Six *less than* the *product* of two numbers	$rs - 6$
Seventy-six percent *of* some number	$0.76z$, or $\dfrac{76}{100}z$
Eight *less than twice* some number	$2x - 8$

8. Eight less than ninety-nine percent of the quotient of two numbers

Do Exercises 4–8.

Answers on page A-2

9. Evaluate $a + b$ when $a = 48$ and $b = 36$.

10. Evaluate $x - y$ when $x = -97$ and $y = 29$.

11. Evaluate a/b when $a = 400$ and $b = -8$.

12. Evaluate $8t$ when $t = 15$.

13. Evaluate $4x + 5y$ when $x = -2$ and $y = 10$.

14. Evaluate $7ab - c$ when $a = -3$, $b = 4$, and $c = 62$.

15. Find the area of a triangle when h is 24 ft and b is 8 ft.

Answers on page A-2

 Evaluating Algebraic Expressions

When we replace a variable with a number, we say that we are **substituting** for the variable. Carrying out the resulting calculation is called **evaluating the expression.**

EXAMPLE 5 Evaluate $x - y$ when $x = 83$ and $y = 49$.

We substitute 83 for x and 49 for y and carry out the subtraction:

$$x - y = 83 - 49 = 34.$$

The number 34 is called the **value** of the expression.

EXAMPLE 6 Evaluate a/b when $a = -63$ and $b = 7$.

We substitute -63 for a and 7 for b and carry out the division:

$$\frac{a}{b} = \frac{-63}{7} = -9.$$

Do Exercises 9–12.

EXAMPLE 7 Evaluate the expression $3xy + z$ when $x = 2$, $y = -5$, and $z = 7$.

We substitute and carry out the calculations according to the rules for order of operations:

$$3xy + z = 3(2)(-5) + 7 = -30 + 7 = -23.$$

Do Exercises 13 and 14.

Geometric formulas must often be evaluated in applied problems. In the next example, we use the formula for the area A of a triangle with a base of length b and a height of length h:

$$A = \tfrac{1}{2}bh.$$

EXAMPLE 8 *Area of a Triangular Sail.* The base of a triangular sail is 8 m and the height is 6.4 m. Find the area of the sail.

We substitute 8 for b and 6.4 for h and multiply:

$$A = \tfrac{1}{2}bh = \tfrac{1}{2} \cdot 8 \cdot 6.4$$
$$= 25.6 \text{ m}^2.$$

Do Exercise 15.

EXAMPLE 9 Evaluate $5 + 2(a - 1)^2$ when $a = 4$.

$$
\begin{aligned}
5 + 2(a - 1)^2 &= 5 + 2(4 - 1)^2 && \text{Substituting} \\
&= 5 + 2(3)^2 && \text{Working within parentheses first} \\
&= 5 + 2(9) && \text{Simplifying } 3^2 \\
&= 5 + 18 && \text{Multiplying} \\
&= 23 && \text{Adding}
\end{aligned}
$$

16. Evaluate $(x - 3)^2$ when $x = 11$.

The rules for order of operations tell us to divide before we multiply when division appears first, reading left to right. Similarly, if subtraction appears before addition, we subtract before we add.

EXAMPLE 10 Evaluate $9 - x^3 + 6 \div 2y^2$ when $x = 2$ and $y = 5$.

$$
\begin{aligned}
9 - x^3 + 6 \div 2y^2 &= 9 - 2^3 + 6 \div 2(5)^2 && \text{Substituting} \\
&= 9 - 8 + 6 \div 2 \cdot 25 && \text{Simplifying } 2^3 \text{ and } 5^2 \\
&= 9 - 8 + 3 \cdot 25 && \text{Dividing} \\
&= 9 - 8 + 75 && \text{Multiplying} \\
&= 1 + 75 && \text{Subtracting} \\
&= 76 && \text{Adding}
\end{aligned}
$$

17. Evaluate $x^2 - 6x + 9$ when $x = 11$.

Do Exercises 16–18.

CALCULATOR CORNER

Evaluating Algebraic Expressions We can evaluate algebraic expressions on a graphing calculator by making the appropriate substitutions, keeping in mind the rules for order of operations, and then carrying out the resulting calculations. To evaluate $5 + 2(a - 1)^2$ when $a = 4$, as in Example 9, we enter $5 + 2(4 - 1)^2$ by pressing ⑤ ⊕ ② ⦅ ④ ⊖ ① ⦆ x² **ENTER**, or ⑤ ⊕ ② ⦅ ④ ⊖ ① ⦆ ⌃ ② **ENTER**. The result is 23.

```
5+2(4−1)²
                           23
5+2(4−1)^2
                           23
```

Exercises: Evaluate.

1. $a + b$, when $a = 5.8$ and $b = 6.4$

2. $b - a$, when $a = 10.3$ and $b = 2.7$

3. $\dfrac{12m}{n}$, when $m = 25$ and $n = -15$

4. $13xy$, when $x = -12.7$ and $y = 100.4$

5. $3a + 2b$, when $a = 3.3$ and $b = 5.7$

6. $2a + 3b$, when $a = -7.3$ and $b = 4.9$

18. Evaluate $8 - x^3 + 10 \div 5y^2$ when $x = 4$ and $y = 6$.

Answers on page A-2

R.4 EXERCISE SET

For Extra Help

MathXL   MyMathLab   InterAct Math   Math Tutor Center   Digital Video Tutor CD 1 Videotape 1   Student's Solutions Manual

a   Translate the phrase to an algebraic expression.

1. 8 more than b

2. 11 more than t

3. 13.4 less than c

4. 0.203 less than d

5. 5 increased by q

6. 18 increased by z

7. b more than a

8. c more than d

9. x divided by y

10. c divided by h

11. x plus w

12. s added to t

13. m subtracted from n

14. p subtracted from q

15. The sum of p and q

16. The sum of a and b

17. Three times q

18. Twice z

19. -18 multiplied by m

20. The product of -6 and t

21. The product of 17% and your salary

22. 48% of the women attending

23. Megan drove at a speed of 75 mph for t hours on an interstate highway in Arizona. How far did Megan travel?

24. Joe had d dollars before spending $19.95 on a DVD of the movie *Pearl Harbor*. How much did Joe have after the purchase?

25. Jennifer had $40 before spending x dollars on a pizza. How much remains?

26. Lance drove his pickup truck at a speed of 65 mph for t hours. How far did he travel?

b   Evaluate.

27. $23z$, when $z = -4$

28. $57y$, when $y = -8$

29. $\dfrac{a}{b}$, when $a = -24$ and $b = -8$

30. $\dfrac{x}{y}$, when $x = 30$ and $y = -6$

31. $\dfrac{m-n}{8}$, when $m = 36$ and $n = 4$

32. $\dfrac{5}{p+q}$, when $p = 20$ and $q = 30$

33. $\dfrac{5z}{y}$, when $z = 9$ and $y = 2$

34. $\dfrac{18m}{n}$, when $m = 7$ and $n = 18$

35. $2c \div 3b$, when $b = 4$ and $c = 6$

36. $4x - y$, when $x = 3$ and $y = -2$

37. $25 - r^2 + s \div r^2$, when $r = 3$ and $s = 27$

38. $n^3 - 2 + p \div n^2$, when $n = 2$ and $p = 12$

39. $m + n(5 + n^2)$, when $m = 15$ and $n = 3$

40. $a^2 - 3(a - b)$, when $a = 10$ and $b = -8$

Simple Interest. The **simple interest** I on a principal of P dollars at interest rate r for t years is given by $I = Prt$.

41. Find the simple interest on a principal of $7345 at 6% for 1 yr.

42. Find the simple interest on a principal of $18,000 at 4.6% for 2 yr. (*Hint*: 4.6% = 0.046.)

43. *Area of a Compact Disc.* The area A of a circle with radius r is given by $A = \pi r^2$. The circumference C of the circle is given by $C = 2\pi r$. The standard compact disc used for software and music has a radius of 6 cm. Find the area and the circumference of such a CD (ignoring the hole in the middle). Use 3.14 for π.

44. *Area of a Parallelogram.* The area A of a parallelogram with base b and height h is given by $A = bh$. Find the area of a parallelogram-shaped horse-feeding area with a height of 24.3 ft and a base of 67.8 ft.

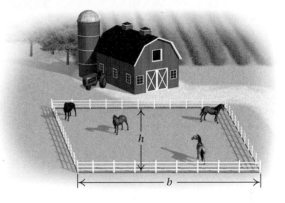

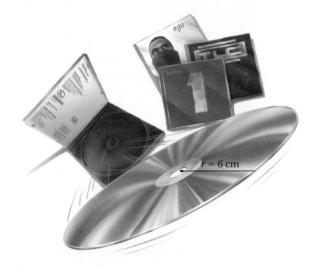

$r = 6$ cm

45. D_W If the base and the height of a triangle are doubled, does its area double? Explain.

46. D_W If the base and the height of a parallelogram are doubled, does its area double? Explain.

SKILL MAINTENANCE

Evaluate. [R.3a]

47. 3^5

48. $(-3)^5$

49. $(-10)^2$

50. $(-10)^4$

51. $(-5.3)^2$

52. $\left(\dfrac{3}{5}\right)^2$

53. $(4.5)^0$

54. $(4.5)^1$

55. $(3x)^1$

56. $(3x)^0$

SYNTHESIS

Translate to an equation.

57. The distance d that a rapid transit train in the Denver airport travels in time t at a speed r is given by speed times time. Write an equation for d.

58. You invest P dollars at 6.7% simple interest. Write an equation for the number of dollars N in the account 1 yr from now.

Evaluate.

59. $\dfrac{x + y}{2} + \dfrac{3y}{2}$, when $x = 2$ and $y = 4$

60. $\dfrac{2.56y}{3.2x}$, when $y = 3$ and $x = 4$

Objectives

a Determine whether two expressions are equivalent by completing a table of values.

b Find equivalent fraction expressions by multiplying by 1, and simplify fraction expressions.

c Use the commutative and the associative laws to find equivalent expressions.

d Use the distributive laws to find equivalent expressions by multiplying and factoring.

a Equivalent Expressions

When solving equations and performing other operations in algebra, we manipulate expressions in various ways. For example, rather than $x + 2x$, we might write $3x$, knowing that the two expressions represent the same number for any allowable replacement of x. In that sense, the expressions $x + 2x$ and $3x$ are **equivalent,** as are $5/x$ and $5x/x^2$ even though 0 is not an allowable replacement because division by 0 is not defined.

EQUIVALENT EXPRESSIONS

Two expressions that have the same value for all *allowable* replacements are called **equivalent expressions.**

Complete the table by evaluating each expression for the given values. Then look for expressions that may be equivalent.

1.

Value	$6x - x$	$5x$	$8x + x$
$x = -2$			
$x = 8$			
$x = 0$			

EXAMPLE 1 Complete the table by evaluating each of the expressions $x + 2x$, $3x$, and $8x - x$ for the given values. Then look for expressions that are equivalent.

Value	$x + 2x$	$3x$	$8x - x$
$x = -2$			
$x = 5$			
$x = 0$			

We substitute and find the value of each expression. For example, for $x = -2$,

$$x + 2x = -2 + 2(-2) = -2 - 4 = -6,$$
$$3x = 3(-2) = -6, \quad \text{and}$$
$$8x - x = 8(-2) - (-2) = -16 + 2 = -14.$$

Value	$x + 2x$	$3x$	$8x - x$
$x = -2$	-6	-6	-14
$x = 5$	15	15	35
$x = 0$	0	0	0

2.

Value	$(x + 3)^2$	$x^2 + 9$
$x = -2$		
$x = 5$		
$x = 4.8$		

Note that the values of $x + 2x$ and $3x$ are the same for the given values of x. Indeed, they are the same for any allowable real-number replacement of x, though we cannot substitute them all to find out. The expressions $x + 2x$ and $3x$ are **equivalent.** But the expressions $x + 2x$ and $8x - x$ are not equivalent, and the expressions $3x$ and $8x - x$ are not equivalent. Although $3x$ and $8x - x$ have the same value for $x = 0$, they are not equivalent since values are not the same for *all x*.

Answers on page A-2

Do Exercises 1 and 2.

b Equivalent Fraction Expressions

For the remainder of this section, we will consider several laws of real numbers that will allow us to find equivalent expressions.

> **THE IDENTITY PROPERTY OF 1**
>
> For any real number a,
>
> $$a \cdot 1 = 1 \cdot a = a.$$
>
> (The number 1 is the **multiplicative identity.**)

We will often refer to the use of the identity property of 1 as "multiplying by 1." We can use multiplying by 1 to change from one fraction expression to an equivalent one with a different denominator.

EXAMPLE 2   Use multiplying by 1 to find an expression equivalent to $\frac{3}{5}$ with a denominator of $10x$.

We multiply by 1, using $2x/(2x)$ as a name for 1:

$$\frac{3}{5} = \frac{3}{5} \cdot 1 = \frac{3}{5} \cdot \frac{2x}{2x} = \frac{3 \cdot 2x}{5 \cdot 2x} = \frac{6x}{10x}.$$

Note that the expressions $3/5$ and $6x/(10x)$ are equivalent. They have the same value for any allowable replacement. Note too that 0 is not an allowable replacement in $6x/(10x)$, but for all nonzero real numbers, the expressions $3/5$ and $6x/(10x)$ have the same value.

Do Exercises 3 and 4.

In algebra, we consider an expression like $3/5$ to be a "simplified" form of $6x/(10x)$. To find such simplified expressions, we reverse the identity property of 1 in order to "remove a factor of 1."

EXAMPLE 3   Simplify: $\dfrac{7x}{9x}$.

We do the reverse of what we did in Example 2:

$$\frac{7x}{9x} = \frac{7 \cdot x}{9 \cdot x} \qquad \text{We factor the numerator and the denominator and then look for the largest common factor of both.}$$

$$= \frac{7}{9} \cdot \frac{x}{x} \qquad \text{Factoring the expression}$$

$$= \frac{7}{9} \cdot 1 \qquad \frac{x}{x} = 1$$

$$= \frac{7}{9}. \qquad \text{Removing a factor of 1 using the identity property of 1 in reverse}$$

EXAMPLE 4   Simplify: $-\dfrac{24y}{16y}$.

$$-\frac{24y}{16y} = -\frac{3 \cdot 8y}{2 \cdot 8y} = -\frac{3}{2} \cdot \frac{8y}{8y} = -\frac{3}{2} \cdot 1 = -\frac{3}{2}$$

Do Exercises 5 and 6.

3. Use multiplying by 1 to find an expression equivalent to $\frac{2}{7}$ with a denominator of $7y$.

4. Use multiplying by 1 to find an expression equivalent to $\frac{2}{11}$ with a denominator of $44x$.

Simplify.

5. $\dfrac{2y}{3y}$

6. $-\dfrac{20m}{12m}$

Answers on page A-2

7. Evaluate $x + y$ and $y + x$ when $x = -3$ and $y = 5$.

C ## The Commutative and the Associative Laws

Let's examine the expressions $x + y$ and $y + x$, as well as xy and yx.

EXAMPLE 5 Evaluate $x + y$ and $y + x$ when $x = 5$ and $y = 8$.

We substitute 5 for x and 8 for y in both expressions:

$$x + y = 5 + 8 = 13; \qquad y + x = 8 + 5 = 13.$$

EXAMPLE 6 Evaluate xy and yx when $x = 26$ and $y = 13$.

We substitute 26 for x and 13 for y in both expressions:

$$xy = 26 \cdot 13 = 338; \qquad yx = 13 \cdot 26 = 338.$$

Do Exercises 7 and 8.

Note that the expressions $x + y$ and $y + x$ have the same values no matter what the variables stand for. Thus they are equivalent. They illustrate that when we add two numbers, the order in which we add does not matter. Similarly, when we multiply two numbers, the order in which we multiply does not matter. Thus the expressions xy and yx are equivalent. They have the same values no matter what the variables stand for. These are examples of general patterns or laws.

THE COMMUTATIVE LAWS

Addition.   For any numbers a and b,

$$a + b = b + a.$$

(We can change the order when adding without affecting the answer.)

Multiplication.   For any numbers a and b,

$$ab = ba.$$

(We can change the order when multiplying without affecting the answer.)

---

8. Evaluate xy and yx when $x = -2$ and $y = 7$.

Using a commutative law, we know that $x + 4$ and $4 + x$ are equivalent. Similarly, $5x$ and $x(5)$ are equivalent. Thus, in an algebraic expression, we can replace one with the other and the result will be equivalent to the original expression.

Now let's examine the expressions $a + (b + c)$ and $(a + b) + c$. Note that these expressions use parentheses as grouping symbols, and they also involve three numbers. Calculations within grouping symbols are to be done first.

EXAMPLE 7 Evaluate $a + (b + c)$ and $(a + b) + c$ when $a = 4$, $b = 8$, and $c = 5$.

$$
\begin{aligned}
a + (b + c) &= 4 + (8 + 5) && \text{Substituting} \\
&= 4 + 13 && \text{Calculating within parentheses first:} \\
&= 17; && \text{adding 8 and 5}
\end{aligned}
$$

$$
\begin{aligned}
(a + b) + c &= (4 + 8) + 5 && \text{Substituting} \\
&= 12 + 5 && \text{Calculating within parentheses first:} \\
&= 17 && \text{adding 4 and 8}
\end{aligned}
$$

Answers on page A-2

EXAMPLE 8 Evaluate $a \cdot (b \cdot c)$ and $(a \cdot b) \cdot c$ when $a = 7$, $b = 4$, and $c = 2$.

$$a \cdot (b \cdot c) = 7 \cdot (4 \cdot 2) = 7 \cdot 8 = 56;$$

$$(a \cdot b) \cdot c = (7 \cdot 4) \cdot 2 = 28 \cdot 2 = 56$$

Do Exercises 9 and 10.

When only addition is involved, grouping symbols can be placed any way we please without affecting the answer. Likewise, when only multiplication is involved, grouping symbols can be placed any way we please without affecting the answer.

THE ASSOCIATIVE LAWS

Addition. For any numbers a, b, and c,

$$a + (b + c) = (a + b) + c.$$

(Numbers can be grouped in any manner for addition.)

Multiplication. For any numbers a, b, and c,

$$a \cdot (b \cdot c) = (a \cdot b) \cdot c.$$

(Numbers can be grouped in any manner for multiplication.)

When only additions or only multiplications are involved, grouping symbols can be placed any way we please. Thus we often omit them. For example,

$$x + (y + 3) \text{ means } x + y + 3, \quad \text{and} \quad l(wh) \text{ means } lwh.$$

EXAMPLE 9 Use the commutative and the associative laws to write at least three expressions equivalent to $(x + 8) + y$.

a) $(x + 8) + y = x + (8 + y)$ Using the associative law first and then the commutative law

$$= x + (y + 8)$$

b) $(x + 8) + y = y + (x + 8)$ Using the commutative law and then the commutative law again

$$= y + (8 + x)$$

c) $(x + 8) + y = (8 + x) + y$ Using the commutative law first and then the associative law

$$= 8 + (x + y)$$

Do Exercises 11 and 12.

d   The Distributive Laws

Let's now examine two laws, each of which involves two operations. The first involves multiplication and addition.

EXAMPLE 10 Evaluate $8(x + y)$ and $8x + 8y$ when $x = 4$ and $y = 5$.

$$8(x + y) = 8(4 + 5) \qquad 8x + 8y = 8 \cdot 4 + 8 \cdot 5$$
$$= 8(9) \qquad\qquad\qquad = 32 + 40$$
$$= 72; \qquad\qquad\qquad = 72$$

9. Evaluate

$$a + (b + c) \quad \text{and} \quad (a + b) + c$$

when $a = 10$, $b = 9$, and $c = 2$.

10. Evaluate

$$a \cdot (b \cdot c) \quad \text{and} \quad (a \cdot b) \cdot c$$

when $a = 11$, $b = 5$, and $c = 8$.

11. Use the commutative laws to write an expression equivalent to each of $y + 5$, ab, and $8 + mn$.

12. Use the commutative and the associative laws to write at least three expressions equivalent to $(2 \cdot x) \cdot y$.

Answers on page A-2

13. Evaluate $10(x + y)$ and $10x + 10y$ when $x = 7$ and $y = 11$.

14. Evaluate $9(a + b)$, $(a + b)9$, and $9a + 9b$ when $a = 5$ and $b = -2$.

The expressions $8(x + y)$ and $8x + 8y$ in Example 10 are equivalent. This fact is the result of a law called *the distributive law of multiplication over addition*.

THE DISTRIBUTIVE LAW OF MULTIPLICATION OVER ADDITION

For any numbers a, b, and c,

$$a(b + c) = ab + ac, \quad \text{or} \quad (b + c)a = ba + ca.$$

(We can add and then multiply, or we can multiply and then add.)

Do Exercises 13 and 14.

The other distributive law involves multiplication and subtraction.

15. Evaluate $5(a - b)$ and $5a - 5b$ when $a = 10$ and $b = 8$.

■ **EXAMPLE 11**   Evaluate $\frac{1}{2}(a - b)$ and $\frac{1}{2}a - \frac{1}{2}b$ when $a = 42$ and $b = 78$.

$$\begin{aligned}\frac{1}{2}(a - b) &= \frac{1}{2}(42 - 78) & \frac{1}{2}a - \frac{1}{2}b &= \frac{1}{2} \cdot 42 - \frac{1}{2} \cdot 78 \\ &= \frac{1}{2}(-36) & &= 21 - 39 \\ &= -18; & &= -18\end{aligned}$$

16. Evaluate $\frac{2}{3}(p - q)$ and $\frac{2}{3}p - \frac{2}{3}q$ when $p = 60$ and $q = 24$.

The expressions $\frac{1}{2}(a - b)$ and $\frac{1}{2}a - \frac{1}{2}b$ in Example 11 are equivalent. This fact is the result of a law called *the distributive law of multiplication over subtraction*.

THE DISTRIBUTIVE LAW OF MULTIPLICATION OVER SUBTRACTION

For any real numbers a, b, and c,

$$a(b - c) = ab - ac, \quad \text{or} \quad (b - c)a = ba - ca.$$

(We can subtract and then multiply, or we can multiply and then subtract.)

Multiply.

17. $8(y - 10)$

We often refer to "*the* distributive law" when we mean *either* or *both* of these laws.

18. $a(x + y - z)$

Do Exercises 15 and 16.

MULTIPLYING EXPRESSIONS WITH VARIABLES

The distributive laws are the basis of multiplication in algebra as well as in arithmetic. In the following examples, note that we multiply each number or letter inside the parentheses by the factor outside.

19. $10\left(4x - 6y + \frac{1}{2}z\right)$

■ **EXAMPLES**   Multiply.

12. $4(x - 2) = 4 \cdot x - 4 \cdot 2 = 4x - 8$

13. $b(s - t + f) = bs - bt + bf$

14. $-3(y + 4) = -3 \cdot y + (-3) \cdot 4 = -3y - 12$

15. $-2x(y - 1) = -2x \cdot y - (-2x) \cdot 1 = -2xy + 2x$

Answers on page A-2

Do Exercises 17–19.

FACTORING EXPRESSIONS WITH VARIABLES

The reverse of multiplying is called **factoring.** Factoring an expression involves factoring its *terms.* **Terms** of algebraic expressions are the parts separated by plus signs.

EXAMPLE 16 List the terms of $3x - 4y - 2z$.

We first find an equivalent expression that uses addition signs:

$3x - 4y - 2z = 3x + (-4y) + (-2z)$.   Using the property $a - b = a + (-b)$

Thus the terms are $3x$, $-4y$, and $-2z$.

Do Exercise 20.

Now we can consider the reverse of multiplying, *factoring.*

FACTORS

To **factor** an expression is to find an equivalent expression that is a product. If $N = ab$, then a and b are **factors** of N.

EXAMPLES Factor.

17. $8x + 8y = 8(x + y)$ 8 and $x + y$ are factors.
18. $cx - cy = c(x - y)$ c and $x - y$ are factors.

The distributive laws tell us that $8(x + y)$ and $8x + 8y$ are equivalent. We consider $8(x + y)$ to be **factored.** The factors are 8 and $x + y$. Whenever the terms of an expression have a factor in common, we can "remove" that factor, or "factor it out," using the distributive laws. We proceed as in Examples 17 and 18, but we may have to factor some of the terms first in order to display the common factor.

Generally, we try to factor out the largest factor common to all the terms. In the following example, we might factor out 3, but there is a larger factor common to the terms, 9. So we factor out the 9.

EXAMPLE 19 Factor: $9x + 27y$.

$9x + 27y = 9 \cdot x + 9 \cdot (3y) = 9(x + 3y)$

We often have to supply a factor of 1 when factoring out a common factor, as in the next example, which is a formula involving simple interest.

EXAMPLE 20 Factor: $P + Prt$.

$P + Prt = P \cdot 1 + P \cdot rt$ Writing P as a product of P and 1
$\quad\quad\quad = P(1 + rt)$ Using the distributive law

You can always check your factoring by multiplying.

Do Exercises 21–25.

20. List the terms of
$$-5x - 7y + 67t - \frac{4}{5}.$$

Factor.

21. $9x + 9y$

22. $ac - ay$

23. $6x - 12$

24. $35x - 25y + 15w + 5$

25. $bs + bt - bw$

Answers on page A-2

a Complete the table by evaluating each expression for the given values. Then look for expressions that are equivalent.

1.

Value	$2x + 3x$	$5x$	$2x - 3x$
$x = -2$			
$x = 5$			
$x = 0$			

2.

Value	$7x + 2x$	$5x$	$7x - 2x$
$x = -2$			
$x = 5$			
$x = 0$			

3.

Value	$4x + 8x$	$4(x + 3x)$	$4(x + 2x)$
$x = -1$			
$x = 3.2$			
$x = 0$			

4.

Value	$5(x - 2)$	$5x - 2$	$5x - 10$
$x = -1$			
$x = 4.6$			
$x = 0$			

b Use multiplying by 1 to find an equivalent expression with the given denominator.

5. $\dfrac{7}{8}$; $8x$

6. $\dfrac{4}{3}$; $3a$

7. $\dfrac{3}{4}$; $8a$

8. $\dfrac{3}{10}$; $50y$

Simplify.

9. $\dfrac{25x}{15x}$

10. $\dfrac{36y}{18y}$

11. $-\dfrac{100a}{25a}$

12. $\dfrac{-625t}{15t}$

c Use a commutative law to find an equivalent expression.

13. $w + 3$

14. $y + 5$

15. rt

16. cd

17. $4 + cd$

18. $pq + 14$

19. $yz + x$

20. $s + qt$

Use an associative law to find an equivalent expression.

21. $m + (n + 2)$

22. $5 \cdot (p \cdot q)$

23. $(7 \cdot x) \cdot y$

24. $(7 + p) + q$

Use the commutative and the associative laws to find three equivalent expressions.

25. $(a + b) + 8$

26. $(4 + x) + y$

27. $7 \cdot (a \cdot b)$

28. $(8 \cdot m) \cdot n$

d Multiply.

29. $4(a + 1)$

30. $3(c + 1)$

31. $8(x - y)$

32. $7(b - c)$

33. $-5(2a + 3b)$

34. $-2(3c + 5d)$

35. $2a(b - c + d)$

36. $5x(y - z + w)$

37. $2\pi r(h + 1)$

38. $P(1 + rt)$

39. $\frac{1}{2}h(a + b)$

40. $\frac{1}{4}\pi r(1 + s)$

List the terms of each of the following.

41. $4a - 5b + 6$

42. $5x - 9y + 12$

43. $2x - 3y - 2z$

44. $5a - 7b - 9c$

Factor.

45. $24x + 24y$

46. $9a + 9b$

47. $7p - 7$

48. $22x - 22$

49. $7x - 21$

50. $6y - 36$

51. $xy + x$

52. $ab + a$

53. $2x - 2y + 2z$

54. $3x + 3y - 3z$

55. $3x + 6y - 3$

56. $4a + 8b - 4$

57. $ab + ac - ad$

58. $xy - xz + xw$

59. $\frac{1}{4}\pi rr + \frac{1}{4}\pi rs$

60. $\frac{1}{2}ah + \frac{1}{2}bh$

61. $\mathbf{D_W}$ If we omit the parentheses in the expression $a(b + c)$, will the result be equivalent to $ab + c$? Explain how you would decide.

62. $\mathbf{D_W}$ Look at the answers to Exercises 1 and 2 and at the expressions. What conclusion might you reach about the expressions $2x + 3x$ and $7x - 2x$? How would you verify it?

SKILL MAINTENANCE

Translate to an algebraic expression.   [R.4a]

63. The square of the sum of two numbers

64. The sum of the squares of two numbers

Subtract.   [R.2c]

65. $-34.2 - 67.8$

66. $-\frac{11}{5} - \left(-\frac{17}{10}\right)$

67. $-\frac{1}{4}\left(-\frac{1}{2}\right)$

68. $0.23(-200)$

SYNTHESIS

Make substitutions to determine whether the pair of expressions is equivalent.

69. $x^2 + y^2$; $(x + y)^2$

70. $(a - b)(a + b)$; $a^2 - b^2$

71. $x^2 \cdot x^3$; x^5

72. $\dfrac{x^8}{x^4}$; x^2

Objectives

a Simplify an expression by collecting like terms.

b Simplify an expression by removing parentheses and collecting like terms.

Collect like terms.

1. $9x + 11x$

2. $5x - 12x$

3. $5x + x$

4. $x - 7x$

5. $22x - 2.5y + 1.4x + 6.4y$

6. $\dfrac{2}{3}x - \dfrac{3}{4}y + \dfrac{4}{5}x - \dfrac{5}{6}y + 23$

Multiply.

7. $-1 \cdot 24$

8. $-1 \cdot 0$

9. $-1 \cdot (-10)$

Answers on page A-3

There are many situations in algebra in which we want to find either an alternative or a simpler expression equivalent to a given one.

a Collecting Like Terms

If two terms have the same letter, or letters, we say that they are **like terms,** or **similar terms.** (If powers, or exponents, are involved, then like terms must have the same letters raised to the same powers. We will consider this in Chapter 4.) If two terms have no letters at all but are just numbers, they are also similar terms. We can simplify by **collecting** or **combining like terms,** using the distributive laws.

EXAMPLES Collect like terms.

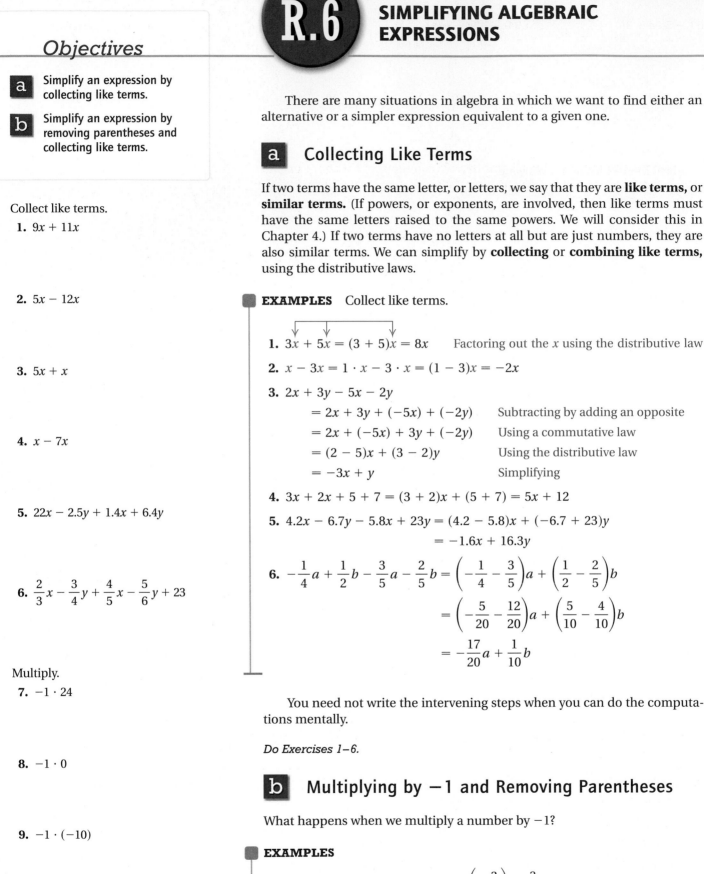

1. $3x + 5x = (3 + 5)x = 8x$ Factoring out the x using the distributive law

2. $x - 3x = 1 \cdot x - 3 \cdot x = (1 - 3)x = -2x$

3. $2x + 3y - 5x - 2y$

$\quad = 2x + 3y + (-5x) + (-2y)$ Subtracting by adding an opposite

$\quad = 2x + (-5x) + 3y + (-2y)$ Using a commutative law

$\quad = (2 - 5)x + (3 - 2)y$ Using the distributive law

$\quad = -3x + y$ Simplifying

4. $3x + 2x + 5 + 7 = (3 + 2)x + (5 + 7) = 5x + 12$

5. $4.2x - 6.7y - 5.8x + 23y = (4.2 - 5.8)x + (-6.7 + 23)y$

$\quad\quad\quad\quad\quad\quad\quad\quad\quad\quad = -1.6x + 16.3y$

6. $-\dfrac{1}{4}a + \dfrac{1}{2}b - \dfrac{3}{5}a - \dfrac{2}{5}b = \left(-\dfrac{1}{4} - \dfrac{3}{5}\right)a + \left(\dfrac{1}{2} - \dfrac{2}{5}\right)b$

$\quad\quad\quad\quad\quad\quad\quad\quad\quad = \left(-\dfrac{5}{20} - \dfrac{12}{20}\right)a + \left(\dfrac{5}{10} - \dfrac{4}{10}\right)b$

$\quad\quad\quad\quad\quad\quad\quad\quad\quad = -\dfrac{17}{20}a + \dfrac{1}{10}b$

You need not write the intervening steps when you can do the computations mentally.

Do Exercises 1–6.

b Multiplying by −1 and Removing Parentheses

What happens when we multiply a number by -1?

EXAMPLES

7. $-1 \cdot 9 = -9$ **8.** $-1 \cdot \left(-\dfrac{3}{5}\right) = \dfrac{3}{5}$ **9.** $-1 \cdot 0 = 0$

Do Exercises 7–9.

THE PROPERTY OF -1

For any number a,

$$-1 \cdot a = -a.$$

(Negative 1 times a is the opposite of a; in other words, changing the sign is the same as multiplying by -1.)

From the property of -1, we know that we can replace $-$ with -1 or the reverse, in any expression. In that way, we can find an equivalent expression for an opposite.

EXAMPLES Find an equivalent expression without parentheses.

10. $-(3x) = -1(3x)$ Replacing $-$ with -1 using the property of -1
$\quad\quad\quad = (-1 \cdot 3)x$ Using an associative law
$\quad\quad\quad = -3x$ Multiplying

11. $-(-9y) = -1(-9y)$ Replacing $-$ with -1
$\quad\quad\quad\quad = [-1(-9)]y$ Using an associative law
$\quad\quad\quad\quad = 9y$ Multiplying

Do Exercises 10 and 11.

EXAMPLES Find an equivalent expression without parentheses.

12. $-(4 + x) = -1(4 + x)$ Replacing $-$ with -1
$\quad\quad\quad\quad = -1 \cdot 4 + (-1) \cdot x$ Multiplying using the distributive law
$\quad\quad\quad\quad = -4 + (-x)$ Replacing $-1 \cdot x$ with $-x$
$\quad\quad\quad\quad = -4 - x$ Adding an opposite is the same as subtracting.

13. $-(3x - 2y + 4) = -1(3x - 2y + 4)$
$\quad\quad\quad\quad\quad = -1 \cdot 3x - (-1)2y + (-1)4$ Using the distributive law
$\quad\quad\quad\quad\quad = -3x - (-2y) + (-4)$ Multiplying
$\quad\quad\quad\quad\quad = -3x + [-(-2y)] + (-4)$ Adding an opposite
$\quad\quad\quad\quad\quad = -3x + 2y - 4$

14. $-(a - b) = -1(a - b)$
$\quad\quad\quad\quad = -1 \cdot a - (-1) \cdot b$
$\quad\quad\quad\quad = -a + [-(-1)b]$
$\quad\quad\quad\quad = -a + b = b - a$

Example 14 illustrates something that you should remember, because it is a convenient shortcut.

THE OPPOSITE OF A DIFFERENCE

For any real numbers a and b,

$$-(a - b) = b - a.$$

(The opposite of $a - b$ is $b - a$.)

Do Exercises 12–16.

Find an equivalent expression without parentheses.

17. $-(-2x - 5z + 24)$

18. $-(3x - 2y)$

19. $-\left(\dfrac{1}{4}t + 41w - 5d - 23\right)$

Remove parentheses and simplify.

20. $6x - (3x + 8)$

21. $6y - 4 - (2y - 5)$

22. $6x - (9y - 4) - (8x + 10)$

23. $7x - (-9y - 4) + (8x - 10)$

Remove parentheses and simplify.

24. $x - 2(y + x)$

25. $3x - 5(2y - 4x)$

26. $(4a - 3b) - \dfrac{1}{4}(4a - 3) + 5$

Answers on page A-3

Examples 10–14 show that we can find an equivalent expression for an opposite by multiplying every term by -1. We could also say that we change the sign of every term inside the parentheses. Thus we can skip some steps.

EXAMPLE 15 Find an equivalent expression without parentheses:
$$-\left(-9t + 7z - \tfrac{1}{4}w\right).$$
We have
$$-\left(-9t + 7z - \tfrac{1}{4}w\right) = 9t - 7z + \tfrac{1}{4}w. \qquad \text{Changing the sign of every term}$$

Do Exercises 17–19.

In some expressions commonly encountered in algebra, there are parentheses preceded by subtraction signs. These parentheses can be removed by changing the sign of *every* term inside. In this way, we simplify by finding a less complicated equivalent expression.

EXAMPLES Remove parentheses and simplify.

16. $6x - (4x + 2) = 6x + [-(4x + 2)]$ Subtracting by adding the opposite
$\qquad\qquad\quad = 6x - 4x - 2$ Changing the sign of every term inside
$\qquad\qquad\quad = 2x - 2$ Collecting like terms

17. $3y - 4 - (9y - 7) = 3y - 4 - 9y + 7$
$\qquad\qquad\qquad\quad = -6y + 3,\ \text{or}\ 3 - 6y$

In Example 16, we see the reason for the word "simplify." The expression $2x - 2$ is equivalent to $6x - (4x + 2)$ but it is shorter.

If parentheses are preceded by an addition sign, *no* signs are changed when they are removed.

EXAMPLES Remove parentheses and simplify.

18. $3y + (3x - 8) - (5 - 12y) = 3y + 3x - 8 - 5 + 12y$
$\qquad\qquad\qquad\qquad\qquad\quad = 15y + 3x - 13$

19. $\tfrac{1}{3}(15x - 4) - (5x + 2y) + 1 = \tfrac{1}{3} \cdot 15x - \tfrac{1}{3} \cdot 4 - 5x - 2y + 1$
$\qquad\qquad\qquad\qquad\qquad\qquad = 5x - \tfrac{4}{3} - 5x - 2y + 1$
$\qquad\qquad\qquad\qquad\qquad\qquad = -2y - \tfrac{1}{3}$

Do Exercises 20–23.

We now consider subtracting an expression consisting of several terms preceded by a number other than -1.

EXAMPLES Remove parentheses and simplify.

20. $x - 3(x + y) = x + [-3(x + y)]$ Subtracting by adding the opposite
$\qquad\qquad\quad = x - 3x - 3y$ Removing parentheses by multiplying $x + y$ by -3
$\qquad\qquad\quad = -2x - 3y$ Collecting like terms

> **Caution!**
>
> A common error is to forget to change this sign. *Remember*: When multiplying by a negative number, change the sign of *every* term inside the parentheses.

21. $3y - 2(4y - 5) = 3y - 8y + 10$ Removing parentheses by multiplying $4y - 5$ by -2

$\qquad\qquad\qquad = -5y + 10$ Collecting like terms

Do Exercises 24–26 on the preceding page.

When expressions with grouping symbols contain variables, we still work from the inside out when simplifying, using the rules for order of operations.

EXAMPLE 22   Simplify: $[2(x + 7) - 4^2] - (2 - x)$.

$[2(x + 7) - 4^2] - (2 - x)$

$\quad = [2x + 14 - 4^2] - (2 - x)$ Multiplying to remove the innermost grouping symbols using the distributive law

$\quad = [2x + 14 - 16] - (2 - x)$ Evaluating the exponential expression

$\quad = [2x - 2] - (2 - x)$ Collecting like terms inside the brackets

$\quad = 2x - 2 - 2 + x$ Multiplying by -1 to remove the parentheses

$\quad = 3x - 4$ Collecting like terms

Do Exercises 27 and 28.

EXAMPLE 23   Simplify: $6y - \{4[3(y - 2) - 4(y + 2)] - 3\}$.

$6y - \{4[3(y - 2) - 4(y + 2)] - 3\}$

$\quad = 6y - \{4[3y - 6 - 4y - 8] - 3\}$ Multiplying to remove the innermost grouping symbols using the distributive law

$\quad = 6y - \{4[-y - 14] - 3\}$ Collecting like terms inside the brackets

$\quad = 6y - \{-4y - 56 - 3\}$ Multiplying to remove the inner brackets using the distributive law

$\quad = 6y - \{-4y - 59\}$ Collecting like terms in the braces

$\quad = 6y + 4y + 59$ Removing braces

$\quad = 10y + 59$ Collecting like terms

Do Exercises 29 and 30.

Simplify.

27. $(3x - 5) - [4(x - 1) + 2]$

28. $[3 - 2(x + 9)] - 4(3^2 - x)$

Simplify.

29. $15x - \{2[2(x - 5) - 6(x + 3)] + 4\}$

30. $9a + \{3a - 2[(a - 4) - (a + 2)]\}$

Answers on page A-3

R.6
EXERCISE SET

For Extra Help

MathXL   MyMathLab   InterAct   Math Tutor   Digital Video   Student's
                       Math   Center   Tutor CD 1   Solutions
                                             Videotape 1   Manual

ⓐ Collect like terms.

1. $7x + 5x$

2. $6a + 9a$

3. $8b - 11b$

4. $9c - 12c$

5. $14y + y$

6. $13x + x$

7. $12a - a$

8. $15x - x$

9. $t - 9t$

10. $x - 6x$

11. $5x - 3x + 8x$

12. $3x - 11x + 2x$

13. $3x - 5y + 8x$

14. $4a - 9b + 10a$

15. $3c + 8d - 7c + 4d$

16. $12a + 3b - 5a + 6b$

17. $4x - 7 + 18x + 25$

18. $13p + 5 - 4p + 7$

19. $1.3x + 1.4y - 0.11x - 0.47y$

20. $0.17a + 1.7b - 12a - 38b$

21. $\dfrac{2}{3}a + \dfrac{5}{6}b - 27 - \dfrac{4}{5}a - \dfrac{7}{6}b$

22. $-\dfrac{1}{4}x - \dfrac{1}{2}x + \dfrac{1}{4}y + \dfrac{1}{2}y - 34$

The **perimeter** of a rectangle is the distance around it. The perimeter P is given by $P = 2l + 2w$.

23. Find an equivalent expression for the perimeter formula $P = 2l + 2w$ by factoring.

24. *Perimeter of a Football Field.* The standard football field has $l = 360$ ft and $w = 160$ ft. Evaluate both expressions in Exercise 23 to find the perimeter.

b For each opposite, find an equivalent expression without parentheses.

25. $-(-2c)$ **26.** $-(-5y)$ **27.** $-(b + 4)$ **28.** $-(a + 9)$

29. $-(b - 3)$ **30.** $-(x - 8)$ **31.** $-(t - y)$ **32.** $-(r - s)$

33. $-(x + y + z)$ **34.** $-(r + s + t)$

35. $-(8x - 6y + 13)$ **36.** $-(9a - 7b + 24)$

37. $-(-2c + 5d - 3e + 4f)$ **38.** $-(-4x + 8y - 5w + 9z)$

39. $-\left(-1.2x + 56.7y - 34z - \dfrac{1}{4}\right)$ **40.** $-\left(-x + 2y - \dfrac{2}{3}z - 56.3w\right)$

Simplify by removing parentheses and collecting like terms.

41. $a + (2a + 5)$ **42.** $x + (5x + 9)$

43. $4m - (3m - 1)$ **44.** $5a - (4a - 3)$

45. $5d - 9 - (7 - 4d)$ **46.** $6x - 7 - (9 - 3x)$

47. $-2(x + 3) - 5(x - 4)$ **48.** $-9(y + 7) - 6(y - 3)$

49. $5x - 7(2x - 3) - 4$

50. $8y - 4(5y - 6) + 9$

51. $8x - (-3y + 7) + (9x - 11)$

52. $-5t + (4t - 12) - 2(3t + 7)$

53. $\dfrac{1}{4}(24x - 8) - \dfrac{1}{2}(-8x + 6) - 14$

54. $-\dfrac{1}{2}(10t - w) + \dfrac{1}{4}(-28t + 4) + 1$

Simplify.

55. $7a - [9 - 3(5a - 2)]$

56. $14b - [7 - 3(9b - 4)]$

57. $5\{-2 + 3[4 - 2(3 + 5)]\}$

58. $7\{-7 + 8[5 - 3(4 + 6)]\}$

59. $[10(x + 3) - 4] + [2(x - 1) + 6]$

60. $[9(x + 5) - 7] + [4(x - 12) + 9]$

61. $[7(x + 5) - 19] - [4(x - 6) + 10]$

62. $[6(x + 4) - 12] - [5(x - 8) + 11]$

63. $3\{[7(x - 2) + 4] - [2(2x - 5) + 6]\}$

64. $4\{[8(x - 3) + 9] - [4(3x - 7) + 2]\}$

65. $4\{[5(x - 3) + 2^2] - 3[2(x + 5) - 9^2]\}$

66. $3\{[6(x - 4) + 5^2] - 2[5(x + 8) - 10^2]\}$

67. $2y + \{8[3(2y - 5) - (8y + 9)] + 6\}$

68. $7b - \{5[4(3b - 8) - (9b + 10)] + 14\}$

69. $^{\mathbf{D}}\mathbf{w}$ Explain in your own words the meaning of the statement $(-x)^2 = x^2$. Determine whether it is true or not. Explain why or why not.

70. $^{\mathbf{D}}\mathbf{w}$ Explain in your own words the meaning of the statement $ab = (-a)(-b)$. Determine whether it is true or not. Explain why or why not.

SKILL MAINTENANCE

Add.   [R.2a]

71. $17 + (-54)$

72. $-17 + (-54)$

73. $-13.78 + (-9.32)$

74. $-\dfrac{2}{3} + \dfrac{7}{8}$

Divide.   [R.2e]

75. $-256 \div 16$

76. $-256 \div (-16)$

77. $256 \div (-16)$

78. $-\dfrac{3}{8} \div \dfrac{9}{4}$

Multiply.   [R.5d]

79. $8(a - b)$

80. $-8(2a - 3b + 4)$

81. $6x(a - b + 2c)$

82. $\dfrac{2}{3}(24x - 12y + 15)$

Factor.   [R.5d]

83. $24a - 24$

84. $24a - 16b$

85. $ab - ac + a$

86. $15p + 45q - 10$

SYNTHESIS

Insert one pair of parentheses to convert the false statement into a true statement.

87. $3 - 8^2 + 9 = 34$

88. $2 \cdot 7 + 3^2 \cdot 5 = 104$

89. $5 \cdot 2^3 \div 3 - 4^4 = 40$

90. $2 - 7 \cdot 2^2 + 9 = -11$

Simplify.

91. $[11(a - 3) + 12a] - \{6[4(3b - 7) - (9b + 10)] + 11\}$

92. $-3[9(x - 4) + 5x] - 8\{3[5(3y + 4)] - 12\}$

93. $z - \{2z + [3z - (4z + 5x) - 6z] + 7z\} - 8z$

94. $\{x + [f - (f + x)] + [x - f]\} + 3x$

95. $x - \{x + 1 - [x + 2 - (x - 3 - \{x + 4 - [x - 5 + (x - 6)]\})]\}$

Objectives

a Use exponential notation in multiplication and division.

b Use exponential notation in raising a power to a power, and in raising a product or a quotient to a power.

c Convert between decimal notation and scientific notation and use scientific notation with multiplication and division.

Multiply and simplify.

1. $8^{-3} \cdot 8^7$

2. $y^7 \cdot y^{-2}$

3. $(9x^{-4})(-2x^7)$

4. $(-3x^{-4})(25x^{-10})$

5. $(-7x^{3n})(6x^{5n})$

6. $(5x^{-3}y^4)(-2x^{-9}y^{-2})$

7. $(4x^{-2}y^4)(15x^2y^{-3})$

Answers on page A-3

We often need to find ways to determine *equivalent exponential expressions*. We do this with several rules or properties regarding exponents.

a **Multiplication and Division**

To see how to multiply, or simplify, in an expression such as $a^3 \cdot a^2$, we use the definition of exponential notation:

$$a^3 \cdot a^2 = \underbrace{a \cdot a \cdot a}_{3 \text{ factors}} \cdot \underbrace{a \cdot a}_{2 \text{ factors}} = a^5$$

The exponent in a^5 is the *sum* of those in $a^3 \cdot a^2$. In general, the exponents are added when we multiply, but note that the base must be the same in all factors. This is true for any integer exponents, even those that may be negative or zero.

> **THE PRODUCT RULE**
>
> For any number a and any integers m and n,
>
> $$a^m \cdot a^n = a^{m+n}.$$
>
> (When multiplying with exponential notation, add the exponents if the bases are the same.)

EXAMPLES Multiply and simplify.

1. $x^4 \cdot x^3 = x^{4+3} = x^7$

2. $4^5 \cdot 4^{-3} = 4^{5+(-3)} = 4^2 = 16$

3. $(-2)^{-3}(-2)^7 = (-2)^{-3+7}$
$= (-2)^4 = 16$

4. $(8x^n)(6x^{2n}) = 8 \cdot 6x^{n+2n}$
$= 48x^{3n}$

5. $(8x^4y^{-2})(-3x^{-3}y) = 8 \cdot (-3) \cdot x^4 \cdot x^{-3} \cdot y^{-2} \cdot y^1$ Using the associative and the commutative laws

$= -24x^{4-3}y^{-2+1}$ Using the product rule

$= -24xy^{-1} = -\dfrac{24x}{y}$ Using $a^{-n} = \dfrac{1}{a^n}$

Note that we give answers using positive exponents. In some situations, this may not be appropriate, but we do so here.

Do Exercises 1–7.

Consider this division:

$$\frac{8^5}{8^3} = \frac{8 \cdot 8 \cdot 8 \cdot 8 \cdot 8}{8 \cdot 8 \cdot 8} = \frac{8 \cdot 8 \cdot 8}{8 \cdot 8 \cdot 8} \cdot 8 \cdot 8 = 8 \cdot 8 = 8^2.$$

We can obtain the result by subtracting exponents. This is always the case, even if exponents are negative or zero.

Divide and simplify.

8. $\dfrac{4^8}{4^5}$

THE QUOTIENT RULE

For any nonzero number a and any integers m and n,

$$\frac{a^m}{a^n} = a^{m-n}.$$

(When dividing with exponential notation, subtract the exponent of the denominator from the exponent of the numerator, if the bases are the same.)

9. $\dfrac{5^4}{5^{-2}}$

EXAMPLES   Divide and simplify.

6. $\dfrac{5^7}{5^3} = 5^{7-3} = 5^4$ Subtracting exponents using the quotient rule

7. $\dfrac{5^7}{5^{-3}} = 5^{7-(-3)} = 5^{7+3} = 5^{10}$ Subtracting exponents (adding an opposite)

8. $\dfrac{9^{-2}}{9^5} = 9^{-2-5} = 9^{-7} = \dfrac{1}{9^7}$

10. $\dfrac{10^{-8}}{10^{-2}}$

9. $\dfrac{7^{-4}}{7^{-5}} = 7^{-4-(-5)} = 7^{-4+5} = 7^1 = 7$

10. $\dfrac{16x^4y^7}{-8x^3y^9} = \dfrac{16}{-8} \cdot \dfrac{x^4}{x^3} \cdot \dfrac{y^7}{y^9} = -2xy^{-2} = -\dfrac{2x}{y^2}$

> The answers $\dfrac{-2x}{y^2}$ or $\dfrac{2x}{-y^2}$ would also be correct here.

11. $\dfrac{40x^{-2n}}{4x^{5n}} = \dfrac{40}{4}x^{-2n-5n} = 10x^{-7n} = \dfrac{10}{x^{7n}}$

11. $\dfrac{45x^{5n}}{-9x^{3n}}$

12. $\dfrac{14x^7y^{-3}}{4x^5y^{-5}} = \dfrac{14}{4} \cdot \dfrac{x^7}{x^5} \cdot \dfrac{y^{-3}}{y^{-5}} = \dfrac{7}{2}x^2y^2$

12. $\dfrac{42y^7x^6}{-21y^{-3}x^{10}}$

In exercises such as Examples 6–12 above, it may help to think as follows: After writing the base, write the top exponent. Then write a subtraction sign. Then write the bottom exponent. Then do the subtraction. For example,

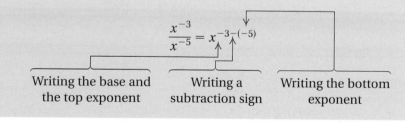

Writing the base and the top exponent   Writing a subtraction sign   Writing the bottom exponent

13. $\dfrac{33a^5b^{-2}}{22a^2b^{-4}}$

Do Exercises 8–13.

Answers on page A-3

Simplify.

14. $(3^7)^6$

15. $(z^{-4})^{-5}$

16. $(t^2)^{-7m}$

b Raising Powers to Powers and Products and Quotients to Powers

When an expression inside parentheses is raised to a power, the inside expression is the base. Consider an expression like $(5^2)^4$. In this case, we are raising 5^2 to the fourth power:

$$(5^2)^4 = (5^2)(5^2)(5^2)(5^2)$$
$$= (5 \cdot 5)(5 \cdot 5)(5 \cdot 5)(5 \cdot 5)$$
$$= 5 \cdot 5 \cdot 5 \cdot 5 \cdot 5 \cdot 5 \cdot 5 \cdot 5 \qquad \text{Using an associative law}$$
$$= 5^8.$$

Note that here we could have multiplied the exponents:

$$(5^2)^4 = 5^{2 \cdot 4} = 5^8.$$

Likewise, $(y^8)^3 = (y^8)(y^8)(y^8) = y^{24}$. Once again, we get the same result if we multiply the exponents:

$$(y^8)^3 = y^{8 \cdot 3} = y^{24}.$$

THE POWER RULE

For any real number a and any integers m and n,

$$(a^m)^n = a^{mn}.$$

(To raise a power to a power, multiply the exponents.)

EXAMPLES Simplify.

13. $(x^5)^7 = x^{5 \cdot 7} \qquad$ Multiply exponents.
$$= x^{35}$$

14. $(y^{-2})^{-2} = y^{(-2)(-2)}$
$$= y^4$$

15. $(x^{-5})^4 = x^{-5 \cdot 4}$
$$= x^{-20} = \frac{1}{x^{20}}$$

16. $(x^4)^{-2t} = x^{4(-2t)}$
$$= x^{-8t} = \frac{1}{x^{8t}}$$

Do Exercises 14–16.

Let's compare $2a^3$ and $(2a)^3$:

$$2a^3 = 2 \cdot a \cdot a \cdot a \qquad \text{The base is } a.$$

and

$$(2a)^3 = (2a)(2a)(2a) \qquad \text{The base is } 2a.$$
$$= (2 \cdot 2 \cdot 2)(a \cdot a \cdot a) \qquad \text{Using the associative law of multiplication}$$
$$= 2^3 a^3 = 8a^3.$$

We see that $2a^3$ and $(2a)^3$ are *not* equivalent. We also see that we can evaluate the power $(2a)^3$ by raising each factor to the power 3. This leads us to the following rule for raising a product to a power.

RAISING A PRODUCT TO A POWER

For any real numbers a and b and any integer n,

$$(ab)^n = a^n b^n.$$

(To raise a product to the nth power, raise each factor to the nth power.)

EXAMPLES Simplify.

17. $(3x^2 y^{-2})^3 = 3^3 (x^2)^3 (y^{-2})^3 = 3^3 x^6 y^{-6} = 27 x^6 y^{-6} = \dfrac{27 x^6}{y^6}$

18. $(5x^3 y^{-5} z^2)^4 = 5^4 (x^3)^4 (y^{-5})^4 (z^2)^4 = 625 x^{12} y^{-20} z^8 = \dfrac{625 x^{12} z^8}{y^{20}}$

Do Exercises 17–20.

There is a similar rule for raising a quotient to a power.

RAISING A QUOTIENT TO A POWER

For any real numbers a and b, and any integer n,

$$\left(\frac{a}{b}\right)^n = \frac{a^n}{b^n}, \, b \neq 0; \quad \text{and} \quad \left(\frac{a}{b}\right)^{-n} = \left(\frac{b}{a}\right)^n = \frac{b^n}{a^n}, \, a \neq 0, b \neq 0.$$

(To raise a quotient to the nth power, raise the numerator to the nth power and divide by the denominator to the nth power.)

EXAMPLES Simplify. Write the answer using positive exponents.

19. $\left(\dfrac{x^2}{y^{-3}}\right)^{-5} = \dfrac{x^{2 \cdot (-5)}}{y^{-3 \cdot (-5)}} = \dfrac{x^{-10}}{y^{15}} = \dfrac{1}{x^{10} y^{15}}$

20. $\left(\dfrac{2x^3 y^{-2}}{3y^4}\right)^5 = \dfrac{(2x^3 y^{-2})^5}{(3y^4)^5} = \dfrac{2^5 (x^3)^5 (y^{-2})^5}{3^5 (y^4)^5} = \dfrac{32 x^{15} y^{-10}}{243 y^{20}}$

$\qquad = \dfrac{32 x^{15} y^{-10-20}}{243} = \dfrac{32 x^{15} y^{-30}}{243} = \dfrac{32 x^{15}}{243 y^{30}}$

21. $\left[\dfrac{-3a^{-5} b^3}{2a^{-2} b^{-4}}\right]^{-2} = \dfrac{(-3a^{-5} b^3)^{-2}}{(2a^{-2} b^{-4})^{-2}} = \dfrac{(-3)^{-2} (a^{-5})^{-2} (b^3)^{-2}}{2^{-2} (a^{-2})^{-2} (b^{-4})^{-2}}$

$\qquad = \dfrac{\frac{1}{(-3)^2} a^{10} b^{-6}}{\frac{1}{2^2} a^4 b^8} = \dfrac{2^2}{(-3)^2} a^{10-4} b^{-6-8}$

$\qquad = \dfrac{4}{9} a^6 b^{-14} = \dfrac{4a^6}{9b^{14}}$

An alternative way to carry out Example 21 is to first write the expression with a positive exponent, as follows:

$$\left[\dfrac{-3a^{-5} b^3}{2a^{-2} b^{-4}}\right]^{-2} = \left[\dfrac{2a^{-2} b^{-4}}{-3a^{-5} b^3}\right]^2 = \dfrac{(2a^{-2} b^{-4})^2}{(-3a^{-5} b^3)^2} = \dfrac{2^2 (a^{-2})^2 (b^{-4})^2}{(-3)^2 (a^{-5})^2 (b^3)^2}$$

$$= \dfrac{4a^{-4} b^{-8}}{9a^{-10} b^6} = \dfrac{4}{9} a^{-4-(-10)} b^{-8-6} = \dfrac{4}{9} a^6 b^{-14} = \dfrac{4a^6}{9b^{14}}.$$

Simplify.

17. $(2xy)^3$

18. $(4x^{-2} y^7)^2$

19. $(-2x^4 y^2)^5$

20. $(10x^{-4} y^7 z^{-2})^3$

Answers on page A-3

R.7 Properties of Exponents and Scientific Notation

Do Exercises 21–23.

c Scientific Notation

There are many kinds of symbolism, or *notation*, for numbers. You are already familiar with fraction notation, decimal notation, and percent notation. Now we study another, **scientific notation,** which is especially useful when calculations involve very large or very small numbers and when estimating.

The following are examples of scientific notation:

- *The distance from the sun to the planet Pluto*:

  3.664×10^9 mi $= 3,664,000,000$ mi

- *The diameter of a helium atom*:

  2.2×10^{-8} cm $= 0.000000022$ cm

- There are 3.778×10^8 pets in the United States. In a recent year, Americans spent $\$3.24 \times 10^{10}$ on pets. This amount includes $\$1.37 \times 10^{10}$ for food and $\$7.9 \times 10^9$ for veterinary care.

  Source: 2003–2004 National Pet Owners Survey, American Pet Products Manufacturers Association

- Elvis Presley's total album sales were 1.175×10^8. The Beatles hold the all-time album sales record of 1.665×10^8.

  Source: Recording Industry Association of America

Simplify.

21. $\left(\dfrac{x^{-3}}{y^4}\right)^{-3}$

22. $\left(\dfrac{3x^2 y^{-3}}{y^5}\right)^2$

23. $\left[\dfrac{-3a^{-5}b^3}{2a^{-2}b^{-4}}\right]^{-3}$

SCIENTIFIC NOTATION

Scientific notation for a number is an expression of the type

$$M \times 10^n,$$

where n is an integer, M is greater than or equal to 1 and less than 10 $(1 \le M < 10)$, and M is expressed in decimal notation. 10^n is also considered to be scientific notation when $M = 1$.

Answers on page A-3

You should try to make conversions to scientific notation mentally as much as possible. Here is a handy mental device.

> A positive exponent in scientific notation indicates a large number (greater than or equal to 10) and a negative exponent indicates a small number (between 0 and 1).

EXAMPLES Convert mentally to scientific notation.

22. Light travels 9,460,000,000,000 km in one year.

$$9,460,000,000,000 = 9.46 \times 10^{12} \qquad 9.460,000,000,000.$$

12 places

Large number, so the exponent is positive.

23. The mass of a grain of sand is 0.0648 g (grams).

$$0.0648 = 6.48 \times 10^{-2} \qquad 0.06.48$$

2 places

Small number, so the exponent is negative.

EXAMPLES Convert mentally to decimal notation.

24. $4.893 \times 10^5 = 489,300 \qquad 4.89300.$

5 places

Positive exponent, so the answer is a large number.

25. $8.7 \times 10^{-8} = 0.000000087 \qquad 0.00000008.7$

8 places

Negative exponent, so the answer is a small number.

Each of the following is *not* scientific notation.

$$13.95 \times 10^{13}, \qquad\qquad 0.468 \times 10^{-8}$$

This number is greater than 10.　　This number is less than 1.

Do Exercises 24–27.

Multiplying and dividing in scientific notation is easy because we can use the properties of exponents.

EXAMPLE 26 Multiply and write scientific notation for the answer: $(3.1 \times 10^5)(4.5 \times 10^{-3})$.

We apply the commutative and the associative laws to get

$$(3.1 \times 10^5)(4.5 \times 10^{-3}) = (3.1 \times 4.5)(10^5 \times 10^{-3}) = 13.95 \times 10^2.$$

To find scientific notation for the result, we convert 13.95 to scientific notation and then simplify:

$$13.95 \times 10^2 = (1.395 \times 10^1) \times 10^2 = 1.395 \times 10^3.$$

Do Exercises 28 and 29.

Convert to scientific notation.

24. Light travels 5,880,000,000,000 mi in one year.

25. 0.000000000257

Convert to decimal notation.

26. 4.567×10^{-13}

27. The distance from the earth to the sun is 9.3×10^7 mi. (This was the answer to a \$1 million question on the TV show "Who Wants to Be a Millionaire?")

Multiply and write scientific notation for the answer.

28. $(9.1 \times 10^{-17})(8.2 \times 10^3)$

29. $(1.12 \times 10^{-8})(5 \times 10^{-7})$

Answers on page A-3

Divide and write scientific notation for the answer.

30. $\dfrac{4.2 \times 10^5}{2.1 \times 10^2}$

31. $\dfrac{1.1 \times 10^{-4}}{2.0 \times 10^{-7}}$

32. Light from the Sun to Earth. The distance from Earth to the sun is about 93,000,000 mi. Light travels 1.86×10^5 mi in 1 sec. About how many seconds does it take light from the sun to reach Earth? Write scientific notation for the answer.

Source: *The Handy Science Answer Book*

33. Mass of Jupiter. The mass of the planet Jupiter is about 318 times the mass of Earth. Write scientific notation for the mass of Jupiter. See Example 29.

EXAMPLE 27 Divide and write scientific notation for the answer:

$$\frac{6.4 \times 10^{-7}}{8.0 \times 10^6}.$$

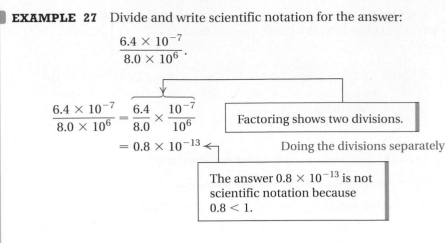

$$\frac{6.4 \times 10^{-7}}{8.0 \times 10^6} = \frac{6.4}{8.0} \times \frac{10^{-7}}{10^6} \quad \boxed{\text{Factoring shows two divisions.}}$$

$$= 0.8 \times 10^{-13} \quad \text{Doing the divisions separately}$$

> The answer 0.8×10^{-13} is not scientific notation because $0.8 < 1$.

$$= (8.0 \times 10^{-1}) \times 10^{-13} \quad \text{Converting 0.8 to scientific notation}$$

$$= 8.0 \times (10^{-1} \times 10^{-13}) \quad \text{Using the associative law of multiplication}$$

$$= 8.0 \times 10^{-14}$$

Do Exercises 30 and 31.

EXAMPLE 28 *Light from the Sun to Neptune.* The planet Neptune is about 2,790,000,000 mi from the sun. Light travels 1.86×10^5 mi in 1 sec. About how many seconds does it take light from the sun to reach Neptune? Write scientific notation for the answer.

Source: *The Handy Science Answer Book*

The time it takes light to travel from the sun to Neptune is

$$\frac{2{,}790{,}000{,}000}{1.86 \times 10^5} = \frac{2.79 \times 10^9}{1.86 \times 10^5} = \frac{2.79}{1.86} \times \frac{10^9}{10^5} = 1.5 \times 10^4 \text{ sec.}$$

EXAMPLE 29 *Mass of the Sun.* The mass of Earth is about 5.98×10^{24} kg. The mass of the sun is about 333,000 times the mass of Earth. Write scientific notation for the mass of the sun.

Source: *The Handy Science Answer Book*

The mass of the sun is 333,000 times the mass of Earth. We convert to scientific notation and multiply:

$$(333{,}000)(5.98 \times 10^{24}) = (3.33 \times 10^5)(5.98 \times 10^{24})$$

$$= (3.33 \times 5.98)(10^5 \times 10^{24})$$

$$= 19.9134 \times 10^{29}$$

$$= (1.99134 \times 10^1) \times 10^{29}$$

$$= 1.99134 \times 10^{30} \text{ kg.}$$

Do Exercises 32 and 33.

Answers on page A-3

Scientific Notation   To enter a number in scientific notation on a graphing calculator, we first type the decimal portion of the number. Then we press **2ND** **EE**. (EE is the second operation associated with the **,** key.) Finally, we type the exponent, which can be at most two digits. For example, to enter 2.36×10^{-8} in scientific notation, we press **2** **.** **3** **6** **2ND** **EE** **(-)** **8** **ENTER**. The decimal portion of the number appears before a small E and the exponent follows the E.

    2.36ᴇ−8
                    2.36ᴇ−8

The graphing calculator can be used to perform computations using scientific notation. To find the product in Example 26 and express the result in scientific notation, we first set the calculator in Scientific mode by pressing **MODE**, positioning the cursor over Sci on the first line, and pressing **ENTER**. Then we press **2ND** **QUIT** to go to the home screen and enter the computation by pressing **3** **.** **1** **2ND** **EE** **5** **×** **4** **.** **5** **2ND** **EE** **(-)** **3** **ENTER**.

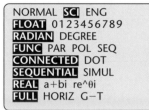

    NORMAL **SCI** ENG
    FLOAT 0123456789
    RADIAN DEGREE
    FUNC PAR POL SEQ
    CONNECTED DOT
    SEQUENTIAL SIMUL
    REAL a+bi re^θi
    FULL HORIZ G−T

    3.1ᴇ5∗4.5ᴇ−3
                    1.395ᴇ3

Exercises: Multiply or divide and express the answer in scientific notation.

1. $(5.13 \times 10^8)(2.4 \times 10^{-13})$

2. $(3.45 \times 10^{-4})(7.1 \times 10^6)$

3. $(7 \times 10^9)(4 \times 10^{-5})$

4. $(6 \times 10^6)(9 \times 10^7)$

5. $\dfrac{4.8 \times 10^6}{1.6 \times 10^{12}}$

6. $\dfrac{7.2 \times 10^{-5}}{1.2 \times 10^{-10}}$

7. $\dfrac{6 \times 10^{-10}}{5 \times 10^4}$

8. $\dfrac{12 \times 10^9}{4 \times 10^{-3}}$

a   Multiply and simplify.

1. $3^6 \cdot 3^3$

2. $8^2 \cdot 8^6$

3. $6^{-6} \cdot 6^2$

4. $9^{-5} \cdot 9^3$

5. $8^{-2} \cdot 8^{-4}$

6. $9^{-1} \cdot 9^{-6}$

7. $b^2 \cdot b^{-5}$

8. $a^4 \cdot a^{-3}$

9. $a^{-3} \cdot a^4 \cdot a^2$

10. $x^{-8} \cdot x^5 \cdot x^3$

11. $(2x)^3 \cdot (3x)^2$

12. $(9y)^2 \cdot (2y)^3$

13. $(14m^2n^3)(-2m^3n^2)$

14. $(6x^5y^{-2})(-3x^2y^3)$

15. $(-2x^{-3})(7x^{-8})$

16. $(6x^{-4}y^3)(-4x^{-8}y^{-2})$

17. $(15x^{4t})(7x^{-6t})$

18. $(9x^{-4n})(-4x^{-8n})$

19. $(2y^{3m})(-4y^{-9m})$

20. $(-3t^{-4a})(-5t^{-a})$

Divide and simplify.

21. $\dfrac{8^9}{8^2}$

22. $\dfrac{7^8}{7^2}$

23. $\dfrac{6^3}{6^{-2}}$

24. $\dfrac{5^{10}}{5^{-3}}$

25. $\dfrac{10^{-3}}{10^6}$

26. $\dfrac{12^{-4}}{12^8}$

27. $\dfrac{9^{-4}}{9^{-6}}$

28. $\dfrac{2^{-7}}{2^{-5}}$

29. $\dfrac{x^{-4n}}{x^{6n}}$

30. $\dfrac{y^{-3t}}{y^{8t}}$

31. $\dfrac{w^{-11q}}{w^{-6q}}$

32. $\dfrac{m^{-7t}}{m^{-5t}}$

33. $\dfrac{a^3}{a^{-2}}$

34. $\dfrac{y^4}{y^{-5}}$

35. $\dfrac{27x^7z^5}{-9x^2z}$

36. $\dfrac{24a^5b^3}{-8a^4b}$

37. $\dfrac{-24x^6y^7}{18x^{-3}y^9}$

38. $\dfrac{14a^4b^{-3}}{-8a^8b^{-5}}$

39. $\dfrac{-18x^{-2}y^3}{-12x^{-5}y^5}$

40. $\dfrac{-14a^{14}b^{-5}}{-18a^{-2}b^{-10}}$

b   Simplify.

41. $(4^3)^2$

42. $(5^4)^5$

43. $(8^4)^{-3}$

44. $(9^3)^{-4}$

45. $(6^{-4})^{-3}$

46. $(7^{-8})^{-5}$

47. $(5a^2b^2)^3$

48. $(2x^3y^4)^5$

49. $(-3x^3y^{-6})^{-2}$

50. $(-3a^2b^{-5})^{-3}$

51. $(-6a^{-2}b^3c)^{-2}$

52. $(-8x^{-4}y^5z^2)^{-4}$

53. $\left(\dfrac{4^{-3}}{3^4}\right)^3$

54. $\left(\dfrac{5^2}{4^{-3}}\right)^{-3}$

55. $\left(\dfrac{2x^3y^{-2}}{3y^{-3}}\right)^3$

56. $\left(\dfrac{-4x^4y^{-2}}{5x^{-1}y^4}\right)^{-4}$

57. $\left(\dfrac{125a^2b^{-3}}{5a^4b^{-2}}\right)^{-5}$

58. $\left(\dfrac{-200x^3y^{-5}}{8x^5y^{-7}}\right)^{-4}$

59. $\left(\dfrac{-6^5y^4z^{-5}}{2^{-2}y^{-2}z^3}\right)^6$

60. $\left(\dfrac{9^{-2}x^{-4}y}{3^{-3}x^{-3}y^2}\right)^8$

61. $[(-2x^{-4}y^{-2})^{-3}]^{-2}$

62. $[(-4a^{-4}b^{-5})^{-3}]^4$

63. $\left(\dfrac{3a^{-2}b}{5a^{-7}b^5}\right)^{-7}$

64. $\left(\dfrac{2x^2y^{-2}}{3x^8y^7}\right)^9$

65. $\dfrac{10^{2a+1}}{10^{a+1}}$

66. $\dfrac{11^{b+2}}{11^{3b-3}}$

67. $\dfrac{9a^{x-2}}{3a^{2x+2}}$

68. $\dfrac{-12x^{a+1}}{4x^{2-a}}$

69. $\dfrac{45x^{2a+4}y^{b+1}}{-9x^{a+3}y^{2+b}}$

70. $\dfrac{-28x^{b+5}y^{4+c}}{7x^{b-5}y^{c-4}}$

71. $(8^x)^{4y}$

72. $(7^{2p})^{3q}$

73. $(12^{3-a})^{2b}$

74. $(x^{a-1})^{3b}$

75. $(5x^{a-1}y^{b+1})^{2c}$

76. $(4x^{3a}y^{2b})^{5c}$

77. $\dfrac{4x^{2a+3}y^{2b-1}}{2x^{a+1}y^{b+1}}$

78. $\dfrac{25x^{a+b}y^{b-a}}{-5x^{a-b}y^{b+a}}$

C   Convert the number to scientific notation.

79. 47,000,000,000

80. 2,600,000,000,000

81. 0.000000016

82. 0.000000263

83. *Population Explosion.*   It has been projected that in 2020, the population of the world will be 7,585,000,000. Write scientific notation for the world population.
Source: U.S. Bureau of the Census, International Database

84. *Cell-Phone Subscribers.*   In 1985, there were 340 thousand cell-phone subscribers. This number increased to 175 million in 2005. Write the number of cell-phone subscribers in 1985 and in 2005 in scientific notation.
Source: *USA Weekend*, December 31, 2004–January 2, 2005

85. *Insect-Eating Lizard.* A gecko is an insect-eating lizard. Its feet will adhere to virtually any surface because they contain millions of miniscule hairs, or setae, that are 200 billionths of a meter wide. Write 200 billionths in scientific notation.

Source: *The Proceedings of the National Academy of Sciences,* Dr. Kellar Autumn and Wendy Hansen of Lewis and Clark College, Portland, Oregon

86. *Firearm Death Rate.* In the United States, the rate of death by firearms is $\dfrac{10.4}{100,000}$ per year. Write scientific notation for the rate of death by firearms.

Source: Centers for Disease Control and Prevention, *National Vital Statistics Reports,* vol 52, no. 3, September 18, 2003

Convert the number to decimal notation.

87. 6.73×10^8

88. 9.24×10^7

89. The wavelength of a certain red light is 6.6×10^{-5} cm.

90. The mass of an electron is 9.11×10^{-28} g.

91. *Reader's Digest* has a total paid circulation of about 1.1×10^7.

Source: Audit Bureau of Circulation, tabulated by Magazine Publishers of America

92. Nicole Kidman reportedly earned $\$3.715 \times 10^6$ for starring in a four-minute commercial for Chanel No. 5.

Source: 2005 *Guinness Book of World Records*

Multiply and write the answer in scientific notation.

93. $(2.3 \times 10^6)(4.2 \times 10^{-11})$

94. $(6.5 \times 10^3)(5.2 \times 10^{-8})$

95. $(2.34 \times 10^{-8})(5.7 \times 10^{-4})$

96. $(3.26 \times 10^{-6})(8.2 \times 10^9)$

Divide and write the answer in scientific notation.

97. $\dfrac{8.5 \times 10^8}{3.4 \times 10^5}$

98. $\dfrac{5.1 \times 10^6}{3.4 \times 10^3}$

99. $\dfrac{4.0 \times 10^{-6}}{8.0 \times 10^{-3}}$

100. $\dfrac{7.5 \times 10^{-9}}{2.5 \times 10^{-4}}$

Write the answers to Exercises 101–110 in scientific notation.

101. Ocean's Twelve *Debut.* The movie *Ocean's Twelve* debuted in 3290 theaters and earned an average of $12,426 per cinema the weekend of the debut. Find the amount that the movie earned in its first weekend.

Source: Exhibitor Relations Co.

102. *Orbit of Venus.* Venus has a nearly circular orbit of the sun. The average distance from the sun to Venus is about 6.71×10^7 mi. How far does Venus travel in one orbit?

103. *Seconds in One Year.* About how many seconds are there in 2000 yr? Assume that there are 365 days in one year.

104. *Popcorn Consumption.* On average, Americans eat 6.5 million gal of popcorn each day. How much popcorn do they eat in one year?

105. *Alpha Centauri.* Other than the sun, the star closest to Earth is Alpha Centauri. Its distance from Earth is about 2.4×10^{13} mi. One light-year = the distance that light travels in one year = 5.88×10^{12} mi. How many light-years is it from Earth to Alpha Centauri?

Source: *The Handy Science Answer Book*

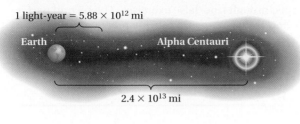

1 light-year = 5.88×10^{12} mi

Earth   Alpha Centauri

2.4×10^{13} mi

106. *Amazon River Water Flow.* The average discharge at the mouth of the Amazon River is 4,200,000 cubic feet per second. How much water is discharged from the Amazon River in one hour? in one year?

107. *Word Knowledge.* There are 300,000 words in the English language. The average person knows about 10,000 of them. What part of the total number of words does the average person know?

108. *Computer Calculations.* An IBM supercomputer has attained a sustained performance of 36.01 trillion calculations per second. How many calculations can be performed in one minute? in one hour?

Source: IBM, *New York Times,* 9/29/04

109. *Printing and Engraving.* A ton of five-dollar bills is worth $4,540,000. How many pounds does a five-dollar bill weigh?

110. *Astronomy.* The brightest star in the night sky, Sirius, is about 4.704×10^{13} mi from Earth. One light-year is 5.88×10^{12} mi. How many light-years is it from Earth to Sirius?

Source: *The Handy Science Answer Book*

111. **D$_\mathbf{W}$** A $20 bill weighs about 2.2×10^{-3} lb. A criminal claims to be carrying $5 million in $20 bills in his suitcase. Is this possible? Why or why not?

112. **D$_\mathbf{W}$** When a calculator indicates that $5^{17} = 7.629394531 \times 10^{11}$, you know that an approximation is being made. How can you tell? (*Hint:* What should the ones digit be?)

SKILL MAINTENANCE

Simplify. [R.3c], [R.6b]

113. $9x - (-4y + 8) + (10x - 12)$

114. $-6t - (5t - 13) + 2(4 - 6t)$

115. $4^2 + 30 \cdot 10 - 7^3 + 16$

116. $5^4 - 38 \cdot 24 - (16 - 4 \cdot 18)$

117. $20 - 5 \cdot 4 - 8$

118. $20 - (5 \cdot 4 - 8)$

SYNTHESIS

Simplify.

119. $\dfrac{(2^{-2})^{-4} \cdot (2^3)^{-2}}{(2^{-2})^2 \cdot (2^5)^{-3}}$

120. $\left[\dfrac{(-3x^{-2}y^5)^{-3}}{(2x^4y^{-8})^{-2}} \right]^2$

121. $\left[\left(\dfrac{a^{-2}}{b^7} \right)^{-3} \cdot \left(\dfrac{a^4}{b^{-3}} \right)^2 \right]^{-1}$

Simplify. Assume that variables in exponents represent integers.

122. $(m^{x-b}n^{x+b})^x(m^b n^{-b})^x$

123. $\left[\dfrac{(2x^a y^b)^3}{(-2x^a y^b)^2} \right]^2$

124. $(x^b y^a \cdot x^a y^b)^c$

The review that follows is meant to prepare you for a chapter exam. It consists of three parts. The first part, Concept Reinforcement, is designed to increase understanding of the concepts through true/false exercises. The second part is a list of important properties and formulas. The third part is the Review Exercises. These provide practice exercises for the exam, together with references to section objectives so you can go back and review. Before beginning, stop and look back over the skills you have obtained. What skills in mathematics do you have now that you did not have before studying this chapter?

CONCEPT REINFORCEMENT

Determine whether the statement is true or false. Answers are given at the back of the book.

_____ **1.** The number 4.6×10^n, where n is an integer, is greater than 0 and less than 1 when $n < 0$.

_____ **2.** For any numbers a and b, $a - b = b - a$.

_____ **3.** Each member of the set of natural numbers is a member of the set of whole numbers.

_____ **4.** The opposite of $-a$ when $a < 0$ is negative.

_____ **5.** Zero is both positive and negative.

_____ **6.** The absolute value of any real number is positive.

_____ **7.** The reciprocal of a negative number is negative.

_____ **8.** If c and d are real numbers and $c + d = 0$, then c and d are additive inverses of each other.

IMPORTANT PROPERTIES AND FORMULAS

Properties of Real Numbers

Commutative Laws:	$a + b = b + a, \quad ab = ba$
Associative Laws:	$a + (b + c) = (a + b) + c, \quad a(bc) = (ab)c$
Distributive Laws:	$a(b + c) = ab + ac, \quad a(b - c) = ab - ac$
Inverses:	$a + (-a) = 0, \quad a \cdot \dfrac{1}{a} = 1$

Identity Property of 0:	$a + 0 = a$
Identity Property of 1:	$1 \cdot a = a$
Property of -1:	$-1 \cdot a = -a$

Properties of Exponents: $a^1 = a, \quad a^0 = 1, \quad a^{-n} = \dfrac{1}{a^n}, \quad \dfrac{1}{a^{-n}} = a^n$

Product Rule:	$a^m \cdot a^n = a^{m+n}$
Power Rule:	$(a^m)^n = a^{mn}$
Quotient Rule:	$\dfrac{a^m}{a^n} = a^{m-n}$
Raising a Product to a Power:	$(ab)^n = a^n b^n$
Raising a Quotient to a Power:	$\left(\dfrac{a}{b}\right)^n = \dfrac{a^n}{b^n}, \quad \left(\dfrac{a}{b}\right)^{-n} = \left(\dfrac{b}{a}\right)^n = \dfrac{b^n}{a^n}$
Scientific Notation:	$M \times 10^n$, or 10^n, where M is such that $1 \le M < 10$.

Review Exercises

PART 1

1. Which of the following numbers are rational? [R.1a]

   $2, \sqrt{3}, -\frac{2}{3}, 0.45\overline{45}, -23.788$

2. Use set-builder notation to name the set of all real numbers less than or equal to 46. [R.1a]

3. Use $<$ or $>$ for $\square$ to write a true sentence: [R.1b]

   $-3.9 \; \square \; 2.9$.

4. Write a different inequality with the same meaning as $19 > x$. [R.1b]

Is each of the following true or false? [R.1b]

5. $-13 \geq 5$

6. $7.01 \leq 7.01$

Graph the inequality on a number line. [R.1c]

7. $x > -4$

8. $x \leq 1$

Find the absolute value. [R.1d]

9. $|-7.23|$

10. $|9 - 9|$

Add, subtract, multiply, or divide, if possible. [R.2a, c, d, e]

11. $6 + (-8)$

12. $-3.8 + (-4.1)$

13. $\frac{3}{4} + \left(-\frac{13}{7}\right)$

14. $-8 - (-3)$

15. $-17.3 - 9.4$

16. $\frac{3}{2} - \left(-\frac{13}{4}\right)$

17. $(-3.8)(-2.7)$

18. $-\frac{2}{3}\left(\frac{9}{14}\right)$

19. $-6(-7)(4)$

20. $-12 \div 3$

21. $\frac{-84}{-4}$

22. $\frac{49}{-7}$

23. $\frac{5}{6} \div \left(-\frac{10}{7}\right)$

24. $-\frac{5}{2} \div \left(-\frac{15}{16}\right)$

25. $\frac{21}{0}$

26. $-108 \div 4.5$

Evaluate $-a$ for each of the following. [R.2b]

27. $a = -7$

28. $a = 2.3$

29. $a = 0$

Write using exponential notation. [R.3a]

30. $a \cdot a \cdot a \cdot a \cdot a$

31. $\left(-\frac{7}{8}\right)\left(-\frac{7}{8}\right)\left(-\frac{7}{8}\right)$

32. Rewrite using a positive exponent: a^{-4}. [R.3b]

33. Rewrite using a negative exponent: $\frac{1}{x^8}$. [R.3b]

Simplify. [R.3c]

34. $2^3 - 3^4 + (13 \cdot 5 + 67)$

35. $64 \div (-4) + (-5)(20)$

PART 2

Translate to an algebraic expression. [R.4a]

36. Five times some number

37. Twenty-eight percent of some number

38. Nine less than t

39. Eight less than the quotient of two numbers

Evaluate. [R.4b]

40. $5x - 7$, when $x = -2$

41. $\dfrac{x - y}{2}$, when $x = 4$ and $y = 20$

42. *Area of a Rug.* The area A of a rectangle is given by the length l times the width w: $A = lw$. Find the area of a rectangular rug that measures 7 ft by 12 ft.

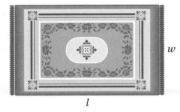

Complete the table by evaluating each expression for the given values. Then look for expressions that are equivalent. [R.5a]

43.

	$x^2 - 5$	$(x + 5)^2$	$(x - 5)^2$	$x^2 + 5$
$x = -1$				
$x = 10$				
$x = 0$				

44.

	$2x - 14$	$2x - 7$	$2(x - 7)$	$2x + 14$
$x = -1$				
$x = 10$				
$x = 0$				

45. Use multiplying by 1 to find an equivalent expression with the given denominator: [R.5b]

$$\frac{7}{3}; \quad 9x.$$

46. Simplify: $\dfrac{-84x}{7x}$. [R.5b]

Use a commutative law to find an equivalent expression. [R.5c]

47. $11 + a$

48. $8y$

Use an associative law to find an equivalent expression. [R.5c]

49. $(9 + a) + b$

50. $8(xy)$

Multiply. [R.5d]

51. $-3(2x - y)$

52. $4ab(2c + 1)$

Factor. [R.5d]

53. $5x + 10y - 5z$

54. $ptr + pts$

Collect like terms. [R.6a]

55. $2x + 6y - 5x - y$

56. $7c - 6 + 9c + 2 - 4c$

57. Find an equivalent expression: $-(-9c + 4d - 3)$. [R.6b]

Simplify. [R.6b]

58. $4(x - 3) - 3(x - 5)$

59. $12x - 3(2x - 5)$

60. $7x - [4 - 5(3x - 2)]$

61. $4m - 3[3(4m - 2) - (5m + 2) + 12]$

Multiply or divide, and simplify. [R.7a]

62. $(2x^4y^{-3})(-5x^3y^{-2})$

63. $\dfrac{-15x^2y^{-5}}{10x^6y^{-8}}$

Simplify. [R.7b]

64. $(-3a^{-4}bc^3)^{-2}$

65. $\left[\dfrac{-2x^4y^{-4}}{3x^{-2}y^6}\right]^{-4}$

Multiply or divide, and write scientific notation for the answer. [R.7c]

66. $\dfrac{2.2 \times 10^7}{3.2 \times 10^{-3}}$

67. $(3.2 \times 10^4)(4.1 \times 10^{-6})$

68. *Volume of a Plastic Sheet.* The volume of a rectangular solid is given by the length l times the width w times the height h: $V = lwh$. A sheet of plastic has a thickness of 0.00015 m. The sheet is 1.2 m by 79 m. Find the volume of the sheet and express the answer in scientific notation. [R.7c]

69. *Finance.* A **mil** is one thousandth of a dollar. The taxation rate in a certain school district is 5.0 mils for every dollar of assessed valuation. The assessed valuation for the district is 13.4 million dollars. How much tax revenue will be raised? [R.7c]

70. $\mathbf{D_W}$ Explain and compare the commutative, associative, and distributive laws. [R.5c, d]

71. $\mathbf{D_W}$ Give two expressions that are equivalent. Give two expressions that are not equivalent and explain why they are not. [R.5a]

SYNTHESIS
---

72. Simplify: $(x^y \cdot x^{3y})^3$. [R.7b]

73. If $a = 2^x$ and $b = 2^{x+5}$, find $a^{-1}b$. [R.7a]

74. Which of the following expressions are equivalent?

a) $3x - 3y$

b) $3x - y$

c) $x^{-2}x^5$

d) x^{-10}

e) x^{-3}

f) $(x^{-2})^5$

g) $x(yz)$

h) $x(y + z)$

i) $3(x - y)$

j) $xy + xz$

R Chapter Test

For Extra Help

Work It Out!
Chapter Test Video
on CD

PART 1

1. Which of the following numbers are irrational?

$$-43, \quad \sqrt{7}, \quad -\frac{2}{3}, \quad 2.3\overline{76}, \quad \pi$$

2. Use set-builder notation to name the set of real numbers greater than 20.

3. Use < or > for ☐ to write a true sentence:

$-4.5 \;\square\; -8.7.$

4. Write a different inequality with the same meaning as $a \le 5$.

Is each of the following true or false?

5. $-6 \ge -6$

6. $-8 \le -6$

7. Graph $x > -2$ on a number line.

$$\xleftarrow{\quad} \overset{}{\underset{-4 \;-3 \;-2 \;-1 \;\; 0 \;\; 1 \;\; 2 \;\; 3 \;\; 4}{\quad\quad\quad\quad\quad\quad}} \xrightarrow{\quad}$$

Find the absolute value.

8. $|0|$

9. $\left|-\dfrac{7}{8}\right|$

Add, subtract, multiply, or divide, if possible.

10. $7 + (-9)$

11. $-5.3 + (-7.8)$

12. $-\dfrac{5}{2} + \left(-\dfrac{7}{2}\right)$

13. $-6 - (-5)$

14. $-18.2 - 11.5$

15. $\dfrac{19}{4} - \left(-\dfrac{3}{2}\right)$

16. $(-4.1)(8.2)$

17. $-\dfrac{4}{5}\left(-\dfrac{15}{16}\right)$

18. $-6(-4)(-11)2$

19. $-75 \div (-5)$

20. $\dfrac{-10}{2}$

21. $-\dfrac{5}{2} \div \left(-\dfrac{15}{16}\right)$

22. $-459.2 \div 5.6$

23. $\dfrac{-3}{0}$

Evaluate $-a$ for each of the following.

24. $a = -13$

25. $a = 0$

26. Write exponential notation: $q \cdot q \cdot q \cdot q$.

27. Rewrite using a negative exponent: $\dfrac{1}{a^9}$.

Simplify.

28. $1 - (2 - 5)^2 + 5 \div 10 \cdot 4^2$

29. $\dfrac{7(5 - 2 \cdot 3) - 3^2}{4^2 - 3^2}$

PART 2

Translate to an algebraic expression.

30. Nine more than t

31. Twelve less than the quotient of two numbers

32. Evaluate $3x - 3y$ when $x = 2$ and $y = -4$.

33. *Area of a Triangular Stamp.* The area A of a triangle is given by $A = \frac{1}{2}bh$. Find the area of a stamp whose base measures 3 cm and whose height measures 2.5 cm.

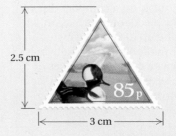

2.5 cm

3 cm

85p

Complete a table by evaluating each expression for $x = -1$, $x = 10$, and $x = 0$. Then determine whether the expressions are equivalent. Answer yes or no.

34. $x(x - 3)$; $x^2 - 3x$

35. $3x + 5x^2$; $8x^2$

36. Use multiplying by 1 to find an equivalent expression with the given denominator.

$$\frac{3}{4};\ 36x$$

37. Simplify:

$$\frac{-54x}{-36x}.$$

Use a commutative law to find an equivalent expression.

38. pq

39. $t + 4$

Use an associative law to find an equivalent expression.

40. $3 + (t + w)$

41. $(4a)b$

Multiply.

42. $-2(3a - 4b)$

43. $3\pi r(s + 1)$

Factor.

44. $ab - ac + 2ad$

45. $2ah + h$

Collect like terms.

46. $6y - 8x + 4y + 3x$

47. $4a - 7 + 17a + 21$

48. Find an equivalent expression without parentheses: $-(-9x + 7y - 22)$.

Simplify.

49. $-3(x + 2) - 4(x - 5)$

50. $4x - [6 - 3(2x - 5)]$

51. $3a - 2[5(2a - 5) - (8a + 4) + 10]$

Multiply or divide, and simplify.

52. $\dfrac{-12x^3y^{-4}}{8x^7y^{-6}}$

53. $(3a^4b^{-2})(-2a^5b^{-3})$

54. $(5a^{4n})(-10a^{5n})$

55. $\dfrac{-60x^{3t}}{12x^{7t}}$

Simplify.

56. $(-3a^{-3}b^2c)^{-4}$

57. $\left[\dfrac{-5a^{-2}b^8}{10a^{10}b^{-4}}\right]^{-4}$

58. Convert to scientific notation: 0.0000437.

Multiply or divide, and write scientific notation for the answer.

59. $(8.7 \times 10^{-9})(4.3 \times 10^{15})$

60. $\dfrac{1.2 \times 10^{-12}}{6.4 \times 10^{-7}}$

61. *Mass of Pluto.* The mass of Earth is 5.98×10^{24} kg. The mass of the planet Pluto is about 0.002 times the mass of Earth. Find the mass of Pluto and express the answer in scientific notation.

SYNTHESIS

62. *CD Memory Storage.* The music on a compact disc is recorded using small pits, or pieces of information, in the disc surface. There can be up to 600 megabytes on a single disc. A megabyte is 1 million bytes and a byte is 8 pieces of information. How many pits, or pieces of information, can be on the surface of a compact disc? Write scientific notation for your answer.

63. Which of the following expressions are equivalent?

a) $x^{-3}x^{-4}$
b) x^{12}
c) x^{-12}
d) $5x + 5$
e) $(x^{-3})^{-4}$
f) $5(x + 1)$
g) $5x$
h) $5 + 5x$
i) $5(xy)$
j) $(5x)y$

Solving Linear Equations and Inequalities

1

Real-World Application

The Iditarod is a 1150-mi dogsled race run from Anchorage to Nome, Alaska, each year in March. The 2005 race was won by Robert Sorlie. If, at one point, Sorlie was four times as far from the end of the course as he was from the starting point, how many miles of the course had he completed?

Source: espn.com

This problem appears as Exercise 2 in Section 1.3.

SOLVING EQUATIONS

Objectives

a Determine whether a given number is a solution of a given equation.

b Solve equations using the addition principle.

c Solve equations using the multiplication principle.

d Solve equations using the addition and multiplication principles together, removing parentheses where appropriate.

Consider the following equations.

a) $3 + 4 = 7$
b) $5 - 1 = 2$
c) $21 + 2 = 24$
d) $x - 5 = 12$
e) $9 - x = x$
f) $13 + 2 = 15$

1. Which equations are true?

2. Which equations are false?

3. Which equations are neither true nor false?

Answers on page A-4

a Equations and Solutions

In order to solve many kinds of problems, we must be able to solve *equations*. Some examples of equations are

$$3 + 5 = 8, \qquad 15 - 10 = 2 + 3,$$
$$x + 8 = 23, \qquad 5x - 2 = 9 - x.$$

EQUATION

An **equation** is a number sentence that says that the expressions on either side of the equals sign, =, represent the same number.

The sentence "$15 - 10 = 2 + 3$" asserts that the expressions $15 - 10$ and $2 + 3$ name the same number. It is a *true* equation because both represent the number 5.

Some equations are true. Some are false. Some are neither true nor false.

EXAMPLES Determine whether the equation is true, false, or neither.

1. $1 + 10 = 11$ Both expressions represent 11. The equation is *true*.

2. $7 - 8 = 9 - 13$ $7 - 8$ represents -1 and $9 - 13$ represents -4. The equation is *false*.

3. $x - 9 = 3$ The equation is *neither* true nor false, because we do not know what number x represents.

Do Exercises 1–3.

If an equation contains a variable, then some replacements or values of the variable may make it true and some may make it false.

SOLUTION OF AN EQUATION

The replacements for the variable that make an equation true are called the **solutions** of the equation. The set of all solutions is called the **solution set** of the equation. When we find all the solutions, we say that we have **solved** the equation.

To determine whether a number is a solution of an equation, we evaluate the algebraic expression on each side of the equals sign by substitution. If the values are the same, then the number is a solution of the equation. If they are not, then the number is not a solution.

EXAMPLE 4 Determine whether 5 is a solution of $x + 6 = 11$.

We have

$$\frac{x + 6 = 11}{5 + 6 \ ? \ 11} \qquad \text{Writing the equation}$$
$$\qquad\qquad \text{Substituting 5 for } x$$
$$11 \ | \qquad \text{TRUE}$$

Since the left-hand and the right-hand sides are the same, 5 is a solution of the equation.

EXAMPLE 5 Determine whether 18 is a solution of $2x - 3 = 5$.

We have

$$
\begin{array}{ll}
2x - 3 = 5 & \text{Writing the equation} \\
\hline
2 \cdot 18 - 3 \;?\; 5 & \text{Substituting 18 for } x \\
36 - 3 & \\
33 & \text{FALSE}
\end{array}
$$

Since the left-hand and the right-hand sides are not the same, 18 is not a solution of the equation.

Do Exercises 4–6.

EQUIVALENT EQUATIONS

Consider the equation

$$x = 5.$$

The solution of this equation is easily "seen" to be 5. If we replace x with 5, we get

$$5 = 5, \quad \text{which is true.}$$

In Example 4, we saw that the solution of the equation $x + 6 = 11$ is also 5, but the fact that 5 is the solution is not so readily apparent. We now consider principles that allow us to start with one equation and end up with an *equivalent equation*, like $x = 5$, in which the variable is alone on one side, and for which the solution is read directly from the equation.

> **EQUIVALENT EQUATIONS**
>
> Equations with the same solutions are called **equivalent equations.**

Do Exercises 7 and 8.

b   The Addition Principle

One of the principles we use in solving equations involves addition. The equation $a = b$ says that a and b represent the same number. Suppose that $a = b$ is true and then add a number c to a. We will get the same result if we add c to b, because a and b are the same number.

> **THE ADDITION PRINCIPLE**
>
> For any real numbers a, b, and c,
>
> $$a = b \quad \text{is equivalent to} \quad a + c = b + c.$$

When we use the addition principle, we sometimes say that we "add the same number on both sides of an equation." We can also "subtract the same number on both sides of an equation," because we can express subtraction as the addition of an opposite. That is,

$$a - c = b - c \quad \text{is equivalent to} \quad a + (-c) = b + (-c).$$

Determine whether the given number is a solution of the given equation.

4. 8; $x + 5 = 13$

5. -4; $7x = 16$

6. 5; $2x + 3 = 13$

7. Determine whether

$$3x + 2 = 11 \quad \text{and} \quad x = 3$$

are equivalent.

8. Determine whether

$$4 - 5x = -11 \quad \text{and} \quad x = -3$$

are equivalent.

Answers on page A-4

EXAMPLE 6 Solve: $x + 6 = 11$.

$$x + 6 = 11$$

$$\left.\begin{array}{l} x + 6 + (-6) = 11 + (-6) \\ x + 6 - 6 = 11 - 6 \end{array}\right\}$$ Using the addition principle: adding -6 on both sides or subtracting 6 on both sides. Note that 6 and -6 are opposites.

$$x + 0 = 5$$ Simplifying

$$x = 5$$ Using the identity property of 0: $x + 0 = x$

Check: $$\begin{array}{c} x + 6 = 11 \\ \hline 5 + 6 \; ? \; 11 \\ 11 \; | \end{array}$$ Substituting 5 for x

$\qquad$ TRUE

The solution is 11.

In Example 6, we wanted to get x alone so that we could readily see the solution, so we added the opposite of 6. This eliminated the 6 on the left, giving us the *additive identity* 0, which when added to x is x. We began with $x + 6 = 11$. Using the addition principle, we derived a simpler equation, $x = 5$. The equations $x + 6 = 11$ and $x = 5$ are *equivalent*.

EXAMPLE 7 Solve: $y - 4.7 = 13.9$.

$$y - 4.7 = 13.9$$

$$y - 4.7 + 4.7 = 13.9 + 4.7$$ Using the addition principle: adding 4.7 on both sides. Note that -4.7 and 4.7 are opposites.

$$y + 0 = 18.6$$ Simplifying

$$y = 18.6$$ Using the identity property of 0: $y + 0 = y$

Check: $$\begin{array}{c} y - 4.7 = 13.9 \\ \hline 18.6 - 4.7 \; ? \; 13.9 \\ 13.9 \; | \end{array}$$ Substituting 18.6 for y

$\qquad$ TRUE

The solution is 18.6.

EXAMPLE 8 Solve: $-\frac{3}{8} + x = -\frac{5}{7}$.

$$-\frac{3}{8} + x = -\frac{5}{7}$$

$$\frac{3}{8} + \left(-\frac{3}{8}\right) + x = \frac{3}{8} + \left(-\frac{5}{7}\right)$$ Using the addition principle: adding $\frac{3}{8}$

$$0 + x = \frac{3}{8} - \frac{5}{7}$$

$$x = \frac{3}{8} \cdot \frac{7}{7} - \frac{5}{7} \cdot \frac{8}{8}$$ Multiplying by 1 to obtain the least common denominator

$$x = \frac{21}{56} - \frac{40}{56}$$

$$x = -\frac{19}{56}$$

Check:

$$-\frac{3}{8} + x = -\frac{5}{7}$$

$$\overline{-\frac{3}{8} + \left(-\frac{19}{56}\right) \;?\; -\frac{5}{7}} \qquad \text{Substituting } -\frac{19}{56} \text{ for } x$$

$$-\frac{3}{8} \cdot \frac{7}{7} + \left(-\frac{19}{56}\right)$$

$$-\frac{21}{56} + \left(-\frac{19}{56}\right)$$

$$-\frac{40}{56}$$

$$-\frac{5}{7} \qquad \text{TRUE}$$

The solution is $-\frac{19}{56}$.

Do Exercises 9–12.

Solve using the addition principle.

9. $x + 9 = 2$

 C The Multiplication Principle

A second principle for solving equations involves multiplication. Suppose that $a = b$ is true and we multiply a by a nonzero number c. We get the same result if we multiply b by c, because a and b are the same number.

THE MULTIPLICATION PRINCIPLE

For any real numbers a, b, and c, $c \neq 0$,

$$a = b \quad \text{is equivalent to} \quad a \cdot c = b \cdot c.$$

---

10. $x + \dfrac{1}{4} = -\dfrac{3}{5}$

EXAMPLE 9 Solve: $\frac{4}{5}x = 22$.

$$\frac{4}{5}x = 22$$

$$\frac{5}{4} \cdot \frac{4}{5}x = \frac{5}{4} \cdot 22 \qquad \text{Multiplying by } \tfrac{5}{4}, \text{ the reciprocal of } \tfrac{4}{5}$$

$$1 \cdot x = \frac{55}{2} \qquad \text{Multiplying and simplifying}$$

$$x = \frac{55}{2} \qquad \text{Using the identity property of 1: } 1 \cdot x = x$$

11. $13 = -25 + y$

Check:

$$\frac{4}{5}x = 22$$

$$\overline{\frac{4}{5} \cdot \frac{55}{2} \;?\; 22}$$

$$22 \Big| \qquad \text{TRUE}$$

The solution is $\frac{55}{2}$.

In Example 9, in order to get x alone, we multiplied by the *multiplicative inverse*, or *reciprocal*, of $\frac{4}{5}$. When we multiplied, we got the *multiplicative identity* 1 times x, or $1 \cdot x$, which simplified to x. This enabled us to eliminate the $\frac{4}{5}$ on the left.

The multiplication principle also tells us that we can "divide both sides by a nonzero number" because division is the same as multiplying by a reciprocal. That is,

$$\frac{a}{c} = \frac{b}{c} \quad \text{is equivalent to} \quad a \cdot \frac{1}{c} = b \cdot \frac{1}{c}, \quad \text{when } c \neq 0.$$

In a product like $\frac{4}{5}x$, the number in front of the variable is called the **coefficient.** When this number is in fraction notation, it is usually most convenient to multiply both sides by its reciprocal. If the coefficient is an integer or is in decimal notation, it is usually more convenient to divide by the coefficient.

12. $y - 61.4 = 78.9$

Answers on page A-4

Solve using the multiplication principle.

13. $8x = 10$

14. $-\dfrac{3}{7}y = 21$

15. $-4x = -\dfrac{6}{7}$

16. Solve: $-12.6 = 4.2y$.

Solve.

17. $-\dfrac{x}{8} = 17$

18. $-x = -5$

Answers on page A-4

EXAMPLE 10 Solve: $4x = 9$.

$$4x = 9$$

$$\dfrac{4x}{4} = \dfrac{9}{4} \qquad \text{Using the multiplication principle: multiplying both sides by } \tfrac{1}{4} \text{ or dividing both sides by the coefficient, 4}$$

$$1 \cdot x = \dfrac{9}{4} \qquad \text{Simplifying}$$

$$x = \dfrac{9}{4} \qquad \text{Using the identity property of 1: } 1 \cdot x = x$$

Check: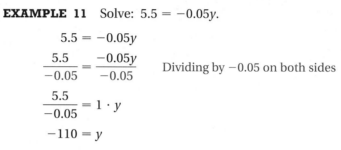

The solution is $\frac{9}{4}$.

Do Exercises 13–15.

EXAMPLE 11 Solve: $5.5 = -0.05y$.

$$5.5 = -0.05y$$

$$\dfrac{5.5}{-0.05} = \dfrac{-0.05y}{-0.05} \qquad \text{Dividing by } -0.05 \text{ on both sides}$$

$$\dfrac{5.5}{-0.05} = 1 \cdot y$$

$$-110 = y$$

The check is left to the student. The solution is -110.

Note that equations are reversible. That is, $a = b$ is equivalent to $b = a$. Thus, $-110 = y$ and $y = -110$ are equivalent, and the solution of both equations is -110.

Do Exercise 16.

EXAMPLE 12 Solve: $-\dfrac{x}{4} = 10$.

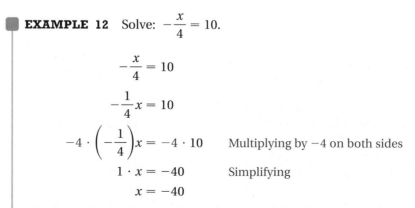

The check is left to the student. The solution is -40.

Do Exercises 17 and 18.

d Using the Principles Together

Let's see how we can use the addition and multiplication principles together.

EXAMPLE 13 Solve: $3x - 4 = 13$.

$$3x - 4 = 13$$

$$3x - 4 + 4 = 13 + 4 \qquad \text{Using the addition principle: adding 4}$$

$$3x = 17 \qquad \text{Simplifying}$$

$$\frac{3x}{3} = \frac{17}{3} \qquad \text{Dividing by 3}$$

$$x = \frac{17}{3} \qquad \text{Simplifying}$$

Check:
$$\begin{array}{c|c} 3x - 4 = 13 \\ \hline 3 \cdot \frac{17}{3} - 4 \;?\; 13 \\ 17 - 4 \\ 13 & \text{TRUE} \end{array}$$

The solution is $\frac{17}{3}$, or $5\frac{2}{3}$.

> In algebra, "improper" fraction notation, such as $\frac{17}{3}$, is quite "proper." We will generally use such notation rather than $5\frac{2}{3}$.

Do Exercise 19.

In a situation such as Example 13, it is easier to first use the addition principle. In a situation in which fractions or decimals are involved, it may be easier to use the multiplication principle first to clear them, but it is not mandatory.

EXAMPLE 14 Clear the fractions and solve: $\frac{3}{16}x + \frac{1}{2} = \frac{11}{8}$.

We multiply both sides by the least common denominator—in this case, 16:

$$\frac{3}{16}x + \frac{1}{2} = \frac{11}{8} \qquad \text{The LCM is 16.}$$

$$16\left(\frac{3}{16}x + \frac{1}{2}\right) = 16\left(\frac{11}{8}\right) \qquad \text{Multiplying by 16}$$

$$16 \cdot \frac{3}{16}x + 16 \cdot \frac{1}{2} = 22 \qquad \begin{array}{l}\text{Carrying out the multiplication. We use} \\ \text{the distributive law on the left, being} \\ \text{careful to multiply } both \text{ terms by 16.}\end{array}$$

$$3x + 8 = 22 \qquad \text{Simplifying. The fractions are cleared.}$$

$$3x + 8 - 8 = 22 - 8 \qquad \text{Subtracting 8}$$

$$3x = 14$$

$$\frac{3x}{3} = \frac{14}{3} \qquad \text{Dividing by 3}$$

$$x = \frac{14}{3}.$$

The number $\frac{14}{3}$ checks and is the solution.

Do Exercise 20.

19. Solve: $-4 + 9x = 8$.

20. Clear the fractions and solve:
$$\frac{2}{3} - \frac{5}{6}y = \frac{1}{3}.$$

Answers on page A-4

21. Clear the decimals and solve:

$$6.3x - 9.1 = 3x.$$

Solve.

22. $\dfrac{5}{2}x + \dfrac{9}{2}x = 21$

23. $1.4x - 0.9x + 0.7 = -2.2$

24. $-4x + 2 + 5x = 3x - 15$

EXAMPLE 15 Clear the decimals and solve: $12.4 - 5.12x = 3.14x$.

We multiply both sides by a power of ten—10, 100, 1000, and so on—to clear the equation of decimals. In this case, we use 10^2, or 100, because the greatest number of decimal places is 2.

$$12.4 - 5.12x = 3.14x$$

$$100(12.4 - 5.12x) = 100(3.14x) \qquad \text{Multiplying by 100}$$

$$100(12.4) - 100(5.12x) = 314x \qquad \begin{array}{l}\text{Carrying out the multiplication. We} \\ \text{use the distributive law on the left.}\end{array}$$

$$1240 - 512x = 314x \qquad \text{Simplifying}$$

$$1240 - 512x + 512x = 314x + 512x \qquad \text{Adding } 512x$$

$$1240 = 826x$$

$$\frac{1240}{826} = \frac{826x}{826} \qquad \text{Dividing by 826}$$

$$x = \frac{1240}{826}, \text{ or } \frac{620}{413}$$

The solution is $\frac{620}{413}$.

Do Exercise 21.

When there are like terms on the same side of an equation, we collect them. If there are like terms on opposite sides of an equation, we use the addition principle to get them on the same side of the equation.

EXAMPLE 16 Solve: $8x + 6 - 2x = -4x - 14$.

$$8x + 6 - 2x = -4x - 14$$

$$6x + 6 = -4x - 14 \qquad \text{Collecting like terms on the left}$$

$$4x + 6x + 6 = 4x - 4x - 14 \qquad \text{Adding } 4x$$

$$10x + 6 = -14 \qquad \text{Collecting like terms}$$

$$10x + 6 - 6 = -14 - 6 \qquad \text{Subtracting 6}$$

$$10x = -20$$

$$\frac{10x}{10} = \frac{-20}{10} \qquad \text{Dividing by 10}$$

$$x = -2$$

Check:

$$\begin{array}{c|c} \hline 8x + 6 - 2x = -4x - 14 \\ \hline 8(-2) + 6 - 2(-2) \overset{?}{} -4(-2) - 14 \\ -16 + 6 + 4 8 - 14 \\ -6 \big| -6 \qquad \text{TRUE} \end{array}$$

The solution is -2.

Do Exercises 22–24.

SPECIAL CASES

Some equations have no solution.

EXAMPLE 17 Solve: $-8x + 5 = 14 - 8x$.

We have

$$-8x + 5 = 14 - 8x$$
$$8x - 8x + 5 = 8x + 14 - 8x \qquad \text{Adding } 8x$$
$$5 = 14. \qquad \text{We get a false equation.}$$

No matter what number we use for x, we get a false sentence. Thus the equation has *no* solution.

There are some equations for which any real number is a solution.

EXAMPLE 18 Solve: $-8x + 5 = 5 - 8x$.

We have

$$-8x + 5 = 5 - 8x$$
$$8x - 8x + 5 = 8x + 5 - 8x \qquad \text{Adding } 8x$$
$$5 = 5. \qquad \text{We get a true equation.}$$

Replacing x with any real number gives a true sentence. Thus any real number is a solution. The equation has *infinitely* many solutions.

Do Exercises 25 and 26.

EQUATIONS CONTAINING PARENTHESES

Equations containing parentheses can often be solved by first multiplying to remove parentheses and then proceeding as before.

EXAMPLE 19 Solve: $30 + 5(x + 3) = -3 + 5x + 48$.

We have

$$30 + 5(x + 3) = -3 + 5x + 48$$
$$30 + 5x + 15 = -3 + 5x + 48 \qquad \text{Multiplying, using the distributive law, to remove parentheses}$$
$$45 + 5x = 45 + 5x \qquad \text{Collecting like terms on each side}$$
$$45 + 5x - 5x = 45 + 5x - 5x \qquad \text{Subtracting } 5x$$
$$45 = 45. \qquad \text{Simplifying. We get a true equation.}$$

All real numbers are solutions.

Do Exercises 27–29.

Solve.

25. $4 + 7x = 7x + 9$

26. $3 + 9x = 9x + 3$

Solve.

27. $7x - 17 = 4 + 7(x - 3)$

28. $3x + 4(x + 2) = 11 + 7x$

29. $3x + 8(x + 2) = 11 + 7x$

Answers on page A-4

Solve.

30. $30 + 7(x - 1) = 3(2x + 7)$

EXAMPLE 20 Solve: $3(7 - 2x) = 14 - 8(x - 1)$.

$$3(7 - 2x) = 14 - 8(x - 1)$$

$21 - 6x = 14 - 8x + 8$	Multiplying, using the distributive law, to remove parentheses
$21 - 6x = 22 - 8x$	Collecting like terms
$21 - 6x + 8x = 22 - 8x + 8x$	Adding $8x$
$21 + 2x = 22$	Collecting like terms
$21 + 2x - 21 = 22 - 21$	Subtracting 21
$2x = 1$	
$\dfrac{2x}{2} = \dfrac{1}{2}$	Dividing by 2
$x = \dfrac{1}{2}$	

Check:
$$\frac{3(7 - 2x) = 14 - 8(x - 1)}{}$$

$$3\left(7 - 2 \cdot \tfrac{1}{2}\right) \;?\; 14 - 8\left(\tfrac{1}{2} - 1\right)$$

$$3(7 - 1) \;\bigg|\; 14 - 8\left(-\tfrac{1}{2}\right)$$

$$3 \cdot 6 \;\bigg|\; 14 + 4$$

$$18 \;\bigg|\; 18 \quad \text{TRUE}$$

The solution is $\frac{1}{2}$.

Do Exercises 30 and 31.

31. $3(y - 1) - 1 = 2 - 5(y + 5)$

AN EQUATION-SOLVING PROCEDURE

1. Clear the equation of fractions or decimals if that is needed.
2. If parentheses occur, multiply to remove them using the distributive law.
3. Collect like terms on each side of the equation, if necessary.
4. Use the addition principle to get all like terms with letters on one side and all other terms on the other side.
5. Collect like terms on each side again, if necessary.
6. Use the multiplication principle to solve for the variable.

The following table describes the kinds of solutions we have considered.

EQUIVALENT EQUATION	NUMBER OF SOLUTIONS	SOLUTION(S)
$x = a$, where a is a real number	One	The number a
A true equation, such as $8 = 8$, $-15 = -15$, or $0 = 0$	Infinitely many	Every real number is a solution.
A false equation, such as $3 = 8$, $-4 = 5$, or $0 = -5$	Zero	There are no solutions.

Answers on page A-4

Checking Possible Solutions   Although a calculator is *not* required for this textbook, the book contains a series of *optional* discussions on using a graphing calculator. The keystrokes for the TI-83 Plus and TI-84 Plus graphing calculators will be shown throughout. For keystrokes for other models of calculators, consult the user's manual for your particular model.

For a brief introduction to the graphing calculator, see the Calculator Corner on p. 9.

Note: If you set your graphing calculator in Sci (scientific) mode to do the exercises in Section R.7, return it to Normal mode now.

To check possible solutions of an equation on a calculator, we can substitute and carry out the calculations on each side of the equation just as we do when we check by hand. If the left- and right-hand sides of the equation have the same value, then the number that was substituted is a solution of the equation. To check the possible solution -2 in the equation $8x + 6 - 2x = -4x - 14$ in Example 16, for instance, we first substitute -2 for x in the expression on the left side of the equation. We press (8) (×) ((-)) (2) (+) (6) (−) (2) (×) ((-)) (2) ⟨ENTER⟩, and get -6. Then we substitute -2 for x in the expression on the right side of the equation. We press ((-)) (4) (×) ((-)) (2) (−) (1) (4) ⟨ENTER⟩. Again, we get -6. Since the two sides of the equation have the same value when x is -2, we know that -2 is the solution of the equation.

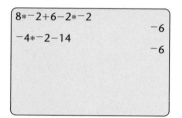

A table can also be used to check possible solutions of equations. First, we press (Y=) to display the equation-editor screen. If an expression for Y1 is currently entered, we place the cursor on it and press ⟨CLEAR⟩ to delete it. We do the same for any other entries that are present.

Next, we position the cursor to the right of Y1 = and enter the left side of the equation by pressing (8) (X,T,θ,n) (+) (6) (−) (2) (X,T,θ,n). Then we position the cursor beside Y2 = and enter the right side of the equation by pressing ((-)) (4) (X,T,θ,n) (−) (1) (4). Now we press ⟨2ND⟩ (TBLSET) to display the Table Setup screen (TBLSET is the second operation associated with the (WINDOW) key.) On the INDPNT line, we position the cursor on "Ask" and press ⟨ENTER⟩ to set up a table in ASK mode. (The settings for TblStart and ΔTbl are irrelevant in ASK mode.)

```
Plot1  Plot2  Plot3
\Y1 ■ 8X+6−2X
\Y2 ■ −4X−14
\Y3 =
\Y4 =
\Y5 =
\Y6 =
\Y7 =
```

```
TABLE SETUP
  TblStart=1
  ΔTbl=1
Indpnt: Auto  [Ask]
Depend: [Auto]  Ask
```

X	Y1	Y2
-2	-6	-6
X=		

We press ⟨2ND⟩ (TABLE) to display the table. (TABLE is the second operation associated with the (GRAPH) key.) We enter the possible solution, -2, by pressing ((-)) (2) ⟨ENTER⟩, and see that Y1 = -6 = Y2 for this value of x. This confirms that the left and right sides of the equation have the same value for $x = -2$, so -2 is the solution of the equation.

Exercises:

1. Use substitution to check the solutions found in Examples 9, 13, and 15.

2. Use a table set in ASK mode to check the solutions found in Margin Exercises 24, 30, and 31.

Remember to review the objectives before doing the exercises.

a   Determine whether the given number is a solution of the given equation.

1. $17;\ x + 23 = 40$

2. $24;\ 47 - x = 23$

3. $-8;\ 2x - 3 = -18$

4. $-10;\ 3x + 14 = -27$

5. $45;\ \dfrac{-x}{9} = -2$

6. $32;\ \dfrac{-x}{8} = -3$

7. $10;\ 2 - 3x = 21$

8. $-11;\ 4 - 5x = 59$

9. $19;\ 5x + 7 = 102$

10. $9;\ 9y + 5 = 86$

11. $-11;\ 7(y - 1) = 84$

12. $-13;\ x + 5 = 5 + x$

b   Solve using the addition principle. Don't forget to check.

13. $y + 6 = 13$

14. $x + 7 = 14$

15. $-20 = x - 12$

16. $-27 = y - 17$

17. $-8 + x = 19$

18. $-8 + r = 17$

19. $-12 + z = -51$

20. $-37 + x = -89$

21. $p - 2.96 = 83.9$

22. $z - 14.9 = -5.73$

23. $-\dfrac{3}{8} + x = -\dfrac{5}{24}$

24. $x + \dfrac{1}{12} = -\dfrac{5}{6}$

c   Solve using the multiplication principle. Don't forget to check.

25. $3x = 18$

26. $5x = 30$

27. $-11y = 44$

28. $-4x = 124$

29. $-\dfrac{x}{7} = 21$

30. $-\dfrac{x}{3} = -25$

31. $-96 = -3z$

32. $-120 = -8y$

33. $4.8y = -28.8$

34. $0.39t = -2.73$

35. $\dfrac{3}{2}t = -\dfrac{1}{4}$

36. $-\dfrac{7}{6}y = -\dfrac{7}{8}$

CHAPTER 1: Solving Linear Equations
and Inequalities

Solve using the principles together. Don't forget to check.

37. $6x - 15 = 45$

38. $4x - 7 = 81$

39. $5x - 10 = 45$

40. $6z - 7 = 11$

41. $9t + 4 = -104$

42. $5x + 7 = -108$

43. $-\dfrac{7}{3}x + \dfrac{2}{3} = -18$

44. $-\dfrac{9}{2}y + 4 = -\dfrac{91}{2}$

45. $\dfrac{6}{5}x + \dfrac{4}{10}x = \dfrac{32}{10}$

46. $\dfrac{9}{5}y + \dfrac{4}{10}y = \dfrac{66}{10}$

47. $0.9y - 0.7y = 4.2$

48. $0.8t - 0.3t = 6.5$

49. $8x + 48 = 3x - 12$

50. $15x + 40 = 8x - 9$

51. $7y - 1 = 27 + 7y$

52. $3x - 15 = 15 + 3x$

53. $3x - 4 = 5 + 12x$

54. $9t - 4 = 14 + 15t$

55. $5 - 4a = a - 13$

56. $6 - 7x = x - 14$

57. $3m - 7 = -7 - 4m - m$

58. $5x - 8 = -8 + 3x - x$

59. $5x + 3 = 11 - 4x + x$

60. $6y + 20 = 10 + 3y + y$

61. $-7 + 9x = 9x - 7$

62. $-3t + 4 = 5 - 3t$

63. $6y - 8 = 9 + 6y$

64. $5 - 2y = -2y + 5$

65. $2(x + 7) = 4x$

66. $3(y + 6) = 9y$

67. $80 = 10(3t + 2)$

68. $27 = 9(5y - 2)$

69. $180(n - 2) = 900$

70. $210(x - 3) = 840$

71. $5y - (2y - 10) = 25$

72. $8x - (3x - 5) = 40$

73. $7(3x + 6) = 11 - (x + 2)$

74. $3(4 - 2x) = 4 - (6x - 8)$

75. $2[9 - 3(-2x - 4)] = 12x + 42$

76. $-40x + 45 = 3[7 - 2(7x - 4)]$

77. $\frac{1}{8}(16y + 8) - 17 = -\frac{1}{4}(8y - 16)$

78. $\frac{1}{6}(12t + 48) - 20 = -\frac{1}{8}(24t - 144)$

79. $3[5 - 3(4 - t)] - 2 = 5[3(5t - 4) + 8] - 26$

80. $6[4(8 - y) - 5(9 + 3y)] - 21 = -7[3(7 + 4y) - 4]$

81. $\frac{2}{3}\left(\frac{7}{8} + 4x\right) - \frac{5}{8} = \frac{3}{8}$

82. $\frac{3}{4}\left(3x - \frac{1}{2}\right) + \frac{2}{3} = \frac{1}{3}$

83. $5(4x - 3) - 2(6 - 8x) + 10(-2x + 7) = -4(9 - 12x)$

84. $9(4x + 7) - 3(5x - 8) = 6(\frac{2}{3} - x) - 5(\frac{3}{5} + 2x)$

To the student and the instructor: The Discussion and Writing exercises are meant to be answered with one or more sentences. They can be discussed and answered collaboratively by the entire class or by small groups. Because of their open-ended nature, the answers to these exercises do not appear at the back of the book. They are denoted by the symbol $\mathbf{D_W}$.

85. $\mathbf{D_W}$ Explain the difference between equivalent expressions and equivalent equations.

86. $\mathbf{D_W}$ Explain how the distributive and commutative laws can be used to express $3x + 6y + 4x + 2y$ as $7x + 8y$.

> **SKILL MAINTENANCE**

This heading indicates that the exercises that follow are *Skill Maintenance* exercises, which review any skill previously studied in the text. You will see them in virtually every exercise set. Answers to *all* skill maintenance exercises are found at the back of the book. If you miss an exercise, restudy the objective shown in red.

Multiply or divide, and simplify. [R.7a]

87. $a^{-9} \cdot a^{23}$

88. $\dfrac{a^{-9}}{a^{23}}$

89. $(6x^5y^{-4})(-3x^{-3}y^{-7})$

90. $\dfrac{6x^5y^{-4}}{-3x^{-3}y^{-7}}$

Multiply. [R.5d]

91. $2(6 - 10x)$

92. $-1(5 - 6x)$

93. $-4(3x - 2y + z)$

94. $5(-2x + 7y - 4)$

Factor. [R.5d]

95. $2x - 6y$

96. $-4x - 24y$

97. $4x - 10y + 2$

98. $-10x + 35y - 20$

99. Name the set consisting of the positive integers less than 10, using both roster notation and set-builder notation. [R.1a]

100. Name the set consisting of the negative integers greater than -9 using both roster notation and set-builder notation. [R.1a]

> **SYNTHESIS**

To the student and the instructor: The Synthesis exercises found at the end of every exercise set challenge students to combine concepts or skills studied in that section or in preceding parts of the text.

Solve. (The symbol ⚏ indicates an exercise designed to be solved with any type of calculator.)

101. ⚏ $4.23x - 17.898 = -1.65x - 42.454$

102. ⚏ $-0.00458y + 1.7787 = 13.002y - 1.005$

103. $\dfrac{3x}{2} + \dfrac{5x}{3} - \dfrac{13x}{6} - \dfrac{2}{3} = \dfrac{5}{6}$

104. $\dfrac{2x - 5}{6} + \dfrac{4 - 7x}{8} = \dfrac{10 + 6x}{3}$

105. $x - \{3x - [2x - (5x - (7x - 1))]\} = x + 7$

106. $23 - 2\{4 + 3(x - 1)\} + 5\{x - 2(x + 3)\} = 7\{x - 2[5 - (2x + 3)]\}$

Objective

a Evaluate formulas and solve a formula for a specified letter.

a Evaluating and Solving Formulas

A **formula** is an equation that represents or models a relationship between two or more quantities. For example, the relationship between the perimeter P of a square and the length of its sides s is given by the formula $P = 4s$. The formula $A = s^2$ represents the relationship between the area A of a square and the length s of its sides.

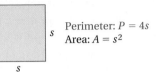

Perimeter: $P = 4s$
Area: $A = s^2$

Other important geometric formulas are $A = \pi r^2$ (for the area A of a circle of radius r), $C = \pi d$ (for the circumference C of a circle of diameter d), and $A = b \cdot h$ (for the area A of a parallelogram of height h and base b). A more complete list of geometric formulas appears at the end of this text.

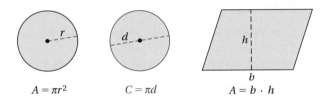

$A = \pi r^2$ $C = \pi d$ $A = b \cdot h$

EXAMPLE 1   *Body Mass Index.*   **Body mass index** I can be used to determine whether an individual has a healthy weight for his or her height. An index between 18.5 and 25 indicates a healthy weight. Body mass index is given by the formula, or model,

$$I = \frac{704.5W}{H^2},$$

where W is weight, in pounds, and H is height, in inches.

Source: National Center for Health Statistics

a) Actor Will Smith is 6'2" and weighs 210 lb as of this writing. What is his body mass index?

b) Professional basketball player Diane Taurasi has a body mass index of 24 and a height of 5'11". What is her weight?

Will Smith,
Body mass index = ?

Diana Taurasi,
Body mass index = 24

Jennifer Aniston,
Body mass index = ?

Marv Bittinger,
Body mass index = 25.4

a) We substitute 210 lb for W and 6'2", or $6 \cdot 12 + 2 = 74$ in., for H. Then we have

$$I = \frac{704.5W}{H^2} = \frac{704.5(210)}{74^2} \approx 27.$$

Thus Will Smith has a body mass index of 27.

b) We substitute 24 for I and 5'11", or $5 \cdot 12 + 11 = 71$ in., for H, and solve for W using the equation-solving principles introduced in Section 1.1.

$$I = \frac{704.5W}{H^2}$$

$$24 = \frac{704.5W}{71^2} \qquad \text{Substituting}$$

$$24 = \frac{704.5W}{5041}$$

$$24 \cdot 5041 = \frac{704.5W}{5041} \cdot 5041 \qquad \text{Multiplying by 5041}$$

$$120{,}984 = 704.5W \qquad \text{Simplifying}$$

$$\frac{120{,}984}{704.5} = \frac{704.5W}{704.5} \qquad \text{Dividing by 704.5}$$

$$172 \approx W.$$

Diana Taurasi weighs approximately 172 lb.

Do Exercise 1.

If we want to make repeated calculations of W, as in Example 1(b), it might be easier to first solve for W, getting it alone on one side of the equation. We "solve" for W as we did above, using the equation-solving principles of Section 1.1.

EXAMPLE 2   Solve for W: $I = \dfrac{704.5W}{H^2}$.

$$I = \frac{704.5W}{H^2} \qquad \text{We want this letter alone.}$$

$$I \cdot H^2 = \frac{704.5W}{H^2} \cdot H^2 \qquad \begin{array}{l}\text{Multiplying by } H^2 \text{ on both} \\ \text{sides to clear the fraction}\end{array}$$

$$IH^2 = 704.5W \qquad \text{Simplifying}$$

$$\frac{IH^2}{704.5} = \frac{704.5W}{704.5} \qquad \text{Dividing by 704.5}$$

$$\frac{IH^2}{704.5} = W$$

Do Exercise 2.

1. Body Mass Index.

a) Actress Jennifer Aniston weighs 110 lb and is 5'6" tall. What is her body mass index?

b) Your author Marv Bittinger has a body mass index of 25.4 and a height of 6'1". What is his weight?

c) Calculate your own body mass index.

2. Solve for m: $F = \dfrac{mv^2}{r}$.

(This is a physics formula.)

Answers on page A-4

3. Solve for m: $H = 2r + 3m$.

EXAMPLE 3 Solve for r: $H = 2r + 3m$.

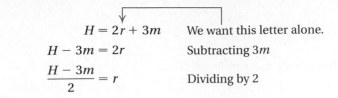

$$H = 2r + 3m \qquad \text{We want this letter alone.}$$
$$H - 3m = 2r \qquad \text{Subtracting } 3m$$
$$\frac{H - 3m}{2} = r \qquad \text{Dividing by 2}$$

Do Exercise 3.

EXAMPLE 4 Solve for b: $A = \frac{5}{2}(b - 20)$.

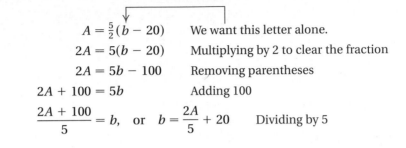

$$A = \frac{5}{2}(b - 20) \qquad \text{We want this letter alone.}$$
$$2A = 5(b - 20) \qquad \text{Multiplying by 2 to clear the fraction}$$
$$2A = 5b - 100 \qquad \text{Removing parentheses}$$
$$2A + 100 = 5b \qquad \text{Adding 100}$$
$$\frac{2A + 100}{5} = b, \quad \text{or} \quad b = \frac{2A}{5} + 20 \qquad \text{Dividing by 5}$$

4. Solve for c: $P = \dfrac{3}{5}(c + 10)$.

Do Exercise 4.

EXAMPLE 5 *Area of a Trapezoid.* Solve for a: $A = \frac{1}{2}h(a + b)$. (To find the area of a trapezoid, take half the product of the height, h, and the sum of the lengths of the parallel sides, a and b.)

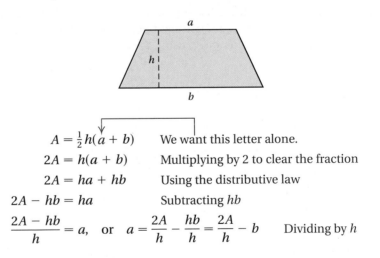

$$A = \frac{1}{2}h(a + b) \qquad \text{We want this letter alone.}$$
$$2A = h(a + b) \qquad \text{Multiplying by 2 to clear the fraction}$$
$$2A = ha + hb \qquad \text{Using the distributive law}$$
$$2A - hb = ha \qquad \text{Subtracting } hb$$
$$\frac{2A - hb}{h} = a, \quad \text{or} \quad a = \frac{2A}{h} - \frac{hb}{h} = \frac{2A}{h} - b \qquad \text{Dividing by } h$$

5. Solve for b: $A = \dfrac{1}{2}h(a + b)$.

Note that there is more than one correct form of the answer to Example 5. Such is a common occurrence in formula solving.

Do Exercise 5.

Answers on page A-4

CHAPTER 1: Solving Linear Equations
and Inequalities

We used the addition and multiplication principles to solve equations in Section 1.1. In a similar manner, we use the same principles in this section to solve a formula for a given letter.

6. Solve for Q: $T = Q + Qiy$.

> To solve a formula for a given letter, identify the letter, and:
>
> **1.** Multiply on both sides to clear the fractions or decimals, if necessary.
> **2.** If parentheses occur, multiply to remove them using the distributive law.
> **3.** Collect like terms on each side, if necessary. This may require factoring if a variable is in more than one term.
> **4.** Using the addition principle, get all terms with the letter to be solved for on one side of the equation and all other terms on the other side.
> **5.** Collect like terms again, if necessary.
> **6.** Solve for the letter in question using the multiplication principle.

As indicated in step (3) above, sometimes we must factor to isolate a letter.

EXAMPLE 6 *Simple Interest.* Solve for P: $A = P + Prt$. (To find the amount A to which principal P, in dollars, will grow at simple interest rate r, in t years, add the principal P to the interest, Prt.)

$$A = P + Prt \qquad \text{We want this letter alone.}$$

$$A = P(1 + rt) \qquad \text{Factoring (or collecting like terms)}$$

$$\frac{A}{1 + rt} = P \qquad \text{Dividing by } 1 + rt \text{ on both sides}$$

Do Exercise 6.

EXAMPLE 7 *Chess Ratings.* The formula

$$R = r + \frac{400(W - L)}{N}$$

is used to establish a chess player's rating R, after he or she has played N games, where W is the number of wins, L is the number of losses, and r is the average rating of the opponents.

Source: U.S. Chess Federation

a) Find the chess rating of a player who plays 8 games in a tournament, winning 5 games and losing 3. The average rating of his or her opponents is 1205.

b) Solve the formula for L.

a) We substitute 8 for N, 5 for W, 3 for L, and 1205 for r in the formula. Then we calculate R:

$$R = r + \frac{400(W - L)}{N} = 1205 + \frac{400(5 - 3)}{8} = 1305.$$

Answer on page A-4

7. Chess Ratings. Use the formula given in Example 7.

a) Find the chess rating of a player who plays 6 games in a tournament, winning 2 games and losing 4. The average rating of his opponents is 1384.

b) Solve the formula for W.

b) We solve as follows:

$$R = r + \frac{400(W - L)}{N} \qquad \text{We want this letter alone.}$$

$$NR = N\left[r + \frac{400(W - L)}{N}\right] \qquad \text{Multiplying by } N \text{ to clear the fraction}$$

$$NR = N \cdot r + N \cdot \frac{400(W - L)}{N} \qquad \text{Multiplying using the distributive law}$$

$$NR = Nr + 400(W - L) \qquad \text{Simplifying}$$

$$NR - Nr = 400(W - L) \qquad \text{Subtracting } Nr$$

$$NR - Nr = 400W - 400L \qquad \text{Using the distributive law}$$

$$NR - Nr - 400W = -400L \qquad \text{Subtracting } 400W$$

$$\frac{NR - Nr - 400W}{-400} = L. \qquad \text{Dividing by } -400$$

Other correct forms of the answer are

$$L = \frac{Nr + 400W - NR}{400} \quad \text{and} \quad L = W - \frac{NR - Nr}{400}.$$

Answers on page A-4

Do Exercise 7.

Study Tips

USING THIS TEXTBOOK

You will find many Study Tips throughout the book. An index of all Study Tips can be found at the front of the book. One of the most important ways to improve your math study skills is to learn the proper use of the textbook. Here we highlight a few points that we consider most helpful.

■ **Be sure to note the special symbols** a , b , c **, and so on, that correspond to the objectives you are to be able to perform.** The first time you see them is in the margin at the beginning of each section; the second time is in the subheadings of each section; and the third time is in the exercise set for the section. You will also find them next to the skill maintenance exercises in each exercise set and in the review exercises at the end of the chapter, as well as in the answers to the chapter tests and the cumulative reviews. These objective symbols allow you to refer to the appropriate place in the text whenever you need to review a topic.

■ **Read and study each step of each example.** The examples include important side comments that explain each step. These carefully chosen examples and notes prepare you for success in the exercise set.

■ **Stop and do the margin exercises as you study a section.** Doing the margin exercises is one of the most effective ways to enhance your ability to learn mathematics from this text. Don't deprive yourself of its benefits!

■ **Note the icons listed at the top of each exercise set.** These refer to the many distinctive multimedia study aids that accompany the book.

■ **Odd-numbered exercises.** Usually an instructor assigns some odd-numbered exercises. When you complete these, you can check your answers at the back of the book. If you miss any, check your work in the *Student's Solutions Manual* or ask your instructor for guidance.

■ **Even-numbered exercises.** Whether or not your instructor assigns the even-numbered exercises, always do some on your own. Remember, there are no answers given for the class tests, so you need to practice doing exercises without answers. Check your answers later with a friend or your instructor.

a Solve for the given letter.

1. *Distance Formula:*
$$d = rt, \quad \text{for } r$$
(Distance d, speed r, time t)

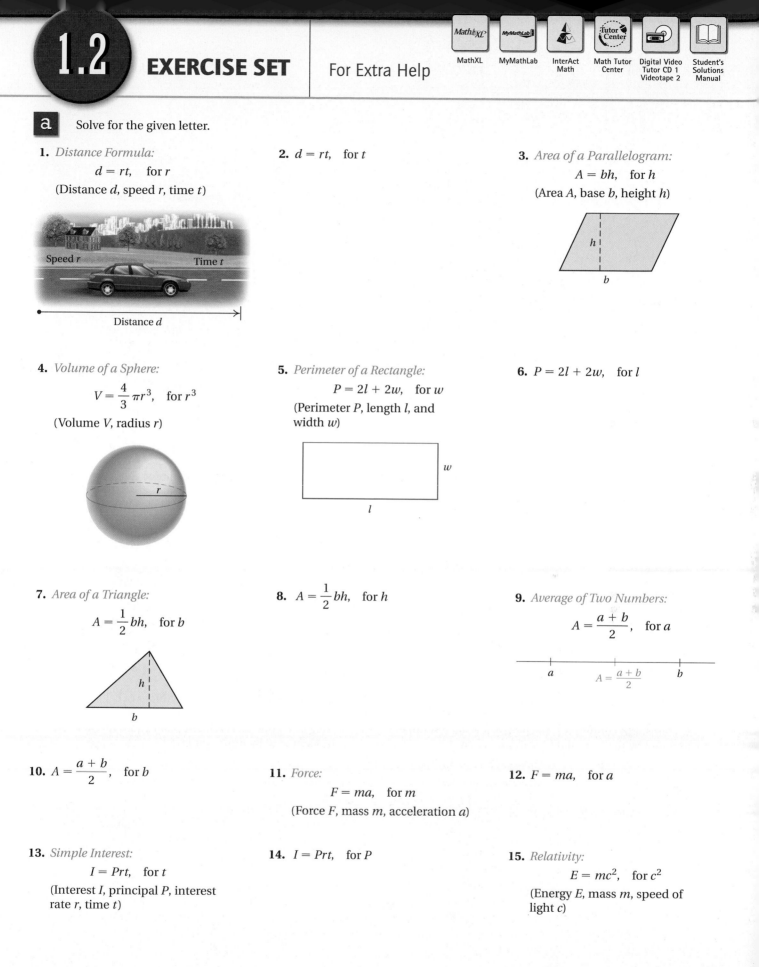

Speed r Time t

Distance d

2. $d = rt$,   for t

3. *Area of a Parallelogram:*
$$A = bh, \quad \text{for } h$$
(Area A, base b, height h)

4. *Volume of a Sphere:*
$$V = \frac{4}{3}\pi r^3, \quad \text{for } r^3$$
(Volume V, radius r)

5. *Perimeter of a Rectangle:*
$$P = 2l + 2w, \quad \text{for } w$$
(Perimeter P, length l, and width w)

6. $P = 2l + 2w$,   for l

7. *Area of a Triangle:*
$$A = \frac{1}{2}bh, \quad \text{for } b$$

8. $A = \frac{1}{2}bh$,   for h

9. *Average of Two Numbers:*
$$A = \frac{a + b}{2}, \quad \text{for } a$$

a $A = \frac{a+b}{2}$ b

10. $A = \dfrac{a + b}{2}$,   for b

11. *Force:*
$$F = ma, \quad \text{for } m$$
(Force F, mass m, acceleration a)

12. $F = ma$,   for a

13. *Simple Interest:*
$$I = Prt, \quad \text{for } t$$
(Interest I, principal P, interest rate r, time t)

14. $I = Prt$,   for P

15. *Relativity:*
$$E = mc^2, \quad \text{for } c^2$$
(Energy E, mass m, speed of light c)

16. $E = mc^2$, for m

17. $Q = \dfrac{p - q}{2}$, for p

18. $Q = \dfrac{p - q}{2}$, for q

19. $Ax + By = c$, for y

20. $Ax + By = c$, for x

21. $I = 1.08\dfrac{T}{N}$, for N

22. $F = \dfrac{mv^2}{r}$, for v^2

23. $C = \dfrac{3}{4}(m + 5)$, for m

24. $A = \dfrac{1}{2}h(a + b)$, for b

25. $n = \dfrac{1}{3}(a + b - c)$, for b

26. $t = \dfrac{1}{6}(x - y + z)$, for z

27. $d = R - Rst$, for R

28. $g = m + mnp$, for m

29. $T = B + Bqt$, for B

30. $Z = Q - Qab$, for Q

31. *Male Caloric Needs.* The number of calories K needed each day by a moderately active man who weighs w kilograms, is h centimeters tall, and is a years old can be estimated by the formula

$$K = 19.18w + 7h - 9.52a + 92.4.$$

Source: M. Parker, *She Does Math.* Mathematical Association of America, p. 96

a) Marv is moderately active, weighs 87 kg, is 185 cm tall, and is 60 yr old. What are his caloric needs?

b) Solve the formula for a, for h, and for w.

32. *Female Caloric Needs.* The number of calories K needed each day by a moderately active woman who weighs w pounds, is h inches tall, and is a years old can be estimated by the formula

$$K = 917 + 6(w + h - a).$$

Source: M. Parker, *She Does Math.* Mathematical Association of America, p. 96

a) Elaine is moderately active, weighs 120 lb, is 67 in. tall, and is 23 yr old. What are her caloric needs?

b) Solve the formula for a, for h, and for w.

Projecting Birth Weight. Ultrasonic images of 29-week-old fetuses can be used to predict weight. One model, or formula, developed by Thurnau, is $P = 9.337da - 299$; a second model, developed by Weiner, is $P = 94.593c + 34.227a - 2134.616$. For both formulas, P is the estimated birth weight, in grams, d is the diameter of the fetal head, in centimeters, c is the circumference of the fetal head, in centimeters, and a is the circumference of the fetal abdomen, in centimeters.

Sources: G. R. Thurnau, R. K. Tamura, R. E. Sabbagha, et al. *Am. J. Obstet Gynecol* 1983; **145**:557; C. P. Weiner, R. E. Sabbagha, N. Vaisrub, et al. *Obstet Gynecol* 1985; **65**:812.

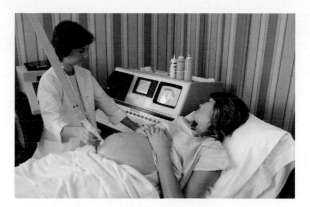

33. a) Use Thurnau's model to estimate the birth weight of a 29-week-old fetus when the diameter of the fetal head is 8.5 cm and the circumference of the fetal abdomen is 24.1 cm.
  b) Solve the formula for a.

34. a) Use Weiner's model to estimate the birth weight of a 29-week-old fetus when the circumference of the fetal head is 26.7 cm and the circumference of the fetal abdomen is 24.1 cm.
  b) Solve the formula for c.

35. *Young's Rule in Medicine.* Young's rule for determining the amount of a medicine dosage for a child is given by

$$c = \frac{ad}{a + 12},$$

where a is the child's age and d is the usual adult dosage, in milligrams. (*Warning!* Do not apply this formula without checking with a physician!)

Source: June Looby Olsen, et al., *Medical Dosage Calculations,* 6th ed. Reading, MA: Addison Wesley Longman, p. A-31

a) The usual adult dosage of a particular medication is 250 mg. Find the dosage for a child of age 3.
b) Solve the formula for d.

36. *Full-Time-Equivalent Students.* Colleges accommodate students who need to take different total-credit-hour loads. They determine the number of "full-time-equivalent" students, F, using the formula

$$F = \frac{n}{15},$$

where n is the total number of credits students enroll in for a given semester.

a) Determine the number of full-time-equivalent students on a campus in which students register for 42,690 credits.
b) Solve the formula for n.

37. **D**_W The equations

$$P = 2l + 2w \quad \text{and} \quad w = \frac{P}{2} - l$$

are equivalent formulas involving the perimeter P, length l, and width w of a rectangle. Devise a problem for which the second of the two formulas would be more useful.

38. **D**_W Devise an application in which it would be useful to solve the equation $d = rt$ for r. (See Exercise 1.)

Divide.   [R.2e]

39. $\dfrac{80}{-16}$

40. $-2000 \div (-8)$

41. $-\dfrac{1}{2} \div \dfrac{1}{4}$

42. $120 \div (-4.8)$

43. $-\dfrac{2}{3} \div \left(-\dfrac{5}{6}\right)$

44. $\dfrac{-90}{-15}$

45. $\dfrac{-90}{15}$

46. $\dfrac{-80}{16}$

Solve.

47. $A = \pi rs + \pi r^2$, for s

48. $s = v_1 t + \frac{1}{2} at^2$, for a; for v_1

49. $\dfrac{P_1 V_1}{T_1} = \dfrac{P_2 V_2}{T_2}$, for V_1; for P_2

50. $\dfrac{P_1 V_1}{T_1} = \dfrac{P_2 V_2}{T_2}$, for T_2; for P_1

51. In Exercise 13, you solved the formula $I = Prt$ for t. Now use it to find how long it will take a deposit of \$75 to earn \$3 interest when invested at 5% simple interest.

52. The area of the shaded triangle ABE is 20 cm^2. Find the area of the trapezoid.

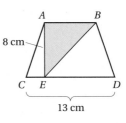

53. *Horsepower of an Engine.*   The horsepower of an engine can be calculated by the formula

$$H = W\left(\dfrac{v}{234}\right)^3,$$

where W is the weight, in pounds, of the car, including the driver, fluids, and fuel, and v is the maximum velocity, or speed, in miles per hour, of the car attained a quarter mile after beginning acceleration.

a) Find the horsepower of a V-6, 2.8-liter engine if $W = 2700$ lb and $v = 83$ mph.

b) Find the horsepower of a 4-cylinder, 2.0-liter engine if $W = 3100$ lb and $v = 73$ mph.

1.3 APPLICATIONS AND PROBLEM SOLVING

a Five Steps for Problem Solving

One very important use of algebra is as a tool for problem solving. The following five-step strategy for solving problems will be used throughout this text.

FIVE STEPS FOR PROBLEM SOLVING

1. *Familiarize* yourself with the problem situation.
2. *Translate* the problem to an equation.
3. *Solve* the equation.
4. *Check* the answer in the original problem.
5. *State* the answer to the problem clearly.

Of the five steps, probably the most important is the first one: becoming familiar with the problem situation. Here are some hints for familiarization.

To familiarize yourself with the problem:

- If a problem is given in words, read it carefully.
- Reread the problem, perhaps aloud. Try to verbalize the problem to yourself. Make notes as you read.
- List the information given and the question to be answered. Choose a variable (or variables) to represent the unknown(s) and clearly state what the variable represents. Be descriptive! For example, let L = length (in meters), d = distance (in miles), and so on.
- Make a drawing and label it with known information. Also, indicate unknown information, using specific units if given.
- Find further information if necessary. Look up a formula at the back of this book or in a reference book. Talk to a reference librarian or look up the topic using an Internet search engine.
- Make a table that lists all the information you have collected. Look for patterns that may help in the translation to an equation.
- Think of a possible answer and check the guess. Observe the manner in which the guess is checked.

EXAMPLE 1 *Installing Seamless Guttering.* Seamless guttering is delivered on a continuous roll and sections are cut from the roll as needed. The Jordans know that they need 127 ft of guttering for six separate sections of their home and that the four shortest sections will be the same size. The longest piece of guttering is 13 ft more than twice the length of the midsize piece. The shortest piece of guttering is 10 ft less than the midsize piece. How long is each piece of guttering?

1. **Familiarize.** All the pieces are described in terms of the midsize piece, so we begin by assigning a variable to that piece and using that variable to describe the lengths of the other pieces.

Longest piece = $2x + 13$

Midsize piece = x

Shortest piece = $x - 10$

DIMENSIONS

Be aware of the dimensions in a problem. They should all be expressed in the same unit—all units of length in feet, for example, or all units of time in hours—when the problem is translated to an equation.

1. **Cutting a Board.** A 106-in. board is cut into three pieces. The shortest length is two-thirds of the midsize length and the longest length is 15 in. less than two times the midsize length. Find the length of each piece.

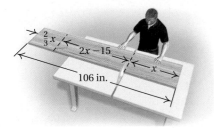

We let x = the length of the midsize piece, in feet, $2x + 13$ = the length of the longest piece, and $x - 10$ = the length of the shortest piece.

2. **Translate.** The sum of the length of the longest piece, plus the length of the midsize piece, plus four times the length of the shortest piece is 127 ft. This gives us the following translation:

Longest piece	plus	Midsize piece	plus	Four times the shortest piece	is	Total length
$(2x + 13)$	$+$	x	$+$	$4(x - 10)$	$=$	$127.$

3. **Solve.** We solve the equation, as follows:

$$(2x + 13) + x + 4(x - 10) = 127$$
$$2x + 13 + x + 4x - 40 = 127 \quad \text{Using the distributive law}$$
$$7x - 27 = 127 \quad \text{Collecting like terms}$$
$$7x - 27 + 27 = 127 + 27 \quad \text{Adding 27}$$
$$7x = 154 \quad \text{Collecting like terms}$$
$$\frac{7x}{7} = \frac{154}{7} \quad \text{Dividing by 7}$$
$$x = 22.$$

4. **Check.** Do we have an answer to the *problem*? If the length of the midsize piece is 22 ft, then the length of the longest piece is

$$2 \cdot 22 + 13, \text{ or } 57 \text{ ft,}$$

and the length of the shortest piece is

$$22 - 10, \text{ or } 12 \text{ ft.}$$

The sum of the lengths of the longest piece, the midsize piece, and four times the shortest piece must be 127 ft:

$$57 \text{ ft} + 22 \text{ ft} + 4(12 \text{ ft}) = 127 \text{ ft.}$$

These lengths check.

5. **State.** The length of the longest piece is 57 ft, the length of the midsize piece is 22 ft, and the length of the shortest piece is 12 ft.

Do Exercise 1.

EXAMPLE 2 *Price of a Mobile Home.* The price of a mobile home was cut 11% to a new price of $48,950. What was the original price?

1. **Familiarize.** We let x = the original price.
2. **Translate.**

$$\underbrace{\text{Original price}} \quad \underbrace{\text{minus}} \quad \underbrace{11\%} \quad \underbrace{\text{of}} \quad \underbrace{\text{Original price}} \quad \underbrace{\text{is}} \quad \underbrace{\text{New price}}$$

$$x \quad - \quad 11\% \quad \cdot \quad x \quad = \quad 48{,}950 \qquad \text{Translation}$$

3. **Solve.** We solve the equation:

$$x - 11\% \cdot x = 48{,}950$$

$$1x - 0.11x = 48{,}950 \qquad \text{Replacing } 11\% \text{ with } 0.11$$

$$\left.\begin{array}{l} (1 - 0.11)x = 48{,}950 \\ 0.89x = 48{,}950 \end{array}\right\} \quad \text{Collecting like terms}$$

$$x = 55{,}000. \qquad \text{Dividing by } 0.89$$

4. **Check.** Note that 11% of $55,000 is $6050. Subtracting this decrease from $55,000 (the original price), we get $48,950 (the new price):

$$11\% \cdot \$55{,}000 = \$6050, \quad \text{and}$$

$$\$55{,}000 - \$6050 = \$48{,}950.$$

The answer checks.

5. **State.** The original price of the mobile home was $55,000.

Do Exercise 2.

EXAMPLE 3 *Real Estate Commission.* The Mendozas negotiated to pay the realtor who sold their house the following commission:

6% for the first $100,000 of the selling price, and

5.5% for the amount that exceeded $100,000.

The realtor received a commission of $10,565 for selling the house. What was the selling price?

1. **Familiarize.** Let's make a guess or estimate to become familiar with the problem. Suppose the house sold for $225,000. The realtor's commission would be

$$6\% \text{ of } \$100{,}000 = 0.06(\$100{,}000) = \$6000$$

plus

$$5.5\% \text{ times } (\$225{,}000 - \$100{,}000) = 0.055(\$125{,}000) = \$6875.$$

The total commission would be $6000 + $6875, or $12,875. Although our guess is not correct, the calculation we performed familiarizes us with the problem, and it also tells us that the house sold for less than $225,000, since $10,565 is less than $12,875. We let S = the selling price of the house.

2. **Translate.** We translate as follows:

$$\underbrace{\begin{array}{c}\text{Commission} \\ \text{on the first} \\ \$100{,}000\end{array}} \quad \underbrace{\text{plus}} \quad \underbrace{\begin{array}{c}\text{Commission on} \\ \text{the amount that} \\ \text{exceeds } \$100{,}000\end{array}} \quad \underbrace{\text{is}} \quad \underbrace{\begin{array}{c}\text{Total} \\ \text{commission}\end{array}}$$

$$6\% \cdot 100{,}000 \quad + \quad 5.5\%(S - 100{,}000) \quad = \quad 10{,}565.$$

2. **Price of a Suit.** A sporting goods store drops the price of a running suit 25% to a sale price of $93. What was the former price?

Answer on page A-5

101

3. Real Estate Commission. The Currys negotiated to pay the realtor who sold their house the following commission:

  7% for the first $100,000 of the selling price

and

  5% for the amount that exceeded $100,000.

The realtor received a commission of $13,400 for selling the house. What was the selling price?

3. Solve. We solve the equation:

$$6\% \cdot 100{,}000 + 5.5\%(S - 100{,}000) = 10{,}565$$
$$0.06 \cdot 100{,}000 + 0.055(S - 100{,}000) = 10{,}565 \qquad \text{Converting to decimal notation}$$
$$6000 + 0.055S - 0.055 \cdot 100{,}000 = 10{,}565 \qquad \text{Using the distributive law}$$
$$6000 + 0.055S - 5500 = 10{,}565 \qquad \text{Simplifying}$$
$$0.055S + 500 = 10{,}565 \qquad \text{Collecting like terms}$$
$$0.055S + 500 - 500 = 10{,}565 - 500 \qquad \text{Subtracting 500}$$
$$0.055S = 10{,}065$$
$$\frac{0.055S}{0.055} = \frac{10{,}065}{0.055} \qquad \text{Dividing by 0.055}$$
$$S = \$183{,}000.$$

4. Check. Performing the check is similar to the sample calculation in the *Familiarize* step. The check is left to the student.

5. State. The selling price of the house was $183,000.

Do Exercise 3.

EXAMPLE 4 *Digital Prints.* The equation

$$y = 0.85x - 0.27$$

can be used to estimate the number of digital camera photographs printed at home, in billions, x years after 2000. That is, $x = 1$ corresponds to 2001, $x = 8$ corresponds to 2008, and so on.

Source: Photo Marketing Association International

a) Estimate the number of digital prints made at home in 2003 and in 2010.

b) In what year were about 4 billion digital prints made at home?

  Since an equation is given, we know that we have the correct translation and thus we will not use the five-step problem-solving strategy in this case.

a) To estimate the number of digital prints made at home in 2003, we first note that 2003 is 3 years after 2000 ($2003 - 2000 = 3$). We substitute 3 for x in the equation:

$$y = 0.85x - 0.27 = 0.85(3) - 0.27 = 2.28.$$

To estimate the number of digital prints that will be made at home in 2010, we substitute 10 for x ($2010 - 2000 = 10$):

$$y = 0.85x - 0.27 = 0.85(10) - 0.27 = 8.23.$$

We estimate that 2.28 billion digital prints were made at home in 2003 and 8.23 billion will be made at home in 2010. (The estimate for 2010 assumes that current trends continue.)

Answer on page A-5

b) To determine the year in which about 4 billion digital prints were made at home, we substitute 4 for y in the equation and solve for x:

$$y = 0.85x - 0.27$$
$$4 = 0.85x - 0.27 \qquad \text{Substituting 4 for } y$$
$$4 + 0.27 = 0.85x - 0.27 + 0.27 \qquad \text{Adding 0.27}$$
$$4.27 = 0.85x$$
$$\frac{4.27}{0.85} = \frac{0.85x}{0.85} \qquad \text{Dividing by 0.85}$$
$$5 \approx x.$$

About 4 billion digital prints were made at home about 5 years after 2000, or in 2005.

Do Exercise 4.

EXAMPLE 5 *Picture Framing.* A piece of wood trim that is 100 in. long is to be cut into two pieces, and each of those pieces is to be cut into four pieces to form a square frame. The length of a side of one square is to be $1\frac{1}{2}$ times the length of a side of the other. How should the wood be cut?

1. **Familiarize.** We first make a drawing of the situation, letting $s =$ the length of a side of the smaller square, in inches, and $1\frac{1}{2}s$, or $\frac{3}{2}s =$ the length of a side of the larger square. Note that the *perimeter* (distance around) of each square is four times the length of a side and that the sum of the two perimeters must be 100 in.

2. **Translate.** We reword the problem and translate.

	Perimeter of smaller square	plus	Perimeter of larger square	is	100 in.
Rewording:					
Translating:	$4s$	$+$	$4\left(\frac{3}{2}s\right)$	$=$	100

3. **Solve.** We solve the equation:

$$4s + 4\left(\tfrac{3}{2}s\right) = 100$$
$$4s + 6s = 100 \qquad \text{Multiplying}$$
$$10s = 100 \qquad \text{Collecting like terms}$$
$$\frac{10s}{10} = \frac{100}{10} \qquad \text{Dividing by 10}$$
$$s = 10.$$

Then $4s = 4 \cdot 10 = 40$ and $4\left(\frac{3}{2}s\right) = 6s = 6 \cdot 10 = 60$. These are the lengths of the pieces into which the wood is to be cut.

4. **Check.** If 10 in. is the length of a side of the smaller square, then $\frac{3}{2}(10)$, or 15 in., is the length of a side of the larger square. The sum of the two perimeters is

$$4 \cdot 10 + 4 \cdot 15 = 40 + 60 = 100,$$

so the lengths check.

5. **State.** The wood should be cut into two pieces, one 40 in. long and the other 60 in. long. Each piece should then be cut into four pieces of the same length in order to form the frames.

Do Exercise 5 on the following page.

4. Digital Prints. Refer to Example 4.

a) Estimate the number of digital prints made at home in 2002 and in 2008.

b) In what year were about 4.8 billion digital prints made at home?

Answers on page A-5

5. Picture Framing. Repeat Example 5, given that the piece of wood is 120 in. long.

6. Angles of a Triangle. The second angle of a triangle is three times as large as the first. The measure of the third angle is 25° greater than that of the first angle. How large are the angles?

Answers on page A-5

■ **EXAMPLE 6** *Triangular Peace Monument.* In Indianapolis, Indiana, children have turned a vacant lot into a monument for peace. This community project brought together neighborhood volunteers, businesses, and government in hopes of showing children positive nonviolent ways of dealing with conflict. A landscape architect used the children's drawings and ideas to design a triangular-shaped peace garden as shown in the figure below. Two sides of the triangle are formed by Indiana Avenue and Senate Avenue. The third side is formed by an apartment building. The second angle of the triangle is 7° more than the first. The third angle is 7° less than twice the first. How large are the angles?

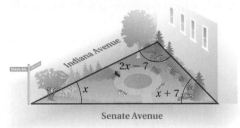

Source: Mentoring in the City Program, Marian College, Indianapolis, Indiana

1. **Familiarize.** We make a drawing as above. Since the second and third angles are described in terms of the first angle, we let the measure of the first angle $= x$. Then the measure of the second angle $= x + 7$ and the measure of the third angle $= 2x - 7$. Recall from geometry that the sum of the measures of the angles of a triangle is 180°.

2. **Translate.** We translate as follows:

$$\underbrace{\text{Measure of first angle}}_{x} \quad \underset{+}{\text{plus}} \quad \underbrace{\text{Measure of second angle}}_{(x+7)} \quad \underset{+}{\text{plus}} \quad \underbrace{\text{Measure of third angle}}_{(2x-7)} \quad \underset{=}{\text{is}} \quad \underset{180}{180°}.$$

3. **Solve.** We solve the equation:

$$x + (x + 7) + (2x - 7) = 180$$
$$4x = 180 \qquad \text{Collecting like terms}$$
$$\frac{4x}{4} = \frac{180}{4} \qquad \text{Dividing by 4}$$
$$x = 45.$$

Possible measures for the angles are as follows:

First angle: $x = 45°$;
Second angle: $x + 7 = 45 + 7 = 52°$;
Third angle: $2x - 7 = 2(45) - 7 = 83°$.

4. **Check.** Consider angles of 45°, 52°, and 83°. The second is 7° more than the first and the third is 7° less than twice the first. The sum is 180°. The answer checks.

5. **State.** The measures of the angles are 45°, 52°, and 83°.

Do Exercise 6.

Sometimes applied problems involve **consecutive integers** like 19, 20, 21, 22 or −34, −33, −32, −31. Consecutive integers can be represented in the form $x, x + 1, x + 2, x + 3$, and so on.

Some examples of **consecutive even integers** are 20, 22, 24, 26 and −34, −32, −30, −28. Consecutive even integers can be represented in the form x, $x + 2, x + 4, x + 6$, and so on, as can **consecutive odd integers** like 19, 21, 23, 25 and −33, −31, −29, −27.

EXAMPLE 7 *Artist's Prints.* Often artists will number in sequence a limited number of prints in order to enhance their value. An artist creates 500 prints and saves three for his children. The numbers of those prints are consecutive integers whose sum is 189. Find the number of each of those prints.

1. **Familiarize.** The numbers of the prints are consecutive integers. Thus we let x = the first integer, $x + 1$ = the second, and $x + 2$ = the third.

2. **Translate.** We translate as follows:

$$\underbrace{\text{First integer}}_{x} + \underbrace{\text{Second integer}}_{(x+1)} + \underbrace{\text{Third integer}}_{(x+2)} = 189$$
$$= 189.$$

3. **Solve.** We solve the equation:

$$x + (x + 1) + (x + 2) = 189$$

$$3x + 3 = 189 \qquad \text{Collecting like terms}$$

$$3x + 3 - 3 = 189 - 3 \qquad \text{Subtracting 3}$$

$$3x = 186$$

$$\frac{3x}{3} = \frac{186}{3} \qquad \text{Dividing by 3}$$

$$x = 62.$$

Then $x + 1 = 62 + 1 = 63$ and $x + 2 = 62 + 2 = 64$.

4. **Check.** The numbers are 62, 63, and 64. These are consecutive integers and their sum is 189. The numbers check.

5. **State.** The numbers of the prints are 62, 63, and 64.

Do Exercise 7.

b Basic Motion Problems

When a problem deals with speed, distance, and time, we can expect to use the following **motion formula.**

THE MOTION FORMULA

Distance = Rate (or speed) · Time

$$d = rt$$
---

7. Artist's Prints. The artist in Example 7 saves three other prints for his own archives. These are also numbered consecutively. The sum of the numbers is 1266. Find the numbers of each of those prints.

Answer on page A-5

4 ft/sec
5 ft/sec

8. Marine Travel. Tim's fishing boat travels 12 km/h in still water. How long will it take him to travel 25 km upstream if the river's current is 3 km/h? 25 km downstream if the river's current is 3 km/h? (*Hint*: To find the boat's speed traveling upstream, subtract the speed of the current from the speed of the boat in still water. To find the boat's speed downstream, add the speed of the current to the speed of the boat in still water.)

■ **EXAMPLE 8** *Moving Sidewalks.* A moving sidewalk in O'Hare Airport is 300 ft long and moves at a speed of 5 ft/sec. If Kate walks at a speed of 4 ft/sec, how long will it take her to travel the 300 ft using the moving sidewalk?

1. **Familiarize.** First read the problem very carefully. This may even involve speaking aloud to a fellow student or to yourself. In this case, organizing the information in a table can be very helpful.

Distance to be traveled	300 ft
Speed of Kate	4 ft/sec
Speed of the moving sidewalk	5 ft/sec
Total speed of Kate on the sidewalk	?
Time required	?

We might now try to determine, possibly with the aid of outside references, what relationships exist among the various quantities in the problem. Since Kate is walking on the sidewalk in the same direction in which it is moving, the two speeds can be added to determine the total speed of Kate on the sidewalk. We can then complete the table, letting $t =$ the time, in seconds, required to travel 300 ft on the moving sidewalk.

Distance to be traveled	300 ft
Speed of Kate	4 ft/sec
Speed of the moving sidewalk	5 ft/sec
Total speed of Kate on the sidewalk	9 ft/sec
Time required	t

2. **Translate.** To translate, we use the motion formula $d = rt$ and substitute 300 ft for d and 9 ft/sec for r:

$$d = rt$$
$$300 = 9 \cdot t.$$

3. **Solve.** We solve the equation:

$$300 = 9t$$
$$\frac{300}{9} = \frac{9t}{9} \qquad \text{Dividing by 9}$$
$$\frac{100}{3} = t.$$

4. **Check.** At a speed of 9 ft/sec and in a time of 100/3, or $33\frac{1}{3}$ sec, Kate would travel $d = 9 \cdot \frac{100}{3} = 300$ ft. This answer checks.

5. **State.** Kate will travel the distance of 300 ft in 100/3, or $33\frac{1}{3}$ sec.

Do Exercise 8.

Answer on page A-5

a. Solve.

1. *Marathon.* A long-distance foot race known as a marathon covers a course of about 26.2 mi. In 1984, the first time women competed in the marathon in the Olympic Games, the race was won by Joan Benoit of the United States. At one point, she was three times as far from the end of the course as she was from the starting point. How many miles had she run at that time?

Source: *The World Almanac* 2005

2. *Iditarod.* The Iditarod is a 1150-mi dogsled race run from Anchorage to Nome, Alaska, each year in March. The 2005 race was won by Robert Sorlie. If, at one point, Sorlie was four times as far from the end of the course as he was from the starting point, how many miles of the course had he completed?

Source: espn.com

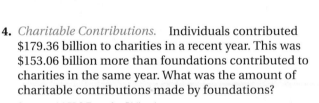

3. *Hurricane Losses.* The estimated insurance loss from Hurricane Andrew, which struck Florida, Louisiana, and Mississippi in 1992, was $20.3 billion. This was $13.5 billion more than the loss from Hurricane Charley, which struck Florida, North Carolina, and South Carolina in 2004. What was the insurance loss from Hurricane Charley?

Sources: Insurance Services Office; Insurance Information Institute

4. *Charitable Contributions.* Individuals contributed $179.36 billion to charities in a recent year. This was $153.06 billion more than foundations contributed to charities in the same year. What was the amount of charitable contributions made by foundations?

Source: AAFRC Trust for Philanthropy

5. *Solar Panel.* The cross section of a support for a solar energy panel is triangular. One angle of the triangle is five times as large as the first angle. The third angle is 2° less than the first angle. How large are the angles?

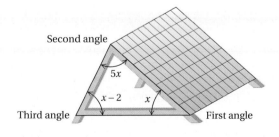

Second angle
$5x$
$x - 2$ x
Third angle   First angle

6. *Cross Section of a Roof.* In a triangular cross section of a roof, the second angle is twice as large as the first. The third angle is 20° greater than the first angle. How large are the angles?

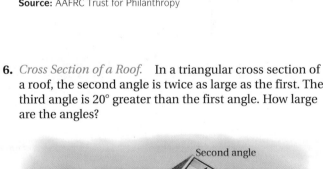

Second angle
$2x$
x $x + 20$
First angle   Third angle

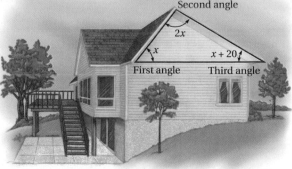

7. *Purchase Price.* Elka pays $1187.20 for a computer. The price includes a 6% sales tax. What is the price of the computer itself?

8. *Pricing.* Media Warehouse prices single general-purpose DVDs by raising the wholesale price of each DVD 50% and adding 25¢. What is the wholesale price of a DVD if it sells for $1.99?

9. *Perimeter of an NBA Court.* The perimeter of an NBA-sized basketball court is 288 ft. The length is 44 ft longer than the width. Find the dimensions of the court.
Source: National Basketball Association

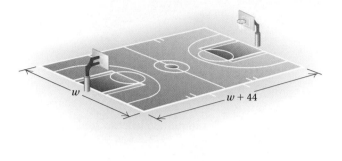

10. *Perimeter of a Tennis Court.* The width of a standard tennis court used for playing doubles is 42 ft less than the length. The perimeter of the court is 228 ft. Find the dimensions of the court.
Source: *Dunlop Illustrated Encyclopedia of Facts*

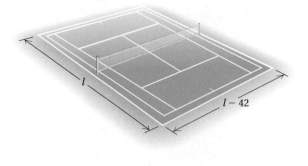

11. *Wire Cutting.* A piece of wire that is 100 cm long is to be cut into two pieces, each to be bent to make a square. The length of a side of one square is to be twice the length of a side of the other. How should the wire be cut?

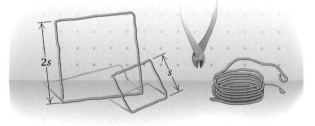

12. *Rope Cutting.* A rope that is 168 ft long is to be cut into three pieces such that the second piece is 6 ft less than three times the first, and the third is 2 ft more than two-thirds of the second. Find the length of the longest piece.

13. *Real Estate Commission.* The Martins negotiated the following real estate commission on the selling price of their house:

    7% for the first $100,000, and

    5% for the amount that exceeds $100,000.

The realtor received a commission of $15,250 for selling the house. What was the selling price?

14. *Real Estate Commission.* The Suzukis negotiated the following real estate commission on the selling price of their house:

    8% for the first $100,000, and

    3% for the amount that exceeds $100,000.

The realtor received a commission of $9200 for selling the house. What was the selling price?

15. *Consecutive Odd Integers.* Find three consecutive odd integers such that the sum of the first, two times the second, and three times the third is 70.

16. *Consecutive Even Integers.* Find three consecutive even integers such that the sum of the first, five times the second, and four times the third is 1226.

17. *Consecutive Post-Office Box Numbers.* The sum of the numbers on two adjacent post-office boxes is 697. What are the numbers?

18. *Consecutive Page Numbers.* The sum of the page numbers on the facing pages of the book *True Success: A New Philosophy of Excellence*, by Tom Morris, is 373. What are the page numbers?

19. *Carpet Cleaning.* A carpet company charges $75 to clean the first 200 sq ft of carpet. There is an additional charge of 25¢ per square foot for any footage that exceeds 200 sq ft and $1.40 per step for any carpeting on a staircase. A customer's cleaning bill was $253.95. This included the cleaning of a staircase with 13 steps. In addition to the staircase, how many square feet of carpet did the customer have cleaned?

20. *School Photos.* Memory Makers prices its school photos as shown here.

BASIC PACKAGE	PRICE
1 8 x 10	
2 5 x 7	$14.95
12 Wallet-size	
1 sheet of 6 extra wallet-sizes	$1.35

The Martinez family purchases the basic package for each of its three children, along with extra wallet-size photos. How many wallet-size photos did they buy in all if their total bill for the photos is $57?

21. *Original Salary.* An editorial assistant receives an 8% raise, bringing her salary to $42,066. What was her salary before the raise?

22. *Original Salary.* After a salesman receives a 5% raise, his new salary is $40,530. What was his old salary?

23. *Coffee Production.* In 2004, Starbucks Corporation announced that it planned to buy 225 million pounds of Arabica coffee beans in 2007. This would be a 650% increase over the amount of beans purchased in 2004. How many pounds of Arabica beans were purchased by Starbucks in 2004?

Source: Starbucks Corporation

24. *Patents Issued.* About 189,600 patents were issued by the U.S. government in 2003. This was a 77% increase over the number of patents issued in 1993. How many patents were issued in 1993?

Source: U.S. Patent and Trademark Office

25. *Driving Fatalities.* The equation

$$y = 19.5x + 17,380.2$$

can be used to estimate the number y of alcohol-related traffic fatalities in the United States x years after 2000. That is, $x = 0$ corresponds to 2000, $x = 5$ corresponds to 2005, and so on.

Source: National Center for Statistics and Analysis

a) Estimate the number of alcohol-related traffic fatalities in 2003 and in 2006.

b) In what year did about 17,458 alcohol-related traffic fatalities occur?

26. *Digital Camera Sales.* The equation

$$y = 3.35x + 3.73$$

can be used to estimate the number y of digital cameras sold, in millions, x years after 2000. That is, $x = 0$ corresponds to 2000, $x = 5$ corresponds to 2005, and so on.

Source: PMA Marketing Research

a) Estimate the number of digital cameras sold in 2002 and in 2007.

b) In what year were about 20.5 million digital cameras sold?

---

b Solve.

27. *Cruising Altitude.* A Boeing 767 has been instructed to climb from its present altitude of 8000 ft to a cruising altitude of 29,000 ft. The plane ascends at a rate of 3500 ft/min. How long will it take the plane to reach the cruising altitude?

28. *Flight into a Headwind.* An airplane traveling 390 mph in still air encounters a 65-mph headwind. How long will it take the plane to travel 725 mi into the wind?

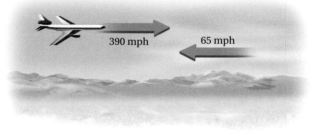

390 mph   65 mph

29. *Boating.* Jen's motorboat travels at a speed of 10 mph in still water. Booth River flows at a speed of 2 mph. How long will it take Jen to travel 15 mi downstream? 15 mi upstream?

30. *Air Travel.* A pilot has been instructed to descend from an altitude of 26,000 ft to 11,000 ft. If the pilot descends at a rate of 2500 ft/min, how long will it take the plane to reach the new altitude?

31. *Boating.* The *Delta Queen* is a paddleboat that tours the Mississippi River near New Orleans, Louisiana. It is not uncommon for the *Delta Queen* to run 7 mph in still water and for the Mississippi to flow at a rate of 3 mph. At these rates, how long will it take the boat to cruise 2 mi upstream?

Source: *Delta Queen* information

32. *Swimming.* Fran swims at a speed of 5 mph in still water. The Lazy River flows at a speed of 2.3 mph. How long will it take Fran to swim 1.8 mi upstream? 1.8 mi downstream?

33. $\mathbf{D_W}$ How can a guess or estimate help prepare you for the *Translate* step when solving problems?

34. $\mathbf{D_W}$ Why is it important to label clearly what a variable represents in an applied problem?

SKILL MAINTENANCE

Simplify. [R.3c]

35. $-44 \cdot 55 - 22$

36. $16 \cdot 8 + 200 \div 25 \cdot 10$

37. $(5 - 12)^2$

38. $5^2 - 12^2$

39. $5^2 - 2 \cdot 5 \cdot 12 + 12^2$

40. $(5 - 12)(5 + 12)$

41. $\dfrac{12|8 - 10| + 9 \cdot 6}{5^4 + 4^5}$

42. $\dfrac{(9 - 4)^2 + (8 - 11)^2}{4^2 + 2^2}$

43. $[-64 \div (-4)] \div (-16)$

44. $-64 \div [-4 \div (-16)]$

45. $2^{13} \div 2^5 \div 2^3$

46. $2^{13} \cdot 2^5 \cdot 2^3$

SYNTHESIS

47. 🖩 *Home Prices.* Real estate prices in Panduski increased 6% from 2002 to 2003 and 2% from 2003 to 2004. From 2004 to 2005, prices dropped 1%. If a house sold for $117,743 in 2005, what was its worth in 2002? (Round to the nearest dollar.)

48. *Adjusted Wages.* Christina's salary is reduced n% during a period of financial difficulty. By what number should her salary be multiplied in order to bring it back to where it was before the reduction?

49. *Novels.* A literature professor owns 400 novels. The number of horror novels is 46% of the number of science fiction novels; the number of science fiction novels is 65% of the number of romance novels; and the number of mystery novels is 17% of the number of horror novels. How many science fiction novels does the professor own? Round to the nearest one.

50. *NBA Shot Clock.* The National Basketball Association operates what is called a 24-second clock. This means that the offensive team has 24 sec in which to attempt a field goal. If it does not make such an attempt, it loses the ball. The NBA arrived at 24 sec by dividing the total number of seconds in a game by the average number of points scored per game. There are 48 min in a game. What is the average number of points scored in a game?

Source: National Basketball Association

51. *Population Change.* The yearly changes in the population census of a city for three consecutive years are, respectively, a 20% increase, a 30% increase, and a 20% decrease. What is the total percent change from the beginning of the first year to the end of the third year, to the nearest percent?

52. *Area of a Triangle.* The height and sides of a triangle are represented by four consecutive integers. The height is the first integer and the base is the third integer. The perimeter of the triangle is 42 in. Find the area of the triangle.

53. *Watch Time.* Your watch loses $1\frac{1}{2}$ sec every hour. You have a friend whose watch gains 1 sec every hour. The watches show the same time now. After how many more seconds will they show the same time again?

54. *Test Scores.* Tico's scores on four tests are 83, 91, 78, and 81. How many points above the average must Tico score on the next test in order to raise his average 2 points?

55. Write a problem for a classmate to solve. Devise the problem so that the solution is, "The material should be cut into two pieces, one 30 cm long and the other 45 cm long."

56. Solve: "The sum of three consecutive integers is 55." Find the integers. Why is it important to check the solution from the *Solve* step in the original problem?

57. *Geometry.* Consider the geometric figure below. Suppose that $L \parallel M$, $m \angle 8 = 5x + 25$, and $m \angle 4 = 8x + 4$. Find $m \angle 2$ and $m \angle 1$.

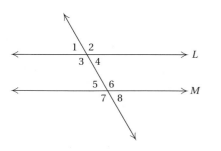

58. *Geometry.* Suppose the figure *ABCD* below is a square. Point *A* is folded onto the midpoint of $\overline{AB}$ and point *D* is folded onto the midpoint of $\overline{DC}$. The perimeter of the smaller figure formed is 25 in. Find the area of the square *ABCD*.

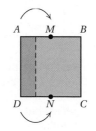

1.4 SETS, INEQUALITIES, AND INTERVAL NOTATION

Objectives

a Determine whether a given number is a solution of an inequality.

b Write interval notation for the solution set or graph of an inequality.

c Solve an inequality using the addition and multiplication principles and then graph the inequality.

d Solve applied problems by translating to inequalities.

a Inequalities

We can extend our equation-solving skills to the solving of inequalities. (See Section R.1 for an introduction to inequalities.)

INEQUALITY

An **inequality** is a sentence containing $<$, $>$, $\leq$, $\geq$, or $\neq$.

Some examples of inequalities are

$$-2 < a, \qquad x > 4, \qquad x + 3 \leq 6, \qquad 6 - 7y \geq 10y - 4, \quad \text{and} \quad 5x \neq 10.$$

SOLUTION OF AN INEQUALITY

Any replacement or value for the variable that makes an inequality true is called a **solution.** The set of all solutions is called the **solution set.** When all the solutions of an inequality have been found, we say that we have **solved** the inequality.

EXAMPLES Determine whether the given number is a solution of the inequality.

1. $x + 3 < 6$;   5

We substitute 5 for x and get $5 + 3 < 6$, or $8 < 6$, a *false* sentence. Therefore, 5 is not a solution.

2. $2x - 3 > -3$;   1

We substitute 1 for x and get $2(1) - 3 > -3$, or $-1 > -3$, a *true* sentence. Therefore, 1 is a solution.

3. $4x - 1 \leq 3x + 2$; -3

We substitute -3 for x and get $4(-3) - 1 \leq 3(-3) + 2$, or $-13 \leq -7$, a *true* sentence. Therefore, -3 is a solution.

Do Exercises 1–3.

b Inequalities and Interval Notation

The **graph** of an inequality is a drawing that represents its solutions. An inequality in one variable can be graphed on the number line.

EXAMPLE 4 Graph $x < 4$ on the number line.

The solutions are all real numbers less than 4, so we shade all numbers less than 4 on the number line. To indicate that 4 is not a solution, we use a right parenthesis ")" at 4.

$$\xleftarrow{\hspace{2cm}} \overset{}{\underset{-7\ -6\ -5\ -4\ -3\ -2\ -1\ \ 0\ \ 1\ \ 2\ \ 3\ \ 4\ \ 5\ \ 6\ \ 7}{\big|\,\big|\,\big|\,\big|\,\big|\,\big|\,\big|\,\big|\,\big|\,\big|\,\big|\,\big|\,}} \xrightarrow{\hspace{1cm}}$$

Determine whether the given number is a solution of the inequality.

1. $3 - x < 2$;   8

2. $3x + 2 > -1$; -2

3. $3x + 2 \leq 4x - 3$;   5

Answers on page A-5

113

Are you aware of all the learning resources that exist for this textbook? Many details are given in the Preface.

- The *Student's Solutions Manual* contains fully worked-out solutions to the odd-numbered exercises in the exercise sets, with the exception of the discussion and writing exercises, as well as solutions to all exercises in Chapter Reviews, Chapter Tests, and Cumulative Reviews. You can order this through the bookstore or by calling 1-800-282-0693.

- An extensive set of *videotapes* supplements this text. These are available on CD-ROM by calling 1-800-282-0693.

- *Tutorial software* called InterAct Math also accompanies this text. If it is not available in the campus learning center, you can order it by calling 1-800-282-0693.

- The Addison-Wesley *Math Tutor Center* is available for help with the odd-numbered exercises. You can order this service by calling 1-800-824-7799.

- Extensive help is available online via MyMathLab and / or MathXL. Ask your instructor for information about these or visit MyMathLab.com and MathXL.com.

We can write the solution set for $x < 4$ using **set-builder notation** (see Section R.1): $\{x \mid x < 4\}$. This is read "The set of all x such that x is less than 4."

Another way to write solutions of an inequality in one variable is to use **interval notation.** Interval notation uses parentheses ( ) and brackets [ ].

If a and b are real numbers such that $a < b$, we define the interval (a, b) as the set of all numbers between but not including a and b—that is, the set of all x for which $a < x < b$. Thus,

$$(a, b) = \{x \mid a < x < b\}.$$

The points a and b are the **endpoints** of the interval. The parentheses indicate that the endpoints are *not* included in the graph.

The interval $[a, b]$ is defined as the set of all numbers x for which $a \le x \le b$. Thus,

$$[a, b] = \{x \mid a \le x \le b\}.$$

The brackets indicate that the endpoints *are* included in the graph.*

Caution!

Do not confuse the *interval* (a, b) with the *ordered pair* (a, b), which denotes a point in the plane, as we will see in Chapter 3. The context in which the notation appears usually makes the meaning clear.

The following intervals include one endpoint and exclude the other:

$$(a, b] = \{x \mid a < x \le b\}. \quad \text{The graph excludes } a \text{ and includes } b.$$

$$[a, b) = \{x \mid a \le x < b\}. \quad \text{The graph includes } a \text{ and excludes } b.$$

Some intervals extend without bound in one or both directions. We use the symbols ∞, read "infinity," and $-\infty$, read "negative infinity," to name these intervals. The notation (a, ∞) represents the set of all numbers greater than a—that is,

$$(a, \infty) = \{x \mid x > a\}.$$

Similarly, the notation $(-\infty, a)$ represents the set of all numbers less than a—that is,

$$(-\infty, a) = \{x \mid x < a\}.$$

---
*Some books use the representations 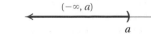 instead of, respectively,

CHAPTER 1: Solving Linear Equations and Inequalities

The notations $[a, \infty)$ and $(-\infty, a]$ are used when we want to include the endpoint a. The interval $(-\infty, \infty)$ names the set of all real numbers.

$$(-\infty, \infty) = \{x \mid x \text{ is a real number}\}$$

Interval notation is summarized in the following table.

INTERVALS: NOTATION AND GRAPHS

INTERVAL NOTATION	SET NOTATION	GRAPH
(a, b)	$\{x \mid a < x < b\}$	
$[a, b]$	$\{x \mid a \leq x \leq b\}$	
$[a, b)$	$\{x \mid a \leq x < b\}$	
$(a, b]$	$\{x \mid a < x \leq b\}$	
(a, ∞)	$\{x \mid x > a\}$	
$[a, \infty)$	$\{x \mid x \geq a\}$	
$(-\infty, b)$	$\{x \mid x < b\}$	
$(-\infty, b]$	$\{x \mid x \leq b\}$	
$(-\infty, \infty)$	$\{x \mid x \text{ is a real number}\}$	

Caution!

Whenever the symbol ∞ is included in interval notation, a right parenthesis ")" is used. Similarly, when $-\infty$ is included, a left parenthesis "(" is used.

EXAMPLES Write interval notation for the given set or graph.

5. $\{x \mid -4 < x < 5\} = (-4, 5)$

6. $\{x \mid x \geq -2\} = [-2, \infty)$

7. $\{x \mid 7 > x \geq 1\} = \{x \mid 1 \leq x < 7\} = [1, 7)$

8.

$(-2, 4]$

$-6\ -5\ -4\ -3\ -2\ -1\ \ 0\ \ 1\ \ 2\ \ 3\ \ 4\ \ 5\ \ 6$

9.

$(-\infty, -1)$

$-6\ -5\ -4\ -3\ -2\ -1\ \ 0\ \ 1\ \ 2\ \ 3\ \ 4\ \ 5\ \ 6$

Do Exercises 4–8.

 Solving Inequalities

Two inequalities are **equivalent** if they have the same solution set. For example, the inequalities $x > 4$ and $4 < x$ are equivalent. Just as the addition principle for equations gives us equivalent equations, the addition principle for inequalities gives us equivalent inequalities.

Write interval notation for the given set or graph.

4. $\{x \mid -4 \leq x < 5\}$

5. $\{x \mid x \leq -2\}$

6. $\{x \mid 6 \geq x > 2\}$

7.

$-40\ -30\ -20\ -10\ \ 0\ \ 10\ \ 20\ \ 30\ \ 40$

8.

$-40\ -30\ -20\ -10\ \ 0\ \ 10\ \ 20\ \ 30\ \ 40$

Answers on page A-5

Solve and graph.

9. $x + 6 > 9$

For any real numbers a, b, and c:

$a < b$ is equivalent to $a + c < b + c$;

$a > b$ is equivalent to $a + c > b + c$.

Similar statements hold for $\leq$ and $\geq$.

Since subtracting c is the same as adding $-c$, there is no need for a separate subtraction principle.

EXAMPLE 10 Solve and graph: $x + 5 > 1$.

We have

$$x + 5 > 1$$

$$x + 5 - 5 > 1 - 5 \qquad \text{Using the addition principle:}$$
$$\text{adding } -5 \text{ or subtracting } 5$$

$$x > -4.$$

We used the addition principle to show that the inequalities $x + 5 > 1$ and $x > -4$ are equivalent. The solution set is $\{x \mid x > -4\}$ and consists of an infinite number of solutions. We cannot possibly check them all. Instead, we can perform a partial check by substituting one member of the solution set (here we use -1) into the original inequality:

$$\frac{x + 5 > 1}{-1 + 5 \; ? \; 1}$$
$$4 \; | \qquad \text{TRUE}$$

Since $4 > 1$ is true, we have our check. The solution set is $\{x \mid x > -4\}$, or $(-4, \infty)$. The graph is as follows:

10. $x + 4 \leq 7$

Do Exercises 9 and 10.

11. Solve and graph:

$$2x - 3 \geq 3x - 1.$$

EXAMPLE 11 Solve and graph: $4x - 1 \geq 5x - 2$.

We have

$$4x - 1 \geq 5x - 2$$

$$4x - 1 + 2 \geq 5x - 2 + 2 \qquad \text{Adding 2}$$

$$4x + 1 \geq 5x \qquad \text{Simplifying}$$

$$4x + 1 - 4x \geq 5x - 4x \qquad \text{Subtracting } 4x$$

$$1 \geq x. \qquad \text{Simplifying}$$

The inequalities $1 \geq x$ and $x \leq 1$ have the same meaning and the same solutions. You can check that any number less than or equal to 1 is a solution. The solution set is $\{x \mid 1 \geq x\}$ or, more commonly, $\{x \mid x \leq 1\}$. Using interval notation, we write that the solution set is $(-\infty, 1]$. The graph is as follows:

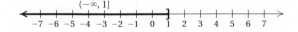

Answers on page A-5

Do Exercise 11.

CHAPTER 1: Solving Linear Equations
and Inequalities

The multiplication principle for inequalities differs from the multiplication principle for equations. Consider the true inequality

$$-4 < 9.$$

If we multiply both numbers by 2, we get another true inequality:

$$-4(2) < 9(2), \quad \text{or} \quad -8 < 18. \qquad \text{True}$$

If we multiply both numbers by -3, we get a false inequality:

$$-4(-3) < 9(-3), \quad \text{or} \quad 12 < -27. \qquad \text{False}$$

However, if we now *reverse* the inequality symbol above, we get a true inequality:

$$12 > -27. \qquad \text{True}$$

**THE MULTIPLICATION PRINCIPLE
FOR INEQUALITIES**

For any real numbers a and b, and any *positive* number c:

$a < b$ is equivalent to $ac < bc$;

$a > b$ is equivalent to $ac > bc$.

For any real numbers a and b, and any *negative* number c:

$a < b$ is equivalent to $ac > bc$;

$a > b$ is equivalent to $ac < bc$.

Similar statements hold for $\leq$ and $\geq$.

Since division by c is the same as multiplication by $1/c$, there is no need for a separate division principle.

The multiplication principle tells us that when we multiply or divide both sides of an inequality by a negative number, we must reverse the inequality symbol to obtain an equivalent inequality.

EXAMPLE 12 Solve and graph: $3y < \frac{3}{4}$.

We have

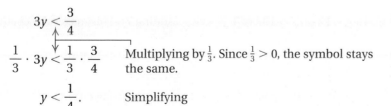

$$\frac{1}{3} \cdot 3y < \frac{1}{3} \cdot \frac{3}{4} \qquad \text{Multiplying by } \tfrac{1}{3}. \text{ Since } \tfrac{1}{3} > 0, \text{ the symbol stays the same.}$$

$$y < \frac{1}{4}. \qquad \text{Simplifying}$$

Any number less than $\frac{1}{4}$ is a solution. The solution set is $\left\{ y \,\middle|\, y < \frac{1}{4} \right\}$, or $\left(-\infty, \frac{1}{4} \right)$. The graph is as follows:

Solve and graph.

12. $5y \leq \dfrac{3}{2}$

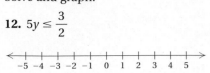

EXAMPLE 13 Solve and graph: $-5x \geq -80$.

We have

$$-5x \geq -80$$

$$\dfrac{-5x}{-5} \leq \dfrac{-80}{-5} \quad \boxed{\begin{array}{l} \text{Dividing by } -5. \text{ Since } -5 < 0, \text{ the} \\ \text{inequality symbol must be reversed.}\end{array}}$$

$$x \leq 16.$$

The solution set is $\{x \,|\, x \leq 16\}$, or $(-\infty, 16]$. The graph is as follows:

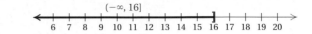

Do Exercises 12–14.

We use the addition and multiplication principles together in solving inequalities in much the same way as in solving equations.

13. $-2y > 10$

EXAMPLE 14 Solve: $16 - 7y \geq 10y - 4$.

We have

$$16 - 7y \geq 10y - 4$$

$$-16 + 16 - 7y \geq -16 + 10y - 4 \qquad \text{Adding } -16$$

$$-7y \geq 10y - 20 \qquad \text{Collecting like terms}$$

$$-10y + (-7y) \geq -10y + 10y - 20 \qquad \text{Adding } -10y$$

$$-17y \geq -20 \qquad \text{Collecting like terms}$$

$$\dfrac{-17y}{-17} \leq \dfrac{-20}{-17} \quad \boxed{\begin{array}{l} \text{Dividing by } -17. \text{ The symbol} \\ \text{must be reversed.}\end{array}}$$

$$y \leq \dfrac{20}{17}. \qquad \text{Simplifying}$$

The solution set is $\left\{ y \,|\, y \leq \dfrac{20}{17} \right\}$, or $\left(-\infty, \dfrac{20}{17} \right]$.

14. $-\dfrac{1}{3}x \leq -4$

In some cases, we can avoid multiplying or dividing by a negative number by using the addition principle in a different way. Let's rework Example 14 by adding $7y$ instead of $-10y$:

$$16 - 7y \geq 10y - 4$$

$$16 - 7y + 7y \geq 10y - 4 + 7y \qquad \begin{array}{l} \text{Adding } 7y. \text{ This makes the coefficient} \\ \text{of the } y\text{-term positive.}\end{array}$$

$$16 \geq 17y - 4 \qquad \text{Collecting like terms}$$

$$16 + 4 \geq 17y - 4 + 4 \qquad \text{Adding } 4$$

$$20 \geq 17y \qquad \text{Collecting like terms}$$

$$\qquad \boxed{\begin{array}{l} \text{Dividing by } 17. \text{ The symbol} \\ \text{stays the same.}\end{array}}$$

$$\dfrac{20}{17} \geq \dfrac{17y}{17}$$

$$\dfrac{20}{17} \geq y.$$

Note that $\dfrac{20}{17} \geq y$ is equivalent to $y \leq \dfrac{20}{17}$.

Answers on page A-5

CHAPTER 1: Solving Linear Equations
and Inequalities

EXAMPLE 15 Solve: $-3(x + 8) - 5x > 4x - 9$.

We have

$$-3(x + 8) - 5x > 4x - 9$$

$-3x - 24 - 5x > 4x - 9$ Using the distributive law

$-24 - 8x > 4x - 9$ Collecting like terms

$-24 - 8x + 8x > 4x - 9 + 8x$ Adding $8x$

$-24 > 12x - 9$ Collecting like terms

$-24 + 9 > 12x - 9 + 9$ Adding 9

$-15 > 12x$

Dividing by 12. The symbol stays the same.

$$\frac{-15}{12} > \frac{12x}{12}$$

$$-\frac{5}{4} > x.$$

The solution set is $\left\{ x \mid -\frac{5}{4} > x \right\}$, or $\left\{ x \mid x < -\frac{5}{4} \right\}$, or $\left(-\infty, -\frac{5}{4} \right)$.

Do Exercises 15–17.

d   Applications and Problem Solving

Many problem-solving and applied situations translate to inequalities. In addition to "is less than" and "is more than," other phrases are commonly used.

IMPORTANT WORDS	SAMPLE SENTENCE	TRANSLATION
is at least	Max is at least 5 years old.	$m \geq 5$
is at most	At most 6 people could fit in the elevator.	$n \leq 6$
cannot exceed	Total weight in the elevator cannot exceed 2000 pounds.	$w \leq 2000$
must exceed	The speed must exceed 15 mph.	$s > 15$
is between	Heather's income is between $23,000 and $35,000.	$23{,}000 < h < 35{,}000$
no more than	Bing weighs no more than 90 pounds.	$w \leq 90$
no less than	Saul would accept no less than $4000 for the piano.	$t \geq 4000$

The following phrases deserve special attention.

TRANSLATING "AT LEAST" AND "AT MOST"

A quantity x is **at least** some amount q: $x \geq q$.
   (If x is at least q, it cannot be less than q.)

A quantity x is **at most** some amount q: $x \leq q$.
   (If x is at most q, it cannot be more than q.)

Do Exercises 18–27.

Solve.

15. $6 - 5y \geq 7$

16. $3x + 5x < 4$

17. $17 - 5(y - 2) \leq 45y + 8(2y - 3) - 39y$

Translate.

18. Emma scored no less than 88 on her Chinese exam.

19. The average credit-card holder is at least $4000 in debt.
   Source: CBS television, "60 Minutes"

20. The price of that PT Cruiser is at most $21,000.

21. The time of the test was between 50 and 60 min.

22. The University of Northern Kentucky is more than 25 miles away.

23. Sarah's weight is less than 110 lb.

24. That number is greater than -8.

25. The costs of production of that DVD model cannot exceed $135,000.

26. At most 11.6% of all deaths in Minnesota are from cancer.

27. Yesterday, at least 37 people got tickets for drunk driving.

Answers on page A-5

28. Cost of Higher Education.
Refer to Example 16. Determine, in terms of an inequality, the years for which the cost of higher education will be more than $14,000.

EXAMPLE 16 *Cost of Higher Education.* The equation

$$C = 587.5t + 8638$$

can be used to estimate the cost of tuition, fees, and room and board at four-year public institutions of higher education, where t is the number of years after 2000. Determine, in terms of an inequality, the years for which the cost will be more than $11,000.

Source: U.S. Department of Education

1. **Familiarize.** We already have a formula. To become more familiar with it, we might make a substitution for t. Suppose we want to know the cost 10 yr after 2000, or in 2010. We substitute 10 for t:

$$C = 587.5(10) + 8638 = \$14{,}513.$$

We see that in 2010 the cost of higher education at four-year public institutions will be more than $11,000. To find all the years in which the cost exceeds $11,000, we could make other guesses less than 10. Instead we will proceed to the next step.

2. **Translate.** The cost C is to be *more than* $11,000. Thus we have

$$C > 11{,}000.$$

We replace C with $587.5t + 8638$ to find the values of t that are solutions of the inequality:

$$587.5t + 8638 > 11{,}000.$$

3. **Solve.** We solve the inequality:

$$587.5t + 8638 > 11{,}000$$
$$587.5t > 2362 \qquad \text{Subtracting 8638}$$
$$t > 4. \qquad \text{Dividing by 587.5 and rounding}$$

4. **Check.** A partial check is to substitute a value for t greater than 4. We did that in the *Familiarize* step and found that the cost was more than $11,000.

5. **State.** The cost of tuition, fees, and room and board at four-year public institutions of higher education will be more than $11,000 for years more than 4 yr after 2000, so we have $\{t \mid t > 4\}$, or years after 2004.

Answer on page A-5

CHAPTER 1: Solving Linear Equations and Inequalities

Do Exercise 28.

EXAMPLE 17 *Salary Plans.* On her new job, Rose can be paid in one of two ways: *Plan A* is a salary of $600 per month, plus a commission of 4% of sales; and *Plan B* is a salary of $800 per month, plus a commission of 6% of sales in excess of $10,000. For what amount of monthly sales is plan A better than plan B, if we assume that sales are always more than $10,000?

1. **Familiarize.** Listing the given information in a table will be helpful.

PLAN A: MONTHLY INCOME	PLAN B: MONTHLY INCOME
$600 salary	$800 salary
4% of sales	6% of sales over $10,000
Total: $600 + 4% of sales	*Total*: $800 + 6% of sales over $10,000

Next, suppose that Rose had sales of $12,000 in one month. Which plan would be better? Under plan A, she would earn $600 plus 4% of $12,000, or

$$600 + 0.04(12,000) = \$1080.$$

Since with plan B commissions are paid only on sales in excess of $10,000, Rose would earn $800 plus 6% of ($12,000 − $10,000), or

$$800 + 0.06(12,000 - 10,000) = \$920.$$

This shows that for monthly sales of $12,000, plan A is better. Similar calculations will show that for sales of $30,000 a month, plan B is better. To determine *all* values for which plan A pays more money, we must solve an inequality that is based on the calculations above.

2. **Translate.** We let S = the amount of monthly sales. If we examine the calculations in the *Familiarize* step, we see that the monthly income from plan A is $600 + 0.04S$ and from plan B is $800 + 0.06(S - 10,000)$. Thus we want to find all values of S for which

$$
\underbrace{600 + 0.04S}_{\text{Income from plan A}} \quad \underbrace{>}_{\text{is greater than}} \quad \underbrace{800 + 0.06(S - 10,000)}_{\text{Income from plan B}}.
$$

3. **Solve.** We solve the inequality:

$$600 + 0.04S > 800 + 0.06(S - 10,000)$$

$600 + 0.04S > 800 + 0.06S - 600$ Using the distributive law

$600 + 0.04S > 200 + 0.06S$ Collecting like terms

$400 > 0.02S$ Subtracting 200 and 0.04S

$20,000 > S$, or $S < 20,000$.   Dividing by 0.02

4. **Check.** For $S = 20,000$, the income from plan A is

$$600 + 4\% \cdot 20,000, \text{ or } \$1400.$$

The income from plan B is

$$800 + 6\% \cdot (20,000 - 10,000), \text{ or } \$1400.$$

This confirms that for sales of $20,000, Rose's pay is the same under either plan.

In the *Familiarize* step, we saw that for sales of $12,000, plan A pays more. Since $12,000 < 20,000$, this is a partial check. Since we cannot check all possible values of S, we will stop here.

5. **State.** For monthly sales of less than $20,000, plan A is better.

Do Exercise 29.

29. Salary Plans. A painter can be paid in one of two ways:

Plan A: $500 plus $4 per hour;

Plan B: Straight $9 per hour.

Suppose that the job takes n hours. For what values of n is plan A better for the painter?

Answer on page A-5

1.4 Sets, Inequalities, and Interval Notation

Translating for Success

1. Consecutive Integers. The sum of two consecutive even integers is 102. Find the integers.

2. Salary Increase. After Susanna earned a 5% raise, her new salary was $25,750. What was her former salary?

3. Dimensions of a Rectangle. The length of a rectangle is 6 in. more than the width. The perimeter of the rectangle is 102 in. Find the length and the width.

4. Population. The population of Doddville is decreasing at a rate of 5% per year. The current population is 25,750. What was the population the previous year?

5. Reading Assignment. Quinn has 6 days to complete a 150-page reading assignment. How many pages must he read the first day so that he has no more than 102 pages left to read on the 5 remaining days?

The goal of these matching questions is to practice step (2), *Translate*, of the five-step problem-solving process. Translate each word problem to an equation or an inequality and select a correct translation from A–O.

A. $0.05(25{,}750) = x$

B. $x + 2x = 102$

C. $2x + 2(x + 6) = 102$

D. $150 - x \leq 102$

E. $x - 0.05x = 25{,}750$

F. $x + (x + 2) = 102$

G. $x + (x + 6) > 102$

H. $x + 5x = 150$

I. $x + 0.05x = 25{,}750$

J. $x + (2x + 6) = 102$

K. $x + (x + 1) = 102$

L. $102 + x > 150$

M. $0.05x = 25{,}750$

N. $102 + 5x > 150$

O. $x + (x + 6) = 102$

Answers on page A-5

6. Numerical Relationship. One number is 6 more than twice another. The sum of the numbers is 102. Find the numbers.

7. DVD Collections. Together Ella and Ken have 102 DVDs. If Ken has 6 more DVDs than Ella, how many does each have?

8. Sales Commissions. Will earns a commission of 5% on his sales. One year he earned commissions totaling $25,750. What were his total sales for the year?

9. Fencing. Jess has 102 ft of fencing that he plans to use to enclose two dog runs. The perimeter of one run is to be twice the perimeter of the other. Into what lengths should the fencing be cut?

10. Quiz Scores. Lupe has a total of 102 points on the first 6 quizzes in her sociology class. How many total points must she earn on the 5 remaining quizzes in order to have more than 150 points for the semester?

a Determine whether the given numbers are solutions of the inequality.

1. $x - 2 \geq 6$; $-4, 0, 4, 8$

2. $3x + 5 \leq -10$; $-5, -10, 0, 27$

3. $t - 8 > 2t - 3$; $0, -8, -9, -3, -\frac{7}{8}$

4. $5y - 7 < 8 - y$; $2, -3, 0, 3, \frac{2}{3}$

b Write interval notation for the given set or graph.

5. $\{x \mid x < 5\}$

6. $\{t \mid t \geq -5\}$

7. $\{x \mid -3 \leq x \leq 3\}$

8. $\{t \mid -10 < t \leq 10\}$

9. $\{x \mid -4 > x > -8\}$

10. $\{x \mid 13 > x \geq 5\}$

11.

12.

13. $-\sqrt{2}$

14.

c Solve and graph.

15. $x + 2 > 1$

16. $x + 8 > 4$

17. $y + 3 < 9$

18. $y + 4 < 10$

19. $a - 9 \leq -31$

20. $a + 6 \leq -14$

21. $t + 13 \geq 9$

22. $x - 8 \leq 17$

23. $y - 8 > -14$

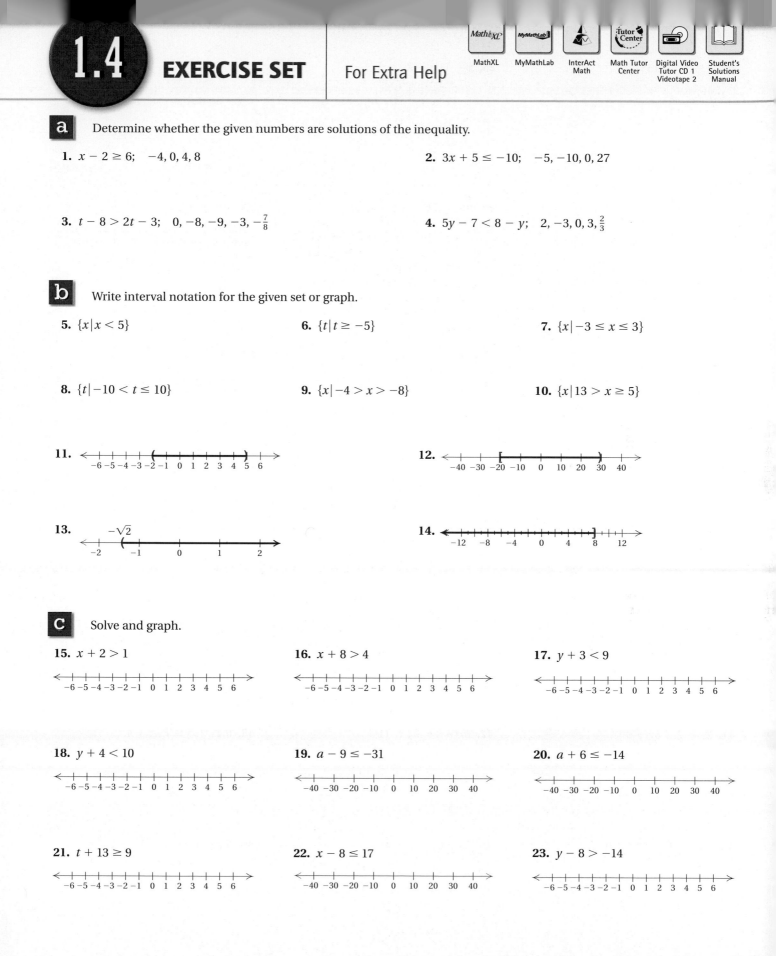

24. $y - 9 > -18$

25. $x - 11 \leq -2$

26. $y - 18 \leq -4$

27. $8x \geq 24$

28. $8t < -56$

29. $0.3x < -18$

30. $0.6x < 30$

31. $\frac{2}{3}x > 2$

32. $\frac{3}{5}x > -3$

Solve.

33. $-9x \geq -8.1$

34. $-5y \leq 3.5$

35. $-\frac{3}{4}x \geq -\frac{5}{8}$

36. $-\frac{1}{8}y \leq -\frac{9}{8}$

37. $2x + 7 < 19$

38. $5y + 13 > 28$

39. $5y + 2y \leq -21$

40. $-9x + 3x \geq -24$

41. $2y - 7 < 5y - 9$

42. $8x - 9 < 3x - 11$

43. $0.4x + 5 \leq 1.2x - 4$

44. $0.2y + 1 > 2.4y - 10$

45. $5x - \frac{1}{12} \leq \frac{5}{12} + 4x$

46. $2x - 3 < \frac{13}{4}x + 10 - 1.25x$

CHAPTER 1: Solving Linear Equations
and Inequalities

47. $4(4y - 3) \geq 9(2y + 7)$

48. $2m + 5 \geq 16(m - 4)$

49. $3(2 - 5x) + 2x < 2(4 + 2x)$

50. $2(0.5 - 3y) + y > (4y - 0.2)8$

51. $5[3m - (m + 4)] > -2(m - 4)$

52. $[8x - 3(3x + 2)] - 5 \geq 3(x + 4) - 2x$

53. $3(r - 6) + 2 > 4(r + 2) - 21$

54. $5(t + 3) + 9 < 3(t - 2) + 6$

55. $19 - (2x + 3) \leq 2(x + 3) + x$

56. $13 - (2c + 2) \geq 2(c + 2) + 3c$

57. $\frac{1}{4}(8y + 4) - 17 < -\frac{1}{2}(4y - 8)$

58. $\frac{1}{3}(6x + 24) - 20 > -\frac{1}{4}(12x - 72)$

59. $2[4 - 2(3 - x)] - 1 \geq 4[2(4x - 3) + 7] - 25$

60. $5[3(7 - t) - 4(8 + 2t)] - 20 \leq -6[2(6 + 3t) - 4]$

61. $\frac{4}{5}(7x - 6) < 40$

62. $\frac{2}{3}(4x - 3) > 30$

63. $\frac{3}{4}(3 + 2x) + 1 \geq 13$

64. $\frac{7}{8}(5 - 4x) - 17 \geq 38$

65. $\frac{3}{4}\left(3x - \frac{1}{2}\right) - \frac{2}{3} < \frac{1}{3}$

66. $\frac{2}{3}\left(\frac{7}{8} - 4x\right) - \frac{5}{8} < \frac{3}{8}$

67. $0.7(3x + 6) \geq 1.1 - (x + 2)$

68. $0.9(2x + 8) < 20 - (x + 5)$

69. $a + (a - 3) \leq (a + 2) - (a + 1)$

70. $0.8 - 4(b - 1) > 0.2 + 3(4 - b)$

---

d Solve.

Body Mass Index. **Body mass index I** can be used to determine whether an individual has a healthy weight for his or her height. An index between 18.5 and 25 indicates a healthy weight. Body mass index is given by the formula, or model,

$$I = \frac{704.5W}{H^2},$$

where W is weight, in pounds, and H is height, in inches. (See Example 1 in Section 1.2.) Use this formula for Exercises 71 and 72.

Source: National Center for Health Statistics

71. *Body Mass Index.* Marv's height is 73 in. Determine, in terms of an inequality, those weights W that will keep his body mass index below 25.

72. *Body Mass Index.* Elaine's height is 67 in. Determine, in terms of an inequality, those weights W that will keep her body mass index below 25.

73. *Grades.* You are taking a European history course in which there will be 4 tests, each worth 100 points. You have scores of 89, 92, and 95 on the first three tests. You must make a total of at least 360 in order to get an A. What scores on the last test will give you an A?

74. *Grades.* You are taking a literature course in which there will be 5 tests, each worth 100 points. You have scores of 94, 90, and 89 on the first three tests. You must make a total of at least 450 in order to get an A. What scores on the fourth test will keep you eligible for an A?

75. *Insurance Claims.* After a serious automobile accident, most insurance companies will replace the damaged car with a new one if repair costs exceed 80% of the N.A.D.A., or "blue-book," value of the car. Miguel's car recently sustained $9200 worth of damage but was not replaced. What was the blue-book value of his car?

76. *Phone Rates.* A long-distance telephone call using Down East Calling costs 10 cents for the first minute and 8 cents for each additional minute. The same call, placed on Long Call Systems, costs 16 cents for the first minute and 6 cents for each additional minute. For what length phone calls is Down East Calling less expensive?

77. *Salary Plans.* Toni can be paid in one of two ways:

    Plan A: A salary of $400 per month plus a commission of 8% of gross sales;

    Plan B: A salary of $610 per month, plus a commission of 5% of gross sales.

For what amount of gross sales should Toni select plan A?

78. *Salary Plans.* Branford can be paid for his masonry work in one of two ways:

    Plan A: $300 plus $9.00 per hour;

    Plan B: Straight $12.50 per hour.

Suppose that the job takes n hours. For what values of n is plan B better for Branford?

79. *Checking-Account Rates.* The Hudson Bank offers two checking-account plans. Their Anywhere plan charges 20¢ per check whereas their Acu-checking plan costs $2 per month plus 12¢ per check. For what numbers of checks per month will the Acu-checking plan cost less?

80. *Insurance Benefits.* Bayside Insurance offers two plans. Under plan A, Giselle would pay the first $50 of her medical bills and 20% of all bills after that. Under plan B, Giselle would pay the first $250 of bills, but only 10% of the rest. For what amount of medical bills will plan B save Giselle money? (Assume that her bills will exceed $250.)

81. *Wedding Costs.* The Arnold Inn offers two plans for wedding parties. Under plan A, the inn charges $30 for each person in attendance. Under plan B, the inn charges $1300 plus $20 for each person in excess of the first 25 who attend. For what size parties will plan B cost less? (Assume that more than 25 guests will attend.)

82. *Investing.* Lillian is about to invest $20,000, part at 6% and the rest at 8%. What is the most that she can invest at 6% and still be guaranteed at least $1500 in interest per year?

83. *Converting Dress Sizes.* The formula

$$I = 2(s + 10)$$

can be used to convert dress sizes s in the United States to dress sizes I in Italy. For what dress sizes in the United States will dress sizes in Italy be larger than 36?

84. *Temperatures of Solids.* The formula

$$C = \frac{5}{9}(F - 32)$$

can be used to convert Fahrenheit temperatures F to Celsius temperatures C.

a) Gold is a solid at Celsius temperatures less than 1063°C. Find the Fahrenheit temperatures for which gold is a solid.

b) Silver is a solid at Celsius temperatures less than 960.8°C. Find the Fahrenheit temperatures for which silver is a solid.

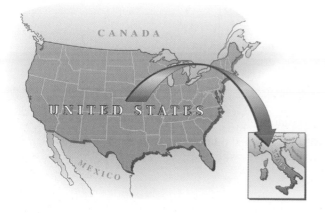

127

85. *Consumption of Bottled Water.* People are consuming more bottled water each year. The number N of gallons per year that each person drinks t years after 1990 is approximated by

$$N = 0.733t + 8.398.$$

a) How many gallons of bottled water did each person drink in 1990 ($t = 0$)? in 1995 ($t = 5$)? in 2000?

b) For what years will the amount of bottled water drunk be at least 15 gal?

86. *Dewpoint Spread.* Pilots use the **dewpoint spread,** or the difference between the current temperature and the dewpoint (the temperature at which dew occurs), to estimate the height of the cloud cover. Each 3° of dewpoint spread corresponds to an increased height of cloud cover of 1000 ft. A plane, flying with limited instruments, must have a cloud cover higher than 3500 ft. What dewpoint spreads will allow the plane to fly?

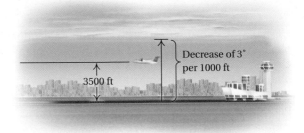

Decrease of 3° per 1000 ft

3500 ft

87. D_W Explain in your own words why the inequality symbol must be reversed when both sides of an inequality are multiplied or divided by a negative number.

88. D_W Graph the solution set of each of the following on a number line and compare:

$$x < 2, \quad x > 2, \quad x = 2, \quad x \le 2, \quad \text{and} \quad x \ge 2.$$

SKILL MAINTENANCE

Simplify.   [R.6b]

89. $3a - 6(2a - 5b)$

90. $2(x - y) + 10(3x - 7y)$

91. $4(a - 2b) - 6(2a - 5b)$

92. $-3(2a - 3b) + 8b$

Factor.   [R.5d]

93. $30x - 70y - 40$

94. $-12a + 30ab$

95. $-8x + 24y - 4$

96. $10n - 45mn + 100m$

Add or subtract.   [R.2a, c]

97. $-2.3 - 8.9$

98. $-2.3 + 8.9$

99. $-2.3 + (-8.9)$

100. $-2.3 - (-8.9)$

SYNTHESIS

101. *Supply and Demand.* The supply S and demand D for a certain product are given by

$$S = 460 + 94p \quad \text{and} \quad D = 2000 - 60p.$$

a) Find those values of p for which supply exceeds demand.
b) Find those values of p for which supply is less than demand.

Determine whether the statement is true or false. If false, give a counterexample.

102. For any real numbers x and y, if $x < y$, then $x^2 < y^2$.

103. For any real numbers a, b, c, and d, if $a < b$ and $c < d$, then $a + c < b + d$.

104. Determine whether the inequalities

$$x < 3 \quad \text{and} \quad 0 \cdot x < 0 \cdot 3$$

are equivalent. Give reasons to support your answer.

Solve.

105. $x + 5 \le 5 + x$

106. $x + 8 < 3 + x$

107. $x^2 + 1 > 0$

1.5 INTERSECTIONS, UNIONS, AND COMPOUND INEQUALITIES

Objectives

a Find the intersection of two sets. Solve and graph conjunctions of inequalities.

b Find the union of two sets. Solve and graph disjunctions of inequalities.

c Solve applied problems involving conjunctions and disjunctions of inequalities.

Cholesterol is a waxy fat that is present in all human beings. One commonly used measure of cholesterol is *total cholesterol*. The following table shows the health-risk levels of this measure of cholesterol.

RISK LEVEL	NORMAL RISK	BORDERLINE HIGH	HIGH RISK
Total cholesterol level	Less than 200	From 200 to 239	240 or higher

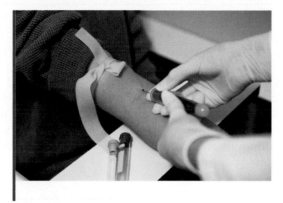

A total-cholesterol level T from 200 to 239 is considered borderline high. We can express this by the sentence

$$200 \leq T \quad and \quad T \leq 239$$

or more simply by

$$200 \leq T \leq 239.$$

This is an example of a *compound inequality*. **Compound inequalities** consist of two or more inequalities joined by the word *and* or the word *or*. We now "solve" such sentences—that is, we find the set of all solutions.

a Intersections of Sets and Conjunctions of Inequalities

INTERSECTION

The **intersection** of two sets A and B is the set of all members that are common to A and B. We denote the intersection of sets A and B as

$$A \cap B.$$

The intersection of two sets is often illustrated as shown at right.

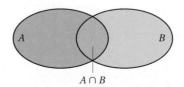

$A \cap B$

129

1. Find the intersection:

$$\{0, 3, 5, 7\} \cap \{0, 1, 3, 11\}.$$

EXAMPLE 1 Find the intersection: $\{1, 2, 3, 4, 5\} \cap \{-2, -1, 0, 1, 2, 3\}$.

The numbers 1, 2, and 3 are common to the two sets, so the intersection is $\{1, 2, 3\}$.

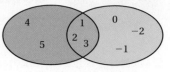

Do Exercises 1 and 2.

> ### CONJUNCTION
>
> When two or more sentences are joined by the word *and* to make a compound sentence, the new sentence is called a **conjunction** of the sentences.

The following is a conjunction of inequalities:

$$-2 < x \quad and \quad x < 1.$$

A number is a solution of a conjunction if it is a solution of *both* inequalities. For example, 0 is a solution of $-2 < x$ *and* $x < 1$ because $-2 < 0$ *and* $0 < 1$. Shown below is the graph of $-2 < x$, followed by the graph of $x < 1$, and then by the graph of the conjunction $-2 < x$ *and* $x < 1$. As the graphs demonstrate, *the solution set of a conjunction is the intersection of the solution sets of the individual inequalities.*

2. Shade the intersection of sets A and B.

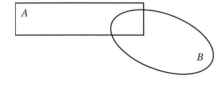

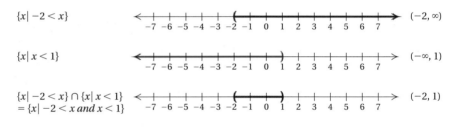

Because there are numbers that are both greater than -2 and less than 1, the conjunction $-2 < x$ *and* $x < 1$ can be abbreviated by $-2 < x < 1$. Thus the interval $(-2, 1)$ can be represented as $\{x \mid -2 < x < 1\}$, the set of all numbers that are *simultaneously* greater than -2 *and* less than 1. Note that, in general, for $a < b$,

$$a < x \quad and \quad x < b \quad \textbf{can be abbreviated} \quad a < x < b;$$

$$\text{and} \quad b > x \quad and \quad x > a \quad \textbf{can be abbreviated} \quad b > x > a.$$

Caution!

"$a > x$ and $x < b$" cannot be abbreviated as "$a > x < b$".

Answers on page A-6

"AND"; "INTERSECTION"

The word **"and"** corresponds to **"intersection"** and to the symbol **"∩"**. In order for a number to be a solution of a conjunction, it must make each part of the conjunction true.

Do Exercise 3.

■ **EXAMPLE 2** Solve and graph: $-1 \leq 2x + 5 < 13$.

This inequality is an abbreviation for the conjunction

$$-1 \leq 2x + 5 \quad and \quad 2x + 5 < 13.$$

The word *and* corresponds to set *intersection*, ∩. To solve the conjunction, we solve each of the two inequalities separately and then find the intersection of the solution sets:

$$-1 \leq 2x + 5 \quad and \quad 2x + 5 < 13$$
$$-6 \leq 2x \qquad and \qquad 2x < 8 \qquad \text{Subtracting 5}$$
$$-3 \leq x \qquad and \qquad x < 4. \qquad \text{Dividing by 2}$$

We now abbreviate the result:

$$-3 \leq x < 4.$$

The solution set is $\{x \mid -3 \leq x < 4\}$, or, in interval notation, $[-3, 4)$. The graph is the intersection of the two separate solution sets.

$\{x \mid -3 \leq x\}$

$[-3, \infty)$

$\{x \mid x < 4\}$

$(-\infty, 4)$

$\{x \mid -3 \leq x\} \cap \{x \mid x < 4\}$
$= \{x \mid -3 \leq x < 4\}$

$[-3, 4)$

The steps above are generally combined as follows:

$$-1 \leq 2x + 5 < 13 \qquad 2x + 5 \text{ appears in both inequalities.}$$
$$-6 \leq 2x < 8 \qquad \text{Subtracting 5}$$
$$-3 \leq x < 4. \qquad \text{Dividing by 2}$$

Such an approach saves some writing and will prove useful in Section 1.6.

Do Exercise 4.

■ **EXAMPLE 3** Solve and graph: $2x - 5 \geq -3 \text{ } and \text{ } 5x + 2 \geq 17$.

We first solve each inequality separately:

$$2x - 5 \geq -3 \quad and \quad 5x + 2 \geq 17$$
$$2x \geq 2 \qquad and \qquad 5x \geq 15$$
$$x \geq 1 \qquad and \qquad x \geq 3.$$

3. Graph and write interval notation:

$$-1 < x \text{ } and \text{ } x < 4.$$

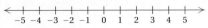

4. Solve and graph:

$$-22 < 3x - 7 \leq 23.$$

Answers on page A-6

131

1.5 Intersections, Unions, and Compound Inequalities

5. Solve and graph:

$3x + 4 < 10$ *and* $2x - 7 < -13$.

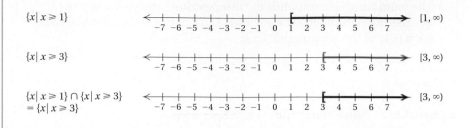

Next, we find the intersection of the two separate solution sets:

$\{x \mid x \geq 1\}$ $[1, \infty)$

$\{x \mid x \geq 3\}$ $[3, \infty)$

$\{x \mid x \geq 1\} \cap \{x \mid x \geq 3\}$
$= \{x \mid x \geq 3\}$ $[3, \infty)$

The numbers common to both sets are those that are greater than or equal to 3. Thus the solution set is $\{x \mid x \geq 3\}$, or, in interval notation, $[3, \infty)$. You should check that any number in $[3, \infty)$ satisfies the conjunction whereas numbers outside $[3, \infty)$ do not.

Do Exercise 5.

EMPTY SET; DISJOINT SETS

Sometimes two sets have no elements in common. In such a case, we say that the intersection of the two sets is the **empty set,** denoted { } or $\varnothing$. Two sets with an empty intersection are said to be **disjoint.**

$A \cap B = \varnothing$ A B

---

6. Solve and graph:

$3x - 7 \leq -13$ *and* $4x + 3 > 8$.

EXAMPLE 4 Solve and graph: $2x - 3 > 1$ *and* $3x - 1 < 2$.

We solve each inequality separately:

$$2x - 3 > 1 \quad and \quad 3x - 1 < 2$$
$$2x > 4 \quad and \quad 3x < 3$$
$$x > 2 \quad and \quad x < 1.$$

The solution set is the intersection of the solution sets of the individual inequalities.

$\{x \mid x > 2\}$ $(2, \infty)$

$\{x \mid x < 1\}$ $(-\infty, 1)$

$\{x \mid x > 2\} \cap \{x \mid x < 1\}$
$= \{x \mid x > 2 \text{ and } x < 1\}$ $\varnothing$
$= \varnothing$

Since no number is both greater than 2 and less than 1, the solution set is the empty set, $\varnothing$.

Do Exercise 6.

Answers on page A-6

CHAPTER 1: Solving Linear Equations
and Inequalities

EXAMPLE 5 Solve: $3 \le 5 - 2x < 7$.

We have

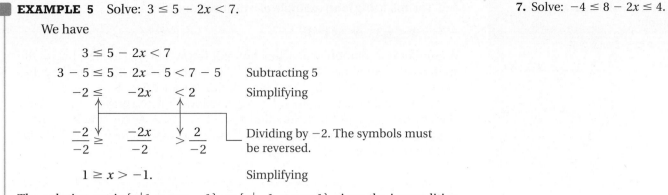

$$3 \le 5 - 2x < 7$$
$$3 - 5 \le 5 - 2x - 5 < 7 - 5 \qquad \text{Subtracting 5}$$
$$-2 \le \quad -2x \quad < 2 \qquad \text{Simplifying}$$

$$\frac{-2}{-2} \ge \frac{-2x}{-2} > \frac{2}{-2} \qquad \begin{array}{l}\text{Dividing by } -2. \text{ The symbols must} \\ \text{be reversed.}\end{array}$$

$$1 \ge x > -1. \qquad \text{Simplifying}$$

The solution set is $\{x \mid 1 \ge x > -1\}$, or $\{x \mid -1 < x \le 1\}$, since the inequalities $1 \ge x > -1$ and $-1 < x \le 1$ are equivalent. The solution, in interval notation, is $(-1, 1]$.

Do Exercise 7.

b Unions of Sets and Disjunctions of Inequalities

The **union** of two sets A and B is the collection of elements belonging to A and/or B. We denote the union of A and B by

$A \cup B.$

The union of two sets is often pictured as shown below.

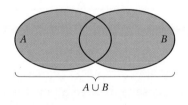

$$A \cup B$$

EXAMPLE 6 Find the union: $\{2, 3, 4\} \cup \{3, 5, 7\}$.

The numbers in either or both sets are 2, 3, 4, 5, and 7, so the union is $\{2, 3, 4, 5, 7\}$. We don't list the number 3 twice.

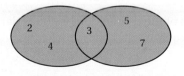

Do Exercises 8 and 9.

When two or more sentences are joined by the word *or* to make a compound sentence, the new sentence is called a **disjunction** of the sentences.

7. Solve: $-4 \le 8 - 2x \le 4$.

8. Find the union:

$$\{0, 1, 3, 4\} \cup \{0, 1, 7, 9\}.$$

9. Shade the union of sets A and B.

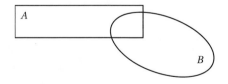

Answers on page A-6

10. Graph and write interval notation:

$$x \le -2 \ or \ x > 4.$$

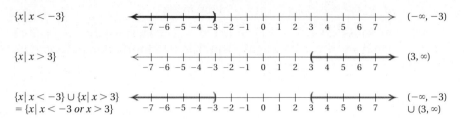

The following is an example of a disjunction:

$$x < -3 \quad or \quad x > 3.$$

A number is a solution of a disjunction if it is a solution of at least one of the individual inequalities. For example, -7 is a solution of $x < -3 \ or \ x > 3$ because $-7 < -3$. Similarly, 5 is also a solution because $5 > 3$.

Shown below is the graph of $x < -3$, followed by the graph of $x > 3$, and then by the graph of the disjunction $x < -3 \ or \ x > 3$. As the graphs demonstrate, *the solution set of a disjunction is the union of the solution sets of the individual sentences.*

$\{x \mid x < -3\}$ $(-\infty, -3)$

$\{x \mid x > 3\}$ $(3, \infty)$

$\{x \mid x < -3\} \cup \{x \mid x > 3\}$ $(-\infty, -3)$
$= \{x \mid x < -3 \ or \ x > 3\}$ $\cup \ (3, \infty)$

The solution set of

$$x < -3 \quad or \quad x > 3$$

is simply written $\{x \mid x < -3 \ or \ x > 3\}$, or, in interval notation, $(-\infty, -3) \cup (3, \infty)$. This cannot be written in a more condensed form.

"OR"; "UNION"

The word **"or"** corresponds to **"union"** and the symbol "$\cup$". In order for a number to be in the solution set of a disjunction, it must be in *at least one* of the solution sets of the individual sentences.

Do Exercise 10.

EXAMPLE 7 Solve and graph: $7 + 2x < -1 \ or \ 13 - 5x \le 3$.

We solve each inequality separately, retaining the word *or*:

$$7 + 2x < -1 \quad or \quad 13 - 5x \le 3$$
$$2x < -8 \quad or \quad -5x \le -10$$

Dividing by -5. The symbol must be reversed.

$$x < -4 \quad or \quad x \ge 2.$$

To find the solution set of the disjunction, we consider the individual graphs. We graph $x < -4$ and then $x \ge 2$. Then we take the union of the graphs.

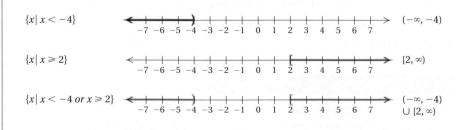

$\{x \mid x < -4\}$ $(-\infty, -4)$

$\{x \mid x \ge 2\}$ $[2, \infty)$

$\{x \mid x < -4 \ or \ x \ge 2\}$ $(-\infty, -4)$ $\cup \ [2, \infty)$

The solution set is $\{x \mid x < -4 \ or \ x \ge 2\}$, or, in interval notation, $(-\infty, -4) \cup [2, \infty)$.

Answer on page A-6

134

A compound inequality like

$$x < -4 \quad or \quad x \geq 2,$$

as in Example 7, *cannot* be expressed as $2 \leq x < -4$ because to do so would be to say that x is *simultaneously* less than -4 and greater than or equal to 2. No number is both less than -4 *and* greater than or equal to 2, but many are less than -4 *or* greater than or equal to 2.

Do Exercises 11 and 12.

EXAMPLE 8 Solve: $-2x - 5 < -2 \, or \, x - 3 < -10$.

We solve the individual inequalities separately, retaining the word *or*:

$$-2x - 5 < -2 \quad or \quad x - 3 < -10$$

$$-2x < 3 \quad or \quad x < -7$$

Reversing the symbol

$$x > -\tfrac{3}{2} \quad or \quad x < -7.$$

Keep the word "or."

The solution set is $\{x \mid x < -7 \, or \, x > -\tfrac{3}{2}\}$, or, in interval notation, $(-\infty, -7) \cup (-\tfrac{3}{2}, \infty)$.

Do Exercise 13.

EXAMPLE 9 Solve: $3x - 11 < 4 \, or \, 4x + 9 \geq 1$.

We solve the individual inequalities separately, retaining the word *or*:

$$3x - 11 < 4 \quad or \quad 4x + 9 \geq 1$$

$$3x < 15 \quad or \quad 4x \geq -8$$

$$x < 5 \quad or \quad x \geq -2.$$

To find the solution set, we first look at the individual graphs.

$\{x \mid x < 5\}$ $(-\infty, 5)$

$\{x \mid x \geq -2\}$ $[-2, \infty)$

$\begin{aligned}\{x \mid x < 5\} \cup \{x \mid x \geq -2\} \\ = \{x \mid x < 5 \, or \, x \geq -2\} \\ = \{x \mid x \text{ is a real number}\}\end{aligned}$ $\begin{aligned}(-\infty, \infty) \\ = \text{The set} \\ \text{of all real} \\ \text{numbers}\end{aligned}$

Since any number is either less than 5 or greater than or equal to -2, the two sets fill the entire number line. Thus the solution set is the set of all real numbers.

Do Exercise 14.

Solve and graph.

11. $x - 4 < -3 \, or \, x - 3 \geq 3$

12. $-2x + 4 \leq -3 \, or \, x + 5 < 3$

13. Solve:

$$-3x - 7 < -1 \, or \, x + 4 < -1.$$

14. Solve and graph:

$$5x - 7 \leq 13 \, or \, 2x - 1 \geq -7.$$

Answers on page A-6

15. Converting Dress Sizes. Refer to Example 10. Which dress sizes in the United States correspond to dress sizes between 36 and 58 in Italy?

Answer on page A-6

Applications and Problem Solving

EXAMPLE 10 *Converting Dress Sizes.* The equation

$$I = 2(s + 10)$$

can be used to convert dress sizes s in the United States to dress sizes I in Italy. Which dress sizes in the United States correspond to dress sizes between 32 and 46 in Italy?

1. **Familiarize.** We have a formula for converting the dress sizes. Thus we can substitute a value into the formula. For a dress of size 6 in the United States, we get the corresponding dress size in Italy as follows:

$$I = 2(6 + 10) = 2 \cdot 16 = 32.$$

This familiarizes us with the formula and also tells us that the United States sizes that we are looking for must be larger than size 6.

2. **Translate.** We want the Italian sizes *between* 32 and 46, so we want to find those values of s for which

$$32 < I < 46 \qquad I \text{ is between 32 and 46}$$

or

$$32 < 2(s + 10) < 46. \qquad \text{Substituting } 2(s + 10) \text{ for } I$$

Thus we have translated the problem to an inequality.

3. **Solve.** We solve the inequality:

$$32 < 2(s + 10) < 46$$
$$\frac{32}{2} < \frac{2(s + 10)}{2} < \frac{46}{2} \qquad \text{Dividing by 2}$$
$$16 < s + 10 < 23$$
$$6 < s < 13. \qquad \text{Subtracting 10}$$

4. **Check.** We substitute some values as we did in the *Familiarize* step.

5. **State.** Dress sizes between 6 and 13 in the United States correspond to dress sizes between 32 and 46 in Italy.

Do Exercise 15.

Study Tips

TAKE THE TIME!

The foundation of all your study skills is *time*! If you invest your time, you increase your likelihood of succeeding.

"Nine-tenths of wisdom is being wise in time."

Theodore Roosevelt

a, **b**   Find the intersection or union.

1. $\{9, 10, 11\} \cap \{9, 11, 13\}$

2. $\{1, 5, 10, 15\} \cap \{5, 15, 20\}$

3. $\{a, b, c, d\} \cap \{b, f, g\}$

4. $\{m, n, o, p\} \cap \{m, o, p\}$

5. $\{9, 10, 11\} \cup \{9, 11, 13\}$

6. $\{1, 5, 10, 15\} \cup \{5, 15, 20\}$

7. $\{a, b, c, d\} \cup \{b, f, g\}$

8. $\{m, n, o, p\} \cup \{m, o, p\}$

9. $\{2, 5, 7, 9\} \cap \{1, 3, 4\}$

10. $\{a, e, i, o, u\} \cap \{m, q, w, s, t\}$

11. $\{3, 5, 7\} \cup \varnothing$

12. $\{3, 5, 7\} \cap \varnothing$

a   Graph and write interval notation.

13. $-4 < a$ *and* $a \le 1$

$$\xleftarrow{\hspace{1em}}\!\!\!\underset{-6\;-5\;-4\;-3\;-2\;-1\;\;0\;\;1\;\;2\;\;3\;\;4\;\;5\;\;6}{\;|\;|\;|\;|\;|\;|\;|\;|\;|\;|\;|\;|\;|\;}\!\!\!\xrightarrow{\hspace{1em}}$$

14. $-\frac{5}{2} \le m$ *and* $m < \frac{3}{2}$

$$\xleftarrow{\hspace{1em}}\!\!\!\underset{-6\;-5\;-4\;-3\;-2\;-1\;\;0\;\;1\;\;2\;\;3\;\;4\;\;5\;\;6}{\;|\;|\;|\;|\;|\;|\;|\;|\;|\;|\;|\;|\;|\;}\!\!\!\xrightarrow{\hspace{1em}}$$

15. $1 < x < 6$

$$\xleftarrow{\hspace{1em}}\!\!\!\underset{-6\;-5\;-4\;-3\;-2\;-1\;\;0\;\;1\;\;2\;\;3\;\;4\;\;5\;\;6}{\;|\;|\;|\;|\;|\;|\;|\;|\;|\;|\;|\;|\;|\;}\!\!\!\xrightarrow{\hspace{1em}}$$

16. $-3 \le y \le 4$

$$\xleftarrow{\hspace{1em}}\!\!\!\underset{-6\;-5\;-4\;-3\;-2\;-1\;\;0\;\;1\;\;2\;\;3\;\;4\;\;5\;\;6}{\;|\;|\;|\;|\;|\;|\;|\;|\;|\;|\;|\;|\;|\;}\!\!\!\xrightarrow{\hspace{1em}}$$

Solve and graph.

17. $-10 \le 3x + 2$ *and* $3x + 2 < 17$

$$\xleftarrow{\hspace{1em}}\!\!\!\underset{-6\;-5\;-4\;-3\;-2\;-1\;\;0\;\;1\;\;2\;\;3\;\;4\;\;5\;\;6}{\;|\;|\;|\;|\;|\;|\;|\;|\;|\;|\;|\;|\;|\;}\!\!\!\xrightarrow{\hspace{1em}}$$

18. $-11 < 4x - 3$ *and* $4x - 3 \le 13$

$$\xleftarrow{\hspace{1em}}\!\!\!\underset{-6\;-5\;-4\;-3\;-2\;-1\;\;0\;\;1\;\;2\;\;3\;\;4\;\;5\;\;6}{\;|\;|\;|\;|\;|\;|\;|\;|\;|\;|\;|\;|\;|\;}\!\!\!\xrightarrow{\hspace{1em}}$$

19. $3x + 7 \ge 4$ *and* $2x - 5 \ge -1$

$$\xleftarrow{\hspace{1em}}\!\!\!\underset{-6\;-5\;-4\;-3\;-2\;-1\;\;0\;\;1\;\;2\;\;3\;\;4\;\;5\;\;6}{\;|\;|\;|\;|\;|\;|\;|\;|\;|\;|\;|\;|\;|\;}\!\!\!\xrightarrow{\hspace{1em}}$$

20. $4x - 7 < 1$ *and* $7 - 3x > -8$

$$\xleftarrow{\hspace{1em}}\!\!\!\underset{-6\;-5\;-4\;-3\;-2\;-1\;\;0\;\;1\;\;2\;\;3\;\;4\;\;5\;\;6}{\;|\;|\;|\;|\;|\;|\;|\;|\;|\;|\;|\;|\;|\;}\!\!\!\xrightarrow{\hspace{1em}}$$

21. $4 - 3x \ge 10$ *and* $5x - 2 > 13$

$$\xleftarrow{\hspace{1em}}\!\!\!\underset{-6\;-5\;-4\;-3\;-2\;-1\;\;0\;\;1\;\;2\;\;3\;\;4\;\;5\;\;6}{\;|\;|\;|\;|\;|\;|\;|\;|\;|\;|\;|\;|\;|\;}\!\!\!\xrightarrow{\hspace{1em}}$$

22. $5 - 7x > 19$ *and* $2 - 3x < -4$

$$\xleftarrow{\hspace{1em}}\!\!\!\underset{-6\;-5\;-4\;-3\;-2\;-1\;\;0\;\;1\;\;2\;\;3\;\;4\;\;5\;\;6}{\;|\;|\;|\;|\;|\;|\;|\;|\;|\;|\;|\;|\;|\;}\!\!\!\xrightarrow{\hspace{1em}}$$

Solve.

23. $-4 < x + 4 < 10$

24. $-6 < x + 6 \le 8$

25. $6 > -x \ge -2$

26. $3 > -x \ge -5$

27. $1 < 3y + 4 \leq 19$

28. $5 \leq 8x + 5 \leq 21$

29. $-10 \leq 3x - 5 \leq -1$

30. $-18 \leq -2x - 7 < 0$

31. $2 < x + 3 \leq 9$

32. $-6 \leq x + 1 < 9$

33. $-6 \leq 2x - 3 < 6$

34. $4 > -3m - 7 \geq 2$

35. $-\dfrac{1}{2} < \dfrac{1}{4}x - 3 \leq \dfrac{1}{2}$

36. $-\dfrac{2}{3} \leq 4 - \dfrac{1}{4}x < \dfrac{2}{3}$

37. $-3 < \dfrac{2x - 5}{4} < 8$

38. $-4 \leq \dfrac{7 - 3x}{5} \leq 4$

b Graph and write interval notation.

39. $x < -2 \ or \ x > 1$

40. $x < -4 \ or \ x > 0$

41. $x \leq -3 \ or \ x > 1$

42. $x \leq -1 \ or \ x > 3$

Solve and graph.

43. $x + 3 < -2 \ or \ x + 3 > 2$

44. $x - 2 < -1 \ or \ x - 2 > 3$

45. $2x - 8 \leq -3 \ or \ x - 1 \geq 3$

46. $x - 5 \leq -4 \ or \ 2x - 7 \geq 3$

47. $7x + 4 \geq -17 \ or \ 6x + 5 \geq -7$

48. $4x - 4 < -8 \ or \ 4x - 4 < 12$

Solve.

49. $7 > -4x + 5 \ or \ 10 \leq -4x + 5$

50. $6 > 2x - 1 \ or \ -4 \leq 2x - 1$

51. $3x - 7 > -10$ *or* $5x + 2 \leq 22$

52. $3x + 2 < 2$ *or* $4 - 2x < 14$

53. $-2x - 2 < -6$ *or* $-2x - 2 > 6$

54. $-3m - 7 < -5$ *or* $-3m - 7 > 5$

55. $\frac{2}{3}x - 14 < -\frac{5}{6}$ *or* $\frac{2}{3}x - 14 > \frac{5}{6}$

56. $\frac{1}{4} - 3x \leq -3.7$ *or* $\frac{1}{4} - 5x \geq 4.8$

57. $\frac{2x - 5}{6} \leq -3$ *or* $\frac{2x - 5}{6} \geq 4$

58. $\frac{7 - 3x}{5} < -4$ *or* $\frac{7 - 3x}{5} > 4$

C Solve.

59. *Pressure at Sea Depth.* The equation

$$P = 1 + \frac{d}{33}$$

gives the pressure P, in atmospheres (atm), at a depth of d feet in the sea. For what depths d is the pressure at least 1 atm and at most 7 atm?

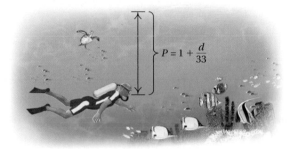

$P = 1 + \frac{d}{33}$

60. *Temperatures of Liquids.* The formula

$$C = \tfrac{5}{9}(F - 32)$$

can be used to convert Fahrenheit temperatures F to Celsius temperatures C.

a) Gold is a liquid for Celsius temperatures C such that $1063° \leq C < 2660°$. Find such an inequality for the corresponding Fahrenheit temperatures.

b) Silver is a liquid for Celsius temperatures C such that $960.8° \leq C < 2180°$. Find such an inequality for the corresponding Fahrenheit temperatures.

61. *Aerobic Exercise.* In order to achieve maximum results from aerobic exercise, one should maintain one's heart rate at a certain level. A 30-year-old woman with a resting heart rate of 60 beats per minute should keep her heart rate between 138 and 162 beats per minute while exercising. She checks her pulse for 10 sec while exercising. What should the number of beats be?

62. *Minimizing Tolls.* A $3.00 toll is charged to cross the bridge from Sanibel Island to mainland Florida. A six-month pass, costing $15.00, reduces the toll to $0.50. A one-year pass, costing $150, allows for free crossings. How many crossings per year does it take, on average, for the two six-month passes to be the most economical choice? Assume a constant number of trips per month.

Source: leewayinfo.com

63. *Body Mass Index.* Refer to Exercises 71 and 72 in Exercise Set 1.4. Marv's height is 73 in. What weights W will allow Marv to keep his body mass index I between 18.5 and 25?

64. *Body Mass Index* Refer to Exercises 71 and 72 in Exercise Set 1.4. Elaine's height is 67 in. What weight W will allow Elaine to keep her body mass index between 18.5 and 25?

65. *Young's Rule in Medicine.* Refer to Exercise 35 in Exercise Set 1.2. The dosage of a medication for an 8-year-old child must stay between 100 mg and 200 mg. Find the equivalent adult dosage.

66. *Young's Rule in Medicine.* Refer to Exercise 35 in Exercise Set 1.2. The dosage of a medication for a 5-year-old child must stay between 50 mg and 100 mg. Find the equivalent adult dosage.

67. $\mathbf{D_W}$ Explain why the conjunction $3 < x$ and $x < 5$ is equivalent to $3 < x < 5$, but the disjunction $3 < x$ or $x < 5$ is not.

68. $\mathbf{D_W}$ Describe the circumstances under which, for intervals, $[a, b] \cup [c, d] = [a, d]$.

SKILL MAINTENANCE

Find the absolute value. [R.1d]

69. $|-3.2|$

70. $|-5| + |7|$

71. $|-5 + 7|$

72. $|7 - 7|$

Simplify. [R.7a, b]

73. $(-2x^{-4}y^6)^5$

74. $(-4a^5b^{-7})(5a^{-12}b^8)$

75. $\dfrac{-4a^5b^{-7}}{5a^{-12}b^8}$

76. $(5p^6q^{11})^2$

77. $\left(\dfrac{56a^5b^{-6}}{28a^7b^{-8}}\right)^{-3}$

78. $\left(\dfrac{125p^{11}q^{12}}{25p^6q^8}\right)^2$

SYNTHESIS

Solve.

79. $x - 10 < 5x + 6 \le x + 10$

80. $4m - 8 > 6m + 5$ or $5m - 8 < -2$

81. $-\frac{2}{15} \le \frac{2}{3}x - \frac{2}{5} \le \frac{2}{15}$

82. $2[5(3 - y) - 2(y - 2)] > y + 4$

83. $3x < 4 - 5x < 5 + 3x$

84. $2x - \frac{3}{4} < -\frac{1}{10}$ or $2x - \frac{3}{4} > \frac{1}{10}$

85. $x + 4 < 2x - 6 \le x + 12$

86. $2x + 3 \le x - 6$ or $3x - 2 \le 4x + 5$

Determine whether the sentence is true or false for all real numbers a, b, and c.

87. If $-b < -a$, then $a < b$.

88. If $a \le c$ and $c \le b$, then $b \ge a$.

89. If $a < c$ and $b < c$, then $a < b$.

90. If $-a < c$ and $-c > b$, then $a > b$.

91. What is the union of the set of all rational numbers with the set of all irrational numbers? the intersection?

140

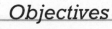

1.6 ABSOLUTE-VALUE EQUATIONS AND INEQUALITIES

a Properties of Absolute Value

We can think of the **absolute value** of a number as its distance from zero on the number line. Recall the formal definition from Section R.2.

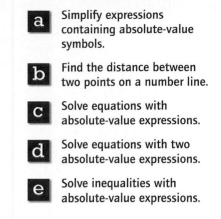

Objectives

a	Simplify expressions containing absolute-value symbols.
b	Find the distance between two points on a number line.
c	Solve equations with absolute-value expressions.
d	Solve equations with two absolute-value expressions.
e	Solve inequalities with absolute-value expressions.

ABSOLUTE VALUE

The **absolute value** of x, denoted $|x|$, is defined as follows:

$$x \geq 0 \rightarrow |x| = x; \qquad x < 0 \rightarrow |x| = -x.$$

This definition tells us that, when x is nonnegative, the absolute value of x is x and, when x is negative, the absolute value of x is the opposite of x. For example, $|3| = 3$ and $|-3| = -(-3) = 3$.

We see that absolute value is never negative. We can also think of a number's absolute value as its distance from 0 on the number line.

Some simple properties of absolute value allow us to manipulate or simplify algebraic expressions.

PROPERTIES OF ABSOLUTE VALUE

a) $|ab| = |a| \cdot |b|$, for any real numbers a and b.

(The absolute value of a product is the product of the absolute values.)

b) $\left|\dfrac{a}{b}\right| = \dfrac{|a|}{|b|}$, for any real numbers a and $b \neq 0$.

(The absolute value of a quotient is the quotient of the absolute values.)

c) $|-a| = |a|$, for any real number a.

(The absolute value of the opposite of a number is the same as the absolute value of the number.)

Simplify, leaving as little as possible inside the absolute-value signs.

1. $|7x|$

2. $|x^8|$

3. $|5a^2b|$

4. $\left|\dfrac{7a}{b^2}\right|$

5. $|-9x|$

■ **EXAMPLES** Simplify, leaving as little as possible inside the absolute-value signs.

1. $|5x| = |5| \cdot |x| = 5|x|$

2. $|-3y| = |-3| \cdot |y| = 3|y|$

3. $|7x^2| = |7| \cdot |x^2| = 7|x^2| = 7x^2$ Since x^2 is never negative for any number x

4. $\left|\dfrac{6x}{-3x^2}\right| = \left|\dfrac{2}{-x}\right| = \dfrac{|2|}{|-x|} = \dfrac{2}{|x|}$

Do Exercises 1–5.

Answers on page A-6

Find the distance between the points.

6. $-6, -35$

7. $19, 14$

8. $0, p$

9. Solve: $|x| = 6$. Then graph on the number line.

10. Solve: $|x| = -6$.

11. Solve: $|p| = 0$.

Answers on page A-6

---

b Distance on the Number Line

The number line below shows that the distance between -3 and 2 is 5.

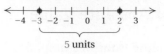

Another way to find the distance between two numbers on the number line is to determine the absolute value of the difference, as follows:

$$|-3 - 2| = |-5| = 5, \quad \text{or} \quad |2 - (-3)| = |5| = 5.$$

Note that the order in which we subtract does not matter because we are taking the absolute value after we have subtracted.

DISTANCE AND ABSOLUTE VALUE
For any real numbers a and b, the **distance** between them is $|a - b|$.

We should note that the distance is also $|b - a|$, because $a - b$ and $b - a$ are opposites and hence have the same absolute value.

EXAMPLE 5 Find the distance between -8 and -92 on the number line.

$$|-8 - (-92)| = |84| = 84, \quad \text{or} \quad |-92 - (-8)| = |-84| = 84$$

EXAMPLE 6 Find the distance between x and 0 on the number line.

$$|x - 0| = |x|$$

Do Exercises 6–8.

c Equations with Absolute Value

EXAMPLE 7 Solve: $|x| = 4$. Then graph on the number line.

Note that $|x| = |x - 0|$, so that $|x - 0|$ is the distance from x to 0. Thus solutions of the equation $|x| = 4$, or $|x - 0| = 4$, are those numbers x whose distance from 0 is 4. Those numbers are -4 and 4. The solution set is $\{-4, 4\}$. The graph consists of just two points, as shown.

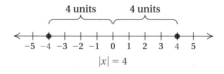

EXAMPLE 8 Solve: $|x| = 0$.

The only number whose absolute value is 0 is 0 itself. Thus the solution is 0. The solution set is $\{0\}$.

EXAMPLE 9 Solve: $|x| = -7$.

The absolute value of a number is always nonnegative. Thus there is no number whose absolute value is -7; consequently, the equation has no solution. The solution set is $\varnothing$.

Examples 7–9 lead us to the following principle for solving linear equations with absolute value.

THE ABSOLUTE-VALUE PRINCIPLE

For any positive number p and any algebraic expression X:

a) The solution of $|X| = p$ are those numbers that satisfy $X = -p$ or $X = p$.

b) The equation $|X| = 0$ is equivalent to the equation $X = 0$.

c) The equation $|X| = -p$ has no solution.

Do Exercises 9–11 on the preceding page.

We can use the absolute-value principle with the addition and multiplication principles to solve equations with absolute value.

EXAMPLE 10 Solve: $2|x| + 5 = 9$.

We first use the addition and multiplication principles to get $|x|$ by itself. Then we use the absolute-value principle.

$$2|x| + 5 = 9$$
$$2|x| = 4 \qquad \text{Subtracting 5}$$
$$|x| = 2 \qquad \text{Dividing by 2}$$
$$x = -2 \quad or \quad x = 2 \qquad \text{Using the absolute-value principle}$$

The solutions are -2 and 2. The solution set is $\{-2, 2\}$.

Do Exercises 12–14.

EXAMPLE 11 Solve: $|x - 2| = 3$.

We can consider solving this equation in two different ways.

METHOD 1 This allows us to see the meaning of the solutions graphically. The solution set consists of those numbers that are 3 units from 2 on the number line.

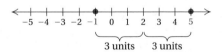

The solutions of $|x - 2| = 3$ are -1 and 5. The solution set is $\{-1, 5\}$.

METHOD 2 This method is more efficient. We use the absolute-value principle, replacing X with $x - 2$ and p with 3. Then we solve each equation separately.

$$|X| = p$$
$$|x - 2| = 3$$
$$x - 2 = -3 \quad or \quad x - 2 = 3 \qquad \text{Absolute-value principle}$$
$$x = -1 \quad or \quad x = 5$$

The solutions are -1 and 5. The solution set is $\{-1, 5\}$.

Do Exercise 15.

Solve.

12. $|3x| = 6$

13. $4|x| + 10 = 27$

14. $3|x| - 2 = 10$

15. Solve: $|x - 4| = 1$. Use two methods as in Example 11.

Answers on page A-6

16. Solve: $|3x - 4| = 17$.

EXAMPLE 12 Solve: $|2x + 5| = 13$.

We use the absolute-value principle, replacing X with $2x + 5$ and p with 13:

$$|X| = p$$
$$|2x + 5| = 13$$
$$2x + 5 = -13 \quad or \quad 2x + 5 = 13 \qquad \text{Absolute-value principle}$$
$$2x = -18 \quad or \qquad 2x = 8$$
$$x = -9 \quad or \qquad x = 4.$$

The solutions are -9 and 4. The solution set is $\{-9, 4\}$.

Do Exercise 16.

17. Solve: $|6 + 2x| = -3$.

EXAMPLE 13 Solve: $|4 - 7x| = -8$.

Since absolute value is always nonnegative, this equation has no solution. The solution set is $\varnothing$.

Do Exercise 17.

d Equations with Two Absolute-Value Expressions

Sometimes equations have two absolute-value expressions. Consider $|a| = |b|$. This means that a and b are the same distance from 0. If a and b are the same distance from 0, then either they are the same number or they are opposites.

Solve.

18. $|5x - 3| = |x + 4|$

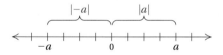

EXAMPLE 14 Solve: $|2x - 3| = |x + 5|$.

Either $2x - 3 = x + 5$ *or* $2x - 3 = -(x + 5)$. We solve each equation:

$$2x - 3 = x + 5 \quad or \quad 2x - 3 = -(x + 5)$$
$$x - 3 = 5 \qquad or \quad 2x - 3 = -x - 5$$
$$x = 8 \qquad or \quad 3x - 3 = -5$$
$$x = 8 \qquad or \qquad 3x = -2$$
$$x = 8 \qquad or \qquad x = -\tfrac{2}{3}.$$

The solutions are 8 and $-\tfrac{2}{3}$. The solution set is $\left\{8, -\tfrac{2}{3}\right\}$.

19. $|x - 3| = |x + 10|$

EXAMPLE 15 Solve: $|x + 8| = |x - 5|$.

$$x + 8 = x - 5 \quad or \quad x + 8 = -(x - 5)$$
$$8 = -5 \qquad or \quad x + 8 = -x + 5$$
$$8 = -5 \qquad or \qquad 2x = -3$$
$$8 = -5 \qquad or \qquad x = -\tfrac{3}{2}$$

The first equation has no solution. The second equation has $-\tfrac{3}{2}$ as a solution. The solution set is $\left\{-\tfrac{3}{2}\right\}$.

Do Exercises 18 and 19.

Answers on page A-6

e Inequalities with Absolute Value

We can extend our methods for solving equations with absolute value to those for solving inequalities with absolute value.

EXAMPLE 16 Solve: $|x| = 4$. Then graph on the number line.

From Example 7, we know that the solutions are -4 and 4. The solution set is $\{-4, 4\}$. The graph consists of just two points, as shown here.

 $\{-4, 4\}$
$|x| = 4$

Do Exercise 20.

EXAMPLE 17 Solve: $|x| < 4$. Then graph.

The solutions of $|x| < 4$ are the solutions of $|x - 0| < 4$ and are those numbers x whose distance from 0 is less than 4. We can check by substituting or by looking at the number line that numbers like $-3, -2, -1, -\frac{1}{2}, -\frac{1}{4}, 0, \frac{1}{4}, \frac{1}{2}, 1, 2,$ and 3 are all solutions. In fact, the solutions are all the real numbers x between -4 and 4. The solution set is $\{x \mid -4 < x < 4\}$ or, in interval notation, $(-4, 4)$. The graph is as follows.

$(-4, 4)$
$|x| < 4$

Do Exercise 21.

EXAMPLE 18 Solve: $|x| \geq 4$. Then graph.

The solutions of $|x| \geq 4$ are solutions of $|x - 0| \geq 4$ and are those numbers whose distance from 0 is greater than or equal to 4—in other words, those numbers x such that $x \leq -4 \ or \ x \geq 4$. The solution set is $\{x \mid x \leq -4 \ or \ x \geq 4\}$, or $(-\infty, -4] \cup [4, \infty)$. The graph is as follows.

$(-\infty, -4] \cup [4, \infty)$
$|x| \geq 4$

Do Exercise 22.

Examples 16–18 illustrate three cases of solving equations and inequalities with absolute value. The following is a general principle for solving equations and inequalities with absolute value.

---

20. Solve: $|x| = 5$. Then graph on the number line.

-8 -6 -4 -2 0 2 4 6 8

21. Solve: $|x| < 5$. Then graph.

-8 -6 -4 -2 0 2 4 6 8

22. Solve: $|x| \geq 5$. Then graph.

-8 -6 -4 -2 0 2 4 6 8

Answers on page A-6

SOLUTIONS OF ABSOLUTE-VALUE EQUATIONS AND INEQUALITIES

For any positive number p and any algebraic expression X:

a) The solutions of $|X| = p$ are those numbers that satisfy $X = -p$ or $X = p$.

As an example, replacing X with $5x - 1$ and p with 8, we see that the solutions of $|5x - 1| = 8$ are those numbers x for which

$$5x - 1 = -8 \quad or \quad 5x - 1 = 8$$
$$5x = -7 \quad or \quad 5x = 9$$
$$x = -\tfrac{7}{5} \quad or \quad x = \tfrac{9}{5}.$$

The solution set is $\left\{-\tfrac{7}{5}, \tfrac{9}{5}\right\}$.

b) The solutions of $|X| < p$ are those numbers that satisfy $-p < X < p$.

As an example, replacing X with $6x + 7$ and p with 5, we see that the solutions of $|6x + 7| < 5$ are those numbers x for which

$$-5 < 6x + 7 < 5$$
$$-12 < 6x < -2$$
$$-2 < x < -\tfrac{1}{3}.$$

The solution set is $\left\{x \mid -2 < x < -\tfrac{1}{3}\right\}$, or $\left(-2, -\tfrac{1}{3}\right)$.

c) The solutions of $|X| > p$ are those numbers that satisfy $X < -p$ or $X > p$.

As an example, replacing X with $2x - 9$ and p with 4, we see that the solutions of $|2x - 9| > 4$ are those numbers x for which

$$2x - 9 < -4 \quad or \quad 2x - 9 > 4$$
$$2x < 5 \quad or \quad 2x > 13$$
$$x < \tfrac{5}{2} \quad or \quad x > \tfrac{13}{2}.$$

The solution set is $\left\{x \mid x < \tfrac{5}{2} \ or \ x > \tfrac{13}{2}\right\}$, or $\left(-\infty, \tfrac{5}{2}\right) \cup \left(\tfrac{13}{2}, \infty\right)$.

CHAPTER 1: Solving Linear Equations
and Inequalities

EXAMPLE 19   Solve: $|3x - 2| < 4$. Then graph.

We use part (b). In this case, X is $3x - 2$ and p is 4:

$$|X| < p$$
$$|3x - 2| < 4 \qquad \text{Replacing } X \text{ with } 3x - 2 \text{ and } p \text{ with } 4$$
$$-4 < 3x - 2 < 4$$
$$-2 < 3x < 6$$
$$-\tfrac{2}{3} < x < 2.$$

The solution set is $\{x \mid -\tfrac{2}{3} < x < 2\}$, or $\left(-\tfrac{2}{3}, 2\right)$. The graph is as follows.

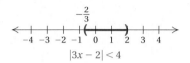

EXAMPLE 20   Solve: $|8 - 4x| \le 5$.

We use part (b). In this case, X is $8 - 4x$ and p is 5:

$$|X| \le p$$
$$|8 - 4x| \le 5 \qquad \text{Replacing } X \text{ with } 8 - 4x \text{ and } p \text{ with } 5$$
$$-5 \le 8 - 4x \le 5$$
$$-13 \le -4x \le -3$$
$$\tfrac{13}{4} \ge x \ge \tfrac{3}{4}. \qquad \begin{array}{l}\text{Dividing by } -4 \text{ and reversing the} \\ \text{inequality symbols}\end{array}$$

The solution set is $\{x \mid \tfrac{13}{4} \ge x \ge \tfrac{3}{4}\}$, or $\{x \mid \tfrac{3}{4} \le x \le \tfrac{13}{4}\}$, or $\left[\tfrac{3}{4}, \tfrac{13}{4}\right]$.

EXAMPLE 21   Solve: $|4x + 2| \ge 6$.

We use part (c). In this case, X is $4x + 2$ and p is 6:

$$|X| \ge p$$
$$|4x + 2| \ge 6 \qquad \text{Replacing } X \text{ with } 4x + 2 \text{ and } p \text{ with } 6$$
$$4x + 2 \le -6 \quad or \quad 4x + 2 \ge 6$$
$$4x \le -8 \quad or \qquad 4x \ge 4$$
$$x \le -2 \quad or \qquad x \ge 1.$$

The solution set is $\{x \mid x \le -2 \ or \ x \ge 1\}$, or $(-\infty, -2] \cup [1, \infty)$.

Do Exercises 23–25.

23. Solve: $|2x - 3| < 7$. Then graph.

←+++++++++++++++++++→
−8 −6 −4 −2   0   2   4   6   8

24. Solve: $|7 - 3x| \le 4$.

25. Solve: $|3x + 2| \ge 5$. Then graph.

←+++++++++++++++++++→
−8 −6 −4 −2   0   2   4   6   8

Answers on page A-6

a Simplify, leaving as little as possible inside absolute-value signs.

1. $|9x|$

2. $|26x|$

3. $|2x^2|$

4. $|8x^2|$

5. $|-2x^2|$

6. $|-20x^2|$

7. $|-6y|$

8. $|-17y|$

9. $\left|\dfrac{-2}{x}\right|$

10. $\left|\dfrac{y}{3}\right|$

11. $\left|\dfrac{x^2}{-y}\right|$

12. $\left|\dfrac{x^4}{-y}\right|$

13. $\left|\dfrac{-8x^2}{2x}\right|$

14. $\left|\dfrac{-9y^2}{3y}\right|$

15. $\left|\dfrac{4y^3}{-12y}\right|$

16. $\left|\dfrac{5x^3}{-25x}\right|$

b Find the distance between the points on a number line.

17. $-8,\ -46$

18. $-7,\ -32$

19. $36,\ 17$

20. $52,\ 18$

21. $-3.9,\ 2.4$

22. $-1.8,\ -3.7$

23. $-5,\ 0$

25. $\dfrac{2}{3},\ -\dfrac{5}{6}$

c Solve.

25. $|x| = 3$

26. $|x| = 5$

27. $|x| = -3$

28. $|x| = -9$

29. $|q| = 0$

30. $|y| = 7.4$

31. $|x - 3| = 12$

32. $|3x - 2| = 6$

33. $|2x - 3| = 4$

34. $|5x + 2| = 3$

35. $|4x - 9| = 14$

36. $|9y - 2| = 17$

37. $|x| + 7 = 18$

38. $|x| - 2 = 6.3$

39. $574 = 283 + |t|$

40. $-562 = -2000 + |x|$

41. $|5x| = 40$

42. $|2y| = 18$

43. $|3x| - 4 = 17$

44. $|6x| + 8 = 32$

45. $7|w| - 3 = 11$

46. $5|x| + 10 = 26$

47. $\left|\dfrac{2x - 1}{3}\right| = 5$

48. $\left|\dfrac{4 - 5x}{6}\right| = 7$

49. $|m + 5| + 9 = 16$

50. $|t - 7| - 5 = 4$

51. $10 - |2x - 1| = 4$

52. $2|2x - 7| + 11 = 25$

53. $|3x - 4| = -2$

54. $|x - 6| = -8$

55. $\left|\dfrac{5}{9} + 3x\right| = \dfrac{1}{6}$

56. $\left|\dfrac{2}{3} - 4x\right| = \dfrac{4}{5}$

d Solve.

57. $|3x + 4| = |x - 7|$

58. $|2x - 8| = |x + 3|$

59. $|x + 3| = |x - 6|$

60. $|x - 15| = |x + 8|$

61. $|2a + 4| = |3a - 1|$

62. $|5p + 7| = |4p + 3|$

63. $|y - 3| = |3 - y|$

64. $|m - 7| = |7 - m|$

65. $|5 - p| = |p + 8|$

66. $|8 - q| = |q + 19|$

67. $\left|\dfrac{2x - 3}{6}\right| = \left|\dfrac{4 - 5x}{8}\right|$

68. $\left|\dfrac{6 - 8x}{5}\right| = \left|\dfrac{7 + 3x}{2}\right|$

69. $\left|\dfrac{1}{2}x - 5\right| = \left|\dfrac{1}{4}x + 3\right|$

70. $\left|2 - \dfrac{2}{3}x\right| = \left|4 + \dfrac{7}{8}x\right|$

71. $|x| < 3$

72. $|x| \leq 5$

73. $|x| \geq 2$

74. $|y| > 12$

75. $|x - 1| < 1$

76. $|x + 4| \leq 9$

77. $5|x + 4| \leq 10$

78. $2|x - 2| > 6$

79. $|2x - 3| \leq 4$

80. $|5x + 2| \leq 3$

81. $|2y - 7| > 10$

82. $|3y - 4| > 8$

83. $|4x - 9| \geq 14$

84. $|9y - 2| \geq 17$

85. $|y - 3| < 12$

86. $|p - 2| < 6$

87. $|2x + 3| \leq 4$

88. $|5x + 2| \leq 13$

89. $|4 - 3y| > 8$

90. $|7 - 2y| > 5$

91. $|9 - 4x| \geq 14$

92. $|2 - 9p| \geq 17$

93. $|3 - 4x| < 21$

94. $|-5 - 7x| \leq 30$

95. $\left|\dfrac{1}{2} + 3x\right| \geq 12$

96. $\left|\dfrac{1}{4}y - 6\right| > 24$

97. $\left|\dfrac{x - 7}{3}\right| < 4$

98. $\left|\dfrac{x + 5}{4}\right| \leq 2$

99. $\left|\dfrac{2 - 5x}{4}\right| \geq \dfrac{2}{3}$

100. $\left|\dfrac{1 + 3x}{5}\right| > \dfrac{7}{8}$

101. $|m + 5| + 9 \leq 16$

102. $|t - 7| + 3 \geq 4$

103. $7 - |3 - 2x| \geq 5$

104. $16 \leq |2x - 3| + 9$

105. $\left|\dfrac{2x - 1}{0.0059}\right| \leq 1$

106. $\left|\dfrac{3x - 2}{5}\right| \geq 1$

CHAPTER 1: Solving Linear Equations
and Inequalities

107. $\mathbf{D_W}$ Explain in your own words why the solutions of the inequality $|x + 5| \le 2$ can be interpreted as "all those numbers x whose distance from -5 is at most 2 units."

108. $\mathbf{D_W}$ Explain in your own words why the interval $[6, \infty)$ is only part of the solution set of $|x| \ge 6$.

↪ VOCABULARY REINFORCEMENT

In each of Exercises 109–116, fill in the blank with the correct term from the given list. Some of the choices may not be used.

109. The _____ of two sets A and B is the collection of elements belonging to A and/or B.   [1.5b]

110. Two sets with an empty intersection are said to be _____ .   [1.5a]

111. The expression $x \ge q$ means x is _____ q.   [1.4d]

112. Interval notation for $\{x \mid a \le x \le b\}$ is _____ .   [1.4b]

113. The _____ of a number is its distance from zero on the number line.   [1.6a]

114. A(n) _____ is a number sentence that says that the expression on either side of the equals sign represents the same number.   [1.1a]

115. Equations with the same solutions are called _____ equations.   [1.1a]

116. A(n) _____ is any sentence containing $<$, $>$, $\le$, $\ge$, or $\ne$.   [1.4a]

$[a, b]$

$[a, b)$

(a, b)

disjoint

union

intersection

equivalent

absolute value

equation

inequality

at least

at most

117. *Motion of a Spring.*   A weighted spring is bouncing up and down so that its distance d above the ground satisfies the inequality $|d - 6 \text{ ft}| \le \frac{1}{2}$ ft. Find all possible distances d.

118. *Container Sizes.*   A container company is manufacturing rectangular boxes of various sizes. The length of any box must exceed the width by at least 3 in., but the perimeter cannot exceed 24 in. What widths are possible?

$$l \ge w + 3,$$
$$2l + 2w \le 24$$

Solve.

119. $|x + 5| = x + 5$

120. $1 - \left|\frac{1}{4}x + 8\right| = \frac{3}{4}$

121. $|7x - 2| = x + 4$

122. $|x - 1| = x - 1$

123. $|x - 6| \le -8$

124. $|3x - 4| > -2$

125. $|x + 5| > x$

126. $\left|\frac{5}{9} + 3x\right| < -\frac{1}{6}$

Find an equivalent inequality with absolute value.

127. $-3 < x < 3$

128. $-5 \le y \le 5$

129. $x \le -6 \text{ or } x \ge 6$

130. $-5 < x < 1$

131. $x < -8 \text{ or } x > 2$

The review that follows is meant to prepare you for a chapter exam. It consists of three parts. The first part, Concept Reinforcement, is designed to increase understanding of the concepts through true/false exercises. The second part is a list of important properties and formulas. The third part is the Review Exercises. These provide practice exercises for the exam, together with references to section objectives so you can go back and review. Before beginning, stop and look back over the skills you have obtained. What skills in mathematics do you have now that you did not have before studying this chapter?

☞ CONCEPT REINFORCEMENT

Determine whether the statement is true or false. Answers are given at the back of the book.

_____ **1.** $2x + 3 = 7$ and $x = 2$ are equivalent equations.

_____ **2.** For any real numbers a, b, and c, $c \neq 0$, $a = b$ is equivalent to $a \cdot c = b \cdot c$.

_____ **3.** For any real numbers a, b, and c, $c \neq 0$, $a \leq b$ is equivalent to $ac \leq bc$.

_____ **4.** If x is negative, $|x| = -x$.

_____ **5.** $|x|$ is always positive.

IMPORTANT PROPERTIES AND FORMULAS

The Addition Principle for Equations:	For any real numbers a, b, and c: $a = b$ is equivalent to $a + c = b + c$.
The Multiplication Principle for Equations:	For any real numbers a, b, and c, $c \neq 0$: $a = b$ is equivalent to $a \cdot c = b \cdot c$.
The Addition Principle for Inequalities:	For any real numbers a, b, and c: $a < b$ is equivalent to $a + c < b + c$; $a > b$ is equivalent to $a + c > b + c$.
The Multiplication Principle for Inequalities:	For any real numbers a and b, and any *positive* number c: $a < b$ is equivalent to $ac < bc$; $a > b$ is equivalent to $ac > bc$.
	For any real numbers a and b, and any *negative* number c: $a < b$ is equivalent to $ac > bc$; $a > b$ is equivalent to $ac < bc$.
	Similar statements hold for $\leq$ and $\geq$.
Set Intersection:	$A \cap B = \{x \mid x \text{ is in } A \text{ and } x \text{ is in } B\}$
Set Union:	$A \cup B = \{x \mid x \text{ is in } A \text{ or in } B, \text{ or both}\}$

"$a < x$ and $x < b$" is equivalent to "$a < x < b$"

Properties of Absolute Value

$$|ab| = |a| \cdot |b|, \qquad \left|\frac{a}{b}\right| = \frac{|a|}{|b|}, \qquad |-a| = |a|, \qquad \text{The distance between } a \text{ and } b \text{ is } |a - b|.$$

Principles for Solving Equations and Inequalities Involving Absolute Value:

For any positive number p and any algebraic expression X:

a) The solutions of $|X| = p$ are those numbers that satisfy $X = -p \text{ or } X = p$.
b) The solutions of $|X| < p$ are those numbers that satisfy $-p < X < p$.
c) The solutions of $|X| > p$ are those numbers that satisfy $X < -p \text{ or } X > p$.

Review Exercises

Solve.   [1.1b, c, d]

1. $-11 + y = -3$

2. $-7x = -3$

3. $-\frac{5}{3}x + \frac{7}{3} = -5$

4. $6(2x - 1) = 3 - (x + 10)$

5. $2.4x + 1.5 = 1.02$

6. $2(3 - x) - 4(x + 1) = 7(1 - x)$

Solve for the indicated letter.   [1.2a]

7. $C = \frac{4}{11}d + 3$, for d

8. $A = 2a - 3b$, for b

9. *Interstate Mile Markers.*   If you are traveling on a U.S. interstate highway, you will notice numbered markers every mile to tell your location in case of an accident or other emergency. In many states, the numbers on the markers increase from west to east. The sum of two consecutive mile markers on I-70 in Utah is 371. Find the numbers on the markers.   [1.3a]

Source: Federal Highway Administration, Ed Rotalewski

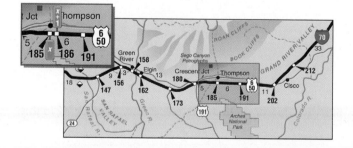

10. *Rope Cutting.*   A piece of rope 27 m long is cut into two pieces so that one piece is four-fifths as long as the other. Find the length of each piece.   [1.3a]

11. *Population Growth.*   The population of Newcastle grew 12% from one year to the next to a total of 179,200. What was the former population?   [1.3a]

12. *Moving Sidewalk.*   A moving sidewalk in an airport is 360 ft long and moves at a speed of 6 ft/sec. If Arnie walks at a speed of 3 ft/sec, how long will it take him to walk the length of the moving sidewalk?   [1.3b]

Write interval notation for the given set or graph.   [1.4b]

13. $\{x \mid -8 \le x < 9\}$

14.

Solve and graph. Write interval notation for the solution set. [1.4c]

15. $x - 2 \le -4$

16. $x + 5 > 6$

Solve.   [1.4c]

17. $a + 7 \le -14$

18. $y - 5 \ge -12$

19. $4y > -16$

20. $-0.3y < 9$

21. $-6x - 5 < 13$

22. $4y + 3 \le -6y - 9$

23. $-\frac{1}{2}x - \frac{1}{4} > \frac{1}{2} - \frac{1}{4}x$

24. $0.3y - 8 < 2.6y + 15$

25. $-2(x - 5) \ge 6(x + 7) - 12$

26. *Moving Costs.*   Musclebound Movers charges $85 plus $40 an hour to move households across town. Champion Moving charges $60 an hour for cross-town moves. For what lengths of time is Champion more expensive?   [1.4d]

27. *Investments.*   You are going to invest $30,000, part at 13% and part at 15%, for one year. What is the most that can be invested at 13% in order to make at least $4300 interest in one year?   [1.4d]

Graph and write interval notation. [1.5a, b]

28. $-2 \le x < 5$

←—+—+—+—+—+—+—+—+—+—+—+—+→
    −6 −5 −4 −3 −2 −1 0 1 2 3 4 5 6

29. $x \le -2 \text{ or } x > 5$

←—+—+—+—+—+—+—+—+—+—+—+—+→
    −6 −5 −4 −3 −2 −1 0 1 2 3 4 5 6

Solve. [1.5a, b]

30. $2x - 5 < -7 \text{ and } 3x + 8 \ge 14$

31. $-4 < x + 3 \le 5$

32. $-15 < -4x - 5 < 0$

33. $3x < -9 \text{ or } -5x < -5$

34. $2x + 5 < -17 \text{ or } -4x + 10 \le 34$

35. $2x + 7 \le -5 \text{ or } x + 7 \ge 15$

Simplify. [1.6a]

36. $\left| -\dfrac{3}{x} \right|$ **37.** $\left| \dfrac{2x}{y^2} \right|$ **38.** $\left| \dfrac{12y}{-3y^2} \right|$

39. Find the distance between -23 and 39. [1.6b]

Solve. [1.6c, d]

40. $|x| = 6$ **41.** $|x - 2| = 7$

42. $|2x + 5| = |x - 9|$ **43.** $|5x + 6| = -8$

Solve. [1.6e]

44. $|2x + 5| < 12$ **45.** $|x| \ge 3.5$

46. $|3x - 4| \ge 15$ **47.** $|x| < 0$

48. Find the intersection: [1.5a]
$$\{1, 2, 5, 6, 9\} \cap \{1, 3, 5, 9\}.$$

49. Find the union: [1.5b]
$$\{1, 2, 5, 6, 9\} \cup \{1, 3, 5, 9\}.$$

50. *Records in the Women's 100-m Dash.*   In 1988, Florence Griffith Joyner set a world record of 10.49 sec in the women's 100-m dash. The equation
$$R = -0.0433t + 10.49$$
can be used to estimate the world record in the women's 100-m dash t years after 1988.
Source: International Amateur Athletic Foundation

a) Estimate the world record in 2008. [1.3a]
b) In what year will the record be 9.5374 sec? [1.3a]
c) For what years was the record between 10.15 and 10.35 sec? [1.5c]

51. $\mathbf{D_W}$ Find the error or errors in each of the following steps: [1.4c]
$$7 - 9x + 6x < -9(x + 2) + 10x$$

$7 - 9x + 6x < -9x + 2 + 10x$	(1)
$7 + 6x > 2 + 10x$	(2)
$-4x > 8$	(3)
$x > -2.$	(4)

52. $\mathbf{D_W}$ How does the word "solve" vary in meaning in this chapter?

SYNTHESIS

53. Solve: $|2x + 5| \le |x + 3|$. [1.6d, e]

CHAPTER 1: Solving Linear Equations
and Inequalities

Solve.

1. $x + 7 = 5$

2. $-12x = -8$

3. $x - \frac{3}{5} = \frac{2}{3}$

4. $3y - 4 = 8$

5. $1.7y - 0.1 = 2.1 - 0.3x$

6. $5(3x + 6) = 6 - (x + 8)$

7. Solve $A = 3B - C$ for B.

8. Solve $m = n - nt$ for n.

Solve.

9. *Room Dimensions.* A rectangular room has a perimeter of 48 ft. The width is two-thirds of the length. What are the dimensions of the room?

10. *Copy Budget.* Copy Solutions rents a copier for $240 per month plus 1.8¢ per copy. A law firm needs to lease a copy machine for use during a special case that they anticipate will take 3 months. They allot a budget of $1500 for copying costs. How many copies can they make and stay within budget?

11. *Population Decrease.* The population of Baytown dropped 12% from one year to the next to a total of 158,400. What was the former population?

12. *Angles in a Triangle.* The measures of the angles of a triangle are three consecutive integers. Find the measures of the angles.

13. *Boating.* A paddleboat moves at a rate of 12 mph in still water. If the river's current moves at a rate of 3 mph, how long will it take the boat to travel 36 mi downstream? 36 mi upstream?

Write interval notation for the given set or graph.

14. $\{x \mid -3 < x \leq 2\}$

15.
$-6\ -5\ -4\ -3\ -2\ -1\ \ 0\ \ 1\ \ 2\ \ 3\ \ 4\ \ 5\ \ 6$

Solve and graph. Write interval notation for the solution set.

16. $x - 2 \leq 4$

$-6\ -5\ -4\ -3\ -2\ -1\ \ 0\ \ 1\ \ 2\ \ 3\ \ 4\ \ 5\ \ 6$

17. $-4y - 3 \geq 5$

$-6\ -5\ -4\ -3\ -2\ -1\ \ 0\ \ 1\ \ 2\ \ 3\ \ 4\ \ 5\ \ 6$

Solve.

18. $x - 4 \geq 6$

19. $-0.6y < 30$

20. $3a - 5 \leq -2a + 6$

21. $-5y - 1 > -9y + 3$

22. $4(5 - x) < 2x + 5$

23. $-8(2x + 3) + 6(4 - 5x) \geq 2(1 - 7x) - 4(4 + 6x)$

Solve.

24. *Moving Costs.* Motivated Movers charges $105 plus $30 an hour to move households across town. Quick-Pak Moving charges $80 an hour for cross-town moves. For what lengths of time is Quick-Pak more expensive?

25. *Pressure at Sea Depth.* The equation

$$P = 1 + \frac{d}{33}$$

gives the pressure P, in atmospheres (atm), at a depth of d feet in the sea. For what depths d is the pressure at least 2 atm and at most 8 atm?

Graph and write interval notation.

26. $-3 \leq x \leq 4$

27. $x < -3 \text{ or } x > 4$

Solve.

28. $5 - 2x \leq 1 \text{ and } 3x + 2 \geq 14$

29. $-3 < x - 2 < 4$

30. $-11 \leq -5x - 2 < 0$

31. $-3x > 12 \text{ or } 4x > -10$

32. $x - 7 \leq -5 \text{ or } x - 7 \geq -10$

33. $3x - 2 < 7 \text{ or } x - 2 > 4$

Simplify.

34. $\left| \dfrac{7}{x} \right|$

35. $\left| \dfrac{-6x^2}{3x} \right|$

36. Find the distance between 4.8 and -3.6.

37. Find the intersection:

$$\{1, 3, 5, 7, 9\} \cap \{3, 5, 11, 13\}.$$

38. Find the union:

$$\{1, 3, 5, 7, 9\} \cup \{3, 5, 11, 13\}.$$

Solve.

39. $|x| = 9$

40. $|x - 3| = 9$

41. $|x + 10| = |x - 12|$

42. $|2 - 5x| = -10$

43. $|4x - 1| < 4.5$

44. $|x| > 3$

45. $\left| \dfrac{6 - x}{7} \right| \leq 15$

46. $|-5x - 3| \geq 10$

Solve.

47. $|3x - 4| \leq -3$

48. $7x < 8 - 3x < 6 + 7x$

Graphs, Functions, and Applications

2

Real-World Application

Reports of mishandled baggage filed with major U.S. airlines have declined slightly in recent years. In 1993, 5.60 reports per 1000 passengers were filed. By 2003, this number had decreased to 4.19 reports per 1000 passengers. Find the rate of change of the number of reports filed per 1000 passengers with respect to time, in years.

Source: U.S. Department of Transportation

This problem appears as Margin Exercise 12 in Section 2.4.

a   Plot points associated with ordered pairs of numbers.

b   Determine whether an ordered pair of numbers is a solution of an equation.

c   Graph linear equations using tables.

d   Graph nonlinear equations using tables.

Graphs display information in a compact way and can provide a visual approach to problem solving. We often see graphs in newspapers and magazines. Examples of bar, circle, and line graphs are shown below.

a   Plotting Ordered Pairs

We have already learned to graph numbers and inequalities in one variable on a line. To graph an equation that contains two variables, we graph pairs of numbers on a plane.

On the number line, each point is the graph of a number. On a plane, each point is the graph of a number pair. To locate points on a plane, we use two perpendicular number lines called **axes.** They cross at a point called the **origin.** The arrows show the positive directions on the axes. Consider the **ordered pair** (2, 3). The numbers in an ordered pair are called **coordinates.** In (2, 3), the **first coordinate** is 2 and the **second coordinate** is 3. (The first coordinate is sometimes called the **abscissa** and the second the **ordinate.**) To plot (2, 3), we start at the origin and move 2 units in the positive horizontal direction (2 units to the right). Then we move 3 units in the positive vertical direction (3 units up) and make a dot.

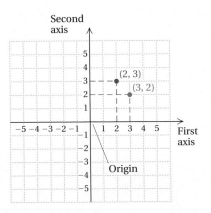

The point (3, 2) is also plotted in the figure. Note that (3, 2) and (2, 3) are different points. The order of the numbers in the pair is indeed important. They are called *ordered pairs* because it makes a difference which number is listed first.

The coordinates of the origin are (0, 0). In general, the first axis is called the *x*-axis and the second axis is called the *y*-axis. We call this the **Cartesian coordinate system** in honor of the great French mathematician and philosopher René Descartes (1596–1650).

EXAMPLE 1 Plot the points $(-4, 3)$, $(-5, -3)$, $(0, 4)$, and $(2.5, 0)$.

To plot $(-4, 3)$, we note that the first number, -4, tells us the distance in the first, or horizontal, direction. We move 4 units in the negative direction, *left.* The second number tells us the distance in the second, or vertical, direction. We move 3 units in the positive direction, *up.* The point $(-4, 3)$ is then marked, or plotted.

The points $(-5, -3)$, $(0, 4)$, and $(2.5, 0)$ are plotted in the same manner.

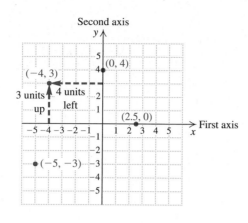

Do Exercises 1–10.

QUADRANTS

The axes divide the plane into four regions called **quadrants,** denoted by Roman numerals and numbered counterclockwise starting at the upper right. In region I (the *first* quadrant), both coordinates of a point are positive. In region II (the *second* quadrant), the first coordinate is negative and the second coordinate is positive. In the *third* quadrant, both coordinates are negative, and in the *fourth* quadrant, the first coordinate is positive and the second is negative.

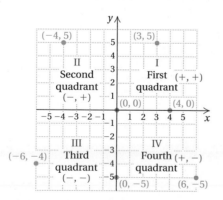

Points with one or more 0's as coordinates, such as $(0, -5)$, $(4, 0)$, and $(0, 0)$, are on axes and *not* in quadrants.

Do Exercises 11 and 12.

b Solutions of Equations

If an equation has two variables, its solutions are pairs of numbers. When such a solution is written as an ordered pair, the first number listed in the pair generally replaces the variable that occurs first alphabetically.

Plot the points on the plane below.

1. $(6, 4)$ 2. $(4, 6)$

3. $(-3, 5)$ 4. $(5, -3)$

5. $(-4, -3)$ 6. $(4, -2)$

7. $(0, 3)$ 8. $(3, 0)$

9. $(0, -4)$ 10. $(-4, 0)$

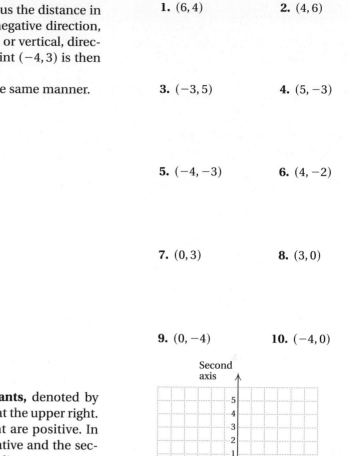

11. What can you say about the coordinates of a point in the third quadrant?

12. What can you say about the coordinates of a point in the fourth quadrant?

Answers on page A-8

159

2.1 Graphs of Equations

13. Determine whether $(2, -4)$ is a solution of $5b - 3a = 34$.

14. Determine whether $(2, -4)$ is a solution of $7p + 5q = -6$.

15. Use the line in Example 3 to find at least two more points that are solutions.

Answers on page A-8

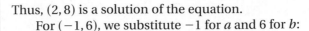

EXAMPLE 2 Determine whether each of the following pairs is a solution of $5b - 3a = 34$: $(2, 8)$ and $(-1, 6)$.

For the pair $(2, 8)$, we substitute 2 for a and 8 for b (alphabetical order of variables):

$$
\begin{array}{c}
5b - 3a = 34 \\
\hline
5 \cdot 8 - 3 \cdot 2 \ ? \ 34 \\
40 - 6 \\
34 \quad \text{TRUE}
\end{array}
$$

Thus, $(2, 8)$ is a solution of the equation.

For $(-1, 6)$, we substitute -1 for a and 6 for b:

$$
\begin{array}{c}
5b - 3a = 34 \\
\hline
5 \cdot 6 - 3 \cdot (-1) \ ? \ 34 \\
30 + 3 \\
33 \quad \text{FALSE}
\end{array}
$$

Thus, $(-1, 6)$ is *not* a solution of the equation.

Do Exercises 13 and 14.

EXAMPLE 3 Show that the pairs $(-4, 3)$, $(0, 1)$, and $(4, -1)$ are solutions of $y = 1 - \frac{1}{2}x$. Then plot the three points and use them to help determine another pair that is a solution.

We replace x with the first coordinate and y with the second coordinate of each pair:

$$
\begin{array}{c}
y = 1 - \frac{1}{2}x \\
\hline
3 \ ? \ 1 - \frac{1}{2} \cdot (-4) \\
1 + 2 \\
3 \quad \text{TRUE}
\end{array}
\qquad
\begin{array}{c}
y = 1 - \frac{1}{2}x \\
\hline
1 \ ? \ 1 - \frac{1}{2} \cdot 0 \\
1 - 0 \\
1 \quad \text{TRUE}
\end{array}
\qquad
\begin{array}{c}
y = 1 - \frac{1}{2}x \\
\hline
-1 \ ? \ 1 - \frac{1}{2} \cdot 4 \\
1 - 2 \\
-1 \quad \text{TRUE}
\end{array}
$$

In each case, the substitution results in a true equation. Thus all the pairs are solutions of the equation.

We plot the points as shown at right. Note that the three points appear to "line up." That is, they appear to be on a straight line. We use a ruler and draw a line passing through $(-4, 3)$, $(0, 1)$, and $(4, -1)$.

The line appears to pass through $(2, 0)$ as well. Let's see if this pair is a solution of $y = 1 - \frac{1}{2}x$:

$$
\begin{array}{c}
y = 1 - \frac{1}{2}x \\
\hline
0 \ ? \ 1 - \frac{1}{2} \cdot 2 \\
1 - 1 \\
0 \quad \text{TRUE}
\end{array}
$$

We see that $(2, 0)$ is another solution of the equation.

Do Exercise 15.

Example 3 leads us to believe that any point on the line that passes through $(-4, 3)$, $(0, 1)$, and $(4, -1)$ represents a solution of $y = 1 - \frac{1}{2}x$. In fact, every solution of $y = 1 - \frac{1}{2}x$ is represented by a point on that line and every point on that line represents a solution. The line is said to be the *graph* of the equation.

GRAPH OF AN EQUATION

The **graph** of an equation is a drawing that represents all its solutions.

C   Graphs of Linear Equations

Equations like $y = 1 - \frac{1}{2}x$ and $2x + 3y = 6$ are said to be **linear** because the graph of their solutions is a line. In general, a linear equation is any equation equivalent to one of the form $y = mx + b$ or $Ax + By = C$, where m, b, A, B, and C are constants (that is, they are numbers, not variables) and A and B are not both 0.

EXAMPLE 4   Graph: $y = 2x$.

We find some ordered pairs that are solutions. This time we list the pairs in a table. To find an ordered pair, we can choose *any* number for x and then determine y. For example, if we choose 3 for x, then $y = 2 \cdot 3 = 6$ (substituting into the equation $y = 2x$). We choose some negative values for x, as well as some positive ones. If a number takes us off the graph paper, we generally do not use it. Next, we plot these points. If we plotted *many* such points, they would appear to make a solid line. We draw the line with a ruler and label it $y = 2x$.

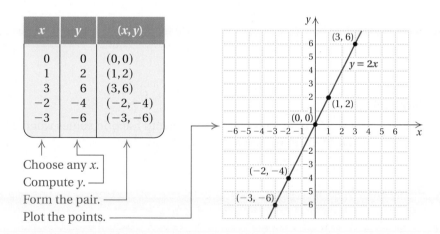

Choose any x.
Compute y.
Form the pair.
Plot the points.

To graph a linear equation:

1. Select a value for one variable and calculate the corresponding value of the other variable. Form an ordered pair using alphabetical order as indicated by the variables.
2. Repeat step (1) to obtain at least two other ordered pairs. Two ordered pairs are essential. A third serves as a check.
3. Plot the ordered pairs and draw a straight line passing through the points.

Do Exercises 16 and 17.

Graph.

16. $y = -2x$

x	y	(x, y)
-3		
-1		
0		
1		
3		

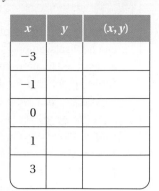

17. $y = \frac{1}{2}x$

x	y	(x, y)
4		
2		
0		
-2		
-4		
-1		

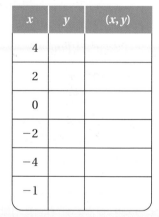

Answers on page A-8

Finding Solutions of Equations   A table of values representing ordered pairs that are solutions of an equation can be displayed on a graphing calculator. To do this for the equation in Example 4, $y = 2x$, we first press ⎡Y=⎤ to access the equation-editor screen. Then we clear any equations that are present. (See the Calculator Corner on p. 85 for the procedure for doing this.) Next, we enter the equation by positioning the cursor beside "Y1 =" and pressing ⎡2⎤ ⎡X,T,θ,n⎤. Now we press ⎡2ND⎤ ⎡TBLSET⎤ to display the table set-up screen. (TBLSET is the second function associated with the ⎡WINDOW⎤ key.) You can choose to supply the x-values yourself or you can set the calculator to supply them. To supply them yourself, follow the procedure for selecting ASK mode on p. 85. To have the calculator supply the x-values, set "Indpnt" to "Auto" by positioning the cursor over "Auto" and pressing ⎡ENTER⎤. "Depend" should also be set to "Auto."

When "Indpnt" is set to "Auto," the graphing calculator will supply values of x, beginning with the value specified as TBLSTART and continuing by adding the value of △TBL to the preceding value for x. Below, we show a table of values that starts with $x = -2$ and adds 1 to the preceding x-value. We press ⎡(-)⎤⎡2⎤⎡▽⎤⎡1⎤ or ⎡(-)⎤⎡2⎤⎡ENTER⎤⎡1⎤ to select a minimum x-value of -2 and an increment of 1. To display the table, we press ⎡2ND⎤ ⎡TABLE⎤. (TABLE is the second operation associated with the ⎡GRAPH⎤ key.) We can use the ⎡△⎤ and ⎡▽⎤ keys to scroll up and down through the table to see other solutions of the equation.

TABLE SETUP
TblStart=−2
△Tbl=1
Indpnt: **Auto** Ask
Depend: **Auto** Ask

X	Y₁	
−2	−4	
−1	−2	
0	0	
1	2	
2	4	
3	6	
4	8	
X = −2		

Exercise: Create a table of ordered pairs that are solutions of the equation.

1. Example 5
2. Example 7

---

18. Graph: $y = 2x + 3$.

x	y	(x, y)

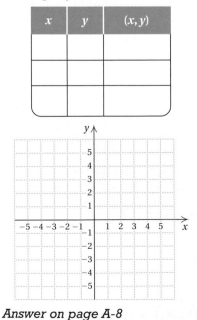

Answer on page A-8

EXAMPLE 5   Graph: $y = -\frac{1}{2}x + 3$.

By choosing even integers for x, we can avoid fraction values when calculating y. For example, if we choose 4 for x, we get

$$y = -\frac{1}{2}x + 3 = -\frac{1}{2}(4) + 3 = -2 + 3 = 1.$$

When x is -6, we get

$$y = -\frac{1}{2}x + 3 = -\frac{1}{2}(-6) + 3 = 3 + 3 = 6,$$

and when x is 0, we get

$$y = -\frac{1}{2}x + 3 = -\frac{1}{2}(0) + 3 = 0 + 3 = 3.$$

CHAPTER 2: Graphs, Functions,
and Applications

We list the results in a table. Then we plot the points corresponding to each pair.

x	y	(x, y)
4	1	(4, 1)
−6	6	(−6, 6)
0	3	(0, 3)

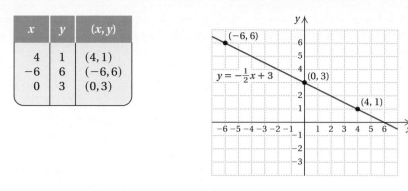

Note that the three points line up. If they did not, we would know that we had made a mistake. When only two points are plotted, an error is harder to detect. We use a ruler or other straightedge to draw a line through the points and then label the graph. Every point on the line represents a solution of $y = -\frac{1}{2}x + 3$.

Do Exercises 18 and 19. (Exercise 18 is on the preceding page.)

Calculating ordered pairs is usually easiest when y is isolated on one side of the equation, as in $y = 2x$ and $y = -\frac{1}{2}x + 3$. To graph an equation in which y is not isolated, we can use the addition and multiplication principles (see Sections 1.1 and 1.2) to first solve for y.

EXAMPLE 6 Graph: $3x + 5y = 10$.

We first solve for y:

$$3x + 5y = 10$$
$$3x + 5y - 3x = 10 - 3x \qquad \text{Subtracting } 3x$$
$$5y = 10 - 3x \qquad \text{Simplifying}$$
$$\tfrac{1}{5} \cdot 5y = \tfrac{1}{5} \cdot (10 - 3x) \qquad \text{Multiplying by } \tfrac{1}{5}, \text{ or dividing by 5}$$
$$y = \tfrac{1}{5} \cdot (10) - \tfrac{1}{5} \cdot (3x) \qquad \text{Using the distributive law}$$
$$y = 2 - \tfrac{3}{5}x$$
$$y = -\tfrac{3}{5}x + 2.$$

Thus the equation $3x + 5y = 10$ is equivalent to $y = -\frac{3}{5}x + 2$. We now find three ordered pairs, using multiples of 5 for x to avoid fractions.

x	y	(x, y)
0	2	(0, 2)
5	−1	(5, −1)
−5	5	(−5, 5)

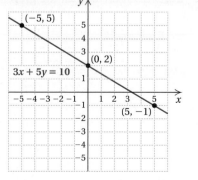

We plot the points, draw the line, and label the graph as shown.

Do Exercises 20 and 21. (Exercise 21 is on the following page.)

19. Graph: $y = -\dfrac{1}{2}x - 3$.

x	y	(x, y)

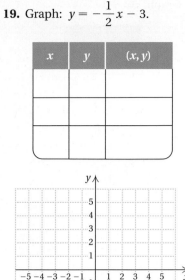

20. Graph: $4y - 3x = -8$.

x	y

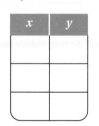

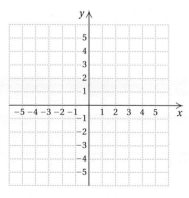

Answers on page A-8

21. Graph: $5x + 2y = 4$.

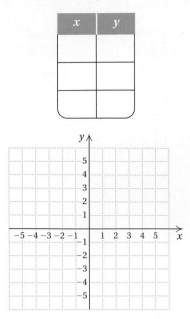

d Graphing Nonlinear Equations

We have seen that equations whose graphs are straight lines are called **linear.** There are many equations whose graphs are not straight lines. Here are some examples.

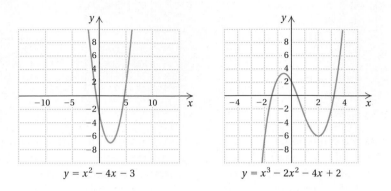

$$y = x^2 - 4x - 3$$

$$y = x^3 - 2x^2 - 4x + 2$$

Let's graph some of these **nonlinear equations.** We usually need to plot more than three points in order to get a good idea of the shape of the graph.

EXAMPLE 7 Graph: $y = x^2 - 5$.

We select numbers for x and find the corresponding values for y. For example, if we choose -2 for x, we get $y = (-2)^2 - 5 = 4 - 5 = -1$. The table lists several ordered pairs.

22. Graph: $y = 4 - x^2$.

x	y
-2	
-1	
0	
1	
2	
3	

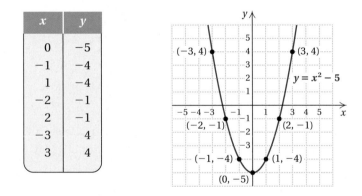

x	y
0	-5
-1	-4
1	-4
-2	-1
2	-1
-3	4
3	4

Next, we plot the points. The more points we plot, the more clearly we see the shape of the graph. Since the value of $x^2 - 5$ grows rapidly as x moves away from the origin, the graph rises steeply on either side of the y-axis.

Do Exercise 22.

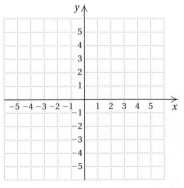

Answers on page A-8

CHAPTER 2: Graphs, Functions, and Applications

EXAMPLE 8 Graph: $y = 1/x$.

We select x-values and find the corresponding y-values. The table lists the ordered pairs $\left(3, \frac{1}{3}\right)$, $\left(2, \frac{1}{2}\right)$, $(1, 1)$, and so on.

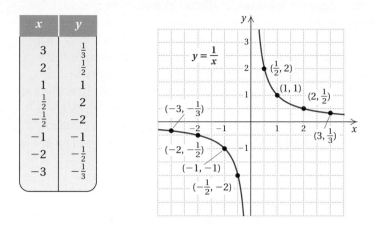

x	y
3	$\frac{1}{3}$
2	$\frac{1}{2}$
1	1
$\frac{1}{2}$	2
$-\frac{1}{2}$	-2
-1	-1
-2	$-\frac{1}{2}$
-3	$-\frac{1}{3}$

We plot these points, noting that each first coordinate is paired with its reciprocal. Since $1/0$ is undefined, we cannot use 0 as a first coordinate. Thus there are two "branches" to this graph—one on each side of the y-axis. Note that for x-values far to the right or far to the left of 0, the graph approaches, but does not touch, the x-axis; and for x-values close to 0, the graph approaches, but does not touch, the y-axis.

Do Exercise 23.

EXAMPLE 9 Graph: $y = |x|$.

We select numbers for x and find the corresponding values for y. For example, if we choose -1 for x, we get $y = |-1| = 1$. Several ordered pairs are listed in the table below.

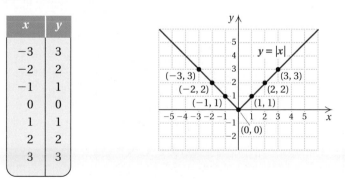

x	y
-3	3
-2	2
-1	1
0	0
1	1
2	2
3	3

We plot these points, noting that the absolute value of a positive number is the same as the absolute value of its opposite. Thus the x-values 3 and -3 both are paired with the y-value 3. Note that the graph is V-shaped and centered at the origin.

Do Exercise 24.

With equations like $y = -\frac{1}{2}x + 3$, $y = x^2 - 5$, and $y = |x|$, which we have graphed in this section, it is understood that y is the **dependent variable** and x is the **independent variable,** since y is expressed in terms of x and consequently y is calculated after first choosing x.

23. Graph: $y = \dfrac{2}{x}$.

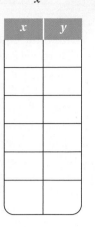

x	y

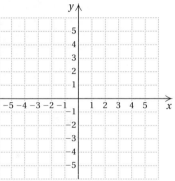

24. Graph: $y = 4 - |x|$.

x	y

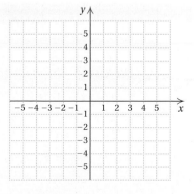

Answers on page A-8

CALCULATOR CORNER

Graphing Equations Graphs of equations are displayed in the **viewing window** of a graphing calculator. The viewing window is the portion of the coordinate plane that appears on the calculator's screen. It is defined by the minimum and maximum values of x and y: Xmin, Xmax, Ymin, and Ymax. The notation [Xmin, Xmax, Ymin, Ymax] is used to represent these window settings or dimensions. For example, $[-12, 12, -8, 8]$ denotes a window that displays the portion of the x-axis from -12 to 12 and the portion of the y-axis from -8 to 8. In addition, the distance between tick marks on the axes is defined by the settings Xscl and Yscl. The Xres setting indicates the pixel resolution. We usually select Xres $= 1$. The window corresponding to the settings $[-20, 30, -12, 20]$, Xscl $= 5$, Yscl $= 2$, Xres $= 1$, is shown on the left below. Press (WINDOW) on the top row of the keypad of your calculator to display the current window settings. The settings for the **standard viewing window** are shown on the right below.

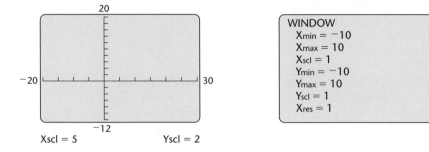

To change a setting, we position the cursor beside the setting we wish to change and enter the new value. For example, to change from the standard settings to $[-20, 30, -12, 20]$, Xscl $= 5$, Yscl $= 2$, on the **WINDOW** screen, we press (-) (2) (0) (ENTER) (3) (0) (ENTER) (5) (ENTER) (-) (1) (2) (ENTER) (2) (0) (ENTER) (2) (ENTER). The ▽ key can be used instead of (ENTER) after typing each window setting. To see the window, we press (GRAPH) on the top row of the keypad. To return quickly to the standard window setting $[-10, 10, -10, 10]$, Xscl $= 1$, Yscl $= 1$, we press (ZOOM) (6).

Equations must be solved for y before they can be graphed on the TI-84 Plus. Consider the equation $3x + 2y = 6$. Solving for y, we enter $y_1 = (6 - 3x)/2$ as described on p. 85. Then we select a window and press (GRAPH) to see the graph of the equation. (Press (ZOOM) (6) to see the graph in the standard window as shown on the right below.)

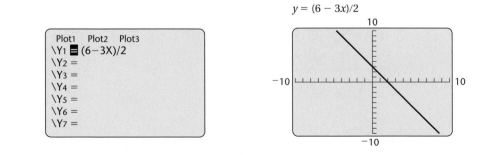

Exercises: Graph the equation in the standard viewing window $[-10, 10, -10, 10]$, with Xscl $= 1$ and Yscl $= 1$.

1. $y = 2x - 1$
2. $3x + y = 2$
3. $y = 5x - 3$
4. $y = -4x + 5$
5. $y = \frac{2}{3}x - 3$
6. $y = -\frac{3}{4}x + 4$
7. $y = 3.104x - 6.21$
8. $2.98x + y = -1.75$

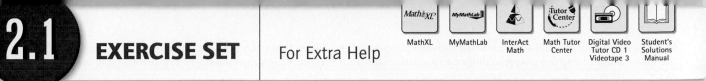

2.1 EXERCISE SET

For Extra Help

a   Plot the following points.

1. $A(4, 1)$, $B(2, 5)$, $C(0, 3)$, $D(0, -5)$, $E(6, 0)$, $F(-3, 0)$, $G(-2, -4)$, $H(-5, 1)$, $J(-6, 6)$

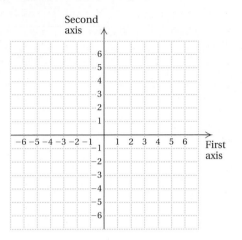

2. $A(-3, -5)$, $B(1, 3)$, $C(0, 7)$, $D(0, -2)$, $E(5, 0)$, $F(-4, 0)$, $G(1, -7)$, $H(-6, 4)$, $J(-3, 3)$

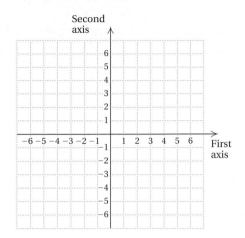

3. Plot the points $M(2, 3)$, $N(5, -3)$, and $P(-2, -3)$. Draw $\overline{MN}$, $\overline{NP}$, and $\overline{MP}$. $\left(\overline{MN}\right.$ means the line segment from M to N.$\left.\right)$ What kind of geometric figure is formed? What is its area?

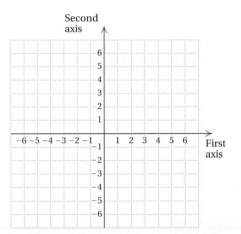

4. Plot the points $Q(-4, 3)$, $R(5, 3)$, $S(2, -1)$, and $T(-7, -1)$. Draw $\overline{QR}$, $\overline{RS}$, $\overline{ST}$, and $\overline{TQ}$. What kind of figure is formed? What is its area?

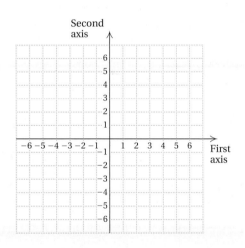

b   Determine whether the given point is a solution of the equation.

5. $(1, -1)$; $y = 2x - 3$

6. $(3, 4)$; $3s + t = 4$

7. $(3, 5)$; $4x - y = 7$

8. $(2, -1)$; $4r + 3s = 5$

9. $\left(0, \dfrac{3}{5}\right)$; $2a + 5b = 7$

10. $(-5, 1)$; $2p - 3q = -13$

In Exercises 11–16, an equation and two ordered pairs are given. Show that each pair is a solution of the equation. Then graph the equation and use the graph to determine another solution. Answers for solutions may vary, but the graphs do not.

11. $y = 4 - x$; $(-1, 5), (3, 1)$

12. $y = x - 3$; $(5, 2), (-1, -4)$

13. $3x + y = 7$; $(2, 1), (4, -5)$

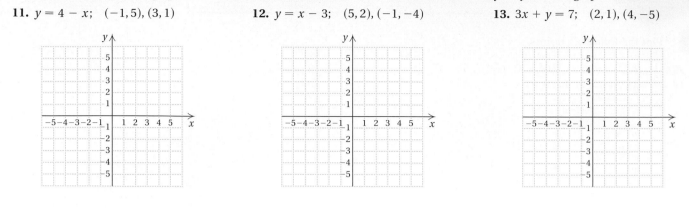

14. $y = \dfrac{1}{2}x + 3$; $(4, 5), (-2, 2)$

15. $6x - 3y = 3$; $(1, 1), (-1, -3)$

16. $4x - 2y = 10$; $(0, -5), (4, 3)$

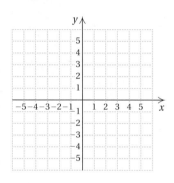

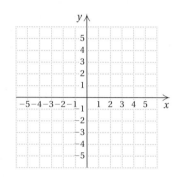

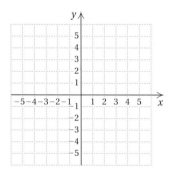

C   Graph.

17. $y = x - 1$

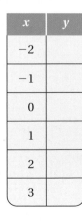

x	y
-2	
-1	
0	
1	
2	
3	

18. $y = x + 1$

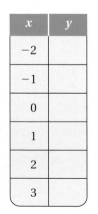

x	y
-2	
-1	
0	
1	
2	
3	

19. $y = x$

x	y
-2	
-1	
0	
1	
2	
3	

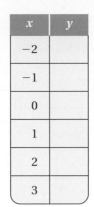

20. $y = -3x$

x	y
-2	
-1	
0	
1	
2	
3	

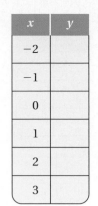

21. $y = \dfrac{1}{4}x$

x	y

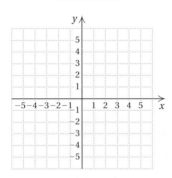

22. $y = \dfrac{1}{3}x$

x	y

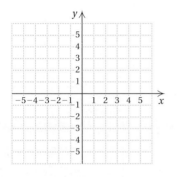

23. $y = 3 - x$

x	y

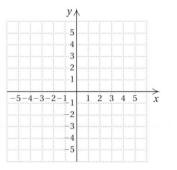

24. $y = x + 3$

x	y

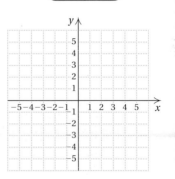

25. $y = 5x - 2$

x	y

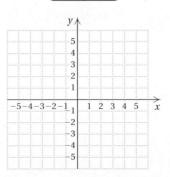

26. $y = \dfrac{1}{4}x + 2$

x	y

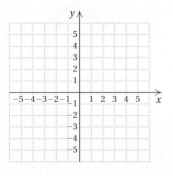

27. $y = \dfrac{1}{2}x + 1$

x	y

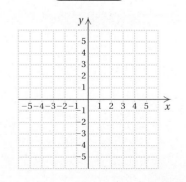

28. $y = \dfrac{1}{3}x - 4$

x	y

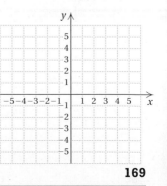

29. $x + y = 5$

x	y

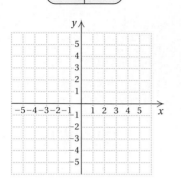

30. $x + y = -4$

x	y

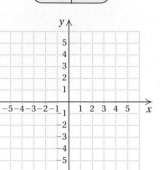

31. $y = -\dfrac{5}{3}x - 2$

x	y

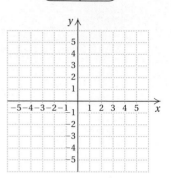

32. $y = -\dfrac{5}{2}x + 3$

x	y

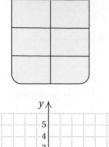

33. $x + 2y = 8$

x	y

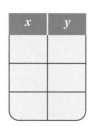

34. $x + 2y = -6$

x	y

35. $y = \dfrac{3}{2}x + 1$

x	y

36. $y = -\dfrac{1}{2}x - 3$

x	y

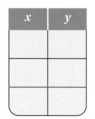

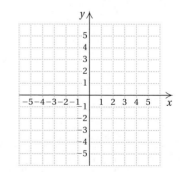
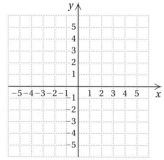
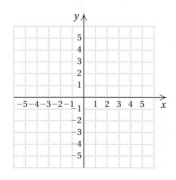

37. $8y + 2x = 4$

x	y

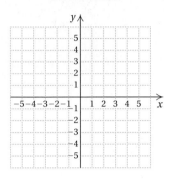

38. $6x - 3y = -9$

x	y

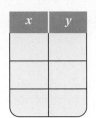

39. $8y + 2x = -4$

x	y

40. $6y + 2x = 8$

x	y

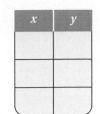

 Graph.

41. $y = x^2$

x	y
-2	
-1	
0	
1	
2	

42. $y = -x^2$
(*Hint:* $-x^2 = -1 \cdot x^2$.)

x	y
-2	
-1	
0	
1	
2	

43. $y = x^2 + 2$

x	y
-2	
-1	
0	
1	
2	

44. $y = 3 - x^2$

x	y
-3	
-2	
-1	
0	
1	
2	

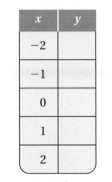

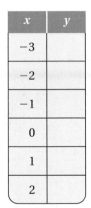

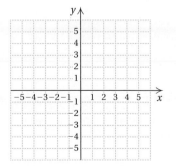

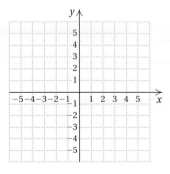

45. $y = x^2 - 3$

x	y
-2	
-1	
0	
1	
2	
3	

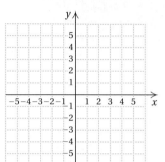

46. $y = x^2 - 3x$

x	y
-1	
0	
1	
2	
3	
4	

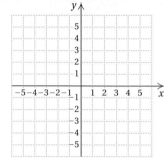

47. $y = -\dfrac{1}{x}$

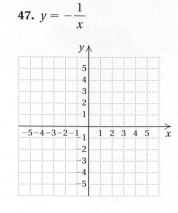

48. $y = \dfrac{3}{x}$

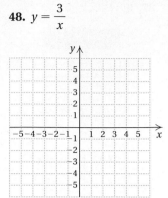

49. $y = |x - 2|$

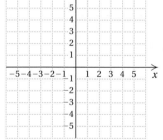

50. $y = |x| + 2$

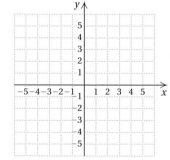

51. $y = x^3$

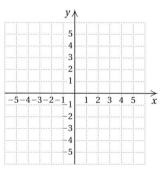

52. $y = x^3 - 2$

53. **D**w Using the equation $y = |x|$, explain why it is "risky" to draw a graph after plotting only two points.

54. **D**w Without making a drawing, how can you tell that the graph of $y = x - 30$ passes through three quadrants?

Solve. [1.5a, b]

55. $-3 < 2x - 5 \le 10$

56. $2x - 5 \ge -10 \; or$
$-4x - 2 < 10$

57. $3x - 5 \le -12 \; or$
$3x - 5 \ge 12$

58. $-13 < 3x + 5 < 23$

Solve. [1.3a]

59. *Landscaping.*   Grass seed is being spread on a triangular traffic island. If the grass seed can cover an area of 200 ft² and the island's base is 16 ft long, how tall a triangle can the seed fill?

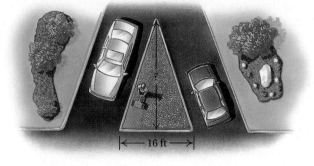

60. *Value of an Office Machine.*   The value of Dupliographic's color copier is given by the equation

$$v = -0.68t + 3.4,$$

where v is the value, in thousands of dollars, t years from the date of purchase.

a) Find the value after 1 yr, 2 yr, 4 yr, and 5 yr.
b) After what amount of time is the value of the copier $1500?

61. *Taxi Fare.*   The fare for a taxi ride from Johnson Street to Elm Street is $5.20. The driver charges $1.00 for the first $\frac{1}{2}$ mi and 30¢ for each additional $\frac{1}{4}$ mi. How far is it from Johnson Street to Elm Street?

62. *Real Estate Commission.*   The Jeffersons negotiated the following real estate commission on the selling price of their house:

7% for the first $100,000   and

4% for the amount that exceeds $100,000.

The realtor received a commission of $16,200 for selling the house. What was the selling price?

   Use a graphing calculator to graph each of the equations in Exercises 63–66. Use a standard viewing window of $[-10, 10, -10, 10]$, with Xscl = 1 and Yscl = 1.

63. $y = x^3 - 3x + 2$

64. $y = x - |x|$

65. $y = \dfrac{1}{x - 2}$

66. $y = \dfrac{1}{x^2}$

In Exercises 67–70, try to find an equation for the given graph.

67.

68.

69.

70.

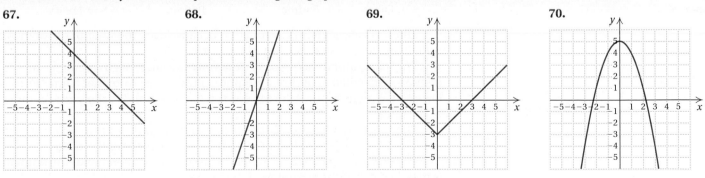

Objectives

2.2   FUNCTIONS AND GRAPHS

a   **Identifying Functions**

Consider the equation $y = 2x - 3$. If we substitute a value for x—say 5—we get a value for y, 7:

$$y = 2x - 3 = 2(5) - 3 = 10 - 3 = 7.$$

The equation $y = 2x - 3$ is an example of a *function*. We now develop the concept of a *function*, one of the most important concepts in mathematics.

In much the same way that ordered pairs form correspondences between first and second coordinates, a *function* is a correspondence from one set to another. For example:

> To each student in a college, there corresponds his or her student ID.
>
> To each item in a store, there corresponds its price.
>
> To each real number, there corresponds the cube of that number.

In each case, the first set is called the **domain** and the second set is called the **range.** This kind of correspondence is called a **function.** Given a member of the domain, there is *just one* member of the range to which it corresponds.

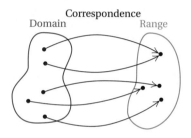

EXAMPLE 1   Determine whether the correspondence is a function.

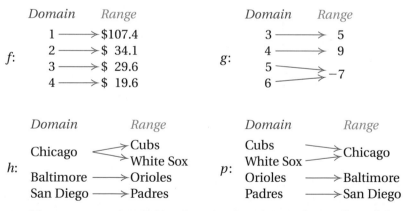

The correspondence f is a function because each member of the domain is matched to *only one* member of the range.

The correspondence g is a function because each member of the domain is matched to *only one* member of the range. Note that a function allows two or more members of the domain to correspond to the same member of the range.

The correspondence h is *not* a function because one member of the domain, Chicago, is matched to *more than one* member of the range.

The correspondence p is a function because each member of the domain is matched to *only one* member of the range.

FUNCTION; DOMAIN; RANGE

A **function** is a correspondence between a first set, called the **domain,** and a second set, called the **range,** such that each member of the domain corresponds to **exactly one** member of the range.

Do Exercises 1–4.

EXAMPLE 2 Determine whether the correspondence is a function.

Domain	*Correspondence*	*Range*
a) The integers	Each number's square	A set of nonnegative integers
b) The set of all states	Each state's members of the U.S. Senate	The set of U.S. Senators
c) The set of U.S. Senators	The state that a Senator represents	The set of all states

a) The correspondence *is* a function because each integer has *only one* square.

b) The correspondence *is not* a function because each state has two U.S. Senators.

c) The correspondence *is* a function because each Senator represents only one state.

Do Exercises 5–7. (Exercise 7 is on the following page.)

When a correspondence between two sets is not a function, it is still an example of a **relation.**

RELATION

A **relation** is a correspondence between a first set, called the **domain,** and a second set, called the **range,** such that each member of the domain corresponds to **at least one** member of the range.

Thus, although the correspondences of Examples 1 and 2 are not all functions, they *are* all relations. A function is a special type of relation—one in which each member of the domain is paired with *exactly one* member of the range.

Determine whether the correspondence is a function.

1. *Domain*   *Range*

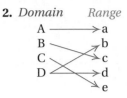

Cheetah ⟶ 70 mph
Human ⟶ 28 mph
Lion ⟶ 50 mph
Chicken ⟶ 9 mph

2. *Domain*   *Range*

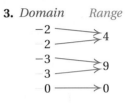

A ⟶ a
B ⟶ b
C ⟶ c
D ⟶ d
⟶ e

3. *Domain*   *Range*

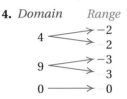

−2 ⟶ 4
2 ⟶ 4
−3 ⟶ 9
3 ⟶ 9
0 ⟶ 0

4. *Domain*   *Range*

4 ⟶ −2
4 ⟶ 2
9 ⟶ −3
9 ⟶ 3
0 ⟶ 0

Determine whether the correspondence is a function.

5. *Domain*
A set of numbers

Correspondence
Square each number and subtract 10.

Range
A set of numbers

6. *Domain*
A set of polygons

Correspondence
Find the perimeter of each polygon.

Range
A set of numbers

Answers on page A-10

175

7. Determine whether the correspondence is a function.

Domain
A set of numbers

Correspondence
The area of a rectangle

Range
A set of rectangles

Answer on page A-10

Answer on page A-10

Study Tips

ASKING QUESTIONS

Don't be afraid to ask questions in class. Most instructors welcome this and encourage students to ask them. Other students probably have the same questions you do.

"Better to ask twice than lose your way once."

Danish proverb

176

---

b   Finding Function Values

Most functions considered in mathematics are described by equations like $y = 2x + 3$ or $y = 4 - x^2$. We graph the function $y = 2x + 3$ by first performing calculations like the following:

for $x = 4$, $y = 2x + 3 = 2 \cdot 4 + 3 = 8 + 3 = 11$;

for $x = -5$, $y = 2x + 3 = 2 \cdot (-5) + 3 = -10 + 3 = -7$;

for $x = 0$, $y = 2x + 3 = 2 \cdot 0 + 3 = 0 + 3 = 3$; and so on.

For $y = 2x + 3$, the **inputs** (members of the domain) are values of x substituted into the equation. The **outputs** (members of the range) are the resulting values of y. If we call the function f, we can use x to represent an arbitrary *input* and $f(x)$—read "f of x," or "f at x," or "the value of f at x"—to represent the corresponding *output*. In this notation, the function given by $y = 2x + 3$ is written as $f(x) = 2x + 3$ and the calculations above can be written more concisely as follows:

$y = f(4) = 2 \cdot 4 + 3 = 8 + 3 = 11$;

$y = f(-5) = 2 \cdot (-5) + 3 = -10 + 3 = -7$;

$y = f(0) = 2 \cdot 0 + 3 = 0 + 3 = 3$; and so on.

Thus instead of writing "when $x = 4$, the value of y is 11," we can simply write "$f(4) = 11$," which can also be read as "f of 4 is 11" or "for the input 4, the output of f is 11."

We can think of a function as a machine. Think of $f(4) = 11$ as putting 4, a member of the domain (an input), into the machine. The machine knows the correspondence $f(x) = 2x + 3$, multiplies 4 by 2 and adds 3, and produces 11, a member of the range (the output).

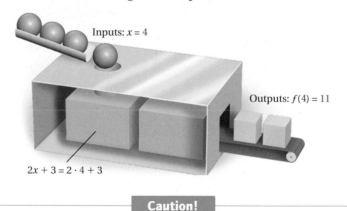

Inputs: $x = 4$

Outputs: $f(4) = 11$

$2x + 3 = 2 \cdot 4 + 3$

> **Caution!**
>
> The notation $f(x)$ *does not mean* "f times x" and should not be read that way.

EXAMPLE 3   A function f is given by $f(x) = 3x^2 - 2x + 8$. Find each of the indicated function values.

a) $f(0)$ **b)** $f(1)$ **c)** $f(-5)$ **d)** $f(7a)$

One way to find function values when a formula is given is to think of the formula with blanks, or placeholders, replacing the variable as follows:

$$f(\square) = 3\,\square^2 - 2\,\square + 8.$$

To find an output for a given input, we think: "Whatever goes in the blank on the left goes in the blank(s) on the right." With this in mind, let's complete the example.

a) $f(0) = 3 \cdot 0^2 - 2 \cdot 0 + 8 = 8$

b) $f(1) = 3 \cdot 1^2 - 2 \cdot 1 + 8 = 3 \cdot 1 - 2 + 8 = 3 - 2 + 8 = 9$

c) $f(-5) = 3(-5)^2 - 2 \cdot (-5) + 8 = 3 \cdot 25 + 10 + 8 = 75 + 10 + 8 = 93$

d) $f(7a) = 3(7a)^2 - 2(7a) + 8 = 3 \cdot 49a^2 - 14a + 8 = 147a^2 - 14a + 8$

Do Exercise 8.

EXAMPLE 4 Find the indicated function value.

a) $f(5)$, for $f(x) = 3x + 2$

b) $g(-2)$, for $g(x) = 7$

c) $F(a + 1)$, for $F(x) = 5x - 8$

d) $f(a + h)$, for $f(x) = -2x + 1$

a) $f(5) = 3 \cdot 5 + 2 = 15 + 2 = 17$

b) For the function given by $g(x) = 7$, all inputs share the same output, 7. Thus, $g(-2) = 7$. The function g is an example of a **constant function.**

c) $F(a + 1) = 5(a + 1) - 8 = 5a + 5 - 8 = 5a - 3$

d) $f(a + h) = -2(a + h) + 1 = -2a - 2h + 1$

Do Exercise 9.

8. Find the indicated function values for the following function:

$$f(x) = 2x^2 + 3x - 4.$$

a) $f(0)$

b) $f(8)$

c) $f(-5)$

d) $f(2a)$

9. Find the indicated function value.

a) $f(-6)$, for $f(x) = 5x - 3$

b) $g(55)$, for $g(x) = -3$

c) $F(a + 2)$, for $F(x) = -5x + 8$

d) $f(a - h)$, for $f(x) = 6x - 7$

Answers on page A-10

CALCULATOR CORNER

Finding Function Values We can find function values on a graphing calculator. One method is to substitute inputs directly into the formula. Consider the function $f(x) = x^2 + 3x - 4$. To find $f(-5)$, we press ⬤ ⬤ ⑤ ⬤ ⓧ² ⊕ ③ ⬤ ⬤ ⑤ ⬤ ⊖ ④ **ENTER**. We find that $f(-5) = 6$.

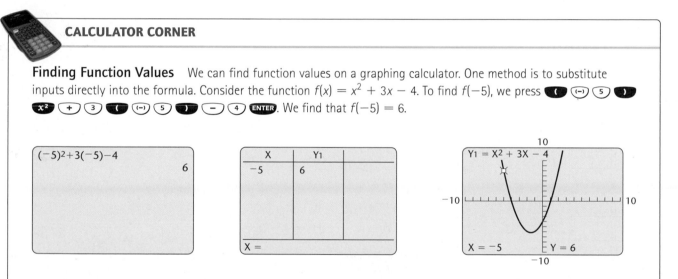

After we have entered the function as $y_1 = x^2 + 3x - 4$ on the equation-editor screen, there are several other methods that we can use to find function values. We can use a table set in ASK mode and enter $x = -5$. (See p. 85.) We see that the function value, y_1, is 6. We can also use the VALUE feature to evaluate the function. To do this, we first graph the function. Then we press **2ND** **CALC** ① to access the VALUE feature. Next, we supply the desired x-value by pressing ⊖ ⑤. Finally, we press **ENTER** to see $X = -5$, $Y = 6$ at the bottom of the screen. Again we see that the function value is 6. Note that, when the VALUE feature is used to find a function value, the x-value must be in the viewing window.

A fourth method for finding function values uses the TRACE feature. With the function graphed in a window that includes the x-value -5, we press **TRACE**. The coordinates of the point where the blinking cursor is positioned on the graph are displayed at the bottom of the screen. To move the cursor to the point with x-coordinate -5, we press ⊖ ⑤ **ENTER**. Now we see $X = -5$, $Y = 6$ displayed at the bottom of the screen. This tells us that $f(-5) = 6$. The final calculator display for this method is the same as the one shown above for the VALUE feature. There are other ways to find function values, but we will not discuss them here.

Exercises: Find the function values.

1. $f(-5.1)$, for $f(x) = 3x + 2$

2. $f(4)$, for $f(x) = -3.6x$

3. $f(-3)$, for $f(x) = x^2 + 5$

4. $f(3)$, for $f(x) = 4x^2 + x - 5$

10. Graph: $f(x) = x - 4$.

x	$f(x)$

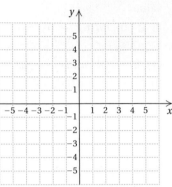

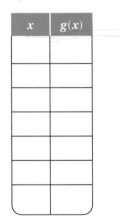

11. Graph: $g(x) = 5 - x^2$.

x	$g(x)$

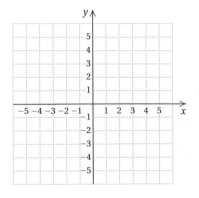

C   Graphs of Functions

To graph a function, we find ordered pairs (x, y) or $(x, f(x))$, plot them, and connect the points. Note that y and $f(x)$ are used interchangeably—that is, $y = f(x)$—when we are working with functions and their graphs.

■ **EXAMPLE 5**   Graph: $f(x) = x + 2$.

A list of some function values is shown in this table. We plot the points and connect them. The graph is a straight line. The "y" on the vertical axis could also be labeled "$f(x)$".

x	$f(x)$
-4	-2
-3	-1
-2	0
-1	1
0	2
1	3
2	4
3	5
4	6

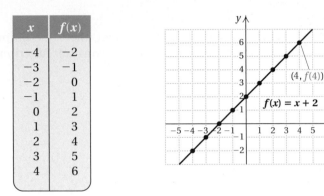

Do Exercise 10.

■ **EXAMPLE 6**   Graph: $g(x) = 4 - x^2$.

We calculate some function values, plot the corresponding points, and draw the curve.

$$g(0) = 4 - 0^2 = 4 - 0 = 4,$$
$$g(-1) = 4 - (-1)^2 = 4 - 1 = 3,$$
$$g(2) = 4 - 2^2 = 4 - 4 = 0,$$
$$g(-3) = 4 - (-3)^2 = 4 - 9 = -5$$

x	$g(x)$
-3	-5
-2	0
-1	3
0	4
1	3
2	0
3	-5

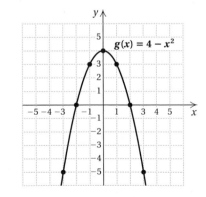

Do Exercise 11.

Answers on page A-10

EXAMPLE 7 Graph: $h(x) = |x|$.

A list of some function values is shown in the following table. We plot the points and connect them. The graph is a V-shaped "curve" that rises on either side of the vertical axis.

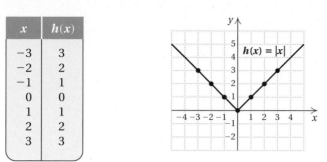

x	$h(x)$
-3	3
-2	2
-1	1
0	0
1	1
2	2
3	3

Do Exercise 12.

d The Vertical-Line Test

Consider the graph of the function f described by $f(x) = x^2 - 5$ shown at right. It is also the graph of the equation $y = x^2 - 5$.

To find a function value, like $f(3)$, from a graph, we locate the input on the horizontal axis, move directly up or down to the graph of the function, and then move left or right to find the output on the vertical axis. Thus, $f(3) = 4$. Keep in mind that members of the domain are found on the horizontal axis, members of the range are found on the vertical axis, and the y on the vertical axis could also be labeled $f(x)$.

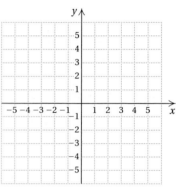

When one member of the domain is paired with two or more different members of the range, the correspondence is not a function. Thus, when a graph contains two or more different points with the same first coordinate, the graph cannot represent a function. Points sharing a common first coordinate are vertically above or below each other. (See the following graph.)

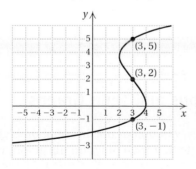

Since 3 is paired with more than one member of the range, the graph does not represent a function.

This observation leads to the *vertical-line test*.

THE VERTICAL-LINE TEST

If it is possible for a vertical line to cross a graph more than once, then the graph is *not* the graph of a function.

12. Graph: $t(x) = 3 - |x|$.

x	$t(x)$

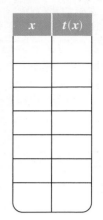

Answer on page A-10

179

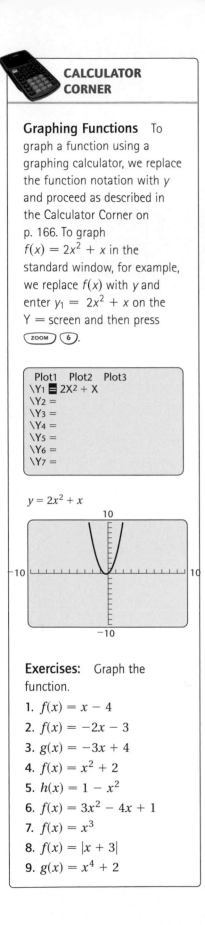
Graphing Functions   To graph a function using a graphing calculator, we replace the function notation with y and proceed as described in the Calculator Corner on p. 166. To graph $f(x) = 2x^2 + x$ in the standard window, for example, we replace $f(x)$ with y and enter $y_1 = 2x^2 + x$ on the $Y =$ screen and then press ZOOM 6.

```
Plot1   Plot2   Plot3
\Y1 ■ 2X² + X
\Y2 =
\Y3 =
\Y4 =
\Y5 =
\Y6 =
\Y7 =
```

$y = 2x^2 + x$

Exercises: Graph the function.

1. $f(x) = x - 4$

2. $f(x) = -2x - 3$

3. $g(x) = -3x + 4$

4. $f(x) = x^2 + 2$

5. $h(x) = 1 - x^2$

6. $f(x) = 3x^2 - 4x + 1$

7. $f(x) = x^3$

8. $f(x) = |x + 3|$

9. $g(x) = x^4 + 2$

EXAMPLE 8   Determine whether each of the following is the graph of a function.

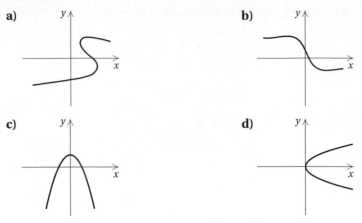

a)

b)

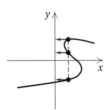

c)

d)

a) The graph *is not* that of a function because a vertical line can cross the graph at more than one point.

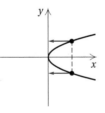

b) The graph *is* that of a function because no vertical line can cross the graph at more than one point. This can be confirmed with a ruler or straightedge.

c) The graph *is* that of a function because no vertical line can cross the graph more than once.

d) The graph *is not* that of a function because a vertical line can cross the graph more than once.

Do Exercises 13–16 on the following page.

e   **Applications of Functions and Their Graphs**

Functions are often described by graphs, whether or not an equation is given. To use a graph in an application, we note that each point on the graph represents a pair of values.

EXAMPLE 9 *Wireless-Phone Subscribers.* The following graph represents the number of wireless-phone subscribers, in millions, in the United States. The number of subscribers is a function *f* of the year. Note that no equation is given for the function.

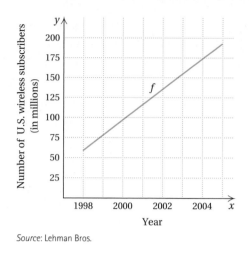

Source: Lehman Bros.

a) How many wireless-phone subscribers were there in 1999? That is, find *f*(1999).

b) How many wireless-phone subscribers were there in 2002? That is, find *f*(2002).

a) To estimate the number of wireless-phone subscribers in 1999, we locate 1999 on the horizontal axis and move directly up until we reach the graph. Then we move across to the vertical axis. We come to a point that is about 78, so we estimate that there were about 78 million wireless-phone subscribers in 1999.

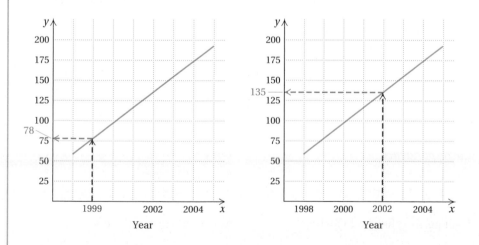

b) To estimate the number of wireless-phone subscribers in 2002, we locate 2002 on the horizontal axis and move directly up until we reach the graph. Then we move across to the vertical axis. We come to a point that is about 135, so we estimate that there were about 135 million wireless-phone subscribers in 2002.

Do Exercises 17 and 18.

Determine whether each of the following is the graph of a function.

13.

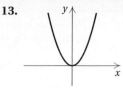

14.

15.

16.

Refer to the graph in Example 9 for Margin Exercises 17 and 18.

17. How many wireless-phone subscribers were there in 2000?

18. How many wireless-phone subscribers were there in 2003?

Answers on page A-10

2.2
EXERCISE SET
For Extra Help

MathXL · MyMathLab · InterAct Math · Math Tutor Center · Digital Video Tutor CD 1 Videotape 3 · Student's Solutions Manual

a Determine whether the correspondence is a function.

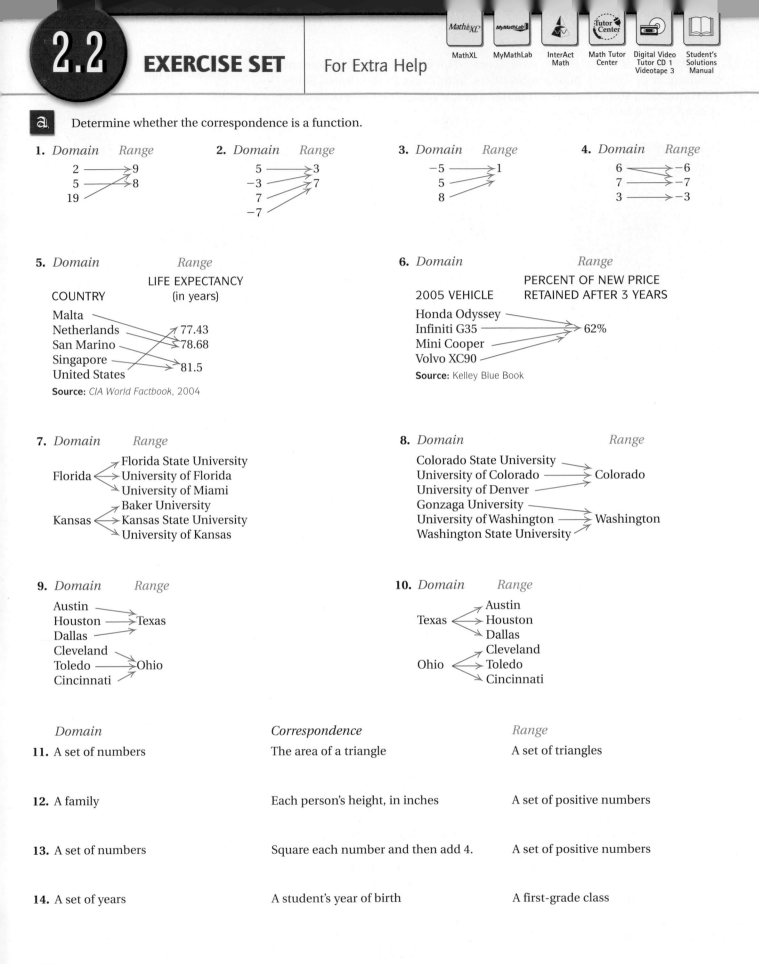

1. Domain   Range

2 ⟶ 9
5 ⟶ 8
19

2. Domain   Range

5 ⟶ 3
−3 ⟶ 7
7
−7

3. Domain   Range

−5 ⟶ 1
5 ⟶
8 ⟶

4. Domain   Range

6 ⟶ −6
7 ⟶ −7
3 ⟶ −3

5. Domain   Range

COUNTRY   LIFE EXPECTANCY (in years)

Malta
Netherlands ⟶ 77.43
San Marino ⟶ 78.68
Singapore
United States ⟶ 81.5

Source: *CIA World Factbook*, 2004

6. Domain   Range

2005 VEHICLE   PERCENT OF NEW PRICE RETAINED AFTER 3 YEARS

Honda Odyssey
Infiniti G35 ⟶ 62%
Mini Cooper
Volvo XC90

Source: Kelley Blue Book

7. Domain   Range

Florida State University
Florida ⟷ University of Florida
University of Miami

Baker University
Kansas ⟷ Kansas State University
University of Kansas

8. Domain   Range

Colorado State University
University of Colorado ⟶ Colorado
University of Denver

Gonzaga University
University of Washington ⟶ Washington
Washington State University

9. Domain   Range

Austin
Houston ⟶ Texas
Dallas
Cleveland
Toledo ⟶ Ohio
Cincinnati

10. Domain   Range

Austin
Texas ⟷ Houston
Dallas
Cleveland
Ohio ⟷ Toledo
Cincinnati

Domain	Correspondence	Range
11. A set of numbers	The area of a triangle	A set of triangles
12. A family	Each person's height, in inches	A set of positive numbers
13. A set of numbers	Square each number and then add 4.	A set of positive numbers
14. A set of years	A student's year of birth	A first-grade class

b Find the function values.

15. $f(x) = x + 5$

    a) $f(4)$ **b)** $f(7)$

    c) $f(-3)$ **d)** $f(0)$

    e) $f(2.4)$ **f)** $f\left(\frac{2}{3}\right)$

16. $g(t) = t - 6$

    a) $g(0)$ **b)** $g(6)$

    c) $g(13)$ **d)** $g(-1)$

    e) $g(-1.08)$ **f)** $g\left(\frac{7}{8}\right)$

17. $h(p) = 3p$

    a) $h(-7)$ **b)** $h(5)$

    c) $h\left(\frac{2}{3}\right)$ **d)** $h(0)$

    e) $h(6a)$ **f)** $h(a + 1)$

18. $f(x) = -4x$

    a) $f(6)$ **b)** $f\left(-\frac{1}{2}\right)$

    c) $f(0)$ **d)** $f(-1)$

    e) $f(3a)$ **f)** $f(a - 1)$

19. $g(s) = 3s + 4$

    a) $g(1)$ **b)** $g(-7)$

    c) $g\left(\frac{2}{3}\right)$ **d)** $g(0)$

    e) $g(a - 2)$ **f)** $g(a + h)$

20. $h(x) = 19$, a constant function

    a) $h(4)$ **b)** $h(-6)$

    c) $h(12.5)$ **d)** $h(0)$

    e) $h\left(\frac{2}{3}\right)$ **f)** $h(a + 3)$

21. $f(x) = 2x^2 - 3x$

    a) $f(0)$ **b)** $f(-1)$

    c) $f(2)$ **d)** $f(10)$

    e) $f(-5)$ **f)** $f(4a)$

22. $f(x) = 3x^2 - 2x + 1$

    a) $f(0)$ **b)** $f(1)$

    c) $f(-1)$ **d)** $f(10)$

    e) $f(-3)$ **f)** $f(2a)$

23. $f(x) = |x| + 1$

    a) $f(0)$ **b)** $f(-2)$

    c) $f(2)$ **d)** $f(-10)$

    e) $f(a - 1)$ **f)** $f(a + h)$

24. $g(t) = |t - 1|$

    a) $g(4)$ **b)** $g(-2)$

    c) $g(-1)$ **d)** $g(100)$

    e) $g(5a)$ **f)** $g(a + 1)$

25. $f(x) = x^3$

    a) $f(0)$ **b)** $f(-1)$

    c) $f(2)$ **d)** $f(10)$

    e) $f(-5)$ **f)** $f(-3a)$

26. $f(x) = x^4 - 3$

    a) $f(1)$ **b)** $f(-1)$

    c) $f(0)$ **d)** $f(2)$

    e) $f(-2)$ **f)** $f(-a)$

27. *Archaeology.* The function H described by

$$H(x) = 2.75x + 71.48$$

can be used to determine the height, in centimeters, of a woman whose *humerus* (the bone from the elbow to the shoulder) is x cm long. If a humerus is known to be from a female, how tall was she if the bone is **(a)** 32 cm long? **(b)** 35 cm long?

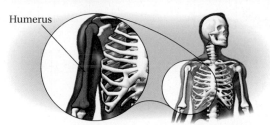

Humerus

28. Refer to Exercise 27. When a humerus is from a male, the function

$$M(x) = 2.89x + 70.64$$

can be used to find the male's height, in centimeters. If a humerus is known to be from a male, how tall was the male if the bone is **(a)** 30 cm long? **(b)** 35 cm long?

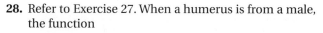

183

29. *Pressure at Sea Depth.* The function $P(d) = 1 + (d/33)$ gives the pressure, in *atmospheres* (atm), at a depth of d feet in the sea. Note that $P(0) = 1$ atm, $P(33) = 2$ atm, and so on. Find the pressure at 20 ft, 30 ft, and 100 ft.

30. *Temperature as a Function of Depth.* The function $T(d) = 10d + 20$ gives the temperature, in degrees Celsius, inside the earth as a function of the depth d, in kilometers. Find the temperature at 5 km, 20 km, and 1000 km.

31. *Melting Snow.* The function $W(d) = 0.112d$ approximates the amount, in centimeters, of water that results from d centimeters of snow melting. Find the amount of water that results from snow melting from depths of 16 cm, 25 cm, and 100 cm.

32. *Temperature Conversions.* The function $C(F) = \frac{5}{9}(F - 32)$ determines the Celsius temperature that corresponds to F degrees Fahrenheit. Find the Celsius temperature that corresponds to 62°F, 77°F, and 23°F.

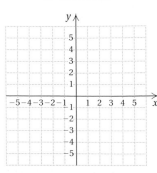

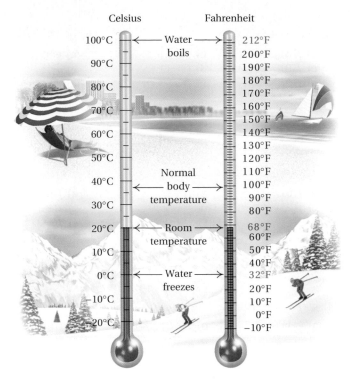

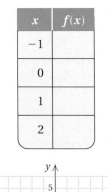
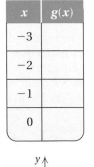
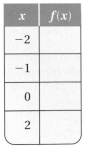
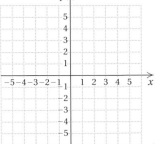
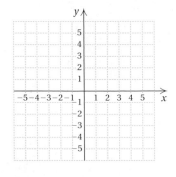
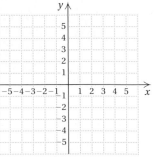

C Graph the function.

33. $f(x) = -2x$

x	f(x)
-2	
-1	
0	
2	

34. $g(x) = 3x$

x	f(x)
-1	
0	
1	
2	

35. $f(x) = 3x - 1$

x	f(x)
-1	
0	
1	
2	

36. $g(x) = 2x + 5$

x	g(x)
-3	
-2	
-1	
0	

37. $g(x) = -2x + 3$

x	g(x)

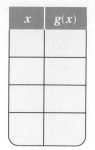

38. $f(x) = -\frac{1}{2}x + 2$

x	f(x)

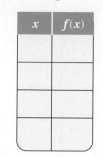

39. $f(x) = \frac{1}{2}x + 1$

x	f(x)

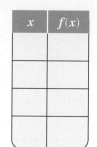

40. $f(x) = -\frac{3}{4}x - 2$

x	f(x)

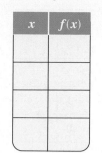

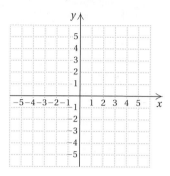

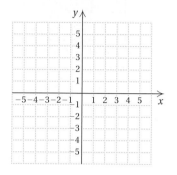

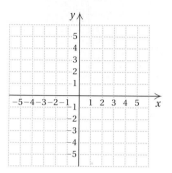

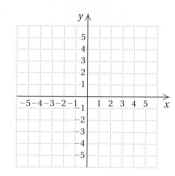

41. $f(x) = 2 - |x|$

x	f(x)
-3	
-2	
-1	
0	
1	
2	
3	

42. $f(x) = |x| - 4$

x	f(x)
-3	
-2	
-1	
0	
1	
2	
3	

43. $g(x) = |x - 1|$

x	g(x)

44. $g(x) = |x + 3|$

x	g(x)

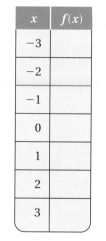

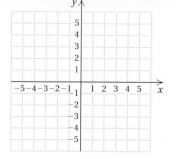

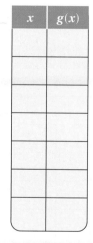

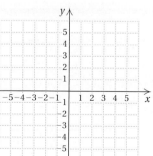

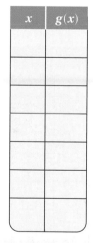

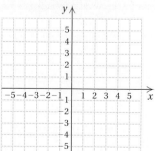

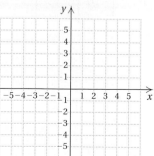

45. $f(x) = x^2$

x	f(x)
-3	
-2	
-1	
0	
1	
2	
3	

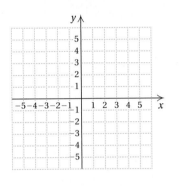

46. $f(x) = x^2 - 1$

x	f(x)
-3	
-2	
-1	
0	
1	
2	
3	

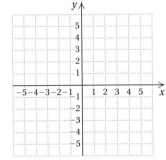

47. $f(x) = x^2 - x - 2$

x	f(x)
-3	
-2	
-1	
0	
1	
2	
3	

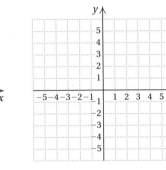
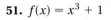

48. $f(x) = x^2 + 6x + 5$

x	f(x)
-5	
-4	
-3	
-2	
-1	
0	
1	

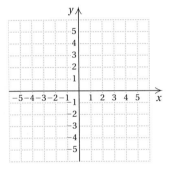

49. $f(x) = 2 - x^2$

x	f(x)

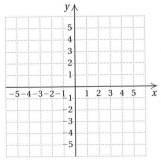

50. $f(x) = 1 - x^2$

x	f(x)

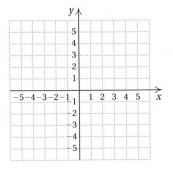

51. $f(x) = x^3 + 1$

x	f(x)
-2	
-1	
0	
1	
2	

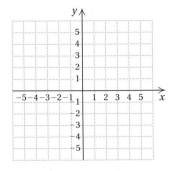

52. $f(x) = x^3 - 2$

x	f(x)
-2	
-1	
0	
1	
2	

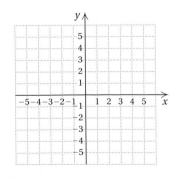

Determine whether each of the following is the graph of a function.

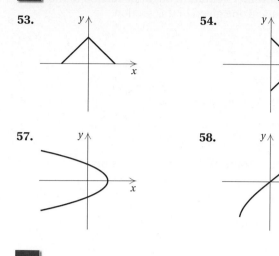

53.

54.

55.

56.

57.

58.

59.

60.

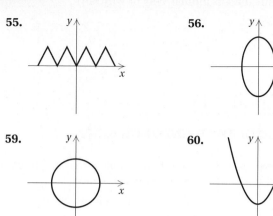

 Solve.

News/Talk Radio Stations. The following graph approximates the number of U.S. commercial radio stations with a news/talk format. The number of stations is a function f of the year x.

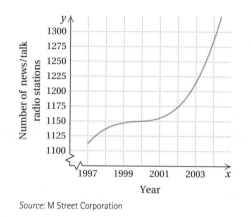

Source: M Street Corporation

61. Approximate the number of news/talk radio stations in 1999. That is, find $f(1999)$.

62. Approximate the number of news/talk radio stations in 2003. That is, find $f(2003)$.

Digital Photographs. The following graph approximates the number of digital photos taken but not printed, in billions. The number of photos that are not printed is a function f of the year x.

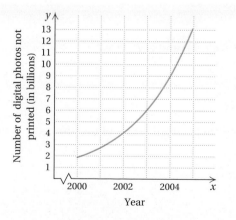

63. Approximate the number of digital photos taken but not printed in 2000. That is, find $f(2000)$.

64. Approximate the number of digital photos taken but not printed in 2002. That is, find $f(2002)$.

65. D_W Is it possible for a function to have more numbers as outputs than as inputs? Why or why not?

66. D_W Look up the word "function" in a dictionary. Explain how that definition might be related to the mathematical one given in this section.

🖎 **VOCABULARY REINFORCEMENT**

In each of Exercises 67–74, fill in the blank with the correct term from the given list. Some of the choices may not be used and some may be used more than once.

67. The axes divide the plane into four regions called _____ . [2.1a]

68. A(n) _____ is a correspondence between two sets such that each member of the first set corresponds to at least one member of the second set. [2.2a]

69. A(n) _____ is a correspondence between a first set, called the _____ , and a second set, called the _____ , such that each member of the _____ corresponds to exactly one member of the _____ . [2.2a]

70. The _____ of an equation is a drawing that represents all of its solutions. [2.1b]

71. Members of the domain of a function are its _____ . [2.2b]

72. The replacements for the variable that make an equation true are its _____ . [1.1a]

73. The _____ states that for any real numbers a, b, and c, $a = b$ is equivalent to $a + c = b + c$. [1.1b]

74. The _____ can be used to determine whether a graph represents a function. [2.2d]

axes

coordinates

quadrants

addition principle

multiplication principle

vertical-line test

graph

domain

range

relation

function

inputs

outputs

solutions

values

For Exercises 75 and 76, let $f(x) = 3x^2 - 1$ and $g(x) = 2x + 5$.

75. Find $f(g(-4))$ and $g(f(-4))$.

76. Find $f(g(-1))$ and $g(f(-1))$.

77. Suppose that a function g is such that $g(-1) = -7$ and $g(3) = 8$. Find a formula for g if $g(x)$ is of the form $g(x) = mx + b$, where m and b are constants.

2.3 FINDING DOMAIN AND RANGE

a Finding Domain and Range

The solutions of an equation in two variables consist of a set of ordered pairs. An arbitrary set of ordered pairs is called a **relation.** When a set of ordered pairs is such that no two different pairs share a common first coordinate, we have a **function.** The **domain** is the set of all first coordinates and the **range** is the set of all second coordinates.

EXAMPLE 1 Find the domain and the range of the function f whose graph is shown below.

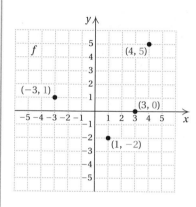

This function contains just four ordered pairs and it can be written as

$$\{(-3, 1), (1, -2), (3, 0), (4, 5)\}.$$

We can determine the domain and the range by reading the x- and y-values directly from the graph.

The domain is the set of all first coordinates, or x-values, $\{-3, 1, 3, 4\}$. The range is the set of all second coordinates, or y-values, $\{1, -2, 0, 5\}$.

Do Exercise 1.

1. Find the domain and the range of the function f whose graph is shown below.

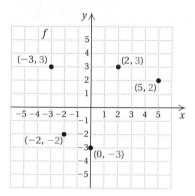

EXAMPLE 2 For the function f whose graph is shown below, determine each of the following.

a) The number in the range that is paired with 1 from the domain. That is, find $f(1)$.

b) The domain of f

c) The numbers in the domain that are paired with 1 from the range. That is, find all x such that $f(x) = 1$.

d) The range of f

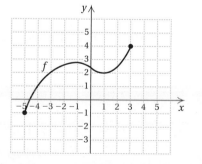

a) To determine which number in the range is paired with 1 in the domain, we locate 1 on the horizontal axis. Next, we find the point on the graph of f for which 1 is the first coordinate. From that point, we can look to the vertical axis to find the corresponding y-coordinate, 2. The input 1 has the output 2—that is, $f(1) = 2$.

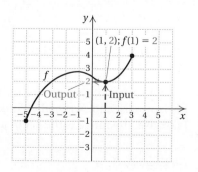

Answer on page A-12

Study Tips

HOMEWORK TIPS

Prepare for your homework assignment by reading the explanations of concepts and following the step-by-step solutions of examples in the text. The time you spend preparing will save valuable time when you do your assignment.

"I will prepare and some day my chance will come."

Abraham Lincoln

2. For the function f whose graph is shown below, determine each of the following.

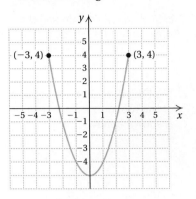

a) The number in the range that is paired with the input 1. That is, find $f(1)$.

b) The domain of f

c) The numbers in the domain that are paired with 4

d) The range of f

b) The domain of the function is the set of all x-values, or inputs, of the points on the graph. These extend from -5 to 3 and can be viewed as the curve's shadow, or projection, onto the x-axis. Thus the domain is the set $\{x \mid -5 \le x \le 3\}$, or, in interval notation, $[-5, 3]$.

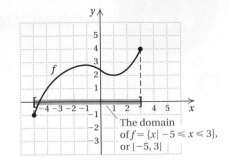

The domain of $f = \{x \mid -5 \le x \le 3\}$, or $[-5, 3]$

c) To determine which numbers in the domain are paired with 1 in the range, we locate 1 on the vertical axis. From there, we look left and right to the graph of f to find any points for which 1 is the second coordinate (output). One such point exists, $(-4, 1)$. For this function, we note that $x = -4$ is the only member of the domain paired with 1. For other functions, there might be more than one member of the domain paired with a member of the range.

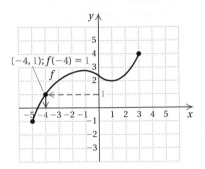

$(-4, 1); f(-4) = 1$

d) The range of the function is the set of all y-values, or outputs, of the points on the graph. These extend from -1 to 4 and can be viewed as the curve's shadow, or projection, onto the y-axis. Thus the range is the set $\{y \mid -1 \le y \le 4\}$, or, in interval notation, $[-1, 4]$.

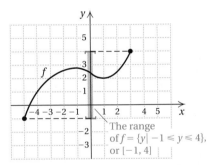

The range of $f = \{y \mid -1 \le y \le 4\}$, or $[-1, 4]$

Do Exercise 2.

3. Find the domain and the range of the function f whose graph is shown below.

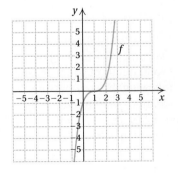

EXAMPLE 3 Find the domain and the range of the function f whose graph is shown below.

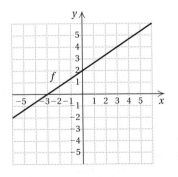

Since no endpoints are indicated, the graph extends indefinitely both horizontally and vertically. Thus the domain is the set of all real numbers. Likewise, the range is the set of all real numbers.

Do Exercise 3.

Answers on page A-12

When a function is given by an equation or formula, the domain is understood to be the largest set of real numbers (inputs) for which function values (outputs) can be calculated. That is, the domain is the set of all possible allowable inputs into the formula. To find the domain, think, "What can we substitute?"

EXAMPLE 4 Find the domain: $f(x) = |x|$.

We ask, "What can we substitute?" Is there any number x for which we cannot calculate $|x|$? The answer is no. Thus the domain of f is the set of all real numbers.

EXAMPLE 5 Find the domain: $f(x) = \dfrac{3}{2x - 5}$.

We ask, "What can we substitute?" Is there any number x for which we cannot calculate $3/(2x - 5)$? Since $3/(2x - 5)$ cannot be calculated when the denominator $2x - 5$ is 0, we solve the following equation to find those real numbers that must be excluded from the domain of f:

$$2x - 5 = 0 \qquad \text{Setting the denominator equal to 0}$$
$$2x = 5 \qquad \text{Adding 5}$$
$$x = \tfrac{5}{2}. \qquad \text{Dividing by 2}$$

Thus $\frac{5}{2}$ is not in the domain, whereas all other real numbers are.

The domain of f is $\left\{x \mid x \text{ is a real number } and \; x \neq \frac{5}{2}\right\}$. In interval notation, the domain is $\left(-\infty, \frac{5}{2}\right) \cup \left(\frac{5}{2}, \infty\right)$.

Do Exercises 4 and 5.

The task of determining the domain and the range of a function is one that we will return to several times as we consider other types of functions in this book.

FUNCTIONS: A REVIEW

The following is a review of the function concepts considered in Sections 2.2 and 2.3.

Function Concepts

- Formula for f: $f(x) = x^2 - 7$
- For every input of f, there is exactly one output.
- When 1 is the input, -6 is the output.
- $f(1) = -6$
- $(1, -6)$ is on the graph.
- Domain = The set of all inputs
  = The set of all real numbers
- Range = The set of all outputs
  = $\{x \mid x \geq 7\}$
  = $[-7, \infty)$

Graph

$f(x) = x^2 - 7$

Find the domain.

4. $f(x) = x^3 - |x|$

5. $f(x) = \dfrac{4}{3x + 2}$

Answers on page A-12

a In Exercises 1–12, the graph is that of a function. Determine for each one **(a)** $f(1)$; **(b)** the domain; **(c)** all x-values such that $f(x) = 2$; and **(d)** the range. An open dot indicates that the point does not belong to the graph.

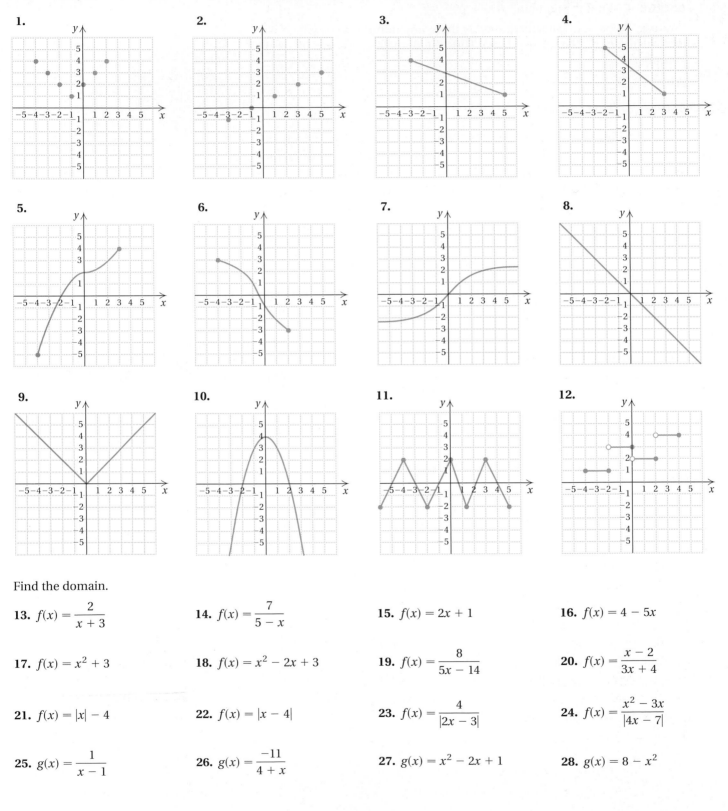

Find the domain.

13. $f(x) = \dfrac{2}{x + 3}$

14. $f(x) = \dfrac{7}{5 - x}$

15. $f(x) = 2x + 1$

16. $f(x) = 4 - 5x$

17. $f(x) = x^2 + 3$

18. $f(x) = x^2 - 2x + 3$

19. $f(x) = \dfrac{8}{5x - 14}$

20. $f(x) = \dfrac{x - 2}{3x + 4}$

21. $f(x) = |x| - 4$

22. $f(x) = |x - 4|$

23. $f(x) = \dfrac{4}{|2x - 3|}$

24. $f(x) = \dfrac{x^2 - 3x}{|4x - 7|}$

25. $g(x) = \dfrac{1}{x - 1}$

26. $g(x) = \dfrac{-11}{4 + x}$

27. $g(x) = x^2 - 2x + 1$

28. $g(x) = 8 - x^2$

29. $g(x) = x^3 - 1$

30. $g(x) = 4x^3 + 5x^2 - 2x$

31. $g(x) = \dfrac{7}{20 - 8x}$

32. $g(x) = \dfrac{2x - 3}{6x - 12}$

33. $g(x) = |x + 7|$

34. $g(x) = |x| + 1$

35. $g(x) = \dfrac{-2}{|4x + 5|}$

36. $g(x) = \dfrac{x^2 + 2x}{|10x - 20|}$

37. For the function f whose graph is shown below, find $f(-1)$, $f(0)$, and $f(1)$.

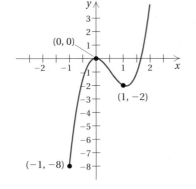

38. For the function g whose graph is shown below, find all the x-values for which $g(x) = 1$.

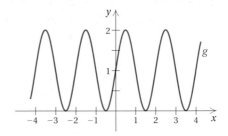

39. $\mathbf{D_W}$ Explain the difference between the domain and the range of a function.

40. $\mathbf{D_W}$ For a given function f, it is known that $f(2) = -3$. Give as many interpretations of this fact as you can.

Solve. [1.4d]

41. On a new job, Anthony can be paid in one of two ways:

    Plan A: A salary of $800 per month, plus a commission of 5% of sales;

    Plan B: A salary of $1000 per month, plus a commission of 7% of sales in excess of $15,000.

For what amount of monthly sales is plan B better than plan A, if we assume that sales are always more than $15,000?

42. *Test Score.* You are taking a business course for which there will be 4 tests, each worth 100 points. You have scores of 92, 90, and 88 on the first three tests. You must make a total of at least 360 points in order to get an A. What scores on the fourth test will keep you eligible for an A?

Solve. [1.6c, d]

43. $|x| = 8$

44. $|x| = -8$

45. $|x - 7| = 11$

46. $|2x + 3| = 13$

47. $|3x - 4| = |x + 2|$

48. $|5x - 6| = |3 - 8x|$

49. $|3x - 8| = -11$

50. $|3x - 8| = 0$

51. Determine the range of each of the functions in Exercises 13, 18, 21, and 22.

52. Determine the range of each of the functions in Exercises 26, 27, 28, and 34.

Objectives

a Find the *y*-intercept of a line from the equation $y = mx + b$ or $f(x) = mx + b$.

b Given two points on a line, find the slope; given a linear equation, derive the equivalent slope–intercept equation and determine the slope and the *y*-intercept.

c Solve applied problems involving slope.

We now turn our attention to functions whose graphs are straight lines. Such functions are called **linear** and can be written in the form $f(x) = mx + b$.

LINEAR FUNCTION

A **linear function** f is any function that can be described by $f(x) = mx + b$.

Compare the two equations $7y + 2x = 11$ and $y = 3x + 5$. Both are linear equations because their graphs are straight lines. Each can be expressed in an equivalent form that is a linear function.

The equation $y = 3x + 5$ can be expressed as $f(x) = mx + b$, where $m = 3$ and $b = 5$.

The equation $7y + 2x = 11$ also has an equivalent form $f(x) = mx + b$. To see this, we solve for y:

$$7y + 2x = 11$$
$$7y + 2x - 2x = -2x + 11 \qquad \text{Subtracting } 2x$$
$$7y = -2x + 11$$
$$\frac{7y}{7} = \frac{-2x + 11}{7} \qquad \text{Dividing by 7}$$
$$y = -\frac{2}{7}x + \frac{11}{7}. \qquad \text{Simplifying}$$

(It might be helpful to review the discussion on solving formulas in Section 1.2.) We now have an equivalent equation in the form

$$f(x) = -\frac{2}{7}x + \frac{11}{7}, \qquad \text{where} \quad m = -\frac{2}{7} \quad \text{and} \quad b = \frac{11}{7}.$$

In this section, we consider the effects of the constants m and b on the graphs of linear functions.

a The Constant b: The y-Intercept

Let's first explore the effect of the constant b.

EXAMPLE 1 Graph $y = 2x$ and $y = 2x + 3$ using the same set of axes. Compare the graphs.

We first make a table of solutions of both equations.

	y	y
x	$y = 2x$	$y = 2x + 3$
0	0	3
1	2	5
−1	−2	1
2	4	7
−2	−4	−1

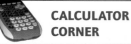

CALCULATOR CORNER

Exploring *b* We can use a graphing calculator to explore the effect of the constant *b* on the graph of a function of the form $f(x) = mx + b$. Graph $y_1 = x$ in the standard $[-10, 10, -10, 10]$ viewing window. Then graph $y_2 = x + 4$, followed by $y_3 = x - 3$, in the same viewing window.

1. Compare the graph of y_2 with the graph of y_1.
2. Compare the graph of y_3 with the graph of y_1.
3. Visualize the graph of $y = x + 8$. Compare it with the graph of y_1.
4. Visualize the graph of $y = x - 5$. Compare it with the graph of y_1.

Next, we plot these points. Drawing a red line for $y = 2x$ and a blue line for $y = 2x + 3$, we note that the graph of $y = 2x + 3$ is simply the graph of $y = 2x$ shifted, or *translated*, up 3 units. The lines are parallel.

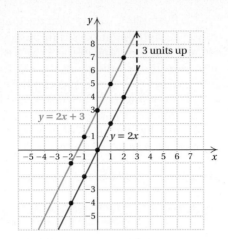

3 units up

$y = 2x + 3$

$y = 2x$

Do Exercises 1 and 2.

EXAMPLE 2 Graph $f(x) = \frac{1}{3}x$ and $g(x) = \frac{1}{3}x - 2$ using the same set of axes. Compare the graphs.

We first make a table of solutions of both equations. By choosing multiples of 3, we can avoid fractions.

	$f(x)$	$g(x)$
x	$f(x) = \frac{1}{3}x$	$g(x) = \frac{1}{3}x - 2$
0	0	-2
3	1	-1
-3	-1	-3
6	2	0

We then plot these points. Drawing a red line for $f(x) = \frac{1}{3}x$ and a blue line for $g(x) = \frac{1}{3}x - 2$, we see that the graph of $g(x) = \frac{1}{3}x - 2$ is simply the graph of $f(x) = \frac{1}{3}x$ shifted, or translated, down 2 units. The lines are parallel.

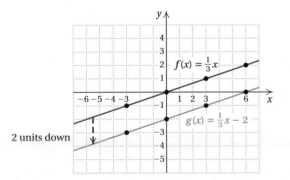

$f(x) = \frac{1}{3}x$

$g(x) = \frac{1}{3}x - 2$

2 units down

1. Graph $y = 3x$ and $y = 3x - 6$ using the same set of axes. Compare the graphs.

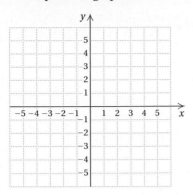

2. Graph $y = -2x$ and $y = -2x + 3$ using the same set of axes. Compare the graphs.

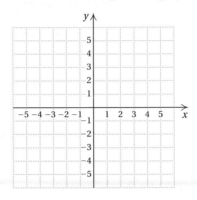

3. Graph $f(x) = \frac{1}{3}x$ and $g(x) = \frac{1}{3}x + 2$ using the same set of axes. Compare the graphs.

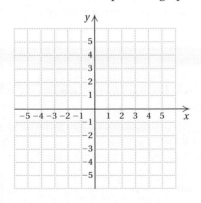

Answers on page A-12

Find the *y*-intercept.

4. $y = 7x + 8$

5. $f(x) = -6x - \frac{2}{3}$

Answers on page A-12

Study Tips

WRITING ALL THE STEPS

Take the time to include all the steps when working your homework problems. Doing so will help you organize your thinking and avoid computational errors. If you find a wrong answer, having all the steps allows easier checking of your work. It will also give you complete, step-by-step solutions of the exercises that can be used to study for an exam.

Writing down all the steps and keeping your work organized may also give you a better chance of getting partial credit.

"Success comes before work only in the dictionary."

Anonymous

In Example 1, we saw that the graph of $y = 2x + 3$ is parallel to the graph of $y = 2x$ and that it passes through the point $(0, 3)$. Similarly, in Example 2, we saw that the graph of $y = \frac{1}{3}x - 2$ is parallel to the graph of $y = \frac{1}{3}x$ and that it passes through the point $(0, -2)$. In general, the graph of $y = mx + b$ is a line parallel to $y = mx$, passing through the point $(0, b)$. The point $(0, b)$ is called the **y-intercept** because it is the point at which the graph crosses the *y*-axis. Often it is convenient to refer to the number *b* as the *y*-intercept. The constant *b* has the effect of moving the graph of $y = mx$ up or down $|b|$ units to obtain the graph of $y = mx + b$.

Do Exercise 3 on the preceding page.

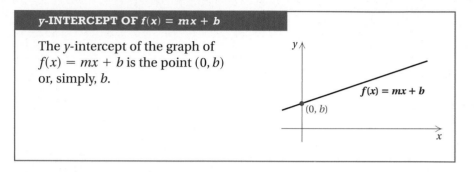

y-INTERCEPT OF $f(x) = mx + b$

The *y*-intercept of the graph of $f(x) = mx + b$ is the point $(0, b)$ or, simply, *b*.

EXAMPLE 3   Find the *y*-intercept: $y = -5x + 4$.

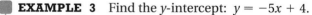

$y = -5x + 4$ $(0, 4)$ is the *y*-intercept.

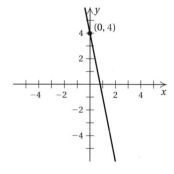

EXAMPLE 4   Find the *y*-intercept: $f(x) = 6.3x - 7.8$.

$f(x) = 6.3x - 7.8$ $(0, -7.8)$ is the *y*-intercept.

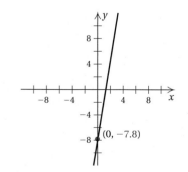

Do Exercises 4 and 5.

b The Constant *m*: Slope

Look again at the graphs in Examples 1 and 2. Note that the slant of each red line seems to match the slant of each blue line. This leads us to believe that the number *m* in the equation $y = mx + b$ is related to the slant of the line. Let's consider some examples.

Graphs with *m* < 0:

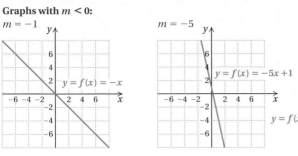

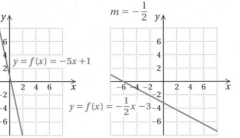

Graphs with *m* = 0:

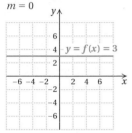

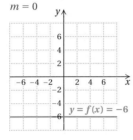

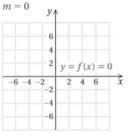

Graphs with *m* > 0:

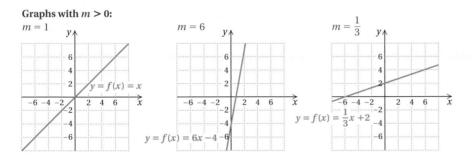

Note that

$m < 0 \longrightarrow$ The graph slants down from left to right;

$m = 0 \longrightarrow$ the graph is horizontal; and

$m > 0 \longrightarrow$ the graph slants up from left to right.

The following definition enables us to visualize the slant and attach a number, a geometric ratio, or *slope*, to the line.

SLOPE

The **slope** of a line containing points (x_1, y_1) and (x_2, y_2) is given by

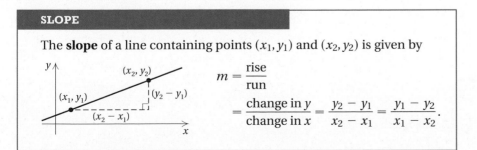

$$m = \frac{\text{rise}}{\text{run}}$$

$$= \frac{\text{change in } y}{\text{change in } x} = \frac{y_2 - y_1}{x_2 - x_1} = \frac{y_1 - y_2}{x_1 - x_2}.$$

Consider a line with two points marked P_1 and P_2, as follows. As we move from P_1 to P_2, the y-coordinate changes from 1 to 3 and the x-coordinate changes from 2 to 7. The change in y is $3 - 1$, or 2. The change in x is $7 - 2$, or 5.

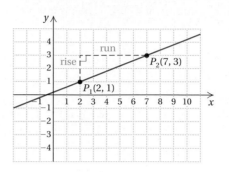

We call the change in y the **rise** and the change in x the **run**. The ratio rise/run is the same for any two points on a line. We call this ratio the **slope**. Slope describes the slant of a line. The slope of the line in the graph above is given by

$$\frac{\text{rise}}{\text{run}}, \quad \text{or} \quad \frac{\text{change in } y}{\text{change in } x}, \quad \text{or} \quad \frac{2}{5}.$$

Whenever x increases by 5 units, y increases by 2 units. Equivalently, whenever x increases by 1 unit, y increases by $\frac{2}{5}$ unit.

EXAMPLE 5 Graph the line containing the points $(-4, 3)$ and $(2, -5)$ and find the slope.

The graph is shown below. Going from $(-4, 3)$ to $(2, -5)$, we see that the change in y, or the rise, is $-5 - 3$, or -8. The change in x, or the run, is $2 - (-4)$, or 6.

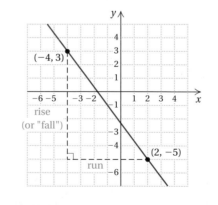

$$\begin{aligned}
\text{Slope} &= \frac{\text{rise}}{\text{run}} = \frac{\text{change in } y}{\text{change in } x} \\
&= \frac{-5 - 3}{2 - (-4)} \\
&= \frac{-8}{6} = -\frac{8}{6}, \text{ or } -\frac{4}{3}
\end{aligned}$$

CALCULATOR CORNER

Visualizing Slope For Exercises 1–4, use the window settings $[-6, 6, -4, 4]$, with $Xscl = 1$ and $Yscl = 1$.

1. Graph $y = x$, $y = 2x$, and $y = 5x$ in the same window. What do you think the graph of $y = 10x$ will look like?

2. Graph $y = x$, $y = \frac{1}{2}x$, and $y = 0.1x$ in the same window. What do you think the graph of $y = 0.005x$ will look like?

3. Graph $y = -x$, $y = -2x$, and $y = -5x$ in the same window. What do you think the graph of $y = -10x$ will look like?

4. Graph $y = -x$, $y = -\frac{1}{2}x$, and $y = -0.1x$ in the same window. What do you think the graph of $y = -0.005x$ will look like?

The formula

$$m = \frac{y_2 - y_1}{x_2 - x_1} = \frac{y_1 - y_2}{x_1 - x_2}$$

tells us that we can subtract in two ways. We must remember, however, to subtract the x-coordinates in the same order that we subtract the y-coordinates.

Let's do Example 5 again:

$$\text{Slope} = \frac{\text{change in } y}{\text{change in } x} = \frac{3 - (-5)}{-4 - 2} = \frac{8}{-6} = -\frac{8}{6} = -\frac{4}{3}.$$

We see that both ways give the same value for the slope.

The slope of a line tells how it slants. A line with positive slope slants up from left to right. The larger the positive number, the steeper the slant. A line with negative slope slants downward from left to right. The smaller the negative number, the steeper the line.

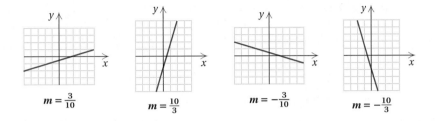

$$m = \frac{3}{10} \qquad m = \frac{10}{3} \qquad m = -\frac{3}{10} \qquad m = -\frac{10}{3}$$

Do Exercises 6 and 7.

How can we find the slope from a given equation? Let's consider the equation $y = 2x + 3$, which is in the form $y = mx + b$. We can find two points by choosing convenient values for x—say 0 and 1—and substituting to find the corresponding y-values.

If $x = 0$, $y = 2 \cdot 0 + 3 = 3$.

If $x = 1$, $y = 2 \cdot 1 + 3 = 5$.

We find two points on the line to be

$(0, 3)$ and $(1, 5)$.

The slope of the line is found as follows, using the definition of slope:

$$m = \frac{\text{change in } y}{\text{change in } x}$$
$$= \frac{5 - 3}{1 - 0} = \frac{2}{1} = 2.$$

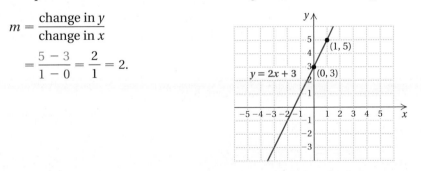

The slope is 2. Note that this is the coefficient of the x-term in the equation $y = 2x + 3$.

If we had chosen different points on the line—say $(-2, -1)$ and $(4, 11)$—the slope would still be 2, as we see in the following calculation:

$$m = \frac{11 - (-1)}{4 - (-2)} = \frac{11 + 1}{4 + 2} = \frac{12}{6} = 2.$$

Do Exercise 8 on the following page.

Graph the line through the given points and find its slope.

6. $(-1, -1)$ and $(2, -4)$

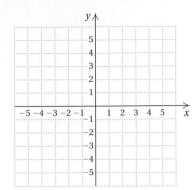

7. $(0, 2)$ and $(3, 1)$

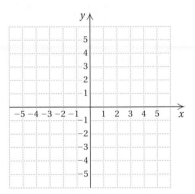

Answers on page A-12

8. Find the slope of the line $f(x) = -\frac{2}{3}x + 1$. Use the points $(9, -5)$ and $(3, -1)$.

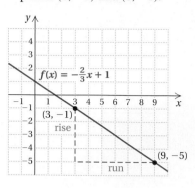

Find the slope and the y-intercept.

9. $f(x) = -8x + 23$

10. $5x - 10y = 25$

We see that the slope of the line $y = mx + b$ is indeed the constant m, the coefficient of x.

SLOPE OF $y = mx + b$

The **slope** of the line $y = mx + b$ is m.

From a linear equation in the form $y = mx + b$, we can read the slope and the y-intercept of the graph directly.

SLOPE–INTERCEPT EQUATION

The equation $y = mx + b$ is called the **slope–intercept equation.** The slope is m and the y-intercept is $(0, b)$.

Note that any graph of an equation $y = mx + b$ passes the vertical-line test and thus represents a function.

EXAMPLE 6 Find the slope and the y-intercept of $y = 5x - 4$.

Since the equation is already in the form $y = mx + b$, we simply read the slope and the y-intercept from the equation:

$$y = 5x - 4.$$

The slope is 5.   The y-intercept is $(0, -4)$.

EXAMPLE 7 Find the slope and the y-intercept of $2x + 3y = 8$.

We first solve for y so we can easily read the slope and the y-intercept:

$$2x + 3y = 8$$
$$3y = -2x + 8 \qquad \text{Subtracting } 2x$$
$$\frac{3y}{3} = \frac{-2x + 8}{3} \qquad \text{Dividing by 3}$$
$$y = -\frac{2}{3}x + \frac{8}{3}. \qquad \text{Finding the form } y = mx + b$$

The slope is $-\frac{2}{3}$.   The y-intercept is $\left(0, \frac{8}{3}\right)$.

Do Exercises 9 and 10.

C   Applications

Slope has many real-world applications. For example, numbers like 2%, 3%, and 6% are often used to represent the *grade* of a road, a measure of how steep a road on a hill or mountain is. A 3% grade $\left(3\% = \frac{3}{100}\right)$ means that for every horizontal distance of 100 ft that the road runs, the road rises 3 ft, and a -3% grade means that for every horizontal distance of 100 ft, the road drops 3 ft. (Normally, the road signs do not include negative signs, since it is obvious whether you are climbing or descending.) An athlete might change the grade of a treadmill during a workout. An escape ramp on an airliner might have a slope of about −0.6.

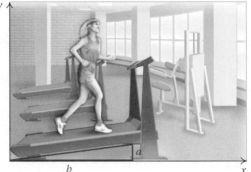

Architects and carpenters use slope when designing and building stairs, ramps, or roof pitches. Another application occurs in hydrology. The strength or force of a river depends on how far the river falls vertically compared to how far it flows horizontally. Slope can also be considered as a **rate of change.**

EXAMPLE 8 *Retail Trade.* Sales at warehouse clubs and superstores have increased steadily in recent years. In 1998, sales of $98.6 billion were registered at these types of stores. By 2003, sales had risen to $217.5 billion. Find the rate of change of sales with respect to time, in years.

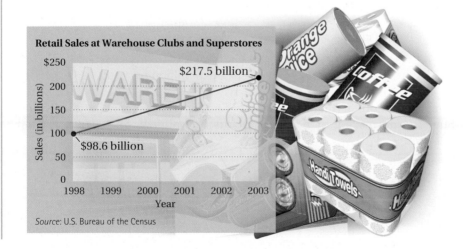

11. Haircutting. The graph below displays data from a day's work at Lee's Barbershop. At 2:00, 8 haircuts had been completed. At 4:00, 14 haircuts had been done. Use the graph to determine the rate of change in the number of haircuts with respect to time.

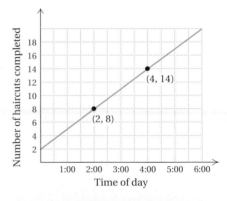

Answer on page A-12

2.4   Linear Functions: Graphs and Slope

12. Mishandled Baggage. Reports of mishandled baggage filed with major U.S. airlines have declined slightly in recent years. The graph below shows the number of reports filed per 1000 passengers for several years. Find the rate of change of the number of reports filed per 1000 passengers.

Reports of Mishandled Baggage

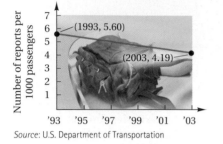

Source: U.S. Department of Transportation

The rate of change of sales with respect to time, in years, is given by

$$\text{Rate of change} = \frac{\$217.5 \text{ billion} - \$98.6 \text{ billion}}{2003 - 1998}$$

$$= \frac{\$118.9 \text{ billion}}{5 \text{ yr}}$$

$$= \$23.78 \text{ billion per year.}$$

Retail sales at warehouse clubs and superstores are increasing at a rate of about $23.78 billion per year.

Do Exercise 11 on the preceding page.

EXAMPLE 9 *Sunday Newspapers.* The circulation of English-language Sunday newspapers in the United States has declined gradually in recent years, as shown by the graph below. Use the graph to find the rate at which this number is decreasing. (Note that the jagged break in the vertical axis is used to avoid including a large portion of unused grid.)

Circulation of English-Language Sunday Newspapers

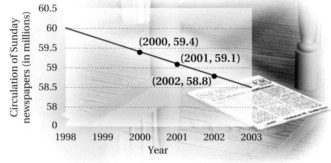

Source: Editor & Publisher International Yearbook

Since the graph is linear, we can use any pair of points to determine the rate of change. We choose the points (2000, 59.4 million) and (2002, 58.8 million), which gives us

$$\text{Rate of change} = \frac{58.8 \text{ million} - 59.4 \text{ million}}{2002 - 2000}$$

$$= \frac{-0.6 \text{ million}}{2 \text{ yr}} = -0.3 \text{ million per year.}$$

The circulation of English-language Sunday newspapers in the United States is decreasing at a rate of about 0.3 million, or 300,000, readers per year.

Do Exercise 12.

Answer on page A-12

, **b**   Find the slope and the *y*-intercept.

1. $y = 4x + 5$

2. $y = -5x + 10$

3. $f(x) = -2x - 6$

4. $g(x) = -5x + 7$

5. $y = -\frac{3}{8}x - \frac{1}{5}$

6. $y = \frac{15}{7}x + \frac{16}{5}$

7. $g(x) = 0.5x - 9$

8. $f(x) = -3.1x + 5$

9. $2x - 3y = 8$

10. $-8x - 7y = 24$

11. $9x = 3y + 6$

12. $9y + 36 - 4x = 0$

13. $3 - \frac{1}{4}y = 2x$

14. $5x = \frac{2}{3}y - 10$

15. $17y + 4x + 3 = 7 + 4x$

16. $3y - 2x = 5 + 9y - 2x$

b   Find the slope of the line.

17.   **18.**   **19.**   **20.**

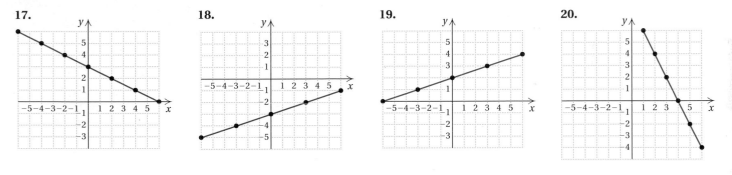

Find the slope of the line containing the given pair of points.

21. $(6, 9)$ and $(4, 5)$

22. $(8, 7)$ and $(2, -1)$

23. $(9, -4)$ and $(3, -8)$

24. $(17, -12)$ and $(-9, -15)$

25. $(-16.3, 12.4)$ and $(-5.2, 8.7)$

26. $(14.4, -7.8)$ and $(-12.5, -17.6)$

c   Find the slope (or rate of change).

27. Find the slope (or grade) of the treadmill.

28. Find the slope (or pitch) of the roof.

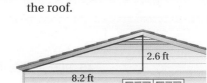

2.6 ft

8.2 ft

29. Find the slope (or head) of the river.

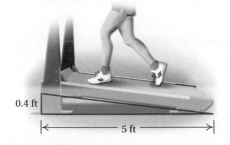

0.4 ft

5 ft

43.33 ft

1238 ft

30. Public buildings regularly include steps with 7-in. risers and 11-in. treads. Find the grade of such a stairway.

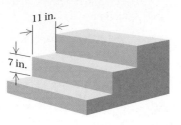

31. *E-mail.* Find the rate of change of the average daily volume of e-mail messages, in billions, in North America, with respect to time, in years.

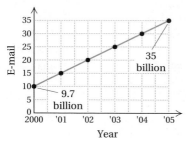

Source: Pitney Bowes Telephone Survey

32. Find the rate of change of the cost of a formal wedding.

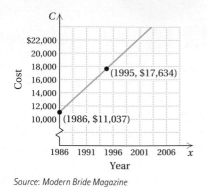

Source: *Modern Bride Magazine*

Find the rate of change.

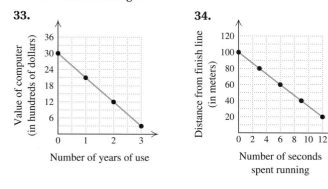

35.

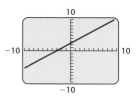

Source: College Board Online

36.

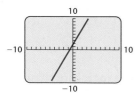

Source: College Board Online

37. **D**_W A student makes a mistake when using a graphing calculator to draw $4x + 5y = 12$ and the following screen appears. Use algebra to show that a mistake has been made. What do you think the mistake was?

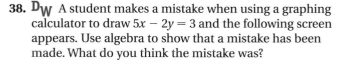

38. **D**_W A student makes a mistake when using a graphing calculator to draw $5x - 2y = 3$ and the following screen appears. Use algebra to show that a mistake has been made. What do you think the mistake was?

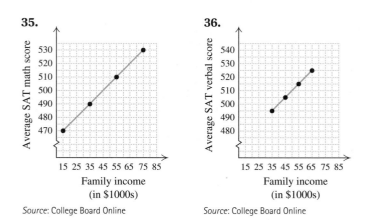

SKILL MAINTENANCE

Simplify. [R.3c], [R.6b]

39. $3^2 - 24 \cdot 56 + 144 \div 12$

40. $9\{2x - 3[5x + 2(-3x + y^0 - 2)]\}$

41. $10\{2x + 3[5x - 2(-3x + y^1 - 2)]\}$

42. $5^4 \div 625 \div 5^2 \cdot 5^7 \div 5^3$

Solve. [1.3a]

43. One side of a square is 5 yd less than a side of an equilateral triangle. If the perimeter of the square is the same as the perimeter of the triangle, what is the length of a side of the square? of the triangle?

Solve. [1.6c, e]

44. $|5x - 8| \geq 32$

45. $|5x - 8| < 32$

46. $|5x - 8| = 32$

47. $|5x - 8| = -32$

2.5 MORE ON GRAPHING LINEAR EQUATIONS

Objectives

a Graph linear equations using intercepts.

b Given a linear equation in slope–intercept form, use the slope and the y-intercept to graph the line.

c Graph linear equations of the form $x = a$ or $y = b$.

d Given the equations of two lines, determine whether their graphs are parallel or whether they are perpendicular.

a Graphing Using Intercepts

The **x-intercept** of the graph of a linear equation or function is the point at which the graph crosses the x-axis. The **y-intercept** is the point at which the graph crosses the y-axis. We know from geometry that only one line can be drawn through two given points. Thus, if we know the intercepts, we can graph the line. To ensure that a computation error has not been made, it is a good idea to calculate a third point as a check.

Many equations of the type $Ax + By = C$ can be graphed conveniently using intercepts.

---

x- AND y-INTERCEPTS

A **y-intercept** is a point $(0, b)$. To find b, let $x = 0$ and solve for y.

An **x-intercept** is a point $(a, 0)$. To find a, let $y = 0$ and solve for x.

EXAMPLE 1   Find the intercepts of $3x + 2y = 12$ and then graph the line.

y-intercept: To find the y-intercept, we let $x = 0$ and solve for y:

$$3x + 2y = 12$$
$$3 \cdot 0 + 2y = 12 \qquad \text{Substituting 0 for } x$$
$$2y = 12$$
$$y = 6.$$

The y-intercept is $(0, 6)$.

x-intercept: To find the x-intercept, we let $y = 0$ and solve for x:

$$3x + 2y = 12$$
$$3x + 2 \cdot 0 = 12 \qquad \text{Substituting 0 for } y$$
$$3x = 12$$
$$x = 4.$$

The x-intercept is $(4, 0)$.

We plot these points and draw the line, using a third point as a check. We choose $x = 6$ and solve for y:

$$3(6) + 2y = 12$$
$$18 + 2y = 12$$
$$2y = -6$$
$$y = -3.$$

We plot $(6, -3)$ and note that it is on the line.

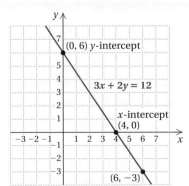

1. Find the intercepts of $4y - 12 = -6x$ and then graph the line.

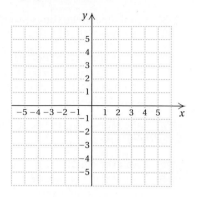

When both the x- and the y-intercepts are $(0, 0)$, as is the case with an equation such as $y = 2x$, whose graph passes through the origin, another point would have to be calculated and a third point used as a check.

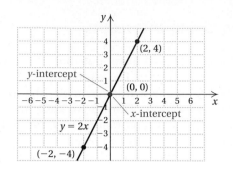

Do Exercise 1.

Answer on page A-13

CALCULATOR CORNER

Viewing the Intercepts   Knowing the intercepts of a linear equation helps us determine a good viewing window for the graph of the equation. For example, when we graph the equation $y = -x + 15$ in the standard window, we see only a small portion of the graph in the upper right-hand corner of the screen, as shown on the left below.

$y = -x + 15$

$y = -x + 15$

Xscl = 5   Yscl = 5

Using algebra, as we did in Example 1, we can find that the intercepts of the graph of this equation are $(0, 15)$ and $(15, 0)$. This tells us that, if we are to see more of the graph than is shown above, both Xmax and Ymax should be greater than 15. We can try different window settings until we find one that suits us. One good choice, shown on the right above, is $[-25, 25, -25, 25]$, with Xscl = 5 and Yscl = 5.

Exercises: Find the intercepts of the equation algebraically. Then graph the equation on a graphing calculator, choosing window settings that allow the intercepts to be seen clearly. (Settings may vary.)

1. $y = -3.2x - 16$

2. $y - 4.25x = 85$

3. $6x + 5y = 90$

4. $5x - 6y = 30$

5. $8x + 3y = 9$

6. $y = 0.4x - 5$

7. $y = 1.2x - 12$

8. $4x - 5y = 2$

b   Graphing Using the Slope and the *y*-Intercept

We can also graph a line using its slope and *y*-intercept.

EXAMPLE 2   Graph: $y = -\frac{2}{3}x + 1$.

This equation is in slope–intercept form, $y = mx + b$. The *y*-intercept is $(0, 1)$. We plot $(0, 1)$. We can think of the slope $\left(m = -\frac{2}{3}\right)$ as $\frac{-2}{3}$.

$$m = \frac{\text{Rise}}{\text{Run}} = \frac{-2}{3} \qquad \begin{array}{l} \text{Move 2 units down.} \\ \text{Move 3 units right.} \end{array}$$

Starting at the *y*-intercept and using the slope, we find another point by moving 2 units down (since the numerator is *negative* and corresponds to the change in *y*) and 3 units to the right (since the denominator is *positive* and corresponds to the change in *x*). We get to a new point, $(3, -1)$. In a similar manner, we can move from the point $(3, -1)$ to find another point, $(6, -3)$.

We could also think of the slope $\left(m = -\frac{2}{3}\right)$ as $\frac{2}{-3}$.

$$m = \frac{\text{Rise}}{\text{Run}} = \frac{2}{-3} \qquad \begin{array}{l} \text{Move 2 units up.} \\ \text{Move 3 units left.} \end{array}$$

Then we can start again at $(0, 1)$, but this time we move 2 units up (since the numerator is *positive* and corresponds to the change in *y*) and 3 units to the left (since the denominator is *negative* and corresponds to the change in *x*). We get another point on the graph, $(-3, 3)$, and from it we can obtain $(-6, 5)$ and others in a similar manner. We plot the points and draw the line.

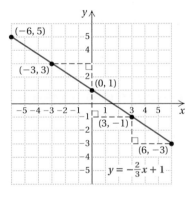

EXAMPLE 3   Graph: $f(x) = \frac{2}{5}x + 4$.

First, we plot the *y*-intercept, $(0, 4)$. We then consider the slope $\frac{2}{5}$. A slope of $\frac{2}{5}$ tells us that, for every 2 units the graph rises, it runs 5 units horizontally in the positive direction, or to the right. Thus, starting at the *y*-intercept and using the slope, we find another point by moving 2 units up (since the numerator is *positive* and corresponds to the change in *y*) and 5 units to the right (since the denominator is *positive* and corresponds to the change in *x*). We get to a new point, $(5, 6)$.

Graph using the slope and the *y*-intercept.

2. $y = \frac{3}{2}x + 1$

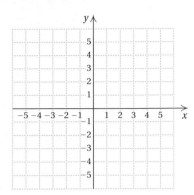

3. $f(x) = \frac{3}{4}x - 2$

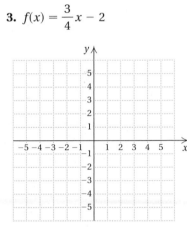

4. $g(x) = -\frac{3}{5}x + 5$

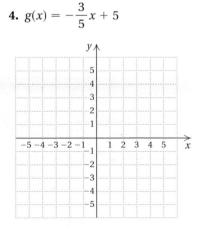

Answers on page A-13

5. $y = -\dfrac{5}{3}x - 4$

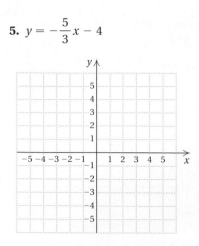

Graph and determine the slope.

6. $f(x) = -4$

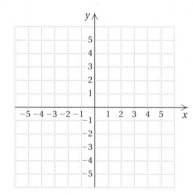

7. $y = 3.6$

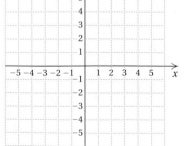

We can also think of the slope $\frac{2}{5}$ as $\frac{-2}{-5}$. A slope of $\frac{-2}{-5}$ tells us that, for every 2 units the graph drops, it runs 5 units horizontally in the negative direction, or to the left. We again start at the y-intercept, $(0, 4)$. We move 2 units down (since the numerator is *negative* and corresponds to the change in y) and 5 units to the left (since the denominator is *negative* and corresponds to the change in x). We get to another new point, $(-5, 2)$. We plot the points and draw the line.

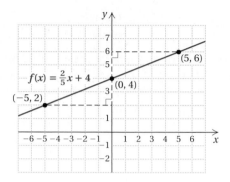

Do Exercises 2–5. (Exercises 2–4 are on the preceding page.)

C Horizontal and Vertical Lines

Some equations have graphs that are parallel to one of the axes. This happens when either A or B is 0 in $Ax + By = C$. These equations have a missing variable; that is, there is only one variable in the equation. In the following example, x is missing.

EXAMPLE 4 Graph: $y = 3$.

Since x is missing, any number for x will do. Thus all ordered pairs $(x, 3)$ are solutions. The graph is a **horizontal line** parallel to the x-axis.

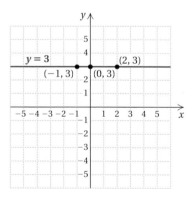

x	y
-1	3
0	3	← y-intercept
2	3

Regardless of x, y must be 3.
Choose *any* number for x.

What about the slope of a horizontal line? In Example 4, consider the points $(-1, 3)$ and $(2, 3)$, which are on the line $y = 3$. The change in y is $3 - 3$, or 0. The change in x is $-1 - 2$, or -3. Thus,

$$m = \frac{3 - 3}{-1 - 2} = \frac{0}{-3} = 0.$$

Any two points on a horizontal line have the same y-coordinate. Thus the change in y is always 0, so the slope is 0.

Answers on page A-13

We can also determine the slope by noting that $y = 3$ can be written in slope–intercept form as $y = 0x + 3$, or $f(x) = 0x + 3$. From this equation, we read that the slope is 0. A function of this type is called a **constant function.** We can express it in the form $y = b$, or $f(x) = b$. Its graph is a horizontal line that crosses the y-axis at $(0, b)$.

Do Exercises 6 and 7 on the preceding page.

In the following example, y is missing and the graph is parallel to the y-axis.

EXAMPLE 5   Graph: $x = -2$.

Since y is missing, any number for y will do. Thus all ordered pairs $(-2, y)$ are solutions. The graph is a **vertical line** parallel to the y-axis.

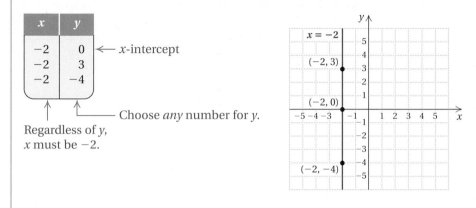

x	y
−2	0	← x-intercept
−2	3
−2	−4

Choose *any* number for y.

Regardless of y, x must be -2.

This graph is not the graph of a function because it fails the vertical-line test. The vertical line itself crosses the graph more than once.

What about the slope of a vertical line? In Example 5, consider the points $(-2, 3)$ and $(-2, -4)$, which are on the line $x = -2$. The change in y is $3 - (-4)$, or 7. The change in x is $-2 - (-2)$, or 0. Thus,

$$m = \frac{3 - (-4)}{-2 - (-2)} = \frac{7}{0}. \qquad \text{Not defined}$$

Since division by 0 is not defined, the slope of this line is not defined. Any two points on a vertical line have the same x-coordinate. Thus the change in x is always 0, so the slope of any vertical line is not defined.

The following summarizes horizontal and vertical lines and their equations.

HORIZONTAL LINE; VERTICAL LINE

The graph of $y = b$, or $f(x) = b$, is a **horizontal line** with y-intercept $(0, b)$. It is the graph of a constant function with slope 0.

The graph of $x = a$ is a **vertical line** through the point $(a, 0)$. The slope is not defined. It is not the graph of a function.

Do Exercises 8–10.

Graph.

8. $x = -5$

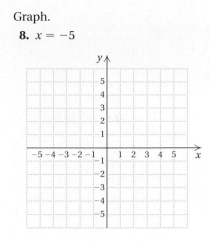

9. $8x - 5 = 19$ (*Hint*: Solve for x.)

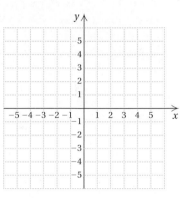

10. Determine, if possible, the slope of each line.

a) $x = -12$

b) $y = 6$

c) $2y + 7 = 11$

d) $x = 0$

e) $y = -\frac{3}{4}$

f) $10 - 5x = 15$

Answers on page A-13

To graph a linear equation:

1. Is the equation of the type $x = a$ or $y = b$? If so, the graph will be a line parallel to an axis; $x = a$ is vertical and $y = b$ is horizontal.
2. If the line is of the type $y = mx$, both intercepts are the origin, $(0, 0)$. Plot $(0, 0)$ and one other point.
3. If the line is of the type $y = mx + b$, plot the y-intercept and one other point.
4. If the equation is of the form $Ax + By = C$, graph using intercepts. If the intercepts are too close together, choose another point farther from the origin.
5. In all cases, use a third point as a check.

We have graphed linear equations in several ways in this chapter. Although, in general, you can use any method that works best for you, we list some guidelines in the margin at left.

Parallel and Perpendicular Lines

PARALLEL LINES

Parallel lines extend indefinitely without intersecting. If two lines are vertical, they are parallel. How can we tell whether nonvertical lines are parallel? We examine their slopes and y-intercepts.

PARALLEL LINES

Two nonvertical lines are **parallel** if they have the *same* slope and *different* y-intercepts.

EXAMPLE 6   Determine whether the graphs of

$$y - 3x = 1 \quad \text{and} \quad 3x + 2y = -2$$

are parallel.

To determine whether lines are parallel, we first find their slopes. To do this, we find the slope–intercept form of each equation by solving for y:

$$y - 3x = 1 \qquad\qquad 3x + 2y = -2$$
$$y = 3x + 1; \qquad\qquad 2y = -3x - 2$$
$$y = \tfrac{1}{2}(-3x - 2)$$
$$= \tfrac{1}{2}(-3x) - \tfrac{1}{2}(2)$$
$$= -\tfrac{3}{2}x - 1.$$

The slopes, 3 and $-\tfrac{3}{2}$, are different. Thus the lines are not parallel, as the graphs at right confirm.

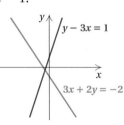

EXAMPLE 7   Determine whether the graphs of

$$3x - y = -5 \quad \text{and} \quad y - 3x = -2$$

are parallel.

We first find the slope–intercept form of each equation by solving for y:

$$3x - y = -5 \qquad\qquad y - 3x = -2$$
$$-y = -3x - 5 \qquad\qquad y = 3x - 2.$$
$$-1(-y) = -1(-3x - 5)$$
$$y = 3x + 5;$$

The slopes, 3, are the same. The y-intercepts, $(0, 5)$ and $(0, -2)$, are different. Thus the lines are parallel, as the graphs appear to confirm.

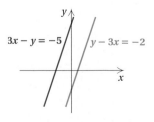

Do Exercises 11–13.

PERPENDICULAR LINES

If one line is vertical and another is horizontal, they are perpendicular. For example, the lines $x = 5$ and $y = -3$ are perpendicular. Otherwise, how can we tell whether two lines are perpendicular?

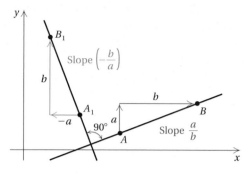

Consider a line $\overleftrightarrow{AB}$, as shown in the figure above, with slope a/b. Then think of rotating the line 90° to get a line $\overleftrightarrow{A_1B_1}$ perpendicular to $\overleftrightarrow{AB}$. For the new line, the rise and the run are interchanged, but the run is now negative. Thus the slope of the new line is $-b/a$, which is the opposite of the reciprocal of the slope of the first line. Also note that when we multiply the slopes, we get

$$\frac{a}{b}\left(\frac{b}{a}\right) = -1.$$

This is the condition under which lines will be perpendicular.

PERPENDICULAR LINES

Two nonvertical lines are **perpendicular** if the product of their slopes is -1. (If one line has slope m, the slope of a line perpendicular to it is $-1/m$. That is, to find the slope of a line perpendicular to a given line, we take the reciprocal of the given slope and change the sign.)

Lines are also perpendicular if one of them is vertical ($x = a$) and one of them is horizontal ($y = b$).

EXAMPLE 8 Determine whether the graphs of $5y = 4x + 10$ and $4y = -5x + 4$ are perpendicular.

To determine whether the lines are perpendicular, we determine whether the product of their slopes is -1. We first find the slope–intercept form of each equation by solving for y.

We have

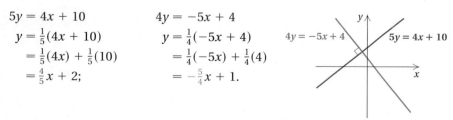

$5y = 4x + 10$

$y = \frac{1}{5}(4x + 10)$

$= \frac{1}{5}(4x) + \frac{1}{5}(10)$

$= \frac{4}{5}x + 2;$

$4y = -5x + 4$

$y = \frac{1}{4}(-5x + 4)$

$= \frac{1}{4}(-5x) + \frac{1}{4}(4)$

$= -\frac{5}{4}x + 1.$

The slope of the first line is $\frac{4}{5}$, and the slope of the second line is $-\frac{5}{4}$. The product of the slopes is $\frac{4}{5} \cdot \left(-\frac{5}{4}\right) = -1$. Thus the lines are perpendicular.

Do Exercises 14 and 15.

Determine whether the graphs of the given pair of lines are parallel.

11. $x + 4 = y,$
$y - x = -3$

12. $y + 4 = 3x,$
$4x - y = -7$

13. $y = 4x + 5,$
$2y = 8x + 10$

Determine whether the graphs of the given pair of lines are perpendicular.

14. $2y - x = 2,$
$y + 2x = 4$

15. $3y = 2x + 15,$
$2y = 3x + 10$

Answers on page A-13

Squaring the Viewing Window; Visualizing Perpendicular Lines Consider the standard $[-10, 10, -10, 10]$ viewing window on the left below. Note that the distance between tick marks is not the same on both axes. In fact, the distance between tick marks on the y-axis is about two-thirds the distance between tick marks on the x-axis. If we change the dimensions of the window to $[-6, 6, -4, 4]$, we get a window in which the distance between tick marks is about the same on both axes, as shown in the window on the right below. Such a window is called a **square viewing window.** In a square window, the length of the portion of the y-axis shown is two-thirds the length of the portion of the x-axis shown. One way to obtain a square window, then, is to choose window dimensions in which the length of the y-axis is two-thirds the length of the x-axis. Other examples of square windows are $[-3, 3, -2, 2]$ and $[-9, 9, -6, 6]$.

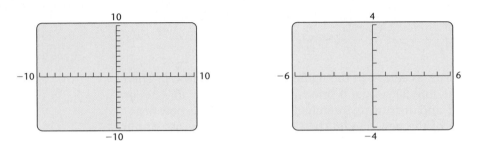

When the viewing window is square, we get an accurate representation of the slope of a line. This is important when we want to visualize perpendicular lines. The graphs of the perpendicular lines $y = x + 1$ and $y = -x + 2$ are shown in the standard window on the left below and in the square window $[-6, 6, -4, 4]$ on the right. Note that the lines do not appear to be perpendicular in the standard window, whereas they do in the square window.

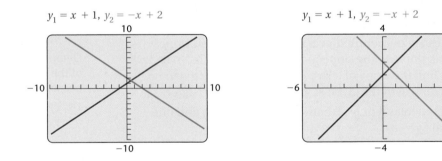

Exercise:

1. Graph each pair of equations in Margin Exercises 14 and 15 using a square viewing window. Note that each equation must be solved for y before it is entered on the equation-editor screen. Check visually whether the lines appear to be perpendicular.

Visualizing for Success

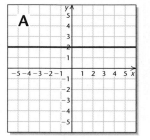

A

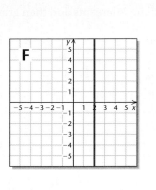

F

Match each equation with its graph.

1. $y = 2 - x$

2. $x - y = 2$

3. $x + 2y = 2$

4. $2x - 3y = 6$

5. $x = 2$

6. $y = 2$

7. $y = |x + 2|$

8. $y = |x| + 2$

9. $y = x^2 - 2$

10. $y = 2 - x^2$

Answers on page A-14

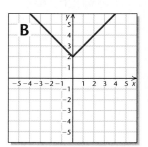

B

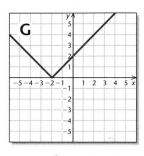

G

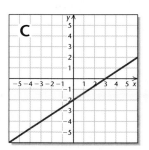

C

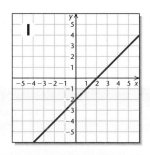

H

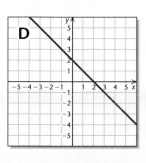

D

I

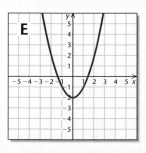

E

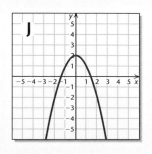

J

MathXL   MyMathLab   InterAct Math   Math Tutor Center   Digital Video Tutor CD 1 Videotape 3   Student's Solutions Manual

Find the intercepts and then graph the line.

1. $x - 2 = y$

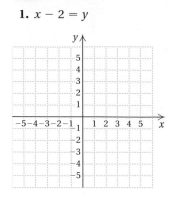

2. $x + 3 = y$

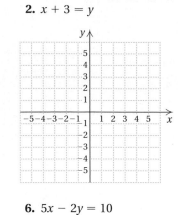

3. $x + 3y = 6$

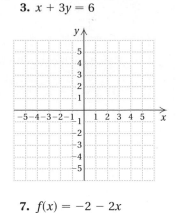

4. $x - 2y = 4$

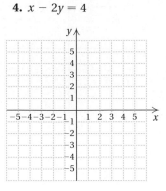

5. $2x + 3y = 6$

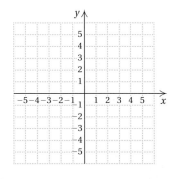

6. $5x - 2y = 10$

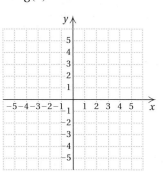

7. $f(x) = -2 - 2x$

8. $g(x) = 5x - 5$

9. $5y = -15 + 3x$

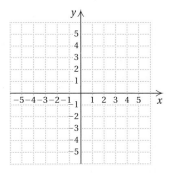

10. $5x - 10 = 5y$

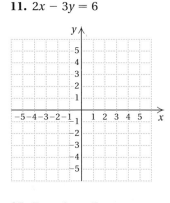

11. $2x - 3y = 6$

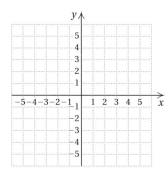

12. $4x + 5y = 20$

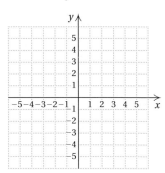

13. $2.8y - 3.5x = -9.8$

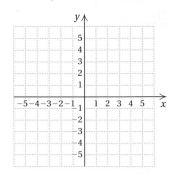

14. $10.8x - 22.68 = 4.2y$

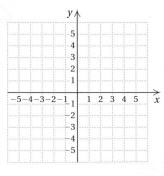

15. $5x + 2y = 7$

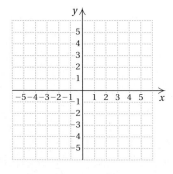

16. $3x - 4y = 10$

Graph using the slope and the y-intercept.

17. $y = \dfrac{5}{2}x + 1$

18. $y = \dfrac{2}{5}x - 4$

19. $f(x) = -\dfrac{5}{2}x - 4$

20. $f(x) = \dfrac{2}{5}x + 3$

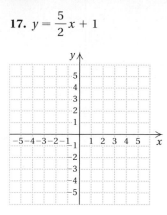
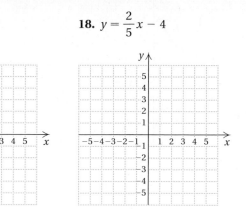
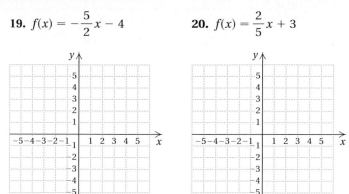

21. $x + 2y = 4$

22. $x - 3y = 6$

23. $4x - 3y = 12$

24. $2x + 6y = 12$

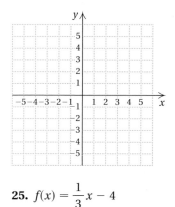
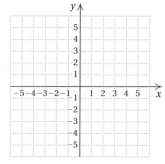
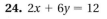
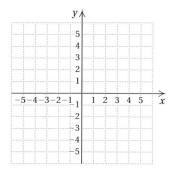

25. $f(x) = \dfrac{1}{3}x - 4$

26. $g(x) = -0.25x + 2$

27. $5x + 4 \cdot f(x) = 4$
(*Hint*: Solve for $f(x)$.)

28. $3 \cdot f(x) = 4x + 6$

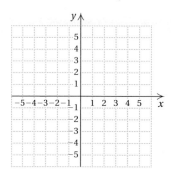
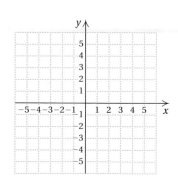

C Graph and, if possible, determine the slope.

29. $x = 1$

30. $x = -4$

31. $y = -1$

32. $y = \dfrac{3}{2}$

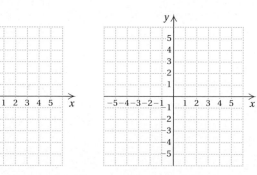

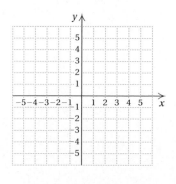
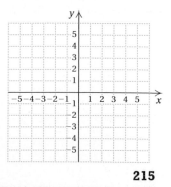

33. $f(x) = -6$

34. $f(x) = 2$

35. $y = 0$

36. $x = 0$

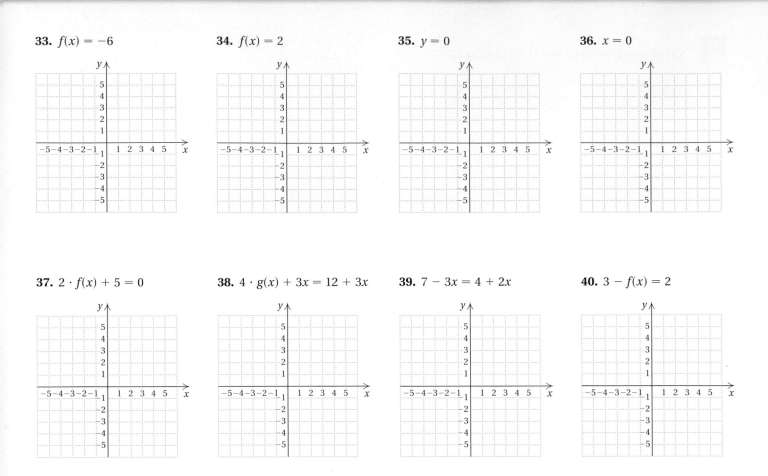

37. $2 \cdot f(x) + 5 = 0$

38. $4 \cdot g(x) + 3x = 12 + 3x$

39. $7 - 3x = 4 + 2x$

40. $3 - f(x) = 2$

d   Determine whether the graphs of the given pair of lines are parallel.

41. $x + 6 = y$,
   $y - x = -2$

42. $2x - 7 = y$,
   $y - 2x = 8$

43. $y + 3 = 5x$,
   $3x - y = -2$

44. $y + 8 = -6x$,
   $-2x + y = 5$

45. $y = 3x + 9$,
   $2y = 6x - 2$

46. $y + 7x = -9$,
   $-3y = 21x + 7$

47. $12x = 3$,
   $-7x = 10$

48. $5y = -2$,
   $\frac{3}{4}x = 16$

Determine whether the graphs of the given pair of lines are perpendicular.

49. $y = 4x - 5$,
   $4y = 8 - x$

50. $2x - 5y = -3$,
   $2x + 5y = 4$

51. $x + 2y = 5$,
   $2x + 4y = 8$

52. $y = -x + 7$,
   $y = x + 3$

53. $2x - 3y = 7$,
   $2y - 3x = 10$

54. $x = y$,
   $y = -x$

55. $2x = 3$,
   $-3y = 6$

56. $-5y = 10$,
   $y = -\frac{4}{9}$

57. D_W Under what conditions will the x- and the y-intercepts of a line be the same? What would the equation for such a line look like?

58. D_W Explain why the slope of a vertical line is not defined but the slope of a horizontal line is 0.

SKILL MAINTENANCE

Write in scientific notation. [R.7c]

59. 53,000,000,000

60. 0.000047

61. 0.018

62. 99,902,000

Write in decimal notation. [R.7c]

63. 2.13×10^{-5}

64. 9.01×10^{8}

65. 2×10^{4}

66. 8.5677×10^{-2}

Factor. [R.5d]

67. $9x - 15y$

68. $12a + 21ab$

69. $21p - 7pq + 14p$

70. $64x - 128y + 256$

SYNTHESIS

71. Find an equation of a horizontal line that passes through the point $(-2, 3)$.

72. Find an equation of a vertical line that passes through the point $(-2, 3)$.

73. Find the value of a such that the graphs of $5y = ax + 5$ and $\frac{1}{4}y = \frac{1}{10}x - 1$ are parallel.

74. Find the value of k such that the graphs of $x + 7y = 70$ and $y + 3 = kx$ are perpendicular.

75. Write an equation of the line that has x-intercept $(-3, 0)$ and y-intercept $\left(0, \frac{2}{5}\right)$.

76. Find the coordinates of the point of intersection of the graphs of the equations $x = -4$ and $y = 5$.

77. Write an equation for the x-axis. Is this equation a function?

78. Write an equation for the y-axis. Is this equation a function?

79. Find the value of m in $y = mx + 3$ so that the x-intercept of its graph will be $(4, 0)$.

80. Find the value of b in $2y = -7x + 3b$ so that the y-intercept of its graph will be $(0, -13)$.

81. Match each sentence with the most appropriate graph.

   a) The rate at which fluids were given intravenously was doubled after 3 hr.

   b) The rate at which fluids were given intravenously was gradually reduced to 0.

   c) The rate at which fluids were given intravenously remained constant for 5 hr.

   d) The rate at which fluids were given intravenously was gradually increased.

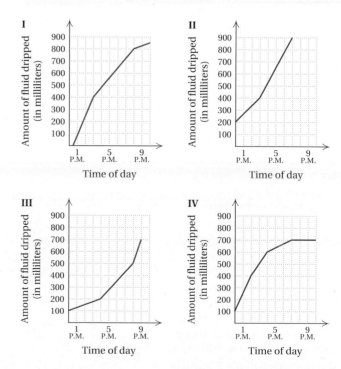

217

Objectives

a Find an equation of a line when the slope and the y-intercept are given.

b Find an equation of a line when the slope and a point are given.

c Find an equation of a line when two points are given.

d Given a line and a point not on the given line, find an equation of the line parallel to the line and containing the point, and find an equation of the line perpendicular to the line and containing the point.

e Solve applied problems involving linear functions.

Study Tips

THE AW MATH TUTOR CENTER

www.aw-bc.com/tutorcenter

The AW Math Tutor Center is staffed by highly qualified mathematics instructors who provide students with tutoring on text examples and odd-numbered exercises. Tutoring is provided free to students who have bought a new textbook with a special access card bound with the book. Tutoring is available by toll-free telephone, toll-free fax, e-mail, and the Internet. White-board technology allows tutors and students to see problems worked while they "talk" live in real time during the tutoring sessions. If you purchased a book without this card, you can purchase an access code through your bookstore using ISBN 0-201-72170-8. (This is also discussed in the Preface.)

In this section, we will learn to find an equation of a line for which we have been given two pieces of information.

a Finding an Equation of a Line When the Slope and the y-Intercept Are Given

If we know the slope and the y-intercept of a line, we can find an equation of the line using the slope–intercept equation $y = mx + b$.

EXAMPLE 1 A line has slope -0.7 and y-intercept $(0, 13)$. Find an equation of the line.

We use the slope–intercept equation and substitute -0.7 for m and 13 for b:

$$y = mx + b$$
$$y = -0.7x + 13.$$

Do Exercise 1 on the following page.

b Finding an Equation of a Line When the Slope and a Point Are Given

Suppose we know the slope of a line and the coordinates of one point on the line. We can use the slope–intercept equation to find an equation of the line. Or, we can use the **point–slope equation.** We first develop a formula for such a line.

Suppose that a line of slope m passes through the point (x_1, y_1). For any other point (x, y) on this line, we must have

$$\frac{y - y_1}{x - x_1} = m.$$

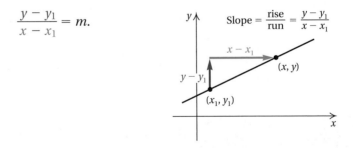

It is tempting to use this last equation as an equation of the line of slope m that passes through (x_1, y_1). The only problem with this form is that when x and y are replaced with x_1 and y_1, we have $\frac{0}{0} = m$, a false equation. To avoid this difficulty, we multiply by $x - x_1$ on both sides and simplify:

$$(x - x_1)\frac{y - y_1}{x - x_1} = m(x - x_1) \qquad \text{Multiplying by } x - x_1 \text{ on both sides}$$

$$y - y_1 = m(x - x_1). \qquad \text{Removing a factor of 1: } \frac{x - x_1}{x - x_1} = 1$$

This is the *point–slope* form of a linear equation.

POINT–SLOPE EQUATION

The **point–slope equation** of a line with slope m, passing through (x_1, y_1), is

$$y - y_1 = m(x - x_1).$$

If we know the slope of a line and a certain point on the line, we can find an equation of the line using either the point–slope equation,

$$y - y_1 = m(x - x_1),$$

or the slope–intercept equation,

$$y = mx + b.$$

Find an equation of the line with the given slope and containing the given point.

2. $m = -5, \ (-4, 2)$

EXAMPLE 2 Find an equation of the line with slope 5 and containing the point $\left(\frac{1}{2}, -1\right)$.

Using the Point–Slope Equation: We consider $\left(\frac{1}{2}, -1\right)$ to be (x_1, y_1) and 5 to be the slope m, and substitute:

$$y - y_1 = m(x - x_1) \qquad \text{Point–slope equation}$$
$$y - (-1) = 5\left(x - \tfrac{1}{2}\right) \qquad \text{Substituting}$$
$$y + 1 = 5x - \tfrac{5}{2} \qquad \text{Simplifying}$$
$$y = 5x - \tfrac{5}{2} - 1$$
$$y = 5x - \tfrac{5}{2} - \tfrac{2}{2}$$
$$y = 5x - \tfrac{7}{2}.$$

3. $m = 3, \ (1, -2)$

Using the Slope–Intercept Equation: The point $\left(\frac{1}{2}, -1\right)$ is on the line, so it is a solution. Thus we can substitute $\frac{1}{2}$ for x and -1 for y in $y = mx + b$. We also substitute 5 for m, the slope. Then we solve for b:

$$y = mx + b \qquad \text{Slope–intercept equation}$$
$$-1 = 5 \cdot \left(\tfrac{1}{2}\right) + b \qquad \text{Substituting}$$
$$-1 = \tfrac{5}{2} + b$$
$$-1 - \tfrac{5}{2} = b$$
$$-\tfrac{2}{2} - \tfrac{5}{2} = b$$
$$-\tfrac{7}{2} = b. \qquad \text{Solving for } b$$

4. $m = 8, \ (3, 5)$

We then use the slope–intercept equation $y = mx + b$ again and substitute 5 for m and $-\frac{7}{2}$ for b:

$$y = 5x - \tfrac{7}{2}.$$

Do Exercises 2–5.

5. $m = -\dfrac{2}{3}, \ (1, 4)$

Answers on page A-15

6. Find an equation of the line containing the points $(4, -3)$ and $(1, 2)$.

Finding an Equation of a Line When Two Points Are Given

We can also use the slope–intercept equation or the point–slope equation to find an equation of a line when two points are given.

■ **EXAMPLE 3** Find an equation of the line containing the points $(2, 3)$ and $(-6, 1)$.

First, we find the slope:

$$m = \frac{3 - 1}{2 - (-6)} = \frac{2}{8}, \text{ or } \frac{1}{4}.$$

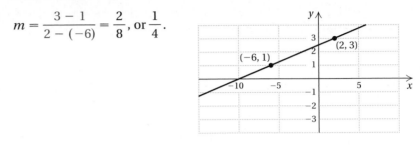

Now we have the slope and two points. We then proceed as we did in Example 2, using either point, and either the point–slope equation or the slope–intercept equation.

Using the Point–Slope Equation: We choose $(2, 3)$ and substitute 2 for x_1, 3 for y_1, and $\frac{1}{4}$ for m:

$$\begin{aligned}
y - y_1 &= m(x - x_1) \quad &&\text{Point–slope equation} \\
y - 3 &= \tfrac{1}{4}(x - 2) \quad &&\text{Substituting} \\
y - 3 &= \tfrac{1}{4}x - \tfrac{1}{2} \\
y &= \tfrac{1}{4}x - \tfrac{1}{2} + 3 \\
y &= \tfrac{1}{4}x - \tfrac{1}{2} + \tfrac{6}{2} \\
y &= \tfrac{1}{4}x + \tfrac{5}{2}.
\end{aligned}$$

7. Find an equation of the line containing the points $(-3, -5)$ and $(-4, 12)$.

Using the Slope–Intercept Equation: We choose $(2, 3)$ and substitute 2 for x, 3 for y, and $\frac{1}{4}$ for m:

$$\begin{aligned}
y &= mx + b \quad &&\text{Slope–intercept equation} \\
3 &= \tfrac{1}{4} \cdot 2 + b \quad &&\text{Substituting} \\
3 &= \tfrac{1}{2} + b \\
3 - \tfrac{1}{2} &= \tfrac{1}{2} + b - \tfrac{1}{2} \\
\tfrac{6}{2} - \tfrac{1}{2} &= b \\
\tfrac{5}{2} &= b. \quad &&\text{Solving for } b
\end{aligned}$$

Finally, we use the slope–intercept equation $y = mx + b$ again and substitute $\frac{1}{4}$ for m and $\frac{5}{2}$ for b:

$$y = \tfrac{1}{4}x + \tfrac{5}{2}.$$

Do Exercises 6 and 7.

Answers on page A-15

d Finding an Equation of a Line Parallel or Perpendicular to a Given Line Through a Point Off the Line

We can also use the methods of Example 2 to find an equation of a line through a point off the line parallel or perpendicular to a given line.

EXAMPLE 4   Find an equation of the line containing the point $(-1, 3)$ and parallel to the line $2x + y = 10$.

A line parallel to the given line $2x + y = 10$ must have the same slope as the given line. To find that slope, we first find the slope–intercept equation by solving for y:

$$2x + y = 10$$
$$y = -2x + 10.$$

Thus the line we want to find through $(-1, 3)$ must also have slope -2.

Using the Point–Slope Equation: We use the point $(-1, 3)$ and the slope -2, substituting -1 for x_1, 3 for y_1, and -2 for m:

$$y - y_1 = m(x - x_1)$$
$$y - 3 = -2(x - (-1)) \quad \text{Substituting}$$
$$y - 3 = -2(x + 1) \quad \text{Simplifying}$$
$$y - 3 = -2x + (-2) \cdot 1$$
$$y - 3 = -2x - 2$$
$$y = -2x - 2 + 3$$
$$y = -2x + 1.$$

Using the Slope–Intercept Equation: We substitute -1 for x and 3 for y in $y = mx + b$, and -2 for m, the slope. Then we solve for b:

$$y = mx + b$$
$$3 = -2(-1) + b \quad \text{Substituting}$$
$$3 = 2 + b$$
$$1 = b. \quad \text{Solving for } b$$

We then use the equation $y = mx + b$ again and substitute -2 for m and 1 for b:

$$y = -2x + 1.$$

The given line $2x + y = 10$, or $y = -2x + 10$, and the line $y = -2x + 1$ have the same slope but different y-intercepts. Thus their graphs are parallel.

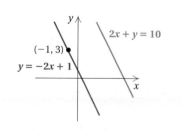

Do Exercise 8.

---

8. Find an equation of the line containing the point $(2, -1)$ and parallel to the line $9x - 3y = 5$.

Answer on page A-15

---

9. Find an equation of the line containing the point $(5, 4)$ and perpendicular to the line $2x - 4y = 9$.

EXAMPLE 5 Find an equation of the line containing the point $(2, -3)$ and perpendicular to the line $4y - x = 20$.

To find the slope of the given line, we first find its slope–intercept form by solving for y:

$$4y - x = 20$$
$$4y = x + 20$$
$$\frac{4y}{4} = \frac{x + 20}{4} \qquad \text{Dividing by 4}$$
$$y = \tfrac{1}{4}x + 5.$$

We know that the slope of the perpendicular line must be the opposite of the reciprocal of $\tfrac{1}{4}$. Thus the new line through $(2, -3)$ must have slope -4.

Using the Point–Slope Equation: We use the point $(2, -3)$ and the slope -4, substituting 2 for x_1, -3 for y_1, and -4 for m:

$$y - y_1 = m(x - x_1)$$
$$y - (-3) = -4(x - 2) \qquad \text{Substituting}$$
$$y + 3 = -4(x - 2) \qquad \text{Simplifying}$$
$$y + 3 = -4x + (-4)(-2)$$
$$y + 3 = -4x + 8$$
$$y = -4x + 8 - 3$$
$$y = -4x + 5.$$

Using the Slope–Intercept Equation: We now substitute 2 for x and -3 for y in $y = mx + b$. We also substitute -4 for m, the slope. Then we solve for b:

$$y = mx + b$$
$$-3 = -4(2) + b \qquad \text{Substituting}$$
$$-3 = -8 + b$$
$$5 = b. \qquad \text{Solving for } b$$

Finally, we use the equation $y = mx + b$ again and substitute -4 for m and 5 for b:

$$y = -4x + 5.$$

The product of the slopes of the lines $4y - x = 20$ and $y = -4x + 5$ is $\tfrac{1}{4} \cdot (-4) = -1$. Thus their graphs are perpendicular.

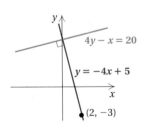

Do Exercise 9.

Answer on page A-15

e   Applications of Linear Functions

When the essential parts of a problem are described in mathematical language, we say that we have a **mathematical model.** We have already studied many kinds of mathematical models in this text—for example, the formulas in Section 1.2 and the functions in Section 2.2. Here we study linear functions as models.

EXAMPLE 6   *Cost of a Service Call.*   For a service call, Williams Plumbing charges a \$45 trip fee and \$80 per hour for labor.

a) Formulate a linear function that models the total cost of a service call $C(t)$, where t is the number of hours of the call.

b) Graph the model.

c) Use the model to determine the cost of a $2\frac{1}{2}$-hr service call.

a) The problem describes a situation in which an hourly fee is charged after an initial charge has been assessed. The total cost of a 1-hr service call is

$$\$45 + \$80 \cdot 1 = \$125.$$

For a 2-hr service call, the total cost is

$$\$45 + \$80 \cdot 2 = \$205.$$

These calculations lead us to generalize that for a service call that lasts t hours, the total cost is given by $C(t) = 45 + 80t$, where $t \geq 0$ (since the number of hours cannot be negative). The notation $C(t)$ indicates that the cost C is a function of time t.

b) Before we draw the graph, we rewrite the model in slope–intercept form:

$$C(t) = 80t + 45.$$

The y-intercept is $(0, 45)$ and the slope, or rate of change, is \$80 per hour, or $\frac{80}{1}$. We first plot $(0, 45)$; from that point, we move up 80 units and right 1 unit to the point $(1, 125)$. We then draw a line through these points. We also calculate a third value as a check:

$$C(3) = 80 \cdot 3 + 45 = 285.$$

The point $(3, 285)$ lines up with the other two points, so the graph is correct.

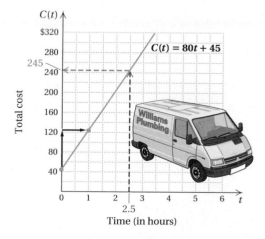

Time (in hours)

HIGHLIGHTING

Reading and highlighting a section before your instructor lectures on it allows you to maximize your learning and understanding during the lecture.

- **Try to keep one section ahead of your syllabus.** If you study ahead of your lectures, you can concentrate on what is being explained in them, rather than trying to write everything down. You can then take notes only of special points or of questions related to what is happening in class.

- **Highlight important points.** You are probably used to highlighting key points as you study. If that works for you, continue to do so. But you will notice many design features throughout this book that already highlight important points. Thus you may not need to highlight as much as you generally do.

- **Highlight points that you do not understand.** Use a unique mark to indicate trouble spots that can lead to questions to be asked during class, in a tutoring session, or when calling or contacting the AW Math Tutor Center.

223

10. Cable TV Service. Clear County Cable TV Service charges a \$50 installation fee and \$40 per month for basic service.

a) Formulate a linear function for cost $C(t)$ for t months of cable TV service.

b) Graph the model.

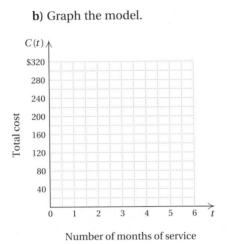

Total cost

$C(t)$

\$320
280
240
200
160
120
80
40

0   1   2   3   4   5   6 t

Number of months of service

c) Use the model to determine the cost of 4 months of service.

c) To determine the cost of a $2\frac{1}{2}$-hr service call, we find $C(2.5)$:

$$C(2.5) = 80(2.5) + 45 = 245.$$

From the graph, we also see that the input 2.5 corresponds to the output 245. Thus we see that a $2\frac{1}{2}$-hr service call costs \$245.

Do Exercise 10.

In the following example, we use two points and find an equation for the linear function through these points. Then we use the equation to make a prediction.

EXAMPLE 7 *Ring Size as a Function of Finger Diameter.* The table below shows data regarding the correspondence between the diameter of a person's finger, in inches, and his or her ring size (U.S.).

FINGER DIAMETER, x (in inches)	RING SIZE, R (in U.S. units)
0.586	4
0.783	10

Source: Oromi, Inc.

a) Assuming a constant rate of change, use the two data points to find a linear function that fits the data.

b) Use the function to determine the ring size of a person whose finger has a diameter of 0.619 in.

c) Cliff's ring size is $11\frac{1}{2}$. What is the diameter of his finger, in inches?

a) We let x = the diameter of a finger, in inches, and R = the U.S. ring size. The table gives us two ordered pairs, $(0.586, 4)$ and $(0.783, 10)$. We use them to find a linear function that fits the data. First, we find the slope:

$$m = \frac{10 - 4}{0.783 - 0.586} = \frac{6}{0.197} \approx 30.4569.$$

Next, we find an equation $R = mx + b$ that fits the data. We can use either the point–slope equation or the slope–intercept equation to do this.

Answers on page A-15

CHAPTER 2: Graphs, Functions, and Applications

Using the Point–Slope Equation: We substitute one of the points—say, $(0.586, 4)$—and the slope in the point–slope equation:

$$R - R_1 = m(x - x_1) \qquad \text{Point–slope equation}$$
$$R - 4 = 30.4569(x - 0.586) \qquad \text{Substituting}$$
$$R - 4 = 30.4569x - 17.8477 \qquad \text{Rounding to four decimal places}$$
$$R = 30.4569x - 13.8477.$$

Using the Slope–Intercept Equation: We choose one of the points—say $(0.586, 4)$—and substitute 4 for R, 0.586 for x, and 30.4569 for m in the slope–intercept equation. Then we solve for b:

$$R = mx + b \qquad \text{Slope–intercept equation}$$
$$4 = 30.4569(0.586) + b \qquad \text{Substituting}$$
$$4 = 17.8477 + b \qquad \text{Rounding to four decimal places}$$
$$-13.8477 = b.$$

Now we use the equation $R = mx + b$ again and substitute 30.4569 for m and -13.8477 for b:

$$R = 30.4569x - 13.8477.$$

Using function notation, we have

$$R(x) = 30.4569x - 13.8477.$$

b) To find the ring size of a person whose finger has a diameter of 0.619 in., we substitute 0.619 for x in the function $R(x) = 30.4569x - 13.8477$:

$$R(x) = 30.4569x - 13.8477$$
$$R(0.619) = 30.4569(0.619) - 13.8477 \qquad \text{Substituting}$$
$$\approx 18.8528 - 13.8477$$
$$\approx 5.$$

The ring size is 5.

c) To find the diameter that corresponds to a ring size of $11\frac{1}{2}$, we substitute 11.5 for $R(x)$ and solve for x:

$$R(x) = 30.4569x - 13.8477$$
$$11.5 = 30.4569x - 13.8477 \qquad \text{Substituting}$$
$$25.3477 = 30.4569x$$
$$0.832 \approx x. \qquad \text{Dividing by 30.4569 and rounding}$$

The diameter of Cliff's finger is about 0.832 in.

Do Exercise 11.

---

11. Hat Size as a Function of Head Circumference. The table below shows data relating hat size to head circumference.

HEAD CIRCUMFERENCE, C (in inches)	HAT SIZE, H
21.2	$6\frac{3}{4}$
22	7

Source: Shushan's New Orleans

a) Assuming a constant rate of change, use the two data points to find a linear function that fits the data.

b) Use the function to determine the hat size of a person whose head has a circumference of 24.8 in.

c) Jerome's hat size is 8. What is the circumference of his head?

Answers on page A-15

225

Linear Regression: Fitting a Linear Equation to a Set of Data

We can use the LINEAR REGRESSION feature on a graphing calculator to fit a linear equation to a set of data. The following table shows the correspondence between the diameter of a person's finger and his or her ring size, in U.S. units.

FINGER DIAMETER (in inches)	RING SIZE (in U.S. Units)
0.487	1
0.528	$2\frac{1}{4}$
0.577	$3\frac{3}{4}$
0.619	5
0.676	$6\frac{3}{4}$
0.693	$7\frac{1}{4}$
0.734	$8\frac{1}{2}$
0.783	10
0.825	$11\frac{1}{4}$

Source: Oromi, Inc.

a) Fit a regression line to the data using the LINEAR REGRESSION feature on a graphing calculator.

b) Use the linear model to determine the ring size of a person with a finger diameter of 0.685 in.

a) We will enter the data on the STAT list editor screen as ordered pairs in which the first coordinate is the finger diameter and the second is the corresponding ring size. First, we clear any existing lists by pressing

STAT 4 **2ND** L_1 **,** **2ND** L_2 **,** **2ND** L_3 **,** **2ND** L_4 **,** **2ND** L_5 **,** **2ND** L_6 **ENTER**. (L_1 through L_6 are the second operations associated with the numeric keys 1 through 6.)

Once the lists have been cleared, we can enter the coordinates of the points. We press **STAT** ① and enter the first coordinates in L_1 and the second coordinates in L_2. We position the cursor at the top of column L_1, below the L_1 heading. To enter 0.487, we press ⊙ ④ ⑧ ⑦ **ENTER**, or ⊙ ④ ⑧ ⑦ ⌄. We then continue typing the first coordinates in order, each followed by **ENTER** or ⌄. Next, we press ▷ to move to the top of column L_2, and type the second coordinates in succession, each followed by **ENTER** or ⌄. (We enter the mixed numerals in decimal notation, as shown in the window below.) Note that the coordinates of each point must be in the same position in both lists.

L1	L2	L3	2
.487	1	------	
.528	2.25		
.577	3.75		
.619	5		
.676	6.75		
.693	7.25		
.734	8.5		

L2 (1) = 1

(continued)

CHAPTER 2: Graphs, Functions, and Applications

Now we will plot the data points to determine whether a linear equation would be a good model for these data. First, we turn on the STAT PLOT feature. To access the STAT PLOT screen, we press **2ND** (STAT PLOT). (STAT PLOT is the second operation associated with the (Y=) key in the upper left-hand corner of the keypad.)

```
STAT PLOTS
1: Plot 1...Off
    ⟋    L1   L2     □
2: Plot 2...Off
    ⋰    L1   L3     □
3: Plot 3...Off
    ⋰    L1   L2     □
4↓Plots Off
```

We will use Plot 1. We access it by highlighting 1 and pressing **ENTER**, or simply by pressing 1. Next, we position the cursor over On and press **ENTER** to turn on Plot 1. The entries Type, Xlist, and Ylist should be as shown below. The last item, Mark, allows us to choose a box, a cross, or a dot for each point. Here we have selected a box. To select Type and Mark, we position the cursor over the appropriate selection and press **ENTER**. We use the L_1 and L_2 keys (associated with the 1 and 2 numeric keys) to select Xlist and Ylist.

```
Plot1  Plot2  Plot3
On   Off
Type:  ⚏   ⟋   ⊞
       ⊞   ⊞   ⟋
Xlist: L1
Ylist: L2
Mark: □  •  ·
```

Note that there should be no equations entered on the "Y = " screen. If there are entries present, they should be cleared before we proceed. (See p. 85.) To see the plotted points, we press (ZOOM) (9). These keystrokes activate the ZOOMSTAT operation, which automatically defines a viewing window that displays all the points.

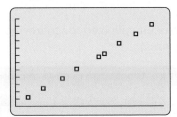

The data points appear to fall on or near a straight line, so we will fit a linear equation to these data. We press **STAT** (▷) (4) **VARS** (▷) (1) (1) **ENTER**. The first three keystrokes select LinReg($ax + b$) from the STAT CALC menu and display the coefficients a and b of the regression equation $y = ax + b$. The keystrokes **VARS** (▷) (1) (1) copy the regression equation to the equation-editor screen as y_1. We see that the regression equation is $y = 30.34160781x - 13.7703335$. If we press (ZOOM) 9 again, we will see the regression equation graphed with the data points.

(continued)

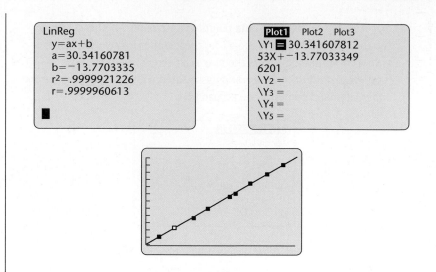

Before graphing other equations, we turn off the STAT PLOT by pressing Y= , positioning the cursor over Plot 1 at the top of the screen, and then pressing ENTER. Plot 1 is now no longer highlighted, indicating that it is not turned on.

b) We can use a table set in ASK mode to determine the ring size of a person with a finger diameter of 0.685 in. (See p. 177.) When $x = 0.685$, $y \approx 7$, so a person with a finger diameter of 0.685 in. would wear a size 7 ring.

X	Y₁	
.685	7.0137	
X =		

Exercises:

1. Find the ring size of a person with a finger diameter of 0.619 in. Compare your answer with the answer you found in Example 7(b).

2. An algebra instructor collected the following data relating study time and test scores.

STUDY TIME (in hours)	TEST SCORES (in percent)
5	73
6	80
8	88
11	92
12	94

a) Fit a regression line to the data using the LINEAR REGRESSION feature on a graphing calculator.
b) Use the linear model to estimate the test score of a student who studies for 10 hr.

MathXL MyMathLab InterAct Math Math Tutor Center Digital Video Tutor CD 1 Videotape 3 Student's Solutions Manual

a Find an equation of the line having the given slope and *y*-intercept.

1. Slope: -8; *y*-intercept: $(0, 4)$

2. Slope: 5; *y*-intercept: $(0, -3)$

3. Slope: 2.3; *y*-intercept: $(0, -1)$

4. Slope: -9.1; *y*-intercept: $(0, 2)$

Find a linear function $f(x) = mx + b$ whose graph has the given slope and *y*-intercept.

5. Slope: $-\frac{7}{3}$; *y*-intercept: $(0, -5)$

6. Slope: $\frac{4}{5}$; *y*-intercept: $(0, 28)$

7. Slope: $\frac{2}{3}$; *y*-intercept: $\left(0, \frac{5}{8}\right)$

8. Slope: $-\frac{7}{8}$; *y*-intercept: $\left(0, -\frac{7}{11}\right)$

b Find an equation of the line having the given slope and containing the given point.

9. $m = 5$, $(4, 3)$

10. $m = 4$, $(5, 2)$

11. $m = -3$, $(9, 6)$

12. $m = -2$, $(2, 8)$

13. $m = 1$, $(-1, -7)$

14. $m = 3$, $(-2, -2)$

15. $m = -2$, $(8, 0)$

16. $m = -3$, $(-2, 0)$

17. $m = 0$, $(0, -7)$

18. $m = 0$, $(0, 4)$

19. $m = \frac{2}{3}$, $(1, -2)$

20. $m = -\frac{4}{5}$, $(2, 3)$

c Find an equation of the line containing the given pair of points.

21. $(1, 4)$ and $(5, 6)$

22. $(2, 5)$ and $(4, 7)$

23. $(-3, -3)$ and $(2, 2)$

24. $(-1, -1)$ and $(9, 9)$

25. $(-4, 0)$ and $(0, 7)$

26. $(0, -5)$ and $(3, 0)$

27. $(-2, -3)$ and $(-4, -6)$

28. $(-4, -7)$ and $(-2, -1)$

29. $(0, 0)$ and $(6, 1)$

30. $(0, 0)$ and $(-4, 7)$

31. $\left(\frac{1}{4}, -\frac{1}{2}\right)$ and $\left(\frac{3}{4}, 6\right)$

32. $\left(\frac{2}{3}, \frac{3}{2}\right)$ and $\left(-3, \frac{5}{6}\right)$

d Write an equation of the line containing the given point and parallel to the given line.

33. $(3, 7)$; $x + 2y = 6$

34. $(0, 3)$; $2x - y = 7$

35. $(2, -1)$; $5x - 7y = 8$

36. $(-4, -5)$; $2x + y = -3$

37. $(-6, 2)$; $3x = 9y + 2$

38. $(-7, 0)$; $2y + 5x = 6$

Write an equation of the line containing the given point and perpendicular to the given line.

39. $(2, 5)$; $2x + y = 3$

40. $(4, 1)$; $x - 3y = 9$

41. $(3, -2)$; $3x + 4y = 5$

42. $(-3, -5)$; $5x - 2y = 4$

43. $(0, 9)$; $2x + 5y = 7$

44. $(-3, -4)$; $6y - 3x = 2$

e Solve.

45. *Moving Costs.* Metro Movers charges $85 plus $40 an hour to move households across town.
   a) Formulate a linear function for total cost $C(t)$ for t hours of moving.
   b) Graph the model.
   c) Use the model to determine the cost of $6\frac{1}{2}$ hr of moving service.

46. *Deluxe Cable TV Service.* Vista Cable TV Service charges a $35 installation fee and $20 per month for basic service.
   a) Formulate a linear function for total cost $C(t)$ for t months of cable TV service.
   b) Graph the model.
   c) Use the model to determine the cost of 9 months of service.

47. *Value of a Fax Machine.* Melton Corporation bought a multifunction fax machine for $750. The value $V(t)$ of the machine depreciates (declines) at a rate of $25 per month.
   a) Formulate a linear function for the value $V(t)$ of the machine after t months.
   b) Graph the model.
   c) Use the model to determine the value of the machine after 13 months.

48. *Value of a Computer.* True Tone Graphics bought a computer for $3800. The value $V(t)$ of the computer depreciates at a rate of $50 per month.
   a) Formulate a linear function for the value $V(t)$ of the computer after t months.
   b) Graph the model.
   c) Use the model to determine the value of the computer after $10\frac{1}{2}$ months.

In Exercises 49–54, assume that a constant rate of change exists for each model formed.

49. *Prescription Drug Sales.* The table below shows data regarding the number of prescriptions filled, in trillions, in 1999 and in 2003.

YEAR	NUMBER OF PRESCRIPTIONS FILLED (in trillions)
1999	2.707
2003	3.215

Source: National Association of Chain Drug Stores

   a) Use the two data points to find a linear function that fits the data. Let $x =$ the number of years since 1999 and $P =$ the number of prescriptions filled, in trillions.
   b) Use the function to estimate the number of prescriptions filled in 2002 and in 2008.

50. *Median Age of Women at First Marriage.* The table below shows data regarding the median age of women at first marriage in 1970 and in 2003.

YEAR	MEDIAN AGE OF WOMEN AT FIRST MARRIAGE
1970	20.8
2003	25.0

Source: U.S. Bureau of the Census

   a) Use the two data points to find a linear function that fits the data. Let $x =$ the number of years since 1970 and $A =$ the median age at first marriage x years from 1970.
   b) Use the function to predict the median age of women at first marriage in 2008 and in 2015.

51. *Records in the 400-Meter Run.* In 1930, the record for the 400-m run was 46.8 sec. In 1970, it was 43.8 sec. Let $R(t) =$ the record in the 400-m run and $t =$ the number of years since 1930.

  a) Find a linear function that fits the data.

  b) Use the function of part (a) to estimate the record in 2003; in 2006.

  c) When will the record be 40 sec?

52. *Recycling.* In 1993, Americans recycled 43.8 million tons of solid waste. In 2001, the figure grew to 68.0 million tons. Let $N(t) =$ the number of tons recycled, in millions, and $t =$ the number of years since 1993.

Source: Franklin Associates

  a) Find a linear function that fits the data.

  b) Use the function of part (a) to estimate the amount recycled in 2005.

53. *Life Expectancy of Males in the United States.* In 1990, the life expectancy of males was 71.8 yr. In 2001, it was 74.4 yr. Let $M(t) =$ life expectancy and $t =$ the number of years since 1990.

Source: U.S. National Center for Health Statistics

  a) Find a linear function that fits the data.

  b) Use the function of part (a) to estimate the life expectancy of males in 2007.

54. *Life Expectancy of Females in the United States.* In 1990, the life expectancy of females was 78.8 yr. In 2001, it was 79.8 yr. Let $F(t) =$ life expectancy and $t =$ the number of years since 1990.

Source: U.S. National Center for Health Statistics

  a) Find a linear function that fits the data.

  b) Use the function of part (a) to estimate the life expectancy of females in 2010.

Determine m and b in each application and explain their meaning.

55. $\mathbf{D_W}$ *Computer Repair.* The cost $R(t)$, in dollars, of computer repair at PC Pros is given by

$$R(t) = 50t + 35,$$

where t is the number of hours that the repair requires.

56. $\mathbf{D_W}$ *Cost of a Taxi Ride.* The cost $C(d)$, in dollars, of a taxi ride in Pelham is given by

$$C(d) = 0.75d + 2,$$

where d is the number of miles traveled.

SKILL MAINTENANCE

Solve. [1.4c], [1.5a], [1.6c, d, e]

57. $2x + 3 > 51$ **58.** $|2x + 3| = 51$ **59.** $2x + 3 \leq 51$ **60.** $2x + 3 \leq 5x - 4$

61. $|2x + 3| \leq 13$ **62.** $|2x + 3| = |x - 4|$ **63.** $|5x - 4| = -8$ **64.** $-12 \leq 2x + 3 < 51$

SYNTHESIS

For Exercises 65 and 66, assume that a linear equation models each situation.

65. *Depreciation of a Computer.* After 6 mos of use, the value of Pearl's computer had dropped to $900. After 8 mos, the value had gone down to $750. How much did the computer cost originally?

66. *Cellular Phone Charges.* The total cost of Mel's cellular phone was $230 after 5 mos of service and $390 after 9 mos. What costs had Mel already incurred when his service just began?

231

The review that follows is meant to prepare you for a chapter exam. It consists of three parts. The first part, Concept Reinforcement, is designed to increase understanding of the concepts through true/false exercises. The second part is a list of important properties and formulas. The third part is the Review Exercises. These provide practice exercises for the exam, together with references to section objectives so you can go back and review. Before beginning, stop and look back over the skills you have obtained. What skills in mathematics do you have now that you did not have before studying this chapter?

⌣ CONCEPT REINFORCEMENT

Determine whether the statement is true or false. Answers are given at the back of the book.

_____ **1.** Every function is a relation.

_____ **2.** It is possible for one input of a function to have two or more outputs.

_____ **3.** It is possible for all the inputs of a function to have the same output.

_____ **4.** If it is possible for a vertical line to cross a graph more than once, the graph is not the graph of a function.

_____ **5.** The slope of a vertical line is 0.

_____ **6.** A line with slope 4 slants up from left to right.

_____ **7.** A line with slope 1 slants more steeply than a line with slope -5.

_____ **8.** Parallel lines have the same slope and y-intercept.

IMPORTANT PROPERTIES AND FORMULAS

$$\text{Slope} = m = \frac{y_2 - y_1}{x_2 - x_1}, \text{ or } \frac{y_1 - y_2}{x_1 - x_2}$$

Equations of Lines and Linear Functions

Horizontal Line:	$f(x) = b$, or $y = b$; slope $= 0$
Vertical Line:	$x = a$, slope is not defined.
Slope–Intercept Equation:	$f(x) = mx + b$, or $y = mx + b$
Point–Slope Equation:	$y - y_1 = m(x - x_1)$
Parallel Lines:	$m_1 = m_2, b_1 \neq b_2$
Perpendicular Lines:	$m_1 = -\dfrac{1}{m_2}, m_1, m_2 \neq 0$

Review Exercises

1. Show that the ordered pairs $(0, -2)$ and $(-1, -5)$ are solutions of the equation $3x - y = 2$. Then use the graph of the two points to determine another solution. Answers may vary. Show your work. [2.1a, b]

Graph. [2.1c, d]

2. $y = -3x + 2$

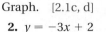

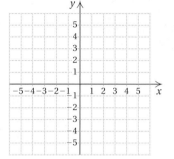

3. $y = \frac{5}{2}x - 3$

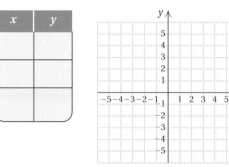

4. $y = |x - 3|$

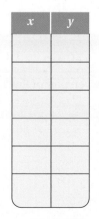

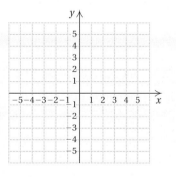

5. $y = 3 - x^2$

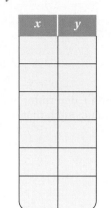

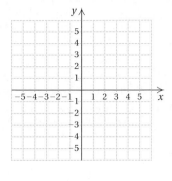

Determine whether the correspondence is a function. [2.2a]

6. $3 \rightarrow a$
   $5 \rightarrow b$
   $7 \rightarrow c$
   $9 \rightarrow d$
   $\rightarrow e$

7. $1 \rightarrow a$
   $2 \rightarrow b$
   $3 \rightarrow c$
   $4 \rightarrow d$
   5

Find the function values. [2.2b]

8. $g(x) = -2x + 5$; $g(0)$ and $g(-1)$

9. $f(x) = 3x^2 - 2x + 7$; $f(0)$ and $f(-1)$

10. *Tuition Cost.* The function $C(t) = 309.2t + 3717.7$ can be used to approximate the average cost of tuition and fees for in-state students at public four-year colleges, where t is the number of years after 2000. Estimate the average cost of tuition and fees in 2010. That is, find $C(10)$. [2.2b]

Source: U.S. National Center for Education Statistics

Determine whether each of the following is the graph of a function. [2.2d]

11.
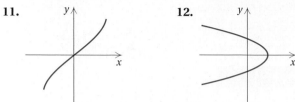

12.

13. For the following graph of a function f, determine **(a)** $f(2)$; **(b)** the domain; **(c)** all x-values such that $f(x) = 2$; and **(d)** the range. [2.3a]

Find the domain. [2.3a]

14. $f(x) = \dfrac{5}{x - 4}$

15. $g(x) = x - x^2$

Find the slope and the y-intercept. [2.4a, b]

16. $y = -3x + 2$

17. $4y + 2x = 8$

18. Find the slope, if it exists, of the line containing the points $(13, 7)$ and $(10, -4)$. [2.4b]

Find the intercepts. Then graph the equation. [2.5a]

19. $2y + x = 4$

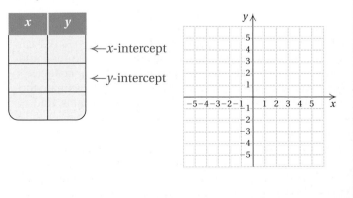

x	y

←x-intercept

←y-intercept

20. $2y = 6 - 3x$

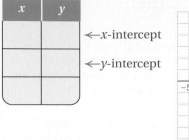

x	y

←x-intercept

←y-intercept

Graph using the slope and the y-intercept. [2.5b]

21. $g(x) = -\dfrac{2}{3}x - 4$

22. $f(x) = \dfrac{5}{2}x + 3$

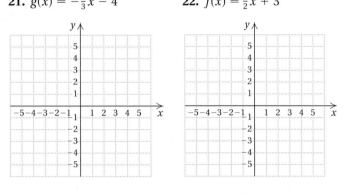

Graph. [2.5c]

23. $x = -3$

24. $f(x) = 4$

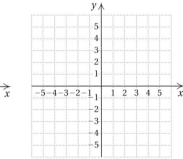

Determine whether the graphs of the given pair of lines are parallel or perpendicular. [2.5d]

25. $y + 5 = -x,$
    $x - y = 2$

26. $3x - 5 = 7y,$
    $7y - 3x = 7$

27. $4y + x = 3,$
    $2x + 8y = 5$

28. $x = 4,$
    $y = -3$

29. Find a linear function $f(x) = mx + b$ whose graph has the given slope and y-intercept: [2.6a]

slope: 4.7; y-intercept: $(0, -23)$.

30. Find an equation of the line having the given slope and containing the given point: [2.6b]

$m = -3$; $(3, -5)$.

31. Find an equation of the line containing the given pair of points: [2.6c]

$(-2, 3)$ and $(-4, 6)$.

32. Find an equation of the line containing the given point and parallel to the given line: [2.6d]

$(14, -1)$; $5x + 7y = 8$.

33. Find an equation of the line containing the given point and perpendicular to the given line: [2.6d]

$(5, 2)$; $3x + y = 5$.

34. *Records in the 400-Meter Run.* The table below shows data regarding the world indoor records in the men's 400-m run. [2.6e]

YEAR	RECORDS IN THE 400-M RUN (in seconds)
1970	46.8
2004	44.63

a) Use the two data points to find a linear function that fits the data. Let x = the number of years since 1970 and R = the world record x years from 1970.

b) Use the function to estimate the world record in the men's 400-m run in 2008 and in 2010.

35. **D**_W Explain the usefulness of the slope concept when describing a line. [2.4b, c], [2.5b], [2.6a, b, c, d]

36. **D**_W Explain the meaning of the notation $f(x)$ when considering a function in as many ways as you can. [2.2b]

SYNTHESIS

37. Homespun Jellies charges \$2.49 for each jar of preserves. Shipping charges are \$3.75 for handling, plus \$0.60 per jar. Find a linear function for determining the cost of buying and shipping x jars of preserves. [2.6e]

Determine whether the given points are solutions of the equation.

1. $(2, -3)$; $y = 5 - 4x$

2. $(2, -3)$; $5b - 7a = 10$

Graph.

3. $y = -2x - 5$

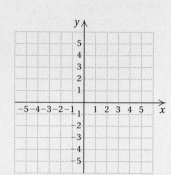

4. $f(x) = -\dfrac{3}{5}x$

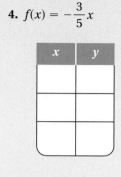

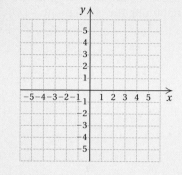

5. $g(x) = 2 - |x|$

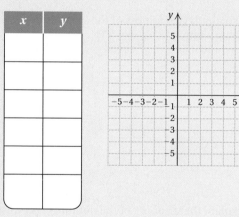

6. $y = \dfrac{4}{x}$

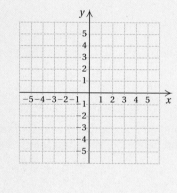

7. *Median Age of Cars.* The function

$$A(t) = 0.233t + 5.87$$

can be used to estimate the median age of cars in the United States t years after 1990. (In this context, we mean that if the median age of cars is 3 yr, then half the cars are older than 3 yr and half are younger.)

Source: The Polk Co.

a) Find the median age of cars in 2002.
b) In what year was the median age of cars 7.734 yr?

Determine whether the correspondence is a function.

8.
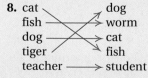

cat
fish
dog
tiger
teacher

dog
worm
cat
fish
student

9. Lake Placid
Oslo
Squaw Valley
Innsbruck

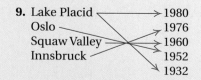

1980
1976
1960
1952
1932

Find the function values.

10. $f(x) = -3x - 4$; $f(0)$ and $f(-2)$

11. $g(x) = x^2 + 7$; $g(0)$ and $g(-1)$

Determine whether each of the following is the graph of a function.

12.

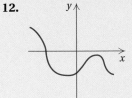

13.

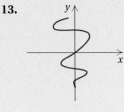

14. *A Graying Society.* From 2008 to 2030, about 76 million baby boomers will retire, with only 48 million new workers available to replace them. The graph below approximates the number of persons 65 and older in the United States for various years.

a) How many persons 65 and older were there in 2000?

b) How many persons 65 and older will there be in 2040?

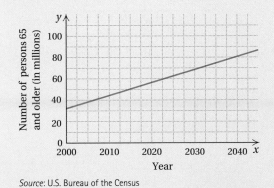

Source: U.S. Bureau of the Census

15. For the following graph of function f, determine **(a)** $f(2)$; **(b)** the domain; **(c)** all x-values such that $f(x) = 2$; and **(d)** the range.

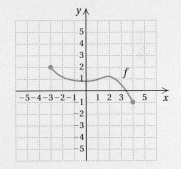

Find the domain.

16. $g(x) = 5 - x^2$

17. $f(x) = \dfrac{8}{2x + 3}$

Find the slope and the y-intercept.

18. $f(x) = -\frac{3}{5}x + 12$

19. $-5y - 2x = 7$

Find the slope, if it exists, of the line containing the following points.

20. $(-2, -2)$ and $(6, 3)$

21. $(-3.1, 5.2)$ and $(-4.4, 5.2)$

22. Find the slope, or rate of change, of the graph at right.

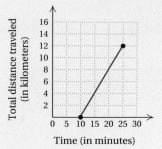

23. Find the intercepts. Then graph the equation.

$$2x + 3y = 6$$

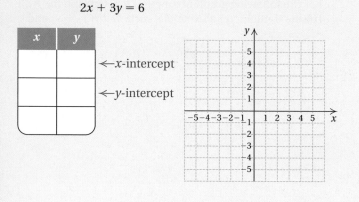

24. Graph using the slope and the *y*-intercept:

$$f(x) = -\frac{2}{3}x - 1.$$

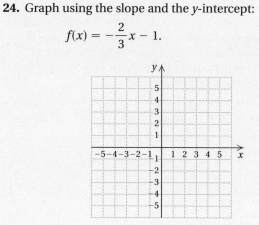

Graph.

25. $y = f(x) = -3$

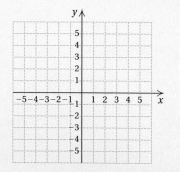

26. $2x = -4$

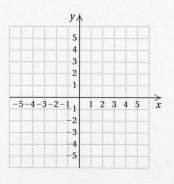

Determine whether the graphs of the given pair of lines are parallel or perpendicular.

27. $4y + 2 = 3x$,
   $-3x + 4y = -12$

28. $y = -2x + 5$,
   $2y - x = 6$

29. Find an equation of the line that has the given characteristics:

slope: -3; y-intercept: $(0, 4.8)$.

30. Find a linear function $f(x) = mx + b$ whose graph has the given slope and y-intercept:

slope: 5.2; y-intercept: $\left(0, -\frac{5}{8}\right)$.

31. Find an equation of the line having the given slope and containing the given point:

$m = -4$; $(1, -2)$.

32. Find an equation of the line containing the given pair of points:

$(4, -6)$ and $(-10, 15)$.

33. Find an equation of the line containing the given point and parallel to the given line:

$(4, -1)$; $x - 2y = 5$.

34. Find an equation of the line containing the given point and perpendicular to the given line:

$(2, 5)$; $x + 3y = 2$.

35. *Median Age of Men at First Marriage.* The table below displays data regarding the median age of men at first marriage in 1970 and in 2003.

YEAR	MEDIAN AGE OF MEN AT FIRST MARRIAGE
1970	23.2
2003	27.0

Source: U.S. Bureau of the Census

a) Use the two data points to find a linear function that fits the data. Let $x =$ the number of years since 1970 and $A =$ the median age at first marriage x years from 1970.

b) Use the function to estimate the median age of men at first marriage in 2008 and in 2015.

SYNTHESIS

36. Find k such that the line $3x + ky = 17$ is perpendicular to the line $8x - 5y = 26$.

37. Find a formula for a function f for which $f(-2) = 3$.

1. *Records in the 1500-Meter Run.* The table below shows data regarding the world indoor records in the men's 1500-m run in 1950 and in 2004.

YEAR	RECORDS IN THE 1500-M RUN (in minutes)
1950	3.85
2004	3.50

a) Use the two data points to find a linear function that fits the data. Let x = the number of years since 1950 and R = the world record x years from 1950.
b) Use the function to estimate the world record in the 1500-m run in 2008 and in 2010.

2. For the graph of function f shown at right, determine **(a)** $f(15)$; **(b)** the domain; **(c)** all x-values such that $f(x) = 14$; and **(d)** the range.

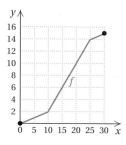

3. Evaluate $a^3 + b^0 - c$ when $a = 3$, $b = 7$, and $c = -3$.

Simplify.

4. $|4.3 - 2.1|$

5. $\left| -\dfrac{2}{3} \right|$

6. $-\dfrac{1}{3} - \left(-\dfrac{5}{6} \right)$

7. $-3.2(-11.4)$

8. $2x - 4(3x - 8)$

9. $(-16x^3 y^{-4})(-2x^5 y^3)$

10. $\dfrac{27x^0 y^3}{-3x^2 y^5}$

11. $3(x - 7) - 4[2 - 5(x + 3)]$

12. $-128 \div 16 + 32 \cdot (-10)$

13. $2^3 - (4 \cdot 2 - 3)^2 + 23^0 \cdot 16^1$

Solve.

14. $x + 9.4 = -12.6$

15. $\dfrac{2}{3}x - \dfrac{1}{4} = -\dfrac{4}{5}x$

16. $-2.4t = -48$

17. $4x + 7 = -14$

18. $3n - (4n - 2) = 7$

19. Solve $W = Ax + By$ for x.

20. Solve $M = A + 4AB$ for A.

Solve.

21. $y - 12 \leq -5$

22. $6x - 7 < 2x - 13$

23. $5(1 - 2x) + x < 2(3 + x)$

24. $x + 3 < -1 \text{ or } x + 9 \geq 1$

25. $-3 < x + 4 \leq 8$

26. $-8 \leq 2x - 4 \leq -1$

27. $|x| = 8$

28. $|y| > 4$

29. $|4x - 1| \leq 7$

30. Find an equation of the line containing the point $(-4, -6)$ and perpendicular to the line whose equation is $4y - x = 3$.

31. Find an equation of the line containing the point $(-4, -6)$ and parallel to the line whose equation is $4y - x = 3$.

Graph on a plane.

32. $y = -2x + 3$

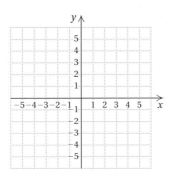

33. $3x = 2y + 6$

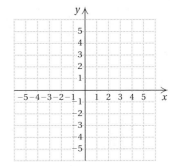

34. $4x + 16 = 0$

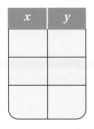
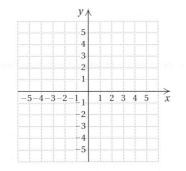

35. $-2y = -6$

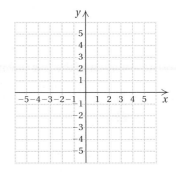

36. $f(x) = \dfrac{2}{3}x + 1$

x	y

37. $g(x) = 5 - |x|$

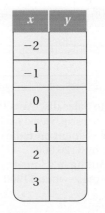

x	y
−2	
−1	
0	
1	
2	
3	

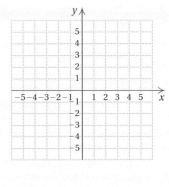

38. Find the slope and the y-intercept of $-4y + 9x = 12$.

39. Find the slope, if it exists, of the line containing the points $(2, 7)$ and $(-1, 3)$.

40. Find an equation of the line with slope -3 and containing the point $(2, -11)$.

41. Find an equation of the line containing the points $(-6, 3)$ and $(4, 2)$.

Solve.

42. *Lot Dimensions.* The perimeter of a lot is 80 m. The length exceeds the width by 6 m. Find the dimensions.

43. *Salary Raise.* After David receives a 20% raise in salary, his new salary is $27,000. What was the old salary?

44. *Matching.* Match each item in the first column with the appropriate item in the second column by drawing connecting lines. Some choices might be used more than once, and some choices might not be used.

$\left(\dfrac{1}{8}\right)^2$ 16

            -16

$\left(\dfrac{1}{8}\right)^{-2}$ 64

            -64

8^{-2}

8^2 $\dfrac{1}{64}$

-8^2 $-\dfrac{1}{64}$

$(-8)^2$

$\left(-\dfrac{1}{8}\right)^{-2}$ $-\dfrac{1}{16}$

$\left(-\dfrac{1}{8}\right)^2$ $\dfrac{1}{16}$

For each of Exercises 45–48, choose the correct answer from the selections given.

45. Find an equation of the line having slope -2 and containing the point $(3, 1)$.

  a) $y - 1 = 2(x - 3)$ **b)** $y - 1 = -2(x - 3)$
  c) $x - 1 = -2(y - 3)$ **d)** $x - 1 = 2(y - 3)$
  e) None of these

46. A 28-ft piece of wire is cut into two pieces so that one piece is one-third as long as the other. Find the length of the shorter piece.

  a) $\dfrac{28}{3}$ ft   **b)** 7 ft
  c) $\dfrac{3}{28}$ ft   **d)** 21 ft
  e) None of these

47. Write exponential notation for a.

  a) a^0 **b)** a^{-1}
  c) $1 \cdot a$ **d)** a^1
  e) None of these

48. Simplify: $[9(2a + 3) - (a - 2)] - [3(2a - 1)]$.

  a) $23a - 22$ **b)** $11a + 28$
  c) $23a + 26$ **d)** $11a + 27$
  e) None of these

SYNTHESIS

49. *Radio Advertising.* Wayside Auto Sales discovers that when $1000 is spent on radio advertising, weekly sales increase by $101,000. When $1250 is spent on radio advertising, weekly sales increase by $126,000. Assuming that sales increase according to a linear equation, by what would sales increase when $1500 is spent on radio advertising?

50. Which pairs of the following four equations represent perpendicular lines?

  (1) $7y - 3x = 21$
  (2) $-3x - 7y = 12$
  (3) $7y + 3x = 21$
  (4) $3y + 7x = 12$

51. Solve: $x + 5 < 3x - 7 \leq x + 13$.

242

Systems of Equations

Real-World Application

White River Gardens in Indianapolis, Indiana, presents a yearly butterfly exhibit. Pupae from Kenya, Tanzania, Costa Rica, and Florida are delivered semiweekly. As part of a recent order, White River Gardens received 63 pupae—*morpho granadensis* at $4.15 per pupa and *battus polydamus* at $1.50 per pupa. The total cost of the two species was $147.50. How many pupae of each species did the exhibit receive?

Source: Lewis J. Birmelin, General Curator of Horticulture, Indianapolis Zoological Society, White River Gardens

This problem appears as Example 1 in Section 3.4.

SYSTEMS OF EQUATIONS IN TWO VARIABLES

Objective

a. Solve a system of two linear equations or two functions by graphing and determine whether a system is consistent or inconsistent and whether the equations in a system are dependent or independent.

We can solve many applied problems more easily by translating to two or more equations in two or more variables than by translating to a single equation. Let's look at such a problem.

Scholastic Aptitude Test. Many high-school students take the Scholastic Aptitude Test. Each student receives two scores, a *verbal* score and a *math* score. In 2003–2004, the average total score of students was 1026, with the average math score exceeding the verbal score by 10 points. What was the average verbal score and what was the average math score?

Source: National Center for Education Statistics

To solve, we first let

x = the average verbal score and

y = the average math score.

The problem contains two statements to translate. First, we look at the total of the two scores:

$$\underbrace{\text{The average verbal score}}_{x} \quad \underbrace{\text{plus}}_{+} \quad \underbrace{\text{The average math score}}_{y} \quad \underbrace{\text{is}}_{=} \quad \underbrace{\text{The total SAT score.}}_{1026}$$

The second statement compares the two scores.

$$\underbrace{\text{The math score}}_{y} \quad \underbrace{\text{is}}_{=} \quad \underbrace{\text{Ten points more than the verbal score.}}_{x + 10}$$

We have now translated the problem to a **pair,** or **system, of equations:**

$$x + y = 1026,$$
$$y = x + 10.$$

We can also write this system in function notation by solving each equation for y:

$$x + y = 1026 \rightarrow y = 1026 - x \rightarrow f(x) = 1026 - x,$$
$$y = x + 10 \rightarrow g(x) = x + 10.$$

To solve this system, whether it is written in function notation or not, we graph each equation and look for the point of intersection of the graphs. (We may need to use detailed graphing paper to determine the point of intersection exactly.)

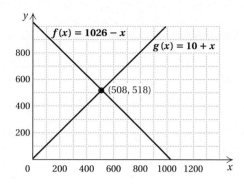

As we see in the graph above, the ordered pair $(508, 518)$ is the intersection and thus the solution—that is, $x = 508$ and $y = 518$. This tells us that the average verbal score was 508 and the average math score was 518.

a Solving Systems of Equations Graphically

ONE SOLUTION

A **solution** of a system of two equations in two variables is an ordered pair that makes *both* equations true. If we graph a system of equations or functions, the point at which the graphs intersect will be a solution of *both* equations or functions.

EXAMPLE 1 Solve this system graphically:

$$y - x = 1,$$
$$y + x = 3.$$

We draw the graph of each equation using any method studied in Chapter 2.

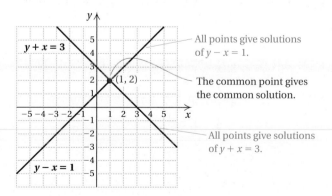

All points give solutions of $y - x = 1$.

The common point gives the common solution.

All points give solutions of $y + x = 3$.

The point of intersection has coordinates that make *both* equations true. The solution seems to be the point $(1, 2)$. However, since graphing alone is not perfectly accurate, solving by graphing may give only approximate answers. Thus we check the pair $(1, 2)$ in both equations as follows.

Solve the system graphically.

1. $-2x + y = 1,$
  $3x + y = 1$

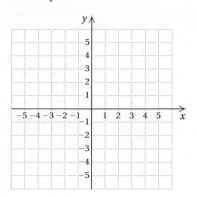

2. $f(x) = \frac{1}{2}x,$
  $g(x) = -\frac{1}{4}x + \frac{3}{2}$

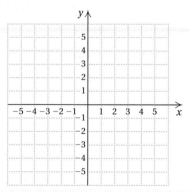

Answers on page A-17

Check:

$y - x = 1$		$y + x = 3$	
$2 - 1 \overset{?}{\mid} 1$		$2 + 1 \overset{?}{\mid} 3$	
$1 \mid$	TRUE	$3 \mid$	TRUE

The solution is $(1, 2)$.

Do Exercises 1 and 2 on the preceding page.

CALCULATOR CORNER

Solving Systems of Equations   We can solve a system of two equations in two variables on a graphing calculator. Consider the system of equations in Example 1:

$$y - x = 1,$$
$$y + x = 3.$$

First, we solve the equations for y, obtaining $y = x + 1$ and $y = -x + 3$. Next, we enter $y_1 = x + 1$ and $y_2 = -x + 3$ on the equation-editor screen and graph the equations. We can use the standard viewing window, $[-10, 10, -10, 10]$. (If you turned on Plot 1 to do the Calculator Corner in Section 2.6 and haven't turned it off yet, do so now. See p. 228 for the procedure.)

   We will use the **INTERSECT** feature to find the coordinates of the point of intersection of the lines. To access this feature, we press **2ND** **CALC** **5**. (CALC is the second operation associated with the **TRACE** key.) The query "First curve?" appears on the graph screen. The blinking cursor is positioned on the graph of y_1. We press **ENTER** to indicate that this is the first curve involved in the intersection. Next, the query "Second curve?" appears and the blinking cursor is positioned on the graph of y_2. We press **ENTER** to indicate that this is the second curve. Now the query "Guess?" appears. We use the ▷ and ◁ keys to move the cursor or enter an x-value close to the point of intersection and press **ENTER**. The coordinates of the point of intersection of the graphs, $x = 1$, $y = 2$, appear at the bottom of the screen. Thus the solution of the system of equations is $(1, 2)$.

$$y_1 = x + 1, \ y_2 = -x + 3$$

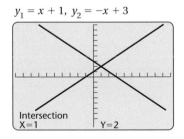

Exercises: Use a graphing calculator to solve the system of equations.

1. $x + y = 5,$
   $\quad y = x + 1$

2. $y = x + 3,$
   $\quad 2x - y = -7$

3. $x - y = -6,$
   $\quad y = 2x + 7$

4. $x + 4y = -1,$
   $\quad x - y = 4$

NO SOLUTION

Sometimes the equations in a system have graphs that are parallel lines.

EXAMPLE 2 Solve graphically:

$$f(x) = -3x + 5,$$
$$g(x) = -3x - 2.$$

We graph the functions. The graphs have the same slope, -3, and different y-intercepts, so they are parallel. There is no point at which they cross, so the system has no solution. No matter what point we try, it will *not* check in *both* equations. The solution set is thus the empty set, denoted $\varnothing$ or { }.

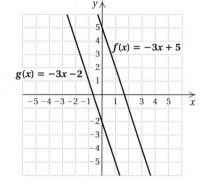

CONSISTENT AND INCONSISTENT SYSTEMS
A system of equations has at least one solution. ➡ It is **consistent.** A system of equations has no solution. ➡ It is **inconsistent.**

The system in Example 1 is consistent. The system in Example 2 is inconsistent.

Do Exercises 3 and 4.

INFINITELY MANY SOLUTIONS

Sometimes the equations in a system have the same graph. In such a case, the equations have an *infinite* number of solutions in common.

EXAMPLE 3 Solve graphically:

$$3y - 2x = 6,$$
$$-12y + 8x = -24.$$

We graph the equations and see that the graphs are the same. Thus any solution of one of the equations is a solution of the other. Each equation has an infinite number of solutions, two of which are shown on the graph.

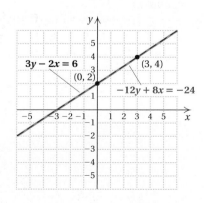

3. Solve graphically:

$$y + 2x = 3,$$
$$y + 2x = -4.$$

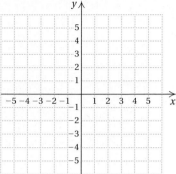

4. Classify each of the systems in Margin Exercises 1–3 as consistent or inconsistent.

Answers on page A-17

5. Solve graphically:

$$2x - 5y = 10,$$
$$-6x + 15y = -30.$$

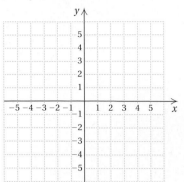

6. Classify the equations in Margin Exercises 1, 2, 3, and 5 as dependent or independent.

7. a) Solve $x + 1 = \frac{2}{3}x$ algebraically.

b) Solve $x + 1 = \frac{2}{3}x$ graphically using Method 1.

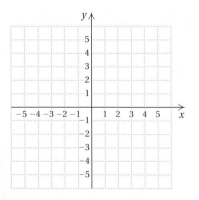

c) Compare your answers to parts (a) and (b).

We check one such solution, $(0, 2)$, which is the y-intercept of each equation.

Check:
$$\begin{array}{c|c} 3y - 2x = 6 \\ \hline 3(2) - 2(0) \;?\; 6 \\ 6 - 0 \\ 6 \end{array} \quad \text{TRUE} \qquad \begin{array}{c|c} -12y + 8x = -24 \\ \hline -12(2) + 8(0) \;?\; -24 \\ -24 + 0 \\ -24 \end{array} \quad \text{TRUE}$$

On your own, check that $(3, 4)$ is a solution of both equations. If $(0, 2)$ and $(3, 4)$ are solutions, then all points on the line containing them will be solutions. The system has an infinite number of solutions.

DEPENDENT AND INDEPENDENT EQUATIONS

A system of two equations in two variables:

has infinitely many solutions. ➡ The equations are **dependent.**

has one solution or no solutions. ➡ The equations are **independent.**

When we graph a system of two equations, one of the following three things can happen.

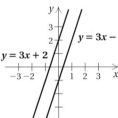 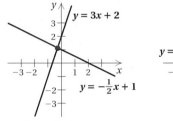 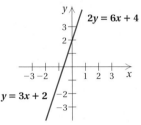

One solution.
Graphs intersect.
The system is consistent
and *the equations are*
independent.

No solution.
Graphs are parallel.
The system is inconsistent
and *the equations are*
independent.

Infinitely many solutions.
Equations have the same
graph. *The system is* consistent
and *the equations are*
dependent.

The equations in Example 1 have one point in common, and that point is the only solution of the system. The equations in Example 3 have an infinite number of solutions in common. Any system that has at least one solution is said to be *consistent*. The systems in Examples 1 and 3 are *consistent*. The system in Example 2 is *inconsistent*.

When one equation in a system can be obtained by multiplying both sides of another equation by a constant, the two equations are said to be *dependent*. The equations in Example 3 are *dependent*, and the equations in Examples 1 and 2 are *independent*.

Do Exercises 5 and 6.

Answers on page A-17

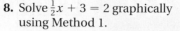

AG ALGEBRAIC–GRAPHICAL CONNECTION

To bring together the concepts of Chapters 1–3, let's look at equation solving from both algebraic and graphical viewpoints.

Consider the equation $-2x + 13 = 4x - 17$. Let's solve it algebraically as we did in Chapter 2:

$$-2x + 13 = 4x - 17$$
$$13 = 6x - 17 \qquad \text{Adding } 2x$$
$$30 = 6x \qquad \text{Adding } 17$$
$$5 = x. \qquad \text{Dividing by } 6$$

Could we also solve the equation graphically? The answer is yes, as we see in the following two methods.

METHOD 1 Solve $-2x + 13 = 4x - 17$ graphically.

We let $f(x) = -2x + 13$ and $g(x) = 4x - 17$. Graphing the system of equations, we get the graph shown at right.

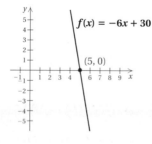

The point of intersection of the two graphs is $(5, 3)$. Note that the x-coordinate of the intersection is 5. This value for x is the solution of the equation $-2x + 13 = 4x - 17$.

Do Exercises 7 and 8. (Exercise 7 is on the preceding page.)

METHOD 2 Solve $-2x + 13 = 4x - 17$ graphically.

Adding $-4x$ and 17 on both sides, we obtain the form

$$-6x + 30 = 0.$$

This time we let $f(x) = -6x + 30$ and $g(x) = 0$. Since $g(x) = 0$, or $y = 0$, is the x-axis, we need only graph $f(x) = -6x + 30$ and see where it crosses the x-axis.

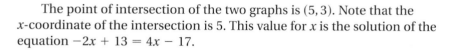

Note that the x-intercept of $f(x) = -6x + 30$ is $(5, 0)$, or just 5. This x-value is the solution of the equation $-2x + 13 = 4x - 17$.

Do Exercise 9.

Let's compare the two methods. Using Method 1, we graph two functions. The solution of the original equation is the x-coordinate of the point of intersection. Using Method 2, we graph one function. The solution of the original equation is the x-coordinate of the x-intercept of the graph.

Do Exercise 10.

8. Solve $\frac{1}{2}x + 3 = 2$ graphically using Method 1.

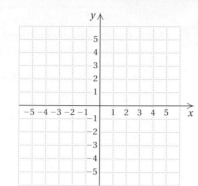

9. a) Solve $x + 1 = \frac{2}{3}x$ graphically using Method 2.

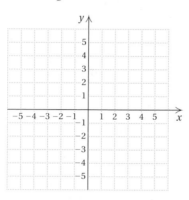

b) Compare your answers to Margin Exercises 7(a), 7(b), and 9(a).

10. Solve $\frac{1}{2}x + 3 = 2$ graphically using Method 2.

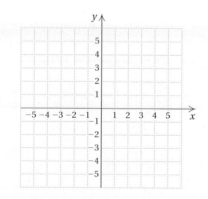

Answers on page A-17

249

3.1 Systems of Equations in
Two Variables

3.1

EXERCISE SET

For Extra Help

MathXL   MyMathLab   InterAct   Math Tutor   Digital Video   Student's
                      Math   Center   Tutor CD 2   Solutions
                                            Videotape 4   Manual

a Solve the system of equations graphically. Then classify the system as consistent or inconsistent and the equations as dependent or independent. Complete the check for Exercises 1–4.

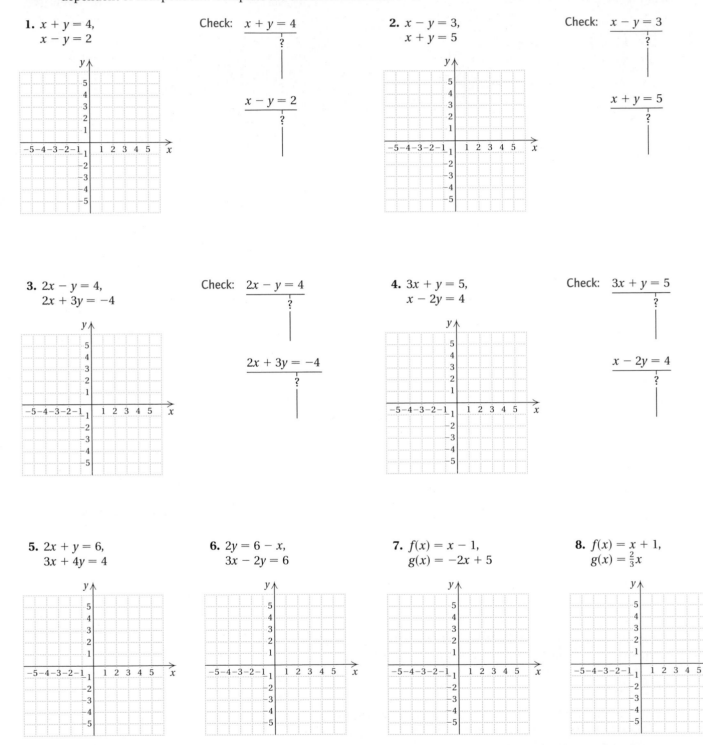

1. $x + y = 4,$
$x - y = 2$

Check: $x + y = 4$
$\dfrac{}{?}$

$x - y = 2$
$\dfrac{}{?}$

2. $x - y = 3,$
$x + y = 5$

Check: $x - y = 3$
$\dfrac{}{?}$

$x + y = 5$
$\dfrac{}{?}$

3. $2x - y = 4,$
$2x + 3y = -4$

Check: $2x - y = 4$
$\dfrac{}{?}$

$2x + 3y = -4$
$\dfrac{}{?}$

4. $3x + y = 5,$
$x - 2y = 4$

Check: $3x + y = 5$
$\dfrac{}{?}$

$x - 2y = 4$
$\dfrac{}{?}$

5. $2x + y = 6,$
$3x + 4y = 4$

6. $2y = 6 - x,$
$3x - 2y = 6$

7. $f(x) = x - 1,$
$g(x) = -2x + 5$

8. $f(x) = x + 1,$
$g(x) = \frac{2}{3}x$

9. $2u + v = 3,$
    $2u = v + 7$

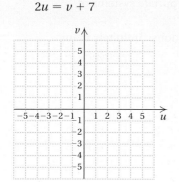

10. $2b + a = 11,$
    $a - b = 5$

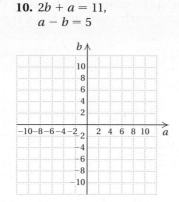

11. $f(x) = -\frac{1}{3}x - 1,$
    $g(x) = \frac{4}{3}x - 6$

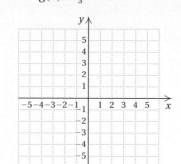

12. $f(x) = -\frac{1}{4}x + 1,$
    $g(x) = \frac{1}{2}x - 2$

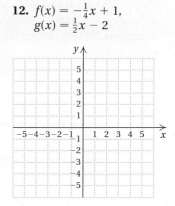

13. $6x - 2y = 2,$
    $9x - 3y = 1$

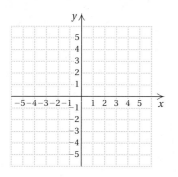

14. $y - x = 5,$
    $2x - 2y = 10$

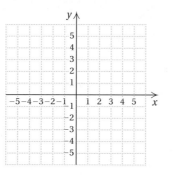

15. $2x - 3y = 6,$
    $3y - 2x = -6$

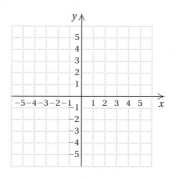

16. $y = 3 - x,$
    $2x + 2y = 6$

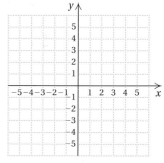

17. $x = 4,$
    $y = -5$

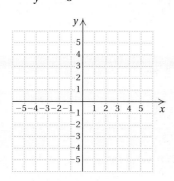

18. $x = -3,$
    $y = 2$

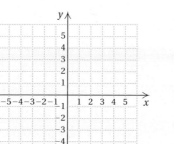

19. $y = -x - 1,$
    $4x - 3y = 17$

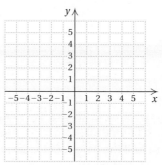

20. $a + 2b = -3,$
    $b - a = 6$

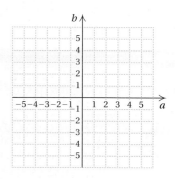

Matching. Each of Exercises 21–26 shows the graph of a system and its solution. First, classify the system as consistent or inconsistent and the equations as dependent or independent. Then match it with one of the appropriate systems of equations (A)–(F), which follow.

21. Solution: $(3, 3)$

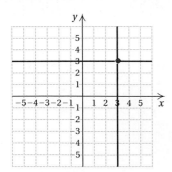

22. Solution: $(1, 1)$

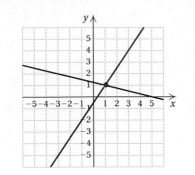

23. Solutions: Infinitely many

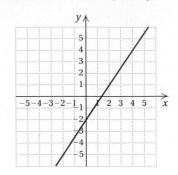

24. Solution: $(4, -3)$

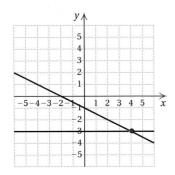

25. Solution: No solution

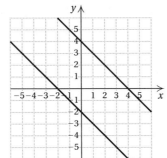

26. Solution: $(-1, 3)$

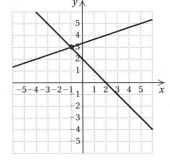

A. $3y - x = 10,$
   $x = -y + 2$

B. $9x - 6y = 12,$
   $y = \frac{3}{2}x - 2$

C. $2y - 3x = -1,$
   $x + 4y = 5$

D. $x + y = 4,$
   $y = -x - 2$

E. $\frac{1}{2}x + y = -1,$
   $y = -3$

F. $x = 3,$
   $y = 3$

27. D_W Explain how to find the solution of $\frac{3}{4}x + 2 = \frac{2}{5}x - 5$ in two ways graphically and in two ways algebraically.

28. D_W Write a system of equations with the given solution. Answers may vary.

a) $(4, -3)$ **b)** No solution
c) Infinitely many solutions

Solve.   [1.1d]

29. $3x + 4 = x - 2$

30. $\frac{3}{4}x + 2 = \frac{2}{5}x - 5$

31. $4x - 5x = 8x - 9 + 11x$

32. $5(10 - 4x) = -3(7x - 4)$

SYNTHESIS

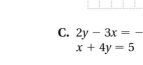 Use a graphing calculator to find the point of intersection of the pair of equations. Round all answers to the nearest hundredth. You may need to solve for y first.

33. $2.18x + 7.81y = 13.78,$
   $5.79x - 3.45y = 8.94$

34. $f(x) = 123.52x + 89.32,$
   $g(x) = -89.22x + 33.76$

Solve graphically.

35. $y = |x|,$
   $x + 4y = 15$

36. $x - y = 0,$
   $y = x^2$

252

3.2 SOLVING BY SUBSTITUTION

Objectives

a Solve systems of equations in two variables by the substitution method.

b Solve applied problems by solving systems of two equations using substitution.

Consider this system of equations:

$$5x + 9y = 2,$$
$$4x - 9y = 10.$$

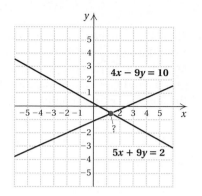

What is the solution? It is rather difficult to tell exactly by graphing. It would appear that fractions are involved. It turns out that the solution is

$$\left(\frac{4}{3}, -\frac{14}{27}\right).$$

Solving by graphing, though useful in many applied situations, is not always fast or accurate in cases where solutions are not integers. We need techniques involving algebra to determine the solution exactly. Because they use algebra, they are called **algebraic methods.**

a The Substitution Method

One nongraphical method for solving systems is known as the **substitution method.**

EXAMPLE 1 Solve this system:

$$x + y = 4, \qquad (1)$$
$$x = y + 1. \qquad (2)$$

Equation (2) says that x and $y + 1$ name the same number. Thus we can substitute $y + 1$ for x in equation (1):

$$x + y = 4 \qquad \text{Equation (1)}$$
$$(y + 1) + y = 4. \qquad \text{Substituting } y + 1 \text{ for } x$$

Since this equation has only one variable, we can solve for y using methods learned earlier:

$$(y + 1) + y = 4$$
$$2y + 1 = 4 \qquad \text{Removing parentheses and collecting like terms}$$
$$2y = 3 \qquad \text{Subtracting 1}$$
$$y = \tfrac{3}{2}. \qquad \text{Dividing by 2}$$

We return to the original pair of equations and substitute $\frac{3}{2}$ for y in *either* equation so that we can solve for x. Calculation will be easier if we choose equation (2) since it is already solved for x:

$$x = y + 1 \qquad \text{Equation (2)}$$
$$= \tfrac{3}{2} + 1 \qquad \text{Substituting } \tfrac{3}{2} \text{ for } y$$
$$= \tfrac{3}{2} + \tfrac{2}{2} = \tfrac{5}{2}.$$

We obtain the ordered pair $\left(\frac{5}{2}, \frac{3}{2}\right)$. Even though we solved for y *first*, it is still the *second* coordinate since x is before y alphabetically. We check to be sure that the ordered pair is a solution.

253

Solve by the substitution method.

1. $x + y = 6,$
$\quad y = x + 2$

2. $y = 7 - x,$
$\quad 2x - y = 8$

(*Caution*: Use parentheses when you substitute, being careful about removing them. Remember to solve for both variables.)

Solve by the substitution method.

3. $2y + x = 1,$
$\quad y - 2x = 8$

4. $8x + 5y = 184,$
$\quad x - y = -3$

Answers on page A-18

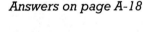

CALCULATOR CORNER

Solving Systems of Equations Use the INTERSECT feature to solve the systems of equations in Margin Exercises 1, 2, and 3. (See the Calculator Corner on p. 246 for the procedure.)

Check:

$$\begin{array}{c|c} x + y = 4 \\ \hline \frac{5}{2} + \frac{3}{2} \;?\; 4 \\ \frac{8}{2} \\ 4 & \text{TRUE} \end{array} \qquad \begin{array}{c|c} x = y + 1 \\ \hline \frac{5}{2} \;?\; \frac{3}{2} + 1 \\ \frac{3}{2} + \frac{2}{2} \\ \frac{5}{2} & \text{TRUE} \end{array}$$

Since $\left(\frac{5}{2}, \frac{3}{2}\right)$ checks, it is the solution. Even though exact fraction solutions are difficult to determine graphically, a graph can help us to visualize whether the solution is reasonable.

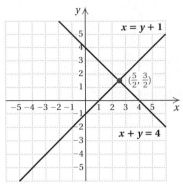

Do Exercises 1 and 2.

Suppose neither equation of a pair has a variable alone on one side. We then solve one equation for one of the variables.

EXAMPLE 2 Solve this system:

$$2x + y = 6, \quad (1)$$
$$3x + 4y = 4. \quad (2)$$

First, we solve one equation for one variable. Since the coefficient of y is 1 in equation (1), it is the easier one to solve for y:

$$y = 6 - 2x. \quad (3)$$

Next, we substitute $6 - 2x$ for y in equation (2) and solve for x:

$$\begin{array}{ll} 3x + 4(6 - 2x) = 4 & \text{Substituting } 6 - 2x \text{ for } y \\ 3x + 24 - 8x = 4 & \text{Multiplying to remove parentheses} \\ 24 - 5x = 4 & \text{Collecting like terms} \\ -5x = -20 & \text{Subtracting 24} \\ x = 4. & \text{Dividing by } -5 \end{array}$$

Caution!

Remember to use parentheses when you substitute. Then remove them carefully.

In order to find y, we return to either of the original equations, (1) or (2), or equation (3), which we solved for y. It is generally easier to use an equation like (3), where we have solved for the specific variable. We substitute 4 for x in equation (3) and solve for y:

$$y = 6 - 2x = 6 - 2(4) = 6 - 8 = -2.$$

We obtain the ordered pair $(4, -2)$.

Check:

$$\begin{array}{c|c} 2x + y = 6 \\ \hline 2(4) + (-2) \;?\; 6 \\ 8 - 2 \\ 6 & \text{TRUE} \end{array} \qquad \begin{array}{c|c} 3x + 4y = 4 \\ \hline 3(4) + 4(-2) \;?\; 4 \\ 12 - 8 \\ 4 & \text{TRUE} \end{array}$$

Since $(4, -2)$ checks, it is the solution.

Do Exercises 3 and 4.

EXAMPLE 3 Solve this system of equations:

$$y = -3x + 5, \quad (1)$$
$$y = -3x - 2. \quad (2)$$

We solved this system graphically in Example 2 of Section 3.1. We found that the graphs are parallel and the system has no solution. Let's try to solve this system algebraically using substitution.

We substitute $-3x - 2$ for y in equation (1):

$$-3x - 2 = -3x + 5 \qquad \text{Substituting } -3x - 2 \text{ for } y$$
$$-2 = 5. \qquad \text{Adding } 3x$$

We have a false equation. The equation has **no solution.** (See also Example 17 of Section 1.1.)

Do Exercise 5.

5. a) Solve this system of equations algebraically using substitution:

$$y + 2x = 3,$$
$$y + 2x = -4.$$

b) Check your answer in part (a) with the one you found graphically in Margin Exercise 3 of Section 3.1.

b Solving Applied Problems Involving Two Equations

Many applied problems are easier to solve if we first translate to a system of two equations rather than to a single equation. Here we will solve a few problems that can be solved using substitution. Section 3.4 is devoted entirely to applied problems.

EXAMPLE 4 *Architecture.* The architects who designed the John Hancock Building in Chicago created a visually appealing building that slants on the sides. Thus the ground floor is a rectangle that is larger than the rectangle formed by the top floor. The ground floor has a perimeter of 860 ft. The length is 100 ft more than the width. Find the length and the width.

1. **Familiarize.** We first make a drawing and label it, using l for length and w for width. We recall, or look up, the formula for perimeter: $P = 2l + 2w$. This formula can be found at the back of this book.

2. **Translate.** We translate as follows:

The perimeter is 860 ft.

$$2l + 2w = 860$$

We then translate the second statement:

The length is 100 ft more than the width.

$$l = w + 100$$

We now have a system of equations:

$$2l + 2w = 860, \quad (1)$$
$$l = w + 100. \quad (2)$$

Answers on page A-18

6. Architecture. The top floor of the John Hancock Building is also a rectangle, but its perimeter is 520 ft. The width is 60 ft less than the length. Find the length and the width.

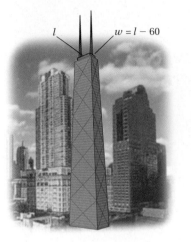

l $w = l - 60$

3. Solve. We substitute $w + 100$ for l in equation (1):

$$2(w + 100) + 2w = 860 \qquad \text{Substituting in equation (1)}$$

$$2w + 200 + 2w = 860 \qquad \begin{array}{l}\text{Multiplying to remove parentheses} \\ \text{on the left}\end{array}$$

$$4w + 200 = 860 \qquad \text{Collecting like terms}$$

$$\left.\begin{array}{l}4w = 660 \\ w = 165.\end{array}\right\} \text{Solving for } w$$

Next, we substitute 165 for w in equation (2) and solve for l:

$$l = 165 + 100 = 265.$$

4. Check. Consider the dimensions 265 ft and 165 ft. The length is 100 ft more than the width. The perimeter is 2(265 ft) + 2(165 ft), or 860 ft. The dimensions 265 ft and 165 ft check in the original problem.

5. State. The length is 265 ft and the width is 165 ft.

Do Exercise 6.

Answer on page A-18

Study Tips

TEST PREPARATION

Success on exams will increase when you put in place a plan of study. Here are some test-preparation and test-taking study tips.

- **Make up your own test questions as you study.** After you have done your homework over a particular objective, write one or two questions on your own that you think might be on a test. You will be amazed at the insight this will provide.

- **Do an overall review of the chapter, focusing on the objectives and the examples.** This should be accompanied by a study of any class notes you may have taken.

- **Do the review exercises at the end of the chapter.** Check your answers at the back of the book. If you have trouble with an exercise, use the objective symbol as a guide to go back and do further study of that objective.

- **Call the AW Math Tutor Center at 1-888-777-0463 if you need extra help.**

- **Do the chapter test at the end of the chapter.** Check the answers and use the objective symbols at the back of the book as a reference for where to review.

- **Ask former students for old exams.** Working such exams can be very helpful and allows you to see what various professors think is important.

- **When taking a test, read each question carefully and try to do all the questions the first time through, but pace yourself.** Answer all the questions, and mark those to recheck if you have time at the end. Very often, your first hunch will be correct.

- **Try to write your test in a neat and orderly manner.** Very often, your instructor tries to give you partial credit when grading an exam. If your test paper is sloppy and disorderly, it is difficult to verify the partial credit. Doing your work neatly can ease such a task for the instructor.

a   Solve the system by the substitution method.

1. $y = 5 - 4x,$
    $2x - 3y = 13$

2. $x = 8 - 4y,$
    $3x + 5y = 3$

3. $2y + x = 9,$
    $x = 3y - 3$

4. $9x - 2y = 3,$
    $3x - 6 = y$

5. $3s - 4t = 14,$
    $5s + t = 8$

6. $m - 2n = 3,$
    $4m + n = 1$

7. $9x - 2y = -6,$
    $7x + 8 = y$

8. $t = 4 - 2s,$
    $t + 2s = 6$

9. $-5s + t = 11,$
    $4s + 12t = 4$

10. $5x + 6y = 14,$
    $-3y + x = 7$

11. $2x + 2y = 2,$
    $3x - y = 1$

12. $4p - 2q = 16,$
    $5p + 7q = 1$

13. $3a - b = 7,$
    $2a + 2b = 5$

14. $5x + 3y = 4,$
    $x - 4y = 3$

15. $2x - 3 = y,$
    $y - 2x = 1$

16. $4x + 13y = 5,$
    $-6x + y = 13$

b   Solve.

17. *Racquetball Court.*   A regulation racquetball court has a perimeter of 120 ft, with a length that is twice the width. Find the length and the width of such a court.

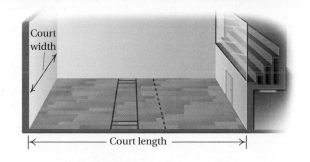

18. *Soccer Field.*   The perimeter of a soccer field is 340 m. The length exceeds the width by 50 m. Find the length and the width.

19. *Supplementary Angles.* **Supplementary angles** are angles whose sum is 180°. Two supplementary angles are such that one angle is 12° less than three times the other. Find the measures of the angles.

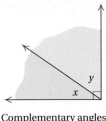

Supplementary angles:
$x + y = 180°$

20. *Complementary Angles.* **Complementary angles** are angles whose sum is 90°. Two complementary angles are such that one angle is 6° more than five times the other. Find the measures of the angles.

Complementary angles:
$x + y = 90°$

21. *Hockey Points.* At one time, hockey teams received two points when they won a game and one point when they tied. One season, a team won a championship with 60 points. They won 9 more games than they tied. How many wins and how many ties did the team have?

22. *Airplane Seating.* An airplane has a total of 152 seats. The number of coach-class seats is 5 more than six times the number of first-class seats. How many of each type of seat are there on the plane?

23. D_W Describe a method that could be used to create inconsistent systems of equations.

24. D_W Write a problem for a classmate to solve that requires writing a system of two equations. Devise the problem so that the solution is "The Lakers made 6 three-point baskets and 31 two-point baskets."

SKILL MAINTENANCE

25. Find the slope of the line $y = 1.3x - 7$. [2.4b]

26. Simplify: $-9(y + 7) - 6(y - 4)$. [R.6b]

27. Solve $A = \dfrac{pq}{7}$ for p. [1.2a]

28. Find the slope of the line containing the points $(-2, 3)$ and $(-5, -4)$. [2.4b]

Solve. [1.1d]

29. $-4x + 5(x - 7) = 8x - 6(x + 2)$

30. $-12(2x - 3) = 16(4x - 5)$

SYNTHESIS

31. For $y = mx + b$, two solutions are $(1, 2)$ and $(-3, 4)$. Find m and b.

32. Solve for x and y in terms of a and b:

$$5x + 2y = a,$$
$$x - y = b.$$

33. *Design.* A piece of posterboard has a perimeter of 156 in. If you cut 6 in. off the width, the length becomes four times the width. What are the dimensions of the original piece of posterboard?

$P = 156$ in.

34. *Nontoxic Scouring Powder.* A nontoxic scouring powder is made up of 4 parts baking soda and 1 part vinegar. How much of each ingredient is needed for a 16-oz mixture?

3.3 SOLVING BY ELIMINATION

Objectives

a Solve systems of equations in two variables by the elimination method.

b Solve applied problems by solving systems of two equations using elimination.

a The Elimination Method

The **elimination method** for solving systems of equations makes use of the *addition principle* for equations. Some systems are much easier to solve using the elimination method rather than the substitution method.

EXAMPLE 1 Solve this system:

$$2x - 3y = 0, \qquad (1)$$
$$-4x + 3y = -1. \qquad (2)$$

The key to the advantage of the elimination method in this case is the $-3y$ in one equation and the $3y$ in the other. These terms are opposites. If we add them, these terms will add to 0, and in effect, the variable y will have been "eliminated."

We will use the addition principle for equations, adding the same number on both sides of the equation. According to equation (2), $-4x + 3y$ and -1 are the same number. Thus we can use a vertical form and add $-4x + 3y$ to the left side of equation (1) and -1 to the right side:

$$
\begin{array}{ll}
2x - 3y = 0 & (1) \\
\underline{-4x + 3y = -1} & (2) \\
-2x + 0y = -1 & \text{Adding} \\
-2x + 0 = -1 & \\
-2x = -1. &
\end{array}
$$

We have eliminated the variable y, which is why we call this the *elimination method*.* We now have an equation with just one variable, which we solve for x:

$$-2x = -1$$
$$x = \tfrac{1}{2}.$$

Next, we substitute $\tfrac{1}{2}$ for x in either equation and solve for y:

$$
\begin{array}{ll}
2 \cdot \tfrac{1}{2} - 3y = 0 & \text{Substituting in equation (1)} \\
1 - 3y = 0 & \\
-3y = -1 & \text{Subtracting 1} \\
y = \tfrac{1}{3}. & \text{Dividing by } -3
\end{array}
$$

We obtain the ordered pair $\left(\tfrac{1}{2}, \tfrac{1}{3}\right)$.

Check:

$$
\begin{array}{c|c}
2x - 3y = 0 & -4x + 3y = -1 \\
\hline
2\left(\tfrac{1}{2}\right) - 3\left(\tfrac{1}{3}\right) \;?\; 0 & -4\left(\tfrac{1}{2}\right) + 3\left(\tfrac{1}{3}\right) \;?\; -1 \\
1 - 1 & -2 + 1 \\
0 \quad \text{TRUE} & -1 \quad \text{TRUE}
\end{array}
$$

---

*Also called the *addition method*.

Solve by the elimination method.

1. $5x + 3y = 17,$
$-5x + 2y = 3$

2. $-3a + 2b = 0,$
$3a - 4b = -1$

3. Solve by the elimination method:

$$2y + 3x = 12,$$
$$-4y + 5x = -2.$$

Answers on page A-18

Since $\left(\frac{1}{2}, \frac{1}{3}\right)$ checks, it is the solution. We can also see this in the graph shown at right.

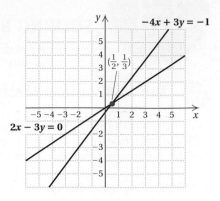

Do Exercises 1 and 2.

In order to eliminate a variable, we sometimes use the multiplication principle to multiply one or both of the equations by a particular number before adding.

EXAMPLE 2   Solve this system:

$$3x + 3y = 15, \quad \textbf{(1)}$$
$$2x + 6y = 22. \quad \textbf{(2)}$$

If we add directly, we get $5x + 9y = 37$, and we have not eliminated a variable. However, note that if the $3y$ in equation (1) were $-6y$, we could eliminate y. Thus we multiply by -2 on both sides of equation (1) and add:

$$
\begin{array}{ll}
-6x - 6y = -30 & \text{Multiplying by } -2 \text{ on both sides of equation (1)} \\
\underline{2x + 6y = 22} & \text{Equation (2)} \\
-4x + 0 = -8 & \text{Adding} \\
-4x = -8 & \\
x = 2. & \text{Solving for } x
\end{array}
$$

Then

$$
\begin{array}{ll}
2 \cdot 2 + 6y = 22 & \text{Substituting 2 for } x \text{ in equation (2)} \\
\left.\begin{array}{l}
4 + 6y = 22 \\
6y = 18 \\
y = 3.
\end{array}\right\} & \text{Solving for } y
\end{array}
$$

We obtain $(2, 3)$, or $x = 2$, $y = 3$. This checks, so it is the solution.

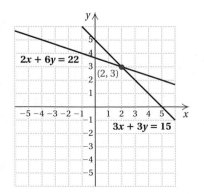

Do Exercise 3.

Sometimes we must multiply twice in order to make two terms opposites.

EXAMPLE 3 Solve this system:

$$2x + 3y = 17, \quad \textbf{(1)}$$
$$5x + 7y = 29. \quad \textbf{(2)}$$

We must first multiply in order to make one pair of terms with the same variable opposites. We decide to do this with the x-terms in each equation. We multiply equation (1) by 5 and equation (2) by -2. Then we get $10x$ and $-10x$, which are opposites.

From equation (1):	$10x + 15y = 85$	Multiplying by 5
From equation (2):	$-10x - 14y = -58$	Multiplying by -2
	$0 + y = 27$	Adding
	$y = 27.$	Solving for y

Then

$$2x + 3 \cdot 27 = 17 \qquad \text{Substituting 27 for } y \text{ in equation (1)}$$
$$\left.\begin{array}{r} 2x + 81 = 17 \\ 2x = -64 \\ x = -32. \end{array}\right\} \quad \text{Solving for } x$$

We check the ordered pair $(-32, 27)$.

Check:

$$\begin{array}{c|c} 2x + 3y = 17 & 5x + 7y = 29 \\ \hline 2(-32) + 3(27) \ ? \ 17 & 5(-32) + 7(27) \ ? \ 29 \\ -64 + 81 & -160 + 189 \\ 17 \ \big| \quad \text{TRUE} & 29 \ \big| \quad \text{TRUE} \end{array}$$

We obtain $(-32, 27)$, or $x = -32$, $y = 27$, as the solution.

Do Exercises 4 and 5.

Some systems have no solution, as we saw graphically in Section 3.1 and algebraically in Example 3 of Section 3.2. How do we recognize such systems if we are solving using elimination?

EXAMPLE 4 Solve this system:

$$y + 3x = 5, \quad \textbf{(1)}$$
$$y + 3x = -2. \quad \textbf{(2)}$$

If we find the slope–intercept equations for this system, we get

$$y = -3x + 5,$$
$$y = -3x - 2.$$

The graphs are parallel lines.
The system has no solution.

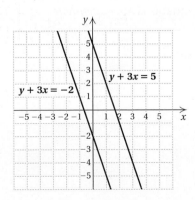

Solve by the elimination method.

4. $4x + 5y = -8,$
$\quad 7x + 9y = 11$

5. $4x - 5y = 38,$
$\quad 7x - 8y = -22$

Answers on page A-18

6. Solve by the elimination method:

$$y + 2x = 3,$$
$$y + 2x = -1.$$

Let's see what happens if we attempt to solve the system by the elimination method. We multiply by -1 on both sides of equation (2) and add:

$$y + 3x = 5 \qquad \text{Equation (1)}$$
$$\underline{-y - 3x = 2} \qquad \text{Multiplying equation (2) by } -1$$
$$0 = 7. \qquad \text{Adding, we obtain a false equation.}$$

The x-terms and the y-terms are eliminated and we end up with a *false* equation. Thus, if we obtain a false equation, such as $0 = 7$, when solving algebraically, we know that the system has **no solution.** The system is inconsistent, and the equations are independent.

Do Exercise 6.

Some systems have infinitely many solutions. How can we recognize such a situation when we are solving systems using an algebraic method?

EXAMPLE 5 Solve this system:

$$3y - 2x = 6, \qquad \textbf{(1)}$$
$$-12y + 8x = -24. \qquad \textbf{(2)}$$

The graphs are the same line. The system has an infinite number of solutions.

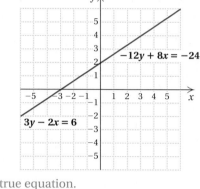

Suppose we try to solve this system by the elimination method:

$$12y - 8x = 24 \qquad \text{Multiplying equation (1) by 4}$$
$$\underline{-12y + 8x = -24} \qquad \text{Equation (2)}$$
$$0 = 0. \qquad \text{Adding, we obtain a true equation.}$$

We have eliminated both variables, and what remains is a true equation, $0 = 0$. It can be expressed as $0 \cdot x + 0 \cdot y = 0$, and is true for all numbers x and y. If an ordered pair is a solution of one of the original equations, then it will be a solution of the other. The system has an **infinite number of solutions.** The system is consistent, and the equations are dependent.

7. Solve by the elimination method:

$$2x - 5y = 10,$$
$$-6x + 15y = -30.$$

SPECIAL CASES

When solving a system of two linear equations in two variables:

1. If a false equation is obtained, such as $0 = 7$, then the system has no solution. The system is *inconsistent,* and the equations are *independent.*
2. If a true equation is obtained, such as $0 = 0$, then the system has an infinite number of solutions. The system is *consistent,* and the equations are *dependent.*

Do Exercise 7.

Answers on page A-18

Solving Systems of Equations

1. Consider the system of equations in Example 4. Before we can enter these equations on a graphing calculator, each must be solved for y. What happens when we do this? What does this indicate about the nature of the solutions of the system of equations?

2. Consider the system of equations in Example 5. Before we can enter these equations on a graphing calculator, each must be solved for y. What happens when we do this? What does this indicate about the nature of the solutions of the system of equations?

When solving a system by the elimination method, it helps to first write the equations in the form $Ax + By = C$. When decimals or fractions occur, it also helps to *clear* before solving.

EXAMPLE 6 Solve this system:

$$0.2x + 0.3y = 1.7,$$
$$\tfrac{1}{7}x + \tfrac{1}{5}y = \tfrac{29}{35}.$$

We have

$$0.2x + 0.3y = 1.7, \xrightarrow[\text{to clear decimals}]{\text{Multiplying by 10}} 2x + 3y = 17,$$
$$\tfrac{1}{7}x + \tfrac{1}{5}y = \tfrac{29}{35} \xrightarrow[\text{to clear fractions}]{\text{Multiplying by 35}} 5x + 7y = 29.$$

We multiplied by 10 to clear the decimals. Multiplication by 35, the least common denominator, clears the fractions. The problem is now identical to Example 3. The solution is $(-32, 27)$, or $x = -32$, $y = 27$.

Do Exercises 8 and 9.

To use the elimination method to solve systems of two equations:

1. Write both equations in the form $Ax + By = C$.
2. Clear any decimals or fractions.
3. Choose a variable to eliminate.
4. Make a chosen variable's terms opposites by multiplying one or both equations by appropriate numbers if necessary.
5. Eliminate a variable by adding the sides of the equations and then solve for the remaining variable.
6. Substitute in either of the original equations to find the value of the other variable.

8. Clear the decimals. Then solve.

$$0.02x + 0.03y = 0.01,$$
$$0.3x - 0.1y = 0.7$$

(*Hint*: Multiply the first equation by 100 and the second one by 10.)

9. Clear the fractions. Then solve.

$$\frac{3}{5}x + \frac{2}{3}y = \frac{1}{3},$$
$$\frac{3}{4}x - \frac{1}{3}y = \frac{1}{4}$$

Answers on page A-18

COMPARING METHODS

When deciding which method to use, consider this table and directions from your instructor. The situation is analogous to having a piece of wood to cut and three different types of saws available. Although all three saws can cut the wood, the "best" choice depends on the particular piece of wood, the type of cut being made, and your level of skill with each saw.

METHOD	STRENGTHS	WEAKNESSES
Graphical	Can "see" solutions.	Inexact when solutions involve numbers that are not integers. Solutions may not appear on the part of the graph drawn.
Substitution	Yields exact solutions. Convenient to use when a variable has a coefficient of 1.	Can introduce extensive computations with fractions. Cannot "see" solutions quickly.
Elimination	Yields exact solutions. Convenient to use when no variable has a coefficient of 1. The preferred method for systems of 3 or more equations in 3 or more variables (see Section 3.5).	Cannot "see" solutions quickly.

b Solving Applied Problems Using Elimination

Let's now work a few applied problems using elimination. Section 3.4 is devoted entirely to applied problems. We revisit Example 4 of Section 3.2, but this time we solve it using elimination.

EXAMPLE 7 *Architecture.* The ground floor of the John Hancock Building has a perimeter of 860 ft. The length is 100 ft more than the width. Find the length and the width of the ground floor of the building.

1., 2. Familiarize and **Translate.** Earlier, we translated the problem to the system of equations

$$2l + 2w = 860, \quad \textbf{(1)}$$
$$l = w + 100. \quad \textbf{(2)}$$

$l = w + 100$ w

3. Solve. To use *elimination*, we first rewrite both equations in the form $Ax + By = C$. Since equation (1) is already in this form, we need only rewrite equation (2):

$$l = w + 100 \quad \text{Equation (2)}$$
$$l - w = 100. \quad \text{Subtracting } w$$

Now we solve the system

$$2l + 2w = 860, \quad \textbf{(1)}$$
$$l - w = 100. \quad \textbf{(2)}$$

We multiply by 2 on both sides of equation (2) and add:

$$2l + 2w = 860 \qquad \text{Equation (1)}$$
$$\underline{2l - 2w = 200} \qquad \text{Multiplying by 2 on both sides of equation (2)}$$
$$4l \qquad = 1060 \qquad \text{Adding}$$
$$l = 265. \qquad \text{Solving for } l$$

Next, we substitute 265 for l in the equation $l = w + 100$ and solve for w:

$$265 = w + 100$$
$$165 = w.$$

Which method, substitution or elimination, do you prefer to use?

4. Check. Consider the dimensions 265 ft and 165 ft. The length is 100 ft more than the width. The perimeter is 2(265 ft) + 2(165 ft), or 860 ft. The dimensions 265 ft and 165 ft check in the original problem.

5. State. The length is 265 ft and the width is 165 ft.

Do Exercise 10.

10. Architecture. The top floor of the John Hancock Building is also a rectangle, but its perimeter is 520 ft. The width is 60 ft less than the length. Find the length and the width.

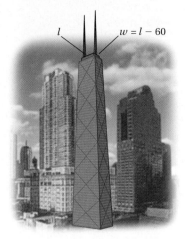

$l \qquad w = l - 60$

a) Use elimination to solve the resulting system.

---

b) Compare the use of the substitution method in Margin Exercise 6 of Section 3.2 with the use of the elimination method here.

Answers on page A-18

a Solve the system by the elimination method.

1. $x + 3y = 7,$
$-x + 4y = 7$

2. $x + y = 9,$
$2x - y = -3$

3. $9x + 5y = 6,$
$2x - 5y = -17$

4. $8x - 3y = 16,$
$8x + 3y = -8$

5. $5x + 3y = 19,$
$2x - 5y = 11$

6. $3x + 2y = 3,$
$9x - 8y = -2$

7. $5r - 3s = 24,$
$3r + 5s = 28$

8. $5x - 7y = -16,$
$2x + 8y = 26$

9. $0.3x - 0.2y = 4,$
$0.2x + 0.3y = 1$

10. $0.7x - 0.3y = 0.5,$
$-0.4x + 0.7y = 1.3$

11. $\frac{1}{2}x + \frac{1}{3}y = 4,$
$\frac{1}{4}x + \frac{1}{3}y = 3$

12. $\frac{2}{3}x + \frac{1}{7}y = -11,$
$\frac{1}{7}x - \frac{1}{3}y = -10$

13. $\frac{2}{5}x + \frac{1}{2}y = 2,$
$\frac{1}{2}x - \frac{1}{6}y = 3$

14. $\frac{1}{3}x + \frac{1}{5}y = 7,$
$\frac{1}{6}x - \frac{2}{5}y = -4$

15. $2x + 3y = 1,$
$4x + 6y = 2$

16. $3x - 2y = 1,$
$-6x + 4y = -2$

17. $2x - 4y = 5,$
$2x - 4y = 6$

18. $3x - 5y = -2,$
$5y - 3x = 7$

19. $5x - 9y = 7,$
$7y - 3x = -5$

20. $a - 2b = 16,$
$b + 3 = 3a$

21. $3(a - b) = 15,$
$4a = b + 1$

22. $10x + y = 306,$
$10y + x = 90$

23. $x - \frac{1}{10}y = 100,$
$y - \frac{1}{10}x = -100$

24. $\frac{1}{8}x + \frac{3}{5}y = \frac{19}{2},$
$-\frac{3}{10}x - \frac{7}{20}y = -1$

25. $0.05x + 0.25y = 22,$
$0.15x + 0.05y = 24$

26. $1.3x - 0.2y = 12,$
$0.4x + 17y = 89$

b   Solve. Use the elimination method when solving the translated system.

27. *Soccer Field.*   The perimeter of a soccer field is 340 m. The length exceeds the width by 50 m. Find the length and the width.

28. *Racquetball Court.*   A regulation racquetball court has a perimeter of 120 ft, with a length that is twice the width. Find the length and the width of such a court.

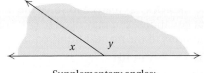
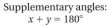

Court width

Court length

29. *Complementary Angles.*   **Complementary angles** are angles whose sum is 90°. Two complementary angles are such that one angle is 6° more than five times the other. Find the measures of the angles.

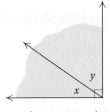

y

x

Complementary angles:
$x + y = 90°$

30. *Supplementary Angles.*   **Supplementary angles** are angles whose sum is 180°. Two supplementary angles are such that one angle is 12° less than three times the other. Find the measures of the angles.

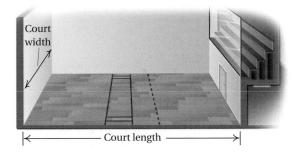

x y

Supplementary angles:
$x + y = 180°$

31. *Airplane Seating.* An airplane has a total of 152 seats. The number of coach-class seats is 5 more than six times the number of first-class seats. How many of each type of seat are there on the plane?

32. *Hockey Points.* At one time, hockey teams received two points when they won a game and one point when they tied. One season, a team won a championship with 60 points. They won 9 more games than they tied. How many wins and how many ties did the team have?

33. D_W Describe a method that could be used to create dependent systems of equations.

34. D_W Write a problem for a classmate to solve that can be translated into a system of two equations. Devise the problem so that the solution is "Shelly gave 9 haircuts and 5 shampoos."

SKILL MAINTENANCE

Given the function $f(x) = 3x^2 - x + 1$, find each of the following function values.   [2.2b]

35. $f(0)$

36. $f(-1)$

37. $f(1)$

38. $f(10)$

39. $f(-2)$

40. $f(2a)$

41. $f(-4)$

42. $f(1.8)$

43. Find the domain of the function: [2.3a]

$$f(x) = \frac{x - 5}{x + 7}.$$

44. Find the domain and the range of the function: [2.3a]

$$g(x) = 5 - x^2.$$

45. Find an equation of the line with slope $-\frac{3}{5}$ and y-intercept $(0, -7)$.   [2.6a]

46. Simplify: $\dfrac{(a^2 b^3)^5}{a^7 b^{16}}$.   [R.7b]

SYNTHESIS

47. Use the INTERSECT feature to solve the following system of equations. You may need to first solve for y. Round answers to the nearest hundredth.

$$3.5x - 2.1y = 106.2,$$
$$4.1x + 16.7y = -106.28$$

48. Solve:

$$\frac{x + y}{2} - \frac{x - y}{5} = 1,$$
$$\frac{x - y}{2} + \frac{x + y}{6} = -2.$$

49. The solution of this system is $(-5, -1)$. Find A and B.

$$Ax - 7y = -3,$$
$$x - By = -1$$

50. Find an equation to pair with $6x + 7y = -4$ such that $(-3, 2)$ is a solution of the system.

51. The points $(0, -3)$ and $\left(-\frac{3}{2}, 6\right)$ are two of the solutions of the equation $px - qy = -1$. Find p and q.

52. Determine a and b for which $(-4, -3)$ will be a solution of the system

$$ax + by = -26,$$
$$bx - ay = 7.$$

3.4 SOLVING APPLIED PROBLEMS: TWO EQUATIONS

Objectives

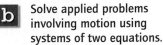

Solve applied problems involving total value and mixture using systems of two equations.

Solve applied problems involving motion using systems of two equations.

a   Total-Value and Mixture Problems

Systems of equations can be a useful tool in solving applied problems. Using systems often makes the *Translate* step easier than using a single equation. The first kind of problem we consider involves quantities of items purchased and the total value, or cost, of the items. We refer to this type of problem as a **total-value problem.**

EXAMPLE 1   *Butterfly Exhibit.*   White River Gardens in Indianapolis, Indiana, presents a yearly butterfly exhibit. Pupae from Kenya, Tanzania, Costa Rica, and Florida are delivered semiweekly. They are stored on rods in a glass case inside a greenhouse until the metamorphosis takes place. As part of a recent order, White River Gardens received 63 pupae—*morpho granadensis* at \$4.15 per pupa and *battus polydamus* at \$1.50 per pupa. The total cost of the two species was \$147.50. How many pupae of each species did the exhibit receive?

Source: Lewis J. Birmelin, General Curator of Horticulture, Indianapolis Zoological Society, White River Gardens

1. **Familiarize.**   To familiarize ourselves with the problem situation, let's guess that the order contained 30 *morpho granadensis* and 33 *battus polydamus*. The total is 63. What is the total cost of these two species? Since *morpho granadensis* costs \$4.15 per pupa and *battus polydamus* \$1.50 per pupa, the total cost would be

$$\underbrace{\text{Cost of } morpho}_{\$4.15(30)} \underbrace{\text{ plus}}_{+} \underbrace{\text{Cost of } battus}_{\$1.50(33)} = \$124.50 + \$49.50$$
$$= \$174.00.$$

Although the total number of pupae is correct, our guess is incorrect because the problem states that the total cost was \$147.50. Since \$174.00 is greater than \$147.50, more of the inexpensive species were received than we had guessed. We could now adjust our guess accordingly. Instead, let's use an algebraic approach that avoids guessing.

   We let p = the number of pupae of *morpho granadensis* and q = the number of pupae of *battus polydamus* received. It helps to organize the information in a table as follows:

	MORPHO GRANADENSIS	BATTUS POLYDAMUS	TOTAL	
Number of pupae received	p	q	63	→ $p + q = 63$
Cost per pupa	\$4.15	\$1.50		
Total cost	$4.15p$	$1.50q$	\$147.50	→ $4.15p + 1.50q$ $= 147.50$

1. Retail Sales of Sweatshirts. Sandy's Sweatshirt Shop sells college sweatshirts. White sweatshirts sell for $18.95 each and red ones sell for $19.50 each. If receipts for the sale of 30 sweatshirts total $572.90, how many of each color did the shop sell? Complete the following table, letting $w =$ the number of white sweatshirts and $r =$ the number of red sweatshirts.

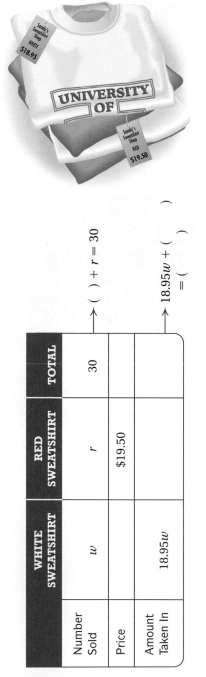

$(\quad) + r = 30$

$18.95w + (\quad) = (\quad)$

	WHITE SWEATSHIRT	RED SWEATSHIRT	TOTAL
Number Sold	w	r	30
Price		$19.50	
Amount Taken In	18.95w		

Answer on page A-18

2. Translate. The first row of the table and the fourth sentence of the problem tell us that a total of 63 pupae was received. Thus we have one equation:

$$p + q = 63.$$

Since each pupa of *morpho granadensis* costs $4.15 and p pupae were received, $4.15p$ is the cost for the *morpho granadensis* species. Similarly, $1.50q$ is the cost of the *battus polydamus* species. From the third row of the table and the information in the statement of the problem, we get a second equation:

$$4.15p + 1.50q = 147.50.$$

We can multiply by 100 on both sides of this equation in order to clear the decimals. This gives us the following system of equations as a translation:

$$p + q = 63, \qquad \textbf{(1)}$$
$$415p + 150q = 14{,}750. \qquad \textbf{(2)}$$

3. Solve. We decide to use the elimination method to solve the system. We eliminate q by multiplying equation (1) by -150 and adding it to equation (2):

$$
\begin{array}{ll}
-150p - 150q = -9450 & \text{Multiplying equation (1) by } -150 \\
\underline{415p + 150q = 14{,}750} & \\
265p = 5300 & \text{Adding} \\
p = 20. & \text{Solving for } p
\end{array}
$$

To find q, we substitute 20 for p in equation (1) and solve for q:

$$
\begin{array}{ll}
p + q = 63 & \text{Equation (1)} \\
20 + q = 63 & \text{Substituting 20 for } p \\
q = 43. & \text{Solving for } q
\end{array}
$$

We obtain $(20, 43)$, or $p = 20$, $q = 43$.

4. Check. We check in the original problem. Remember that p is the number of pupae of *morpho granadensis* and q is the number of pupae of *battus polydamus*.

Number of pupae: $p + q = 20 + 43 = 63$

Cost of morpho granadensis: $\$4.15p = 4.15(20) = \$83.00$

Cost of battus polydamus: $\$1.50q = 1.50(43) = \underline{\$64.50}$

$$\text{Total} = \$147.50$$

The numbers check.

5. State. The exhibit received 20 pupae of *morpho granadensis* and 43 pupae of *battus polydamus*.

Do Exercise 1.

The following problem, similar to Example 1, is called a **mixture problem.**

EXAMPLE 2 *Blending Flower Seeds.* Tara's website, <u>Garden Edibles</u>, specializes in the sale of herbs and flowers for colorful meals and garnishes. Tara sells packets of nasturtium seeds for $0.95 each and packets of Johnny-jump-up seeds for $1.43 each. She decides to offer a 16-packet spring-garden combination, combining packets of both types of seeds at $1.10 per packet. How many packets of each type of seed should be put in her garden mix?

1. **Familiarize.** To familiarize ourselves with the problem situation, we make a guess and do some calculations. The total number of packets of seed is 16. Let's try 12 packets of nasturtiums and 4 packets of Johnny-jump-ups.

The sum of the number of packets is 12 + 4, or 16.

The value of these seed packets is found by multiplying the cost per packet by the number of packets and adding:

$0.95(12) + $1.43(4), or $17.12.

The desired cost is $1.10 per packet. If we multiply $1.10 by 16, we get 16($1.10), or $17.60. This does not agree with $17.12, but these calculations give us a basis for understanding how to translate.

We let $a =$ the number of packets of nasturtium seeds and $b =$ the number of packets of Johnny-jump-up seeds. Next, we organize the information in a table, as follows.

	NASTURTIUM	JOHNNY-JUMP-UP	SPRING	
Number of Packets	a	b	16	→ $a + b = 16$
Price per Packet	$0.95	$1.43	$1.10	
Value of Packets	$0.95a$	$1.43b$	$16 \cdot 1.10$, or 17.60	→ $0.95a + 1.43b$ = 17.60

2. Blending Coffees. The Coffee Counter charges $9.00 per pound for Kenyan French Roast coffee and $8.00 per pound for Sumatran coffee. How much of each type should be used to make a 20-lb blend that sells for $8.40 per pound?

2. Translate. The total number of packets is 16, so we have one equation:
$$a + b = 16.$$

The value of the nasturtium seeds is $0.95a$ and the value of the Johnny-jump-up seeds is $1.43b$. These amounts are in dollars. Since the total value is to be $16(\$1.10)$, or 17.60, we have
$$0.95a + 1.43b = 17.60.$$

We can multiply by 100 on both sides of this equation in order to clear the decimals. Thus we have the translation, a system of equations:

$$a + b = 16, \qquad \textbf{(1)}$$
$$95a + 143b = 1760. \qquad \textbf{(2)}$$

3. Solve. We decide to use substitution, although elimination could be used as we did in Example 1. When equation (1) is solved for b, we get $b = 16 - a$. Substituting $16 - a$ for b in equation (2) and solving gives us

$95a + 143(16 - a) = 1760$	Substituting
$95a + 2288 - 143a = 1760$	Using the distributive law
$-48a = -528$	Subtracting 2288 and collecting like terms
$a = 11.$	

We have $a = 11$. Substituting this value in the equation $b = 16 - a$, we obtain $b = 16 - 11$, or 5.

4. Check. We check in a manner similar to our guess in the *Familiarize* step. The total number of packets is $11 + 5$, or 16. The value of the packet mixture is

$$\$0.95(11) + \$1.43(5), \text{ or } \$17.60.$$

Thus the numbers of packets check.

5. State. The spring garden mixture can be made by combining 11 packets of nasturtium seeds with 5 packets of Johnny-jump-up seeds.

Do Exercise 2.

EXAMPLE 3 *Student Loans.* Jed's student loans totaled $16,200. Part was a Perkins loan made at 5% interest and the rest was a Stafford loan made at 4% interest. After one year, Jed's loans accumulated $715 in interest. What was the amount of each loan?

1. Familiarize. Listing the given information in a table will help. The columns in the table come from the formula for simple interest: $I = Prt$. We let $x =$ the number of dollars in the Perkins loan and $y =$ the number of dollars in the Stafford loan.

	PERKINS LOAN	STAFFORD LOAN	TOTAL	
Principal	x	y	$16,200	$\rightarrow x + y = 16,200$
Rate of Interest	5%	4%		
Time	1 yr	1 yr		
Interest	$0.05x$	$0.04y$	$715	$\rightarrow 0.05x + 0.04y = 715$

2. Translate. The total of the amounts of the loans is found in the first row of the table. This gives us one equation:

$$x + y = 16{,}200.$$

Look at the last row of the table. The interest totals $715. This gives us a second equation:

$$5\%x + 4\%y = 715, \quad \text{or} \quad 0.05x + 0.04y = 715.$$

After we multiply on both sides to clear the decimals, we have

$$5x + 4y = 71{,}500.$$

3. Solve. Using either elimination or substitution, we solve the resulting system:

$$x + y = 16{,}200,$$
$$5x + 4y = 71{,}500.$$

We find that $x = 6700$ and $y = 9500$.

4. Check. The sum is $6700 + $9500, or $16,200. The interest from $6700 at 5% for one year is 5%($6700), or $335. The interest from $9500 at 4% for one year is 4%($9500), or $380. The total interest is $335 + $380, or $715. The numbers check in the problem.

5. State. The Perkins loan was for $6700 and the Stafford loan was for $9500.

Do Exercise 3.

EXAMPLE 4 *Mixing Fertilizers.* Yardbird Gardening carries two kinds of fertilizer containing nitrogen and water. "Gently Green" is 5% nitrogen and "Sun Saver" is 15% nitrogen. Yardbird Gardening needs to combine the two types of solution to make 90 L of a solution that is 12% nitrogen. How much of each brand should be used?

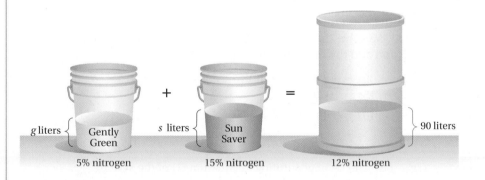

1. Familiarize. We first make a drawing and a guess to become familiar with the problem.

We choose two numbers that total 90 L—say, 40 L of Gently Green and 50 L of Sun Saver—for the amounts of each fertilizer. Will the resulting mixture have the correct percentage of nitrogen? To find out, we multiply as follows:

$$5\%(40\,\text{L}) = 2\,\text{L of nitrogen} \quad \text{and} \quad 15\%(50\,\text{L}) = 7.5\,\text{L of nitrogen}.$$

Thus the total amount of nitrogen in the mixture is 2 L + 7.5 L, or 9.5 L.

3. Client Investments. Kaufman Financial Corporation makes investments for corporate clients. It makes an investment of $3700 for one year at simple interest, yielding $297. Part of the money is invested at 7% and the rest at 9%. How much was invested at each rate?

Do the *Familiarize* and *Translate* steps by completing the following table. Let $x =$ the number of dollars invested at 7% and $y =$ the number of dollars invested at 9%.

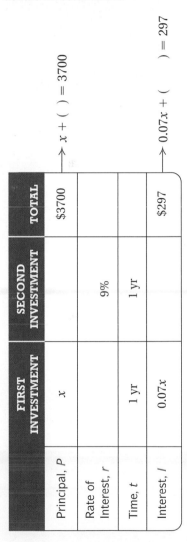

	FIRST INVESTMENT	SECOND INVESTMENT	TOTAL
Principal, P	x		$3700
Rate of Interest, r		9%	
Time, t	1 yr	1 yr	
Interest, I	$0.07x$		$297

$x + (\ \ ) = 3700$

$0.07x + (\ \ ) = 297$

Answer on page A-18

The final mixture of 90 L is supposed to be 12% nitrogen. Now

$$12\%(90 \text{ L}) = 10.8 \text{ L}.$$

Since 9.5 L and 10.8 L are not the same, our guess is incorrect. But these calculations help us to become familiar with the problem and to make the translation.

We let g = the number of liters of Gently Green and s = the number of liters of Sun Saver.

The information can be organized in a table, as follows.

	GENTLY GREEN	SUN SAVER	MIXTURE	
Number of Liters	g	s	90	→ $g + s = 90$
Percent of Nitrogen	5%	15%	12%	
Amount of Nitrogen	$0.05g$	$0.15s$	0.12×90, or 10.8 liters	→ $0.05g + 0.15s = 10.8$

2. **Translate.** If we add g and s in the first row, we get 90, and this gives us one equation:

$$g + s = 90.$$

If we add the amounts of nitrogen listed in the third row, we get 10.8, and this gives us another equation:

$$5\%g + 15\%s = 10.8, \quad \text{or} \quad 0.05g + 0.15s = 10.8.$$

After clearing the decimals, we have the following system:

$$g + s = 90, \qquad \textbf{(1)}$$
$$5g + 15s = 1080. \qquad \textbf{(2)}$$

3. **Solve.** We solve the system using elimination. We multiply equation (1) by -5 and add the result to equation (2):

$$
\begin{aligned}
-5g - 5s &= -450 \qquad \text{Multiplying equation (1) by } -5 \\
\underline{5g + 15s} &= \underline{1080} \\
10s &= 630 \qquad \text{Adding} \\
s &= 63; \qquad \text{Dividing by 10}
\end{aligned}
$$

$$
\begin{aligned}
g + 63 &= 90 \qquad \text{Substituting in equation (1) of the system} \\
g &= 27. \qquad \text{Solving for } g
\end{aligned}
$$

4. Check. Remember that g is the number of liters of Gently Green, with 5% nitrogen, and s is the number of liters of Sun Saver, with 15% nitrogen.

Total number of liters of mixture: $g + s = 27 + 63 = 90$

Amount of nitrogen: $5\%(27) + 15\%(63) = 1.35 + 9.45 = 10.8$ L

Percentage of nitrogen in mixture: $\dfrac{10.8}{90} = 0.12 = 12\%$

The numbers check in the original problem.

5. State. Yardbird Gardening should mix 27 L of Gently Green and 63 L of Sun Saver.

Do Exercise 4.

b   Motion Problems

When a problem deals with speed, distance, and time, we can expect to use the following *motion formula*.

THE MOTION FORMULA

Distance = Rate (or speed) · Time

$$d = rt$$

TIPS FOR SOLVING MOTION PROBLEMS

1. Draw a diagram using an arrow or arrows to represent distance and the direction of each object in motion.
2. Organize the information in a table or chart.
3. Look for as many things as you can that are the same, so you can write equations.

EXAMPLE 5 *Auto Travel.* Your brother leaves on a trip, forgetting his suitcase. You know that he normally drives at a speed of 55 mph. You do not discover the suitcase until 1 hr after he has left. If you follow him at a speed of 65 mph, how long will it take you to catch up with him?

1. Familiarize. We first make a drawing. From the drawing, we see that when you catch up with your brother, the distances from home are the same. We let d = the distance, in miles. If we let t = the time, in hours, for you to catch your brother, then $t + 1$ = the time traveled by your brother at a slower speed.

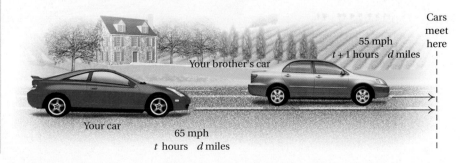

4. Mixing Cleaning Solutions. King's Service Station uses two kinds of cleaning solution containing acid and water. "Attack" is 2% acid and "Blast" is 6% acid. They want to mix the two to get 60 qt of a solution that is 5% acid. How many quarts of each should they use?

Do the *Familiarize* and *Translate* steps by completing the following table. Let a = the number of quarts of Attack and b = the number of quarts of Blast.

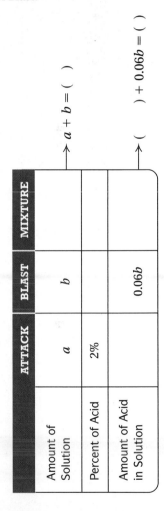

	ATTACK	BLAST	MIXTURE
Amount of Solution	a	b	$a + b = (\quad)$
Percent of Acid	2%		
Amount of Acid in Solution		$0.06b$	$(\quad) + 0.06b = (\quad)$

Answer on page A-18

5. Train Travel. A train leaves Barstow traveling east at 35 km/h. One hour later, a faster train leaves Barstow, also traveling east on a parallel track at 40 km/h. How far from Barstow will the faster train catch up with the slower one?

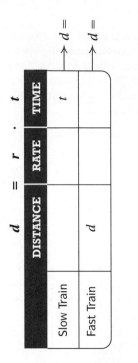

We organize the information in a table as follows.

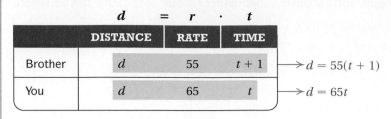

	DISTANCE	RATE	TIME	
Brother	d	55	$t + 1$	→ $d = 55(t + 1)$
You	d	65	t	→ $d = 65t$

2. Translate. Using $d = rt$ in each row of the table, we get an equation. Thus we have a system of equations:

$$d = 55(t + 1), \quad \textbf{(1)}$$
$$d = 65t. \quad \textbf{(2)}$$

3. Solve. We solve the system using the substitution method:

$65t = 55(t + 1)$ Substituting $65t$ for d in equation (1)

$65t = 55t + 55$ Multiplying to remove parentheses on the right

$10t = 55$
$t = 5.5.$ Solving for t

Your time is 5.5 hr, which means that your brother's time is $5.5 + 1$, or 6.5 hr.

4. Check. At 65 mph, you will travel $65 \cdot 5.5$, or 357.5 mi, in 5.5 hr. At 55 mph, your brother will travel $55 \cdot 6.5$, or the same 357.5 mi, in 6.5 hr. The numbers check.

5. State. You will catch up with your brother in 5.5 hr.

Do Exercise 5.

EXAMPLE 6 *Marine Travel.* A Coast-Guard patrol boat travels 4 hr on a trip downstream with a 6-mph current. The return trip against the same current takes 5 hr. Find the speed of the boat in still water.

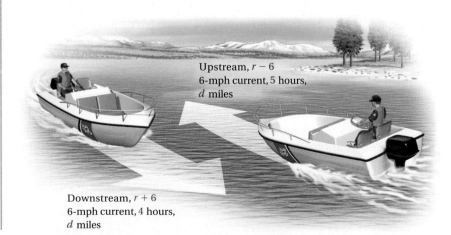

Upstream, $r - 6$
6-mph current, 5 hours,
d miles

Downstream, $r + 6$
6-mph current, 4 hours,
d miles

Answer on page A-18

1. **Familiarize.** We first make a drawing. From the drawing, we see that the distances are the same. We let $d =$ the distance, in miles, and $r =$ the speed of the boat in still water, in miles per hour. Then, when the boat is traveling downstream, its speed is $r + 6$ (the current helps the boat along). When it is traveling upstream, its speed is $r - 6$ (the current holds the boat back). We can organize the information in a table. We use the formula $d = rt$.

	d =	r ·	t	
	DISTANCE	**RATE**	**TIME**	
Downstream	d	$r + 6$	4	→ $d = (r + 6)4$
Upstream	d	$r - 6$	5	→ $d = (r - 6)5$

2. **Translate.** From each row of the table, we get an equation, $d = rt$:

$$d = 4r + 24, \quad (1)$$
$$d = 5r - 30. \quad (2)$$

3. **Solve.** We solve the system by the substitution method:

$$4r + 24 = 5r - 30 \quad \text{Substituting } 4r + 24 \text{ for } d \text{ in equation (2)}$$
$$\left.\begin{aligned} 24 &= r - 30 \\ 54 &= r. \end{aligned}\right\} \quad \text{Solving for } r$$

4. **Check.** If $r = 54$, then $r + 6 = 60$; and $60 \cdot 4 = 240$, the distance traveled downstream. If $r = 54$, then $r - 6 = 48$; and $48 \cdot 5 = 240$, the distance traveled upstream. The distances are the same. In this type of problem, a problem-solving tip to keep in mind is "Have I found what the problem asked for?" We could solve for a certain variable but still have not answered the question of the original problem. For example, we might have found speed when the problem wanted distance. In this problem, we want the speed of the boat in still water, and that is r.

5. **State.** The speed in still water is 54 mph.

Do Exercise 6.

6. **Air Travel.** An airplane flew for 4 hr with a 20-mph tailwind. The return flight against the same wind took 5 hr. Find the speed of the plane in still air.

	d =	r ·	t	
	DISTANCE	**RATE**	**TIME**	
With Wind		$r + 20$		→ $d =$
Against Wind	d			→ $d =$

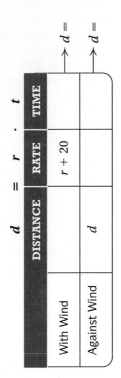

Answer on page A-18

Translating for Success

1. *Office Expense.* The monthly phone expense for an office is $1094 less than the janitorial expense. Three times the janitorial expense minus four times the phone expense is $248. What is the total of the two expenses?

2. *Dimensions of a Triangle.* The sum of the base and the height of a triangle is 192 in. The height is twice the base. Find the base and the height.

3. *Supplementary Angles.* Two supplementary angles are such that twice one angle is 7° more than the other. Find the measures of the angles.

4. *SAT Scores.* The total of Megan's verbal and math scores on the SAT was 1094. Her math score was 248 points higher than her verbal score. What were her math and verbal SAT scores?

5. *Sightseeing Boat.* A sightseeing boat travels 3 hr on a trip downstream with a 2.5-mph current. The return trip against the same current takes 3.5 hr. Find the speed of the boat in still water.

The goal of these matching questions is to practice step (2), *Translate,* of the five-step problem-solving process. Translate each word problem to a system of equations and select a correct translation from systems A–J.

A. $x = y + 248$,
$x + y = 1094$

B. $5x = 2y - 3$,
$y = \frac{2}{3}x + 5$

C. $y = \frac{1}{2}x$,
$2x + 2y = 192$

D. $2x = 7 + y$,
$x + y = 180$

E. $x + y = 192$,
$x = 2y$

F. $x + y = 180$,
$x = 2y + 7$

G. $x - 1094 = y$,
$3x - 4y = 248$

H. $3\%x + 2.5\%y = 97.50$,
$x + y = 2500$

I. $2x = 5 + \frac{2}{3}y$,
$3y = 15x - 4$

J. $x(y + 2.5) \cdot 3$,
$3.5(y - 2.5) = x$

Answers on page A-19

6. *Running Distances.* Each day Tricia runs 5 miles more than two-thirds the distance that Chris runs. Five times the distance that Chris runs is 3 mi less than twice the distance that Tricia runs. How far does Tricia run daily?

7. *Dimensions of a Rectangle.* The perimeter of a rectangle is 192 in. The width is half the length. Find the length and the width.

8. *Mystery Numbers.* Teka asked her students to determine the two numbers that she placed in a sealed envelope. Twice the smaller number is 5 more than two-thirds of the larger number. Three times the larger number is 4 less than fifteen times the smaller. Find the numbers.

9. *Supplementary Angles.* Two supplementary angles are such that one angle is 7° more than twice the other. Find the measures of the angles.

10. *Student Loans.* Brandt's student loans totaled $2500. Part was made at 3% interest and the rest at 2.5%. After one year, Brandt had accumulated $97.50 in interest. What was the amount of each loan?

a   Solve.

1. *Retail Sales.*   Paint Town sold 45 paintbrushes, one kind at $8.50 each and another at $9.75 each. In all, $398.75 was taken in for the brushes. How many of each kind were sold?

$8.50 $9.75

2. *Retail Sales.*   Mountainside Fleece sold 40 neckwarmers. Solid-color neckwarmers sold for $9.90 each and print ones sold for $12.75 each. In all, $421.65 was taken in for the neckwarmers. How many of each type were sold?

3. *Sales of Pharmaceuticals.*   In 2004, the Diabetic Express charged $27.06 for a vial of Humulin insulin and $34.39 for a vial of Novolin Velosulin insulin. If a total of $1565.57 was collected for 50 vials of insulin, how many vials of each type were sold?

4. *Fundraising.*   The St. Mark's Community Barbecue served 250 dinners. A child's plate cost $3.50 and an adult's plate cost $7.00. A total of $1347.50 was collected. How many of each type of plate was served?

5. *Radio Airplay.*   Rudy must play 12 commercials during his 1-hr radio show. Each commercial is either 30 sec or 60 sec long. If the total commercial time during that hour is 10 min, how many commercials of each type does Rudy play?

6. *Nontoxic Floor Wax.*   A nontoxic floor wax can be made by combining lemon juice and food-grade linseed oil. The amount of oil should be twice the amount of lemon juice. How much of each ingredient is needed in order to make 32 oz of floor wax? (The mix should be spread with a rag and buffed when dry.)

7. *Catering.*   Stella's Catering is planning a wedding reception. The bride and groom would like to serve a nut mixture containing 25% peanuts. Stella has available mixtures that are either 40% or 10% peanuts. How much of each type should be mixed to get a 10-lb mixture that is 25% peanuts?

8. *Blending Granola.*   Deep Thought Granola is 25% nuts and dried fruit. Oat Dream Granola is 10% nuts and dried fruit. How much of Deep Thought and how much of Oat Dream should be mixed to form a 20-lb batch of granola that is 19% nuts and dried fruit?

9. *Ink Remover.*   Etch Clean Graphics uses one cleanser that is 25% acid and a second that is 50% acid. How many liters of each should be mixed to get 10 L of a solution that is 40% acid?

10. *Livestock Feed.*   Soybean meal is 16% protein and corn meal is 9% protein. How many pounds of each should be mixed to get a 350-lb mixture that is 12% protein?

11. *Dry Cleaners.* Claudio, a banking vice-president, took 17 neckties to Milto Cleaners. The rate for non-silk ties is $3.25 per tie and for silk ties is $3.60 per tie. His total bill was $58.75. How many silk ties did he have dry cleaned?

12. *Laundry.* While on a four-week hiking trip in the mountains, the Tryon family washed 11 loads of clothes at The Mountain View Laundry. The 20-lb capacity washing machine costs $1.50 per load while the 30-lb costs $2.50. Their total laundry expense was $20.50. How many loads were laundered in each size washing machine?

13. *Student Loans.* Sarah's two student loans totaled $12,000. One of her loans was at 6% simple interest and the other at 9%. After one year, Sarah owed $855 in interest. What was the amount of each loan?

14. *Investments.* An executive nearing retirement made two investments totaling $45,000. In one year, these investments yielded $2430 in simple interest. Part of the money was invested at 4% and the rest at 6%. How much was invested at each rate?

15. *Food Science.* The following bar graph shows the milk fat percentages in three dairy products. How many pounds each of whole milk and cream should be mixed to form 200 lb of milk for cream cheese?

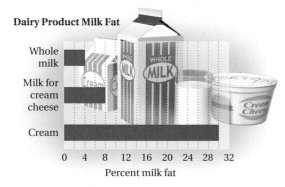

Dairy Product Milk Fat

Whole milk

Milk for cream cheese

Cream

0 4 8 12 16 20 24 28 32

Percent milk fat

16. *Automotive Maintenance.* "Arctic Antifreeze" is 18% alcohol and "Frost No-More" is 10% alcohol. How many liters of Arctic Antifreeze should be mixed with 7.5 L of Frost No-More in order to get a mixture that is 15% alcohol?

17. *Teller Work.* Juan goes to a bank and gets change for a $50 bill consisting of all $5 bills and $1 bills. There are 22 bills in all. How many of each kind are there?

18. *Making Change.* Christina makes a $9.25 purchase at a bookstore in Reno with a $20 bill. The store has no bills and gives her the change in quarters and fifty-cent pieces. There are 30 coins in all. How many of each kind are there?

19. *Investments.* William opened two investment accounts for his grandson's college fund. The first year, these investments, which totaled $18,000, yielded $831 in simple interest. Part of the money was invested at 5.5% and the rest at 4%. How much was invested at each rate?

20. *Student Loans.* Cole's two student loans totaled $31,000. One of his loans was at 2.8% simple interest and the other at 4.5%. After one year, Cole owed $1024.40 in interest. What was the amount of each loan?

b Solve.

21. *Train Travel.* A train leaves Danville Junction and travels north at a speed of 75 mph. Two hours later, a second train leaves on a parallel track and travels north at 125 mph. How far from the station will they meet?

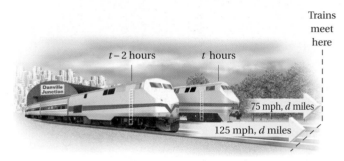

Trains meet here

$t - 2$ hours t hours

Danville Junction

75 mph, d miles

125 mph, d miles

22. *Car Travel.* Two cars leave Denver traveling in opposite directions. One car travels at a speed of 80 km/h and the other at 96 km/h. In how many hours will they be 528 km apart?

23. *Canoeing.* Darren paddled for 4 hr with a 6-km/h current to reach a campsite. The return trip against the same current took 10 hr. Find the speed of Darren's canoe in still water.

24. *Boating.* Mia's motorboat took 3 hr to make a trip downstream with a 6-mph current. The return trip against the same current took 5 hr. Find the speed of the boat in still water.

25. *Car Travel.* Donna is late for a sales meeting after traveling from one town to another at a speed of 32 mph. If she had traveled 4 mph faster, she could have made the trip in $\frac{1}{2}$ hr less time. How far apart are the towns?

26. *Air Travel.* Rod is a pilot for Crossland Airways. He computes his flight time against a headwind for a trip of 2900 mi at 5 hr. The flight would take 4 hr and 50 min if the headwind were half as great. Find the headwind and the plane's air speed.

27. *Air Travel.* Two planes travel toward each other from cities that are 780 km apart at rates of 190 km/h and 200 km/h. They started at the same time. In how many hours will they meet?

28. *Motorcycle Travel.* Sally and Rocky travel on motorcycles toward each other from Chicago and Indianapolis, which are about 350 km apart, and they are biking at rates of 110 km/h and 90 km/h. They started at the same time. In how many hours will they meet?

29. *Air Travel.* Two airplanes start at the same time and fly toward each other from points 1000 km apart at rates of 420 km/h and 330 km/h. After how many hours will they meet?

30. *Truck and Car Travel.* A truck and a car leave a service station at the same time and travel in the same direction. The truck travels at 55 mph and the car at 40 mph. They can maintain CB radio contact within a range of 10 mi. When will they lose contact?

31. ▦ *Point of No Return.* A plane flying the 3458-mi trip from New York City to London has a 50-mph tailwind. The flight's *point of no return* is the point at which the flight time required to return to New York is the same as the time required to continue to London. If the speed of the plane in still air is 360 mph, how far is New York from the point of no return?

32. ▦ *Point of No Return.* A plane is flying the 2553-mi trip from Los Angeles to Honolulu into a 60-mph headwind. If the speed of the plane in still air is 310 mph, how far from Los Angeles is the plane's point of no return? (See Exercise 31.)

33. $\mathbf{D_W}$ List three or four study tips for someone beginning this exercise set.

34. $\mathbf{D_W}$ Write a problem similar to Margin Exercise 1 for a classmate to solve. Design the problem so the answer is "The florist sold 14 hanging plants and 9 flats of petunias."

SKILL MAINTENANCE

Given the function $f(x) = 4x - 7$, find each of the following function values. [2.2b]

35. $f(0)$ **36.** $f(-1)$ **37.** $f(1)$ **38.** $f(10)$

39. $f(-2)$ **40.** $f(2a)$ **41.** $f(-4)$ **42.** $f(1.8)$

43. $f\left(\frac{3}{4}\right)$ **44.** $f(-2.5)$ **45.** $f(-3h)$ **46.** $f(1000)$

SYNTHESIS

47. *Automotive Maintenance.* The radiator in Michelle's car contains 16 L of antifreeze and water. This mixture is 30% antifreeze. How much of this mixture should she drain and replace with pure antifreeze so that there will be a mixture of 50% antifreeze?

48. *Physical Exercise.* Natalie jogs and walks to school each day. She averages 4 km/h walking and 8 km/h jogging. The distance from home to school is 6 km and Natalie makes the trip in 1 hr. How far does she jog in a trip?

49. *Fuel Economy.* Sally Cline's SUV gets 18 miles per gallon (mpg) in city driving and 24 mpg in highway driving. The SUV is driven 465 mi on 23 gal of gasoline. How many miles were driven in the city and how many were driven on the highway?

50. *Gender.* Phil and Phyllis are siblings. Phyllis has twice as many brothers as she has sisters. Phil has the same number of brothers as sisters. How many girls and how many boys are in the family?

51. *Wood Stains.* Bennet Custom Flooring has 0.5 gal of stain that is 20% brown and 80% neutral. A customer orders 1.5 gal of a stain that is 60% brown and 40% neutral. How much pure brown stain and how much neutral stain should be added to the original 0.5 gal in order to make up the order?

52. ◣◢ See Exercise 51. Let x = the amount of pure brown stain added to the original 0.5 gal. Find a function $P(x)$ that can be used to determine the percentage of brown stain in the 1.5-gal mixture. On a graphing calculator, draw the graph of P and use ZOOM and TRACE or the TABLE feature to confirm the answer to Exercise 51.

3.5

SYSTEMS OF EQUATIONS IN THREE VARIABLES

a | Solving Systems in Three Variables

A **linear equation in three variables** is an equation equivalent to one of the type $Ax + By + Cz = D$. A **solution** of a system of three equations in three variables is an ordered triple (x, y, z) that makes *all three* equations true.

The substitution method can be used to solve systems of three equations, but it is not efficient unless a variable has already been eliminated from one or more of the equations. Therefore, we will use only the elimination method—essentially the same procedure for systems of three equations as for systems of two equations.* The first step is to eliminate a variable and obtain a system of two equations in two variables.

EXAMPLE 1 Solve the following system of equations:

$$
\begin{aligned}
x + y + z &= 4, & \textbf{(1)} \\
x - 2y - z &= 1, & \textbf{(2)} \\
2x - y - 2z &= -1. & \textbf{(3)}
\end{aligned}
$$

a) We first use *any* two of the three equations to get an equation in two variables. In this case, let's use equations (1) and (2) and add to eliminate z:

$$
\begin{aligned}
x + y + z &= 4 & \textbf{(1)} \\
\underline{x - 2y - z} &= \underline{1} & \textbf{(2)} \\
2x - y &= 5. & \textbf{(4)} \quad \text{Adding to eliminate } z
\end{aligned}
$$

b) We use a *different* pair of equations and eliminate the **same variable** that we did in part (a). Let's use equations (1) and (3) and again eliminate z.

> **Caution!**
>
> A common error is to eliminate a different variable the second time.

$$
\begin{aligned}
x + y + z &= 4, & \textbf{(1)} \\
2x - y - 2z &= -1; & \textbf{(3)}
\end{aligned}
$$

$$
\begin{aligned}
2x + 2y + 2z &= 8 & & \text{Multiplying equation (1) by 2} \\
\underline{2x - y - 2z} &= \underline{-1} & \textbf{(3)} \\
4x + y &= 7 & \textbf{(5)} & \quad \text{Adding to eliminate } z
\end{aligned}
$$

c) Now we solve the resulting system of equations, (4) and (5). That solution will give us two of the numbers. Note that we now have two equations in two variables. Had we eliminated two *different* variables in parts (a) and (b), this would not be the case.

$$
\begin{aligned}
2x - y &= 5 & \textbf{(4)} \\
\underline{4x + y} &= \underline{7} & \textbf{(5)} \\
6x &= 12 & \text{Adding} \\
x &= 2
\end{aligned}
$$

*Other methods for solving systems of equations are considered in Appendixes B and C.

Study Tips

BEING A TUTOR

Try being a tutor for a fellow student. You can maximize your understanding and retention of concepts if you explain the material to someone else.

1. Solve. Don't forget to check.

$$4x - y + z = 6,$$
$$-3x + 2y - z = -3,$$
$$2x + y + 2z = 3$$

We can use either equation (4) or (5) to find y. We choose equation (5):

$$4x + y = 7 \qquad \textbf{(5)}$$
$$4(2) + y = 7 \qquad \text{Substituting 2 for } x$$
$$8 + y = 7$$
$$y = -1.$$

d) We now have $x = 2$ and $y = -1$. To find the value for z, we use any of the original three equations and substitute to find the third number, z. Let's use equation (1) and substitute our two numbers in it:

$$x + y + z = 4 \qquad \textbf{(1)}$$
$$2 + (-1) + z = 4 \qquad \text{Substituting 2 for } x \text{ and } -1 \text{ for } y$$
$$\left. \begin{array}{l} 1 + z = 4 \\ z = 3. \end{array} \right\} \quad \text{Solving for } z$$

We have obtained the ordered triple $(2, -1, 3)$. We check as follows, substituting $(2, -1, 3)$ into each of the three equations using alphabetical order.

Check:

$$\frac{x + y + z = 4}{2 + (-1) + 3 \; ? \; 4}$$
$$\qquad\qquad 4 \;\Big|\qquad \text{TRUE}$$

$$\frac{x - 2y - z = 1}{2 - 2(-1) - 3 \; ? \; 1}$$
$$\qquad 2 + 2 - 3 \;\Big|$$
$$\qquad\qquad 1 \;\Big|\qquad \text{TRUE}$$

$$\frac{2x - y - 2z = -1}{2(2) - (-1) - 2 \cdot 3 \; ? \; -1}$$
$$\qquad\quad 4 + 1 - 6 \;\Big|$$
$$\qquad\qquad\quad -1 \;\Big|\qquad \text{TRUE}$$

The triple $(2, -1, 3)$ checks and is the solution.

To use the elimination method to solve systems of three equations:

1. Write all equations in the standard form $Ax + By + Cz = D$.
2. Clear any decimals or fractions.
3. Choose a variable to eliminate. Then use *any* two of the three equations to eliminate that variable, getting an equation in two variables.
4. Next, use a different pair of equations and get another equation in *the same two variables*. That is, eliminate the same variable that you did in step (3).
5. Solve the resulting system (pair) of equations. That will give two of the numbers.
6. Then use any of the original three equations to find the third number.

Do Exercise 1.

Answer on page A-19

EXAMPLE 2 Solve this system:

$$4x - 2y - 3z = 5, \qquad \textbf{(1)}$$
$$-8x - y + z = -5, \qquad \textbf{(2)}$$
$$2x + y + 2z = 5. \qquad \textbf{(3)}$$

a) The equations are in standard form and do not contain decimals or fractions.

b) We decide to eliminate the variable y since the y-terms are opposites in equations (2) and (3). We add:

$$-8x - y + z = -5 \qquad \textbf{(2)}$$
$$\underline{2x + y + 2z = 5} \qquad \textbf{(3)}$$
$$-6x + 3z = 0. \qquad \textbf{(4)} \qquad \text{Adding}$$

c) We use another pair of equations to get an equation in the same two variables, x and z. That is, we eliminate the same variable y that we did in step (b). We use equations (1) and (3) and eliminate y:

$$4x - 2y - 3z = 5, \qquad \textbf{(1)}$$
$$2x + y + 2z = 5; \qquad \textbf{(3)}$$

$$4x - 2y - 3z = 5 \qquad \textbf{(1)}$$
$$\underline{4x + 2y + 4z = 10} \qquad \qquad \text{Multiplying equation (3) by 2}$$
$$8x + z = 15. \qquad \textbf{(5)} \qquad \text{Adding}$$

d) Now we solve the resulting system of equations (4) and (5). That will give us two of the numbers:

$$-6x + 3z = 0, \qquad \textbf{(4)}$$
$$8x + z = 15. \qquad \textbf{(5)}$$

We multiply equation (5) by -3. $\left(\text{We could also have multiplied equation (4) by } -\frac{1}{3}.\right)$

$$-6x + 3z = 0 \qquad \textbf{(4)}$$
$$\underline{-24x - 3z = -45} \qquad \text{Multiplying equation (5) by } -3$$
$$-30x = -45 \qquad \text{Adding}$$
$$x = \frac{-45}{-30} = \frac{3}{2}$$

We now use equation (5) to find z:

$$8x + z = 15 \qquad \textbf{(5)}$$
$$8\left(\tfrac{3}{2}\right) + z = 15 \qquad \text{Substituting } \tfrac{3}{2} \text{ for } x$$
$$\left.\begin{array}{c} 12 + z = 15 \\ z = 3. \end{array}\right\} \quad \text{Solving for } z$$

2. Solve. Don't forget to check.

$$2x + y - 4z = 0,$$
$$x - y + 2z = 5,$$
$$3x + 2y + 2z = 3$$

e) Next, we use any of the original equations and substitute to find the third number, y. We choose equation (3) since the coefficient of y there is 1:

$$2x + y + 2z = 5 \qquad \textbf{(3)}$$
$$2\left(\tfrac{3}{2}\right) + y + 2(3) = 5 \qquad \text{Substituting } \tfrac{3}{2} \text{ for } x \text{ and 3 for } z$$
$$\left.\begin{array}{r} 3 + y + 6 = 5 \\ y + 9 = 5 \\ y = -4. \end{array}\right\} \quad \text{Solving for } y$$

The solution is $\left(\tfrac{3}{2}, -4, 3\right)$. The check is as follows.

Check:

$$\begin{array}{c} \underline{4x - 2y - 3z = 5} \\ 4 \cdot \tfrac{3}{2} - 2(-4) - 3(3) \ ? \ 5 \\ 6 + 8 - 9 \ \bigg| \\ 5 \ \bigg| \quad \text{TRUE} \end{array}$$

$$\begin{array}{c} \underline{-8x - y + z = -5} \\ -8 \cdot \tfrac{3}{2} - (-4) + 3 \ ? \ -5 \\ -12 + 4 + 3 \ \bigg| \\ -5 \ \bigg| \quad \text{TRUE} \end{array}$$

$$\begin{array}{c} \underline{2x + y + 2z = 5} \\ 2 \cdot \tfrac{3}{2} + (-4) + 2(3) \ ? \ 5 \\ 3 - 4 + 6 \ \bigg| \\ 5 \ \bigg| \quad \text{TRUE} \end{array}$$

Do Exercise 2.

In Example 3, two of the equations have a missing variable.

EXAMPLE 3 Solve this system:

$$x + y + z = 180, \qquad \textbf{(1)}$$
$$x \qquad - z = -70, \qquad \textbf{(2)}$$
$$2y - z = 0. \qquad \textbf{(3)}$$

We note that there is no y in equation (2). In order to have a system of two equations in the variables x and z, we need to find another equation without a y. We use equations (1) and (3) to eliminate y:

$$x + y + z = 180, \qquad \textbf{(1)}$$
$$2y - z = 0; \qquad \textbf{(3)}$$

$$\begin{array}{rl} -2x - 2y - 2z = -360 & \text{Multiplying equation (1) by } -2 \\ \underline{2y - z = \quad 0} & \textbf{(3)} \\ -2x \quad - 3z = -360. & \textbf{(4)} \quad \text{Adding} \end{array}$$

Now we solve the resulting system of equations (2) and (4):

$$x - z = -70, \qquad \textbf{(2)}$$
$$-2x - 3z = -360; \qquad \textbf{(4)}$$

$$\begin{array}{rl} 2x - 2z = -140 & \text{Multiplying equation (2) by 2} \\ \underline{-2x - 3z = -360} & \textbf{(4)} \\ -5z = -500 & \text{Adding} \\ z = 100. \end{array}$$

Answer on page A-19

CHAPTER 3: Systems of Equations

To find x, we substitute 100 for z in equation (2) and solve for x:

$$x - z = -70$$
$$x - 100 = -70$$
$$x = 30.$$

To find y, we substitute 100 for z in equation (3) and solve for y:

$$2y - z = 0$$
$$2y - 100 = 0$$
$$2y = 100$$
$$y = 50.$$

The triple $(30, 50, 100)$ is the solution. The check is left to the student.

Do Exercise 3.

It is possible for a system of three equations to have no solution, that is, to be inconsistent. An example is the system

$$x + y + z = 14,$$
$$x + y + z = 11,$$
$$2x - 3y + 4z = -3.$$

Note the first two equations. It is not possible for a sum of three numbers to be both 14 and 11. Thus the system has no solution. We will not consider such systems here, nor will we consider systems with infinitely many solutions, which also exist.

3. Solve. Don't forget to check.

$$x + y + z = 100,$$
$$x - y \quad\quad = -10,$$
$$x \quad\quad - z = -30$$

Answer on page A-19

Study Tips

TIME MANAGEMENT (PART 1)

Time is the most critical factor in your success in learning mathematics. Have reasonable expectations about the time you need to study math. (See also the Study Tips on time management in Sections 5.6 and 6.1.)

■ **Juggling time.** Working 40 hours per week and taking 12 credit hours is equivalent to working two full-time jobs. Can you handle such a load? Your ratio of number of work hours to number of credit hours should be about 40/3, 30/6, 20/9, 10/12, or 5/14.

■ **A rule of thumb on study time.** Budget about 2–3 hours for homework and study per week for every hour of class time.

■ **Scheduling your time.** Make an hour-by-hour schedule of your typical week. Include work, school, home, sleep, study, and leisure times. Try to schedule time for study when you are most alert. Choose a setting that will enable you to maximize your concentration. Plan for success and it will happen!

"You cannot increase the quality or quantity of your achievement or performance except to the degree in which you increase your ability to use time effectively."

Brian Tracy, motivational speaker

a Solve.

1. $x + y + z = 2,$
$2x - y + 5z = -5,$
$-x + 2y + 2z = 1$

2. $2x - y - 4z = -12,$
$2x + y + z = 1,$
$x + 2y + 4z = 10$

3. $2x - y + z = 5,$
$6x + 3y - 2z = 10,$
$x - 2y + 3z = 5$

4. $x - y + z = 4,$
$3x + 2y + 3z = 7,$
$2x + 9y + 6z = 5$

5. $2x - 3y + z = 5,$
$x + 3y + 8z = 22,$
$3x - y + 2z = 12$

6. $6x - 4y + 5z = 31,$
$5x + 2y + 2z = 13,$
$x + y + z = 2$

7. $3a - 2b + 7c = 13,$
$a + 8b - 6c = -47,$
$7a - 9b - 9c = -3$

8. $x + y + z = 0,$
$2x + 3y + 2z = -3,$
$-x + 2y - 3z = -1$

9. $2x + 3y + z = 17,$
$x - 3y + 2z = -8,$
$5x - 2y + 3z = 5$

10. $2x + y - 3z = -4,$
$4x - 2y + z = 9,$
$3x + 5y - 2z = 5$

11. $2x + y + z = -2,$
$2x - y + 3z = 6,$
$3x - 5y + 4z = 7$

12. $2x + y + 2z = 11,$
$3x + 2y + 2z = 8,$
$x + 4y + 3z = 0$

13. $x - y + z = 4,$
$5x + 2y - 3z = 2,$
$3x - 7y + 4z = 8$

14. $2x + y + 2z = 3,$
$x + 6y + 3z = 4,$
$3x - 2y + z = 0$

15. $4x - y - z = 4,$
$2x + y + z = -1,$
$6x - 3y - 2z = 3$

16. $a + 2b + c = 1,$
$7a + 3b - c = -2,$
$a + 5b + 3c = 2$

17. $2r + 3s + 12t = 4,$
$4r - 6s + 6t = 1,$
$r + s + t = 1$

18. $10x + 6y + z = 7,$
$5x - 9y - 2z = 3,$
$15x - 12y + 2z = -5$

19. $4a + 9b = 8,$
$8a + 6c = -1,$
$6b + 6c = -1$

20. $3p + 2r = 11,$
$q - 7r = 4,$
$p - 6q = 1$

21. $x + y + z = 57,$
$-2x + y = 3,$
$x - z = 6$

22. $x + y + z = 105,$
$10y - z = 11,$
$2x - 3y = 7$

23. $r + s = 5,$
$3s + 2t = -1,$
$4r + t = 14$

24. $a - 5c = 17,$
$b + 2c = -1,$
$4a - b - 3c = 12$

25. D_W Explain a procedure that could be used to solve a system of four equations in four variables.

26. D_W Is it possible for a system of three equations to have exactly two ordered triples in its solution set? Why or why not?

SKILL MAINTENANCE

Solve for the indicated letter. [1.2a]

27. $F = 3ab$, for a

28. $Q = 4(a + b)$, for a

29. $F = \frac{1}{2}t(c - d)$, for c

30. $F = \frac{1}{2}t(c - d)$, for d

31. $Ax - By = c$, for y

32. $Ax + By = c$, for y

Find the slope and the y-intercept. [2.4b]

33. $y = -\frac{2}{3}x - \frac{5}{4}$

34. $y = 5 - 4x$

35. $2x - 5y = 10$

36. $7x - 6.4y = 20$

SYNTHESIS

Solve.

37. $w + x + y + z = 2,$
$w + 2x + 2y + 4z = 1,$
$w - x + y + z = 6,$
$w - 3x - y + z = 2$

38. $w + x - y + z = 0,$
$w - 2x - 2y - z = -5,$
$w - 3x - y + z = 4,$
$2w - x - y + 3z = 7$

Objective

a Solve applied problems using systems of three equations.

---

1. **Triangle Measures.** One angle of a triangle is twice as large as a second angle. The remaining angle is 20° greater than the first angle. Find the measure of each angle.

3.6 SOLVING APPLIED PROBLEMS: THREE EQUATIONS

a Using Systems of Three Equations

Solving systems of three or more equations is important in many applications occurring in the natural and social sciences, business, and engineering.

EXAMPLE 1 *Architecture.* In a triangular cross-section of a roof, the largest angle is 70° greater than the smallest angle. The largest angle is twice as large as the remaining angle. Find the measure of each angle.

1. **Familiarize.** We first make a drawing. Since we do not know the size of any angle, we use x, y, and z for the measures of the angles. We let $x =$ the smallest angle, $z =$ the largest angle, and $y =$ the remaining angle.

2. **Translate.** In order to translate the problem, we need to make use of a geometric fact—that is, the sum of the measures of the angles of a triangle is 180°. This fact about triangles gives us one equation:

$$x + y + z = 180.$$

There are two statements in the problem that we can translate directly.

The largest angle	is	70°	greater than	the smallest angle.
↓	↓	↓	↓	↓
z	$=$	70	$+$	x

The largest angle	is	twice as large as the remaining angle.
↓	↓	↓
z	$=$	$2y$

We now have a system of three equations:

$$x + y + z = 180, \qquad x + y + z = 180,$$
$$x + 70 = z, \qquad \text{or} \quad x \qquad -z = -70,$$
$$2y = z; \qquad\qquad\qquad 2y - z = 0.$$

3. **Solve.** The system was solved in Example 3 of Section 3.5. The solution is $(30, 50, 100)$.

4. **Check.** The sum of the numbers is 180. The largest angle measures 100° and the smallest measures 30°. The largest angle is 70° greater than the smallest. The remaining angle measures 50°. The largest angle is twice as large as the remaining angle. We do have an answer to the problem.

5. **State.** The measures of the angles of the triangle are 30°, 50°, and 100°.

Do Exercise 1.

Answer on page A-19

EXAMPLE 2 *Cholesterol Levels.* Americans have become very conscious of their cholesterol levels. Recent studies indicate that a child's intake of cholesterol should be no more than 300 mg per day. By eating 1 egg, 1 cupcake, and 1 slice of pizza, a child consumes 302 mg of cholesterol. If the child eats 2 cupcakes and 3 slices of pizza, he or she takes in 65 mg of cholesterol. By eating 2 eggs and 1 cupcake, a child consumes 567 mg of cholesterol. How much cholesterol is in each item?

1. **Familiarize.** After we have read the problem a few times, it becomes clear that an egg contains considerably more cholesterol than the other foods. Let's guess that one egg contains 200 mg of cholesterol and one cupcake contains 50 mg. Because of the third sentence in the problem, it would follow that a slice of pizza contains 52 mg of cholesterol since $200 + 50 + 52 = 302$.

   To see if our guess satisfies the other statements in the problem, we find the amount of cholesterol that 2 cupcakes and 3 slices of pizza would contain: $2 \cdot 50 + 3 \cdot 52 = 256$. Since this does not match the 65 mg listed in the fourth sentence of the problem, our guess was incorrect. Rather than guess again, we examine how we checked our guess and let e, c, and s = the number of milligrams of cholesterol in an egg, a cupcake, and a slice of pizza, respectively.

2. **Translate.** By rewording some of the sentences in the problem, we can translate it into three equations.

$$
\underbrace{\text{The amount of}\atop \text{cholesterol in}\atop \text{1 egg}}_{e} \quad \underbrace{\text{plus}}_{+} \quad \underbrace{\text{the amount of}\atop \text{cholesterol in}\atop \text{1 cupcake}}_{c} \quad \underbrace{\text{plus}}_{+} \quad \underbrace{\text{the amount of}\atop \text{cholesterol in}\atop \text{1 slice of pizza}}_{s} \quad \underbrace{\text{is}}_{=} \quad \underbrace{\text{302 mg.}}_{302}
$$

$$
\underbrace{\text{The amount of cholesterol}\atop \text{in 2 cupcakes}}_{2c} \quad \underbrace{\text{plus}}_{+} \quad \underbrace{\text{the amount of cholesterol}\atop \text{in 3 slices of pizza}}_{3s} \quad \underbrace{\text{is}}_{=} \quad \underbrace{\text{65 mg.}}_{65}
$$

$$
\underbrace{\text{The amount of cholesterol}\atop \text{in 2 eggs}}_{2e} \quad \underbrace{\text{plus}}_{+} \quad \underbrace{\text{the amount of cholesterol}\atop \text{in 1 cupcake}}_{c} \quad \underbrace{\text{is}}_{=} \quad \underbrace{\text{567 mg.}}_{567}
$$

We now have a system of three equations:

$$
\begin{aligned}
e + c + s &= 302, & \textbf{(1)} \\
2c + 3s &= 65, & \textbf{(2)} \\
2e + c &= 567. & \textbf{(3)}
\end{aligned}
$$

2. Client Investments. Kaufman Financial Corporation makes investments for corporate clients. One year, a client receives $1620 in simple interest from three investments that total $25,000. Part is invested at 5%, part at 6%, and part at 7%. There is $11,000 more invested at 7% than at 6%. How much was invested at each rate?

3. Solve. To solve, we first note that the variable e does not appear in equation (2). In order to have a system of two equations in the variables c and s, we need to find another equation without the variable e. We use equations (1) and (3) to eliminate e:

$$e + c + s = 302, \qquad (1)$$
$$2e + c = 567; \qquad (3)$$

$$\begin{array}{ll} -2e - 2c - 2s = -604 & \text{Multiplying equation (1) by } -2 \\ \underline{2e + c = 567} & (3) \\ -c - 2s = -37. & (4) \qquad \text{Adding} \end{array}$$

Next, we solve the resulting system of equations (2) and (4):

$$2c + 3s = 65, \qquad (2)$$
$$-c - 2s = -37; \qquad (4)$$

$$\begin{array}{ll} 2c + 3s = 65 & (2) \\ \underline{-2c - 4s = -74} & \text{Multiplying equation (4) by 2} \\ -s = -9 & \text{Adding} \\ s = 9. \end{array}$$

To find c, we substitute 9 for s in equation (4) and solve for c:

$$\begin{array}{ll} -c - 2s = -37 & (4) \\ -c - 2(9) = -37 & \text{Substituting} \\ -c - 18 = -37 \\ -c = -19 \\ c = 19. \end{array}$$

To find e, we substitute 19 for c in equation (3) and solve for e:

$$\begin{array}{ll} 2e + c = 567 & (3) \\ 2e + 19 = 567 & \text{Substituting} \\ 2e = 548 \\ e = 274. \end{array}$$

The solution is $c = 19$, $e = 274$, $s = 9$, or $(19, 274, 9)$.

4. Check. The sum of 19, 274, and 9 is 302 so the total cholesterol in 1 cupcake, 1 egg, and 1 slice of pizza checks. Two cupcakes and three slices of pizza would contain $2 \cdot 19 + 3 \cdot 9$, or 65 mg, while two eggs and one cupcake would contain $2 \cdot 274 + 19$, or 567 mg of cholesterol. The answer checks.

5. State. A cupcake contains 19 mg of cholesterol, an egg contains 274 mg of cholesterol, and a slice of pizza contains 9 mg of cholesterol.

Do Exercise 2.

Answer on page A-19

3.6

EXERCISE SET

For Extra Help

Math XL   MyMathLab   InterAct   Math Tutor   Digital Video   Student's
                        Math   Center   Tutor CD 2   Solutions
                                              Videotape 4   Manual

a   Solve.

1. *Low-Fat Fruit Drinks.*   Smoothie King®, Nutritional Lifestyle Center ™, recently sold 20-oz low-fat fruit Smoothies for $3.25, 32-oz Smoothies for $4.75, and 40-oz Smoothies for $5.75. One hot summer afternoon, Jake sold 34 Smoothies for a total of $153. The number of 20-oz and 40-oz Smoothies, combined, was 4 more than the number of 32-oz Smoothies. How many of each size were sold?
Source: Smoothie King®

2. *Cappuccinos.*   Starbucks® sells cappuccinos in three sizes: tall for $2.65, grande for $3.20, and vente for $3.50. After an outdoor rally for the college football team, Bryce served 50 cappuccinos. The number of tall and vente cappuccinos, combined, was 2 fewer than the number of grande cappuccinos. If he collected a total of $157, how many cappuccinos of each size did he serve?
Source: Starbucks® Corporation

3. *Triangle Measures.*   In triangle *ABC,* the measure of angle *B* is three times that of angle *A*. The measure of angle *C* is 20° more than that of angle *A*. Find the measure of each angle.

4. *Triangle Measures.*   In triangle *ABC*, the measure of angle *B* is twice the measure of angle *A*. The measure of angle *C* is 80° more than that of angle *A*. Find the measure of each angle.

5. The sum of three numbers is 55. The difference of the largest and the smallest is 49, and the sum of the two smaller is 13. Find the numbers.

6. The sum of three numbers is −30. The largest minus twice the smallest is 45, and the largest is 20 more than the middle number. Find the numbers.

7. *Automobile Pricing.* A recent basic model of a particular automobile had a price of $12,685. The basic model with the added features of automatic transmission and power door locks was $14,070. The basic model with air conditioning (AC) and power door locks was $13,580. The basic model with AC and automatic transmission was $13,925. What was the individual cost of each of the three options?

8. *Telemarketing.* Sven, Tillie, and Isaiah can process 740 telephone orders per day. Sven and Tillie together can process 470 orders, while Tillie and Isaiah together can process 520 orders per day. How many orders can each person process alone?

9. *Lens Production.* When Sight-Rite's three polishing machines, A, B, and C, are all working, 5700 lenses can be polished in one week. When only A and B are working, 3400 lenses can be polished in one week. When only B and C are working, 4200 lenses can be polished in one week. How many lenses can be polished in a week by each machine alone?

10. *Welding Rates.* Elrod, Dot, and Wendy can weld 74 linear feet per hour when working together. Elrod and Dot together can weld 44 linear feet per hour, while Elrod and Wendy can weld 50 linear feet per hour. How many linear feet per hour can each weld alone?

11. *Investments.* A business class divided an imaginary investment of $80,000 among three mutual funds. The first fund grew by 10%, the second by 6%, and the third by 15%. Total earnings were $8850. The earnings from the first fund were $750 more than the earnings from the third. How much was invested in each fund?

12. *Advertising.* In a recent year, companies spent a total of $84.8 billion on newspaper, television, and radio ads. The total amount spent on television and radio ads was only $2.6 billion more than the amount spent on newspaper ads alone. The amount spent on newspaper ads was $5.1 billion more than what was spent on television ads. How much was spent on each form of advertising? (*Hint*: Let the variables represent numbers of billions of dollars.)

13. *Twin Births.* In the United States, the highest incidence of fraternal twin births occurs among Asian–Americans, then African–Americans, and then Caucasians. Of every 15,400 births, the total number of fraternal twin births for all three is 739, where there are 185 more for Asian–Americans than African–Americans and 231 more for Asian–Americans than Caucasians. How many births of fraternal twins are there for each group out of every 15,400 births?

14. *Crying Rate.* The sum of the average number of times a man, a woman, and a one-year-old child cry each month is 71.7. A one-year-old cries 46.4 more times than a man. The average number of times a one-year-old cries per month is 28.3 more than the average number of times combined that a man and a woman cry. What is the average number of times per month that each cries?

15. *Nutrition.* A dietician in a hospital prepares meals under the guidance of a physician. Suppose that for a particular patient a physician prescribes a meal to have 800 calories, 55 g of protein, and 220 mg of vitamin C. The dietician prepares a meal of roast beef, baked potato, and broccoli according to the data in the following table.

	CALORIES	PROTEIN (in grams)	VITAMIN C (in milligrams)
Roast Beef, 3 oz	300	20	0
Baked Potato	100	5	20
Broccoli, 156 g	50	5	100

How many servings of each food are needed in order to satisfy the doctor's orders?

16. *Nutrition.* Repeat Exercise 15 but replace the broccoli with asparagus, for which one 180-g serving contains 50 calories, 5 g of protein, and 44 mg of vitamin C. Which meal would you prefer eating?

17. *Golf.* On an 18-hole golf course, there are par-3 holes, par-4 holes, and par-5 holes. A golfer who shoots par on every hole has a total of 70. There are twice as many par-4 holes as there are par-5 holes. How many of each type of hole are there on the golf course?

18. *Golf.* On an 18-hole golf course, there are par-3 holes, par-4 holes, and par-5 holes. A golfer who shoots par on every hole has a total of 72. The sum of the number of par-3 holes and the number of par-5 holes is 8. How many of each type of hole are there on the golf course?

19. *Basketball Scoring.* The New York Knicks recently scored a total of 92 points on a combination of 2-point field goals, 3-point field goals, and 1-point foul shots. Altogether, the Knicks made 50 baskets and 19 more 2-pointers than foul shots. How many shots of each kind were made?

20. *History.* Find the year in which the first U.S. transcontinental railroad was completed. The following are some facts about the number. The sum of the digits in the year is 24. The ones digit is 1 more than the hundreds digit. Both the tens and the ones digits are multiples of 3.

21. $\mathbf{D_W}$ Exercise 10 can be solved mentally after a careful reading of the problem. How is this possible?

22. $\mathbf{D_W}$ *Ticket Revenue.* A pops-concert audience of 100 people consists of adults, senior citizens, and children. The ticket prices are $10 each for adults, $3 each for senior citizens, and $0.50 each for children. The total amount of money taken in is $100. How many adults, senior citizens, and children are in attendance? Does there seem to be some information missing? Do some careful reasoning and explain.

23. **Dw** Consider Exercise 19. Suppose 92 points were scored with 50 baskets but no foul shots. Would there still be a solution? Why or why not?

24. **Dw** Consider Exercise 1. Suppose Jake collected $150. Could the problem still be solved? Why or why not?

VOCABULARY REINFORCEMENT

In each of Exercises 25–32, fill in the blank with the correct term from the given list. Some of the choices may not be used.

25. The expression $x \leq q$ means x is _____ q. [1.4d]

26. When two sets have no elements in common, the intersection of the two sets is called the _____ . [1.5a]

27. The graph of a(n) _____ equation is a line. [2.1c]

28. When the slope of a line is _____ , the graph of the line slants down from left to right. [2.4b]

29. A(n) _____ system of equations has at least one solution. [3.1a]

30. Two lines are _____ if the product of their slopes is −1. [2.5d]

31. The _____ of the graph of $f(x) = mx + b$ is the point $(0, b)$. [2.4a]

32. When the slope of a line is zero, the graph of the line is _____ . [2.4b]

parallel

perpendicular

union

empty set

consistent

inconsistent

linear

x-intercept

y-intercept

positive

zero

negative

vertical

horizontal

at least

at most

33. Find the sum of the angle measures at the tips of the star in this figure.

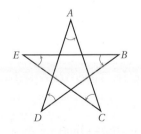

34. *Sharing Raffle Tickets.* Hal gives Tom as many raffle tickets as Tom has and Gary as many as Gary has. In like manner, Tom then gives Hal and Gary as many tickets as each then has. Similarly, Gary gives Hal and Tom as many tickets as each then has. If each finally has 40 tickets, with how many tickets does Tom begin?

35. *Digits.* Find a three-digit positive integer such that the sum of all three digits is 14, the tens digit is 2 more than the ones digit, and if the digits are reversed, the number is unchanged.

36. *Ages.* Tammy's age is the sum of the ages of Carmen and Dennis. Carmen's age is 2 more than the sum of the ages of Dennis and Mark. Dennis's age is four times Mark's age. The sum of all four ages is 42. How old is Tammy?

Objectives

a Determine whether an ordered pair of numbers is a solution of an inequality in two variables.

b Graph linear inequalities in two variables.

c Graph systems of linear inequalities and find coordinates of any vertices.

A **graph** of an inequality is a drawing that represents its solutions. An inequality in one variable can be graphed on the number line (see Section 1.4). An inequality in two variables can be graphed on a coordinate plane.

A **linear inequality** is one that we can get from a related linear equation by changing the equals symbol to an inequality symbol. The graph of a linear inequality is a region on one side of a line. This region is called a **half-plane.** The graph sometimes includes the graph of the related line at the boundary of the half-plane.

1. Determine whether $(1, -4)$ is a solution of $4x - 5y < 12$.

$$\frac{4x - 5y < 12}{?}$$

a Solutions of Inequalities in Two Variables

The solutions of an inequality in two variables are ordered pairs.

EXAMPLES Determine whether the ordered pair is a solution of the inequality $5x - 4y > 13$.

1. $(-3, 2)$

We have

$$\frac{5x - 4y > 13}{5(-3) - 4 \cdot 2 \; ? \; 13}$$
$$-15 - 8$$
$$-23 \quad \text{FALSE}$$

We use alphabetical order to replace x with -3 and y with 2.

Since $-23 > 13$ is false, $(-3, 2)$ is not a solution.

2. $(4, -3)$

We have

$$\frac{5x - 4y > 13}{5(4) - 4(-3) \; ? \; 13}$$
$$20 + 12$$
$$32 \quad \text{TRUE}$$

Replacing x with 4 and y with -3

Since $32 > 13$ is true, $(4, -3)$ is a solution.

2. Determine whether $(4, -3)$ is a solution of $3y - 2x \leq 6$.

$$\frac{3y - 2x \leq 6}{?}$$

Do Exercises 1 and 2.

b Graphing Inequalities in Two Variables

Let's visualize the results of Examples 1 and 2. The equation $5x - 4y = 13$ is represented by the dashed line in the graphs below. The solutions of the inequality $5x - 4y > 13$ are shaded below that dashed line. The pair $(-3, 2)$ is not a solution of the inequality $5x - 4y > 13$ and is not in the shaded region. See the graph on the left below.

Answers on page A-19

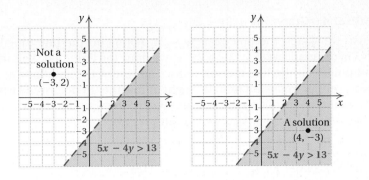

The pair $(4, -3)$ is a solution of the inequality $5x - 4y > 13$ and is in the shaded region. See the graph on the right above.

We now consider how to graph inequalities.

EXAMPLE 3   Graph: $y < x$.

We first graph the line $y = x$. Every solution of $y = x$ is an ordered pair like $(3, 3)$, where the first and second coordinates are the same. The graph of $y = x$ is shown on the left below. We draw it dashed because these points are *not* solutions of $y < x$.

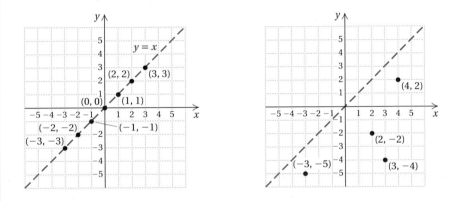

Now look at the graph on the right above. Several ordered pairs are plotted on the half-plane below $y = x$. Each is a solution of $y < x$. We can check the pair $(4, 2)$ as follows:

$$\frac{y < x}{2 \ ? \ 4} \quad \text{TRUE}$$

It turns out that any point on the same side of $y = x$ as $(4, 2)$ is also a solution. Thus, *if you know that one point in a half-plane is a solution, then all points in that half-plane are solutions.* In this text, we will usually indicate this by color shading. We shade the half-plane below $y = x$.

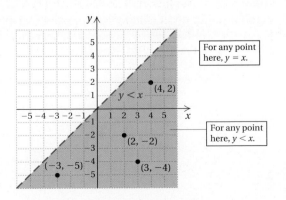

EXAMPLE 4 Graph: $8x + 3y \geq 24$.

First, we sketch the line $8x + 3y = 24$. Points on the line $8x + 3y = 24$ are also in the graph of $8x + 3y \geq 24$, so we draw the line solid. This indicates that all points on the line are solutions. The rest of the solutions are in the half-plane either to the left or to the right of the line. To determine which, we select a point that is not on the line and determine whether it is a solution of $8x + 3y \geq 24$. We try $(-3, 4)$ as a test point:

$$
\begin{array}{c|c}
\multicolumn{2}{c}{8x + 3y \geq 24} \\
\hline
8(-3) + 3(4) \ ? \ 24 & \\
-24 + 12 & \\
-12 & \text{FALSE}
\end{array}
$$

We see that $-12 \geq 24$ is *false*. Since $(-3, 4)$ is not a solution, none of the points in the half-plane containing $(-3, 4)$ is a solution. Thus the points in the opposite half-plane are solutions. We shade that half-plane and obtain the graph shown below.

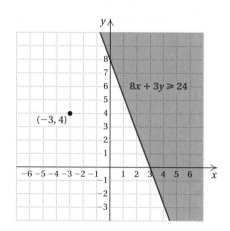

To graph an inequality in two variables:

1. Replace the inequality symbol with an equals sign and graph this related equation. This separates points that represent solutions from those that do not.
2. If the inequality symbol is $<$ or $>$, draw the line dashed. If the inequality symbol is $\leq$ or $\geq$, draw the line solid.
3. The graph consists of a half-plane that is either above or below or to the left or right of the line and, if the line is solid, the line as well. To determine which half-plane to shade, choose a point not on the line as a test point. If the line does not go through the origin, $(0, 0)$ is an easy point to use. Substitute to determine whether that point is a solution. If so, shade the half-plane containing that point. If not, shade the opposite half-plane.

EXAMPLE 5 Graph: $6x - 2y < 12$.

1. We first graph the related equation $6x - 2y = 12$.

2. Since the inequality uses the symbol $<$, points on the line are not solutions of the inequality, so we draw a dashed line.

3. To determine which half-plane to shade, we consider a test point *not* on the line. We try $(0, 0)$ and substitute:

$$\begin{array}{c|l} 6x - 2y < 12 \\ \hline 6(0) - 2(0) \text{ ? } 12 \\ 0 - 0 \\ \quad 0 & \text{TRUE} \end{array}$$

Since the inequality $0 < 12$ is *true*, the point $(0, 0)$ is a solution; each point in the half-plane containing $(0, 0)$ is a solution. Thus each point in the opposite half-plane is *not* a solution. The graph is shown below.

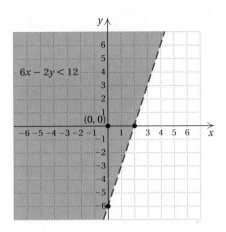

Do Exercises 3 and 4.

EXAMPLE 6 Graph $x > -3$ on a plane.

There is a missing variable in this inequality. If we graph the inequality on the number line, its graph is as follows:

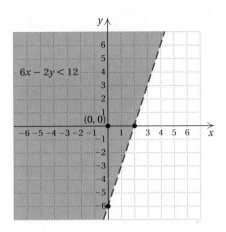

However, we can also write this inequality as $x + 0y > -3$ and consider graphing it in the plane. We are, in effect, determining which ordered pairs have x-values greater than -3. We use the same technique that we have used with the other examples. We first graph the related equation $x = -3$ in the plane. We draw the boundary with a dashed line. The rest of the graph is a half-plane to the right or left of the line $x = -3$. To determine which, we consider a test point, $(2, 5)$:

$$\begin{array}{c|l} x + 0y > -3 \\ \hline 2 + 0(5) \text{ ? } -3 \\ \quad 2 & \text{TRUE} \end{array}$$

Graph.

3. $6x - 3y < 18$

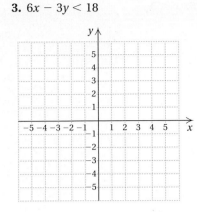

4. $4x + 3y \geq 12$

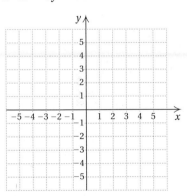

Answers on page A-19

301

3.7 Systems of Inequalities
in Two Variables

Graph on a plane.

5. $x < 3$

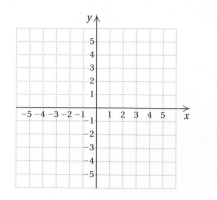

Since $(2,5)$ is a solution, all the points in the half-plane containing $(2,5)$ are solutions. We shade that half-plane.

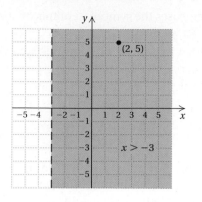

EXAMPLE 7 Graph $y \leq 4$ on a plane.

We first graph $y = 4$ using a solid line. We then use $(-2, 5)$ as a test point and substitute in $0x + y \leq 4$:

$$\begin{array}{c|c} 0x + y \leq 4 \\ \hline 0(-2) + 5 \; ? \; 4 \\ 0 + 5 \\ 5 & \text{FALSE} \end{array}$$

We see that $(-2, 5)$ is *not* a solution, so all the points in the half-plane containing $(-2, 5)$ are not solutions. Thus each point in the opposite half-plane is a solution.

6. $y \geq -4$

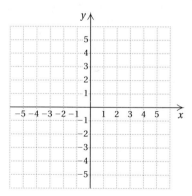

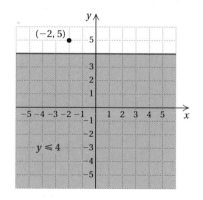

Do Exercises 5 and 6.

C Systems of Linear Inequalities

The following is an example of a system of two linear inequalities in two variables:

$$x + y \leq 4,$$
$$x - y < 4.$$

A **solution** of a system of linear inequalities is an ordered pair that is a solution of *both* inequalities. We now graph solutions of systems of linear inequalities. To do so, we graph each inequality and determine where the graphs overlap, or intersect. That will be a region in which the ordered pairs are solutions of both inequalities.

Answers on page A-19

EXAMPLE 8 Graph the solutions of the system

$$x + y \leq 4,$$
$$x - y < 4.$$

We graph the inequality $x + y \leq 4$ by first graphing the equation $x + y = 4$ using a solid red line. We consider $(0, 0)$ as a test point and find that it is a solution, so we shade all points on that side of the line using red shading. (See the graph on the left below.) The arrows at the ends of the line also indicate the half-plane, or region, that contains the solutions.

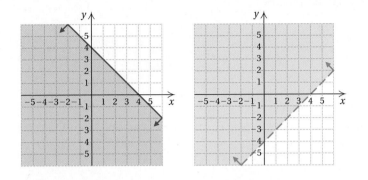

Next, we graph $x - y < 4$. We begin by graphing the equation $x - y = 4$ using a dashed blue line and consider $(0, 0)$ as a test point. Again, $(0, 0)$ is a solution so we shade that side of the line using blue shading. (See the graph on the right above.) The solution set of the system is the region that is shaded both red and blue and part of the line $x + y = 4$. (See the graph below.)

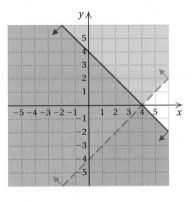

Do Exercise 7.

7. Graph:

$$x + y \geq 1,$$
$$y - x \geq 2.$$

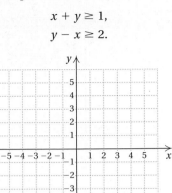

Answer on page A-19

3.7 Systems of Inequalities
in Two Variables

8. Graph: $-3 \leq y < 4$.

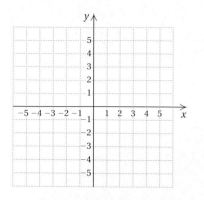

EXAMPLE 9 Graph: $-2 < x \leq 5$.

This is actually a system of inequalities:

$$-2 < x,$$
$$x \leq 5.$$

We graph the equation $-2 = x$ and see that the graph of the first inequality is the half-plane to the right of the line $-2 = x$. (See the graph on the left below.)

Next, we graph the second inequality, starting with the line $x = 5$, and find that its graph is the line and also the half-plane to the left of it. (See the graph on the right below.)

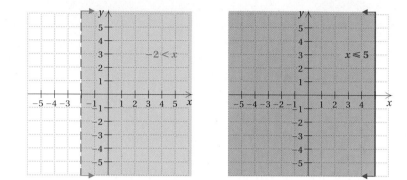

We shade the intersection of these graphs.

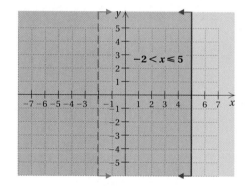

Do Exercise 8.

Answer on page A-19

A system of inequalities may have a graph that consists of a polygon and its interior. In *linear programming*, which is a topic rich in application that you may study in a later course, it is important to be able to find the vertices of such a polygon.

EXAMPLE 10 Graph the following system of inequalities. Find the coordinates of any vertices formed.

$$6x - 2y \leq 12, \quad \textbf{(1)}$$
$$y - 3 \leq 0, \quad \textbf{(2)}$$
$$x + y \geq 0 \quad \textbf{(3)}$$

We graph the lines $6x - 2y = 12$, $y - 3 = 0$, and $x + y = 0$ using solid lines. The regions for each inequality are indicated by the arrows at the ends of the lines. We then note where the regions overlap and shade the region of solutions using one color.

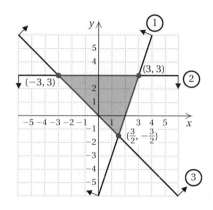

To find the vertices, we solve three different systems of equations. The system of equations from inequalities (1) and (2) is

$$6x - 2y = 12, \quad \textbf{(1)}$$
$$y - 3 = 0. \quad \textbf{(2)}$$

Solving, we obtain the vertex $(3, 3)$.

The system of equations from inequalities (1) and (3) is

$$6x - 2y = 12, \quad \textbf{(1)}$$
$$x + y = 0. \quad \textbf{(3)}$$

Solving, we obtain the vertex $\left(\frac{3}{2}, -\frac{3}{2}\right)$.

The system of equations from inequalities (2) and (3) is

$$y - 3 = 0, \quad \textbf{(2)}$$
$$x + y = 0. \quad \textbf{(3)}$$

Solving, we obtain the vertex $(-3, 3)$.

Do Exercise 9.

9. Graph the system of inequalities. Find the coordinates of any vertices formed.

$$5x + 6y \leq 30,$$
$$0 \leq y \leq 3,$$
$$0 \leq x \leq 4$$

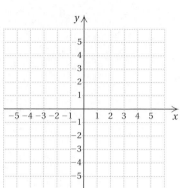

Answer on page A-19

10. Graph the system of inequalities. Find the coordinates of any vertices formed.

$$2x + 4y \leq 8,$$
$$x + y \leq 3,$$
$$x \geq 0,$$
$$y \geq 0$$

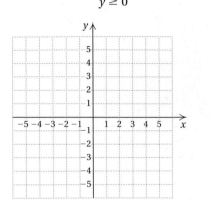

EXAMPLE 11 Graph the following system of inequalities. Find the coordinates of any vertices formed.

$$x + y \leq 16, \quad \textbf{(1)}$$
$$3x + 6y \leq 60, \quad \textbf{(2)}$$
$$x \geq 0, \quad \textbf{(3)}$$
$$y \geq 0 \quad \textbf{(4)}$$

We graph each inequality using solid lines. The regions for each inequality are indicated by the arrows at the ends of the lines. We then note where the regions overlap and shade the region of solutions using one color.

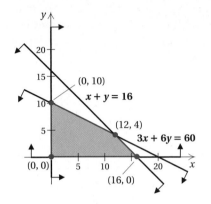

To find the vertices, we solve four different systems of equations. The system of equations from inequalities (1) and (2) is

$$x + y = 16, \quad \textbf{(1)}$$
$$3x + 6y = 60. \quad \textbf{(2)}$$

Solving, we obtain the vertex $(12, 4)$.
The system of equations from inequalities (1) and (4) is

$$x + y = 16, \quad \textbf{(1)}$$
$$y = 0. \quad \textbf{(4)}$$

Solving, we obtain the vertex $(16, 0)$.
The system of equations from inequalities (3) and (4) is

$$x = 0, \quad \textbf{(3)}$$
$$y = 0. \quad \textbf{(4)}$$

The vertex is $(0, 0)$.
The system of equations from inequalities (2) and (3) is

$$3x + 6y = 60, \quad \textbf{(2)}$$
$$x = 0. \quad \textbf{(3)}$$

Solving, we obtain the vertex $(0, 10)$.

Do Exercise 10.

Answer on page A-19

CALCULATOR CORNER

Graphs of Inequalities   We can graph inequalities on a graphing calculator, shading the region of the solution set. To graph the inequality in Example 4, $8x + 3y \geq 24$, we first graph the line $8x + 3y = 24$. Solving for y, we get $y = (24 - 8x)/3$, or $y = (-8/3)x + 8$. We enter this equation on the equation-editor screen. After determining algebraically that the solution set consists of all points above the line, we use the calculator's "shade above" GraphStyle to shade this region. To do this, we position the cursor over the graphstyle icon on the equation-editor screen to the left of the equation. Then we press **ENTER** repeatedly, until the "shade above" icon appears. Then we select a window and press **GRAPH** to display the graph of the inequality.

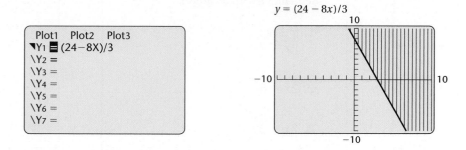

$y = (24 - 8x)/3$

Note that we cannot use the GraphStyle option to graph an inequality like the one in Example 6, $x > -3$, because the related equation has no y-term and thus cannot be entered in "$y =$" form.

We can also graph a system of inequalities by shading the solution set of each inequality in the system with a different pattern. When the "shade above" or "shade below" GraphStyle options are selected, the TI-84 Plus rotates through four shading patterns. For example, to graph the solution set of the system of inequalities

$$x + y \leq 4,$$
$$x - y \geq 2,$$

we first graph the equation $x + y = 4$, entering it in the form $y_1 = -x + 4$. We determine that the solution set of $x + y \leq 4$ consists of all points below the line $x + y = 4$, or $y_1 = -x + 4$, so we select the "shade below" GraphStyle for this function. Next, we graph $x - y = 2$, entering it in the form $y_2 = x - 2$. The solution set of $x - y \geq 2$ consists of all points below the line $x - y = 2$, or $y_2 = x - 2$, so we also choose the "shade below" GraphStyle for this function. Now we select a window and press **GRAPH** to display the solution sets of each inequality in the system and the region where they overlap. The region of overlap is the solution set of the system of inequalities. We could also use the **INTERSECT** feature to find the coordinates of the vertex formed by this system.

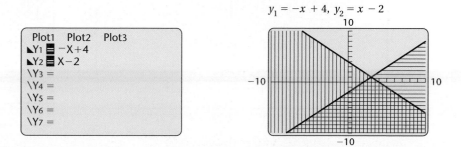

$y_1 = -x + 4, \ y_2 = x - 2$

The TI-84 Plus has an **INEQUALZ** application that allows us to enter a linear inequality directly on the equation-editor screen. Then, when the **GRAPH** key is pressed, the calculator determines the region to be shaded and graphs the inequality accordingly. This application is accessed by pressing the **APPS** key.

Exercises:

1. Use a graphing calculator to graph the inequalities in Margin Exercises 4 and 6.

2. Use a graphing calculator to graph the system of inequalities in Margin Exercise 7.

Visualizing for Success

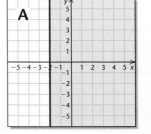

A

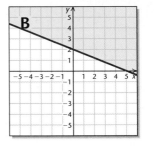

B

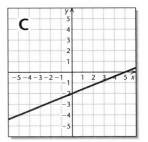

C

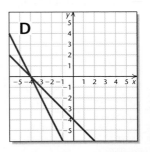

D

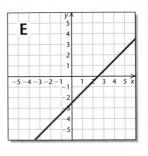

E

Match the equation, inequality, system of equations, or system of inequalities with its graph.

1. $x + y = -4,$
   $2x + y = -8$

2. $2x + 5y \geq 10$

3. $2x - 2y = 5$

4. $2x - 5y = 10$

5. $-2y < 8$

6. $5x - 2y = 10$

7. $2x = 10$

8. $5x + 2y < 10,$
   $2x - 5y > 10$

9. $5x \geq -10$

10. $y - 2x < 8$

Answers on page A-20

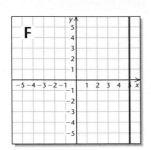

F

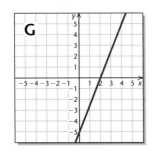

G

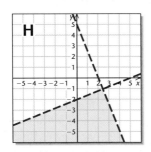

H

I

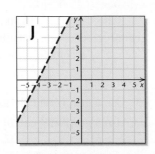

J

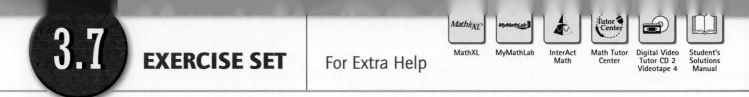

3.7 EXERCISE SET

For Extra Help

a Determine whether the given ordered pair is a solution of the given inequality.

1. $(-3, 3);$ $3x + y < -5$

2. $(6, -8);$ $4x + 3y \geq 0$

3. $(5, 9);$ $2x - y > -1$

4. $(5, -2);$ $6y - x > 2$

b Graph the inequality on a plane.

5. $y > 2x$

6. $y < 3x$

7. $y < x + 1$

8. $y \leq x - 3$

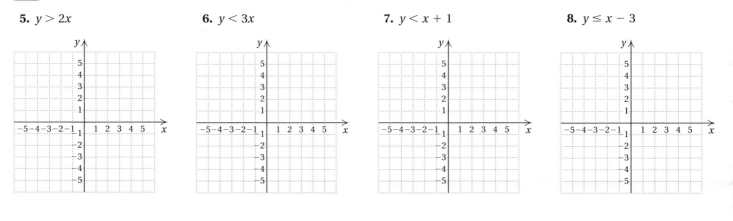

9. $y > x - 2$

10. $y \geq x + 4$

11. $x + y < 4$

12. $x - y \geq 3$

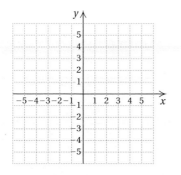
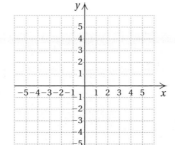
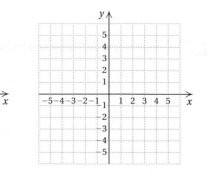

13. $3x + 4y \leq 12$

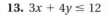

14. $2x + 3y < 6$

15. $2y - 3x > 6$

16. $2y - x \leq 4$

17. $3x - 2 \leq 5x + y$

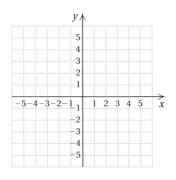

18. $2x - 2y \geq 8 + 2y$

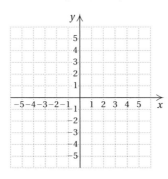

19. $x < 5$

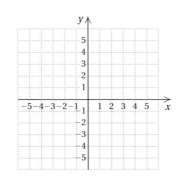

20. $y \geq -2$

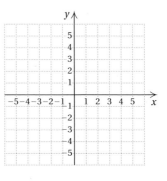

21. $y > 2$

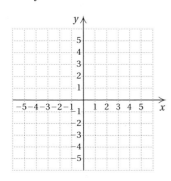

22. $x \leq -4$

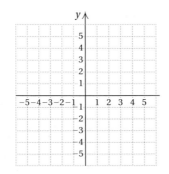

23. $2x + 3y \leq 6$

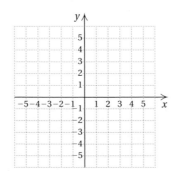

24. $7x + 2y \geq 21$

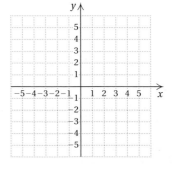

CHAPTER 3: Systems of Equations

Matching. Each of Exercises 25–30 shows the graph of an inequality. Match the graph with one of the appropriate inequalities (A)–(F) that follow.

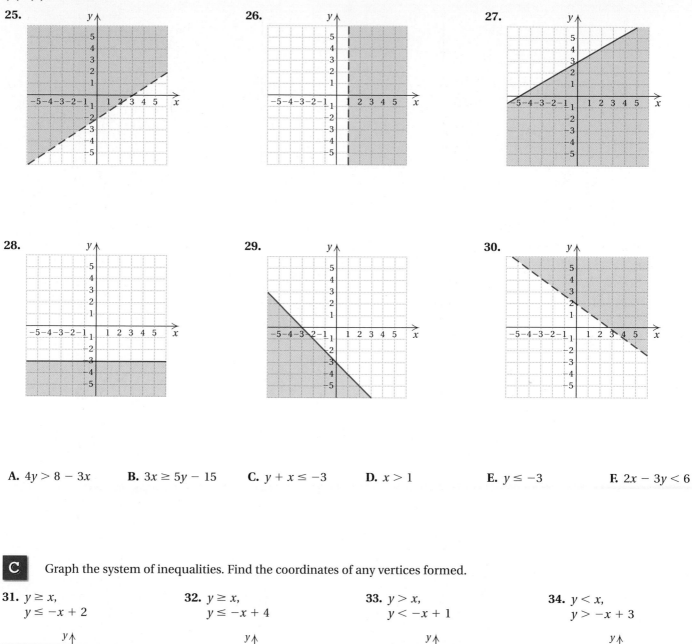

25.

26.

27.

28.

29.

30.

A. $4y > 8 - 3x$ **B.** $3x \geq 5y - 15$ **C.** $y + x \leq -3$ **D.** $x > 1$ **E.** $y \leq -3$ **F.** $2x - 3y < 6$

C   Graph the system of inequalities. Find the coordinates of any vertices formed.

31. $y \geq x,$
    $y \leq -x + 2$

32. $y \geq x,$
    $y \leq -x + 4$

33. $y > x,$
    $y < -x + 1$

34. $y < x,$
    $y > -x + 3$

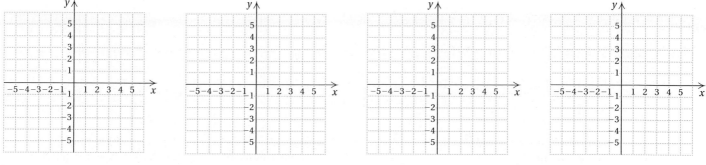

35. $y \geq -2$,
$x \geq 1$

36. $y \leq -2$,
$x \geq 2$

37. $x \leq 3$,
$y \geq -3x + 2$

38. $x \geq -2$,
$y \leq -2x + 3$

39. $y \leq 2x + 1$,
$y \geq -2x + 1$,
$x \leq 2$

40. $x - y \leq 2$,
$x + 2y \geq 8$,
$y \leq 4$

41. $x + y \leq 1$,
$x - y \leq 2$

42. $x + y \leq 3$,
$x - y \leq 4$

43. $x + 2y \leq 12$,
$2x + y \leq 12$,
$x \geq 0$,
$y \geq 0$

44. $4y - 3x \geq -12$,
$4y + 3x \geq -36$,
$y \leq 0$,
$x \leq 0$

45. $8x + 5y \leq 40$,
$x + 2y \leq 8$,
$x \geq 0$,
$y \geq 0$

46. $y - x \geq 1$,
$y - x \leq 3$,
$2 \leq x \leq 5$

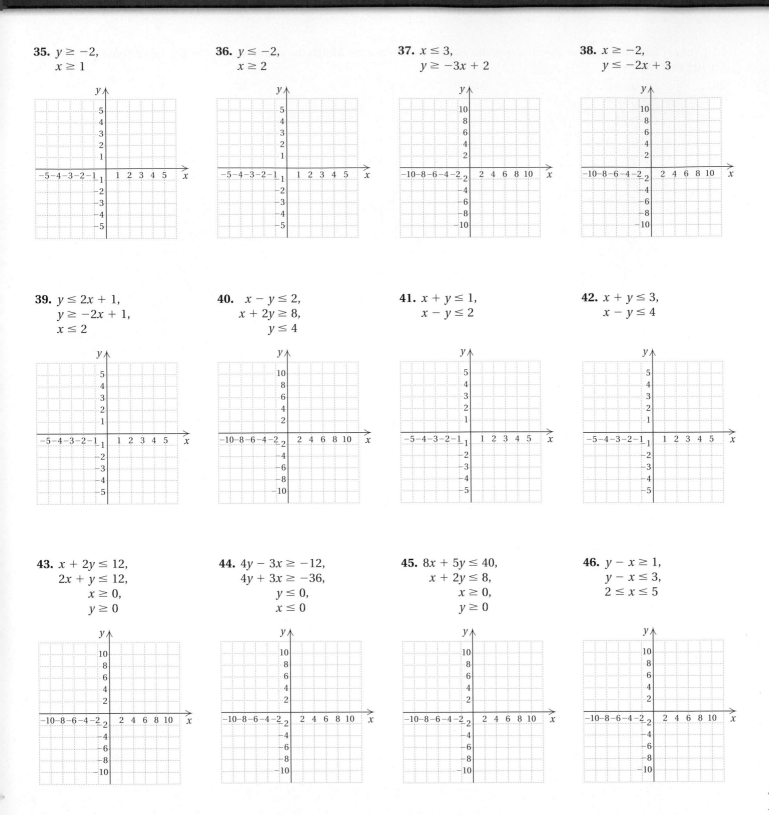

47. $\mathbf{D_W}$ Do all systems of linear inequalities have solutions? Why or why not?

48. $\mathbf{D_W}$ When graphing linear inequalities, Ron always shades above the line when he sees a $\geq$ symbol. Is this wise? Why or why not?

Solve. [1.1d]

49. $5(3x - 4) = -2(x + 5)$

50. $4(3x + 4) = 2 - x$

51. $2(x - 1) + 3(x - 2) - 4(x - 5) = 10$

52. $10x - 8(3x - 7) = 2(4x - 1)$

53. $5x + 7x = -144$

54. $0.5x - 2.34 + 2.4x = 7.8x - 9$

Given the function $f(x) = |2 - x|$, find each of the following function values. [2.2b]

55. $f(0)$

56. $f(-1)$

57. $f(1)$

58. $f(10)$

59. $f(-2)$

60. $f(2a)$

61. $f(-4)$

62. $f(1.8)$

63. *Luggage Size.* Unless an additional fee is paid, most major airlines will not check any luggage for which the sum of the item's length, width, and height exceeds 62 in. The U.S. Postal Service will ship a package only if the sum of the package's length and girth (distance around its midsection) does not exceed 130 in. Video Promotions is ordering several 30-in. long cases that will be both mailed and checked as luggage. Using w and h for width and height (in inches), respectively, write and graph an inequality that represents all acceptable combinations of width and height.

Source: U.S. Postal Service

64. *Exercise Danger Zone.* It is dangerous to exercise when the weather is hot and humid. The solutions of the following system of inequalities give a "danger zone" for which it is dangerous to exercise intensely:

$$4H - 3F < 70,$$
$$F + H > 160,$$
$$2F + 3H > 390,$$

where F is the temperature, in degrees Fahrenheit, and H is the humidity.

a) Draw the danger zone by graphing the system of inequalities.

b) Is it dangerous to exercise when $F = 80°$ and $H = 80\%$?

In Exercises 65–68, use a graphing calculator with a SHADE feature to graph the inequality.

65. $3x + 6y > 2$

66. $x - 5y \leq 10$

67. $13x - 25y + 10 \leq 0$

68. $2x + 5y > 0$

69. Use a graphing calculator with a SHADE feature to check your answers to Exercises 31 and 32. Then use the INTERSECT feature to determine any point(s) of intersection.

BUSINESS AND ECONOMICS APPLICATIONS

Objectives

a Given total-cost and total-revenue functions, find the total-profit function and the break-even point.

b Given supply and demand functions, find the equilibrium point.

a Break-Even Analysis

When a company manufactures x units of a product, it invests money. This is **total cost** and can be thought of as a function C, where $C(x)$ is the total cost of producing x units. When the company sells x units of the product, it takes in money. This is **total revenue** and can be thought of as a function R, where $R(x)$ is the total revenue from the sale of x units. **Total profit** is the money taken in less the money spent, or total revenue minus total cost. Total profit from the production and sale of x units is a function P given by

$$\textbf{Profit} = \textbf{Revenue} - \textbf{Cost}, \quad \text{or} \quad P(x) = R(x) - C(x).$$

If $R(x)$ is greater than $C(x)$, the company has a profit. If $C(x)$ is greater than $R(x)$, the company has a loss. When $R(x) = C(x)$, the company breaks even.

There are two kinds of costs. First, there are costs like rent, insurance, machinery, and so on. These costs, which must be paid whether a product is produced or not, are called **fixed costs.** When a product is being produced, there are costs for labor, materials, marketing, and so on. These are called **variable costs,** because they vary according to the amount of the product being produced. The sum of the fixed costs and the variable costs gives the **total cost** of producing a product.

EXAMPLE 1 *Manufacturing Radios.* Ergs, Inc., is planning to make a new kind of radio. Fixed costs will be $90,000, and it will cost $15 to produce each radio (variable costs). Each radio sells for $26.

a) Find the total cost $C(x)$ of producing x radios.

b) Find the total revenue $R(x)$ from the sale of x radios.

c) Find the total profit $P(x)$ from the production and sale of x radios.

d) What profit or loss will the company realize from the production and sale of 3000 radios? of 14,000 radios?

e) Graph the total-cost, total-revenue, and total-profit functions using the same set of axes. Determine the break-even point.

a) Total cost is given by

$$C(x) = \text{(Fixed costs) plus (Variable costs)},$$

$$\text{or} \quad C(x) = 90{,}000 + 15x,$$

where x is the number of radios produced.

b) Total revenue is given by

$$R(x) = 26x.$$ $26 times the number of radios sold. We assume that every radio produced is sold.

c) Total profit is given by

$$P(x) = R(x) - C(x)$$
$$= 26x - (90{,}000 + 15x)$$
$$= 11x - 90{,}000.$$

d) Profits will be

$$P(3000) = 11 \cdot 3000 - 90{,}000 = -\$57{,}000$$

when 3000 radios are produced and sold, and

$$P(14{,}000) = 11 \cdot 14{,}000 - 90{,}000 = \$64{,}000$$

when 14,000 radios are produced and sold. Thus the company loses $57,000 if only 3000 radios are sold, but makes $64,000 if 14,000 are sold.

e) The graphs of each of the three functions are shown below:

$$C(x) = 90{,}000 + 15x,\quad \textbf{(1)}$$
$$R(x) = 26x,\quad \textbf{(2)}$$
$$P(x) = 11x - 90{,}000.\quad \textbf{(3)}$$

$R(x)$, $C(x)$, and $P(x)$ are all in dollars.

Equation (2) has a graph that goes through the origin and has a slope of 26. Equation (1) has an intercept on the y-axis of 90,000 and has a slope of 15. Equation (3) has an intercept on the y-axis of $-90{,}000$ and has a slope of 11. It is shown by the dashed line. The red dashed line shows a "negative" profit, which is a loss. (That is what is known as "being in the red.") The black dashed line shows a "positive" profit, or gain. (That is what is known as "being in the black.")

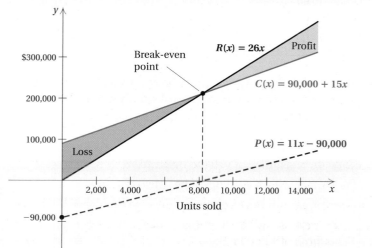

Profits occur when the revenue is greater than the cost. Losses occur when the revenue is less than the cost. The **break-even point** occurs where the graphs of R and C cross. Thus to find the break-even point, we solve a system:

$$C(x) = 90{,}000 + 15x,$$
$$R(x) = 26x.$$

1. Manufacturing Radios. Refer to Example 1. Suppose that fixed costs are $80,000, and it costs $20 to produce each radio. Each radio sells for $36.

a) Find the total cost $C(x)$ of producing x radios.

b) Find the total revenue $R(x)$ from the sale of x radios.

c) Find the total profit $P(x)$ from the production and sale of x radios.

d) What profit or loss will the company realize from the production and sale of 4000 radios? of 16,000 radios?

e) Graph the total-cost, total-revenue, and total-profit functions using the same set of axes. Determine the break-even point.

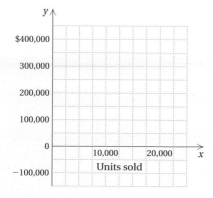

Answers on page A-21

Since both revenue and cost are in *dollars* and they are equal at the break-even point, the system can be rewritten as

$$d = 90{,}000 + 15x, \qquad (1)$$
$$d = 26x \qquad\qquad\qquad (2)$$

and solved using substitution:

$$26x = 90{,}000 + 15x \qquad \text{Substituting } 26x \text{ for } d \text{ in equation (1)}$$
$$11x = 90{,}000$$
$$x \approx 8181.8.$$

The firm will break even if it produces and sells about 8182 radios (8181 will yield a tiny loss and 8182 a tiny gain), and takes in a total of $R(8182) = 26 \cdot 8182 = \$212{,}732$ in revenue. Note that the x-coordinate of the break-even point is also the x-coordinate of the x-intercept of the profit function. It can also be found by solving $P(x) = 0$.

Do Exercise 1 on the preceding page.

b   Supply and Demand

As the price of coffee varies, the amount sold varies. The table and the graph below both show that consumer demand goes down as the price goes up and the demand goes up as the price goes down.

DEMAND FUNCTION, D

PRICE, p, PER KILOGRAM	QUANTITY, D(p) (in millions of kilograms)
$ 8.00	25
9.00	20
10.00	15
11.00	10
12.00	5

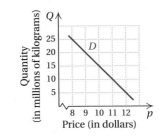

As the price of coffee varies, the amount available varies. The table and the graph below both show that sellers will supply less as the price goes down, but will supply more as the price goes up.

SUPPLY FUNCTION, S

PRICE, p, PER KILOGRAM	QUANTITY, S(p) (in millions of kilograms)
$ 9.00	5
9.50	10
10.00	15
10.50	20
11.00	25

Let's look at the above graphs together. We see that as price increases, demand decreases. As price increases, supply increases. The point of intersection of the demand and supply functions is called the **equilibrium point.** At the equilibrium point, the amount that the seller will supply is the same amount that the consumer will buy. The situation is analogous to a buyer and a seller negotiating the price of an item. The equilibrium point is the price and quantity on which they finally agree.

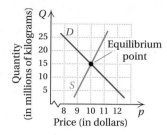

Any ordered pair of coordinates from the graph is (price, quantity), because the horizontal axis is the price axis and the vertical axis is the quantity axis. If D is a demand function and S is a supply function, then the equilibrium point is where demand equals supply:

$$D(p) = S(p).$$

EXAMPLE 2 Find the equilibrium point for the following demand and supply functions:

$$D(p) = 1000 - 60p, \quad \textbf{(1)}$$
$$S(p) = 200 + 4p. \quad \textbf{(2)}$$

Since both demand and supply are *quantities* and they are equal at the equilibrium point, we rewrite the system as

$$q = 1000 - 60p, \quad \textbf{(1)}$$
$$q = 200 + 4p. \quad \textbf{(2)}$$

We substitute $200 + 4p$ for q in equation (1) and solve:

$$200 + 4p = 1000 - 60p$$
$$200 + 64p = 1000 \qquad \text{Adding } 60p \text{ on both sides}$$
$$64p = 800 \qquad \text{Subtracting 200 on both sides}$$
$$p = \frac{800}{64} = 12.5.$$

Thus the equilibrium price is $12.50 per unit.

To find the equilibrium quantity, we substitute $12.50 into either $D(p)$ or $S(p)$. We use $S(p)$:

$$S(12.5) = 200 + 4(12.5)$$
$$= 200 + 50 = 250.$$

Thus the equilibrium quantity is 250 units, and the equilibrium point is ($12.50, 250$).

Do Exercise 2.

2. Find the equilibrium point for the following supply and demand functions:

$$D(p) = 1000 - 46p,$$
$$S(p) = 300 + 4p.$$

Answer on page A-21

3.8

EXERCISE SET

For Extra Help

Math XL   MyMathLab   InterAct   Math Tutor   Digital Video   Student's
MathXL   MyMathLab   Math   Center   Tutor CD 2   Solutions
                                                  Videotape 4   Manual

a. For each of the following pairs of total-cost and total-revenue functions, find **(a)** the total-profit function and **(b)** the break-even point.

1. $C(x) = 25x + 270{,}000$;
$R(x) = 70x$

2. $C(x) = 45x + 300{,}000$;
$R(x) = 65x$

3. $C(x) = 10x + 120{,}000$;
$R(x) = 60x$

4. $C(x) = 30x + 49{,}500$;
$R(x) = 85x$

5. $C(x) = 20x + 10{,}000$;
$R(x) = 100x$

6. $C(x) = 40x + 22{,}500$;
$R(x) = 85x$

7. $C(x) = 22x + 16{,}000$;
$R(x) = 40x$

8. $C(x) = 15x + 75{,}000$;
$R(x) = 55x$

9. $C(x) = 50x + 195{,}000$;
$R(x) = 125x$

10. $C(x) = 34x + 928{,}000$;
$R(x) = 128x$

Solve.

11. *Manufacturing Lamps.* City Lights is planning to manufacture a new type of lamp. For the first year, the fixed costs for setting up production are $22,500. The variable costs for producing each lamp are $40. The revenue from each lamp is $85. Find the following.

a) The total cost $C(x)$ of producing x lamps
b) The total revenue $R(x)$ from the sale of x lamps
c) The total profit $P(x)$ from the production and sale of x lamps
d) The profit or loss from the production and sale of 3000 lamps; of 400 lamps
e) The break-even point

12. *Computer Manufacturing.* Sky View Electronics is planning to introduce a new line of computers. For the first year, the fixed costs for setting up production are $125,100. The variable costs for producing each computer are $750. The revenue from each computer is $1050. Find the following.

a) The total cost $C(x)$ of producing x computers
b) The total revenue $R(x)$ from the sale of x computers
c) The total profit $P(x)$ from the production and sale of x computers
d) The profit or loss from the production and sale of 400 computers; of 700 computers
e) The break-even point

Solve.

13. *Manufacturing Caps.* Martina's Custom Printing is planning on adding painter's caps to its product line. For the first year, the fixed costs for setting up production are $16,404. The variable costs for producing a dozen caps are $6.00. The revenue on each dozen caps is $18.00. Find the following.

   a) The total cost $C(x)$ of producing x dozen caps
   b) The total revenue $R(x)$ from the sale of x dozen caps
   c) The total profit $P(x)$ from the production and sale of x dozen caps
   d) The profit or loss from the production and sale of 3000 dozen caps; of 1000 dozen caps
   e) The break-even point

14. *Sport Coat Production.* Sarducci's is planning a new line of sport coats. For the first year, the fixed costs for setting up production are $10,000. The variable costs for producing each coat are $20. The revenue from each coat is $100. Find the following.

   a) The total cost $C(x)$ of producing x coats
   b) The total revenue $R(x)$ from the sale of x coats
   c) The total profit $P(x)$ from the production and sale of x coats
   d) The profit or loss from the production and sale of 2000 coats; of 50 coats
   e) The break-even point

b   Find the equilibrium point for each of the following pairs of demand and supply functions.

15. $D(p) = 1000 - 10p;$
    $S(p) = 230 + p$

16. $D(p) = 2000 - 60p;$
    $S(p) = 460 + 94p$

17. $D(p) = 760 - 13p;$
    $S(p) = 430 + 2p$

18. $D(p) = 800 - 43p;$
    $S(p) = 210 + 16p$

19. $D(p) = 7500 - 25p;$
    $S(p) = 6000 + 5p$

20. $D(p) = 8800 - 30p;$
    $S(p) = 7000 + 15p$

21. $D(p) = 1600 - 53p;$
    $S(p) = 320 + 75p$

22. $D(p) = 5500 - 40p;$
    $S(p) = 1000 + 85p$

23. D_W Variable costs and fixed costs are often compared to the slope and the y-intercept, respectively, of an equation of a line. Explain why this analogy is valid.

24. D_W In this section, we examined supply and demand functions for coffee. Does it seem realistic to you for the graph of D to have a constant slope? Why or why not?

Find the slope and the y-intercept.   [2.4b]

25. $5y - 3x = 8$

26. $6x + 7y - 9 = 4$

27. $2y = 3.4x + 98$

28. $\dfrac{x}{3} + \dfrac{y}{4} = 1$

319

The review that follows is meant to prepare you for a chapter exam. It consists of two parts. The first part, Concept Reinforcement, is designed to increase understanding of the concepts through true/false exercises. The second part is the Review Exercises. These provide practice exercises for the exam, together with references to section objectives so you can go back and review. Before beginning, stop and look back over the skills you have obtained. What skills in mathematics do you have now that you did not have before studying this chapter?

CONCEPT REINFORCEMENT

Determine whether the statement is true or false. Answers are given at the back of the book.

_____ **1.** When $(0, b)$ is a solution of each equation in a system of two equations, the graphs of the two equations have the same y-intercept.

_____ **2.** If, when solving a system of two linear equations in two variables, a false equation is obtained, the system has infinitely many solutions.

_____ **3.** Every system of equations has at least one solution.

_____ **4.** If the graphs of two linear equations intersect, then the system is consistent.

_____ **5.** The intersection of the graphs of the lines $x = a$ and $y = b$ is (a, b).

Review Exercises

Solve graphically. Then classify the system as consistent or inconsistent and the equations as dependent or independent. [3.1a]

1. $4x - y = -9,$
$\quad x - y = -3$

2. $15x + 10y = -20,$
$\quad 3x + 2y = -4$

3. $y - 2x = 4,$
$\quad y - 2x = 5$

Solve by the substitution method. [3.2a]

4. $7x - 4y = 6,$
$\quad y - 3x = -2$

5. $y = x + 2,$
$\quad y - x = 8$

6. $9x - 6y = 2,$
$\quad x = 4y + 5$

Solve by the elimination method. [3.3a]

7. $8x - 2y = 10,$
$\quad -4y - 3x = -17$

8. $4x - 7y = 18,$
$\quad 9x + 14y = 40$

9. $3x - 5y = -4,$
$\quad 5x - 3y = 4$

10. $1.5x - 3 = -2y,$
$\quad 3x + 4y = 6$

11. _Music Spending._ Sean has $37 to spend. He can spend all of it on two compact discs and a cassette, or he can buy one CD and two cassettes and have $5.00 left over. What is the price of a CD? of a cassette? [3.4a]

12. _Orange Drink Mixtures._ "Orange Thirst" is 15% orange juice and "Quencho" is 5% orange juice. How many liters of each should be combined in order to get 10 L of a mixture that is 10% orange juice? [3.4a]

13. *Train Travel.* A train leaves Watsonville at noon traveling north at 44 mph. One hour later, another train, going 52 mph, travels north on a parallel track. How many hours will the second train travel before it overtakes the first train?   [3.4b]

Trains meet here

$t - 1$ hours t hours

Watsonville

44 mph

52 mph

Solve.   [3.5a]

14. $x + 2y + z = 10,$
$\quad 2x - y + z = 8,$
$\quad 3x + y + 4z = 2$

15. $3x + 2y + z = 3,$
$\quad 6x - 4y - 2z = -34,$
$\quad -x + 3y - 3z = 14$

16. $2x - 5y - 2z = -4,$
$\quad 7x + 2y - 5z = -6,$
$\quad -2x + 3y + 2z = 4$

17. $x + y + 2z = 1,$
$\quad x - y + z = 1,$
$\quad x + 2y + z = 2$

18. *Triangle Measure.* In triangle ABC, the measure of angle A is four times the measure of angle C, and the measure of angle B is 45° more than the measure of angle C. What are the measures of the angles of the triangle?   [3.6a]

19. *Money Mixtures.* Elaine has $194, consisting of $20, $5, and $1 bills. The number of $1 bills is 1 less than the total number of $20 and $5 bills. If she has 39 bills in her purse, how many of each denomination does she have?   [3.6a]

Graph.   [3.7b]

20. $2x + 3y < 12$

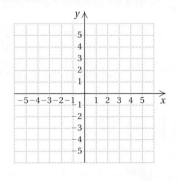

21. $y \leq 0$

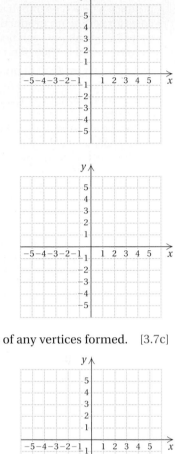

22. $x + y \geq 1$

Graph. Find the coordinates of any vertices formed.   [3.7c]

23. $y \geq -3,$
$\quad x \geq 2$

24. $x + 3y \geq -1,$
$\quad x + 3y \leq 4$

25. $x - y \leq 3,$
$\quad x + y \geq -1,$
$\quad y \leq 2$

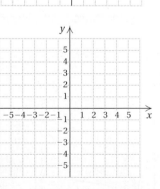

26. *Bed Manufacturing.* Kregel Furniture is planning to produce a new type of bed. For the first year, the fixed costs for setting up production are $35,000. The variable costs for producing each bed are $175. The revenue from each bed is $300. Find the following. [3.8a]

a) The total cost $C(x)$ of producing x beds
b) The total revenue $R(x)$ from the sale of x beds
c) The total profit from the production and sale of x beds
d) The profit or loss from the production and sale of 1200 beds; of 200 beds
e) The break-even point

27. Find the equilibrium point for the following demand and supply functions: [3.8b]

$$D(p) = 120 - 13p,$$
$$S(p) = 60 + 7p.$$

28. $\mathbf{D_W}$ Briefly compare the strengths and the weaknesses of the graphical, substitution, and elimination methods as applied to the solution of two equations in two variables. [3.1a], [3.2a], [3.3a]

29. $\mathbf{D_W}$ Explain the advantages of using a system of equations to solve an applied problem. [3.2b], [3.3b], [3.4a, b]

30. *Peanuts.* Help Peppermint Patty with her dilemma. Translate Peppermint Patty's problem to a system of equations. Then solve the problem. [3.3b]

Source: PEANUTS reprinted by permission of United Feature Syndicate, Inc.

31. Solve graphically:

$$y = x + 2,$$
$$y = x^2 + 2.$$

[2.1d], [3.1a]

32. *Height Estimation in Anthropology.* An anthropologist can use linear functions to estimate the height of a male or female, given the length of certain bones. The *femur* is the large bone from the hip to the knee. Let $x =$ the length of the femur, in centimeters. Then the height, in centimeters, of a male with a femur of length x is given by the function

$$M(x) = 1.88x + 81.31.$$

The height, in centimeters, of a female with a femur of length x is given by the function

$$F(x) = 1.95x + 72.85.$$

[2.2b], [3.1a]

A 45-cm femur was uncovered at an anthropological dig.

a) If we assume that it was from a male, how tall was he?
b) If we assume that it was from a female, how tall was she?
c) Graph each equation and find the point of intersection of the graphs of the equations.
d) For what length of a male femur and a female femur, if any, would the height be the same?

Solve graphically. Then classify the system as consistent or inconsistent and the equations as dependent or independent.

1. $y = 3x + 7,$
$3x + 2y = -4$

2. $y = 3x + 4,$
$y = 3x - 2$

3. $y - 3x = 6,$
$6x - 2y = -12$

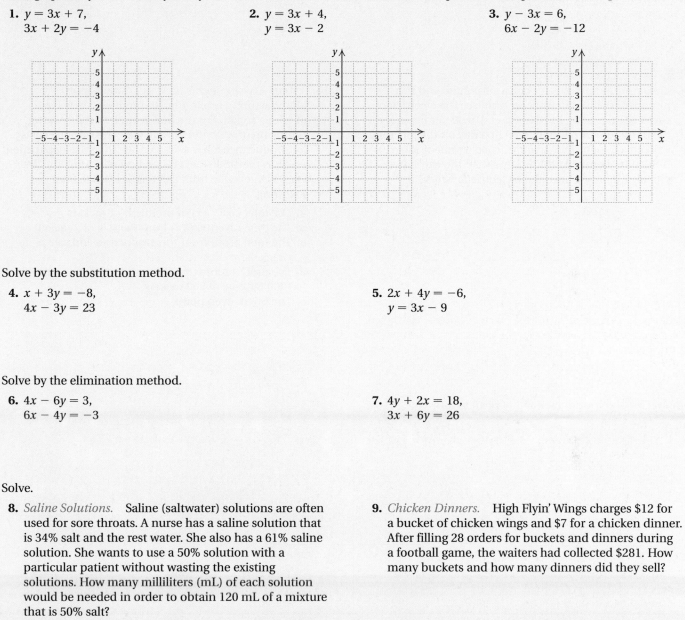

Solve by the substitution method.

4. $x + 3y = -8,$
$4x - 3y = 23$

5. $2x + 4y = -6,$
$y = 3x - 9$

Solve by the elimination method.

6. $4x - 6y = 3,$
$6x - 4y = -3$

7. $4y + 2x = 18,$
$3x + 6y = 26$

Solve.

8. *Saline Solutions.* Saline (saltwater) solutions are often used for sore throats. A nurse has a saline solution that is 34% salt and the rest water. She also has a 61% saline solution. She wants to use a 50% solution with a particular patient without wasting the existing solutions. How many milliliters (mL) of each solution would be needed in order to obtain 120 mL of a mixture that is 50% salt?

9. *Chicken Dinners.* High Flyin' Wings charges $12 for a bucket of chicken wings and $7 for a chicken dinner. After filling 28 orders for buckets and dinners during a football game, the waiters had collected $281. How many buckets and how many dinners did they sell?

10. *Air Travel.* An airplane flew for 5 hr with a 20-km/h tailwind and returned in 7 hr against the same wind. Find the speed of the plane in still air.

11. *Tennis Court.* The perimeter of a standard tennis court used for playing doubles is 288 ft. The width of the court is 42 ft less than the length. Find the length and the width.

12. Solve:

$$6x + 2y - 4z = 15,$$
$$-3x - 4y + 2z = -6,$$
$$4x - 6y + 3z = 8.$$

13. Find the equilibrium point for the following demand and supply functions:

$$D(p) = 79 - 8p,$$
$$S(p) = 37 + 6p.$$

Solve.

14. *Repair Rates.*   An electrician, a carpenter, and a plumber are hired to work on a house. The electrician earns $21 per hour, the carpenter $19.50 per hour, and the plumber $24 per hour. The first day on the job, they worked a total of 21.5 hr and earned a total of $469.50. If the plumber worked 2 hr more than the carpenter did, how many hours did the electrician work?

15. *Manufacturing Tennis Rackets.*   Sweet Spot Manufacturing is planning to produce a new type of tennis racket. For the first year, the fixed costs for setting up production are $40,000. The variable costs for producing each racket are $30. The sales department predicts that 1500 rackets can be sold during the first year. The revenue from each racket is $80. Find the following.

a) The total cost $C(x)$ of producing x rackets
b) The total revenue $R(x)$ from the sale of x rackets
c) The total profit from the production and sale of x rackets
d) The profit or loss from the production and sale of 1200 rackets; of 200 rackets
e) The break-even point

Graph. Find the coordinates of any vertices formed.

16. $x - 6y < -6$

17. $x + y \geq 3,$
$\quad\ \ x - y \geq 5$

18. $2y - x \geq -4,$
$\quad\ \ 2y + 3x \leq -6,$
$\quad\ \ y \leq 0,$
$\quad\ \ x \leq 0$

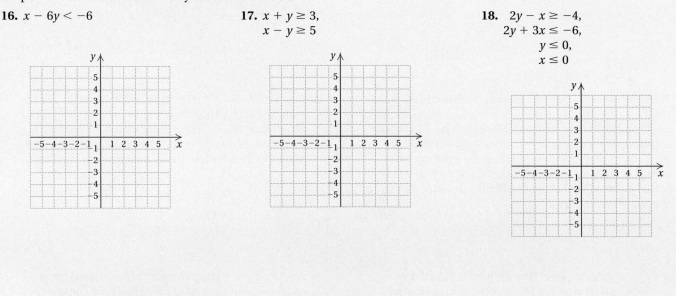

SYNTHESIS

19. The graph of the function $f(x) = mx + b$ contains the points $(-1, 3)$ and $(-2, -4)$. Find m and b.

CHAPTER 3: Systems of Equations

Polynomials and Polynomial Functions

4

Real-World Application

Suppose that a bottle rocket is launched upward with an initial velocity of 96 ft/sec and from a height of 880 ft. Its height h after t seconds is given by the function

$$h(t) = -16t^2 + 96t + 880.$$

a) What is the height of the bottle rocket after 0 sec? 1 sec? 3 sec? 8 sec? 10 sec?

b) Find an equivalent expression for $h(t)$ by factoring.

This problem appears as Exercise 52 in Section 4.5.

Objectives

a Identify the degree of each term and the degree of a polynomial; identify terms, coefficients, monomials, binomials, and trinomials; arrange polynomials in ascending or descending order; and identify the leading term, the leading coefficient, and the constant term.

b Evaluate a polynomial function for given inputs.

c Collect like terms in a polynomial and add polynomials.

d Find the opposite of a polynomial and subtract polynomials.

A **polynomial** is a particular type of algebraic expression. Let's examine an application before we consider definitions and manipulations involving polynomials.

Stacks of Golf Balls. Each stack of golf balls pictured below is formed by square layers of golf balls. The number N of balls in the stack is given by the polynomial function

$$N(x) = \tfrac{1}{3}x^3 + \tfrac{1}{2}x^2 + \tfrac{1}{6}x,$$

where x is the number of layers. The graph of the function is shown below.

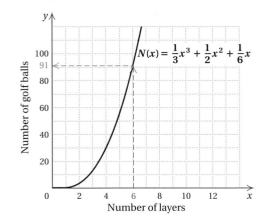

For a stack with 6 layers, there is a total of 91 golf balls, as we can see from the graph and from substituting 6 for x in the polynomial function:

$$N(x) = \tfrac{1}{3}x^3 + \tfrac{1}{2}x^2 + \tfrac{1}{6}x$$
$$N(6) = \tfrac{1}{3} \cdot 6^3 + \tfrac{1}{2} \cdot 6^2 + \tfrac{1}{6} \cdot 6 = 72 + 18 + 1 = 91.$$

Although we will not be considering graphs of polynomial functions in detail in this chapter (other than in Calculator Corners), this situation gives us an idea of how polynomial functions can occur in applied problems.

a Polynomial Expressions

The following are examples of *monomials*:

$$0, \quad -3, \quad z, \quad 8x, \quad -7y^2, \quad 4a^2b^3, \quad 1.3p^4q^5r^7.$$

MONOMIAL

A **monomial** is a one-term expression like $ax^n y^m z^q$, where a is a real number and n, m, and q are nonnegative integers. More specifically, a monomial is a constant or a constant times some variable or variables raised to powers that are nonnegative integers.

Expressions like these are called **polynomials in one variable:**

$$5x^2, \quad 8a, \quad 2, \quad 2x + 3,$$
$$-7x + 5, \quad 2y^2 + 5y - 3,$$
$$5a^4 - 3a^2 + \tfrac{1}{4}a - 8, \quad b^6 + 3b^5 - 8b + 7b^4 + \tfrac{1}{2}.$$

Expressions like these are called **polynomials in several variables:**

$$15x^3y^2,$$
$$5a - ab^2 + 7b + 2,$$
$$9xy^2z - 4x^3z - 14x^4y^2 + 9.$$

POLYNOMIAL

A **polynomial** is a monomial or a combination of sums and/or differences of monomials.

The following are algebraic expressions that are not polynomials:

$$(1) \ \frac{y^2 - 3}{y^2 + 4}, \qquad (2) \ 8x^4 - 2x^3 + \frac{1}{x}, \qquad (3) \ \frac{2xy}{x^3 - y^3}.$$

Expressions (1) and (3) are not polynomials because they represent quotients. Expression (2) is not a polynomial because

$$\frac{1}{x} = x^{-1};$$

this is not a monomial because the exponent is negative.

The polynomial $5x^3y - 7xy^2 - y^3 + 2$ has four **terms:**

$$5x^3y, \qquad -7xy^2, \qquad -y^3, \quad \text{and} \quad 2.$$

The **coefficients** of the terms are 5, -7, -1, and 2. The term 2 is called a **constant term.**

The **degree of a term** is the sum of the exponents of the variables, if there are variables. For example,

the degree of the term $9x^5$ is 5 and

the degree of the term $0.6a^2b^7$ is 9.

The degree of a nonzero constant term, such as 2, is 0. We can express 2 as $2x^0$. Mathematicians agree that the polynomial 0 has *no* degree. This is because we can express 0 as

$$0 = 0x^5 = 0x^{12},$$

and so on, using any exponent we wish.

The **degree of a polynomial** is the same as the degree of its term of highest degree. For example,

the degree of the polynomial $4 - x^3 + 5x^2 - x^6$ is 6.

The **leading term** of a polynomial is the term of highest degree. Its coefficient is called the **leading coefficient.** For example,

the leading term of $9x^2 - 5x^3 + x - 10$ is $-5x^3$ and

the leading coefficient is -5.

1. Identify the terms and the leading term.

$$-92x^5 - 8x^4 + x^2 + 5$$

2. Identify the coefficient of each term and the leading coefficient.

$$5x^3y - 4xy^2 - 2x^3 + xy - y - 5$$

3. Identify the terms, the degree of each term, and the degree of the polynomial. Then identify the leading term, the leading coefficient, and the constant term.

a) $6x^2 - 5x^3 + 2x - 7$

b) $2y - 4 - 5x + 7x^2y^3z^2 + 5xy^2$

4. Consider the following polynomials.

a) $3x^2 - 2$

b) $5x^3 + 9x - 3$

c) $4x^2$

d) $-7y$

e) -3

f) $8x^3 - 2x^2$

g) $-4y^2 - 5 - 5y$

h) $5 - 3x$

Identify the monomials, the binomials, and the trinomials.

5. a) Arrange in ascending order:

$$5 - 6x^2 + 7x^3 - x^4 + 10x.$$

b) Arrange in descending order:

$$5 - 6x^2 + 7x^3 - x^4 + 10x.$$

Answers on page A-22

EXAMPLE 1 Identify the terms, the degree of each term, and the degree of the polynomial. Then identify the leading term, the leading coefficient, and the constant term.

$$2x^3 + 8x^2 - 17x - 3$$

Term		$2x^3$	$8x^2$	$-17x$	-3
Degree of Term		3	2	1	0
Degree of Polynomial		3			
Leading Term		$2x^3$			
Leading Coefficient		2			
Constant Term		-3			

EXAMPLE 2 Identify the terms, the degree of each term, and the degree of the polynomial. Then identify the leading term, the leading coefficient, and the constant term.

$$6x^2 + 8x^2y^3 - 17xy - 24xy^2z^4 + 2y + 3$$

Term	$6x^2$	$8x^2y^3$	$-17xy$	$-24xy^2z^4$	$2y$	3
Degree of Term	2	5	2	7	1	0
Degree of Polynomial	7					
Leading Term	$-24xy^2z^4$					
Leading Coefficient	-24					
Constant Term	3					

Do Exercises 1–3.

The following are some names for certain types of polynomials.

TYPE	DEFINITION: POLYNOMIAL OF	EXAMPLES
Monomial	One term	4, $\quad -3p$, $\quad -7a^2b^3$, $\quad 0$, $\quad xyz$
Binomial	Two terms	$2x + 7$, $\quad a^2 - 3b$, $\quad 5x^3 + 8x$
Trinomial	Three terms	$x^2 - 7x + 12$, $\quad 4a^2 + 2ab + b^2$

Do Exercise 4.

We generally arrange polynomials in one variable so that the exponents *decrease* from left to right, which is **descending order.** Sometimes they may be written so that the exponents *increase* from left to right, which is **ascending order.** In general, if an exercise is written in a particular order, we write the answer in that same order.

EXAMPLE 3 Consider $12 + x^2 - 7x$. Arrange in descending order and then in ascending order.

$$\text{Descending order:} \quad x^2 - 7x + 12 \qquad \text{Ascending order:} \quad 12 - 7x + x^2$$

Do Exercise 5 on the preceding page.

EXAMPLE 4 Consider $x^4 + 2 - 5x^2 + 3x^3y + 7xy^2$. Arrange in descending powers of x and then in ascending powers of x.

$$\text{Descending powers of } x: \quad x^4 + 3x^3y - 5x^2 + 7xy^2 + 2$$
$$\text{Ascending powers of } x: \quad 2 + 7xy^2 - 5x^2 + 3x^3y + x^4$$

Do Exercise 6.

b Evaluating Polynomial Functions

A polynomial function is one like

$$P(x) = 5x^7 + 3x^5 - 4x^2 - 5,$$

where the algebraic expression used to describe the function is a polynomial. To find the outputs of a polynomial function for a given input, we substitute the input for each occurrence of the variable as we did in Section 2.2.

EXAMPLE 5 For the polynomial function $P(x) = -x^2 + 4x - 1$, find $P(2)$, $P(10)$, and $P(-10)$.

$$P(2) = -2^2 + 4(2) - 1 = -4 + 8 - 1 = 3;$$

$$P(10) = -10^2 + 4(10) - 1 = -100 + 40 - 1 = -61;$$

$$P(-10) = -(-10)^2 + 4(-10) - 1 = -100 - 40 - 1 = -141$$

Do Exercise 7.

EXAMPLE 6 *Veterinary Medicine.* Gentamicin is an antibiotic frequently used by veterinarians. The concentration, in micrograms per milliliter (mcg/mL), of Gentamicin in a horse's bloodstream t hours after injection can be approximated by the polynomial function

$$C(t) = -0.005t^4 + 0.003t^3 + 0.35t^2 + 0.5t.$$

a) Evaluate $C(2)$ to find the concentration 2 hr after injection.

b) Use only the graph below to estimate $C(4)$.

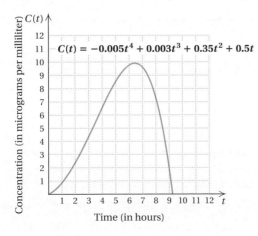

6. a) Arrange in ascending powers of y:

$$5x^4y - 3y^2 + 3x^2y^3 + x^3 - 5.$$

b) Arrange in descending powers of y:

$$5x^4y - 3y^2 + 3x^2y^3 + x^3 - 5.$$

7. For the polynomial function

$$P(x) = x^2 - 2x + 5,$$

find $P(0)$, $P(4)$, and $P(-2)$.

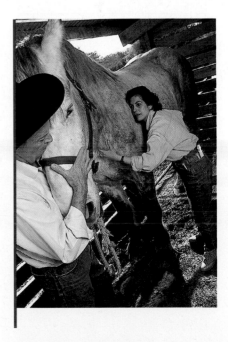

Answers on page A-22

8. Veterinary Medicine. Refer to the function and the graph of Example 6.

a) Evaluate $C(3)$ to find the concentration 3 hr after injection.

b) Use only the graph at right to estimate $C(9)$.

Collect like terms.

9. $3y - 4x + 6xy^2 - 2xy^2$

10. $3xy^3 + 2x^3y + 5xy^3 - 8x + 15 - 3x^2y - 6x^2y + 11x - 8$

Add.

11. $(3x^3 + 4x^2 - 7x - 2) + (-7x^3 - 2x^2 + 3x + 4)$

12. $(7y^5 - 5) + (3y^5 - 4y^2 + 10)$

13. $(5p^2q^4 - 2p^2q^2 - 3q) + (-6p^2q^2 + 3q + 5)$

Answers on page A-22

a) We evaluate the function when $t = 2$:

$$C(2) = -0.005(2)^4 + 0.003(2)^3 + 0.35(2)^2 + 0.5(2)$$
$$= -0.005(16) + 0.003(8) + 0.35(4) + 0.5(2)$$
$$= -0.08 + 0.024 + 1.4 + 1 = 2.344.$$

We carry out the calculation using the rules for order of operations.

The concentration after 2 hr is about 2.344 mcg/mL.

b) To estimate $C(4)$, the concentration after 4 hr, we locate 4 on the horizontal axis. From there we move vertically to the graph of the function and then horizontally to the $C(t)$-axis. This locates a value of about 6.5. Thus,

$$C(4) \approx 6.5.$$

The concentration after 4 hr is about 6.5 mcg/mL.

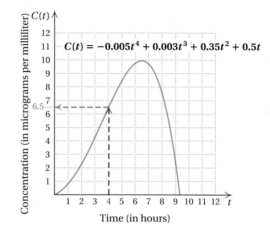

Do Exercise 8.

C Adding Polynomials

When two terms have the same variable(s) raised to the same power(s), they are called **like terms,** or **similar terms,** and they can be "collected," or "combined," using the distributive laws, adding or subtracting the coefficients as follows.

EXAMPLES Collect like terms.

7. $3x^2 - 4y + 2x^2 = 3x^2 + 2x^2 - 4y$ Rearranging using the commutative law for addition

$$= (3 + 2)x^2 - 4y$$ Using the distributive law
$$= 5x^2 - 4y$$

8. $9x^3 + 5x - 4x^2 - 2x^3 + 5x^2 = 7x^3 + x^2 + 5x$

9. $3x^2y + 5xy^2 - 3x^2y - xy^2 = 4xy^2$

Do Exercises 9 and 10.

The sum of two polynomials can be found by writing a plus sign between them and then collecting like terms to simplify the expression.

EXAMPLE 10 Add: $(-3x^3 + 2x - 4) + (4x^3 + 3x^2 + 2)$.

$$(-3x^3 + 2x - 4) + (4x^3 + 3x^2 + 2) = x^3 + 3x^2 + 2x - 2$$

EXAMPLE 11 Add: $13x^3y + 3x^2y - 5y$ and $x^3y + 4x^2y - 3xy$.

$$(13x^3y + 3x^2y - 5y) + (x^3y + 4x^2y - 3xy) = 14x^3y + 7x^2y - 3xy - 5y$$

Do Exercises 11–13 on the preceding page.

Using columns to add is sometimes helpful. To do so, we write the polynomials one under the other, listing like terms under one another and leaving spaces for missing terms.

EXAMPLE 12 Add: $4ax^2 + 4bx - 5$ and $-6ax^2 + 8$.

$$\begin{array}{r} 4ax^2 + 4bx - 5 \\ -6ax^2 \qquad\;\; + 8 \\ \hline -2ax^2 + 4bx + 3 \end{array}$$

d Subtracting Polynomials

If the sum of two polynomials is 0, they are called **opposites,** or **additive inverses,** of each other. For example,

$$(3x^2 - 5x + 2) + (-3x^2 + 5x - 2) = 0,$$

so the opposite of $3x^2 - 5x + 2$ is $-3x^2 + 5x - 2$. We can say the same thing using algebraic symbolism, as follows:

The opposite of $(3x^2 - 5x + 2)$ is $(-3x^2 + 5x - 2)$.

$$-\qquad (3x^2 - 5x + 2) \quad = \quad -3x^2 + 5x - 2$$

Thus, $-(3x^2 - 5x + 2)$ and $-3x^2 + 5x - 2$ are equivalent.

The *opposite* of a polynomial P can be symbolized by $-P$ or by replacing each term with its opposite. The two expressions for the opposite are equivalent.

EXAMPLE 13 Write two equivalent expressions for the opposite of

$7xy^2 - 6xy - 4y + 3$.

First expression: $-(7xy^2 - 6xy - 4y + 3)$ Writing an inverse sign in front

Second expression: $-7xy^2 + 6xy + 4y - 3$ Writing the opposite of each term (see also Section R.6)

Do Exercises 14–16.

To subtract a polynomial, we add its opposite.

EXAMPLE 14 Subtract: $(-5x^2 + 4) - (2x^2 + 3x - 1)$.

We have

$(-5x^2 + 4) - (2x^2 + 3x - 1)$

$= (-5x^2 + 4) + [-(2x^2 + 3x - 1)]$ Adding the opposite

$= (-5x^2 + 4) + (-2x^2 - 3x + 1)$ $-2x^2 - 3x + 1$ is equivalent to $-(2x^2 + 3x - 1)$.

$= -7x^2 - 3x + 5.$ Adding

Write two equivalent expressions for the opposite, or additive inverse.

14. $4x^3 - 5x^2 + \dfrac{1}{4}x - 10$

15. $8xy^2 - 4x^3y^2 - 9x - \dfrac{1}{5}$

16. $-9y^5 - 8y^4 + \dfrac{1}{2}y^3 - y^2 + y - 1$

Subtract.

17. $(6x^2 + 4) - (3x^2 - 1)$

18. $(9y^3 - 2y - 4) - (-5y^3 - 8)$

19. $(-3p^2 + 5p - 4) - (-4p^2 + 11p - 2)$

Answers on page A-22

Subtract.

20.

$(2y^5 - y^4 + 3y^3 - y^2 - y - 7) -$
$(-y^5 + 2y^4 - 2y^3 + y^2 - y - 4)$

21. $(4p^4q - 5p^3q^2 + p^2q^3 + 2q^4) -$
$(-5p^4q + 5p^3q^2 - 3p^2q^3 - 7q^4)$

22. $\left(\dfrac{3}{2}y^3 - \dfrac{1}{2}y^2 + 0.3\right) -$
$\left(\dfrac{1}{2}y^3 + \dfrac{1}{2}y^2 - \dfrac{4}{3}y + 0.2\right)$

Answers on page A-22

With practice, you may find that you can skip some steps, by mentally taking the opposite of each term and then combining like terms. Eventually, all you will write is the answer.

$(-5x^2 + 4) - (2x^2 + 3x - 1)$ *Think*:
$= -7x^2 - 3x + 5$

$-5x^2 - 2x^2 = -5x^2 + (-2x^2) = -7x^2,$
$0x - 3x = 0x + (-3x) = -3x,$
$4 - (-1) = 4 + 1 = 5.$

Do Exercises 17–19 on the preceding page.

To use columns for subtraction, we mentally change the signs of the terms being subtracted.

EXAMPLE 15 Subtract:

$$(4x^2y - 6x^3y^2 + x^2y^2) - (4x^2y + x^3y^2 + 3x^2y^3 - 8x^2y^2).$$

Write: (Subtract)   *Think*: (Add)

$4x^2y - 6x^3y^2 \qquad + x^2y^2$
$-(4x^2y + x^3y^2 + 3x^2y^3 - 8x^2y^2)$ ⟵

$4x^2y - 6x^3y^2 \qquad + x^2y^2$
$-4x^2y - x^3y^2 - 3x^2y^3 + 8x^2y^2$
$\qquad\qquad - 7x^3y^2 - 3x^2y^3 + 9x^2y^2$

Take the opposite of each term mentally and add.

Do Exercises 20–22.

CALCULATOR CORNER

Checking Addition and Subtraction of Polynomials A table set in AUTO mode can be used to perform a partial check that polynomials in a single variable have been added or subtracted correctly. To check Example 10, we enter $y_1 = (-3x^3 + 2x - 4) + (4x^3 + 3x^2 + 2)$ and $y_2 = x^3 + 3x^2 + 2x - 2$. If the addition has been done correctly, the values of y_1 and y_2 will be the same regardless of the table settings used.

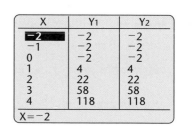

X	Y₁	Y₂
-2	-2	-2
-1	-2	-2
0	-2	-2
1	4	4
2	22	22
3	58	58
4	118	118

X=-2

Graphs can also be used to check addition and subtraction. See the Calculator Corner on p. 341 for the procedure. Keep in mind that these procedures provide only a partial check since we can neither view all possible values of *x* in a table nor see the entire graph.

Exercises: Use a table to determine whether the sum or difference is correct.

1. $(x^3 - 2x^2 + 3x - 7) + (3x^2 - 4x + 5) = x^3 + x^2 - x - 2$

2. $(2x^2 + 3x - 6) + (5x^2 - 7x + 4) = 7x^2 + 4x - 2$

3. $(4x^3 + 3x^2 + 2) + (-3x^3 + 2x - 4) = x^3 + 3x^2 + 2x - 2$

4. $(7x^5 + 2x^4 - 5x) - (-x^5 - 2x^4 + 3) = 8x^5 + 4x^4 - 5x - 3$

5. $(-2x^3 + 3x^2 - 4x + 5) - (3x^2 + 2x - 8) = -2x^3 - 6x - 3$

6. $(3x^4 - 2x^2 - 1) - (2x^4 - 3x^2 - 4) = x^4 + x^2 - 5$

a   Identify the terms, the degree of each term, and the degree of the polynomial. Then identify the leading term, the leading coefficient, and the constant term.

1. $-9x^4 - x^3 + 7x^2 + 6x - 8$

2. $y^3 - 5y^2 + y + 1$

3. $t^3 + 4t^7 + s^2t^4 - 2$

4. $a^2 + 9b^5 - a^4b^3 - 11$

5. $u^7 + 8u^2v^6 + 3uv + 4u - 1$

6. $2p^6 + 5p^4w^4 - 13p^3w + 7p^2 - 10$

Arrange in descending powers of y.

7. $23 - 4y^3 + 7y - 6y^2$

8. $5 - 8y + 6y^2 + 11y^3 - 18y^4$

9. $x^2y^2 + x^3y - xy^3 + 1$

10. $x^3y - x^2y^2 + xy^3 + 6$

11. $2by - 9b^5y^5 - 8b^2y^3$

12. $dy^6 - 2d^7y^2 + 3cy^5 - 7y - 2d$

Arrange in ascending powers of x.

13. $12x + 5 + 8x^5 - 4x^3$

14. $-3x^2 + 8x + 2$

15. $-9x^3y + 3xy^3 + x^2y^2 + 2x^4$

16. $5x^2y^2 - 9xy + 8x^3y^2 - 5x^4$

17. $4ax - 7ab + 4x^6 - 7ax^2$

18. $5xy^8 - 3ax^5 + 4ax^3 - 12a + 5x^5$

b   Evaluate the polynomial function for the given values of the variable.

19. $P(x) = 3x^2 - 2x + 5$; $P(4), P(-2), P(0)$

20. $f(x) = -7x^3 + 10x^2 - 13$; $f(4), f(-1), f(0)$

21. $p(x) = 9x^3 + 8x^2 - 4x - 9$; $p(-3), p(0), p(1), p\left(\frac{1}{2}\right)$

22. $Q(x) = 6x^3 - 11x - 4$; $Q(-2), Q\left(\frac{1}{3}\right), Q(0), Q(10)$

23. *Golf Ball Stacks.*   Each stack of golf balls is formed by square layers of golf balls. The number N of balls in the stack is given by the polynomial function
$$N(x) = \tfrac{1}{3}x^3 + \tfrac{1}{2}x^2 + \tfrac{1}{6}x,$$
where x is the number of layers. Look at the photograph at the beginning of this section. How many golf balls are in each of the stacks?

24. *Falling Distance.*   The distance $s(t)$, in feet, traveled by an object falling freely from rest in t seconds is approximated by the function given by
$$s(t) = 16t^2.$$

a) A paintbrush falls from a scaffold and takes 3 sec to hit the ground. How high is the scaffold?

b) A stone is dropped from a cliff and takes 8 sec to hit the ground. How high is the cliff?

$s(t) = 16t^2$

25. *Medicine.* Ibuprofen is a medication used to relieve pain. The polynomial function

$$M(t) = 0.5t^4 + 3.45t^3 - 96.65t^2 + 347.7t,$$

$$0 \leq t \leq 6,$$

can be used to estimate the number of milligrams of ibuprofen in the bloodstream t hours after 400 mg of the medication has been swallowed.

Source: Based on data from Dr. P. Carey, Burlington, VT

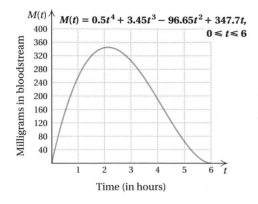

a) Use the graph above to estimate the number of milligrams of ibuprofen in the bloodstream 2 hr after 400 mg has been swallowed.

b) Use the graph above to estimate the number of milligrams of ibuprofen in the bloodstream 4 hr after 400 mg has been swallowed.

c) Approximate $M(5)$.

d) Approximate $M(3)$.

26. *Median Income by Age.* The polynomial function

$$I(x) = -0.0560x^4 + 7.9980x^3 - 436.1840x^2$$

$$+ 11,627.8376x - 90,625.0001,$$

$$13 \leq x \leq 65,$$

can be used to approximate the median income I by age x of a person living in the United States. The graph is shown below.

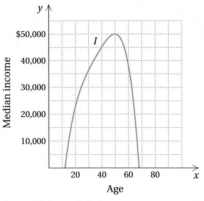

Source: U.S. Bureau of the Census;
The Conference Board: Simmons Bureau of Labor Statistics

a) Evaluate $I(22)$ to estimate the median income of a 22-year-old.

b) Use only the graph to estimate $I(40)$.

27. *Total Revenue.* An office machine firm is marketing a new kind of inkjet color printer. The firm determines that when it sells x printers, its total revenue is

$$R(x) = 280x - 0.4x^2 \text{ dollars.}$$

a) What is the total revenue from the sale of 75 printers?

b) What is the total revenue from the sale of 100 printers?

28. *Total Cost.* The office machine firm determines that the total cost, in dollars, of producing x printers is given by

$$C(x) = 5000 + 0.6x^2.$$

a) What is the total cost of producing 75 printers?

b) What is the total cost of producing 100 printers?

Total Profit. **Total profit P** is defined as total revenue R minus total cost C, and is given by the function

$$P(x) = R(x) - C(x).$$

For each of the following, find the total profit $P(x)$.

29. $R(x) = 280x - 0.4x^2, \quad C(x) = 7000 + 0.6x^2$

30. $R(x) = 280x - 0.7x^2, \quad C(x) = 8000 + 0.5x^2$

Magic Number. In a recent season, the Arizona Diamondbacks were leading the San Francisco Giants for the Western Division championship of the National League. In the table below, the number in parentheses, 18, was the **magic number.** It means that any combination of Diamondbacks wins and Giants losses that totals 18 would ensure the championship for the Diamondbacks. The magic number M is given by the polynomial

$$M = G - W_1 - L_2 + 1,$$

where W_1 is the number of wins for the first-place team, L_2 is the number of losses for the second-place team, and G is the total number of games in the season, which is 162 in the major leagues. When the magic number reaches 1, a tie for the championship is clinched. When the magic number reaches 0, the championship is clinched. For the situation shown below, $G = 162$, $W_1 = 81$, and $L_2 = 64$. Then the magic number is

$$
\begin{aligned}
M &= G - W_1 - L_2 + 1 \\
&= 162 - 81 - 64 + 1 \\
&= 18.
\end{aligned}
$$

WEST	W	L	Pct.	GB
Arizona (18)	81	62	.566	—
San Francisco	80	64	.556	1½
Los Angeles	78	65	.545	3
San Diego	70	73	.490	11
Colorado	62	80	.437	18½

Magic number in parentheses

31. Compute the magic number for Atlanta.

EAST	W	L	PCT.	GB
Atlanta (?)	78	64	.549	—
Philadelphia	75	68	.524	$3\frac{1}{2}$
New York	71	73	.493	8
Florida	66	77	.462	$12\frac{1}{2}$
Montreal	61	82	.427	$17\frac{1}{2}$

32. Compute the magic number for Houston.

CENTRAL	W	L	PCT.	GB
Houston (?)	84	59	.587	—
St. Louis	78	64	.549	$5\frac{1}{2}$
Chicago	78	65	.545	6
Milwaukee	63	80	.441	21
Cincinnati	58	86	.403	$26\frac{1}{2}$
Pittsburgh	55	88	.385	29

33. Compute the magic number for New York.

EAST	W	L	PCT.	GB
New York (?)	86	57	.601	—
Boston	72	69	.511	13
Toronto	70	73	.490	16
Baltimore	55	87	.387	$30\frac{1}{2}$
Tampa Bay	50	93	.350	36

34. Compute the magic number for Cleveland.

CENTRAL	W	L	PCT.	GB
Cleveland (?)	82	62	.569	—
Minnesota	76	68	.528	6
Chicago	74	70	.514	8
Detroit	57	86	.399	$24\frac{1}{2}$
Kansas City	57	86	.399	$24\frac{1}{2}$

C Collect like terms.

35. $6x^2 - 7x^2 + 3x^2$

36. $-2y^2 - 7y^2 + 5y^2$

37. $7x - 2y - 4x + 6y$

38. $a - 8b - 5a + 7b$

39. $3a + 9 - 2 + 8a - 4a + 7$

40. $13x + 14 - 6 - 7x + 3x + 5$

41. $3a^2b + 4b^2 - 9a^2b - 6b^2$

42. $5x^2y^2 + 4x^3 - 8x^2y^2 - 12x^3$

43. $8x^2 - 3xy + 12y^2 + x^2 - y^2 + 5xy + 4y^2$

44. $a^2 - 2ab + b^2 + 9a^2 + 5ab - 4b^2 + a^2$

45. $4x^2y - 3y + 2xy^2 - 5x^2y + 7y + 7xy^2$

46. $3xy^2 + 4xy - 7xy^2 + 7xy + x^2y$

Add.

47. $(3x^2 + 5y^2 + 6) + (2x^2 - 3y^2 - 1)$

48. $(11y^2 + 6y - 3) + (9y^2 - 2y + 9)$

49. $(2a - c + 3b) + (4a - 2b + 2c)$

50. $(8x + z - 7y) + (5x + 10y - 4z)$

51. $(a^2 - 3b^2 + 4c^2) + (-5a^2 + 2b^2 - c^2)$

52. $(x^2 - 5y^2 - 9z^2) + (-6x^2 + 9y^2 - 2z^2)$

53. $(x^2 + 3x - 2xy - 3) + (-4x^2 - x + 3xy + 2)$

54. $(5a^2 - 3b + ab + 6) + (-a^2 + 8b - 8ab - 4)$

55. $(7x^2y - 3xy^2 + 4xy) + (-2x^2y - xy^2 + xy)$

56. $(7ab - 3ac + 5bc) + (13ab - 15ac - 8bc)$

57. $(2r^2 + 12r - 11) + (6r^2 - 2r + 4) + (r^2 - r - 2)$

58. $(5x^2 + 19x - 23) + (-7x^2 - 11x + 12) + (-x^2 - 9x + 8)$

59. $\left(\frac{2}{3}xy + \frac{5}{6}xy^2 + 5.1x^2y\right) + \left(-\frac{4}{5}xy + \frac{3}{4}xy^2 - 3.4x^2y\right)$

60. $\left(\frac{1}{8}xy - \frac{3}{5}x^3y^2 + 4.3y^3\right) + \left(-\frac{1}{3}xy - \frac{3}{4}x^3y^2 - 2.9y^3\right)$

d   Write two equivalent expressions for the opposite of the polynomial.

61. $5x^3 - 7x^2 + 3x - 6$

62. $-8y^4 - 18y^3 + 4y - 9$

63. $-13y^2 + 6ay^4 - 5by^2$

64. $9ax^5y^3 - 8by^5 - abx - 16ay$

Subtract.

65. $(7x - 2) - (-4x + 5)$

66. $(8y + 1) - (-5y - 2)$

67. $(-3x^2 + 2x + 9) - (x^2 + 5x - 4)$

68. $(-9y^2 + 4y + 8) - (4y^2 + 2y - 3)$

69. $(5a + c - 2b) - (3a + 2b - 2c)$

70. $(z + 8x - 4y) - (4x + 6y - 3z)$

71. $(3x^2 - 2x - x^3) - (5x^2 - x^3 - 8x)$

72. $(8y^2 - 4y^3 - 3y) - (3y^2 - 9y - 7y^3)$

73. $(5a^2 + 4ab - 3b^2) - (9a^2 - 4ab + 2b^2)$

74. $(9y^2 - 14yz - 8z^2) - (12y^2 - 8yz + 4z^2)$

75. $(6ab - 4a^2b + 6ab^2) - (3ab^2 - 10ab - 12a^2b)$

76. $(10xy - 4x^2y^2 - 3y^3) - (-9x^2y^2 + 4y^3 - 7xy)$

77. $(0.09y^4 - 0.052y^3 + 0.93) - (0.03y^4 - 0.084y^3 + 0.94y^2)$

78. $(1.23x^4 - 3.122x^3 + 1.11x) - (0.79x^4 - 8.734x^3 + 0.04x^2 + 6.71x)$

79. $\left(\frac{5}{8}x^4 - \frac{1}{4}x^2 - \frac{1}{2}\right) - \left(-\frac{3}{8}x^4 + \frac{3}{4}x^2 + \frac{1}{2}\right)$

80. $\left(\frac{5}{6}y^4 - \frac{1}{2}y^2 - 7.8y + \frac{1}{3}\right) - \left(-\frac{3}{8}y^4 + \frac{3}{4}y^2 + 3.4y - \frac{1}{5}\right)$

81. $\mathbf{D_W}$ Is the sum of two binomials always a binomial? Why or why not?

82. $\mathbf{D_W}$ Annie claims that she can add any two polynomials but finds subtraction difficult. What advice would you offer her?

SKILL MAINTENANCE

Graph.   [2.1c, d]

83. $f(x) = \frac{2}{3}x - 1$

84. $g(x) = |x| - 1$

85. $g(x) = \dfrac{4}{x - 3}$

86. $f(x) = 1 - x^2$

Multiply.   [R.5d]

87. $3(y - 2)$

88. $-10(x + 2y - 7)$

89. $-14(3p - 2q - 10)$

90. $\frac{2}{3}(12w - 9t + 30)$

Graph using the slope and the y-intercept.   [2.5b]

91. $y = \frac{4}{3}x + 2$

92. $y = -0.4x + 1$

93. $y = 0.4x - 3$

94. $y = -\frac{2}{3}x - 4$

SYNTHESIS

95. *Triangular Layers.*   The number of spheres in a triangular pyramid with x triangular layers is given by the function

$$N(x) = \frac{1}{6}x^3 + \frac{1}{2}x^2 + \frac{1}{3}x.$$

The volume of a sphere of radius r is given by the function

$$V(r) = \frac{4}{3}\pi r^3,$$

where π can be approximated as 3.14.

Chocolate Heaven has a window display of truffles piled in a triangular pyramid formation 5 layers deep. If the diameter of each truffle is 3 cm, find the volume of chocolate in the display.

Truffle Balls
$5.95 per lb.

96. *Surface Area.*   Find a polynomial function that gives the outside surface area of a box like this one, with an open top and dimensions as shown.

Perform the indicated operations. Assume that the exponents are natural numbers.

97. $(47x^{4a} + 3x^{3a} + 22x^{2a} + x^a + 1) + (37x^{3a} + 8x^{2a} + 3)$

98. $(3x^{6a} - 5x^{5a} + 4x^{3a} + 8) - (2x^{6a} + 4x^{4a} + 3x^{3a} + 2x^{2a})$

99. 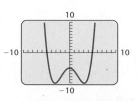 Use the TABLE and GRAPH features of a graphing calculator to check your answers to Exercises 57, 65, and 67.

100. ⊞ A student who is trying to graph $f(x) = 0.05x^4 - x^2 + 5$ gets the following screen. How can the student tell at a glance that a mistake has been made?

Objectives

a Multiply any two polynomials.

b Use the FOIL method to multiply two binomials.

c Use a rule to square a binomial.

d Use a rule to multiply a sum and a difference of the same two terms.

e For functions f described by second-degree polynomials, find and simplify notation like

$$f(a + h)$$

and

$$f(a + h) - f(a).$$

Multiply.

1. $(9y^2)(-2y)$

2. $(4x^3y)(6x^5y^2)$

3. $(-5xy^7z^4)(18x^3y^2z^8)$

Multiply.

4. $(-3y)(2y + 6)$

5. $(2xy)(4y^2 - 5)$

Answers on page A-23

4.2 MULTIPLICATION OF POLYNOMIALS

a Multiplication of Any Two Polynomials

MULTIPLYING MONOMIALS

Monomials are expressions like $10x^2$, $8x^5$, and $-7a^2b^3$. To multiply monomials, we first multiply their coefficients. Then we multiply the variables using the commutative and associative laws and the rules for exponents that we studied in Chapter R.

EXAMPLES Multiply and simplify.

1. $(10x^2)(8x^5) = (10 \cdot 8)(x^2 \cdot x^5)$
$$= 80x^{2+5} \quad \text{Adding exponents}$$
$$= 80x^7$$

2. $(-8x^4y^7)(5x^3y^2) = (-8 \cdot 5)(x^4 \cdot x^3)(y^7 \cdot y^2)$
$$= -40x^{4+3}y^{7+2} \quad \text{Adding exponents}$$
$$= -40x^7y^9$$

Do Exercises 1–3.

MULTIPLYING MONOMIALS AND BINOMIALS

The distributive law is the basis for multiplying polynomials other than monomials. We first multiply a monomial and a binomial.

EXAMPLE 3 Multiply: $2x(3x - 5)$.

$$2x \cdot (3x - 5) = 2x \cdot 3x - 2x \cdot 5 \quad \text{Using the distributive law}$$
$$= 6x^2 - 10x \quad \text{Multiplying monomials}$$

EXAMPLE 4 Multiply: $3a^2b(a^2 - b^2)$.

$$3a^2b \cdot (a^2 - b^2) = 3a^2b \cdot a^2 - 3a^2b \cdot b^2 \quad \text{Using the distributive law}$$
$$= 3a^4b - 3a^2b^3$$

Do Exercises 4 and 5.

MULTIPLYING BINOMIALS

Next, we multiply two binomials. To do so, we use the distributive law twice, first considering one of the binomials as a single expression and multiplying it by each term of the other binomial.

EXAMPLE 5 Multiply: $(3y^2 + 4)(y^2 - 2)$.

$$(3y^2 + 4)(y^2 - 2) = (3y^2 + 4) \cdot y^2 - (3y^2 + 4) \cdot 2 \qquad \text{Using the distributive law}$$

$$= [3y^2 \cdot y^2 + 4 \cdot y^2] - [3y^2 \cdot 2 + 4 \cdot 2] \qquad \text{Using the distributive law}$$

$$= 3y^2 \cdot y^2 + 4 \cdot y^2 - 3y^2 \cdot 2 - 4 \cdot 2 \qquad \text{Removing parentheses}$$

$$= 3y^4 + 4y^2 - 6y^2 - 8 \qquad \text{Multiplying the monomials}$$

$$= 3y^4 - 2y^2 - 8 \qquad \text{Collecting like terms}$$

Do Exercises 6 and 7.

MULTIPLYING ANY TWO POLYNOMIALS

To find a quick way to multiply any two polynomials, let's consider another example.

EXAMPLE 6 Multiply: $(p + 2)(p^4 - 2p^3 + 3)$.

By the distributive law, we have

$$(\ p + 2\ )(p^4 - 2p^3 + 3)$$

$$= (\ p + 2\ )(p^4) - (\ p + 2\ )(2p^3) + (\ p + 2\ )(3)$$

$$= p(p^4) + 2(p^4) - p(2p^3) - 2(2p^3) + p(3) + 2(3)$$

$$= p^5 + 2p^4 - 2p^4 - 4p^3 + 3p + 6$$

$$= p^5 - 4p^3 + 3p + 6. \qquad \text{Collecting like terms}$$

Do Exercises 8 and 9.

From the preceding examples, we can see how to multiply any two polynomials.

> **PRODUCT OF TWO POLYNOMIALS**
>
> To multiply two polynomials P and Q, select one of the polynomials, say P. Then multiply each term of P by every term of Q and collect like terms.

We can use columns to save time doing long multiplications. We multiply each term at the top by every term at the bottom, keeping like terms in columns and *adding spaces for missing terms*. Then we add.

EXAMPLE 7 Multiply: $(5x^3 + 3x^2 + x - 4)(-2x^2 + 3x + 6)$.

$$
\begin{array}{r}
5x^3 + 3x^2 + x - 4 \\
-2x^2 + 3x + 6 \\
\hline
30x^3 + 18x^2 + 6x - 24 \qquad \text{Multiplying by } 6 \\
15x^4 + 9x^3 + 3x^2 - 12x \qquad \text{Multiplying by } 3x \\
-10x^5 - 6x^4 - 2x^3 + 8x^2 \qquad \text{Multiplying by } -2x^2 \\
\hline
-10x^5 + 9x^4 + 37x^3 + 29x^2 - 6x - 24
\end{array}
$$

Multiply.

6. $(5x^2 - 4)(x + 3)$

7. $(2y + 3)(3y - 4)$

Multiply.

8. $(p - 3)(p^3 + 4p^2 - 5)$

9. $(2x^3 + 4x - 5)(x - 4)$

Answers on page A-23

Multiply. Use columns.

10. $(-4x^3 + 5x^2 - 2x + 1) \times (-2x^2 - 3x + 6)$

11. $(-4x^3 - 2x + 1) \times (-2x^2 - 3x + 6)$

12. $(a^2 - 2ab + b^2) \times (a^3 + 3ab - b^2)$

A visualization of
$(x + 7)(x + 4)$ using areas

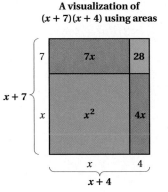

EXAMPLE 8 Multiply: $(5x^3 + x - 4)(-2x^2 + 3x + 6)$.

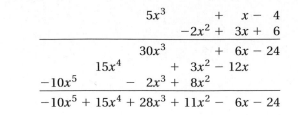

$$
\begin{array}{r}
5x^3 \qquad\quad + x - 4 \\
-2x^2 + 3x + 6 \\
\hline
30x^3 \qquad + 6x - 24 \\
15x^4 \qquad + 3x^2 - 12x \\
-10x^5 \qquad - 2x^3 + 8x^2 \\
\hline
-10x^5 + 15x^4 + 28x^3 + 11x^2 - 6x - 24
\end{array}
$$

Multiplying by 6
Multiplying by $3x$
Multiplying by $-2x^2$

Do Exercises 10–12.

b   Product of Two Binomials Using the FOIL Method

We now consider some **special products.** There are rules for faster multiplication in certain situations.

Let's find a faster special-product rule for the product of two binomials. Consider $(x + 7)(x + 4)$. We multiply each term of $(x + 7)$ by each term of $(x + 4)$:

$$(x + 7)(x + 4) = x \cdot x + x \cdot 4 + 7 \cdot x + 7 \cdot 4.$$

This multiplication illustrates a pattern that occurs whenever two binomials are multiplied:

First terms   Outside terms   Inside terms   Last terms

$$(x + 7)(x + 4) = x \cdot x + 4x + 7x + 7(4) = x^2 + 11x + 28.$$

This special method of multiplying is called the **FOIL method.** Keep in mind that this method is based on the distributive law.

> **THE FOIL METHOD**
>
> To multiply two binomials, $A + B$ and $C + D$, multiply the **F**irst terms AC, the **O**utside terms AD, the **I**nside terms BC, and then the **L**ast terms BD. Then collect like terms, if possible.
>
> $$(A + B)(C + D) = AC + AD + BC + BD$$
>
> **1.** Multiply First terms: AC.
> **2.** Multiply Outside terms: AD.
> **3.** Multiply Inside terms: BC.
> **4.** Multiply Last terms: BD.
>
> $\downarrow$
> **FOIL**
>
>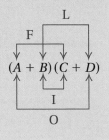

Answers on page A-23

EXAMPLES Multiply.

$$\overset{\text{F} \quad \text{O} \quad \text{I} \quad \text{L}}{}$$

9. $(x + 5)(x - 8) = x^2 - 8x + 5x - 40$
$$= x^2 - 3x - 40 \qquad \text{Collecting like terms}$$

We write the result in descending order since the original binomials are in descending order.

$$\overset{\text{F} \quad \text{O} \quad \text{I} \quad \text{L}}{}$$

10. $(3xy + 2x)(x^2 + 2xy^2) = 3x^3y + 6x^2y^3 + 2x^3 + 4x^2y^2$

11. $(2x - 3)(y + 2) = 2xy + 4x - 3y - 6$

12. $(2x + 3y)(x - 4y) = 2x^2 - 8xy + 3xy - 12y^2$
$$= 2x^2 - 5xy - 12y^2 \qquad \text{Collecting like terms}$$

Do Exercises 13–15.

Multiply.

13. $(y - 4)(y + 10)$

14. $(p + 5q)(2p - 3q)$

15. $(x^2y + 2x)(xy^2 + y^2)$

Answers on page A-23

CALCULATOR CORNER

Checking Multiplication of Polynomials A partial check of multiplication of polynomials can be performed graphically. Consider the product $(x + 3)(x - 2) = x^2 + x - 6$. We will use two graph styles to determine whether this product is correct. First, we press **MODE** to determine if **SEQUENTIAL** mode is selected. If it is not, we position the blinking cursor over **SEQUENTIAL** and then press **ENTER**. Next, on the Y = screen, we enter $y_1 = (x + 3)(x - 2)$ and $y_2 = x^2 + x - 6$. We will select the line-graph style for y_1 and the path style for y_2. To select these graph styles, we use ◁ to position the cursor over the icon to the left of the equation and press **ENTER** repeatedly until the desired style of icon appears, as shown below.

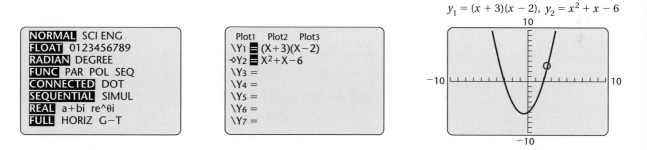

The graphing calculator will graph y_1 first as a solid line. Then it will graph y_2 as the circular cursor traces the leading edge of the graph, allowing us to determine visually whether the graphs coincide. In this case, the graphs appear to coincide, so the factorization is probably correct.

A table can also be used to perform a partial check of a product. See the Calculator Corner on p. 332 for the procedure. Remember that these procedures provide only a partial check since we can neither see the entire graph nor view all possible values of x in a table.

Exercises: Determine graphically whether the product is correct.

1. $(x + 4)(x + 3) = x^2 + 7x + 12$

2. $(3x + 2)(x - 1) = 3x^2 + x - 2$

3. $(4x - 1)(x - 5) = 4x^2 - 21x + 5$

4. $(2x - 1)(3x - 4) = 6x^2 - 11x - 4$

5. $(x - 1)(x - 1) = x^2 + 1$

6. $(x - 2)(x + 2) = x^2 - 4$

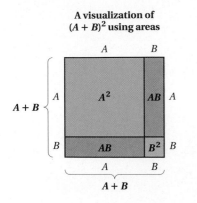

A visualization of $(A + B)^2$ using areas

c | Squares of Binomials

We can use the FOIL method to develop special products for the square of a binomial:

$$(A + B)^2 = (A + B)(A + B)$$
$$= A^2 + AB + AB + B^2$$
$$= A^2 + 2AB + B^2;$$

$$(A - B)^2 = (A - B)(A - B)$$
$$= A^2 - AB - AB + B^2$$
$$= A^2 - 2AB + B^2.$$

SQUARE OF A BINOMIAL

The **square of a binomial** is the square of the first term, plus twice the product of the two terms, plus the square of the last term.

$$(A + B)^2 = A^2 + 2AB + B^2;$$
$$(A - B)^2 = A^2 - 2AB + B^2$$

Multiply.

16. $(a - b)^2$

Caution!

In general, it is true that

$$(AB)^2 = A^2 B^2, \quad \text{but} \quad (A + B)^2 \neq A^2 + B^2.$$

---

17. $(x + 8)^2$

■ **EXAMPLES** Multiply.

$$(A - B)^2 = A^2 - 2\ A\ B\ +\ B^2$$

13. $(y - 5)^2 = y^2 - 2(y)(5) + 5^2$
$$= y^2 - 10y + 25$$

$$(A + B)^2 = A^2 + 2\ A\ B\ +\ B^2$$

14. $(2x + 3y)^2 = (2x)^2 + 2(2x)(3y) + (3y)^2$
$$= 4x^2 + 12xy + 9y^2$$

18. $(3x - 7)^2$

15. $(3x^2 + 5xy^2)^2 = (3x^2)^2 + 2(3x^2)(5xy^2) + (5xy^2)^2$
$$= 9x^4 + 30x^3 y^2 + 25x^2 y^4$$

16. $\left(\frac{1}{2}a^2 - b^3\right)^2 = \left(\frac{1}{2}a^2\right)^2 - 2\left(\frac{1}{2}a^2\right)(b^3) + (b^3)^2$
$$= \frac{1}{4}a^4 - a^2 b^3 + b^6$$

Do Exercises 16–19.

19. $\left(m^3 + \dfrac{1}{4}n\right)^2$

d | Products of Sums and Differences

Another special case of a product of two binomials is the product of a sum and a difference. Note the following:

$$\begin{array}{cccc} \text{F} & \text{O} & \text{I} & \text{L} \\ \downarrow & \downarrow & \downarrow & \downarrow \end{array}$$
$$(A + B)(A - B) = A^2 - AB + AB - B^2 = A^2 - B^2$$

Answers on page A-23

PRODUCT OF A SUM AND A DIFFERENCE

The product of the sum and the difference of the same two terms is the square of the first term minus the square of the second term (the difference of their squares.)

$$(A + B)(A - B) = A^2 - B^2$$ This is called a **difference of squares.**

EXAMPLES Multiply. (Say the rule as you work.)

$$(A + B)(A - B) = A^2 - B^2$$

17. $(y + 5)(y - 5) = y^2 - 5^2 = y^2 - 25$

18. $(2xy^2 + 3x)(2xy^2 - 3x) = (2xy^2)^2 - (3x)^2 = 4x^2y^4 - 9x^2$

19. $(0.2t - 1.4m)(0.2t + 1.4m) = (0.2t)^2 - (1.4m)^2 = 0.04t^2 - 1.96m^2$

20. $\left(\frac{2}{3}n - m^2\right)\left(\frac{2}{3}n + m^2\right) = \left(\frac{2}{3}n\right)^2 - (m^2)^2 = \frac{4}{9}n^2 - m^4$

Do Exercises 20–23.

EXAMPLES Multiply.

21. $(5y + 4 + 3x)(5y + 4 - 3x) = (5y + 4)^2 - (3x)^2$

$$= 25y^2 + 40y + 16 - 9x^2$$

Here we treat the binomial $5y + 4$ as the first expression, A, and $3x$ as the second, B.

22. $(3xy^2 + 4y)(-3xy^2 + 4y) = (4y + 3xy^2)(4y - 3xy^2)$

$$= (4y)^2 - (3xy^2)^2$$
$$= 16y^2 - 9x^2y^4$$

Do Exercises 24 and 25.

Try to multiply polynomials mentally, even when several types are mixed. First check to see what types of polynomials are to be multiplied. Then use the quickest method. Sometimes we might use more than one method. Remember that FOIL *always* works for multiplying binomials!

EXAMPLE 23 Multiply: $(a - 5b)(a + 5b)(a^2 - 25b^2)$.

We first note that $a - 5b$ and $a + 5b$ can be multiplied using the rule $(A - B)(A + B) = A^2 - B^2$. Then we square, using $(A - B)^2 = A^2 - 2AB + B^2$.

$(a - 5b)(a + 5b)(a^2 - 25b^2)$

$= (a^2 - 25b^2)(a^2 - 25b^2)$ Using $(A - B)(A + B) = A^2 - B^2$

$= (a^2 - 25b^2)^2$

$= (a^2)^2 - 2(a^2)(25b^2) + (25b^2)^2$ Using $(A - B)^2 = A^2 - 2AB + B^2$

$= a^4 - 50a^2b^2 + 625b^4$

Do Exercise 26.

Multiply.

20. $(x + 8)(x - 8)$

21. $(4y - 7)(4y + 7)$

22. $(2.8a + 4.1b)(2.8a - 4.1b)$

23. $\left(3w - \frac{3}{5}q^2\right)\left(3w + \frac{3}{5}q^2\right)$

Multiply.

24. $(7x^2y + 2y)(-2y + 7x^2y)$

25. $(2x + 3 - 5y)(2x + 3 + 5y)$

26. Multiply:

$(3x + 2y)(3x - 2y)(9x^2 + 4y^2)$.

Answers on page A-23

343

27. Given $f(x) = x^2 + 2x - 7$, find and simplify $f(a + 1)$ and $f(a + h) - f(a)$.

e Using Function Notation

AG *ALGEBRAIC–GRAPHICAL CONNECTION*

Let's stop for a moment and look back at what we have done in this section. We have shown, for example, that

$$(x - 2)(x + 2) = x^2 - 4,$$

that is, $x^2 - 4$ and $(x - 2)(x + 2)$ are equivalent expressions.

From the viewpoint of functions, if

$$f(x) = (x - 2)(x + 2)$$

and

$$g(x) = x^2 - 4,$$

then for any given input x, the outputs $f(x)$ and $g(x)$ are identical. Thus the graphs of these functions are identical and we say that f and g represent the same function. Functions like these are graphed in detail in Chapter 7.

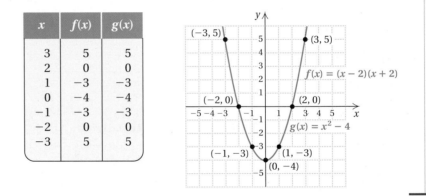

x	$f(x)$	$g(x)$
3	5	5
2	0	0
1	-3	-3
0	-4	-4
-1	-3	-3
-2	0	0
-3	5	5

Our work with multiplying can be used when manipulating functions.

EXAMPLE 24 Given $f(x) = x^2 - 4x + 5$, find and simplify $f(a + 3)$ and $f(a + h) - f(a)$.

To find $f(a + 3)$, we replace x with $a + 3$. Then we simplify:

$$f(a + 3) = (a + 3)^2 - 4(a + 3) + 5$$
$$= a^2 + 6a + 9 - 4a - 12 + 5 = a^2 + 2a + 2.$$

To find $f(a + h) - f(a)$, we replace x with $a + h$ for $f(a + h)$ and x with a for $f(a)$. Then we simplify:

$$f(a + h) - f(a) = [(a + h)^2 - 4(a + h) + 5] - [a^2 - 4a + 5]$$
$$= a^2 + 2ah + h^2 - 4a - 4h + 5 - a^2 + 4a - 5$$
$$= 2ah + h^2 - 4h.$$

Do Exercise 27.

Answer on page A-23

Study Tips

MEMORIZING FORMULAS

Memorizing can be a very helpful tool in the study of mathematics. Don't underestimate its power as you consider the special products. Consider putting the rules, in words and in math symbols, on index cards and go over them many times.

a   Multiply.

1. $8y^2 \cdot 3y$

2. $-5x^2 \cdot 6xy$

3. $2x(-10x^2y)$

4. $-7ab^2(4a^2b^2)$

5. $(5x^5y^4)(-2xy^3)$

6. $(2a^2bc^2)(-3ab^5c^4)$

7. $2z(7 - x)$

8. $4a(a^2 - 3a)$

9. $6ab(a + b)$

10. $2xy(2x - 3y)$

11. $5cd(3c^2d - 5cd^2)$

12. $a^2(2a^2 - 5a^3)$

13. $(5x + 2)(3x - 1)$

14. $(2a - 3b)(4a - b)$

15. $(s + 3t)(s - 3t)$

16. $(y + 4)(y - 4)$

17. $(x - y)(x - y)$

18. $(a + 2b)(a + 2b)$

19. $(x^3 + 8)(x^3 - 5)$

20. $(2x^4 - 7)(3x^3 + 5)$

21. $(a^2 - 2b^2)(a^2 - 3b^2)$

22. $(2m^2 - n^2)(3m^2 - 5n^2)$

23. $(x - 4)(x^2 + 4x + 16)$

24. $(y + 3)(y^2 - 3y + 9)$

25. $(x + y)(x^2 - xy + y^2)$

26. $(a - b)(a^2 + ab + b^2)$

27. $(a^2 + a - 1)(a^2 + 4a - 5)$

28. $(x^2 - 2x + 1)(x^2 + x + 2)$

29. $(4a^2b - 2ab + 3b^2)(ab - 2b + a)$

30. $(2x^2 + y^2 - 2xy)(x^2 - 2y^2 - xy)$

31. $\left(x + \frac{1}{4}\right)\left(x + \frac{1}{4}\right)$

32. $\left(b - \frac{1}{3}\right)\left(b - \frac{1}{3}\right)$

33. $\left(\frac{1}{2}x - \frac{2}{3}\right)\left(\frac{1}{4}x + \frac{1}{3}\right)$

34. $\left(\frac{2}{3}a + \frac{1}{6}b\right)\left(\frac{1}{3}a - \frac{5}{6}b\right)$

35. $(1.3x - 4y)(2.5x + 7y)$

36. $(40a - 0.24b)(0.3a + 10b)$

37. $(a + 8)(a + 5)$

38. $(x + 2)(x + 3)$

39. $(y + 7)(y - 4)$

40. $(y - 2)(y + 3)$

41. $\left(3a + \frac{1}{2}\right)^2$

42. $\left(2x - \frac{1}{3}\right)^2$

43. $(x - 2y)^2$

44. $(2s + 3t)^2$

45. $\left(b - \frac{1}{3}\right)\left(b - \frac{1}{2}\right)$

46. $\left(x - \frac{1}{2}\right)\left(x - \frac{1}{4}\right)$

47. $(2x + 9)(x + 2)$

48. $(3b + 2)(2b - 5)$

49. $(20a - 0.16b)^2$

50. $(10p^2 + 2.3q)^2$

51. $(2x - 3y)(2x + y)$

52. $(2a - 3b)(2a - b)$

53. $(x^3 + 2)^2$

54. $(y^4 - 7)^2$

55. $(2x^2 - 3y^2)^2$

56. $(3s^2 + 4t^2)^2$

57. $(a^3b^2 + 1)^2$

58. $(x^2y - xy^3)^2$

59. $(0.1a^2 - 5b)^2$

60. $(6p + 0.45q^2)^2$

61. *Compound Interest.* Suppose that P dollars is invested in a savings account at interest rate i, compounded annually, for 2 yr. The amount A in the account after 2 yr is given by

$$A = P(1 + i)^2.$$

Find an equivalent expression for A.

62. *Compound Interest.* Suppose that P dollars is invested in a savings account at interest rate i, compounded semiannually, for 1 yr. The amount A in the account after 1 yr is given by

$$A = P\left(1 + \frac{i}{2}\right)^2.$$

Find an equivalent expression for A.

d Multiply.

63. $(d + 8)(d - 8)$

64. $(y - 3)(y + 3)$

65. $(2c + 3)(2c - 3)$

66. $(1 - 2x)(1 + 2x)$

67. $(6m - 5n)(6m + 5n)$

68. $(3x + 7y)(3x - 7y)$

69. $(x^2 + yz)(x^2 - yz)$

70. $(2a^2 + 5ab)(2a^2 - 5ab)$

71. $(-mn + m^2)(mn + m^2)$

72. $(1.6 + pq)(-1.6 + pq)$

73. $(-3pq + 4p^2)(4p^2 + 3pq)$

74. $(-10xy + 5x^2)(5x^2 + 10xy)$

75. $\left(\frac{1}{2}p - \frac{2}{3}q\right)\left(\frac{1}{2}p + \frac{2}{3}q\right)$

76. $\left(\frac{3}{5}ab + 4c\right)\left(\frac{3}{5}ab - 4c\right)$

77. $(x + 1)(x - 1)(x^2 + 1)$

78. $(y - 2)(y + 2)(y^2 + 4)$

79. $(a - b)(a + b)(a^2 - b^2)$

80. $(2x - y)(2x + y)(4x^2 - y^2)$

81. $(a + b + 1)(a + b - 1)$

82. $(m + n + 2)(m + n - 2)$

83. $(2x + 3y + 4)(2x + 3y - 4)$

84. $(3a - 2b + c)(3a - 2b - c)$

e For each of the following functions, find $f(t - 1)$, $f(p + 1)$, $f(a + h) - f(a)$, $f(t - 2) + c$, and $f(a) + 5$.

85. $f(x) = 5x + x^2$

86. $f(x) = 4x + 2x^2$

87. $f(x) = 3x^2 - 7x + 8$

88. $f(x) = 3x^2 - 4x + 7$

89. $f(x) = 5x - x^2$

90. $f(x) = 4x - 2x^2$

91. $f(x) = 4 + 3x - x^2$

92. $f(x) = 2 - 4x - 3x^2$

93. $\mathbf{D_W}$ Find two binomials whose product is $x^2 - 9$ and explain how you decided on those two binomials.

94. $\mathbf{D_W}$ Find two binomials whose product is $x^2 - 6x + 9$ and explain how you decided on those two binomials.

SKILL MAINTENANCE

Solve.   [3.4b]

95. *Auto Travel.*   Rachel leaves on a business trip, forgetting her briefcase. Her sister discovers Rachel's briefcase 2 hr later, and knowing that Rachel needs its contents for her sales presentation and that Rachel normally travels at a speed of 55 mph, she decides to follow her at a speed of 75 mph. After how long will Rachel's sister catch up with her?

96. *Air Travel.*   An airplane flew for 5 hr against a 20-mph headwind. The return trip with the wind took 4 hr. Find the speed of the plane in still air.

Solve.   [3.2a], [3.3a]

97. $5x + 9y = 2$,
$4x - 9y = 10$

98. $x + 4y = 13$,
$5x - 7y = -16$

99. $2x - 3y = 1$,
$4x - 6y = 2$

100. $9x - 8y = -2$,
$3x + 2y = 3$

SYNTHESIS

101. Use the TABLE and GRAPH features of a graphing calculator to check your answers to Exercises 28, 40, and 77.

102. Use the TABLE and GRAPH features of a graphing calculator to determine whether each of the following is correct.
   a) $(x - 1)^2 = x^2 - 1$
   b) $(x - 2)(x + 3) = x^2 + x - 6$
   c) $(x - 1)^3 = x^3 - 3x^2 + 3x - 1$
   d) $(x + 1)^4 = x^4 + 1$

Multiply. Assume that variables in exponents represent natural numbers.

103. $(z^{n^2})^{n^3}(z^{4n^3})^{n^2}$

104. $y^3 z^n (y^{3n} z^3 - 4yz^{2n})$

105. $(r^2 + s^2)^2 (r^2 + 2rs + s^2)(r^2 - 2rs + s^2)$

106. $(y - 1)^6 (y + 1)^6$

107. $\left(3x^5 - \frac{5}{11}\right)^2$

108. $(4x^2 + 2xy + y^2)(4x^2 - 2xy + y^2)$

109. $(x^a + y^b)(x^a - y^b)(x^{2a} + y^{2b})$

110. $\left(x - \frac{1}{7}\right)\left(x^2 + \frac{1}{7}x + \frac{1}{49}\right)$

111. $(x - 1)(x^2 + x + 1)(x^3 + 1)$

112. $(x^{a-b})^{a+b}$

4.3 INTRODUCTION TO FACTORING

Objectives

a Factor polynomials whose terms have a common factor.

b Factor certain polynomials with four terms by grouping.

Factoring is the reverse of multiplication. To **factor** an expression is to find an equivalent expression that is a product. For example, reversing a type of multiplication we have considered, we know that

$$x^2 - 9 = (x + 3)(x - 3).$$

We say that $x + 3$ and $x - 3$ are **factors** of $x^2 - 9$ and that $(x + 3)(x - 3)$ is a **factorization.**

1. Consider

$$x^2 - 4x - 5 = (x - 5)(x + 1).$$

a) What are the factors of $x^2 - 4x - 5$?

FACTOR

To **factor** a polynomial is to express it as a product.

A **factor** of a polynomial P is a polynomial that can be used to express P as a product.

FACTORIZATION

A **factorization** of a polynomial P is an expression that names P as a product of factors.

---

b) What are the terms of $x^2 - 4x - 5$?

Caution!

Be careful not to confuse terms with factors! The terms of $x^2 - 9$ are x^2 and -9. Terms are used to form sums. Factors of $x^2 - 9$ are $x - 3$ and $x + 3$. Factors are used to form products.

Factor.

2. $3x^2 - 6$

Do Exercise 1.

a Terms with Common Factors

To multiply a monomial and a polynomial with more than one term, we multiply each term by the monomial using the distributive laws. To factor, we do the reverse. We express a polynomial as a product using the distributive laws in reverse. Compare.

Multiply

$5x(x^2 - 3x + 1)$

$= 5x \cdot x^2 - 5x \cdot 3x + 5x \cdot 1$

$= 5x^3 - 15x^2 + 5x$

Factor

$5x^3 - 15x^2 + 5x$

$= 5x \cdot x^2 - 5x \cdot 3x + 5x \cdot 1$

$= 5x(x^2 - 3x + 1)$

3. $4x^5 - 8x^3$

4. $9y^4 - 15y^3 + 3y^2$

EXAMPLE 1 Factor: $4y^2 - 8$.

$4y^2 - 8 = 4 \cdot y^2 - 4 \cdot 2$ 4 is the largest common factor.

$= 4(y^2 - 2)$ Factoring out the common factor 4

5. $6x^2y - 21x^3y^2 + 3x^2y^3$

Answers on page A-23

Factor out a common factor with a
negative coefficient.

6. $-8x + 32$

In some cases, there is more than one common factor. In Example 2 below, for instance, 5 is a common factor, x^3 is a common factor, and $5x^3$ is a common factor. If there is more than one common factor, we generally choose the one with the largest coefficient and the largest exponent.

EXAMPLES Factor.

2. $5x^4 - 20x^3 = 5x^3 \cdot x - 5x^3 \cdot 4$
$= 5x^3(x - 4)$ Multiply mentally to check your answer.

3. $12x^2y - 20x^3y = 4x^2y(3 - 5x)$

EXAMPLE 4 Factor: $10p^6q^2 - 4p^5q^3 + 2p^4q^4$.

First, we look for the greatest positive common factor in the coefficients:

$10, -4, 2$ ⟹ Greatest common factor $= 2$.

Second, we look for the greatest common factor in the powers of p:

p^6, p^5, p^4 ⟹ Greatest common factor $= p^4$.

Third, we look for the greatest common factor in the powers of q:

q^2, q^3, q^4 ⟹ Greatest common factor $= q^2$.

Thus, $2p^4q^2$ is the greatest common factor of the given polynomial. Then

$$10p^6q^2 - 4p^5q^3 + 2p^4q^4 = 2p^4q^2 \cdot 5p^2 - 2p^4q^2 \cdot 2pq + 2p^4q^2 \cdot q^2$$
$$= 2p^4q^2(5p^2 - 2pq + q^2).$$

7. $-3x^2 - 15x + 9$

The polynomials in Examples 1–4 have been **factored completely.** They cannot be factored further. The factors in the resulting factorization are said to be **prime polynomials.**

Do Exercises 2–5 on the preceding page.

When the leading coefficient is a negative number, we generally factor out the negative coefficient.

EXAMPLES Factor out a common factor with a negative coefficient.

5. $-4x - 24 = -4(x + 6)$
6. $-2x^2 + 6x - 10 = -2(x^2 - 3x + 5)$

Do Exercises 6 and 7.

EXAMPLE 7 *Height of a Thrown Object.* Suppose that a softball is thrown upward with an initial velocity of 64 ft/sec. Its height h, in feet, after t seconds is given by the function

$$h(t) = -16t^2 + 64t.$$

a) Find an equivalent expression for $h(t)$ by factoring out a common factor with a negative coefficient.

b) Check your factoring by evaluating both expressions for $h(t)$ at $t = 1$.

Answers on page A-23

8. Height of a Rocket. A model rocket is launched upward with an initial velocity of 96 ft/sec. Its height h, in feet, after t seconds is given by the function

$$h(t) = -16t^2 + 96t.$$

a) Find an equivalent expression for $h(t)$ by factoring out a common factor with a negative coefficient.

a) We factor out $-16t$ as follows:

$$h(t) = -16t^2 + 64t = -16t(t - 4).$$

b) We check as follows:

$$h(1) = -16 \cdot 1^2 + 64 \cdot 1 = 48;$$
$$h(1) = -16 \cdot 1(1 - 4) = 48. \qquad \text{Using the factorization}$$

Do Exercise 8.

b) Check your factoring by evaluating both expressions for $h(t)$ at $t = 2$.

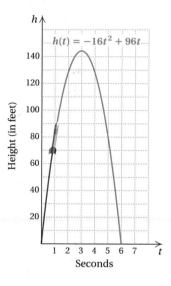

b Factoring by Grouping

In expressions of four or more terms, there may be a *common binomial factor*. We proceed as in the following examples.

EXAMPLE 8 Factor: $(a - b)(x + 5) + (a - b)(x - y^2)$.

$$(a - b)(x + 5) + (a - b)(x - y^2) = (a - b)[(x + 5) + (x - y^2)]$$
$$= (a - b)(2x + 5 - y^2)$$

Do Exercises 9 and 10.

In Example 9, we factor two parts of the expression. Then we factor as in Example 8.

EXAMPLE 9 Factor: $y^3 + 3y^2 + 4y + 12$.

$$y^3 + 3y^2 + 4y + 12 = (y^3 + 3y^2) + (4y + 12) \qquad \text{Grouping}$$
$$= y^2(y + 3) + 4(y + 3) \qquad \text{Factoring each binomial}$$
$$= (y^2 + 4)(y + 3) \qquad \text{Factoring out the common factor } y + 3$$

Factor.

9. $(p + q)(x + 2) + (p + q)(x + y)$

EXAMPLE 10 Factor: $3x^3 - 6x^2 - x + 2$.

First, we consider the first two terms and factor out the greatest common factor:

$$3x^3 - 6x^2 = 3x^2(x - 2).$$

Next, we look at the third and fourth terms to see if we can factor them in order to have $x - 2$ as a factor. We see that if we factor out -1, we get $x - 2$:

$$-x + 2 = -1 \cdot (x - 2).$$

10. $(y + 3)(y - 21) + (y + 3)(y + 10)$

Answers on page A-23

4.3 Introduction to Factoring

Factor by grouping.

11. $5y^3 + 2y^2 - 10y - 4$

Finally, we factor out the common factor $x - 2$:

$$3x^3 - 6x^2 - x + 2 = (3x^3 - 6x^2) + (-x + 2)$$
$$= 3x^2(x - 2) + (-x + 2)$$
$$= 3x^2(x - 2) - 1(x - 2) \qquad Check: -1(x - 2) = -x + 2$$
$$= (3x^2 - 1)(x - 2). \qquad \text{Factoring out the common factor } x - 2$$

EXAMPLE 11 Factor: $4x^3 - 15 + 20x^2 - 3x$.

$$4x^3 - 15 + 20x^2 - 3x = 4x^3 + 20x^2 - 3x - 15 \qquad \text{Rearranging}$$
$$= 4x^2(x + 5) - 3(x + 5) \qquad Check:$$
$$\qquad\qquad -3(x + 5) = -3x - 15$$
$$= (4x^2 - 3)(x + 5) \qquad \text{Factoring out } x + 5$$

12. $x^3 + 5x^2 + 4x - 20$

Not all polynomials with four terms can be factored by grouping. An example is

$$x^3 + x^2 + 3x - 3.$$

Note that in a grouping like $x^2(x + 1) + 3(x - 1)$, the expressions $x + 1$ and $x - 1$ are not the same. No other grouping allows us to factor out a common binomial.

Answers on page A-23

Do Exercises 11 and 12.

CALCULATOR CORNER

Checking Factorizations A partial check of a factorization can be performed using a table or a graph. To check the factorization $3x^3 - 6x^2 - x + 2 = (3x^2 - 1)(x - 2)$, for example, we enter $y_1 = 3x^3 - 6x^2 - x + 2$ and $y_2 = (3x^2 - 1)(x - 2)$ on the equation-editor screen (see p. 85). Then we set up a table in AUTO mode (see p. 162). If the factorization is correct, the values of y_1 and y_2 will be the same regardless of the table settings used. We can also graph $y_1 = 3x^3 - 6x^2 - x + 2$ and $y_2 = (3x^2 - 1)(x - 2)$. If the graphs appear to coincide, the factorization is probably correct. Keep in mind that these procedures provide only a partial check since we cannot view all possible values of x in a table nor can we see the entire graph.

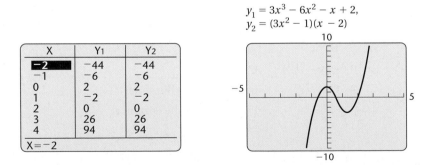

$$y_1 = 3x^3 - 6x^2 - x + 2,$$
$$y_2 = (3x^2 - 1)(x - 2)$$

X	Y₁	Y₂
−2	−44	−44
−1	−6	−6
0	2	2
1	−2	−2
2	0	0
3	26	26
4	94	94
X = −2		

Exercises: Use a table or a graph to determine whether the factorization is correct.

1. $18x^2 + 3x - 6 = 3(2x - 1)(3x + 2)$

2. $3x^2 - 11x - 20 = (3x + 4)(x - 5)$

3. $2x^2 + 5x - 12 = (2x + 3)(x - 4)$

4. $20x^2 - 13x - 2 = (4x + 1)(5x - 2)$

5. $6x^2 + 13x + 6 = (6x + 1)(x + 6)$

6. $6x^2 + 13x + 6 = (3x + 2)(2x + 3)$

7. $x^2 + 16 = (x - 4)(x - 4)$

8. $x^2 - 16 = (x + 4)(x - 4)$

EXERCISE SET   For Extra Help

a   Factor.

1. $6a^2 + 3a$

2. $4x^2 + 2x$

3. $x^3 + 9x^2$

4. $y^3 + 8y^2$

5. $8x^2 - 4x^4$

6. $6x^2 + 3x^4$

7. $4x^2y - 12xy^2$

8. $5x^2y^3 + 15x^3y^2$

9. $3y^2 - 3y - 9$

10. $5x^2 - 5x + 15$

11. $4ab - 6ac + 12ad$

12. $8xy + 10xz - 14xw$

13. $10a^4 + 15a^2 - 25a - 30$

14. $12t^5 - 20t^4 + 8t^2 - 16$

15. $15x^2y^5z^3 - 12x^4y^4z^7$

16. $21a^3b^5c^7 - 14a^7b^6c^2$

17. $14a^4b^3c^5 + 21a^3b^5c^4 - 35a^4b^4c^3$

18. $9x^3y^6z^2 - 12x^4y^4z^4 + 15x^2y^5z^3$

Factor out a common factor with a negative coefficient.

19. $-5x - 45$

20. $-3t + 18$

21. $-6a - 84$

22. $-8t + 40$

23. $-2x^2 + 2x - 24$

24. $-2x^2 + 16x - 20$

25. $-3y^2 + 24y$

26. $-7x^2 - 56y$

27. $-a^4 + 2a^3 - 13a^2 - 1$

28. $-m^3 - m^2 + m - 2$

29. $-3y^3 + 12y^2 - 15y + 24$

30. $-4m^4 - 32m^3 + 64m - 12$

31. *Golf Ball Stacks.*   Each stack of golf balls is formed by square layers of golf balls. The number N of balls in the stack is given by the polynomial function
$$N(x) = \tfrac{1}{3}x^3 + \tfrac{1}{2}x^2 + \tfrac{1}{6}x,$$
where x is the number of layers. Find an equivalent expression for $N(x)$ by factoring out a common factor.

32. *Surface Area of a Silo.*   A silo is a structure that is shaped like a right circular cylinder with a half sphere on top. The surface area of a silo of height h and radius r (including the area of the base) is given by the polynomial $2\pi rh + \pi r^2$. Find an equivalent expression by factoring out a common factor.

33. *Height of a Baseball.*   A baseball is popped up with an upward velocity of 72 ft/sec. Its height h, in feet, after t seconds is given by

$$h(t) = -16t^2 + 72t.$$

a) Find an equivalent expression for $h(t)$ by factoring out a common factor with a negative coefficient.
b) Perform a partial check of part (a) by evaluating both expressions for $h(t)$ at $t = 2$.

34. *Number of Diagonals.*   The number of diagonals of a polygon having n sides is given by the polynomial function

$$P(n) = \tfrac{1}{2}n^2 - \tfrac{3}{2}n.$$

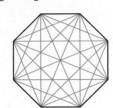

Find an equivalent expression for $P(n)$ by factoring out a common factor.

35. *Total Revenue.*   Household Sound is marketing a new kind of home theater projector. The firm determines that when it sells x projectors, the total revenue R is given by the polynomial function

$$R(x) = 280x + 0.4x^2 \text{ dollars.}$$

Find an equivalent expression for $R(x)$ by factoring out $0.4x$.

36. *Total Cost.*   Household Sound determines that the total cost C of producing x home theater projectors is given by the polynomial function

$$C(x) = 0.18x + 0.6x^2.$$

Find an equivalent expression for $C(x)$ by factoring out $0.6x$.

b   Factor.

37. $a(b - 2) + c(b - 2)$

38. $a(x^2 - 3) - 2(x^2 - 3)$

39. $(x - 2)(x + 5) + (x - 2)(x + 8)$

40. $(m - 4)(m + 3) + (m - 4)(m - 3)$

41. $a^2(x - y) + a^2(x - y)$

42. $3x^2(x - 6) + 3x^2(x - 6)$

43. $ac + ad + bc + bd$

44. $xy + xz + wy + wz$

45. $b^3 - b^2 + 2b - 2$

46. $y^3 - y^2 + 3y - 3$

47. $y^3 - 8y^2 + y - 8$

48. $t^3 + 6t^2 - 2t - 12$

49. $24x^3 - 36x^2 + 72x - 108$

50. $10a^4 + 15a^2 - 25a - 30$

51. $a^4 - a^3 + a^2 + a$

52. $p^6 + p^5 - p^3 + p^2$

53. $2y^4 + 6y^2 + 5y^2 + 15$

54. $2xy + x^2y - 6 - 3x$

55. D_W Explain in your own words why $-(a - b) = b - a$.

56. D_W Is it true that if a polynomial's coefficients and exponents are all prime numbers, then the polynomial itself is prime? Why or why not?

VOCABULARY REINFORCEMENT

In each of Exercises 57–64, fill in the blank with the correct term from the given list. Some of the choices may not be used.

57. The equation $y = mx + b$ is called the _____ equation of the line with slope m and y-intercept $(0, b)$. [2.4b]

58. Equations with the same solutions are called _____ equations. [1.1a]

59. If the slope of a line is less than 0, the graph slants _____ from left to right. [2.4b]

60. A(n) _____ system of equations has no solution. [3.1a]

61. The equation $y - y_1 = m(x - x_1)$, where m is the slope of the line and (x_1, y_1) is a point on the line, is called the _____ equation. [2.6b]

62. _____ angles are angles whose sum is $180°$. [3.2b]

63. When the terms of a polynomial are written such that the exponents increase from left to right, we say the polynomial is written in _____ order. [4.1a]

64. The function $h(x) = 5$ is an example of a(n) _____ function. [2.2b]

point–slope

slope–intercept

complementary

supplementary

consistent

inconsistent

equivalent

ascending

descending

up

down

constant

increasing

decreasing

Complete each of the following.

65. $x^5y^4 + \underline{\quad} = x^3y(\underline{\quad} + xy^5)$

66. $a^3b^7 - \underline{\quad} = \underline{\quad}(ab^4 - c^2)$

Factor.

67. $rx^2 - rx + 5r + sx^2 - sx + 5s$

68. $3a^2 + 6a + 30 + 7a^2b + 14ab + 70b$

69. $a^4x^4 + a^4x^2 + 5a^4 + a^2x^4 + a^2x^2 + 5a^2 + 5x^4 + 5x^2 + 25$ (*Hint*: Use three groups of three.)

Factor out the smallest power of x in each of the following.

70. $x^{1/2} + 5x^{3/2}$

71. $x^{1/3} - 7x^{4/3}$

72. $x^{3/4} + x^{1/2} - x^{1/4}$

73. $x^{1/3} - 5x^{1/2} + 3x^{3/4}$

Factor. Assume that all exponents are natural numbers.

74. $2x^{3a} + 8x^a + 4x^{2a}$

75. $3a^{n+1} + 6a^n - 15a^{n+2}$

76. $4x^{a+b} + 7x^{a-b}$

77. $7y^{2a+b} - 5y^{a+b} + 3y^{a+2b}$

Objective

Factor. Check by multiplying.

1. $x^2 + 5x + 6$

2. $y^2 + 7y + 10$

a Factoring Trinomials: $x^2 + bx + c$

We now consider factoring trinomials of the type $x^2 + bx + c$. We use a re-fined trial-and-error process that is based on the FOIL method.

CONSTANT TERM POSITIVE

Recall the FOIL method of multiplying two binomials:

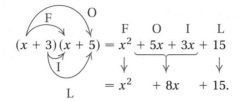

The product is a trinomial. In this example, the leading term has a coefficient of 1. The constant term is positive. To factor $x^2 + 8x + 15$, we think of FOIL in reverse. We multiplied x times x to get the first term of the trinomial. Thus the first term of each binomial factor is x. We want to find numbers p and q such that

$$x^2 + 8x + 15 = (x + p)(x + q).$$

To get the middle term and the last term of the trinomial, we look for two numbers whose product is 15 and whose sum is 8. Those numbers are 3 and 5. Thus the factorization is

$$(x + 3)(x + 5), \quad \text{or} \quad (x + 5)(x + 3)$$

by the commutative law of multiplication. In general,

$$(x + p)(x + q) = x^2 + (p + q)x + pq.$$

To factor, we can use this equation in reverse.

EXAMPLE 1   Factor: $x^2 + 9x + 8$.

Think of FOIL in reverse. The first term of each factor is x. We are looking for numbers p and q such that

$$x^2 + 9x + 8 = (x + p)(x + q) = x^2 + (p + q)x + pq.$$

We look for two numbers p and q whose product is 8 and whose sum is 9. Since both 8 and 9 are positive, we need consider only positive factors.

PAIRS OF FACTORS	SUMS OF FACTORS
2, 4	6
1, 8	9 ←

The numbers we need are 1 and 8.

The factorization is $(x + 1)(x + 8)$. We can check by multiplying:

$$(x + 1)(x + 8) = x^2 + 9x + 8.$$

Do Exercises 1 and 2.

When the constant term of a trinomial is positive, we look for two factors with the same sign (both positive or both negative). The sign is that of the middle term.

EXAMPLE 2 Factor: $y^2 - 9y + 20$.

Since the constant term, 20, is positive and the coefficient of the middle term, -9, is negative, we look for a factorization of 20 in which both factors are negative. Their sum must be -9.

PAIRS OF FACTORS	SUMS OF FACTORS
$-1, -20$	-21
$-2, -10$	-12
$-4, \ -5$	-9 ←

The numbers we need are -4 and -5.

The factorization is $(y - 4)(y - 5)$.

Do Exercises 3 and 4.

CONSTANT TERM NEGATIVE

When the constant term of a trinomial is negative, we look for two factors whose product is negative. One of them must be positive and the other negative. Their sum must be the coefficient of the middle term.

EXAMPLE 3 Factor: $x^3 - x^2 - 30x$.

Always look first for the largest common factor. This time x is the common factor. We first factor it out:

$$x^3 - x^2 - 30x = x(x^2 - x - 30).$$

Now consider $x^2 - x - 30$. Since the constant term, -30, is negative, we look for a factorization of -30 in which one factor is positive and one factor is negative. The sum of the factors must be -1, the coefficient of the middle term, so the negative factor must have the larger absolute value. Thus we consider only pairs of factors in which the negative factor has the larger absolute value.

PAIRS OF FACTORS	SUMS OF FACTORS
$1, -30$	-29
$2, -15$	-13
$3, -10$	-7
$5, \ -6$	-1 ←

The numbers we want are 5 and -6.

The factorization of $x^2 - x - 30$ is $(x + 5)(x - 6)$. But do not forget the common factor! The factorization of the original trinomial is

$$x(x + 5)(x - 6).$$

Do Exercises 5–7.

Factor.

3. $m^2 - 8m + 12$

4. $24 - 11t + t^2$

5. a) Factor $x^2 - x - 20$.

b) Explain why you would not consider these pairs of factors in factoring $x^2 - x - 20$.

PAIRS OF FACTORS	PRODUCTS OF FACTORS
$1, \quad 20$	
$2, \quad 10$	
$4, \quad 5$	
$-1, -20$	
$-2, -10$	
$-4, \ -5$	

Factor.

6. $x^3 - 3x^2 - 54x$

7. $2x^3 - 2x^2 - 84x$

Answers on page A-24

Factor.

8. $x^3 + 4x^2 - 12x$

9. $y^2 - 4y - 12$

10. $x^2 - 110 - x$

11. Factor: $x^2 + x - 5$.

Factor.

12. $x^2 - 5xy + 6y^2$

13. $p^2 - 6pq - 16q^2$

Answers on page A-24

CHAPTER 4: Polynomials and
Polynomial Functions

EXAMPLE 4 Factor: $x^2 + 17x - 110$.

Since the constant term, -110, is negative, we look for a factorization of -110 in which one factor is positive and one factor is negative. Their sum must be 17, so the positive factor must have the larger absolute value.

PAIRS OF FACTORS	SUMS OF FACTORS
$-1, 110$	109
$-2, 55$	53
$-5, 22$	17 ←
$-10, 11$	1

We consider only pairs of factors in which the positive term has the larger absolute value.

The numbers we need are -5 and 22.

The factorization is $(x - 5)(x + 22)$.

Do Exercises 8–10.

Some trinomials are not factorable.

EXAMPLE 5 Factor: $x^2 - x - 7$.

There are no factors of -7 whose sum is -1. This trinomial is *not* factorable into binomials.

Do Exercise 11.

To factor $x^2 + bx + c$:

1. First arrange in descending order.
2. Use a trial-and-error procedure that looks for factors of c whose sum is b.

- If c is positive, then the signs of the factors are the same as the sign of b.
- If c is negative, then one factor is positive and the other is negative. (If the sum of the two factors is the opposite of b, changing the signs of each factor will give the desired factors whose sum is b.)

3. Check your result by multiplying.

The procedure considered here can also be applied to a trinomial with more than one variable.

EXAMPLE 6 Factor: $x^2 - 2xy - 48y^2$.

We look for numbers p and q such that

$$x^2 - 2xy - 48y^2 = (x + py)(x + qy).$$

Our thinking is much the same as if we were factoring $x^2 - 2x - 48$. We look for factors of -48 whose sum is -2. Those factors are 6 and -8. Then

$$x^2 - 2xy - 48y^2 = (x + 6y)(x - 8y).$$

We can check by multiplying.

Do Exercises 12 and 13.

Sometimes a trinomial like $x^4 + 2x^2 - 15$ can be factored using the following method. We can first think of the trinomial as $(x^2)^2 + 2x^2 - 15$, or we can make a substitution (perhaps just mentally), letting $u = x^2$. Then the trinomial becomes

$$u^2 + 2u - 15.$$

As we see in Example 7, we factor this trinomial and if a factorization is found, we replace each occurrence of u with x^2.

EXAMPLE 7   Factor: $x^4 + 2x^2 - 15$.

We let $u = x^2$. Then consider $u^2 + 2u - 15$. The constant term is negative and the middle term is positive. Thus we look for pairs of factors of -15, one positive and one negative, such that the positive factor has the larger absolute value and the sum of the factors is 2.

PAIRS OF FACTORS	SUMS OF FACTORS
$-1, 15$	14
$-3, \ 5$	2 ←

The numbers we need are -3 and 5.

The desired factorization of $u^2 + 2u - 15$ is

$$(u - 3)(u + 5).$$

Replacing u with x^2, we obtain the following factorization of the original trinomial:

$$(x^2 - 3)(x^2 + 5).$$

Do Exercises 14 and 15.

LEADING COEFFICIENT OF −1

EXAMPLE 8   Factor: $14 + 5x - x^2$.

Note that this trinomial is written in ascending order. When we rewrite it in descending order, we get

$$-x^2 + 5x + 14,$$

which has a leading coefficient of -1. Before factoring, in such a case, we can factor out a -1:

$$-x^2 + 5x + 14 = -1(x^2 - 5x - 14).$$

Then we proceed to factor $x^2 - 5x - 14$. We get

$$-x^2 + 5x + 14 = -1(x^2 - 5x - 14) = -1(x - 7)(x + 2).$$

We can also express this answer two other ways by multiplying through either binomial by -1. Thus each of the following is a correct answer:

$$-x^2 + 5x + 14 = -1(x - 7)(x + 2);$$
$$= (-x + 7)(x + 2); \quad \text{Multiplying } x - 7 \text{ by } -1$$
$$= (x - 7)(-x - 2). \quad \text{Multiplying } x + 2 \text{ by } -1$$

Do Exercises 16 and 17.

Factor.
14. $x^4 - 9x^2 + 14$

15. $p^6 + p^3 - 6$

Factor.
16. $10 - 3x - x^2$

17. $-x^2 + 8x - 16$

Answers on page A-24

a   Factor.

1. $x^2 + 13x + 36$

2. $x^2 + 9x + 18$

3. $t^2 - 8t + 15$

4. $y^2 - 10y + 21$

5. $x^2 - 8x - 33$

6. $t^2 - 15 - 2t$

7. $2y^2 - 16y + 32$

8. $2a^2 - 20a + 50$

9. $p^2 + 3p - 54$

10. $m^2 + m - 72$

11. $12x + x^2 + 27$

12. $10y + y^2 + 24$

13. $y^2 - \dfrac{2}{3}y + \dfrac{1}{9}$

14. $p^2 + \dfrac{2}{5}p + \dfrac{1}{25}$

15. $t^2 - 4t + 3$

16. $y^2 - 14y + 45$

17. $5x + x^2 - 14$

18. $x + x^2 - 90$

19. $x^2 + 5x + 6$

20. $y^2 + 8y + 7$

21. $56 + x - x^2$

22. $32 + 4y - y^2$

23. $32y + 4y^2 - y^3$

24. $56x + x^2 - x^3$

25. $x^4 + 11x^2 - 80$

26. $y^4 + 5y^2 - 84$

27. $x^2 - 3x + 7$

28. $x^2 + 12x + 13$

29. $x^2 + 12xy + 27y^2$

30. $p^2 - 5pq - 24q^2$

CHAPTER 4: Polynomials and
Polynomial Functions

31. $45 + 4x - x^2$

32. $30 + 7x - x^2$

33. $-z^2 + 36 - 9z$

34. $24 - a^2 - 10a$

35. $x^4 + 50x^2 + 49$

36. $p^4 + 80p^2 + 79$

37. $x^6 + 11x^3 + 18$

38. $x^6 - x^3 - 42$

39. $x^8 - 11x^4 + 24$

40. $x^8 - 7x^4 + 10$

41. $\mathbf{D_W}$ Under what conditions would it be easier to evaluate a polynomial function after it has been factored?

42. $\mathbf{D_W}$ Checking the factorization of a second-degree polynomial by making a single replacement is only a *partial* check. Write an *incorrect* factorization and explain how evaluating both the polynomial and the factorization might not catch a possible error.

SKILL MAINTENANCE

Solve.   [3.4a]

43. *Mixing Rice.*   Countryside Rice is 90% white rice and 10% wild rice. Mystic Rice is 50% wild rice. How much of each type should be used to create a 25-lb batch of rice that is 35% wild rice?

44. *Wages.*   Takako worked a total of 17 days last month at her father's restaurant. She earned $50 a day during the week and $60 a day during the weekend. Last month Takako earned $940. How many weekdays did she work?

Determine whether each of the following is the graph of a function.   [2.2d]

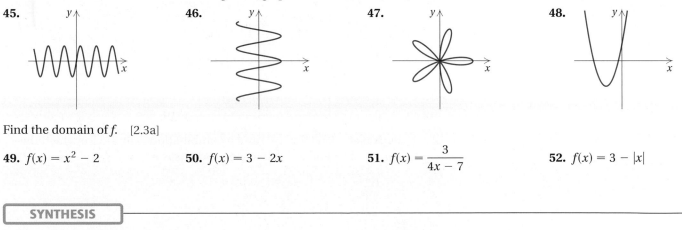

45.

46.

47.

48.

Find the domain of f.   [2.3a]

49. $f(x) = x^2 - 2$

50. $f(x) = 3 - 2x$

51. $f(x) = \dfrac{3}{4x - 7}$

52. $f(x) = 3 - |x|$

SYNTHESIS

53. Find all integers m for which $x^2 + mx + 75$ can be factored.

54. Find all integers q for which $x^2 + qx - 32$ can be factored.

55. One of the factors of $x^2 - 345x - 7300$ is $x + 20$. Find the other factor.

56. Use the TABLE and GRAPH features of a graphing calculator to check your answers to Exercises 1–6.

361

Objectives

a Factor trinomials of the type $ax^2 + bx + c, a \neq 1$, by the FOIL method.

b Factor trinomials of the type $ax^2 + bx + c, a \neq 1$, by the ac-method.

Factor by the FOIL method.

1. $3x^2 - 13x - 56$

2. $3x^2 + 5x + 2$

Now we learn to factor trinomials of the type $ax^2 + bx + c, a \neq 1$. We use two methods: the FOIL method and the ac-method.* Although one is discussed before the other, this should not be taken as a recommendation of one form over the other.

a The FOIL Method

We first consider the **FOIL method** for factoring trinomials of the type $ax^2 + bx + c, a \neq 1$. Consider the following multiplication.

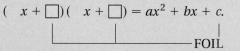

$$(3x + 2)(4x + 5) = 12x^2 + \underbrace{15x + 8x}_{} + 10$$
$$= 12x^2 + 23x + 10$$

To factor $12x^2 + 23x + 10$, we must reverse what we just did. We look for two binomials whose product is this trinomial. The product of the First terms must be $12x^2$. The product of the Outside terms plus the product of the Inside terms must be $23x$. The product of the Last terms must be 10. We know from the preceding discussion that the answer is $(3x + 2)(4x + 5)$. In general, however, finding such an answer involves trial and error. We use the following method.

THE FOIL METHOD

To factor trinomials of the type $ax^2 + bx + c, a \neq 1$, using the **FOIL method:**

1. Factor out the largest common factor.
2. Find two First terms whose product is ax^2:

$$(\square x + \quad)(\square x + \quad) = ax^2 + bx + c.$$
$$\text{FOIL}$$

3. Find two Last terms whose product is c:

$$(\quad x + \square)(\quad x + \square) = ax^2 + bx + c.$$
$$\text{FOIL}$$

4. Repeat steps (2) and (3) until a combination is found for which the sum of the Outside and Inside products is bx:

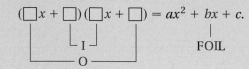

$$(\square x + \square)(\square x + \square) = ax^2 + bx + c.$$
$$\text{FOIL}$$

5. Always check by multiplying.

To the instructor: Here we present two ways to factor general trinomials: the FOIL method and the ac-method. You can teach both methods and let the student use the one he or she prefers or you can select just one for the student.

EXAMPLE 1 Factor: $3x^2 + 10x - 8$.

1. First, we factor out the largest common factor, if any. There is none (other than 1 or -1).

2. Next, we factor the first term, $3x^2$. The only possibility is $3x \cdot x$. The desired factorization is then of the form $(3x + \boxed{})(x + \boxed{})$.

3. We then factor the last term, -8, which is negative. The possibilities are $(-8)(1)$, $8(-1)$, $2(-4)$, and $(-2)(4)$. They can be written in either order.

4. We look for combinations of factors from steps (2) and (3) such that the sum of the outside and the inside products is the middle term, $10x$:

$$\overset{\displaystyle 3x}{(3x - 8)(x + 1)} = 3x^2 - 5x - 8; \qquad \overset{\displaystyle -3x}{(3x + 8)(x - 1)} = 3x^2 + 5x - 8;$$
$$\underset{-8x \quad \text{Wrong middle term}}{} \qquad \underset{8x \quad \text{Wrong middle term}}{}$$

$$\overset{\displaystyle -12x}{(3x + 2)(x - 4)} = 3x^2 - 10x - 8; \qquad \overset{\displaystyle 12x}{(3x - 2)(x + 4)} = 3x^2 + 10x - 8$$
$$\underset{2x \quad \text{Wrong middle term}}{} \qquad \underset{-2x \quad \text{Correct middle term!}}{}$$

There are four other possibilities that we could try, but we have a factorization: $(3x - 2)(x + 4)$.

5. *Check:* $(3x - 2)(x + 4) = 3x^2 + 10x - 8$.

Do Exercises 1 and 2 on the preceding page.

EXAMPLE 2 Factor: $18x^6 - 57x^5 + 30x^4$.

1. First, we factor out the largest common factor, if any. The expression $3x^4$ is common to all terms, so we factor it out: $3x^4(6x^2 - 19x + 10)$.

2. Next, we factor the trinomial $6x^2 - 19x + 10$. We factor the first term, $6x^2$, and get $6x \cdot x$, or $3x \cdot 2x$. We then have these as possibilities for factorizations: $(3x + \boxed{})(2x + \boxed{})$ or $(6x + \boxed{})(x + \boxed{})$.

3. We then factor the last term, 10, which is positive. The possibilities are $(10)(1)$, $(-10)(-1)$, $(5)(2)$, and $(-5)(-2)$. They can be written in either order.

4. We look for combinations of factors from steps (2) and (3) such that the sum of the outside and the inside products is the middle term, $-19x$. The sign of the middle term is negative, but the sign of the last term, 10, is positive. Thus the signs of both factors of the last term, 10, must be negative. From our list of factors in step (3), we can use only -10, -1 and -5, -2 as possibilities. This reduces the possibilities for factorizations by half. We begin by using these factors with $(3x + \boxed{})(2x + \boxed{})$. Should we not find the correct factorization, we will consider $(6x + \boxed{})(x + \boxed{})$.

363

Factor.

3. $24y^2 - 46y + 10$

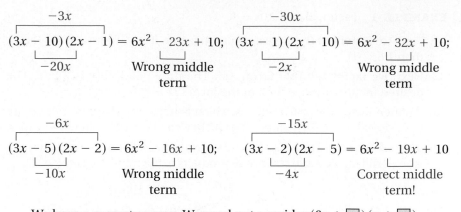

We have a correct answer. We need not consider $(6x + \square)(x + \square)$.

Look again at the possibility $(3x - 1)(2x - 10)$. Without multiplying, we can reject such a possibility, noting that

$$(3x - 1)(2x - 10) = 2(3x - 1)(x - 5).$$

The expression $2x - 10$ has a common factor, 2. But we removed the largest common factor before we began. If this expression were a factorization, then 2 would have to be a common factor along with $3x^4$. Thus, as we saw when we multiplied, $(3x - 1)(2x - 10)$ cannot be part of the factorization of the original trinomial. Given that we factored out the largest common factor at the outset, we can now eliminate factorizations that have a common factor.

The factorization of $6x^2 - 19x + 10$ is $(3x - 2)(2x - 5)$. But do not forget the common factor! We must include it in order to get a complete factorization of the original trinomial:

$$18x^6 - 57x^5 + 30x^4 = 3x^4(3x - 2)(2x - 5).$$

4. $20x^5 - 46x^4 + 24x^3$

5. *Check*: $3x^4(3x - 2)(2x - 5) = 3x^4(6x^2 - 19x + 10)$

$$= 18x^6 - 57x^5 + 30x^4.$$

Here is another tip that might speed up your factoring. Suppose in Example 2 that we considered the possibility

$$(3x + 2)(2x + 5) = 6x^2 + 19x + 10.$$

We might have tried this before noting that using all plus signs would give us a plus sign for the middle term. If we change *both* signs, however, we get the correct answer before including the common factor:

$$(3x - 2)(2x - 5) = 6x^2 - 19x + 10.$$

Do Exercises 3 and 4.

Answers on page A-24

TIPS FOR FACTORING $ax^2 + bx + c$, $a \neq 1$, USING THE FOIL METHOD

1. If the largest common factor has been factored out of the original trinomial, then no binomial factor can have a common factor (other than 1 or -1).
2. **a)** If the signs of all the terms are positive, then the signs of all the terms of the binomial factors are positive.
   b) If a and c are positive and b is negative, then the signs of the factors of c are negative.
   c) If a is positive and c is negative, then the factors of c will have opposite signs.
3. Be systematic about your trials. Keep track of those you have tried and those you have not.
4. Changing the signs of the factors of c will change the sign of the middle term.

Keep in mind that this method of factoring trinomials of the type $ax^2 + bx + c$ involves trial and error. As you practice, you will find that you will need fewer trials to arrive at the factorization.

Do Exercises 5 and 6.

The procedure considered here can also be applied to a trinomial with more than one variable.

EXAMPLE 3   Factor: $30m^2 + 23mn - 11n^2$.

1. First, we factor out the largest common factor, if any. In this polynomial, there is no common factor (other than 1 or -1).
2. Next, we factor the first term, $30m^2$, and get the following possibilities:

   $$30m \cdot m, \quad 15m \cdot 2m, \quad 10m \cdot 3m, \quad \text{and} \quad 6m \cdot 5m.$$

   We then have these as possibilities for factorizations:

   $(30m + \square)(m + \square), \qquad (15m + \square)(2m + \square),$
   $(10m + \square)(3m + \square), \qquad (6m + \square)(5m + \square).$

3. We then factor the last term, $-11n^2$, which is negative. The possibilities are $-11n \cdot n$ and $11n \cdot (-n)$.
4. We look for combinations of factors from steps (2) and (3) such that the sum of the outside and the inside products is the middle term, $23mn$. Since the coefficient of the middle term is positive, let's begin our search using $11n \cdot (-n)$. Should we not find the correct factorization, we will consider $-11n \cdot n$.

   $(30m + 11n)(m - n) = 30m^2 - 19mn - 11n^2;$ ⎫ Note that changing
   $(30m - n)(m + 11n) = 30m^2 + 329mn - 11n^2;$ ⎭ the order of $11n$ and $-n$ changes the middle term.
   $(15m + 11n)(2m - n) = 30m^2 + 7mn - 11n^2;$
   $(15m - n)(2m + 11n) = 30m^2 + 163mn - 11n^2;$
   $(10m + 11n)(3m - n) = 30m^2 + 23mn - 11n^2 \longleftarrow$ Correct middle term

Answers on page A-24

Factor.

7. $21x^2 - 5xy - 4y^2$

We have a correct answer: $30m^2 + 23mn - 11n^2$. The factorization of $30m^2 + 23mn - 11n^2$ is $(10m + 11n)(3m - n)$.

5. *Check:* $(10m + 11n)(3m - n) = (30m^2 + 23mn - 11n^2)$.

Do Exercises 7 and 8.

b The *ac*-Method

The second method of factoring trinomials of the type $ax^2 + bx + c$, $a \neq 1$, is known as the **ac-method,** or the **grouping method.** It involves not only trial and error and FOIL, but also factoring by grouping. This method can cut down on the guesswork of the trials.

We can factor $x^2 + 7x + 10$ by "splitting" the middle term, $7x$, and using factoring by grouping:

$$x^2 + 7x + 10 = x^2 + 2x + 5x + 10$$
$$= x(x + 2) + 5(x + 2)$$
$$= (x + 5)(x + 2).$$

If the leading coefficient is not 1, as in $6x^2 + 23x + 20$, we use a method for factoring similar to what we just did with $x^2 + 7x + 10$.

THE *ac*-METHOD

To factor $ax^2 + bx + c$, $a \neq 1$, using the *ac*-method:

1. Factor out the largest common factor.
2. Multiply the leading coefficient a and the constant c.
3. Try to factor the product ac so that the sum of the factors is b. That is, find integers p and q such that $pq = ac$ and $p + q = b$.
4. Split the middle term. That is, write it as a sum using the factors found in step (3).
5. Factor by grouping.
6. Always check by multiplying.

---

8. $60a^2 + 123ab - 27b^2$

EXAMPLE 4 Factor: $6x^2 + 23x + 20$.

1. First, factor out a common factor, if any. There is none (other than 1 or -1).

2. Multiply the leading coefficient, 6, and the constant, 20: $6 \cdot 20 = 120$.

3. Then look for a factorization of 120 in which the sum of the factors is the coefficient of the middle term, 23. Since both 120 and 23 are positive, we need consider only positive factors of 120.

PAIRS OF FACTORS	SUMS OF FACTORS		PAIRS OF FACTORS	SUMS OF FACTORS
1, 120	121		5, 24	29
2, 60	62		6, 20	26
3, 40	43		8, 15	23
4, 30	34		10, 12	22

Answers on page A-24

4. Next, split the middle term as a sum or a difference using the factors found in step (3):

$$6x^2 + 23x + 20 = 6x^2 + 8x + 15x + 20.$$

5. Factor by grouping as follows:

$$6x^2 + 23x + 20 = 6x^2 + 8x + 15x + 20$$
$$= 2x(3x + 4) + 5(3x + 4) \qquad \text{Factoring by grouping;}$$
$$\text{see Section 4.3}$$
$$= (2x + 5)(3x + 4).$$

We could also split the middle term as $15x + 8x$. We still get the same factorization, although the factors are in a different order:

$$6x^2 + 23x + 20 = 6x^2 + 15x + 8x + 20$$
$$= 3x(2x + 5) + 4(2x + 5)$$
$$= (3x + 4)(2x + 5).$$

6. *Check*: $(3x + 4)(2x + 5) = 6x^2 + 23x + 20.$

Do Exercises 9 and 10.

EXAMPLE 5 Factor: $6x^4 - 116x^3 - 80x^2$.

1. First, factor out the largest common factor, if any. The expression $2x^2$ is common to all three terms: $2x^2(3x^2 - 58x - 40)$.

2. Now, factor the trinomial $3x^2 - 58x - 40$. Multiply the leading coefficient, 3, and the constant, -40: $3(-40) = -120$.

3. Next, try to factor -120 so that the sum of the factors is -58. Since the coefficient of the middle term, -58, is negative, the negative factor of -120 must have the larger absolute value.

PAIRS OF FACTORS	SUMS OF FACTORS	PAIRS OF FACTORS	SUMS OF FACTORS
1, −120	−119	5, −24	−19
2, −60	−58	6, −20	−14
3, −40	−37	8, −15	−7
4, −30	−26	10, −12	−2

4. Split the middle term, $-58x$, as follows: $-58x = 2x - 60x$.

5. Factor by grouping:

$$3x^2 - 58x - 40 = 3x^2 + 2x - 60x - 40 \qquad \text{Substituting } 2x - 60x$$
$$\text{for } -58x$$
$$= x(3x + 2) - 20(3x + 2) \qquad \text{Factoring by grouping}$$
$$= (x - 20)(3x + 2).$$

The factorization of $3x^2 - 58x - 40$ is $(x - 20)(3x + 2)$. But don't forget the common factor! We must include it to get a factorization of the original trinomial:

$$6x^4 - 116x^3 - 80x^2 = 2x^2(x - 20)(3x + 2).$$

6. *Check*: $2x^2(x - 20)(3x + 2) = 2x^2(3x^2 - 58x - 40)$
$$= 6x^4 - 116x^3 - 80x^2.$$

Do Exercises 11 and 12.

Factor by the *ac*-method.

9. $4x^2 + 4x - 3$

10. $4x^2 + 37x + 9$

Factor by the *ac*-method.

11. $10y^4 - 7y^3 - 12y^2$

12. $36a^3 + 21a^2 + a$

Answers on page A-24

By now you have probably encountered certain topics that gave you more difficulty than others. It is important to know that this happens to every person who studies mathematics. Unfortunately, frustration is often part of the learning process and it is important not to give up when difficulty arises.

One source of frustration for many students is not being able to set aside sufficient time for studying. Family commitments, work schedules, and extracurricular activities are just a few of the time demands that many students face. Couple these demands with a math lesson that seems to require a greater than usual amount of study time, and it is no wonder that many students often feel frustrated. Below are some study tips that might be useful if and when troubles arise.

■ **Realize that everyone—even your instructor—has been stumped at times when studying math.** You are not the first person, nor will you be the last, to encounter a "roadblock."

■ **Whether working alone or with a classmate, try to allow enough study time so that you won't need to constantly glance at a clock.** Difficult material is best mastered when your mind is completely focused on the subject matter. Thus, if you are tired, it is usually best to study early the next morning or to take a ten-minute "power-nap" in order to make the most productive use of your time. Consider redoing the weekly planner on the Student Organizer in the Preface. You may need to adjust your schedule frequently. PLAN FOR SUCCESS with extra study time!

■ **Talk about your trouble spot with a classmate.** It is possible that he or she is also having difficulty with the same material. If that is the case, perhaps the majority of your class is confused and your instructor's coverage of the topic is not yet finished. If your classmate *does* understand the topic that is troubling you, patiently allow him or her to explain it to you. By verbalizing the math in question, your classmate may help clarify the material for both of you. Perhaps you will be able to return the favor for your classmate when he or she is struggling with a topic that you understand.

■ **Try to study in a "controlled" environment.** What we mean by this is that you can often put yourself in a setting that will enable you to maximize your powers of concentration. For example, some students may succeed in studying at home or in a dorm room, but for many these settings are filled with distractions. Consider a trip to a library, classroom building, or perhaps the attic or basement if such a setting is more conducive to studying. If you plan on working with a classmate, try to find a location in which conversation will not be bothersome to others.

■ **When working on difficult material, it is often helpful to first "back up" and review the most recent material that did make sense.** This can build your confidence and create a momentum that can often carry you through the roadblock. Sometimes a small piece of information that appeared in a previous section is all that is needed for your problem spot to disappear. When the difficult material is finally mastered, try to make use of what is fresh in your mind by taking a "sneak preview" of what your next topic for study will be.

4.5

EXERCISE SET

For Extra Help

MathXL · MyMathLab · InterAct Math · Math Tutor Center · Digital Video Tutor CD 2 Videotape 5 · Student's Solutions Manual

a , **b** Factor.

1. $3x^2 - 14x - 5$

2. $8x^2 - 6x - 9$

3. $10y^3 + y^2 - 21y$

4. $6x^3 + x^2 - 12x$

5. $3c^2 - 20c + 32$

6. $12b^2 - 8b + 1$

7. $35y^2 + 34y + 8$

8. $9a^2 + 18a + 8$

9. $4t + 10t^2 - 6$

10. $8x + 30x^2 - 6$

11. $8x^2 - 16 - 28x$

12. $18x^2 - 24 - 6x$

13. $18a^2 - 51a + 15$

14. $30a^2 - 85a + 25$

15. $30t^2 + 85t + 25$

16. $18y^2 + 51y + 15$

17. $12x^3 - 31x^2 + 20x$

18. $15x^3 - 19x^2 - 10x$

19. $14x^4 - 19x^3 - 3x^2$

20. $70x^4 - 68x^3 + 16x^2$

21. $3a^2 - a - 4$

22. $6a^2 - 7a - 10$

23. $9x^2 + 15x + 4$

24. $6y^2 - y - 2$

25. $3 + 35z - 12z^2$

26. $8 - 6a - 9a^2$

27. $-4t^2 - 4t + 15$

28. $-12a^2 + 7a - 1$

29. $3x^3 - 5x^2 - 2x$

30. $18y^3 - 3y^2 - 10y$

31. $24x^2 - 2 - 47x$

32. $15y^2 - 10 - 15y$

33. $-8t^3 - 8t^2 + 30t$

34. $-36a^3 + 21a^2 - 3a$

35. $-24x^3 + 2x + 47x^2$

36. $-15y^3 + 10y + 47y^2$

37. $21x^2 + 37x + 12$

38. $10y^2 + 23y + 12$

39. $40x^4 + 16x^2 - 12$

40. $24y^4 + 2y^2 - 15$

41. $12a^2 - 17ab + 6b^2$

42. $20p^2 - 23pq + 6q^2$

43. $2x^2 + xy - 6y^2$

44. $8m^2 - 6mn - 9n^2$

45. $12x^2 - 58xy + 56y^2$

46. $30p^2 + 21pq - 36q^2$

47. $9x^2 - 30xy + 25y^2$

48. $4p^2 + 12pq + 9q^2$

49. $3x^6 + 4x^3 - 4$

50. $2p^8 + 11p^4 + 15$

51. *Height of a Thrown Baseball.* Suppose that a baseball is thrown upward with an initial velocity of 80 ft/sec from a height of 224 ft. Its height h after t seconds is given by the function

$$h(t) = -16t^2 + 80t + 224.$$

a) What is the height of the ball after 0 sec? 1 sec? 3 sec? 4 sec? 6 sec?

b) Find an equivalent expression for $h(t)$ by factoring.

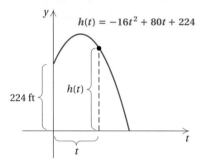

52. *Fireworks.* Suppose that a bottle rocket is launched upward with an initial velocity of 96 ft/sec and from a height of 880 ft. Its height h after t seconds is given by the function

$$h(t) = -16t^2 + 96t + 880.$$

a) What is the height of the bottle rocket after 0 sec? 1 sec? 3 sec? 8 sec? 10 sec?

b) Find an equivalent expression for $h(t)$ by factoring.

53. $\mathbf{D_W}$ Explain how to use the *ac*-method to factor trinomials of the type $ax^2 + bx + c, a \neq 1$.

54. $\mathbf{D_W}$ Explain how to use the FOIL method to factor trinomials of the type $ax^2 + bx + c, a \neq 1$.

SKILL MAINTENANCE

Solve. [3.5a]

55. $x + 2y - z = 0,$
$4x + 2y + 5z = 6,$
$2x - y + z = 5$

56. $2x + y + 2z = 5,$
$4x - 2y - 3z = 5,$
$-8x - y + z = -5$

57. $2x + 9y + 6z = 5,$
$x - y + z = 4,$
$3x + 2y + 3z = 7$

58. $x - 3y + 2z = -8,$
$2x + 3y + z = 17,$
$5x - 2y + 3z = 5$

Determine whether the graphs of the given pairs of lines are parallel or perpendicular. [2.5d]

59. $y - 2x = 18,$
$2x - 7 = y$

60. $21x + 7 = -3y,$
$y + 7x = -9$

61. $2x + 5y = 4,$
$2x - 5y = -3$

62. $y + x = 7,$
$y - x = 3$

Find an equation of the line containing the given pair of points. [2.6c]

63. $(-2, -3)$ and $(5, -4)$

64. $(2, -3)$ and $(5, -4)$

65. $(-10, 3)$ and $(7, -4)$

66. $\left(-\frac{2}{3}, 1\right)$ and $\left(\frac{4}{3}, -4\right)$

SYNTHESIS

67. Use the TABLE and GRAPH features of a graphing calculator to check your answers to Exercises 2, 17, and 28.

68. Use the TABLE and GRAPH features of a graphing calculator to check your answers to Exercises 4, 11, and 32.

Factor. Assume that variables in exponents represent positive integers.

69. $p^2q^2 + 7pq + 12$

70. $2x^4y^6 - 3x^2y^3 - 20$

71. $x^2 - \frac{4}{25} + \frac{3}{5}x$

72. $y^2 - \frac{8}{49} + \frac{2}{7}y$

73. $y^2 + 0.4y - 0.05$

74. $t^2 + 0.6t - 0.27$

75. $7a^2b^2 + 6 + 13ab$

76. $9x^2y^2 - 4 + 5xy$

77. $3x^2 + 12x - 495$

78. $15t^3 - 60t^2 - 315t$

79. $216x + 78x^2 + 6x^3$

80. $\frac{1}{4}p^2 - \frac{2}{5}p + \frac{4}{25}$

81. $x^{2a} + 5x^a - 24$

82. $4x^{2a} - 4x^a - 3$

4.6 SPECIAL FACTORING

Objectives

a Factor trinomial squares.

b Factor differences of squares.

c Factor certain polynomials with four terms by grouping and possibly using the factoring of a trinomial square or the difference of squares.

d Factor sums and differences of cubes.

In this section, we consider some special factoring methods. When we recognize certain types of polynomials, we can factor more quickly using these special methods. Most of them are the reverse of the methods of special multiplication.

a Trinomial Squares

Consider the trinomial $x^2 + 6x + 9$. To factor it, we can use the method considered in Section 4.4. We look for factors of 9 whose sum is 6. We see that these factors are 3 and 3 and the factorization is

$$x^2 + 6x + 9 = (x + 3)(x + 3) = (x + 3)^2.$$

Note that the result is the square of a binomial. We also call $x^2 + 6x + 9$ a **trinomial square,** or **perfect-square trinomial.** We can certainly use the procedures of Sections 4.4 and 4.5 to factor trinomial squares, but we want to develop an even faster procedure. In order to do so, we must first be able to recognize when a trinomial is a square.

> **How to recognize a trinomial square:**
>
> **a)** The two expressions A^2 and B^2 must be squares.
> **b)** There must be no minus sign before either A^2 or B^2.
> **c)** Multiplying A and B (expressions whose squares are A^2 and B^2) and doubling the result gives either the remaining term, $2AB$, or its opposite, $-2AB$.

EXAMPLES Determine whether the polynomial is a trinomial square.

1. $x^2 + 10x + 25$

 a) Two terms are squares: x^2 and 25.

 b) There is no minus sign before either x^2 or 25.

 c) If we multiply the expressions whose squares are x^2 and 25, x and 5, and double the product, we get $10x$, the remaining term.

 Thus this is a trinomial square.

2. $4x + 16 + 3x^2$

 a) Only one term, 16, is a square ($3x^2$ is not a square because 3 is not a perfect square and $4x$ is not a square because x is not a square).

 Thus this is not a trinomial square.

3. $100y^2 + 81 - 180y$

 (It can help to first write this in descending order: $100y^2 - 180y + 81$.)

 a) Two of the terms, $100y^2$ and 81, are squares.

 b) There is no minus sign before either $100y^2$ or 81.

 c) If we multiply the expressions whose squares are $100y^2$ and 81, $10y$ and 9, and double the product, we get the opposite of the remaining term: $2(10y)(9) = 180y$, which is the opposite of $-180y$.

 Thus this is a trinomial square.

1. Which of the following are trinomial squares?

 a) $x^2 + 6x + 9$

 b) $x^2 - 8x + 16$

 c) $x^2 + 6x + 11$

 d) $4x^2 + 25 - 20x$

 e) $16x^2 - 20x + 25$

 f) $16 + 14x + 5x^2$

 g) $x^2 + 8x - 16$

 h) $x^2 - 8x - 16$

Answer on page A-24

Factor.

2. $x^2 + 14x + 49$

3. $9y^2 - 30y + 25$

4. $16x^2 + 72xy + 81y^2$

5. $16x^4 - 40x^2y^3 + 25y^6$

Factor.

6. $-8a^2 + 24ab - 18b^2$

7. $3a^2 - 30ab + 75b^2$

Do Exercise 1 on the preceding page.

The factors of a trinomial square are two identical binomials. We use the following equations.

TRINOMIAL SQUARES
$A^2 + 2AB + B^2 = (A + B)^2$; $A^2 - 2AB + B^2 = (A - B)^2$

EXAMPLE 4   Factor: $x^2 - 10x + 25$.

$$x^2 - 10x + 25 = (x - 5)^2$$

Note the sign!

We find the square terms and write their square roots with a minus sign between them.

EXAMPLE 5   Factor: $16y^2 + 49 + 56y$.

$$16y^2 + 49 + 56y = 16y^2 + 56y + 49 \qquad \text{Rewriting in descending order}$$
$$= (4y + 7)^2$$

We find the square terms and write their square roots with a plus sign between them.

EXAMPLE 6   Factor: $-20xy + 4y^2 + 25x^2$.

We have

$$-20xy + 4y^2 + 25x^2 = 4y^2 - 20xy + 25x^2 \qquad \text{Writing descending order in } y$$
$$= (2y - 5x)^2.$$

This square can also be expressed as

$$25x^2 - 20xy + 4y^2 = (5x - 2y)^2.$$

Do Exercises 2–5.

In factoring, we must always remember to look *first* for the largest factor common to all the terms.

EXAMPLE 7   Factor: $2x^2 - 12xy + 18y^2$.

Always remember to look first for a common factor. This time the largest common factor is 2.

$$2x^2 - 12xy + 18y^2 = 2(x^2 - 6xy + 9y^2) \qquad \text{Removing the common factor 2}$$
$$= 2(x - 3y)^2 \qquad \text{Factoring the trinomial square}$$

EXAMPLE 8   Factor: $-4y^2 - 144y^8 + 48y^5$.

$$-4y^2 - 144y^8 + 48y^5$$
$$= -4y^2(1 + 36y^6 - 12y^3) \qquad \text{Removing the common factor } -4y^2$$
$$= -4y^2(1 - 12y^3 + 36y^6) \qquad \text{Changing order}$$
$$= -4y^2(1 - 6y^3)^2 \qquad \text{Factoring the trinomial square}$$

Do Exercises 6 and 7.

Answers on page A-24

CHAPTER 4: Polynomials and
Polynomial Functions

b Differences of Squares

The following are *differences of squares*:

$$x^2 - 9, \qquad 49 - 4y^2, \qquad a^2 - 49b^2.$$

To factor a difference of two expressions that are squares, we can use a pattern for multiplying a sum and a difference that we used earlier.

DIFFERENCE OF TWO SQUARES

To factor a difference of two squares, write the square root of the first expression *plus* the square root of the second *times* the square root of the first *minus* the square root of the second.

$$A^2 - B^2 = (A + B)(A - B)$$

You should memorize this rule, and then say it as you work.

EXAMPLE 9 Factor: $x^2 - 9$.

$$\begin{array}{c} A^2 - B^2 = (A + B)(A - B) \\ \downarrow \quad \downarrow \quad \quad \downarrow \quad \downarrow \quad \downarrow \quad \downarrow \\ x^2 - 9 = x^2 - 3^2 = (x + 3)(x - 3) \end{array}$$

EXAMPLE 10 Factor: $25y^6 - 49x^2$.

$$\begin{array}{c} A^2 \quad - \quad B^2 \quad = (\ A \ + \ B)(A \ - \ B) \\ \downarrow \quad\quad \downarrow \quad\quad \downarrow \quad \downarrow \quad \downarrow \quad\quad \downarrow \\ 25y^6 - 49x^2 = (5y^3)^2 - (7x)^2 = (5y^3 + 7x)(5y^3 - 7x) \end{array}$$

EXAMPLE 11 Factor: $x^2 - \frac{1}{16}$.

$$x^2 - \tfrac{1}{16} = x^2 - \left(\tfrac{1}{4}\right)^2 = \left(x + \tfrac{1}{4}\right)\left(x - \tfrac{1}{4}\right)$$

Do Exercises 8–10.

Common factors should always be removed. Removing common factors actually eases the factoring process because the type of factoring to be done becomes clearer.

EXAMPLE 12 Factor: $5 - 5x^2y^6$.

There is a common factor, 5.

$5 - 5x^2y^6 = 5(1 - x^2y^6)$	Removing the common factor 5
$= 5[1^2 - (xy^3)^2]$	Recognizing the difference of squares; $x^2y^6 = (x^1y^3)^2 = (xy^3)^2$
$= 5(1 + xy^3)(1 - xy^3)$	Factoring the difference of squares

EXAMPLE 13 Factor: $2x^4 - 8y^4$.

There is a common factor, 2.

$2x^4 - 8y^4 = 2(x^4 - 4y^4)$	Removing the common factor 2
$= 2[(x^2)^2 - (2y^2)^2]$	Recognizing the difference of squares
$= 2(x^2 + 2y^2)(x^2 - 2y^2)$	Factoring the difference of squares

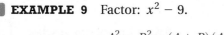

Factor.

8. $y^2 - 4$

9. $49x^4 - 25y^{10}$

10. $m^2 - \dfrac{1}{9}$

Answers on page A-24

Factor.

11. $25x^2y^2 - 4a^2$

12. $9x^2 - 16y^2$

13. $20x^2 - 5y^2$

14. $81x^4y^2 - 16y^2$

15. Factor: $a^3 + a^2 - 16a - 16$.

EXAMPLE 14   Factor: $16x^4y - 81y$.

There is a common factor, y.

$$
\begin{aligned}
16x^4y - 81y &= y(16x^4 - 81) && \text{Removing the common factor } y \\
&= y[(4x^2)^2 - 9^2] \\
&= y(4x^2 + 9)(4x^2 - 9) && \text{Factoring the difference of squares} \\
&= y(4x^2 + 9)(2x + 3)(2x - 3) && \text{Factoring } 4x^2 - 9, \text{ which is} \\
& && \text{also a difference of squares}
\end{aligned}
$$

In Example 14, it may be tempting to try to factor $4x^2 + 9$. Note that it is a sum of two expressions that are squares, but it cannot be factored further. Also note that one of the factors, $4x^2 - 9$, could be factored further. Whenever that is possible, you should do so. That way you will be factoring *completely*.

Caution!

If the greatest common factor has been removed, then you cannot factor a sum of squares further. In particular,

$$A^2 + B^2 \neq (A + B)^2.$$

Consider $25x^2 + 225$. This is a case in which we have a sum of squares, but there is a common factor, 25. Factoring, we get $25(x^2 + 9)$. Now $x^2 + 9$ cannot be factored further.

Do Exercises 11–14.

C More Factoring by Grouping

Sometimes when factoring a polynomial with four terms completely, we might get a factor that can be factored further using other methods we have learned.

EXAMPLE 15   Factor completely: $x^3 + 3x^2 - 4x - 12$.

$$
\begin{aligned}
x^3 + 3x^2 - 4x - 12 &= x^2(x + 3) - 4(x + 3) \\
&= (x^2 - 4)(x + 3) \\
&= (x + 2)(x - 2)(x + 3)
\end{aligned}
$$

Do Exercise 15.

A difference of squares can have more than two terms. For example, one of the squares may be a trinomial. We can factor by a type of grouping.

EXAMPLE 16   Factor completely: $x^2 + 6x + 9 - y^2$.

$$
\begin{aligned}
x^2 + 6x + 9 - y^2 &= (x^2 + 6x + 9) - y^2 && \text{Grouping as a} \\
& && \text{trinomial minus } y^2 \\
& && \text{to show a difference} \\
& && \text{of squares} \\
&= (x + 3)^2 - y^2 \\
&= (x + 3 + y)(x + 3 - y)
\end{aligned}
$$

Answers on page A-24

Factor completely.

16. $x^2 + 2x + 1 - p^2$

d   Sums or Differences of Cubes

We can factor the sum or the difference of two expressions that are cubes.
    Consider the following products:

$$(A + B)(A^2 - AB + B^2) = A(A^2 - AB + B^2) + B(A^2 - AB + B^2)$$
$$= A^3 - A^2B + AB^2 + A^2B - AB^2 + B^3$$
$$= A^3 + B^3$$

17. $y^2 - 8y + 16 - 9m^2$

and $(A - B)(A^2 + AB + B^2) = A(A^2 + AB + B^2) - B(A^2 + AB + B^2)$
$$= A^3 + A^2B + AB^2 - A^2B - AB^2 - B^3$$
$$= A^3 - B^3.$$

The above equations (reversed) show how we can factor a sum or a difference of two cubes.

18. $x^2 + 8x + 16 - 100t^2$

SUM OR DIFFERENCE OF CUBES

$A^3 + B^3 = (A + B)(A^2 - AB + B^2);$
$A^3 - B^3 = (A - B)(A^2 + AB + B^2)$

---

19. $64p^2 - (x^2 + 8x + 16)$

    Note that what we are considering here is a sum or a difference of cubes. We are not cubing a binomial. For example, $(A + B)^3$ is *not* the same as $A^3 + B^3$. The table of cubes in the margin is helpful.

EXAMPLE 17   Factor: $x^3 - 27$.
    We have

Answers on page A-24

$$\overset{\displaystyle A^3 \quad\; B^3}{\underset{\displaystyle x^3 - 27 = x^3 - 3^3.}{\downarrow \quad\;\; \downarrow}}$$

In one set of parentheses, we write the cube root of the first term, x. Then we write the cube root of the second term, -3. This gives us the expression $x - 3$:

$$(x - 3)(\qquad\qquad).$$

To get the next factor, we think of $x - 3$ and do the following:

     Square the first term: $x \cdot x = x^2$.
     Multiply the terms, $x(-3) = -3x$, and then change the sign: $3x$.
     Square the second term: $(-3)^2 = 9$.

$$(x - 3)(x^2 + 3x + 9).$$
$$(A - B)(A^2 + AB + B^2)$$

Note that we cannot factor $x^2 + 3x + 9$. It is not a trinomial square nor can it be factored by trial and error. Check this on your own.

N	N^3
0.2	0.008
0.1	0.001
0	0
1	1
2	8
3	27
4	64
5	125
6	216
7	343
8	512
9	729
10	1000

Factor.

20. $x^3 - 8$

21. $64 - y^3$

Factor.

22. $27x^3 + y^3$

23. $8y^3 + z^3$

Do Exercises 20 and 21.

> ■ **EXAMPLE 18** Factor: $125x^3 + y^3$.
>
> We have
>
> $$125x^3 + y^3 = (5x)^3 + y^3.$$
>
> In one set of parentheses, we write the cube root of the first term, $5x$. Then we write a plus sign, and then the cube root of the second term, y:
>
> $$(5x + y)(\qquad\qquad).$$
>
> To get the next factor, we think of $5x + y$ and do the following:
>
> Square the first term: $(5x)(5x) = 25x^2$.
>
> Multiply the terms, $5x \cdot y = 5xy$, and then change the sign: $-5xy$.
>
> Square the second term: $y \cdot y = y^2$.
>
> $$(5x + y)(25x^2 - 5xy + y^2).$$
> $$(A + B)(A^2 - AB + B^2)$$

Do Exercises 22 and 23.

> ■ **EXAMPLE 19** Factor: $128y^7 - 250x^6y$.
>
> We first look for the largest common factor:
>
> $$\begin{aligned} 128y^7 - 250x^6y &= 2y(64y^6 - 125x^6) \\ &= 2y[(4y^2)^3 - (5x^2)^3] \\ &= 2y(4y^2 - 5x^2)(16y^4 + 20x^2y^2 + 25x^4). \end{aligned}$$

> ■ **EXAMPLE 20** Factor: $a^6 - b^6$.
>
> We can express this polynomial as a difference of squares:
>
> $$a^6 - b^6 = (a^3)^2 - (b^3)^2.$$
>
> We factor as follows:
>
> $$a^6 - b^6 = (a^3 + b^3)(a^3 - b^3).$$
>
> One factor is a sum of two cubes, and the other factor is a difference of two cubes. We factor them:
>
> $$a^6 - b^6 = (a + b)(a^2 - ab + b^2)(a - b)(a^2 + ab + b^2).$$
>
> We have now factored completely.

In Example 20, had we thought of factoring first as a difference of two cubes, we would have had

$$\begin{aligned} (a^2)^3 - (b^2)^3 &= (a^2 - b^2)(a^4 + a^2b^2 + b^4) \\ &= (a + b)(a - b)(a^4 + a^2b^2 + b^4). \end{aligned}$$

In this case, we might have missed some factors; $a^4 + a^2b^2 + b^4$ can be factored as $(a^2 - ab + b^2)(a^2 + ab + b^2)$, but we probably would not have known to do such factoring.

Answers on page A-24

EXAMPLE 21 Factor: $64a^6 - 729b^6$.

We have

$$64a^6 - 729b^6 = (8a^3)^2 - (27b^3)^2$$

$$= (8a^3 - 27b^3)(8a^3 + 27b^3) \quad \text{Factoring a difference of squares}$$

$$= [(2a)^3 - (3b)^3][(2a)^3 + (3b)^3].$$

Each factor is a sum or a difference of cubes. We factor each:

$$= (2a - 3b)(4a^2 + 6ab + 9b^2)(2a + 3b)(4a^2 - 6ab + 9b^2).$$

FACTORING SUMMARY

Sum of cubes:	$A^3 + B^3 = (A + B)(A^2 - AB + B^2)$;
Difference of cubes:	$A^3 - B^3 = (A - B)(A^2 + AB + B^2)$;
Difference of squares:	$A^2 - B^2 = (A + B)(A - B)$;
Sum of squares:	$A^2 + B^2$ cannot be factored using real numbers if the largest common factor has been removed.

Do Exercises 24–27.

Factor.

24. $m^6 - n^6$

25. $16x^7y + 54xy^7$

26. $729x^6 - 64y^6$

27. $x^3 - 0.027$

Answers on page A-24

Study Tips

BETTER TEST TAKING

How often do you make the following statement after taking a test: "I was able to do the homework, but I froze during the test"? This can be an excuse for poor study habits. Here are two tips to help you with this difficulty. Both are intended to make test taking less stressful by getting you to practice good test-taking habits on a daily basis.

■ **Treat every homework exercise as if it were a test question.** If you had to work a problem at your job with no backup answer provided, what would you do? You would probably work it very deliberately, checking and rechecking every step. You might work it more than one time, or you might try to work it another way to check the result. Try to use this approach when doing your homework. Treat every exercise as though it were a test question with no answer at the back of the book.

■ **Be sure that you do questions without answers as part of every homework assignment whether or not the instructor has assigned them!** One reason a test may seem such a different task is that questions on a test lack answers. That is the reason for taking a test: to see if you can do the questions without assistance. As part of your test preparation, be sure you do some exercises for which you do not have the answers. Thus when you take a test, you are doing a more familiar task.

The purpose of doing your homework using these approaches is to give you more test-taking practice beforehand. Let's use a sports analogy: At a basketball game, the players take lots of practice shots before the game. They play the first half, go to the locker room, and come out for the second half. What do they do before the second half, even though they have just played 20 minutes of basketball? They shoot baskets again! We suggest the same approach here. Create more and more situations in which you practice taking test questions by treating each homework exercise like a test question and by doing exercises for which you have no answers. Good luck!

"He who does not venture has no luck."

Mexican proverb

Visualizing for Success

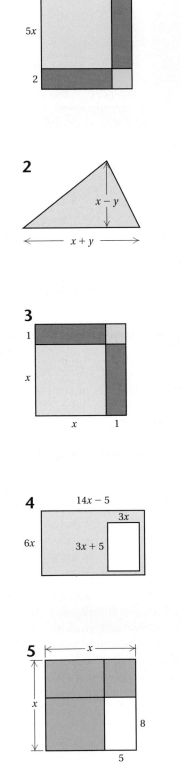

1
5x   2
5x
2

2
x − y
x + y

3
1
x
x   1

4
14x − 5
3x
6x   3x + 5

5
x
x
8
5

In each of Exercises 1–10, find two algebraic expressions from the list below for the shaded area of the figure.

A. $(5x + 2)^2$

B. $13x$

C. $400 − 4x^2$

D. $x^2 − (x − 2y)^2$

E. $25x^2 + 20x + 4$

F. $\frac{1}{2}(x^2 − y^2)$

G. $(x + 1)^2$

H. $4y(x − y)$

I. $4(10 − x)(10 + x)$

J. $\frac{1}{2}(x − y)(x + y)$

K. $x^2 + 2x + 1$

L. $6x(14x − 5) − 3x(3x + 5)$

M. $x^2 + 9x + 20$

N. $(x − 2)^2$

O. $(x + 4)(x + 5)$

P. $8(x − 5) + (x − 5)(x − 8) + 5(x − 8)$

Q. $x^2 − 40$

R. $5x + 8x$

S. $15x(5x − 3)$

T. $x^2 − 4x + 4$

Answers on page A-24

6
20
x
x
20
x
x

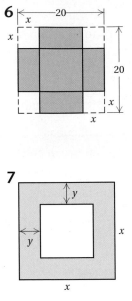

7
y
x
y
x
x

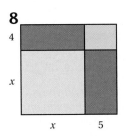

8
4
x
x   5

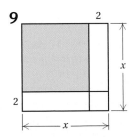

9
2
x
2
x

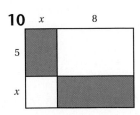

10
x   8
5
x

4.6

EXERCISE SET

For Extra Help

MathXL MyMathLab InterAct   Math Tutor   Digital Video   Student's
                    Math   Center   Tutor CD 2   Solutions
                                           Videotape 5   Manual

a Factor.

1. $x^2 - 4x + 4$

2. $y^2 - 16y + 64$

3. $y^2 + 18y + 81$

4. $x^2 + 8x + 16$

5. $x^2 + 1 + 2x$

6. $x^2 + 1 - 2x$

7. $9y^2 + 12y + 4$

8. $25x^2 - 60x + 36$

9. $-18y^2 + y^3 + 81y$

10. $24a^2 + a^3 + 144a$

11. $12a^2 + 36a + 27$

12. $20y^2 + 100y + 125$

13. $2x^2 - 40x + 200$

14. $32x^2 + 48x + 18$

15. $1 - 8d + 16d^2$

16. $64 + 25y^2 - 80y$

17. $y^4 - 8y^2 + 16$

18. $y^4 - 18y^2 + 81$

19. $0.25x^2 + 0.30x + 0.09$

20. $0.04x^2 - 0.28x + 0.49$

21. $p^2 - 2pq + q^2$

22. $m^2 + 2mn + n^2$

23. $a^2 + 4ab + 4b^2$

24. $49p^2 - 14pq + q^2$

25. $25a^2 - 30ab + 9b^2$

26. $49p^2 - 84pq + 36q^2$

27. $y^6 + 26y^3 + 169$

28. $p^6 - 10p^3 + 25$

29. $16x^{10} - 8x^5 + 1$

30. $9x^{10} + 12x^5 + 4$

31. $x^4 + 2x^2y^2 + y^4$

32. $p^6 - 2p^3q^4 + q^8$

b Factor.

33. $x^2 - 16$

34. $y^2 - 9$

35. $p^2 - 49$

36. $m^2 - 64$

37. $p^2q^2 - 25$

38. $a^2b^2 - 81$

39. $6x^2 - 6y^2$

40. $8x^2 - 8y^2$

41. $4xy^4 - 4xz^4$

42. $25ab^4 - 25az^4$

43. $4a^3 - 49a$

44. $9x^3 - 25x$

45. $3x^8 - 3y^8$

46. $2a^9 - 32a$

47. $9a^4 - 25a^2b^4$

48. $16x^6 - 121x^2y^4$

49. $\frac{1}{36} - z^2$

50. $\frac{1}{100} - y^2$

51. $0.04x^2 - 0.09y^2$

52. $0.01x^2 - 0.04y^2$

c   Factor.

53. $m^3 - 7m^2 - 4m + 28$

54. $x^3 + 8x^2 - x - 8$

55. $a^3 - ab^2 - 2a^2 + 2b^2$

56. $p^2q - 25q + 3p^2 - 75$

57. $(a + b)^2 - 100$

58. $(p - 7)^2 - 144$

59. $144 - (p - 8)^2$

60. $100 - (x - 4)^2$

61. $a^2 + 2ab + b^2 - 9$

62. $x^2 - 2xy + y^2 - 25$

63. $r^2 - 2r + 1 - 4s^2$

64. $c^2 + 4cd + 4d^2 - 9p^2$

65. $2m^2 + 4mn + 2n^2 - 50b^2$

66. $12x^2 + 12x + 3 - 3y^2$

67. $9 - (a^2 + 2ab + b^2)$

68. $16 - (x^2 - 2xy + y^2)$

d   Factor.

69. $z^3 + 27$

70. $a^3 + 8$

71. $x^3 - 1$

72. $c^3 - 64$

73. $y^3 + 125$

74. $x^3 + 1$

75. $8a^3 + 1$

76. $27x^3 + 1$

CHAPTER 4: Polynomials and
Polynomial Functions

77. $y^3 - 8$

78. $p^3 - 27$

79. $8 - 27b^3$

80. $64 - 125x^3$

81. $64y^3 + 1$

82. $125x^3 + 1$

83. $8x^3 + 27$

84. $27y^3 + 64$

85. $a^3 - b^3$

86. $x^3 - y^3$

87. $a^3 + \frac{1}{8}$

88. $b^3 + \frac{1}{27}$

89. $2y^3 - 128$

90. $3z^3 - 3$

91. $24a^3 + 3$

92. $54x^3 + 2$

93. $rs^3 + 64r$

94. $ab^3 + 125a$

95. $5x^3 - 40z^3$

96. $2y^3 - 54z^3$

97. $x^3 + 0.001$

98. $y^3 + 0.125$

99. $64x^6 - 8t^6$

100. $125c^6 - 8d^6$

101. $2y^4 - 128y$

102. $3z^5 - 3z^2$

103. $z^6 - 1$

104. $t^6 + 1$

105. $t^6 + 64y^6$

106. $p^6 - q^6$

107. $\mathbf{D_W}$ Under what conditions, if any, can the sum of two squares be factored? Explain.

108. $\mathbf{D_W}$ Explain how you could use factoring or graphing to explain why $x^3 - 8 \neq (x - 2)^3$.

Solve.   [3.2a], [3.3a]

109. $7x - 2y = -11,$
$2x + 7y = 18$

110. $y = 3x - 8,$
$4x - 6y = 100$

111. $x - y = -12,$
$x + y = 14$

112. $7x - 2y = -11,$
$2y - 7x = -18$

Graph the given system of inequalities and determine coordinates of any vertices formed.   [3.7c]

113. $x - y \le 5,$
$x + y \ge 3$

114. $x - y \le 5,$
$x + y \ge 3,$
$x \le 6$

115. $x - y \ge 5,$
$x + y \le 3,$
$x \ge 1$

116. $x - y \ge 5,$
$x + y \le 3$

Given the line and a point not on the line, find an equation through the point parallel to the given line, and find an equation through the point perpendicular to the given line.   [2.6d]

117. $x - y = 5; (-2, -4)$

118. $2x - 3y = 6; (1, -7)$

119. $y = -\frac{1}{2}x + 3; (4, 5)$

120. $x - 4y = -10; (6, 0)$

121. Given that $P(x) = x^3$, use factoring to simplify $P(a + h) - P(a)$.

122. Given that $P(x) = x^4$, use factoring to simplify $P(a + h) - P(a)$.

123. Show how the geometric model below can be used to verify the formula for factoring $a^3 - b^3$.

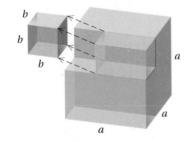

124. *Volume of Carpeting.*   The volume of a carpet that is rolled up can be estimated by the polynomial $\pi R^2 h - \pi r^2 h$.

a) Factor the polynomial.
b) Use both the original and the factored forms to find the volume of a roll for which $R = 50$ cm, $r = 10$ cm, and $h = 4$ m. Use 3.14 for π.

Factor. Assume that variables in exponents represent positive integers.

125. $5c^{100} - 80d^{100}$

126. $9x^{2n} - 6x^n + 1$

127. $x^{6a} + y^{3b}$

128. $a^3x^3 - b^3y^3$

129. $3x^{3a} + 24y^{3b}$

130. $\frac{8}{27}x^3 + \frac{1}{64}y^3$

131. $\frac{1}{24}x^3y^3 + \frac{1}{3}z^3$

132. $7x^3 - \frac{7}{8}$

133. $(x + y)^3 - x^3$

134. $(1 - x)^3 + (x - 1)^6$

135. $(a + 2)^3 - (a - 2)^3$

136. $y^4 - 8y^3 - y + 8$

4.7

FACTORING: A GENERAL STRATEGY

Objective

a Factor polynomials completely using any of the methods considered in this chapter.

a   A General Factoring Strategy

Factoring is an important algebraic skill, used for solving equations and many other manipulations of algebraic symbolism. We now consider polynomials of many types and learn to use a general strategy for factoring. The key is to recognize the type of polynomial to be factored.

> **A STRATEGY FOR FACTORING**
>
> **a)** Always look for a *common factor* (other than 1 or −1). If there are any, factor out the largest one.
>
> **b)** Then look at the number of terms.
>
> > *Two terms*: Try factoring as a difference of squares first. Next, try factoring as a sum or a difference of cubes. Do *not* try to factor a *sum* of squares: $A^2 + B^2$.
> >
> > *Three terms*: Determine whether the expression is a trinomial square. If it is, you know how to factor. If not, try the trial-and-error method or the *ac*-method.
> >
> > *Four or more terms*: Try factoring by grouping and removing a common binomial factor. Next, try grouping into a difference of squares, one of which is a trinomial.
>
> **c)** Always *factor completely*. If a factor with more than one term can be factored, you should factor it.
>
> **d)** Always *check* by multiplying.

EXAMPLE 1   Factor: $10a^2x - 40b^2x$.

a) We look first for a common factor:

$$10x(a^2 - 4b^2). \qquad \text{Factoring out the largest common factor}$$

b) The factor $a^2 - 4b^2$ has only two terms. It is a difference of squares. We factor it, keeping the common factor: $10x(a + 2b)(a - 2b)$.

c) Have we factored completely? Yes, because none of the factors with more than one term can be factored further using polynomials of smaller degree.

d) *Check*: $10x(a + 2b)(a - 2b) = 10x(a^2 - 4b^2) = 10xa^2 - 40xb^2$, or $10a^2x - 40b^2x$

EXAMPLE 2   Factor: $x^6 - y^6$.

a) We look for a common factor. There isn't one (other than 1 or −1).

b) There are only two terms. It is a difference of squares: $(x^3)^2 - (y^3)^2$. We factor it: $(x^3 + y^3)(x^3 - y^3)$. One factor is a sum of two cubes, and the other factor is a difference of two cubes. We factor them:

$$x^6 - y^6 = \underbrace{(x^3 + y^3)}\,\underbrace{(x^3 - y^3)}$$

$$= \overbrace{(x + y)(x^2 - xy + y^2)}\,\overbrace{(x - y)(x^2 + xy + y^2)}.$$

Factor completely.

1. $3y^3 - 12x^2y$

2. $7a^3 - 7$

3. $64x^6 - 729y^6$

c) We have factored completely because none of the factors can be factored further using polynomials of smaller degree.

d) *Check:* $(x + y)(x^2 - xy + y^2)(x - y)(x^2 + xy + y^2)$
$$= (x^3 + y^3)(x^3 - y^3)$$
$$= (x^3)^2 - (y^3)^2 = x^6 - y^6.$$

Do Exercises 1–3.

 EXAMPLE 3 Factor: $10x^6 + 40y^2$.

a) We remove the largest common factor: $10(x^6 + 4y^2)$.

b) In the parentheses, there are two terms, a sum of squares, which cannot be factored.

c) We have factored $10x^6 + 40y^2$ completely as $10(x^6 + 4y^2)$.

d) *Check:* $10(x^6 + 4y^2) = 10x^6 + 40y^2$.

 EXAMPLE 4 Factor: $2x^2 + 50a^2 - 20ax$.

a) We remove the largest common factor: $2(x^2 + 25a^2 - 10ax)$.

b) In the parentheses, there are three terms. The trinomial is a square. We factor it: $2(x - 5a)^2$.

c) None of the factors with more than one term can be factored further.

d) *Check:* $2(x - 5a)^2 = 2(x^2 - 10ax + 25a^2) = 2x^2 - 20ax + 50a^2$, or $2x^2 + 50a^2 - 20ax$.

 EXAMPLE 5 Factor: $6x^2 - 20x - 16$.

a) We remove the largest common factor: $2(3x^2 - 10x - 8)$.

b) In the parentheses, there are three terms. The trinomial is not a square. We factor: $2(x - 4)(3x + 2)$.

c) We cannot factor further.

d) *Check:* $2(x - 4)(3x + 2) = 2(3x^2 - 10x - 8) = 6x^2 - 20x - 16$.

 EXAMPLE 6 Factor: $3x + 12 + ax^2 + 4ax$.

a) There is no common factor (other than 1 or -1).

b) There are four terms. We try grouping to remove a common binomial factor:

$$3(x + 4) + ax(x + 4) \qquad \text{Factoring two grouped binomials}$$
$$= (3 + ax)(x + 4). \qquad \text{Removing the common binomial factor}$$

c) None of the factors with more than one term can be factored further.

d) *Check:* $(3 + ax)(x + 4) = 3x + 12 + ax^2 + 4ax$.

Answers on page A-25

EXAMPLE 7   Factor: $y^2 - 9a^2 + 12y + 36$.

a) There is no common factor (other than 1 or -1).

b) There are four terms. We try grouping to remove a common binomial factor, but that is not possible. We try grouping as a difference of squares:

$$(y^2 + 12y + 36) - 9a^2 = (y + 6)^2 - (3a)^2$$
$$= (y + 6 + 3a)(y + 6 - 3a). \quad \text{Factoring the difference of squares}$$

c) No factor with more than one term can be factored further.

d) *Check*: $(y + 6 + 3a)(y + 6 - 3a) = [(y + 6) + 3a][(y + 6) - 3a]$
$$= (y + 6)^2 - (3a)^2$$
$$= y^2 + 12y + 36 - 9a^2, \text{ or}$$
$$y^2 - 9a^2 + 12y + 36.$$

EXAMPLE 8   Factor: $x^3 - xy^2 + x^2y - y^3$.

a) There is no common factor (other than 1 or -1).

b) There are four terms. We try grouping to remove a common binomial factor:

$$x(x^2 - y^2) + y(x^2 - y^2) \quad \text{Factoring two grouped binomials}$$
$$= (x + y)(x^2 - y^2). \quad \text{Removing the common binomial factor}$$

c) The factor $x^2 - y^2$ can be factored further, giving

$$(x + y)(x + y)(x - y). \quad \text{Factoring a difference of squares}$$

None of the factors with more than one term can be factored further, so we have factored completely.

d) *Check*: $(x + y)(x + y)(x - y) = (x + y)(x^2 - y^2)$
$$= x^3 - xy^2 + x^2y - y^3.$$

Do Exercises 4–9.

Factor.

4. $3x - 6 - bx^2 + 2bx$

5. $5y^4 + 20x^6$

6. $6x^2 - 3x - 18$

7. $a^3 - ab^2 - a^2b + b^3$

8. $3x^2 + 18ax + 27a^2$

9. $2x^2 - 20x + 50 - 18b^2$

Answers on page A-25

Study Tips

SPECIAL VIDEOTAPES

In addition to the videotaped lectures for the books, there is a special VHS tape, *Math Problem Solving in the Real World*. Check with your instructor to see whether this tape is available on your campus.

There is also a special Math Study Skills for Students Video on CD that is designed to help you make better use of your math study time and improve your retention of concepts and procedures taught in classes from basic mathematics through intermediate algebra. (See the Preface for more information.)

a Factor completely.

1. $y^2 - 225$

2. $x^2 - 400$

3. $2x^2 + 11x + 12$

4. $8a^2 + 18a - 5$

5. $5x^4 - 20$

6. $3xy^2 - 75x$

7. $p^2 + 36 + 12p$

8. $a^2 + 49 + 14a$

9. $2x^2 - 10x - 132$

10. $3y^2 - 15y - 252$

11. $9x^2 - 25y^2$

12. $16a^2 - 81b^2$

13. $4m^4 - 100$

14. $2x^2 - 288$

15. $3x^2 + 15x - 252$

16. $2y^2 + 10y - 132$

17. $2xy^2 - 50x$

18. $3a^3b - 108ab$

19. $225 - (a - 3)^2$

20. $625 - (t - 10)^2$

21. $m^6 - 1$

22. $64t^6 - 1$

23. $x^2 + 6x - y^2 + 9$

24. $t^2 + 10t - p^2 + 25$

25. $250x^3 - 128y^3$

26. $27a^3 - 343b^3$

27. $8m^3 + m^6 - 20$

28. $-37x^2 + x^4 + 36$

29. $ac + cd - ab - bd$

30. $xw - yw + xz - yz$

31. $50b^2 - 5ab - a^2$

32. $9c^2 + 12cd - 5d^2$

33. $-7x^2 + 2x^3 + 4x - 14$

34. $9m^2 + 3m^3 + 8m + 24$

35. $2x^3 + 6x^2 - 8x - 24$

36. $3x^3 + 6x^2 - 27x - 54$

37. $16x^3 + 54y^3$

38. $250a^3 + 54b^3$

39. $36y^2 - 35 + 12y$

40. $2b - 28a^2b + 10ab$

41. $a^8 - b^8$

42. $2x^4 - 32$

43. $a^3b - 16ab^3$

44. $x^3y - 25xy^3$

45. $\frac{1}{16}x^2 - \frac{1}{6}xy^2 + \frac{1}{9}y^4$ **46.** $36x^2 + 15x + \frac{25}{16}$ **47.** $5x^3 - 5x^2y - 5xy^2 + 5y^3$ **48.** $a^3 - ab^2 + a^2b - b^3$

49. $42ab + 27a^2b^2 + 8$ **50.** $-23xy + 20x^2y^2 + 6$ **51.** $8y^4 - 125y$ **52.** $64p^4 - p$

53. $a^2 - b^2 - 6b - 9$ **54.** $m^2 - n^2 - 8n - 16$

55. $\mathbf{D_W}$ In your own words, outline a procedure that can be used to try to factor a polynomial.

56. $\mathbf{D_W}$ Emily has factored a particular polynomial as $(a - b)(x - y)$. George factors the same polynomial and gets $(b - a)(y - x)$. Who is correct and why?

SKILL MAINTENANCE

Solve.   [3.2b]

57. *Exam Scores.* There are 75 questions on a college entrance examination. Two points are awarded for each correct answer, and one half point is deducted for each incorrect answer. A score of 100 indicates how many correct and how many incorrect answers, assuming that all questions are answered?

58. *Perimeter.* A pentagon with all five sides the same length has the same perimeter as an octagon in which all eight sides are the same length. One side of the pentagon is 2 less than three times the length of one side of the octagon. Find the perimeters.

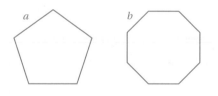

SYNTHESIS

Factor. Assume that variables in exponents represent natural numbers.

59. $30y^4 - 97xy^2 + 60x^2$

60. $3x^2y^2z + 25xyz^2 + 28z^3$

61. $5x^3 - \frac{5}{27}$

62. $x^4 - 50x^2 + 49$

63. $(x - p)^2 - p^2$

64. $s^6 - 729t^6$

65. $(y - 1)^4 - (y - 1)^2$

66. $27x^{6s} + 64y^{3t}$

67. $4x^2 + 4xy + y^2 - r^2 + 6rs - 9s^2$

68. $c^4d^4 - a^{16}$

69. $c^{2w+1} + 2c^{w+1} + c$

70. $24x^{2a} - 6$

71. $3(x + 1)^2 + 9(x + 1) - 12$

72. $8(a - 3)^2 - 64(a - 3) + 128$

73. $x^6 - 2x^5 + x^4 - x^2 + 2x - 1$

74. $1 - \dfrac{x^{27}}{1000}$

75. $y^9 - y$

76. $(m - 1)^3 - (m + 1)^3$

APPLICATIONS OF POLYNOMIAL EQUATIONS AND FUNCTIONS

Objectives

a Solve quadratic and other polynomial equations by first factoring and then using the principle of zero products.

b Solve applied problems involving quadratic and other polynomial equations that can be solved by factoring.

Whenever two polynomials are set equal to each other, we have a **polynomial equation.** Some examples of polynomial equations are

$$4x^3 + x^2 + 5x = 6x - 3,$$
$$x^2 - x = 6,$$

and $3y^4 + 2y^2 + 2 = 0$.

A second-degree polynomial equation in one variable is often called a **quadratic equation.** Of the equations listed above, only $x^2 - x = 6$ is a quadratic equation.

Polynomial equations, and quadratic equations in particular, occur frequently in applications, so the ability to solve them is an important skill. One way of solving certain polynomial equations involves factoring.

a The Principle of Zero Products

When we multiply two or more numbers, if either factor is 0, then the product is 0. Conversely, if a product is 0, then at least one of the factors must be 0. This property of 0 gives us a new principle for solving equations.

THE PRINCIPLE OF ZERO PRODUCTS

For any real numbers a and b:

If $ab = 0$, then $a = 0$ or $b = 0$ (or both).

If $a = 0$ or $b = 0$, then $ab = 0$.

To solve an equation using the principle of zero products, we first write it in *standard form*: with 0 on one side of the equation and the leading coefficient positive.

EXAMPLE 1 Solve: $x^2 - x = 6$.

In order to use the principle of zero products, we must have 0 on one side of the equation, so we subtract 6 on both sides:

$$x^2 - x - 6 = 0. \qquad \text{Getting 0 on one side}$$

We need a factorization on the other side, so we factor the polynomial:

$$(x - 3)(x + 2) = 0. \qquad \text{Factoring}$$

We now have two expressions, $x - 3$ and $x + 2$, whose product is 0. Using the principle of zero products, we set each expression or factor equal to 0:

$$x - 3 = 0 \quad or \quad x + 2 = 0. \qquad \text{Using the principle of zero products}$$

This gives us two simple linear equations. We solve them separately,

$$x = 3 \quad or \quad x = -2,$$

and check in the original equation as follows.

Check:
$$\begin{array}{c|c} x^2 - x = 6 \\ \hline 3^2 - 3 \;?\; 6 \\ 9 - 3 \\ 6 \end{array} \quad \text{TRUE}$$

$$\begin{array}{c|c} x^2 - x = 6 \\ \hline (-2)^2 - (-2) \;?\; 6 \\ 4 + 2 \\ 6 \end{array} \quad \text{TRUE}$$

The numbers 3 and -2 are both solutions.

1. Solve: $x^2 + 8 = 6x$.

> To solve an equation using the principle of zero products:
>
> **1.** Obtain a 0 on one side of the equation.
> **2.** Factor the other side.
> **3.** Set each factor equal to 0.
> **4.** Solve the resulting equations.

Do Exercise 1.

When you solve an equation using the principle of zero products, you may wish to check by substitution as we did in Example 1. Such a check will detect errors in solving.

Caution!

When using the principle of zero products, it is important that there is a 0 on one side of the equation. If neither side of the equation is 0, the procedure will not work.

For example, consider $x^2 - x = 6$ in Example 1 as

$$x(x - 1) = 6.$$

Suppose we reasoned as follows, setting factors equal to 6:

$$x = 6 \quad or \quad x - 1 = 6 \qquad \text{This step is incorrect!}$$
$$x = 7.$$

Neither 6 nor 7 checks, as shown below:

$$\begin{array}{c|c} x(x - 1) = 6 \\ \hline 6(6 - 1) & 6 \\ 6(5) \\ 30 \end{array} \quad \text{FALSE}$$

$$\begin{array}{c|c} x(x - 1) = 6 \\ \hline 7(7 - 1) & 6 \\ 7(6) \\ 42 \end{array} \quad \text{FALSE}$$

2. Solve: $5y + 2y^2 = 3$.

EXAMPLE 2 Solve: $7y + 3y^2 = -2$.

Since there must be a 0 on one side of the equation, we add 2 to get 0 on the right-hand side and arrange in descending order. Then we factor and use the principle of zero products.

$$7y + 3y^2 = -2$$
$$3y^2 + 7y + 2 = 0 \qquad \text{Getting 0 on one side}$$
$$(3y + 1)(y + 2) = 0 \qquad \text{Factoring}$$
$$3y + 1 = 0 \quad or \quad y + 2 = 0 \qquad \begin{array}{l}\text{Using the principle of} \\ \text{zero products}\end{array}$$
$$y = -\tfrac{1}{3} \quad or \qquad y = -2$$

The solutions are $-\tfrac{1}{3}$ and -2.

Do Exercise 2.

Answers on page A-25

3. Solve: $8b^2 = 16b$.

EXAMPLE 3 Solve: $5b^2 = 10b$.

$$5b^2 = 10b$$
$$5b^2 - 10b = 0 \qquad \text{Getting 0 on one side}$$
$$5b(b - 2) = 0 \qquad \text{Factoring}$$
$$5b = 0 \quad or \quad b - 2 = 0 \qquad \text{Using the principle of zero products}$$
$$b = 0 \quad or \qquad b = 2$$

The solutions are 0 and 2.

Do Exercise 3.

4. Solve: $25 + x^2 = -10x$.

EXAMPLE 4 Solve: $x^2 - 6x + 9 = 0$.

$$x^2 - 6x + 9 = 0 \qquad \text{Getting 0 on one side}$$
$$(x - 3)(x - 3) = 0 \qquad \text{Factoring}$$
$$x - 3 = 0 \quad or \quad x - 3 = 0 \qquad \text{Using the principle of zero products}$$
$$x = 3 \quad or \qquad x = 3$$

There is only one solution, 3.

Do Exercise 4.

5. Solve: $x^3 + x^2 = 6x$.

EXAMPLE 5 Solve: $3x^3 - 9x^2 = 30x$.

$$3x^3 - 9x^2 = 30x$$
$$3x^3 - 9x^2 - 30x = 0 \qquad \text{Getting 0 on one side}$$
$$3x(x^2 - 3x - 10) = 0 \qquad \text{Factoring out a common factor}$$
$$3x(x + 2)(x - 5) = 0 \qquad \text{Factoring the trinomial}$$
$$3x = 0 \quad or \quad x + 2 = 0 \quad or \quad x - 5 = 0 \qquad \text{Using the principle of zero products}$$
$$x = 0 \quad or \qquad x = -2 \quad or \qquad x = 5$$

The solutions are 0, -2, and 5.

Do Exercise 5.

6. Given that $f(x) = 10x^2 + 13x$, find all values of x for which $f(x) = 3$.

EXAMPLE 6 Given that $f(x) = 3x^2 - 4x$, find all values of x for which $f(x) = 4$.

We want all numbers x for which $f(x) = 4$. Since $f(x) = 3x^2 - 4x$, we must have

$$3x^2 - 4x = 4 \qquad \text{Setting } f(x) \text{ equal to 4}$$
$$3x^2 - 4x - 4 = 0 \qquad \text{Getting 0 on one side}$$
$$(3x + 2)(x - 2) = 0 \qquad \text{Factoring}$$
$$3x + 2 = 0 \quad or \quad x - 2 = 0$$
$$x = -\tfrac{2}{3} \quad or \qquad x = 2.$$

We can check as follows.

$$f\left(-\tfrac{2}{3}\right) = 3\left(-\tfrac{2}{3}\right)^2 - 4\left(-\tfrac{2}{3}\right) = 3 \cdot \tfrac{4}{9} + \tfrac{8}{3} = \tfrac{4}{3} + \tfrac{8}{3} = \tfrac{12}{3} = 4;$$
$$f(2) = 3(2)^2 - 4(2) = 3 \cdot 4 - 8 = 12 - 8 = 4.$$

To have $f(x) = 4$, we must have $x = -\tfrac{2}{3}$ or $x = 2$.

Do Exercise 6.

Answers on page A-25

EXAMPLE 7 Find the domain of F if $F(x) = \dfrac{x-2}{x^2 + 2x - 15}$.

The domain of F is the set of all values for which

$$\frac{x-2}{x^2 + 2x - 15}$$

is a real number. Since division by 0 is undefined, $F(x)$ cannot be calculated for any x-value for which the denominator, $x^2 + 2x - 15$, is 0. To make sure these values are *excluded*, we solve:

$$x^2 + 2x - 15 = 0 \qquad \text{Setting the denominator equal to 0}$$
$$(x-3)(x+5) = 0 \qquad \text{Factoring}$$
$$x - 3 = 0 \quad or \quad x + 5 = 0$$
$$x = 3 \quad or \qquad x = -5. \qquad \text{These are the values to \textit{exclude}.}$$

The domain of F is $\{x \,|\, x$ is a real number *and* $x \neq -5$ *and* $x \neq 3\}$.

Do Exercise 7.

AG **ALGEBRAIC–GRAPHICAL CONNECTION**

We now consider graphical connections with the algebraic equation-solving concepts.

In Chapter 2, we briefly considered the graph of a quadratic function

$$f(x) = ax^2 + bx + c, \quad a \neq 0.$$

For example, the graph of the function $f(x) = x^2 + 6x + 8$ and its x-intercepts are shown below.

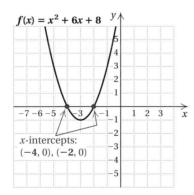

The x-intercepts are $(-4, 0)$ and $(-2, 0)$. These pairs are also the points of intersection of the graphs of $f(x) = x^2 + 6x + 8$ and $g(x) = 0$ (the x-axis).

In this section, we began studying how to solve quadratic equations like $x^2 + 6x + 8 = 0$ using factoring:

$$x^2 + 6x + 8 = 0$$
$$(x+4)(x+2) = 0 \qquad\qquad \text{Factoring}$$
$$x + 4 = 0 \quad or \quad x + 2 = 0 \qquad \text{Principle of zero products}$$
$$x = -4 \quad or \qquad x = -2.$$

We see that the solutions of $0 = x^2 + 6x + 8$, -4 and -2, are the first coordinates of the x-intercepts, $(-4, 0)$ and $(-2, 0)$, of the graph of $f(x) = x^2 + 6x + 8$.

Do Exercise 8.

7. Find the domain of the function G if

$$G(x) = \frac{2x-9}{x^2 - 3x - 28}.$$

8. Consider solving the equation

$$x^2 - 6x + 8 = 0$$

graphically.

a) Below is the graph of

$$f(x) = x^2 - 6x + 8.$$

Use *only* the graph to find the x-intercepts of the graph.

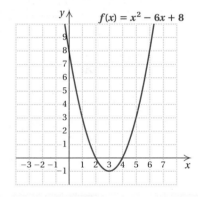

b) Use *only* the graph to find the solutions of $x^2 - 6x + 8 = 0$.

c) Compare your answers to parts (a) and (b).

Answers on page A-25

4.8   Applications of Polynomial
Equations and Functions

CALCULATOR CORNER

Solving Quadratic Equations   We can solve quadratic equations graphically. Consider the equation $x^2 - x = 6$. First, we must write the equation with 0 on one side. To do this, we subtract 6 on both sides of the equation. We get $x^2 - x - 6 = 0$. Next, we graph $y = x^2 - x - 6$ in a window that shows the x-intercepts. The standard window works well in this case.

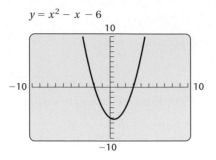

$$y = x^2 - x - 6$$

   The solutions of the equation are the values of x for which $x^2 - x - 6 = 0$. These are also the first coordinates of the x-intercepts of the graph. We use the **ZERO** feature from the **CALC** menu to find these numbers. To find the solution corresponding to the leftmost x-intercept, we first press **2ND** **CALC** **2** to select the **ZERO** feature. The prompt "Left Bound?" appears. We use the ◁ or the ▷ key to move the cursor to the left of the intercept and press **ENTER** . Now, the prompt "Right Bound?" appears. We move the cursor to the right of the intercept and press **ENTER** . Next, the prompt "Guess?" appears. We move the cursor close to the intercept and press **ENTER** again. We now see the cursor positioned at the leftmost x-intercept and the coordinates of that point, $x = -2$, $y = 0$, are displayed. Thus, $x^2 - x - 6 = 0$ when $x = -2$. This is one solution of the equation.

   Repeat this procedure to find the first coordinate of the other x-intercept. We see that $x = 3$ at that point. Thus the solutions of the equation are -2 and 3.

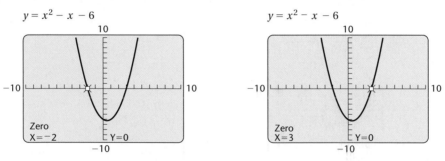

$$y = x^2 - x - 6 \qquad\qquad y = x^2 - x - 6$$

   This equation could also be solved by entering $y_1 = x^2 - x$ and $y_2 = 6$ and finding the first coordinate of the points of intersection using the **INTERSECT** feature as described on p. 246.

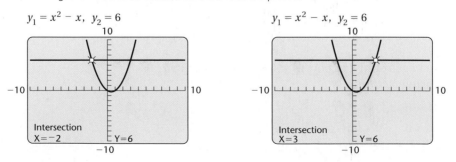

$$y_1 = x^2 - x, \ y_2 = 6 \qquad\qquad y_1 = x^2 - x, \ y_2 = 6$$

Exercise:

1. Solve the equations in Examples 2–5 graphically. Note that, regardless of the variable used in an example, each equation should be entered on the equation-editor screen in terms of x.

b Applications and Problem Solving

Some problems can be translated to quadratic equations. The problem-solving process is the same one we use for other kinds of applied problems.

EXAMPLE 8 *Prize Tee Shirts.* During intermission at sporting events, team mascots commonly use a powerful slingshot to launch tightly rolled tee shirts into the stands.

The height $h(t)$, in feet, of an airborne tee shirt t seconds after being launched can be approximated by

$$h(t) = -15t^2 + 75t + 10.$$

After peaking, a rolled-up tee shirt is caught by a fan 70 ft above ground level. How long was the tee shirt in the air?

1. **Familiarize.** We make a drawing and label it, using the information provided (see the figure). We could evaluate $h(t)$ for a few values of t. Note that t cannot be negative, since it represents time from launch.

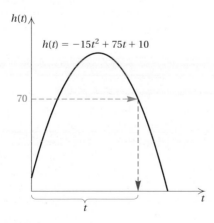

2. **Translate.** The function is given. Since we are asked to determine how long it will take for the shirt to reach someone 70 ft above ground level, we are interested in the value of t for which $h(t) = 70$:

$$-15t^2 + 75t + 10 = 70.$$

9. Fireworks. Suppose that a bottle rocket is launched upward with an initial velocity of 96 ft/sec and from a height of 880 ft. Its height h, in feet, after t seconds is given by

$$h(t) = -16t^2 + 96t + 880.$$

After how long will the rocket reach the ground?

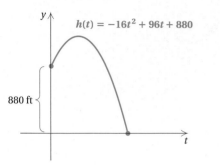

3. Solve. We solve by factoring:

$$-15t^2 + 75t + 10 = 70$$
$$-15t^2 + 75t - 60 = 0 \qquad \text{Subtracting 70}$$
$$\left. \begin{array}{l} -15(t^2 - 5t + 4) = 0 \\ -15(t - 4)(t - 1) = 0 \end{array} \right\} \text{Factoring}$$
$$t - 4 = 0 \quad or \quad t - 1 = 0$$
$$t = 4 \quad or \qquad t = 1.$$

The solutions appear to be 4 and 1.

4. Check. We have

$$h(4) = -15 \cdot 4^2 + 75 \cdot 4 + 10 = -240 + 300 + 10 = 70 \text{ ft};$$
$$h(1) = -15 \cdot 1^2 + 75 \cdot 1 + 10 = -15 + 75 + 10 = 70 \text{ ft}.$$

Both 1 and 4 check, as we can also see from the graph below.

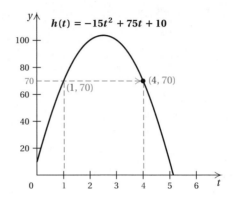

However, the problem states that the tee shirt is caught on the way down from its peak height. Thus we reject the solution 1 since that would indicate when the height of the tee shirt was 70 ft on the way up.

5. State. The tee shirt was in the air for 4 sec.

Do Exercise 9.

The following example involves the **Pythagorean theorem,** which relates the lengths of the sides of a right triangle. A **right triangle** has a 90°, or right, angle, which is denoted by a symbol like ⌐. The longest side, opposite the 90° angle, is called the **hypotenuse.** The other sides, called **legs,** form the two sides of the right angle.

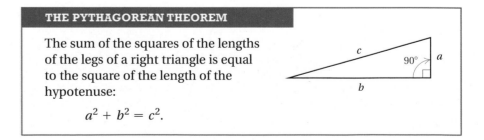

THE PYTHAGOREAN THEOREM

The sum of the squares of the lengths of the legs of a right triangle is equal to the square of the length of the hypotenuse:

$$a^2 + b^2 = c^2.$$

Answer on page A-25

EXAMPLE 9 *Carpentry.* In order to build a deck at a right angle to his house, master carpenter Dick Bonewitz decides to plant a stake in the ground a precise distance from the back wall of his house. This stake will combine with two marks on the house to form a right triangle. From a course in geometry, Dick remembers that there are three consecutive integers that can work as sides of a right triangle. Find the measurements of that triangle.

1. **Familiarize.** Recall that x, $x + 1$, and $x + 2$ can be used to represent three unknown consecutive integers. Since $x + 2$ is the largest number, it must represent the hypotenuse. The legs serve as the sides of the right angle, so one leg must be formed by the marks on the house. We make a drawing in which

   $x =$ the distance between the marks on the house,

   $x + 1 =$ the length of the other leg,

   and

   $x + 2 =$ the length of the hypotenuse.

2. **Translate.** Applying the Pythagorean theorem, we translate as follows:

   $$a^2 + b^2 = c^2$$
   $$x^2 + (x + 1)^2 = (x + 2)^2.$$

3. **Solve.** We solve the equation as follows:

$x^2 + (x^2 + 2x + 1) = x^2 + 4x + 4$	Squaring the binomials
$2x^2 + 2x + 1 = x^2 + 4x + 4$	Collecting like terms
$x^2 - 2x - 3 = 0$	Subtracting $x^2 + 4x + 4$
$(x - 3)(x + 1) = 0$	Factoring

   $$x - 3 = 0 \quad or \quad x + 1 = 0$$
   $$x = 3 \quad or \quad x = -1.$$

4. **Check.** The integer -1 cannot be a length of a side because it is negative. For $x = 3$, we have $x + 1 = 4$, and $x + 2 = 5$. Since $3^2 + 4^2 = 5^2$, the lengths 3, 4, and 5 determine a right triangle. Thus, 3, 4, and 5 check.

5. **State.** Dick should use a triangle with sides having a ratio of $3:4:5$. Thus, if the marks on the house are 3 yd apart, he should locate the stake at the point in the yard that is precisely 4 yd from one mark and 5 yd from the other mark.

Do Exercise 10.

10. Child's Block. The lengths of the sides of a right triangle formed by a child's wooden block are such that one leg has length 5 cm. The lengths of the other sides are consecutive integers. Find the lengths of the other sides of the triangle.

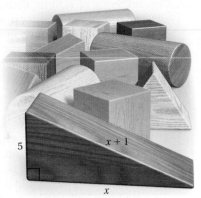

Answer on page A-25

Translating
for Success

1. *Car Travel.* Two cars leave town at the same time going in different directions. One travels 50 mph and the other travels 55 mph. In how many hours will they be 200 mi apart?

2. *Mixture of Solutions.* Solution A is 27% alcohol and solution B is 55% alcohol. How much of each should be used in order to make 10 L of a solution that is 48% alcohol?

3. *Triangle Dimensions.* The base of a triangle is 3 cm less than the height. The area is 27 cm². Find the height and the base.

4. *Three Numbers.* The sum of three numbers is 38. The first number is 3 less than twice the second number. The second number minus the third number is −7. What are the numbers?

5. *Supplementary Angles.* Two angles are supplementary. One angle measures 27° more than three times the measure of the other. Find the measure of each angle.

Translate each word problem to an equation or a system of equations and select a correct translation from equations A–Q.

A. $x + y + z = 38,$
$x = 2y - 3,$
$y - z = -7$

B. $\frac{1}{2}x(x - 3) = 27$

C. $x + y = 180,$
$x = 3y - 27$

D. $x^2 + 36 = (x + 4)^2$

E. $x^2 + (x + 4)^2 = 36$

F. $x + y = 10,$
$0.27x + 0.55y = 4.8$

G. $x + y = 45,$
$10x - 7y = 402$

H. $x + y + z = 180,$
$y - 3x - 38 = 0,$
$x - z = 7$

I. $x + y = 90,$
$x = 3y + 10$

J. $x + 29.3\%x = 77.2$

K. $x + y + z = 38,$
$x - 2y = 3,$
$x - z = -7$

L. $x + y = 10,$
$27x + 55y = 4.8$

M. $55x - 50x = 200$

N. $x^2 - 3x = 27$

O. $x + y = 45,$
$7x + 10y = 402$

P. $x + y = 180,$
$x = 3y + 27$

Q. $50x + 55x = 200$

Answers on page A-25

6. *Triangle Dimensions.* The length of one leg of a right triangle is 6 m. The length of the hypotenuse is 4 m longer than the length of the other leg. Find the lengths of the hypotenuse and the other leg.

7. *Pizza Sales.* Todd's fraternity sold 45 pizzas over a football weekend. Small pizzas sold for $7 each and large pizzas for $10 each. The total amount of the sales was $402. How many of each size pizza were sold?

8. *Angle Measures.* The second angle of a triangle measures 38° more than three times the measure of the first. The measure of the third angle is 7° less than the first. Find the measures of each angle of the triangle.

9. *Complementary Angles.* Two angles are complementary. One angle measures 10° more than three times the measure of the other. Find the measure of each angle.

10. *Life Expectancy.* Life expectancy in the United States was 77.2 yr in 2002. This was a 29.3% increase in life expectancy in 1930. What was the life expectancy in 1930?

Source: National Center for Health Statistics

a   Solve.

1. $x^2 + 3x = 28$

2. $y^2 - 4y = 45$

3. $y^2 + 9 = 6y$

4. $r^2 + 4 = 4r$

5. $x^2 + 20x + 100 = 0$

6. $y^2 + 10y + 25 = 0$

7. $9x + x^2 + 20 = 0$

8. $8y + y^2 + 15 = 0$

9. $x^2 + 8x = 0$

10. $t^2 + 9t = 0$

11. $x^2 - 25 = 0$

12. $p^2 - 49 = 0$

13. $z^2 = 144$

14. $y^2 = 64$

15. $y^2 + 2y = 63$

16. $a^2 + 3a = 40$

17. $32 + 4x - x^2 = 0$

18. $27 + 6t - t^2 = 0$

19. $3b^2 + 8b + 4 = 0$

20. $9y^2 + 15y + 4 = 0$

21. $8y^2 - 10y + 3 = 0$

22. $4x^2 + 11x + 6 = 0$

23. $6z - z^2 = 0$

24. $8y - y^2 = 0$

25. $12z^2 + z = 6$

26. $6x^2 - 7x = 10$

27. $7x^2 - 7 = 0$

28. $4y^2 - 36 = 0$

29. $10 - r - 21r^2 = 0$

30. $28 + 5a - 12a^2 = 0$

31. $15y^2 = 3y$

32. $18x^2 = 9x$

33. $14 = x(x - 5)$

34. $x(x - 5) = 24$

35. $2x^3 - 2x^2 = 12x$

36. $50y + 5y^3 = 35y^2$

37. $2x^3 = 128x$

38. $147y = 3y^3$

39. $t^4 - 26t^2 + 25 = 0$

40. $x^4 - 13x^2 + 36 = 0$

41. $(a - 4)(a + 4) = 20$

42. $(t - 6)(t + 6) = 45$

43. $x(5 + 12x) = 28$

44. $a(1 + 21a) = 10$

45. Given that $f(x) = x^2 + 12x + 40$, find all values of x such that $f(x) = 8$.

46. Given that $f(x) = x^2 + 14x + 50$, find all values of x such that $f(x) = 5$.

47. Given that $g(x) = 2x^2 + 5x$, find all values of x such that $g(x) = 12$.

48. Given that $g(x) = 2x^2 - 15x$, find all values of x such that $g(x) = -7$.

49. Given that $h(x) = 12x + x^2$, find all values of x such that $h(x) = -27$.

50. Given that $h(x) = 4x - x^2$, find all values of x such that $h(x) = -32$.

Find the domain of the function f given by each of the following.

51. $f(x) = \dfrac{3}{x^2 - 4x - 5}$

52. $f(x) = \dfrac{2}{x^2 - 7x + 6}$

53. $f(x) = \dfrac{x}{6x^2 - 54}$

54. $f(x) = \dfrac{2x}{5x^2 - 20}$

55. $f(x) = \dfrac{x - 5}{25x^2 - 10x + 1}$

56. $f(x) = \dfrac{1 + x}{9x^2 + 30x + 25}$

57. $f(x) = \dfrac{7}{5x^3 - 35x^2 + 50x}$

58. $f(x) = \dfrac{3}{2x^3 - 2x^2 - 12x}$

In each of Exercises 59–62, an equation $ax^2 + bx + c = 0$ is given. Use *only* the graph of $f(x) = ax^2 + bx + c$ to find the x-intercepts of the graph and the solutions of the equation $ax^2 + bx + c = 0$.

59. $x^2 - 4x - 45 = 0$

60. $-x^2 - 3x + 40 = 0$

61. $32 + 4x - x^2 = 0$

62. $3x^2 - 12x = 0$

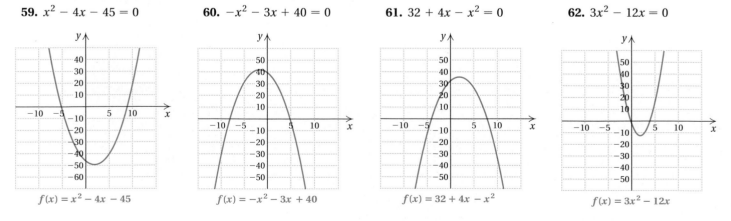

$f(x) = x^2 - 4x - 45$ | $f(x) = -x^2 - 3x + 40$ | $f(x) = 32 + 4x - x^2$ | $f(x) = 3x^2 - 12x$

Solve.

63. *Area of an Envelope.* An envelope is 4 cm longer than it is wide. The area is 96 cm². Find the length and the width.

64. *Book Area.* A book is 5 cm longer than it is wide. Find the length and the width if the area is 84 cm².

65. *Tent Design.* The triangular entrance to a tent is 2 ft taller than it is wide. The area of the entrance is 12 ft². Find the height and the base.

66. *Sailing.* A triangular sail is 9 m taller than it is wide. The area is 56 m². Find the height and the base of the sail.

Area = 56 m²

67. *Flower Bed Design.* A rectangular flower bed is to be 3 m longer than it is wide. The flower bed will have an area of 108 m². What will its dimensions be?

68. *Workbench Design.* The length of the top of a workbench is 4 ft greater than the width. The area is 96 ft². Find the length and the width.

69. *Antenna Wires.* A wire is stretched from the ground to the top of an antenna tower, as shown. The wire is 20 ft long. The height of the tower is 4 ft greater than the distance *d* from the tower's base to the end of the wire. Find the distance *d* and the height of the tower.

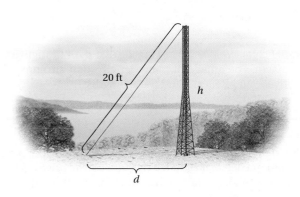

20 ft

h

d

70. *Ladder Location.* The foot of an extension ladder is 9 ft from a wall. The height that the ladder reaches on the wall and the length of the ladder are consecutive integers. How long is the ladder?

←9 ft→

71. *Consecutive Even Integers.* Three consecutive even integers are such that the square of the third is 76 more than the square of the second. Find the three integers.

72. *Consecutive Even Integers.* Three consecutive even integers are such that the square of the first plus the square of the third is 136. Find the three integers.

73. *Geometry.* If each of the sides of a square is lengthened by 6 cm, the area becomes 144 cm². Find the length of a side of the original square.

74. *Geometry.* If each of the sides of a square is lengthened by 4 m, the area becomes 49 m². Find the length of a side of the original square.

75. *Framing a Picture.* A picture frame measures 12 cm by 20 cm, and 84 cm² of picture shows. Find the width of the frame.

76. *Enclosure Dimensions.* The area of the square base of a walled enclosure is 12 more than its perimeter. Find the length of a side.

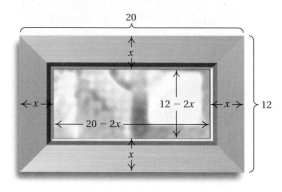

77. *Parking Lot Design.* A rectangular parking lot is 50 ft longer than it is wide. Determine the dimensions of the parking lot if it measures 250 ft diagonally.

78. *Framing a Picture.* A picture frame measures 14 cm by 20 cm, and 160 cm² of picture shows. Find the width of the frame.

79. *Triangle Dimensions.* One leg of a right triangle has length 9 m. The other sides have lengths that are consecutive integers. Find these lengths.

80. *Triangle Dimensions.* One leg of a right triangle is 10 cm. The other sides have lengths that are consecutive even integers. Find these lengths.

81. *Triangle Dimensions.* The lengths of the hypotenuse and one leg of a right triangle are consecutive integers. The length of the other leg is 7 ft. Find the missing lengths.

82. *Triangle Dimensions.* The lengths of the hypotenuse and one leg of a right triangle are consecutive odd integers. The length of the other leg is 8 ft. Find the missing lengths.

83. *Safety Flares.* Suppose that a flare is launched upward with an initial velocity of 80 ft/sec and from a height of 224 ft. Its height h, in feet, after t seconds is given by

$$h(t) = -16t^2 + 80t + 224.$$

After how long will the flare reach the ground?

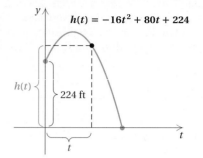

84. *Camcorder Production.* Suppose that the cost of making x video cameras is $C(x) = \frac{1}{9}x^2 + 2x + 1$, where $C(x)$ is in thousands of dollars. If the revenue from the sale of x video cameras is given by $R(x) = \frac{5}{36}x^2 + 2x$, where $R(x)$ is in thousands of dollars, how many cameras must be sold in order for the firm to break even? (See Section 3.8.)

85. $\mathbf{D_W}$ Explain how one could write a quadratic equation that has 5 and -3 as solutions. Can the number of solutions of a quadratic equation exceed two? Why or why not?

86. $\mathbf{D_W}$ Suppose that you are given a detailed graph of $y = P(x)$, where $P(x)$ is a polynomial. How could you use the graph to solve the equation $P(x) = 0$? $P(x) = 4$?

SKILL MAINTENANCE

Find the distance between the given pair of points on the number line. [1.6b]

87. $-3, 4$ **88.** $-3, -4$ **89.** $3, -4$ **90.** $-7.8, -10.3$

91. $3.6, 4.9$ **92.** $-\frac{3}{5}, \frac{2}{3}$ **93.** $-123, 568$ **94.** $0, -1023$

Find an equation of the line containing the given pair of points. [2.6c]

95. $(-2, 7)$ and $(-8, -4)$ **96.** $(-2, 7)$ and $(8, -4)$ **97.** $(-2, 7)$ and $(8, 4)$ **98.** $(-24, 10)$ and $(-86, -42)$

SYNTHESIS

99. Following is the graph of $f(x) = -x^2 - 2x + 3$. Use *only* the graph to solve $-x^2 - 2x + 3 = 0$ and $-x^2 - 2x + 3 \geq -5$.

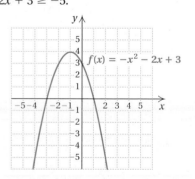

100. Following is the graph of $f(x) = x^4 - 3x^3$. Use *only* the graph to solve $x^4 - 3x^3 = 0$, $x^4 - 3x^3 \leq 0$, and $x^4 - 3x^3 > 0$.

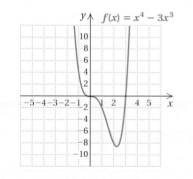

101. 📟 Use the TABLE feature of a graphing calculator to check that -5 and 3 are not in the domain of F, as shown in Example 7.

102. 📟 Use the TABLE feature of a graphing calculator to check your answers to Exercises 51, 54, and 57.

103. 📟 Use a graphing calculator to solve each equation.
   a) $x^4 - 3x^3 - x^2 + 5 = 0$
   b) $x^4 - 3x^3 - x^2 + 5 = 5$
   c) $x^4 - 3x^3 - x^2 + 5 = -8$
   d) $x^4 = 1 + 3x^3 + x^2$

104. Solve each of the following equations.
   a) $(8x + 11)(12x^2 - 5x - 2) = 0$
   b) $(3x^2 - 7x - 20)(x - 5) = 0$
   c) $3x^3 + 6x^2 - 27x - 54 = 0$
      (*Hint*: Factor by grouping.)
   d) $2x^3 + 6x^2 = 8x + 24$

The review that follows is meant to prepare you for a chapter exam. It consists of three parts. The first part, Concept Reinforcement, is designed to increase understanding of the concepts through true/false exercises. The second part is a list of important properties and formulas. The third part is the Review Exercises. These provide practice exercises for the exam, together with references to section objectives so you can go back and review. Before beginning, stop and look back over the skills you have obtained. What skills in mathematics do you have now that you did not have before studying this chapter?

✎ CONCEPT REINFORCEMENT

Determine whether the statement is true or false. Answers are given at the back of the book.

_____ 1. The polynomial $5x + 2x^2 - 4x^3$ can be factored.

_____ 2. The binomial $27 - t^3$ is a difference of cubes.

_____ 3. According to the principle of zero products, if $ab = 0$, then $a = 0$ and $b = 0$.

_____ 4. The expression $3x^{-1}$ is a monomial.

_____ 5. The expression $(3x - 12y)(2x + 5)$ is a complete factorization of $6x^2 - 9xy - 60y^2$.

_____ 6. The expression $(2a - 3)^2$ is a complete factorization of $4a^2 - 12a + 9$.

_____ 7. $a^3 - 8 = (a - 2)(a^2 - 4a + 4)$

_____ 8. $(5x - 7y)^2 = 25x^2 - 49y^2$

IMPORTANT PROPERTIES AND FORMULAS

Factoring Formulas: $\quad A^2 - B^2 = (A + B)(A - B), \quad A^2 + 2AB + B^2 = (A + B)^2,$
$\qquad\qquad\qquad\quad A^2 - 2AB + B^2 = (A - B)^2, \quad A^3 + B^3 = (A + B)(A^2 - AB + B^2),$
$\qquad\qquad\qquad\quad A^3 - B^3 = (A - B)(A^2 + AB + B^2)$

The Principle of Zero Products: For any real numbers a and b:
$\qquad\qquad\qquad$ If $ab = 0$, then $a = 0$ or $b = 0$.
$\qquad\qquad\qquad$ If $a = 0$ or $b = 0$, then $ab = 0$.

Review Exercises

1. Given the polynomial   [4.1a]
$$3x^6y - 7x^8y^3 + 2x^3 - 3x^2:$$

a) Identify the degree of each term and the degree of the polynomial.
b) Identify the leading term and the leading coefficient.
c) Arrange in ascending powers of x.
d) Arrange in descending powers of y.

Evaluate the polynomial function for the given values. [4.1b]

2. $P(x) = x^3 - x^2 + 4x;$ $P(0)$ and $P(-1)$

3. $P(x) = 4 - 2x - x^2;$ $P(-2)$ and $P(5)$

4. *Emergency-Room Visits.* The number E, in thousands, of hospital emergency-room visits involving narcotic painkillers can be estimated by the polynomial function given by

$$E(t) = 1.55x^2 + 2.71x + 47.04,$$

where t is the number of years since 1996.   [4.1b]

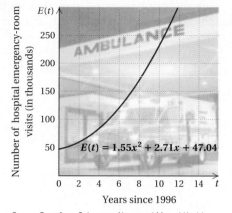

Source: Data from Substance Abuse and Mental Health Services Administration, Drug Abuse Warning Network

a) Use this graph to predict the number of hospital emergency visits in 2006.
b) Use the function to predict the number of hospital emergency visits in 2010.

Collect like terms.   [4.1c]

5. $4x^2y - 3xy^2 - 5x^2y + xy^2$

6. $3ab - 10 + 5ab^2 - 2ab + 7ab^2 + 14$

Add, subtract, or multiply.   [4.1c, d], [4.2a, b, c, d]

7. $(-6x^3 - 4x^2 + 3x + 1) + (5x^3 + 2x + 6x^2 + 1)$

8. $(4x^3 - 2x^2 - 7x + 5) + (8x^2 - 3x^3 - 9 + 6x)$

9. $(-9xy^2 - xy + 6x^2y) + (-5x^2y - xy + 4xy^2) + (12x^2y - 3xy^2 + 6xy)$

10. $(3x - 5) - (-6x + 2)$

11. $(4a - b + 3c) - (6a - 7b - 4c)$

12. $(9p^2 - 4p + 4) - (-7p^2 + 4p + 4)$

13. $(6x^2 - 4xy + y^2) - (2y^2 + 3xy - 2y^2)$

14. $(3x^2y)(-6xy^3)$

15. $(x^4 - 2x^2 + 3)(x^4 + x^2 - 1)$

16. $(4ab + 3c)(2ab - c)$

17. $(2x + 5y)(2x - 5y)$

18. $(2x - 5y)^2$

19. $(5x^2 - 7x + 3)(4x^2 + 2x - 9)$

20. $(x^2 + 4y^3)^2$

21. $(x - 5)(x^2 + 5x + 25)$

22. $\left(x - \frac{1}{3}\right)\left(x - \frac{1}{6}\right)$

23. Given that $f(x) = x^2 - 2x - 7$, find and simplify $f(a - 1)$ and $f(a + h) - f(a)$.   [4.2e]

Factor.   [4.3a, b], [4.4a], [4.5a, b], [4.6a, b, c, d], [4.7a]
24. $9y^4 - 3y^2$

25. $15x^4 - 18x^3 + 21x^2 - 9x$

26. $a^2 - 12a + 27$

27. $3m^2 + 14m + 8$

28. $25x^2 + 20x + 4$

29. $4y^2 - 16$

30. $ax + 2bx - ay - 2by$

31. $4x^4 + 4x^2 + 20$

32. $27x^3 - 8$

33. $0.064b^3 - 0.125c^3$

34. $y^5 - y$

35. $2z^8 - 16z^6$

36. $54x^6y - 2y$

37. $1 + a^3$

38. $36x^2 - 120x + 100$

39. $6t^2 + 17pt + 5p^2$

40. $x^3 + 2x^2 - 9x - 18$

41. $a^2 - 2ab + b^2 - 4t^2$

Solve. [4.8a]

42. $x^2 - 20x = -100$

43. $6b^2 - 13b + 6 = 0$

44. $8y^2 = 14y$

45. $r^2 = 16$

46. Given that $f(x) = x^2 - 7x - 40$, find all values of x such that $f(x) = 4$.

47. Find the domain of the function f given by
$$f(x) = \frac{x - 3}{3x^2 + 19x - 14}.$$

Solve. [4.8b]

48. *Photograph Dimensions.* A photograph is 3 in. longer than it is wide. When a 2-in. matte border is placed around the photograph, the total area of the photograph and the border is 108 in². Find the dimensions of the photograph.

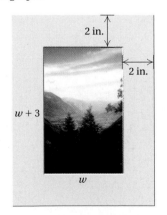

49. The sum of the squares of three consecutive odd integers is 83. Find the integers.

50. *Area.* The area of a square is 7 more than six times the length of a side. What is the length of a side of the square?

51. $\mathbf{D_W}$ In this chapter, we learned to solve equations that we could not have solved before. Describe these new equations and the way we go about solving them. How is the procedure different from those we have used before now? [4.8a]

52. $\mathbf{D_W}$ Explain the error in each of the following.
  a) $(a + 3)^2 = a^2 + 9$ [4.2c]
  b) $a^3 + b^3 = (a + b)(a^2 - 2ab + b^2)$ [4.6d]
  c) $(a - b)(a - b) = a^2 - b^2$ [4.2c]
  d) $(x + 3)(x - 4) = x^2 - 12$ [4.2b]
  e) $(p + 7)(p - 7) = p^2 + 49$ [4.2d]
  f) $(t - 3)^2 = t^2 - 9$ [4.2c]

SYNTHESIS

Factor. [4.6d]

53. $128x^6 - 2y^6$

54. $(x + 1)^3 - (x - 1)^3$

55. Multiply: $[a - (b - 1)][(b - 1)^2 + a(b - 1) + a^2]$. [4.6d]

56. Solve: $64x^3 = x$. [4.8a]

1. Given the polynomial
$$3xy^3 - 4x^2y + 5x^5y^4 - 2x^4y:$$
   a) Identify the degree of each term and the degree of the polynomial.
   b) Identify the leading term and the leading coefficient.
   c) Arrange in ascending powers of x.
   d) Arrange in descending powers of y.

2. Given that $P(x) = 2x^3 + 3x^2 - x + 4$, find $P(0)$ and $P(-2)$.

3. *Video-Game Sales.* Projected sales S of video games, in billions of dollars, can be estimated by the polynomial function given by
$$S(t) = 0.0496t^4 - 0.6705t^3 + 2.6367t^2 - 2.3880t + 1.6123,$$
where t is the number of years since 2004.
Source: Jupiter Research

   a) Use the graph to predict the sales of video games, in billions of dollars, in 2010.
   b) Use the function to predict the sales of video games, in billions of dollars, in 2009.

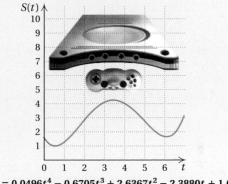

$S(t) = 0.0496t^4 - 0.6705t^3 + 2.6367t^2 - 2.3880t + 1.6123$

4. Collect like terms: $5xy - 2xy^2 - 2xy + 5xy^2$.

Add, subtract, or multiply.

5. $(-6x^3 + 3x^2 - 4y) + (3x^3 - 2y - 7y^2)$

6. $(4a^3 - 2a^2 + 6a - 5) + (3a^3 - 3a + 2 - 4a^2)$

7. $(5m^3 - 4m^2n - 6mn^2 - 3n^3) +$
   $(9mn^2 - 4n^3 + 2m^3 + 6m^2n)$

8. $(9a - 4b) - (3a + 4b)$

9. $(4x^2 - 3x + 7) - (-3x^2 + 4x - 6)$

10. $(6y^2 - 2y - 5y^3) - (4y^2 - 7y - 6y^3)$

11. $(-4x^2y)(-16xy^2)$

12. $(6a - 5b)(2a + b)$

13. $(x - y)(x^2 - xy - y^2)$

14. $(3m^2 + 4m - 2)(-m^2 - 3m + 5)$

15. $(4y - 9)^2$

16. $(x - 2y)(x + 2y)$

17. Given that $f(x) = x^2 - 5x$, find and simplify $f(a + 10)$ and $f(a + h) - f(a)$.

Factor.

18. $9x^2 + 7x$

19. $24y^3 + 16y^2$

20. $y^3 + 5y^2 - 4y - 20$

21. $p^2 - 12p - 28$

22. $12m^2 + 20m + 3$

23. $9y^2 - 25$

24. $3r^3 - 3$

25. $9x^2 + 25 - 30x$

26. $(z + 1)^2 - b^2$

27. $x^8 - y^8$

28. $y^2 + 8y + 16 - 100t^2$

29. $20a^2 - 5b^2$

30. $24x^2 - 46x + 10$

31. $16a^7b + 54ab^7$

Solve.

32. $x^2 - 18 = 3x$

33. $5y^2 - 125 = 0$

34. $2x^2 + 21 = -17x$

35. Given that $f(x) = 3x^2 - 15x + 11$, find all values of x such that $f(x) = 11$.

36. Find the domain of the function f given by
$$f(x) = \frac{3 - x}{x^2 + 2x + 1}.$$

Solve.

37. *Photograph Dimensions.* A photograph is 3 cm longer than it is wide. Its area is 40 cm². Find its length and its width.

38. *Ladder Location.* The foot of an extension ladder is 10 ft from a wall. The ladder is 2 ft longer than the distance that it reaches up the wall. How far up the wall does the ladder reach?

39. *Area.* The area of a square is 5 more than four times the length of a side. What is the length of a side of the square?

40. *Number of Games in a League.* If there are n teams in a league and each team plays every other team once, the total number of games played is given by the polynomial function $f(n) = \frac{1}{2}n^2 - \frac{1}{2}n$. Find an equivalent expression for $f(n)$ by factoring out the largest common factor.

SYNTHESIS

41. Factor: $6x^{2n} - 7x^n - 20$.

42. If $pq = 5$ and $(p + q)^2 = 29$, find the value of $p^2 + q^2$.

CHAPTER 4: Polynomials and
Polynomial Functions

Cumulative Review

1. *Scoville Scale.* In 1912, the pharmacist Wilbur Scoville created a rating scale for hot peppers. Ratings for various peppers are given in the table. Complete the table.

   Source: Kathy Wollard, *How Come Planet Earth*? New York: Workman, 1999

Hot pepper	Pepper rating: scientific notation (Scoville units)	Hot pepper rating: standard notation (Scoville units)
Green pepper	0	0
Jalapeno	4 x 10³	
Tabasco		40,000
Oranhe habanero	3 x 10⁵	
Red savina		577,000

   Source: Kathy Wollard, *How Come Planet Earth?*

2. Evaluate $\dfrac{2m - n}{4}$ when $m = 3$ and $n = 2$.

Simplify.

3. $|0|$

4. $-8 - (-4)$

5. $-3a + (6a - 1) - 4(a + 1)$

Simplify.

6. $2[5(3x + 4) - 2x] - [7(3x + 4) - 8]$

7. $[(-3a^{-6}b^2)^5]^{-2}$

8. $(x^2 + 4x - xy - 9) + (-3x^2 - 3x + 8)$

9. $(6x^2 - 3x + 2x^3) - (8x^2 - 9x + 2x^3)$

10. $(a^2 - a - 3) \cdot (a^2 + 2a - 3)$

11. $(x + 4)(x + 9)$

Solve.

12. $8 - 3x = 6x - 10$

13. $\frac{1}{2}x - 3 = \frac{7}{2}$

14. $A = \frac{1}{2}h(a + b)$, for b

15. $6x - 1 \le 3(5x + 2)$

16. $4x - 3 < 2$ *or* $x - 3 > 1$

17. $|2x - 3| < 7$

18. $x + y + z = -5,$
    $x - z = 10,$
    $y - z = 12$

19. $2x + 5y = -2,$
    $5x + 3y = 14$

20. $3x - y = 7,$
    $2x + 2y = 5$

21. $x + 2y - z = 0,$
    $3x + y - 2z = -1,$
    $x - 4y + z = -2$

22. $2x + 3y = 2,$
    $-4x + 3y = -7$

23. $-3a + 2b = 0,$
    $3a - 4b = -1$

24. $2x + 2y - 4z = 1,$
$-2x - 4y + 8z = -1,$
$4x + 4y + 4z = 5$

25. $11x + x^2 + 24 = 0$

26. $2x^2 - 15x = -7$

27. Given that $f(x) = 3x^2 + 4x$, find all values of x such that $f(x) = 4$.

28. Find the domain of the function F given by
$$F(x) = \frac{x + 7}{x^2 - 2x - 15}.$$

Factor.

29. $3x^3 - 12x^2$

30. $2x^4 + x^3 + 2x + 1$

31. $x^2 + 5x - 14$

32. $20a^2 - 23a + 6$

33. $4x^2 - 25$

34. $2x^2 - 28x + 98$

35. $a^3 + 64$

36. $8x^3 - 1$

37. $4a^3 + a^6 - 12$

38. $4x^4y^2 - x^2y^4$

Graph.

39. $x < 1 \text{ or } x \geq 2$

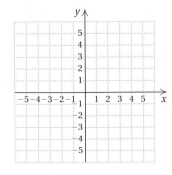

40. $y = -2x$

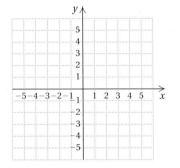

41. $y = \frac{1}{2}x$

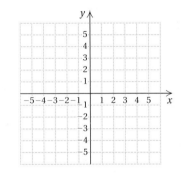

42. $4y + 3x = 12 + 3x$

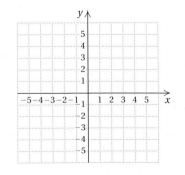

43. $6y + 24 = 0$

44. $y > x + 6$

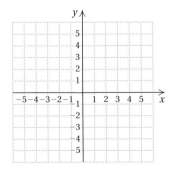

45. $2x + y \leq 2$

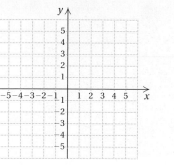

46. $f(x) = x^2 - 3$

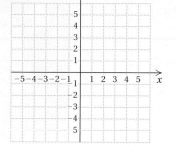

47. $g(x) = 4 - |x|$

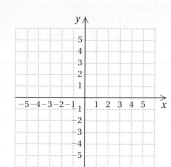

Graph. List the vertices.

48. $y \geq -x,$
$\quad y \leq 2x + 1$

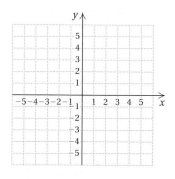

49. $2x + 3y \leq 6,$
$\quad 5x - 5y \leq 15,$
$\quad\quad x \geq 0$

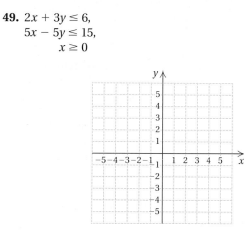

50. Find an equation of the line containing the point $(3, 7)$ and parallel to the line $x + 2y = 6$.

51. Find an equation of the line containing the point $(3, -2)$ and perpendicular to the line $3x + 4y = 5$.

52. Find an equation of the line containing the points $(-1, 4)$ and $(-2, 0)$.

53. Find an equation of the line with slope -3 and through the point $(2, 1)$.

Solve.

54. *Lens Polishing.* In a factory there are three polishing machines, A, B, and C. When all three of them are working, 5700 lenses can be polished in one week. When only A and B are working, 3400 lenses can be polished in one week. When only B and C are working, 4200 lenses can be polished in one week. How many lenses can be polished in a week by each machine?

55. *Percentage of Food Dollars Spent Away From Home.*
Americans are eating out more and more often. The percentage P of the total amount spent for food that is spent eating out is given by the linear function

$$P(t) = 0.505t + 46.84,$$

where t is the number of years since 2000.

Sources: U.S. Bureau of Labor Statistics; National Restaurant Association

a) What percentage of food dollars will be spent eating away from home in 2010?
b) In what year will the percentage of food dollars spent eating away from home be 60%?
c) In what years will the percentage of food dollars spent eating away from home be between 50% and 70%?

56. *Games in a Sports League.* In a sports league of n teams in which each team plays every other team twice, the total number N of games to be played is given by the function

$$N(n) = n^2 - n.$$

a) A women's college volleyball league has 6 teams. If we assume that each team plays every other team twice, what is the total number of games to be played?
b) Another volleyball league plays a total of 72 games. If we assume that each team plays every other team twice, how many teams are in the league?

57. *Display of a Sports Card.* A valuable sports card is 4 cm wide and 5 cm long. The card is to be sandwiched by two pieces of Lucite, each of which is $5\frac{1}{2}$ times the area of the card. Determine the dimensions of the Lucite that will ensure a uniform border.

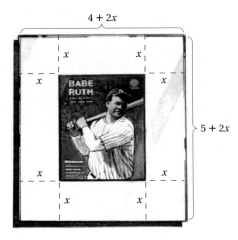

In each of Exercises 58–61, choose the correct answer from the selections given.

58. Find the slope, if it exists, of the line through the points $(-8, -5)$ and $(-7, 1)$.

a) $\dfrac{2}{5}$ **b)** $\dfrac{1}{6}$ **c)** $\dfrac{5}{2}$
d) 6 **e)** None of these

59. Find an equation of the line containing $(3, 5)$ and parallel to $-2x + y = -5$.

a) $y = -2x + 11$ **b)** $y = 2x - 1$
c) $y = -2x + 13$ **d)** $y = 2x - 7$
e) None of these

60. Which, if any, of the following systems has $(-2, 3)$ as a solution?

a) $5x + 4y = 2,$
    $x + 3y = 6$
b) $5x + 4y = 2,$
    $x + 3y = 7$
c) $5x - 4y = 2,$
    $x + 3y = 7$
d) $x + y = 1,$
    $x - y = 5$
e) None of these

61. Which of the following is a factor of $t^3 - 64$?

a) $t - 8$ **b)** $t^2 - 4t + 16$
c) $t^2 + 8t + 16$ **d)** $t + 4$
e) None of these

SYNTHESIS

62. Solve: $|x + 1| \le |x - 3|$.

63. Bert and Sally Ng are mathematics professors at a state university. Together they have 46 years of service. Two years ago, Bert had taught 2.5 times as long as Sally. How long has each taught at the university?

Rational Expressions, Equations, and Functions

5

Real-World Application

Wild horses still exist in at least ten states of the United States. To estimate the number in California, a forest ranger catches 616 wild horses, tags them, and releases them. Later, 244 horses are caught, and it is found that 49 of them are tagged. Estimate how many wild horses there are in California.

Source: U.S. Bureau of Land Management

This problem appears as Example 4 in Section 5.6.

Objectives

a Find all numbers for which a rational expression is not defined or that are not in the domain of a rational function.

b Multiply a rational expression by 1, using an expression like A/A.

c Simplify rational expressions.

d Multiply rational expressions and simplify.

e Divide rational expressions and simplify.

1. Find all numbers for which the rational expression

$$\frac{x^2 - 4x + 9}{2x + 5}$$

is not defined.

2. Find the domain of f if

$$f(x) = \frac{x^2 - 4x + 9}{2x + 5}.$$

Write both set-builder and interval notation for the answer.

a Rational Expressions and Functions

An expression that consists of the quotient of two polynomials, where the polynomial in the denominator is nonzero, is called a **rational expression.** The following are examples of rational expressions:

$$\frac{7}{8}, \quad \frac{z}{-6}, \quad \frac{a}{b}, \quad \frac{8}{y + 5}, \quad \frac{t^4 - 5t}{t^2 - 3t - 28}, \quad \frac{x^2 + 7xy - 4}{x^3 - y^3}.$$

Note that every rational number is a rational expression.

Rational expressions indicate division. Thus we cannot make a replacement of the variable that allows a denominator to be 0. (For a discussion of why we exclude division by 0, see Section R.2.)

EXAMPLE 1 Find all numbers for which the rational expression

$$\frac{2x + 1}{x - 3}$$

is not defined.

When x is replaced with 3, the denominator is 0, and the rational expression is not defined:

$$\frac{2x + 1}{x - 3} = \frac{2 \cdot 3 + 1}{3 - 3} = \frac{7}{0}. \leftarrow \text{Division by 0 is not defined.}$$

You can check some replacements other than 3 to see that it appears that 3 is the only replacement that is not allowable. Thus the rational expression is not defined for the number 3.

You may have noticed that the procedure in Example 1 is similar to one we have performed when finding the domain of a function.

EXAMPLE 2 Find the domain of f if $f(x) = \dfrac{2x + 1}{x - 3}$.

The domain is the set of all replacements for which the rational expression is defined (see Section 2.3). We begin by determining the replacements that make the denominator 0. We can do this by setting the denominator equal to 0 and solving for x:

$$x - 3 = 0$$
$$x = 3.$$

The domain of f is $\{x \mid x \text{ is a real number } and\ x \neq 3\}$, or, in interval notation, $(-\infty, 3) \cup (3, \infty)$.

Do Exercises 1 and 2.

Answers on page A-27

CHAPTER 5: Rational Expressions, Equations, and Functions

Let's make a visual check of Example 2 by looking at the following graph.

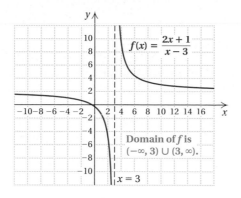

Note that the graph consists of two unconnected "branches." If a vertical line were drawn at $x = 3$, shown dashed here, it would not touch the graph of f. Thus 3 is not in the domain of f.

EXAMPLE 3 Find all numbers for which the rational expression

$$\frac{t^4 - 5t}{t^2 - 3t - 28}$$

is not defined.

The rational expression is not defined for a replacement that makes the denominator 0. To determine those replacements to exclude, we set the denominator equal to 0 and solve:

$$t^2 - 3t - 28 = 0 \qquad \text{Setting the denominator equal to 0}$$
$$(t - 7)(t + 4) = 0 \qquad \text{Factoring}$$
$$t - 7 = 0 \quad or \quad t + 4 = 0 \qquad \text{Using the principle of zero products}$$
$$t = 7 \quad or \qquad t = -4.$$

Thus the expression is not defined for the replacements 7 and -4.

EXAMPLE 4 Find the domain of g if

$$g(t) = \frac{t^4 - 5t}{t^2 - 3t - 28}.$$

We proceed as we did in Example 3. The expression is not defined for the replacements 7 and -4. Thus the domain is $\{t \mid t \text{ is a real number } and \ t \neq 7 \ and \ t \neq -4\}$, or, in interval notation, $(-\infty, -4) \cup (-4, 7) \cup (7, \infty)$.

Do Exercises 3 and 4.

3. Find all numbers for which the rational expression

$$\frac{t^2 - 9}{t^2 - 7t + 10}$$

is not defined.

4. Find the domain of g if

$$g(t) = \frac{t^2 - 9}{t^2 - 7t + 10}.$$

Write both set-builder and interval notation for the answer.

Answers on page A-27

Multiply.

5. $\dfrac{3x + 2y}{5x + 4y} \cdot \dfrac{x}{x}$

6. $\dfrac{2x^2 - y}{3x + 4} \cdot \dfrac{3x + 2}{3x + 2}$

7. $\dfrac{-1}{-1} \cdot \dfrac{2a - 5}{a - b}$

b   Finding Equivalent Rational Expressions

Calculations with rational expressions are similar to those with rational numbers.

> **MULTIPLYING RATIONAL EXPRESSIONS**
>
> To multiply rational expressions, multiply numerators and multiply denominators:
>
> $$\frac{A}{B} \cdot \frac{C}{D} = \frac{AC}{BD}.$$

For example, we have the following:

$$\frac{3}{5} \cdot \frac{2}{7} = \frac{3 \cdot 2}{5 \cdot 7} = \frac{6}{35}, \qquad \frac{3x}{4} \cdot \frac{5x}{7} = \frac{(3x)(5x)}{4 \cdot 7} = \frac{15x^2}{28},$$

and $\dfrac{x + 3}{y - 4} \cdot \dfrac{x^3}{y + 5} = \dfrac{(x + 3)x^3}{(y - 4)(y + 5)}.$ Multiplying numerators and multiplying denominators

For purposes of our work in this chapter, it is better in the example above to leave the numerator $(x + 3)x^3$ and the denominator $(y - 4)(y + 5)$ in factored form because it is easier to simplify if we do not multiply.

To simplify, we first consider multiplying by 1.

> Any rational expression with the same numerator and denominator is a symbol for 1:
>
> $$\frac{73}{73} = 1, \qquad \frac{x - y}{x - y} = 1, \qquad \frac{4x^2 - 5}{4x^2 - 5} = 1, \qquad \frac{-1}{-1} = 1, \qquad \frac{x + 5}{x + 5} = 1.$$

We can multiply by 1 to get equivalent expressions—for example,

$$\frac{7}{9} \cdot \frac{4}{4} = \frac{7 \cdot 4}{9 \cdot 4} = \frac{28}{36} \quad \text{and} \quad \frac{5}{6} \cdot \frac{x}{x} = \frac{5 \cdot x}{6 \cdot x} = \frac{5x}{6x}.$$

As another example, let's multiply $(x + y)/5$ by 1, using the symbol $(x - y)/(x - y)$:

$$\frac{x + y}{5} \cdot \frac{x - y}{x - y} = \frac{(x + y)(x - y)}{5(x - y)}. \qquad \text{Multiplying by } \frac{x - y}{x - y}, \text{ which is 1}$$

We know that the expressions

$$\frac{x + y}{5} \quad \text{and} \quad \frac{(x + y)(x - y)}{5(x - y)}$$

are equivalent. This means that they will name the same number for all replacements that do not make a denominator 0.

EXAMPLES Multiply to obtain an equivalent expression.

5. $\dfrac{x^2 + 3}{x - 1} \cdot 1 = \dfrac{x^2 + 3}{x - 1} \cdot \dfrac{x + 1}{x + 1} = \dfrac{(x^2 + 3)(x + 1)}{(x - 1)(x + 1)}$ Using $\dfrac{x + 1}{x + 1}$ for 1

6. $1 \cdot \dfrac{x - 4}{x - y} = \dfrac{-1}{-1} \cdot \dfrac{x - 4}{x - y} = \dfrac{-1 \cdot (x - 4)}{-1 \cdot (x - y)}$ Using $\dfrac{-1}{-1}$ for 1

$\qquad\qquad\quad = \dfrac{-x + 4}{-x + y} = \dfrac{4 - x}{y - x}$

Do Exercises 5–7 on the preceding page.

C   Simplifying Rational Expressions

We simplify rational expressions using the identity property of 1 (see Section R.5b) in reverse. That is, we "remove" factors that are equal to 1. We first factor the numerator and the denominator and then factor the rational expression, so that a factor is equal to 1. We also say, accordingly, that we "remove a factor of 1."

EXAMPLE 7   Simplify: $\dfrac{120}{320}$.

$\dfrac{120}{320} = \dfrac{40 \cdot 3}{40 \cdot 8}$ Factoring the numerator and the denominator, looking for common factors

$\qquad = \dfrac{40}{40} \cdot \dfrac{3}{8}$ Factoring the rational expression; $\dfrac{40}{40}$ is a factor of 1

$\qquad = 1 \cdot \dfrac{3}{8}$ $\dfrac{40}{40} = 1$

$\qquad = \dfrac{3}{8}$ Removing a factor of 1

Do Exercise 8.

EXAMPLES   Simplify.

8. $\dfrac{5x^2}{x} = \dfrac{5x \cdot x}{1 \cdot x}$ Factoring the numerator and the denominator

$\qquad = \dfrac{5x}{1} \cdot \dfrac{x}{x}$ Factoring the rational expression; $\dfrac{x}{x}$ is a factor of 1

$\qquad = 5x \cdot 1$ $\dfrac{x}{x} = 1$

$\qquad = 5x$ Removing a factor of 1

In this example, we supplied a 1 in the denominator. This can always be done, but it is not necessary.

9. $\dfrac{4a + 8}{2} = \dfrac{2(2a + 4)}{2 \cdot 1}$ Factoring the numerator and the denominator

$\qquad = \dfrac{2}{2} \cdot \dfrac{2a + 4}{1}$ Factoring the rational expression; $\dfrac{2}{2}$ is a factor of 1

$\qquad = \dfrac{2a + 4}{1}$ Removing a factor of 1

$\qquad = 2a + 4$

Do Exercises 9 and 10.

8. Simplify: $\dfrac{128}{160}$.

Simplify.

9. $\dfrac{7x^2}{x}$

10. $\dfrac{6a + 9}{3}$

Answers on page A-27

EXAMPLES Simplify.

10. $\dfrac{2x^2 + 4x}{6x^2 + 2x} = \dfrac{2x(x + 2)}{2x(3x + 1)}$ Factoring the numerator and the denominator

$\qquad\qquad = \dfrac{2x}{2x} \cdot \dfrac{x + 2}{3x + 1}$ Factoring the rational expression

$\qquad\qquad = \dfrac{x + 2}{3x + 1}$ Removing a factor of 1

11. $\dfrac{x^2 - 1}{2x^2 - x - 1} = \dfrac{(x - 1)(x + 1)}{(2x + 1)(x - 1)}$ Factoring the numerator and the denominator

$\qquad\qquad = \dfrac{x + 1}{2x + 1} \cdot \dfrac{x - 1}{x - 1}$ Factoring the rational expression

$\qquad\qquad = \dfrac{x + 1}{2x + 1}$ Removing a factor of 1

12. $\dfrac{9x^2 + 6xy - 3y^2}{12x^2 - 12y^2} = \dfrac{3(x + y)(3x - y)}{3(4)(x + y)(x - y)}$ Factoring the numerator and the denominator

$\qquad\qquad = \dfrac{3(x + y)}{3(x + y)} \cdot \dfrac{3x - y}{4(x - y)}$ Factoring the rational expression

$\qquad\qquad = \dfrac{3x - y}{4(x - y)}$ Removing a factor of 1

For purposes of later work, we generally do not multiply out the numerator and the denominator after simplifying rational expressions.

CANCELING

Canceling is a shortcut that you may have used for removing a factor of 1 when working with fraction notation or rational expressions. With great concern, we mention it here as a possible way to speed up your work. **Canceling may be done for removing factors of 1 only in products.** It *cannot* be done in sums or when adding expressions together. Our concern is that canceling be done with care and understanding. Example 12 might have been done faster as follows:

$$\dfrac{9x^2 + 6xy - 3y^2}{12x^2 - 12y^2} = \dfrac{\cancel{3}\cancel{(x + y)}(3x - y)}{\cancel{3}(4)\cancel{(x + y)}(x - y)}$$ When a factor of 1 is noted, it is "canceled" as shown.

$$\qquad\qquad = \dfrac{3x - y}{4(x - y)}.$$ Removing a factor of 1: $\dfrac{3(x + y)}{3(x + y)} = 1$

Caution!

The difficulty with canceling is that it can be applied incorrectly in situations such as the following:

$$\dfrac{\cancel{2} + 3}{\cancel{2}} = 3, \qquad \dfrac{\cancel{4} + 1}{\cancel{4} + 2} = \dfrac{1}{2}, \qquad \dfrac{1\cancel{5}}{\cancel{5}4} = \dfrac{1}{4}.$$

Wrong!   Wrong!   Wrong!

In each of these situations, the expressions canceled are *not* factors of 1. Factors are parts of products. For example, in $2 \cdot 3$, 2 and 3 are factors, but in $2 + 3$, 2 and 3 are *not* factors. **If you can't factor, you can't cancel! If in doubt, don't cancel!**

Do Exercises 11–13.

OPPOSITES IN RATIONAL EXPRESSIONS

Expressions of the form $a - b$ and $b - a$ are opposites, or additive inverses, of each other. When either of these binomials is multiplied by -1, the result is the other binomial:

$$\left.\begin{array}{l} -1(a - b) = -a + b = b - a; \\ -1(b - a) = -b + a = a - b. \end{array}\right\}$$ Multiplication by -1 reverses the order in which subtraction occurs.

Consider

$$\frac{x - 8}{8 - x}.$$

At first glance, the numerator and the denominator do not appear to have any common factors other than 1. But $x - 8$ and $8 - x$ are opposites of each other. Therefore, we can rewrite one as the opposite of the other by factoring out a -1.

EXAMPLE 13 Simplify: $\dfrac{x - 8}{8 - x}$.

$$\begin{aligned} \frac{x - 8}{8 - x} &= \frac{x - 8}{-(x - 8)} && \text{Rewriting } 8 - x \text{ as } -(x - 8). \text{ See Section R.6.} \\ &= \frac{1(x - 8)}{-1(x - 8)} \\ &= \frac{1}{-1} \cdot \frac{x - 8}{x - 8} \\ &= -1 \cdot 1 && \text{Note that } \frac{1}{-1} = -1, \text{ not } 1. \\ &= -1 \end{aligned}$$

Do Exercises 14–16.

d Multiplying and Simplifying

After multiplying, we generally simplify, if possible. That is one reason why we leave the numerator and the denominator in factored form. Even so, we might need to factor them further in order to simplify.

EXAMPLES Multiply and simplify.

14.
$$\begin{aligned} \frac{x + 2}{x - 3} \cdot \frac{x^2 - 4}{x^2 + x - 2} &= \frac{(x + 2)(x^2 - 4)}{(x - 3)(x^2 + x - 2)} && \begin{array}{l}\text{Multiplying the} \\ \text{numerators and the} \\ \text{denominators}\end{array} \\ &= \frac{(x + 2)(x + 2)(x - 2)}{(x - 3)(x + 2)(x - 1)} && \begin{array}{l}\text{Factoring the numerator} \\ \text{and the denominator}\end{array} \\ &= \frac{(x + 2)\cancel{(x + 2)}(x - 2)}{(x - 3)\cancel{(x + 2)}(x - 1)} && \begin{array}{l}\text{Removing a factor of 1:} \\ \dfrac{x + 2}{x + 2} = 1\end{array} \\ &= \frac{(x + 2)(x - 2)}{(x - 3)(x - 1)} && \text{Simplifying} \end{aligned}$$

Simplify.

11. $\dfrac{6x^2 + 4x}{4x^2 + 8x}$

12. $\dfrac{2y^2 + 6y + 4}{y^2 - 1}$

13. $\dfrac{20a^2 - 80b^2}{16a^2 - 64ab + 64b^2}$

Simplify.

14. $\dfrac{y - 3}{3 - y}$

15. $\dfrac{p - q}{q - p}$

16. $\dfrac{t + 8}{-t - 8}$

Answers on page A-27

Multiply and simplify.

17. $\dfrac{(x-y)^3}{x+y} \cdot \dfrac{3x+3y}{x^2-y^2}$

18. $\dfrac{a^3+b^3}{a^2-b^2} \cdot \dfrac{a^2-2ab+b^2}{a^2-ab+b^2}$

15. $\dfrac{a^3-b^3}{a^2-b^2} \cdot \dfrac{a^2+2ab+b^2}{a^2+ab+b^2}$

$$= \frac{(a^3-b^3)(a^2+2ab+b^2)}{(a^2-b^2)(a^2+ab+b^2)}$$

$$= \frac{(a-b)(a^2+ab+b^2)(a+b)(a+b)}{(a-b)(a+b)(a^2+ab+b^2)\cdot 1}$$ Factoring the numerator and the denominator

$$= \frac{\cancel{(a-b)}\cancel{(a^2+ab+b^2)}\cancel{(a+b)}(a+b)}{\cancel{(a-b)}\cancel{(a+b)}\cancel{(a^2+ab+b^2)}\cdot 1}$$ Removing a factor of 1: $\dfrac{(a-b)(a^2+ab+b^2)(a+b)}{(a-b)(a^2+ab+b^2)(a+b)}=1$

$$= \frac{a+b}{1}$$ Simplifying

$$= a+b$$

Do Exercises 17 and 18.

Answers on page A-27

CALCULATOR CORNER

Checking Multiplication and Simplification   We can use the TABLE feature as a partial check that rational expressions have been multiplied and/or simplified correctly. To check the simplification in Example 11,

$$\frac{x^2-1}{2x^2-x-1} = \frac{x+1}{2x+1},$$

we first enter $y_1 = (x^2-1)/(2x^2-x-1)$ and $y_2 = (x+1)/(2x+1)$. Then, using AUTO mode, we look at a table of values of y_1 and y_2. (See p. 162.) If the simplification is correct, the values should be the same for all replacements for which the rational expression is defined. The ERROR messages indicate that -0.5 and 1 are replacements in the first rational expression for which the expression is not defined and -0.5 is a replacement in the second rational expression for which the expression is not defined. For all other numbers, we see that y_1 and y_2 are the same, so the simplification appears to be correct. Remember, this is only a partial check since we cannot check all possible values of x.

X	Y₁	Y₂
-1.5	.25	.25
-1	0	0
-.5	ERR:	ERR:
0	1	1
.5	.75	.75
1	ERR:	.66667
1.5	.625	.625

X=-1.5

Exercises: Use the TABLE feature to determine whether each of the following is correct.

1. $\dfrac{5x^2}{x} = 5x$

2. $\dfrac{2x^2+4x}{6x^2+2x} = \dfrac{x+2}{3x+1}$

3. $\dfrac{x+5}{x} = 5$

4. $\dfrac{x^2-3x+2}{x^2-1} = \dfrac{x+2}{x-1}$

5. $\dfrac{x^2-16}{x^2-4} = 4$

6. $\dfrac{x+2}{x-3} \cdot \dfrac{x^2-4}{x^2+x-2} = \dfrac{(x+2)(x-2)}{(x-3)(x-1)}$

7. $\dfrac{x^2+4x+4}{x^2-9} \cdot \dfrac{x-3}{x+2} = \dfrac{x+2}{x+3}$

8. $\dfrac{x^2-5x}{x^2} \cdot \dfrac{4}{x^2-25} = \dfrac{4}{x(x+5)}$

---

e · Dividing and Simplifying

Two expressions are reciprocals (or multiplicative inverses) of each other if their product is 1. To find the reciprocal of a rational expression, we interchange the numerator and the denominator.

The reciprocal of $\dfrac{3}{7}$ is $\dfrac{7}{3}$.

The reciprocal of $\dfrac{x + 2y}{x + y - 1}$ is $\dfrac{x + y - 1}{x + 2y}$.

The reciprocal of $y - 8$ is $\dfrac{1}{y - 8}$.

Do Exercises 19–21.

We divide rational expressions in the same way that we divide fraction notation in arithmetic. For a review, see Section R.2.

DIVIDING RATIONAL EXPRESSIONS

To divide by a rational expression, multiply by its reciprocal:

$$\frac{A}{B} \div \frac{C}{D} = \frac{A}{B} \cdot \frac{D}{C} = \frac{AD}{BC}.$$

Then factor and simplify if possible.

For example,

$$\frac{2}{3} \div \frac{4}{5} = \frac{2}{3} \cdot \frac{5}{4} = \frac{2 \cdot 5}{3 \cdot 2 \cdot 2} = \frac{5}{3 \cdot 2} \cdot \frac{2}{2} = \frac{5}{6} \cdot 1 = \frac{5}{6}.$$

EXAMPLES Divide and simplify.

16. $\dfrac{x - 2}{x + 1} \div \dfrac{x + 5}{x - 3} = \dfrac{x - 2}{x + 1} \cdot \dfrac{x - 3}{x + 5}$ Multiplying by the reciprocal of the divisor

$$= \frac{(x - 2)(x - 3)}{(x + 1)(x + 5)}$$

17. $\dfrac{a^2 - 1}{a - 1} \div \dfrac{a^2 - 2a + 1}{a + 1}$

$$= \frac{a^2 - 1}{a - 1} \cdot \frac{a + 1}{a^2 - 2a + 1}$$ Multiplying by the reciprocal of the divisor

$$= \frac{(a^2 - 1)(a + 1)}{(a - 1)(a^2 - 2a + 1)}$$ Multiplying the numerators and the denominators

$$= \frac{(a + 1)(a - 1)(a + 1)}{(a - 1)(a - 1)(a - 1)}$$ Factoring the numerator and the denominator

$$= \frac{(a + 1)\cancel{(a - 1)}(a + 1)}{(a - 1)\cancel{(a - 1)}(a - 1)}$$ Removing a factor of 1: $\dfrac{a - 1}{a - 1} = 1$

$$= \frac{(a + 1)(a + 1)}{(a - 1)(a - 1)}$$ Simplifying

Find the reciprocal.

19. $\dfrac{x + 3}{x - 5}$

20. $x + 7$

21. $\dfrac{1}{y^3 - 9}$

Answers on page A-27

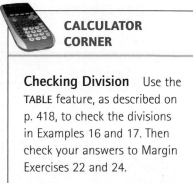

CALCULATOR CORNER

Checking Division Use the TABLE feature, as described on p. 418, to check the divisions in Examples 16 and 17. Then check your answers to Margin Exercises 22 and 24.

419

Divide and simplify.

22. $\dfrac{x^2 + 7x + 10}{2x - 4} \div \dfrac{x^2 - 3x - 10}{x - 2}$

23. $\dfrac{a^2 - b^2}{ab} \div \dfrac{a^2 - 2ab + b^2}{2a^2b^2}$

24. Perform the indicated operations and simplify:

$\dfrac{a^3 + 8}{a - 2} \div (a^2 - 2a + 4) \cdot (a - 2)^2.$

Do Exercises 22 and 23.

EXAMPLE 18 Perform the indicated operations and simplify:

$$\frac{c^3 - d^3}{(c + d)^2} \div (c - d) \cdot (c + d).$$

Using the rules for order of operations, we do the division first:

$$\frac{c^3 - d^3}{(c + d)^2} \div (c - d) \cdot (c + d)$$

$$= \frac{c^3 - d^3}{(c + d)^2} \cdot \frac{1}{c - d} \cdot (c + d)$$

$$= \frac{(c - d)(c^2 + cd + d^2)(c + d)}{(c + d)(c + d)(c - d)}$$

$$= \frac{\cancel{(c - d)}(c^2 + cd + d^2)\cancel{(c + d)}}{(c + d)\cancel{(c + d)}\cancel{(c - d)}} \qquad \begin{array}{l} \text{Removing a factor of 1:} \\ \dfrac{(c - d)(c + d)}{(c - d)(c + d)} = 1 \end{array}$$

$$= \frac{c^2 + cd + d^2}{c + d}.$$

Do Exercise 24.

Answers on page A-27

Study Tips

WORKING WITH RATIONAL EXPRESSIONS

The procedures covered in this chapter are by their nature rather long. It may help to write out lots of steps as you do the problems. If you have difficulty, consider taking a clean sheet of paper and starting over. Don't squeeze your work into a small amount of space. When using lined paper, consider using two spaces at a time, with the paper's line representing the fraction bar.

a   Find all numbers for which the rational expression is not defined.

1. $\dfrac{5t^2 - 64}{3t + 17}$

2. $\dfrac{x^2 + x + 105}{5x - 45}$

3. $\dfrac{x^3 - x^2 + x + 2}{x^2 + 12x + 35}$

4. $\dfrac{x^2 - 3x - 4}{x^2 - 18x + 77}$

Find the domain. Write both set-builder and interval notation for the answer.

5. $f(t) = \dfrac{5t^2 - 64}{3t + 17}$

6. $f(x) = \dfrac{x^2 + x + 105}{5x - 45}$

7. $f(x) = \dfrac{x^3 - x^2 + x + 2}{x^2 + 12x + 35}$

8. $f(x) = \dfrac{x^2 - 3x - 4}{x^2 - 18x + 77}$

b   Multiply to obtain an equivalent expression. Do not simplify.

9. $\dfrac{7x}{7x} \cdot \dfrac{x + 2}{x + 8}$

10. $\dfrac{2 - y^2}{8 - y} \cdot \dfrac{-1}{-1}$

11. $\dfrac{q - 5}{q + 3} \cdot \dfrac{q + 5}{q + 5}$

12. $\dfrac{p + 1}{p + 4} \cdot \dfrac{p - 4}{p - 4}$

c   Simplify.

13. $\dfrac{15y^5}{5y^4}$

14. $\dfrac{7w^3}{28w^2}$

15. $\dfrac{16p^3}{24p^7}$

16. $\dfrac{48t^5}{56t^{11}}$

17. $\dfrac{9a - 27}{9}$

18. $\dfrac{6a - 30}{6}$

19. $\dfrac{12x - 15}{21}$

20. $\dfrac{18a - 2}{22}$

21. $\dfrac{4y - 12}{4y + 12}$

22. $\dfrac{8x + 16}{8x - 16}$

23. $\dfrac{t^2 - 16}{t^2 - 8t + 16}$

24. $\dfrac{p^2 - 25}{p^2 + 10p + 25}$

25. $\dfrac{x^2 - 9x + 8}{x^2 + 3x - 4}$

26. $\dfrac{y^2 + 8y - 9}{y^2 - 5y + 4}$

27. $\dfrac{w^3 - z^3}{w^2 - z^2}$

28. $\dfrac{a^2 - b^2}{a^3 + b^3}$

d Multiply and simplify.

29. $\dfrac{x^4}{3x + 6} \cdot \dfrac{5x + 10}{5x^7}$

30. $\dfrac{10t}{6t - 12} \cdot \dfrac{20t - 40}{30t^3}$

31. $\dfrac{x^2 - 16}{x^2} \cdot \dfrac{x^2 - 4x}{x^2 - x - 12}$

32. $\dfrac{y^2 + 10y + 25}{y^2 - 9} \cdot \dfrac{y^2 - 3y}{y + 5}$

33. $\dfrac{y^2 - 16}{2y + 6} \cdot \dfrac{y + 3}{y - 4}$

34. $\dfrac{m^2 - n^2}{4m + 4n} \cdot \dfrac{m + n}{m - n}$

35. $\dfrac{x^2 - 2x - 35}{2x^3 - 3x^2} \cdot \dfrac{4x^3 - 9x}{7x - 49}$

36. $\dfrac{y^2 - 10y + 9}{y^2 - 1} \cdot \dfrac{y + 4}{y^2 - 5y - 36}$

37. $\dfrac{c^3 + 8}{c^2 - 4} \cdot \dfrac{c^2 - 4c + 4}{c^2 - 2c + 4}$

38. $\dfrac{x^3 - 27}{x^2 - 9} \cdot \dfrac{x^2 - 6x + 9}{x^2 + 3x + 9}$

39. $\dfrac{x^2 - y^2}{x^3 - y^3} \cdot \dfrac{x^2 + xy + y^2}{x^2 + 2xy + y^2}$

40. $\dfrac{4x^2 - 9y^2}{8x^3 - 27y^3} \cdot \dfrac{4x^2 + 6xy + 9y^2}{4x^2 + 12xy + 9y^2}$

Divide and simplify.

41. $\dfrac{12x^8}{3y^4} \div \dfrac{16x^3}{6y}$

42. $\dfrac{9a^7}{8b^2} \div \dfrac{12a^2}{24b^7}$

43. $\dfrac{3y + 15}{y} \div \dfrac{y + 5}{y}$

44. $\dfrac{6x + 12}{x} \div \dfrac{x + 2}{x^3}$

45. $\dfrac{y^2 - 9}{y} \div \dfrac{y + 3}{y + 2}$

46. $\dfrac{x^2 - 4}{x} \div \dfrac{x - 2}{x + 4}$

47. $\dfrac{4a^2 - 1}{a^2 - 4} \div \dfrac{2a - 1}{a - 2}$

48. $\dfrac{25x^2 - 4}{x^2 - 9} \div \dfrac{5x - 2}{x + 3}$

49. $\dfrac{x^2 - 16}{x^2 - 10x + 25} \div \dfrac{3x - 12}{x^2 - 3x - 10}$

50. $\dfrac{y^2 - 36}{y^2 - 8y + 16} \div \dfrac{3y - 18}{y^2 - y - 12}$

51. $\dfrac{y^3 + 3y}{y^2 - 9} \div \dfrac{y^2 + 5y - 14}{y^2 + 4y - 21}$

52. $\dfrac{a^3 + 4a}{a^2 - 16} \div \dfrac{a^2 + 8a + 15}{a^2 + a - 20}$

53. $\dfrac{x^3 - 64}{x^3 + 64} \div \dfrac{x^2 - 16}{x^2 - 4x + 16}$

54. $\dfrac{8y^3 + 27}{64y^3 - 1} \div \dfrac{4y^2 - 9}{16y^2 + 4y + 1}$

55. $\dfrac{8x^3y^3 + 27x^3}{64x^3y^3 - x^3} \div \dfrac{4x^2y^2 - 9x^2}{16x^2y^2 + 4x^2y + x^2}$

56. $\dfrac{x^3y - 64y}{x^3y + 64y} \div \dfrac{x^2y^2 - 16y^2}{x^2y^2 - 4xy^2 + 16y^2}$

Perform the indicated operations and simplify.

57. $\dfrac{r^2 - 4s^2}{r + 2s} \div (r + 2s) \cdot \dfrac{2s}{r - 2s}$

58. $\dfrac{d^2 - d}{d^2 - 6d + 8} \cdot \dfrac{d - 2}{d^2 + 5d} \div \dfrac{5d}{d^2 - 9d + 20}$

59. $\mathbf{D_W}$ Is it possible to understand how to simplify rational expressions without first understanding how to multiply? Why or why not?

60. $\mathbf{D_W}$ Nancy *incorrectly* simplifies $(x + 2)/x$ as follows:

$$\frac{x + 2}{x} = \frac{\not{x} + 2}{\not{x}} = 1 + 2 = 3.$$

She insists that this is correct because when x is replaced with 1, her answer checks. Explain her error.

SKILL MAINTENANCE

In Exercises 61–64, the graph is that of a function. Determine the domain and the range.   [2.3a]

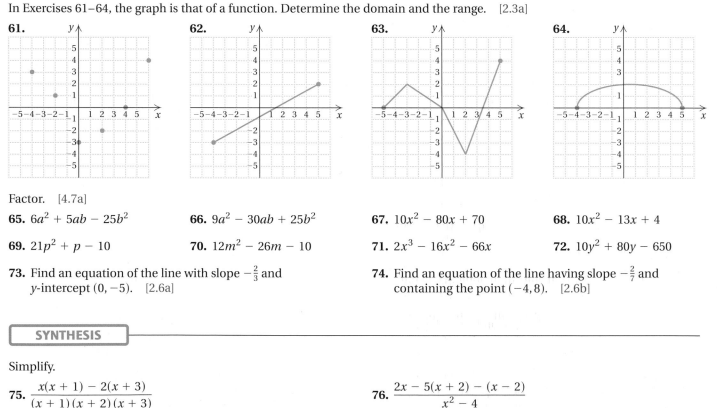

61.   **62.**   **63.**   **64.**

Factor.   [4.7a]

65. $6a^2 + 5ab - 25b^2$

66. $9a^2 - 30ab + 25b^2$

67. $10x^2 - 80x + 70$

68. $10x^2 - 13x + 4$

69. $21p^2 + p - 10$

70. $12m^2 - 26m - 10$

71. $2x^3 - 16x^2 - 66x$

72. $10y^2 + 80y - 650$

73. Find an equation of the line with slope $-\frac{2}{3}$ and y-intercept $(0, -5)$.   [2.6a]

74. Find an equation of the line having slope $-\frac{2}{7}$ and containing the point $(-4, 8)$.   [2.6b]

SYNTHESIS

Simplify.

75. $\dfrac{x(x + 1) - 2(x + 3)}{(x + 1)(x + 2)(x + 3)}$

76. $\dfrac{2x - 5(x + 2) - (x - 2)}{x^2 - 4}$

77. $\dfrac{m^2 - t^2}{m^2 + t^2 + m + t + 2mt}$

78. $\dfrac{a^3 - 2a^2 + 2a - 4}{a^3 - 2a^2 - 3a + 6}$

79. Let

$$g(x) = \frac{2x + 3}{4x - 1}.$$

Find $g(5)$, $g(0)$, $g\left(\frac{1}{4}\right)$, and $g(a + h)$.

CHAPTER 5: Rational Expressions,
Equations, and Functions

5.2

LCMS, LCDS, ADDITION, AND SUBTRACTION

a   Finding LCMs by Factoring

To add rational expressions when denominators are different, we first find a common denominator. Let's review the procedure used in arithmetic first. To do the addition

$$\frac{5}{42} + \frac{7}{12},$$

we find a common denominator. We look for the least common multiple (LCM) of 42 and 12. That number becomes the least common denominator (LCD).

To find the LCM, we factor both numbers completely (into primes).

$$42 = 2 \cdot 3 \cdot 7 \longleftarrow \boxed{\text{Any multiple of 42 has these factors.}}$$

$$12 = 2 \cdot 2 \cdot 3 \longleftarrow \boxed{\text{Any multiple of 12 has these factors.}}$$

The LCM is the number that has 2 as a factor twice, 3 as a factor once, and 7 as a factor once: LCM = $2 \cdot 2 \cdot 3 \cdot 7$, or 84.

> **FINDING LCMS**
>
> To find the LCM, use each factor the greatest number of times that it occurs in any one prime factorization.

EXAMPLE 1   Find the LCM of 18 and 24.

$$\left. \begin{array}{l} 18 = \boxed{3 \cdot 3} \cdot 2 \\ \\ 24 = \boxed{2 \cdot 2 \cdot 2} \cdot 3 \end{array} \right\} \quad \text{The LCM is } \boxed{3 \cdot 3} \cdot \boxed{2 \cdot 2 \cdot 2} \text{, or 72.}$$

Do Exercises 1 and 2.

Now let's return to adding $\frac{5}{42}$ and $\frac{7}{12}$:

$$\frac{5}{42} + \frac{7}{12} = \frac{5}{2 \cdot 3 \cdot 7} + \frac{7}{2 \cdot 2 \cdot 3}. \quad \text{Factoring the denominators}$$

The LCD is the LCM of the denominators, $2 \cdot 2 \cdot 3 \cdot 7$. To get this LCD in the first denominator, we need a factor of 2. In the second denominator, we need a factor of 7. We multiply by 1, as follows:

$$\frac{5}{2 \cdot 3 \cdot 7} \cdot \frac{2}{2} + \frac{7}{2 \cdot 2 \cdot 3} \cdot \frac{7}{7} = \frac{10}{2 \cdot 2 \cdot 3 \cdot 7} + \frac{49}{2 \cdot 2 \cdot 3 \cdot 7}$$

$$= \frac{59}{2 \cdot 2 \cdot 3 \cdot 7} = \frac{59}{84}.$$

Objectives

a   Find the LCM of several algebraic expressions by factoring.

b   Add and subtract rational expressions.

c   Simplify combined additions and subtractions of rational expressions.

Find the LCM by factoring.

1. 18, 30

2. 12, 18, 24

Answers on page A-28

Add, first finding the LCD of the denominators.

3. $\dfrac{5}{12} + \dfrac{11}{30}$

4. $\dfrac{7}{12} + \dfrac{13}{18} + \dfrac{1}{24}$

Multiplying the first fraction by $\frac{2}{2}$ gave us an equivalent fraction with a denominator that is the LCD. Multiplying the second fraction by $\frac{7}{7}$ also gave us an equivalent fraction with a denominator that is the LCD. Once we had a common denominator, we added the numerators.

Do Exercises 3 and 4.

We find the LCM of algebraic expressions in the same way that we find the LCM of natural numbers.

Our reasoning for learning how to find LCMs is so that we will be able to add rational expressions. For example, to do the addition

$$\frac{7}{12xy^2} + \frac{8}{15x^3y},$$

we first need to find the LCM of $12xy^2$ and $15x^3y$, which is $60x^3y^2$.

EXAMPLES

2. Find the LCM of $12xy^2$ and $15x^3y$.

We factor each expression completely. To find the LCM, we use each factor the greatest number of times that it occurs in any one prime factorization.

$$12xy^2 = \boxed{2 \cdot 2 \cdot 3} \cdot x \cdot \boxed{y \cdot y} \;;$$

$$15x^3y = 3 \cdot \boxed{5 \cdot x \cdot x \cdot x} \cdot y$$ Factoring

$12xy^2$ is a factor.

$$\text{LCM} = 2 \cdot 2 \cdot 3 \cdot 5 \cdot x \cdot x \cdot x \cdot y \cdot y = 60x^3y^2$$

$15x^3y$ is a factor.

3. Find the LCM of $x^2 + 2x + 1$, $5x^2 - 5x$, and $x^2 - 1$.

$$x^2 + 2x + 1 = \boxed{(x + 1)(x + 1)} \;;$$

$$5x^2 - 5x = \boxed{5x(x - 1)} \;;$$ Factoring

$$x^2 - 1 = (x + 1)(x - 1)$$ Both factors of $x^2 - 1$ are already present in the previous factorizations.

$$\text{LCM} = 5x(x + 1)(x + 1)(x - 1)$$

4. Find the LCM of $x^2 - y^2$, $x^3 + y^3$, and $x^2 + 2xy + y^2$.

$$x^2 - y^2 = \boxed{(x - y)} \,(x + y);$$

$$x^3 + y^3 = (x + y)\boxed{(x^2 - xy + y^2)} \;;$$ Factoring

$$x^2 + 2xy + y^2 = (x + y)(x + y) = \boxed{(x + y)^2}$$

$$\text{LCM} = (x - y)(x + y)^2(x^2 - xy + y^2)$$

Answers on page A-28

Recall that $-(x - 3) = -1(x - 3) = 3 - x$. If $(x - 3)(x + 2)$ is an LCM, then $-1(x - 3)(x + 2) = (3 - x)(x + 2)$ is also an LCM.

If, when we are finding LCMs, factors that are opposites occur, we do not use both of them. For example, if $a - b$ occurs in one factorization and $b - a$ occurs in another, we do not use both, since they are opposites.

EXAMPLE 5 Find the LCM of $x^2 - y^2$ and $3y - 3x$.

$$x^2 - y^2 = \boxed{(x + y)(x - y)} \longleftarrow$$

> We can use $(x - y)$ or $(y - x)$, but we do not use both.

$$3y - 3x = \boxed{3}\ (y - x),\ \text{or}\ -3(x - y)$$

$$\text{LCM} = 3(x + y)(x - y),\ \text{or}\ 3(x + y)(y - x),\ \text{or}\ -3(x + y)(x - y)$$

In most cases, we would use the form LCM $= 3(x + y)(x - y)$.

Do Exercises 5–8.

b Adding and Subtracting Rational Expressions

ADDITION AND SUBTRACTION WITH LIKE DENOMINATORS

To add or subtract when denominators are the same, add or subtract the numerators and keep the same denominator.

$$\frac{A}{C} + \frac{B}{C} = \frac{A + B}{C} \quad \text{and} \quad \frac{A}{C} - \frac{B}{C} = \frac{A - B}{C}, \quad \text{where } C \neq 0.$$

Then factor and simplify if possible.

EXAMPLE 6 Add: $\dfrac{3 + x}{x} + \dfrac{4}{x}$.

$$\frac{3 + x}{x} + \frac{4}{x} = \frac{3 + x + 4}{x} \qquad \text{Adding numerators and keeping the same denominator}$$

Caution!

$$= \frac{7 + x}{x} \longleftarrow$$

This expression does *not* simplify to 7: $\dfrac{7 + x}{x} \neq 7$.

Example 6 shows that

$$\frac{3 + x}{x} + \frac{4}{x} \quad \text{and} \quad \frac{7 + x}{x}$$

are equivalent expressions. They name the same number for all replacements for which the rational expressions are defined.

Find the LCM.

5. $a^2 b^2,\ 5a^3 b$

6. $y^2 + 7y + 12,\ y^2 + 8y + 16,$
$y + 4$

7. $x^2 - 9,\ x^3 - x^2 - 6x,\ 2x^2$

8. $a^2 - b^2,\ 2b - 2a$

Answers on page A-28

Add.

9. $\dfrac{5 + y}{y} + \dfrac{7}{y}$

EXAMPLE 7   Add: $\dfrac{4x^2 - 5xy}{x^2 - y^2} + \dfrac{2xy - y^2}{x^2 - y^2}$.

$$\dfrac{4x^2 - 5xy}{x^2 - y^2} + \dfrac{2xy - y^2}{x^2 - y^2} = \dfrac{4x^2 - 3xy - y^2}{x^2 - y^2} \qquad \text{Adding the numerators}$$

$$= \dfrac{(4x + y)(x - y)}{(x + y)(x - y)} \qquad \begin{array}{l}\text{Factoring the numerator and}\\\text{the denominator}\end{array}$$

$$= \dfrac{(4x + y)\cancel{(x - y)}}{(x + y)\cancel{(x - y)}} \qquad \text{Removing a factor of 1: } \dfrac{x - y}{x - y} = 1$$

$$= \dfrac{4x + y}{x + y}$$

Do Exercises 9 and 10.

10. $\dfrac{2x^2 + 5x - 9}{x - 5} + \dfrac{x^2 - 19x + 4}{x - 5}$

EXAMPLE 8   Subtract: $\dfrac{4x + 5}{x + 3} - \dfrac{x - 2}{x + 3}$.

$$\dfrac{4x + 5}{x + 3} - \dfrac{x - 2}{x + 3} = \dfrac{4x + 5 - (x - 2)}{x + 3} \qquad \text{Subtracting numerators}$$

$$= \dfrac{4x + 5 - x + 2}{x + 3}$$

$$= \dfrac{3x + 7}{x + 3}$$

> A common error: forgetting these parentheses. If you forget them, you will be subtracting only *part* of the numerator, $x - 2$.

Do Exercises 11 and 12.

Subtract.

11. $\dfrac{a}{b + 2} - \dfrac{b}{b + 2}$

When denominators are different, we find the least common denominator, LCD. The procedure we will use is as follows.

> **ADDITION AND SUBTRACTION WITH DIFFERENT DENOMINATORS**
>
> To add or subtract rational expressions with different denominators:
>
> 1. Find the LCM of the denominators. This is the least common denominator (LCD).
> 2. For each rational expression, find an equivalent expression with the LCD. To do so, multiply by 1 using an expression for 1 made up of factors of the LCD that are missing from the original denominator.
> 3. Add or subtract the numerators. Write the result over the LCD.
> 4. Simplify, if possible.

12. $\dfrac{4y + 7}{x^2 + y^2} - \dfrac{3y - 5}{x^2 + y^2}$

EXAMPLE 9   Add: $\dfrac{2a}{5} + \dfrac{3b}{2a}$.

We first find the LCD: $\left.\begin{array}{l}5\\2a\end{array}\right\}$ LCD $= 5 \cdot 2a$, or $10a$.

Now we multiply each expression by 1. We choose symbols for 1 that will give us the LCD in each denominator. In this case, we use $2a/(2a)$ and $5/5$:

$$\dfrac{2a}{5} \cdot \dfrac{2a}{2a} + \dfrac{3b}{2a} \cdot \dfrac{5}{5} = \dfrac{4a^2}{10a} + \dfrac{15b}{10a} = \dfrac{4a^2 + 15b}{10a}.$$

Multiplying the first term by $2a/(2a)$ gave us a denominator of $10a$. Multiplying the second term by $\frac{5}{5}$ also gave us a denominator of $10a$.

EXAMPLE 10 Add: $\dfrac{3x^2 + 3xy}{x^2 - y^2} + \dfrac{2 - 3x}{x - y}$.

We first find the LCD of the denominators. To do so, we first factor:

$$\left. \begin{array}{l} x^2 - y^2 = (x + y)(x - y) \\[4pt] x - y = x - y \end{array} \right\} \quad \text{LCD} = (x + y)(x - y).$$

We now multiply by 1 to get the LCD in the second expression. Then we add and simplify if possible.

$$\dfrac{3x^2 + 3xy}{(x + y)(x - y)} + \dfrac{2 - 3x}{x - y} \cdot \dfrac{x + y}{x + y} \qquad \begin{array}{l}\text{Multiplying by 1 to get}\\ \text{the LCD}\end{array}$$

$$= \dfrac{3x^2 + 3xy}{(x + y)(x - y)} + \dfrac{(2 - 3x)(x + y)}{(x - y)(x + y)}$$

$$= \dfrac{3x^2 + 3xy}{(x + y)(x - y)} + \dfrac{2x + 2y - 3x^2 - 3xy}{(x - y)(x + y)} \qquad \text{Multiplying in the numerator}$$

$$= \dfrac{3x^2 + 3xy + 2x + 2y - 3x^2 - 3xy}{(x + y)(x - y)} \qquad \text{Adding the numerators}$$

$$= \dfrac{2x + 2y}{(x + y)(x - y)} \qquad \text{Combining like terms}$$

$$= \dfrac{2(x + y)}{(x + y)(x - y)} \qquad \text{Factoring the numerator}$$

$$= \dfrac{2\cancel{(x + y)}}{\cancel{(x + y)}(x - y)} \qquad \text{Removing a factor of 1: } \dfrac{x + y}{x + y} = 1$$

$$= \dfrac{2}{x - y}$$

Do Exercises 13 and 14.

EXAMPLE 11 Subtract: $\dfrac{2y + 1}{y^2 - 7y + 6} - \dfrac{y + 3}{y^2 - 5y - 6}$.

$$\dfrac{2y + 1}{y^2 - 7y + 6} - \dfrac{y + 3}{y^2 - 5y - 6}$$

$$= \dfrac{2y + 1}{(y - 6)(y - 1)} - \dfrac{y + 3}{(y - 6)(y + 1)} \qquad \text{LCD} = (y - 6)(y - 1)(y + 1)$$

$$= \dfrac{2y + 1}{(y - 6)(y - 1)} \cdot \dfrac{y + 1}{y + 1} - \dfrac{y + 3}{(y - 6)(y + 1)} \cdot \dfrac{y - 1}{y - 1} \qquad \begin{array}{l}\text{Multiplying by 1}\\ \text{to get the LCD}\end{array}$$

$$= \dfrac{(2y + 1)(y + 1) - (y + 3)(y - 1)}{(y - 6)(y - 1)(y + 1)} \qquad \text{Subtracting the numerators}$$

$$= \dfrac{(2y^2 + 3y + 1) - (y^2 + 2y - 3)}{(y - 6)(y - 1)(y + 1)} \qquad \begin{array}{l}\text{Multiplying. Note the use}\\ \text{of parentheses.}\end{array}$$

$$= \dfrac{2y^2 + 3y + 1 - y^2 - 2y + 3}{(y - 6)(y - 1)(y + 1)}$$

$$= \dfrac{y^2 + y + 4}{(y - 6)(y - 1)(y + 1)} \qquad \begin{array}{l}\text{The numerator cannot be factored.}\\ \text{The rational expression is simplified.}\end{array}$$

We generally do not multiply out a numerator or a denominator if it has three or more factors (other than monomials). This will be helpful when we solve equations.

Do Exercises 15 and 16.

Add.

13. $\dfrac{3x}{7} + \dfrac{4y}{3x}$

14. $\dfrac{2xy - 2x^2}{x^2 - y^2} + \dfrac{2x + 3}{x + y}$

Subtract.

15. $\dfrac{a}{a + 3} - \dfrac{a - 4}{a}$

16. $\dfrac{4y - 5}{y^2 - 7y + 12} - \dfrac{y + 7}{y^2 + 2y - 15}$

Answers on page A-28

Add.

17. $\dfrac{b}{3b} + \dfrac{b^3}{-3b}$

18. $\dfrac{3x^2 + 4}{x - 5} + \dfrac{x^2 - 7}{5 - x}$

Subtract.

19. $\dfrac{3}{4y} - \dfrac{7x}{-4y}$

20. $\dfrac{4x^2}{2x - y} - \dfrac{7x^2}{y - 2x}$

Answers on page A-28

CALCULATOR CORNER

Checking Addition and Subtraction   Use the TABLE feature, as described on p. 418, to check the sums and differences in Examples 6, 8, 11, and 15. Then check your answers to Margin Exercises 15 and 21.

DENOMINATORS THAT ARE OPPOSITES

When one denominator is the opposite of the other, we can first multiply either expression by 1 using $-1/-1$.

EXAMPLE 12   Add: $\dfrac{a}{2a} + \dfrac{a^3}{-2a}$.

$$\dfrac{a}{2a} + \dfrac{a^3}{-2a} = \dfrac{a}{2a} + \dfrac{a^3}{-2a} \cdot \dfrac{-1}{-1} \qquad \text{Multiplying by 1, using } \dfrac{-1}{-1} \leftarrow$$

$\boxed{\text{This is equal to 1 (not } -1\text{).}}$

$$= \dfrac{a}{2a} + \dfrac{-a^3}{2a}$$

$$= \dfrac{a - a^3}{2a} \qquad \text{Adding numerators}$$

$$= \dfrac{a(1 - a)^2}{2a} \qquad \text{Factoring}$$

$$= \dfrac{\cancel{a}(1 - a^2)}{2\cancel{a}} \qquad \text{Removing a factor of 1: } \dfrac{a}{a} = 1$$

$$= \dfrac{1 - a^2}{2}$$

EXAMPLE 13   Subtract: $\dfrac{x^2}{5y} - \dfrac{x^3}{-5y}$.

$$\dfrac{x^2}{5y} - \dfrac{x^3}{-5y} = \dfrac{x^2}{5y} - \dfrac{x^3}{-5y} \cdot \dfrac{-1}{-1} \qquad \text{Multiplying by } \dfrac{-1}{-1}$$

$$= \dfrac{x^2}{5y} - \dfrac{-x^3}{5y}$$

$$= \dfrac{x^2 - (-x^3)}{5y} \leftarrow \boxed{\text{Don't forget these parentheses!}}$$

$$= \dfrac{x^2 + x^3}{5y}$$

EXAMPLE 14   Subtract: $\dfrac{5x}{x - 2y} - \dfrac{3y - 7}{2y - x}$.

$$\dfrac{5x}{x - 2y} - \dfrac{3y - 7}{2y - x} = \dfrac{5x}{x - 2y} - \dfrac{3y - 7}{2y - x} \cdot \dfrac{-1}{-1}$$

$$= \dfrac{5x}{x - 2y} - \dfrac{-3y + 7}{x - 2y} \leftarrow \boxed{\begin{array}{l}\text{Remember: } (2y - x)(-1) = \\ -2y + x = x - 2y.\end{array}}$$

$$= \dfrac{5x - (-3y + 7)}{x - 2y} \qquad \text{Subtracting numerators}$$

$$= \dfrac{5x + 3y - 7}{x - 2y}$$

Do Exercises 17–20.

C | Combined Additions and Subtractions

EXAMPLE 15 Perform the indicated operations and simplify.

$$\frac{2x}{x^2 - 4} + \frac{5}{2 - x} - \frac{1}{2 + x}$$

$$= \frac{2x}{(x - 2)(x + 2)} + \frac{5}{2 - x} - \frac{1}{2 + x}$$

$$= \frac{2x}{(x - 2)(x + 2)} + \frac{5}{2 - x} \cdot \frac{-1}{-1} - \frac{1}{x + 2} \qquad \text{Multiplying by } \frac{-1}{-1}$$

$$= \frac{2x}{(x - 2)(x + 2)} + \frac{-5}{x - 2} - \frac{1}{x + 2} \qquad \text{LCD} = (x - 2)(x + 2)$$

$$= \frac{2x}{(x - 2)(x + 2)} + \frac{-5}{x - 2} \cdot \frac{x + 2}{x + 2} - \frac{1}{x + 2} \cdot \frac{x - 2}{x - 2} \qquad \begin{array}{l}\text{Multiplying by 1} \\ \text{to get the LCD}\end{array}$$

$$= \frac{2x - 5(x + 2) - (x - 2)}{(x - 2)(x + 2)} \qquad \text{Adding and subtracting the numerators}$$

$$= \frac{2x - 5x - 10 - x + 2}{(x - 2)(x + 2)} \qquad \text{Removing parentheses}$$

$$= \frac{-4x - 8}{(x - 2)(x + 2)}$$

$$= \frac{-4\cancel{(x + 2)}}{(x - 2)\cancel{(x + 2)}} \qquad \text{Removing a factor of 1: } \frac{x + 2}{x + 2} = 1$$

$$= \frac{-4}{x - 2}, \text{ or } -\frac{4}{x - 2}$$

Another correct form of the answer is $4/(2 - x)$. It is found by multiplying by $-1/-1$.

Do Exercise 21.

21. Perform the indicated operations and simplify:

$$\frac{8x}{x^2 - 1} + \frac{2}{1 - x} - \frac{4}{x + 1}.$$

Answer on page A-28

5.2

EXERCISE SET

For Extra Help

MathXL | MyMathLab | InterAct Math | Math Tutor Center | Digital Video Tutor CD 2 Videotape 6 | Student's Solutions Manual

a   Find the LCM by factoring.

1. 15, 40   **2.** 12, 32   **3.** 18, 48   **4.** 45, 54

5. 30, 105   **6.** 24, 60   **7.** 9, 15, 5   **8.** 27, 35, 63

Add. Find the LCD first.

9. $\dfrac{5}{6} + \dfrac{4}{15}$ **10.** $\dfrac{5}{12} + \dfrac{13}{18}$ **11.** $\dfrac{7}{36} + \dfrac{1}{24}$

12. $\dfrac{11}{30} + \dfrac{19}{75}$ **13.** $\dfrac{3}{4} + \dfrac{7}{30} + \dfrac{1}{16}$ **14.** $\dfrac{5}{8} + \dfrac{7}{12} + \dfrac{11}{40}$

Find the LCM.

15. $21x^2y,\ 7xy$ **16.** $18a^2b,\ 50ab^3$ **17.** $y^2 - 100,\ 10y + 100$

18. $r^2 - s^2,\ rs + s^2$ **19.** $15ab^2,\ 3ab,\ 10a^3b$ **20.** $6x^2y^2,\ 9x^3y,\ 15y^3$

21. $5y - 15,\ y^2 - 6y + 9$ **22.** $x^2 + 10x + 25,\ x^2 + 2x - 15$

23. $y^2 - 25,\ 5 - y$ **24.** $x^2 - 36,\ 6 - x$

25. $2r^2 - 5r - 12,\ 3r^2 - 13r + 4,\ r^2 - 16$ **26.** $2x^2 - 5x - 3,\ 2x^2 - x - 1,\ x^2 - 6x + 9$

27. $x^5 + 4x^3,\ x^3 - 4x^2 + 4x$ **28.** $9x^3 + 9x^2 - 18x,\ 6x^5 + 24x^4 + 24x^3$

29. $x^5 - 2x^4 + x^3,\ 2x^3 + 2x,\ 5x + 5$ **30.** $x^5 - 4x^4 + 4x^3,\ 3x^2 - 12,\ 2x + 4$

b   Add or subtract. Then simplify. If a denominator has three or more factors (other than monomials), leave it factored.

31. $\dfrac{x - 2y}{x + y} + \dfrac{x + 9y}{x + y}$

32. $\dfrac{a - 8b}{a + b} + \dfrac{a + 13b}{a + b}$

33. $\dfrac{4y + 3}{y - 2} - \dfrac{y - 2}{y - 2}$

34. $\dfrac{3t + 2}{t - 4} - \dfrac{t - 4}{t - 4}$

35. $\dfrac{a^2}{a - b} + \dfrac{b^2}{b - a}$

36. $\dfrac{r^2}{r - s} + \dfrac{s^2}{s - r}$

37. $\dfrac{6}{y} - \dfrac{7}{-y}$

38. $\dfrac{4}{x} - \dfrac{9}{-x}$

39. $\dfrac{4a - 2}{a^2 - 49} + \dfrac{5 + 3a}{49 - a^2}$

40. $\dfrac{2y - 3}{y^2 - 1} - \dfrac{4 - y}{1 - y^2}$

41. $\dfrac{a^3}{a - b} + \dfrac{b^3}{b - a}$

42. $\dfrac{x^3}{x^2 - y^2} + \dfrac{y^3}{y^2 - x^2}$

43. $\dfrac{y - 2}{y + 4} + \dfrac{y + 3}{y - 5}$

44. $\dfrac{x - 2}{x + 3} + \dfrac{x + 2}{x - 4}$

45. $\dfrac{4xy}{x^2 - y^2} + \dfrac{x - y}{x + y}$

46. $\dfrac{5ab}{a^2 - b^2} + \dfrac{a + b}{a - b}$

47. $\dfrac{9x + 2}{3x^2 - 2x - 8} + \dfrac{7}{3x^2 + x - 4}$

48. $\dfrac{3y + 2}{2y^2 - y - 10} + \dfrac{8}{2y^2 - 7y + 5}$

49. $\dfrac{4}{x + 1} + \dfrac{x + 2}{x^2 - 1} + \dfrac{3}{x - 1}$

50. $\dfrac{-2}{y + 2} + \dfrac{5}{y - 2} + \dfrac{y + 3}{y^2 - 4}$

51. $\dfrac{x-1}{3x+15} - \dfrac{x+3}{5x+25}$

52. $\dfrac{y-2}{4y+8} - \dfrac{y+6}{5y+10}$

53. $\dfrac{5ab}{a^2-b^2} - \dfrac{a-b}{a+b}$

54. $\dfrac{6xy}{x^2-y^2} - \dfrac{x+y}{x-y}$

55. $\dfrac{3y}{y^2-7y+10} - \dfrac{2y}{y^2-8y+15}$

56. $\dfrac{5x}{x^2-6x+8} - \dfrac{3x}{x^2-x-12}$

57. $\dfrac{y}{y^2-y-20} + \dfrac{2}{y+4}$

58. $\dfrac{6}{y^2+6y+9} + \dfrac{5}{y^2-9}$

59. $\dfrac{3y+2}{y^2+5y-24} + \dfrac{7}{y^2+4y-32}$

60. $\dfrac{3y+2}{y^2-7y+10} + \dfrac{2y}{y^2-8y+15}$

61. $\dfrac{3x-1}{x^2+2x-3} - \dfrac{x+4}{x^2-9}$

62. $\dfrac{3p-2}{p^2+2p-24} - \dfrac{p-3}{p^2-16}$

C Perform the indicated operations and simplify.

63. $\dfrac{1}{x+1} - \dfrac{x}{x-2} + \dfrac{x^2+2}{x^2-x-2}$

64. $\dfrac{2}{y+3} - \dfrac{y}{y-1} + \dfrac{y^2+2}{y^2+2y-3}$

65. $\dfrac{x-1}{x-2} - \dfrac{x+1}{x+2} + \dfrac{x-6}{x^2-4}$

66. $\dfrac{y-3}{y-4} - \dfrac{y+2}{y+4} + \dfrac{y-7}{y^2-16}$

67. $\dfrac{y+2}{y+4} + \dfrac{y-7}{y^2-16} - \dfrac{y-3}{y-4}$

68. $\dfrac{x-6}{x^2-4} - \dfrac{x-1}{x-2} - \dfrac{x+1}{x+2}$

69. $\dfrac{4x}{x^2 - 1} + \dfrac{3x}{1 - x} - \dfrac{4}{x - 1}$

70. $\dfrac{5y}{1 - 2y} - \dfrac{2y}{2y + 1} + \dfrac{3}{4y^2 - 1}$

71. $\dfrac{1}{x + y} + \dfrac{1}{y - x} - \dfrac{2x}{x^2 - y^2}$

72. $\dfrac{1}{b - a} + \dfrac{1}{a + b} - \dfrac{2b}{a^2 - b^2}$

73. $\mathbf{D_W}$ Janine found that the sum of two rational expressions was $(3 - x)/(x - 5)$, but the answer at the back of the book was $(x - 3)/(5 - x)$. Was Janine's answer correct? Why or why not?

74. $\mathbf{D_W}$ Many students make the mistake of always multiplying the denominators when looking for a least common denominator. Use Example 11 to explain why this approach can yield results that are more difficult to simplify.

SKILL MAINTENANCE

Graph.   [3.7b]

75. $2x - 3y > 6$

76. $y - x > 3$

77. $5x + 3y \le 15$

78. $5x - 3y \le 15$

Factor.   [4.6d]

79. $t^3 - 8$

80. $q^3 + 125$

81. $23x^4 + 23x$

82. $64a^3 - 27b^3$

83. Simplify: $\dfrac{15x^{-7}y^{12}z^4}{35x^{-2}y^6z^{-3}}$.   [R.7a]

84. Find an equation of the line that passes through the point $(-2, 3)$ and is perpendicular to the line $5y + 4x = 7$.   [2.6d]

SYNTHESIS

85. Determine the domain and the range of the function graphed below.

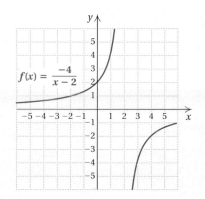

$$f(x) = \dfrac{-4}{x - 2}$$

Perform the indicated operations and simplify.

89. $\dfrac{x + y + 1}{y - (x + 1)} + \dfrac{x + y - 1}{x - (y - 1)} - \dfrac{x - y - 1}{1 - (y - x)}$

91. $\dfrac{x}{x^4 - y^4} - \dfrac{1}{x^2 + 2xy + y^2}$

Find the LCM.

86. 18, 42, 82, 120, 300, 700

87. $x^8 - x^4, \ x^5 - x^2, \ x^5 - x^3, \ x^5 + x^2$

88. The LCM of two expressions is $8a^4b^7$. One of the expressions is $2a^3b^7$. List all possibilities for the other expression.

90. $\dfrac{b - c}{a - (b - c)} - \dfrac{b - a}{(b - a) - c}$

92. $\dfrac{x^2}{3x^2 - 5x - 2} - \dfrac{2x}{3x + 1} \cdot \dfrac{1}{x - 2}$

435

Objectives

a Divide a polynomial by a monomial.

b Divide a polynomial by a divisor that is not a monomial, and if there is a remainder, express the result in two ways.

c Use synthetic division to divide a polynomial by a binomial of the type $x - a$.

1. Divide: $\dfrac{x^3 + 16x^2 + 6x}{2x}$.

Divide.

2. $(15y^5 - 6y^4 + 18y^3) \div (3y^2)$

3. $(x^4y^3 + 10x^3y^2 + 16x^2y) \div (2x^2y)$

Answers on page A-28

5.3 DIVISION OF POLYNOMIALS

A rational expression represents division. "Long" division of polynomials, like division of real numbers, relies on our multiplication and subtraction skills.

a Divisor a Monomial

We first consider division by a monomial (a term like $45x^{10}$ or $48a^2b^5$). When we are dividing a monomial by a monomial, we can use the rules of exponents and subtract exponents when the bases are the same. (We studied this in Section R.7.) For example,

$$\frac{45x^{10}}{3x^4} = \frac{45}{3}x^{10-4} = 15x^6 \quad \text{and} \quad \frac{48a^2b^5}{-3ab^2} = \frac{48}{-3}a^{2-1}b^{5-2} = -16ab^3.$$

When we divide a polynomial by a monomial, we break up the division into a sum of quotients of monomials. To do so, we reverse the rule for adding fractions. That is, since

$$\frac{A}{C} + \frac{B}{C} = \frac{A+B}{C}, \quad \text{we know that} \quad \frac{A+B}{C} = \frac{A}{C} + \frac{B}{C}.$$

EXAMPLE 1 Divide $12x^3 + 8x^2 + x + 4$ by $4x$.

$$\frac{12x^3 + 8x^2 + x + 4}{4x} \qquad \text{Writing a fraction expression}$$

$$= \frac{12x^3}{4x} + \frac{8x^2}{4x} + \frac{x}{4x} + \frac{4}{4x} \qquad \begin{array}{l}\text{Dividing each term of the numerator} \\ \text{by the monomial}\end{array}$$

$$= 3x^2 + 2x + \frac{1}{4} + \frac{1}{x} \qquad \text{Doing the four indicated divisions}$$

Do Exercise 1.

EXAMPLE 2 Divide: $(8x^4y^5 - 3x^3y^4 + 5x^2y^3) \div (x^2y^3)$.

$$\frac{8x^4y^5 - 3x^3y^4 + 5x^2y^3}{x^2y^3} = \frac{8x^4y^5}{x^2y^3} - \frac{3x^3y^4}{x^2y^3} + \frac{5x^2y^3}{x^2y^3}$$

$$= 8x^2y^2 - 3xy + 5$$

> **DIVIDING BY A MONOMIAL**
>
> To divide a polynomial by a monomial, divide each term by the monomial.

Do Exercises 2 and 3.

b Divisor Not a Monomial

When the divisor is not a monomial, we use a procedure very much like long division in arithmetic.

EXAMPLE 3 Divide $x^2 + 5x + 8$ by $x + 3$.

We have

$$
\begin{array}{r}
x \\
x + 3\overline{\smash{)}x^2 + 5x + 8} \\
\end{array}
$$

— Divide the first term by the first term: $x^2/x = x$.

$$x^2 + 3x$$ — Multiply x above by the divisor, $x + 3$.

$$2x$$ — Subtract: $(x^2 + 5x) - (x^2 + 3x) = x^2 + 5x - x^2 - 3x$
$$= 2x.$$

We now "bring down" the other terms of the dividend—in this case, 8.

$$
\begin{array}{r}
x + 2 \\
x + 3\overline{\smash{)}x^2 + 5x + 8} \\
x^2 + 3x \\
\end{array}
$$

— Divide the first term by the first term: $2x/x = 2$.

$$2x + 8$$ — The 8 has been "brought down."

$$2x + 6$$ — Multiply 2 above by the divisor, $x + 3$.

$$2$$ — Subtract: $(2x + 8) - (2x + 6) = 2x + 8 - 2x - 6 = 2.$

The answer is $x + 2$, R 2; or

$$x + 2 + \frac{2}{x + 3}.$$

This expression is the remainder over the divisor.

Note that the answer is not a polynomial unless the remainder is 0.

To check, we multiply the quotient by the divisor and add the remainder to see if we get the dividend:

Divisor	Quotient	Remainder	Dividend
$(x + 3)$	$\cdot \quad (x + 2)$	$+ \quad 2$	$= \quad (x^2 + 5x + 6) + 2$
			$= \quad x^2 + 5x + 8$

The answer checks.

EXAMPLE 4 Divide: $(5x^4 + x^3 - 3x^2 - 6x - 8) \div (x - 1)$.

$$
\begin{array}{r}
5x^3 + 6x^2 + 3x - 3 \\
x - 1\overline{\smash{)}5x^4 + x^3 - 3x^2 - 6x - 8} \\
\underline{5x^4 - 5x^3} \\
6x^3 - 3x^2 \\
\underline{6x^3 - 6x^2} \\
3x^2 - 6x \\
\underline{3x^2 - 3x} \\
-3x - 8 \\
\underline{-3x + 3} \\
-11
\end{array}
$$

Subtract:
$(5x^4 + x^3) - (5x^4 - 5x^3) = 6x^3.$

Subtract:
$(6x^3 - 3x^2) - (6x^3 - 6x^2) = 3x^2.$

Subtract:
$(3x^2 - 6x) - (3x^2 - 3x) = -3x.$

Subtract:
$(-3x - 8) - (-3x + 3) = -11.$

The answer is $5x^3 + 6x^2 + 3x - 3$, R -11; or

$$5x^3 + 6x^2 + 3x - 3 + \frac{-11}{x - 1}.$$

Do Exercises 4 and 5.

4. Divide and check:

$$x - 2\overline{\smash{)}x^2 + 3x - 10}.$$

5. Divide and check:

$(2x^4 + 3x^3 - x^2 - 7x + 9) \div (x + 4).$

Answers on page A-29

6. Divide and check:

$$(9y^4 + 14y^2 - 8) \div (3y + 2).$$

Always remember when dividing polynomials to arrange the polynomials in descending order. In a polynomial division, if there are *missing* terms in the dividend, either write them with 0 coefficients or leave space for them. For example, in $125y^3 - 8$, we say that "the y^2- and y-terms are **missing**." We could write them in as follows: $125y^3 + 0y^2 + 0y - 8$.

EXAMPLE 5   Divide: $(125y^3 - 8) \div (5y - 2)$.

$$
\begin{array}{r}
25y^2 + 10y + 4 \\
5y - 2\overline{)125y^3 + 0y^2 + 0y - 8} \\
\underline{125y^3 - 50y^2} \\
50y^2 + 0y \\
\underline{50y^2 - 20y} \\
20y - 8 \\
\underline{20y - 8} \\
0
\end{array}
$$

When there are missing terms, we can write them in.

Subtract: $125y^3 - (125y^3 - 50y^2) = 50y^2$.

Subtract: $50y^2 - (50y^2 - 20y) = 20y$.

Subtract: $(20y - 8) - (20y - 8) = 0$.

The answer is $25y^2 + 10y + 4$.

Do Exercise 6.

Divide and check.

7. $(y^3 - 11y^2 + 6) \div (y - 3)$

Another way to deal with missing terms is to leave space for them, as we see in Example 6.

EXAMPLE 6   Divide: $(x^4 - 9x^2 - 5) \div (x - 2)$.

Note that the x^3- and x-terms are missing in the dividend.

$$
\begin{array}{r}
x^3 + 2x^2 - 5x - 10 \\
x - 2\overline{)x^4 - 9x^2 - 5} \\
\underline{x^4 - 2x^3} \\
2x^3 - 9x^2 \\
\underline{2x^3 - 4x^2} \\
- 5x^2 \\
\underline{- 5x^2 + 10x} \\
- 10x - 5 \\
\underline{- 10x + 20} \\
- 25
\end{array}
$$

← We leave spaces for missing terms.

Subtract: $x^4 - (x^4 - 2x^3) = 2x^3$.

Subtract: $(2x^3 - 9x^2) - (2x^3 - 4x^2) = -5x^2$.

Subtract: $-5x^2 - (-5x^2 + 10x) = -10x$.

Subtract: $(-10x - 5) - (-10x + 20) = -25$.

8. $(x^3 + 9x^2 - 5) \div (x - 1)$

The answer is $x^3 + 2x^2 - 5x - 10$, R -25; or

$$x^3 + 2x^2 - 5x - 10 + \frac{-25}{x - 2}.$$

Do Exercises 7 and 8.

Answers on page A-29

When dividing, we may "come out even" (have a remainder of 0) or we may not. If not, how long should we keep working? We continue until the degree of the remainder is less than the degree of the divisor, as in the next example.

9. Divide and check:

$$(y^3 - 11y^2 + 6) \div (y^2 - 3).$$

EXAMPLE 7 Divide: $(6x^3 + 9x^2 - 5) \div (x^2 - 2x)$.

$$
\begin{array}{r}
6x\ + 21 \\
x^2 - 2x\overline{)6x^3 + \ 9x^2 + \ \ 0x - 5} \\
\underline{6x^3 - 12x^2} \\
21x^2 + \ \ 0x \\
\underline{21x^2 - 42x} \\
42x - 5
\end{array}
$$

We have a missing term.
We can write it in.

The degree of the remainder, 1, is less than the degree of the divisor, 2, so we are finished.

The answer is $6x + 21$, R $42x - 5$; or

$$6x + 21 + \frac{42x - 5}{x^2 - 2x}.$$

Do Exercise 9.

Answer on page A-29

C Synthetic Division

To divide a polynomial by a binomial of the type $x - a$, we can streamline the general procedure by a process called **synthetic division.**

Compare the following. In **A,** we perform a division. In **B,** we also divide but we do not write the variables.

A.
$$
\begin{array}{r}
4x^2 + 5x + 11 \\
x - 2\overline{)4x^3 - 3x^2 + \ \ x + \ 7} \\
\underline{4x^3 - 8x^2} \\
5x^2 + \ \ x \\
\underline{5x^2 - 10x} \\
11x + \ 7 \\
\underline{11x - 22} \\
29
\end{array}
$$

B.
$$
\begin{array}{r}
4 + 5 + 11 \\
1 - 2\overline{)4 - 3 + \ 1 + \ 7} \\
\underline{4 - 8} \\
5 + \ 1 \\
\underline{5 - 10} \\
11 + \ 7 \\
\underline{11 - 22} \\
29
\end{array}
$$

In **B,** there is still some duplication of writing. Also, since we can subtract by adding the opposite, we can use 2 instead of -2 and then add instead of subtracting.

Study Tips

AUDIO RECORDINGS

Your instructor can request a complete set of audio recordings designed to help lead you through each section of this textbook. If you have difficulty reading or if you want extra review, these recordings explain solution steps for examples, caution you about errors, give instructions to work margin exercises, and then review the solutions to the margin exercises. These recordings are ideal for use outside of class. To obtain these audio recordings, consult with your instructor and refer to the Preface of this text. Recordings are available as MP3 files within MyMathLab.

439

10. Use synthetic division to divide:

$(2x^3 - 4x^2 + 8x - 8) \div (x - 3).$

C. Synthetic Division

a) $\underline{2\rfloor 4 \quad -3 \quad 1 \quad 7}$

$\qquad\qquad 4$

Write the 2, the opposite of -2 in the divisor $x - 2$, and the coefficients of the dividend.

Bring down the first coefficient.

b)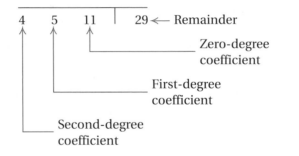

Multiply 4 by 2 to get 8. Add 8 and -3.

c) $\begin{array}{r} 2\rfloor 4 \quad -3 \quad 1 \quad 7 \\ 8 \quad 10 \\ \hline 4 \quad 5 \quad 11 \end{array}$

Multiply 5 by 2 to get 10. Add 10 and 1.

d) $\begin{array}{r} 2\rfloor 4 \quad -3 \quad 1 \quad 7 \\ 8 \quad 10 \quad 22 \\ \hline 4 \quad 5 \quad 11 \mid 29 \end{array}$

Multiply 11 by 2 to get 22. Add 22 and 7.

$\underbrace{\qquad\qquad}_{\text{Quotient}} \quad \text{Remainder}$

The last number, 29, is the remainder. The other numbers are the coefficients of the quotient with that of the term of highest degree first, as follows. Note that the degree of the term of highest degree is 1 less than the degree of the dividend.

$$4 \qquad 5 \qquad 11 \quad \mid \quad 29 \leftarrow \text{Remainder}$$

Zero-degree coefficient

First-degree coefficient

Second-degree coefficient

The answer is $4x^2 + 5x + 11$, R 29; or $4x^2 + 5x + 11 + \dfrac{29}{x - 2}$.

> It is important to remember that in order for synthetic division to work, the divisor must be of the form $x - a$, that is, a variable minus a constant. The coefficient of the variable must be 1.

EXAMPLE 8 Use synthetic division to divide:

$$(x^3 + 6x^2 - x - 30) \div (x - 2).$$

We have

$$\begin{array}{r} 2\rfloor 1 \quad 6 \quad -1 \quad -30 \\ 2 \quad 16 \quad 30 \\ \hline 1 \quad 8 \quad 15 \mid 0 \end{array}$$

The answer is $x^2 + 8x + 15$, R 0; or just $x^2 + 8x + 15$.

Answer on page A-29

Do Exercise 10.

CHAPTER 5: Rational Expressions,
Equations, and Functions

When there are missing terms, be sure to write 0's for their coefficients.

EXAMPLES Use synthetic division to divide.

9. $(2x^3 + 7x^2 - 5) \div (x + 3)$

There is no x-term, so we must write a 0 for its coefficient. Note that $x + 3 = x - (-3)$, so we write -3 at the left.

$$
\begin{array}{r|rrrr}
-3 & 2 & 7 & 0 & -5 \\
   &   & -6 & -3 & 9 \\
\hline
   & 2 & 1 & -3 & \,|\, 4
\end{array}
$$

The answer is $2x^2 + x - 3$, R 4; or $2x^2 + x - 3 + \dfrac{4}{x + 3}$.

10. $(x^3 + 4x^2 - x - 4) \div (x + 4)$

Note that $x + 4 = x - (-4)$, so we write -4 at the left.

$$
\begin{array}{r|rrrr}
-4 & 1 & 4 & -1 & -4 \\
   &   & -4 & 0 & 4 \\
\hline
   & 1 & 0 & -1 & \,|\, 0
\end{array}
$$

The answer is $x^2 - 1$.

11. $(x^4 - 1) \div (x - 1)$

The divisor is $x - 1$, so we write 1 at the left.

$$
\begin{array}{r|rrrrr}
1 & 1 & 0 & 0 & 0 & -1 \\
  &   & 1 & 1 & 1 & 1 \\
\hline
  & 1 & 1 & 1 & 1 & \,|\, 0
\end{array}
$$

The answer is $x^3 + x^2 + x + 1$.

12. $(8x^5 - 6x^3 + x - 8) \div (x + 2)$

Note that $x + 2 = x - (-2)$, so we write -2 at the left.

$$
\begin{array}{r|rrrrrr}
-2 & 8 & 0 & -6 & 0 & 1 & -8 \\
   &   & -16 & 32 & -52 & 104 & -210 \\
\hline
   & 8 & -16 & 26 & -52 & 105 & \,|\, -218
\end{array}
$$

The answer is $8x^4 - 16x^3 + 26x^2 - 52x + 105$, R -218; or

$$8x^4 - 16x^3 + 26x^2 - 52x + 105 + \dfrac{-218}{x + 2}.$$

Do Exercises 11 and 12.

Use synthetic division to divide.

11. $(x^3 - 2x^2 + 5x - 4) \div (x + 2)$

12. $(y^3 + 1) \div (y + 1)$

Answers on page A-29

a   Divide and check.

1. $\dfrac{24x^6 + 18x^5 - 36x^2}{6x^2}$

2. $\dfrac{30y^8 - 15y^6 + 40y^4}{5y^4}$

3. $\dfrac{45y^7 - 20y^4 + 15y^2}{5y^2}$

4. $\dfrac{60x^8 + 44x^5 - 28x^3}{4x^3}$

5. $(32a^4b^3 + 14a^3b^2 - 22a^2b) \div (2a^2b)$

6. $(7x^3y^4 - 21x^2y^3 + 28xy^2) \div (7xy)$

b   Divide.

7. $(x^2 + 10x + 21) \div (x + 3)$

8. $(y^2 - 8y + 16) \div (y - 4)$

9. $(a^2 - 8a - 16) \div (a + 4)$

10. $(y^2 - 10y - 25) \div (y - 5)$

11. $(x^2 + 7x + 14) \div (x + 5)$

12. $(t^2 - 7t - 9) \div (t - 3)$

13. $(4y^3 + 6y^2 + 14) \div (2y + 4)$

14. $(6x^3 - x^2 - 10) \div (3x + 4)$

15. $(10y^3 + 6y^2 - 9y + 10) \div (5y - 2)$

16. $(6x^3 - 11x^2 + 11x - 2) \div (2x - 3)$

17. $(2x^4 - x^3 - 5x^2 + x - 6) \div (x^2 + 2)$

18. $(3x^4 + 2x^3 - 11x^2 - 2x + 5) \div (x^2 - 2)$

19. $(2x^5 - x^4 + 2x^3 - x) \div (x^2 - 3x)$

20. $(2x^5 + 3x^3 + x^2 - 4) \div (x^2 + x)$

c   Use synthetic division to divide.

21. $(x^3 - 2x^2 + 2x - 5) \div (x - 1)$

22. $(x^3 - 2x^2 + 2x - 5) \div (x + 1)$

23. $(a^2 + 11a - 19) \div (a + 4)$

24. $(a^2 + 11a - 19) \div (a - 4)$

25. $(x^3 - 7x^2 - 13x + 3) \div (x - 2)$

26. $(x^3 - 7x^2 - 13x + 3) \div (x + 2)$

27. $(3x^3 + 7x^2 - 4x + 3) \div (x + 3)$

28. $(3x^3 + 7x^2 - 4x + 3) \div (x - 3)$

29. $(y^3 - 3y + 10) \div (y - 2)$

30. $(x^3 - 2x^2 + 8) \div (x + 2)$

31. $(3x^4 - 25x^2 - 18) \div (x - 3)$

32. $(6y^4 + 15y^3 + 28y + 6) \div (y + 3)$

33. $(x^3 - 8) \div (x - 2)$

34. $(y^3 + 125) \div (y + 5)$

35. $(y^4 - 16) \div (y - 2)$

36. $(x^5 - 32) \div (x - 2)$

37. $\mathbf{D_W}$ Do addition, subtraction, multiplication, and division of polynomials always result in a polynomial? Why or why not?

38. $\mathbf{D_W}$ Explain how synthetic division can be useful when factoring a polynomial.

SKILL MAINTENANCE

Graph. [3.7b]

39. $2x - 3y < 6$

40. $5x + 3y \le 15$

41. $y > 4$

42. $x \le -2$

Graph. [2.2c]

43. $f(x) = x^2$

44. $g(x) = x^2 - 3$

45. $f(x) = 3 - x^2$

46. $f(x) = x^2 + 6x + 6$

Solve. [4.8a]

47. $x^2 - 5x = 0$

48. $25y^2 = 64$

49. $12x^2 = 17x + 5$

50. $12x^2 + 11x + 2 = 0$

SYNTHESIS

51. Let $f(x) = 4x^3 + 16x^2 - 3x - 45$. Find $f(-3)$ and then solve $f(x) = 0$.

52. Let $f(x) = 6x^3 - 13x^2 - 79x + 140$. Find $f(4)$ and then solve $f(x) = 0$.

53. When $x^2 - 3x + 2k$ is divided by $x + 2$, the remainder is 7. Find k.

54. Find k such that when $x^3 - kx^2 + 3x + 7k$ is divided by $x + 2$, the remainder is 0.

Divide.

55. $(4a^3b + 5a^2b^2 + a^4 + 2ab^3) \div (a^2 + 2b^2 + 3ab)$

56. $(a^7 + b^7) \div (a + b)$

Objective

[a] Simplify complex rational expressions.

[a]   A **complex rational expression** is a rational expression that contains rational expressions within its numerator and/or its denominator. Here are some examples:

$$\dfrac{\dfrac{2}{3}}{\dfrac{4}{5}}, \quad \dfrac{1 + \dfrac{5}{x}}{4x}, \quad \dfrac{\dfrac{x - y}{x + y}}{\dfrac{2x - y}{3x + y}}, \quad \dfrac{\dfrac{3x}{5} - \dfrac{2}{x}}{\dfrac{4x}{3} + \dfrac{7}{6x}}.$$

The rational expressions within each complex rational expression are red.

There are two methods that can be used to simplify complex rational expressions. We will consider both of them.

METHOD 1: MULTIPLYING BY THE LCM OF ALL THE DENOMINATORS

Method 1. To simplify a complex rational expression:

1. First, find the LCM of all the denominators of all the rational expressions occurring within both the numerator and the denominator of the (original) complex rational expression.
2. Multiply by 1 using LCM/LCM.
3. If possible, simplify.

EXAMPLE 1   Simplify: $\dfrac{x + \dfrac{1}{5}}{x - \dfrac{1}{3}}$.

We first find the LCM of all the denominators of all the rational expressions occurring in both the numerator and the denominator of the complex rational expression. The denominators are 3 and 5. The LCM of these denominators is $3 \cdot 5$, or 15. We multiply by 15/15.

$$\dfrac{x + \dfrac{1}{5}}{x - \dfrac{1}{3}} = \left(\dfrac{x + \dfrac{1}{5}}{x - \dfrac{1}{3}} \right) \cdot \dfrac{15}{15} \qquad \text{Multiplying by 1}$$

$$= \dfrac{\left(x + \dfrac{1}{5} \right) \cdot 15}{\left(x - \dfrac{1}{3} \right) \cdot 15} \qquad \begin{array}{l}\text{Multiplying the numerators}\\\text{and the denominators}\end{array}$$

$$= \dfrac{15x + \dfrac{1}{5} \cdot 15}{15x - \dfrac{1}{3} \cdot 15} \qquad \begin{array}{l}\text{Carrying out the multiplications}\\\text{using the distributive laws}\end{array}$$

$$= \dfrac{15x + 3}{15x - 5} \qquad \text{No further simplification is possible.}$$

To the instructor and the student: Students can be instructed to try both methods and then choose the one that works best for them, or one method can be chosen by the instructor.

In Example 1, if you feel more comfortable doing so, you can always write denominators of 1 where there are no denominators. In this case, you could start out by writing

$$\frac{\dfrac{x}{1} + \dfrac{1}{5}}{\dfrac{x}{1} - \dfrac{1}{3}}.$$

Do Exercise 1.

EXAMPLE 2 Simplify: $\dfrac{1 + \dfrac{1}{x}}{1 - \dfrac{1}{x^2}}$.

We first find the LCM of all the denominators of all the rational expressions occurring in both the numerator and the denominator of the complex rational expression. The denominators are x and x^2. The LCM of these denominators is x^2. We multiply by x^2/x^2.

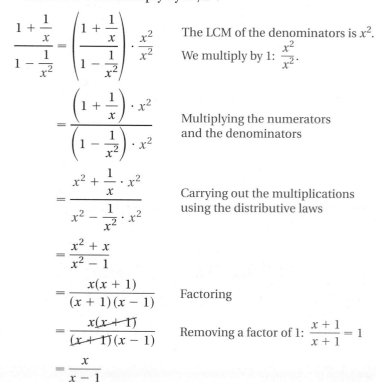

$\dfrac{1 + \dfrac{1}{x}}{1 - \dfrac{1}{x^2}} = \left(\dfrac{1 + \dfrac{1}{x}}{1 - \dfrac{1}{x^2}}\right) \cdot \dfrac{x^2}{x^2}$ The LCM of the denominators is x^2.
We multiply by 1: $\dfrac{x^2}{x^2}$.

$= \dfrac{\left(1 + \dfrac{1}{x}\right) \cdot x^2}{\left(1 - \dfrac{1}{x^2}\right) \cdot x^2}$ Multiplying the numerators and the denominators

$= \dfrac{x^2 + \dfrac{1}{x} \cdot x^2}{x^2 - \dfrac{1}{x^2} \cdot x^2}$ Carrying out the multiplications using the distributive laws

$= \dfrac{x^2 + x}{x^2 - 1}$

$= \dfrac{x(x + 1)}{(x + 1)(x - 1)}$ Factoring

$= \dfrac{x\cancel{(x + 1)}}{\cancel{(x + 1)}(x - 1)}$ Removing a factor of 1: $\dfrac{x + 1}{x + 1} = 1$

$= \dfrac{x}{x - 1}$

Do Exercise 2.

1. Simplify. Use method 1.

$$\frac{y + \dfrac{1}{2}}{y - \dfrac{1}{7}}$$

2. Simplify. Use method 1.

$$\frac{1 - \dfrac{1}{x}}{1 - \dfrac{1}{x^2}}$$

Answers on page A-29

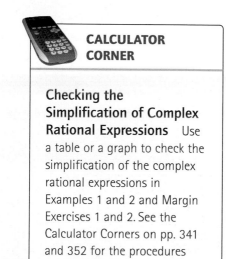

CALCULATOR CORNER

Checking the Simplification of Complex Rational Expressions   Use a table or a graph to check the simplification of the complex rational expressions in Examples 1 and 2 and Margin Exercises 1 and 2. See the Calculator Corners on pp. 341 and 352 for the procedures to follow.

Simplify. Use method 1.

3. $\dfrac{\dfrac{1}{a}+\dfrac{1}{b}}{\dfrac{1}{a}-\dfrac{1}{b}}$

EXAMPLE 3 Simplify: $\dfrac{\dfrac{1}{a}+\dfrac{1}{b}}{\dfrac{1}{a^3}+\dfrac{1}{b^3}}$.

The denominators are a, b, a^3, and b^3. The LCM of these denominators is a^3b^3. We multiply by a^3b^3/a^3b^3.

$$\dfrac{\dfrac{1}{a}+\dfrac{1}{b}}{\dfrac{1}{a^3}+\dfrac{1}{b^3}}=\left(\dfrac{\dfrac{1}{a}+\dfrac{1}{b}}{\dfrac{1}{a^3}+\dfrac{1}{b^3}}\right)\cdot\dfrac{a^3b^3}{a^3b^3}$$

The LCM of the denominators is a^3b^3. We multiply by 1: $\dfrac{a^3b^3}{a^3b^3}$.

$$=\dfrac{\left(\dfrac{1}{a}+\dfrac{1}{b}\right)\cdot a^3b^3}{\left(\dfrac{1}{a^3}+\dfrac{1}{b^3}\right)\cdot a^3b^3}$$

Multiplying the numerators and the denominators

$$=\dfrac{\dfrac{1}{a}\cdot a^3b^3+\dfrac{1}{b}\cdot a^3b^3}{\dfrac{1}{a^3}\cdot a^3b^3+\dfrac{1}{b^3}\cdot a^3b^3}$$

Carrying out the multiplications using a distributive law

$$=\dfrac{a^2b^3+a^3b^2}{b^3+a^3}=\dfrac{a^2b^2(b+a)}{(b+a)(b^2-ba+a^2)}$$

Factoring

$$=\dfrac{a^2b^2\cancel{(b+a)}}{\cancel{(b+a)}(b^2-ba+a^2)}$$

Removing a factor of 1: $\dfrac{b+a}{b+a}=1$

$$=\dfrac{a^2b^2}{b^2-ba+a^2}.$$

Do Exercises 3 and 4.

4. $\dfrac{\dfrac{1}{a}-\dfrac{1}{b}}{\dfrac{1}{a^3}-\dfrac{1}{b^3}}$

METHOD 2: ADDING OR SUBTRACTING IN THE NUMERATOR AND THE DENOMINATOR

Method 2. To simplify a complex rational expression:

1. Add or subtract, as necessary, to get a single rational expression in the numerator.
2. Add or subtract, as necessary, to get a single rational expression in the denominator.
3. Divide the numerator by the denominator.
4. If possible, simplify.

We will redo Examples 1–3 using this method.

Answers on page A-29

CHAPTER 5: Rational Expressions, Equations, and Functions

EXAMPLE 4 Simplify: $\dfrac{x + \dfrac{1}{5}}{x - \dfrac{1}{3}}$.

$\dfrac{x + \dfrac{1}{5}}{x - \dfrac{1}{3}} = \dfrac{x \cdot \dfrac{5}{5} + \dfrac{1}{5}}{x - \dfrac{1}{3}} = \dfrac{\dfrac{5x + 1}{5}}{x - \dfrac{1}{3}}$

To get a single rational expression in the numerator, we note that the LCM in the numerator is 5. We multiply by 1 and add.

$= \dfrac{\dfrac{5x + 1}{5}}{x \cdot \dfrac{3}{3} - \dfrac{1}{3}} = \dfrac{\dfrac{5x + 1}{5}}{\dfrac{3x - 1}{3}}$

To get a single rational expression in the denominator, we note that the LCM in the denominator is 3. We multiply by 1 and subtract.

$= \dfrac{5x + 1}{5} \cdot \dfrac{3}{3x - 1}$

Multiplying by the reciprocal of the denominator

$= \dfrac{15x + 3}{15x - 5}$

No further simplification is possible.

Simplify. Use method 2.

5. $\dfrac{y + \dfrac{1}{2}}{y - \dfrac{1}{7}}$

EXAMPLE 5 Simplify: $\dfrac{1 + \dfrac{1}{x}}{1 - \dfrac{1}{x^2}}$.

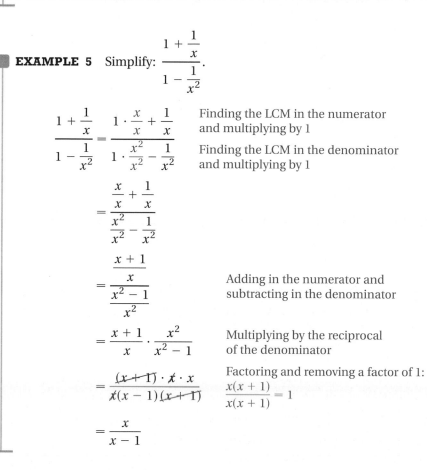

$\dfrac{1 + \dfrac{1}{x}}{1 - \dfrac{1}{x^2}} = \dfrac{1 \cdot \dfrac{x}{x} + \dfrac{1}{x}}{1 \cdot \dfrac{x^2}{x^2} - \dfrac{1}{x^2}}$

Finding the LCM in the numerator and multiplying by 1

Finding the LCM in the denominator and multiplying by 1

$= \dfrac{\dfrac{x}{x} + \dfrac{1}{x}}{\dfrac{x^2}{x^2} - \dfrac{1}{x^2}}$

$= \dfrac{\dfrac{x + 1}{x}}{\dfrac{x^2 - 1}{x^2}}$

Adding in the numerator and subtracting in the denominator

$= \dfrac{x + 1}{x} \cdot \dfrac{x^2}{x^2 - 1}$

Multiplying by the reciprocal of the denominator

$= \dfrac{(x + 1) \cdot x \cdot x}{x(x - 1)(x + 1)}$

Factoring and removing a factor of 1: $\dfrac{x(x + 1)}{x(x + 1)} = 1$

$= \dfrac{x}{x - 1}$

6. $\dfrac{1 - \dfrac{1}{x}}{1 - \dfrac{1}{x^2}}$

Do Exercises 5 and 6.

Answers on page A-29

Simplify. Use method 2.

7. $\dfrac{\dfrac{1}{a} + \dfrac{1}{b}}{\dfrac{1}{a} - \dfrac{1}{b}}$

8. $\dfrac{\dfrac{1}{a} - \dfrac{1}{b}}{\dfrac{1}{a^3} - \dfrac{1}{b^3}}$

Answers on page A-29

 EXAMPLE 6 Simplify: $\dfrac{\dfrac{1}{a} + \dfrac{1}{b}}{\dfrac{1}{a^3} + \dfrac{1}{b^3}}$.

The LCM in the numerator is ab, and the LCM in the denominator is a^3b^3.

$$\dfrac{\dfrac{1}{a} + \dfrac{1}{b}}{\dfrac{1}{a^3} + \dfrac{1}{b^3}} = \dfrac{\dfrac{1}{a} \cdot \dfrac{b}{b} + \dfrac{1}{b} \cdot \dfrac{a}{a}}{\dfrac{1}{a^3} \cdot \dfrac{b^3}{b^3} + \dfrac{1}{b^3} \cdot \dfrac{a^3}{a^3}}$$

$$= \dfrac{\dfrac{b}{ab} + \dfrac{a}{ab}}{\dfrac{b^3}{a^3b^3} + \dfrac{a^3}{a^3b^3}}$$

$$= \dfrac{\dfrac{b + a}{ab}}{\dfrac{b^3 + a^3}{a^3b^3}}$$ Adding in the numerator and the denominator

$$= \dfrac{b + a}{ab} \cdot \dfrac{a^3b^3}{b^3 + a^3}$$ Multiplying by the reciprocal of the denominator

$$= \dfrac{(b + a)a^3b^3}{ab(b^3 + a^3)}$$

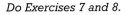

$$= \dfrac{(b + a) \cdot ab \cdot a^2b^2}{ab(b + a)(b^2 - ba + a^2)}$$ Factoring and removing a factor of 1: $\dfrac{ab(b + a)}{ab(b + a)} = 1$

$$= \dfrac{a^2b^2}{b^2 - ba + a^2}$$

Do Exercises 7 and 8.

CHAPTER 5: Rational Expressions, Equations, and Functions

a Simplify.

1. $\dfrac{2 + \dfrac{3}{5}}{4 - \dfrac{1}{2}}$

2. $\dfrac{\dfrac{3}{8} - 5}{\dfrac{2}{3} + 6}$

3. $\dfrac{\dfrac{2}{3} + \dfrac{4}{5}}{\dfrac{3}{4} - \dfrac{1}{2}}$

4. $\dfrac{\dfrac{5}{8} - \dfrac{2}{3}}{\dfrac{3}{4} + \dfrac{5}{6}}$

5. $\dfrac{\dfrac{x}{y^2}}{\dfrac{y^3}{x^2}}$

6. $\dfrac{\dfrac{a^3}{b^5}}{\dfrac{a^4}{b^2}}$

7. $\dfrac{\dfrac{9x^2 - y^2}{xy}}{\dfrac{3x - y}{y}}$

8. $\dfrac{\dfrac{a^2 - 16b^2}{ab}}{\dfrac{a + 4b}{b}}$

9. $\dfrac{\dfrac{1}{a} + 2}{\dfrac{1}{a} - 1}$

10. $\dfrac{\dfrac{1}{t} + 6}{\dfrac{1}{t} - 5}$

11. $\dfrac{x - \dfrac{1}{x}}{x + \dfrac{1}{x}}$

12. $\dfrac{y + \dfrac{1}{y}}{y - \dfrac{1}{y}}$

13. $\dfrac{\dfrac{3}{x} + \dfrac{4}{y}}{\dfrac{4}{x} - \dfrac{3}{y}}$

14. $\dfrac{\dfrac{2}{y} + \dfrac{5}{z}}{\dfrac{1}{y} - \dfrac{4}{z}}$

15. $\dfrac{a - \dfrac{3a}{b}}{b - \dfrac{b}{a}}$

16. $\dfrac{1 - \dfrac{2}{3x}}{x - \dfrac{4}{9x}}$

17. $\dfrac{\dfrac{1}{a} + \dfrac{1}{b}}{\dfrac{a^2 - b^2}{ab}}$

18. $\dfrac{\dfrac{1}{x} - \dfrac{1}{y}}{\dfrac{x^2 - y^2}{xy}}$

19. $\dfrac{\dfrac{1}{x + h} - \dfrac{1}{x}}{h}$

20. $\dfrac{\dfrac{1}{a - h} - \dfrac{1}{a}}{h}$

It may help you to write h as $\dfrac{h}{1}$.

21. $\dfrac{\dfrac{x^2 - x - 12}{x^2 - 2x - 15}}{\dfrac{x^2 + 8x + 12}{x^2 - 5x - 14}}$

22. $\dfrac{\dfrac{y^2 - y - 6}{y^2 - 5y - 14}}{\dfrac{y^2 + 6y + 5}{y^2 - 6y - 7}}$

23. $\dfrac{\dfrac{1}{x + 2} + \dfrac{4}{x - 3}}{\dfrac{2}{x - 3} - \dfrac{7}{x + 2}}$

24. $\dfrac{\dfrac{1}{y - 4} + \dfrac{1}{y + 5}}{\dfrac{6}{y + 5} + \dfrac{2}{y - 4}}$

25. $\dfrac{\dfrac{6}{x^2-4}-\dfrac{5}{x+2}}{\dfrac{7}{x^2-4}-\dfrac{4}{x-2}}$

26. $\dfrac{\dfrac{1}{x^2-1}+\dfrac{5}{x^2-5x+4}}{\dfrac{1}{x^2-1}+\dfrac{2}{x^2+3x+2}}$

27. $\dfrac{\dfrac{1}{z^2}-\dfrac{1}{w^2}}{\dfrac{1}{z^3}+\dfrac{1}{w^3}}$

28. $\dfrac{\dfrac{1}{b^2}-\dfrac{1}{c^2}}{\dfrac{1}{b^3}-\dfrac{1}{c^3}}$

29. $\dfrac{\dfrac{3}{x^2+2x-3}-\dfrac{1}{x^2-3x-10}}{\dfrac{3}{x^2-6x+5}-\dfrac{1}{x^2+5x+6}}$

30. $\dfrac{\dfrac{1}{a^2+7a+12}+\dfrac{1}{a^2+a-6}}{\dfrac{1}{a^2+2a-8}+\dfrac{1}{a^2+5a+4}}$

31. $\mathbf{D_W}$ In arithmetic, we are taught that
$$\frac{a}{b} \div \frac{c}{d} = \frac{a}{b} \cdot \frac{d}{c}.$$
(To divide using fraction notation, we invert and multiply.) Use method 1 to explain why we can do this.

32. $\mathbf{D_W}$ Explain why it is easier to use method 1 than method 2 to simplify the following expression.
$$\frac{\dfrac{a}{b}+\dfrac{c}{d}}{\dfrac{a}{b}-\dfrac{c}{d}}$$

SKILL MAINTENANCE

Factor.   [4.3a], [4.4a], [4.6d]

33. $4x^3 + 20x^2 + 6x$

34. $y^3 + 8$

35. $y^3 - 8$

36. $2x^3 - 32x^2 + 126x$

37. $1000x^3 + 1$

38. $1 - 1000a^3$

39. $y^3 - 64x^3$

40. $\frac{1}{8}a^3 - 343$

41. Solve for s: $T = \dfrac{r+s}{3}$.   [1.2a]

42. Graph: $f(x) = -3x + 2$.   [2.2c]

43. Given that $f(x) = x^2 - 3$, find $f(-5)$.   [2.2b]

44. Solve: $|2x - 5| = 7$.   [1.6c]

SYNTHESIS

For each function in Exercises 45–48, find and simplify $\dfrac{f(a+h)-f(a)}{h}$.

45. $f(x) = \dfrac{3}{x^2}$

46. $f(x) = \dfrac{5}{x}$

47. $f(x) = \dfrac{1}{1-x}$

48. $f(x) = \dfrac{x}{1+x}$

Simplify.

49. $\dfrac{5x^{-1} - 5y^{-1} + 10x^{-1}y^{-1}}{6x^{-1} - 6y^{-1} + 12x^{-1}y^{-1}}$

50. $\left[\dfrac{\dfrac{x+3}{x-3}+1}{\dfrac{x+3}{x-3}-1}\right]^8$

Find the reciprocal and simplify.

51. $x^2 - \dfrac{1}{x}$

52. $\dfrac{1-\dfrac{1}{a}}{a-1}$

53. $\dfrac{a^3 + b^3}{a+b}$

54. $x^2 + x + 1 + \dfrac{1}{x} + \dfrac{1}{x^2}$

5.5

SOLVING RATIONAL EQUATIONS

[a] Rational Equations

In Sections 5.1–5.4, we studied operations with *rational expressions*. These expressions do not have equals signs. Although we can perform the operations and simplify, we cannot solve them. Note the following examples:

$$\frac{x^2 - 6x + 9}{x^2 - 4} \cdot \frac{x - 2}{x - 3}, \qquad \frac{x + y}{x - y} \div \frac{x^2 + y}{x^2 - y^2}, \quad \text{and} \quad \frac{a + 7}{a^2 - 16} + \frac{5}{5a - 15}.$$

Operation signs occur. There are no equals signs!

Most often, the result of our calculation is another rational expression that is not cleared of fractions.

Equations *do have* equals signs, and we can clear them of fractions as we did in Section 1.1. A **rational,** or **fraction, equation** is an equation containing one or more rational expressions. Here are some examples:

$$\frac{2}{3} - \frac{5}{6} = \frac{1}{x}, \qquad x + \frac{6}{x} = 5, \quad \text{and} \quad \frac{2x}{x - 3} - \frac{6}{x} = \frac{18}{x^2 - 3x}.$$

There are equals signs as well as operation signs.

> **SOLVING RATIONAL EQUATIONS**
>
> To solve a rational equation, the first step is to clear the equation of fractions. To do this, multiply all terms on both sides of the equation by the LCM of all the denominators. Then carry out the equation-solving process as discussed in Chapters 1 and 4.

EXAMPLE 1 Solve: $\dfrac{2}{3} - \dfrac{5}{6} = \dfrac{1}{x}$.

The LCM of all the denominators is $6x$, or $2 \cdot 3 \cdot x$. Using the multiplication principle of Chapter 1, we multiply all terms on both sides of the equation by the LCM.

$$(2 \cdot 3 \cdot x) \cdot \left(\frac{2}{3} - \frac{5}{6} \right) = (2 \cdot 3 \cdot x) \cdot \frac{1}{x} \qquad \text{Multiplying both sides by the LCM}$$

$$2 \cdot 3 \cdot x \cdot \frac{2}{3} - 2 \cdot 3 \cdot x \cdot \frac{5}{6} = 2 \cdot 3 \cdot x \cdot \frac{1}{x} \qquad \text{Multiplying to remove parentheses}$$

> When clearing fractions, be sure to multiply *every* term in the equation by the LCM.

$$2 \cdot x \cdot 2 - x \cdot 5 = 2 \cdot 3$$
$$4x - 5x = 6$$
$$-x = 6$$
$$-1 \cdot x = 6$$
$$x = -6$$

1. Solve: $\dfrac{2}{3} + \dfrac{5}{6} = \dfrac{1}{x}$.

Check:

$$\dfrac{2}{3} - \dfrac{5}{6} = \dfrac{1}{x}$$

$$\dfrac{\dfrac{2}{3} - \dfrac{5}{6} \quad ? \quad \dfrac{1}{-6}}{}$$

$$\dfrac{4}{6} - \dfrac{5}{6} \quad \Big| \quad -\dfrac{1}{6}$$

$$-\dfrac{1}{6} \quad \Big| \qquad \text{TRUE}$$

The solution is -6.

Do Exercise 1.

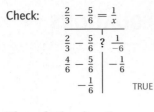 **EXAMPLE 2** Solve: $\dfrac{x+1}{2} - \dfrac{x-3}{3} = 3$.

The LCM of all the denominators is $2 \cdot 3$, or 6. We multiply all terms on both sides of the equation by the LCM.

$$2 \cdot 3 \cdot \left(\dfrac{x+1}{2} - \dfrac{x-3}{3} \right) = 2 \cdot 3 \cdot 3 \qquad \text{Multiplying both sides by the LCM}$$

$$2 \cdot 3 \cdot \dfrac{x+1}{2} - 2 \cdot 3 \cdot \dfrac{x-3}{3} = 2 \cdot 3 \cdot 3 \qquad \text{Multiplying to remove parentheses}$$

$$3(x+1) - 2(x-3) = 18 \qquad \text{Simplifying}$$

$$\left. \begin{array}{r} 3x + 3 - 2x + 6 = 18 \\ x + 9 = 18 \end{array} \right\} \quad \begin{array}{l} \text{Multiplying and collecting} \\ \text{like terms} \end{array}$$

$$x = 9$$

2. Solve: $\dfrac{y-4}{5} - \dfrac{y+7}{2} = 5$.

Check:

$$\dfrac{x+1}{2} - \dfrac{x-3}{3} = 3$$

$$\dfrac{\dfrac{9+1}{2} - \dfrac{9-3}{3} \quad ? \quad 3}{}$$

$$5 - 2 \quad \Big|$$

$$3 \quad \Big| \qquad \text{TRUE}$$

The solution is 9.

> **Caution!**
>
> *Clearing fractions* is a valid procedure only when solving equations, *not* when adding, subtracting, multiplying, or dividing rational expressions.

Do Exercise 2.

CHECKING POSSIBLE SOLUTIONS

When we multiply all terms on both sides of an equation by the LCM, the resulting equation might yield numbers that are *not* solutions of the original equation. Thus we must *always* check possible solutions in the original equation.

1. If you have carried out all algebraic procedures correctly, you need only check to see whether a number makes a denominator 0 in the original equation. If it does, it is not a solution.
2. To be sure that no computational errors have been made and that you indeed have a solution, a complete check is necessary, as we did in Chapter 1.

The next example illustrates the importance of checking all possible solutions.

EXAMPLE 3 Solve: $\dfrac{2x}{x-3} - \dfrac{6}{x} = \dfrac{18}{x^2-3x}$.

The LCM of the denominators is $x(x-3)$. We multiply all terms on both sides by $x(x-3)$.

$$x(x-3)\left(\dfrac{2x}{x-3} - \dfrac{6}{x}\right) = x(x-3)\left(\dfrac{18}{x^2-3x}\right)$$ Multiplying both sides by the LCM

$$x(x-3)\cdot\dfrac{2x}{x-3} - x(x-3)\cdot\dfrac{6}{x} = x(x-3)\left(\dfrac{18}{x^2-3x}\right)$$ Multiplying to remove parentheses

$$2x^2 - 6(x-3) = 18$$ Simplifying

$$2x^2 - 6x + 18 = 18$$

$$2x^2 - 6x = 0$$

$$2x(x-3) = 0$$ Factoring

$$2x = 0 \quad or \quad x-3 = 0$$ Using the principle of zero products

$$x = 0 \quad or \quad x = 3$$

3. Solve:

$$\dfrac{4x}{x+5} + \dfrac{20}{x} = \dfrac{100}{x^2+5x}.$$

The numbers 0 and 3 are possible solutions. We look at the original equation and see that each makes a denominator 0, so neither is a solution. We can carry out a check, as follows.

Check:

For 0:

$$\dfrac{2x}{x-3} - \dfrac{6}{x} = \dfrac{18}{x^2-3x}$$

$$\dfrac{2(0)}{0-3} - \dfrac{6}{0} \;\overset{?}{\vert}\; \dfrac{18}{0^2-3(0)}$$

$$0 - \dfrac{6}{0} \;\Big\vert\; \dfrac{18}{0}$$ NOT DEFINED

For 3:

$$\dfrac{2x}{x-3} - \dfrac{6}{x} = \dfrac{18}{x^2-3x}$$

$$\dfrac{2(3)}{3-3} - \dfrac{6}{3} \;\overset{?}{\vert}\; \dfrac{18}{3^2-3(3)}$$

$$\dfrac{6}{0} - 2 \;\Big\vert\; \dfrac{18}{0}$$ NOT DEFINED

The equation has *no solution*.

Do Exercise 3.

EXAMPLE 4 Solve: $\dfrac{x^2}{x-2} = \dfrac{4}{x-2}$.

The LCM of the denominators is $x-2$. We multiply all terms on both sides by $x-2$.

$$(x-2)\cdot\dfrac{x^2}{x-2} = (x-2)\cdot\dfrac{4}{x-2}$$

$$x^2 = 4$$ Simplifying

$$x^2 - 4 = 0$$

$$(x+2)(x-2) = 0$$

$$x = -2 \quad or \quad x = 2$$ Using the principle of zero products

Answer on page A-30

4. Solve: $\dfrac{x^2}{x - 3} = \dfrac{9}{x - 3}$.

Check: For 2:

$$\frac{x^2}{x - 2} = \frac{4}{x - 2}$$

$$\frac{2^2}{2 - 2} \overset{?}{\vrule} \frac{4}{2 - 2}$$

$$\frac{4}{0} \ \vrule \ \frac{4}{0} \qquad \text{NOT DEFINED}$$

For -2:

$$\frac{x^2}{x - 2} = \frac{4}{x - 2}$$

$$\frac{(-2)^2}{-2 - 2} \overset{?}{\vrule} \frac{4}{-2 - 2}$$

$$\frac{4}{-4} \ \vrule \ \frac{4}{-4}$$

$$-1 \ \vrule \ -1 \qquad \text{TRUE}$$

The number -2 is a solution, but 2 is not (it results in division by 0).

Do Exercise 4.

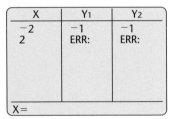

CALCULATOR CORNER

Checking Solutions of Rational Equations We can use a table to check possible solutions of rational equations. Consider the equation in Example 4,

$$\frac{x^2}{x - 2} = \frac{4}{x - 2},$$

and the possible solutions that were found, -2 and 2. To check these solutions, we enter $y_1 = x^2/(x - 2)$ and $y_2 = 4/(x - 2)$ on the equation-editor screen. Then, with a table set in ASK mode, we enter $x = -2$. (See p. 85.) Since y_1 and y_2 have the same value, we know that the equation is true when $x = -2$, and thus -2 is a solution. Now we enter $x = 2$. The ERROR messages indicate that 2 is not a solution because it is not an allowable replacement for x in the equation.

X	Y₁	Y₂
-2	-1	-1
2	ERR:	ERR:
X=		

Exercises:

1. Use a graphing calculator to check the possible solutions found in Examples 1, 2, and 3.

2. Use a graphing calculator to check the possible solutions you found in Margin Exercises 1–4.

Answer on page A-30

CHAPTER 5: Rational Expressions,
Equations, and Functions

EXAMPLE 5 Solve: $\dfrac{2}{x-1} = \dfrac{3}{x+1}$.

The LCM of the denominators is $(x-1)(x+1)$. We multiply all terms on both sides by $(x-1)(x+1)$.

$$(x-1)(x+1) \cdot \dfrac{2}{x-1} = (x-1)(x+1) \cdot \dfrac{3}{x+1} \quad \text{Multiplying}$$

$$2(x+1) = 3(x-1) \quad \text{Simplifying}$$

$$2x + 2 = 3x - 3$$

$$5 = x$$

The check is left to the student. The number 5 checks and is the solution.

EXAMPLE 6 Solve: $\dfrac{2}{x+5} + \dfrac{1}{x-5} = \dfrac{16}{x^2 - 25}$.

The LCM of the denominators is $(x+5)(x-5)$. We multiply all terms on both sides by $(x+5)(x-5)$.

$$(x+5)(x-5) \cdot \left[\dfrac{2}{x+5} + \dfrac{1}{x-5} \right] = (x+5)(x-5) \cdot \dfrac{16}{x^2-25}$$

$$(x+5)(x-5) \cdot \dfrac{2}{x+5} + (x+5)(x-5) \cdot \dfrac{1}{x-5} = (x+5)(x-5) \cdot \dfrac{16}{x^2-25}$$

$$2(x-5) + (x+5) = 16$$

$$2x - 10 + x + 5 = 16$$

$$3x - 5 = 16$$

$$3x = 21$$

$$x = 7$$

Check:

$$\dfrac{2}{x+5} + \dfrac{1}{x-5} = \dfrac{16}{x^2 - 25}$$

$$\dfrac{2}{7+5} + \dfrac{1}{7-5} \;\overset{?}{\vert}\; \dfrac{16}{7^2 - 25}$$

$$\dfrac{2}{12} + \dfrac{1}{2} \;\vert\; \dfrac{16}{49 - 25}$$

$$\dfrac{8}{12} \;\vert\; \dfrac{16}{24}$$

$$\dfrac{2}{3} \;\vert\; \dfrac{2}{3} \quad \text{TRUE}$$

The solution is 7.

Do Exercises 5 and 6.

EXAMPLE 7 Given that $f(x) = x + 6/x$, find all values of x for which $f(x) = 5$.

Since $f(x) = x + 6/x$, we want to find all values of x for which

$$x + \dfrac{6}{x} = 5.$$

Solve.

5. $\dfrac{2}{x-1} = \dfrac{3}{x+2}$

6. $\dfrac{2}{x^2 - 9} + \dfrac{5}{x-3} = \dfrac{3}{x+3}$

Answers on page A-30

7. Given that $f(x) = x - 12/x$, find all values of x for which $f(x) = 1$.

The LCM of the denominators is x. We multiply all terms on both sides by x:

$$x\left(x + \frac{6}{x}\right) = x \cdot 5 \qquad \text{Multiplying both sides by } x$$

$$x \cdot x + x \cdot \frac{6}{x} = 5x$$

$$x^2 + 6 = 5x \qquad \text{Simplifying}$$

$$x^2 - 5x + 6 = 0 \qquad \text{Getting 0 on one side}$$

$$(x - 3)(x - 2) = 0 \qquad \text{Factoring}$$

$$x = 3 \quad or \quad x = 2. \qquad \text{Using the principle of zero products}$$

Check: For $x = 3$, $f(3) = 3 + \dfrac{6}{3} = 3 + 2 = 5$.

For $x = 2$, $f(2) = 2 + \dfrac{6}{2} = 2 + 3 = 5$.

The solutions are 2 and 3.

Do Exercise 7.

AG **ALGEBRAIC–GRAPHICAL CONNECTION**

Let's make a visual check of Example 7 by looking at a graph. We can think of the equation

$$x + \frac{6}{x} = 5$$

as the intersection of the graphs of

$$f(x) = x + \frac{6}{x} \quad \text{and} \quad g(x) = 5.$$

We see in the graph that there are two points of intersection, at $x = 2$ and at $x = 3$.

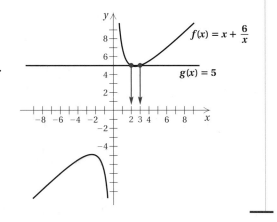

| Caution! |

In this section, we have introduced a new use of the LCM. Before, you used the LCM in adding or subtracting rational expressions. Now we are working with equations. There are equals signs. We clear the fractions by multiplying all terms on both sides of the equation by the LCM. This eliminates the denominators. *Do not* make the mistake of trying to "clear the fractions" when you do not have an equation!

Answer on page A-30

ARE YOU CALCULATING OR SOLVING?

One of the common difficulties with this chapter is knowing for sure the task at hand. Are you combining expressions using operations to get another *rational expression*, or are you solving equations for which the results are numbers that are *solutions* of an equation? To learn to make these decisions, complete the following list by writing in the blank the type of answer you should get: "Rational expression" or "Solutions." You need not complete the mathematical operations. Answers can be found at the back of the book.

TASK	ANSWER (Just write "Rational expression" or "Solutions.")
1. Add: $\dfrac{4}{x-2} + \dfrac{1}{x+2}$.	
2. Solve: $\dfrac{4}{x-2} = \dfrac{1}{x+2}$.	
3. Subtract: $\dfrac{4}{x-2} - \dfrac{1}{x+2}$.	
4. Multiply: $\dfrac{4}{x-2} \cdot \dfrac{1}{x+2}$.	
5. Divide: $\dfrac{4}{x-2} \div \dfrac{1}{x+2}$.	
6. Solve: $\dfrac{4}{x-2} + \dfrac{1}{x+2} = \dfrac{26}{x^2-4}$.	
7. Perform the indicated operations and simplify: $\dfrac{4}{x-2} + \dfrac{1}{x+2} - \dfrac{26}{x^2-4}$.	
8. Solve: $\dfrac{x^2}{x-1} = \dfrac{1}{x-1}$.	
9. Solve: $\dfrac{2}{y^2-25} = \dfrac{3}{y-5} + \dfrac{1}{y-5}$.	
10. Solve: $\dfrac{x}{x+4} - \dfrac{4}{x-4} = \dfrac{x^2+16}{x^2-16}$.	
11. Perform the indicated operations and simplify: $\dfrac{x}{x+4} - \dfrac{4}{x-4} - \dfrac{x^2+16}{x^2-16}$.	
12. Solve: $\dfrac{5}{y-3} - \dfrac{30}{y^2-9} = 1$.	
13. Add: $\dfrac{5}{y-3} + \dfrac{30}{y^2-9} + 1$.	

a   Solve. Don't forget to check!

1. $\dfrac{y}{10} = \dfrac{2}{5} + \dfrac{3}{8}$

2. $\dfrac{3}{8} + \dfrac{1}{3} = \dfrac{t}{12}$

3. $\dfrac{1}{4} - \dfrac{5}{6} = \dfrac{1}{a}$

4. $\dfrac{5}{8} - \dfrac{2}{5} = \dfrac{1}{y}$

5. $\dfrac{x}{3} - \dfrac{x}{4} = 12$

6. $\dfrac{y}{5} - \dfrac{y}{3} = 15$

7. $x + \dfrac{8}{x} = -9$

8. $y + \dfrac{22}{y} = -13$

9. $\dfrac{3}{y} + \dfrac{7}{y} = 5$

10. $\dfrac{4}{3y} - \dfrac{3}{y} = \dfrac{10}{3}$

11. $\dfrac{1}{2} = \dfrac{z-5}{z+1}$

12. $\dfrac{x-6}{x+9} = \dfrac{2}{7}$

13. $\dfrac{3}{y+1} = \dfrac{2}{y-3}$

14. $\dfrac{4}{x-1} = \dfrac{3}{x+2}$

15. $\dfrac{y-1}{y-3} = \dfrac{2}{y-3}$

16. $\dfrac{x-2}{x-4} = \dfrac{2}{x-4}$

17. $\dfrac{x+1}{x} = \dfrac{3}{2}$

18. $\dfrac{y+2}{y} = \dfrac{5}{3}$

19. $\dfrac{1}{2} - \dfrac{4}{9x} = \dfrac{4}{9} - \dfrac{1}{6x}$

20. $-\dfrac{1}{3} - \dfrac{5}{4y} = \dfrac{3}{4} - \dfrac{1}{6y}$

21. $\dfrac{60}{x} - \dfrac{60}{x-5} = \dfrac{2}{x}$

22. $\dfrac{50}{y} - \dfrac{50}{y-2} = \dfrac{4}{y}$

23. $\dfrac{7}{5x-2} = \dfrac{5}{4x}$

24. $\dfrac{5}{y+4} = \dfrac{3}{y-2}$

25. $\dfrac{x}{x-2} + \dfrac{x}{x^2-4} = \dfrac{x+3}{x+2}$

26. $\dfrac{3}{y-2} + \dfrac{2y}{4-y^2} = \dfrac{5}{y+2}$

27. $\dfrac{6}{x^2-4x+3} - \dfrac{1}{x-3} = \dfrac{1}{4x-4}$

28. $\dfrac{1}{2x+10} = \dfrac{8}{x^2-25} - \dfrac{2}{x-5}$

29. $\dfrac{5}{y+3} = \dfrac{1}{4y^2-36} + \dfrac{2}{y-3}$

30. $\dfrac{7}{x-2} - \dfrac{8}{x+5} = \dfrac{1}{2x^2+6x-20}$

31. $\dfrac{a}{2a-6} - \dfrac{3}{a^2-6a+9} = \dfrac{a-2}{3a-9}$

32. $\dfrac{1}{x-2} = \dfrac{2}{x+4} + \dfrac{2x-1}{x^2+2x-8}$

33. $\dfrac{2x+3}{x-1} = \dfrac{10}{x^2-1} + \dfrac{2x-3}{x+1}$

34. $\dfrac{y}{y+1} + \dfrac{3y+5}{y^2+4y+3} = \dfrac{2}{y+3}$

35. $\dfrac{4}{x+3} + \dfrac{7}{x^2-3x+9} = \dfrac{108}{x^3+27}$

36. $\dfrac{3x}{x+2} + \dfrac{72}{x^3+8} = \dfrac{24}{x^2-2x+4}$

37. $\dfrac{5x}{x-7} - \dfrac{35}{x+7} = \dfrac{490}{x^2-49}$

38. $\dfrac{3x}{x+2} + \dfrac{6}{x} + 4 = \dfrac{12}{x^2+2x}$

39. $\dfrac{x-1}{3} + \dfrac{6x+1}{15} + \dfrac{2(x-2)}{13-7x} = \dfrac{2(x+2)}{5}$

For the given rational function f, find all values of x for which $f(x)$ has the indicated value.

40. $f(x) = 2x - \dfrac{15}{x}; \ f(x) = 1$

41. $f(x) = 2x - \dfrac{6}{x}; \ f(x) = 1$

42. $f(x) = \dfrac{x-5}{x+1}; \ f(x) = \dfrac{3}{5}$

43. $f(x) = \dfrac{x-3}{x+2}; \ f(x) = \dfrac{1}{5}$

44. $f(x) = \dfrac{12}{x} - \dfrac{12}{2x}; \ f(x) = 8$

45. $f(x) = \dfrac{6}{x} - \dfrac{6}{2x}; \ f(x) = 5$

46. $\mathbf{D_W}$ Explain how you might easily create rational equations for which there is no solution. (See Example 4 for a hint.)

47. $\mathbf{D_W}$ Explain why it is sufficient, when checking a possible solution of a rational equation, to verify that the number in question does not make a denominator 0.

SKILL MAINTENANCE

Factor. [4.6d]

48. $4t^3 + 500$

49. $1 - t^6$

50. $a^3 + 8b^3$

51. $a^3 - 8b^3$

Solve. [4.8a]

52. $x^2 - 6x + 9 = 0$

53. $(x-3)(x+4) = 0$

54. $x^2 - 49 = 0$

55. $12x^2 - 11x + 2 = 0$

SYNTHESIS

56.
a) Use the INTERSECT feature of a graphing calculator to find the points of intersection of the graphs of
$$f(x) = \dfrac{1}{1+x} + \dfrac{x}{1-x} \quad \text{and} \quad g(x) = \dfrac{1}{1-x} - \dfrac{x}{1+x}.$$
b) Use the algebraic methods of this section to check your answers to part (a).
c) Explain which procedure you prefer.

57.
a) Use the INTERSECT feature of a graphing calculator to find the points of intersection of the graphs of
$$f(x) = \dfrac{x+3}{x+2} - \dfrac{x+4}{x+3} \quad \text{and} \quad g(x) = \dfrac{x+5}{x+4} - \dfrac{x+6}{x+5}.$$
b) Use the algebraic methods of this section to check your answers to part (a).
c) Explain which procedure you prefer.

Objectives

a) Solve work problems and certain basic problems using rational equations.

b) Solve applied problems involving proportions.

c) Solve motion problems using rational equations.

a Work Problems

■ **EXAMPLE 1**   *Cake Decorating.*   Sara and Jeff work at Gourmet Perfection, Inc. Sara can decorate a wedding cake in 4 hr. Jeff can decorate the same cake in 5 hr. How long would it take them, working together, to decorate the cake?

1. **Familiarize.**   We familiarize ourselves with the problem by considering two *incorrect* ways of translating the problem to mathematical language.

   a) A common *incorrect* way to translate the problem is to add the two times: 4 hr + 5 hr = 9 hr. Let's think about this. Sara can do the job alone in 4 hr. If Sara and Jeff work together, the time it takes them should be *less* than 4 hr. Thus we reject 9 hr as a solution, but we do have a partial check on any answer we get. The answer should be less than 4 hr.

   b) Another *incorrect* way to translate the problem is as follows. Suppose the two people split up the decorating job in such a way that Sara does half the decorating and Jeff does the other half. Then

   Sara decorates $\frac{1}{2}$ of the cake in $\frac{1}{2}(4 \text{ hr})$, or 2 hr,

   and

   Jeff decorates $\frac{1}{2}$ of the cake in $\frac{1}{2}(5 \text{ hr})$, or $2\frac{1}{2}$ hr.

   But time is wasted since Sara would finish her part $\frac{1}{2}$ hr earlier than Jeff. In effect, they have not worked together to complete the job as fast as possible. If Sara helps Jeff after completing her half, the entire job could be finished in a time somewhere between 2 hr and $2\frac{1}{2}$ hr.

   We proceed to a translation by considering how much of the job is finished in 1 hr, 2 hr, 3 hr, and so on. It takes Sara 4 hr to do the decorating job alone. Then in 1 hr, she can do $\frac{1}{4}$ of the job. It takes Jeff 5 hr to do the job alone. Then in 1 hr, he can do $\frac{1}{5}$ of the job. Working together, they can do

   $\frac{1}{4} + \frac{1}{5}$, or $\frac{9}{20}$ of the job in 1 hr.

In 2 hr, Sara can do $2\left(\frac{1}{4}\right)$ of the job and Jeff can do $2\left(\frac{1}{5}\right)$ of the job. Working together, they can do

$2\left(\frac{1}{4}\right) + 2\left(\frac{1}{5}\right)$, or $\frac{9}{10}$ of the job in 2 hr.

Continuing this reasoning, we can form a table like the following one.

TIME	FRACTION OF THE JOB COMPLETED		
	SARAH	JEFF	TOGETHER
1 hr	$\frac{1}{4}$	$\frac{1}{5}$	$\frac{1}{4} + \frac{1}{5}$, or $\frac{9}{20}$
2 hr	$2\left(\frac{1}{4}\right)$	$2\left(\frac{1}{5}\right)$	$2\left(\frac{1}{4}\right) + 2\left(\frac{1}{5}\right)$, or $\frac{9}{10}$
3 hr	$3\left(\frac{1}{4}\right)$	$3\left(\frac{1}{5}\right)$	$3\left(\frac{1}{4}\right) + 3\left(\frac{1}{5}\right)$, or $1\frac{7}{20}$
t hr	$t\left(\frac{1}{4}\right)$	$t\left(\frac{1}{5}\right)$	$t\left(\frac{1}{4}\right) + t\left(\frac{1}{5}\right)$

From the table, we see that if they work 3 hr, the fraction of the job that they will finish is $1\frac{7}{20}$, which is more of the job than needs to be done. We also see that the answer is somewhere between 2 hr and 3 hr. What we want is a number t such that the fraction of the job that is completed is 1; that is, the job is just completed—not more $\left(1\frac{7}{20}\right)$ and not less $\left(\frac{9}{10}\right)$.

2. **Translate.** From the table, we see that the time we want is some number t for which

$$t\left(\frac{1}{4}\right) + t\left(\frac{1}{5}\right) = 1, \quad \text{or} \quad \frac{t}{4} + \frac{t}{5} = 1,$$

where 1 represents the idea that the entire job is completed in time t.

3. **Solve.** We solve the equation:

$$\frac{t}{4} + \frac{t}{5} = 1$$

$$20\left(\frac{t}{4} + \frac{t}{5}\right) = 20 \cdot 1 \qquad \text{The LCM is } 2 \cdot 2 \cdot 5, \text{ or } 20. \\ \text{We multiply by 20.}$$

$$20 \cdot \frac{t}{4} + 20 \cdot \frac{t}{5} = 20 \qquad \text{Using the distributive law}$$

$$5t + 4t = 20 \qquad \text{Simplifying}$$

$$9t = 20$$

$$t = \frac{20}{9}, \text{ or } 2\frac{2}{9} \text{ hr.}$$

4. **Check.** The check can be done by using $\frac{20}{9}$ for t and substituting into the original equation:

$$\frac{20}{9}\left(\frac{1}{4}\right) + \frac{20}{9}\left(\frac{1}{5}\right) = \frac{5}{9} + \frac{4}{9} = \frac{9}{9} = 1.$$

We also have a partial check in what we learned from the *Familiarize* step. The answer, $2\frac{2}{9}$ hr, is between 2 hr and 3 hr (see the table), and it is less than 4 hr, the time it takes Sara working alone.

5. **State.** It takes $2\frac{2}{9}$ hr for them to complete the job working together.

1. **Trimming Shrubbery.** Alex can trim the shrubbery at Beecher Community College in 6 hr. Tanya can do the same job in 4 hr. How long would it take them, working together, to do the same trimming job?

THE WORK PRINCIPLE

Suppose that a is the time it takes A to do a job, b is the time it takes B to do the same job, and t is the time it takes them to do the job working together. Then

$$\frac{t}{a} + \frac{t}{b} = 1.$$

Do Exercise 1.

EXAMPLE 2 *Planning an Intertribal Dance.* Hannah Two Rivers and Feather McBride are planning the Intertribal Dance Event for September's Council Tree Pow Wow and Cultural Festival. Less experienced, Feather estimates that if each were working alone, it would take her 9 hr longer than Hannah to complete the job. Working together, the two women finish in 20 hr. How long would it take each, working alone, to complete the job?

1. **Familiarize.** Comparing this problem to Example 1, we note that we do not know the times required by each person to complete the tasks had each worked alone. We let

> h = the amount of time, in hours, that it would take Hannah working alone.

Then

> $h + 9$ = the amount of time, in hours, that it would take Feather working alone.

We also know that $t = 20\,\text{hr}$ = total time. Thus,

> $\dfrac{20}{h}$ = the fraction of the job that Hannah could finish in 20 hr

and

> $\dfrac{20}{h + 9}$ = the fraction of the job that Feather could finish in 20 hr.

Answer on page A-30

2. Translate. Using the work principle, we know that

$$\frac{t}{a} + \frac{t}{b} = 1 \qquad \text{Using the work principle}$$

$$\frac{20}{h} + \frac{20}{h+9} = 1. \qquad \text{Substituting } \frac{20}{h} \text{ for } \frac{t}{a} \text{ and } \frac{20}{h+9} \text{ for } \frac{t}{b}$$

3. Solve. We solve the equation:

$$\frac{20}{h} + \frac{20}{h+9} = 1$$

$$h(h+9)\left(\frac{20}{h} + \frac{20}{h+9}\right) = h(h+9) \cdot 1 \qquad \begin{array}{l}\text{We multiply by the LCM,}\\ \text{which is } h(h+9).\end{array}$$

$$h(h+9) \cdot \frac{20}{h} + h(h+9) \cdot \frac{20}{h+9} = h(h+9) \qquad \text{Using the distributive law}$$

$$(h+9) \cdot 20 + h \cdot 20 = h^2 + 9h \qquad \text{Simplifying}$$

$$20h + 180 + 20h = h^2 + 9h$$

$$0 = h^2 - 31h - 180 \qquad \text{Getting 0 on one side}$$

$$0 = (h-36)(h+5) \qquad \text{Factoring}$$

$$h - 36 = 0 \quad or \quad h + 5 = 0 \qquad \begin{array}{l}\text{Using the principle}\\ \text{of zero products}\end{array}$$

$$h = 36 \quad or \qquad h = -5.$$

4. Check. Since negative time has no meaning in the problem, we reject -5 as a solution to the original problem. The number 36 checks since if Hannah takes 36 hr alone and Feather takes $36 + 9$, or 45 hr alone, in 20 hr, working together, they would have completed

$$\frac{20}{36} + \frac{20}{45} = \frac{5}{9} + \frac{4}{9}, \text{ or 1 job.}$$

5. State. It would take Hannah 36 hr and Feather 45 hr to complete the tasks alone.

Do Exercise 2.

2. Filling a Water Tank. Two pipes carry water to the same tank. Pipe A, working alone, can fill the tank three times as fast as pipe B. Together, the pipes can fill the tank in 24 hr. Find the time it would take each pipe to fill the tank alone.

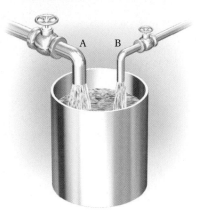

b Applications Involving Proportions

Any rational expression a/b represents a **ratio.** Percent can be considered a ratio. For example, 67% is the ratio of 67 to 100, or 67/100. The ratio of two different kinds of measure is called a **rate.** Speed is an example of a rate. Florence Griffith Joyner set a world record in a recent Olympics with a time of 10.49 sec in the 100-m dash. Her speed, or rate, was

$$\frac{100 \text{ m}}{10.49 \text{ sec}}, \quad or \quad 9.5 \frac{\text{m}}{\text{sec}}. \qquad \text{Rounded to the nearest tenth}$$

PROPORTION

An equality of ratios, $A/B = C/D$, read "*A* is to *B* as *C* is to *D*," is called a **proportion.** The numbers named in a true proportion are said to be **proportional** to each other.

Answer on page A-30

We can use proportions to solve applied problems by expressing a ratio in two ways, as shown below. For example, suppose that it takes 8 gal of gas to drive for 120 mi, and we want to determine how much will be required to drive for 550 mi. If we assume that the car uses gas at the same rate throughout the trip, the ratios are the same, and we can write a proportion.

$$\text{Miles} \longrightarrow \frac{120}{8} = \frac{550}{x} \longleftarrow \text{Miles}$$
$$\text{Gallons} \longrightarrow \qquad\qquad \longleftarrow \text{Gallons}$$

To solve this proportion, we note that the LCM is $8x$. Thus we multiply by $8x$.

$$8x \cdot \frac{120}{8} = 8x \cdot \frac{550}{x} \qquad \text{Multiplying by } 8x$$

$$x \cdot 120 = 8 \cdot 550 \qquad \text{Simplifying}$$

$$120x = 8 \cdot 550$$

$$x = \frac{8 \cdot 550}{120} \qquad \text{Dividing by 120}$$

$$x \approx 36.67$$

Thus, 36.67 gal will be required to drive for 550 mi.

It is common to use **cross products** to solve proportions, as follows:

$$\frac{120}{8} \bowtie \frac{550}{x} \qquad \text{If } \frac{A}{B} = \frac{C}{D}, \text{ then } AD = BC.$$

$$120 \cdot x = 8 \cdot 550 \qquad 120 \cdot x \text{ and } 8 \cdot 550 \text{ are called } cross\ products. \text{ Note that this is the equation above that results from clearing fractions.}$$

$$x = \frac{8 \cdot 550}{120}$$

$$x \approx 36.67.$$

EXAMPLE 3 *Calories Burned.* Your author generally exercises every day. The readout on a StairMaster machine tells him that if he exercises for 24 min, he will burn 356 calories. How many calories will he burn if he exercises for 30 min?

1. **Familiarize.** We let $c =$ the number of calories burned in 30 min.

2. **Translate.** Next, we translate to a proportion. We make each side the ratio of minutes to number of calories, with minutes in the numerator and number of calories in the denominator.

$$\text{Minutes} \longrightarrow \frac{24}{356} = \frac{30}{c} \longleftarrow \text{Minutes}$$
$$\text{Calories} \longrightarrow \qquad\qquad \longleftarrow \text{Calories}$$

3. **Solve.** We solve the proportion:

$$\frac{24}{356} = \frac{30}{c}$$

$$24c = 356 \cdot 30 \qquad \text{Equating cross products}$$

$$c = \frac{356 \cdot 30}{24} \qquad \text{Dividing by 24}$$

$$= 445. \qquad \text{Multiplying and dividing}$$

4. Check. We substitute into the proportion and check cross products:

$$\frac{24}{356} = \frac{30}{445};$$

$$24 \cdot 445 = 10{,}680; \qquad 356 \cdot 30 = 10{,}680.$$

Since the cross products are the same, the answer checks.

5. State. In 30 min, he will burn 445 calories.

Do Exercise 3.

EXAMPLE 4 *Estimating Wild Horse Population in California.* Wild horses still exist in at least ten states of the United States. To estimate the number in California, a forest ranger catches 616 wild horses, tags them, and releases them. Later, 244 horses are caught, and it is found that 49 of them are tagged. Estimate how many wild horses there are in California.

Source: U.S. Bureau of Land Management

1. **Familiarize.** We let $H =$ the number of wild horses in California. For the purposes of this example, we assume that the tagged horses mix freely with others in the state. We also assume that when some horses have been captured, the ratio of those tagged to the total number captured is the same as the ratio of horses originally tagged to the total number of wild horses in the state. For example, given that 1 of every 3 horses captured later is tagged, we assume that 1 of every 3 horses in the state was originally tagged.

2. **Translate.** We translate to a proportion, as follows:

Horses tagged originally $\longrightarrow \dfrac{616}{H} = \dfrac{49}{244} \begin{array}{l} \longleftarrow \text{Tagged horses caught} \\ \longleftarrow \text{Horses caught} \end{array}$

Horses in California $\longrightarrow$

3. **Solve.** We solve the proportion:

$$616 \cdot 244 = H \cdot 49 \qquad \text{Equating cross products}$$

$$\frac{616 \cdot 244}{49} = H \qquad \text{Dividing by 49}$$

$$3067 \approx H. \qquad \text{Multiplying and dividing and approximating}$$

4. **Check.** We substitute into the proportion and check cross products:

$$\frac{616}{3067} = \frac{49}{244};$$

$$616 \cdot 244 = 150{,}304; \qquad 3067 \cdot 49 = 150{,}283.$$

The cross products are close but not exact because we rounded the total.

5. **State.** We estimate that there are about 3067 wild horses in California.

Do Exercise 4.

---

3. Determining Medication Dosage. To control a fever, a doctor suggests that a child who weighs 28 kg be given 420 mg of Tylenol. If dosage is proportional to the child's weight, how much Tylenol would be recommended for a child who weighs 35 kg?

4. Estimating Wild Horse Population in Utah. To estimate the number of wild horses in Utah, a forest ranger catches 620 wild horses, tags them, and releases them. Later, 122 horses are caught, and it is found that 31 of them are tagged. Estimate how many wild horses there are in Utah.

Source: U.S. Bureau of Land Management

Answers on page A-30

C Motion Problems

We considered motion problems earlier in Sections 1.3 and 3.4. To translate them, we know that we can use either the basic motion formula, $d = rt$, or either of two formulas $r = d/t$, or $t = d/r$, which can be derived from $d = rt$.

MOTION FORMULAS

The following are the formulas for motion problems:

$d = rt \longrightarrow$ Distance = Rate · Time;

$r = \dfrac{d}{t} \longrightarrow$ Rate = $\dfrac{\text{Distance}}{\text{Time}}$;

$t = \dfrac{d}{r} \longrightarrow$ Time = $\dfrac{\text{Distance}}{\text{Rate}}$.

EXAMPLE 5 *Bicycling.* A racer is bicycling 15 km/h faster than a person on a mountain bike. In the time it takes the racer to travel 80 km, the person on the mountain bike has gone 50 km. Find the speed of each bicyclist.

1. **Familiarize.** Let's guess that the person on the mountain bike is going 10 km/h. The racer would then be traveling $10 + 15$, or 25 km/h. At 25 km/h, the racer will travel 80 km in $\frac{80}{25} = 3.2$ hr. Going 10 km/h, the mountain bike will cover 50 km in $\frac{50}{10} = 5$ hr. Since $3.2 \neq 5$, our guess is wrong, but we can see that if r is the rate, in kilometers per hour, of the slower bike, then the rate of the racer, who is traveling 15 km/h faster, is $r + 15$.

    Making a drawing and organizing the facts in a chart can be helpful.

50 km
r km/h

80 km
$r + 15$ km/h

	DISTANCE	SPEED	TIME
Mountain Bike	50	r	t
Racing Bike	80	$r + 15$	t

$\rightarrow 50 = rt \rightarrow t = \dfrac{50}{r}$

$\rightarrow 80 = (r + 15)t \rightarrow t = \dfrac{80}{r + 15}$

2. **Translate.** The time is the same for both bikes. Using the formula $d = rt$ and then $t = d/r$ across both rows of the table, we find two expressions for time and can equate them as

$$\frac{50}{r} = \frac{80}{r + 15}.$$

3. Solve. We solve the equation:

$$\frac{50}{r} = \frac{80}{r + 15}$$

$$r(r + 15) \cdot \frac{50}{r} = r(r + 15) \cdot \frac{80}{r + 15} \qquad \text{The LCM is } r(r + 15).$$
$$\text{We multiply by } r(r + 15).$$

$$(r + 15) \cdot 50 = r \cdot 80 \qquad \text{Simplifying. We can also obtain}$$
$$\text{this by equating cross products.}$$

$$50r + 750 = 80r \qquad \text{Using the distributive law}$$

$$750 = 30r \qquad \text{Subtracting } 50r$$

$$\frac{750}{30} = r \qquad \text{Dividing by 30}$$

$$25 = r.$$

4. Check. If our answer checks, the mountain bike is going 25 km/h and the racing bike is going 25 + 15, or 40 km/h.

Traveling 80 km at 40 km/h, the racer is riding for $\frac{80}{40} = 2$ hr. Traveling 50 km at 25 km/h, the person on the mountain bike is riding for $\frac{50}{25} = 2$ hr. Our answer checks since the two times are the same.

5. State. The speed of the racer is 40 km/h, and the speed of the person on the mountain bike is 25 km/h.

Do Exercise 5.

EXAMPLE 6 *Air Travel.* An airplane flies 1062 mi with the wind. In the same amount of time, it can fly 738 mi against the wind. The speed of the plane in still air is 200 mph. Find the speed of the wind.

1. Familiarize. We first make a drawing. We let w = the speed of the wind and t = the time, and then organize the facts in a chart.

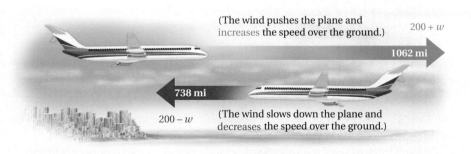

(The wind pushes the plane and increases the speed over the ground.) $200 + w$

1062 mi

738 mi

$200 - w$

(The wind slows down the plane and decreases the speed over the ground.)

	DISTANCE	SPEED	TIME
With Wind	1062	$200 + w$	t
Against Wind	738	$200 - w$	t

$$\rightarrow 1062 = (200 + w)t \rightarrow t = \frac{1062}{200 + w}$$
$$\rightarrow 738 = (200 - w)t \rightarrow t = \frac{738}{200 - w}$$

2. Translate. Using the formula $t = d/r$ across both rows of the table, we find two expressions for time and can equate them as

$$\frac{1062}{200 + w} = \frac{738}{200 - w}.$$

5. Four-Wheeler Travel. Jaime's four-wheeler travels 8 km/h faster than Mara's. Jaime travels 69 km in the same time it takes Mara to travel 45 km. Find the speed of each person's four-wheeler.

Answer on page A-30

6. River Travel. The *Delta Queen* is a riverboat that travels the Ohio River. Suppose that it travels 246 mi downstream in the same time that it takes to travel 180 mi upstream. The speed of the current in the river is 5.5 mph. Find the speed of the boat in still water.

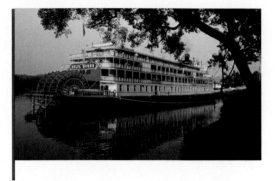

3. Solve. We solve the equation:

$$\frac{1062}{200 + w} = \frac{738}{200 - w}$$

$$(200 + w)(200 - w)\left(\frac{1062}{200 + w}\right) = (200 + w)(200 - w)\left(\frac{738}{200 - w}\right)$$

$$(200 - w)1062 = (200 + w)738$$ We can also obtain this by equating cross products.

$$212{,}400 - 1062w = 147{,}600 + 738w$$
$$64{,}800 = 1800w$$
$$w = 36.$$

4. Check. With the wind, the speed of the plane is $200 + 36$, or 236 mph. Dividing the distance, 1062 mi, by the speed, 236 mph, we get 4.5 hr. Against the wind, the speed of the plane is $200 - 36$, or 164 mph. Dividing the distance, 738 mi, by the speed, 164 mph, we get 4.5 hr. Since the times are the same, the answer checks.

5. State. The speed of the wind is 36 mph.

Do Exercise 6.

Answer on page A-30

Study Tips

TIME MANAGEMENT (PART 2)

Here are some additional tips to help you with time management. (See also the Study Tips on time management in Sections 3.5 and 6.1.)

■ **Avoid "time killers."** We live in a media age, and the Internet, e-mail, television, and movies are all time killers. Allow yourself a break to enjoy some college and outside activities. But keep track of the time you spend on such activities and compare it to the time you spend studying.

■ **Prioritize your tasks.** Be careful about taking on too many college activities that fall outside of academics. Examples of such activities are decorating a homecoming float, joining a fraternity or sorority, and participating on a student council committee. Any of these is important but keep them to a minimum to be sure that you have enough time for your studies.

■ **Be aggressive about your study tasks.** Instead of worrying over your math homework or test preparation, do something to get yourself started. Work a problem here and a problem there, and before long you will accomplish the task at hand. If the task is large, break it down into smaller parts, and do one at a time. You will be surprised at how quickly the large task can then be completed.

"Time is more valuable than money. You can get more money, but you can't get more time."

Jim Rohn, motivational speaker

Translating for Success

1. *Sums of Squares.* The sum of the squares of two consecutive odd integers is 650. Find the integers.

2. *Estimating Fish Population.* To determine the number of fish in a lake, a conservationist catches 225 of them, tags them, and releases them back into the lake. Later, 108 fish are caught, and it is found that 15 of them are tagged. Estimate how many fish are in the lake.

3. *Consecutive Integers.* The sum of two consecutive even integers is 650. Find the integers.

4. *Sums of Squares.* The sum of the squares of two consecutive integers is 685. Find the integers.

5. *Hockey Results.* A hockey team played 81 games in a season. They won 1 fewer game than three times the number of ties and lost 8 fewer games than they won. How many games did they win? lose? tie?

Translate each word problem to an equation or a system of equations and select a correct translation from equations A–O.

A. $x + (x + 2) = 650$

B. $\dfrac{225}{x} = \dfrac{15}{108}$

C. $x^2 + (x + 1)^2 = 685$

D. $\dfrac{30}{x + 3} = \dfrac{40}{x}$

E. $x + y + z = 81,$
$x = 3y - 1,$
$z = x - 8$

F. $x + y + z = 81,$
$x - 1 = 3y,$
$z = x - 8$

G. $x^2 + (x + 5)^2 = 650$

H. $x + y + z = 650,$
$x + y = 480,$
$y + z = 685$

I. $\dfrac{40}{x + 3} = \dfrac{30}{x}$

J. $\dfrac{15}{x} = \dfrac{108}{225}$

K. $x + y + z = 685,$
$x + y = 480,$
$y + z = 650$

L. $x^2 + (x + 5)^2 = 685$

M. $\dfrac{1}{3} + \dfrac{1}{8} = \dfrac{1}{x}$

N. $x^2 + (x + 2)^2 = 650$

O. $x = y + 3,$
$2x + 2y = 81$

Answers on page A-30

---

6. *Sides of a Square.* If the sides of a square are increased by 5 ft, the area of the original square plus the area of the enlarged square is 650 ft^2. Find the length of a side of the original square.

7. *Bicycling.* The speed of one mountain biker is 3 km/h faster than the speed of another biker. The first biker travels 40 mi in the same amount of time that it takes the second to travel 30 mi. Find the speed of each biker.

8. *PDQ Shopping Network.* Sarah, Claire, and Maggie can process 685 telephone orders per day for PDQ shopping network. Sarah and Claire together can process 480 orders, while Claire and Maggie can process 650 orders per day. How many orders can each process alone?

9. *Filling Time.* A spa can be filled in 3 hr by hose A alone and in 8 hr by hose B alone. How long would it take to fill the tank if both hoses are working?

10. *Rectangle Dimensions.* The length of a rectangle is 3 ft longer than its width. Find the dimensions of the rectangle such that the perimeter of the rectangle is 81 ft.

5.6

EXERCISE SET

For Extra Help

Math XL MyMathLab InterAct Math Tutor Digital Video Student's
MathXL MyMathLab Math Center Tutor CD 2 Solutions
Videotape 6 Manual

a   Solve.

1. *Mail Order.*   Jason, an experienced shipping clerk, can fill a certain order in 5 hr. Jessica, a new clerk, needs 9 hr to do the same job. Working together, how long will it take them to fill the order?

2. *Painting.*   Alexander can paint a room in 4 hr. Alexandra can paint the same room in 3 hr. Working together, how long will it take them to paint the room?

3. *Filling a Pool.*   An in-ground backyard pool can be filled in 12 hr if water enters through a pipe alone, or in 30 hr if water enters through a hose alone. If water is entering through both the pipe and the hose, how long will it take to fill the pool?

4. *Filling a Tank.*   A tank can be filled in 18 hr by pipe A alone and in 22 hr by pipe B alone. How long will it take to fill the tank if both pipes are working?

5. *Printing.*   One printing press can print an order of advertising brochures in 4.5 hr. Another press can do the same job in 5.5 hr. How long will it take if both presses are used? (*Hint*: You may find that multiplying by $\frac{1}{10}$ on both sides of the equation will clear the decimals.)

6. *Wood Cutting.*   Hannah can clear a lot in 5.5 hr. Damon can do the same job in 7.5 hr. How long will it take them to clear the lot working together?

7. *Painting.*   Juan can paint the neighbor's house four times as fast as Ariel. The year they worked together it took them 8 days. How long would it take each to paint the house alone?

8. *Newspaper Delivery.*   Matt can deliver papers three times as fast as his father. If they work together, it takes them 1 hr. How long would it take each to deliver the papers alone?

9. *Cutting Firewood.* Jake can cut and split a cord of firewood in 6 fewer hr than Skyler can. When they work together, it takes them 4 hr. How long would it take each of them to do the job alone?

10. *Painting.* Sara takes 3 hr longer to paint a floor than Kate does. When they work together, it takes them 2 hr. How long would each take to do the job alone?

b Solve.

11. *Women's Hip Measurements.* For improved health, it is recommended that a woman's waist-to-hip ratio be 0.85 (or lower). Marta's hip measurement is 40 in. To meet the recommendation, what should Marta's waist measurement be?

Source: David Schmidt, "Lifting Weight Myths," *Nutrition Action Newsletter,* 20, no. **4,** October 1993

12. *Men's Hip Measurements.* It is recommended that a man's waist-to-hip ratio be 0.95 (or lower). Malcolm's hip measurement is 40 in. To meet the recommendation, what should Malcolm's waist measurement be?

13. *Home-Run Pace.* After 45 games of the 2005 Major League Baseball season, Alex Rodriguez of the New York Yankees led the American League with 14 home runs. Assuming that he would continue to hit home runs at the same rate, how many would he hit in the 162-game season?

Source: Major League Baseball

14. *Batting Average and Hits.* After 40 games of the 2005 Major League Baseball season, Miguel Cabrera of the Florida Marlins led the National League with 30 hits in 152 at-bats. Assuming that he would continue to get hits at the same rate, how many would he get in the entire season if he had 650 at-bats?

Source: Major League Baseball

15. *Coffee Consumption.* Coffee beans from 14 trees are required to produce 7.7 kg of coffee. (This is the average amount that each person in the United States drinks each year.) The beans from how many trees are required to produce 638 kg of coffee?

16. *Human Blood.* 10 cm^3 of a normal specimen of human blood contains 1.2 g of hemoglobin. How many grams does 32 cm^3 of the same blood contain?

17. *Estimating Wildlife Populations.* To determine the number of deer in a game preserve, a conservationist catches 318 deer, tags them, and lets them loose. Later, 168 deer are caught; 56 of them are tagged. How many deer are in the preserve?

18. *Estimating Wildlife Populations.* To determine the number of trout in a lake, a conservationist catches 112 trout, tags them, and releases them back into the lake. Later, 82 trout are caught; 32 of them are tagged. How many trout are in the lake?

19. *Weight on Mars.* The ratio of the weight of an object on Mars to the weight of an object on Earth is 0.4 to 1.

a) How much will a 12-T rocket weigh on Mars?
b) How much will a 120-lb astronaut weigh on Mars?

20. *Weight on Moon.* The ratio of the weight of an object on the moon to the weight of an object on Earth is 0.16 to 1.

a) How much will a 12-T rocket weigh on the moon?
b) How much will a 180-lb astronaut weigh on the moon?

21. Consider the numbers 1, 2, 3, and 5. If the same number is added to each of the numbers, it is found that the ratio of the first new number to the second is the same as the ratio of the third new number to the fourth. Find the number.

22. *Rope Cutting.* A rope is 28 ft long. How can the rope be cut in such a way that the ratio of the resulting two segments is 3 to 5?

C Solve.

23. *Boating.* The current in the Lazy River moves at a rate of 4 mph. Ken's dinghy motors 6 mi upstream in the same time that it takes to motor 12 mi downstream. What is the speed of the dinghy in still water?

24. *Kayaking.* The speed of the current in Catamount Creek is 3 mph. Jaci's kayak can travel 4 mi upstream in the same time that it takes to travel 10 mi downstream. What is the speed of Jaci's kayak in still water?

25. *Marine Travel.* Sandy's tugboat moves at a rate of 10 mph in still water. It travels 24 mi upstream and 24 mi downstream in a total time of 5 hr. What is the speed of the current?

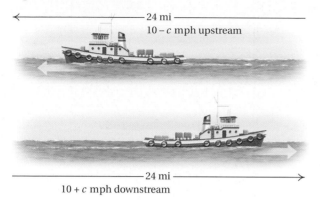

← ——————— 24 mi ———————
10 – c mph upstream

—————————— 24 mi ——————————→
10 + c mph downstream

26. *Shipping.* A barge moves at a rate of 7 km/h in still water. It travels 45 km upriver and 45 km downriver in a total time of 14 hr. What is the speed of the current?

27. *Moving Sidewalks.* A moving sidewalk at an airport moves at a rate of 1.8 ft/sec. Walking on the moving sidewalk, Camille travels 105 ft forward in the time it takes to travel 51 ft in the opposite direction. How fast would Camille be walking on a nonmoving sidewalk?

28. *Moving Sidewalks.* Newark Airport's moving sidewalk moves at a rate of 1.7 ft/sec. Walking on the moving sidewalk, Benny can travel 120 ft forward in the same time it takes to travel 52 ft in the opposite direction. How fast would Benny be walking on a nonmoving sidewalk?

29. *Walking.* Rosanna walks 2 mph slower than Simone. In the time it takes Simone to walk 8 mi, Rosanna walks 5 mi. Find the speed of each person.

30. *Bus Travel.* A local bus travels 7 mph slower than the express. The express travels 90 mi in the time it takes the local to travel 75 mi. Find the speed of each bus.

31. **DW** The photograph at right shows a picture of Corona Arch in Moab, Utah, one of the favorite hiking places of your author Marv Bittinger. He appears at the bottom of the photograph.

a) Given that Marv is 6 ft 1 in. tall, discuss how you might use the photograph to estimate the actual height *H* of the arch.

b) Assume that an $8\frac{1}{2}$-in. by 11-in. photograph has been printed from a digital file and that in that photo Marv is $\frac{11}{32}$, or 0.34375 in. tall, and the height of the arch in the photo is $7\frac{5}{8}$, or 7.625 in. Find the actual height *H* of the arch.

H

73 in.

In Exercises 32–35, the graph is that of a function. Determine the domain and the range.   [2.3a]

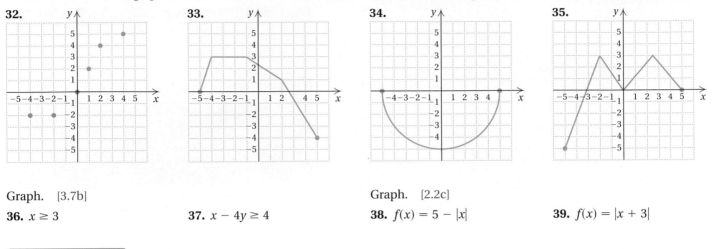

32.

33.

34.

35.

Graph.   [3.7b]

36. $x \geq 3$

37. $x - 4y \geq 4$

Graph.   [2.2c]

38. $f(x) = 5 - |x|$

39. $f(x) = |x + 3|$

40. *Escalators.*   Together, a 100-cm-wide escalator and a 60-cm-wide escalator can empty a 1575-person auditorium in 14 min. The wider escalator moves twice as many people as the narrower one does. How many people per hour does the 60-cm-wide escalator move?

Source: McGraw-Hill *Encyclopedia of Science and Technology*

41. *Travel by Car.*   Melissa drives to work at 50 mph and arrives 1 min late. She drives to work at 60 mph and arrives 5 min early. How far does Melissa live from work?

42. *Filling a Tank.*   A tank can be filled in 9 hr and drained in 11 hr. How long will it take to fill the tank if the drain is left open?

43. *Gas Mileage.*   An automobile gets 22.5 miles per gallon (mpg) in city driving and 30 mpg in highway driving. The car is driven 465 mi on a full tank of 18.4 gal of gasoline. How many miles were driven in the city and how many were driven on the highway?

44. Three trucks, A, B, and C, working together, can move a load of sand in t hours. When working alone, it takes A 1 extra hour to move the sand, B 6 extra hours, and C t extra hours. Find t.

45. *Clock Hands.*   At what time after 4:00 will the minute hand and the hour hand of a clock first be in the same position?

5.7

FORMULAS AND APPLICATIONS

a Formulas

Formulas occur frequently as mathematical models. Here we consider rational formulas. The procedure for solving a rational formula for a given letter is as follows.

> To solve a rational formula for a given letter, identify the letter, and:
>
> 1. Multiply on both sides to clear fractions or decimals, if that is needed.
> 2. Multiply if necessary to remove parentheses.
> 3. Get all terms with the letter to be solved for on one side of the equation and all other terms on the other side, using the addition principle.
> 4. Factor out the unknown.
> 5. Solve for the letter in question, using the multiplication principle.

EXAMPLE 1 *Optics.* The formula $f = L/d$ defines a camera's "f-stop," where L is the *focal length* (the distance from the lens to the film) and d is the *aperture* (the diameter of the lens). Solve the formula for d.

We solve this equation as we did the rational equations in Section 5.5:

$$f = \frac{L}{d}$$ We want the letter d alone.

$$d \cdot f = d \cdot \frac{L}{d}$$ The LCM is d. We multiply by d.

$$df = L$$ Simplifying

$$d = \frac{L}{f}.$$ Dividing by f

The formula $d = L/f$ can now be used to find the aperture if we know the focal length and the f-stop.

Do Exercise 1.

EXAMPLE 2 *Astronomy.* The formula $L = \dfrac{dR}{D - d}$, where D is the diameter of the sun, d is the diameter of the earth, R is the earth's distance from the sun, and L is some fixed distance, is used to calculate when lunar eclipses occur. Solve the formula for D.

$$L = \frac{dR}{D - d}$$ We want the letter D alone.

$$(D - d) \cdot L = (D - d) \cdot \frac{dR}{D - d}$$ The LCM is $D - d$. We multiply by $D - d$.

$$(D - d)L = dR$$ Simplifying

$$DL - dL = dR$$

$$DL = dR + dL$$ Adding dL

$$D = \frac{dR + dL}{L}$$ Dividing by L

1. **Combined Gas Law.** The formula

$$\frac{PV}{T} = k$$

relates the pressure P, the volume V, and the temperature T of a gas. Solve the formula for T. (*Hint*: Begin by clearing the fraction.)

2. Solve the formula

$$I = \frac{pT}{M + pn}$$

for n.

Answers on page A-30

475

3. The Doppler Effect. The formula

$$F = \frac{sg}{s + v}$$

is used to determine the frequency F of a sound that is moving at velocity v toward a listener who hears the sound as frequency g. Here s is the speed of sound in a particular medium. Solve the formula for s.

Since D appears by itself on one side and not on the other, we have solved for D.

Do Exercise 2 on the preceding page.

EXAMPLE 3 Solve the formula $L = \dfrac{dR}{D - d}$ for d.

We proceed as we did in Example 2 until we reach the equation

$$DL - dL = dR. \qquad \text{We want } d \text{ alone.}$$

We must get all terms containing d alone on one side:

$$DL - dL = dR$$
$$DL = dR + dL \qquad \text{Adding } dL$$
$$DL = d(R + L) \qquad \text{Factoring out the letter } d$$
$$\frac{DL}{R + L} = d. \qquad \text{Dividing by } R + L$$

We now have d alone on one side, so we have solved the formula for d.

Caution!

If, when you are solving an equation for a letter, the letter appears on both sides of the equation, you know the answer is wrong. The letter must be alone on one side and *not* occur on the other.

Do Exercise 3.

4. Work Formula. The formula

$$\frac{t}{a} + \frac{t}{b} = 1$$

involves the total time t for some work to be done by two workers whose individual times are a and b. Solve the formula for t.

EXAMPLE 4 *Resistance.* The formula

$$\frac{1}{R} = \frac{1}{r_1} + \frac{1}{r_2}$$

involves the resistance R of two resistors r_1 and r_2 connected in parallel.* Solve the formula for r_1.

We multiply by the LCM, which is Rr_1r_2:

$$Rr_1r_2 \cdot \frac{1}{R} = Rr_1r_2 \cdot \left(\frac{1}{r_1} + \frac{1}{r_2}\right) \qquad \text{Multiplying by the LCM}$$

$$Rr_1r_2 \cdot \frac{1}{R} = Rr_1r_2 \cdot \frac{1}{r_1} + Rr_1r_2 \cdot \frac{1}{r_2} \qquad \begin{array}{l}\text{Multiplying to remove}\\\text{parentheses}\end{array}$$

$$r_1r_2 = Rr_2 + Rr_1. \qquad \text{Simplifying by removing factors of 1}$$

We might be tempted at this point to multiply by $1/r_2$ to get r_1 alone on the left, *but* note that there is an r_1 on the right. We must get all the terms involving r_1 on the *same side* of the equation.

$$r_1r_2 - Rr_1 = Rr_2 \qquad \text{Subtracting } Rr_1$$
$$r_1(r_2 - R) = Rr_2 \qquad \text{Factoring out } r_1$$
$$r_1 = \frac{Rr_2}{r_2 - R} \qquad \text{Dividing by } r_2 - R \text{ to get } r_1 \text{ alone}$$

Do Exercise 4.

*Note that R, r_1, and r_2 are all different variables. It is common to use subscripts, as in r_1 (read "r sub 1") and r_2, to distinguish variables.

5.7

EXERCISE SET

For Extra Help

a   Solve.

1. $\dfrac{W_1}{W_2} = \dfrac{d_1}{d_2}$, for W_2

2. $\dfrac{W_1}{W_2} = \dfrac{d_1}{d_2}$, for d_1

3. $\dfrac{1}{R} = \dfrac{1}{r_1} + \dfrac{1}{r_2}$, for r_2
(Electricity formula)

4. $\dfrac{1}{R} = \dfrac{1}{r_1} + \dfrac{1}{r_2}$, for R
(Electricity formula)

5. $s = \dfrac{(v_1 + v_2)t}{2}$, for t

6. $s = \dfrac{(v_1 + v_2)t}{2}$, for v_1

7. $R = \dfrac{gs}{g + s}$, for s

8. $I = \dfrac{2V}{V + 2r}$, for V

9. $\dfrac{1}{p} + \dfrac{1}{q} = \dfrac{1}{f}$, for p
(An optics formula)

10. $\dfrac{1}{p} + \dfrac{1}{q} = \dfrac{1}{f}$, for f
(Optics formula)

11. $\dfrac{t}{a} + \dfrac{t}{b} = 1$, for a
(Work formula)

12. $\dfrac{t}{a} + \dfrac{t}{b} = 1$, for b
(Work formula)

13. $I = \dfrac{nE}{E + nr}$, for E

14. $I = \dfrac{nE}{E + nr}$, for n

15. $I = \dfrac{704.5W}{H^2}$, for H^2

16. $S = \dfrac{H}{m(t_1 - t_2)}$, for t_1

17. $\dfrac{E}{e} = \dfrac{R + r}{r}$, for r

18. $\dfrac{E}{e} = \dfrac{R + r}{r}$, for e

19. $V = \dfrac{1}{3}\pi h^2(3R - h)$, for R

20. $A = P(1 + rt)$, for r
(Interest formula)

21. *Interest.* The formula

$$P = \frac{A}{1 + r}$$

is used to determine what amount of principal P should be invested for one year at simple interest r in order to have A dollars after a year. Solve the formula for r.

22. *Average Speed.* The formula

$$v = \frac{d_2 - d_1}{t_2 - t_1}$$

gives an object's average speed v when that object has traveled d_1 miles in t_1 hours and d_2 miles in t_2 hours. Solve the formula for t_2.

23. *Escape Velocity.* The formula

$$\frac{V^2}{R^2} = \frac{2g}{R + h}$$

is used to find a satellite's *escape velocity* V, where R is a planet's radius, h is the satellite's height above the planet, and g is the planet's acceleration due to gravity. Solve the formula for h.

24. *Earned Run Average.* The formula

$$A = 9 \cdot \frac{R}{I}$$

gives a pitcher's *earned run average*, where A is the earned run average, R is the number of earned runs, and I is the number of innings pitched. How many earned runs were given up if a pitcher's earned run average is 2.4 after 45 innings? Solve the formula for I.

25. *Semester Average.* The formula

$$A = \frac{2Tt + Qq}{2T + Q}$$

gives a student's average A after T tests and Q quizzes, where each test counts as 2 quizzes, t is the test average, and q is the quiz average. Solve the formula for Q.

26. $\mathbf{D_W}$ Which is easier to solve for x? Explain why.

$$\frac{1}{38} + \frac{1}{47} = \frac{1}{x} \quad \text{or} \quad \frac{1}{a} + \frac{1}{b} = \frac{1}{x}$$

27. $\mathbf{D_W}$ Describe a situation in which the result of Margin Exercise 4,

$$t = \frac{ab}{a + b},$$

would be especially useful.

SKILL MAINTENANCE

Solve. [3.4a], [3.6a]

28. *Coin Value.* There are 50 dimes in a roll of dimes, 40 nickels in a roll of nickels, and 40 quarters in a roll of quarters. Rob has 12 rolls of coins with a total value of $70.00. He has 3 more rolls of nickels than dimes. How many of each roll of coin does he have?

29. *Audiotapes.* Esther wants to buy tapes for her work at the campus radio station. She needs some 30-min tapes and some 60-min tapes. She buys 12 tapes with a total recording time of 10 hr. How many of each length did she buy?

Given that $f(x) = x^3 - x$, find each of the following. [2.2b]

30. $f(-2)$ **31.** $f(2)$ **32.** $f(0)$ **33.** $f(2a)$

34. Find the slope of the line containing the points $(-2, 5)$ and $(8, -3)$. [2.4b]

35. Find an equation of the line containing the points $(-2, 5)$ and $(8, -3)$. [2.6c]

SYNTHESIS

36. *Escape Velocity.* (Refer to Exercise 23.) A satellite's escape velocity is 6.5 mi/sec, the radius of the earth is 3960 mi, and the acceleration due to gravity is 32.2 ft/sec². How far is the satellite from the surface of the earth?

5.8 VARIATION AND APPLICATIONS

Objectives

a Find an equation of direct variation given a pair of values of the variables.

b Solve applied problems involving direct variation.

c Find an equation of inverse variation given a pair of values of the variables.

d Solve applied problems involving inverse variation.

e Find equations of other kinds of variation given values of the variables.

f Solve applied problems involving other kinds of variation.

We now extend our study of formulas and functions by considering applications involving variation.

a Equations of Direct Variation

An electrician earns $21 per hour. In 1 hr, $21 is earned; in 2 hr, $42 is earned; in 3 hr, $63 is earned; and so on. We plot this information on a graph, using the number of hours as the first coordinate and the amount earned as the second coordinate to form a set of ordered pairs:

$(1, 21)$, $(2, 42)$,

$(3, 63)$, $(4, 84)$,

and so on.

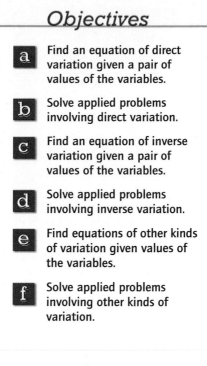

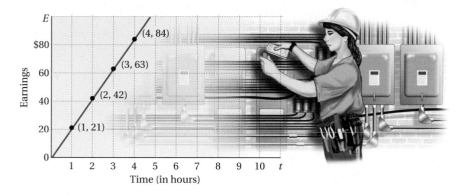

Note that the ratio of the second coordinate to the first is the same number for each point:

$$\frac{21}{1} = 21, \qquad \frac{42}{2} = 21,$$

$$\frac{63}{3} = 21, \qquad \frac{84}{4} = 21,$$

and so on.

Whenever a situation produces pairs of numbers in which the *ratio is constant,* we say that there is **direct variation.** Here the amount earned varies directly as the time:

$$\frac{E}{t} = 21 \text{ (a constant)}, \quad \text{or} \quad E = 21t,$$

or, using function notation, $E(t) = 21t$. The equation is an equation of **direct variation.** The coefficient, 21 in the situation above, is called the **variation constant.** In this case, it is the rate of change of earnings with respect to time.

DIRECT VARIATION

If a situation gives rise to a linear function $f(x) = kx$, or $y = kx$, where k is a positive constant, we say that we have **direct variation,** or that **y varies directly as x,** or that **y is directly proportional to x.** The number k is called the **variation constant,** or **constant of proportionality.**

479

1. Find the variation constant and an equation of variation in which y varies directly as x, and $y = 8$ when $x = 20$.

EXAMPLE 1 Find the variation constant and an equation of variation in which y varies directly as x, and $y = 32$ when $x = 2$.

We know that $(2, 32)$ is a solution of $y = kx$. Thus,

$$y = kx$$
$$32 = k \cdot 2 \qquad \text{Substituting}$$
$$\frac{32}{2} = k, \text{ or } k = 16. \qquad \text{Solving for } k$$

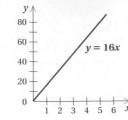

The variation constant, 16, is the rate of change of y with respect to x. The equation of variation is $y = 16x$.

2. Find the variation constant and an equation of variation in which y varies directly as x, and $y = 5.6$ when $x = 8$.

The graph of $y = kx$, $k > 0$, always goes through the origin and rises from left to right. Note that as x increases, y increases. The constant k is also the slope of the line.

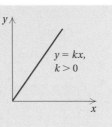

Do Exercises 1 and 2.

b   Applications of Direct Variation

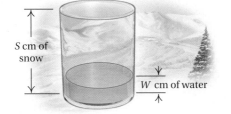

S cm of snow

W cm of water

EXAMPLE 2 *Water from Melting Snow.*   The number of centimeters W of water produced from melting snow varies directly as S, the number of centimeters of snow. Meteorologists have found that 150 cm of snow will melt to 16.8 cm of water. To how many centimeters of water will 200 cm of snow melt?

We first find the variation constant using the data and then find an equation of variation:

$$W = kS \qquad W \text{ varies directly as } S.$$
$$16.8 = k \cdot 150 \qquad \text{Substituting}$$
$$\frac{16.8}{150} = k \qquad \text{Solving for } k$$
$$0.112 = k. \qquad \text{This is the variation constant.}$$

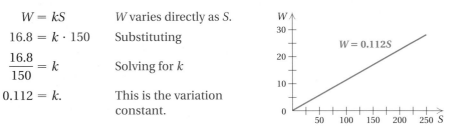

The equation of variation is $W = 0.112S$.

Next, we use the equation to find how many centimeters of water will result from melting 200 cm of snow:

$$W = 0.112S$$
$$W = 0.112(200) \qquad \text{Substituting}$$
$$W = 22.4.$$

Thus, 200 cm of snow will melt to 22.4 cm of water.

3. Ohm's Law.   Ohm's Law states that the voltage V in an electric circuit varies directly as the number of amperes I of electric current in the circuit. If the voltage is 10 volts when the current is 3 amperes, what is the voltage when the current is 15 amperes?

Do Exercises 3 and 4. (Exercise 4 is on the following page.)

Answers on page A-31

CHAPTER 5: Rational Expressions, Equations, and Functions

C Equations of Inverse Variation

A bus is traveling a distance of 20 mi. At a speed of 5 mph, the trip will take 4 hr; at 20 mph, it will take 1 hr; at 40 mph, it will take $\frac{1}{2}$ hr; and so on. We plot this information on a graph, using speed as the first coordinate and time as the second coordinate to determine a set of ordered pairs:

$(5, 4)$, $(20, 1)$,

$\left(40, \frac{1}{2}\right)$,

and so on.

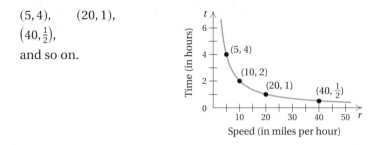

Note that the products of the coordinates are all the same number:

$$5 \cdot 4 = 20, \quad 20 \cdot 1 = 20, \quad 40 \cdot \frac{1}{2} = 20, \quad \text{and so on.}$$

Whenever a situation produces pairs of numbers in which the *product is constant,* we say that there is **inverse variation.** Here the time varies inversely as the speed:

$$rt = 20 \,(\text{a constant}), \quad \text{or} \quad t = \frac{20}{r}.$$

The equation is an equation of **inverse variation.** The coefficient, 20 in the situation above, is called the **variation constant.** Note that as the first number (speed) increases, the second number (time) decreases.

INVERSE VARIATION

If a situation gives rise to a function $f(x) = k/x$, or $y = k/x$, where k is a positive constant, we say that we have **inverse variation,** or that y **varies inversely as x,** or that y **is inversely proportional to x.** The number k is called the **variation constant,** or **constant of proportionality.**

EXAMPLE 3 Find the variation constant and an equation of variation in which y varies inversely as x, and $y = 32$ when $x = 0.2$.

We know that $(0.2, 32)$ is a solution of $y = k/x$. We substitute:

$$y = \frac{k}{x}$$

$$32 = \frac{k}{0.2} \qquad \text{Substituting}$$

$$(0.2)32 = k \qquad \text{Solving for } k$$

$$6.4 = k.$$

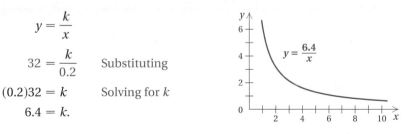

The variation constant is 6.4. The equation of variation is $y = \dfrac{6.4}{x}$.

---

4. An Ecology Problem. The amount of garbage G produced in the United States varies directly as the number of people N who produce the garbage. It is known that 50 tons of garbage is produced by 200 people in 1 year. The population of the San Francisco–Oakland–San Jose area is 6,300,000. How much garbage is produced by this area in 1 year?

Answer on page A-31

5. Find the variation constant and an equation of variation in which y varies inversely as x, and $y = 0.012$ when $x = 50$.

It is helpful to look at the graph of $y = k/x$, $k > 0$. The graph is like the one shown at right for positive values of x. Note that as x increases, y decreases.

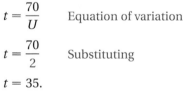

$y = \dfrac{k}{x}$, $k > 0$

Do Exercise 5.

6. Building a Shed. The time t required to do a job varies inversely as the number of people P who work on the job (assuming that all work at the same rate). It takes 4 hr for 12 people to build a woodshed. How long would it take 3 people to complete the same job?

d   Applications of Inverse Variation

EXAMPLE 4   *Ultraviolet Index.*   The ultraviolet index, UV, is a measure issued daily by the National Weather Service. It indicates the strength of the sun rays in a particular locale. For those people whose skin is quite sensitive, a UV rating of 7 will cause a sunburn after 10 min. Given that the number of minutes it takes to burn t varies inversely as the UV rating U, how long will it take a person with highly sensitive skin to burn on a day with a UV rating of 2?
Source: *Los Angeles Times, 3/24/98*

We first find the variation constant using the data given and then find an equation of variation:

$$t = \frac{k}{U} \qquad t \text{ varies inversely as } U.$$

$$10 = \frac{k}{7} \qquad \text{Substituting}$$

$$70 = k. \qquad \text{Solving for } k, \text{ the variation constant}$$

The equation of variation is $t = \dfrac{70}{U}$.

Next, we use the equation to find the time it would take a person with highly sensitive skin to burn on a day with a UV rating of 2:

$$t = \frac{70}{U} \qquad \text{Equation of variation}$$

$$t = \frac{70}{2} \qquad \text{Substituting}$$

$$t = 35.$$

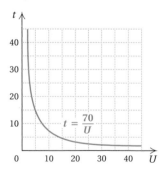

$t = \dfrac{70}{U}$

It would take 35 min for a person with highly sensitive skin to burn on a day when the UV rating is 2.

Do Exercise 6.

Answers on page A-31

e Other Kinds of Variation

We now look at other kinds of variation. Consider the equation for the area of a circle, in which A and r are variables and π is a constant:

$$A = \pi r^2, \quad \text{or, as a function,} \quad A(r) = \pi r^2.$$

We say that the area *varies directly* as the square of the radius.

> y varies directly as the nth power of x if there is some positive constant k such that $y = kx^n$.

EXAMPLE 5 Find an equation of variation in which y varies directly as the square of x, and $y = 12$ when $x = 2$.

We write an equation of variation and find k:

$$y = kx^2$$
$$12 = k \cdot 2^2$$
$$12 = k \cdot 4$$
$$3 = k.$$

Thus, $y = 3x^2$.

Do Exercise 7.

From the law of gravity, we know that the weight W of an object *varies inversely* as the square of its distance d from the center of the earth:

$$W = \frac{k}{d^2}.$$

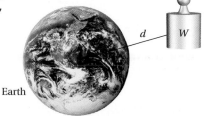
Earth

> y varies inversely as the nth power of x if there is some positive constant k such that
> $$y = \frac{k}{x^n}.$$

7. Find an equation of variation in which y varies directly as the square of x, and $y = 175$ when $x = 5$.

Answer on page A-31

8. Find an equation of variation in which y varies inversely as the square of x, and $y = \frac{1}{4}$ when $x = 6$.

EXAMPLE 6 Find an equation of variation in which W varies inversely as the square of d, and $W = 3$ when $d = 5$.

$$W = \frac{k}{d^2}$$

$$3 = \frac{k}{5^2} \qquad \text{Substituting}$$

$$3 = \frac{k}{25}$$

$$75 = k$$

Thus, $W = \dfrac{75}{d^2}$.

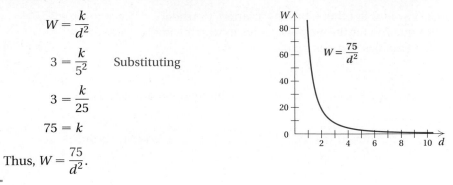

Do Exercise 8.

Consider the equation for the area A of a triangle with height h and base b: $A = \frac{1}{2}bh$. We say that the area *varies jointly* as the height and the base.

> y varies jointly as x and z if there is some positive constant k such that
>
> $y = kxz$.

9. Find an equation of variation in which y varies jointly as x and z, and $y = 65$ when $x = 10$ and $z = 13$.

EXAMPLE 7 Find an equation of variation in which y varies jointly as x and z, and $y = 42$ when $x = 2$ and $z = 3$.

$$y = kxz$$

$$42 = k \cdot 2 \cdot 3 \qquad \text{Substituting}$$

$$42 = k \cdot 6$$

$$7 = k$$

Thus, $y = 7xz$.

Do Exercise 9.

10. Find an equation of variation in which y varies jointly as x and the square of z and inversely as w, and $y = 80$ when $x = 4$, $z = 10$, and $w = 25$.

The equation

$$y = k \cdot \frac{xz^2}{w}$$

asserts that y varies jointly as x and the square of z, and inversely as w.

EXAMPLE 8 Find an equation of variation in which y varies jointly as x and z and inversely as the square of w, and $y = 105$ when $x = 3$, $z = 20$, and $w = 2$.

$$y = k \cdot \frac{xz}{w^2}$$

$$105 = k \cdot \frac{3 \cdot 20}{2^2} \qquad \text{Substituting}$$

$$105 = k \cdot 15$$

$$7 = k$$

Thus, $y = 7 \cdot \dfrac{xz}{w^2}$.

Answers on page A-31

Do Exercise 10.

f   Other Applications of Variation

Many problem situations can be described with equations of variation.

EXAMPLE 9 *Volume of a Tree.*   The volume of wood V in a tree varies jointly as the height h and the square of the girth g (girth is distance around). If the volume of a redwood tree is 216 m^3 when the height is 30 m and the girth is 1.5 m, what is the height of a tree whose volume is 960 m^3 and girth is 2 m?

We first find k using the first set of data. Then we solve for h using the second set of data.

$$V = khg^2$$
$$216 = k \cdot 30 \cdot 1.5^2$$
$$3.2 = k$$

Then the equation of variation is $V = 3.2hg^2$. We substitute the second set of data into the equation:

$$960 = 3.2 \cdot h \cdot 2^2$$
$$75 = h.$$

Therefore, the height of the tree is 75 m.

EXAMPLE 10 *TV Signal.*   The intensity I of a TV signal varies inversely as the square of the distance d from the transmitter. If the intensity is 23 watts per square meter (W/m^2) at a distance of 2 km, what is the intensity at a distance of 6 km?

We first find k using the first set of data. Then we solve for I using the second set of data.

$$I = \frac{k}{d^2}$$
$$23 = \frac{k}{2^2}$$
$$92 = k$$

$$I = \frac{92}{d^2}$$

Then the equation of variation is $I = 92/d^2$. We substitute the second distance into the equation:

$$I = \frac{92}{d^2} = \frac{92}{6^2} \approx 2.56. \qquad \text{Rounded to the nearest hundredth}$$

Therefore, at 6 km, the intensity is about 2.56 W/m^2.

Do Exercises 11 and 12.

---

11. Distance of a Dropped Object.
The distance s that an object falls when dropped from some point above the ground varies directly as the square of the time t that it falls. If the object falls 19.6 m in 2 sec, how far will the object fall in 10 sec?

12. Electrical Resistance.   At a fixed temperature, the resistance R of a wire varies directly as the length l and inversely as the square of its diameter d. If the resistance is 0.1 ohm when the diameter is 1 mm and the length is 50 cm, what is the resistance when the length is 2000 cm and the diameter is 2 mm?

Answers on page A-31

a   Find the variation constant and an equation of variation in which y varies directly as x and the following are true.

1. $y = 40$ when $x = 8$

2. $y = 54$ when $x = 12$

3. $y = 4$ when $x = 30$

4. $y = 3$ when $x = 33$

5. $y = 0.9$ when $x = 0.4$

6. $y = 0.8$ when $x = 0.2$

b   Solve.

7. *Aluminum Usage.*   The number N of aluminum cans used each year varies directly as the number of people using the cans. If 250 people use 60,000 cans in one year, how many cans are used each year in Dallas, which has a population of 1,189,000?

8. *Weekly Allowance.*   The average weekly allowance A of children varies directly as their grade level G. It is known that the average allowance of a 9th-grade student is $9.66 per week. What then is the average weekly allowance of a 4th-grade student?

Source: Fidelity Investments *Investment Vision Magazine*

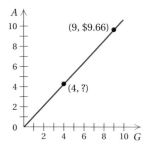

9. *Hooke's Law.*   Hooke's law states that the distance d that a spring is stretched by a hanging object varies directly as the weight w of the object. If a spring is stretched 40 cm by a 3-kg barbell, what is the distance stretched by a 5-kg barbell?

10. *Lead Pollution.*   The average U.S. community of population 12,500 released about 385 tons of lead into the environment in a recent year. How many tons were released nationally? Use 295,000,000 as the U.S. population.

Source: *Conservation Matters,* Autumn 1995 issue. Boston: Conservation Law Foundation, p. 30

11. *Fat Intake.*   The maximum number of grams of fat that should be in a diet varies directly as a person's weight. A person weighing 120 lb should have no more than 60 g of fat per day. What is the maximum daily fat intake for a person weighing 180 lb?

12. *Relative Aperture.*   The relative aperture, or f-stop, of a 23.5-mm diameter lens is directly proportional to the focal length F of the lens. If a 150-mm focal length has an f-stop of 6.3, find the f-stop of a 23.5-mm diameter lens with a focal length of 80 mm.

13. *Mass of Water in Body.*   The number of kilograms W of water in a human body varies directly as the mass of the body. A 96-kg person contains 64 kg of water. How many kilograms of water are in a 60-kg person?

14. *Weight on Mars.*   The weight M of an object on Mars varies directly as its weight E on Earth. A person who weighs 95 lb on Earth weighs 38 lb on Mars. How much would a 100-lb person weigh on Mars?

c   Find the variation constant and an equation of variation in which *y* varies inversely as *x* and the following are true.

15. *y* = 14 when *x* = 7

16. *y* = 1 when *x* = 8

17. *y* = 3 when *x* = 12

18. *y* = 12 when *x* = 5

19. *y* = 0.1 when *x* = 0.5

20. *y* = 1.8 when *x* = 0.3

d   Solve.

21. *Work Rate.*   The time *T* required to do a job varies inversely as the number of people *P* working. It takes 5 hr for 7 bricklayers to build a park wall. How long will it take 10 bricklayers to complete the job?

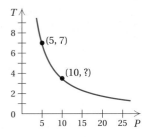

22. *Pumping Rate.*   The time *t* required to empty a tank varies inversely as the rate *r* of pumping. If a pump can empty a tank in 45 min at the rate of 600 kL/min, how long will it take the pump to empty the same tank at the rate of 1000 kL/min?

23. *Current and Resistance.*   The current *I* in an electrical conductor varies inversely as the resistance *R* of the conductor. If the current is $\frac{1}{2}$ ampere when the resistance is 240 ohms, what is the current when the resistance is 540 ohms?

24. *Wavelength and Frequency.*   The wavelength *W* of a radio wave varies inversely as its frequency *F*. A wave with a frequency of 1200 kilohertz has a length of 300 meters. What is the length of a wave with a frequency of 800 kilohertz?

25. *Musical Pitch.*   The pitch *P* of a musical tone varies inversely as its wavelength *W*. One tone has a pitch of 330 vibrations per second and a wavelength of 3.2 ft. Find the wavelength of another tone that has a pitch of 550 vibrations per second.

26. *Beam Weight.*   The weight *W* that a horizontal beam can support varies inversely as the length *L* of the beam. Suppose that an 8-m beam can support 1200 kg. How many kilograms can a 14-m beam support?

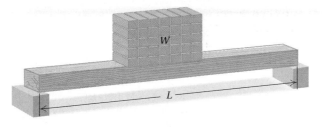

27. *Volume and Pressure.*   The volume *V* of a gas varies inversely as the pressure *P* upon it. The volume of a gas is 200 cm³ under a pressure of 32 kg/cm². What will be its volume under a pressure of 40 kg/cm²?

28. *Rate of Travel.*   The time *t* required to drive a fixed distance varies inversely as the speed *r*. It takes 5 hr at a speed of 80 km/h to drive a fixed distance. How long will it take to drive the same distance at a speed of 70 km/h?

e   Find an equation of variation in which the following are true.

29. *y* varies directly as the square of *x*, and *y* = 0.15 when *x* = 0.1

30. *y* varies directly as the square of *x*, and *y* = 6 when *x* = 3

31. y varies inversely as the square of x, and $y = 0.15$ when $x = 0.1$

32. y varies inversely as the square of x, and $y = 6$ when $x = 3$

33. y varies jointly as x and z, and $y = 56$ when $x = 7$ and $z = 8$

34. y varies directly as x and inversely as z, and $y = 4$ when $x = 12$ and $z = 15$

35. y varies jointly as x and the square of z, and $y = 105$ when $x = 14$ and $z = 5$

36. y varies jointly as x and z and inversely as w, and $y = \frac{3}{2}$ when $x = 2$, $z = 3$, and $w = 4$

37. y varies jointly as x and z and inversely as the product of w and p, and $y = \frac{3}{28}$ when $x = 3$, $z = 10$, $w = 7$, and $p = 8$

38. y varies jointly as x and z and inversely as the square of w, and $y = \frac{12}{5}$ when $x = 16$, $z = 3$, and $w = 5$

f   Solve.

39. *Stopping Distance of a Car.* The stopping distance d of a car after the brakes have been applied varies directly as the square of the speed r. If a car traveling 60 mph can stop in 200 ft, how fast can a car travel and still stop in 72 ft?

40. *Combined Gas Law.* The volume V of a given mass of a gas varies directly as the temperature T and inversely as the pressure P. If $V = 231$ cm^3 when $T = 42°$ and $P = 20$ kg/cm^2, what is the volume when $T = 30°$ and $P = 15$ kg/cm^2?

41. *Intensity of Light.* The intensity I of light from a light bulb varies inversely as the square of the distance d from the bulb. Suppose that I is 90 W/m^2 (watts per square meter) when the distance is 5 m. How much *further* would it be to a point where the intensity is 40 W/m^2?

42. *Weight of an Astronaut.* The weight W of an object varies inversely as the square of the distance d from the center of the earth. At sea level (3978 mi from the center of the earth), an astronaut weighs 220 lb. Find his weight when he is 200 mi above the surface of the earth and the spacecraft is not in motion.

43. *Earned-Run Average.* A pitcher's earned-run average E varies directly as the number R of earned runs allowed and inversely as the number I of innings pitched. In 2005, Roger Clemens of the Houston Astros had an earned-run average of 1.87. He gave up 44 earned runs in $211\frac{1}{3}$ innings. How many earned runs would he have given up had he pitched 240 innings with the same average? Round to the nearest whole number.

Source: Major League Baseball

44. *Atmospheric Drag.* Wind resistance, or atmospheric drag, tends to slow down moving objects. Atmospheric drag varies jointly as an object's surface area A and velocity v. If a car traveling at a speed of 40 mph with a surface area of 37.8 ft^2 experiences a drag of 222 N (Newtons), how fast must a car with 51 ft^2 of surface area travel in order to experience a drag force of 430 N?

45. *Water Flow.* The amount Q of water emptied by a pipe varies directly as the square of the diameter d. A pipe 5 in. in diameter will empty 225 gal of water over a fixed time period. If we assume the same kind of flow, how many gallons of water are emptied in the same amount of time by a pipe that is 9 in. in diameter?

46. *Weight of a Sphere.* The weight W of a sphere of a given material varies directly as its volume V, and its volume V varies directly as the cube of its diameter.
   a) Find an equation of variation relating the weight W to the diameter d.
   b) An iron ball that is 5 in. in diameter is known to weigh 25 lb. Find the weight of an iron ball that is 8 in. in diameter.

47. $\mathbf{D_W}$ Write a variation problem for a classmate to solve. Design the problem so that the answer is "When Simone studies for 8 hr a week, her quiz score is 92."

48. $\mathbf{D_W}$ If y varies directly as x and x varies inversely as z, how does y vary with regard to z? Why?

| SKILL MAINTENANCE |

✎ VOCABULARY REINFORCEMENT

In each of Exercises 49–56, fill in the blank with the correct term from the given list. Some of the choices may not be used.

49. When two terms have the same variable(s) raised to the same power(s), they are called _____ terms. [4.1c]

50. _____ angles are angles whose sum is 90°. [3.2b]

51. If the sum of two polynomials is 0, they are called _____ , or _____ inverses of each other. [4.1d]

52. The graph of $x = a$ is a(n) _____ line through the point $(a, 0)$. [2.5c]

53. The _____ of two sets A and B is the set of all members that are common to A and B. [1.5a]

54. A(n) _____ function f is any function that can be described by $f(x) = mx + b$. [2.4a]

55. The _____ states that for any real numbers a, b, and c, $c \neq 0$, $a = b$ is equivalent to $a \cdot c = b \cdot c$. [1.1c]

56. A y-intercept is a point _____ . [2.5a]

multiplicative

additive

intersection

union

linear

addition principle

multiplication principle

like

opposite(s)

supplementary

complementary

horizontal

vertical

$(a, 0)$

$(0, a)$

| SYNTHESIS |

57. *Area of a Circle.* The area of a circle varies directly as the square of the length of a diameter. What is the variation constant?

58. In each of the following equations, state whether y varies directly as x, inversely as x, or neither directly nor inversely as x.
   a) $7xy = 14$
   b) $x - 2y = 12$
   c) $-2x + 3y = 0$
   d) $x = \dfrac{3}{4}y$
   e) $\dfrac{x}{y} = 2$

Describe, in words, the variation given by the equation.

59. $Q = \dfrac{kp^2}{q^3}$

60. $W = \dfrac{km_1 M_1}{d^2}$

61. *Volume and Cost.* A peanut butter jar in the shape of a right circular cylinder is 4 in. high and 3 in. in diameter and sells for $1.20. If we assume that cost is proportional to volume, how much should a jar 6 in. high and 6 in. in diameter cost?

The review that follows is meant to prepare you for a chapter exam. It consists of three parts. The first part, Concept Reinforcement, is designed to increase understanding of the concepts through true/false exercises. The second part is a list of important properties and formulas. The third part is the Review Exercises. These provide practice exercises for the exam, together with references to section objectives so you can go back and review. Before beginning, stop and look back over the skills you have obtained. What skills in mathematics do you have now that you did not have before studying this chapter?

☞ CONCEPT REINFORCEMENT

Determine whether the statement is true or false. Answers are given at the back of the book.

_____ 1. The expressions $a - b$ and $-(b - a)$ are opposites, or additive inverses, of each other.

_____ 2. For synthetic division, the divisor must be in the form $x - a$.

_____ 3. If y is inversely proportional to x, then the rational function $f(x) = \dfrac{k}{x}$ can model the situation.

_____ 4. Clearing fractions is a valid procedure only when solving equations, not when adding, subtracting, multiplying, or dividing rational expressions.

_____ 5. The sum of two rational expressions is the sum of the numerators over the sum of the denominators.

_____ 6. The domain of $f(x) = \dfrac{(x - 5)(x + 4)}{x - 4}$ is $\{x \mid x \neq 5 \text{ and } x \neq 4 \text{ and } x \neq -4\}$.

IMPORTANT PROPERTIES AND FORMULAS

Direct Variation: $y = kx$

Inverse Variation: $y = \dfrac{k}{x}$

Joint Variation: $y = kxz$

Work Principle: $\dfrac{t}{a} + \dfrac{t}{b} = 1$, where a is the time needed for A to complete the job alone, b is the time needed for B to complete the job alone, and t is the time needed for A and B to complete the job working together.

Review Exercises

1. Find all numbers for which the rational expression
$$\frac{x^2 - 3x + 2}{x^2 - 9}$$
is not defined.   [5.1a]

2. Find the domain of f where   [5.1a]
$$f(x) = \frac{x^2 - 3x + 2}{x^2 - 9}.$$

Simplify.   [5.1c]

3. $\dfrac{4x^2 - 7x - 2}{12x^2 + 11x + 2}$

4. $\dfrac{a^2 + 2a + 4}{a^3 - 8}$

Find the LCM.   [5.2a]

5. $6x^3, \ 16x^2$

6. $x^2 - 49, \ 3x + 1$

7. $x^2 + x - 20, \ x^2 + 3x - 10$

Perform the indicated operations and simplify.   [5.1d, e], [5.2b, c]

8. $\dfrac{y^2 - 64}{2y + 10} \cdot \dfrac{y + 5}{y + 8}$

9. $\dfrac{x^3 - 8}{x^2 - 25} \cdot \dfrac{x^2 + 10x + 25}{x^2 + 2x + 4}$

10. $\dfrac{9a^2 - 1}{a^2 - 9} \div \dfrac{3a + 1}{a + 3}$

11. $\dfrac{x^3 - 64}{x^2 - 16} \div \dfrac{x^2 + 5x + 6}{x^2 - 3x - 18}$

12. $\dfrac{x}{x^2 + 5x + 6} - \dfrac{2}{x^2 + 3x + 2}$

13. $\dfrac{2x^2}{x - y} + \dfrac{2y^2}{x + y}$

14. $\dfrac{3}{y + 4} - \dfrac{y}{y - 1} + \dfrac{y^2 + 3}{y^2 + 3y - 4}$

Divide.

15. $(16ab^3c - 10ab^2c^2 + 12a^2b^2c) \div (4ab)$ [5.3a]

16. $(y^2 - 20y + 64) \div (y - 6)$ [5.3b]

17. $(6x^4 + 3x^2 + 5x + 4) \div (x^2 + 2)$ [5.3b]

Divide using synthetic division. Show your work.   [5.3c]

18. $(x^3 + 5x^2 + 4x - 7) \div (x - 4)$

19. $(3x^4 - 5x^3 + 2x - 7) \div (x + 1)$

Simplify.   [5.4a]

20. $\dfrac{3 + \dfrac{3}{y}}{4 + \dfrac{4}{y}}$

21. $\dfrac{\dfrac{2}{a} + \dfrac{2}{b}}{\dfrac{4}{a^3} + \dfrac{4}{b^3}}$

22. $\dfrac{\dfrac{x^2 - 5x - 36}{x^2 - 36}}{\dfrac{x^2 + x - 12}{x^2 - 12x + 36}}$

23. $\dfrac{\dfrac{4}{x + 3} - \dfrac{2}{x^2 - 3x + 2}}{\dfrac{3}{x - 2} + \dfrac{1}{x^2 + 2x - 3}}$

Solve.   [5.5a]

24. $\dfrac{x}{4} + \dfrac{x}{7} = 1$

25. $\dfrac{5}{3x + 2} = \dfrac{3}{2x}$

26. $\dfrac{4x}{x + 1} + \dfrac{4}{x} + 9 = \dfrac{4}{x^2 + x}$

27. $\dfrac{90}{x^2 - 3x + 9} - \dfrac{5x}{x + 3} = \dfrac{405}{x^3 + 27}$

28. $\dfrac{2}{x - 3} + \dfrac{1}{4x + 20} = \dfrac{1}{x^2 + 2x - 15}$

29. Given that
$$f(x) = \frac{6}{x} + \frac{4}{x},$$
find all x for which $f(x) = 5$.

30. *House Painting.* David can paint the outside of a house in 12 hr. Bill can paint the same house in 9 hr. How long would it take them working together to paint the house? [5.6a]

31. *Boat Travel.* The current of the Gold River is 6 mph. A boat travels 50 mi downstream in the same time that it takes to travel 30 mi upstream. Fill in the table below and then find the speed of the boat in still water. [5.6c]

	DISTANCE	SPEED	TIME
Downstream			
Upstream			

32. *Travel Distance.* Fred operates a potato-chip delivery route. He drives 800 mi in 3 days. How far will he travel in 15 days? [5.6b]

Solve for the indicated letter. [5.7a]

33. $W = \dfrac{cd}{c + d}$, for d; for c

34. $S = \dfrac{p}{a} + \dfrac{t}{b}$, for b; for t

35. Find an equation of variation in which y varies directly as x, and $y = 100$ when $x = 25$. [5.8a]

36. Find an equation of variation in which y varies inversely as x, and $y = 100$ when $x = 25$. [5.8c]

37. *Pumping Time.* The time t required to empty a tank varies inversely as the rate r of pumping. If a pump can empty a tank in 35 min at the rate of 800 kL per minute, how long will it take the pump to empty the same tank at the rate of 1400 kL per minute? [5.8d]

38. *Test Score.* The score N on a test varies directly as the number of correct responses a. Ellen answers 28 questions correctly and earns a score of 87. What would Ellen's score have been if she had answered 25 questions correctly? [5.8b]

39. *Power of Electric Current.* The power P expended by heat in an electric circuit of fixed resistance varies directly as the square of the current C in the circuit. A circuit expends 180 watts when a current of 6 amperes is flowing. What is the amount of heat expended when the current is 10 amperes? [5.8f]

40. **D**_W Discuss at least three different uses of the LCM studied in this chapter. [5.2b], [5.4a], [5.5a]

41. **D**_W You have learned to solve a new kind of equation in this chapter. Explain how this type differs from those you have studied previously and how the equation-solving process differs. [5.5a]

SYNTHESIS

42. Find the reciprocal and simplify: $\dfrac{a - b}{a^3 - b^3}$. [5.1c, e]

43. Solve: $\dfrac{5}{x - 13} - \dfrac{5}{x} = \dfrac{65}{x^2 - 13x}$. [5.5a]

1. Find all numbers for which the rational expression

$$\frac{x^2 - 16}{x^2 - 3x + 2}$$

is not defined.

2. Find the domain of f where

$$f(x) = \frac{x^2 - 16}{x^2 - 3x + 2}.$$

Simplify.

3. $\dfrac{12x^2 + 11x + 2}{4x^2 - 7x - 2}$

4. $\dfrac{p^3 + 1}{p^2 - p - 2}$

5. Find the LCM of $x^2 + x - 6$ and $x^2 + 8x + 15$.

Perform the indicated operations and simplify.

6. $\dfrac{2x^2 + 20x + 50}{x^2 - 4} \cdot \dfrac{x + 2}{x + 5}$

7. $\dfrac{x}{x^2 + 11x + 30} - \dfrac{5}{x^2 + 9x + 20}$

8. $\dfrac{y^2 - 16}{2y + 6} \div \dfrac{y - 4}{y + 3}$

9. $\dfrac{x^2}{x - y} + \dfrac{y^2}{y - x}$

10. $\dfrac{1}{x + 1} - \dfrac{x + 2}{x^2 - 1} + \dfrac{3}{x - 1}$

11. $\dfrac{a}{a - b} + \dfrac{b}{a^2 + ab + b^2} - \dfrac{2}{a^3 - b^3}$

Divide.

12. $(20r^2s^3 + 15r^2s^2 - 10r^3s^3) \div (5r^2s)$

13. $(y^3 + 125) \div (y + 5)$

14. $(4x^4 + 3x^3 - 5x - 2) \div (x^2 + 1)$

Divide using synthetic division. Show your work.

15. $(x^3 + 3x^2 + 2x - 6) \div (x - 3)$

16. $(4x^3 - 6x^2 - 9) \div (x + 5)$

Simplify.

17. $\dfrac{1 - \dfrac{1}{x^2}}{1 - \dfrac{1}{x}}$

18. $\dfrac{\dfrac{1}{a^3} + \dfrac{1}{b^3}}{\dfrac{1}{a} + \dfrac{1}{b}}$

19. Given that

$$f(x) = \frac{2}{x - 1} + \frac{2}{x + 2},$$

find all x for which $f(x) = 1$.

Solve.

20. $\dfrac{2}{x-1} = \dfrac{3}{x+3}$

21. $\dfrac{7x}{x+3} + \dfrac{21}{x-3} = \dfrac{126}{x^2-9}$

22. $\dfrac{2x}{x+7} = \dfrac{5}{x+1}$

23. $\dfrac{1}{3x-6} - \dfrac{1}{x^2-4} = \dfrac{3}{x+2}$

24. *Completing a Puzzle.* Working together, Bella and Sam can complete a jigsaw puzzle in 1.5 hr. Bella takes 4 hr longer than Sam does when working alone. How long would it take Sam to complete the puzzle?

25. *Bicycle Travel.* Jody can bicycle at a rate of 12 mph when there is no wind. Against the wind, Jody bikes 8 mi in the same time that it takes to bike 14 mi with the wind. What is the speed of the wind?

26. *Predicting Paint Needs.* Lowell and Chris run a summer painting company to defray their college expenses. They need 4 gal of paint to paint 1700 ft² of clapboard. How much paint would they need for a building with 6000 ft² of clapboard?

Solve for the indicated letter.

27. $T = \dfrac{ab}{a-b}$, for a; for b

28. $Q = \dfrac{2}{a} - \dfrac{t}{b}$, for a

29. Find an equation of variation in which Q varies jointly as x and y, and $Q = 25$ when $x = 2$ and $y = 5$.

30. Find an equation of variation in which y varies inversely as x, and $y = 10$ when $x = 25$.

31. *Income vs Time.* Dean's income I varies directly as the time t worked. He gets a job that pays $275 for 40 hr of work. What is he paid for working 72 hr, assuming that there is no change in pay scale for overtime?

32. *Time and Speed.* The time t required to drive a fixed distance varies inversely as the speed r. It takes 5 hr at 60 km/h to drive a fixed distance. How long would it take to drive that same distance at 40 km/h?

33. *Area of a Balloon.* The surface area of a balloon varies directly as the square of its radius. The area is 314 cm² when the radius is 5 cm. What is the area when the radius is 7 cm?

<table>
<tr><td>SYNTHESIS</td></tr>
</table>

34. Solve: $\dfrac{6}{x-15} - \dfrac{6}{x} = \dfrac{90}{x^2-15x}$.

35. Find the LCM of $1 - t^6$ and $1 + t^6$.

36. Find the x- and y-intercepts of the function f given by

$$f(x) = \dfrac{\dfrac{5}{x+4} - \dfrac{3}{x-2}}{\dfrac{2}{x-3} + \dfrac{1}{x+4}}.$$

494

CHAPTER 5: Rational Expressions,
Equations, and Functions

Radical Expressions, Equations, and Functions

6

Real-World Application

An observation deck near the top of the Sears Tower in Chicago is 1353 ft high. How far can a tourist see to the horizon from this deck?

This problem appears as Exercise 45 in Section 6.6.

Objectives

 a Find principal square roots and their opposites, approximate square roots, find outputs of square-root functions, graph square-root functions, and find the domains of square-root functions.

b Simplify radical expressions with perfect-square radicands.

c Find cube roots, simplifying certain expressions, and find outputs of cube-root functions.

d Simplify expressions involving odd and even roots.

Find the square roots.

1. 9

2. 36

3. 121

Simplify.

4. $\sqrt{1}$ **5.** $\sqrt{36}$

6. $\sqrt{\dfrac{81}{100}}$ **7.** $\sqrt{0.0064}$

Answers on page A-32

In this section, we consider roots, such as square roots and cube roots. We define the symbolism and consider methods of manipulating symbols to get equivalent expressions.

a Square Roots and Square-Root Functions

When we raise a number to the second power, we say that we have **squared** the number. Sometimes we may need to find the number that was squared. We call this process **finding a square root** of a number.

SQUARE ROOT

The number c is a **square root** of a if $c^2 = a$.

For example:

5 is a *square root* of 25 because $5^2 = 5 \cdot 5 = 25$;

-5 is a *square root* of 25 because $(-5)^2 = (-5)(-5) = 25$.

The number -4 does not have a real-number square root because there is no real number c such that $c^2 = -4$.

PROPERTIES OF SQUARE ROOTS

Every positive real number has two real-number square roots.

The number 0 has just one square root, 0 itself.

Negative numbers do not have real-number square roots.*

EXAMPLE 1 Find the two square roots of 64.

The square roots of 64 are 8 and -8 because $8^2 = 64$ and $(-8)^2 = 64$.

Do Exercises 1–3.

PRINCIPAL SQUARE ROOT

The **principal square root** of a nonnegative number is its nonnegative square root. The symbol $\sqrt{a}$ represents the principal square root of a. To name the negative square root of a, we can write $-\sqrt{a}$.

EXAMPLES Simplify.

2. $\sqrt{25} = 5$ Remember: $\sqrt{}$ indicates the principal (nonnegative) square root.

3. $-\sqrt{25} = -5$

---

*In Section 6.8, we will consider a number system in which negative numbers do have square roots.

4. $\sqrt{\dfrac{81}{64}} = \dfrac{9}{8}$

5. $\sqrt{0.0049} = 0.07$

6. $-\sqrt{0.000001} = -0.001$

7. $\sqrt{0} = 0$

8. $\sqrt{-25}$ Does not exist as a real number. Negative numbers do not have real-number square roots.

Do Exercises 4–13. (Exercises 4–7 are on the preceding page.)

We found exact square roots in Examples 1–8. We often need to use rational numbers to *approximate* square roots that are irrational (see Section R.1). Such expressions can be found using a calculator with a square-root key.

EXAMPLES Use a calculator to approximate each of the following.

Number	*Using a calculator with a 10-digit readout*	*Rounded to three decimal places*
9. $\sqrt{11}$	3.316624790	3.317
10. $\sqrt{487}$	22.06807649	22.068
11. $-\sqrt{7297.8}$	-85.42716196	-85.427
12. $\sqrt{\dfrac{463}{557}}$	.9117229728	0.912

Do Exercises 14–19.

RADICAL; RADICAL EXPRESSION; RADICAND

The symbol $\sqrt{}$ is called a **radical.**

An expression written with a radical is called a **radical expression.**

The expression written under the radical is called the **radicand.**

These are radical expressions:

$$\sqrt{5}, \qquad \sqrt{a}, \qquad -\sqrt{5x}, \qquad \sqrt{y^2 + 7}.$$

The radicands in these expressions are 5, a, $5x$, and $y^2 + 7$, respectively.

EXAMPLE 13 Identify the radicand in $\sqrt{x^2 - 9}$.

The radicand in $\sqrt{x^2 - 9}$ is $x^2 - 9$.

Do Exercises 20 and 21 on the following page.

Find the following.

8. a) $\sqrt{16}$

b) $-\sqrt{16}$

c) $\sqrt{-16}$

9. a) $\sqrt{49}$

b) $-\sqrt{49}$

c) $\sqrt{-49}$

10. a) $\sqrt{144}$

b) $-\sqrt{144}$

c) $\sqrt{-144}$

11. $\sqrt{\dfrac{25}{64}}$

12. $-\sqrt{0.81}$

13. $\sqrt{1.44}$

It would be helpful to memorize the following table of exact square roots.

TABLE OF COMMON SQUARE ROOTS	
$\sqrt{1} = 1$	$\sqrt{196} = 14$
$\sqrt{4} = 2$	$\sqrt{225} = 15$
$\sqrt{9} = 3$	$\sqrt{256} = 16$
$\sqrt{16} = 4$	$\sqrt{289} = 17$
$\sqrt{25} = 5$	$\sqrt{324} = 18$
$\sqrt{36} = 6$	$\sqrt{361} = 19$
$\sqrt{49} = 7$	$\sqrt{400} = 20$
$\sqrt{64} = 8$	$\sqrt{441} = 21$
$\sqrt{81} = 9$	$\sqrt{484} = 22$
$\sqrt{100} = 10$	$\sqrt{529} = 23$
$\sqrt{121} = 11$	$\sqrt{576} = 24$
$\sqrt{144} = 12$	$\sqrt{625} = 25$
$\sqrt{169} = 13$	

Use a calculator to approximate the square root to three decimal places.

14. $\sqrt{17}$

15. $\sqrt{40}$

16. $\sqrt{1138}$

17. $-\sqrt{867.6}$

18. $\sqrt{\dfrac{22}{35}}$

19. $-\sqrt{\dfrac{2103.4}{67.82}}$

Answers on page A-32

Identify the radicand.

20. $\sqrt{28 + x}$

21. $\sqrt{\dfrac{y}{y + 3}}$

For the given function, find the indicated function values.

22. $g(x) = \sqrt{6x + 4}$; $g(0)$, $g(3)$, and $g(-5)$

23. $f(x) = -\sqrt{x}$; $f(4)$, $f(7)$, and $f(-3)$

Since each nonnegative real number x has exactly one principal square root, the symbol $\sqrt{x}$ represents exactly one real number and thus can be used to define a square-root function:

$$f(x) = \sqrt{x}.$$

The domain of this function is the set of nonnegative real numbers. In interval notation, the domain is $[0, \infty)$. This function will be discussed further in Example 16.

EXAMPLE 14   For the given function, find the indicated function values:

$$f(x) = \sqrt{3x - 2};   f(1), f(5), \text{ and } f(0).$$

We have

$f(1) = \sqrt{3 \cdot 1 - 2}$ Substituting 1 for x

$\quad = \sqrt{3 - 2} = \sqrt{1} = 1$;   Simplifying and taking the square root

$f(5) = \sqrt{3 \cdot 5 - 2}$ Substituting 5 for x

$\quad = \sqrt{13} \approx 3.606$;   Simplifying and approximating

$f(0) = \sqrt{3 \cdot 0 - 2}$ Substituting 0 for x

$\quad = \sqrt{-2}$.   Negative radicand. No real-number function value exists; 0 is not in the domain of f.

Do Exercises 22 and 23.

EXAMPLE 15   Find the domain of $g(x) = \sqrt{x + 2}$.

The expression $\sqrt{x + 2}$ is a real number only when $x + 2$ is nonnegative. Thus the domain of $g(x) = \sqrt{x + 2}$ is the set of all x-values for which $x + 2 \geq 0$. We solve as follows:

$x + 2 \geq 0$

$\quad x \geq -2$.   Adding -2

The domain of $g = \{x \mid x \geq -2\} = [-2, \infty)$.

EXAMPLE 16   Graph: **(a)** $f(x) = \sqrt{x}$; **(b)** $g(x) = \sqrt{x + 2}$.

We first find outputs as we did in Example 14. We can either select inputs that have exact outputs or use a calculator to make approximations. Once ordered pairs have been calculated, a smooth curve can be drawn.

a)

x	$f(x) = \sqrt{x}$	$(x, f(x))$
0	0	$(0, 0)$
1	1	$(1, 1)$
3	1.7	$(3, 1.7)$
4	2	$(4, 2)$
7	2.6	$(7, 2.6)$
9	3	$(9, 3)$

We can see from the table and the graph that the domain is $[0, \infty)$. The range is also the set of nonnegative real numbers $[0, \infty)$.

Answers on page A-32

b)

x	$g(x) = \sqrt{x+2}$	$(x, g(x))$
-2	0	$(-2, 0)$
-1	1	$(-1, 1)$
0	1.4	$(0, 1.4)$
3	2.2	$(3, 2.2)$
5	2.6	$(5, 2.6)$
10	3.5	$(10, 3.5)$

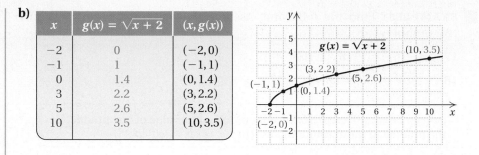

We can see from the table, the graph, and Example 15 that the domain is $[-2, \infty)$. The range is the set of nonnegative real numbers $[0, \infty)$.

Do Exercises 24–27.

b   Finding $\sqrt{a^2}$

In the expression $\sqrt{a^2}$, the radicand is a perfect square. It is tempting to think that $\sqrt{a^2} = a$, but we see below that this is not the case.

Suppose $a = 5$. Then we have $\sqrt{5^2}$, which is $\sqrt{25}$, or 5.

Suppose $a = -5$. Then we have $\sqrt{(-5)^2}$, which is $\sqrt{25}$, or 5.

Suppose $a = 0$. Then we have $\sqrt{0^2}$, which is $\sqrt{0}$, or 0.

The symbol $\sqrt{a^2}$ never represents a negative number. It represents the principal square root of a^2. Note the following.

SIMPLIFYING $\sqrt{a^2}$

$a \geq 0 \longrightarrow \sqrt{a^2} = a$

If a is positive or 0, the principal square root of a^2 is a.

$a < 0 \longrightarrow \sqrt{a^2} = -a$

If a is negative, the principal square root of a^2 is the opposite of a.

In all cases, the radical expression represents the absolute value of a.

PRINCIPAL SQUARE ROOT OF a^2

For any real number a, $\sqrt{a^2} = |a|$. The principal (nonnegative) square root of a^2 is the absolute value of a.

The absolute value is used to ensure that the principal square root is nonnegative, which is as it is defined.

Find the domain of the function.

24. $f(x) = \sqrt{x - 5}$

25. $g(x) = \sqrt{2x + 3}$

Graph.

26. $g(x) = -\sqrt{x}$

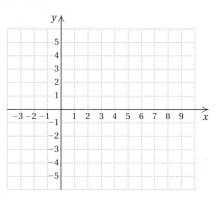

27. $f(x) = 2\sqrt{x} + 3$

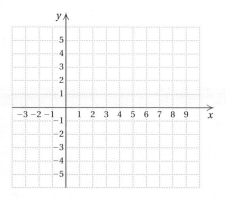

Answers on page A-32

Find the following. Assume that letters can represent *any* real number.

28. $\sqrt{y^2}$

29. $\sqrt{(-24)^2}$

30. $\sqrt{(5y)^2}$

31. $\sqrt{16y^2}$

32. $\sqrt{(x+7)^2}$

33. $\sqrt{4(x-2)^2}$

34. $\sqrt{49(y+5)^2}$

35. $\sqrt{x^2 - 6x + 9}$

Find the following.

36. $\sqrt[3]{-64}$

37. $\sqrt[3]{27y^3}$

38. $\sqrt[3]{8(x+2)^3}$

39. $\sqrt[3]{-\dfrac{343}{64}}$

Answers on page A-32

EXAMPLES Find the following. Assume that letters can represent any real number.

17. $\sqrt{(-16)^2} = |-16|$, or 16

18. $\sqrt{(3b)^2} = |3b| = |3| \cdot |b| = 3|b|$

> $|3b|$ can be simplified to $3|b|$ because the absolute value of any product is the product of the absolute values. That is, $|a \cdot b| = |a| \cdot |b|$.

19. $\sqrt{(x-1)^2} = |x-1|$

20. $\sqrt{x^2 + 8x + 16} = \sqrt{(x+4)^2}$
    $= |x+4|$

> **Caution!**
> $|x+4|$ is *not* the same as $|x| + 4$.

Do Exercises 28–35.

C Cube Roots

> **CUBE ROOT**
>
> The number c is the **cube root** of a, written $\sqrt[3]{a}$, if the third power of c is a—that is, if $c^3 = a$, then $\sqrt[3]{a} = c$.

For example:

> 2 is the *cube root* of 8 because $2^3 = 2 \cdot 2 \cdot 2 = 8$;
>
> -4 is the *cube root* of -64 because $(-4)^3 = (-4)(-4)(-4) = -64$.

We talk about *the* cube root of a number rather than *a* cube root because of the following.

> Every real number has exactly one cube root in the system of real numbers. The symbol $\sqrt[3]{a}$ represents *the* cube root of a.

EXAMPLES Find the following.

21. $\sqrt[3]{8} = 2$ because $2^3 = 8$.

22. $\sqrt[3]{-27} = -3$

23. $\sqrt[3]{-\dfrac{216}{125}} = -\dfrac{6}{5}$

24. $\sqrt[3]{0.001} = 0.1$

25. $\sqrt[3]{x^3} = x$

26. $\sqrt[3]{-8} = -2$

27. $\sqrt[3]{0} = 0$

28. $\sqrt[3]{-8y^3} = \sqrt[3]{(-2y)^3} = -2y$

When we are determining a cube root, no absolute-value signs are needed because a real number has just one cube root. The real-number cube root of a positive number is positive. The real-number cube root of a negative number is negative. The cube root of 0 is 0. That is, $\sqrt[3]{a^3} = a$ whether $a > 0$, $a < 0$, or $a = 0$.

Do Exercises 36–39.

Since the symbol $\sqrt[3]{x}$ represents exactly one real number, it can be used to define a cube-root function: $f(x) = \sqrt[3]{x}$.

EXAMPLE 29 For the given function, find the indicated function values:

$$f(x) = \sqrt[3]{x}; \quad f(125), \ f(0), \ f(-8), \text{ and } f(-10).$$

We have

$$f(125) = \sqrt[3]{125} = 5;$$
$$f(0) = \sqrt[3]{0} = 0;$$
$$f(-8) = \sqrt[3]{-8} = -2;$$
$$f(-10) = \sqrt[3]{-10} \approx -2.1544.$$

For calculator instructions for finding higher roots, see the Calculator Corner on p. 503.

Do Exercise 40.

The graph of $f(x) = \sqrt[3]{x}$ is shown below for reference. Note that the domain and the range *each* consists of the entire set of real numbers, $(-\infty, \infty)$.

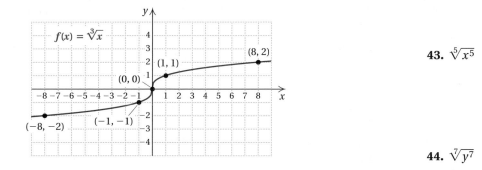

d   Odd and Even *k*th Roots

In the expression $\sqrt[k]{a}$, we call k the **index** and assume $k \geq 2$.

ODD ROOTS

The 5th root of a number a is the number c for which $c^5 = a$. There are also 7th roots, 9th roots, and so on. Whenever the number k in $\sqrt[k]{\ }$ is an odd number, we say that we are taking an **odd root.**

Every number has just one real-number odd root. For example, $\sqrt[3]{8} = 2$, $\sqrt[3]{-8} = -2$, and $\sqrt[3]{0} = 0$. If the number is positive, then the root is positive. If the number is negative, then the root is negative. If the number is 0, then the root is 0. Absolute-value signs are *not* needed when we are finding odd roots.

If k is an *odd* natural number, then for any real number a,

$$\sqrt[k]{a^k} = a.$$

---

40. For the given function, find the indicated function values:

$$g(x) = \sqrt[3]{x - 4}; \quad g(-23),$$
$$g(4), g(-1), \text{ and } g(11).$$

Find the following.

41. $\sqrt[5]{243}$

42. $\sqrt[5]{-243}$

43. $\sqrt[5]{x^5}$

44. $\sqrt[7]{y^7}$

45. $\sqrt[5]{0}$

46. $\sqrt[5]{-32x^5}$

47. $\sqrt[7]{(3x + 2)^7}$

Answers on page A-32

Find the following. Assume that letters can represent any real number.

48. $\sqrt[4]{81}$

EXAMPLES Find the following.

30. $\sqrt[5]{32} = 2$

31. $\sqrt[5]{-32} = -2$

32. $-\sqrt[5]{32} = -2$

33. $-\sqrt[5]{-32} = -(-2) = 2$

34. $\sqrt[7]{x^7} = x$

35. $\sqrt[7]{128} = 2$

36. $\sqrt[7]{-128} = -2$

37. $\sqrt[7]{0} = 0$

38. $\sqrt[5]{a^5} = a$

39. $\sqrt[9]{(x-1)^9} = x - 1$

49. $-\sqrt[4]{81}$

Do Exercises 41–47 on the preceding page.

EVEN ROOTS

When the index k in $\sqrt[k]{}$ is an even number, we say that we are taking an **even root.** When the index is 2, we do not write it. Every positive real number has two real-number kth roots when k is even. One of those roots is positive and one is negative. Negative real numbers do not have real-number kth roots when k is even. When we are finding even kth roots, absolute-value signs are sometimes necessary, as they are with square roots. For example,

50. $\sqrt[4]{-81}$

$$\sqrt{64} = 8, \qquad \sqrt[6]{64} = 2, \qquad -\sqrt[6]{64} = -2, \qquad \sqrt[6]{64x^6} = \sqrt[6]{(2x)^6} = |2x| = 2|x|.$$

Note that in $\sqrt[6]{64x^6}$, we need absolute-value signs because a variable is involved.

51. $\sqrt[4]{0}$

EXAMPLES Find the following. Assume that letters can represent any real number.

40. $\sqrt[4]{16} = 2$

41. $-\sqrt[4]{16} = -2$

42. $\sqrt[4]{-16}$ Does not exist as a real number.

52. $\sqrt[4]{16(x-2)^4}$

43. $\sqrt[4]{81x^4} = \sqrt[4]{(3x)^4} = 3|x|$

44. $\sqrt[6]{(y+7)^6} = |y+7|$

53. $\sqrt[6]{x^6}$

45. $\sqrt{81y^2} = \sqrt{(9y)^2} = 9|y|$

The following is a summary of how absolute value is used when we are taking even or odd roots.

54. $\sqrt[8]{(x+3)^8}$

SIMPLIFYING $\sqrt[k]{a^k}$

For any real number a:

a) $\sqrt[k]{a^k} = |a|$ when k is an *even* natural number. We use absolute value when k is even unless a is nonnegative.

55. $\sqrt[7]{(x+3)^7}$

b) $\sqrt[k]{a^k} = a$ when k is an *odd* natural number greater than 1. We do not use absolute value when k is odd.

Do Exercises 48–56.

56. $\sqrt[5]{243x^5}$

Answers on page A-32

Approximating Roots   We can use a graphing calculator to approximate square roots, cube roots, and higher roots of real numbers. To approximate $\sqrt{21}$, for example, we press ⟨2ND⟩ ⟨√⟩ ⟨2⟩ ⟨1⟩ ⟨)⟩ ⟨ENTER⟩. ($\sqrt{}$ is the second operation associated with the ⟨x²⟩ key.) To approximate $-\sqrt{6.95}$, we press ⟨(-)⟩ ⟨2ND⟩ ⟨√⟩ ⟨6⟩ ⟨.⟩ ⟨9⟩ ⟨5⟩ ⟨)⟩ ⟨ENTER⟩. Although it is not necessary to include the right parenthesis in either of these entries, we do so here in order to close the set of parentheses that are opened when the calculator displays "$\sqrt{}$(". We see that $\sqrt{21} \approx 4.583$ and $-\sqrt{6.95} \approx -2.636$.

We can also find higher roots on a graphing calculator. To find $\sqrt[3]{-71}$, we will use the cube-root operation from the MATH menu. We press ⟨MATH⟩ ⟨4⟩ to select this operation. Then we press ⟨(-)⟩ ⟨7⟩ ⟨1⟩ ⟨)⟩ ⟨ENTER⟩ to enter the radicand and display the result. As with square roots, we choose to close the parentheses although it is not necessary for this calculation. To find fourth, fifth, or higher roots, we use the xth-root operation from the MATH menu. To find $\sqrt[6]{178.4}$, we first press ⟨6⟩ to indicate that we are finding a sixth root. Then we press ⟨MATH⟩ ⟨5⟩ to select the xth-root operation. Finally, we press ⟨1⟩ ⟨7⟩ ⟨8⟩ ⟨.⟩ ⟨4⟩ ⟨ENTER⟩ to enter the radicand and display the result. Note that since this operation does not supply a left parenthesis, we do not enter a right parenthesis at the end. We see that $\sqrt[3]{-71} \approx -4.141$ and $\sqrt[6]{178.4} \approx 2.373$.

```
√(21)
            4.582575695
-√(6.95)
            -2.636285265
```

```
³√(-71)
            -4.140817749
6ˣ√178.4
            2.372643426
```

Exercises: Use a graphing calculator to approximate each of the following to three decimal places.

1. $\sqrt{43}$

2. $\sqrt{10,467}$

3. $-\sqrt{9406}$

4. $-\sqrt{11/17}$

5. $\sqrt[3]{416.73}$

6. $-\sqrt[3]{-800}$

7. $\sqrt[4]{16.4}$

8. $\sqrt[7]{-1389.7}$

9. $\sqrt{3/5}$

10. $\sqrt{3}/\sqrt{5}$

11. $\sqrt{3+7}$

12. $\sqrt{3} + \sqrt{7}$

Study Tips

TIME MANAGEMENT (PART 3)

Here are some additional tips to help you with time management. (See also the Study Tips on time management in Sections 3.5 and 5.6.)

■ **Are you a morning or an evening person?**   If you are an evening person, it might be best to avoid scheduling early-morning classes. If you are a morning person, do the opposite, but go to bed earlier to compensate. Nothing can drain your study time and effectiveness like fatigue.

■ **Keep on schedule.**   Your course syllabus provides a plan for the semester's schedule. Use a write-on calendar, daily planner, laptop computer, or personal digital assistant to outline your time for the semester. Be sure to note deadlines involving term papers and exams so you can begin a task early, breaking it down into smaller segments that can be accomplished more easily.

■ **Balance your class schedule.**   You may be someone who prefers large blocks of time for study on the off days. In that case, it might be advantageous for you to take courses that meet only three days a week. Keep in mind, however, that this might be a problem when tests in more than one course are scheduled for the same day.

"Time is our most important asset, yet we tend to waste it, kill it, and spend it rather than invest it."

Jim Rohn, motivational speaker

6.1

EXERCISE SET

For Extra Help

MathXL MyMathLab InterAct Math Tutor Center Math Tutor Center Digital Video Tutor CD 3 Videotape 7 Student's Solutions Manual

a Find the square roots.

1. 16

2. 225

3. 144

4. 9

5. 400

6. 81

Simplify.

7. $-\sqrt{\dfrac{49}{36}}$

8. $-\sqrt{\dfrac{361}{9}}$

9. $\sqrt{196}$

10. $\sqrt{441}$

11. $\sqrt{0.0036}$

12. $\sqrt{0.04}$

13. $\sqrt{-225}$

14. $\sqrt{-64}$

Use a calculator to approximate to three decimal places.

15. $\sqrt{347}$

16. $-\sqrt{1839.2}$

17. $\sqrt{\dfrac{285}{74}}$

18. $\sqrt{\dfrac{839.4}{19.7}}$

Identify the radicand.

19. $9\sqrt{y^2 + 16}$

20. $-3\sqrt{p^2 - 10}$

21. $x^4 y^5 \sqrt{\dfrac{x}{y - 1}}$

22. $a^2 b^2 \sqrt{\dfrac{a^2 - b}{b}}$

For the given function, find the indicated function values.

23. $f(x) = \sqrt{5x - 10}$; $f(6), f(2), f(1)$, and $f(-1)$

24. $t(x) = -\sqrt{2x + 1}$; $t(4), t(0), t(-1)$, and $t\!\left(-\tfrac{1}{2}\right)$

25. $g(x) = \sqrt{x^2 - 25}$; $g(-6), g(3), g(6)$, and $g(13)$

26. $F(x) = \sqrt{x^2 + 1}$; $F(0), F(-1)$, and $F(-10)$

27. Find the domain of the function f in Exercise 23.

28. Find the domain of the function t in Exercise 24.

29. *Speed of a Skidding Car.* How do police determine how fast a car had been traveling after an accident has occurred? The function

$$S(x) = 2\sqrt{5x}$$

can be used to approximate the speed S, in miles per hour, of a car that has left a skid mark of length x, in feet. What was the speed of a car that left skid marks of length 30 ft? 150 ft?

30. *Parking-Lot Arrival Spaces.* The attendants at a parking lot park cars in temporary spaces before the cars are taken to permanent parking stalls. The number N of such spaces needed is approximated by the function

$$N(a) = 2.5\sqrt{a},$$

where a is the average number of arrivals in peak hours. What is the number of spaces needed when the average number of arrivals is 66? 100?

Graph.

31. $f(x) = 2\sqrt{x}$

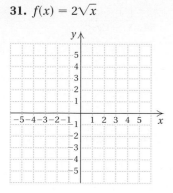

32. $g(x) = 3 - \sqrt{x}$

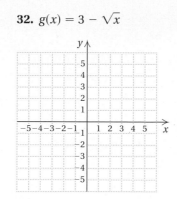

33. $F(x) = -3\sqrt{x}$

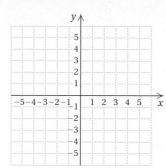

34. $f(x) = 2 + \sqrt{x - 1}$

35. $f(x) = \sqrt{x}$

36. $g(x) = -\sqrt{x}$

37. $f(x) = \sqrt{x - 2}$

38. $g(x) = \sqrt{x + 3}$

39. $f(x) = \sqrt{12 - 3x}$

40. $g(x) = \sqrt{8 - 4x}$

41. $g(x) = \sqrt{3x + 9}$

42. $f(x) = \sqrt{3x - 6}$

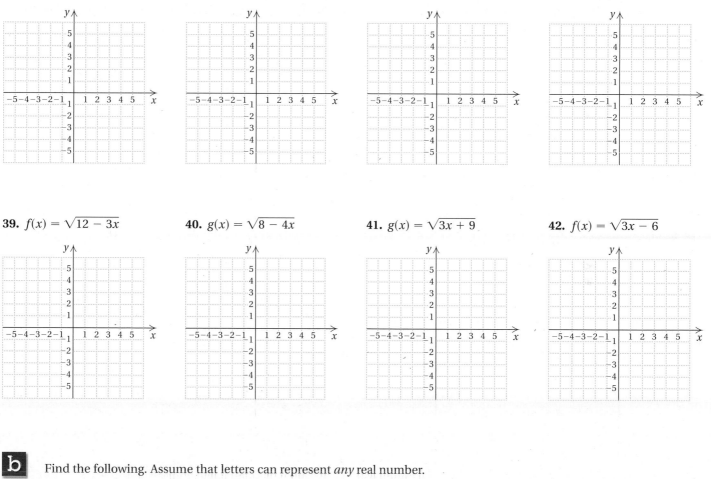

b Find the following. Assume that letters can represent *any* real number.

43. $\sqrt{16x^2}$

44. $\sqrt{25t^2}$

45. $\sqrt{(-12c)^2}$

46. $\sqrt{(-9d)^2}$

47. $\sqrt{(p + 3)^2}$

48. $\sqrt{(2 - x)^2}$

49. $\sqrt{x^2 - 4x + 4}$

50. $\sqrt{9t^2 - 30t + 25}$

C Simplify.

51. $\sqrt[3]{27}$

52. $-\sqrt[3]{64}$

53. $\sqrt[3]{-64x^3}$

54. $\sqrt[3]{-125y^3}$

55. $\sqrt[3]{-216}$

56. $-\sqrt[3]{-1000}$

57. $\sqrt[3]{0.343(x+1)^3}$

58. $\sqrt[3]{0.000008(y-2)^3}$

For the given function, find the indicated function values.

59. $f(x) = \sqrt[3]{x+1}$; $f(7)$, $f(26)$, $f(-9)$, and $f(-65)$

60. $g(x) = -\sqrt[3]{2x-1}$; $g(-62)$, $g(0)$, $g(-13)$, and $g(63)$

61. $f(x) = -\sqrt[3]{3x+1}$; $f(0)$, $f(-7)$, $f(21)$, and $f(333)$

62. $g(t) = \sqrt[3]{t-3}$; $g(30)$, $g(-5)$, $g(1)$, and $g(67)$

d Find the following. Assume that letters can represent *any* real number.

63. $-\sqrt[4]{625}$

64. $-\sqrt[4]{256}$

65. $\sqrt[5]{-1}$

66. $\sqrt[5]{-32}$

67. $\sqrt[5]{-\dfrac{32}{243}}$

68. $\sqrt[5]{-\dfrac{1}{32}}$

69. $\sqrt[6]{x^6}$

70. $\sqrt[8]{y^8}$

71. $\sqrt[4]{(5a)^4}$

72. $\sqrt[4]{(7b)^4}$

73. $\sqrt[10]{(-6)^{10}}$

74. $\sqrt[12]{(-10)^{12}}$

75. $\sqrt[414]{(a+b)^{414}}$

76. $\sqrt[1999]{(2a+b)^{1999}}$

77. $\sqrt[7]{y^7}$

78. $\sqrt[3]{(-6)^3}$

79. $\sqrt[5]{(x-2)^5}$

80. $\sqrt[9]{(2xy)^9}$

81. $\mathbf{D_W}$ Does the nth root of x^2 always exist? Why or why not?

82. $\mathbf{D_W}$ Explain how to formulate a radical expression that can be used to define a function f with a domain of $\{x \mid x \le 5\}$.

SKILL MAINTENANCE

Solve. [4.8a]

83. $x^2 + x - 2 = 0$

84. $x^2 + x = 0$

85. $4x^2 - 49 = 0$

86. $2x^2 - 26x + 72 = 0$

87. $3x^2 + x = 10$

88. $4x^2 - 20x + 25 = 0$

89. $4x^3 - 20x^2 + 25x = 0$

90. $x^3 - x^2 = 0$

Simplify. [R.7a, b]

91. $(a^3 b^2 c^5)^3$

92. $(5a^7 b^8)(2a^3 b)$

SYNTHESIS

93. Find the domain of
$$f(x) = \frac{\sqrt{x+3}}{\sqrt{2-x}}.$$

94. Use a graphing calculator to check your answers to Exercises 35, 39, and 41.

95. Use only the graph of $f(x) = \sqrt{x}$, shown below, to approximate $\sqrt{3}$, $\sqrt{5}$, and $\sqrt{10}$. Answers may vary.

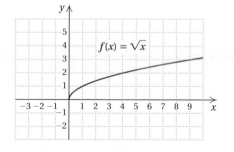

96. Use only the graph of $f(x) = \sqrt[3]{x}$, shown below, to approximate $\sqrt[3]{4}$, $\sqrt[3]{6}$, and $\sqrt[3]{-5}$. Answers may vary.

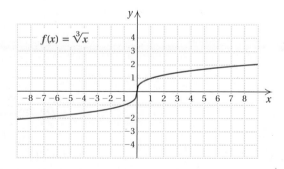

97. Use the TABLE, TRACE, and GRAPH features of a graphing calculator to find the domain and the range of each of the following functions.

a) $f(x) = \sqrt[3]{x}$

b) $g(x) = \sqrt[3]{4x - 5}$

c) $q(x) = 2 - \sqrt{x + 3}$

d) $h(x) = \sqrt[4]{x}$

e) $t(x) = \sqrt[4]{x - 3}$

Objectives

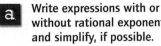

a Write expressions with or without rational exponents, and simplify, if possible.

b Write expressions without negative exponents, and simplify, if possible.

c Use the laws of exponents with rational exponents.

d Use rational exponents to simplify radical expressions.

In this section, we give meaning to expressions such as $a^{1/3}$, $7^{-1/2}$, and $(3x)^{0.84}$, which have rational numbers as exponents. We will see that using such notation can help simplify certain radical expressions.

a Rational Exponents

Expressions like $a^{1/2}$, $5^{-1/4}$, and $(2y)^{4/5}$ have not yet been defined. We will define such expressions so that the general properties of exponents hold.

Consider $a^{1/2} \cdot a^{1/2}$. If we want to multiply by adding exponents, it must follow that $a^{1/2} \cdot a^{1/2} = a^{1/2+1/2}$, or a^1. Thus we should define $a^{1/2}$ to be a square root of a. Similarly, $a^{1/3} \cdot a^{1/3} \cdot a^{1/3} = a^{1/3+1/3+1/3}$, or a^1, so $a^{1/3}$ should be defined to mean $\sqrt[3]{a}$.

$a^{1/n}$

For any *nonnegative* real number a and any natural number index n ($n \neq 1$),

$$a^{1/n} \quad \text{means} \quad \sqrt[n]{a} \quad \text{(the nonnegative } n\text{th root of } a\text{)}.$$

Whenever we use rational exponents, we assume that the bases are nonnegative.

EXAMPLES Rewrite without rational exponents, and simplify, if possible.

1. $27^{1/3} = \sqrt[3]{27} = 3$

2. $(abc)^{1/5} = \sqrt[5]{abc}$

3. $x^{1/2} = \sqrt{x}$ An index of 2 is not written.

Do Exercises 1–5.

EXAMPLES Rewrite with rational exponents.

4. $\sqrt[5]{7xy} = (7xy)^{1/5}$ We need parentheses around the radicand here.

5. $8\sqrt[3]{xy} = 8(xy)^{1/3}$

6. $\sqrt[7]{\dfrac{x^3 y}{9}} = \left(\dfrac{x^3 y}{9}\right)^{1/7}$

Do Exercises 6–9.

How should we define $a^{2/3}$? If the general properties of exponents are to hold, we have $a^{2/3} = (a^{1/3})^2$, or $(a^2)^{1/3}$, or $(\sqrt[3]{a})^2$, or $\sqrt[3]{a^2}$. We define this accordingly.

$a^{m/n}$

For any natural numbers m and n ($n \neq 1$) and any nonnegative real number a,

$$a^{m/n} \quad \text{means} \quad \sqrt[n]{a^m}, \quad \text{or} \quad (\sqrt[n]{a})^m.$$

Rewrite without rational exponents, and simplify, if possible.

1. $y^{1/4}$

2. $(3a)^{1/2}$

3. $16^{1/4}$

4. $(125)^{1/3}$

5. $(a^3 b^2 c)^{1/5}$

Rewrite with rational exponents.

6. $\sqrt[3]{19ab}$

7. $19\sqrt[3]{ab}$

8. $\sqrt[5]{\dfrac{x^2 y}{16}}$

9. $7\sqrt[4]{2ab}$

Answers on page A-32

EXAMPLES Rewrite without rational exponents, and simplify, if possible.

7. $(27)^{2/3} = \sqrt[3]{27^2}$
$= \left(\sqrt[3]{27}\right)^2$
$= 3^2$
$= 9$

8. $4^{3/2} = \sqrt[2]{4^3}$
$= \left(\sqrt[2]{4}\right)^3$
$= 2^3$
$= 8$

Do Exercises 10–12.

EXAMPLES Rewrite with rational exponents.

The index becomes the denominator of the rational exponent.

9. $\sqrt[3]{9^4} = 9^{4/3}$

10. $\left(\sqrt[4]{7xy}\right)^5 = (7xy)^{5/4}$

Do Exercises 13 and 14.

b Negative Rational Exponents

Negative rational exponents have a meaning similar to that of negative integer exponents.

> $a^{-m/n}$
>
> For any rational number m/n and any positive real number a,
>
> $$a^{-m/n} \text{ means } \frac{1}{a^{m/n}},$$
>
> that is, $a^{m/n}$ and $a^{-m/n}$ are reciprocals.

EXAMPLES Rewrite with positive exponents, and simplify, if possible.

11. $9^{-1/2} = \frac{1}{9^{1/2}} = \frac{1}{\sqrt{9}} = \frac{1}{3}$

12. $(5xy)^{-4/5} = \frac{1}{(5xy)^{4/5}}$

13. $64^{-2/3} = \frac{1}{64^{2/3}} = \frac{1}{\left(\sqrt[3]{64}\right)^2} = \frac{1}{4^2} = \frac{1}{16}$

14. $4x^{-2/3}y^{1/5} = 4 \cdot \frac{1}{x^{2/3}} \cdot y^{1/5} = \frac{4y^{1/5}}{x^{2/3}}$

15. $\left(\frac{3r}{7s}\right)^{-5/2} = \left(\frac{7s}{3r}\right)^{5/2}$ Since $\left(\frac{a}{b}\right)^{-n} = \left(\frac{b}{a}\right)^n$

Do Exercises 15–19.

Rewrite without rational exponents, and simplify, if possible.

10. $x^{3/5}$ **11.** $8^{2/3}$

12. $4^{5/2}$

Rewrite with rational exponents.

13. $\left(\sqrt[3]{7abc}\right)^4$ **14.** $\sqrt[5]{6^7}$

Rewrite with positive exponents, and simplify, if possible.

15. $16^{-1/4}$ **16.** $(3xy)^{-7/8}$

17. $81^{-3/4}$ **18.** $7p^{3/4}q^{-6/5}$

19. $\left(\frac{11m}{7n}\right)^{-2/3}$

Answers on page A-33

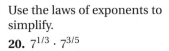

CALCULATOR CORNER

Rational Exponents We can use a graphing calculator to approximate rational roots of real numbers. To approximate $7^{2/3}$, we press ⑦ ⌃ ⦅ ② ÷ ③ ⦆ **ENTER** . Note that the parentheses around the exponent are necessary. If they are not used, the calculator will read the expression as $7^2 \div 3$. To approximate $14^{-1.9}$, we press ① ④ ⌃ ⦵ ① ⦁ ⑨ **ENTER** . Parentheses are not required when a rational exponent is expressed in a single decimal number. The display indicates that $7^{2/3} \approx 3.659$ and $14^{-1.9} \approx 0.007$.

```
7^(2/3)
                    3.65930571
14^-1.9
                    .006642885
```

Exercises: Approximate each of the following.

1. $5^{3/4}$

2. $8^{4/7}$

3. $29^{-3/8}$

4. $73^{0.56}$

5. $34^{-2.78}$

6. $32^{0.2}$

Use the laws of exponents to simplify.

20. $7^{1/3} \cdot 7^{3/5}$

21. $\dfrac{5^{7/6}}{5^{5/6}}$

22. $(9^{3/5})^{2/3}$

23. $(p^{-2/3}q^{1/4})^{1/2}$

C Laws of Exponents

The same laws hold for rational-number exponents as for integer exponents. We list them for review.

For any real number a and any rational exponents m and n:

1. $a^m \cdot a^n = a^{m+n}$ In multiplying, we can add exponents if the bases are the same.

2. $\dfrac{a^m}{a^n} = a^{m-n}$ In dividing, we can subtract exponents if the bases are the same.

3. $(a^m)^n = a^{m \cdot n}$ To raise a power to a power, we can multiply the exponents.

4. $(ab)^m = a^m b^m$ To raise a product to a power, we can raise each factor to the power.

5. $\left(\dfrac{a}{b}\right)^n = \dfrac{a^n}{b^n}$ To raise a quotient to a power, we can raise both the numerator and the denominator to the power.

EXAMPLES Use the laws of exponents to simplify.

16. $3^{1/5} \cdot 3^{3/5} = 3^{1/5+3/5} = 3^{4/5}$ Adding exponents

17. $\dfrac{7^{1/4}}{7^{1/2}} = 7^{1/4-1/2} = 7^{1/4-2/4} = 7^{-1/4} = \dfrac{1}{7^{1/4}}$ Subtracting exponents

18. $(7.2^{2/3})^{3/4} = 7.2^{2/3 \cdot 3/4} = 7.2^{6/12} = 7.2^{1/2}$ Multiplying exponents

19. $(a^{-1/3}b^{2/5})^{1/2} = a^{-1/3 \cdot 1/2} \cdot b^{2/5 \cdot 1/2}$ Raising a product to a power and multiplying exponents

$$= a^{-1/6}b^{1/5} = \dfrac{b^{1/5}}{a^{1/6}}$$

Answers on page A-33

Do Exercises 20–23.

CHAPTER 6: Radical Expressions, Equations, and Functions

d Simplifying Radical Expressions

Rational exponents can be used to simplify some radical expressions. The procedure is as follows.

> **SIMPLIFYING RADICAL EXPRESSIONS**
>
> 1. Convert radical expressions to exponential expressions.
> 2. Use arithmetic and the laws of exponents to simplify.
> 3. Convert back to radical notation when appropriate.
>
> *Important*: This procedure works only when all expressions under radicals are *nonnegative* since rational exponents are not defined otherwise. With this assumption, no absolute-value signs will be needed.

EXAMPLES Use rational exponents to simplify.

20. $\sqrt[6]{x^3} = x^{3/6}$ Converting to an exponential expression

   $\quad = x^{1/2}$ Simplifying the exponent

   $\quad = \sqrt{x}$ Converting back to radical notation

21. $\sqrt[6]{4} = 4^{1/6}$ Converting to exponential notation

   $\quad = (2^2)^{1/6}$ Renaming 4 as 2^2

   $\quad = 2^{2/6}$ Using $(a^m)^n = a^{mn}$; multiplying exponents

   $\quad = 2^{1/3}$ Simplifying the exponent

   $\quad = \sqrt[3]{2}$ Converting back to radical notation

Do Exercises 24–26.

EXAMPLE 22   Use rational exponents to simplify: $\sqrt[8]{a^2b^4}$.

   $\sqrt[8]{a^2b^4} = (a^2b^4)^{1/8}$ Converting to exponential notation

   $\quad = a^{2/8} \cdot b^{4/8}$ Using $(ab)^n = a^n b^n$

   $\quad = a^{1/4} \cdot b^{1/2}$ Simplifying the exponents

   $\quad = a^{1/4} \cdot b^{2/4}$ Rewriting $\frac{1}{2}$ with a denominator of 4

   $\quad = (ab^2)^{1/4}$ Using $a^n b^n = (ab)^n$

   $\quad = \sqrt[4]{ab^2}$ Converting back to radical notation

Do Exercises 27–29.

We can use properties of rational exponents to write a single radical expression for a product or a quotient.

EXAMPLE 23   Use rational exponents to write a single radical expression for $\sqrt[3]{5} \cdot \sqrt{2}$.

   $\sqrt[3]{5} \cdot \sqrt{2} = 5^{1/3} \cdot 2^{1/2}$ Converting to exponential notation

   $\quad = 5^{2/6} \cdot 2^{3/6}$ Rewriting so that exponents have a common denominator

   $\quad = (5^2 \cdot 2^3)^{1/6}$ Using $a^n b^n = (ab)^n$

   $\quad = \sqrt[6]{5^2 \cdot 2^3}$ Converting back to radical notation

   $\quad = \sqrt[6]{200}$ Multiplying under the radical

Use rational exponents to simplify.

24. $\sqrt[4]{a^2}$

25. $\sqrt[4]{x^4}$

26. $\sqrt[6]{8}$

Use rational exponents to simplify.

27. $\sqrt[12]{x^3y^6}$

28. $\sqrt[6]{a^{12}b^3}$

29. $\sqrt[5]{a^5b^{10}}$

30. Use rational exponents to write a single radical expression:

   $\sqrt[4]{7} \cdot \sqrt{3}$.

Answers on page A-33

Write a single radical expression.

31. $x^{2/3}y^{1/2}z^{5/6}$

Do Exercise 30 on the preceding page.

EXAMPLE 24 Write a single radical expression for $a^{1/2}b^{-1/2}c^{5/6}$.

$$a^{1/2}b^{-1/2}c^{5/6} = a^{3/6}b^{-3/6}c^{5/6}$$ Rewriting so that exponents have a common denominator

$$= (a^3b^{-3}c^5)^{1/6}$$ Using $a^nb^n = (ab)^n$

$$= \sqrt[6]{a^3b^{-3}c^5}$$ Converting to radical notation

32. $\dfrac{a^{1/2}b^{3/8}}{a^{1/4}b^{1/8}}$

EXAMPLE 25 Write a single radical expression for $\dfrac{x^{5/6} \cdot y^{3/8}}{x^{4/9} \cdot y^{1/4}}$.

$$\frac{x^{5/6} \cdot y^{3/8}}{x^{4/9} \cdot y^{1/4}} = x^{5/6-4/9} \cdot y^{3/8-1/4}$$ Subtracting exponents

$$= x^{15/18-8/18} \cdot y^{3/8-2/8}$$ Finding common denominators so that exponents can be subtracted

$$= x^{7/18} \cdot y^{1/8}$$ Carrying out the subtraction of exponents

$$= x^{28/72} \cdot y^{9/72}$$ Rewriting so that all exponents have a common denominator

$$= \sqrt[72]{x^{28}y^9}$$ Converting to radical notation

Use rational exponents to simplify.

33. $\sqrt[14]{(5m)^2}$

Do Exercises 31 and 32.

EXAMPLES Use rational exponents to simplify.

26. $\sqrt[6]{(5x)^3} = (5x)^{3/6}$ Converting to exponential notation

$\qquad = (5x)^{1/2}$ Simplifying the exponent

$\qquad = \sqrt{5x}$ Converting back to radical notation

27. $\sqrt[5]{t^{20}} = t^{20/5}$ Converting to exponential notation

$\qquad = t^4$ Simplifying the exponent

34. $\sqrt[18]{m^3}$

28. $\left(\sqrt[3]{pq^2c}\right)^{12} = (pq^2c)^{12/3}$ Converting to exponential notation

$\qquad = (pq^2c)^4$ Simplifying the exponent

$\qquad = p^4q^8c^4$ Using $(ab)^n = a^nb^n$

29. $\sqrt{\sqrt[3]{x}} = \sqrt{x^{1/3}}$ Converting the radicand to exponential notation

$\qquad = (x^{1/3})^{1/2}$ Try to go directly to this step.

35. $\left(\sqrt[6]{a^5b^3c}\right)^{24}$

$\qquad = x^{1/6}$ Multiplying exponents

$\qquad = \sqrt[6]{x}$ Converting back to radical notation

Do Exercises 33–36.

36. $\sqrt[5]{\sqrt{x}}$

Answers on page A-33

6.2

EXERCISE SET

For Extra Help

a Rewrite without rational exponents, and simplify, if possible.

1. $y^{1/7}$

2. $x^{1/6}$

3. $8^{1/3}$

4. $16^{1/2}$

5. $(a^3b^3)^{1/5}$

6. $(x^2y^2)^{1/3}$

7. $16^{3/4}$

8. $4^{7/2}$

9. $49^{3/2}$

10. $27^{4/3}$

Rewrite with rational exponents.

11. $\sqrt{17}$

12. $\sqrt{x^3}$

13. $\sqrt[3]{18}$

14. $\sqrt[3]{23}$

15. $\sqrt[5]{xy^2z}$

16. $\sqrt[7]{x^3y^2z^2}$

17. $\left(\sqrt{3mn}\right)^3$

18. $\left(\sqrt[3]{7xy}\right)^4$

19. $\left(\sqrt[7]{8x^2y}\right)^5$

20. $\left(\sqrt[6]{2a^5b}\right)^7$

b Rewrite with positive exponents, and simplify, if possible.

21. $27^{-1/3}$

22. $100^{-1/2}$

23. $100^{-3/2}$

24. $16^{-3/4}$

25. $3x^{-1/4}$

26. $8y^{-1/7}$

27. $(2rs)^{-3/4}$

28. $(5xy)^{-5/6}$

29. $2a^{3/4}b^{-1/2}c^{2/3}$

30. $5x^{-2/3}y^{4/5}z$

31. $\left(\dfrac{7x}{8yz}\right)^{-3/5}$

32. $\left(\dfrac{2ab}{3c}\right)^{-5/6}$

33. $\dfrac{1}{x^{-2/3}}$

34. $\dfrac{1}{a^{-7/8}}$

35. $2^{-1/3}x^4y^{-2/7}$

36. $3^{-5/2}a^3b^{-7/3}$

37. $\dfrac{7x}{\sqrt[3]{z}}$

38. $\dfrac{6a}{\sqrt[4]{b}}$

39. $\dfrac{5a}{3c^{-1/2}}$

40. $\dfrac{2z}{5x^{-1/3}}$

c Use the laws of exponents to simplify. Write the answers with positive exponents.

41. $5^{3/4} \cdot 5^{1/8}$

42. $11^{2/3} \cdot 11^{1/2}$

43. $\dfrac{7^{5/8}}{7^{3/8}}$

44. $\dfrac{3^{5/8}}{3^{-1/8}}$

45. $\dfrac{4.9^{-1/6}}{4.9^{-2/3}}$

46. $\dfrac{2.3^{-3/10}}{2.3^{-1/5}}$

47. $(6^{3/8})^{2/7}$

48. $(3^{2/9})^{3/5}$

49. $a^{2/3} \cdot a^{5/4}$

50. $x^{3/4} \cdot x^{2/3}$

51. $(a^{2/3} \cdot b^{5/8})^4$ **52.** $(x^{-1/3} \cdot y^{-2/5})^{-15}$ **53.** $(x^{2/3})^{-3/7}$ **54.** $(a^{-3/2})^{2/9}$

d Use rational exponents to simplify. Write the answer in radical notation if appropriate.

55. $\sqrt[6]{a^2}$ **56.** $\sqrt[6]{t^4}$ **57.** $\sqrt[3]{x^{15}}$ **58.** $\sqrt[4]{a^{12}}$ **59.** $\sqrt[6]{x^{-18}}$

60. $\sqrt[5]{a^{-10}}$ **61.** $(\sqrt[3]{ab})^{15}$ **62.** $(\sqrt[7]{cd})^{14}$ **63.** $\sqrt[14]{128}$ **64.** $\sqrt[6]{81}$

65. $\sqrt[6]{4x^2}$ **66.** $\sqrt[3]{8y^6}$ **67.** $\sqrt{x^4y^6}$ **68.** $\sqrt[4]{16x^4y^2}$ **69.** $\sqrt[5]{32c^{10}d^{15}}$

Use rational exponents to write a single radical expression.

70. $\sqrt[3]{3}\,\sqrt{3}$ **71.** $\sqrt[3]{7} \cdot \sqrt[4]{5}$ **72.** $\sqrt[7]{11} \cdot \sqrt[6]{13}$ **73.** $\sqrt[4]{5} \cdot \sqrt[5]{7}$ **74.** $\sqrt[3]{y}\,\sqrt[5]{3y}$

75. $\sqrt{x}\,\sqrt[3]{2x}$ **76.** $(\sqrt[3]{x^2y^5})^{12}$ **77.** $(\sqrt[5]{a^2b^4})^{15}$ **78.** $\sqrt[4]{\sqrt{x}}$ **79.** $\sqrt[3]{\sqrt[6]{m}}$

80. $a^{2/3} \cdot b^{3/4}$ **81.** $x^{1/3} \cdot y^{1/4} \cdot z^{1/6}$ **82.** $\dfrac{x^{8/15} \cdot y^{7/5}}{x^{1/3} \cdot y^{-1/5}}$ **83.** $\left(\dfrac{c^{-4/5}d^{5/9}}{c^{3/10}d^{1/6}}\right)^3$ **84.** $\sqrt[3]{\sqrt[4]{xy}}$

85. $\mathbf{D_W}$ Find the domain of
$$f(x) = (x + 5)^{1/2}(x + 7)^{-1/2}$$
and explain how you found your answer.

86. $\mathbf{D_W}$ Explain why $\sqrt[3]{x^6} = x^2$ for any value of x, but $\sqrt{x^6} = x^3$ only when $x \geq 0$.

SKILL MAINTENANCE

Solve. [5.7a]

87. $A = \dfrac{ab}{a + b}$, for a **88.** $Q = \dfrac{st}{s - t}$, for s **89.** $Q = \dfrac{st}{s - t}$, for t **90.** $\dfrac{1}{t} = \dfrac{1}{a} - \dfrac{1}{b}$, for b

SYNTHESIS

91. Use the SIMULTANEOUS mode to graph
$$y_1 = x^{1/2}, \quad y_2 = 3x^{2/5}, \quad y_3 = x^{4/7}, \quad y_4 = \tfrac{1}{5}x^{3/4}.$$
Then, looking only at coordinates, match each graph with its equation.

92. Simplify:
$$\left(\sqrt[10]{\sqrt[5]{x^{15}}}\right)^5 \left(\sqrt[5]{\sqrt[10]{x^{15}}}\right)^5.$$

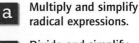

6.3 SIMPLIFYING RADICAL EXPRESSIONS

Objectives

a Multiply and simplify radical expressions.

b Divide and simplify radical expressions.

a Multiplying and Simplifying Radical Expressions

Note that $\sqrt{4}\ \sqrt{25} = 2 \cdot 5 = 10$. Also $\sqrt{4 \cdot 25} = \sqrt{100} = 10$. Likewise,

$$\sqrt[3]{27}\ \sqrt[3]{8} = 3 \cdot 2 = 6 \quad \text{and} \quad \sqrt[3]{27 \cdot 8} = \sqrt[3]{216} = 6.$$

These examples suggest the following.

THE PRODUCT RULE FOR RADICALS

For any nonnegative real numbers a and b and any index k,

$$\sqrt[k]{a} \cdot \sqrt[k]{b} = \sqrt[k]{a \cdot b}, \quad \text{or} \quad a^{1/k} \cdot b^{1/k} = (ab)^{1/k}.$$

The index must be the same throughout.

(To multiply, multiply the radicands.)

EXAMPLES Multiply.

1. $\sqrt{3} \cdot \sqrt{5} = \sqrt{3 \cdot 5} = \sqrt{15}$

2. $\sqrt{5a}\ \sqrt{2b} = \sqrt{5a \cdot 2b} = \sqrt{10ab}$

3. $\sqrt[3]{4}\ \sqrt[3]{5} = \sqrt[3]{4 \cdot 5} = \sqrt[3]{20}$

4. $\sqrt[4]{\dfrac{y}{5}}\ \sqrt[4]{\dfrac{7}{x}} = \sqrt[4]{\dfrac{y}{5} \cdot \dfrac{7}{x}} = \sqrt[4]{\dfrac{7y}{5x}}$

Caution!
A common error is to omit the index in the answer.

Do Exercises 1–4.

Keep in mind that the product rule can be used only when the indexes are the same. When indexes differ, we can use rational exponents as we did in Examples 23 and 24 of Section 6.2.

EXAMPLE 5 Multiply: $\sqrt{5x} \cdot \sqrt[4]{3y}$.

$$\begin{aligned}
\sqrt{5x} \cdot \sqrt[4]{3y} &= (5x)^{1/2}(3y)^{1/4} && \text{Converting to exponential notation} \\
&= (5x)^{2/4}(3y)^{1/4} && \text{Rewriting so that exponents have a common denominator} \\
&= [(5x)^2(3y)]^{1/4} && \text{Using } a^n b^n = (ab)^n \\
&= [(25x^2)(3y)]^{1/4} && \text{Squaring } 5x \\
&= \sqrt[4]{(25x^2)(3y)} && \text{Converting back to radical notation} \\
&= \sqrt[4]{75x^2y} && \text{Multiplying under the radical}
\end{aligned}$$

Do Exercises 5 and 6.

Multiply.

1. $\sqrt{19}\ \sqrt{7}$

2. $\sqrt{3p}\ \sqrt{7q}$

3. $\sqrt[4]{403}\ \sqrt[4]{7}$

4. $\sqrt[3]{\dfrac{5}{p}} \cdot \sqrt[3]{\dfrac{2}{q}}$

Multiply.

5. $\sqrt{5}\ \sqrt[3]{2}$

6. $\sqrt{x}\ \sqrt[3]{5y}$

Answers on page A-33

Simplify by factoring.

7. $\sqrt{32}$

8. $\sqrt[3]{80}$

Answers on page A-33

Study Tips

TEST TAKING: MORE ON DOING EVEN-NUMBERED EXERCISES

In an earlier study tip (p. 377), as a way to improve your test-taking skills, we encouraged you to build some even-numbered exercises into your homework. Here we explore this issue further.

Working a test is different from working your homework, when the answers are provided. When taking the test, you are "on your own," so to speak. Keep the following tips in mind when taking your next test or quiz.

1. Work a bit slower and deliberately, taking a fresh piece of paper to redo the problem. Check your work against your previous work to see if there is a difference and why. This is especially helpful if you finish the test early and have extra time.

2. Use estimation techniques to solve the problem as a check.

3. Do the checks to applied problems that we so often discuss in the book.

CHAPTER 6: Radical Expressions, Equations, and Functions

We can reverse the product rule to simplify a product. We simplify the root of a product by taking the root of each factor separately.

FACTORING RADICAL EXPRESSIONS

For any nonnegative real numbers a and b and any index k,

$$\sqrt[k]{ab} = \sqrt[k]{a} \cdot \sqrt[k]{b}, \quad \text{or} \quad (ab)^{1/k} = a^{1/k} \cdot b^{1/k}.$$

(Take the kth root of each factor separately.)

Compare the following:

$$\sqrt{50} = \sqrt{10 \cdot 5} = \sqrt{10}\,\sqrt{5};$$
$$\sqrt{50} = \sqrt{25 \cdot 2} = \sqrt{25}\,\sqrt{2} = 5\sqrt{2}.$$

In the second case, the radicand has the perfect-square factor 25. If you do not recognize perfect-square factors, try factoring the radicand into its prime factors. For example,

$$\sqrt{50} = \sqrt{2 \cdot \underset{\uparrow}{5 \cdot 5}} = 5\sqrt{2}.$$

Perfect square (a pair of the same numbers)

Square-root radical expressions in which the radicand has no perfect-square factors, such as $5\sqrt{2}$, are considered to be in simplest form. A procedure for simplifying kth roots follows.

SIMPLIFYING kth ROOTS

To simplify a radical expression by factoring:

1. Look for the largest factors of the radicand that are perfect kth powers (where k is the index).
2. Then take the kth root of the resulting factors.
3. A radical expression, with index k, is *simplified* when its radicand has no factors that are perfect kth powers.

EXAMPLES Simplify by factoring.

6. $\sqrt{50} = \sqrt{25 \cdot 2} = \sqrt{25} \cdot \sqrt{2} = \sqrt{5 \cdot 5} \cdot \sqrt{2} = 5\sqrt{2}$

This factor is a perfect square.

7. $\sqrt[3]{32} = \sqrt[3]{8 \cdot 4} = \sqrt[3]{8} \cdot \sqrt[3]{4} = \sqrt[3]{2 \cdot 2 \cdot 2} \cdot \sqrt[3]{2 \cdot 2} = 2\sqrt[3]{4}$

This factor is a perfect cube (third power).

8. $\sqrt[4]{48} = \sqrt[4]{16 \cdot 3} = \sqrt[4]{16} \cdot \sqrt[4]{3} = \sqrt[4]{2 \cdot 2 \cdot 2 \cdot 2} \cdot \sqrt[4]{3} = 2\sqrt[4]{3}$

This factor is a perfect fourth power.

Do Exercises 7 and 8.

In many situations, expressions under radicals never represent negative numbers. In such cases, absolute-value notation is not necessary. For this reason, we will henceforth assume that *all expressions under radicals are nonnegative.*

EXAMPLES Simplify by factoring. Assume that all expressions under radicals represent nonnegative numbers.

9. $\sqrt{5x^2} = \sqrt{5 \cdot x^2}$ Factoring the radicand

$\phantom{\sqrt{5x^2}} = \sqrt{5} \cdot \sqrt{x^2}$ Factoring into two radicals

$\phantom{\sqrt{5x^2}} = \sqrt{5} \cdot \underset{\smile}{x}$ Taking the square root of x^2

> Absolute-value notation is not needed because the expression under the radical is not negative.

10. $\sqrt{18x^2y} = \sqrt{9 \cdot 2 \cdot x^2 \cdot y}$ Factoring the radicand and looking for perfect-square factors

$\phantom{\sqrt{18x^2y}} = \sqrt{9 \cdot x^2 \cdot 2 \cdot y}$

$\phantom{\sqrt{18x^2y}} = \sqrt{9} \cdot \sqrt{x^2} \cdot \sqrt{2} \cdot \sqrt{y}$ Factoring into several radicals

$\phantom{\sqrt{18x^2y}} = 3x\sqrt{2y}$ Taking square roots

11. $\sqrt{216x^5y^3} = \sqrt{36 \cdot 6 \cdot x^4 \cdot x \cdot y^2 \cdot y}$ Factoring the radicand and looking for perfect-square factors

$\phantom{\sqrt{216x^5y^3}} = \sqrt{36 \cdot x^4 \cdot y^2 \cdot 6 \cdot x \cdot y}$

$\phantom{\sqrt{216x^5y^3}} = \sqrt{36}\,\sqrt{x^4}\,\sqrt{y^2}\,\sqrt{6xy}$ Factoring into several radicals

$\phantom{\sqrt{216x^5y^3}} = 6x^2y\sqrt{6xy}$ Taking square roots

Let's look at this example another way. We do a complete factorization and look for pairs of factors. Each pair of factors makes a square:

$\sqrt{216x^5y^3} = \sqrt{2 \cdot 2 \cdot 2 \cdot 3 \cdot 3 \cdot 3 \cdot x \cdot x \cdot x \cdot x \cdot x \cdot y \cdot y \cdot y}$ Each pair of factors makes a perfect square.

$\phantom{\sqrt{216x^5y^3}} = 2 \cdot 3 \cdot x \cdot x \cdot y \cdot \sqrt{2 \cdot 3 \cdot x \cdot y}$

$\phantom{\sqrt{216x^5y^3}} = 6x^2y\sqrt{6xy}.$

12. $\sqrt[3]{16a^7b^{11}} = \sqrt[3]{8 \cdot 2 \cdot a^6 \cdot a \cdot b^9 \cdot b^2}$ Factoring the radicand. The index is 3, so we look for the largest powers that are multiples of 3 because these are perfect cubes.

$\phantom{\sqrt[3]{16a^7b^{11}}} = \sqrt[3]{8} \cdot \sqrt[3]{a^6} \cdot \sqrt[3]{b^9} \cdot \sqrt[3]{2ab^2}$ Factoring into radicals

$\phantom{\sqrt[3]{16a^7b^{11}}} = 2a^2b^3\sqrt[3]{2ab^2}$ Taking cube roots

Let's look at this example another way. We do a complete factorization and look for triples of factors. Each triple of factors makes a cube:

$\sqrt[3]{16a^7b^{11}}$

$= \sqrt[3]{2 \cdot 2 \cdot 2 \cdot 2 \cdot a \cdot a \cdot a \cdot a \cdot a \cdot a \cdot a \cdot b \cdot b \cdot b \cdot b \cdot b \cdot b \cdot b \cdot b \cdot b \cdot b \cdot b}$

                         Each triple of factors makes a cube.

$= 2 \cdot a \cdot a \cdot b \cdot b \cdot b \cdot \sqrt[3]{2 \cdot a \cdot b \cdot b}$

$= 2a^2b^3\sqrt[3]{2ab^2}.$

Do Exercises 9–14.

Simplify by factoring. Assume that all expressions under radicals represent nonnegative numbers.

9. $\sqrt{300}$

10. $\sqrt{36y^2}$

11. $\sqrt{12a^2b}$

12. $\sqrt{12ab^3c^2}$

13. $\sqrt[3]{16}$

14. $\sqrt[3]{81x^4y^8}$

Answers on page A-33

Multiply and simplify. Assume that all expressions under radicals represent nonnegative numbers.

15. $\sqrt{3}\,\sqrt{6}$

16. $\sqrt{18y}\,\sqrt{14y}$

17. $\sqrt[3]{3x^2y}\,\sqrt[3]{36x}$

18. $\sqrt{7a}\,\sqrt{21b}$

Divide and simplify. Assume that all expressions under radicals represent positive numbers.

19. $\dfrac{\sqrt{75}}{\sqrt{3}}$

20. $\dfrac{14\sqrt{128xy}}{2\sqrt{2}}$

21. $\dfrac{\sqrt{50a^3}}{\sqrt{2a}}$

22. $\dfrac{4\sqrt[3]{250}}{7\sqrt[3]{2}}$

Sometimes after we have multiplied, we can then simplify by factoring.

EXAMPLES   Multiply and simplify. Assume that all expressions under radicals represent nonnegative numbers.

13. $\sqrt{20}\sqrt{8} = \sqrt{20 \cdot 8} = \sqrt{4 \cdot 5 \cdot 4 \cdot 2} = 4\sqrt{10}$

14. $3\sqrt[3]{25} \cdot 2\sqrt[3]{5} = 6 \cdot \sqrt[3]{25 \cdot 5}$
$$= 6 \cdot \sqrt[3]{5 \cdot 5 \cdot 5}$$
$$= 6 \cdot 5 = 30$$

15. $\sqrt[3]{18y^3}\,\sqrt[3]{4x^2} = \sqrt[3]{18y^3 \cdot 4x^2}$ Multiplying radicands
$$= \sqrt[3]{2 \cdot 3 \cdot 3 \cdot y \cdot y \cdot y \cdot 2 \cdot 2 \cdot x \cdot x}$$
$$= 2 \cdot y \cdot \sqrt[3]{3 \cdot 3 \cdot x \cdot x}$$
$$= 2y\sqrt[3]{9x^2}.$$

Do Exercises 15–18.

b   Dividing and Simplifying Radical Expressions

Note that $\dfrac{\sqrt[3]{27}}{\sqrt[3]{8}} = \dfrac{3}{2}$ and that $\sqrt[3]{\dfrac{27}{8}} = \dfrac{3}{2}$. This example suggests the following.

THE QUOTIENT RULE FOR RADICALS

For any nonnegative number a, any positive number b, and any index k,

$$\frac{\sqrt[k]{a}}{\sqrt[k]{b}} = \sqrt[k]{\frac{a}{b}}, \quad \text{or} \quad \frac{a^{1/k}}{b^{1/k}} = \left(\frac{a}{b}\right)^{1/k}.$$

(To divide, divide the radicands. After doing this, you can sometimes simplify by taking roots.)

EXAMPLES   Divide and simplify. Assume that all expressions under radicals represent positive numbers.

16. $\dfrac{\sqrt{80}}{\sqrt{5}} = \sqrt{\dfrac{80}{5}} = \sqrt{16} = 4$ [We divide the radicands.]

17. $\dfrac{5\sqrt[3]{32}}{\sqrt[3]{2}} = 5\sqrt[3]{\dfrac{32}{2}} = 5\sqrt[3]{16} = 5\sqrt[3]{8 \cdot 2} = 5\sqrt[3]{8}\,\sqrt[3]{2} = 5 \cdot 2\sqrt[3]{2} = 10\sqrt[3]{2}$

18. $\dfrac{\sqrt{72xy}}{2\sqrt{2}} = \dfrac{1}{2} \dfrac{\sqrt{72xy}}{\sqrt{2}} = \dfrac{1}{2}\sqrt{\dfrac{72xy}{2}} = \dfrac{1}{2}\sqrt{36xy} = \dfrac{1}{2}\sqrt{36}\,\sqrt{xy}$
$$= \dfrac{1}{2} \cdot 6\sqrt{xy} = 3\sqrt{xy}$$

Do Exercises 19–22.

Answers on page A-33

We can reverse the quotient rule to simplify a quotient. We simplify the root of a quotient by taking the roots of the numerator and of the denominator separately.

Simplify by taking the roots of the numerator and the denominator. Assume that all expressions under radicals represent positive numbers.

23. $\sqrt{\dfrac{25}{36}}$

---

*k*th ROOTS OF QUOTIENTS

For any nonnegative number a, any positive number b, and any index k,

$$\sqrt[k]{\dfrac{a}{b}} = \dfrac{\sqrt[k]{a}}{\sqrt[k]{b}}, \quad \text{or} \quad \left(\dfrac{a}{b}\right)^{1/k} = \dfrac{a^{1/k}}{b^{1/k}}.$$

(Take the kth roots of the numerator and of the denominator separately.)

EXAMPLES Simplify by taking the roots of the numerator and the denominator. Assume that all expressions under radicals represent positive numbers.

24. $\sqrt{\dfrac{x^2}{100}}$

19. $\sqrt[3]{\dfrac{27}{125}} = \dfrac{\sqrt[3]{27}}{\sqrt[3]{125}} = \dfrac{3}{5}$ We take the cube root of the numerator and of the denominator.

20. $\sqrt{\dfrac{25}{y^2}} = \dfrac{\sqrt{25}}{\sqrt{y^2}} = \dfrac{5}{y}$ We take the square root of the numerator and of the denominator.

21. $\sqrt{\dfrac{16x^3}{y^4}} = \dfrac{\sqrt{16x^3}}{\sqrt{y^4}} = \dfrac{\sqrt{16x^2 \cdot x}}{\sqrt{y^4}} = \dfrac{\sqrt{16x^2} \cdot \sqrt{x}}{\sqrt{y^4}} = \dfrac{4x\sqrt{x}}{y^2}$

22. $\sqrt[3]{\dfrac{27y^5}{343x^3}} = \dfrac{\sqrt[3]{27y^5}}{\sqrt[3]{343x^3}} = \dfrac{\sqrt[3]{27y^3 \cdot y^2}}{\sqrt[3]{343x^3}} = \dfrac{\sqrt[3]{27y^3} \cdot \sqrt[3]{y^2}}{\sqrt[3]{343x^3}} = \dfrac{3y\sqrt[3]{y^2}}{7x}$

25. $\sqrt[3]{\dfrac{54x^5}{125}}$

We are assuming here that no expression represents 0 or a negative number. Thus we need not be concerned about zero denominators.

Do Exercises 23–25.

When indexes differ, we can use rational exponents.

EXAMPLE 23 Divide and simplify: $\dfrac{\sqrt[3]{a^2b^4}}{\sqrt{ab}}$.

26. Divide and simplify:

$$\dfrac{\sqrt[4]{x^3y^2}}{\sqrt[3]{x^2y}}.$$

$\dfrac{\sqrt[3]{a^2b^4}}{\sqrt{ab}} = \dfrac{(a^2b^4)^{1/3}}{(ab)^{1/2}}$ Converting to exponential notation

$= \dfrac{a^{2/3}b^{4/3}}{a^{1/2}b^{1/2}}$ Using the product and power rules

$= a^{2/3-1/2}b^{4/3-1/2}$ Subtracting exponents

$= a^{4/6-3/6}b^{8/6-3/6}$ Finding common denominators so exponents can be subtracted

$= a^{1/6}b^{5/6}$

$= (ab^5)^{1/6}$ Using $a^n b^n = (ab)^n$

$= \sqrt[6]{ab^5}$ Converting back to radical notation

Do Exercise 26.

Answers on page A-33

6.3

EXERCISE SET

For Extra Help

MathXL   MyMathLab   InterAct   Math Tutor   Digital Video   Student's
                      Math   Center   Tutor CD 3   Solutions
                                             Videotape 7   Manual

a   Simplify by factoring. Assume that all expressions under radicals represent nonnegative numbers.

1. $\sqrt{24}$

2. $\sqrt{20}$

3. $\sqrt{90}$

4. $\sqrt{18}$

5. $\sqrt[3]{250}$

6. $\sqrt[3]{108}$

7. $\sqrt{180x^4}$

8. $\sqrt{175y^6}$

9. $\sqrt[3]{54x^8}$

10. $\sqrt[3]{40y^3}$

11. $\sqrt[3]{80t^8}$

12. $\sqrt[3]{108x^5}$

13. $\sqrt[4]{80}$

14. $\sqrt[4]{32}$

15. $\sqrt{32a^2b}$

16. $\sqrt{75p^3q^4}$

17. $\sqrt[4]{243x^8y^{10}}$

18. $\sqrt[4]{162c^4d^6}$

19. $\sqrt[5]{96x^7y^{15}}$

20. $\sqrt[5]{p^{14}q^9r^{23}}$

Multiply and simplify. Assume that all expressions under radicals represent nonnegative numbers.

21. $\sqrt{10}\,\sqrt{5}$

22. $\sqrt{6}\,\sqrt{3}$

23. $\sqrt{15}\,\sqrt{6}$

24. $\sqrt{2}\,\sqrt{32}$

25. $\sqrt[3]{2}\,\sqrt[3]{4}$

26. $\sqrt[3]{9}\,\sqrt[3]{3}$

27. $\sqrt{45}\,\sqrt{60}$

28. $\sqrt{24}\,\sqrt{75}$

29. $\sqrt{3x^3}\,\sqrt{6x^5}$

30. $\sqrt{5a^7}\,\sqrt{15a^3}$

31. $\sqrt{5b^3}\,\sqrt{10c^4}$

32. $\sqrt{2x^3y}\,\sqrt{12xy}$

33. $\sqrt[3]{5a^2}\ \sqrt[3]{2a}$

34. $\sqrt[3]{7x}\ \sqrt[3]{3x^2}$

35. $\sqrt[3]{y^4}\ \sqrt[3]{16y^5}$

36. $\sqrt[3]{s^2t^4}\ \sqrt[3]{s^4t^6}$

37. $\sqrt[4]{16}\ \sqrt[4]{64}$

38. $\sqrt[5]{64}\ \sqrt[5]{16}$

39. $\sqrt{12a^3b}\ \sqrt{8a^4b^2}$

40. $\sqrt{30x^3y^4}\ \sqrt{18x^2y^5}$

41. $\sqrt{2}\ \sqrt[3]{5}$

42. $\sqrt{6}\ \sqrt[3]{5}$

43. $\sqrt[4]{3}\ \sqrt{2}$

44. $\sqrt[3]{5}\ \sqrt[4]{2}$

45. $\sqrt{a}\ \sqrt[4]{a^3}$

46. $\sqrt[3]{x^2}\ \sqrt[6]{x^5}$

47. $\sqrt[5]{b^2}\ \sqrt{b^3}$

48. $\sqrt[4]{a^3}\ \sqrt[3]{a^2}$

49. $\sqrt{xy^3}\ \sqrt[3]{x^2y}$

50. $\sqrt[4]{9ab^3}\ \sqrt{3a^4b}$

b   Divide and simplify. Assume that all expressions under radicals represent positive numbers.

51. $\dfrac{\sqrt{90}}{\sqrt{5}}$

52. $\dfrac{\sqrt{98}}{\sqrt{2}}$

53. $\dfrac{\sqrt{35q}}{\sqrt{7q}}$

54. $\dfrac{\sqrt{30x}}{\sqrt{10x}}$

55. $\dfrac{\sqrt[3]{54}}{\sqrt[3]{2}}$

56. $\dfrac{\sqrt[3]{40}}{\sqrt[3]{5}}$

57. $\dfrac{\sqrt{56xy^3}}{\sqrt{8x}}$

58. $\dfrac{\sqrt{52ab^3}}{\sqrt{13a}}$

59. $\dfrac{\sqrt[3]{96a^4b^2}}{\sqrt[3]{12a^2b}}$

60. $\dfrac{\sqrt[3]{189x^5y^7}}{\sqrt[3]{7x^2y^2}}$

61. $\dfrac{\sqrt{128xy}}{2\sqrt{2}}$

62. $\dfrac{\sqrt{48ab}}{2\sqrt{3}}$

63. $\dfrac{\sqrt[4]{48x^9y^{13}}}{\sqrt[4]{3xy^5}}$

64. $\dfrac{\sqrt[5]{64a^{11}b^{28}}}{\sqrt[5]{2ab^2}}$

65. $\dfrac{\sqrt[3]{a}}{\sqrt{a}}$

66. $\dfrac{\sqrt{x}}{\sqrt[4]{x}}$

67. $\dfrac{\sqrt[3]{a^2}}{\sqrt[4]{a}}$

68. $\dfrac{\sqrt[3]{x^2}}{\sqrt[5]{x}}$

69. $\dfrac{\sqrt[4]{x^2y^3}}{\sqrt[3]{xy}}$

70. $\dfrac{\sqrt[5]{a^4b^2}}{\sqrt[3]{ab^2}}$

Simplify.

71. $\sqrt{\dfrac{25}{36}}$

72. $\sqrt{\dfrac{49}{64}}$

73. $\sqrt{\dfrac{16}{49}}$

74. $\sqrt{\dfrac{100}{81}}$

75. $\sqrt[3]{\dfrac{125}{27}}$

76. $\sqrt[3]{\dfrac{343}{1000}}$

77. $\sqrt{\dfrac{49}{y^2}}$

78. $\sqrt{\dfrac{121}{x^2}}$

79. $\sqrt{\dfrac{25y^3}{x^4}}$

80. $\sqrt{\dfrac{36a^5}{b^6}}$

81. $\sqrt[3]{\dfrac{27a^4}{8b^3}}$

82. $\sqrt[3]{\dfrac{64x^7}{216y^6}}$

CHAPTER 6: Radical Expressions,
Equations, and Functions

83. $\sqrt[4]{\dfrac{81x^4}{16}}$

84. $\sqrt[4]{\dfrac{81x^4}{y^8z^4}}$

85. $\sqrt[5]{\dfrac{32x^8}{y^{10}}}$

86. $\sqrt[5]{\dfrac{32b^{10}}{243a^{20}}}$

87. $\sqrt[6]{\dfrac{x^{13}}{y^6z^{12}}}$

88. $\sqrt[6]{\dfrac{p^9q^{24}}{r^{18}}}$

89. $\mathbf{D_W}$ Is the quotient of two irrational numbers always an irrational number? Why or why not?

90. $\mathbf{D_W}$ 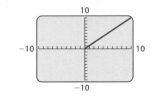 Ron is puzzled. When he uses a graphing calculator to graph $y = \sqrt{x} \cdot \sqrt{x}$, he gets the following screen. Explain why Ron did not get the complete line $y_1 = x$.

SKILL MAINTENANCE

Solve.

91. *Boating.* A paddleboat moves at a rate of 14 km/h in still water. If the river's current moves at a rate of 7 km/h, how long will it take the boat to travel 56 km downstream? 56 km upstream? [1.3b]

92. *Triangle Dimensions.* The base of a triangle is 2 in. longer than the height. The area is 12 in². Find the height and the base. [4.8b]

Solve. [5.5a]

93. $\dfrac{12x}{x-4} - \dfrac{3x^2}{x+4} = \dfrac{384}{x^2-16}$

94. $\dfrac{2}{3} + \dfrac{1}{t} = \dfrac{4}{5}$

95. $\dfrac{18}{x^2-3x} = \dfrac{2x}{x-3} - \dfrac{6}{x}$

96. $\dfrac{4x}{x+5} + \dfrac{20}{x} = \dfrac{100}{x^2+5x}$

SYNTHESIS

---

97. *Pendulums.* The **period** of a pendulum is the time it takes to complete one cycle, swinging to and fro. For a pendulum that is L centimeters long, the period T is given by the function

$$T(L) = 2\pi\sqrt{\dfrac{L}{980}},$$

where T is in seconds. Find, to the nearest hundredth of a second, the period of a pendulum of length **(a)** 65 cm; **(b)** 98 cm; **(c)** 120 cm. Use a calculator's $\boxed{\pi}$ key if possible.

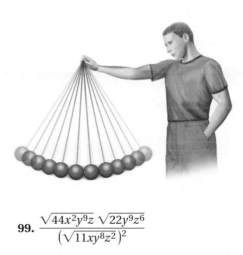

Simplify.

98. $\dfrac{\sqrt[3]{x^3 - y^3}}{\sqrt[3]{x-y}}$

99. $\dfrac{\sqrt{44x^2y^9z}\ \sqrt{22y^9z^6}}{\left(\sqrt{11xy^8z^2}\right)^2}$

100. 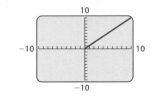 Use a graphing calculator to check your answers to Exercises 7, 12, 30, and 54.

Objectives

a   Add or subtract with radical notation and simplify.

b   Multiply expressions involving radicals in which some factors contain more than one term.

Add or subtract. Simplify by collecting like radical terms, if possible.

1. $5\sqrt{2} + 8\sqrt{2}$

2. $7\sqrt[4]{5x} + 3\sqrt[4]{5x} - \sqrt{7}$

Add or subtract. Simplify by collecting like radical terms, if possible.

3. $7\sqrt{45} - 2\sqrt{5}$

4. $3\sqrt[3]{y^5} + 4\sqrt[3]{y^2} + \sqrt[3]{8y^6}$

5. $\sqrt{25x - 25} - \sqrt{9x - 9}$

Answers on page A-33

---

a   Addition and Subtraction

Any two real numbers can be added. For example, the sum of 7 and $\sqrt{3}$ can be expressed as $7 + \sqrt{3}$. We cannot simplify this sum. However, when we have **like radicals** (radicals having the same index and radicand), we can use the distributive laws to simplify by collecting like radical terms. For example,

$$7\sqrt{3} + \sqrt{3} = 7\sqrt{3} + 1 \cdot \sqrt{3} = (7 + 1)\sqrt{3} = 8\sqrt{3}.$$

EXAMPLES   Add or subtract. Simplify by collecting like radical terms, if possible.

1. $6\sqrt{7} + 4\sqrt{7} = (6 + 4)\sqrt{7}$ Using a distributive law (factoring out $\sqrt{7}$)

$\qquad\qquad\qquad = 10\sqrt{7}$

2. $8\sqrt[3]{2} - 7x\sqrt[3]{2} + 5\sqrt[3]{2} = (8 - 7x + 5)\sqrt[3]{2}$ Factoring out $\sqrt[3]{2}$

$\qquad\qquad\qquad\qquad\qquad = (13 - 7x)\sqrt[3]{2}$

> These parentheses *are* necessary!

3. $6\sqrt[5]{4x} + 4\sqrt[5]{4x} - \sqrt[3]{4x} = (6 + 4)\sqrt[5]{4x} - \sqrt[3]{4x}$

$\qquad\qquad\qquad\qquad\qquad = 10\sqrt[5]{4x} - \sqrt[3]{4x}$

> Note that these expressions have the same *radicand,* but they are not like radicals because they do not have the same *index.*

Do Exercises 1 and 2.

Sometimes we need to simplify radicals by factoring in order to obtain terms with like radicals.

EXAMPLES   Add or subtract. Simplify by collecting like radical terms, if possible.

4. $3\sqrt{8} - 5\sqrt{2} = 3(\sqrt{4 \cdot 2}) - 5\sqrt{2}$ Factoring 8

$\qquad\qquad = 3\sqrt{4} \cdot \sqrt{2} - 5\sqrt{2}$ Factoring $\sqrt{4 \cdot 2}$ into two radicals

$\qquad\qquad = 3 \cdot 2\sqrt{2} - 5\sqrt{2}$ Taking the square root of 4

$\qquad\qquad = 6\sqrt{2} - 5\sqrt{2}$

$\qquad\qquad = (6 - 5)\sqrt{2}$ Collecting like radical terms

$\qquad\qquad = \sqrt{2}$

5. $5\sqrt{2} - 4\sqrt{3}$ No simplification possible

6. $5\sqrt[3]{16y^4} + 7\sqrt[3]{2y} = 5\sqrt[3]{8y^3 \cdot 2y} + 7\sqrt[3]{2y}$ ⎫
   Factoring the first

$\qquad\qquad\qquad\quad = 5\sqrt[3]{8y^3} \cdot \sqrt[3]{2y} + 7\sqrt[3]{2y}$ ⎬   radical

$\qquad\qquad\qquad\quad = 5 \cdot 2y \cdot \sqrt[3]{2y} + 7\sqrt[3]{2y}$ Taking the cube root of $8y^3$

$\qquad\qquad\qquad\quad = 10y\sqrt[3]{2y} + 7\sqrt[3]{2y}$

$\qquad\qquad\qquad\quad = (10y + 7)\sqrt[3]{2y}$ Collecting like radical terms

Do Exercises 3–5.

b More Multiplication

To multiply expressions in which some factors contain more than one term, we use the procedures for multiplying polynomials.

EXAMPLES Multiply.

7. $\sqrt{3}(x - \sqrt{5}) = \sqrt{3} \cdot x - \sqrt{3} \cdot \sqrt{5}$ Using a distributive law

$\qquad = x\sqrt{3} - \sqrt{15}$ Multiplying radicals

8. $\sqrt[3]{y}(\sqrt[3]{y^2} + \sqrt[3]{2}) = \sqrt[3]{y} \cdot \sqrt[3]{y^2} + \sqrt[3]{y} \cdot \sqrt[3]{2}$ Using a distributive law

$\qquad = \sqrt[3]{y^3} + \sqrt[3]{2y}$ Multiplying radicals

$\qquad = y + \sqrt[3]{2y}$ Simplifying $\sqrt[3]{y^3}$

Do Exercises 6 and 7.

EXAMPLE 9 Multiply: $(4\sqrt{3} + \sqrt{2})(\sqrt{3} - 5\sqrt{2})$.

$$\qquad\qquad\quad \overset{F}{} \quad \overset{O}{} \quad \overset{I}{} \quad \overset{L}{}$$

$$(4\sqrt{3} + \sqrt{2})(\sqrt{3} - 5\sqrt{2}) = 4(\sqrt{3})^2 - 20\sqrt{3} \cdot \sqrt{2} + \sqrt{2} \cdot \sqrt{3} - 5(\sqrt{2})^2$$

$$= 4 \cdot 3 - 20\sqrt{6} + \sqrt{6} - 5 \cdot 2$$

$$= 12 - 20\sqrt{6} + \sqrt{6} - 10$$

$$= 2 - 19\sqrt{6} \quad \text{Collecting like terms}$$

EXAMPLE 10 Multiply: $(\sqrt{a} + \sqrt{3})(\sqrt{b} + \sqrt{3})$. Assume that all expressions under radicals represent nonnegative numbers.

$$(\sqrt{a} + \sqrt{3})(\sqrt{b} + \sqrt{3}) = \sqrt{a}\sqrt{b} + \sqrt{a}\sqrt{3} + \sqrt{3}\sqrt{b} + \sqrt{3}\sqrt{3}$$

$$= \sqrt{ab} + \sqrt{3a} + \sqrt{3b} + 3$$

EXAMPLE 11 Multiply: $(\sqrt{5} + \sqrt{7})(\sqrt{5} - \sqrt{7})$.

$$(\sqrt{5} + \sqrt{7})(\sqrt{5} - \sqrt{7}) = (\sqrt{5})^2 - (\sqrt{7})^2 \quad \begin{array}{l}\text{This is now a difference} \\ \text{of two squares:} \\ (A - B)(A + B) = A^2 - B^2.\end{array}$$

$$= 5 - 7 = -2$$

EXAMPLE 12 Multiply: $(\sqrt{a} + \sqrt{b})(\sqrt{a} - \sqrt{b})$. Assume that all expressions under radicals represent nonnegative numbers.

$$(\sqrt{a} + \sqrt{b})(\sqrt{a} - \sqrt{b}) = (\sqrt{a})^2 - (\sqrt{b})^2$$

$$= a - b \longleftarrow \boxed{\text{No radicals}}$$

Expressions of the form $\sqrt{a} + \sqrt{b}$ and $\sqrt{a} - \sqrt{b}$ are called **conjugates.** Their product is always an expression that has no radicals.

Do Exercises 8–11.

EXAMPLE 13 Multiply: $(\sqrt{3} + x)^2$.

$$(\sqrt{3} + x)^2 = (\sqrt{3})^2 + 2x\sqrt{3} + x^2 \quad \text{Squaring a binomial}$$

$$= 3 + 2x\sqrt{3} + x^2$$

Do Exercises 12 and 13.

Multiply. Assume that all expressions under radicals represent nonnegative numbers.

6. $\sqrt{2}(5\sqrt{3} + 3\sqrt{7})$

7. $\sqrt[3]{a^2}(\sqrt[3]{3a} - \sqrt[3]{2})$

Multiply. Assume that all expressions under radicals represent nonnegative numbers.

8. $(\sqrt{3} - 5\sqrt{2})(2\sqrt{3} + \sqrt{2})$

9. $(\sqrt{a} + 2\sqrt{3})(3\sqrt{b} - 4\sqrt{3})$

10. $(\sqrt{2} + \sqrt{5})(\sqrt{2} - \sqrt{5})$

11. $(\sqrt{p} - \sqrt{q})(\sqrt{p} + \sqrt{q})$

Multiply.

12. $(2\sqrt{5} - y)^2$

13. $(3\sqrt{6} + 2)^2$

Answer on page A-33

MathXL MyMathLab InterAct Math Math Tutor Center Digital Video Tutor CD 3 Videotape 7 Student's Solutions Manual

a Add or subtract. Then simplify by collecting like radical terms, if possible. Assume that all expressions under radicals represent nonnegative numbers.

1. $7\sqrt{5} + 4\sqrt{5}$

2. $2\sqrt{3} + 9\sqrt{3}$

3. $6\sqrt[3]{7} - 5\sqrt[3]{7}$

4. $13\sqrt[5]{3} - 8\sqrt[5]{3}$

5. $4\sqrt[3]{y} + 9\sqrt[3]{y}$

6. $6\sqrt[4]{t} - 3\sqrt[4]{t}$

7. $5\sqrt{6} - 9\sqrt{6} - 4\sqrt{6}$

8. $3\sqrt{10} - 8\sqrt{10} + 7\sqrt{10}$

9. $4\sqrt[3]{3} - \sqrt{5} + 2\sqrt[3]{3} + \sqrt{5}$

10. $5\sqrt{7} - 8\sqrt[4]{11} + \sqrt{7} + 9\sqrt[4]{11}$

11. $8\sqrt{27} - 3\sqrt{3}$

12. $9\sqrt{50} - 4\sqrt{2}$

13. $8\sqrt{45} + 7\sqrt{20}$

14. $9\sqrt{12} + 16\sqrt{27}$

15. $18\sqrt{72} + 2\sqrt{98}$

16. $12\sqrt{45} - 8\sqrt{80}$

17. $3\sqrt[3]{16} + \sqrt[3]{54}$

18. $\sqrt[3]{27} - 5\sqrt[3]{8}$

19. $2\sqrt{128} - \sqrt{18} + 4\sqrt{32}$

20. $5\sqrt{50} - 2\sqrt{18} + 9\sqrt{32}$

21. $\sqrt{5a} + 2\sqrt{45a^3}$

22. $4\sqrt{3x^3} - \sqrt{12x}$

23. $\sqrt[3]{24x} - \sqrt[3]{3x^4}$

24. $\sqrt[3]{54x} - \sqrt[3]{2x^4}$

25. $7\sqrt{27x^3} + \sqrt{3x}$

26. $2\sqrt{45x^3} - \sqrt{5x}$

27. $\sqrt{4} + \sqrt{18}$

28. $\sqrt[3]{8} - \sqrt[3]{24}$

29. $5\sqrt[3]{32} - \sqrt[3]{108} + 2\sqrt[3]{256}$

30. $3\sqrt[3]{8x} - 4\sqrt[3]{27x} + 2\sqrt[3]{64x}$

31. $\sqrt[3]{6x^4} + \sqrt[3]{48x} - \sqrt[3]{6x}$

32. $\sqrt[4]{80x^5} - \sqrt[4]{405x^9} + \sqrt[4]{5x}$

33. $\sqrt{4a - 4} + \sqrt{a - 1}$

34. $\sqrt{9y + 27} + \sqrt{y + 3}$

35. $\sqrt{x^3 - x^2} + \sqrt{9x - 9}$

36. $\sqrt{4x - 4} + \sqrt{x^3 - x^2}$

b   Multiply.

37. $\sqrt{5}\left(4 - 2\sqrt{5}\right)$

38. $\sqrt{6}\left(2 + \sqrt{6}\right)$

39. $\sqrt{3}\left(\sqrt{2} - \sqrt{7}\right)$

40. $\sqrt{2}\left(\sqrt{5} - \sqrt{2}\right)$

41. $\sqrt{3}\left(-4\sqrt{3} + 6\right)$

42. $\sqrt{2}\left(-5\sqrt{2} - 7\right)$

43. $\sqrt{3}\left(2\sqrt{5} - 3\sqrt{4}\right)$

44. $\sqrt{2}\left(3\sqrt{10} - 2\sqrt{2}\right)$

45. $\sqrt[3]{2}\left(\sqrt[3]{4} - 2\sqrt[3]{32}\right)$

46. $\sqrt[3]{3}\left(\sqrt[3]{9} - 4\sqrt[3]{21}\right)$

47. $\sqrt[3]{a}\left(\sqrt[3]{2a^2} + \sqrt[3]{16a^2}\right)$

48. $\sqrt[3]{x}\left(\sqrt[3]{3x^2} - \sqrt[3]{81x^2}\right)$

49. $\left(\sqrt{3} - \sqrt{2}\right)\left(\sqrt{3} + \sqrt{2}\right)$

50. $\left(\sqrt{5} + \sqrt{6}\right)\left(\sqrt{5} - \sqrt{6}\right)$

51. $\left(\sqrt{8} + 2\sqrt{5}\right)\left(\sqrt{8} - 2\sqrt{5}\right)$

52. $\left(\sqrt{18} + 3\sqrt{7}\right)\left(\sqrt{18} - 3\sqrt{7}\right)$

53. $\left(7 + \sqrt{5}\right)\left(7 - \sqrt{5}\right)$

54. $\left(4 - \sqrt{3}\right)\left(4 + \sqrt{3}\right)$

55. $\left(2 - \sqrt{3}\right)\left(2 + \sqrt{3}\right)$

56. $\left(11 - \sqrt{2}\right)\left(11 + \sqrt{2}\right)$

57. $\left(\sqrt{8} + \sqrt{5}\right)\left(\sqrt{8} - \sqrt{5}\right)$

58. $\left(\sqrt{6} - \sqrt{7}\right)\left(\sqrt{6} + \sqrt{7}\right)$

59. $\left(3 + 2\sqrt{7}\right)\left(3 - 2\sqrt{7}\right)$

60. $\left(6 - 3\sqrt{2}\right)\left(6 + 3\sqrt{2}\right)$

For the following exercises, assume that all expressions under radicals represent nonnegative numbers.

61. $\left(\sqrt{a} + \sqrt{b}\right)\left(\sqrt{a} - \sqrt{b}\right)$

62. $\left(\sqrt{x} - \sqrt{y}\right)\left(\sqrt{x} + \sqrt{y}\right)$

63. $\left(3 - \sqrt{5}\right)\left(2 + \sqrt{5}\right)$

64. $\left(2 + \sqrt{6}\right)\left(4 - \sqrt{6}\right)$

65. $\left(\sqrt{3} + 1\right)\left(2\sqrt{3} + 1\right)$

66. $\left(4\sqrt{3} + 5\right)\left(\sqrt{3} - 2\right)$

67. $\left(2\sqrt{7} - 4\sqrt{2}\right)\left(3\sqrt{7} + 6\sqrt{2}\right)$

68. $\left(4\sqrt{5} + 3\sqrt{3}\right)\left(3\sqrt{5} - 4\sqrt{3}\right)$

69. $\left(\sqrt{a} + \sqrt{2}\right)\left(\sqrt{a} + \sqrt{3}\right)$

70. $\left(2 - \sqrt{x}\right)\left(1 - \sqrt{x}\right)$

71. $\left(2\sqrt[3]{3} + \sqrt[3]{2}\right)\left(\sqrt[3]{3} - 2\sqrt[3]{2}\right)$

72. $\left(3\sqrt[4]{7} + \sqrt[4]{6}\right)\left(2\sqrt[4]{9} - 3\sqrt[4]{6}\right)$

73. $\left(2 + \sqrt{3}\right)^2$

74. $\left(\sqrt{5} + 1\right)^2$

75. $\left(\sqrt[5]{9} - \sqrt[5]{3}\right)\left(\sqrt[5]{8} + \sqrt[5]{27}\right)$

76. $\left(\sqrt[3]{8x} - \sqrt[3]{5y}\right)^2$

77. $\mathbf{D_W}$ Why do we need to know how to simplify radical expressions before we learn to add them?

78. $\mathbf{D_W}$ In what way(s) is collecting like radical terms the same as collecting like monomial terms?

SKILL MAINTENANCE

Multiply or divide and simplify. [5.1d, e]

79. $\dfrac{x^3 + 4x}{x^2 - 16} \div \dfrac{x^2 + 8x + 15}{x^2 + x - 20}$

80. $\dfrac{a^2 - 4}{a} \div \dfrac{a - 2}{a + 4}$

81. $\dfrac{a^3 + 8}{a^2 - 4} \cdot \dfrac{a^2 - 4a + 4}{a^2 - 2a + 4}$

82. $\dfrac{y^3 - 27}{y^2 - 9} \cdot \dfrac{y^2 - 6y + 9}{y^2 + 3y + 9}$

Simplify. [5.4a]

83. $\dfrac{x - \dfrac{1}{3}}{x + \dfrac{1}{4}}$

84. $\dfrac{1 - \dfrac{1}{x}}{1 - \dfrac{1}{x^2}}$

85. $\dfrac{\dfrac{1}{p} - \dfrac{1}{q}}{\dfrac{1}{p^2} - \dfrac{1}{q^2}}$

86. $\dfrac{\dfrac{1}{a} + \dfrac{1}{b}}{\dfrac{1}{a^3} + \dfrac{1}{b^3}}$

Solve. [1.6c, d, e]

87. $|3x + 7| = 22$

88. $|3x + 7| < 22$

89. $|3x + 7| \geq 22$

90. $|3x + 7| = |2x - 5|$

SYNTHESIS

91. 📈 Graph the function $f(x) = \sqrt{(x - 2)^2}$. What is the domain?

92. 📈 Use a graphing calculator to check your answers to Exercises 5, 22, and 70.

Multiply and simplify.

93. $\sqrt{9 + 3\sqrt{5}}\,\sqrt{9 - 3\sqrt{5}}$

94. $\left(\sqrt{x + 2} - \sqrt{x - 2}\right)^2$

95. $\left(\sqrt{3} + \sqrt{5} - \sqrt{6}\right)^2$

96. $\sqrt[3]{y}\left(1 - \sqrt[3]{y}\right)\left(1 + \sqrt[3]{y}\right)$

97. $\left(\sqrt[3]{9} - 2\right)\left(\sqrt[3]{9} + 4\right)$

98. $\left[\sqrt{3} + \sqrt{2} + \sqrt{1}\right]^4$

529

Objectives

Rationalize the denominator of a radical expression having one term in the denominator.

Rationalize the denominator of a radical expression having two terms in the denominator.

1. Rationalize the denominator:

$$\sqrt{\frac{2}{5}}.$$

2. Rationalize the denominator:

$$\sqrt[3]{\frac{5}{4}}.$$

6.5 MORE ON DIVISION OF RADICAL EXPRESSIONS

 Rationalizing Denominators

Sometimes in mathematics it is useful to find an equivalent expression without a radical in the denominator. This provides a standard notation for expressing results. The procedure for finding such an expression is called **rationalizing the denominator.** We carry this out by multiplying by 1.

EXAMPLE 1 Rationalize the denominator: $\sqrt{\dfrac{7}{3}}$.

We multiply by 1, using $\sqrt{3}/\sqrt{3}$. We do this so that the denominator of the radicand will be a perfect square.

$$\sqrt{\frac{7}{3}} = \frac{\sqrt{7}}{\sqrt{3}} \cdot \frac{\sqrt{3}}{\sqrt{3}}$$

$$= \frac{\sqrt{7} \cdot \sqrt{3}}{\sqrt{3} \cdot \sqrt{3}}$$

$$= \frac{\sqrt{21}}{\sqrt{3^2}} = \frac{\sqrt{21}}{3}$$

The radicand is a perfect square.

Do Exercise 1.

EXAMPLE 2 Rationalize the denominator: $\sqrt[3]{\dfrac{7}{25}}$.

We first factor the denominator:

$$\sqrt[3]{\frac{7}{25}} = \sqrt[3]{\frac{7}{5 \cdot 5}}.$$

To get a perfect cube in the denominator, we consider the index 3 and the factors. We have 2 factors of 5, and we need 3 factors of 5. We achieve this by multiplying by 1, using $\sqrt[3]{5}/\sqrt[3]{5}$.

$$\sqrt[3]{\frac{7}{25}} = \sqrt[3]{\frac{7}{5 \cdot 5}} \cdot \frac{\sqrt[3]{5}}{\sqrt[3]{5}}$$

Multiplying by $\dfrac{\sqrt[3]{5}}{\sqrt[3]{5}}$ to make the denominator of the radicand a perfect cube

$$= \frac{\sqrt[3]{7} \cdot \sqrt[3]{5}}{\sqrt[3]{5 \cdot 5} \cdot \sqrt[3]{5}}$$

$$= \frac{\sqrt[3]{35}}{\sqrt[3]{5^3}}$$

The radicand is a perfect cube.

$$= \frac{\sqrt[3]{35}}{5}$$

Do Exercise 2.

EXAMPLE 3 Rationalize the denominator: $\sqrt{\dfrac{2a}{5b}}$. Assume that all expressions under radicals represent positive numbers.

$$\sqrt{\frac{2a}{5b}} = \frac{\sqrt{2a}}{\sqrt{5b}} \qquad \text{Converting to a quotient of radicals}$$

$$= \frac{\sqrt{2a}}{\sqrt{5b}} \cdot \frac{\sqrt{5b}}{\sqrt{5b}} \qquad \text{Multiplying by 1}$$

$$= \frac{\sqrt{10ab}}{\sqrt{5^2 b^2}} \qquad \begin{array}{l}\text{The radicand in the denominator}\\ \text{is a perfect square.}\end{array}$$

$$= \frac{\sqrt{10ab}}{5b}$$

Do Exercise 3.

EXAMPLE 4 Rationalize the denominator: $\dfrac{\sqrt[3]{a}}{\sqrt[3]{9x}}$.

We factor the denominator:

$$\frac{\sqrt[3]{a}}{\sqrt[3]{9x}} = \frac{\sqrt[3]{a}}{\sqrt[3]{3 \cdot 3 \cdot x}}.$$

To choose the symbol for 1, we look at $3 \cdot 3 \cdot x$. To make it a cube, we need another 3 and two more x's. Thus we multiply by 1, using $\sqrt[3]{3x^2}/\sqrt[3]{3x^2}$:

$$\frac{\sqrt[3]{a}}{\sqrt[3]{9x}} = \frac{\sqrt[3]{a}}{\sqrt[3]{3 \cdot 3 \cdot x}} \cdot \frac{\sqrt[3]{3x^2}}{\sqrt[3]{3x^2}} \qquad \text{Multiplying by 1}$$

$$= \frac{\sqrt[3]{3ax^2}}{\sqrt[3]{3^3 x^3}} \qquad \begin{array}{l}\text{The radicand in the denominator}\\ \text{is a perfect cube.}\end{array}$$

$$= \frac{\sqrt[3]{3ax^2}}{3x}.$$

Do Exercises 4 and 5.

EXAMPLE 5 Rationalize the denominator: $\dfrac{3x}{\sqrt[5]{2x^2y^3}}$.

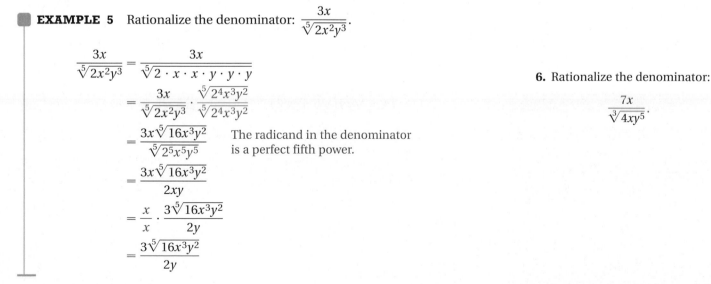

$$\frac{3x}{\sqrt[5]{2x^2y^3}} = \frac{3x}{\sqrt[5]{2 \cdot x \cdot x \cdot y \cdot y \cdot y}}$$

$$= \frac{3x}{\sqrt[5]{2x^2y^3}} \cdot \frac{\sqrt[5]{2^4 x^3 y^2}}{\sqrt[5]{2^4 x^3 y^2}}$$

$$= \frac{3x\sqrt[5]{16x^3y^2}}{\sqrt[5]{2^5 x^5 y^5}} \qquad \begin{array}{l}\text{The radicand in the denominator}\\ \text{is a perfect fifth power.}\end{array}$$

$$= \frac{3x\sqrt[5]{16x^3y^2}}{2xy}$$

$$= \frac{x}{x} \cdot \frac{3\sqrt[5]{16x^3y^2}}{2y}$$

$$= \frac{3\sqrt[5]{16x^3y^2}}{2y}$$

Do Exercise 6.

3. Rationalize the denominator:
$$\sqrt{\frac{4a}{3b}}.$$

Rationalize the denominator.

4. $\dfrac{\sqrt[4]{7}}{\sqrt[4]{2}}$

5. $\sqrt[3]{\dfrac{3x^5}{2y}}$

6. Rationalize the denominator:
$$\frac{7x}{\sqrt[3]{4xy^5}}.$$

Answers on page A-33

Multiply.

7. $(c - \sqrt{b})(c + \sqrt{b})$

8. $(\sqrt{a} + \sqrt{b})(\sqrt{a} - \sqrt{b})$

What symbol for 1 would you use to rationalize the denominator?

9. $\dfrac{\sqrt{5} + 1}{\sqrt{3} - y}$

10. $\dfrac{1}{\sqrt{2} + \sqrt{3}}$

Rationalize the denominator.

11. $\dfrac{14}{3 + \sqrt{2}}$

12. $\dfrac{5 + \sqrt{2}}{1 - \sqrt{2}}$

b   Rationalizing When There Are Two Terms

Do Exercises 7 and 8.

Certain pairs of expressions containing square roots, such as $c - \sqrt{b}$, $c + \sqrt{b}$ and $\sqrt{a} - \sqrt{b}$, $\sqrt{a} + \sqrt{b}$, are called **conjugates.** The product of such a pair of conjugates has no radicals in it. (See Example 12 of Section 6.4.) Thus when we wish to rationalize a denominator that has two terms and one or more of them involves a square-root radical, we multiply by 1 using the conjugate of the denominator to write a symbol for 1.

EXAMPLES What symbol for 1 would you use to rationalize the denominator?

Expression   *Symbol for 1*

6. $\dfrac{3}{x + \sqrt{7}}$ $\dfrac{x - \sqrt{7}}{x - \sqrt{7}}$

> Change the operation sign in the denominator to obtain the conjugate. Use the conjugate for the numerator and denominator of the symbol for 1.

7. $\dfrac{\sqrt{7} + 4}{3 - 2\sqrt{5}}$ $\dfrac{3 + 2\sqrt{5}}{3 + 2\sqrt{5}}$

Do Exercises 9 and 10.

EXAMPLE 8 Rationalize the denominator: $\dfrac{4}{\sqrt{3} + x}$.

$$\dfrac{4}{\sqrt{3} + x} = \dfrac{4}{\sqrt{3} + x} \cdot \dfrac{\sqrt{3} - x}{\sqrt{3} - x}$$

$$= \dfrac{4(\sqrt{3} - x)}{(\sqrt{3} + x)(\sqrt{3} - x)}$$

$$= \dfrac{4\sqrt{3} - 4x}{3 - x^2}$$

EXAMPLE 9 Rationalize the denominator: $\dfrac{4 + \sqrt{2}}{\sqrt{5} - \sqrt{2}}$.

$$\dfrac{4 + \sqrt{2}}{\sqrt{5} - \sqrt{2}} = \dfrac{4 + \sqrt{2}}{\sqrt{5} - \sqrt{2}} \cdot \dfrac{\sqrt{5} + \sqrt{2}}{\sqrt{5} + \sqrt{2}}$$ Multiplying by 1, using the conjugate of $\sqrt{5} - \sqrt{2}$, which is $\sqrt{5} + \sqrt{2}$

$$= \dfrac{(4 + \sqrt{2})(\sqrt{5} + \sqrt{2})}{(\sqrt{5} - \sqrt{2})(\sqrt{5} + \sqrt{2})}$$ Multiplying numerators and denominators

$$= \dfrac{4\sqrt{5} + 4\sqrt{2} + \sqrt{2}\,\sqrt{5} + (\sqrt{2})^2}{(\sqrt{5})^2 - (\sqrt{2})^2}$$ Using $(A - B)(A + B) = A^2 - B^2$ in the denominator

$$= \dfrac{4\sqrt{5} + 4\sqrt{2} + \sqrt{10} + 2}{5 - 2}$$

$$= \dfrac{4\sqrt{5} + 4\sqrt{2} + \sqrt{10} + 2}{3}$$

Do Exercises 11 and 12.

MathXL MyMathLab InterAct Math Math Tutor Center Digital Video Tutor CD 3 Videotape 7 Student's Solutions Manual

a Rationalize the denominator. Assume that all expressions under radicals represent positive numbers.

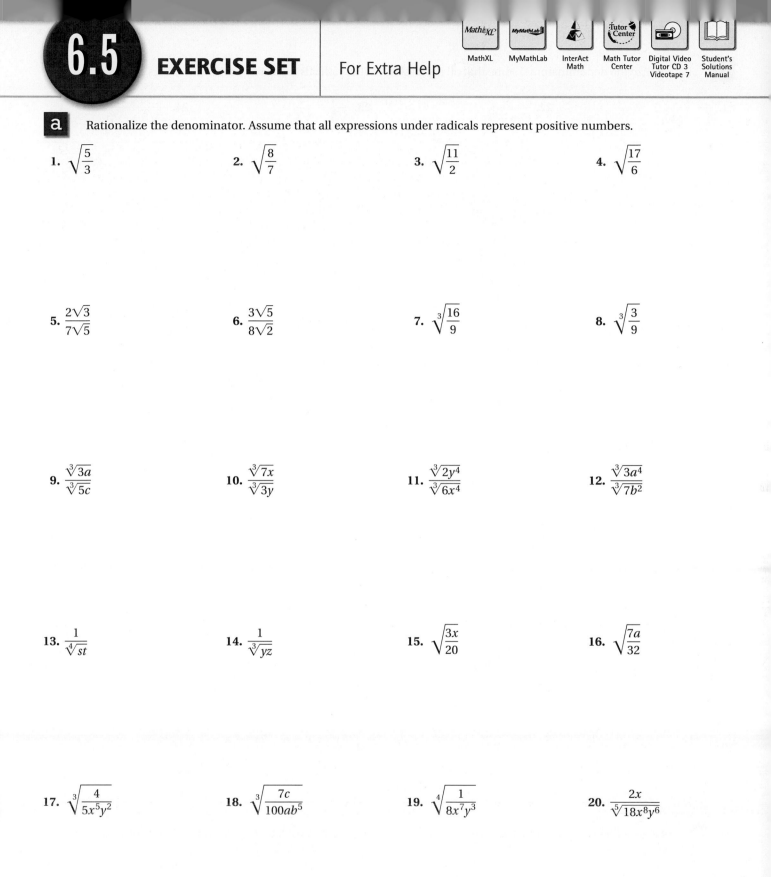

1. $\sqrt{\dfrac{5}{3}}$

2. $\sqrt{\dfrac{8}{7}}$

3. $\sqrt{\dfrac{11}{2}}$

4. $\sqrt{\dfrac{17}{6}}$

5. $\dfrac{2\sqrt{3}}{7\sqrt{5}}$

6. $\dfrac{3\sqrt{5}}{8\sqrt{2}}$

7. $\sqrt[3]{\dfrac{16}{9}}$

8. $\sqrt[3]{\dfrac{3}{9}}$

9. $\dfrac{\sqrt[3]{3a}}{\sqrt[3]{5c}}$

10. $\dfrac{\sqrt[3]{7x}}{\sqrt[3]{3y}}$

11. $\dfrac{\sqrt[3]{2y^4}}{\sqrt[3]{6x^4}}$

12. $\dfrac{\sqrt[3]{3a^4}}{\sqrt[3]{7b^2}}$

13. $\dfrac{1}{\sqrt[4]{st}}$

14. $\dfrac{1}{\sqrt[3]{yz}}$

15. $\sqrt{\dfrac{3x}{20}}$

16. $\sqrt{\dfrac{7a}{32}}$

17. $\sqrt[3]{\dfrac{4}{5x^5y^2}}$

18. $\sqrt[3]{\dfrac{7c}{100ab^5}}$

19. $\sqrt[4]{\dfrac{1}{8x^7y^3}}$

20. $\dfrac{2x}{\sqrt[5]{18x^8y^6}}$

b   Rationalize the denominator. Assume that all expressions under radicals represent positive numbers.

21. $\dfrac{9}{6 - \sqrt{10}}$

22. $\dfrac{3}{8 + \sqrt{5}}$

23. $\dfrac{-4\sqrt{7}}{\sqrt{5} - \sqrt{3}}$

24. $\dfrac{34\sqrt{5}}{2\sqrt{5} - \sqrt{3}}$

25. $\dfrac{\sqrt{5} - 2\sqrt{6}}{\sqrt{3} - 4\sqrt{5}}$

26. $\dfrac{\sqrt{6} - 3\sqrt{5}}{\sqrt{3} - 2\sqrt{7}}$

27. $\dfrac{2 - \sqrt{a}}{3 + \sqrt{a}}$

28. $\dfrac{5 + \sqrt{x}}{8 - \sqrt{x}}$

29. $\dfrac{5\sqrt{3} - 3\sqrt{2}}{3\sqrt{2} - 2\sqrt{3}}$

30. $\dfrac{7\sqrt{2} + 4\sqrt{3}}{4\sqrt{3} - 3\sqrt{2}}$

31. $\dfrac{\sqrt{x} - \sqrt{y}}{\sqrt{x} + \sqrt{y}}$

32. $\dfrac{\sqrt{a} + \sqrt{b}}{\sqrt{a} - \sqrt{b}}$

33. $\mathbf{D_W}$ A student *incorrectly* claims that

$$\frac{5 + \sqrt{2}}{\sqrt{18}} = \frac{5 + \sqrt{1}}{\sqrt{9}} = \frac{5 + 1}{3} = 2.$$

How could you convince the student that a mistake has been made? How would you explain the correct way of rationalizing the denominator?

34. $\mathbf{D_W}$ A student considers the radical expression

$$\frac{11}{\sqrt[3]{4} - \sqrt[3]{5}}$$

and tries to rationalize the denominator by multiplying by

$$\frac{\sqrt[3]{4} + \sqrt[3]{5}}{\sqrt[3]{4} + \sqrt[3]{5}}.$$

Discuss the difficulties of such a plan.

SKILL MAINTENANCE

Solve.   [5.5a]

35. $\dfrac{1}{2} - \dfrac{1}{3} = \dfrac{5}{t}$

36. $\dfrac{5}{x - 1} + \dfrac{9}{x^2 + x + 1} = \dfrac{15}{x^3 - 1}$

Divide and simplify.   [5.1e]

37. $\dfrac{1}{x^3 - y^3} \div \dfrac{1}{(x - y)(x^2 + xy + y^2)}$

38. $\dfrac{2x^2 - x - 6}{x^2 + 4x + 3} \div \dfrac{2x^2 + x - 3}{x^2 - 1}$

SYNTHESIS

39. ▨ Use a graphing calculator to check your answers to Exercises 15 and 16.

40. Express each of the following as the product of two radical expressions.

    a) $x - 5$ **b)** $x - a$

Simplify. (*Hint*: Rationalize the denominator.)

41. $\sqrt{a^2 - 3} - \dfrac{a^2}{\sqrt{a^2 - 3}}$

42. $\dfrac{1}{4 + \sqrt{3}} + \dfrac{1}{\sqrt{3}} + \dfrac{1}{\sqrt{3} - 4}$

534

CHAPTER 6: Radical Expressions,
Equations, and Functions

Copyright © 2007 Pearson Education, Inc.

6.6

SOLVING RADICAL EQUATIONS

Objectives

a | Solve radical equations with one radical term.

b | Solve radical equations with two radical terms.

c | Solve applied problems involving radical equations.

a | The Principle of Powers

A **radical equation** has variables in one or more radicands—for example,

$$\sqrt[3]{2x} + 1 = 5, \qquad \sqrt{x} + \sqrt{4x - 2} = 7.$$

To solve such an equation, we need a new equation-solving principle. Suppose that an equation $a = b$ is true. If we square both sides, we get another true equation: $a^2 = b^2$. This can be generalized.

THE PRINCIPLE OF POWERS

For any natural number n, if an equation $a = b$ is true, then $a^n = b^n$ is true.

However, if an equation $a^n = b^n$ is true, it *may not* be true that $a = b$, if n is even. For example, $3^2 = (-3)^2$ is true, but $3 = -3$ is not true. Thus we *must* make a check when we solve an equation using the principle of powers.

EXAMPLE 1 Solve: $\sqrt{x} - 3 = 4$.

We have

$$\sqrt{x} - 3 = 4$$
$$\sqrt{x} = 7 \qquad \text{Adding to isolate the radical}$$
$$(\sqrt{x})^2 = 7^2 \qquad \text{Using the principle of powers (squaring)}$$
$$x = 49. \qquad \sqrt{x} \cdot \sqrt{x} = x$$

The number 49 is a possible solution. But we *must* make a check in order to be sure!

Check: $\qquad \dfrac{\sqrt{x} - 3 = 4}{\sqrt{49} - 3 \;?\; 4}$
$\qquad\qquad\quad 7 - 3 \;\Big|$
$\qquad\qquad\qquad\quad 4 \;\Big|\;$ TRUE

The solution is 49.

Caution!

The principle of powers does not always give equivalent equations. For this reason, a check is a must!

Study Tips

BEGINNING TO STUDY FOR THE FINAL EXAM (PART 1)

It is never too soon to begin to study for the final examination. Take a few minutes each week to review the highlighted information, such as formulas, properties, and procedures. Make special use of the Summary and Reviews, Chapter Tests, and Cumulative Reviews, as well as the supplements such as the Work It Out! Chapter Test Video on CD, the Interact Math Tutorial Web site, and MathXL. The Cumulative Review/Final Examination for Chapters R–9 is a sample final exam.

"Practice does not make perfect; practice makes permanent."

Dr. Richard Chase, former president, Wheaton College

Solve.

1. $\sqrt{x} - 7 = 3$

2. $\sqrt{x} = -2$

Solve.

3. $x + 2 = \sqrt{2x + 7}$

4. $x + 1 = 3\sqrt{x - 1}$

Answers on page A-34

536

CHAPTER 6: Radical Expressions,
Equations, and Functions

EXAMPLE 2 Solve: $\sqrt{x} = -3$.

We might observe at the outset that this equation has no solution because the principal square root of a number is never negative. Let's continue as above for comparison.

$$\sqrt{x} = -3$$
$$(\sqrt{x})^2 = (-3)^2$$
$$x = 9$$

Check: $\sqrt{x} = -3$
$\sqrt{9} \ ? \ -3$
$3 \ | \quad$ FALSE

The number 9 does *not* check. Thus the equation $\sqrt{x} = -3$ has no real-number solution. Note that the equation $x = 9$ has solution 9, but that $\sqrt{x} = -3$ has *no* solution. Thus the equations $x = 9$ and $\sqrt{x} = -3$ are *not* equivalent. That is, $\sqrt{9} \neq -3$.

Do Exercises 1 and 2.

To solve an equation with a radical term, we first isolate the radical term on one side of the equation. Then we use the principle of powers.

EXAMPLE 3 Solve: $x - 7 = 2\sqrt{x + 1}$.

The radical term is already isolated. We proceed with the principle of powers:

$$x - 7 = 2\sqrt{x + 1}$$
$$(x - 7)^2 = (2\sqrt{x + 1})^2 \qquad \text{Using the principle of powers (squaring)}$$
$$(x - 7) \cdot (x - 7) = (2\sqrt{x + 1})(2\sqrt{x + 1})$$
$$x^2 - 14x + 49 = 2^2(\sqrt{x + 1})^2$$
$$x^2 - 14x + 49 = 4(x + 1)$$
$$x^2 - 14x + 49 = 4x + 4$$
$$x^2 - 18x + 45 = 0$$
$$(x - 3)(x - 15) = 0 \qquad \text{Factoring}$$
$$x - 3 = 0 \quad or \quad x - 15 = 0 \qquad \text{Using the principle of zero products}$$
$$x = 3 \quad or \qquad x = 15.$$

The possible solutions are 3 and 5. We check.

For 3:

$x - 7 = 2\sqrt{x + 1}$
$3 - 7 \ ? \ 2\sqrt{3 + 1}$
$-4 \ | \ 2\sqrt{4}$
$| \ 2(2)$
$| \ 4 \qquad$ FALSE

For 15:

$x - 7 = 2\sqrt{x + 1}$
$15 - 7 \ ? \ 2\sqrt{15 + 1}$
$8 \ | \ 2\sqrt{16}$
$| \ 2(4)$
$| \ 8 \qquad$ TRUE

The number 3 does *not* check, but the number 15 does check. The solution is 15.

The number 3 in Example 3 is what is sometimes called an *extraneous solution,* but such terminology is risky to use at best because the number 3 is in *no way* a solution of the original equation.

Do Exercises 3 and 4.

ALGEBRAIC–GRAPHICAL CONNECTION

We can visualize or check the solutions of a radical equation graphically. Consider the equation of Example 3: $x - 7 = 2\sqrt{x + 1}$. We can examine the solutions by graphing the equations

$$y = x - 7 \quad \text{and} \quad y = 2\sqrt{x + 1}$$

using the same set of axes. A hand-drawn graph of $y = 2\sqrt{x + 1}$ would involve approximating square roots on a calculator.

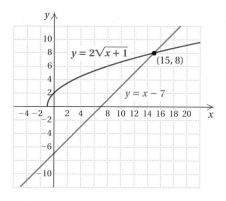

It appears from the graph that when $x = 15$, the values of $y = x - 7$ and $y = 2\sqrt{x + 1}$ are the same, 8. We can check this as we did in Example 3. Note too that the graphs *do not* intersect at $x = 3$, the "extraneous" solution.

CALCULATOR CORNER

Solving Radical Equations We can solve radical equations graphically. Consider the equation in Example 3,

$$x - 7 = 2\sqrt{x + 1}.$$

We first graph each side of the equation. We enter $y_1 = x - 7$ and $y_2 = 2\sqrt{x + 1}$ on the equation-editor screen and graph the equations using the window $[-5, 20, -10, 10]$. Note that there is one point of intersection. We use the **INTERSECT** feature to find its coordinates. (See the Calculator Corner on p. 246 for the procedure.) The first coordinate, 15, is the value of x for which $y_1 = y_2$, or $x - 7 = 2\sqrt{x + 1}$. It is the solution of the equation. Note that the graph shows a single solution whereas the algebraic solution in Example 3 yields two possible solutions, 3 and 15, that must be checked. The algebraic check shows that 15 is the only solution.

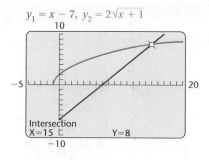

Exercises:

1. Solve the equations in Examples 1 and 4 graphically.

2. Solve the equations in Margin Exercises 1, 3, and 4 graphically.

Solve.

5. $x = \sqrt{x + 5} + 1$

6. $\sqrt[4]{x - 1} - 2 = 0$

EXAMPLE 4   Solve: $x = \sqrt{x + 7} + 5$.

We have

$$x = \sqrt{x + 7} + 5$$

$$x - 5 = \sqrt{x + 7} \qquad \text{Subtracting 5 to isolate the radical term}$$

$$(x - 5)^2 = \left(\sqrt{x + 7}\right)^2 \qquad \text{Using the principle of powers (squaring both sides)}$$

$$x^2 - 10x + 25 = x + 7$$

$$x^2 - 11x + 18 = 0$$

$$(x - 9)(x - 2) = 0 \qquad \text{Factoring}$$

$$x = 9 \quad or \quad x = 2. \qquad \text{Using the principle of zero products}$$

The possible solutions are 9 and 2. Let's check.

For 9:

$$x = \sqrt{x + 7} + 5$$

$$\overline{9 \; ? \; \sqrt{9 + 7} + 5}$$

$$| \; 9 \qquad \text{TRUE}$$

For 2:

$$x = \sqrt{x + 7} + 5$$

$$\overline{2 \; ? \; \sqrt{2 + 7} + 5}$$

$$| \; 8 \qquad \text{FALSE}$$

Since 9 checks but 2 does not, the solution is 9.

EXAMPLE 5   Solve: $\sqrt[3]{2x + 1} + 5 = 0$.

We have

$$\sqrt[3]{2x + 1} + 5 = 0$$

$$\sqrt[3]{2x + 1} = -5 \qquad \text{Subtracting 5. This isolates the radical term.}$$

$$\left(\sqrt[3]{2x + 1}\right)^3 = (-5)^3 \qquad \text{Using the principle of powers (raising to the third power)}$$

$$2x + 1 = -125$$

$$2x = -126 \qquad \text{Subtracting 1}$$

$$x = -63.$$

Check:

$$\sqrt[3]{2x + 1} + 5 = 0$$

$$\overline{\sqrt[3]{2 \cdot (-63) + 1} + 5 \; ? \; 0}$$

$$\sqrt[3]{-125} + 5 \;\;|$$

$$-5 + 5 \;\;|$$

$$0 \;\;| \qquad \text{TRUE}$$

The solution is -63.

Do Exercises 5 and 6.

Answers on page A-34

CHAPTER 6: Radical Expressions,
Equations, and Functions

b Equations with Two Radical Terms

A general strategy for solving radical equations, including those with two radical terms, is as follows.

> **SOLVING RADICAL EQUATIONS**
>
> To solve radical equations:
>
> 1. Isolate one of the radical terms.
> 2. Use the principle of powers.
> 3. If a radical remains, perform steps (1) and (2) again.
> 4. Check possible solutions.

EXAMPLE 6 Solve: $\sqrt{x-3} + \sqrt{x+5} = 4$.

$$\sqrt{x-3} + \sqrt{x+5} = 4$$

$$\sqrt{x-3} = 4 - \sqrt{x+5}$$ Subtracting $\sqrt{x+5}$. This isolates one of the radical terms.

$$\left(\sqrt{x-3}\right)^2 = \left(4 - \sqrt{x+5}\right)^2$$ Using the principle of powers (squaring both sides)

$$x-3 = 16 - 8\sqrt{x+5} + (x+5)$$ Using $(A-B)^2 = A^2 - 2AB + B^2$. See this rule in Section 4.2.

$$-3 = 21 - 8\sqrt{x+5}$$ Subtracting x and collecting like terms

$$-24 = -8\sqrt{x+5}$$ Isolating the remaining radical term

$$3 = \sqrt{x+5}$$ Dividing by -8

$$3^2 = \left(\sqrt{x+5}\right)^2$$ Squaring

$$9 = x+5$$

$$4 = x$$

The number 4 checks and is the solution.

EXAMPLE 7 Solve: $\sqrt{2x-5} = 1 + \sqrt{x-3}$.

$$\sqrt{2x-5} = 1 + \sqrt{x-3}$$

$$\left(\sqrt{2x-5}\right)^2 = \left(1 + \sqrt{x-3}\right)^2$$ One radical is already isolated. We square both sides.

$$2x-5 = 1 + 2\sqrt{x-3} + \left(\sqrt{x-3}\right)^2$$

$$2x-5 = 1 + 2\sqrt{x-3} + (x-3)$$

$$x-3 = 2\sqrt{x-3}$$ Isolating the remaining radical term

$$(x-3)^2 = \left(2\sqrt{x-3}\right)^2$$ Squaring both sides

$$x^2 - 6x + 9 = 4(x-3)$$

$$x^2 - 6x + 9 = 4x - 12$$

$$x^2 - 10x + 21 = 0$$

$$(x-7)(x-3) = 0$$ Factoring

$$x = 7 \quad or \quad x = 3$$ Using the principle of zero products

539

Solve.

7. $\sqrt{x} - \sqrt{x-5} = 1$

The possible solutions are 7 and 3. We check.

For 7:
$$\sqrt{2x-5} = 1 + \sqrt{x-3}$$

$\sqrt{2x-5} = 1 + \sqrt{x-3}$	
$\sqrt{2(7)-5}$? $1 + \sqrt{7-3}$	
$\sqrt{14-5}$	$1 + \sqrt{4}$
$\sqrt{9}$	$1 + 2$
3	3 TRUE

For 3:
$$\sqrt{2x-5} = 1 + \sqrt{x-3}$$

$\sqrt{2x-5} = 1 + \sqrt{x-3}$	
$\sqrt{2(3)-5}$? $1 + \sqrt{3-3}$	
$\sqrt{6-5}$	$1 + \sqrt{0}$
$\sqrt{1}$	$1 + 0$
1	1 TRUE

The numbers 7 and 3 check and are the solutions.

Do Exercises 7 and 8.

EXAMPLE 8 Solve: $\sqrt{x+2} - \sqrt{2x+2} + 1 = 0$.

We first isolate one radical.

8. $\sqrt{2x-5} - 2 = \sqrt{x-2}$

$$\sqrt{x+2} - \sqrt{2x+2} + 1 = 0$$

$\qquad \sqrt{x+2} + 1 = \sqrt{2x+2}$ Adding $\sqrt{2x+2}$ to isolate a radical expression

$\qquad (\sqrt{x+2} + 1)^2 = (\sqrt{2x+2})^2$ Squaring both sides

$\qquad x + 2 + 2\sqrt{x+2} + 1 = 2x + 2$

$\qquad\qquad 2\sqrt{x+2} = x - 1$

$\qquad\qquad (2\sqrt{x+2})^2 = (x-1)^2$

$\qquad\qquad 4(x+2) = x^2 - 2x + 1$

$\qquad\qquad 4x + 8 = x^2 - 2x + 1$

$\qquad\qquad 0 = x^2 - 6x - 7$

$\qquad\qquad 0 = (x-7)(x+1)$ Factoring

$\qquad x - 7 = 0 \quad or \quad x + 1 = 0$ Using the principle of zero products

$\qquad\qquad x = 7 \quad or \qquad x = -1$

The possible solutions are 7 and -1. We check.

For 7:

$$\sqrt{x+2} - \sqrt{2x+2} + 1 = 0$$

$\sqrt{x+2} - \sqrt{2x+2} + 1 = 0$	
$\sqrt{7+2} - \sqrt{2 \cdot 7 + 2} + 1$? 0	
$\sqrt{9} - \sqrt{16} + 1$	
$3 - 4 + 1$	
0	TRUE

9. Solve:

$\sqrt{3x+1} - 1 - \sqrt{x+4} = 0$.

For -1:

$$\sqrt{x+2} - \sqrt{2x+2} + 1 = 0$$

$\sqrt{x+2} - \sqrt{2x+2} + 1 = 0$	
$\sqrt{-1+2} - \sqrt{2 \cdot (-1) + 2} + 1$? 0	
$\sqrt{1} - \sqrt{0} + 1$	
$1 - 0 + 1$	
2	FALSE

The number 7 checks, but -1 does not. The solution is 7.

Do Exercise 9.

Answers on page A-34

C Applications

Speed of Sound. Many applications translate to radical equations. For example, at a temperature of t degrees Fahrenheit, sound travels S feet per second, where

$$S = 21.9\sqrt{5t + 2457}. \quad (1)$$

EXAMPLE 9 *Orchestra Practice.* During orchestra practice, the temperature of a room was 72°F. How fast was the sound of the orchestra traveling through the room?

We substitute 72 for t in equation (1) and find an approximation using a calculator:

$$\begin{aligned}
S &= 21.9\sqrt{5t + 2457} \\
&= 21.9\sqrt{5(72) + 2457} \\
&= 21.9\sqrt{360 + 2457} \\
&= 21.9\sqrt{2817} \\
&\approx 1162.4 \text{ ft/sec.}
\end{aligned}$$

Do Exercise 10.

EXAMPLE 10 *Musical Performances.* The group *NSYNC regularly performed outdoors for large audiences. A scientific instrument at one of their concerts determined that the sound of the group was traveling at a rate of 1170 ft/sec. What was the air temperature at the concert?

We substitute 1170 for S in the formula $S = 21.9\sqrt{5t + 2457}$:

$$1170 = 21.9\sqrt{5t + 2457}.$$

Then we solve the equation for t:

$$\begin{aligned}
1170 &= 21.9\sqrt{5t + 2457} \\
\frac{1170}{21.9} &= \sqrt{5t + 2457} && \text{Dividing by 21.9} \\
\left(\frac{1170}{21.9}\right)^2 &= \left(\sqrt{5t + 2457}\right)^2 && \text{Squaring both sides} \\
2854.2 &\approx 5t + 2457 && \text{Simplifying} \\
397.2 &\approx 5t && \text{Subtracting 2457} \\
79 &\approx t. && \text{Dividing by 5}
\end{aligned}$$

The temperature at the concert was about 79°F.

Do Exercise 11.

10. **Music During Surgery.** It has been shown that playing music during surgery may hasten a patient's recovery after surgery. During a heart bypass surgery, doctors played music in the operating room. The temperature of the room was kept at 60°F. How fast did the music travel through the room?

Source: *Acta Anaesthesiologica Scandinavica,* 2001; **45:** 812–817

11. **Musical Performances.** During an outdoor concert given by LeAnn Rimes, the speed of sound from the music was measured by a scientific instrument to be 1179 ft/sec. What was the air temperature at the concert?

Answers on page A-34

a Solve.

1. $\sqrt{2x-3}=4$

2. $\sqrt{5x+2}=7$

3. $\sqrt{6x}+1=8$

4. $\sqrt{3x}-4=6$

5. $\sqrt{y+7}-4=4$

6. $\sqrt{x-1}-3=9$

7. $\sqrt{5y+8}=10$

8. $\sqrt{2y+9}=5$

9. $\sqrt[3]{x}=-1$

10. $\sqrt[3]{y}=-2$

11. $\sqrt{x+2}=-4$

12. $\sqrt{y-3}=-2$

13. $\sqrt[3]{x+5}=2$

14. $\sqrt[3]{x-2}=3$

15. $\sqrt[4]{y-3}=2$

16. $\sqrt[4]{x+3}=3$

17. $\sqrt[3]{6x+9}+8=5$

18. $\sqrt[3]{3y+6}+2=3$

19. $8=\dfrac{1}{\sqrt{x}}$

20. $\dfrac{1}{\sqrt{y}}=3$

21. $x-7=\sqrt{x-5}$

22. $x-5=\sqrt{x+7}$

23. $2\sqrt{x+1}+7=x$

24. $\sqrt{2x+7}-2=x$

25. $3\sqrt{x-1}-1=x$

26. $x-1=\sqrt{x+5}$

27. $x-3=\sqrt{27-3x}$

28. $x-1=\sqrt{1-x}$

Solve.

29. $\sqrt{3y + 1} = \sqrt{2y + 6}$

30. $\sqrt{5x - 3} = \sqrt{2x + 3}$

31. $\sqrt{y - 5} + \sqrt{y} = 5$

32. $\sqrt{x - 9} + \sqrt{x} = 1$

33. $3 + \sqrt{z - 6} = \sqrt{z + 9}$

34. $\sqrt{4x - 3} = 2 + \sqrt{2x - 5}$

35. $\sqrt{20 - x} + 8 = \sqrt{9 - x} + 11$

36. $4 + \sqrt{10 - x} = 6 + \sqrt{4 - x}$

37. $\sqrt{4y + 1} - \sqrt{y - 2} = 3$

38. $\sqrt{y + 15} - \sqrt{2y + 7} = 1$

39. $\sqrt{x + 2} + \sqrt{3x + 4} = 2$

40. $\sqrt{6x + 7} - \sqrt{3x + 3} = 1$

41. $\sqrt{3x - 5} + \sqrt{2x + 3} + 1 = 0$

42. $\sqrt{2m - 3} + 2 - \sqrt{m + 7} = 0$

43. $2\sqrt{t - 1} - \sqrt{3t - 1} = 0$

44. $3\sqrt{2y + 3} - \sqrt{y + 10} = 0$

Sighting to the Horizon.   How far can you see to the horizon from a given height? The function

$$D = 1.2\sqrt{h}$$

can be used to approximate the distance D, in miles, that a person can see to the horizon from a height h, in feet.

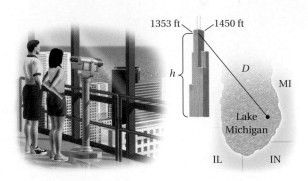

45. An observation deck near the top of the Sears Tower in Chicago is 1353 ft high. How far can a tourist see to the horizon from this deck?

46. The roof of the Sears Tower is 1450 ft high. How far can a worker see to the horizon from the top of the Sears Tower?

47. Elaine can see 31.3 mi to the horizon from the top of a cliff. What is the height of Elaine's eyes?

48. A technician can see 30.4 mi to the horizon from the top of a radio tower. How high is the tower?

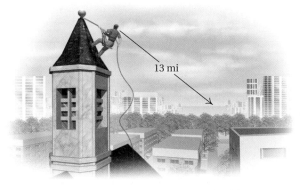

49. A steeplejack can see 13 mi to the horizon from the top of a building. What is the height of the steeplejack's eyes?

50. A person can see 230 mi to the horizon from an airplane window. How high is the airplane?

51. How far can a sailor see to the horizon from the top of a mast that is 13 ft high?

52. How far can you see to the horizon through an airplane window at a height of 32,000 ft?

Speed of a Skidding Car. After an accident, how do police determine the speed at which the car had been traveling? The formula

$$r = 2\sqrt{5L}$$

can be used to approximate the speed r, in miles per hour, of a car that has left a skid mark of length L, in feet. Use this formula for Exercises 53 and 54.

53. How far will a car skid at 55 mph? at 75 mph?

54. How far will a car skid at 65 mph? at 100 mph?

Temperature and the Speed of Sound. Solve Exercises 55 and 56 using the formula $S = 21.9\sqrt{5t + 2457}$ from Example 9.

55. During blasting for avalanche control in Utah's Wasatch Mountains, sound traveled at a rate of 1113 ft/sec. What was the temperature at the time?

56. At a recent concert by the Dave Matthews Band, sound traveled at a rate of 1176 ft/sec. What was the temperature at the time?

Period of a Swinging Pendulum. The formula $T = 2\pi\sqrt{L/32}$ can be used to find the period T, in seconds, of a pendulum of length L, in feet.

57. What is the length of a pendulum that has a period of 1.0 sec? Use 3.14 for π.

58. What is the length of a pendulum that has a period of 2.0 sec? Use 3.14 for π.

59. **D**w The principle of powers contains an "if–then" statement that becomes false when the parts are interchanged. Find another mathematical example of such an "if–then" statement.

60. **D**w Is checking necessary when the principle of powers is used with an odd power n? Why or why not?

Solve. [5.6a]

61. *Painting a Room.* Julia can paint a room in 8 hr. George can paint the same room in 10 hr. How long will it take them, working together, to paint the same room?

62. *Delivering Leaflets.* Jeff can drop leaflets in mailboxes three times as fast as Grace can. If they work together, it takes them 1 hr to complete the job. How long would it take each to deliver the leaflets alone?

Solve. [5.6b]

63. *Bicycle Travel.* A cyclist traveled 702 mi in 14 days. At this same ratio, how far would the cyclist have traveled in 56 days?

64. *Earnings.* Dharma earned $696.64 working for 56 hr at a fruit stand. How many hours must she work in order to earn $1044.96?

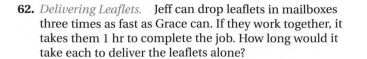

Solve. [4.8a]

65. $x^2 + 2.8x = 0$

66. $3x^2 - 5x = 0$

67. $x^2 - 64 = 0$

68. $2x^2 = x + 21$

For each of the following functions, find and simplify $f(a + h) - f(a)$. [4.2e]

69. $f(x) = x^2$

70. $f(x) = x^2 - x$

71. $f(x) = 2x^2 - 3x$

72. $f(x) = 2x^2 + 3x - 7$

73. Use a graphing calculator to check your answers to Exercises 4, 9, 33, and 38.

74. Consider the equation
$$\sqrt{2x + 1} + \sqrt{5x - 4} = \sqrt{10x + 9}.$$
a) Use a graphing calculator to solve the equation.
b) Solve the equation algebraically.
c) Explain the advantages and disadvantages of using each method. Which do you prefer?

Solve.

75. $\sqrt[3]{\dfrac{z}{4}} - 10 = 2$

76. $\sqrt[4]{z^2 + 17} = 3$

77. $\sqrt{\sqrt{y + 49}} - \sqrt{y} = \sqrt{7}$

78. $\sqrt[3]{x^2 + x + 15} - 3 = 0$

79. $\sqrt{\sqrt{x^2 + 9x + 34}} = 2$

80. $\sqrt{8 - b} = b\sqrt{8 - b}$

81. $\sqrt{x - 2} - \sqrt{x + 2} + 2 = 0$

82. $6\sqrt{y} + 6y^{-1/2} = 37$

83. $\sqrt{a^2 + 30a} = a + \sqrt{5a}$

84. $\sqrt{\sqrt{x + 4}} = \sqrt{x} - 2$

85. $\dfrac{x - 1}{\sqrt{x^2 + 3x + 6}} = \dfrac{1}{4}$

86. $\sqrt{x + 1} - \dfrac{2}{\sqrt{x + 1}} = 1$

87. $\sqrt{y^2 + 6} + y - 3 = 0$

88. $2\sqrt{x - 1} - \sqrt{3x - 5} = \sqrt{x - 9}$

89. $\sqrt{y + 1} - \sqrt{2y - 5} = \sqrt{y - 2}$

90. Evaluate: $\sqrt{7 + 4\sqrt{3}} - \sqrt{7 - 4\sqrt{3}}$.

a **Applications**

a   Solve applied problems involving the Pythagorean theorem and powers and roots.

There are many kinds of applied problems that involve powers and roots. Many also make use of right triangles and the Pythagorean theorem: $a^2 + b^2 = c^2$.

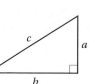

1. Find the length of the hypotenuse of this right triangle. Give an exact answer and an approximation to three decimal places.

EXAMPLE 1 *Vegetable Garden.* Benito and Dominique are planting a vegetable garden in the backyard. They decide that it will be a 30-ft by 40-ft rectangle and begin to lay it out using string. They soon realize that it is difficult to form the right angles and that it would be helpful to know the length of a diagonal. Find the length of a diagonal.

Using the Pythagorean theorem, $a^2 + b^2 = c^2$, we substitute 30 for a and 40 for b and then solve for c:

$$a^2 + b^2 = c^2$$
$$30^2 + 40^2 = c^2 \quad \text{Substituting}$$
$$900 + 1600 = c^2$$
$$2500 = c^2$$
$$\sqrt{2500} = c$$

We consider only the positive root since length cannot be negative.

$$50 = c.$$

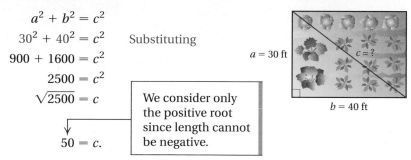

The length of the hypotenuse, or the diagonal, is 50 ft. Knowing this measurement would help in laying out the garden. Construction workers often use a procedure like this to lay out a right angle.

EXAMPLE 2 Find the length of the hypotenuse of this right triangle. Give an exact answer and an approximation to three decimal places.

$$7^2 + 4^2 = c^2 \quad \text{Substituting}$$
$$49 + 16 = c^2$$
$$65 = c^2$$

Exact answer: $\quad c = \sqrt{65}$
Approximation: $\quad c \approx 8.062 \quad$ Using a calculator

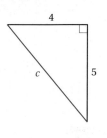

2. Find the length of the leg of this right triangle. Give an exact answer and an approximation to three decimal places.

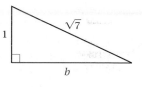

3. Find the length of the hypotenuse of this right triangle. Give an exact answer and an approximation to three decimal places.

EXAMPLE 3 Find the missing length b in this right triangle. Give an exact answer and an approximation to three decimal places.

$$1^2 + b^2 = \left(\sqrt{11}\right)^2 \quad \text{Substituting}$$
$$1 + b^2 = 11$$
$$b^2 = 10$$

Exact answer: $\quad b = \sqrt{10}$
Approximation: $\quad b \approx 3.162 \quad$ Using a calculator

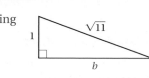

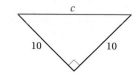

Do Exercises 1–3.

Answers on page A-34

4. Baseball Diamond. A baseball diamond is actually a square 90 ft on a side. Suppose a catcher fields a bunt along the third-base line 10 ft from home plate. How far would the catcher have to throw the ball to first base? Give an exact answer and an approximation to three decimal places.

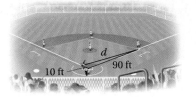

EXAMPLE 4 *Ramps for the Disabled.* Laws regarding access ramps for the disabled state that a ramp must be in the form of a right triangle, where every vertical length (leg) of 1 ft has a horizontal length (leg) of 12 ft. What is the length of a ramp with a 12-ft horizontal leg and a 1-ft vertical leg? Give an exact answer and an approximation to three decimal places.

We make a drawing and let h = the length of the ramp. It is the length of the hypotenuse of a right triangle whose legs are 12 ft and 1 ft. We substitute these values into the Pythagorean theorem to find h.

$$h^2 = 12^2 + 1^2$$
$$h^2 = 144 + 1$$
$$h^2 = 145$$

Exact answer: $h = \sqrt{145}$ ft

Approximation: $h \approx 12.042$ ft   Using a calculator

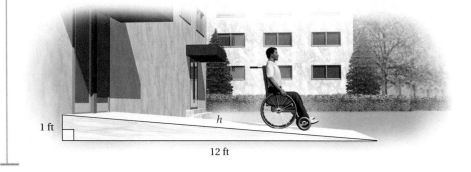

Do Exercise 4.

5. Referring to Example 5, find L given that $h = 3$ ft and $d = 180$ ft. You will need a calculator with an exponential key $\boxed{y^x}$, or ⬛.

EXAMPLE 5 *Road-Pavement Messages.* In a psychological study, it was determined that the proper length L of the letters of a word painted on pavement is given by

$$L = \frac{0.000169d^{2.27}}{h},$$

where d is the distance of a car from the lettering and h is the height of the eye above the road. All units are in feet. For a person h feet above the road, a message d feet away will be the most readable if the length of the letters is L.

Find L, given that $h = 4$ ft and $d = 180$ ft.

We substitute 4 for h and 180 for d and calculate L using a calculator with an exponential key $\boxed{y^x}$, or ⬛:

$$L = \frac{0.000169(180)^{2.27}}{4} \approx 5.6 \text{ ft.}$$

Do Exercise 5.

Answers on page A-34

CHAPTER 6: Radical Expressions, Equations, and Functions

Translating
for Success

1. *Angles of a Triangle.*
The second angle of a triangle is four times as large as the first. The third is 27° less than the sum of the other angles. Find the measures of the angles.

2. *Lengths of a Rectangle.*
The area of a rectangle is 180 ft^2. The length is 26 ft greater than the width. Find the length and the width.

3. *Boat Travel.* The speed of a river is 3 mph. A boat can go 72 mi upstream and 24 mi downstream in a total time of 16 hr. Find the speed of the boat in still water.

4. *Coin Mixture.* A collection of nickels and quarters is worth $13.85. There are 85 coins in all. How many of each coin are there?

5. *Perimeter.* The perimeter of a rectangle is 180 ft. The length is 26 ft greater than the width. Find the length and the width.

Translate each word problem to an equation or a system of equations and select a correct translation from equations A–O.

A. $12^2 + 12^2 = x^2$

B. $x(x + 26) = 180$

C. $10,311 + 5\%x = x$

D. $x + y = 85,$
$5x + 25y = 13.85$

E. $x^2 + 4^2 = 12^2$

F. $\dfrac{240}{x - 18} = \dfrac{384}{x}$

G. $x + 5\%x = 10,311$

H. $\dfrac{x}{65} + 1 = \dfrac{x}{85}$

I. $\dfrac{x}{65} + \dfrac{x}{85} = 1$

J. $x + y + z = 180,$
$y = 4x,$
$z = x + y - 27$

K. $2x + 2(x + 26) = 180$

L. $\dfrac{384}{x - 18} = \dfrac{240}{x}$

M. $x + y = 85,$
$0.05x + 0.25y = 13.85$

N. $2x + 2(x + 24) = 240$

O. $\dfrac{72}{x + 3} + \dfrac{24}{x - 3} = 16$

Answers on page A-34

6. *Shoveling Time.* It takes Marv 65 min to shovel 4 in. of snow from his driveway. It takes Elaine 85 min to do the same job. How long would it take if they worked together?

7. *Money Borrowed.* Claire borrows some money at 5% simple interest. After 1 yr, $10,311 pays off her loan. How much did she originally borrow?

8. *Plank Height.* A 12-ft plank is leaning against a shed. The bottom of the plank is 4 ft from the building. How high up the side of the shed is the top of the plank?

9. *Train Speeds.* The speed of train A is 18 mph slower than the speed of train B. Train A travels 240 mi in the same time that it takes train B to travel 384 mi. Find the speed of train A.

10. *Diagonal of a Square.* Find the length of a diagonal of a square swimming pool whose sides are 12 ft long.

6.7 EXERCISE SET

For Extra Help

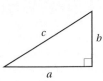

MathXL   MyMathLab   InterAct Math   Math Tutor Center   Digital Video Tutor CD 3 Videotape 7   Student's Solutions Manual

a In a right triangle, find the length of the side not given. Give an exact answer and an approximation to three decimal places.

1. $a = 3$, $b = 5$

2. $a = 8$, $b = 10$

3. $a = 15$, $b = 15$

4. $a = 8$, $b = 8$

5. $b = 12$, $c = 13$

6. $a = 5$, $c = 12$

7. $c = 7$, $a = \sqrt{6}$

8. $c = 10$, $a = 4\sqrt{5}$

9. $b = 1$, $c = \sqrt{13}$

10. $a = 1$, $c = \sqrt{12}$

11. $a = 1$, $c = \sqrt{n}$

12. $c = 2$, $a = \sqrt{n}$

In the following problems, give an exact answer and, where appropriate, an approximation (using a calculator) to three decimal places.

13. *Guy Wire.*   How long is a guy wire reaching from the top of a 10-ft pole to a point on the ground 4 ft from the pole?

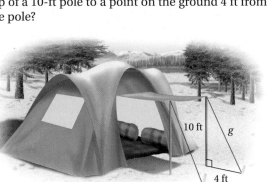

10 ft g

4 ft

14. *Softball Diamond.*   A slow-pitch softball diamond is actually a square 65 ft on a side. How far is it from home to second base?

15. *Road-Pavement Messages.*   Using the formula of Example 5, find the length L of a road-pavement message when $h = 4$ ft and $d = 200$ ft.

16. *Road-Pavement Messages.*   Using the formula of Example 5, find the length L of a road-pavement message when $h = 8$ ft and $d = 300$ ft.

17. *Bridge Expansion.* During the summer heat, a 2-mi bridge expands 2 ft in length. If we assume that the bulge occurs straight up the middle, how high is the bulge? (The answer may surprise you. In reality, bridges are built with expansion spaces to avoid such buckling.)

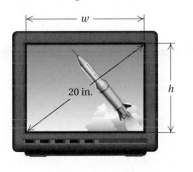

18. *Triangle Areas.* Triangle *ABC* has sides of lengths 25 ft, 25 ft, and 30 ft. Triangle *PQR* has sides of lengths 25 ft, 25 ft, and 40 ft. Which triangle has the greater area and by how much?

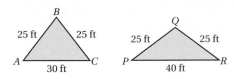

19. Each side of a regular octagon has length *s*. Find a formula for the distance *d* between the parallel sides of the octagon.

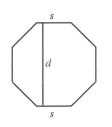

20. The two equal sides of an isosceles right triangle are of length *s*. Find a formula for the length of the hypotenuse.

21. The length and the width of a rectangle are given by consecutive integers. The area of the rectangle is 90 cm^2. Find the length of a diagonal of the rectangle.

22. The diagonal of a square has length $8\sqrt{2}$ ft. Find the length of a side of the square.

23. *Television Sets.* What does it mean to refer to a 20-in. TV set or a 25-in. TV set? Such units refer to the diagonal of the screen. A 20-in. TV set also has a width of 16 in. What is its height?

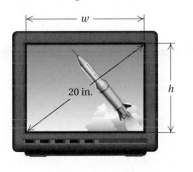

24. *Television Sets.* A 25-in. TV set has a screen with a height of 15 in. What is its width?

25. Find all ordered pairs on the *x*-axis of a Cartesian coordinate system that are 5 units from the point (0, 4).

26. Find all ordered pairs on the *y*-axis of a Cartesian coordinate system that are 5 units from the point (3, 0).

27. *Speaker Placement.* A stereo receiver is in a corner of a 12-ft by 14-ft room. Speaker wire will run under a rug, diagonally, to a speaker in the far corner. If 4 ft of slack is required on each end, how long a piece of wire should be purchased?

28. *Distance Over Water.* To determine the width of a pond, a surveyor locates two stakes at either end of the pond and uses instrumentation to place a third stake so that the distance across the pond is the length of a hypotenuse. If the third stake is 90 m from one stake and 70 m from the other, how wide is the pond?

90 m
70 m
?

29. *Plumbing.* Plumbers use the Pythagorean theorem to calculate pipe length. If a pipe is to be offset, as shown in the figure, the *travel*, or length, of the pipe, is calculated using the lengths of the *advance* and *offset*. Find the travel if the offset is 17.75 in. and the advance is 10.25 in.

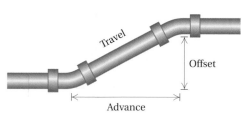

Travel
Offset
Advance

30. *Carpentry.* Darla is laying out the footer of a house. To see if the corner is square, she measures 16 ft from the corner along one wall and 12 ft from the corner along the other wall. How long should the diagonal be between those two points if the corner is a right angle?

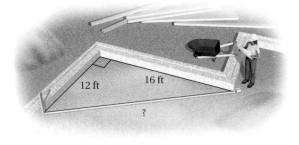

12 ft
16 ft
?

31. **D**_W Write a problem for a classmate to solve in which the solution is "The height of the tepee is $5\sqrt{3}$ yd."

32. **D**_W Write a problem for a classmate to solve in which the solution is "The height of the window is $15\sqrt{3}$ yd."

Solve.   [5.6c]

33. *Commuter Travel.* The speed of the Zionsville Flash commuter train is 14 mph faster than that of the Carmel Crawler. The Flash travels 290 mi in the same time that it takes the Crawler to travel 230 mi. Find the speed of each train.

34. *Marine Travel.* A motor boat travels three times as fast as the current in the Saskatee River. A trip up the river and back takes 10 hr, and the total distance of the trip is 100 mi. Find the speed of the current.

CHAPTER 6: Radical Expressions,
Equations, and Functions

Solve. [4.8a], [5.5a]

35. $2x^2 + 11x - 21 = 0$

36. $x^2 + 24 = 11x$

37. $\dfrac{x+2}{x+3} = \dfrac{x-4}{x-5}$

38. $3x^2 - 12 = 0$

39. $\dfrac{x-5}{x-7} = \dfrac{4}{3}$

40. $\dfrac{x-1}{x-3} = \dfrac{6}{x-3}$

SYNTHESIS

41. *Roofing.* Kit's cottage, which is 24 ft wide and 32 ft long, needs a new roof. By counting clapboards that are 4 in. apart, Kit determines that the peak of the roof is 6 ft higher than the sides. If one packet of shingles covers $33\frac{1}{3}$ square feet, how many packets will the job require?

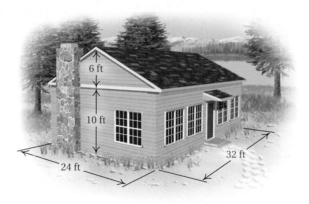

42. *Painting.* (Refer to Exercise 41.) A gallon of paint covers about 275 square feet. If Kit's first floor is 10 ft high, how many gallons of paint should be bought to paint the house? What assumption(s) is made in your answer?

43. *Cube Diagonal.* A cube measures 5 cm on each side. How long is the diagonal that connects two opposite corners of the cube? Give an exact answer.

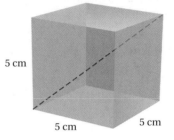

5 cm

5 cm

5 cm

44. *Wind Chill Temperature.* Because wind enhances the loss of heat from the skin, we feel colder when there is wind than when there is not. The *wind chill temperature* is what the temperature would have to be with no wind in order to give the same chilling effect as with the wind. A formula for finding the wind chill temperature, T_w, is

$$T_w = 91.4 - (91.4 - T)\left(0.478 + 0.301\sqrt{v} - 0.02v\right),$$

where T is the actual temperature given by a thermometer, in degrees Fahrenheit, and v is the wind speed, in miles per hour.* Use a calculator to find the wind chill temperature in each case. Round to the nearest degree.

a) $T = 40°F,$
$v = 25$ mph

b) $T = 20°F,$
$v = 25$ mph

c) $T = 10°F,$
$v = 20$ mph

d) $T = 10°F,$
$v = 40$ mph

e) $T = -5°F,$
$v = 35$ mph

f) $T = -16°F,$
$v = 35$ mph

---

*This formula can be used only when the wind speed is *above* 4 mph.

Objectives

Express in terms of i.

1. $\sqrt{-5}$

2. $\sqrt{-25}$

3. $-\sqrt{-11}$

4. $-\sqrt{-36}$

5. $\sqrt{-54}$

a Imaginary and Complex Numbers

Negative numbers do not have square roots in the real-number system. However, mathematicians have described a larger number system that contains the real-number system, such that negative numbers have square roots. That system is called the **complex-number system.** We begin by defining a number that is a square root of -1. We call this new number i.

THE COMPLEX NUMBER i

We define the number i to be $\sqrt{-1}$. That is,
$$i = \sqrt{-1} \quad \text{and} \quad i^2 = -1.$$

To express roots of negative numbers in terms of i, we can use the fact that in the complex numbers, $\sqrt{-p} = \sqrt{-1 \cdot p} = \sqrt{-1}\,\sqrt{p}$ when p is a positive real number.

EXAMPLES Express in terms of i.

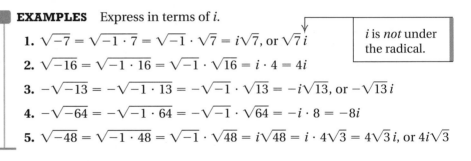

1. $\sqrt{-7} = \sqrt{-1 \cdot 7} = \sqrt{-1} \cdot \sqrt{7} = i\sqrt{7}$, or $\sqrt{7}\,i$

> i is *not* under the radical.

2. $\sqrt{-16} = \sqrt{-1 \cdot 16} = \sqrt{-1} \cdot \sqrt{16} = i \cdot 4 = 4i$

3. $-\sqrt{-13} = -\sqrt{-1 \cdot 13} = -\sqrt{-1} \cdot \sqrt{13} = -i\sqrt{13}$, or $-\sqrt{13}\,i$

4. $-\sqrt{-64} = -\sqrt{-1 \cdot 64} = -\sqrt{-1} \cdot \sqrt{64} = -i \cdot 8 = -8i$

5. $\sqrt{-48} = \sqrt{-1 \cdot 48} = \sqrt{-1} \cdot \sqrt{48} = i\sqrt{48} = i \cdot 4\sqrt{3} = 4\sqrt{3}\,i$, or $4i\sqrt{3}$

Do Exercises 1–5.

IMAGINARY NUMBER

An **imaginary* number** is a number that can be named

$$bi,$$

where b is some real number and $b \neq 0$.

To form the system of **complex numbers,** we take the imaginary numbers and the real numbers and all possible sums of real and imaginary numbers. These are complex numbers:

$$7 - 4i, \quad -\pi + 19i, \quad 37, \quad i\sqrt{8}.$$

---
*Don't let the name "imaginary" fool you. The imaginary numbers are very important in such fields as engineering and the physical sciences.

COMPLEX NUMBER

A **complex number** is any number that can be named

$$a + bi,$$

where a and b are any real numbers. (Note that either a or b or both can be 0.)

Since $0 + bi = bi$, every imaginary number is a complex number. Similarly, $a + 0i = a$, so every real number is a complex number. The relationships among various real and complex numbers are shown in the following diagram.

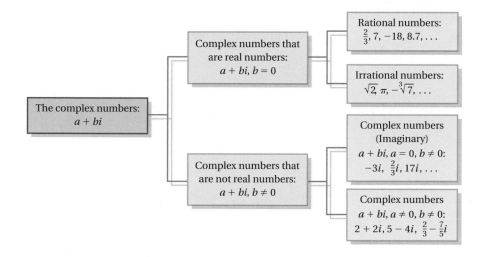

It is important to keep in mind some comparisons between numbers that have real-number roots and those that have complex-number roots that are not real. For example, $\sqrt{-48}$ is a complex number that is not a real number because we are taking the square root of a negative number. *But*, $\sqrt[3]{-125}$ is a real number because we are taking the cube root of a negative number and *any* real number has a cube root that is a real number.

b Addition and Subtraction

The complex numbers follow the commutative and associative laws of addition. Thus we can add and subtract them as we do binomials with real-number coefficients, that is, we collect like terms.

 EXAMPLES Add or subtract.

6. $(8 + 6i) + (3 + 2i) = (8 + 3) + (6 + 2)i = 11 + 8i$

7. $(3 + 2i) - (5 - 2i) = (3 - 5) + [2 - (-2)]i = -2 + 4i$

Do Exercises 6–9.

c Multiplication

The complex numbers obey the commutative, associative, and distributive laws. But although the property $\sqrt{a}\,\sqrt{b} = \sqrt{ab}$ does *not* hold for complex numbers in general, it does hold when $a = -1$ and b is a positive real number.

Add or subtract.

6. $(7 + 4i) + (8 - 7i)$

7. $(-5 - 6i) + (-7 + 12i)$

8. $(8 + 3i) - (5 + 8i)$

9. $(5 - 4i) - (-7 + 3i)$

Answers on page A-34

Multiply.

10. $\sqrt{-25} \cdot \sqrt{-4}$

11. $\sqrt{-2} \cdot \sqrt{-17}$

12. $-6i \cdot 7i$

13. $-3i(4 - 3i)$

14. $5i(-5 + 7i)$

15. $(1 + 3i)(1 + 5i)$

16. $(3 - 2i)(1 + 4i)$

17. $(3 + 2i)^2$

To multiply square roots of negative real numbers, we first express them in terms of i. For example,

$$\sqrt{-2} \cdot \sqrt{-5} = \sqrt{-1} \cdot \sqrt{2} \cdot \sqrt{-1} \cdot \sqrt{5} = i\sqrt{2} \cdot i\sqrt{5}$$
$$= i^2\sqrt{10} = -\sqrt{10} \quad \text{is correct!}$$

But $\sqrt{-2} \cdot \sqrt{-5} = \sqrt{(-2)(-5)} = \sqrt{10}$ is wrong!

Keeping this and the fact that $i^2 = -1$ in mind, we multiply in much the same way that we do with real numbers.

Caution!

The rule $\sqrt{a} \sqrt{b} = \sqrt{ab}$ holds only for nonnegative real numbers.

EXAMPLES Multiply.

8. $\sqrt{-49} \cdot \sqrt{-16} = \sqrt{-1} \cdot \sqrt{49} \cdot \sqrt{-1} \cdot \sqrt{16}$
$$= i \cdot 7 \cdot i \cdot 4$$
$$= i^2(28)$$
$$= (-1)(28) \quad i^2 = -1$$
$$= -28$$

9. $\sqrt{-3} \cdot \sqrt{-7} = \sqrt{-1} \cdot \sqrt{3} \cdot \sqrt{-1} \cdot \sqrt{7}$
$$= i \cdot \sqrt{3} \cdot i \cdot \sqrt{7}$$
$$= i^2(\sqrt{21})$$
$$= (-1)\sqrt{21} \quad i^2 = -1$$
$$= -\sqrt{21}$$

10. $-2i \cdot 5i = -10 \cdot i^2$
$$= (-10)(-1) \quad i^2 = -1$$
$$= 10$$

11. $(-4i)(3 - 5i) = (-4i) \cdot 3 - (-4i)(5i)$ Using a distributive law
$$= -12i + 20i^2$$
$$= -12i + 20(-1) \qquad i^2 = -1$$
$$= -12i - 20$$
$$= -20 - 12i$$

12. $(1 + 2i)(1 + 3i) = 1 + 3i + 2i + 6i^2$ Multiplying each term of one number by every term of the other (FOIL)
$$= 1 + 3i + 2i + 6(-1) \qquad i^2 = -1$$
$$= 1 + 3i + 2i - 6$$
$$= -5 + 5i \quad \text{Collecting like terms}$$

13. $(3 - 2i)^2 = 3^2 - 2(3)(2i) + (2i)^2$ Squaring the binomial
$$= 9 - 12i + 4i^2$$
$$= 9 - 12i + 4(-1) \qquad i^2 = -1$$
$$= 9 - 12i - 4$$
$$= 5 - 12i$$

Do Exercises 10–17.

Answers on page A-34

d | Powers of *i*

We now want to simplify certain expressions involving powers of *i*. To do so, we first see how to simplify powers of *i*. Simplifying powers of *i* can be done by using the fact that $i^2 = -1$ and expressing the given power of *i* in terms of even powers, and then in terms of powers of i^2. Consider the following:

$$i,$$
$$i^2 = -1,$$
$$i^3 = i^2 \cdot i = (-1)i = -i,$$
$$i^4 = (i^2)^2 = (-1)^2 = 1,$$
$$i^5 = i^4 \cdot i = (i^2)^2 \cdot i = (-1)^2 \cdot i = i,$$
$$i^6 = (i^2)^3 = (-1)^3 = -1.$$

Note that the powers of *i* cycle themselves through the values i, -1, $-i$, and 1.

EXAMPLES Simplify.

14. $i^{37} = i^{36} \cdot i = (i^2)^{18} \cdot i = (-1)^{18} \cdot i = 1 \cdot i = i$

15. $i^{58} = (i^2)^{29} = (-1)^{29} = -1$

16. $i^{75} = i^{74} \cdot i = (i^2)^{37} \cdot i = (-1)^{37} \cdot i = -1 \cdot i = -i$

17. $i^{80} = (i^2)^{40} = (-1)^{40} = 1$

Do Exercises 18–21.

Now let's simplify other expressions.

EXAMPLES Simplify to the form $a + bi$.

18. $8 - i^2 = 8 - (-1) = 8 + 1 = 9$

19. $17 + 6i^3 = 17 + 6 \cdot i^2 \cdot i = 17 + 6(-1)i = 17 - 6i$

20. $i^{22} - 67i^2 = (i^2)^{11} - 67(-1) = (-1)^{11} + 67 = -1 + 67 = 66$

21. $i^{23} + i^{48} = (i^{22}) \cdot i + (i^2)^{24} = (i^2)^{11} \cdot i + (-1)^{24} = (-1)^{11} \cdot i + (-1)^{24}$
$$= -i + 1 = 1 - i$$

Do Exercises 22–25.

e | Conjugates and Division

Conjugates of complex numbers are defined as follows.

CONJUGATE

The **conjugate** of a complex number $a + bi$ is $a - bi$, and the **conjugate** of $a - bi$ is $a + bi$.

Simplify.

18. i^{47}

19. i^{68}

20. i^{85}

21. i^{90}

Simplify.

22. $8 - i^5$

23. $7 + 4i^2$

24. $6i^{11} + 7i^{14}$

25. $i^{34} - i^{55}$

Answers on page A-34

Find the conjugate.

26. $6 + 3i$

EXAMPLES Find the conjugate.

22. $5 + 7i$ The conjugate is $5 - 7i$.

23. $14 - 3i$ The conjugate is $14 + 3i$.

24. $-3 - 9i$ The conjugate is $-3 + 9i$.

25. $4i$ The conjugate is $-4i$.

Do Exercises 26–28.

27. $-9 - 5i$

> When we multiply a complex number by its conjugate, we get a real number.

EXAMPLES Multiply.

26. $(5 + 7i)(5 - 7i) = 5^2 - (7i)^2$ Using $(A + B)(A - B) = A^2 - B^2$

$\qquad\qquad\qquad\qquad = 25 - 49i^2$

$\qquad\qquad\qquad\qquad = 25 - 49(-1)$ $i^2 = -1$

$\qquad\qquad\qquad\qquad = 25 + 49$

$\qquad\qquad\qquad\qquad = 74$

28. $\pi - \dfrac{1}{4}i$

27. $(2 - 3i)(2 + 3i) = 2^2 - (3i)^2$

$\qquad\qquad\qquad\qquad = 4 - 9i^2$

$\qquad\qquad\qquad\qquad = 4 - 9(-1)$ $i^2 = -1$

$\qquad\qquad\qquad\qquad = 4 + 9$

$\qquad\qquad\qquad\qquad = 13$

Do Exercises 29 and 30.

We use conjugates in dividing complex numbers.

Multiply.

29. $(7 - 2i)(7 + 2i)$

EXAMPLE 28 Divide and simplify to the form $a + bi$: $\dfrac{-5 + 9i}{1 - 2i}$.

$$\frac{-5 + 9i}{1 - 2i} \cdot \frac{1 + 2i}{1 + 2i} = \frac{(-5 + 9i)(1 + 2i)}{(1 - 2i)(1 + 2i)}$$

Multiplying by 1 using the conjugate of the denominator in the symbol for 1

$$= \frac{-5 - 10i + 9i + 18i^2}{1^2 - 4i^2}$$

$$= \frac{-5 - i + 18(-1)}{1 - 4(-1)} \qquad i^2 = -1$$

$$= \frac{-5 - i - 18}{1 + 4}$$

$$= \frac{-23 - i}{5}$$

$$= -\frac{23}{5} - \frac{1}{5}i$$

30. $(-3 - i)(-3 + i)$

Note the similarity between the preceding example and rationalizing denominators. In both cases, we used the conjugate of the denominator to write another name for 1. In Example 28, the symbol for the number 1 was chosen using the conjugate of the divisor, $1 - 2i$.

Answers on page A-34

EXAMPLE 29 What symbol for 1 would you use to divide?

Division to be done   *Symbol for 1*

$$\frac{3 + 5i}{4 + 3i}$$ $$\frac{4 - 3i}{4 - 3i}$$

Divide and simplify to the form $a + bi$.

31. $\frac{6 + 2i}{1 - 3i}$

EXAMPLE 30 Divide and simplify to the form $a + bi$: $\frac{3 + 5i}{4 + 3i}$.

$$\frac{3 + 5i}{4 + 3i} \cdot \frac{4 - 3i}{4 - 3i} = \frac{(3 + 5i)(4 - 3i)}{(4 + 3i)(4 - 3i)}$$ Multiplying by 1

$$= \frac{12 - 9i + 20i - 15i^2}{4^2 - 9i^2}$$

$$= \frac{12 + 11i - 15(-1)}{16 - 9(-1)}$$ $i^2 = -1$

$$= \frac{27 + 11i}{25} = \frac{27}{25} + \frac{11}{25}i$$

32. $\frac{2 + 3i}{-1 + 4i}$

Do Exercises 31 and 32.

Answers on page A-34

CALCULATOR CORNER

Complex Numbers We can perform operations on complex numbers on a graphing calculator. To do so, we first set the calculator in complex, or $a + bi$, mode by pressing **MODE**, using the ⬇ and ⬇ keys to position the blinking cursor over $a + bi$, and then pressing **ENTER**. We press **2ND** **QUIT** to go to the home screen. Now we can add, subtract, multiply, and divide complex numbers.

To find $(3 + 4i) - (7 - i)$, for example, we press **(** **3** **+** **4** **2ND** *i* **)** **−** **(** **7** **−** **2ND** *i* **)** **ENTER**. (i is the second operation associated with the **.** key.) Note that although the parentheses around $3 + 4i$ are optional, those around $7 - i$ are necessary to ensure that both parts of the second complex number are subtracted from the first number.

To find $\frac{5 - 2i}{-1 + 3i}$ and display the result using fraction notation, we press **(** **5** **−** **2** **2ND** *i* **)** **÷** **(** **(−)** **1** **+** **3** **2ND** *i* **)** **MATH** **1** **ENTER**. Since the fraction bar acts as a grouping symbol in the original expression, the parentheses must be used to group the numerator and the denominator when the expression is entered in the calculator. To find $\sqrt{-4} \cdot \sqrt{-9}$, we press **2ND** **√** **(−)** **4** **)** **×** **2ND** **√** **(−)** **9** **)** **ENTER**. Note that the calculator supplies the left parenthesis in each radicand and we supply the right parenthesis. The results of these operations are shown below.

```
(3+4i)−(7−i)
                       −4+5i
(5−2i)/(−1+3i) ▶Frac
                  −11/10−13/10i
√(−4)*√(−9)
                          −6
```

Exercises: Carry out each operation.

1. $(9 + 4i) + (-11 - 13i)$

2. $(9 + 4i) - (-11 - 13i)$

3. $(9 + 4i) \cdot (-11 - 13i)$

4. $(9 + 4i) \div (-11 - 13i)$

5. $\sqrt{-16} \cdot \sqrt{-25}$

6. $\sqrt{-23} \cdot \sqrt{-35}$

7. $\frac{4 - 5i}{-6 + 8i}$

8. $(-3i)^4$

9. $(1 - i)^3 - (2 + 3i)^4$

10. $\frac{(1 - i)^3}{(2 + 3i)^2}$

33. Determine whether $-i$ is a solution of $x^2 + 1 = 0$.

$$x^2 + 1 = 0$$

$$\overline{}$$
$$?$$

The equation $x^2 + 1 = 0$ has no real-number solution, but it has *two* nonreal complex solutions.

EXAMPLE 31   Determine whether i is a solution of the equation $x^2 + 1 = 0$.

We substitute i for x in the equation.

$$x^2 + 1 = 0$$

$$\overline{}$$
$$i^2 + 1 \;?\; 0$$
$$-1 + 1 \;\Big|$$
$$0 \;\Big| \quad \text{TRUE}$$

The number i is a solution.

Do Exercise 33.

Any equation consisting of a polynomial in one variable on one side and 0 on the other has complex-number solutions (some may be real). It is not always easy to find the solutions, but they always exist.

EXAMPLE 32   Determine whether $1 + i$ is a solution of the equation $x^2 - 2x + 2 = 0$.

We substitute $1 + i$ for x in the equation.

$$x^2 - 2x + 2 = 0$$

$$\overline{}$$
$$(1 + i)^2 - 2(1 + i) + 2 \;?\; 0$$
$$1 + 2i + i^2 - 2 - 2i + 2 \;\Big|$$
$$1 + 2i - 1 - 2 - 2i + 2 \;\Big|$$
$$(1 - 1 - 2 + 2) + (2 - 2)i \;\Big|$$
$$0 + 0i \;\Big|$$
$$0 \;\Big| \quad \text{TRUE}$$

The number $1 + i$ is a solution.

34. Determine whether $1 - i$ is a solution of $x^2 - 2x + 2 = 0$.

$$x^2 - 2x + 2 = 0$$

$$\overline{}$$
$$?$$

EXAMPLE 33   Determine whether $2i$ is a solution of $x^2 + 3x - 4 = 0$.

$$x^2 + 3x - 4 = 0$$

$$\overline{}$$
$$(2i)^2 + 3(2i) - 4 \;?\; 0$$
$$4i^2 + 6i - 4 \;\Big|$$
$$-4 + 6i - 4 \;\Big|$$
$$-8 + 6i \;\Big| \quad \text{FALSE}$$

The number $2i$ is not a solution.

Do Exercise 34.

Answers on page A-34

6.8

EXERCISE SET

For Extra Help

Math XL   MyMathLab   InterAct   Math Tutor   Digital Video   Student's
                        Math   Center   Tutor CD 3   Solutions
MathXL   MyMathLab   Math Tutor   Videotape 7   Manual
                              Center

a Express in terms of i.

1. $\sqrt{-35}$

2. $\sqrt{-21}$

3. $\sqrt{-16}$

4. $\sqrt{-36}$

5. $-\sqrt{-12}$

6. $-\sqrt{-20}$

7. $\sqrt{-3}$

8. $\sqrt{-4}$

9. $\sqrt{-81}$

10. $\sqrt{-27}$

11. $\sqrt{-98}$

12. $-\sqrt{-18}$

13. $-\sqrt{-49}$

14. $-\sqrt{-125}$

15. $4 - \sqrt{-60}$

16. $6 - \sqrt{-84}$

17. $\sqrt{-4} + \sqrt{-12}$

18. $-\sqrt{-76} + \sqrt{-125}$

b Add or subtract and simplify.

19. $(7 + 2i) + (5 - 6i)$

20. $(-4 + 5i) + (7 + 3i)$

21. $(4 - 3i) + (5 - 2i)$

22. $(-2 - 5i) + (1 - 3i)$

23. $(9 - i) + (-2 + 5i)$

24. $(6 + 4i) + (2 - 3i)$

25. $(6 - i) - (10 + 3i)$

26. $(-4 + 3i) - (7 + 4i)$

27. $(4 - 2i) - (5 - 3i)$

28. $(-2 - 3i) - (1 - 5i)$

29. $(9 + 5i) - (-2 - i)$

30. $(6 - 3i) - (2 + 4i)$

c Multiply.

31. $\sqrt{-36} \cdot \sqrt{-9}$

32. $\sqrt{-16} \cdot \sqrt{-64}$

33. $\sqrt{-7} \cdot \sqrt{-2}$

34. $\sqrt{-11} \cdot \sqrt{-3}$

35. $-3i \cdot 7i$

36. $8i \cdot 5i$

37. $-3i(-8 - 2i)$

38. $4i(5 - 7i)$

39. $(3 + 2i)(1 + i)$

40. $(4 + 3i)(2 + 5i)$

41. $(2 + 3i)(6 - 2i)$

42. $(5 + 6i)(2 - i)$

43. $(6 - 5i)(3 + 4i)$

44. $(5 - 6i)(2 + 5i)$

45. $(7 - 2i)(2 - 6i)$

46. $(-4 + 5i)(3 - 4i)$

47. $(3 - 2i)^2$

48. $(5 - 2i)^2$

49. $(1 + 5i)^2$

50. $(6 + 2i)^2$

51. $(-2 + 3i)^2$

52. $(-5 - 2i)^2$

d Simplify.

53. i^7

54. i^{11}

55. i^{24}

56. i^{35}

57. i^{42}

58. i^{64}

59. i^9

60. $(-i)^{71}$

61. i^6

62. $(-i)^4$

63. $(5i)^3$

64. $(-3i)^5$

CHAPTER 6: Radical Expressions,
Equations, and Functions

Simplify to the form $a + bi$.

65. $7 + i^4$

66. $-18 + i^3$

67. $i^{28} - 23i$

68. $i^{29} + 33i$

69. $i^2 + i^4$

70. $5i^5 + 4i^3$

71. $i^5 + i^7$

72. $i^{84} - i^{100}$

73. $1 + i + i^2 + i^3 + i^4$

74. $i - i^2 + i^3 - i^4 + i^5$

75. $5 - \sqrt{-64}$

76. $\sqrt{-12} + 36i$

77. $\dfrac{8 - \sqrt{-24}}{4}$

78. $\dfrac{9 + \sqrt{-9}}{3}$

e   Divide and simplify to the form $a + bi$.

79. $\dfrac{4 + 3i}{3 - i}$

80. $\dfrac{5 + 2i}{2 + i}$

81. $\dfrac{3 - 2i}{2 + 3i}$

82. $\dfrac{6 - 2i}{7 + 3i}$

83. $\dfrac{8 - 3i}{7i}$

84. $\dfrac{3 + 8i}{5i}$

85. $\dfrac{4}{3 + i}$

86. $\dfrac{6}{2 - i}$

87. $\dfrac{2i}{5 - 4i}$

88. $\dfrac{8i}{6 + 3i}$

89. $\dfrac{4}{3i}$

90. $\dfrac{5}{6i}$

91. $\dfrac{2 - 4i}{8i}$

92. $\dfrac{5 + 3i}{i}$

93. $\dfrac{6 + 3i}{6 - 3i}$

94. $\dfrac{4 - 5i}{4 + 5i}$

f   Determine whether the complex number is a solution of the equation.

95. $1 - 2i$;
$x^2 - 2x + 5 = 0$
  ?

96. $1 + 2i$;
$x^2 - 2x + 5 = 0$
  ?

97. $2 + i$;
$x^2 - 4x - 5 = 0$
  ?

98. $1 - i$;
$x^2 + 2x + 2 = 0$
  ?

99. $\mathbf{D_W}$ How are conjugates of complex numbers similar to the conjugates used in Section 6.5?

100. $\mathbf{D_W}$ Is every real number a complex number? Why or why not?

VOCABULARY REINFORCEMENT

In each of Exercises 101–108, fill in the blank with the correct term from the given list. Some of the choices may not be used.

101. An expression that consists of the quotient of two polynomials, where the polynomial in the denominator is nonzero, is called a(n) ————————— expression.   [5.1a]

102. In the equation $(A + B)(A - B) = A^2 - B^2$, the expression $A^2 - B^2$ is called a(n) —————————.   [4.2c]

103. When being graphed, the numbers in an ordered pair are called —————————.   [2.1a]

104. Every ————————— real number has two real-number square roots.   [6.1a]

105. An equality of ratios, $A/B = C/D$, read "A is to B as C is to D" is called a(n) —————————.   [5.6b]

106. A(n) ————————— is a polynomial that can be expressed as a binomial square.   [4.6a]

107. ————————— numbers do not have real-number square roots.   [6.1a]

108. The principle of ————————— states that if $ab = 0$, then $a = 0$ or $b = 0$ (or both).   [4.8a]

coordinates

intercepts

trinomial square

positive

negative

rational

irrational

proportion

zero products

difference of squares

cross product

SYNTHESIS

109. A complex function g is given by
$$g(z) = \frac{z^4 - z^2}{z - 1}.$$
Find $g(2i)$, $g(1 + i)$, and $g(-1 + 2i)$.

110. Evaluate $\dfrac{1}{w - w^2}$ when $w = \dfrac{1 - i}{10}$.

Express in terms of i.

111. $\dfrac{1}{8}\left(-24 - \sqrt{-1024}\right)$

112. $12\sqrt{-\dfrac{1}{32}}$

113. $7\sqrt{-64} - 9\sqrt{-256}$

Simplify.

114. $\dfrac{i^5 + i^6 + i^7 + i^8}{(1 - i)^4}$

115. $(1 - i)^3(1 + i)^3$

116. $\dfrac{5 - \sqrt{5}\,i}{\sqrt{5}\,i}$

117. $\dfrac{6}{1 + \dfrac{3}{i}}$

118. $\left(\dfrac{1}{2} - \dfrac{1}{3}i\right)^2 - \left(\dfrac{1}{2} + \dfrac{1}{3}i\right)^2$

119. $\dfrac{i - i^{38}}{1 + i}$

120. Find all numbers a for which the opposite of a is the same as the reciprocal of a.

The review that follows is meant to prepare you for a chapter exam. It consists of three parts. The first part, Concept Reinforcement, is designed to increase understanding of the concepts through true/false exercises. The second part is a list of important properties and formulas. The third part is the Review Exercises. These provide practice exercises for the exam, together with references to section objectives so you can go back and review. Before beginning, stop and look back over the skills you have obtained. What skills in mathematics do you have now that you did not have before studying this chapter?

CONCEPT REINFORCEMENT

Determine whether the statement is true or false. Answers are given at the back of the book.

_____ 1. If radical expressions are to be multiplied, their indexes must be the same.

_____ 2. For any real numbers $\sqrt[m]{a}$ and $\sqrt[n]{b}$, $\sqrt[m]{a} \cdot \sqrt[n]{b} = \sqrt[mn]{ab}$.

_____ 3. The square of a complex number is always a real number.

_____ 4. Every imaginary number is a complex number, but not every complex number is imaginary.

_____ 5. Every real number has two real-number square roots.

_____ 6. If $\sqrt[3]{q}$ is negative, then q is negative.

IMPORTANT PROPERTIES AND FORMULAS

$\sqrt{a^2} = |a|$;

$\sqrt[k]{a^k} = |a|$, when k is even; $\qquad \sqrt[k]{a^k} = a$, when k is odd;

$\sqrt[k]{ab} = \sqrt[k]{a} \cdot \sqrt[k]{b}$; $\qquad \sqrt[k]{\dfrac{a}{b}} = \dfrac{\sqrt[k]{a}}{\sqrt[k]{b}}$;

$a^{1/n} = \sqrt[n]{a}$;

$a^{m/n} = \sqrt[n]{a^m} = (\sqrt[n]{a})^m$; $\qquad a^{-m/n} = \dfrac{1}{a^{m/n}}$

Principle of Powers: If $a = b$ is true, then $a^n = b^n$ is true.

Pythagorean Theorem: $a^2 + b^2 = c^2$, in a right triangle.

$i = \sqrt{-1}, \qquad i^2 = -1, \qquad i^3 = -i, \qquad i^4 = 1$

Imaginary Numbers: $bi, i^2 = -1, b \neq 0$

Complex Numbers: $a + bi, i^2 = -1$

Conjugates: $a + bi, a - bi$

Review Exercises

Use a calculator to approximate to three decimal places. [6.1a]

1. $\sqrt{778}$

2. $\sqrt{\dfrac{963.2}{23.68}}$

3. For the given function, find the indicated function values. [6.1a]
$$f(x) = \sqrt{3x - 16}; \quad f(0), f(-1), f(1), \text{ and } f\left(\tfrac{41}{3}\right)$$

4. Find the domain of the function f in Exercise 3. [6.1a]

Simplify. Assume that letters represent *any* real number. [6.1b]

5. $\sqrt{81a^2}$

6. $\sqrt{(-7z)^2}$

7. $\sqrt{(c-3)^2}$

8. $\sqrt{x^2 - 6x + 9}$

Simplify. [6.1c]

9. $\sqrt[3]{-1000}$

10. $\sqrt[3]{-\dfrac{1}{27}}$

11. For the given function, find the indicated function values. [6.1c]
$$f(x) = \sqrt[3]{x + 2}; \quad f(6), f(-10), \text{ and } f(25)$$

Simplify. Assume that letters represent *any* real number. [6.1d]

12. $\sqrt[10]{x^{10}}$

13. $-\sqrt[13]{(-3)^{13}}$

Rewrite without rational exponents, and simplify, if possible. [6.2a]

14. $a^{1/5}$

15. $64^{3/2}$

Rewrite with rational exponents. [6.2a]

16. $\sqrt{31}$

17. $\sqrt[5]{a^2b^3}$

Rewrite with positive exponents, and simplify, if possible. [6.2b]

18. $49^{-1/2}$

19. $(8xy)^{-2/3}$

20. $5a^{-3/4}b^{1/2}c^{-2/3}$

21. $\dfrac{3a}{\sqrt[4]{t}}$

Use the laws of exponents to simplify. Write answers with positive exponents. [6.2c]

22. $(x^{-2/3})^{3/5}$

23. $\dfrac{7^{-1/3}}{7^{-1/2}}$

Use rational exponents to simplify. Write the answer in radical notation if appropriate. [6.2d]

24. $\sqrt[3]{x^{21}}$

25. $\sqrt[3]{27x^6}$

Use rational exponents to write a single radical expression. [6.2d]

26. $x^{1/3}y^{1/4}$

27. $\sqrt[4]{x}\,\sqrt[3]{x}$

Simplify by factoring. Assume that all expressions under radicals represent nonnegative numbers. [6.3a]

28. $\sqrt{245}$

29. $\sqrt[3]{-108}$

30. $\sqrt[3]{250a^2b^6}$

Simplify. Assume that all expressions under radicals represent positive numbers. [6.3b]

31. $\sqrt{\dfrac{49}{36}}$

32. $\sqrt[3]{\dfrac{64x^6}{27}}$

33. $\sqrt[4]{\dfrac{16x^8}{81y^{12}}}$

Perform the indicated operations and simplify. Assume that all expressions under radicals represent positive numbers. [6.3a, b]

34. $\sqrt{5x}\,\sqrt{3y}$

35. $\sqrt[3]{a^5b}\,\sqrt[3]{27b}$

36. $\sqrt[3]{a}\,\sqrt[5]{b^3}$

37. $\dfrac{\sqrt[3]{60xy^3}}{\sqrt[3]{10x}}$

38. $\dfrac{\sqrt{75x}}{2\sqrt{3}}$

39. $\dfrac{\sqrt[3]{x^2}}{\sqrt[4]{x}}$

Add or subtract. Assume that all expressions under radicals represent nonnegative numbers. [6.4a]

40. $5\sqrt[3]{x} + 2\sqrt[3]{x}$

41. $2\sqrt{75} - 7\sqrt{3}$

42. $\sqrt[3]{8x^4} + \sqrt[3]{xy^6}$

43. $\sqrt{50} + 2\sqrt{18} + \sqrt{32}$

CHAPTER 6: Radical Expressions, Equations, and Functions

Multiply. [6.4b]
44. $(\sqrt{5} - 3\sqrt{8})(\sqrt{5} + 2\sqrt{8})$

45. $(1 - \sqrt{7})^2$

46. $(\sqrt[3]{27} - \sqrt[3]{2})(\sqrt[3]{27} + \sqrt[3]{2})$

Rationalize the denominator. [6.5a, b]

47. $\sqrt{\dfrac{8}{3}}$ **48.** $\dfrac{2}{\sqrt{a} + \sqrt{b}}$

Solve. [6.6a, b]
49. $\sqrt[4]{x + 3} = 2$ **50.** $1 + \sqrt{x} = \sqrt{3x - 3}$

51. $x - 3 = \sqrt{5 - x}$

52. *Length of a Side of a Square.* The diagonal of a square has length $9\sqrt{2}$ cm. Find the length of a side of the square. [6.7a]

53. *Bookcase Width.* A bookcase is 5 ft tall and has a 7-ft diagonal brace, as shown. How wide is the bookcase? [6.7a]

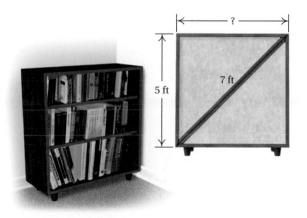

Automotive Repair. For an engine with a displacement of 2.8 L, the function given by
$$d(n) = 0.75\sqrt{2.8n}$$
can be used to determine the diameter size of the carburetor's opening, $d(n)$, in millimeters, where n is the number of rpm's at which the engine achieves peak performance. [6.6c]

Source: macdizzy.com

54. ▦ If a carburetor's opening is 81 mm, for what number of rpm's will the engine produce peak power?

55. ▦ If a carburetor's opening is 84 mm, for what number of rpm's will the engine produce peak power?

In a right triangle, find the length of the side not given. Give an exact answer and an answer to three decimal places. [6.7a]

56. $a = 7, \quad b = 24$ **57.** $a = 2, \quad c = 5\sqrt{2}$

58. Express in terms of i: $\sqrt{-25} + \sqrt{-8}$. [6.8a]

Add or subtract. [6.8b]
59. $(-4 + 3i) + (2 - 12i)$ **60.** $(4 - 7i) - (3 - 8i)$

Multiply. [6.8c, d]
61. $(2 + 5i)(2 - 5i)$ **62.** i^{13}

63. $(6 - 3i)(2 - i)$

Divide. [6.8e]
64. $\dfrac{-3 + 2i}{5i}$ **65.** $\dfrac{6 - 3i}{2 - i}$

66. Determine whether $1 + i$ is a solution of $x^2 + x + 2 = 0$. [6.8f]
$$\dfrac{x^2 + x + 2 = 0}{?}$$

67. Graph: $f(x) = \sqrt{x}$. [6.1a]

68. **D**$_W$ We learned a new method of equation solving in this chapter. Explain how this procedure differs from others we have used. [6.6a, b]

SYNTHESIS
---

69. Simplify: $i \cdot i^2 \cdot i^3 \cdots i^{99} \cdot i^{100}$. [6.8c, d]

70. Solve: $\sqrt{11x + \sqrt{6 + x}} = 6$. [6.6a]

1. Use a calculator to approximate $\sqrt{148}$ to three decimal places.

2. For the given function, find the indicated function values.
$$f(x) = \sqrt{8 - 4x}; \quad f(1) \text{ and } f(3)$$

3. Find the domain of the function f in Question 2.

Simplify. Assume that letters represent *any* real number.

4. $\sqrt{(-3q)^2}$

5. $\sqrt{x^2 + 10x + 25}$

6. $\sqrt[3]{-\dfrac{1}{1000}}$

7. $\sqrt[5]{x^5}$

8. $\sqrt[10]{(-4)^{10}}$

Rewrite without rational exponents, and simplify, if possible.

9. $a^{2/3}$

10. $32^{3/5}$

Rewrite with rational exponents.

11. $\sqrt{37}$

12. $\left(\sqrt{5xy^2}\right)^5$

Rewrite with positive exponents, and simplify, if possible.

13. $1000^{-1/3}$

14. $8a^{3/4}b^{-3/2}c^{-2/5}$

Use the laws of exponents to simplify. Write answers with positive exponents.

15. $(x^{2/3}y^{-3/4})^{12/5}$

16. $\dfrac{2.9^{-5/8}}{2.9^{2/3}}$

Use rational exponents to simplify. Write the answer in radical notation if appropriate. Assume that all expressions under radicals represent nonnegative numbers.

17. $\sqrt[8]{x^2}$

18. $\sqrt[4]{16x^6}$

Use rational exponents to write a single radical expression.

19. $a^{2/5}b^{1/3}$

20. $\sqrt[4]{2y}\,\sqrt[3]{y}$

Simplify by factoring. Assume that all expressions under radicals represent nonnegative numbers.

21. $\sqrt{148}$

22. $\sqrt[4]{80}$

23. $\sqrt[3]{24a^{11}b^{13}}$

Simplify. Assume that all expressions under radicals represent positive numbers.

24. $\sqrt[3]{\dfrac{16x^5}{y^7}}$

25. $\sqrt{\dfrac{25x^2}{36y^4}}$

Perform the indicated operations and simplify. Assume that all expressions under radicals represent positive numbers.

26. $\sqrt[3]{2x}\ \sqrt[3]{5y^2}$

27. $\sqrt[4]{x^3y^2}\ \sqrt[4]{xy}$

28. $\dfrac{\sqrt[5]{x^3y^4}}{\sqrt[5]{xy^2}}$

29. $\dfrac{\sqrt{300a}}{5\sqrt{3}}$

30. Add: $3\sqrt{128} + 2\sqrt{18} + 2\sqrt{32}$.

Multiply.

31. $\left(\sqrt{20} + 2\sqrt{5}\right)\left(\sqrt{20} - 3\sqrt{5}\right)$

32. $\left(3 + \sqrt{x}\right)^2$

33. Rationalize the denominator: $\dfrac{1 + \sqrt{2}}{3 - 5\sqrt{2}}$.

Solve.

34. $\sqrt[5]{x - 3} = 2$

35. $\sqrt{x - 6} = \sqrt{x + 9} - 3$

36. $\sqrt{x - 1} + 3 = x$

37. *Length of a Side of a Square.* The diagonal of a square has length $7\sqrt{2}$ ft. Find the length of a side of the square.

38. *Sighting to the Horizon.* A person can see 72 mi to the horizon from an airplane window. How high is the airplane? Use the formula $V = 3.5\sqrt{h}$.

In a right triangle, find the length of the side not given. Give an exact answer and an answer to three decimal places.

39. $a = 7, \quad b = 7$

40. $a = 1, \quad c = \sqrt{5}$

41. Express in terms of i: $\sqrt{-9} + \sqrt{-64}$.

42. Subtract: $(5 + 8i) - (-2 + 3i)$.

Multiply.

43. $(3 - 4i)(3 + 7i)$

44. i^{95}

45. Divide: $\dfrac{-7 + 14i}{6 - 8i}$.

46. Determine whether $1 + 2i$ is a solution of $x^2 + 2x + 5 = 0$.

47. Simplify: $\dfrac{1 - 4i}{4i(1 + 4i)^{-1}}$.

48. Solve: $\sqrt{2x - 2} + \sqrt{7x + 4} = \sqrt{13x + 10}$.

Games in a Sports League. In a sports league of n teams in which each team plays every other team twice, the total number N of games to be played is given by the function

$$N(n) = n^2 - n.$$

Use this function for Questions 1 and 2.

1. A men's rugby league has 6 teams. If we assume that each team plays every other team twice, what is the total number of games to be played?

2. Another rugby league plays a total of 72 games. If we assume that each team plays every other team twice, how many teams are in the league?

Simplify. Assume that all expressions under radicals represent nonnegative numbers.

3. $-\dfrac{3}{5}\left(-\dfrac{7}{10}\right)$

4. $36.2 - 73.4$

5. $7c - [4 - 3(6c - 9)]$

6. $6^2 - 3^3 \cdot 2 + 3 \cdot 8^2$

7. $\left(\dfrac{4x^4y^{-5}}{3x^{-3}}\right)^3$

8. $(2x^2 - 3x + 1) + (6x - 3x^3 + 7x^2 - 4)$

9. $(2x^2 - y)^2$

10. $(5x^2 - 2x + 1)(3x^2 + x - 2)$

11. $\dfrac{x^3 + 64}{x^2 - 49} \cdot \dfrac{x^2 - 14x + 49}{x^2 - 4x + 16}$

12. $\dfrac{x}{x + 2} + \dfrac{1}{x - 3} - \dfrac{x^2 - 2}{x^2 - x - 6}$

13. $\dfrac{\dfrac{y^2 - 5y - 6}{y^2 - 7y - 18}}{\dfrac{y^2 + 3y + 2}{y^2 + 4y + 4}}$

14. $(y^3 + 3y^2 - 5) \div (y + 2)$

15. $\sqrt[3]{-8x^3}$

16. $\sqrt{16x^2 - 32x + 16}$

17. $9\sqrt{75} + 6\sqrt{12}$

18. $\sqrt{2xy^2} \cdot \sqrt{8xy^3}$

19. $\dfrac{3\sqrt{5}}{\sqrt{6} - \sqrt{3}}$

20. $\sqrt[6]{\dfrac{m^{12}n^{24}}{64}}$

21. $6^{2/9} \cdot 6^{2/3}$

22. $(6 + i) - (3 - 4i)$

23. $\dfrac{2 - i}{6 + 5i}$

Solve.

24. $\dfrac{1}{5} + \dfrac{3}{10}x = \dfrac{4}{5}$

25. $M = \dfrac{1}{8}(c - 3)$, for c

26. $3a - 4 < 10 + 5a$

27. $-8 < x + 2 < 15$

28. $|3x - 6| = 2$

29. $3x + 5y = 30,$
    $5x + 3y = 34$

30. $3x + 2y - z = -7,$
$-x + y + 2z = 9,$
$5x + 5y + z = -1$

31. $625 = 49y^2$

32. $\dfrac{6x}{x-5} - \dfrac{300}{x^2 + 5x + 25} = \dfrac{2250}{x^3 - 125}$

33. $\dfrac{3x^2}{x+2} + \dfrac{5x-22}{x-2} = \dfrac{-48}{x^2-4}$

34. $I = \dfrac{nE}{R+nr}$, for R

35. $\sqrt{4x+1} - 2 = 3$

36. $2\sqrt{1-x} = \sqrt{5}$

37. $13 - x = 5 + \sqrt{x+4}$

Graph.

38. $f(x) = -\dfrac{2}{3}x + 2$

39. $4x - 2y = 8$

40. $4x \geq 5y + 20$

41. $y \geq -3,$
$y \leq 2x + 3$

42. $g(x) = x^2 - x - 2$

43. $f(x) = |x + 4|$

44. $g(x) = \dfrac{4}{x-3}$

45. $f(x) = 2 - \sqrt{x}$

Factor.

46. $12x^2y^2 - 30xy^3$

47. $3x^2 - 17x - 28$

48. $y^2 - y - 132$

49. $27y^3 + 8$

50. $4x^2 - 625$

Find the domain and the range of the function.

51.

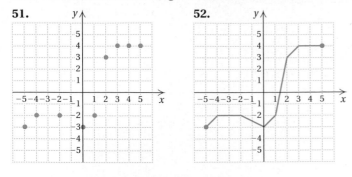

52.

53.

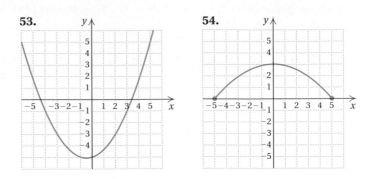

54.

55. Find the slope and the y-intercept of the line $3x - 2y = 8$.

56. Find an equation for the line perpendicular to the line $3x - y = 5$ and passing through $(1, 4)$.

Solve.

57. *Harvesting Time.* One combine can harvest a field in 3 hr. Another combine can harvest the same field in 1.5 hr. How long should it take them to harvest the field together?

58. *Triangle Area.* The height h of triangles of fixed area varies inversely as the base b. Suppose the height is 100 ft when the base is 20 ft. Find the height when the base is 16 ft. What is the fixed area?

59. *Heart Transplants.* In a recent year, the University of Pittsburgh Medical Center and the Medical College of Virginia had performed a total of 669 heart transplants. The University of Pittsburgh had performed 33 more than twice the number of transplants at the Medical College of Virginia. How many transplants had each hospital performed?

60. *Warning Dye.* A warning dye is used by people in lifeboats to aid search planes. The volume V of the dye used varies directly as the square of the diameter d of the circular patch of water formed by the dye. If 4 L of dye is required for a 10-m wide circle, how much dye is needed for a 40-m wide circle?

In each of Questions 61–64, choose the correct answer from the selections given.

61. Rewrite with rational exponents: $\sqrt[5]{xy^4}$.

a) $\dfrac{1}{(xy^4)^5}$

b) $(xy^4)^5$

c) $(xy)^{4/5}$

d) $(xy^4)^{1/5}$

e) None of these

62. A grain bin can be filled in 3 hr if the grain enters through spout A alone or in 15 hr if the grain enters through spout B alone. If grain is entering through both spouts at the same time, how many hours will it take to fill the bin?

a) $\frac{5}{2}$ hr

b) 9 hr

c) $22\frac{1}{2}$ hr

d) $10\frac{1}{2}$ hr

e) None of these

63. Divide: $(x^3 - x^2 + 2x + 4) \div (x - 3)$.

a) $x^2 + 2x + 8$, R -20

b) $x^2 + 2x - 4$, R -8

c) $x^2 - 4x - 10$, R -26

d) $x^2 - 4x + 14$, R 46

e) None of these

64. Solve: $2x + 6 = 8 + \sqrt{5x + 1}$.

a) $\frac{1}{4}$

b) 3

c) $3, \frac{1}{4}$

d) 4, 3

e) None of these

SYNTHESIS

65. *The Pythagorean Formula in Three Dimensions.* The length d of a diagonal of a rectangular box is given by $d = \sqrt{a^2 + b^2 + c^2}$, where a, b, and c are the lengths of the sides. Find the length of a diagonal of a box whose sides have lengths 2 ft, 4 ft, and 5 ft.

66. Solve: $\dfrac{x + \sqrt{x + 1}}{x - \sqrt{x + 1}} = \dfrac{5}{11}$.

CHAPTER 6: Radical Expressions, Equations, and Functions

Quadratic Equations and Functions

7

Real-World Application

The base of a triangular sail is 9 m less than its height. The area is 56 m². Find the base and the height of the sail.

This problem appears as Exercise 5 in Section 7.3.

Objectives

a Solve quadratic equations using the principle of square roots and find the x-intercepts of the graph of a related function.

b Solve quadratic equations by completing the square.

c Solve applied problems using quadratic equations.

AG *ALGEBRAIC–GRAPHICAL CONNECTION*

Let's reexamine the graphical connections to the algebraic equation-solving concepts we have studied before.

In Chapter 2, we introduced the graph of a quadratic function:

$$f(x) = ax^2 + bx + c, \quad a \neq 0.$$

For example, the graph of the function $f(x) = x^2 + 6x + 8$ and its x-intercepts are shown below.

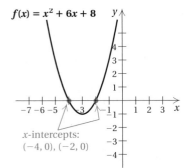

The x-intercepts are $(-4, 0)$ and $(-2, 0)$. These pairs are also the points of intersection of the graphs of $f(x) = x^2 + 6x + 8$ and $g(x) = 0$ (the x-axis). We will analyze the graphs of quadratic functions in greater detail in Sections 7.5–7.7.

In Chapter 4, we solved quadratic equations like $x^2 + 6x + 8 = 0$ using factoring, as here:

$$x^2 + 6x + 8 = 0$$
$$(x + 4)(x + 2) = 0 \qquad \text{Factoring}$$
$$x + 4 = 0 \quad or \quad x + 2 = 0 \qquad \text{Using the principle of zero products}$$
$$x = -4 \quad or \qquad x = -2.$$

We see that the solutions of $0 = x^2 + 6x + 8$, -4 and -2, are the first coordinates of the x-intercepts, $(-4, 0)$ and $(-2, 0)$, of the graph of $f(x) = x^2 + 6x + 8$.

Do Exercise 1.

We now extend our ability to solve quadratic equations.

a **The Principle of Square Roots**

The quadratic equation

$$5x^2 + 8x - 2 = 0$$

is said to be in **standard form.** The quadratic equation

$$5x^2 = 2 - 8x$$

is equivalent to the preceding, but it is *not* in standard form.

---

1. Consider solving the equation

$$x^2 - 6x + 8 = 0.$$

Below is the graph of

$$f(x) = x^2 - 6x + 8.$$

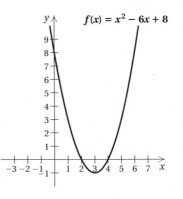

a) What are the x-intercepts of the graph?

b) What are the solutions of $x^2 - 6x + 8 = 0$?

c) What relationship exists between the answers to parts (a) and (b)?

Answers on page A-36

An equation of the type $ax^2 + bx + c = 0$, where a, b, and c are real-number constants and $a > 0$, is called the **standard form of a quadratic equation.**

To find the standard form of the quadratic equation $-5x^2 + 4x - 7 = 0$, we find an equivalent equation by multiplying by -1 on both sides:

$$-1(-5x^2 + 4x - 7) = -1(0)$$
$$5x^2 - 4x + 7 = 0. \qquad \text{Writing in standard form}$$

In Section 4.8, we studied the use of factoring and the principle of zero products to solve certain quadratic equations. Let's review that procedure and introduce a new one.

EXAMPLE 1

a) Solve: $x^2 = 25$.

b) Find the x-intercepts of $f(x) = x^2 - 25$.

a) We first find standard form and then factor:

$$x^2 - 25 = 0 \qquad \text{Subtracting 25}$$
$$(x - 5)(x + 5) = 0 \qquad \text{Factoring}$$
$$x - 5 = 0 \quad or \quad x + 5 = 0 \qquad \begin{array}{l}\text{Using the principle of}\\ \text{zero products}\end{array}$$
$$x = 5 \quad or \qquad x = -5.$$

The solutions are 5 and -5.

b) The x-intercepts of $f(x) = x^2 - 25$ are $(-5, 0)$ and $(5, 0)$. The solutions of the equation $x^2 = 25$ are the first coordinates of the x-intercepts of the graph of $f(x) = x^2 - 25$.

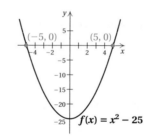

EXAMPLE 2 Solve: $6x^2 - 15x = 0$.

We factor and use the principle of zero products:

$$6x^2 - 15x = 0$$
$$3x(2x - 5) = 0$$
$$3x = 0 \quad or \quad 2x - 5 = 0$$
$$x = 0 \quad or \qquad x = \frac{5}{2}.$$

The solutions are 0 and $\frac{5}{2}$. The check is left to the student.

Do Exercises 2 and 3.

---

2. a) Solve: $x^2 = 16$.

b) Find the x-intercepts of $f(x) = x^2 - 16$.

3. a) Solve: $4x^2 + 14x = 0$.

b) Find the x-intercepts of $f(x) = 4x^2 + 14x$.

Answers on page A-36

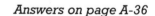

CALCULATOR CORNER

Solving Quadratic Equations Use the ZERO feature to solve the equations in Example 2 and Margin Exercise 3. See the Calculator Corner on p. 392 to review the procedure.

4. a) Solve: $5x^2 = 8x - 3$.

EXAMPLE 3

a) Solve: $3x^2 = 2 - x$.

b) Find the x-intercepts of $f(x) = 3x^2 + x - 2$.

a) We first find standard form. Then we factor and use the principle of zero products.

$$3x^2 = 2 - x$$

$$3x^2 + x - 2 = 0 \qquad \text{Adding } x \text{ and subtracting 2 to get the standard form}$$

$$(3x - 2)(x + 1) = 0 \qquad \text{Factoring}$$

$$3x - 2 = 0 \quad or \quad x + 1 = 0 \qquad \text{Using the principle of zero products}$$

$$3x = 2 \quad or \qquad x = -1$$

$$x = \tfrac{2}{3} \quad or \qquad x = -1$$

Check: For $\tfrac{2}{3}$:

$$\frac{3x^2 = 2 - x}{3\left(\tfrac{2}{3}\right)^2 \ ? \ 2 - \left(\tfrac{2}{3}\right)}$$

$$\begin{array}{c|c} 3 \cdot \tfrac{4}{9} & \tfrac{6}{3} - \tfrac{2}{3} \\ \tfrac{4}{3} & \tfrac{4}{3} \end{array} \quad \text{TRUE}$$

For -1:

$$\frac{3x^2 = 2 - x}{3(-1)^2 \ ? \ 2 - (-1)}$$

$$\begin{array}{c|c} 3 \cdot 1 & 2 + 1 \\ 3 & 3 \end{array} \quad \text{TRUE}$$

b) Find the x-intercepts of $f(x) = 5x^2 - 8x + 3$.

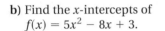
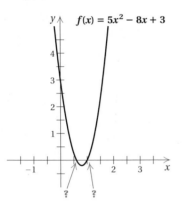

The solutions are -1 and $\tfrac{2}{3}$.

b) The x-intercepts of $f(x) = 3x^2 + x - 2$ are $(-1, 0)$ and $\left(\tfrac{2}{3}, 0\right)$. The solutions of the equation $3x^2 = 2 - x$ are the first coordinates of the x-intercepts of the graph of $f(x) = 3x^2 + x - 2$.

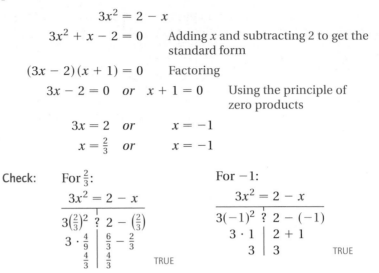
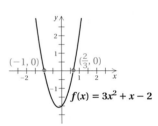

Do Exercise 4.

SOLVING EQUATIONS OF THE TYPE $x^2 = d$

Consider the equation $x^2 = 25$ again. We know from Chapter 6 that the number 25 has two real-number square roots, namely, 5 and -5. Note that these are the solutions of the equation in Example 1. This exemplifies the principle of square roots, which provides a quick method for solving equations of the type $x^2 = d$.

THE PRINCIPLE OF SQUARE ROOTS

The solutions of the equation $x^2 = d$ are $\sqrt{d}$ and $-\sqrt{d}$.

When $d > 0$, the solutions are two real numbers.

When $d = 0$, the only solution is 0.

When $d < 0$, the solutions are two imaginary numbers.

Answers on page A-36

EXAMPLE 4 Solve: $3x^2 = 6$. Give the exact solutions and approximate the solutions to three decimal places.

We have

$$3x^2 = 6$$
$$x^2 = 2$$
$$x = \sqrt{2} \quad or \quad x = -\sqrt{2}.$$

We often use the symbol $\pm\sqrt{2}$ to represent both of the solutions.

$(-\sqrt{2}, 0)$ $(\sqrt{2}, 0)$
$f(x) = 3x^2 - 6$

Check: For $\sqrt{2}$:

$$\frac{3x^2 = 6}{3(\sqrt{2})^2 \;?\; 6}$$
$$3 \cdot 2 \;\Big|$$
$$6 \;\Big| \quad \text{TRUE}$$

For $-\sqrt{2}$:

$$\frac{3x^2 = 6}{3(-\sqrt{2})^2 \;?\; 6}$$
$$3 \cdot 2 \;\Big|$$
$$6 \;\Big| \quad \text{TRUE}$$

The solutions are $\sqrt{2}$ and $-\sqrt{2}$, or $\pm\sqrt{2}$, which are about 1.414 and -1.414 when rounded to three decimal places.

Do Exercise 5.

Sometimes we rationalize denominators to simplify answers.

EXAMPLE 5 Solve: $-5x^2 + 2 = 0$. Give the exact solutions and approximate the solutions to three decimal places.

$$-5x^2 + 2 = 0$$
$$x^2 = \frac{2}{5} \qquad \text{Subtracting 2 and dividing by } -5$$
$$x = \sqrt{\frac{2}{5}} \quad or \quad x = -\sqrt{\frac{2}{5}} \qquad \text{Using the principle of square roots}$$
$$x = \sqrt{\frac{2}{5} \cdot \frac{5}{5}} \quad or \quad x = -\sqrt{\frac{2}{5} \cdot \frac{5}{5}} \qquad \text{Rationalizing the denominators}$$
$$x = \frac{\sqrt{10}}{5} \quad or \quad x = -\frac{\sqrt{10}}{5}$$

Check: We check both numbers at once, since there is no x-term in the equation. We could have checked both numbers at once in Example 4 as well.

$$\frac{-5x^2 + 2 = 0}{-5\left(\pm\frac{\sqrt{10}}{5}\right)^2 + 2 \;?\; 0}$$
$$-5\left(\frac{10}{25}\right) + 2$$
$$-2 + 2$$
$$0 \;\Big| \quad \text{TRUE}$$

$\left(-\dfrac{\sqrt{10}}{5}, 0\right)$ $\left(\dfrac{\sqrt{10}}{5}, 0\right)$
$f(x) = -5x^2 + 2$

The solutions are $\dfrac{\sqrt{10}}{5}$ and $-\dfrac{\sqrt{10}}{5}$, or $\pm\dfrac{\sqrt{10}}{5}$. We can use a calculator for approximations:

$$\pm\frac{\sqrt{10}}{5} \approx \pm 0.632.$$

5. Solve: $5x^2 = 15$. Give the exact solution and approximate the solutions to three decimal places.

6. Solve: $-3x^2 + 8 = 0$. Give the exact solution and approximate the solutions to three decimal places.

Answers on page A-36

CALCULATOR CORNER

Imaginary Solutions of Quadratic Equations

What happens when you use the ZERO feature to solve the equation in Example 6? Explain why this happens.

7. Solve: $2x^2 + 1 = 0$.

8. a) Solve: $(x - 1)^2 = 5$.

b) Find the x-intercepts of $f(x) = (x - 1)^2 - 5$.

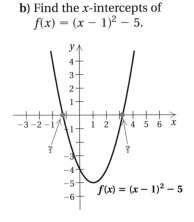

Sometimes we get solutions that are imaginary numbers.

EXAMPLE 6 Solve: $4x^2 + 9 = 0$.

$$4x^2 + 9 = 0$$

$$x^2 = -\frac{9}{4} \qquad \text{Subtracting 9 and dividing by 4}$$

$$x = \sqrt{-\frac{9}{4}} \quad or \quad x = -\sqrt{-\frac{9}{4}} \qquad \text{Using the principle of square roots}$$

$$x = \frac{3}{2}i \quad or \quad x = -\frac{3}{2}i \qquad \text{Simplifying}$$

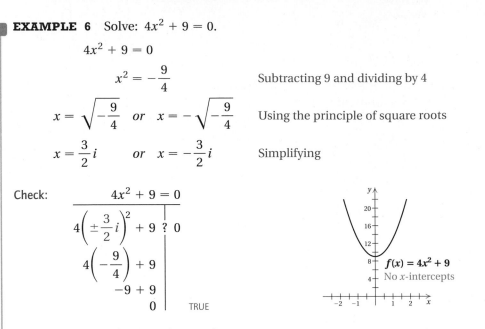

The solutions are $\frac{3}{2}i$ and $-\frac{3}{2}i$, or $\pm\frac{3}{2}i$.

We see that the graph of $f(x) = 4x^2 + 9$ does not cross the x-axis. This is true because the equation $4x^2 + 9 = 0$ has *imaginary* complex-number solutions.

Do Exercise 7.

SOLVING EQUATIONS OF THE TYPE $(x + c)^2 = d$

The equation $(x - 2)^2 = 7$ can also be solved using the principle of square roots.

EXAMPLE 7

a) Solve: $(x - 2)^2 = 7$.

b) Find the x-intercepts of $f(x) = (x - 2)^2 - 7$.

a) We have

$$(x - 2)^2 = 7$$

$$x - 2 = \sqrt{7} \qquad or \quad x - 2 = -\sqrt{7} \qquad \text{Using the principle of square roots}$$

$$x = 2 + \sqrt{7} \quad or \qquad x = 2 - \sqrt{7}.$$

The solutions are $2 + \sqrt{7}$ and $2 - \sqrt{7}$, or $2 \pm \sqrt{7}$.

b) The x-intercepts of $f(x) = (x - 2)^2 - 7$ are $\left(2 - \sqrt{7}, 0\right)$ and $\left(2 + \sqrt{7}, 0\right)$.

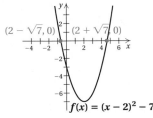

Do Exercise 8.

If we can express the left side of an equation as the square of a binomial, we can proceed as we did in Example 7.

9. Solve: $x^2 + 16x + 64 = 11$.

EXAMPLE 8 Solve: $x^2 + 6x + 9 = 2$.

We have

$$x^2 + 6x + 9 = 2 \quad \text{The left side is the square of a binomial.}$$
$$(x + 3)^2 = 2$$
$$x + 3 = \sqrt{2} \quad or \quad x + 3 = -\sqrt{2} \quad \text{Using the principle of square roots}$$
$$x = -3 + \sqrt{2} \quad or \quad x = -3 - \sqrt{2}.$$

The solutions are $-3 + \sqrt{2}$ and $-3 - \sqrt{2}$, or $-3 \pm \sqrt{2}$.

Do Exercise 9.

b Completing the Square

We can solve quadratic equations like $3x^2 = 6$ and $(x - 2)^2 = 7$ by using the principle of square roots. We can also solve an equation such as $x^2 + 6x + 9 = 2$ in like manner because the expression on the left side is the square of a binomial, $(x + 3)^2$. This second procedure is the basis for a method called **completing the square.** *It can be used to solve any quadratic equation.*

Suppose we have the following quadratic equation:

$$x^2 + 14x = 4.$$

If we could add on both sides of the equation a constant that would make the expression on the left the square of a binomial, we could then solve the equation using the principle of square roots.

How can we determine what to add to $x^2 + 14x$ to construct the square of a binomial? We want to find a number a such that the following equation is satisfied:

$$x^2 + 14x + a^2 = (x + a)(x + a) = x^2 + 2ax + a^2.$$

Thus a is such that $2a = 14$. Solving, we get $a = 7$. That is, a is half of the coefficient of x in $x^2 + 14x$. Since $a^2 = \left(\frac{14}{2}\right)^2 = 7^2 = 49$, we add 49 to our original expression:

$$x^2 + 14x + 49 \text{ is the square of } x + 7;$$

that is,

$$x^2 + 14x + 49 = (x + 7)^2.$$

> **COMPLETING THE SQUARE**
>
> When solving an equation, to **complete the square** of an expression like $x^2 + bx$, we take half the x-coefficient, which is $b/2$, and square it. Then we add that number, $(b/2)^2$, on both sides of the equation.

Answer on page A-36

Solve.

10. $x^2 + 6x + 8 = 0$

Returning to solving our original equation, we first add 49 on *both* sides to *complete the square* on the left. Then we solve:

$$x^2 + 14x = 4 \qquad \text{Original equation}$$
$$x^2 + 14x + 49 = 4 + 49 \qquad \text{Adding 49: } \left(\tfrac{14}{2}\right)^2 = 7^2 = 49$$
$$(x + 7)^2 = 53$$
$$x + 7 = \sqrt{53} \qquad or \quad x + 7 = -\sqrt{53} \qquad \text{Using the principle of square roots}$$
$$x = -7 + \sqrt{53} \quad or \qquad x = -7 - \sqrt{53}.$$

The solutions are $-7 \pm \sqrt{53}$.

We have seen that a quadratic equation $(x + c)^2 = d$ can be solved using the principle of square roots. Any equation, such as $x^2 - 6x + 8 = 0$, can be put in this form by completing the square. Then we can solve as before.

11. $x^2 - 8x - 20 = 0$

EXAMPLE 9 Solve: $x^2 - 6x + 8 = 0$.

We have

$$x^2 - 6x + 8 = 0$$
$$x^2 - 6x = -8. \qquad \text{Subtracting 8}$$

We take half of -6 and square it, to get 9. Then we add 9 on *both* sides of the equation. This makes the left side the square of a binomial, $x - 3$. We have now *completed the square*.

$$x^2 - 6x + 9 = -8 + 9 \qquad \text{Adding 9: } \left(\tfrac{-6}{2}\right)^2 = (-3)^2 = 9$$
$$(x - 3)^2 = 1$$
$$x - 3 = 1 \quad or \quad x - 3 = -1 \qquad \text{Using the principle of square roots}$$
$$x = 4 \quad or \qquad x = 2$$

The solutions are 2 and 4.

Do Exercises 10 and 11.

12. Solve by completing the square:

$$x^2 + 6x - 1 = 0.$$

EXAMPLE 10 Solve $x^2 + 4x - 7 = 0$ by completing the square.

We have

$$x^2 + 4x - 7 = 0$$
$$x^2 + 4x = 7 \qquad \text{Adding 7}$$
$$x^2 + 4x + 4 = 7 + 4 \qquad \text{Adding 4: } \left(\tfrac{4}{2}\right)^2 = (2)^2 = 4$$
$$(x + 2)^2 = 11$$
$$x + 2 = \sqrt{11} \qquad or \quad x + 2 = -\sqrt{11} \qquad \text{Using the principle of square roots}$$
$$x = -2 + \sqrt{11} \quad or \qquad x = -2 - \sqrt{11}.$$

The solutions are $-2 \pm \sqrt{11}$.

Do Exercise 12.

Answers on page A-36

When the coefficient of x^2 is not 1, we can make it 1, as shown in the following example.

EXAMPLE 11 Solve $3x^2 + 7x = 2$ by completing the square.

We have

$$3x^2 + 7x = 2$$

$$\frac{1}{3}(3x^2 + 7x) = \frac{1}{3} \cdot 2 \qquad \text{Multiplying by } \tfrac{1}{3} \text{ to make the } x^2\text{-coefficient 1}$$

$$x^2 + \frac{7}{3}x = \frac{2}{3} \qquad \text{Multiplying and simplifying}$$

$$x^2 + \frac{7}{3}x + \frac{49}{36} = \frac{2}{3} + \frac{49}{36} \qquad \text{Adding } \frac{49}{36}: \left[\frac{1}{2} \cdot \frac{7}{3}\right]^2 = \frac{49}{36}$$

$$\left(x + \frac{7}{6}\right)^2 = \frac{24}{36} + \frac{49}{36} \qquad \text{Finding a common denominator}$$

$$\left(x + \frac{7}{6}\right)^2 = \frac{73}{36}$$

$$x + \frac{7}{6} = \sqrt{\frac{73}{36}} \qquad or \quad x + \frac{7}{6} = -\sqrt{\frac{73}{36}} \qquad \begin{array}{l}\text{Using the principle of} \\ \text{square roots}\end{array}$$

$$x + \frac{7}{6} = \frac{\sqrt{73}}{6} \qquad or \quad x + \frac{7}{6} = -\frac{\sqrt{73}}{6}$$

$$x = -\frac{7}{6} + \frac{\sqrt{73}}{6} \qquad or \qquad x = -\frac{7}{6} - \frac{\sqrt{73}}{6}.$$

The solutions are $-\dfrac{7}{6} \pm \dfrac{\sqrt{73}}{6}$.

$f(x) = 3x^2 + 7x - 2$

$\left(-\dfrac{7}{6} - \dfrac{\sqrt{73}}{6}, 0\right)$ $\left(-\dfrac{7}{6} + \dfrac{\sqrt{73}}{6}, 0\right)$

Do Exercises 13 and 14.

Solve by completing the square.

13. $2x^2 + 6x = 5$

14. $3x^2 - 2x = 7$

Answers on page A-36

15. Solve by completing the square:
$$3x^2 = 2x - 1.$$

EXAMPLE 12 Solve $2x^2 = 3x - 7$ by completing the square.

$$2x^2 = 3x - 7$$

$$2x^2 - 3x = -7 \qquad \text{Subtracting } 3x$$

$$\frac{1}{2}(2x^2 - 3x) = \frac{1}{2} \cdot (-7) \qquad \begin{array}{l}\text{Multiplying by } \frac{1}{2} \text{ to make the} \\ x^2\text{-coefficient } 1\end{array}$$

$$x^2 - \frac{3}{2}x = -\frac{7}{2} \qquad \text{Multiplying and simplifying}$$

$$x^2 - \frac{3}{2}x + \frac{9}{16} = -\frac{7}{2} + \frac{9}{16} \qquad \text{Adding } \frac{9}{16}: \left[\frac{1}{2}\left(-\frac{3}{2}\right)\right]^2 = \left[-\frac{3}{4}\right]^2 = \frac{9}{16}$$

$$\left(x - \frac{3}{4}\right)^2 = -\frac{56}{16} + \frac{9}{16} \qquad \text{Finding a common denominator}$$

$$\left(x - \frac{3}{4}\right)^2 = -\frac{47}{16}$$

$$x - \frac{3}{4} = \sqrt{-\frac{47}{16}} \quad or \quad x - \frac{3}{4} = -\sqrt{-\frac{47}{16}} \qquad \begin{array}{l}\text{Using the principle} \\ \text{of square roots}\end{array}$$

$$x - \frac{3}{4} = i\sqrt{\frac{47}{16}} \quad or \quad x - \frac{3}{4} = -i\sqrt{\frac{47}{16}} \qquad \sqrt{-1} = i$$

$$x = \frac{3}{4} + \frac{i\sqrt{47}}{4} \quad or \quad x = \frac{3}{4} - \frac{i\sqrt{47}}{4}$$

The solutions are $\dfrac{3}{4} \pm i\dfrac{\sqrt{47}}{4}$.

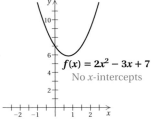

$$f(x) = 2x^2 - 3x + 7$$
No x-intercepts

We see that the graph of $f(x) = 2x^2 - 3x + 7$ does not cross the x-axis. This is true because the equation $2x^2 = 3x - 7$ has nonreal complex-number solutions.

Do Exercise 15.

SOLVING BY COMPLETING THE SQUARE

To solve an equation $ax^2 + bx + c = 0$ by completing the square:

1. If $a \neq 1$, multiply by $1/a$ so that the x^2-coefficient is 1.
2. If the x^2-coefficient is 1, add or subtract so that the equation is in the form

$$x^2 + bx = -c, \quad or \quad x^2 + \frac{b}{a}x = -\frac{c}{a} \text{ if step (1) has been applied.}$$

3. Take half of the x-coefficient and square it. Add the result on both sides of the equation.
4. Express the side with the variables as the square of a binomial.
5. Use the principle of square roots and complete the solution.

Answer on page A-36

CHAPTER 7: Quadratic Equations
and Functions

C Applications and Problem Solving

EXAMPLE 13 *Hang Time.* One of the most exciting plays in basketball is the dunk shot. The amount of time T that passes from the moment a player leaves the ground, goes up, makes the shot, and arrives back on the ground is called the *hang time*. A function relating an athlete's vertical leap V, in inches, to hang time T, in seconds, is given by

$$V(T) = 48T^2.$$

a) Hall-of-Famer Michael Jordan has a hang time of about 0.889 sec. What is his vertical leap?

b) Although his height is only 5 ft 7 in., Spud Webb, formerly of the Sacramento Kings, had a vertical leap of about 44 in. What is his hang time?

Source: Peter Brancazio, "The Mechanics of a Slam Dunk," *Popular Mechanics*, November 1991. Courtesy of Professor Peter Brancazio, Brooklyn College.

a) To find Jordan's vertical leap, we substitute 0.889 for T in the function and compute V:

$$V(0.889) = 48(0.889)^2 \approx 37.9 \text{ in.}$$

Jordan's vertical leap is about 37.9 in. Surprisingly, Jordan does not have the vertical leap most fans would expect.

b) To find Webb's hang time, we substitute 44 for V and solve for T:

$$44 = 48T^2 \qquad \text{Substituting 44 for } V$$

$$\frac{44}{48} = T^2 \qquad \text{Solving for } T^2$$

$$0.91\overline{6} = T^2$$

$$\sqrt{0.91\overline{6}} = T \qquad \text{Hang time is positive.}$$

$$0.957 \approx T. \qquad \text{Using a calculator}$$

Webb's hang time is 0.957 sec. Note that his hang time is greater than Jordan's.

Do Exercises 16 and 17.

Completing the square provides a base for proving the quadratic formula in Section 7.2 and for our work with conic sections in Chapter 9.

---

16. Hang Time. Carl Landry, of Purdue University, has a hang time of about 0.835 sec. What is his vertical leap?

17. Hang Time. The record for the greatest vertical leap in the NBA is held by Darryl Griffith of the Utah Jazz. It was 48 in. What was his hang time?

Answers on page A-36

a

1. a) Solve:
$6x^2 = 30$.
b) Find the x-intercepts of $f(x) = 6x^2 - 30$.

$f(x) = 6x^2 - 30$

2. a) Solve:
$5x^2 = 35$.
b) Find the x-intercepts of $f(x) = 5x^2 - 35$.

$f(x) = 5x^2 - 35$

3. a) Solve:
$9x^2 + 25 = 0$.
b) Find the x-intercepts of $f(x) = 9x^2 + 25$.

$f(x) = 9x^2 + 25$

4. a) Solve:
$36x^2 + 49 = 0$.
b) Find the x-intercepts of $f(x) = 36x^2 + 49$.

$f(x) = 36x^2 + 49$

Solve. Give the exact solution and approximate solutions to three decimal places, when appropriate.

5. $2x^2 - 3 = 0$

6. $3x^2 - 7 = 0$

7. $(x + 2)^2 = 49$

8. $(x - 1)^2 = 6$

9. $(x - 4)^2 = 16$

10. $(x + 3)^2 = 9$

11. $(x - 11)^2 = 7$

12. $(x - 9)^2 = 34$

13. $(x - 7)^2 = -4$

14. $(x + 1)^2 = -9$

15. $(x - 9)^2 = 81$

16. $(t - 2)^2 = 25$

17. $\left(x - \frac{3}{2}\right)^2 = \frac{7}{2}$

18. $\left(y + \frac{3}{4}\right)^2 = \frac{17}{16}$

19. $x^2 + 6x + 9 = 64$

20. $x^2 + 10x + 25 = 100$

21. $y^2 - 14y + 49 = 4$

22. $p^2 - 8p + 16 = 1$

b

Solve by completing the square. Show your work.

23. $x^2 + 4x = 2$

24. $x^2 + 2x = 5$

25. $x^2 - 22x = 11$

26. $x^2 - 18x = 10$

27. $x^2 + x = 1$

28. $x^2 - x = 3$

29. $t^2 - 5t = 7$

30. $y^2 + 9y = 8$

31. $x^2 + \frac{3}{2}x = 3$

32. $x^2 - \frac{4}{3}x = \frac{2}{3}$

33. $m^2 - \frac{9}{2}m = \frac{3}{2}$

34. $r^2 + \frac{2}{5}r = \frac{4}{5}$

35. $x^2 + 6x - 16 = 0$

36. $x^2 - 8x + 15 = 0$

37. $x^2 + 22x + 102 = 0$ **38.** $x^2 + 18x + 74 = 0$ **39.** $x^2 - 10x - 4 = 0$ **40.** $x^2 + 10x - 4 = 0$

41. a) Solve:
$x^2 + 7x - 2 = 0$.
b) Find the
x-intercepts of
$f(x) = x^2 + 7x - 2$.

42. a) Solve:
$x^2 - 7x - 2 = 0$.
b) Find the
x-intercepts of
$f(x) = x^2 - 7x - 2$.

43. a) Solve:
$2x^2 - 5x + 8 = 0$.
b) Find the
x-intercepts of
$f(x) = 2x^2 - 5x + 8$.

44. a) Solve:
$2x^2 - 3x + 9 = 0$.
b) Find the
x-intercepts of
$f(x) = 2x^2 - 3x + 9$.

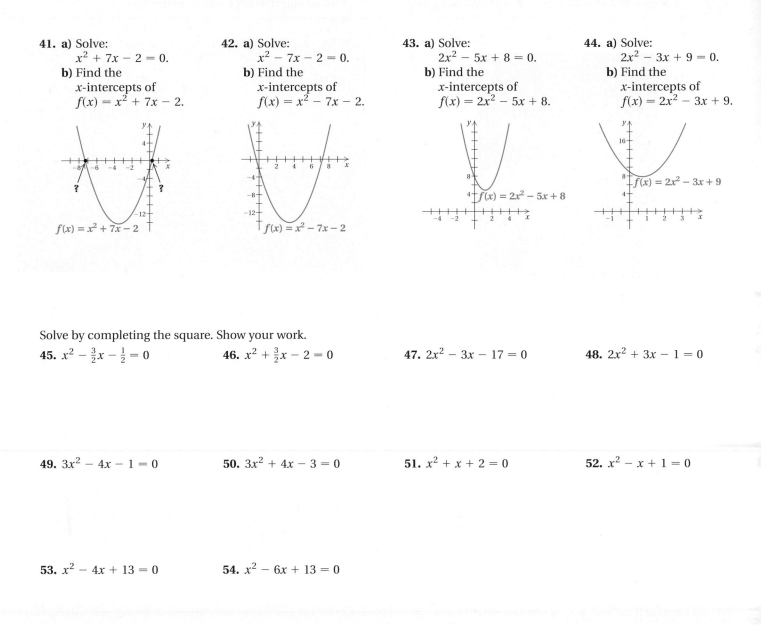

Solve by completing the square. Show your work.

45. $x^2 - \frac{3}{2}x - \frac{1}{2} = 0$ **46.** $x^2 + \frac{3}{2}x - 2 = 0$ **47.** $2x^2 - 3x - 17 = 0$ **48.** $2x^2 + 3x - 1 = 0$

49. $3x^2 - 4x - 1 = 0$ **50.** $3x^2 + 4x - 3 = 0$ **51.** $x^2 + x + 2 = 0$ **52.** $x^2 - x + 1 = 0$

53. $x^2 - 4x + 13 = 0$ **54.** $x^2 - 6x + 13 = 0$

    Hang Time. For Exercises 55 and 56, use the hang-time function $V(T) = 48T^2$, relating vertical leap to hang time.

55. The NBA's Vince Carter, of the New Jersey Nets, has a vertical leap of about 36 in. What is his hang time?

56. Tracy McGrady, of the Houston Rockets, has a vertical leap of 40 in. What is his hang time?

Free-Falling Objects. The function $s(t) = 16t^2$ is used to approximate the distance s, in feet, that an object falls freely from rest in t seconds. Use the formula for Exercises 57–60.

57. Suspended 1053 ft above the water, the bridge over Colorado's Royal Gorge is the world's highest bridge. How long would it take an object to fall freely from the bridge?

Source: *The Guinness Book of Records*

58. The CN Tower in Toronto, at 1815 ft, is the world's tallest self-supporting tower (no guy wires). How long would it take an object to fall freely from the top?

Source: *The Guinness Book of Records*

59. Reaching 745 ft above the water, the towers of California's Golden Gate Bridge are the world's tallest bridge towers. How long would it take an object to fall freely from the top?

Source: *The Guinness Book of Records*

60. The Gateway Arch in St. Louis is 640 ft high. How long would it take an object to fall freely from the top?

61. **D**_W Explain in your own words a sequence of steps that you might follow to solve any quadratic equation.

62. **D**_W Write a problem involving hang time for a classmate to solve. (See Example 13.) Devise the problem so that the solution is "The player's hang time is about 0.935 sec."

SKILL MAINTENANCE

63. *Tattoo Removal.* The following table lists data regarding the number of people who visited a doctor for tattoo removal in 1996 and in 2000. [2.6e]

NUMBER OF YEARS SINCE 1995	NUMBER OF PEOPLE WHO VISITED A DOCTOR FOR TATTOO REMOVAL (in thousands)
1	275
5	410

Source: *Star-Tribune*, Minneapolis–St. Paul

a) Use the two data points to find a linear function $T(t) = mt + b$ that fits the data.
b) Use the function to estimate the number of people who will visit a doctor in 2008 to have a tattoo removed.
c) In what year will 1,085,000 people visit a doctor for tattoo removal?

Graph.   [2.2c], [2.5a]

64. $f(x) = 5 - 2x^2$

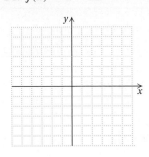

65. $f(x) = 5 - 2x$

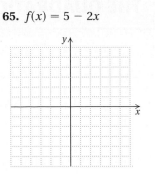

66. $2x - 5y = 10$

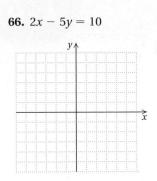

67. $f(x) = |5 - 2x|$

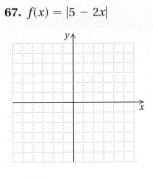

68. Simplify: $\sqrt{88}$.   [6.3a]

69. Rationalize the denominator: $\sqrt{\dfrac{2}{5}}$.   [6.5a]

Solve.   [6.6a, b]

70. $\sqrt{5x - 4} + \sqrt{13 - x} = 7$

71. $\sqrt{4x - 4} = \sqrt{x + 4} + 1$

72. $\sqrt{7x - 5} = \sqrt{4x + 7}$

73. $-35 = \sqrt{2x + 5}$

SYNTHESIS

74. Use a graphing calculator to solve each of the following equations.

a) $25.55x^2 - 1635.2 = 0$
b) $-0.0644x^2 + 0.0936x + 4.56 = 0$
c) $2.101x + 3.121 = 0.97x^2$

75. Problems such as those in Exercises 17, 21, and 25 can be solved without first finding standard form by using the INTERSECT feature on a graphing calculator. We let y_1 = the left side of the equation and y_2 = the right side. Use a graphing calculator to solve Exercises 17, 21, and 25 in this manner.

Find b such that the trinomial is a square.

76. $x^2 + bx + 75$

77. $x^2 + bx + 64$

Solve.

78. $\left(x - \frac{1}{3}\right)\left(x - \frac{1}{3}\right) + \left(x - \frac{1}{3}\right)\left(x + \frac{2}{9}\right) = 0$

79. $x(2x^2 + 9x - 56)(3x + 10) = 0$

80. *Boating.*   A barge and a fishing boat leave a dock at the same time, traveling at right angles to each other. The barge travels 7 km/h slower than the fishing boat. After 4 hr, the boats are 68 km apart. Find the speed of each vessel.

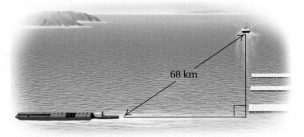

68 km

Objective

a Solve quadratic equations using the quadratic formula, and approximate solutions using a calculator.

There are at least two reasons for learning to complete the square. One is to enhance your ability to graph certain equations that are needed to solve problems in Section 7.7. The other is to prove a general formula for solving quadratic equations.

a Solving Using the Quadratic Formula

Each time you solve by completing the square, the procedure is the same. When we do the same kind of procedure many times, we look for a formula to speed up our work. Consider

$$ax^2 + bx + c = 0, \quad a > 0.$$

Note that if $a < 0$, we can get an equivalent form with $a > 0$ by first multiplying by -1.

Let's solve by *completing the square*. As we carry out the steps, compare them with Example 12 in the preceding section.

$$x^2 + \frac{b}{a}x + \frac{c}{a} = 0 \qquad \text{Multiplying by } \frac{1}{a}$$

$$x^2 + \frac{b}{a}x \qquad = -\frac{c}{a} \qquad \text{Subtracting } \frac{c}{a}$$

Half of $\frac{b}{a}$ is $\frac{b}{2a}$. The square is $\frac{b^2}{4a^2}$. We add $\frac{b^2}{4a^2}$ on both sides:

$$x^2 + \frac{b}{a}x + \frac{b^2}{4a^2} = -\frac{c}{a} + \frac{b^2}{4a^2} \qquad \text{Adding } \frac{b^2}{4a^2}$$

$$\left(x + \frac{b}{2a}\right)^2 = -\frac{4ac}{4a^2} + \frac{b^2}{4a^2} \qquad \begin{array}{l}\text{Factoring the left side and finding a}\\ \text{common denominator on the right}\end{array}$$

$$\left(x + \frac{b}{2a}\right)^2 = \frac{b^2 - 4ac}{4a^2}$$

$$x + \frac{b}{2a} = \sqrt{\frac{b^2 - 4ac}{4a^2}} \quad or \quad x + \frac{b}{2a} = -\sqrt{\frac{b^2 - 4ac}{4a^2}}. \qquad \begin{array}{l}\text{Using the principle}\\ \text{of square roots}\end{array}$$

Since $a > 0$, $\sqrt{4a^2} = 2a$, so we can simplify as follows:

$$x + \frac{b}{2a} = \frac{\sqrt{b^2 - 4ac}}{2a} \quad or \quad x + \frac{b}{2a} = -\frac{\sqrt{b^2 - 4ac}}{2a}.$$

Thus,

$$x = -\frac{b}{2a} \pm \frac{\sqrt{b^2 - 4ac}}{2a}, \quad or \quad x = \frac{-b \pm \sqrt{b^2 - 4ac}}{2a}.$$

We now have the following.

THE QUADRATIC FORMULA

The solutions of $ax^2 + bx + c = 0$ are given by

$$x = \frac{-b \pm \sqrt{b^2 - 4ac}}{2a}.$$

The formula also holds when $a < 0$. A similar proof would show this, but we will not consider it here.

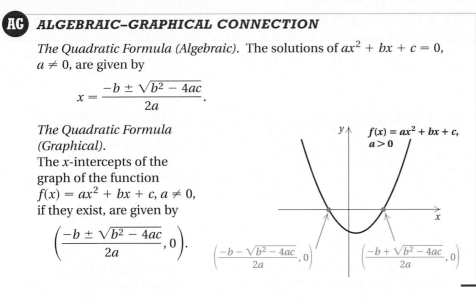

AG **ALGEBRAIC–GRAPHICAL CONNECTION**

The Quadratic Formula (Algebraic). The solutions of $ax^2 + bx + c = 0$, $a \neq 0$, are given by

$$x = \frac{-b \pm \sqrt{b^2 - 4ac}}{2a}.$$

The Quadratic Formula (Graphical).
The x-intercepts of the graph of the function $f(x) = ax^2 + bx + c, a \neq 0$, if they exist, are given by

$$\left(\frac{-b \pm \sqrt{b^2 - 4ac}}{2a}, 0 \right).$$

$f(x) = ax^2 + bx + c,$
$a > 0$

$$\left(\frac{-b - \sqrt{b^2 - 4ac}}{2a}, 0 \right)$$

$$\left(\frac{-b + \sqrt{b^2 - 4ac}}{2a}, 0 \right)$$

EXAMPLE 1 Solve $5x^2 + 8x = -3$ using the quadratic formula.

We first find standard form and determine a, b, and c:

$$5x^2 + 8x + 3 = 0;$$

$$a = 5, \quad b = 8, \quad c = 3.$$

We then use the quadratic formula:

$$x = \frac{-b \pm \sqrt{b^2 - 4ac}}{2a}$$

$$x = \frac{-8 \pm \sqrt{8^2 - 4 \cdot 5 \cdot 3}}{2 \cdot 5} \quad \text{Substituting}$$

$$x = \frac{-8 \pm \sqrt{64 - 60}}{10}$$

Be sure to write the fraction bar all the way across.

$$x = \frac{-8 \pm \sqrt{4}}{10}$$

$$x = \frac{-8 \pm 2}{10}$$

$$x = \frac{-8 + 2}{10} \quad or \quad x = \frac{-8 - 2}{10}$$

$$x = \frac{-6}{10} \quad or \quad x = \frac{-10}{10}$$

$$x = -\frac{3}{5} \quad or \quad x = -1.$$

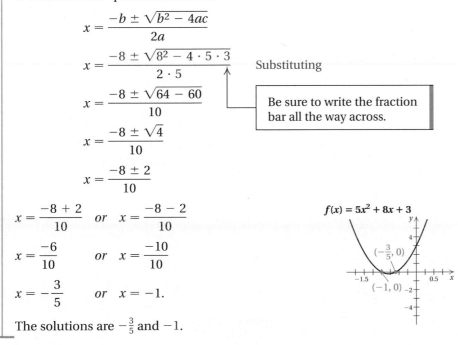

$f(x) = 5x^2 + 8x + 3$

$\left(-\frac{3}{5}, 0\right)$

$(-1, 0)$

The solutions are $-\frac{3}{5}$ and -1.

589

1. Consider the equation

$$2x^2 = 4 + 7x.$$

a) Solve using the quadratic formula.

b) Solve by factoring.

Answers on page A-37

Answers on page A-37

CALCULATOR CORNER

Approximating Solutions of Quadratic Equations   In Example 2, we found that the solutions of the equation $5x^2 - 8x = 3$ are $\dfrac{4 + \sqrt{31}}{5}$ and $\dfrac{4 - \sqrt{31}}{5}$. We can use a calculator to approximate these solutions. To approximate $\dfrac{4 + \sqrt{31}}{5}$, we press `(` `4` `+` `2ND` `√` `3` `1` `)` `)` `÷` `5` `ENTER`. To approximate $\dfrac{4 - \sqrt{31}}{5}$, we press `(` `4` `−` `2ND` `√` `3` `1` `)` `)` `÷` `5` `ENTER`. We see that the solutions are approximately 1.914 and −0.314.

```
(4+√(31))/5
              1.913552873
(4-√(31))/5
              -.3135528726
```

Exercises: Use a calculator to approximate the solutions in each of the following. Round to three decimal places.

1. Example 4

2. Margin Exercise 2

3. Margin Exercise 4

It turns out that we could have solved the equation in Example 1 more easily by factoring as follows:

$$5x^2 + 8x + 3 = 0$$
$$(5x + 3)(x + 1) = 0$$
$$5x + 3 = 0 \quad \textit{or} \quad x + 1 = 0$$
$$5x = -3 \quad \textit{or} \qquad x = -1$$
$$x = -\tfrac{3}{5} \quad \textit{or} \qquad x = -1.$$

To solve a quadratic equation:

1. Check for the form $x^2 = d$ or $(x + c)^2 = d$. If it is in this form, use the principle of square roots as in Section 7.1.
2. If it is not in the form of step (1), write it in standard form $ax^2 + bx + c = 0$ with a and b nonzero.
3. Then try factoring.
4. If it is not possible to factor or if factoring seems difficult, use the quadratic formula.

The solutions of a quadratic equation cannot always be found by factoring. They can *always* be found using the quadratic formula.

The solutions to all the exercises in this section could also be found by completing the square. However, the quadratic formula is the preferred method when solving applied problems because it is faster.

Do Exercise 1.

We will see in Example 2 that we cannot always rely on factoring.

EXAMPLE 2   Solve: $5x^2 - 8x = 3$. Give the exact solution and approximate the solutions to three decimal places.

We first find standard form and determine a, b, and c:

$$5x^2 - 8x - 3 = 0;$$
$$a = 5, \quad b = -8, \quad c = -3.$$

We then use the quadratic formula, $x = \dfrac{-b \pm \sqrt{b^2 - 4ac}}{2a}$:

$$x = \frac{-(-8) \pm \sqrt{(-8)^2 - 4 \cdot 5 \cdot (-3)}}{2 \cdot 5} \quad \text{Substituting}$$

$$= \frac{8 \pm \sqrt{64 + 60}}{10} = \frac{8 \pm \sqrt{124}}{10} = \frac{8 \pm \sqrt{4 \cdot 31}}{10}$$

$$= \frac{8 \pm 2\sqrt{31}}{10} = \frac{2(4 \pm \sqrt{31})}{2 \cdot 5} = \frac{2}{2} \cdot \frac{4 \pm \sqrt{31}}{5} = \frac{4 \pm \sqrt{31}}{5}.$$

Caution!

To avoid a common error in simplifying, remember to *factor the numerator and the denominator* and then remove a factor of 1.

We can use a calculator to approximate the solutions:

$$\frac{4 + \sqrt{31}}{5} \approx 1.914; \qquad \frac{4 - \sqrt{31}}{5} \approx -0.314.$$

CHAPTER 7: Quadratic Equations and Functions

Check: Checking the exact solutions $(4 \pm \sqrt{31})/5$ can be quite cumbersome. It could be done on a calculator or by using the approximations. Here we check 1.914; the check for -0.314 is left to the student.

For 1.914:

$$5x^2 - 8x = 3$$

$$\overline{\begin{array}{l} 5(1.914)^2 - 8(1.914) \;?\; 3 \\ 5(3.663396) - 15.312 \\ 3.00498 \end{array}}$$ APPROXIMATELY TRUE

We do not have a perfect check due to the rounding error. But our check seems to confirm the solutions.

Do Exercise 2.

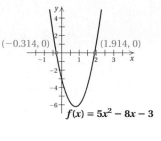

$(-0.314, 0)$ $(1.914, 0)$

$f(x) = 5x^2 - 8x - 3$

2. Solve using the quadratic formula:

$$3x^2 + 2x = 7.$$

Give the exact solution and approximate solutions to three decimal places.

Some quadratic equations have solutions that are nonreal complex numbers.

EXAMPLE 3   Solve: $x^2 + x + 1 = 0$.

We have $a = 1$, $b = 1$, $c = 1$. We use the quadratic formula:

$$x = \frac{-1 \pm \sqrt{1^2 - 4 \cdot 1 \cdot 1}}{2 \cdot 1}$$

$$= \frac{-1 \pm \sqrt{1 - 4}}{2}$$

$$= \frac{-1 \pm \sqrt{-3}}{2}$$

$$= \frac{-1 \pm i\sqrt{3}}{2}.$$

$f(x) = x^2 + x + 1$

No x-intercepts

3. Solve: $x^2 - x + 2 = 0$.

The solutions are

$$\frac{-1 + i\sqrt{3}}{2} \quad \text{and} \quad \frac{-1 - i\sqrt{3}}{2}.$$

The solutions can also be expressed in the form

$$-\frac{1}{2} + i\frac{\sqrt{3}}{2} \quad \text{and} \quad -\frac{1}{2} - i\frac{\sqrt{3}}{2}.$$

Do Exercise 3.

EXAMPLE 4   Solve: $2 + \dfrac{7}{x} = \dfrac{5}{x^2}$. Give the exact solution and approximate solutions to three decimal places.

We first find standard form:

$$x^2\left(2 + \frac{7}{x}\right) = x^2 \cdot \frac{5}{x^2} \qquad \text{Multiplying by } x^2 \text{ to clear fractions, noting that } x \neq 0$$

$$2x^2 + 7x = 5$$

$$2x^2 + 7x - 5 = 0. \qquad \text{Subtracting 5}$$

Answers on page A-37

4. Solve:

$$3 = \frac{5}{x} + \frac{4}{x^2}.$$

Give the exact solution and approximate solutions to three decimal places.

Then

$$a = 2, \quad b = 7, \quad c = -5$$

$$x = \frac{-7 \pm \sqrt{7^2 - 4 \cdot 2 \cdot (-5)}}{2 \cdot 2} \quad \text{Substituting}$$

$$x = \frac{-7 \pm \sqrt{49 + 40}}{4} = \frac{-7 \pm \sqrt{89}}{4}$$

$$x = \frac{-7 + \sqrt{89}}{4} \quad or \quad x = \frac{-7 - \sqrt{89}}{4}.$$

The quadratic formula always gives correct results when we begin with the standard form. In such cases, we need check only to detect errors in substitution and computation. In this case, since we began with a rational equation, which was *not* in standard form, we *do* need to check. We cleared the fractions before obtaining standard form, and this step could introduce numbers that do not check in the original equation. At the very least, we need to show that neither of the numbers makes a denominator 0. Since neither of them does, the solutions are

$$\frac{-7 + \sqrt{89}}{4} \quad \text{and} \quad \frac{-7 - \sqrt{89}}{4}.$$

We can use a calculator to approximate the solutions:

$$\frac{-7 + \sqrt{89}}{4} \approx 0.608;$$

$$\frac{-7 - \sqrt{89}}{4} \approx -4.108.$$

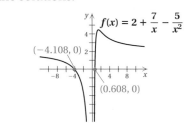

Do Exercise 4.

Answer on page A-37

CALCULATOR CORNER

Visualizing Solutions of Quadratic Equations The graph of $f(x) = 5x^2 - 8x - 3$ in Example 2 has two x-intercepts. This indicates that the equation $5x^2 - 8x - 3 = 0$ (or, equivalently, $5x^2 - 8x = 3$) has two real-number solutions. The graph of $f(x) = x^2 + x + 1$ in Example 3 has no x-intercepts. This indicates that the equation $x^2 + x + 1 = 0$ has no real-number solutions.

Exercises:

1. Explain how the graph of $y = x^2 - x + 2$ shows that the equation $x^2 - x + 2 = 0$ has no real-number solutions.

2. Use a graph to determine whether the equation $x^2 + x = 1$ has real-number solutions. If so, how many are there?

3. Use a graph to determine whether the equation $x^2 - x - 2 = 0$ has real-number solutions. If so, how many are there?

4. Use a graph to determine whether the equation $4x^2 + 1 = 4x$ has real-number solutions. If so, how many are there?

Solving Quadratic Equations   A quadratic equation written with 0 on one side of the equals sign can be solved using the ZERO feature of a graphing calculator. See the Calculator Corner on p. 392 for the procedure.

We can also use the INTERSECT feature to solve a quadratic equation. Consider the equation in Exercise 19 in Exercise Set 7.2: $4x(x - 2) - 5x(x - 1) = 2$. First, we enter $y_1 = 4x(x - 2) - 5x(x - 1)$ and $y_2 = 2$ on the equation-editor screen and graph the equations in a window that shows the point(s) of intersection of the graphs. We use the window $[-5, 3, -2, 4]$. We see that there are two points of intersection, so the equation has two solutions.

We use the INTERSECT feature to find the coordinates of the left-hand point of intersection. (See the Calculator Corner on p. 246 for the procedure.) The first coordinate of this point, -2, is one solution of the equation. We use the INTERSECT feature again to find the other solution, -1.

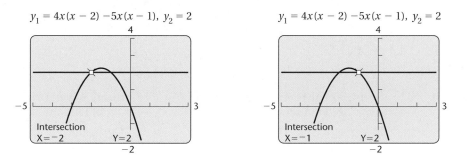

Note that we could use the ZERO feature to solve this equation if we first write it with 0 on one side: $4x(x - 2) - 5x(x - 1) - 2 = 0$.

Exercises: Solve.

1. $5x^2 = -11x + 12$

2. $2x^2 - 15 = 7x$

3. $6(x - 3) = (x - 3)(x - 2)$

4. $(x + 1)(x - 4) = 3(x - 4)$

Study Tips

BEGINNING TO STUDY FOR THE FINAL EXAM (PART 2)

The best scenario for preparing for a final exam is to do so over a period of at least two weeks. Work in a diligent, disciplined manner, doing some final-exam preparation each day. Here is a detailed plan that many find useful.

1. **Begin by browsing through each chapter, reviewing the highlighted or boxed information regarding important formulas in both the text and the Summary and Review.**   There may be some formulas that you will need to memorize.

2. **Retake each chapter test that you took in class, assuming your instructor has returned it. Otherwise, use the chapter test in the book.**   Restudy the objectives in the text that correspond to each question you missed.

3. **If you are still missing questions, use the supplements for extra review.**   For example, you might check out the videotapes or audio recordings, the *Student's Solutions Manual,* the InterAct Math Tutorial Web site, or MathXL.

4. **For remaining difficulties, see your instructor, go to a tutoring session, or participate in a study group.**

5. **Take the Final Examination in the text during the last couple of days before the final.**   Set aside the same amount of time that you will have for the final. See how much of the final exam you can complete under test-like conditions.

"The door of opportunity won't open unless you do some pushing."

Anonymous

a   Solve.

1. $x^2 + 8x + 2 = 0$

2. $x^2 - 6x - 4 = 0$

3. $3p^2 = -8p - 1$

4. $3u^2 = 18u - 6$

5. $x^2 - x + 1 = 0$

6. $x^2 + x + 2 = 0$

7. $x^2 + 13 = 4x$

8. $x^2 + 13 = 6x$

9. $r^2 + 3r = 8$

10. $h^2 + 4 = 6h$

11. $1 + \dfrac{2}{x} + \dfrac{5}{x^2} = 0$

12. $1 + \dfrac{5}{x^2} = \dfrac{2}{x}$

13. a) Solve: $3x + x(x - 2) = 0$.
  b) Find the x-intercepts of $f(x) = 3x + x(x - 2)$.

14. a) Solve: $4x + x(x - 3) = 0$.
  b) Find the x-intercepts of $f(x) = 4x + x(x - 3)$.

15. a) Solve: $11x^2 - 3x - 5 = 0$.
  b) Find the x-intercepts of $f(x) = 11x^2 - 3x - 5$.

16. a) Solve: $7x^2 + 8x = -2$.
  b) Find the x-intercepts of $f(x) = 7x^2 + 8x + 2$.

17. a) Solve: $25x^2 = 20x - 4$.
  b) Find the x-intercepts of $f(x) = 25x^2 - 20x + 4$.

18. a) Solve: $49x^2 - 14x + 1 = 0$.
  b) Find the x-intercepts of $f(x) = 49x^2 - 14x + 1$.

Solve.

19. $4x(x - 2) - 5x(x - 1) = 2$

20. $3x(x + 1) - 7x(x + 2) = 6$

21. $14(x - 4) - (x + 2) = (x + 2)(x - 4)$

22. $11(x - 2) + (x - 5) = (x + 2)(x - 6)$

23. $5x^2 = 17x - 2$

24. $15x = 2x^2 + 16$

25. $x^2 + 5 = 4x$

26. $x^2 + 5 = 2x$

27. $x + \dfrac{1}{x} = \dfrac{13}{6}$

28. $\dfrac{3}{x} + \dfrac{x}{3} = \dfrac{5}{2}$

29. $\dfrac{1}{y} + \dfrac{1}{y + 2} = \dfrac{1}{3}$

30. $\dfrac{1}{x} + \dfrac{1}{x + 4} = \dfrac{1}{7}$

31. $(2t - 3)^2 + 17t = 15$

32. $2y^2 - (y + 2)(y - 3) = 12$

33. $(x - 2)^2 + (x + 1)^2 = 0$

34. $(x + 3)^2 + (x - 1)^2 = 0$

35. $x^3 - 1 = 0$
(*Hint*: Factor the difference of cubes. Then use the quadratic formula.)

36. $x^3 + 27 = 0$

Solve. Give the exact solution and approximate solutions to three decimal places.

37. $x^2 + 6x + 4 = 0$

38. $x^2 + 4x - 7 = 0$

39. $x^2 - 6x + 4 = 0$

40. $x^2 - 4x + 1 = 0$

41. $2x^2 - 3x - 7 = 0$

42. $3x^2 - 3x - 2 = 0$

43. $5x^2 = 3 + 8x$

44. $2y^2 + 2y - 3 = 0$

45. D_W The list of steps on p. 590 does not mention completing the square as a method of solving quadratic equations. Why not?

46. D_W Given the solutions of a quadratic equation, is it possible to reconstruct the original equation? Why or why not?

SKILL MAINTENANCE

Solve. [6.6a, b]

47. $x = \sqrt{x + 2}$

48. $x = \sqrt{15 - 2x}$

49. $\sqrt{x + 2} = \sqrt{2x - 8}$

50. $\sqrt{x + 1} + 2 = \sqrt{3x + 1}$

51. $\sqrt{x + 5} = -7$

52. $\sqrt{2x - 6} + 11 = 2$

53. $\sqrt[3]{4x - 7} = 2$

54. $\sqrt[4]{3x - 1} = 2$

SYNTHESIS

55. Use a graphing calculator to solve the equations in Exercises 3, 16, 17, and 37 using the INTERSECT feature, letting $y_1 = $ the left side and $y_2 = $ the right side. Then solve $2.2x^2 + 0.5x - 1 = 0$.

56. Use a graphing calculator to solve the equations in Exercises 9, 27, and 30. Then solve $5.33x^2 = 8.23x + 3.24$.

Solve.

57. $2x^2 - x - \sqrt{5} = 0$

58. $\dfrac{5}{x} + \dfrac{x}{4} = \dfrac{11}{7}$

59. $ix^2 - x - 1 = 0$

60. $\sqrt{3}x^2 + 6x + \sqrt{3} = 0$

61. $\dfrac{x}{x + 1} = 4 + \dfrac{1}{3x^2 - 3}$

62. $(1 + \sqrt{3})x^2 - (3 + 2\sqrt{3})x + 3 = 0$

63. Let $f(x) = (x - 3)^2$. Find all inputs x such that $f(x) = 13$.

64. Let $f(x) = x^2 + 14x + 49$. Find all inputs x such that $f(x) = 36$.

Objectives

a Solve applied problems involving quadratic equations.

b Solve a formula for a given letter.

1. Landscaping. A rectangular garden is 60 ft by 80 ft. Part of the garden is torn up to install a sidewalk of uniform width around it. The area of the new garden is $\frac{1}{2}$ of the old area. How wide is the sidewalk?

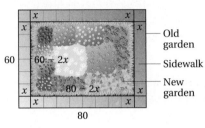

Answer on page A-37

7.3 APPLICATIONS INVOLVING QUADRATIC EQUATIONS

a Applications and Problem Solving

Sometimes when we translate a problem to mathematical language, the result is a quadratic equation.

EXAMPLE 1 *Box Construction.* An open box is to be made from a 10-ft by 20-ft rectangular piece of cardboard by cutting a square from each corner. The area of the bottom of the box is to be 96 ft^2. What is the length of the sides of the squares that are cut from the corners?

1. Familiarize. We first make a drawing and label it with the known information. Since we do not know the length of the sides of the squares that are cut from the corners, we have called that length x.

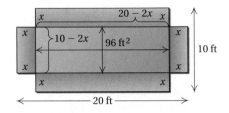

2. Translate. Remember, the area of a rectangle is lw (length times width). We find two expressions for the area of the bottom of the box. First, we know from the statement of the problem that

the area of the bottom of the box is 96 ft^2.

Second, the width of the bottom of the box is 10 ft minus twice the length of the squares cut from the corners, or $10 - 2x$. The length of the bottom of the box is 20 ft minus twice the length of the squares cut from the corners, or $20 - 2x$. Then we can multiply the expressions to find another expression for the area of the bottom of the box:

$$(10 - 2x)(20 - 2x).$$

These expressions give us the equation $(10 - 2x)(20 - 2x) = 96$.

3. Solve. We solve the equation:

$$\begin{aligned} (10 - 2x)(20 - 2x) &= 96 \\ 200 - 20x - 40x + 4x^2 &= 96 &\text{Using FOIL on the left} \\ 4x^2 - 60x + 104 &= 0 &\text{Obtaining 0 on the right and collecting like terms} \\ x^2 - 15x + 26 &= 0 &\text{Dividing by 4} \\ (x - 2)(x - 13) &= 0 &\text{Factoring} \\ x = 2 \quad or \quad x &= 13. &\text{Using the principle of zero products} \end{aligned}$$

4. Check. We check in the original problem. We see that 13 is not a solution because 13 is longer than the width of the piece of cardboard.

If the length of a side of the squares is 2, then the bottom of the box will have width $10 - 2 \cdot 2$, or 6 ft. The length will be $20 - 2 \cdot 2$, or 16 ft. The area is $6 \cdot 16$, or 96 ft^2.

5. State. The length of the sides of the squares is 2 ft.

Do Exercise 1.

EXAMPLE 2 *Town Planning.* Three towns A, B, and C are situated as shown. The roads at A form a right angle. The distance from A to B is 2 mi less than the distance from A to C. The distance from B to C is 10 mi. Find the distance from A to B and the distance from A to C.

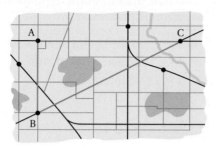

1. **Familiarize.** We first make a drawing and label it. We let d = the distance from A to C. Then the distance from A to B is $d - 2$.

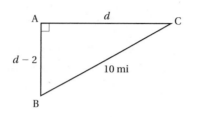

2. **Translate.** We see that a right triangle is formed. We can use the Pythagorean equation, which we studied in Chapter 6: $c^2 = a^2 + b^2$. In this problem, we have

$$10^2 = d^2 + (d - 2)^2.$$

3. **Solve.** We solve the equation:

$$10^2 = d^2 + (d - 2)^2$$
$$100 = d^2 + d^2 - 4d + 4 \qquad \text{Squaring}$$
$$2d^2 - 4d - 96 = 0 \qquad \text{Finding standard form}$$
$$d^2 - 2d - 48 = 0 \qquad \text{Multiplying by } \tfrac{1}{2}, \text{ or dividing by 2}$$
$$(d - 8)(d + 6) = 0 \qquad \text{Factoring}$$
$$d - 8 = 0 \quad or \quad d + 6 = 0 \qquad \text{Using the principle of zero products}$$
$$d = 8 \quad or \qquad d = -6.$$

4. **Check.** We know that -6 cannot be a solution because distances are not negative. If $d = 8$, then $d - 2 = 6$, and

$$d^2 + (d - 2)^2 = 8^2 + 6^2 = 64 + 36 = 100.$$

Since $10^2 = 100$, the distance 8 mi checks.

5. **State.** The distance from A to C is 8 mi, and the distance from A to B is 6 mi.

Do Exercise 2.

EXAMPLE 3 *Town Planning.* Three towns A, B, and C are situated as shown in Example 2. The roads at A form a right angle. The distance from A to B is 2 mi less than the distance from A to C. The distance from B to C is 8 mi. Find the distance from A to B and the distance from A to C. Find exact and approximate answers to the nearest hundredth of a mile.

2. Ladder Location. A ladder leans against a building, as shown below. The ladder is 20 ft long. The distance to the top of the ladder is 4 ft greater than the distance d from the building. Find the distance d and the distance to the top of the ladder.

Answer on page A-37

3. Ladder Location. Refer to Margin Exercise 2. Suppose that the ladder has length 10 ft. Find the distance d and the distance $d + 4$.

Using the same reasoning that we did in Example 2, we translate the problem to the equation

$$8^2 = d^2 + (d - 2)^2.$$

We solve as follows. Note that the quadratic equation we get is not easily factored, so we use the quadratic formula:

$$64 = d^2 + d^2 - 4d + 4 \qquad \text{Squaring}$$
$$2d^2 - 4d - 60 = 0 \qquad \text{Finding standard form}$$
$$d^2 - 2d - 30 = 0. \qquad \text{Multiplying by } \tfrac{1}{2}, \text{ or dividing by 2}$$

Then

$$d = \frac{-b \pm \sqrt{b^2 - 4ac}}{2a}$$

$$= \frac{-(-2) \pm \sqrt{(-2)^2 - 4(1)(-30)}}{2(1)} \qquad \begin{array}{l}\text{Substituting 1 for } a, -2 \text{ for } b, \\ \text{and } -30 \text{ for } c\end{array}$$

$$= \frac{2 \pm \sqrt{124}}{2} = \frac{2 \pm \sqrt{4(31)}}{2} = \frac{2 \pm 2\sqrt{31}}{2} = 1 \pm \sqrt{31}.$$

Since $1 - \sqrt{31} < 0$ and $1 + \sqrt{31} > 0$, it follows that $d = 1 + \sqrt{31}$. Using a calculator, we find that $d = 1 + \sqrt{31} \approx 6.57$ mi, and that $d - 2 \approx 4.57$ mi. Thus the distance from A to C is about 6.57 mi, and the distance from A to B is about 4.57 mi.

Do Exercise 3.

EXAMPLE 4 *Motorcycle Travel.* Karin's motorcycle traveled 300 mi at a certain speed. Had she gone 10 mph faster, she could have made the trip in 1 hr less time. Find her speed.

1. **Familiarize.** We first make a drawing, labeling it with known and unknown information. We can also organize the information, in a table, as we did in Section 5.6. We let $r =$ the speed, in miles per hour, and $t =$ the time, in hours.

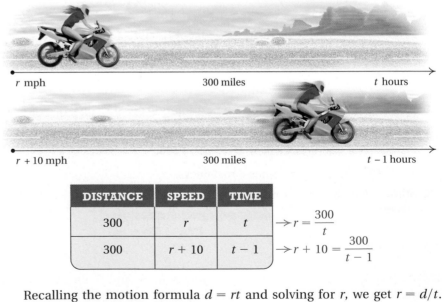

DISTANCE	SPEED	TIME
300	r	t
300	$r + 10$	$t - 1$

$$\rightarrow r = \frac{300}{t}$$
$$\rightarrow r + 10 = \frac{300}{t - 1}$$

Recalling the motion formula $d = rt$ and solving for r, we get $r = d/t$. From the rows of the table, we obtain

$$r = \frac{300}{t} \quad \text{and} \quad r + 10 = \frac{300}{t - 1}.$$

2. Translate. We substitute for r from the first equation into the second and get a translation:

$$\frac{300}{t} + 10 = \frac{300}{t-1}.$$

3. Solve. We solve as follows:

$$\frac{300}{t} + 10 = \frac{300}{t-1}$$

$$t(t-1)\left[\frac{300}{t} + 10\right] = t(t-1)\cdot\frac{300}{t-1} \qquad \text{Multiplying by the LCM}$$

$$t(t-1)\cdot\frac{300}{t} + t(t-1)\cdot 10 = t(t-1)\cdot\frac{300}{t-1}$$

$$300(t-1) + 10(t^2 - t) = 300t$$

$$10t^2 - 10t - 300 = 0 \qquad \text{Standard form}$$

$$t^2 - t - 30 = 0 \qquad \text{Dividing by 10}$$

$$(t-6)(t+5) = 0 \qquad \text{Factoring}$$

$$t = 6 \quad or \quad t = -5. \qquad \text{Using the principle of zero products}$$

4. Check. Since negative time has no meaning in this problem, we try 6 hr. Remembering that $r = d/t$, we get $r = 300/6 = 50$ mph.

To check, we take the speed 10 mph faster, which is 60 mph, and see how long the trip would have taken at that speed:

$$t = \frac{d}{r} = \frac{300}{60} = 5 \text{ hr.}$$

This is 1 hr less than the trip actually took, so we have an answer.

5. State. Karin's speed was 50 mph.

Do Exercise 4.

b   Solving Formulas

Recall that to solve a formula for a certain letter, we use the principles for solving equations to get that letter alone on one side.

EXAMPLE 5 *Period of a Pendulum.*
The time T required for a pendulum of length L to swing back and forth (complete one period) is given by the formula $T = 2\pi\sqrt{L/g}$, where g is the gravitational constant. Solve for L.

$$T = 2\pi\sqrt{\frac{L}{g}} \qquad \text{This is a radical equation (see Section 6.6).}$$

$$T^2 = \left(2\pi\sqrt{\frac{L}{g}}\right)^2 \qquad \text{Principle of powers (squaring)}$$

$$T^2 = 2^2\pi^2\frac{L}{g}$$

$$gT^2 = 4\pi^2 L \qquad \text{Clearing fractions}$$

$$\frac{gT^2}{4\pi^2} = L. \qquad \text{Multiplying by } \frac{1}{4\pi^2}$$

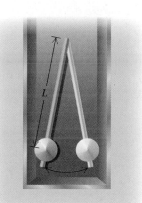

4. Marine Travel. Two ships make the same voyage of 3000 nautical miles. The faster ship travels 10 knots faster than the slower one (a *knot* is 1 nautical mile per hour). The faster ship makes the voyage in 50 hr less time than the slower one. Find the speeds of the two ships.

Complete this table to help with the familiarization.

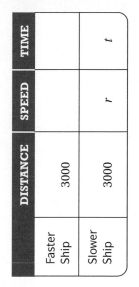

	DISTANCE	SPEED	TIME
Faster Ship	3000	r	t
Slower Ship	3000		

5. Solve $A = \sqrt{\dfrac{w_1}{w_2}}$ for w_2.

Answers on page A-37

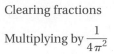

6. Solve $V = \pi r^2 h$ for r.
(Volume of a right circular cylinder)

We now have L alone on one side and L does not appear on the other side, so the formula is solved for L.

Do Exercise 5 on the preceding page.

In most formulas, variables represent nonnegative numbers, so we do not need to use absolute-value signs when taking square roots.

EXAMPLE 6 *Hang Time.* An athlete's *hang time* is the amount of time that the athlete can remain airborne when jumping. A formula relating an athlete's vertical leap V, in inches, to hang time T, in seconds, is $V = 48T^2$. (See Example 13 in Section 7.1.) Solve for T.

We have

$$48T^2 = V$$

$$T^2 = \frac{V}{48} \qquad \text{Multiplying by } \tfrac{1}{48} \text{ to get } T^2 \text{ alone}$$

$$T = \sqrt{\frac{V}{48}} \qquad \text{Using the principle of square roots; note that } T \geq 0.$$

$$T = \sqrt{\frac{V}{2 \cdot 2 \cdot 2 \cdot 2 \cdot 3} \cdot \frac{3}{3}} = \frac{\sqrt{3V}}{2 \cdot 2 \cdot 3} = \frac{\sqrt{3V}}{12}.$$

Do Exercise 6.

7. Solve $s = gt + 16t^2$ for t.

EXAMPLE 7 *Falling Distance.* An object that is tossed downward with an initial speed (velocity) of v_0 will travel a distance of s meters, where $s = 4.9t^2 + v_0 t$ and t is measured in seconds. Solve for t.

Answers on page A-37

CHAPTER 7: Quadratic Equations
and Functions

Since t is squared in one term and raised to the first power in the other term, the equation is quadratic in t. The variable is t; v_0 and s are treated as constants.

We have

$$4.9t^2 + v_0 t = s$$

$$4.9t^2 + v_0 t - s = 0 \qquad \text{Writing standard form}$$

$$a = 4.9, \quad b = v_0, \quad c = -s$$

$$t = \frac{-v_0 \pm \sqrt{v_0^2 - 4(4.9)(-s)}}{2(4.9)} \qquad \begin{array}{l} \text{Using the quadratic formula:} \\ t = \dfrac{-b \pm \sqrt{b^2 - 4ac}}{2a} \end{array}$$

$$t = \frac{-v_0 \pm \sqrt{v_0^2 + 19.6s}}{9.8}.$$

Since the negative square root would yield a negative value for t, we use only the positive root:

$$t = \frac{-v_0 + \sqrt{v_0^2 + 19.6s}}{9.8}.$$

The steps listed in the margin should help you when solving formulas for a given letter. Try to remember that when solving a formula, you do the same things you would do to solve any equation.

Do Exercise 7 on the preceding page.

EXAMPLE 8 Solve $t = \dfrac{a}{\sqrt{a^2 + b^2}}$ for a.

In this case, we could either clear the fractions first or use the principle of powers first. Let's clear the fractions. Multiplying by $\sqrt{a^2 + b^2}$, we have

$$t\sqrt{a^2 + b^2} = a.$$

Now we square both sides and then continue:

$$\left(t\sqrt{a^2 + b^2}\right)^2 = a^2 \qquad \text{Squaring}$$

> **Caution!**
> Don't forget to square both t and $\sqrt{a^2 + b^2}$.

$$t^2(a^2 + b^2) = a^2$$

$$t^2 a^2 + t^2 b^2 = a^2$$

$$t^2 b^2 = a^2 - t^2 a^2 \qquad \text{Getting all } a^2\text{-terms together}$$

$$t^2 b^2 = a^2(1 - t^2) \qquad \text{Factoring out } a^2$$

$$\frac{t^2 b^2}{1 - t^2} = a^2 \qquad \text{Dividing by } 1 - t^2$$

$$\sqrt{\frac{t^2 b^2}{1 - t^2}} = a \qquad \text{Taking the square root}$$

$$\frac{tb}{\sqrt{1 - t^2}} = a. \qquad \text{Simplifying}$$

You need not rationalize denominators in situations such as this.

Do Exercise 8.

To solve a formula for a letter, say, b:

1. Clear the fractions and use the principle of powers, as needed, until b does not appear in any radicand or denominator. (In some cases, you may clear the fractions first, and in some cases you may use the principle of powers first.)
2. Collect all terms with b^2 in them. Also collect all terms with b in them.
3. If b^2 does not appear, you can finish by using just the addition and multiplication principles.
4. If b^2 appears but b does not, solve the equation for b^2. Then take square roots on both sides.
5. If there are terms containing both b and b^2, write the equation in standard form and use the quadratic formula.

8. Solve $\dfrac{b}{\sqrt{a^2 - b^2}} = t$ for b.

Answer on page A-37

Translating
for Success

1. *Car Travel.* Sarah drove her car 800 mi to see her friend. The return trip was 2 hr faster at a speed that was 10 mph more. Find her return speed.

2. *Coin Mixture.* A collection of dimes and quarters is worth $26.95. There are 117 coins in all. How many of each coin are there?

3. *Wire Cutting.* A 537-in. wire is cut into three pieces. The second piece is 7 in. shorter than the first. The third is half as long as the first. How long is each piece?

4. *Marine Travel.* The Columbia River flows at a rate of 2 mph for the length of a popular boating route. In order for a motorized dinghy to travel 3 mi upriver and return in a total of 4 hr, how fast must the boat be able to travel in still water?

5. *Locker Numbers.* The numbers on three adjoining lockers are consecutive integers whose sum is 537. Find the integers.

Translate each word problem to an equation or a system of equations and select a correct translation from equations A–O.

A. $(80 - 2x)(100 - 2x) = \frac{1}{3} \cdot 80 \cdot 100$

B. $\dfrac{800}{x} + 10 = \dfrac{800}{x - 2}$

C. $x + 18\% \cdot x = 2.24$

D. $x + 25y = 26.95,$
$x + y = 117$

E. $2x + 2(x - 7) = 537$

F. $x + (x - 7) + \frac{1}{2}x = 537$

G. $0.10x + 0.25y = 26.95,$
$x + y = 117$

H. $2.24 - 18\% \cdot 2.24 = x$

I. $\dfrac{4}{x + 2} + \dfrac{4}{x - 2} = 3$

J. $x^2 + (x + 1)^2 = 7^2$

K. $75^2 + x^2 = 78^2$

L. $\dfrac{3}{x + 2} + \dfrac{3}{x - 2} = 4$

M. $75^2 + 78^2 = x^2$

N. $x + (x + 1) + (x + 2) = 537$

O. $\dfrac{800}{x} + \dfrac{800}{x - 2} = 10$

Answers on page A-37

6. *Gasoline Prices.* One day the price of gasoline was increased 18% to a new price of $2.24 per gallon. What was the original price?

7. *Triangle Dimensions.* The hypotenuse of a right triangle is 7 ft. The length of one leg is 1 ft longer than the other. Find the lengths of the legs.

8. *Rectangle Dimensions.* The perimeter of a rectangle is 537 ft. The width of the rectangle is 7 ft shorter than the length. Find the length and the width.

9. *Guy Wire.* A guy wire is 78 ft long. It is attached to the top of a 75-ft cellphone tower. How far is it from the base of the pole to the point where the wire is attached to the ground?

10. *Landscaping.* A rectangular garden is 80 ft by 100 ft. Part of the garden is torn up to install a sidewalk of uniform width around it. The area of the new garden is $\frac{1}{3}$ of the old area. How wide is the sidewalk?

7.3

EXERCISE SET

For Extra Help

MathXL   MyMathLab   InterAct   Math Tutor   Digital Video   Student's
                       Math   Center   Tutor CD 3   Solutions
                                                Videotape 8   Manual

a Solve.

1. *Flower Bed.* The width of a rectangular flower bed is 7 ft less than the length. The area is 18 ft². Find the length and the width.

2. *Feed Lot.* The width of a rectangular feed lot is 8 m less than the length. The area is 20 m². Find the length and the width.

3. *Parking Lot.* The length of a rectangular parking lot is twice the width. The area is 162 yd². Find the length and the width.

4. *Computer Part.* The length of a rectangular computer part is twice the width. The area is 242 cm². Find the length and the width.

5. *Sailing.* The base of a triangular sail is 9 m less than its height. The area is 56 m². Find the base and the height of the sail.

6. *Parking Lot.* The width of a rectangular parking lot is 50 ft less than its length. Determine the dimensions of the parking lot if it measures 250 ft diagonally.

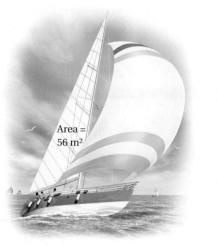

Area = 56 m²

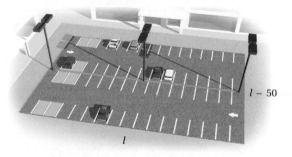

$l - 50$

l

7. *Sailing.* The base of a triangular sail is 8 ft less than its height. The area is 56 ft². Find the base and the height of the sail.

8. *Parking Lot.* The width of a rectangular parking lot is 51 ft less than its length. Determine the dimensions of the parking lot if it measures 250 ft diagonally.

9. *Picture Framing.* The outside of a picture frame measures 12 cm by 20 cm; 84 cm² of picture shows. Find the width of the frame.

10. *Picture Framing.* The outside of a picture frame measures 14 in. by 20 in.; 160 in² of picture shows. Find the width of the frame.

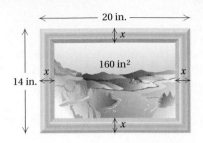

11. *Landscaping.* A landscaper is designing a flower garden in the shape of a right triangle. She wants 10 ft of a perennial border to form the hypotenuse of the triangle, and one leg is to be 2 ft longer than the other. Find the lengths of the legs.

12. The hypotenuse of a right triangle is 25 m long. The length of one leg is 17 m less than the other. Find the lengths of the legs.

13. *Page Numbers.* A student opens a literature book to two facing pages. The product of the page numbers is 812. Find the page numbers.

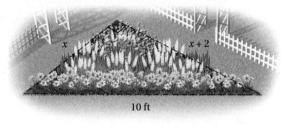

14. *Page Numbers.* A student opens a mathematics book to two facing pages. The product of the page numbers is 1980. Find the page numbers.

Solve. Find exact and approximate answers rounded to three decimal places.

15. The width of a rectangle is 4 ft less than the length. The area is 10 ft². Find the length and the width.

16. The length of a rectangle is twice the width. The area is 328 cm². Find the length and the width.

17. *Page Dimensions.* The outside of an oversized book page measures 14 in. by 20 in.; 100 in² of printed text shows. Find the width of the margin.

18. *Picture Framing.* The outside of a picture frame measures 13 cm by 20 cm; 80 cm² of picture shows. Find the width of the frame.

19. The hypotenuse of a right triangle is 24 ft long. The length of one leg is 14 ft more than the other. Find the lengths of the legs.

20. The hypotenuse of a right triangle is 22 m long. The length of one leg is 10 m less than the other. Find the lengths of the legs.

21. *Car Trips.* During the first part of a trip, Meira's Honda traveled 120 mi at a certain speed. Meira then drove another 100 mi at a speed that was 10 mph slower. If the total time for Meira's trip was 4 hr, what was her speed on each part of the trip?

DISTANCE	SPEED	TIME

22. *Canoeing.* During the first part of a canoe trip, Tim covered 60 km at a certain speed. He then traveled 24 km at a speed that was 4 km/h slower. If the total time for Tim's trip was 8 hr, what was his speed on each part of the trip?

DISTANCE	SPEED	TIME

23. *Car Trips.* Petra's Plymouth travels 200 mi at a certain speed. If the car had gone 10 mph faster, the trip would have taken 1 hr less. Find Petra's speed.

24. *Car Trips.* Sandi's Subaru travels 280 mi at a certain speed. If the car had gone 5 mph faster, the trip would have taken 1 hr less. Find Sandi's speed.

25. *Air Travel.* A Cessna flies 600 mi at a certain speed. A Beechcraft flies 1000 mi at a speed that is 50 mph faster, but takes 1 hr longer. Find the speed of each plane.

26. *Air Travel.* A turbo-jet flies 50 mph faster than a super-prop plane. If a turbo-jet goes 2000 mi in 3 hr less time than it takes the super-prop to go 2800 mi, find the speed of each plane.

27. *Bicycling.* Naoki bikes the 40 mi to Hillsboro at a certain speed. The return trip is made at a speed that is 6 mph slower. Total time for the round trip is 14 hr. Find Naoki's speed on each part of the trip.

28. *Car Speed.* On a sales trip, Gail drives the 600 mi to Richmond at a certain speed. The return trip is made at a speed that is 10 mph slower. Total time for the round trip is 22 hr. How fast did Gail travel on each part of the trip?

29. *Navigation.* The current in a typical Mississippi River shipping route flows at a rate of 4 mph. In order for a barge to travel 24 mi upriver and then return in a total of 5 hr, approximately how fast must the barge be able to travel in still water?

30. *Navigation.* The Hudson River flows at a rate of 3 mph. A patrol boat travels 60 mi upriver and returns in a total time of 9 hr. What is the speed of the boat in still water?

Solve the formula for the given letter. Assume that all variables represent nonnegative numbers.

31. $A = 6s^2$, for s
(Surface area of a cube)

32. $A = 4\pi r^2$, for r
(Surface area of a sphere)

33. $F = \dfrac{Gm_1m_2}{r^2}$, for r

34. $N = \dfrac{kQ_1Q_2}{s^2}$, for s
(Number of phone calls between two cities)

35. $E = mc^2$, for c
(Einstein's energy–mass relationship)

36. $V = \frac{1}{3}s^2h$, for s
(Volume of a pyramid)

37. $a^2 + b^2 = c^2$, for b
(Pythagorean formula in two dimensions)

38. $a^2 + b^2 + c^2 = d^2$, for c
(Pythagorean formula in three dimensions)

39. $N = \dfrac{k^2 - 3k}{2}$, for k
(Number of diagonals of a polygon of k sides)

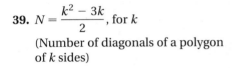

40. $s = v_0t + \dfrac{gt^2}{2}$, for t
(A motion formula)

41. $A = 2\pi r^2 + 2\pi rh$, for r
(Surface area of a cylinder)

42. $A = \pi r^2 + \pi rs$, for r
(Surface area of a cone)

43. $T = 2\pi\sqrt{\dfrac{L}{g}}$, for g
(A pendulum formula)

44. $W = \sqrt{\dfrac{1}{LC}}$, for L
(An electricity formula)

45. $I = \dfrac{704.5W}{H^2}$, for H
(Body mass index; see Example 1 of Section 1.2)

46. $N + p = \dfrac{6.2A^2}{pR^2}$, for R

47. $m = \dfrac{m_0}{\sqrt{1 - \dfrac{v^2}{c^2}}}$, for v

(A relativity formula)

48. Solve the formula given in Exercise 47 for c.

49. $\mathbf{D_W}$  Explain how Exercises 1–30 can be solved using a calculator but without using factoring, completing the square, or the quadratic formula.

50. $\mathbf{D_W}$ Explain how the quadratic formula can be used to factor a quadratic polynomial into two binomials. Use it to factor $5x^2 + 8x - 3$.

SKILL MAINTENANCE

Add or subtract.　[5.2b, c]

51. $\dfrac{1}{x - 1} + \dfrac{1}{x^2 - 3x + 2}$

52. $\dfrac{x + 1}{x - 1} - \dfrac{x + 1}{x^2 + x + 1}$

53. $\dfrac{2}{x + 3} - \dfrac{x}{x - 1} + \dfrac{x^2 + 2}{x^2 + 2x - 3}$

54. Multiply and simplify: $\sqrt{3x^2}\,\sqrt{3x^3}$.　[6.3a]

55. Express in terms of i: $\sqrt{-20}$.　[6.8a]

Simplify.　[5.4a]

56. $\dfrac{\dfrac{3}{x - 1}}{\dfrac{1}{x + 1} + \dfrac{2}{x - 1}}$

57. $\dfrac{\dfrac{4}{a^2 b}}{\dfrac{3}{a} - \dfrac{4}{b^2}}$

SYNTHESIS

58. Solve: $\dfrac{4}{2x + i} - \dfrac{1}{x - i} = \dfrac{2}{x + i}$.

59. Find a when the reciprocal of $a - 1$ is $a + 1$.

60. *Bungee Jumping.*　Jesse is tied to one end of a 40-m elasticized (bungee) cord. The other end of the cord is tied to the middle of a train trestle. If Jesse jumps off the bridge, for how long will he fall before the cord begins to stretch? (See Example 7 and let $v_0 = 0$.)

40 m

61. *Surface Area.*　A sphere is inscribed in a cube as shown in the figure below. Express the surface area of the sphere as a function of the surface area S of the cube.

62. *Pizza Crusts.*　At Pizza Perfect, Ron can make 100 large pizza crusts in 1.2 hr less than Chad. Together they can do the job in 1.8 hr. How long does it take each to do the job alone?

63. *The Golden Rectangle.*　For over 2000 yr, the proportions of a "golden" rectangle have been considered visually appealing. A rectangle of width w and length l is considered "golden" if

$$\frac{w}{l} = \frac{l}{w + l}.$$

Solve for l.

MORE ON QUADRATIC EQUATIONS

Objectives

a Determine the nature of the solutions of a quadratic equation.

b Write a quadratic equation having two numbers specified as solutions.

c Solve equations that are quadratic in form.

a The Discriminant

From the quadratic formula, we know that the solutions x_1 and x_2 of a quadratic equation are given by

$$x_1 = \frac{-b + \sqrt{b^2 - 4ac}}{2a} \quad \text{and} \quad x_2 = \frac{-b - \sqrt{b^2 - 4ac}}{2a}.$$

The expression $b^2 - 4ac$ is called the **discriminant.** When using the quadratic formula, it is helpful to compute the discriminant first. If it is 0, there will be just one real solution. If it is positive, there will be two real solutions. If it is negative, we will be taking the square root of a negative number; hence there will be two nonreal complex-number solutions, and they will be complex conjugates.

DISCRIMINANT $b^2 - 4ac$	NATURE OF SOLUTIONS	x-INTERCEPTS
0	Only one solution; it is a real number	Only one
Positive	Two different real-number solutions	Two different
Negative	Two different nonreal complex-number solutions (complex conjugates)	None

If the discriminant is a perfect square, we can solve the equation by factoring, not needing the quadratic formula.

EXAMPLE 1 Determine the nature of the solutions of $9x^2 - 12x + 4 = 0$.

We have

$$a = 9, \quad b = -12, \quad c = 4.$$

We compute the discriminant:

$$b^2 - 4ac = (-12)^2 - 4 \cdot 9 \cdot 4$$
$$= 144 - 144$$
$$= 0.$$

There is just one solution, and it is a real number. Since 0 is a perfect square, the equation can be solved by factoring.

EXAMPLE 2 Determine the nature of the solutions of $x^2 + 5x + 8 = 0$.

We have

$$a = 1, \quad b = 5, \quad c = 8.$$

We compute the discriminant:

$$b^2 - 4ac = 5^2 - 4 \cdot 1 \cdot 8$$
$$= 25 - 32$$
$$= -7.$$

Since the discriminant is negative, there are two nonreal complex-number solutions. The equation cannot be solved by factoring because -7 is not a perfect square.

EXAMPLE 3 Determine the nature of the solutions of $x^2 + 5x + 6 = 0$.

We have

$$a = 1, \quad b = 5, \quad c = 6;$$
$$b^2 - 4ac = 5^2 - 4 \cdot 1 \cdot 6 = 1.$$

Determine the nature of the solutions without solving.

1. $x^2 + 5x - 3 = 0$

Since the discriminant is positive, there are two solutions, and they are real numbers. The equation can be solved by factoring since the discriminant is a perfect square.

$f(x) = x^2 + 5x + 6$
Two x-intercepts

The discriminant, $b^2 - 4ac$, tells us how many real-number solutions the equation $ax^2 + bx + c = 0$ has, so it also indicates how many x-intercepts the graph of $f(x) = ax^2 + bx + c$ has. Compare the following.

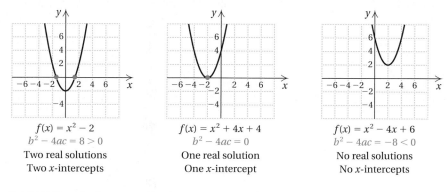

$f(x) = x^2 - 2$
$b^2 - 4ac = 8 > 0$
Two real solutions
Two x-intercepts

$f(x) = x^2 + 4x + 4$
$b^2 - 4ac = 0$
One real solution
One x-intercept

$f(x) = x^2 - 4x + 6$
$b^2 - 4ac = -8 < 0$
No real solutions
No x-intercepts

2. $9x^2 - 6x + 1 = 0$

Do Exercises 1–3.

b   Writing Equations from Solutions

We know by the principle of zero products that $(x - 2)(x + 3) = 0$ has solutions 2 and -3. If we know the solutions of an equation, we can write the equation, using this principle in reverse.

EXAMPLE 4 Find a quadratic equation whose solutions are 3 and $-\frac{2}{5}$.

We have

3. $3x^2 - 2x + 1 = 0$

$$x = 3 \quad or \quad x = -\frac{2}{5}$$
$$x - 3 = 0 \quad or \quad x + \frac{2}{5} = 0 \qquad \text{Getting the 0's on one side}$$
$$x - 3 = 0 \quad or \quad 5x + 2 = 0 \qquad \text{Clearing the fraction}$$
$$(x - 3)(5x + 2) = 0 \qquad \text{Using the principle of zero products in reverse}$$
$$5x^2 - 13x - 6 = 0. \qquad \text{Using FOIL}$$

Answers on page A-38

Find a quadratic equation having the following solutions.

4. 7 and −2

5. −4 and $\dfrac{5}{3}$

6. $5i$ and $-5i$

7. $-2\sqrt{2}$ and $\sqrt{2}$

EXAMPLE 5 Write a quadratic equation whose solutions are $2i$ and $-2i$.

We have

$$x = 2i \quad or \quad x = -2i$$
$$x - 2i = 0 \quad or \quad x + 2i = 0 \qquad \text{Getting the 0's on one side}$$
$$(x - 2i)(x + 2i) = 0 \qquad \text{Using the principle of zero products in reverse}$$
$$x^2 - (2i)^2 = 0 \qquad \text{Using } (A - B)(A + B) = A^2 - B^2$$
$$x^2 - 4i^2 = 0$$
$$x^2 - 4(-1) = 0$$
$$x^2 + 4 = 0.$$

EXAMPLE 6 Write a quadratic equation whose solutions are $\sqrt{3}$ and $-2\sqrt{3}$.

We have

$$x = \sqrt{3} \quad or \quad x = -2\sqrt{3}$$
$$x - \sqrt{3} = 0 \quad or \quad x + 2\sqrt{3} = 0 \qquad \text{Getting the 0's on one side}$$
$$\left(x - \sqrt{3}\right)\left(x + 2\sqrt{3}\right) = 0 \qquad \text{Using the principle of zero products}$$
$$x^2 + 2\sqrt{3}\,x - \sqrt{3}\,x - 2\left(\sqrt{3}\right)^2 = 0 \qquad \text{Using FOIL}$$
$$x^2 + \sqrt{3}\,x - 6 = 0. \qquad \text{Collecting like terms}$$

Do Exercises 4–7.

C Equations Quadratic in Form

Certain equations that are not really quadratic can still be solved as quadratic. Consider this fourth-degree equation.

$$x^4 \quad - 9x^2 \quad + 8 = 0$$
$$\downarrow \qquad \downarrow \qquad \downarrow \quad \downarrow$$
$$(x^2)^2 - 9(x^2) + 8 = 0 \qquad \text{Thinking of } x^4 \text{ as } (x^2)^2$$
$$\downarrow \qquad \downarrow \qquad \downarrow \quad \downarrow$$
$$u^2 \quad - 9u \quad + 8 = 0 \qquad \text{To make this clearer, write } u \text{ instead of } x^2.$$

The equation $u^2 - 9u + 8 = 0$ can be solved by factoring or by the quadratic formula. After that, we can find x by remembering that $x^2 = u$. Equations that can be solved like this are said to be **quadratic in form,** or **reducible to quadratic.**

EXAMPLE 7 Solve: $x^4 - 9x^2 + 8 = 0$.

Let $u = x^2$. Then we solve the equation found by substituting u for x^2:

$$u^2 - 9u + 8 = 0$$
$$(u - 8)(u - 1) = 0 \qquad \text{Factoring}$$
$$u - 8 = 0 \quad or \quad u - 1 = 0 \qquad \text{Using the principle of zero products}$$
$$u = 8 \quad or \qquad u = 1.$$

Next, we substitute x^2 for u and solve these equations:

$$x^2 = 8 \quad or \quad x^2 = 1$$
$$x = \pm\sqrt{8} \quad or \quad x = \pm 1$$
$$x = \pm 2\sqrt{2} \quad or \quad x = \pm 1.$$

Note that when a number and its opposite are raised to an even power, the results are the same. Thus we can make one check for $\pm 2\sqrt{2}$ and one for ± 1.

Check:

For $\pm 2\sqrt{2}$:

$$\frac{x^4 - 9x^2 + 8 = 0}{\left(\pm 2\sqrt{2}\right)^4 - 9\left(\pm 2\sqrt{2}\right)^2 + 8 \; ? \; 0}$$
$$64 - 9 \cdot 8 + 8 \quad\Big|$$
$$0 \quad\Big| \quad \text{TRUE}$$

For ± 1:

$$\frac{x^4 - 9x^2 + 8 = 0}{\left(\pm 1\right)^4 - 9\left(\pm 1\right)^2 + 8 \; ? \; 0}$$
$$1 - 9 + 8 \quad\Big|$$
$$0 \quad\Big| \quad \text{TRUE}$$

The solutions are 1, -1, $2\sqrt{2}$, and $-2\sqrt{2}$.

Caution!

A common error is to solve for u and then forget to solve for x. Remember that you *must* find values for the *original* variable!

Do Exercise 8.

Solving equations quadratic in form can sometimes introduce numbers that are not solutions of the original equation. Thus a check by substitution in the original equation is necessary.

EXAMPLE 8 Solve: $x - 3\sqrt{x} - 4 = 0$.

Let $u = \sqrt{x}$. Then we solve the equation found by substituting u for $\sqrt{x}$ and u^2 for x.

$$u^2 - 3u - 4 = 0$$
$$(u - 4)(u + 1) = 0$$
$$u = 4 \quad or \quad u = -1$$

Next, we substitute $\sqrt{x}$ for u and solve these equations:

$$\sqrt{x} = 4 \quad or \quad \sqrt{x} = -1.$$

Squaring the first equation, we get $x = 16$. Squaring the second equation, we get $x = 1$. We check both solutions.

Check:

For 16:

$$\frac{x - 3\sqrt{x} - 4 = 0}{16 - 3\sqrt{16} - 4 \; ? \; 0}$$
$$16 - 3 \cdot 4 - 4 \quad\Big|$$
$$16 - 12 - 4 \quad\Big|$$
$$0 \quad\Big| \quad \text{TRUE}$$

For 1:

$$\frac{x - 3\sqrt{x} - 4 = 0}{1 - 3\sqrt{1} - 4 \; ? \; 0}$$
$$1 - 3 \cdot 1 - 4 \quad\Big|$$
$$-6 \quad\Big| \quad \text{FALSE}$$

The solution is 16.

Do Exercise 9.

8. Solve: $x^4 - 10x^2 + 9 = 0$.

9. Solve $x + 3\sqrt{x} - 10 = 0$. Be sure to check.

Answers on page A-38

10. Solve: $x^{-2} + x^{-1} - 6 = 0$.

EXAMPLE 9 Solve: $y^{-2} - y^{-1} - 2 = 0$.

Let $u = y^{-1}$. Then we solve the equation found by substituting u for y^{-1} and u^2 for y^{-2}:

$$u^2 - u - 2 = 0$$
$$(u - 2)(u + 1) = 0$$
$$u = 2 \quad or \quad u = -1.$$

Next, we substitute y^{-1} or $1/y$ for u and solve these equations:

$$\frac{1}{y} = 2 \quad or \quad \frac{1}{y} = -1.$$

Solving, we get

$$y = \frac{1}{2} \quad or \quad y = \frac{1}{(-1)} = -1.$$

The numbers $\frac{1}{2}$ and -1 both check. They are the solutions.

Do Exercise 10.

11. Find the x-intercepts of
$f(x) = (x^2 - x)^2 - 14(x^2 - x) + 24$.

EXAMPLE 10 Find the x-intercepts of the graph of

$$f(x) = (x^2 - 1)^2 - (x^2 - 1) - 2.$$

The x-intercepts occur where $f(x) = 0$ so we must have

$$(x^2 - 1)^2 - (x^2 - 1) - 2 = 0.$$

Let $u = x^2 - 1$. Then we solve the equation found by substituting u for $x^2 - 1$:

$$u^2 - u - 2 = 0$$
$$(u - 2)(u + 1) = 0$$
$$u = 2 \quad or \quad u = -1.$$

Next, we substitute $x^2 - 1$ for u and solve these equations:

$$x^2 - 1 = 2 \quad or \quad x^2 - 1 = -1$$
$$x^2 = 3 \quad or \quad x^2 = 0$$
$$x = \pm\sqrt{3} \quad or \quad x = 0.$$

The numbers $\sqrt{3}$, $-\sqrt{3}$, and 0 check. They are the solutions of $(x^2 - 1)^2 - (x^2 - 1) - 2 = 0$. Thus the x-intercepts of the graph of $f(x)$ are $\left(-\sqrt{3}, 0\right)$, $(0, 0)$, and $\left(\sqrt{3}, 0\right)$.

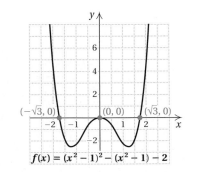

Do Exercise 11.

Answers on page A-38

CHAPTER 7: Quadratic Equations
and Functions

a Determine the nature of the solutions of the equation.

1. $x^2 - 8x + 16 = 0$

2. $x^2 + 12x + 36 = 0$

3. $x^2 + 1 = 0$

4. $x^2 + 6 = 0$

5. $x^2 - 6 = 0$

6. $x^2 - 3 = 0$

7. $4x^2 - 12x + 9 = 0$

8. $4x^2 + 8x - 5 = 0$

9. $x^2 - 2x + 4 = 0$

10. $x^2 + 3x + 4 = 0$

11. $9t^2 - 3t = 0$

12. $4m^2 + 7m = 0$

13. $y^2 = \dfrac{1}{2}y + \dfrac{3}{5}$

14. $y^2 + \dfrac{9}{4} = 4y$

15. $4x^2 - 4\sqrt{3}x + 3 = 0$

16. $6y^2 - 2\sqrt{3}\,y - 1 = 0$

b Write a quadratic equation having the given numbers as solutions.

17. -4 and 4

18. -11 and 9

19. -2 and -7

20. 3 and 10

21. 8, only solution

[*Hint*: It must be a double solution, that is, $(x - 8)(x - 8) = 0$.]

22. -3, only solution

23. $-\dfrac{2}{5}$ and $\dfrac{6}{5}$

24. $-\dfrac{1}{4}$ and $-\dfrac{1}{2}$

25. $\dfrac{k}{3}$ and $\dfrac{m}{4}$

26. $\dfrac{c}{2}$ and $\dfrac{d}{2}$

27. $-\sqrt{3}$ and $2\sqrt{3}$

28. $\sqrt{2}$ and $3\sqrt{2}$

C   Solve.

29. $x^4 - 6x^2 + 9 = 0$

30. $x^4 - 7x^2 + 12 = 0$

31. $x - 10\sqrt{x} + 9 = 0$

32. $2x - 9\sqrt{x} + 4 = 0$

33. $(x^2 - 6x)^2 - 2(x^2 - 6x) - 35 = 0$

34. $(x^2 + 5x)^2 + 2(x^2 + 5x) - 24 = 0$

35. $x^{-2} - 5x^{-1} - 36 = 0$

36. $3x^{-2} - x^{-1} - 14 = 0$

37. $\left(1 + \sqrt{x}\right)^2 + \left(1 + \sqrt{x}\right) - 6 = 0$

38. $\left(2 + \sqrt{x}\right)^2 - 3\left(2 + \sqrt{x}\right) - 10 = 0$

39. $(y^2 - 5y)^2 - 2(y^2 - 5y) - 24 = 0$

40. $(2t^2 + t)^2 - 4(2t^2 + t) + 3 = 0$

41. $t^4 - 10t^2 + 9 = 0$

42. $w^4 - 29w^2 + 100 = 0$

43. $2x^{-2} + x^{-1} - 1 = 0$

44. $m^{-2} + 9m^{-1} - 10 = 0$

45. $6x^4 - 19x^2 + 15 = 0$

46. $6x^4 - 17x^2 + 5 = 0$

47. $x^{2/3} - 4x^{1/3} - 5 = 0$

48. $x^{2/3} + 2x^{1/3} - 8 = 0$

49. $\left(\dfrac{x-4}{x+1}\right)^2 - 2\left(\dfrac{x-4}{x+1}\right) - 35 = 0$

50. $\left(\dfrac{x+3}{x-3}\right)^2 - \left(\dfrac{x+3}{x-3}\right) - 6 = 0$

51. $9\left(\dfrac{x+2}{x+3}\right)^2 - 6\left(\dfrac{x+2}{x+3}\right) + 1 = 0$

52. $16\left(\dfrac{x-1}{x-8}\right)^2 + 8\left(\dfrac{x-1}{x-8}\right) + 1 = 0$

53. $\left(\dfrac{x^2-2}{x}\right)^2 - 7\left(\dfrac{x^2-2}{x}\right) - 18 = 0$

54. $\left(\dfrac{y^2-1}{y}\right)^2 - 4\left(\dfrac{y^2-1}{y}\right) - 12 = 0$

Find the x-intercepts of the function.

55. $f(x) = 5x + 13\sqrt{x} - 6$

56. $f(x) = 3x + 10\sqrt{x} - 8$

57. $f(x) = (x^2 - 3x)^2 - 10(x^2 - 3x) + 24$

58. $f(x) = x^{2/5} + x^{1/5} - 6$

59. $\mathbf{D_W}$ Describe a procedure that could be used to write an equation having the first seven natural numbers as solutions.

60. $\mathbf{D_W}$ Describe a procedure that could be used to write an equation that is quadratic in $3x^2 + 1$ and has real-number solutions.

Solve. [3.4a]

61. *Coffee Beans.* Twin Cities Roasters has Kenyan coffee worth $6.75 per pound and Peruvian coffee worth $11.25 per pound. How many pounds of each kind should be mixed in order to obtain a 50-lb mixture that is worth $8.55 per pound?

62. *Solution Mixtures.* Solution A is 18% alcohol and solution B is 45% alcohol. How many liters of each should be mixed in order to get 12 L of a solution that is 36% alcohol?

Multiply and simplify. Assume that all expressions under radicals represent nonnegative numbers. [6.3a]

63. $\sqrt{8x}\,\sqrt{2x}$

64. $\sqrt[3]{x^2}\,\sqrt[3]{27x^4}$

65. $\sqrt[4]{9a^2}\,\sqrt[4]{18a^3}$

66. $\sqrt[5]{16}\,\sqrt[5]{64}$

Graph. [2.2c], [2.5a, c]

67. $f(x) = -\frac{3}{5}x + 4$

68. $5x - 2y = 8$

69. $y = 4$

70. $f(x) = -x - 3$

71. Use a graphing calculator to check your answers to Exercises 30, 32, 34, and 37.

72. Use a graphing calculator to solve each of the following equations.
a) $6.75x - 35\sqrt{x} - 5.26 = 0$
b) $\pi x^4 - \pi^2 x^2 = \sqrt{99.3}$
c) $x^4 - x^3 - 13x^2 + x + 12 = 0$

For each equation under the given condition, **(a)** find k and **(b)** find the other solution.

73. $kx^2 - 2x + k = 0$; one solution is -3.

74. $kx^2 - 17x + 33 = 0$; one solution is 3.

75. Find a quadratic equation for which the sum of the solutions is $\sqrt{3}$ and the product is 8.

76. Find k given that $kx^2 - 4x + (2k - 1) = 0$ and the product of the solutions is 3.

77. The graph of a function of the form
$$f(x) = ax^2 + bx + c$$
is a curve similar to the one shown below. Determine a, b, and c from the information given.

78. While solving a quadratic equation of the form $ax^2 + bx + c = 0$ with a graphing calculator, Shawn-Marie gets the following screen.

How could the discriminant help her check the graph?

Solve.

79. $\dfrac{x}{x-1} - 6\sqrt{\dfrac{x}{x-1}} - 40 = 0$

80. $\dfrac{x}{x-3} - 24 = 10\sqrt{\dfrac{x}{x-3}}$

81. $\sqrt{x-3} - \sqrt[4]{x-3} = 12$

82. $a^3 - 26a^{3/2} - 27 = 0$

83. $x^6 - 28x^3 + 27 = 0$

84. $x^6 + 7x^3 - 8 = 0$

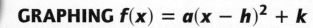

In this section and the next, we develop techniques for graphing quadratic functions.

a   Graphs of $f(x) = ax^2$

The most basic quadratic function is $f(x) = x^2$.

EXAMPLE 1   Graph: $f(x) = x^2$.

We choose some values for x and compute $f(x)$ for each. Then we plot the ordered pairs and connect them with a smooth curve.

x	$f(x) = x^2$	$(x, f(x))$
-3	9	$(-3, 9)$
-2	4	$(-2, 4)$
-1	1	$(-1, 1)$
0	0	$(0, 0)$
1	1	$(1, 1)$
2	4	$(2, 4)$
3	9	$(3, 9)$

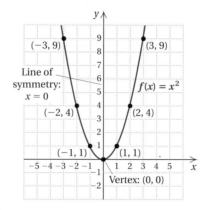

All quadratic functions have graphs similar to the one in Example 1. Such curves are called **parabolas.** They are cup-shaped curves that are symmetric with respect to a vertical line known as the parabola's **line of symmetry,** or **axis of symmetry.** In the graph of $f(x) = x^2$, shown above, the y-axis (or the line $x = 0$) is the line of symmetry. If the paper were to be folded on this line, the two halves of the curve would match. The point $(0, 0)$ is the **vertex** of this parabola.

Let's compare the graphs of $g(x) = \frac{1}{2}x^2$ and $h(x) = 2x^2$ with the graph of $f(x) = x^2$. We choose x-values and plot points for both functions.

x	$g(x) = \frac{1}{2}x^2$
-3	$\frac{9}{2}$
-2	2
-1	$\frac{1}{2}$
0	0
1	$\frac{1}{2}$
2	2
3	$\frac{9}{2}$

x	$h(x) = 2x^2$
-3	18
-2	8
-1	2
0	0
1	2
2	8
3	18

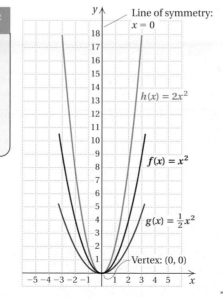

Note the symmetry: For equal increments to the left and right of the vertex, the y-values are the same.

Objectives

a Graph quadratic functions of the type $f(x) = ax^2$ and then label the vertex and the line of symmetry.

b Graph quadratic functions of the type $f(x) = a(x - h)^2$ and then label the vertex and the line of symmetry.

b Graph quadratic functions of the type $f(x) = a(x - h)^2 + k$, finding the vertex, the line of symmetry, and the maximum or minimum y-value.

CALCULATOR CORNER

Graphing Quadratic Functions Use a graphing calculator to make a table of values for $f(x) = x^2$. (See the Calculator Corner on p. 162 for the procedure.)

Graph.

1. $f(x) = -\frac{1}{3}x^2$

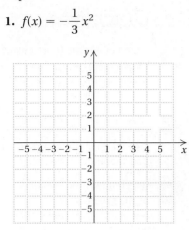

2. $f(x) = 3x^2$

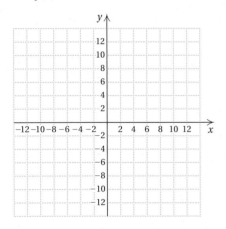

3. $f(x) = -2x^2$

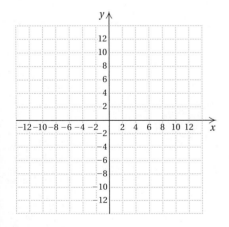

Note that the graph of $g(x) = \frac{1}{2}x^2$ is a wider parabola than the graph of $f(x) = x^2$, and the graph of $h(x) = 2x^2$ is narrower. The vertex and the line of symmetry, however, remain $(0, 0)$ and $x = 0$, respectively.

CALCULATOR CORNER

Graphs of Quadratic Functions

1. Graph each of the following equations in the viewing window $[-5, 5, -10, 10]$:

$$y_1 = x^2, \qquad y_2 = 3x^2, \qquad y_3 = \frac{1}{3}x^2.$$

Determine a rule that describes the effect of the constant a on the graph of $y = ax^2$, where $a > 1$ and where $0 < a < 1$. Graph some other equations to test your rule.

2. Graph each of the following equations in the viewing window $[-5, 5 - 10, 10]$:

$$y_1 = -x^2, \qquad y_2 = -4x^2, \qquad y_3 = -\frac{2}{3}x^2.$$

Determine a rule that describes the effect of the constant a on the graph of $y = ax^2$, where $a < -1$ and where $-1 < a < 0$. Graph some other equations to test your rule.

When we consider the graph of $k(x) = -\frac{1}{2}x^2$, we see that the parabola opens down and is the same shape as the graph of $g(x) = \frac{1}{2}x^2$.

x	$k(x) = -\frac{1}{2}x^2$
-3	$-\frac{9}{2}$
-2	-2
-1	$-\frac{1}{2}$
0	0
1	$-\frac{1}{2}$
2	-2
3	$-\frac{9}{2}$

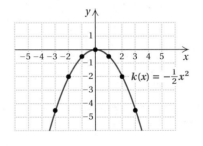

GRAPHS OF $f(x) = ax^2$

The graph of $f(x) = ax^2$, or $y = ax^2$, is a parabola with $x = 0$ as its line of symmetry; its vertex is the origin.

For $a > 0$, the parabola opens up; for $a < 0$, the parabola opens down.

If $|a|$ is greater than 1, the parabola is narrower than $y = x^2$.

If $|a|$ is between 0 and 1, the parabola is wider than $y = x^2$.

Do Exercises 1–3.

Answers on page A-38

b Graphs of $f(x) = a(x - h)^2$

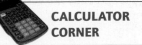

CALCULATOR CORNER

It would seem logical now to consider functions of the type

$$f(x) = ax^2 + bx + c.$$

We are heading in that direction, but it is convenient to first consider graphs of $f(x) = a(x - h)^2$ and then $f(x) = a(x - h)^2 + k$, where a, h, and k are constants.

Graphing Quadratic Functions   Use a graphing calculator to graph the equations in Margin Exercises 1–3.

EXAMPLE 2 Graph: $g(x) = (x - 3)^2$.

We choose some values for x and compute $g(x)$. Then we plot the points and draw the curve.

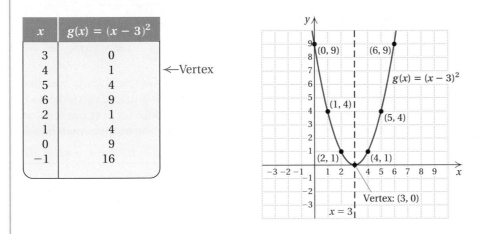

x	$g(x) = (x - 3)^2$
3	0
4	1
5	4
6	9
2	1
1	4
0	9
−1	16

←Vertex

First note that for an x-value of 3, $g(3) = (3 - 3)^2 = 0$. As we increase x-values from 3, note that the corresponding y-values increase. Then as we decrease x-values from 3, note that the corresponding y-values increase again. The line $x = 3$ is a line of symmetry. Equal distances of x-values to the left and right of the vertex produce the same y-values.

CALCULATOR CORNER

Graphs of Quadratic Functions   Graph each of the following equations in the viewing window $[-5, 5, -5, 5]$:

$$y_1 = 7x^2,$$
$$y_2 = 7(x - 1)^2,$$
$$y_3 = 7(x + 2)^2.$$

Determine a rule that describes the effect of the constant h on the graph of $y = a(x - h)^2$. Graph some other equations to test your rule.

EXAMPLE 3 Graph: $t(x) = (x + 3)^2$.

We choose some values for x and compute $t(x)$. Then we plot the points and draw the curve.

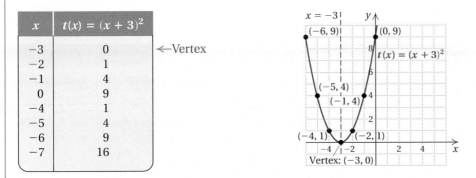

x	$t(x) = (x + 3)^2$
−3	0
−2	1
−1	4
0	9
−4	1
−5	4
−6	9
−7	16

←Vertex

First note that for an x-value of -3, $t(-3) = (-3 + 3)^2 = 0$. As we increase x-values from -3, note that the corresponding y-values increase. Then as we decrease x-values from -3, note that the y-values increase again. The line $x = -3$ is a line of symmetry.

Graph. Find and label the vertex and the line of symmetry.

4. $f(x) = \dfrac{1}{2}(x - 4)^2$

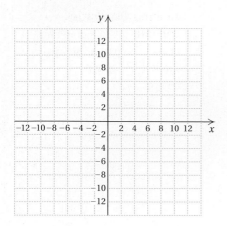

5. $f(x) = -\dfrac{1}{2}(x - 4)^2$

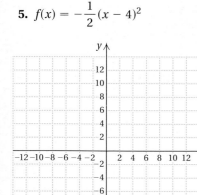

The graph of $g(x) = (x - 3)^2$ in Example 2 looks just like the graph of $f(x) = x^2$ in Example 1, except that it is moved, or translated, 3 units to the right. Comparing the pairs for $g(x)$ with those for $f(x)$, we see that when an input for $g(x)$ is 3 more than an input for $f(x)$, the outputs match.

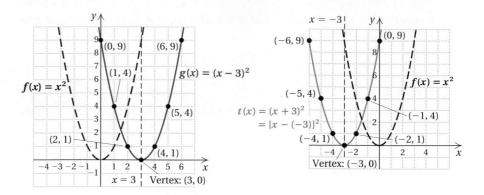

The graph of $t(x) = (x + 3)^2 = [x - (-3)]^2$ in Example 3 looks just like the graph of $f(x) = x^2$ in Example 1, except that it is moved, or translated, 3 units to the left. Comparing the pairs for $t(x)$ with those for $f(x)$, we see that when an input for $t(x)$ is 3 less than an input for $f(x)$, the outputs match.

GRAPHS OF $f(x) = a(x - h)^2$

The graph of $f(x) = a(x - h)^2$ has the same shape as the graph of $y = ax^2$.

If h is positive, the graph of $y = ax^2$ is shifted h units to the right.

If h is negative, the graph of $y = ax^2$ is shifted $|h|$ units to the left.

The vertex is $(h, 0)$ and the axis of symmetry is $x = h$.

EXAMPLE 4 Graph: $f(x) = -2(x + 3)^2$.

We first rewrite the equation as $f(x) = -2[x - (-3)]^2$. In this case, $a = -2$ and $h = -3$, so the graph looks like that of $g(x) = 2x^2$ translated 3 units to the left and, since $-2 < 0$, the graph opens down. The vertex is $(-3, 0)$, and the line of symmetry is $x = -3$. Plotting points as needed, we obtain the graph shown below.

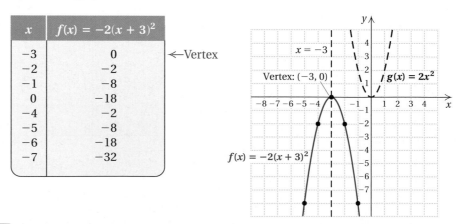

x	$f(x) = -2(x + 3)^2$	
-3	0	←Vertex
-2	-2	
-1	-8	
0	-18	
-4	-2	
-5	-8	
-6	-18	
-7	-32	

Answers on page A-38

Do Exercises 4 and 5.

C Graphs of $f(x) = a(x - h)^2 + k$

Given a graph of $f(x) = a(x - h)^2$, what happens if we add a constant k? Suppose that we add 2. This increases each function value $f(x)$ by 2, so the curve is moved up. If k is negative, the curve is moved down. The line of symmetry for the parabola remains $x = h$, but the vertex will be at (h, k), or equivalently, $(h, f(h))$.

Note that if a parabola opens up $(a > 0)$, the function value, or y-value, at the vertex is a least, or **minimum,** value. That is, it is less than the y-value at any other point on the graph. If the parabola opens down $(a < 0)$, the function value at the vertex is a greatest, or **maximum,** value.

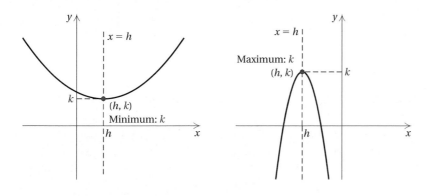

CALCULATOR CORNER

Graphs of Quadratic Functions Graph each of the following equations in the viewing window $[-5, 5, -5, 5]$:

$$y_1 = 7(x - 1)^2,$$
$$y_2 = 7(x - 1)^2 + 2,$$
$$y_3 = 7(x - 1)^2 - 3.$$

Determine a rule that describes the effect of the constant k on the graph of $y = a(x - h)^2 + k$. Graph some other equations to test your rule.

GRAPHS OF $f(x) = a(x - h)^2 + k$

The graph of $f(x) = a(x - h)^2 + k$ has the same shape as the graph of $y = a(x - h)^2$.

If k is positive, the graph of $y = a(x - h)^2$ is shifted k units up.

If k is negative, the graph of $y = a(x - h)^2$ is shifted $|k|$ units down.

The vertex is (h, k), and the line of symmetry is $x = h$.

For $a > 0$, k is the minimum function value. For $a < 0$, k is the maximum function value.

EXAMPLE 5 Graph $f(x) = (x - 3)^2 - 5$, and find the minimum function value.

The graph will look like that of $g(x) = (x - 3)^2$ (see Example 2) but translated 5 units down. You can confirm this by plotting some points. For instance,

$$f(4) = (4 - 3)^2 - 5 = -4,$$

whereas in Example 2,

$$g(4) = (4 - 3)^2 = 1.$$

Note that the vertex is $(h, k) = (3, -5)$, so we begin calculating points on both sides of $x = 3$. The line of symmetry is $x = 3$, and the minimum function value is -5.

Graph. Find the vertex, the line of symmetry, and the maximum or minimum y-value.

6. $f(x) = \frac{1}{2}(x + 2)^2 - 4$

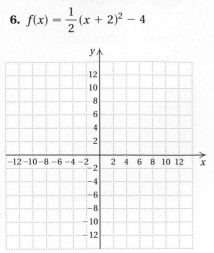

7. $f(x) = -2(x - 5)^2 + 3$

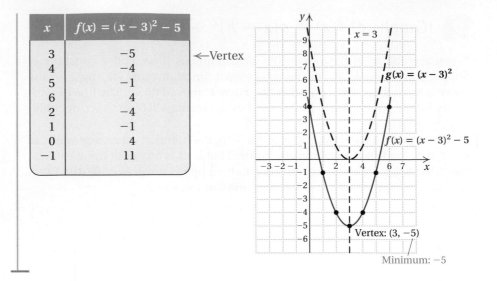

x	$f(x) = (x - 3)^2 - 5$
3	-5
4	-4
5	-1
6	4
2	-4
1	-1
0	4
-1	11

←Vertex

EXAMPLE 6 Graph $t(x) = \frac{1}{2}(x - 3)^2 + 5$, and find the minimum function value.

The graph looks just like that of $f(x) = \frac{1}{2}x^2$ but moved 3 units to the right and 5 units up. The vertex is $(3, 5)$, and the line of symmetry is $x = 3$. We draw $f(x) = \frac{1}{2}x^2$ and then shift the curve over and up. The minimum function value is 5. By plotting some points, we have a check.

x	$t(x) = \frac{1}{2}(x - 3)^2 + 5$
3	5
4	$5\frac{1}{2}$
5	7
6	$9\frac{1}{2}$
2	$5\frac{1}{2}$
1	7
0	$9\frac{1}{2}$

←Vertex

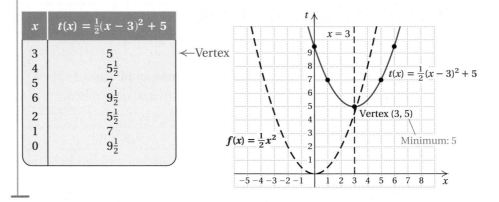

EXAMPLE 7 Graph $f(x) = -2(x + 3)^2 + 5$. Find the vertex, the line of symmetry, and the maximum or minimum value.

We first express the equation in the equivalent form

$$f(x) = -2[x - (-3)]^2 + 5.$$

The graph looks like that of $g(x) = -2x^2$ translated 3 units to the left and 5 units up. The vertex is $(-3, 5)$, and the line of symmetry is $x = -3$. Since $-2 < 0$, we know that the graph opens down so 5, the second coordinate of the vertex, is the maximum y-value.

We compute a few points as needed. The graph is shown on the following page.

x	$f(x) = -2(x + 3)^2 + 5$
-3	5
-2	3
-1	-3
-4	3
-5	-3

←Vertex

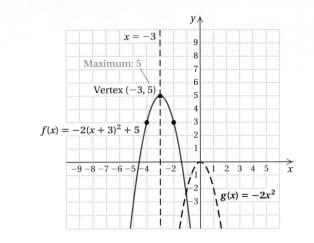

Do Exercises 6 and 7 on the preceding page.

Study Tips

THE 20 BEST JOBS

JOB	ANNUAL EARNINGS	GROWTH RATE THROUGH 2010	ANNUAL OPENINGS
1. Computer software engineers, applications	$70,210	100.0%	28,000
2. Computer systems analysts	61,990	59.7	34,000
3. Computer and information systems managers	82,480	47.9	28,000
4. Teachers, postsecondary	52,115	23.5	184,000
5. Management analysts	57,970	28.9	50,000
6. Registered nurses	46,670	25.6	140,000
7. Computer software engineers, systems software	73,280	89.7	23,000
8. Medical and health services managers	59,220	32.3	27,000
9. Sales agents, financial services	59,690	22.3	55,000
10. Sales agents, securities and commodities	59,690	22.3	55,000
11. Securities, commodities, and financial services sales agents	59,690	22.3	55,000
12. Computer support specialists	38,560	97.0	40,000
13. Sales managers	71,620	32.8	21,000
14. Computer security specialists	53,770	81.9	18,000
15. Network and computer systems administrators	53,770	81.9	18,000
16. Financial managers	70,210	18.5	53,000
17. Financial managers, branch or department	70,210	18.5	53,000
18. Treasurers, comptrollers, and chief financial officers	70,210	18.5	53,000
19. Accountants	45,380	18.5	100,000
20. Accountants and auditors	45,380	18.5	100,000

Source: Based on "The 20 Best Jobs," from Best Jobs for the 21st Century, 3rd ed. Indianapolis: Jist Works, 2004, p. 16.

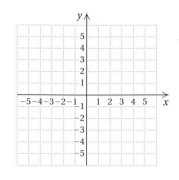
 Graph. Find and label the vertex and the line of symmetry.

1. $f(x) = 4x^2$

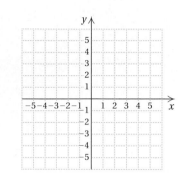

x	$f(x)$
0	
1	
2	
-1	
-2	

Vertex: (____, ____)
Line of symmetry: $x =$ ____

2. $f(x) = 5x^2$

x	$f(x)$
0	
1	
2	
-1	
-2	

Vertex: (____, ____)
Line of symmetry: $x =$ ____

3. $f(x) = \frac{1}{3}x^2$

x	$f(x)$
0	
1	
2	
-1	
-2	

Vertex: (____, ____)
Line of symmetry: $x =$ ____

4. $f(x) = \frac{1}{4}x^2$

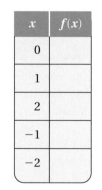

x	$f(x)$
0	
1	
2	
-1	
-2	

Vertex: (____, ____)
Line of symmetry: $x =$ ____

5. $f(x) = -\frac{1}{2}x^2$

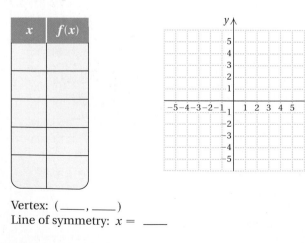

x	$f(x)$

Vertex: (____, ____)
Line of symmetry: $x =$ ____

6. $f(x) = -\frac{1}{4}x^2$

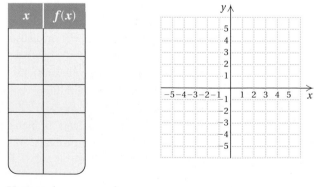

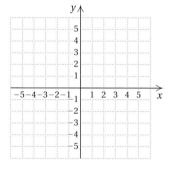

x	$f(x)$

Vertex: (____, ____)
Line of symmetry: $x =$ ____

Copyright © 2007 Pearson Education, Inc.

7. $f(x) = -4x^2$

x	$f(x)$

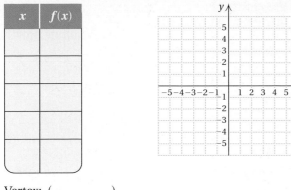

Vertex: (____ , ____)
Line of symmetry: $x =$ ____

8. $f(x) = -3x^2$

x	$f(x)$

Vertex: (____ , ____)
Line of symmetry: $x =$ ____

9. $f(x) = (x + 3)^2$

x	$f(x)$
-3	
-2	
-1	
-4	
-5	

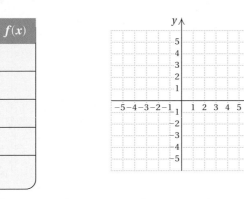

Vertex: (____ , ____)
Line of symmetry: $x =$ ____

10. $f(x) = (x + 1)^2$

x	$f(x)$
-1	
0	
1	
-2	
-3	

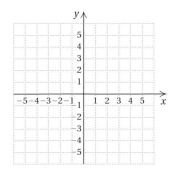

Vertex: (____ , ____)
Line of symmetry: $x =$ ____

11. $f(x) = 2(x - 4)^2$

x	$f(x)$

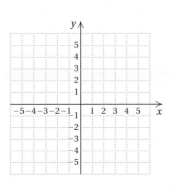

Vertex: (____ , ____)
Line of symmetry: $x =$ ____

12. $f(x) = 4(x - 1)^2$

x	$f(x)$

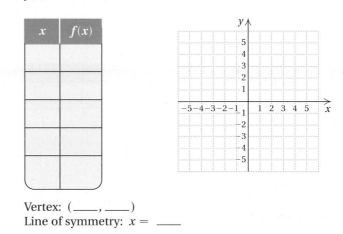

Vertex: (____ , ____)
Line of symmetry: $x =$ ____

625

13. $f(x) = -2(x + 2)^2$

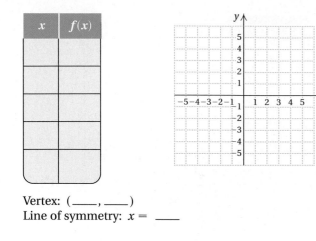

x	$f(x)$

Vertex: (____ , ____)
Line of symmetry: $x =$ ____

14. $f(x) = -2(x + 4)^2$

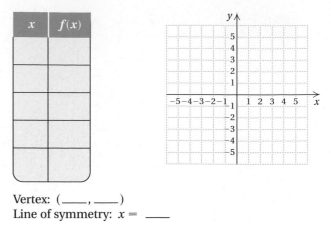

x	$f(x)$

Vertex: (____ , ____)
Line of symmetry: $x =$ ____

15. $f(x) = 3(x - 1)^2$

16. $f(x) = 4(x - 2)^2$

17. $f(x) = -\frac{3}{2}(x + 2)^2$

18. $f(x) = -\frac{5}{2}(x + 3)^2$

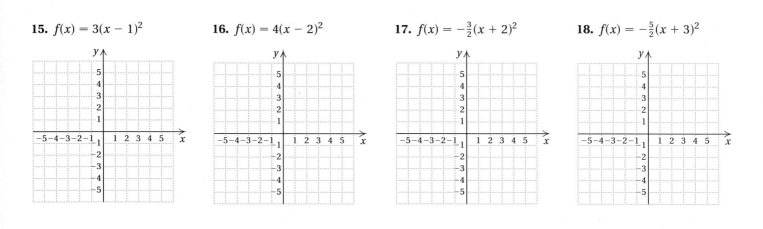

C Graph. Find and label the vertex and the line of symmetry. Find the maximum or minimum value.

19. $f(x) = (x - 3)^2 + 1$

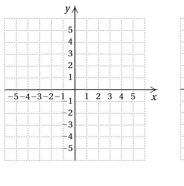

Vertex: (____ , ____)
Line of symmetry: $x =$ ____
Minimum value: ____

20. $f(x) = (x + 2)^2 - 3$

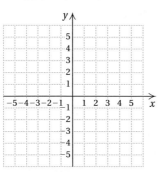

Vertex: (____ , ____)
Line of symmetry: $x =$ ____
Minimum value: ____

21. $f(x) = -3(x + 4)^2 + 1$

Vertex: (____ , ____)
Line of symmetry: $x =$ ____
Maximum value: ____

22. $f(x) = -\frac{1}{2}(x - 1)^2 - 3$

Vertex: (____ , ____)
Line of symmetry: $x =$ ____
Maximum value: ____

23. $f(x) = \frac{1}{2}(x + 1)^2 + 4$

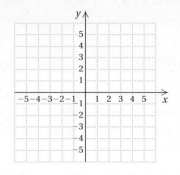

Vertex: (__ , __)
Line of symmetry: $x =$ ____
_____ value: _____

24. $f(x) = -2(x - 5)^2 - 3$

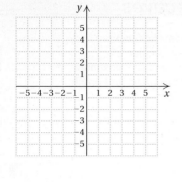

Vertex: (__ , __)
Line of symmetry: $x =$ ____
_____ value: _____

25. $f(x) = -(x + 1)^2 - 2$

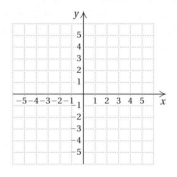

Vertex: (__ , __)
Line of symmetry: $x =$ ____
_____ value: _____

26. $f(x) = 3(x - 4)^2 + 2$

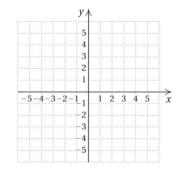

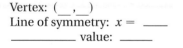

Vertex: (__ , __)
Line of symmetry: $x =$ ____
_____ value: _____

27. D_W Explain, without plotting points, why the graph of $f(x) = (x + 3)^2$ looks like the graph of $f(x) = x^2$ translated 3 units to the left.

28. D_W Explain, without plotting points, why the graph of $f(x) = (x + 3)^2 - 4$ looks like the graph of $f(x) = x^2$ translated 3 units to the left and 4 units down.

SKILL MAINTENANCE

Solve. [6.6a, b]

29. $x - 5 = \sqrt{x + 7}$

30. $\sqrt{2x + 7} = \sqrt{5x - 4}$

31. $\sqrt{x + 4} = -11$

32. $x = 7 + 2\sqrt{x + 1}$

Multiply and simplify. Assume that all expressions under radicals represent nonnegative numbers. [6.3a]

33. $\sqrt[4]{5x^3y^5} \, \sqrt[4]{125x^2y^3}$

34. $\sqrt{9a^3} \, \sqrt{16ab^4}$

SYNTHESIS

35. Use the TRACE and/or TABLE features on a graphing calculator to confirm the maximum or minimum values given in Exercises 23, 24, and 26.

Objectives

a For a quadratic function, find the vertex, the line of symmetry, and the maximum or minimum value, and graph the function.

b Find the intercepts of a quadratic function.

1. For $f(x) = x^2 - 4x + 7$, find the vertex, the line of symmetry, and the minimum value. Then graph.

x	f(x)

Vertex: (_____, _____)
Line of symmetry: $x =$ _____
Minimum value: _____

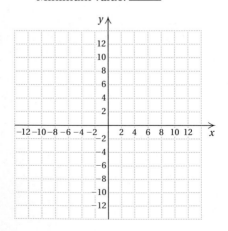

Answer on page A-40

628

CHAPTER 7: Quadratic Equations and Functions

---

a Graphing and Analyzing $f(x) = ax^2 + bx + c$

By *completing the square*, we can begin with any quadratic polynomial $ax^2 + bx + c$ and find an equivalent expression $a(x - h)^2 + k$. This allows us to combine the skills of Sections 7.1 and 7.5 to graph and analyze any quadratic function $f(x) = ax^2 + bx + c$.

EXAMPLE 1 For $f(x) = x^2 - 6x + 4$, find the vertex, the line of symmetry, and the minimum value. Then graph.

We first find the vertex and the line of symmetry. To do so, we find the equivalent form $a(x - h)^2 + k$ by completing the square, beginning as follows:

$$f(x) = x^2 - 6x + 4 = (x^2 - 6x \quad) + 4.$$

We complete the square inside the parentheses, but in a different manner than we did before. We take half the x-coefficient, $-6/2 = -3$, and square it: $(-3)^2 = 9$. Then we add 0, or $9 - 9$, inside the parentheses. (Because we are using function notation, instead of adding $(b/2)^2$ on both sides of an equation, we add and subtract it on the same side, effectively adding 0 and not changing the value of the expression.)

$$
\begin{aligned}
f(x) &= (x^2 - 6x + 0) + 4 & &\text{Adding 0} \\
&= (x^2 - 6x + 9 - 9) + 4 & &\text{Substituting } 9 - 9 \text{ for 0} \\
&= (x^2 - 6x + 9) + (-9 + 4) & &\text{Using the associative law of addition to regroup} \\
&= (x - 3)^2 - 5. & &\text{Factoring and simplifying}
\end{aligned}
$$

(This equation was graphed in Example 5 of Section 7.5.) The vertex is $(3, -5)$, and the line of symmetry is $x = 3$. The coefficient of x^2 is 1, which is positive, so the graph opens up. This tells us that -5 is a minimum. We plot the vertex and draw the line of symmetry. We choose some x-values on both sides of the vertex and graph the parabola. Suppose we compute the pair $(5, -1)$:

$$f(5) = 5^2 - 6(5) + 4 = 25 - 30 + 4 = -1.$$

We note that it is 2 units to the right of the line of symmetry. There will also be a pair with the same y-coordinate on the graph 2 units to the *left* of the line of symmetry. Thus we get a second point, $(1, -1)$, without making another calculation.

x	f(x)	
3	−5	←Vertex
4	−4	
5	−1	
6	4	
2	−4	
1	−1	
0	4	

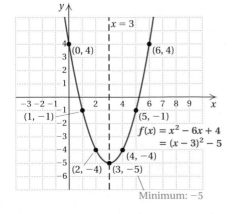

$$f(x) = x^2 - 6x + 4 = (x - 3)^2 - 5$$

Minimum: −5

Do Exercise 1 on the preceding page.

EXAMPLE 2 For $f(x) = 3x^2 + 12x + 13$, find the vertex, the line of symmetry, and the minimum value. Then graph.

Since the coefficient of x^2 is not 1, we factor out 3 from only the *first two* terms of the expression. Remember that we want to get to the form $f(x) = a(x - h)^2 + k$:

$$f(x) = 3x^2 + 12x + 13$$
$$= 3(x^2 + 4x) + 13. \quad \text{Factoring 3 out of the first two terms}$$

Next, we complete the square inside the parentheses:

$$f(x) = 3(x^2 + 4x \qquad) + 13.$$

We take half the x-coefficient, $\frac{1}{2} \cdot 4 = 2$, and square it: $2^2 = 4$. Then we add 0, or $4 - 4$, inside the parentheses:

$$f(x) = 3(x^2 + 4x + 0) + 13 \qquad \text{Adding 0}$$
$$= 3(x^2 + 4x + 4 - 4) + 13 \qquad \text{Substituting } 4 - 4 \text{ for } 0$$
$$\left.\begin{array}{l} = 3(x^2 + 4x + 4 - 4) + 13 \\[6pt] = 3(x^2 + 4x + 4) + 3(-4) + 13 \end{array}\right\} \quad \begin{array}{l}\text{Using the distributive} \\ \text{law to separate } -4 \\ \text{from the trinomial}\end{array}$$
$$= 3(x + 2)^2 + 1. \qquad \text{Factoring and simplifying}$$

The vertex is $(-2, 1)$, and the line of symmetry is $x = -2$. The coefficient of x^2 is 3, so the graph is narrow and opens up. This tells us that 1 is a minimum. We choose a few x-values on one side of the line of symmetry, compute y-values, and use the resulting coordinates to find more points on the other side of the line of symmetry. We plot points and graph the parabola.

x	$f(x)$	
-2	1	←Vertex
-1	4	
-3	4	
0	13	
-4	13	

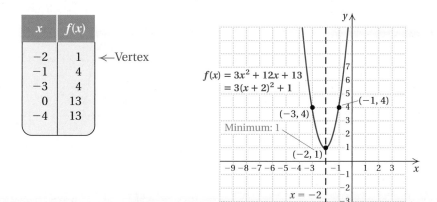

Do Exercise 2.

2. For $f(x) = 3x^2 - 24x + 43$, find the vertex, the line of symmetry, and the minimum value. Then graph.

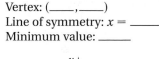

Vertex: (____, ____)
Line of symmetry: $x = $ _____
Minimum value: _____

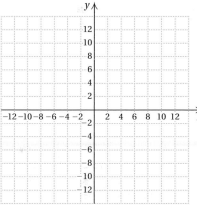

Answer on page A-40

629

7.6 Graphing $f(x) = ax^2 + bx + c$

3. For $f(x) = -4x^2 + 12x - 5$, find the vertex, the line of symmetry, and the maximum value. Then graph.

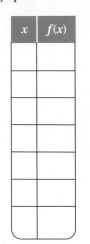

x	$f(x)$

Vertex: (____,____)
Line of symmetry: $x =$ _____
Maximum value: _____

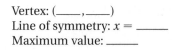

EXAMPLE 3 For $f(x) = -2x^2 + 10x - 7$, find the vertex, the line of symmetry, and the maximum value. Then graph.

Again, the coefficient of x^2 is not 1. We factor out -2 from only the *first two* terms of the expression. This makes the coefficient of x^2 inside the parentheses 1:

$$f(x) = -2x^2 + 10x - 7$$
$$= -2(x^2 - 5x) - 7.$$

Next, we complete the square as before:

$$f(x) = -2(x^2 - 5x \qquad) - 7.$$

We take half the x-coefficient, $\frac{1}{2}(-5) = -\frac{5}{2}$, and square it: $\left(-\frac{5}{2}\right)^2 = \frac{25}{4}$. Then we add 0, or $\frac{25}{4} - \frac{25}{4}$, inside the parentheses:

$$f(x) = -2\left(x^2 - 5x + \frac{25}{4} - \frac{25}{4}\right) - 7 \qquad \text{Adding 0, or } \frac{25}{4} - \frac{25}{4}$$
$$= -2\left(x^2 - 5x + \frac{25}{4} - \frac{25}{4}\right) - 7$$

Using the distributive law to separate the $-\frac{25}{4}$ from the trinomial

$$= -2\left(x^2 - 5x + \frac{25}{4}\right) + (-2)\left(-\frac{25}{4}\right) - 7$$
$$= -2\left(x^2 - 5x + \frac{25}{4}\right) + \frac{25}{2} - 7$$
$$= -2\left(x - \frac{5}{2}\right)^2 + \frac{11}{2}. \qquad \text{Factoring and simplifying}$$

The vertex is $\left(\frac{5}{2}, \frac{11}{2}\right)$, and the line of symmetry is $x = \frac{5}{2}$. The coefficient of x^2 is -2, so the graph is narrow and opens down. This tells us that $\frac{11}{2}$ is a maximum. We choose a few x-values on one side of the line of symmetry, compute y-values, and use the resulting coordinates to find more points on the other side of the line of symmetry. We plot points and graph the parabola.

x	$f(x)$
$\frac{5}{2}$	$\frac{11}{2}$, or $5\frac{1}{2}$ ←Vertex
3	5
4	1
5	-7
2	5
1	1
0	-7

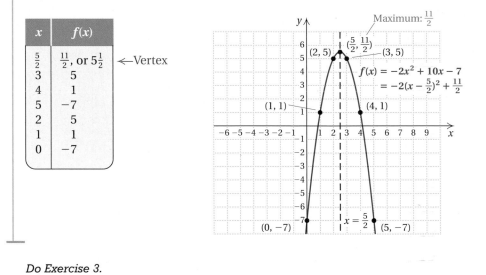

Do Exercise 3.

Answer on page A-40

CHAPTER 7: Quadratic Equations and Functions

The method used in Examples 1–3 can be generalized to find a formula for locating the vertex. We complete the square as follows:

$$f(x) = ax^2 + bx + c$$

$$= a\left(x^2 + \frac{b}{a}x\right) + c. \qquad \text{Factoring } a \text{ out of the first two terms.}$$
$$\text{Check by multiplying.}$$

Half of the x-coefficient, $\dfrac{b}{a}$, is $\dfrac{b}{2a}$. We square it to get $\dfrac{b^2}{4a^2}$ and add $\dfrac{b^2}{4a^2} - \dfrac{b^2}{4a^2}$ inside the parentheses. Then we distribute the a:

$$f(x) = a\left(x^2 + \frac{b}{a}x + \frac{b^2}{4a^2} - \frac{b^2}{4a^2}\right) + c$$

$$\left.\begin{array}{l} \\ \\ \\ = a\left(x^2 + \dfrac{b}{a}x + \dfrac{b^2}{4a^2}\right) + a\left(-\dfrac{b^2}{4a^2}\right) + c \\ \\ = a\left(x + \dfrac{b}{2a}\right)^2 + \dfrac{-b^2}{4a} + \dfrac{4ac}{4a} \\ \\ = a\left[x - \left(-\dfrac{b}{2a}\right)\right]^2 + \dfrac{4ac - b^2}{4a}. \end{array}\right\}$$

Using the distributive law

Factoring and finding a common denominator

Thus we have the following.

VERTEX; LINE OF SYMMETRY

The **vertex** of a parabola given by $f(x) = ax^2 + bx + c$ is

$$\left(-\frac{b}{2a}, \frac{4ac - b^2}{4a}\right), \quad \text{or} \quad \left(-\frac{b}{2a}, f\left(-\frac{b}{2a}\right)\right).$$

The x-coordinate of the vertex is $-b/(2a)$. The **line of symmetry** is $x = -b/(2a)$. The second coordinate of the vertex is easiest to find by computing $f\left(-\dfrac{b}{2a}\right)$.

Let's reexamine Example 3 to see how we could have found the vertex directly. From the formula above,

$$\text{the } x\text{-coordinate of the vertex is} -\frac{b}{2a} = -\frac{10}{2(-2)} = \frac{5}{2}.$$

Substituting $\frac{5}{2}$ into $f(x) = -2x^2 + 10x - 7$, we find the second coordinate of the vertex:

$$f\left(\tfrac{5}{2}\right) = -2\left(\tfrac{5}{2}\right)^2 + 10\left(\tfrac{5}{2}\right) - 7$$
$$= -2\left(\tfrac{25}{4}\right) + 25 - 7$$
$$= -\tfrac{25}{2} + 18 = -\tfrac{25}{2} + \tfrac{36}{2} = \tfrac{11}{2}.$$

The vertex is $\left(\frac{5}{2}, \frac{11}{2}\right)$. The line of symmetry is $x = \frac{5}{2}$.

We have developed two methods for finding the vertex. One is by completing the square and the other is by using a formula. You should check with your instructor about which method to use.

Do Exercises 4–6.

Find the vertex of the parabola using the formula.

4. $f(x) = x^2 - 6x + 4$

5. $f(x) = 3x^2 - 24x + 43$

6. $f(x) = -4x^2 + 12x - 5$

Answers on page A-40

Find the intercepts.

7. $f(x) = x^2 + 2x - 3$

b Finding the Intercepts of a Quadratic Function

The points at which a graph crosses an axis are called **intercepts.** We determine the y-intercept by finding $f(0)$. For $f(x) = ax^2 + bx + c$, the y-intercept is $(0, c)$.

To find the x-intercepts, we look for values of x for which $f(x) = 0$. For $f(x) = ax^2 + bx + c$, we solve

$$0 = ax^2 + bx + c.$$

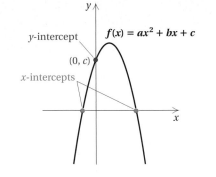

8. $f(x) = x^2 + 8x + 16$

EXAMPLE 4 Find the intercepts of $f(x) = x^2 - 2x - 2$.

The y-intercept is $(0, f(0))$. Since $f(0) = 0^2 - 2 \cdot 0 - 2 = -2$, the y-intercept is $(0, -2)$. To find the x-intercepts, we solve

$$0 = x^2 - 2x - 2.$$

Using the quadratic formula gives us $x = 1 \pm \sqrt{3}$. Thus the x-intercepts are $\left(1 - \sqrt{3}, 0\right)$ and $\left(1 + \sqrt{3}, 0\right)$, or, approximately, $(-0.732, 0)$ and $(2.732, 0)$.

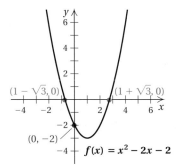

9. $f(x) = x^2 - 4x + 1$

Do Exercises 7–9.

Answers on page A-40

CHAPTER 7: Quadratic Equations
and Functions

Visualizing
for Success

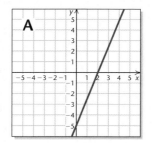

A

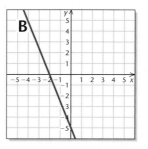

B

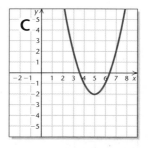

C

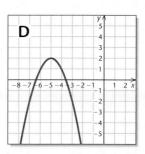

D

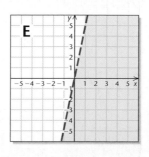

E

Match each equation or inequality with its graph.

1. $y = -(x - 5)^2 + 2$

2. $2x + 5y = 10$

3. $5x - 2y = 10$

4. $2x - 5y = 10$

5. $y = (x - 5)^2 - 2$

6. $y = x^2 - 5$

7. $5x + 2y \geq 10$

8. $5x + 2y = -10$

9. $y < 5x$

10. $y = -(x + 5)^2 + 2$

Answers on page A-40

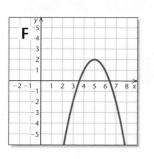

F

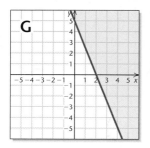

G

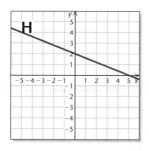

H

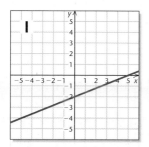

I

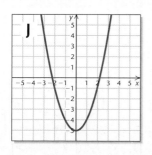

J

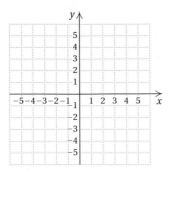

a For each quadratic function, find **(a)** the vertex, **(b)** the line of symmetry, and **(c)** the maximum or minimum value. Then **(d)** graph the function.

1. $f(x) = x^2 - 2x - 3$

x	$f(x)$

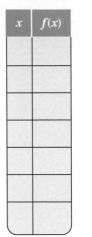

Vertex: (____,____)
Line of symmetry: $x =$ ____
_____ value: _____

2. $f(x) = x^2 + 2x - 5$

x	$f(x)$

Vertex: (____,____)
Line of symmetry: $x =$ ____
_____ value: _____

3. $f(x) = -x^2 - 4x - 2$

x	$f(x)$

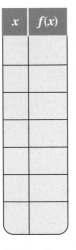

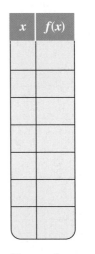

Vertex: (____,____)
Line of symmetry: $x =$ ____
_____ value: _____

4. $f(x) = -x^2 + 4x + 1$

x	$f(x)$

Vertex: (____,____)
Line of symmetry: $x =$ ____
_____ value: _____

5. $f(x) = 3x^2 - 24x + 50$

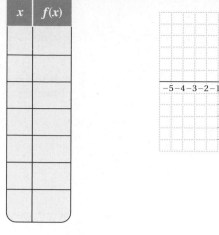

x	$f(x)$

Vertex: (_____, _____)
Line of symmetry: $x =$ _____
_____ value: _____

6. $f(x) = 4x^2 + 8x + 1$

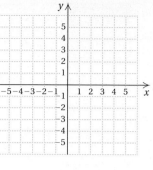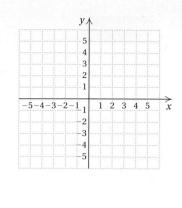

x	$f(x)$

Vertex: (_____, _____)
Line of symmetry: $x =$ _____
_____ value: _____

7. $f(x) = -2x^2 - 2x + 3$

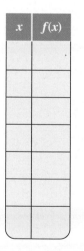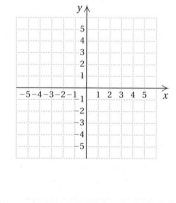

x	$f(x)$

Vertex: (_____, _____)
Line of symmetry: $x =$ _____
_____ value: _____

8. $f(x) = -2x^2 + 2x + 1$

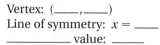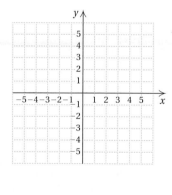

x	$f(x)$

Vertex: (_____, _____),
Line of symmetry: $x =$ _____
_____ value: _____

9. $f(x) = 5 - x^2$

x	$f(x)$

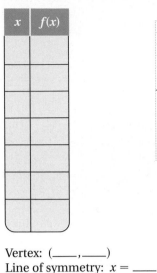

Vertex: (____,____)
Line of symmetry: $x =$ ____
_____ value: ____

10. $f(x) = x^2 - 3x$

x	$f(x)$

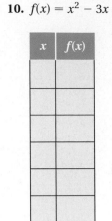

Vertex: (____,____)
Line of symmetry: $x =$ ____
_____ value: ____

11. $f(x) = 2x^2 + 5x - 2$

x	$f(x)$

Vertex: (____,____)
Line of symmetry: $x =$ ____
_____ value: ____

12. $f(x) = -4x^2 - 7x + 2$

x	$f(x)$

Vertex: (____,____)
Line of symmetry: $x =$ ____
_____ value: ____

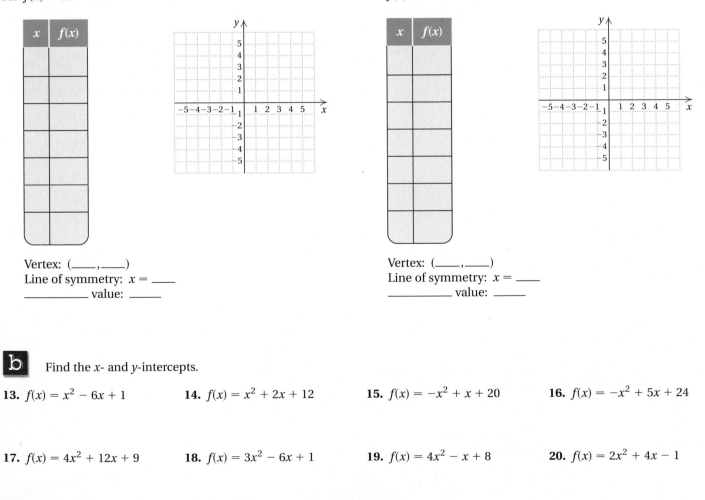

b Find the x- and y-intercepts.

13. $f(x) = x^2 - 6x + 1$

14. $f(x) = x^2 + 2x + 12$

15. $f(x) = -x^2 + x + 20$

16. $f(x) = -x^2 + 5x + 24$

17. $f(x) = 4x^2 + 12x + 9$

18. $f(x) = 3x^2 - 6x + 1$

19. $f(x) = 4x^2 - x + 8$

20. $f(x) = 2x^2 + 4x - 1$

21. D_W Does the graph of every quadratic function have a y-intercept? Why or why not?

22. D_W Is it possible for the graph of a quadratic function to have only one x-intercept if the vertex is off the x-axis? Why or why not?

SKILL MAINTENANCE

Solve. [5.8a, b]

23. *Determining Medication Dosage.* A child's dosage D, in milligrams, of a medication varies directly as the child's weight w, in kilograms. To control a fever, a doctor suggests that a child who weighs 28 kg be given 420 mg of Tylenol.

  a) Find an equation of variation.
  b) How much Tylenol would be recommended for a child who weighs 42 kg?

24. *Calories Burned.* The number C of calories burned while exercising varies directly as the time t, in minutes, spent exercising. Harold exercises for 24 min on a StairMaster and burns 356 calories.

  a) Find an equation of variation.
  b) How many calories would he burn if he were to exercise for 48 min?

Find the variation constant and an equation of variation in which y varies inversely as x and the following are true. [5.8c]

25. $y = 125$ when $x = 2$

26. $y = 2$ when $x = 125$

Find the variation constant and an equation of variation in which y varies directly as x and the following are true. [5.8a]

27. $y = 125$ when $x = 2$

28. $y = 2$ when $x = 125$

SYNTHESIS

29. Use the TRACE and/or TABLE features of a graphing calculator to estimate the maximum or minimum values of the following functions.

  a) $f(x) = 2.31x^2 - 3.135x - 5.89$
  b) $f(x) = -18.8x^2 + 7.92x + 6.18$

30. Use the TRACE and/or TABLE features of a graphing calculator to confirm the maximum or minimum values given in Exercises 8, 11, and 12.

Graph.

31. $f(x) = |x^2 - 1|$

32. $f(x) = |x^2 + 6x + 4|$

33. $f(x) = |x^2 - 3x - 4|$

34. $f(x) = |2(x - 3)^2 - 5|$

35. A quadratic function has $(-1, 0)$ as one of its intercepts and $(3, -5)$ as its vertex. Find an equation for the function.

36. A quadratic function has $(4, 0)$ as one of its intercepts and $(-1, 7)$ as its vertex. Find an equation for the function.

37. Consider

$$f(x) = \frac{x^2}{8} + \frac{x}{4} - \frac{3}{8}.$$

Find the vertex, the line of symmetry, and the maximum or minimum value. Then draw the graph.

38. Use only the graph in Exercise 37 to approximate the solutions of each of the following equations.

  a) $\dfrac{x^2}{8} + \dfrac{x}{4} - \dfrac{3}{8} = 0$
  b) $\dfrac{x^2}{8} + \dfrac{x}{4} - \dfrac{3}{8} = 1$
  c) $\dfrac{x^2}{8} + \dfrac{x}{4} - \dfrac{3}{8} = 2$

 Use the INTERSECT feature of a graphing calculator to find the points of intersection of the graphs of each pair of functions.

39. $f(x) = x^2 - 4x + 2$, $g(x) = 2 + x$

40. $f(x) = x^2 + 2x + 1$, $g(x) = -2x^2 - 4x + 1$

Objectives

a Solve maximum–minimum problems involving quadratic functions.

b Fit a quadratic function to a set of data to form a mathematical model, and solve related applied problems.

7.7 MATHEMATICAL MODELING WITH QUADRATIC FUNCTIONS

We now consider some of the many situations in which quadratic functions can serve as mathematical models.

a Maximum–Minimum Problems

We have seen that for any quadratic function $f(x) = ax^2 + bx + c$, the value of $f(x)$ at the vertex is either a maximum or a minimum, meaning that either all outputs are smaller than that value for a maximum or larger than that value for a minimum.

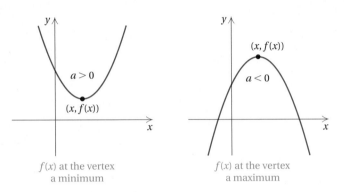

$f(x)$ at the vertex a minimum

$f(x)$ at the vertex a maximum

There are many types of applied problems in which we want to find a maximum or minimum value of a quantity. If a quadratic function can be used as a model, we can find such maximums or minimums by finding coordinates of the vertex.

EXAMPLE 1 *Fenced-In Land.* A farmer has 64 yd of fencing. What are the dimensions of the largest rectangular pen that the farmer can enclose?

1. **Familiarize.** We first make a drawing and label it. We let l = the length of the pen and w = the width. Recall the following formulas:

Perimeter: $2l + 2w$;

Area: $l \cdot w$.

l	w	A
22	10	220
20	12	240
18	14	252
18.5	13.5	249.75
12.4	19.6	243.04
15	17	255

To become familiar with the problem, let's choose some dimensions (shown at left) for which $2l + 2w = 64$ and then calculate the corresponding areas. What choice of l and w will maximize A?

638

CHAPTER 7: Quadratic Equations and Functions

2. Translate. We have two equations, one for perimeter and one for area:

$$2l + 2w = 64,$$
$$A = l \cdot w.$$

Let's use them to express A as a function of l or w, but not both. To express A in terms of w, for example, we solve for l in the first equation:

$$2l + 2w = 64$$
$$2l = 64 - 2w$$
$$l = \frac{64 - 2w}{2}$$
$$= 32 - w.$$

Substituting $32 - w$ for l, we get a quadratic function $A(w)$, or just A:

$$A = lw = (32 - w)w = 32w - w^2 = -w^2 + 32w.$$

3. Carry out. Note here that we are altering the third step of our five-step problem-solving strategy to "carry out" some kind of mathematical manipulation, because we are going to find the vertex rather than solve an equation. To do so, we complete the square as in Section 7.6:

$A = -w^2 + 32w$ This is a parabola opening down, so a maximum exists.

$= -1(w^2 - 32w)$ Factoring out -1

$= -1(w^2 - 32w + 256 - 256)$ $\frac{1}{2}(-32) = -16;\ (-16)^2 = 256.$
We add 0, or $256 - 256$.

$= -1(w^2 - 32w + 256) + (-1)(-256)$ Using the distributive law

$= -(w - 16)^2 + 256.$

The vertex is $(16, 256)$. Thus the maximum value is 256. It occurs when $w = 16$ and $l = 32 - w = 32 - 16 = 16$.

4. Check. We note that 256 is larger than any of the values found in the *Familiarize* step. To be more certain, we could make more calculations. We leave this to the student. We can also use the graph of the function to check the maximum value.

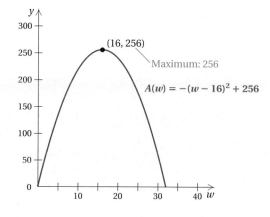

5. State. The largest rectangular pen that can be enclosed is 16 yd by 16 yd; that is, a square.

Do Exercise 1.

---

1. Fenced-In Land. A farmer has 100 yd of fencing. What are the dimensions of the largest rectangular pen that the farmer can enclose?

To familiarize yourself with the problem, complete the following table.

l	w	A
12	38	456
15	35	
24	26	
25	25	
26.2	23.8	

Answer on page A-40

CALCULATOR CORNER

Maximum and Minimum Values We can use a graphing calculator to find the maximum or minimum value of a quadratic function. Consider the quadratic function in Example 1, $A = -w^2 + 32w$. First, we replace w with x and A with y and graph the function in a window that displays the vertex of the graph. We choose $[0, 40, 0, 300]$, with $Xscl = 5$ and $Yscl = 20$. Now, we press **2ND** (CALC) (4) or **2ND** (CALC) ⌄ ⌄ ⌄ **ENTER** to select the MAXIMUM feature from the CALC menu. We are prompted to select a left bound for the maximum point. This means that we must choose an x-value that is to the left of the x-value of the point where the maximum occurs. This can be done by using the left- and right-arrow keys to move the cursor to a point to the left of the maximum point or by keying in an appropriate value. Once this is done, we press **ENTER**. Now, we are prompted to select a right bound. We move the cursor to a point to the right of the maximum point or key in an appropriate value.

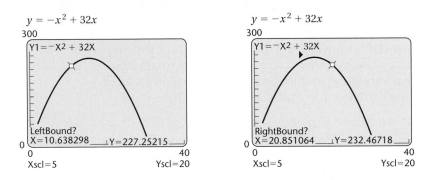

We press **ENTER** again. Finally, we are prompted to guess the x-value at which the maximum occurs. We move the cursor close to the maximum or key in an x-value. We press **ENTER** a third time and see that the maximum function value of 256 occurs when $x = 16$. (One or both coordinates of the maximum point might be approximations of the actual values, as shown with the x-value below, because of the method the calculator uses to find these values.)

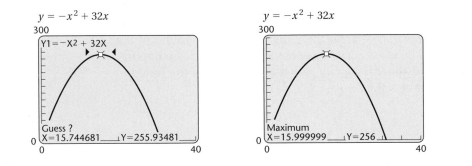

To find a minimum value, we select item 3, "minimum," from the CALC menu by pressing **2ND** (CALC) (3) or **2ND** (CALC) ⌄ ⌄ **ENTER**.

Exercises: Use the maximum or minimum feature on a graphing calculator to find the maximum or minimum value of the function.

1. $y = 3x^2 - 6x + 4$
2. $y = 2x^2 + x + 5$
3. $y = -x^2 + 4x + 2$
4. $y = -4x^2 + 5x - 1$

b   Fitting Quadratic Functions to Data

As we move through our study of mathematics, we develop a library of functions. These functions can serve as models for many applications. Some of them are graphed below. We have not considered the cubic or quartic functions in detail other than in the Calculator Corners (we leave that discussion to a later course), but we show them here for reference.

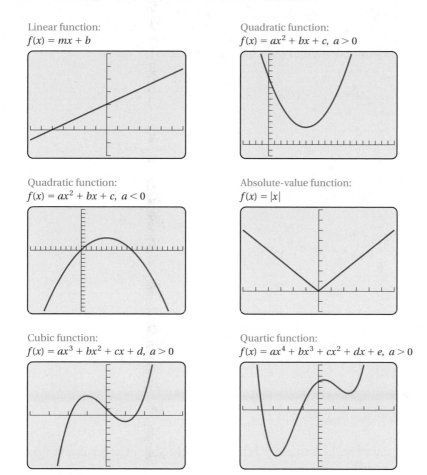

Linear function:
$$f(x) = mx + b$$

Quadratic function:
$$f(x) = ax^2 + bx + c, \ a > 0$$

Quadratic function:
$$f(x) = ax^2 + bx + c, \ a < 0$$

Absolute-value function:
$$f(x) = |x|$$

Cubic function:
$$f(x) = ax^3 + bx^2 + cx + d, \ a > 0$$

Quartic function:
$$f(x) = ax^4 + bx^3 + cx^2 + dx + e, \ a > 0$$

Now let's consider some real-world data. How can we decide which type of function might fit the data of a particular application? One simple way is to graph the data and look for a pattern resembling one of the graphs above. For example, data might be modeled by a linear function if the graph resembles a straight line. The data might be modeled by a quadratic function if the graph rises and then falls, or falls and then rises, in a curved manner resembling a parabola. For a quadratic, it might also just rise or fall in a curved manner as if following only one part of the parabola.

Choosing Models. For the scatterplots and graphs in Margin Exercises 2–5, determine which, if any, of the following functions might be used as a model for the data.

Linear, $f(x) = mx + b$;

Quadratic, $f(x) = ax^2 + bx + c$, $a > 0$;

Quadratic, $f(x) = ax^2 + bx + c$, $a < 0$;

Polynomial, neither quadratic nor linear

2.

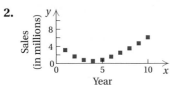

3.

4.

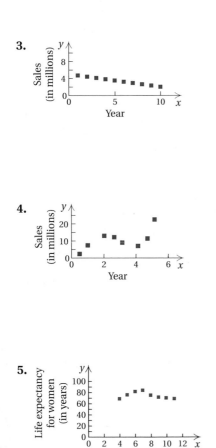

5.

Source: Orthopedic Quarterly

Answers on page A-40

Let's now use our library of functions to see which, if any, might fit certain data situations.

EXAMPLES *Choosing Models.* For the scatterplots and graphs below, determine which, if any, of the following functions might be used as a model for the data.

Linear, $f(x) = mx + b$;

Quadratic, $f(x) = ax^2 + bx + c$, $a > 0$;

Quadratic, $f(x) = ax^2 + bx + c$, $a < 0$;

Polynomial, neither quadratic nor linear

2.

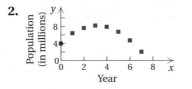

The data rise and then fall in a curved manner fitting a quadratic function $f(x) = ax^2 + bx + c$, $a < 0$.

3.

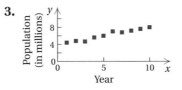

The data seem to fit a linear function $f(x) = mx + b$.

4.

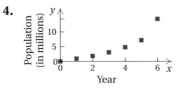

The data rise in a manner fitting the right side of a quadratic function $f(x) = ax^2 + bx + c$, $a > 0$.

5. Driver Fatalities by Age

Number of licensed drivers per 100,000 who died in motor vehicle accidents in 1990. The fatality rates for both the 70–79 group and the 80+ age group were lower than for the 15- to 24-year olds.

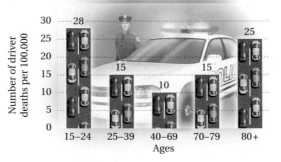

Source: National Highway Traffic Safety Administration

The data fall and then rise in a curved manner fitting a quadratic function $f(x) = ax^2 + bx + c$, $a > 0$.

6.

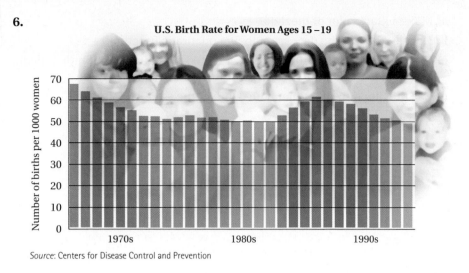

U.S. Birth Rate for Women Ages 15–19

Source: Centers for Disease Control and Prevention

The data fall, then rise, then fall again. They do not appear to fit a linear or quadratic function but might fit a polynomial function that is neither quadratic nor linear.

Do Exercises 2–5 on the preceding page.

Whenever a quadratic function seems to fit a data situation, that function can be determined if at least three inputs and their outputs are known.

EXAMPLE 7 *River Depth.* The drawing below shows the cross section of a river. Typically rivers are deepest in the middle, with the depth decreasing to 0 at the edges. A hydrologist measures the depths D, in feet, of a river at distances x, in feet, from one bank. The results are listed in the table at right.

DISTANCE x FROM THE RIVERBANK (in feet)	DEPTH D OF THE RIVER (in feet)
0	0
15	10.2
25	17
50	20
90	7.2
100	0

x = distance from left bank (in feet)

$D(x)$ = depth of river (in feet)

6. Ticket Profits. Valley Community College is presenting a play. The profit P, in dollars, after x days is given in the following table. (Profit can be negative when costs exceed revenue. See Section 3.8.)

DAYS x	PROFIT P
0	$\$-100$
90	560
180	872
270	870
360	548
450	-100

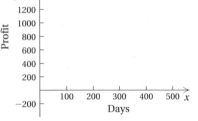

a) Make a scatterplot of the data.

b) Decide whether the data can be modeled by a quadratic function.

c) Use the data points $(0, -100)$, $(180, 872)$, and $(360, 548)$ to find a quadratic function that fits the data.

d) Use the function to estimate the profits after 225 days.

a) Make a scatterplot of the data.

b) Decide whether the data seem to fit a quadratic function.

c) Use the data points $(0, 0)$, $(50, 20)$, and $(100, 0)$ to find a quadratic function that fits the data.

d) Use the function to estimate the depth of the river at 75 ft.

a) The scatterplot is as follows.

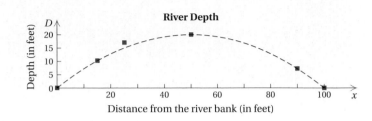

b) The data seem to rise and fall in a manner similar to a quadratic function. The dashed black line in the graph represents a sample quadratic function of fit. Note that it may not necessarily go through each point.

c) We are looking for a quadratic function

$$D(x) = ax^2 + bx + c.$$

We need to determine the constants a, b, and c. We use the three data points $(0, 0)$, $(50, 20)$, and $(100, 0)$ and substitute as follows:

$$0 = a \cdot 0^2 + b \cdot 0 + c,$$
$$20 = a \cdot 50^2 + b \cdot 50 + c,$$
$$0 = a \cdot 100^2 + b \cdot 100 + c.$$

After simplifying, we see that we need to solve the system

$$0 = c,$$
$$20 = 2{,}500a + 50b + c,$$
$$0 = 10{,}000a + 100b + c.$$

Since $c = 0$, the system reduces to a system of two equations in two variables:

$$20 = 2{,}500a + 50b, \qquad \textbf{(1)}$$
$$0 = 10{,}000a + 100b. \qquad \textbf{(2)}$$

We multiply equation (1) by -2, add, and solve for a (see Section 3.3):

$$-40 = -5{,}000a - 100b,$$
$$\underline{ 0 = 10{,}000a + 100b}$$
$$-40 = 5000a \qquad \text{Adding}$$
$$\frac{-40}{5000} = a \qquad \text{Solving for } a$$
$$-0.008 = a.$$

Answers on page A-41

Next, we substitute -0.008 for a in equation (2) and solve for b:

$$0 = 10,000(-0.008) + 100b$$
$$0 = -80 + 100b$$
$$80 = 100b$$
$$0.8 = b.$$

This gives us the quadratic function:

$$D(x) = -0.008x^2 + 0.8x.$$

d) To find the depth 75 ft from the riverbank, we substitute:

$$D(75) = -0.008(75)^2 + 0.8(75) = 15.$$

At a distance of 75 ft from the riverbank, the depth of the river is 15 ft.

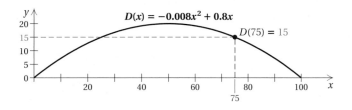

Do Exercise 6 on the preceding page.

CALCULATOR CORNER

Mathematical Modeling: Fitting a Quadratic Function to Data   We can use the quadratic regression feature on a graphing calculator to fit a quadratic function to a set of data. The following table shows the average number of live births for women of various ages.

AGE	AVERAGE NUMBER OF LIVE BIRTHS PER 1000 WOMEN
16	34
18.5	86.5
22	111.1
27	113.9
32	84.5
37	35.4
42	6.8

a) Make a scatterplot of the data and verify that the data can be modeled with a quadratic function.
b) Fit a quadratic function to the data using the quadratic regression feature on a graphing calculator.
c) Use the function to estimate the average number of live births per 1000 women of age 20 and of age 30.

(*continued*)

a) We enter the data on the STAT list editor screen, turn on Plot 1, and plot the points as described in the Calculator Corner on p. 226. The data points rise and then fall in a manner consistent with a quadratic function.

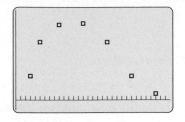

b) To fit a quadratic function to the data, we press ⬛STAT⬛ ⬛▷⬛ ⬛5⬛ ⬛VARS⬛ ⬛▷⬛ ⬛1⬛ ⬛1⬛ ⬛ENTER⬛. The first three keystrokes select QuadReg from the STAT CALC menu and display the coefficients a, b, and c of the regression equation $y = ax^2 + bx + c$. The keystrokes ⬛VARS⬛ ⬛▷⬛ ⬛1⬛ ⬛1⬛ copy the regression equation to the equation-editor screen as y_1. We see that the regression equation is $y = -0.4868465035x^2 + 25.94985182x - 238.4892193$. We can press ⬛ZOOM⬛ ⬛9⬛ to see the regression equation graphed with the data points.

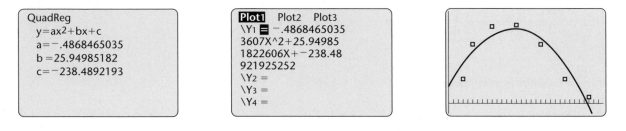

c) Since the equation is entered on the equation-editor screen, we can use a table set in ASK mode to estimate the average number of live births per 1000 women of age 20 and of age 30. We see that when $x = 20$, $y \approx 85.8$, and when $x = 30$, $y \approx 101.8$, so we estimate that there are about 85.8 live births per 1000 women of age 20 and about 101.8 live births per 1000 women of age 30.

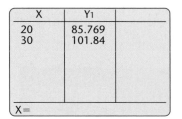

Remember to turn off the STAT PLOT as described on p. 228 before you graph other equations.

Exercise:

1. Consider the data in the table in Example 7.

   a) Use a graphing calculator to make a scatterplot of the data.
   b) Use a graphing calculator to fit a quadratic function to the data. Compare this function with the one found in Example 7.
   c) Graph the quadratic function with the scatterplot.
   d) Use the function found in part (b) to estimate the depth of the river 75 ft from the riverbank. Compare this estimate with the one found in Example 7.

7.7

EXERCISE SET

For Extra Help

MathXL   MyMathLab   InterAct   Math Tutor   Digital Video   Student's
                        Math   Center   Tutor CD 3   Solutions
                                                  Videotape 8   Manual

a   Solve.

1. *Architecture.*   An architect is designing the floors for a hotel. Each floor is to be rectangular and is allotted 720 ft of security piping around walls outside the rooms. What dimensions of the atrium will allow an atrium at the bottom to have maximum area?

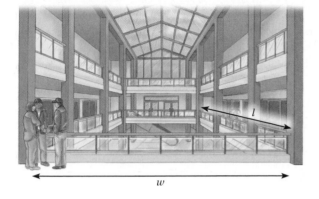

2. *Stained-Glass Window Design.*   An artist is designing a rectangular stained-glass window with a perimeter of 84 in. What dimensions will yield the maximum area?

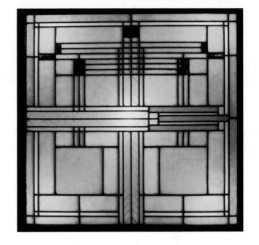

3. *Molding Plastics.*   Economite Plastics plans to produce a one-compartment vertical file by bending the long side of an 8-in. by 14-in. sheet of plastic along two lines to form a U shape. How tall should the file be in order to maximize the volume that the file can hold?

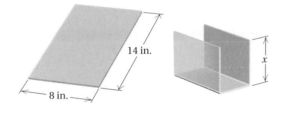

4. *Patio Design.*   A stone mason has enough stones to enclose a rectangular patio with 60 ft of perimeter, assuming that the attached house forms one side of the rectangle. What is the maximum area that the mason can enclose? What should the dimensions of the patio be in order to yield this area?

5. *Minimizing Cost.*   Aki's Bicycle Designs has determined that when x hundred bicycles are built, the average cost per bicycle is given by

$$C(x) = 0.1x^2 - 0.7x + 2.425,$$

where $C(x)$ is in hundreds of dollars. How many bicycles should the shop build in order to minimize the average cost per bicycle?

6. *Corral Design.*   A rancher needs to enclose two adjacent rectangular corrals, one for sheep and one for cattle. If a river forms one side of the corrals and 180 yd of fencing is available, what is the largest total area that can be enclosed?

7. *Garden Design.* A farmer decides to enclose a rectangular garden, using the side of a barn as one side of the rectangle. What is the maximum area that the farmer can enclose with 40 ft of fence? What should the dimensions of the garden be in order to yield this area?

8. *Composting.* A rectangular compost container is to be formed in a corner of a fenced yard, with 8 ft of chicken wire completing the other two sides of the rectangle. If the chicken wire is 3 ft high, what dimensions of the base will maximize the volume of the container?

9. *Ticket Sales.* The number of tickets sold each day for an upcoming performance of Handel's *Messiah* is given by

$$N(x) = -0.4x^2 + 9x + 11,$$

where x is the number of days since the concert was first announced. When will daily ticket sales peak and how many tickets will be sold that day?

10. *Stock Prices.* The value of a share of R. P. Mugahti, in dollars, can be represented by $V(x) = x^2 - 6x + 13$, where x is the number of months after January 2003. What is the lowest value $V(x)$ will reach, and when did that occur?

Maximizing Profit. Recall (Section 3.8) that total profit P is the difference between total revenue R and total cost C. Given the following total-revenue and total-cost functions, find the total profit, the maximum value of the total profit, and the value of x at which it occurs.

11. $R(x) = 1000x - x^2$,
  $C(x) = 3000 + 20x$

12. $R(x) = 200x - x^2$,
  $C(x) = 5000 + 8x$

13. What is the maximum product of two numbers whose sum is 22? What numbers yield this product?

14. What is the maximum product of two numbers whose sum is 45? What numbers yield this product?

15. What is the minimum product of two numbers whose difference is 4? What are the numbers?

16. What is the minimum product of two numbers whose difference is 6? What are the numbers?

17. What is the maximum product of two numbers that add to -12? What numbers yield this product?

18. What is the minimum product of two numbers that differ by 9? What are the numbers?

b *Choosing Models.* For the scatterplots and graphs in Exercises 19–26, determine which, if any, of the following functions might be used as a model for the data: Linear, $f(x) = mx + b$; quadratic, $f(x) = ax^2 + bx + c, a > 0$; quadratic, $f(x) = ax^2 + bx + c, a < 0$; polynomial, neither quadratic nor linear.

19.

Media Usage

Hours per day vs. Years since 1990

20.

Growth of World Wide Web Sites

Number (in millions) vs. Years since 1995

21.

Valley Community College

Average class size vs. Year

22.

Valley Community College

Average class size vs. Year

23.

Valley Community College

Average class size vs. Year

24.

Demand for Earphones

Price per unit vs. Quantity

25. **Average Number of Live Births per 1000 Women**

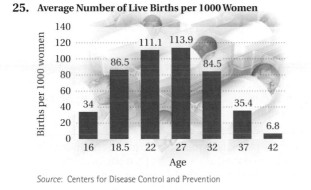

Births per 1000 women vs. Age

16: 34; 18.5: 86.5; 22: 111.1; 27: 113.9; 32: 84.5; 37: 35.4; 42: 6.8

Source: Centers for Disease Control and Prevention

26. **Sony Electronics, Inc.**

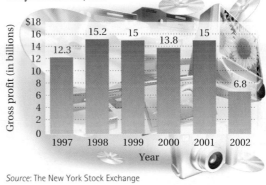

Gross profit (in billions) vs. Year

1997: 12.3; 1998: 15.2; 1999: 15; 2000: 13.8; 2001: 15; 2002: 6.8

Source: The New York Stock Exchange

Find a quadratic function that fits the set of data points.

27. $(1, 4), (-1, -2), (2, 13)$

28. $(1, 4), (-1, 6), (-2, 16)$

29. $(2, 0), (4, 3), (12, -5)$

30. $(-3, -30), (3, 0), (6, 6)$

31. *Nighttime Accidents.*

  a) Find a quadratic function that fits the following data.

TRAVEL SPEED (in kilometers per hour)	NUMBER OF NIGHTTIME ACCIDENTS (for every 200 million kilometers driven)
60	400
80	250
100	250

  b) Use the function to estimate the number of nighttime accidents that occur at 50 km/h.

32. *Daytime Accidents.*

  a) Find a quadratic function that fits the following data.

TRAVEL SPEED (in kilometers per hour)	NUMBER OF DAYTIME ACCIDENTS (for every 200 million kilometers driven)
60	100
80	130
100	200

  b) Use the function to estimate the number of daytime accidents that occur at 50 km/h.

33. *Archery.* The Olympic flame tower at the 1992 Summer Olympics was lit at a height of about 27 m by a flaming arrow that was launched about 63 m from the base of the tower. If the arrow landed about 63 m beyond the tower, find a quadratic function that expresses the height h of the arrow as a function of the distance d that it traveled horizontally.

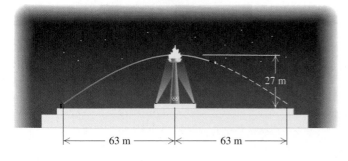

34. *Pizza Prices.* Pizza Unlimited has the following prices for pizzas.

DIAMETER	PRICE
8 in.	$ 6.00
12 in.	$ 8.50
16 in.	$11.50

Is price a quadratic function of diameter? It probably should be, because the price should be proportional to the area, and the area is a quadratic function of the diameter. (The area of a circular region is given by $A = \pi r^2$ or $(\pi/4) \cdot d^2$.)

  a) Express price as a quadratic function of diameter using the data points $(8, 6), (12, 8.50),$ and $(16, 11.50)$.

  b) Use the function to find the price of a 14-in. pizza.

35. D_W Explain the restrictions that should be placed on the domains of the quadratic functions found in Exercises 11 and 31 and why such restrictions are needed.

36. D_W Explain how the leading coefficient of a quadratic function can be used to determine whether a maximum or minimum function value exists.

SKILL MAINTENANCE

VOCABULARY REINFORCEMENT

In each of Exercises 37–44, fill in the blank with the correct word(s) from the given list. Some of the choices may not be used.

37. In the expression $5\sqrt{2x - 9} + 3$, the symbol $\sqrt{}$ is called a(n) _____ and $2x - 9$ is called the _____ . [6.1a]

38. When a system of equations has infinitely many solutions, the equations are _____ . [3.1a]

39. The degree of a term of a polynomial is the _____ of the exponents of the variables. [4.1a]

40. A consistent system of equations has _____ solution. [3.1a]

41. The equation $y = k/x$, where k is a positive constant, is an equation of _____ variation. [5.8c]

42. When a system of equations has one solution or no solutions, the equations are _____ . [3.1a]

43. If the exponents in a polynomial decrease from left to right, the polynomial is written in _____ order. [4.1a]

44. A(n) _____ is a point $(a, 0)$. [2.5a]

at least one	inverse
no	sum
dependent	product
independent	x-intercept
ascending	y-intercept
descending	radical
direct	radicand

SYNTHESIS

45. *Sony Electronics, Inc.* Use the REGRESSION feature on your graphing calculator to fit a quartic function to the data in Exercise 26.

46. The sum of the base and the height of a triangle is 38 cm. Find the dimensions for which the area is a maximum, and find the maximum area.

Objectives

1. Solve by graphing:
$$x^2 + 2x - 3 > 0.$$

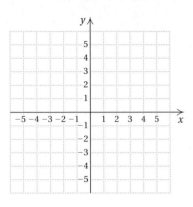

a Quadratic and Other Polynomial Inequalities

Inequalities like the following are called **quadratic inequalities:**

$$x^2 + 3x - 10 < 0, \qquad 5x^2 - 3x + 2 \geq 0.$$

In each case, we have a polynomial of degree 2 on the left. We will solve such inequalities in two ways. The first method provides understanding and the second yields the more efficient method.

The first method for solving a quadratic inequality, such as $ax^2 + bx + c > 0$, is by considering the graph of a related function, $f(x) = ax^2 + bx + c$.

EXAMPLE 1 Solve: $x^2 + 3x - 10 > 0$.

Consider the function $f(x) = x^2 + 3x - 10$ and its graph. The graph opens up since the leading coefficient ($a = 1$) is positive. We find the x-intercepts by setting the polynomial equal to 0 and solving:

$$x^2 + 3x - 10 = 0$$
$$(x + 5)(x - 2) = 0$$
$$x + 5 = 0 \quad or \quad x - 2 = 0$$
$$x = -5 \quad or \qquad x = 2.$$

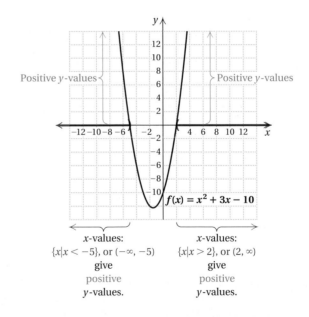

Positive y-values

Positive y-values

$f(x) = x^2 + 3x - 10$

x-values: $\{x \mid x < -5\}$, or $(-\infty, -5)$ give positive y-values.

x-values: $\{x \mid x > 2\}$, or $(2, \infty)$ give positive y-values.

Values of y will be positive to the left and right of the intercepts, as shown. Thus the solution set of the inequality is

$$\{x \mid x < -5 \ or \ x > 2\}, \quad or \quad (-\infty, -5) \cup (2, \infty).$$

Do Exercise 1.

We can solve any inequality by considering the graph of a related function and finding x-intercepts as in Example 1. In some cases, we may need to use the quadratic formula to find the intercepts.

Answer on page A-41

EXAMPLE 2 Solve: $x^2 + 3x - 10 < 0$.

Looking again at the graph of $f(x) = x^2 + 3x - 10$ or at least visualizing it tells us that y-values are negative for those x-values between -5 and 2.

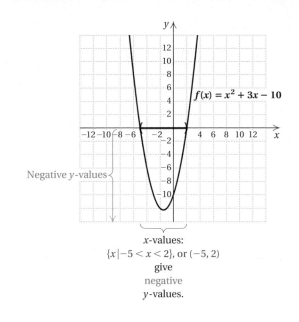

$f(x) = x^2 + 3x - 10$

Negative y-values

x-values:
$\{x | -5 < x < 2\}$, or $(-5, 2)$
give
negative
y-values.

That is, the solution set is $\{x | -5 < x < 2\}$, or $(-5, 2)$.

When an inequality contains $\leq$ or $\geq$, the x-values of the x-intercepts must be included. Thus the solution set of the inequality $x^2 + 3x - 10 \leq 0$ is $\{x | -5 \leq x \leq 2\}$, or $[-5, 2]$.

Do Exercises 2 and 3. (Exercise 3 is on the following page.)

Answer on page A-41

2. Solve by graphing:

$$x^2 + 2x - 3 < 0.$$

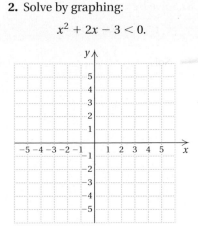

CALCULATOR CORNER

Solving Polynomial Inequalities We can solve polynomial inequalities graphically. Consider the inequality in Example 2, $x^2 + 3x - 10 < 0$. We first graph the function $f(x) = x^2 + 3x - 10$. Then we use the ZERO feature to find the solutions of the equation $f(x) = 0$, or $x^2 + 3x - 10 = 0$. The solutions, -5 and 2, divide the number line into three intervals, $(-\infty, -5)$, $(-5, 2)$, and $(2, \infty)$.

Since we want to find the values of x for which $f(x) < 0$, we look for the interval(s) on which the function values are negative. That is, we note where the graph lies below the x-axis. This occurs in the interval $(-5, 2)$, so the solution set is $\{x | -5 < x < 2\}$, or $(-5, 2)$.

If we were solving the inequality $x^2 + 3x - 10 > 0$, we would look for the intervals on which the graph lies above the x-axis. We can see that $x^2 + 3x - 10 > 0$ for $\{x | x < -5 \ or \ x > 2\}$, or $(-\infty, -5) \cup (2, \infty)$. If the inequality symbol were $\leq$ or $\geq$, we would include the endpoints of the intervals as well.

$y = x^2 + 3x - 10$

Exercises: Solve graphically.

1. $x^2 + 3x - 4 > 0$

2. $x^2 - x - 6 < 0$

3. $6x^3 + 9x^2 - 6x \leq 0$

4. $x^3 - 16x \geq 0$

---

3. Solve by graphing:

$$x^2 + 2x - 3 \le 0.$$

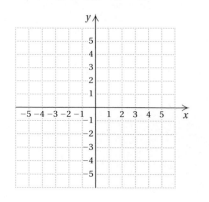

We now consider a more efficient method for solving polynomial inequalities. The preceding discussion provides the understanding for this method. In Examples 1 and 2, we see that the x-intercepts divide the number line into intervals.

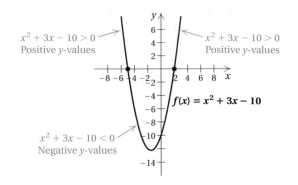

If a function has a positive output for one number in an interval, it will be positive for all the numbers in the interval. The same is true for negative outputs. Thus we can merely make a test substitution in each interval to solve the inequality. This is very similar to our method of using test points to graph a linear inequality in a plane.

EXAMPLE 3 Solve: $x^2 + 3x - 10 < 0$.

We set the polynomial equal to 0 and solve. The solutions of $x^2 + 3x - 10 = 0$, or $(x + 5)(x - 2) = 0$, are -5 and 2. We then locate them on a number line as follows. Note that the numbers divide the number line into three intervals, which we will call A, B, and C.

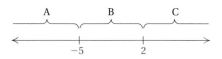

We choose a test number in interval A, say -7, and substitute -7 for x in the function $f(x) = x^2 + 3x - 10$:

$$f(-7) = (-7)^2 + 3(-7) - 10$$
$$= 49 - 21 - 10 = 18. \quad \text{Thus, } f(-7) > 0.$$

Note that $18 > 0$, so the function values will be positive for any number in interval A.

Next, we try a test number in interval B, say 1, and find the corresponding function value:

$$f(1) = 1^2 + 3(1) - 10$$
$$= 1 + 3 - 10 = -6. \quad \text{Thus, } f(1) < 0.$$

Note that $-6 < 0$, so the function values will be negative for any number in interval B.

Answer on page A-41

Next, we try a test number in interval C, say 4, and find the corresponding function value:

$$f(4) = 4^2 + 3(4) - 10$$
$$= 16 + 12 - 10 = 18. \quad \text{Thus, } f(4) > 0.$$

Note that $18 > 0$, so the function values will be positive for any number in interval C.

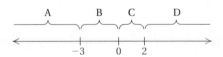

We are looking for numbers x for which $f(x) = x^2 + 3x - 10 < 0$. Thus any number x in interval B is a solution. If the inequality had been $\le$, it would have been necessary to include the intercepts -5 and 2 in the solution set as well. The solution set is $\{x \mid -5 < x < 2\}$, or the interval $(-5, 2)$.

To solve a polynomial inequality:

1. Get 0 on one side, set the expression on the other side equal to 0, and solve to find the x-intercepts.
2. Use the numbers found in step (1) to divide the number line into intervals.
3. Substitute a number from each interval into the related function. If the function value is positive, then the expression will be positive for all numbers in the interval. If the function value is negative, then the expression will be negative for all numbers in the interval.
4. Select the intervals for which the inequality is satisfied and write set-builder or interval notation for the solution set.

Do Exercises 4 and 5.

EXAMPLE 4 Solve: $5x(x + 3)(x - 2) \ge 0$.

The solutions of $f(x) = 0$, or $5x(x + 3)(x - 2) = 0$, are -3, 0, and 2. They divide the real-number line into four intervals, as shown below.

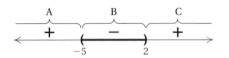

We try test numbers in each interval:

A: Test -5, $f(-5) = 5(-5)(-5 + 3)(-5 - 2) = -350 < 0.$
B: Test -2, $f(-2) = 5(-2)(-2 + 3)(-2 - 2) = 40 > 0.$
C: Test 1, $f(1) = 5(1)(1 + 3)(1 - 2) = -20 < 0.$
D: Test 3, $f(3) = 5(3)(3 + 3)(3 - 2) = 90 > 0.$

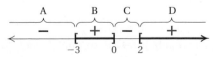

Solve using the method of Example 3.

4. $x^2 + 3x > 4$

5. $x^2 + 3x \le 4$

Answers on page A-41

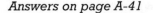

6. Solve: $6x(x + 1)(x - 1) < 0$.

The expression is positive for values of x in intervals B and D. Since the inequality symbol is $\geq$, we need to include the x-intercepts. The solution set of the inequality is

$$\{x \mid -3 \leq x \leq 0 \ or \ 2 \leq x\}, \quad \text{or} \quad [-3, 0] \cup [2, \infty).$$

We visualize this with the graph below.

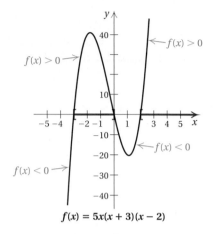

$$f(x) = 5x(x + 3)(x - 2)$$

Do Exercise 6.

 b Rational Inequalities

We adapt the preceding method when an inequality involves rational expressions. We call these **rational inequalities.**

EXAMPLE 5 Solve: $\dfrac{x - 3}{x + 4} \geq 2$.

We write a related equation by changing the $\geq$ symbol to $=$:

$$\frac{x - 3}{x + 4} = 2.$$

Then we solve this related equation. First, we multiply both sides of the equation by the LCM, which is $x + 4$:

$$(x + 4) \cdot \frac{x - 3}{x + 4} = (x + 4) \cdot 2$$
$$x - 3 = 2x + 8$$
$$-11 = x.$$

With rational inequalities, we also need to determine those numbers for which the rational expression is not defined—that is, those numbers that make the denominator 0. We set the denominator equal to 0 and solve: $x + 4 = 0$, or $x = -4$. Next, we use the numbers -11 and -4 to divide the number line into intervals, as shown below.

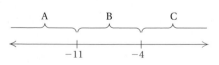

We try test numbers in each interval to see if each satisfies the original inequality.

Answer on page A-41

A: Test -15, $\dfrac{x-3}{x+4} \geq 2$

$$\dfrac{-15-3}{-15+4} \; ? \; 2$$

$$\dfrac{18}{11} \quad \text{FALSE}$$

Since the inequality is false for $x = -15$, the number -15 is not a solution of the inequality. Interval A is *not* part of the solution set.

B: Test -8, $\dfrac{x-3}{x+4} \geq 2$

$$\dfrac{-8-3}{-8+4} \; ? \; 2$$

$$\dfrac{11}{4} \quad \text{TRUE}$$

Since the inequality is true for $x = -8$, the number -8 is a solution of the inequality. Interval B *is* part of the solution set.

C: Test 1, $\dfrac{x-3}{x+4} \geq 2$

$$\dfrac{1-3}{1+4} \; ? \; 2$$

$$-\dfrac{2}{5} \quad \text{FALSE}$$

Since the inequality is false for $x = 1$, the number 1 is not a solution of the inequality. Interval C is *not* part of the solution set.

The solution set includes the interval B. The number -11 is also included since the inequality symbol is $\geq$ and -11 is a solution of the related equation. The number -4 is not included; it is not an allowable replacement because it results in division by 0. Thus the solution set of the original inequality is

$$\{x \mid -11 \leq x < -4\}, \quad \text{or} \quad [-11, -4).$$

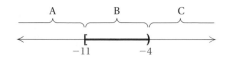

To solve a rational inequality:

1. Change the inequality symbol to an equals sign and solve the related equation.
2. Find the numbers for which any rational expression in the inequality is not defined.
3. Use the numbers found in steps (1) and (2) to divide the number line into intervals.
4. Substitute a number from each interval into the inequality. If the number is a solution, then the interval to which it belongs is part of the solution set.
5. Select the intervals for which the inequality is satisfied and write set-builder or interval notation for the solution set.

Do Exercises 7 and 8.

Solve.

7. $\dfrac{x+1}{x-2} \geq 3$

8. $\dfrac{x}{x-5} < 2$

Answers on page A-41

a Solve algebraically and verify results from the graph.

1. $(x - 6)(x + 2) > 0$

2. $(x - 5)(x + 1) > 0$

3. $4 - x^2 \geq 0$

4. $9 - x^2 \leq 0$

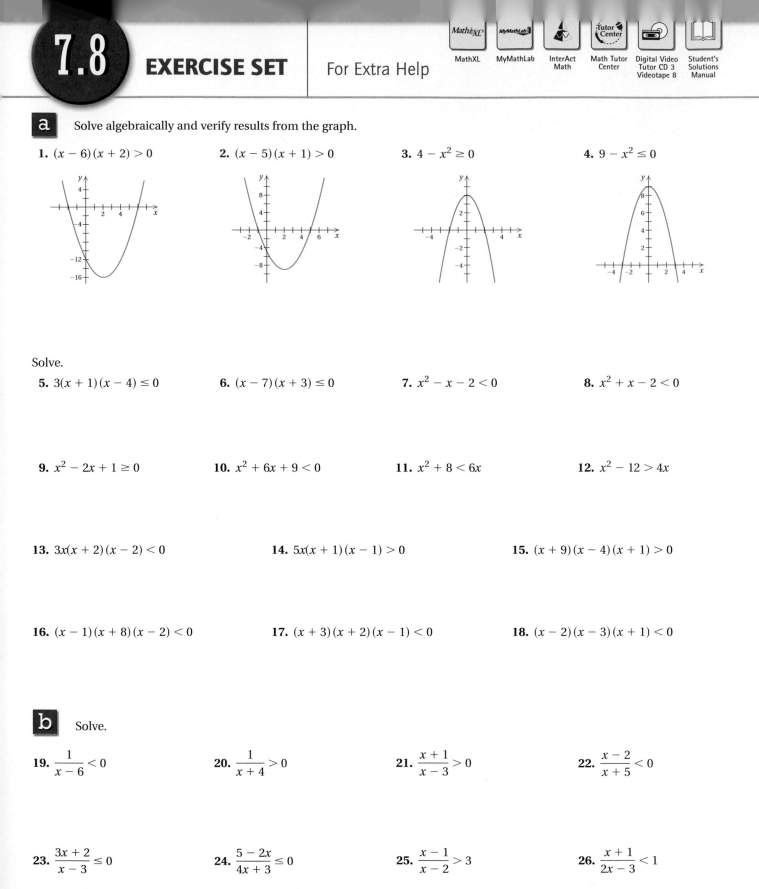

Solve.

5. $3(x + 1)(x - 4) \leq 0$

6. $(x - 7)(x + 3) \leq 0$

7. $x^2 - x - 2 < 0$

8. $x^2 + x - 2 < 0$

9. $x^2 - 2x + 1 \geq 0$

10. $x^2 + 6x + 9 < 0$

11. $x^2 + 8 < 6x$

12. $x^2 - 12 > 4x$

13. $3x(x + 2)(x - 2) < 0$

14. $5x(x + 1)(x - 1) > 0$

15. $(x + 9)(x - 4)(x + 1) > 0$

16. $(x - 1)(x + 8)(x - 2) < 0$

17. $(x + 3)(x + 2)(x - 1) < 0$

18. $(x - 2)(x - 3)(x + 1) < 0$

b Solve.

19. $\dfrac{1}{x - 6} < 0$

20. $\dfrac{1}{x + 4} > 0$

21. $\dfrac{x + 1}{x - 3} > 0$

22. $\dfrac{x - 2}{x + 5} < 0$

23. $\dfrac{3x + 2}{x - 3} \leq 0$

24. $\dfrac{5 - 2x}{4x + 3} \leq 0$

25. $\dfrac{x - 1}{x - 2} > 3$

26. $\dfrac{x + 1}{2x - 3} < 1$

27. $\dfrac{(x-2)(x+1)}{x-5} < 0$

28. $\dfrac{(x+4)(x-1)}{x+3} > 0$

29. $\dfrac{x+3}{x} \le 0$

30. $\dfrac{x}{x-2} \ge 0$

31. $\dfrac{x}{x-1} > 2$

32. $\dfrac{x-5}{x} < 1$

33. $\dfrac{x-1}{(x-3)(x+4)} < 0$

34. $\dfrac{x+2}{(x-2)(x+7)} > 0$

35. $3 < \dfrac{1}{x}$

36. $\dfrac{1}{x} \le 2$

37. $\dfrac{(x-1)(x+2)}{(x+3)(x-4)} > 0$

38. $\dfrac{x^2-11x+30}{x^2-8x-9} \ge 0$

39. $\mathbf{D_W}$ Describe a method that could be used to create quadratic inequalities that have no solution.

40. $\mathbf{D_W}$ Describe a method that could be used to create quadratic inequalities that have all real numbers as solutions.

SKILL MAINTENANCE

Simplify. [6.3b]

41. $\sqrt[3]{\dfrac{125}{27}}$

42. $\sqrt{\dfrac{25}{4a^2}}$

43. $\sqrt{\dfrac{16a^3}{b^4}}$

44. $\sqrt[3]{\dfrac{27c^5}{343d^3}}$

Add or subtract. [6.4a]

45. $3\sqrt{8} - 5\sqrt{2}$

46. $7\sqrt{45} - 2\sqrt{20}$

47. $5\sqrt[3]{16a^4} + 7\sqrt[3]{2a}$

48. $3\sqrt{10} + 8\sqrt{20} - 5\sqrt{80}$

SYNTHESIS

49. 📈 Use a graphing calculator to solve Exercises 11, 22, and 25 by graphing two curves, one for each side of an inequality.

50. 📈 Use a graphing calculator to solve each of the following.

a) $x + \dfrac{1}{x} < 0$

b) $x - \sqrt{x} \ge 0$

c) $\frac{1}{3}x^3 - x + \frac{2}{3} \le 0$

Solve.

51. $x^2 - 2x \le 2$

52. $x^2 + 2x > 4$

53. $x^4 + 2x^2 > 0$

54. $x^4 + 3x^2 \le 0$

55. $\left|\dfrac{x+2}{x-1}\right| < 3$

56. *Total Profit.* A company determines that its total profit on the production and sale of x units of a product is given by

$$P(x) = -x^2 + 812x - 9600.$$

a) A company makes a profit for those nonnegative values of x for which $P(x) > 0$. Find the values of x for which the company makes a profit.

b) A company loses money for those nonnegative values of x for which $P(x) < 0$. Find the values of x for which the company loses money.

57. *Height of a Thrown Object.* The function

$$H(t) = -16t^2 + 32t + 1920$$

gives the height H of an object thrown from a cliff 1920 ft high, after time t seconds.

a) For what times is the height greater than 1920 ft?

b) For what times is the height less than 640 ft?

Summary and Review

The review that follows is meant to prepare you for a chapter exam. It consists of three parts. The first part, Concept Reinforcement, is designed to increase understanding of the concepts through true/false exercises. The second part is a list of important properties and formulas. The third part is the Review Exercises. These provide practice exercises for the exam, together with references to section objectives so you can go back and review. Before beginning, stop and look back over the skills you have obtained. What skills in mathematics do you have now that you did not have before studying this chapter?

↩ CONCEPT REINFORCEMENT

Determine whether the statement is true or false. Answers are given at the back of the book.

_____ 1. Every quadratic equation has exactly two real-number solutions.

_____ 2. The quadratic formula can be used to find all the solutions of any quadratic equation.

_____ 3. The graph of $f(x) = -(-x^2 - 8x - 3)$ opens downward.

_____ 4. The graph of $f(x) = -x^2 - 8x - 3$ is a translation downward of the graph of $g(x) = -x^2 - 8x$.

_____ 5. The graph of $f(x) = -3(x + 2)^2 - 5$ is a translation to the right of the graph of $f(x) = -3x^2 - 5$.

_____ 6. If the graph of a quadratic equation crosses the x-axis, then it has two real-number solutions.

IMPORTANT PROPERTIES AND FORMULAS

Principle of Square Roots: $x^2 = d$ has solutions $\sqrt{d}$ and $-\sqrt{d}$.

Quadratic Formula: $x = \dfrac{-b \pm \sqrt{b^2 - 4ac}}{2a}$;   *Discriminant:* $b^2 - 4ac$

The *vertex* of the graph of $f(x) = ax^2 + bx + c$ is $\left(-\dfrac{b}{2a}, \dfrac{4ac - b^2}{4a}\right)$, or $\left(-\dfrac{b}{2a}, f\left(-\dfrac{b}{2a}\right)\right)$.

The *line of symmetry* of the graph of $f(x) = ax^2 + bx + c$ is $x = -\dfrac{b}{2a}$.

Review Exercises

1. a) Solve: $2x^2 - 7 = 0$.   [7.1a]
   b) Find the x-intercepts of $f(x) = 2x^2 - 7$.

Solve.   [7.2a]

2. $14x^2 + 5x = 0$ **3.** $x^2 - 12x + 27 = 0$

4. $4x^2 + 3x + 1 = 0$

5. $x^2 - 7x + 13 = 0$

6. $4x(x - 1) + 15 = x(3x + 4)$

7. $x^2 + 4x + 1 = 0$. Give exact solutions and approximate solutions to three decimal places.

8. $\dfrac{x}{x - 2} + \dfrac{4}{x - 6} = 0$

9. $\dfrac{x}{4} - \dfrac{4}{x} = 2$

10. $15 = \dfrac{8}{x+2} - \dfrac{6}{x-2}$

11. Solve $x^2 + 4x + 1 = 0$ by completing the square. Show your work.   [7.1b]

12. *Hang Time.*   Use the function $V(T) = 48T^2$. A basketball player has a vertical leap of 39 in. What is his hang time?   [7.1c]

13. *DVD Player Screen.*   The width of a rectangular screen on a portable DVD player is 5 cm less than the length. The area is 126 cm². Find the length and the width. [7.3a]

14. *Picture Matting.*   A picture mat measures 12 in. by 16 in.; 140 in² of picture shows. Find the width of the mat.   [7.3a]

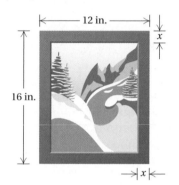

16 in. | 12 in. | x

15. *Motorcycle Travel.*   During the first part of a trip, a motorcyclist travels 50 mi at a certain speed. The rider travels 80 mi on the second part of the trip at a speed that is 10 mph slower. The total time for the trip is 3 hr. What is the speed on each part of the trip?   [7.3a]

Determine the nature of the solutions of the equation. [7.4a]

16. $x^2 + 3x - 6 = 0$

17. $x^2 + 2x + 5 = 0$

Write a quadratic equation having the given solutions. [7.4b]

18. $\dfrac{1}{5}, -\dfrac{3}{5}$

19. -4, only solution

Solve for the indicated letter.   [7.3b]

20. $N = 3\pi\sqrt{\dfrac{1}{p}}$, for p

21. $2A = \dfrac{3B}{T^2}$, for T

Solve.   [7.4c]

22. $x^4 - 13x^2 + 36 = 0$

23. $15x^{-2} - 2x^{-1} - 1 = 0$

24. $(x^2 - 4)^2 - (x^2 - 4) - 6 = 0$

25. $x - 13\sqrt{x} + 36 = 0$

For each quadratic function in Exercises 26–28, find and label **(a)** the vertex, **(b)** the line of symmetry, and **(c)** the maximum or minimum value. Then **(d)** graph the function. [7.5c], [7.6a]

26. $f(x) = -\dfrac{1}{2}(x - 1)^2 + 3$

x	$f(x)$

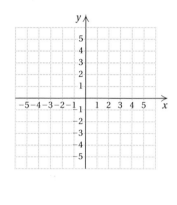

Vertex: (____, ____)
Line of symmetry: $x =$ ____
_____ value: _____

27. $f(x) = x^2 - x + 6$

x	$f(x)$

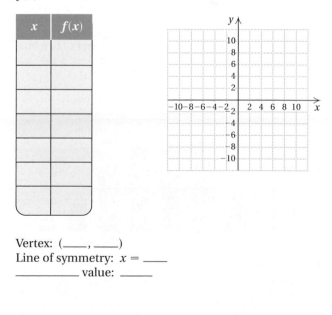

Vertex: (____, ____)
Line of symmetry: $x =$ ____
_____ value: _____

28. $f(x) = -3x^2 - 12x - 8$

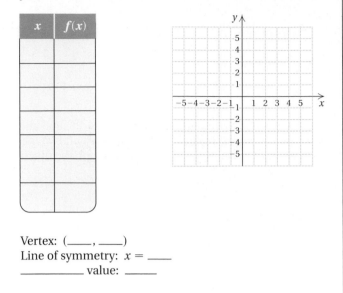

x	$f(x)$

Vertex: (____ , ____)
Line of symmetry: $x =$ ____
_____ value: ____

29. Find the x- and y-intercepts: [7.6b]

$$f(x) = x^2 - 9x + 14.$$

30. What is the minimum product of two numbers whose difference is 22? What numbers yield this product? [7.7a]

31. Find the quadratic function that fits the data points $(0, -2)$, $(1, 3)$, and $(3, 7)$. [7.7b]

32. *Live Births by Age.* The average number of live births per 1000 women rises and falls according to age, as seen in the following bar graph. [7.7b]

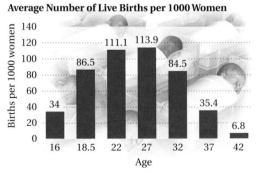

Average Number of Live Births per 1000 Women

Source: Centers for Disease Control and Prevention

a) Use the data points $(16, 34)$, $(27, 113.9)$, and $(37, 35.4)$ to fit a quadratic function to the data.

b) Use the quadratic function to estimate the number of live births per 1000 women of age 30.

Solve. [7.8a, b]

33. $(x + 2)(x - 1)(x - 2) > 0$

34. $\dfrac{(x + 4)(x - 1)}{(x + 2)} < 0$

35. $\mathbf{D_W}$ Explain as many characteristics as you can of the graph of the quadratic function $f(x) = ax^2 + bx + c$. [7.5a, b, c], [7.6a, b]

36. $\mathbf{D_W}$ Explain how the x-intercepts of a quadratic function can be used to help find the vertex of the function. What piece of information would still be missing? [7.6a, b]

SYNTHESIS

---

37. A quadratic function has x-intercepts $(-3, 0)$ and $(5, 0)$ and y-intercept $(0, -7)$. Find an equation for the function. What is its maximum or minimum value? [7.6a, b]

38. Find h and k such that $3x^2 - hx + 4k = 0$, the sum of the solutions is 20, and the product of the solutions is 80. [7.2a], [7.7b]

39. The average of two numbers is 171. One of the numbers is the square root of the other. Find the numbers. [7.3a]

1. **a)** Solve: $3x^2 - 4 = 0$.
   b) Find the x-intercepts of $f(x) = 3x^2 - 4$.

Solve.

2. $x^2 + x + 1 = 0$

3. $x - 8\sqrt{x} + 7 = 0$

4. $4x(x - 2) - 3x(x + 1) = -18$

5. $x^4 - 5x^2 + 5 = 0$

6. $x^2 + 4x = 2$. Give exact solutions and approximate solutions to three decimal places.

7. $\dfrac{1}{4 - x} + \dfrac{1}{2 + x} = \dfrac{3}{4}$

8. Solve $x^2 - 4x + 1 = 0$ by completing the square. Show your work.

9. *Free-Falling Objects.* The Peachtree Plaza in Atlanta, Georgia, is 723 ft tall. Use the function $s(t) = 16t^2$ to approximate how long it would take an object to fall from the top.

10. *Marine Travel.* The Columbia River flows at a rate of 2 mph for the length of a popular boating route. In order for a motorized dinghy to travel 3 mi upriver and then return in a total of 4 hr, how fast must the boat be able to travel in still water?

11. *Memory Board.* A computer-parts company wants to make a rectangular memory board that has a perimeter of 28 cm. What dimensions will allow the board to have a maximum area?

12. *Hang Time.* Use the function $V(T) = 48T^2$. Anfernee Hardaway of the New York Knicks has a vertical leap of 36 in. What is his hang time?

13. Determine the nature of the solutions of the equation $x^2 + 5x + 17 = 0$.

14. Write a quadratic equation having the solutions $\sqrt{3}$ and $3\sqrt{3}$.

15. Solve $V = 48T^2$ for T.

For each quadratic function in Exercises 16 and 17, find and label (a) the vertex, (b) the line of symmetry, and (c) the maximum or minimum value. Then (d) graph the function.

16. $f(x) = -x^2 - 2x$

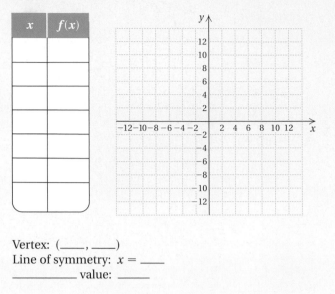

x	$f(x)$

Vertex: (____, ____)
Line of symmetry: $x = $ ____
_____ value: ____

17. $f(x) = 4x^2 - 24x + 41$

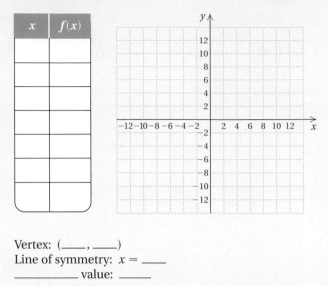

x	$f(x)$

Vertex: (____, ____)
Line of symmetry: $x = $ ____
_____ value: ____

18. Find the x- and y-intercepts:
$$f(x) = -x^2 + 4x - 1.$$

19. What is the minimum product of two numbers whose difference is 8? What numbers yield this product?

20. Find the quadratic function that fits the data points $(0, 0)$, $(3, 0)$, and $(5, 2)$.

21. *Consumer Credit.* The graph at right shows outstanding consumer credit in the United States for various years. It appears that the graph might be fit by one part or side of a quadratic function.

Source: Board of Governors of the Federal Reserve System, Federal Reserve Bulletin (monthly); www.federalreserve.gov

a) Use the data points $(0, 349)$, $(10, 780)$, and $(20, 1692)$ to fit a quadratic function $C(t) = at^2 + bt + c$ to the data, where C is the outstanding consumer credit t years after 1980 and $t = 0$ corresponds to 1980.

b) Use the quadratic function to predict the outstanding consumer credit in 2007 and in 2010.

Outstanding Consumer Credit, 1980–2003 (in billions of dollars)

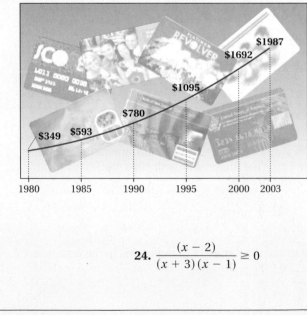

Solve.

22. $x^2 < 6x + 7$

23. $\dfrac{x - 5}{x + 3} < 0$

24. $\dfrac{(x - 2)}{(x + 3)(x - 1)} \geq 0$

25. A quadratic function has x-intercepts $(-2, 0)$ and $(7, 0)$ and y-intercept $(0, 8)$. Find an equation for the function. What is its maximum or minimum value?

26. One solution of $kx^2 + 3x - k = 0$ is -2. Find the other solution.

27. Solve: $x^8 - 20x^4 + 64 = 0$.

CHAPTER 7: Quadratic Equations and Functions

Exponential and Logarithmic Functions

8

Real-World Application

In many colleges and universities, a course in American Sign Language will fulfill the foreign language requirement. The number N of students enrolled in a course in sign language t years after 1990 can be approximated by

$$N(t) = 1602(1.3539)^t,$$

where $t = 0$ corresponds to 1990.

Source: Modern Language Association

a) How many students fulfilled their foreign language requirement by taking American Sign Language in 1990? in 1995? in 2000? in 2005?

b) Graph the function.

This problem appears as Exercise 30 in Section 8.1.

Objectives

a   Graph exponential equations and functions.

b   Graph exponential equations in which *x* and *y* have been interchanged.

c   Solve applied problems involving applications of exponential functions and their graphs.

The rapidly rising graph shown below approximates the graph of an *exponential function*. We will consider such functions and some of their applications.

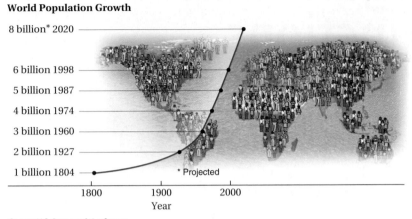

World Population Growth

Source: U.S. Bureau of the Census

a   Graphing Exponential Functions

In Chapter 6, we gave meaning to exponential expressions with rational-number exponents such as

$$8^{1/4}, \quad 3^{-3/4}, \quad 7^{2.34}, \quad 5^{1.73}.$$

For example, $5^{1.73}$, or $5^{173/100}$, or $\sqrt[100]{5^{173}}$, means to raise 5 to the 173rd power and then take the 100th root. We now develop the meaning of exponential expressions with irrational exponents. Examples of expressions with irrational exponents are

$$5^{\sqrt{3}}, \quad 7^{\pi}, \quad 9^{-\sqrt{2}}.$$

Since we can approximate irrational numbers with decimal approximations, we can also approximate expressions with irrational exponents. For example, consider $5^{\sqrt{3}}$. We know that $5^{\sqrt{3}} \approx 5^{1.73} = \sqrt[100]{5^{173}}$. As rational values of r get close to $\sqrt{3}$, 5^r gets close to some real number. Note the following:

r closes in on $\sqrt{3}$.	5^r closes in on some real number p.
r	5^r
$1 < \sqrt{3} < 2$	$5 = 5^1 < p < 5^2 = 25$
$1.7 < \sqrt{3} < 1.8$	$15.426 \approx 5^{1.7} < p < 5^{1.8} \approx 18.119$
$1.73 < \sqrt{3} < 1.74$	$16.189 \approx 5^{1.73} < p < 5^{1.74} \approx 16.452$
$1.732 < \sqrt{3} < 1.733$	$16.241 \approx 5^{1.732} < p < 5^{1.733} \approx 16.267$

As r closes in on $\sqrt{3}$, 5^r closes in on some real number p. We define $5^{\sqrt{3}}$ to be that number p. To seven decimal places, we have

$$5^{\sqrt{3}} \approx 16.2424508.$$

Any positive irrational exponent can be defined in a similar way. Negative irrational exponents are then defined in the same way as negative integer exponents. Then the expression a^x has meaning for any real number x. The general laws of exponents still hold, but we will not prove that here.

We now define exponential functions.

EXPONENTIAL FUNCTION

The function $f(x) = a^x$, where a is a positive constant different from 1, is called an **exponential function,** base a.

We restrict the base a to being positive to avoid the possibility of taking even roots of negative numbers such as the square root of -1, $(-1)^{1/2}$, which is not a real number. We restrict the base from being 1 because for $a = 1$, $t(x) = 1^x = 1$, which is a constant. The following are examples of exponential functions:

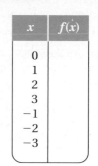

$$f(x) = 2^x, \qquad f(x) = \left(\tfrac{1}{2}\right)^x, \qquad f(x) = (0.4)^x.$$

Note that in contrast to polynomial functions like $f(x) = x^2$ and $f(x) = x^3$, the variable is *in the exponent.* Let's consider graphs of exponential functions.

EXAMPLE 1 Graph the exponential function $f(x) = 2^x$.

We compute some function values and list the results in a table. It is a good idea to begin by letting $x = 0$.

$$f(0) = 2^0 = 1;$$
$$f(1) = 2^1 = 2;$$
$$f(2) = 2^2 = 4;$$
$$f(3) = 2^3 = 8;$$
$$f(-1) = 2^{-1} = \frac{1}{2^1} = \frac{1}{2};$$
$$f(-2) = 2^{-2} = \frac{1}{2^2} = \frac{1}{4};$$
$$f(-3) = 2^{-3} = \frac{1}{2^3} = \frac{1}{8}.$$

x	$f(x)$
0	1
1	2
2	4
3	8
-1	$\frac{1}{2}$
-2	$\frac{1}{4}$
-3	$\frac{1}{8}$

Next, we plot these points and connect them with a smooth curve.

In graphing, be sure to plot enough points to determine how steeply the curve rises.

The curve comes very close to the x-axis, but does not touch or cross it.

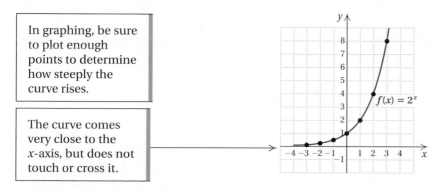

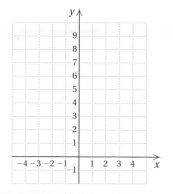

1. Graph: $f(x) = 3^x$.

   a) Complete this table of solutions.

x	$f(x)$
0	
1	
2	
3	
-1	
-2	
-3	

   b) Plot the points from the table and connect them with a smooth curve.

Answers on page A-43

2. Graph: $f(x) = \left(\dfrac{1}{3}\right)^x$.

a) Complete this table of solutions.

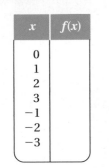

x	$f(x)$
0	
1	
2	
3	
−1	
−2	
−3	

b) Plot the points from the table and connect them with a smooth curve.

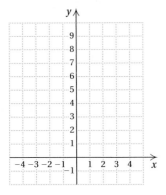

Note that as x increases, the function values increase indefinitely. As x decreases, the function values decrease, getting very close to 0. The x-axis, or the line $y = 0$, is an *asymptote*, meaning here that as x gets very small, the curve comes very close to but never touches the axis.

Do Exercise 1 on the preceding page.

EXAMPLE 2 Graph the exponential function $f(x) = \left(\dfrac{1}{2}\right)^x$.

We compute some function values and list the results in a table. Before we do so, note that

$$f(x) = \left(\tfrac{1}{2}\right)^x = (2^{-1})^x = 2^{-x}.$$

Then we have

$$f(0) = 2^{-0} = 1;$$

$$f(1) = 2^{-1} = \frac{1}{2^1} = \frac{1}{2};$$

$$f(2) = 2^{-2} = \frac{1}{2^2} = \frac{1}{4};$$

$$f(3) = 2^{-3} = \frac{1}{2^3} = \frac{1}{8};$$

$$f(-1) = 2^{-(-1)} = 2^1 = 2;$$

$$f(-2) = 2^{-(-2)} = 2^2 = 4;$$

$$f(-3) = 2^{-(-3)} = 2^3 = 8.$$

x	$f(x)$
0	1
1	$\frac{1}{2}$
2	$\frac{1}{4}$
3	$\frac{1}{8}$
−1	2
−2	4
−3	8

Next, we plot these points and draw the curve. Note that this graph is a reflection across the y-axis of the graph in Example 1. The line $y = 0$ is again an asymptote.

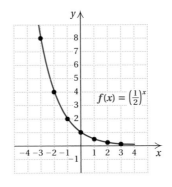

Do Exercise 2.

The preceding examples illustrate exponential functions with various bases. Let's list some of their characteristics. Keep in mind that the definition of an exponential function, $f(x) = a^x$, requires that the base be positive and different from 1.

Answers on page A-43

When $a > 1$, the function $f(x) = a^x$ increases from left to right. The greater the value of a, the steeper the curve. As x gets smaller and smaller, the curve gets closer to the line $y = 0$: It is an asymptote.

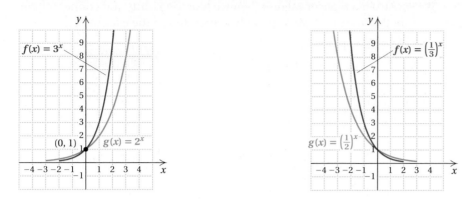

When $0 < a < 1$, the function $f(x) = a^x$ decreases from left to right. As a approaches 1, the curve becomes less steep. As x gets larger and larger, the curve gets closer to the line $y = 0$: It is an asymptote.

y-INTERCEPT OF AN EXPONENTIAL FUNCTION

All functions $f(x) = a^x$ go through the point $(0, 1)$. That is, the y-intercept is $(0, 1)$.

Do Exercises 3 and 4.

EXAMPLE 3 Graph: $f(x) = 2^{x-2}$.

We construct a table of values. Then we plot the points and connect them with a smooth curve. Be sure to note that $x - 2$ is the *exponent*.

$$f(0) = 2^{0-2} = 2^{-2} = \frac{1}{2^2} = \frac{1}{4};$$

$$f(1) = 2^{1-2} = 2^{-1} = \frac{1}{2^1} = \frac{1}{2};$$

$$f(2) = 2^{2-2} = 2^0 = 1;$$

$$f(3) = 2^{3-2} = 2^1 = 2;$$

$$f(4) = 2^{4-2} = 2^2 = 4;$$

$$f(-1) = 2^{-1-2} = 2^{-3} = \frac{1}{2^3} = \frac{1}{8};$$

$$f(-2) = 2^{-2-2} = 2^{-4} = \frac{1}{2^4} = \frac{1}{16}$$

x	$f(x)$
0	$\frac{1}{4}$
1	$\frac{1}{2}$
2	1
3	2
4	4
-1	$\frac{1}{8}$
-2	$\frac{1}{16}$

The graph looks just like the graph of $g(x) = 2^x$, but it is translated 2 units to the right.

The y-intercept of $g(x) = 2^x$ is $(0, 1)$. The y-intercept of $f(x) = 2^{x-2}$ is $\left(0, \frac{1}{4}\right)$. The line $y = 0$ is still an asymptote.

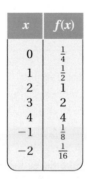

Graph.

3. $f(x) = 4^x$

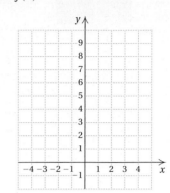

4. $f(x) = \left(\frac{1}{4}\right)^x$

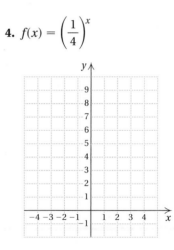

5. Graph: $f(x) = 2^{x+2}$.

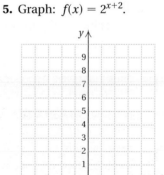

Answers on page A-43

669

6. Graph: $f(x) = 2^x - 4$.

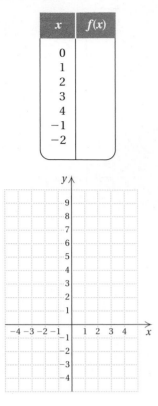

x	$f(x)$
0	
1	
2	
3	
4	
−1	
−2	

Do Exercise 5 on the preceding page.

 **EXAMPLE 4** Graph: $f(x) = 2^x - 3$.

We construct a table of values. Then we plot the points and connect them with a smooth curve. Note that the only expression in the exponent is x.

$$f(0) = 2^0 - 3 = 1 - 3 = -2;$$
$$f(1) = 2^1 - 3 = 2 - 3 = -1;$$
$$f(2) = 2^2 - 3 = 4 - 3 = 1;$$
$$f(3) = 2^3 - 3 = 8 - 3 = 5;$$
$$f(4) = 2^4 - 3 = 16 - 3 = 13;$$
$$f(-1) = 2^{-1} - 3 = \frac{1}{2} - 3 = -\frac{5}{2};$$
$$f(-2) = 2^{-2} - 3 = \frac{1}{4} - 3 = -\frac{11}{4}.$$

x	$f(x)$
0	−2
1	−1
2	1
3	5
4	13
−1	$-\frac{5}{2}$
−2	$-\frac{11}{4}$

The graph looks just like the graph of $g(x) = 2^x$, but it is translated down 3 units. The y-intercept is $(0, -2)$. The line $y = -3$ is an asymptote. The curve gets closer to this line as x gets smaller and smaller.

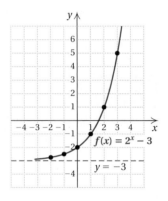

Answer on page A-43

Do Exercise 6.

CALCULATOR CORNER

Graphing Exponential Functions We can use a graphing calculator to graph exponential functions. It might be necessary to try several sets of window dimensions in order to find the ones that give a good view of the curve.

To graph $f(x) = 3^x - 1$, we enter the equation as y_1 by pressing ③ ⌃ Ⓧ,T,θ,n ⊖ ① . We can begin graphing with the standard window $[-10, 10, -10, 10]$ by pressing ⓏOOM ⑥ . Although this window gives a good view of the curve, we might want to adjust it to show more of the curve in the first quadrant. Changing the dimensions to $[-10, 10, -5, 15]$ accomplishes this.

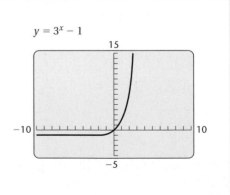

$y = 3^x - 1$

Exercises:

1. Use a graphing calculator to graph the functions in Examples 1–4.

2. Use a graphing calculator to graph the functions in Margin Exercises 1–6.

CHAPTER 8: Exponential and
Logarithmic Functions

b Equations with x and y Interchanged

It will be helpful in later work to be able to graph an equation in which the x and the y in $y = a^x$ are interchanged.

EXAMPLE 5 Graph: $x = 2^y$.

Note that x is alone on one side of the equation. We can find ordered pairs that are solutions more easily by choosing values for y and then computing the x-values.

For $y = 0$, $x = 2^0 = 1$.

For $y = 1$, $x = 2^1 = 2$.

For $y = 2$, $x = 2^2 = 4$.

For $y = 3$, $x = 2^3 = 8$.

For $y = -1$, $x = 2^{-1} = \dfrac{1}{2^1} = \dfrac{1}{2}$.

For $y = -2$, $x = 2^{-2} = \dfrac{1}{2^2} = \dfrac{1}{4}$.

For $y = -3$, $x = 2^{-3} = \dfrac{1}{2^3} = \dfrac{1}{8}$.

x	y
1	0
2	1
4	2
8	3
$\frac{1}{2}$	-1
$\frac{1}{4}$	-2
$\frac{1}{8}$	-3

(1) Choose values for y.
(2) Compute values for x.

We plot the points and connect them with a smooth curve. What happens as y-values become smaller?

This curve does not touch or cross the y-axis.

Note that this curve $x = 2^y$ looks just like the graph of $y = 2^x$, except that it is reflected, or flipped, across the line $y = x$, as shown below.

$y = 2^x$	
x	y
0	1
1	2
2	4
3	8
-1	$\frac{1}{2}$
-2	$\frac{1}{4}$
-3	$\frac{1}{8}$

$x = 2^y$	
x	y
1	0
2	1
4	2
8	3
$\frac{1}{2}$	-1
$\frac{1}{4}$	-2
$\frac{1}{8}$	-3

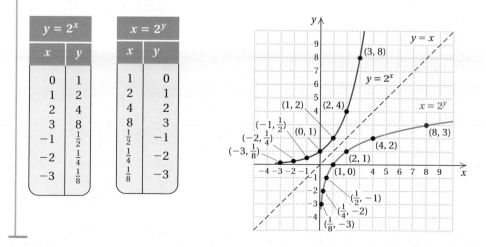

7. Graph: $x = 3^y$.

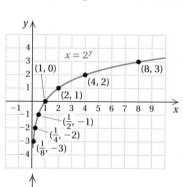

Do Exercise 7.

Answer on page A-43

8. **Interest Compounded Annually.** Find the amount in an account after 1 yr and after 2 yr if $40,000 is invested at 5%, compounded annually.

C Applications of Exponential Functions

When interest is paid on interest, we call it **compound interest.** This is the type of interest paid on investments and loans. Suppose you have $100,000 in a savings account at an interest rate of 8%. This means that in 1 yr, the account will contain the original $100,000 plus 8% of $100,000. Thus the total in the account after 1 yr will be

$100,000   plus $100,000 $\times$ 0.08.

This can also be expressed as

$100,000 + $100,000 $\times$ 0.08 = $100,000 $\times$ 1 + $100,000 $\times$ 0.08

$$= \$100{,}000(1 + 0.08) \quad \text{Factoring out } \$100{,}000 \text{ using the distributive law}$$

$$= \$100{,}000(1.08)$$

$$= \$108{,}000.$$

Now suppose that the total of $108,000 remains in the account for another year. At the end of the second year, the account will contain the $108,000 plus 8% of $108,000. The total in the account will be

$108,000   plus $108,000 $\times$ 0.08,

or

$$\$108{,}000(1.08) = \$100{,}000(1.08)^2 = \$116{,}640.$$

Note that in the second year, interest is earned on the first year's interest as well as the original amount. When this happens, we say the interest is **compounded annually.** If the original amount of $100,000 earned only simple interest for 2 yr, the interest would be

$100,000 $\times$ 0.08 $\times$ 2,   or $16,000,

and the amount in the account would be

$100,000 + $16,000 = $116,000,

less than the $116,640 when interest is compounded annually.

Do Exercise 8.

The following table shows how the computation continues over 4 yr.

$100,000 IN AN ACCOUNT

YEAR	INTEREST COMPOUNDED ANNUALLY	SIMPLE INTEREST
Beginning of 1st yr End of 1st yr	$100,000 $108,000 = $100,000(1.08)^1	$108,000
Beginning of 2nd yr End of 2nd yr	$108,000 $116,640 = $100,000(1.08)^2	$116,000
Beginning of 3rd yr End of 3rd yr	$116,640 $125,971.20 = $100,000(1.08)^3	$124,000
Beginning of 4th yr End of 4th yr	$125,971.20 $136,048.90 $\approx$ $100,000(1.08)^4	$132,000

Answer on page A-43

We can express interest compounded annually using an exponential function.

EXAMPLE 6 *Interest Compounded Annually.* The amount of money A that a principal P will grow to after t years at interest rate r, compounded annually, is given by the formula

$$A = P(1 + r)^t.$$

Suppose that $100,000 is invested at 8% interest, compounded annually.

a) Find a function for the amount in the account after t years.

b) Find the amount of money in the account at $t = 0$, $t = 4$, $t = 8$, and $t = 10$.

c) Graph the function.

a) If $P = \$100,000$ and $r = 8\% = 0.08$, we can substitute these values and form the following function:

$$A(t) = \$100,000(1 + 0.08)^t = \$100,000(1.08)^t.$$

b) To find the function values, you might find a calculator with a power key helpful.

$$A(0) = \$100,000(1.08)^0 = \$100,000;$$
$$A(4) = \$100,000(1.08)^4 \approx \$136,048.90;$$
$$A(8) = \$100,000(1.08)^8 \approx \$185,093.02;$$
$$A(10) = \$100,000(1.08)^{10} \approx \$215,892.50$$

c) We use the function values computed in (b) with others, if we wish, to draw the graph as follows. Note that the axes are scaled differently because of the large numbers.

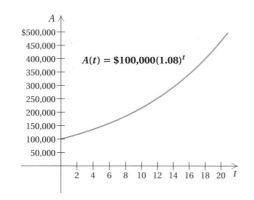

Do Exercise 9.

Suppose the principal of $100,000 we just considered were **compounded semiannually**—that is, every half year. Interest would then be calculated twice a year at a rate of 8% ÷ 2, or 4%, each time. The computations are as follows:

After the first $\frac{1}{2}$ year, the account will contain 104% of $100,000:

$$\$100,000 \times 1.04 = \$104,000.$$

After a second $\frac{1}{2}$ year (1 full year), the account will contain 104% of $104,000:

$$\$104,000 \times 1.04 = \$100,000 \times (1.04)^2 = \$108,160.$$

After a third $\frac{1}{2}$ year $\left(1\frac{1}{2}\text{ full years}\right)$, the account will contain 104% of $108,160:

$$\$108,160 \times 1.04 = \$100,000 \times (1.04)^3 = \$112,486.40.$$

9. Interest Compounded Annually. Suppose that $40,000 is invested at 5% interest, compounded annually.

a) Find a function for the amount in the account after t years.

b) Find the amount of money in the account at $t = 0$, $t = 4$, $t = 8$, and $t = 10$.

c) Graph the function.

10. A couple invests $7000 in an account paying 6.4%, compounded quarterly. Find the amount in the account after $5\frac{1}{2}$ yr.

Answers on page A-43

The Compound-Interest Formula If $1000 is invested at 8%, compounded quarterly, how much is in the account at the end of 2 yr?

We use the compound-interest formula, substituting 1000 for P, 0.08 for r, 4 for n (compounding quarterly), and 2 for t. Then we get

$$A = P\left(1 + \frac{r}{n}\right)^{n \cdot t}$$

$$= 1000\left(1 + \frac{0.08}{4}\right)^{4 \cdot 2}.$$

To do this computation on a calculator, we press

(1)(0)(0)(0)(( )(1)(+)

(.)(0)(8)(÷)(4)( ))

(^)(( )(4)(×)(2)( ))

ENTER. The result is approximately $1171.66.

Exercises:

1. If $1000 is invested at 6%, compounded semiannually, how much is in the account at the end of 2 yr?

2. If $1000 is invested at 6%, compounded monthly, how much is in the account at the end of 2 yr?

3. If $20,000 is invested at 4.2%, compounded quarterly, how much is in the account at the end of 10 yr?

4. If $10,000 is invested at 5.4%, how much is in the account at the end of 1 yr, if interest is compounded **(a)** annually? **(b)** semiannually? **(c)** quarterly? **(d)** daily? **(e)** hourly?

After a fourth $\frac{1}{2}$ year (2 full years), the account will contain 104% of $112,486.40:

$$\$112{,}486.40 \times 1.04 = \$100{,}000 \times (1.04)^4$$

$$\approx \$116{,}985.86. \quad \text{Rounded to the nearest cent}$$

Comparing these results with those in the table on p. 672, we can see that by having more compounding periods, we increase the amount in the account. We have illustrated the following result.

> **COMPOUND-INTEREST FORMULA**
>
> If a principal P has been invested at interest rate r, compounded n times a year, in t years it will grow to an amount A given by
>
> $$A = P \cdot \left(1 + \frac{r}{n}\right)^{n \cdot t}.$$

EXAMPLE 7 The Ibsens invest $4000 in an account paying $8\frac{5}{8}\%$, compounded quarterly. Find the amount in the account after $2\frac{1}{2}$ yr.

The compounding is quarterly—that is, four times a year—so in $2\frac{1}{2}$ yr, there are ten $\frac{1}{4}$-yr periods. We substitute $4000 for P, $8\frac{5}{8}\%$, or 0.08625, for r, 4 for n, and $2\frac{1}{2}$, or $\frac{5}{2}$, for t and compute A:

$$A = P \cdot \left(1 + \frac{r}{n}\right)^{n \cdot t}$$

$$= 4000 \cdot \left(1 + \frac{8\frac{5}{8}\%}{4}\right)^{4 \cdot \frac{5}{2}}$$

$$= 4000 \cdot \left(1 + \frac{0.08625}{4}\right)^{10}$$

$$= 4000(1.0215625)^{10} \quad \text{Using a calculator}$$

$$\approx \$4951.19.$$

The amount in the account after $2\frac{1}{2}$ yr is $4951.19.

Do Exercise 10 on the preceding page.

a   Graph.

1. $f(x) = 2^x$

x	$f(x)$
0	
1	
2	
3	
−1	
−2	
−3	

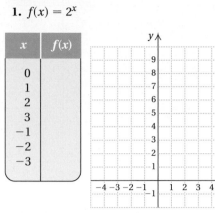

2. $f(x) = 3^x$

x	$f(x)$
0	
1	
2	
3	
−1	
−2	
−3	

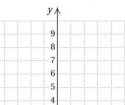

3. $f(x) = 5^x$

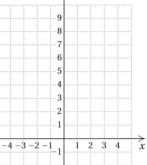

4. $f(x) = 6^x$

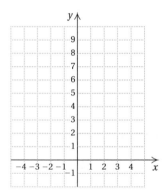

5. $f(x) = 2^{x+1}$

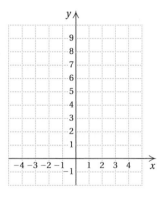

6. $f(x) = 2^{x-1}$

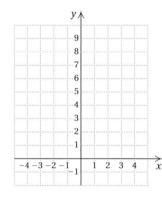

7. $f(x) = 3^{x-2}$

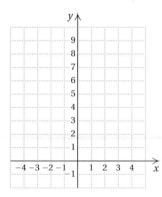

8. $f(x) = 3^{x+2}$

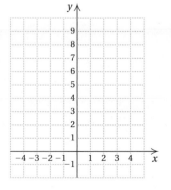

9. $f(x) = 2^x - 3$

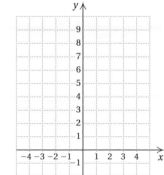

10. $f(x) = 2^x + 1$

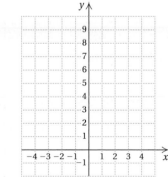

11. $f(x) = 5^{x+3}$

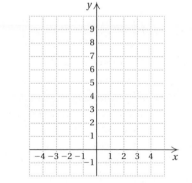

12. $f(x) = 6^{x-4}$

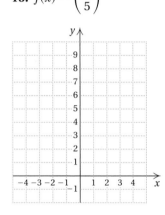

13. $f(x) = \left(\dfrac{1}{2}\right)^x$

x	$f(x)$
0	
1	
2	
3	
−1	
−2	
−3	

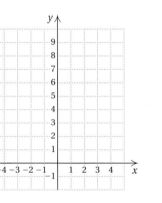

14. $f(x) = \left(\dfrac{1}{3}\right)^x$

x	$f(x)$
0	
1	
2	
3	
−1	
−2	
−3	

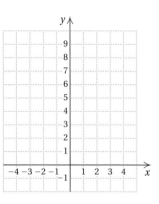

15. $f(x) = \left(\dfrac{1}{5}\right)^x$

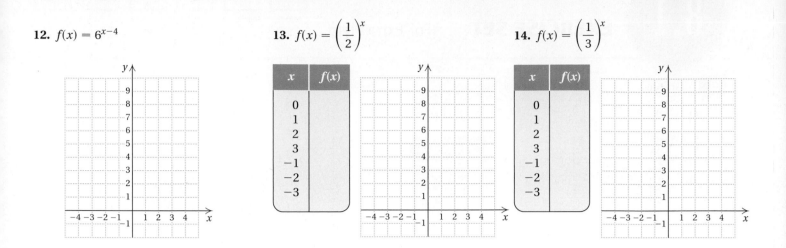

16. $f(x) = \left(\dfrac{1}{4}\right)^x$

17. $f(x) = 2^{2x-1}$

18. $f(x) = 3^{3-x}$

b   Graph.

19. $x = 2^y$

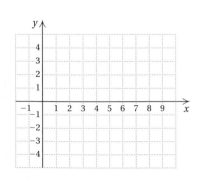

20. $x = 6^y$

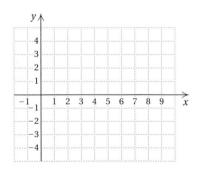

21. $x = \left(\dfrac{1}{2}\right)^y$

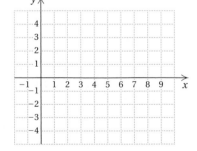

22. $x = \left(\dfrac{1}{3}\right)^y$

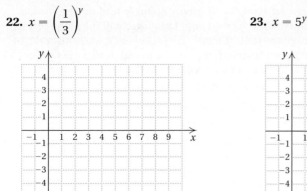

23. $x = 5^y$

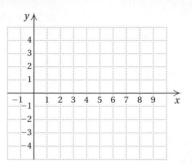

24. $x = \left(\dfrac{2}{3}\right)^y$

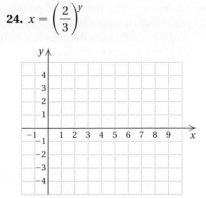

Graph both equations using the same set of axes.

25. $y = 2^x$, $x = 2^y$

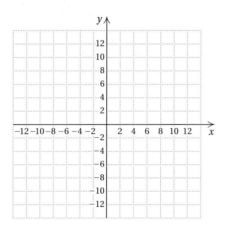

26. $y = \left(\dfrac{1}{2}\right)^x$, $x = \left(\dfrac{1}{2}\right)^y$

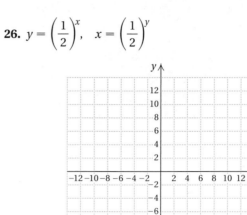

C   Solve.

27. *Interest Compounded Annually.* Suppose that $50,000 is invested at 6% interest, compounded annually.

  a) Find a function A for the amount in the account after t years.

  b) Complete the following table of function values.

t	$A(t)$
0	
1	
2	
4	
8	
10	
20	

  c) Graph the function.

28. *Interest Compounded Annually.* Suppose that $50,000 is invested at 8% interest, compounded annually.

  a) Find a function A for the amount in the account after t years.

  b) Complete the following table of function values.

t	$A(t)$
0	
1	
2	
4	
8	
10	
20	

  c) Graph the function.

677

29. *Diabetes.* The number of people in the United States who have diabetes is growing exponentially. The number N of cases of diabetes, in millions, t years after 2000 can be approximated by

$$N(t) = 17.7(1.018)^t,$$

where $t = 0$ corresponds to 2000.

a) How many cases, in millions, of diabetes were/will there be in 2000? in 2007? in 2010? in 2030?

b) Graph the function.

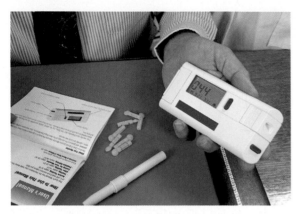

30. *Sign Language.* In many colleges and universities, a course in American Sign Language will fulfill the foreign language requirement. The number N, in thousands, of students enrolled in a course in sign language t years after 1990 can be approximated by

$$N(t) = 1.602(1.3539)^t,$$

where $t = 0$ corresponds to 1990.

Source: Modern Language Association

a) How many students fulfilled their foreign language requirement by taking American Sign Language in 1990? in 1995? in 2000? in 2005?

b) Graph the function.

31. *Flat-Panel TVs.* As more and more plasma flat-screen TVs are sold, the average price is declining exponentially. The average price A of a plasma TV t years after 2000 can be approximated by

$$A(t) = 9809(0.815)^t,$$

where $t = 0$ corresponds to 2000.

a) What is the average price of a plasma TV in 2000? in 2005? in 2008?

b) Graph the function.

32. *Marine Biology.* Due to excessive whaling before the mid-1970s, the humpback whale is considered an endangered species. The worldwide population of humpbacks $P(t)$, in thousands, t years after 1900 ($t < 70$) can be approximated by

$$P(t) = 150(0.960)^t.$$

Source: American Cetecean Society, 2001; ASK Archive, 1998

a) How many humpback whales were alive in 1930? in 1960?

b) Graph the function.

33. *Growth of Bacteria.* Bladder infections are often caused when the bacteria *Escherichia coli* reach the human bladder. Suppose that 3000 of the bacteria are present at time $t = 0$. Then t minutes later, the number of bacteria present will be

$$N(t) = 3000(2)^{t/20}.$$

Source: Chris Hayes, "Detecting a Human Health Risk: *E. coli*," *Laboratory Medicine* 29, no. 6, June 1998: 347–355

a) How many bacteria will be present after 10 min? 20 min? 30 min? 40 min? 60 min?

b) Graph the function.

34. *Salvage Value.* An office machine is purchased for $5200. Its value each year is about 80% of the value the preceding year. Its value after t years is given by the exponential function

$$V(t) = \$5200(0.8)^t.$$

a) Find the value of the machine after 0 yr, 1 yr, 2 yr, 5 yr, and 10 yr.

b) Graph the function.

35. D_W Why was it necessary to discuss irrational exponents before graphing exponential functions?

36. D_W Suppose that $1000 is invested for 5 yr at 6% interest, compounded annually. In what year will the greatest amount of interest be earned? Why?

SKILL MAINTENANCE

37. Multiply and simplify: $x^{-5} \cdot x^3$. [R.7a]

38. Simplify: $(x^{-3})^4$. [R.7b]

Simplify. [R.3a]

39. 9^0

40. $\left(\frac{2}{3}\right)^0$

41. $\left(\frac{2}{3}\right)^1$

42. 2.7^1

Divide and simplify. [R.7a]

43. $\dfrac{x^{-3}}{x^4}$

44. $\dfrac{x}{x^{11}}$

45. $\dfrac{x}{x^0}$

46. $\dfrac{x^{-3}}{x^{-4}}$

SYNTHESIS

47. Simplify: $\left(5^{\sqrt{2}}\right)^{2\sqrt{2}}$.

48. Which is larger: $\pi^{\sqrt{2}}$ or $\left(\sqrt{2}\right)^\pi$?

Graph.

49. $y = 2^x + 2^{-x}$

50. $y = |2^x - 2|$

51. $y = \left|\left(\frac{1}{2}\right)^x - 1\right|$

52. $y = 2^{-x^2}$

Graph both equations using the same set of axes.

53. $y = 3^{-(x-1)}, \quad x = 3^{-(y-1)}$

54. $y = 1^x, \quad x = 1^y$

55. ▱ Use a graphing calculator to graph each of the equations in Exercises 49–52.

679

Objectives

a Find the inverse of a relation if it is described as a set of ordered pairs or as an equation.

b Given a function, determine whether it is one-to-one and has an inverse that is a function.

c Find a formula for the inverse of a function, if it exists, and graph inverse relations and functions.

d Find the composition of functions and express certain functions as a composition of functions.

e Determine whether a function is an inverse by checking its composition with the original function.

When we go from an output of a function back to its input or inputs, we get an *inverse relation*. When that relation is a function, we have an *inverse function*. We now study such inverse functions and how to find formulas when the original function has a formula. We do so to understand the relationships among the special functions that we study in this chapter.

a Inverses

A set of ordered pairs is called a **relation**. When we consider the graph of a function, we are thinking of a set of ordered pairs. Thus a function can be thought of as a special kind of relation, in which to each first coordinate there corresponds one and only one second coordinate.

Consider the relation h given as follows:

$$h = \{(-7, 4), (3, -1), (-6, 5), (0, 2)\}.$$

Suppose we *interchange* the first and second coordinates. The relation we obtain is called the **inverse** of the relation h and is given as follows:

Inverse of $h = \{(4, -7), (-1, 3), (5, -6), (2, 0)\}.$

INVERSE RELATION

Interchanging the coordinates of the ordered pairs in a relation produces the **inverse relation.**

---

1. Consider the relation g given by

$$g = \{(2, 5), (-1, 4), (-2, 1)\}.$$

The graph of the relation is shown below in red. Find the inverse and draw its graph in blue.

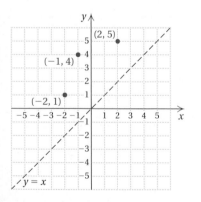

EXAMPLE 1 Consider the relation g given by

$$g = \{(2, 4), (-1, 3), (-2, 0)\}.$$

In the figure below, the relation g is shown in red. The inverse of the relation is

$$\{(4, 2), (3, -1), (0, -2)\}$$

and is shown in blue.

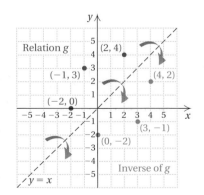

Do Exercise 1.

Answer on page A-45

INVERSE RELATION

If a relation is defined by an equation, interchanging the variables produces an equation of the **inverse relation.**

EXAMPLE 2 Find an equation of the inverse of the relation

$$y = 3x - 4.$$

Then graph both the relation and its inverse.

We interchange x and y and obtain an equation of the inverse:

$$x = 3y - 4.$$

Relation: $y = 3x - 4$ ⟶ *Inverse*: $x = 3y - 4$

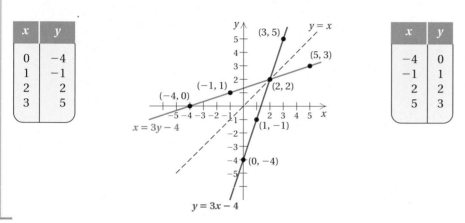

x	y
0	−4
1	−1
2	2
3	5

x	y
−4	0
−1	1
2	2
5	3

Note in Example 2 that the relation $y = 3x - 4$ is a function and its inverse relation $x = 3y - 4$ is also a function. Each graph passes the vertical-line test (see Section 2.2).

EXAMPLE 3 Find an equation of the inverse of the relation

$$y = 6x - x^2.$$

Then graph both the original relation and its inverse.

We interchange x and y and obtain an equation of the inverse:

$$x = 6y - y^2.$$

Relation: $y = 6x - x^2$ ⟶ *Inverse*: $x = 6y - y^2$

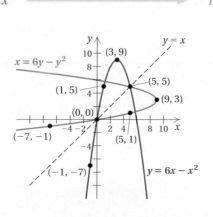

x	y
−1	−7
0	0
1	5
3	9
5	5

x	y
−7	−1
0	0
5	1
9	3
5	5

---

2. Find an equation of the inverse relation. Then complete the table and graph both the original relation and its inverse.

Relation:
$y = 6 - 2x$

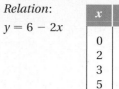

x	y
0	6
2	2
3	0
5	−4

Inverse:
_____

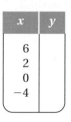

x	y
6	
2	
0	
−4	

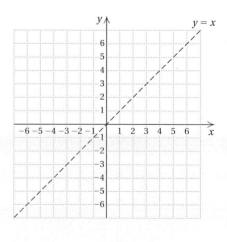

Answers on page A-45

3. Find an equation of the inverse relation. Then complete the table and graph both the original relation and its inverse.

Relation:

$y = x^2 - 4x + 7$

x	y
0	7
1	4
2	3
3	4
4	7

Inverse:

_____

x	y
7	
4	
3	
4	
7	

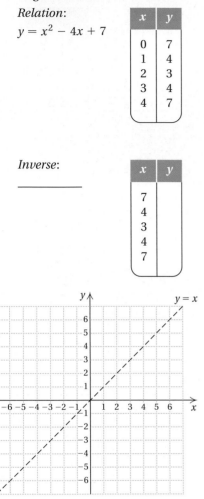

4. Determine whether the function is one-to-one and thus has an inverse that is also a function.

$$f(x) = 4 - x$$

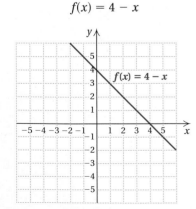

Answers on page A-45

Note in Example 3 that the relation $y = 6x - x^2$ is a function because it passes the vertical-line test. However, its inverse relation $x = 6y - y^2$ is not a function because its graph fails the vertical-line test. Therefore, the inverse of a function is *not* always a function.

Do Exercises 2 and 3. (Exercise 2 is on the preceding page.)

b   Inverses and One-To-One Functions

Let's consider the following two functions.

NUMBER (Domain)	CUBE (Range)
−3 ⟶	−27
−2 ⟶	−8
−1 ⟶	−1
0 ⟶	0
1 ⟶	1
2 ⟶	8
3 ⟶	27

YEAR (Domain)	FIRST-CLASS POSTAGE COST, IN CENTS (Range)
1983 ⟶	20
1984 ⟶	
1989 ⟶	25
1991 ⟶	29
1995 ⟶	32
1999 ⟶	33
2001 ⟶	34
2005 ⟶	37

Source: U.S. Postal Service

Suppose we reverse the arrows. Are these inverse relations functions?

CUBE ROOT (Range)	NUMBER (Domain)
−3 ⟵	−27
−2 ⟵	−8
−1 ⟵	−1
0 ⟵	0
1 ⟵	1
2 ⟵	8
3 ⟵	27

YEAR (Range)	FIRST-CLASS POSTAGE COST, IN CENTS (Domain)
1983 ⟵	20
1984 ⟵	
1989 ⟵	25
1991 ⟵	29
1995 ⟵	32
1999 ⟵	33
2001 ⟵	34
2005 ⟵	37

We see that the inverse of the cubing function is a function. The inverse of the postage function is not a function, however, because the input 20 has *two* outputs, 1983 and 1984. Recall that for a function, each input has exactly one output. However, it can happen that the same output comes from two or more different inputs. If this is the case, the inverse cannot be a function. When this possibility is excluded, the inverse is also a function.

In the cubing function, different inputs have different outputs. Thus its inverse is also a function. The cubing function is what is called a **one-to-one function.** If the inverse of a function f is also a function, it is named f^{-1} (read "f-inverse").

Caution!

The -1 in f^{-1} is *not* an exponent and f^{-1} does not represent a reciprocal!

ONE-TO-ONE FUNCTION AND INVERSES

A function f is **one-to-one** if different inputs have different outputs— that is,

$$\text{if} \quad a \neq b, \quad \text{then} \quad f(a) \neq f(b). \quad \text{Or,}$$

A function f is **one-to-one** if when the outputs are the same, the inputs are the same—that is,

$$\text{if} \quad f(a) = f(b), \quad \text{then} \quad a = b.$$

If a function is one-to-one, then its inverse is a function.

The domain of a one-to-one function f is the range of the inverse f^{-1}.

The range of a one-to-one function f is the domain of the inverse f^{-1}.

How can we tell graphically whether a function is one-to-one and thus has an inverse that is a function?

EXAMPLE 4 The graph of the exponential function $f(x) = 2^x$, or $y = 2^x$, is shown on the left below. The graph of the inverse $x = 2^y$ is shown on the right. How can we tell by examining only the graph on the left whether it has an inverse that is a function?

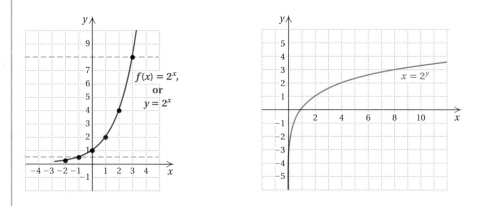

We see that the graph on the right passes the vertical-line test, so we know it is that of a function. However, if we look only at the graph on the left, we think as follows:

A function is one-to-one if different inputs have different outputs. In other words, no two x-values will have the same y-value. For this function, we cannot find two x-values that have the same y-value. Note also that no horizontal line can be drawn that will cross the graph more than once. The function is thus one-to-one and its inverse is a function.

THE HORIZONTAL-LINE TEST

If it is possible for a horizontal line to intersect the graph of a function more than once, then the function is not one-to-one and therefore its inverse is not a function.

Determine whether the function is one-to-one and thus has an inverse that is also a function.

5. $f(x) = x^2 - 1$

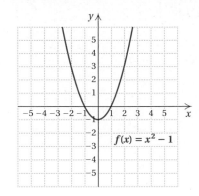

6. $f(x) = 4^x$
(Sketch this graph yourself.)

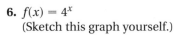
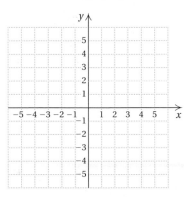

7. $f(x) = |x| - 3$
(Sketch this graph yourself.)

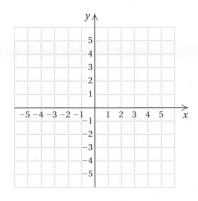

Answers on page A-45

A graph is that of a function if no vertical line crosses the graph more than once. A function has an inverse that is also a function if no horizontal line crosses the graph more than once.

■ **EXAMPLE 5** Determine whether the function $f(x) = x^2$ is one-to-one and has an inverse that is also a function.

The graph of $f(x) = x^2$, or $y = x^2$, is shown on the left below. There are many horizontal lines that cross the graph more than once, so this function is not one-to-one and does not have an inverse that is a function.

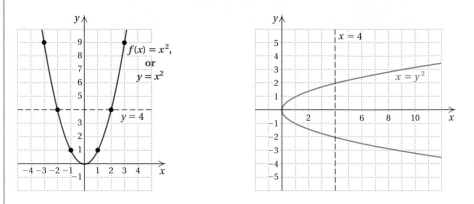

The inverse of the function $y = x^2$ is the relation $x = y^2$. The graph of $x = y^2$ is shown on the right above. It fails the vertical-line test and is not a function.

Do Exercises 4–7 on the preceding pages.

c **Inverse Formulas and Graphs**

Suppose that a function is described by a formula. If it has an inverse that is a function, how do we find a formula for the inverse function? If for any equation with two variables such as x and y we interchange the variables, we obtain an equation of the inverse relation. We proceed as follows to find a formula for f^{-1}.

> If a function f is one-to-one, a formula for its inverse f^{-1} can be found as follows:
>
> 1. Replace $f(x)$ with y.
> 2. Interchange x and y. (This gives the inverse relation.)
> 3. Solve for y.
> 4. Replace y with $f^{-1}(x)$.

■ **EXAMPLE 6** Given $f(x) = x + 1$:

a) Determine whether the function is one-to-one.

b) If it is one-to-one, find a formula for $f^{-1}(x)$.

c) Graph the inverse function, if it exists.

a) The graph of $f(x) = x + 1$ is shown below. It passes the horizontal-line test, so it is one-to-one. Thus its inverse is a function.

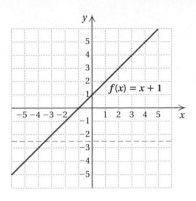

b) 1. Replace $f(x)$ with y: $y = x + 1$.

    2. Interchange x and y: $x = y + 1$.   This gives the inverse relation.

    3. Solve for y: $x - 1 = y$.

    4. Replace y with $f^{-1}(x)$: $f^{-1}(x) = x - 1$.

c) We graph $f^{-1}(x) = x - 1$, or $y = x - 1$. The graph is shown below.

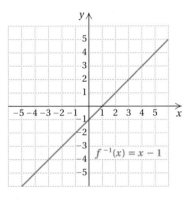

EXAMPLE 7   Given $f(x) = 2x - 3$:

a) Determine whether the function is one-to-one.

b) If it is one-to-one, find a formula for $f^{-1}(x)$.

c) Graph the inverse function, if it exists.

a) The graph of $f(x) = 2x - 3$ is shown below. It passes the horizontal-line test and is one-to-one.

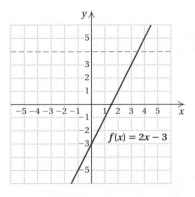

Given each function:

a) Determine whether it is one-to-one.

b) If it is one-to-one, find a formula for the inverse.

c) Graph the inverse function, if it exists.

8. $f(x) = 3 - x$

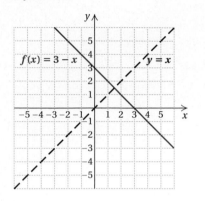

b) 1. Replace $f(x)$ with y: $y = 2x - 3$.

2. Interchange x and y: $x = 2y - 3$.

3. Solve for y: $x + 3 = 2y$

$$\frac{x + 3}{2} = y.$$

4. Replace y with $f^{-1}(x)$: $f^{-1}(x) = \dfrac{x + 3}{2}$.

c) We graph

$$f^{-1}(x) = \frac{x + 3}{2}, \quad \text{or}$$

$$y = \frac{1}{2}x + \frac{3}{2}.$$

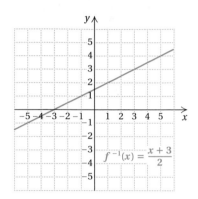

Do Exercises 8 and 9.

Let's now consider inverses of functions in terms of a function machine. Suppose that a one-to-one function f is programmed into a machine. If the machine has a reverse switch, when the switch is thrown, the machine performs the inverse function f^{-1}. Inputs then enter at the opposite end, and the entire process is reversed.

9. $g(x) = 3x - 2$

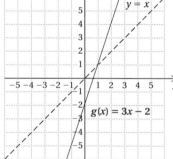

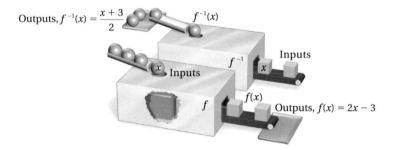

Consider $f(x) = 2x - 3$ and $f^{-1}(x) = (x + 3)/2$ from Example 7. For the input 5,

$$f(5) = 2 \cdot 5 - 3 = 10 - 3 = 7.$$

The output is 7. Now we use 7 for the input in the inverse:

$$f^{-1}(7) = \frac{7 + 3}{2} = \frac{10}{2} = 5.$$

The function f takes 5 to 7. The inverse function f^{-1} takes the number 7 back to 5.

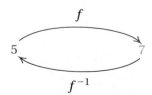

Answers on page A-45

How do the graphs of a function and its inverse compare?

EXAMPLE 8 Graph $f(x) = 2x - 3$ and $f^{-1}(x) = (x + 3)/2$ using the same set of axes. Then compare.

The graph of each function follows. Note that the graph of f^{-1} can be drawn by reflecting the graph of f across the line $y = x$. That is, if we graph $f(x) = 2x - 3$ in wet ink and fold the paper along the line $y = x$, the graph of $f^{-1}(x) = (x + 3)/2$ will appear as the impression made by f.

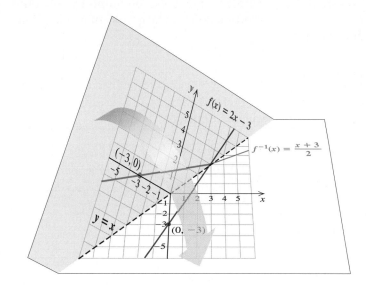

When x and y are interchanged to find a formula for the inverse, we are, in effect, flipping the graph of $f(x) = 2x - 3$ over the line $y = x$. For example, when the coordinates of the y-intercept of the graph of f, $(0, -3)$, are reversed, we get the x-intercept of the graph of f^{-1}, $(-3, 0)$.

> The graph of f^{-1} is a reflection of the graph of f across the line $y = x$.

Do Exercise 10.

EXAMPLE 9 Consider $g(x) = x^3 + 2$.

a) Determine whether the function is one-to-one.

b) If it is one-to-one, find a formula for its inverse.

c) Graph the inverse, if it exists.

a) The graph of $g(x) = x^3 + 2$ is shown at right in red. It passes the horizontal-line test and thus is one-to-one.

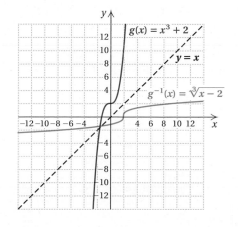

10. Graph $g(x) = 3x - 2$ and $g^{-1}(x) = (x + 2)/3$ using the same set of axes.

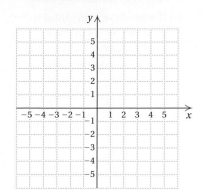

Answer on page A-45

687

11. Given $f(x) = x^3 + 1$:

a) Determine whether the function is one-to-one.

b) If it is one-to-one, find a formula for its inverse.

c) Graph the function and its inverse using the same set of axes.

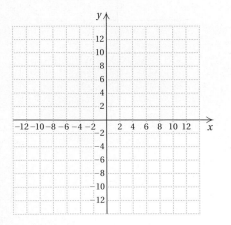

Answers on page A-45

b) **1.** Replace $g(x)$ with y: $y = x^3 + 2$.

     2. Interchange x and y: $x = y^3 + 2$.

     3. Solve for y: $x - 2 = y^3$

                         $\sqrt[3]{x - 2} = y$.   Since a number has only one cube root, we can solve for y.

     4. Replace y with $g^{-1}(x)$: $g^{-1}(x) = \sqrt[3]{x - 2}$.

c) To find the graph, we reflect the graph of $g(x) = x^3 + 2$ across the line $y = x$, as we did in Example 8. It can also be found by substituting into $g^{-1}(x) = \sqrt[3]{x - 2}$ and plotting points. The graphs of g and g^{-1} are shown together above.

Do Exercise 11.

    We can now see why we exclude 1 as a base for an exponential function. Consider

$$f(x) = a^x = 1^x = 1.$$

The graph of f is the horizontal line $y = 1$. The graph is not one-to-one. The function does not have an inverse that is a function. All other positive bases yield exponential functions that are one-to-one.

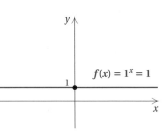

CALCULATOR CORNER

Graphing an Inverse Function   The DRAWINV operation can be used to graph a function and its inverse on the same screen. A formula for the inverse function need not be found in order to do this. The graphing calculator must be set in FUNC mode when this operation is used.

    To graph $f(x) = 2x - 3$ and $f^{-1}(x)$ using the same set of axes, we first clear any existing equations on the equation-editor screen and then enter $y_1 = 2x - 3$. Now, we press **2ND** (DRAW) **8** to select the DRAWINV operation. (DRAW is the second operation associated with the **PRGM** key.) We press **VARS** ▶ **1** **1** to indicate that we want to graph the inverse of y_1. Then we press **ENTER** to see the graph of the function and its inverse. The graphs are shown here in a squared window.

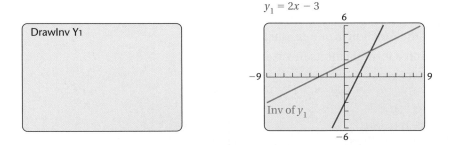

Exercises: Use the DRAWINV operation on a graphing calculator to graph each function with its inverse on the same screen.

1. $f(x) = x - 5$

2. $f(x) = \frac{2}{3}x$

3. $f(x) = x^2 + 2$

4. $f(x) = x^3 - 3$

---

d Composite Functions

In the real world, functions frequently occur in which some quantity depends on a variable that, in turn, depends on another variable. For instance, the number of employees hired by a firm may depend on the firm's profits, which may in turn depend on the number of items the firm produces. Such functions are called **composite functions.**

For example, the function g that gives a correspondence between women's shoe sizes in the United States and those in Italy is given by $g(x) = 2x + 24$, where x is the U.S. size and $g(x)$ is the Italian size. Thus a U.S. size 4 corresponds to a shoe size of $g(4) = 2 \cdot 4 + 24$, or 32, in Italy.

There is also a function that gives a correspondence between women's shoe sizes in Italy and those in Britain. The function is given by $f(x) = \frac{1}{2}x - 14$, where x is the Italian size and $f(x)$ is the corresponding British size. Thus an Italian size 32 corresponds to a British size $f(32) = \frac{1}{2}(32) - 14$, or 2.

It seems reasonable to conclude that a shoe size of 4 in the United States corresponds to a size of 2 in Britain and that some function h describes this correspondence. Can we find a formula for h? If we look at the following tables, we might guess that such a formula is $h(x) = x - 2$, and that is indeed correct. But, for more complicated formulas, we would need to use algebra.

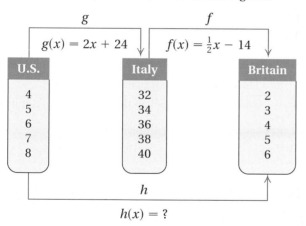

A shoe size x in the United States corresponds to a shoe size $g(x)$ in Italy, where

$$g(x) = 2x + 24.$$

Now $2x + 24$ is a shoe size in Italy. If we replace x in $f(x) = \frac{1}{2}x - 14$ with $g(x)$, or $2x + 24$, we can find the corresponding shoe size in Britain:

$$f(x) = \tfrac{1}{2}x - 14$$
$$f(g(x)) = f(2x + 24) = \tfrac{1}{2}[2x + 24] - 14$$
$$= x + 12 - 14 = x - 2.$$

This gives a formula for h:

$$h(x) = x - 2.$$

Thus a shoe size of 4 in the United States corresponds to a shoe size of $h(4) = 4 - 2$, or 2, in Britain. The function h is the **composition** of f and g, symbolized by $f \circ g$. To find $f \circ g(x)$, we substitute $g(x)$ for x in $f(x)$.

12. Given $f(x) = x + 5$ and $g(x) = x^2 - 1$, find $f \circ g(x)$ and $g \circ f(x)$.

COMPOSITE FUNCTION

The **composite function** $f \circ g$, the **composition** of f and g, is defined as

$$f \circ g(x) = f(g(x)), \quad \text{or} \quad (f \circ g)(x) = f[g(x)].$$

We can visualize the composition of functions as follows.

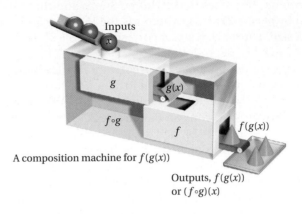

A composition machine for $f(g(x))$

Outputs, $f(g(x))$ or $(f \circ g)(x)$

EXAMPLE 10 Given $f(x) = 3x$ and $g(x) = 1 + x^2$:

a) Find $f \circ g(5)$ and $g \circ f(5)$.

b) Find $f \circ g(x)$ and $g \circ f(x)$.

We consider each function separately:

$f(x) = 3x$ This function multiplies each input by 3.

and $g(x) = 1 + x^2$.   This function adds 1 to the square of each input.

a) $f \circ g(5) = f(g(5)) = f(1 + 5^2) = f(26) = 3(26) = 78;$

$g \circ f(5) = g(f(5)) = g(3 \cdot 5) = g(15) = 1 + 15^2 = 1 + 225 = 226$

b) $f \circ g(x) = f(g(x))$

$\qquad\qquad = f(1 + x^2)$ Substituting $1 + x^2$ for x

$\qquad\qquad = 3(1 + x^2)$

$\qquad\qquad = 3 + 3x^2;$

$g \circ f(x) = g(f(x))$

$\qquad\qquad = g(3x)$ Substituting $3x$ for x

$\qquad\qquad = 1 + (3x)^2$

$\qquad\qquad = 1 + 9x^2$

We can check the values in part (a) with the formulas found in part (b):

$f \circ g(x) = 3 + 3x^2$	$g \circ f(x) = 1 + 9x^2$
$f \circ g(5) = 3 + 3 \cdot 5^2$	$g \circ f(5) = 1 + 9 \cdot 5^2$
$\qquad = 3 + 3 \cdot 25$	$\qquad = 1 + 9 \cdot 25$
$\qquad = 3 + 75$	$\qquad = 1 + 225$
$\qquad = 78;$	$\qquad = 226.$

Example 10 shows that $f \circ g(5) \neq g \circ f(5)$ and, in general, $f \circ g(x) \neq g \circ f(x)$.

Do Exercise 12.

Answer on page A-45

EXAMPLE 11 Given $f(x) = \sqrt{x}$ and $g(x) = x - 1$, find $f \circ g(x)$ and $g \circ f(x)$.

$$f \circ g(x) = f(g(x)) = f(x - 1) = \sqrt{x - 1};$$
$$g \circ f(x) = g(f(x)) = g(\sqrt{x}) = \sqrt{x} - 1$$

Do Exercise 13.

It is important to be able to recognize how a function can be expressed, or "broken down," as a composition. Such a situation can occur in a study of calculus.

EXAMPLE 12 Find $f(x)$ and $g(x)$ such that $h(x) = f \circ g(x)$:

$$h(x) = (7x + 3)^2.$$

This is $7x + 3$ to the 2nd power. Two functions that can be used for the composition are $f(x) = x^2$ and $g(x) = 7x + 3$. We can check by forming the composition:

$$h(x) = f \circ g(x) = f(g(x)) = f(7x + 3) = (7x + 3)^2.$$

This is the most "obvious" answer to the question. There can be other less obvious answers. For example, if

$$f(x) = (x - 1)^2 \quad \text{and} \quad g(x) = 7x + 4,$$

then $\quad h(x) = f \circ g(x) = f(g(x)) = f(7x + 4) = (7x + 4 - 1)^2 = (7x + 3)^2.$

Do Exercise 14.

e Inverse Functions and Composition

Suppose that we used some input x for the function f and found its output, $f(x)$. The function f^{-1} would then take that output back to x. Similarly, if we began with an input x for the function f^{-1} and found its output, $f^{-1}(x)$, the original function f would then take that output back to x.

> If a function f is one-to-one, then f^{-1} is the unique function for which
> $$f^{-1} \circ f(x) = x \quad \text{and} \quad f \circ f^{-1}(x) = x.$$

EXAMPLE 13 Let $f(x) = 2x - 3$. Use composition to show that

$$f^{-1}(x) = \frac{x + 3}{2}. \qquad \text{(See Example 7.)}$$

We find $f^{-1} \circ f(x)$ and $f \circ f^{-1}(x)$ and check to see that each is x.

$$\begin{aligned} f^{-1} \circ f(x) &= f^{-1}(f(x)) \\ &= f^{-1}(2x - 3) \\ &= \frac{(2x - 3) + 3}{2} \\ &= \frac{2x}{2} \\ &= x; \end{aligned} \qquad \begin{aligned} f \circ f^{-1}(x) &= f(f^{-1}(x)) \\ &= f\left(\frac{x + 3}{2}\right) \\ &= 2 \cdot \frac{x + 3}{2} - 3 \\ &= x + 3 - 3 \\ &= x \end{aligned}$$

Do Exercise 15.

13. Given $f(x) = 4x + 5$ and $g(x) = \sqrt[3]{x}$, find $f \circ g(x)$ and $g \circ f(x)$.

14. Find $f(x)$ and $g(x)$ such that $h(x) = f \circ g(x)$. Answers may vary.

a) $h(x) = \sqrt[3]{x^2 + 1}$

b) $h(x) = \dfrac{1}{(x + 5)^4}$

15. Let $f(x) = \frac{2}{3}x - 4$. Use composition to show that

$$f^{-1}(x) = \frac{3x + 12}{2}.$$

Answers on page A-45

Composition of Functions   We can evaluate composite functions on a graphing calculator. For example, given that $f(x) = 2x - 5$ and $g(x) = x^2 - 3x + 8$, we can find $f \circ g(7)$ and $g \circ f(7)$.

On the equation-editor screen, we enter $y_1 = 2x - 5$ and $y_2 = x^2 - 3x + 8$. Then $f \circ g(7) = y_1 \circ y_2(7)$, or $y_1(y_2(7))$ and $g \circ f(7) = y_2 \circ y_1(7)$, or $y_2(y_1(7))$. To find these function values, we press ⟨2ND⟩ ⟨QUIT⟩ to go to the home screen. Then we enter $y_1(y_2(7))$ by pressing ⟨VARS⟩ ⟨▷⟩ ⟨1⟩ ⟨1⟩ ⟨(⟩ ⟨VARS⟩ ⟨▷⟩ ⟨1⟩ ⟨2⟩ ⟨(⟩ ⟨7⟩ ⟨)⟩ ⟨)⟩ ⟨ENTER⟩. Now we enter $y_2(y_1(7))$ by pressing ⟨VARS⟩ ⟨▷⟩ ⟨1⟩ ⟨2⟩ ⟨(⟩ ⟨VARS⟩ ⟨▷⟩ ⟨1⟩ ⟨1⟩ ⟨(⟩ ⟨7⟩ ⟨)⟩ ⟨)⟩ ⟨ENTER⟩. We see that $y_1(y_2(7))$, or $f \circ g(7)$, is 67 and $y_2(y_1(7))$, or $g \circ f(7)$, is 62.

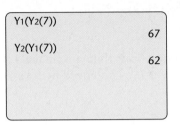

```
Y₁(Y₂(7))
                    67
Y₂(Y₁(7))
                    62
```

Exercises:

1. Use a graphing calculator to check Example 10(a).

2. Given $f(x) = x + 3$ and $g(x) = 2x^2 - 1$, use a graphing calculator to find $f \circ g(-1)$ and $g \circ f(-1)$.

3. Given $f(x) = \sqrt{x}$ and $g(x) = x^2$, use a graphing calculator to find $f \circ g(9)$ and $g \circ f(9)$.

4. Given $f(x) = x^3$ and $g(x) = x - 2$, use a graphing calculator to find $f \circ g(3)$ and $g \circ f(3)$.

Study Tips

BEGINNING TO STUDY FOR THE FINAL EXAM (PART 3): THREE DAYS TO TWO WEEKS OF STUDY TIME

1. **Begin by browsing through each chapter, reviewing the highlighted or boxed information regarding important formulas in both the text and the Summary and Review.**   There may be some formulas that you will need to memorize. Summarize them on an index card and quiz yourself frequently.

2. **Retake each chapter test that you took in class, assuming your instructor has returned it. Otherwise, use the chapter test in the book.**   Restudy the objectives in the text that correspond to each question you missed.

3. **Work the Cumulative Review/Final Examination during the last couple of days before the final.**   Set aside the same amount of time that you will have for the final. See how much of the final exam you can complete under test-like conditions. Be careful to avoid any questions corresponding to objectives not covered. Again, restudy the objectives in the text that correspond to each question you missed.

4. **For remaining difficulties, see your instructor, go to a tutoring session, or participate in a study group.**

"It is a great piece of skill to know how to guide your luck, even while waiting for it."

Baltasar Gracian,
seventeenth-century Spanish philosopher and writer

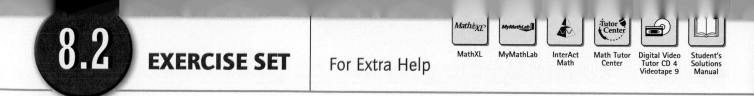
a Find the inverse of the relation. Graph the original relation in red and then graph the inverse relation in blue.

1. $\{(1, 2), (6, -3), (-3, -5)\}$

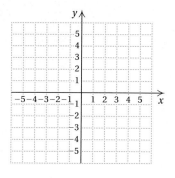

2. $\{(3, -1), (5, 2), (5, -3), (2, 0)\}$

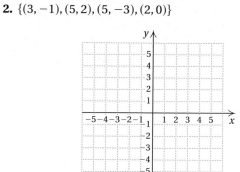

Find an equation of the inverse of the relation. Then complete the second table and graph both the original relation and its inverse.

3. $y = 2x + 6$

x	y
−1	4
0	6
1	8
2	10
3	12

x	y
4	
6	
8	
10	
12	

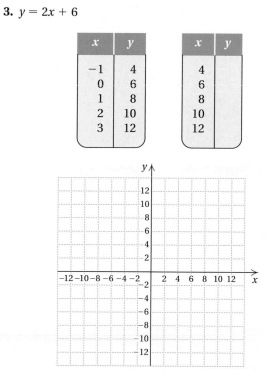

4. $y = \frac{1}{2}x^2 - 8$

x	y
−4	0
−2	−6
0	−8
2	−6
4	0

x	y
0	
−6	
−8	
−6	
0	

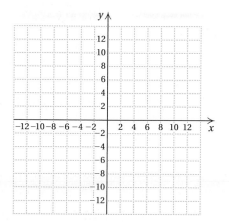

b Determine whether the function is one-to-one.

5. $f(x) = x - 5$

6. $f(x) = 3 - 6x$

7. $f(x) = x^2 - 2$

8. $f(x) = 4 - x^2$

9. $f(x) = |x| - 3$

10. $f(x) = |x - 2|$

11. $f(x) = 3^x$

12. $f(x) = \left(\frac{1}{2}\right)^x$

C Determine whether the function is one-to-one. If it is, find a formula for its inverse.

13. $f(x) = 5x - 2$

14. $f(x) = 4 + 7x$

15. $f(x) = \dfrac{-2}{x}$

16. $f(x) = \dfrac{1}{x}$

17. $f(x) = \frac{4}{3}x + 7$

18. $f(x) = -\frac{7}{8}x + 2$

19. $f(x) = \dfrac{2}{x + 5}$

20. $f(x) = \dfrac{1}{x - 8}$

21. $f(x) = 5$

22. $f(x) = -2$

23. $f(x) = \dfrac{2x + 1}{5x + 3}$

24. $f(x) = \dfrac{2x - 1}{5x + 3}$

25. $f(x) = x^3 - 1$

26. $f(x) = x^3 + 5$

27. $f(x) = \sqrt[3]{x}$

28. $f(x) = \sqrt[3]{x - 4}$

Graph the function and its inverse using the same set of axes.

29. $f(x) = \frac{1}{2}x - 3$,

$f^{-1}(x) =$ _____

x	$f(x)$

x	$f^{-1}(x)$

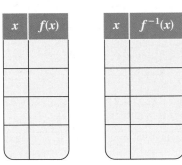

30. $g(x) = x + 4$,

$g^{-1}(x) =$ _____

x	$g(x)$

x	$g^{-1}(x)$

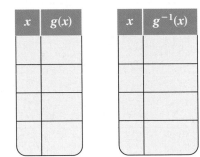

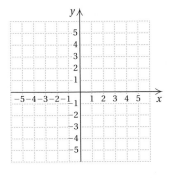

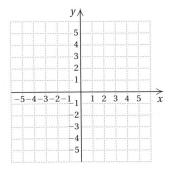

31. $f(x) = x^3$,

$f^{-1}(x) = $ _____

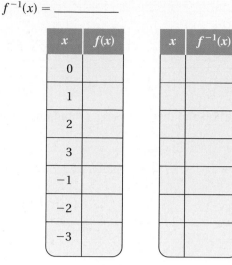

x	$f(x)$
0	
1	
2	
3	
−1	
−2	
−3	

x	$f^{-1}(x)$

32. $f(x) = x^3 - 1$,

$f^{-1}(x) = $ _____

x	$f(x)$
0	
1	
2	
3	
−1	
−2	
−3	

x	$f^{-1}(x)$

d Find $f \circ g(x)$ and $g \circ f(x)$.

33. $f(x) = 2x - 3$,
$g(x) = 6 - 4x$

34. $f(x) = 9 - 6x$,
$g(x) = 0.37x + 4$

35. $f(x) = 3x^2 + 2$,
$g(x) = 2x - 1$

36. $f(x) = 4x + 3$,
$g(x) = 2x^2 - 5$

37. $f(x) = 4x^2 - 1$,
$g(x) = \dfrac{2}{x}$

38. $f(x) = \dfrac{3}{x}$,
$g(x) = 2x^2 + 3$

39. $f(x) = x^2 + 5$,
$g(x) = x^2 - 5$

40. $f(x) = \dfrac{1}{x^2}$,
$g(x) = x - 1$

Find $f(x)$ and $g(x)$ such that $h(x) = f \circ g(x)$. Answers may vary.

41. $h(x) = (5 - 3x)^2$

42. $h(x) = 4(3x - 1)^2 + 9$

43. $h(x) = \sqrt{5x + 2}$

44. $h(x) = (3x^2 - 7)^5$

45. $h(x) = \dfrac{1}{x - 1}$

46. $h(x) = \dfrac{3}{x} + 4$

47. $h(x) = \dfrac{1}{\sqrt{7x + 2}}$

48. $h(x) = \sqrt{x - 7} - 3$

49. $h(x) = \left(\sqrt{x} + 5\right)^4$

50. $h(x) = \dfrac{x^3 + 1}{x^3 - 1}$

For each function, use composition to show that the inverse is correct.

51. $f(x) = \frac{4}{5}x$,
$f^{-1}(x) = \frac{5}{4}x$

52. $f(x) = \frac{x+7}{2}$,
$f^{-1}(x) = 2x - 7$

53. $f(x) = \frac{1-x}{x}$,
$f^{-1}(x) = \frac{1}{x+1}$

54. $f(x) = x^3 - 5$,
$f^{-1}(x) = \sqrt[3]{x+5}$

Find the inverse of the given function by thinking about the operations of the function and then reversing, or undoing, them. Then use composition to show whether the inverse is correct.

Function	*Inverse*
55. $f(x) = 3x$	$f^{-1}(x) = $ _____

Function	*Inverse*
56. $f(x) = \frac{1}{4}x + 7$	$f^{-1}(x) = $ _____

57. $f(x) = -x$ $f^{-1}(x) = $ _____

58. $f(x) = \sqrt[3]{x} - 5$ $f^{-1}(x) = $ _____

59. $f(x) = \sqrt[3]{x-5}$ $f^{-1}(x) = $ _____

60. $f(x) = x^{-1}$ $f^{-1}(x) = $ _____

61. *Dress Sizes in the United States and France.* A size-6 dress in the United States is size 38 in France. A function that converts dress sizes in the United States to those in France is

$$f(x) = x + 32.$$

a) Find the dress sizes in France that correspond to sizes of 8, 10, 14, and 18 in the United States.
b) Determine whether this function has an inverse that is a function. If so, find a formula for the inverse.
c) Use the inverse function to find dress sizes in the United States that correspond to sizes of 40, 42, 46, and 50 in France.

62. *Dress Sizes in the United States and Italy.* A size-6 dress in the United States is size 36 in Italy. A function that converts dress sizes in the United States to those in Italy is

$$f(x) = 2(x + 12).$$

a) Find the dress sizes in Italy that correspond to sizes of 8, 10, 14, and 18 in the United States.
b) Determine whether this function has an inverse that is a function. If so, find a formula for the inverse.
c) Use the inverse function to find dress sizes in the United States that correspond to sizes of 40, 44, 52, and 60 in Italy.

63. **D**_W The function $V(t) = 750(1.2)^t$ is used to predict the value V of a certain rare stamp t years from 1999. Do not calculate $V^{-1}(t)$ but explain how V^{-1} could be used.

64. **D**_W An organization determines that the cost per person of chartering a bus is given by the function

$$C(x) = \frac{100 + 5x}{x},$$

where x is the number of people in the group and $C(x)$ is in dollars. Determine $C^{-1}(x)$ and explain how this inverse function could be used.

Use rational exponents to simplify. [6.2d]

65. $\sqrt[6]{a^2}$　　　　**66.** $\sqrt[6]{x^4}$　　　　**67.** $\sqrt{a^4b^6}$　　　　**68.** $\sqrt[3]{8t^6}$

69. $\sqrt[8]{81}$　　　　**70.** $\sqrt[4]{32}$　　　　**71.** $\sqrt[12]{64x^6y^6}$　　　　**72.** $\sqrt[8]{p^4t^2}$

73. $\sqrt[5]{32a^{15}b^{40}}$　　　　**74.** $\sqrt[3]{1000x^9y^{18}}$　　　　**75.** $\sqrt[4]{81a^8b^8}$　　　　**76.** $\sqrt[3]{27p^3q^9}$

In Exercises 77–80, use a graphing calculator to help determine whether or not the given functions are inverses of each other.

77. $f(x) = 0.75x^2 + 2; \quad g(x) = \sqrt{\dfrac{4(x-2)}{3}}$

78. $f(x) = 1.4x^3 + 3.2; \quad g(x) = \sqrt[3]{\dfrac{x - 3.2}{1.4}}$

79. $f(x) = \sqrt{2.5x + 9.25}; \quad g(x) = 0.4x^2 - 3.7, x \geq 0$

80. $f(x) = 0.8x^{1/2} + 5.23; \quad g(x) = 1.25(x^2 - 5.23), x \geq 0$

81. Use a graphing calculator to help match each function in Column A with its inverse from Column B.

Column A

(1) $y = 5x^3 + 10$

(2) $y = (5x + 10)^3$

(3) $y = 5(x + 10)^3$

(4) $y = (5x)^3 + 10$

Column B

A. $y = \dfrac{\sqrt[3]{x} - 10}{5}$

B. $y = \sqrt[3]{\dfrac{x}{5}} - 10$

C. $y = \sqrt[3]{\dfrac{x - 10}{5}}$

D. $y = \dfrac{\sqrt[3]{x - 10}}{5}$

In Exercises 82 and 83, graph the inverse of f.

82.

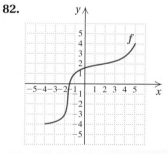

83.

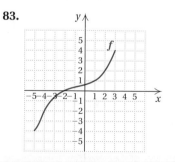

84. Examine the following table. Does it appear that f and g could be inverses of each other? Why or why not?

x	$f(x)$	$g(x)$
6	6	6
7	6.5	8
8	7	10
9	7.5	12
10	8	14
11	8.5	16
12	9	18

85. Assume in Exercise 84 that f and g are both linear functions. Find equations for $f(x)$ and $g(x)$. Are f and g inverses of each other?

Objectives

a Graph logarithmic functions.

b Convert from exponential equations to logarithmic equations and from logarithmic equations to exponential equations.

c Solve logarithmic equations.

d Find common logarithms on a calculator.

---

1. Write the meaning of $\log_2 64$. Then find $\log_2 64$.

We are now ready to study inverses of exponential functions. These functions have many applications and are referred to as *logarithm*, or *logarithmic, functions*.

a Graphing Logarithmic Functions

Consider the exponential function $f(x) = 2^x$. Like all exponential functions, f is one-to-one. Can a formula for f^{-1} be found? To answer this, we use the method of Section 8.2:

1. Replace $f(x)$ with y: $y = 2^x$.

2. Interchange x and y: $x = 2^y$.

3. Solve for y: $y =$ the power to which we raise 2 to get x.

4. Replace y with $f^{-1}(x)$: $f^{-1}(x) =$ the power to which we raise 2 to get x.

We now define a new symbol to replace the words "the power to which we raise 2 to get x."

MEANING OF LOGARITHMS

$\log_2 x$, read "the logarithm, base 2, of x," or "log, base 2, of x," means "the power to which we raise 2 to get x."

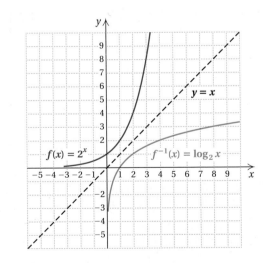

Thus if $f(x) = 2^x$, then $f^{-1}(x) = \log_2 x$. Note that $f^{-1}(8) = \log_2 8 = 3$, because 3 is the *power to which we raise 2 to get* 8; that is, $2^3 = 8$.

Although expressions like $\log_2 13$ can only be approximated, remember that $\log_2 13$ represents the **power** *to which we raise* 2 *to get* 13. That is, $2^{\log_2 13} = 13$.

Answer on page A-47

Do Exercise 1.

For any exponential function $f(x) = a^x$, the inverse is called a **logarithmic function, base a.** The graph of the inverse can, of course, be drawn by reflecting the graph of $f(x) = a^x$ across the line $y = x$. It will be helpful to remember that the inverse of $f(x) = a^x$ is given by $f^{-1}(x) = \log_a x$. Normally, we use a number a that is greater than 1 for the logarithm base.

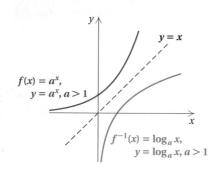

LOGARITHMS

The inverse of $f(x) = a^x$ is given by

$$f^{-1}(x) = \log_a x.$$

We read "$\log_a x$" as "the logarithm, base a, of x." We define $y = \log_a x$ as that number y such that $a^y = x$, where $x > 0$ and a is a positive constant other than 1.

It is helpful in dealing with logarithmic functions to remember that the logarithm of a number is an **exponent.** It is the exponent y in $x = a^y$. Keep thinking, "The logarithm, base a, of a number x is the power to which a must be raised in order to get x."

EXPONENTIAL FUNCTION	LOGARITHMIC FUNCTION
$y = a^x$	$x = a^y$
$f(x) = a^x$	$f^{-1}(x) = \log_a x$
$a > 0, a \neq 1$	$a > 0, a \neq 1$
Domain = The set of real numbers	Range = The set of real numbers
Range = The set of positive numbers	Domain = The set of positive numbers

Why do we exclude 1 from being a logarithm base? See the graph below. If we allow 1 as a logarithm base, the graph of the relation $y = \log_1 x$, or $x = 1^y = 1$, is a vertical line, which is not a function and therefore not a logarithmic function.

2. Graph: $y = f(x) = \log_3 x$.

EXAMPLE 1 Graph: $y = f(x) = \log_5 x$.

The equation $y = \log_5 x$ is equivalent to $5^y = x$. We can find ordered pairs that are solutions by choosing values for y and computing the corresponding x-values.

For $y = 0$, $x = 5^0 = 1$.

For $y = 1$, $x = 5^1 = 5$.

For $y = 2$, $x = 5^2 = 25$.

For $y = 3$, $x = 5^3 = 125$.

For $y = -1$, $x = 5^{-1} = \dfrac{1}{5}$.

For $y = -2$, $x = 5^{-2} = \dfrac{1}{25}$.

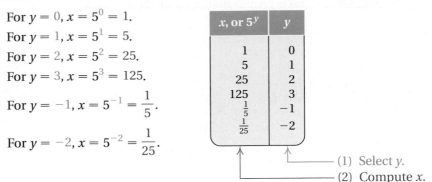

x, or 5^y	y
1	0
5	1
25	2
125	3
$\frac{1}{5}$	-1
$\frac{1}{25}$	-2

(1) Select y.

(2) Compute x.

The table shows the following:

$$\left.\begin{array}{l} \log_5 1 = 0; \\ \log_5 5 = 1; \\ \log_5 25 = 2; \\ \log_5 \frac{1}{5} = -1; \\ \log_5 \frac{1}{25} = -2. \end{array}\right\}$$ These can all be checked using the equations above.

We plot the ordered pairs and connect them with a smooth curve. The graph of $y = 5^x$ has been shown only for reference.

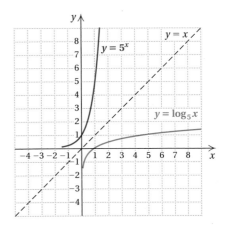

Do Exercise 2.

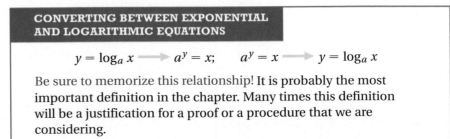

b   Converting Between Exponential Equations and Logarithmic Equations

We use the definition of logarithms to convert from exponential equations to logarithmic equations.

> **CONVERTING BETWEEN EXPONENTIAL AND LOGARITHMIC EQUATIONS**
>
> $$y = \log_a x \longrightarrow a^y = x; \qquad a^y = x \longrightarrow y = \log_a x$$
>
> Be sure to memorize this relationship! It is probably the most important definition in the chapter. Many times this definition will be a justification for a proof or a procedure that we are considering.

Answer on page A-47

EXAMPLES Convert to a logarithmic equation.

2. $8 = 2^x \longrightarrow x = \log_2 8$ The exponent is the logarithm.

The base remains the same.

3. $y^{-1} = 4 \longrightarrow -1 = \log_y 4$

4. $a^b = c \longrightarrow b = \log_a c$

Do Exercises 3–6.

We also use the definition of logarithms to convert from logarithmic equations to exponential equations.

EXAMPLES Convert to an exponential equation.

5. $y = \log_3 5 \longrightarrow 3^y = 5$ The logarithm is the exponent.

The base does not change.

6. $-2 = \log_a 7 \longrightarrow a^{-2} = 7$

7. $a = \log_b d \longrightarrow b^a = d$

Do Exercises 7–10.

C Solving Certain Logarithmic Equations

Certain equations involving logarithms can be solved by first converting to exponential equations. We will solve more complicated equations later.

EXAMPLE 8 Solve: $\log_2 x = -3$.

$$\log_2 x = -3$$
$$2^{-3} = x \qquad \text{Converting to an exponential equation}$$
$$\frac{1}{2^3} = x$$
$$\frac{1}{8} = x$$

Check: $\log_2 \frac{1}{8}$ is the exponent to which we raise 2 to get $\frac{1}{8}$. Since $2^{-3} = \frac{1}{8}$, we know that $\frac{1}{8}$ checks and is the solution.

EXAMPLE 9 Solve: $\log_x 16 = 2$.

$$\log_x 16 = 2$$
$$x^2 = 16 \qquad \text{Converting to an exponential equation}$$
$$x = 4 \quad \text{or} \quad x = -4 \qquad \text{Using the principle of square roots}$$

Check: $\log_4 16 = 2$ because $4^2 = 16$. Thus, 4 is a solution. Since all logarithm bases must be positive, $\log_{-4} 16$ is not defined. Therefore, -4 is not a solution.

Do Exercises 11–13 on the following page.

Convert to a logarithmic equation.

3. $6^0 = 1$

4. $10^{-3} = 0.001$

5. $16^{0.25} = 2$

6. $m^T = P$

Convert to an exponential equation.

7. $\log_2 32 = 5$

8. $\log_{10} 1000 = 3$

9. $\log_a Q = 7$

10. $\log_t M = x$

Answers on page A-47

Solve.

11. $\log_{10} x = 4$

12. $\log_x 81 = 4$

13. $\log_2 x = -2$

Find each of the following.

14. $\log_{10} 10,000$

15. $\log_{10} 0.0001$

16. $\log_7 1$

Answers on page A-47

To think of finding logarithms as solving equations may help in some cases.

EXAMPLE 10   Find $\log_{10} 1000$.

METHOD 1.   Let $\log_{10} 1000 = x$. Then

$$10^x = 1000 \qquad \text{Converting to an exponential equation}$$
$$10^x = 10^3$$
$$x = 3. \qquad \text{The exponents are the same.}$$

Therefore, $\log_{10} 1000 = 3$.

METHOD 2.   Think of the meaning of $\log_{10} 1000$. It is the exponent to which we raise 10 to get 1000. That exponent is 3. Therefore, $\log_{10} 1000 = 3$.

EXAMPLE 11   Find $\log_{10} 0.01$.

METHOD 1.   Let $\log_{10} 0.01 = x$. Then

$$10^x = 0.01 \qquad \text{Converting to an exponential equation}$$
$$10^x = \frac{1}{100}$$
$$10^x = 10^{-2}$$
$$x = -2. \qquad \text{The exponents are the same.}$$

Therefore, $\log_{10} 0.01 = -2$.

METHOD 2. $\log_{10} 0.01$ is the exponent to which we raise 10 to get 0.01. Noting that

$$0.01 = \frac{1}{100} = \frac{1}{10^2} = 10^{-2},$$

we see that the exponent is -2. Therefore, $\log_{10} 0.01 = -2$.

EXAMPLE 12   Find $\log_5 1$.

METHOD 1.   Let $\log_5 1 = x$. Then

$$5^x = 1 \qquad \text{Converting to an exponential equation}$$
$$5^x = 5^0$$
$$x = 0. \qquad \text{The exponents are the same.}$$

Therefore, $\log_5 1 = 0$.

METHOD 2. $\log_5 1$ is the exponent to which we raise 5 to get 1. That exponent is 0. Therefore, $\log_5 1 = 0$.

Do Exercises 14–16.

THE LOGARITHM OF 1
For any base a,
$\qquad \log_a 1 = 0$.
The logarithm, base a, of 1 is always 0.

The proof follows from the fact that $a^0 = 1$. This is equivalent to the logarithmic equation $\log_a 1 = 0$.

Another property follows similarly. We know that $a^1 = a$ for any real number a. In particular, it holds for any positive number a. This is equivalent to the logarithmic equation $\log_a a = 1$.

THE LOGARITHM, BASE a, OF a

For any base a,
$$\log_a a = 1.$$

 EXAMPLE 13 Simplify: $\log_m 1$ and $\log_t t$.

$$\log_m 1 = 0; \qquad \log_t t = 1$$

Do Exercises 17–20.

d | Finding Common Logarithms on a Calculator

Base-10 logarithms are called **common logarithms.** Before calculators became so widely available, common logarithms were used extensively to do complicated calculations. In fact, that is why logarithms were invented. The abbreviation **log,** with no base written, is used for the common logarithm, base-10. Thus,

$$\log 29 \quad \text{means} \quad \log_{10} 29.$$

> Be sure to memorize $\log a = \log_{10} a$.

We can approximate $\log 29$. Note the following:

$$\left.\begin{array}{l} \log 100 = \log_{10} 100 = 2; \\ \\ \qquad\qquad \log 29 = ?; \\ \\ \log 10 = \log_{10} 10 = 1. \end{array}\right\} \quad \begin{array}{l} \text{It seems reasonable that } \log 29 \\ \text{is between 1 and 2.} \end{array}$$

On a scientific or graphing calculator, the key for common logarithms is generally marked **LOG** . We find that

$$\log 29 \approx 1.462397998 \approx 1.4624,$$

rounded to four decimal places. This also tells us that $10^{1.4624} \approx 29$.

On some scientific calculators, the keystrokes for doing such a calculation might be

(2) (9) **LOG** =. The display would then read 1.462398.

Using a graphing calculator, the keystrokes might be

LOG (2) (9) **ENTER** . The display would then read 1.462397998.

 EXAMPLES Find each common logarithm, to four decimal places, on a scientific or graphing calculator.

Function Value	Readout	Rounded
14. log 287,523	5.458672591	5.4587
15. log 0.000486	−3.313363731	−3.3134
16. log (−5)	NONREAL ANS	NONREAL ANS

Simplify.

17. $\log_3 1$

18. $\log_3 3$

19. $\log_c c$

20. $\log_c 1$

Find the common logarithm, to four decimal places, on a scientific or graphing calculator.

21. log 78,235.4

22. log 0.0000309

23. log (−3)

24. Find

log 1000 and log 10,000

without using a calculator. Between what two whole numbers is log 9874? Then on a calculator approximate log 9874, rounded to four decimal places.

Answers on page A-47

25. Complete the following table to express each number in the first column as a power of 10. Round each exponent to the nearest ten-thousandth.

NUMBER	EXPRESSED AS A POWER OF 10
8	
947	
634,567	
0.00708	
0.000778899	
18,600,000	
1860	

26. Find $10^{4.8934}$ using a calculator. (Compare your computation to that of Margin Exercise 21.)

CHAPTER 8: Exponential and Logarithmic Functions

In Example 16, log (−5) does not exist as a real number because there is no real-number power to which we can raise 10 to get −5. The number 10 raised to any power is nonnegative. The logarithm of a negative number does not exist as a real number (though it can be defined as a complex number).

Do Exercises 21–24 on the preceding page.

We can use common logarithms to express any positive number as a power of 10. We simply find the common logarithm of the number on a calculator. Considering very large or very small numbers as powers of 10 might be a helpful way to compare those numbers.

EXAMPLE 17 Complete the following table to express each number in the first column as a power of 10. Round each exponent to the nearest ten-thousandth.

We simply find the common logarithm of the number using a calculator.

NUMBER	EXPRESSED AS A POWER OF 10
4	$4 = 10^{0.6021}$
625	$625 = 10^{2.7959}$
134,567	$134,567 = 10^{5.1289}$
0.00567	$0.00567 = 10^{-2.2464}$
0.000374859	$0.000374859 = 10^{-3.4261}$
186,000	$186,000 = 10^{5.2695}$
186,000,000	$186,000,000 = 10^{8.2695}$

Do Exercise 25.

The inverse of a logarithmic function is an exponential function. Thus, if $f(x) = \log x$, then $f^{-1}(x) = 10^x$. Because of this, on many calculators, the **LOG** key doubles as the $\boxed{10^x}$ key after a **2ND** or $\boxed{\text{SHIFT}}$ key has been pressed. To find $10^{5.4587}$ on a scientific calculator, we might enter 5.4587 and press $\boxed{10^x}$. On many graphing calculators, we press **2ND** $\boxed{10^x}$, followed by 5.4587. In either case, we get the approximation

$$10^{5.4587} \approx 287,541.1465.$$

Compare this computation to Example 14. Note that, apart from the rounding error, $10^{5.4587}$ takes us back to about 287,523.

Do Exercise 26.

Using the scientific keys on a calculator would allow us to construct a graph of $f(x) = \log_{10} x = \log x$ by finding function values directly, rather than converting to exponential form as we did in Example 1.

x	$f(x)$
0.5	−0.3010
1	0
2	0.3010
3	0.4771
5	0.6990
9	0.9542
10	1

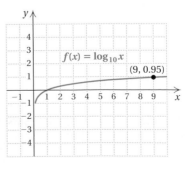

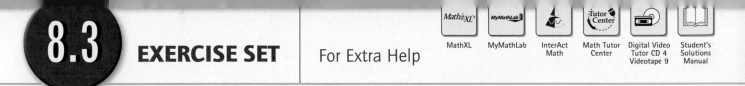

8.3

EXERCISE SET

For Extra Help

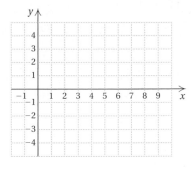

a Graph.

1. $f(x) = \log_2 x$, or $y = \log_2 x$

$y = \log_2 x \longrightarrow x = $ _____

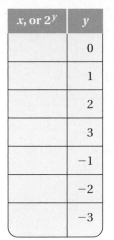

x, or 2^y	y
	0
	1
	2
	3
	-1
	-2
	-3

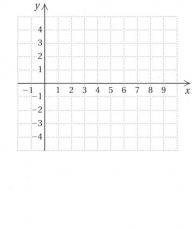

2. $f(x) = \log_{10} x$, or $y = \log_{10} x$

$y = \log_{10} x \longrightarrow x = $ _____

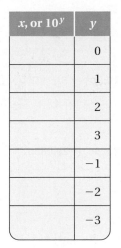

x, or 10^y	y
	0
	1
	2
	3
	-1
	-2
	-3

3. $f(x) = \log_{1/3} x$

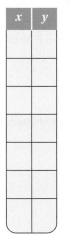

x	y

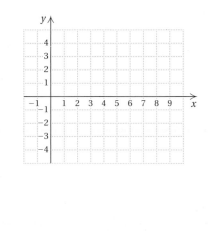

4. $f(x) = \log_{1/2} x$

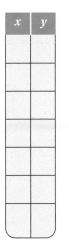

x	y

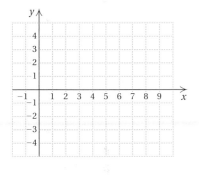

Graph both functions using the same set of axes.

5. $f(x) = 3^x$, $\quad f^{-1}(x) = \log_3 x$

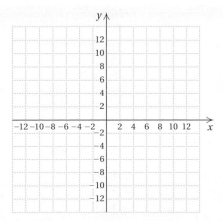

6. $f(x) = 4^x$, $\quad f^{-1}(x) = \log_4 x$

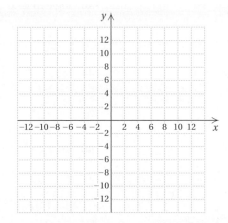

Convert to a logarithmic equation.

7. $10^3 = 1000$

8. $10^2 = 100$

9. $5^{-3} = \dfrac{1}{125}$

10. $4^{-5} = \dfrac{1}{1024}$

11. $8^{1/3} = 2$

12. $16^{1/4} = 2$

13. $10^{0.3010} = 2$

14. $10^{0.4771} = 3$

15. $e^2 = t$

16. $p^k = 3$

17. $Q^t = x$

18. $P^m = V$

19. $e^2 = 7.3891$

20. $e^3 = 20.0855$

21. $e^{-2} = 0.1353$

22. $e^{-4} = 0.0183$

Convert to an exponential equation.

23. $w = \log_4 10$

24. $t = \log_5 9$

25. $\log_6 36 = 2$

26. $\log_7 7 = 1$

27. $\log_{10} 0.01 = -2$

28. $\log_{10} 0.001 = -3$

29. $\log_{10} 8 = 0.9031$

30. $\log_{10} 2 = 0.3010$

31. $\log_e 100 = 4.6052$

32. $\log_e 10 = 2.3026$

33. $\log_t Q = k$

34. $\log_m P = a$

Solve.

35. $\log_3 x = 2$

36. $\log_4 x = 3$

37. $\log_x 16 = 2$

38. $\log_x 64 = 3$

39. $\log_2 16 = x$

40. $\log_5 25 = x$

41. $\log_3 27 = x$

42. $\log_4 16 = x$

43. $\log_x 25 = 1$

44. $\log_x 9 = 1$

45. $\log_3 x = 0$

46. $\log_2 x = 0$

47. $\log_2 x = -1$

48. $\log_3 x = -2$

49. $\log_8 x = \dfrac{1}{3}$

50. $\log_{32} x = \dfrac{1}{5}$

CHAPTER 8: Exponential and
Logarithmic Functions

Find each of the following.

51. $\log_{10} 100$

52. $\log_{10} 100{,}000$

53. $\log_{10} 0.1$

54. $\log_{10} 0.001$

55. $\log_{10} 1$

56. $\log_{10} 10$

57. $\log_5 625$

58. $\log_2 64$

59. $\log_7 49$

60. $\log_5 125$

61. $\log_2 8$

62. $\log_8 64$

63. $\log_9 \dfrac{1}{81}$

64. $\log_5 \dfrac{1}{125}$

65. $\log_8 1$

66. $\log_6 6$

67. $\log_e e$

68. $\log_e 1$

69. $\log_{27} 9$

70. $\log_8 2$

d   Find the common logarithm, to four decimal places, on a calculator.

71. $\log 78{,}889.2$

72. $\log 9{,}043{,}788$

73. $\log 0.67$

74. $\log 0.0067$

75. $\log (-97)$

76. $\log 0$

77. $\log \left(\dfrac{289}{32.7} \right)$

78. $\log \left(\dfrac{23}{86.2} \right)$

79. Complete the following table to express each number in the first column as a power of 10. Round each exponent to the nearest ten-thousandth.

NUMBER	EXPRESSED AS A POWER OF 10
6	
84	
987,606	
0.00987606	
98,760.6	
70,000,000	
7000	

80. Complete the following table to express each number in the first column as a power of 10. Round each exponent to the nearest ten-thousandth.

NUMBER	EXPRESSED AS A POWER OF 10
7	
314	
31.4	
31,400,000	
0.000314	
3.14	
0.0314	

81. D_W Explain in your own words what is meant by $\log_a b = c$.

82. D_W John Napier (1550–1617) of Scotland is credited by mathematicians as the inventor of logarithms. Make a report on Napier and his work.

VOCABULARY REINFORCEMENT

In each of Exercises 83–90, fill in the blank with the correct term from the given list. Some of the choices may not be used.

83. The _____ of a complex number $a + bi$ is $a - bi$.
[6.8e]

84. If a situation gives rise to a linear function $f(x) = kx$, where k is a positive constant, the situation is an example of _____ variation.   [5.8a]

85. In the polynomial $6x^5 - 2x^2 + 4$, $6x^5$ is called the _____.   [4.1a]

86. The expression $b^2 - 4ac$ in the _____ formula is called the _____.   [7.4a]

87. A system of equations that has no solution is called a(n) _____ system.   [3.1a]

88. Graphs of quadratic functions are called _____.
[7.5a]

89. For the graph of $f(x) = (x - 2)^2 + 4$, the line $x = 2$ is called the _____.   [7.5a]

90. A complex number is any number that can be named _____, where a and b are any real numbers.
[6.8a]

direct

indirect

$b^2 - 4ac$

$a + bi$

discriminant

radical

quadratic

parabolas

polynomials

line of symmetry

conjugate

leading term

leading coefficient

consistent

inconsistent

Graph.

91. $f(x) = \log_3 |x + 1|$

92. $f(x) = \log_2 (x - 1)$

Solve.

93. $\log_{125} x = \frac{2}{3}$

94. $|\log_3 x| = 3$

95. $\log_{128} x = \frac{5}{7}$

96. $\log_4 (3x - 2) = 2$

97. $\log_8 (2x + 1) = -1$

98. $\log_{10} (x^2 + 21x) = 2$

Simplify.

99. $\log_{1/4} \frac{1}{64}$

100. $\log_{81} 3 \cdot \log_3 81$

101. $\log_{10} (\log_4 (\log_3 81))$

102. $\log_2 (\log_2 (\log_4 256))$

103. $\log_{1/5} 25$

8.4 PROPERTIES OF LOGARITHMIC FUNCTIONS

Objectives

a   Express the logarithm of a product as a sum of logarithms, and conversely.

b   Express the logarithm of a power as a product.

c   Express the logarithm of a quotient as a difference of logarithms, and conversely.

d   Convert from logarithms of products, quotients, and powers to expressions in terms of individual logarithms, and conversely.

e   Simplify expressions of the type $\log_a a^k$.

The ability to manipulate logarithmic expressions is important in many applications and in more advanced mathematics. We now establish some basic properties that are useful in manipulating logarithmic expressions.

a   Logarithms of Products

PROPERTY 1: THE PRODUCT RULE

For any positive numbers M and N,

$$\log_a (M \cdot N) = \log_a M + \log_a N.$$

(The logarithm of a product is the sum of the logarithms of the factors. The number a can be any logarithm base.)

EXAMPLE 1   Express as a sum of logarithms: $\log_2 (4 \cdot 16)$.

$$\log_2 (4 \cdot 16) = \log_2 4 + \log_2 16 \qquad \text{By Property 1}$$

EXAMPLE 2   Express as a single logarithm: $\log_{10} 0.01 + \log_{10} 1000$.

$$\log_{10} 0.01 + \log_{10} 1000 = \log_{10} (0.01 \times 1000) \qquad \text{By Property 1}$$
$$= \log_{10} 10$$

Do Exercises 1–4.

A PROOF OF PROPERTY 1 (*OPTIONAL*).   Let $\log_a M = x$ and $\log_a N = y$. Converting to exponential equations, we have $a^x = M$ and $a^y = N$. Then we multiply to obtain

$$M \cdot N = a^x \cdot a^y = a^{x+y}.$$

Converting $M \cdot N = a^{x+y}$ back to a logarithmic equation, we get

$$\log_a (M \cdot N) = x + y.$$

Remembering what x and y represent, we get

$$\log_a (M \cdot N) = \log_a M + \log_a N.$$

b   Logarithms of Powers

PROPERTY 2: THE POWER RULE

For any positive number M and any real number k,

$$\log_a M^k = k \cdot \log_a M.$$

(The logarithm of a power of M is the exponent times the logarithm of M. The number a can be any logarithm base.)

Express as a sum of logarithms.

1. $\log_5 (25 \cdot 5)$

2. $\log_b PQ$

Express as a single logarithm.

3. $\log_3 7 + \log_3 5$

4. $\log_a J + \log_a A + \log_a M$

Answers on page A-47

Express as a product.

5. $\log_7 4^5$

6. $\log_a \sqrt{5}$

7. Express as a difference of logarithms:

$$\log_b \frac{P}{Q}.$$

8. Express as a single logarithm:

$$\log_2 125 - \log_2 25.$$

Answers on page A-47

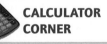
EXAMPLES Express as a product.

3. $\log_a 9^{-5} = -5 \log_a 9$ By Property 2

4. $\log_a \sqrt[4]{5} = \log_a 5^{1/4}$ Writing exponential notation

          $= \frac{1}{4} \log_a 5$ By Property 2

Do Exercises 5 and 6.

A PROOF OF PROPERTY 2 (*OPTIONAL*). Let $x = \log_a M$. Then we convert to an exponential equation to get $a^x = M$. Raising both sides to the kth power, we obtain

$$(a^x)^k = M^k, \quad \text{or} \quad a^{xk} = M^k.$$

Converting back to a logarithmic equation with base a, we get $\log_a M^k = xk$. But $x = \log_a M$, so

$$\log_a M^k = (\log_a M)k = k \cdot \log_a M.$$

C   Logarithms of Quotients

> **PROPERTY 3: THE QUOTIENT RULE**
>
> For any positive numbers M and N,
>
> $$\log_a \frac{M}{N} = \log_a M - \log_a N.$$
>
> (The logarithm of a quotient is the logarithm of the numerator minus the logarithm of the denominator. The number a can be any logarithm base.)

EXAMPLE 5 Express as a difference of logarithms: $\log_t \dfrac{6}{U}$.

$$\log_t \frac{6}{U} = \log_t 6 - \log_t U \qquad \text{By Property 3}$$

EXAMPLE 6 Express as a single logarithm: $\log_b 17 - \log_b 27$.

$$\log_b 17 - \log_b 27 = \log_b \frac{17}{27} \qquad \text{By Property 3}$$

EXAMPLE 7 Express as a single logarithm: $\log_{10} 10{,}000 - \log_{10} 100$.

$$\log_{10} 10{,}000 - \log_{10} 100 = \log_{10} \frac{10{,}000}{100} = \log_{10} 100$$

Do Exercises 7 and 8.

A PROOF OF PROPERTY 3 (*OPTIONAL*). The proof makes use of Property 1 and Property 2.

$$\log_a \frac{M}{N} = \log_a MN^{-1}$$

$$= \log_a M + \log_a N^{-1} \qquad \text{By Property 1}$$

$$= \log_a M + (-1)\log_a N \qquad \text{By Property 2}$$

$$= \log_a M - \log_a N$$

CHAPTER 8: Exponential and
Logarithmic Functions

d Using the Properties Together

Express in terms of logarithms of x, y, z, and w.

EXAMPLES Express in terms of logarithms.

8. $\log_a \dfrac{x^2 y^3}{z^4} = \log_a (x^2 y^3) - \log_a z^4$ Using Property 3

$\qquad\qquad = \log_a x^2 + \log_a y^3 - \log_a z^4$ Using Property 1

$\qquad\qquad = 2 \log_a x + 3 \log_a y - 4 \log_a z$ Using Property 2

9. $\log_a \sqrt{\dfrac{z^3}{xy}}$

9. $\log_a \sqrt[4]{\dfrac{xy}{z^3}} = \log_a \left(\dfrac{xy}{z^3}\right)^{1/4}$ Writing exponential notation

$\qquad\qquad = \tfrac{1}{4} \log_a \dfrac{xy}{z^3}$ Using Property 2

$\qquad\qquad = \tfrac{1}{4}(\log_a xy - \log_a z^3)$ Using Property 3 (note the parentheses)

$\qquad\qquad = \tfrac{1}{4}(\log_a x + \log_a y - 3 \log_a z)$ Using Properties 1 and 2

$\qquad\qquad = \tfrac{1}{4} \log_a x + \tfrac{1}{4} \log_a y - \tfrac{3}{4} \log_a z$ Distributive law

10. $\log_a \dfrac{x^2}{y^3 z}$

10. $\log_b \dfrac{xy}{m^3 n^4} = \log_b xy - \log_b m^3 n^4$ Using Property 3

$\qquad\qquad = (\log_b x + \log_b y) - (\log_b m^3 + \log_b n^4)$ Using Property 1

$\qquad\qquad = \log_b x + \log_b y - \log_b m^3 - \log_b n^4$ Removing parentheses

$\qquad\qquad = \log_b x + \log_b y - 3 \log_b m - 4 \log_b n$ Using Property 2

11. $\log_a \dfrac{x^3 y^4}{z^5 w^9}$

Do Exercises 9–11.

EXAMPLES Express as a single logarithm.

11. $\dfrac{1}{2} \log_a x - 7 \log_a y + \log_a z$

$\qquad = \log_a x^{1/2} - \log_a y^7 + \log_a z$ Using Property 2

$\qquad = \log_a \dfrac{\sqrt{x}}{y^7} + \log_a z$ Using Property 3

$\qquad = \log_a \dfrac{z\sqrt{x}}{y^7}$ Using Property 1

Express as a single logarithm.

12. $5 \log_a x - \log_a y + \dfrac{1}{4} \log_a z$

12. $\log_a \dfrac{b}{\sqrt{x}} + \log_a \sqrt{bx}$

$\qquad = \log_a b - \log_a \sqrt{x} + \log_a \sqrt{bx}$ Using Property 3

$\qquad = \log_a b - \tfrac{1}{2} \log_a x + \tfrac{1}{2} \log_a (bx)$ Using Property 2

$\qquad = \log_a b - \tfrac{1}{2} \log_a x + \tfrac{1}{2}(\log_a b + \log_a x)$ Using Property 1

$\qquad = \log_a b - \tfrac{1}{2} \log_a x + \tfrac{1}{2} \log_a b + \tfrac{1}{2} \log_a x$

$\qquad = \tfrac{3}{2} \log_a b$ Collecting like terms

$\qquad = \log_a b^{3/2}$ Using Property 2

13. $\log_a \dfrac{\sqrt{x}}{b} - \log_a \sqrt{bx}$

Example 12 could also be done as follows:

$\log_a \dfrac{b}{\sqrt{x}} + \log_a \sqrt{bx} = \log_a \dfrac{b}{\sqrt{x}} \sqrt{bx}$ Using Property 1

$\qquad\qquad\qquad = \log_a \dfrac{b}{\sqrt{x}} \cdot \sqrt{b} \cdot \sqrt{x}$

$\qquad\qquad\qquad = \log_a b\sqrt{b}, \text{ or } \log_a b^{3/2}.$

Answers on page A-47

Given

$$\log_a 2 = 0.301,$$
$$\log_a 5 = 0.699,$$

find each of the following.

14. $\log_a 4$ **15.** $\log_a 10$

16. $\log_a \dfrac{2}{5}$ **17.** $\log_a \dfrac{5}{2}$

18. $\log_a \dfrac{1}{5}$ **19.** $\log_a \sqrt{a^3}$

20. $\log_a 5a$ **21.** $\log_a 16$

Simplify.

22. $\log_2 2^6$

23. $\log_{10} 10^{3.2}$

24. $\log_e e^{12}$

Answers on page A-47

CHAPTER 8: Exponential and
Logarithmic Functions

Do Exercises 12 and 13 on the preceding page.

EXAMPLES Given $\log_a 2 = 0.301$ and $\log_a 3 = 0.477$, find each of the following.

13. $\log_a 6 = \log_a (2 \cdot 3) = \log_a 2 + \log_a 3$ Property 1
$$= 0.301 + 0.477 = 0.778$$

14. $\log_a \dfrac{2}{3} = \log_a 2 - \log_a 3$ Property 3

$$= 0.301 - 0.477 = -0.176$$

15. $\log_a 81 = \log_a 3^4 = 4 \log_a 3$ Property 2
$$= 4(0.477) = 1.908$$

16. $\log_a \dfrac{1}{3} = \log_a 1 - \log_a 3$ Property 3

$$= 0 - 0.477 = -0.477$$

17. $\log_a \sqrt{a} = \log_a a^{1/2} = \dfrac{1}{2} \log_a a = \dfrac{1}{2} \cdot 1 = \dfrac{1}{2}$ Property 2

18. $\log_a 2a = \log_a 2 + \log_a a$ Property 1
$$= 0.301 + 1 = 1.301$$

19. $\log_a 5$ No way to find using these properties
           $(\log_a 5 \neq \log_a 2 + \log_a 3)$

20. $\dfrac{\log_a 3}{\log_a 2} = \dfrac{0.477}{0.301} \approx 1.58$ We simply divide the logarithms, not using any property.

Do Exercises 14–21.

e   The Logarithm of the Base to a Power

> **PROPERTY 4**
>
> For any base a,
>
> $$\log_a a^k = k.$$
>
> (The logarithm, base a, of a to a power is the power.)

A PROOF OF PROPERTY 4 (OPTIONAL). The proof involves Property 2 and the fact that $\log_a a = 1$:

$$\log_a a^k = k(\log_a a) \qquad \text{Using Property 2}$$
$$= k \cdot 1 \qquad \text{Using } \log_a a = 1$$
$$= k.$$

EXAMPLES Simplify.

21. $\log_3 3^7 = 7$

22. $\log_{10} 10^{5.6} = 5.6$

23. $\log_e e^{-t} = -t$

Do Exercises 22–24.

> **Caution!**
>
> Keep in mind that, in general,
>
> $$\log_a (M + N) \neq \log_a M + \log_a N,$$
> $$\log_a (M - N) \neq \log_a M - \log_a N,$$
> $$\log_a (MN) \neq (\log_a M)(\log_a N),$$
> and
> $$\log_a (M/N) \neq (\log_a M) \div (\log_a N).$$

a Express as a sum of logarithms.

1. $\log_2 (32 \cdot 8)$

2. $\log_3 (27 \cdot 81)$

3. $\log_4 (64 \cdot 16)$

4. $\log_5 (25 \cdot 125)$

5. $\log_a Qx$

6. $\log_r 8Z$

Express as a single logarithm.

7. $\log_b 3 + \log_b 84$

8. $\log_a 75 + \log_a 5$

9. $\log_c K + \log_c y$

10. $\log_t H + \log_t M$

b Express as a product.

11. $\log_c y^4$

12. $\log_a x^3$

13. $\log_b t^6$

14. $\log_{10} y^7$

15. $\log_b C^{-3}$

16. $\log_c M^{-5}$

c Express as a difference of logarithms.

17. $\log_a \dfrac{67}{5}$

18. $\log_t \dfrac{T}{7}$

19. $\log_b \dfrac{2}{5}$

20. $\log_a \dfrac{z}{y}$

Express as a single logarithm.

21. $\log_c 22 - \log_c 3$

22. $\log_d 54 - \log_d 9$

d Express in terms of logarithms.

23. $\log_a x^2 y^3 z$

24. $\log_a 5xy^4 z^3$

25. $\log_b \dfrac{xy^2}{z^3}$

26. $\log_b \dfrac{p^2 q^5}{m^4 n^7}$

27. $\log_c \sqrt[3]{\dfrac{x^4}{y^3 z^2}}$

28. $\log_a \sqrt{\dfrac{x^6}{p^5 q^8}}$

29. $\log_a \sqrt[4]{\dfrac{m^8 n^{12}}{a^3 b^5}}$

30. $\log_a \sqrt{\dfrac{a^6 b^8}{a^2 b^5}}$

Express as a single logarithm and, if possible, simplify.

31. $\dfrac{2}{3} \log_a x - \dfrac{1}{2} \log_a y$

32. $\dfrac{1}{2} \log_a x + 3 \log_a y - 2 \log_a x$

33. $\log_a 2x + 3(\log_a x - \log_a y)$

34. $\log_a x^2 - 2 \log_a \sqrt{x}$

35. $\log_a \dfrac{a}{\sqrt{x}} - \log_a \sqrt{ax}$

36. $\log_a (x^2 - 4) - \log_a (x - 2)$

Given $\log_b 3 = 1.099$ and $\log_b 5 = 1.609$, find each of the following.

37. $\log_b 15$

38. $\log_b \dfrac{3}{5}$

39. $\log_b \dfrac{5}{3}$

40. $\log_b \dfrac{1}{3}$

41. $\log_b \dfrac{1}{5}$

42. $\log_b \sqrt{b}$

43. $\log_b \sqrt{b^3}$

44. $\log_b 3b$

45. $\log_b 5b$

46. $\log_b 9$

e   Simplify.

47. $\log_e e^t$

48. $\log_w w^8$

49. $\log_p p^5$

50. $\log_Y Y^{-4}$

Solve for x.

51. $\log_2 2^7 = x$

52. $\log_9 9^4 = x$

53. $\log_e e^x = -7$

54. $\log_a a^x = 2.7$

55. D_W Find a way to express $\log_a (x/5)$ as a difference of logarithms without using the quotient rule. Explain your work.

56. D_W A student incorrectly reasons that

$$\log_a \frac{1}{x} = \log_a \frac{x}{x \cdot x}$$
$$= \log_a x - \log_a x + \log_a x$$
$$= \log_a x.$$

What mistake has the student made? Explain what the answer should be.

SKILL MAINTENANCE

Compute and simplify. Express answers in the form $a + bi$, where $i^2 = -1$.   [6.8b, c, d, e]

57. i^{29}

58. i^{34}

59. $(2 + i)(2 - i)$

60. $\dfrac{2 + i}{2 - i}$

61. $(7 - 8i) - (-16 + 10i)$

62. $2i^2 \cdot 5i^3$

63. $(8 + 3i)(-5 - 2i)$

64. $(2 - i)^2$

SYNTHESIS

65. 📈 Use the TABLE and GRAPH features to show that
$$\log x^2 \neq (\log x)(\log x).$$

66. 📈 Use the TABLE and GRAPH features to show that
$$\frac{\log x}{\log 4} \neq \log x - \log 4.$$

Express as a single logarithm and, if possible, simplify.

67. $\log_a (x^8 - y^8) - \log_a (x^2 + y^2)$

68. $\log_a (x + y) + \log_a (x^2 - xy + y^2)$

Express as a sum or a difference of logarithms.

69. $\log_a \sqrt{1 - s^2}$

70. $\log_a \dfrac{c - d}{\sqrt{c^2 - d^2}}$

Determine whether each is true or false.

71. $\dfrac{\log_a P}{\log_a Q} = \log_a \dfrac{P}{Q}$

72. $\dfrac{\log_a P}{\log_a Q} = \log_a P - \log_a Q$

73. $\log_a 3x = \log_a 3 + \log_a x$

74. $\log_a 3x = 3 \log_a x$

75. $\log_a (P + Q) = \log_a P + \log_a Q$

76. $\log_a x^2 = 2 \log_a x$

714

8.5 NATURAL LOGARITHMIC FUNCTIONS

Objectives

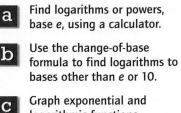

a Find logarithms or powers, base e, using a calculator.

b Use the change-of-base formula to find logarithms to bases other than e or 10.

c Graph exponential and logarithmic functions, base e.

Any positive number other than 1 can serve as the base of a logarithmic function. Common, or base-10, logarithms, which were introduced in Section 8.3, are useful because they have the same base as our "commonly" used decimal system of naming numbers.

Today, another base is widely used. It is an irrational number named e. We now consider e and **natural logarithms,** or logarithms base e.

a The Base e and Natural Logarithms

When interest is computed n times per year, the compound-interest formula is

$$A = P\left(1 + \frac{r}{n}\right)^{nt},$$

where A is the amount that an initial investment P will grow to after t years at interest rate r. Suppose that \$1 could be invested at 100% interest for 1 year. (In reality, no financial institution would pay such an interest rate.) The preceding formula becomes a function A defined in terms of the number of compounding periods n:

$$A(n) = \left(1 + \frac{1}{n}\right)^n. \quad \text{(See Section 8.1.)}$$

Let's find some function values, using a calculator and rounding to six decimal places. The numbers in the table at right approach a very important number called e. It is an irrational number, so its decimal representation neither terminates nor repeats.

n	$A(n) = \left(1 + \frac{1}{n}\right)^n$
1 (compounded annually)	\$2.00
2 (compounded semiannually)	\$2.25
3	\$2.370370
4 (compounded quarterly)	\$2.441406
5	\$2.488320
100	\$2.704814
365 (compounded daily)	\$2.714567
8760 (compounded hourly)	\$2.718127

THE NUMBER e

$e \approx 2.7182818284\ldots$

Logarithms, base e, are called **natural logarithms,** or **Naperian logarithms,** in honor of John Napier (1550–1617), a Scotsman who invented logarithms.

The abbreviation **ln** is commonly used with natural logarithms. Thus,

ln 29   means $\log_e 29$.   Be sure to memorize $\ln a = \log_e a$.

We usually read "ln 29" as "the natural log of 29," or simply "el en of 29."

On a calculator, the key for natural logarithms is generally marked . Using that key, we find that

$\ln 29 \approx 3.36729583 \approx 3.3673,$

rounded to four decimal places. This also tells us that $e^{3.3673} \approx 29$.

On some scientific calculators, the keystrokes for doing such a calculation might be

②⑨ LN ⊟ .

The display would then read 3.3672958.

Find the natural logarithm, to four decimal places, on a calculator.

1. ln 78,235.4

2. ln 0.0000309

3. ln (−3)

4. ln 0

5. ln 10

6. Find $e^{11.2675}$ using a calculator. (Compare this computation to that of Margin Exercise 1.)

7. Find e^{-2} using a calculator.

8. a) Find $\log_6 7$ using common logarithms.

b) Find $\log_6 7$ using natural logarithms.

9. Find $\log_2 46$ using natural logarithms.

Answers on page A-47

If we were to use a graphing calculator, the keystrokes might be

LN ② ⑨ **ENTER** .

The display would then read 3.36729583.

EXAMPLES Find each natural logarithm, to four decimal places, on a calculator.

Function Value	Readout	Rounded
1. ln 287,523	12.56905814	12.5691
2. ln 0.000486	−7.629301934	−7.6293
3. ln (−5)	NONREAL ANS	NONREAL ANS
4. ln (e)	1	1
5. ln 1	0	0

Do Exercises 1–5.

The inverse of a logarithmic function is an exponential function. Thus, if $f(x) = \ln x$, then $f^{-1}(x) = e^x$. Because of this, on many calculators, the **LN** key doubles as the $\boxed{e^x}$ key after a **2ND** or $\boxed{\text{SHIFT}}$ key has been pressed.

EXAMPLE 6 Find $e^{12.5691}$ using a calculator.

On a scientific calculator, we might enter 12.5691 and press $\boxed{e^x}$. On a graphing calculator, we might press **2ND** $\boxed{e^x}$, followed by 12.5691. In either case, we get the approximation

$$e^{12.5691} \approx 287{,}535.0371.$$

Compare this computation to Example 1. Note that, apart from the rounding error, $e^{12.5691}$ takes us back to about 287,523.

EXAMPLE 7 Find $e^{-1.524}$ using a calculator.

On a scientific calculator, we might enter −1.524 and press $\boxed{e^x}$. On a graphing calculator, we might press **2ND** $\boxed{e^x}$, followed by −1.524. In either case, we get the approximation

$$e^{-1.524} \approx 0.2178.$$

Do Exercises 6 and 7.

 Changing Logarithm Bases

Most calculators give the values of both common logarithms and natural logarithms. To find a logarithm with some other base, we can use the following conversion formula.

THE CHANGE-OF-BASE FORMULA

For any logarithm bases a and b and any positive number M,

$$\log_b M = \frac{\log_a M}{\log_a b}.$$

A PROOF OF THE CHANGE-OF-BASE FORMULA (*OPTIONAL*).   Let $x = \log_b M$. Then, writing an equivalent exponential equation, we have $b^x = M$. Next, we take the logarithm base a on both sides. This gives us

$$\log_a b^x = \log_a M.$$

By Property 2, the Power Rule,

$$x \log_a b = \log_a M,$$

and solving for x, we obtain

$$x = \frac{\log_a M}{\log_a b}.$$

But $x = \log_b M$, so we have

$$\log_b M = \frac{\log_a M}{\log_a b},$$

which is the change-of-base formula.

EXAMPLE 8   Find $\log_4 7$ using common logarithms.

Let $a = 10$, $b = 4$, and $M = 7$. Then we substitute into the change-of-base formula:

$$\log_b M = \frac{\log_a M}{\log_a b}$$

$$\log_4 7 = \frac{\log_{10} 7}{\log_{10} 4} \qquad \text{Substituting 10 for } a, \\ \qquad\qquad\qquad\quad 4 \text{ for } b, \text{ and } 7 \text{ for } M$$

$$= \frac{\log 7}{\log 4}$$

$$\approx 1.4037.$$

To check, we use a calculator with a power key $\boxed{y^x}$ or $\boxed{\wedge}$ to verify that

$$4^{1.4037} \approx 7.$$

We can also use base e for a conversion.

EXAMPLE 9   Find $\log_4 7$ using natural logarithms.

$$\log_b M = \frac{\log_a M}{\log_a b}$$

$$\log_4 7 = \frac{\log_e 7}{\log_e 4} \qquad \text{Substituting } e \text{ for } a, \\ \qquad\qquad\qquad 4 \text{ for } b, \text{ and } 7 \text{ for } M$$

$$= \frac{\ln 7}{\ln 4}$$

$$\approx 1.4037 \qquad \text{Note that this is the same answer as Example 8.}$$

EXAMPLE 10   Find $\log_5 29$ using natural logarithms.

Substituting e for a, 5 for b, and 29 for M, we have

$$\log_5 29 = \frac{\log_e 29}{\log_e 5} \qquad \text{Using the change-of-base formula}$$

$$= \frac{\ln 29}{\ln 5} \approx 2.0922.$$

Do Exercises 8 and 9 on the preceding page.

Graph.

10. $f(x) = e^{2x}$

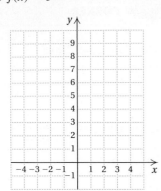

11. $g(x) = \frac{1}{2}e^{-x}$

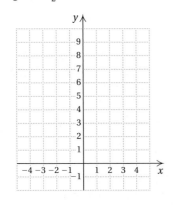

Graph.

12. $f(x) = 2 \ln x$

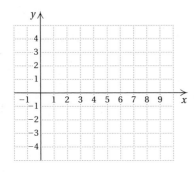

13. $g(x) = \ln (x - 2)$

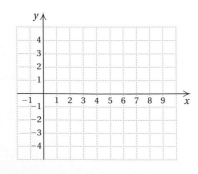

Answers on page A-48

■ **EXAMPLE 11** Graph $f(x) = e^x$ and $g(x) = e^{-x}$.

We use a calculator with an ⌐e^x⌐ key to find approximate values of e^x and e^{-x}. Using these values, we can graph the functions.

x	e^x	e^{-x}
0	1	1
1	2.7	0.4
2	7.4	0.1
−1	0.4	2.7
−2	0.1	7.4

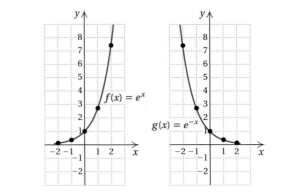

Note that each graph is the image of the other reflected across the *y*-axis.

■ **EXAMPLE 12** Graph: $f(x) = e^{-0.5x}$.

We find some solutions with a calculator, plot them, and then draw the graph. For example, $f(2) = e^{-0.5(2)} = e^{-1} \approx 0.4$.

x	$e^{-0.5x}$
0	1
1	0.6
2	0.4
3	0.2
−1	1.6
−2	2.7
−3	4.5

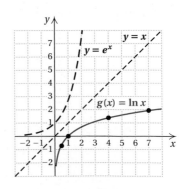

Do Exercises 10 and 11.

■ **EXAMPLE 13** Graph: $g(x) = \ln x$.

We find some solutions with a calculator and then draw the graph. As expected, the graph is a reflection across the line $y = x$ of the graph of $y = e^x$.

x	$\ln x$
1	0
4	1.4
7	1.9
0.5	−0.7

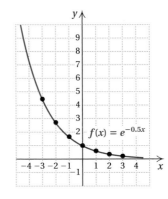

EXAMPLE 14 Graph: $f(x) = \ln (x + 3)$.

We find some solutions with a calculator, plot them, and then draw the graph.

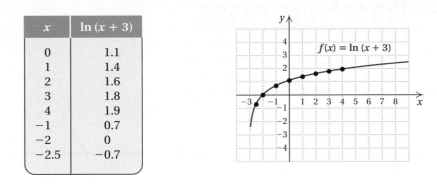

x	$\ln (x + 3)$
0	1.1
1	1.4
2	1.6
3	1.8
4	1.9
-1	0.7
-2	0
-2.5	-0.7

The graph of $y = \ln (x + 3)$ is the graph of $y = \ln x$ translated 3 units to the left.

Do Exercises 12 and 13 on the preceding page.

CALCULATOR CORNER

Graphing Logarithmic Functions   We can graph logarithmic functions with base 10 or base e by entering the function on the equation-editor screen using the **LOG** or **LN** key. To graph a logarithmic function with a base other than 10 or e, we must first use the change-of-base formula to change the base to 10 or e.

In Example 1 of Section 8.3, we graphed the function $y = \log_5 x$ by finding a table of x- and y-values and plotting points. We will now graph this function on a graphing calculator by first changing the base to e. We let $a = e$, $b = 5$, and $M = x$ and substitute in the change-of-base formula. We enter $y_1 = \dfrac{\ln x}{\ln 5}$ on the equation-editor screen, select a window, and press **GRAPH**. The TI-84+ forces the use of parentheses with the ln function, so the parentheses in the numerator must be closed: ln (x)/ln (5). The right parenthesis following the 5 is optional but we include it for completeness.

We could have let $a = 10$ and used base-10 logarithms to graph this function as well.

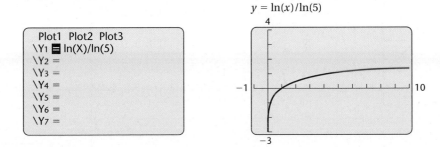

$$y = \ln(x)/\ln(5)$$

Exercises: Graph each of the following on a graphing calculator.

1. $y = \log_2 x$

2. $y = \log_3 x$

3. $y = \log_{1/2} x$

4. $y = \log_{2/3} x$

Visualizing for Success

1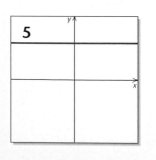

2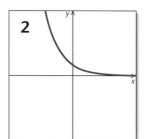

3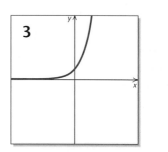

4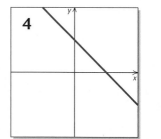

5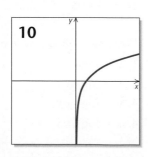

Match each graph with its function.

A. $f(x) = ax^2 + bx + c,$
$a < 0, c < 0$

B. $f(x) = a^x, 0 < a < 1$

C. $f(x) = a^x, a < 0$

D. $f(x) = \log_a x, 0 < a < 1$

E. $f(x) = \log_a x, a < 0$

F. $f(x) = mx + b, m > 0, b < 0$

G. $f(x) = mx + b, m < 0, b > 0$

H. $f(x) = mx + b, m < 0, b < 0$

I. $f(x) = ax^2 + bx + c,$
$a > 0, c < 0$

J. $f(x) = ax^2 + bx + c,$
$a < 0, c > 0$

K. $f(x) = \log_a x, a > 1$

L. $f(x) = ax^2 + bx + c,$
$a > 0, c > 0$

M. $f(x) = mx + b, m < 0, b = 0$

N. $f(x) = mx + b, m = 0, b > 0$

O. $f(x) = a^x, a > 1$

Answers on page A-48

6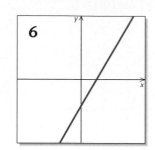

7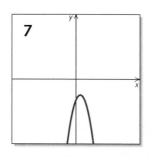

8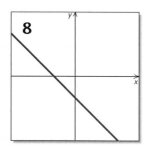

9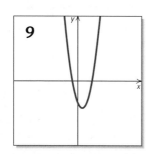

10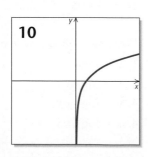

a Find each of the following logarithms or powers, base e, using a calculator. Round answers to four decimal places.

1. $\ln 2$ **2.** $\ln 5$ **3.** $\ln 62$ **4.** $\ln 30$ **5.** $\ln 4365$ **6.** $\ln 901.2$

7. $\ln 0.0062$ **8.** $\ln 0.00073$ **9.** $\ln 0.2$ **10.** $\ln 0.04$ **11.** $\ln 0$ **12.** $\ln (-4)$

13. $\ln \left(\dfrac{97.4}{558} \right)$ **14.** $\ln \left(\dfrac{786.2}{77.2} \right)$ **15.** $\ln e$ **16.** $\ln e^2$ **17.** $e^{2.71}$ **18.** $e^{3.06}$

19. $e^{-3.49}$ **20.** $e^{-2.64}$ **21.** $e^{4.7}$ **22.** $e^{1.23}$ **23.** $\ln e^5$ **24.** $e^{\ln 7}$

b Find each of the following logarithms using the change-of-base formula.

25. $\log_6 100$ **26.** $\log_3 100$ **27.** $\log_2 100$ **28.** $\log_7 100$ **29.** $\log_7 65$ **30.** $\log_5 42$

31. $\log_{0.5} 5$ **32.** $\log_{0.1} 3$ **33.** $\log_2 0.2$ **34.** $\log_2 0.08$ **35.** $\log_\pi 200$ **36.** $\log_\pi \pi$

c Graph.

37. $f(x) = e^x$ **38.** $f(x) = e^{0.5x}$ **39.** $f(x) = e^{-0.5x}$ **40.** $f(x) = e^{-x}$

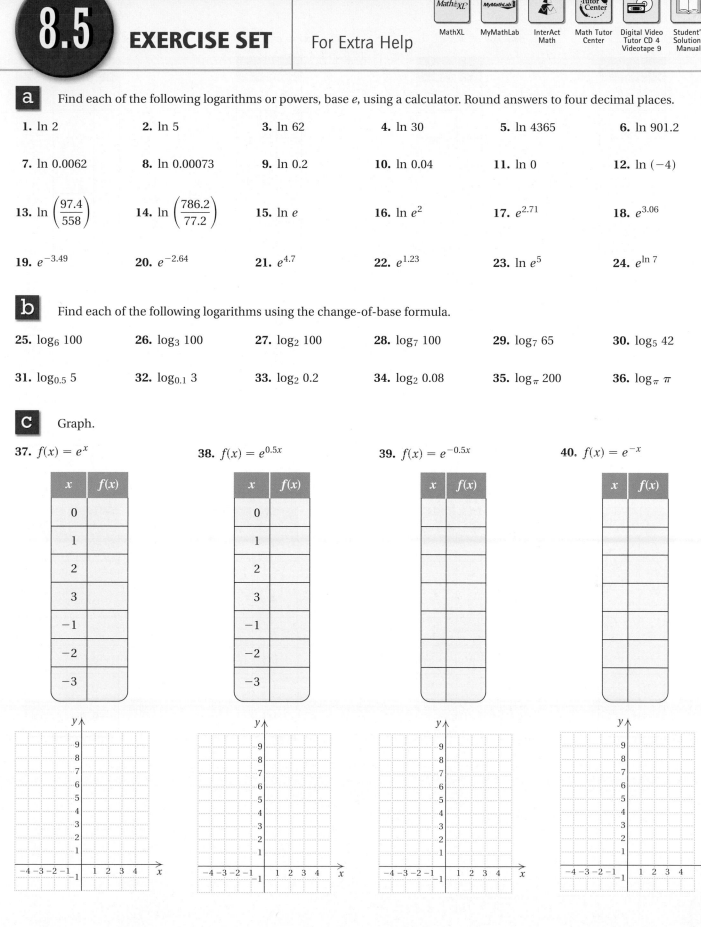

x	$f(x)$
0	
1	
2	
3	
-1	
-2	
-3	

x	$f(x)$
0	
1	
2	
3	
-1	
-2	
-3	

x	$f(x)$

x	$f(x)$

41. $f(x) = e^{x-1}$

x	$f(x)$

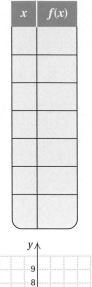

42. $f(x) = e^{-x} + 3$

x	$f(x)$

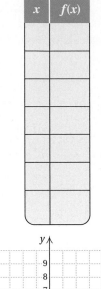

43. $f(x) = e^{x+2}$

x	$f(x)$

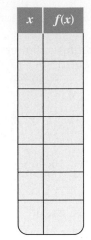

44. $f(x) = e^{x-2}$

x	$f(x)$

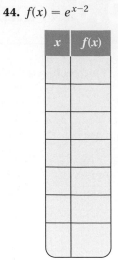

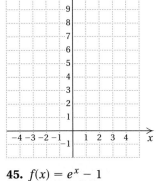

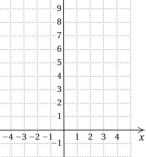

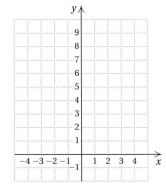

45. $f(x) = e^x - 1$

x	$f(x)$

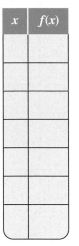

46. $f(x) = 2e^{0.5x}$

x	$f(x)$

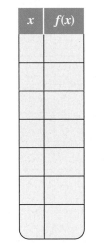

47. $f(x) = \ln(x + 2)$

x	$f(x)$
0	
1	
2	
3	
−0.5	
−1	
−1.5	

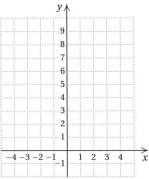

48. $f(x) = \ln(x + 1)$

x	$f(x)$
0	
1	
2	
3	
4	
−0.5	
−0.75	

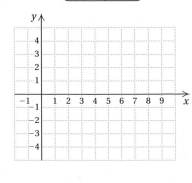

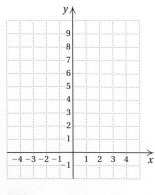

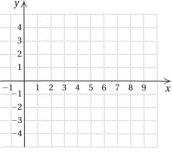

49. $f(x) = \ln(x - 3)$

x	$f(x)$

50. $f(x) = 2\ln(x - 2)$

x	$f(x)$

51. $f(x) = 2\ln x$

x	$f(x)$

52. $f(x) = \ln x - 3$

x	$f(x)$

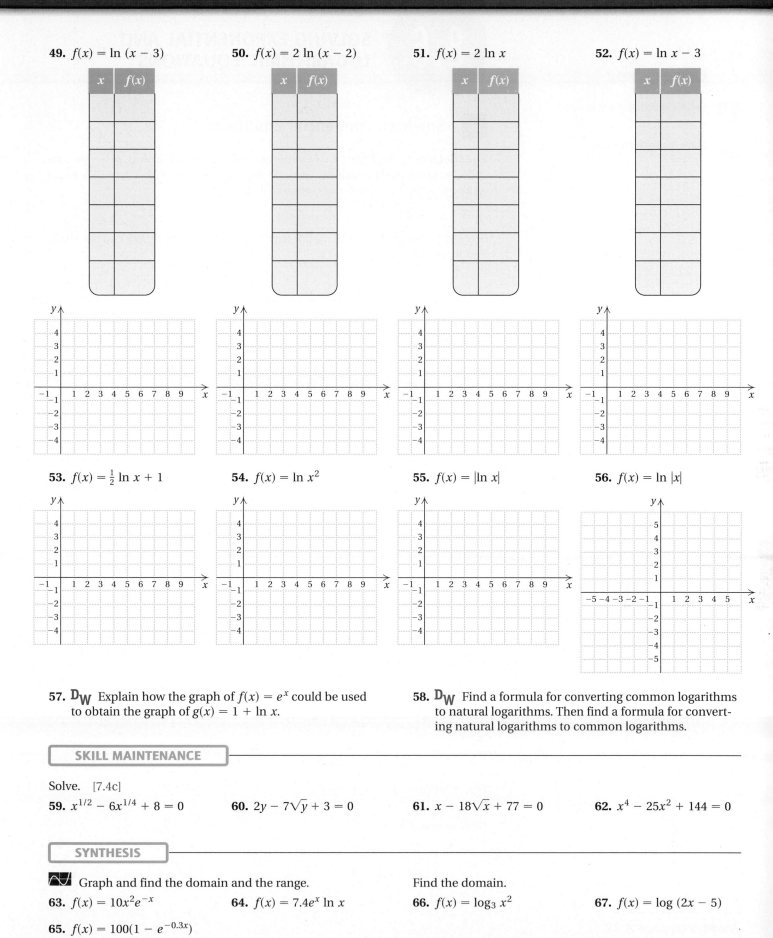

53. $f(x) = \frac{1}{2}\ln x + 1$

54. $f(x) = \ln x^2$

55. $f(x) = |\ln x|$

56. $f(x) = \ln |x|$

57. **D**_W Explain how the graph of $f(x) = e^x$ could be used to obtain the graph of $g(x) = 1 + \ln x$.

58. **D**_W Find a formula for converting common logarithms to natural logarithms. Then find a formula for converting natural logarithms to common logarithms.

SKILL MAINTENANCE

Solve. [7.4c]

59. $x^{1/2} - 6x^{1/4} + 8 = 0$

60. $2y - 7\sqrt{y} + 3 = 0$

61. $x - 18\sqrt{x} + 77 = 0$

62. $x^4 - 25x^2 + 144 = 0$

SYNTHESIS

Graph and find the domain and the range.

63. $f(x) = 10x^2 e^{-x}$

64. $f(x) = 7.4e^x \ln x$

65. $f(x) = 100(1 - e^{-0.3x})$

Find the domain.

66. $f(x) = \log_3 x^2$

67. $f(x) = \log(2x - 5)$

Objectives

 a Solve exponential equations.

b Solve logarithmic equations.

Solve.

1. $3^{2x} = 9$

2. $4^{2x-3} = 64$

a Solving Exponential Equations

Equations with variables in exponents, such as $5^x = 12$ and $2^{7x} = 64$, are called **exponential equations.** Sometimes, as is the case with $2^{7x} = 64$, we can write each side as a power of the *same* number:

$$2^{7x} = 2^6.$$

Since the base is the same, 2, the exponents are the same. We can set them equal and solve:

$$7x = 6$$
$$x = \frac{6}{7}.$$

We use the following property, which is true because exponential functions are one-to-one.

PRINCIPLE OF EXPONENTIAL EQUALITY

For any $a > 0$, $a \neq 1$,

$$a^x = a^y \longrightarrow x = y.$$

(When powers are equal, the exponents are equal.)

EXAMPLE 1 Solve: $2^{3x-5} = 16$.

Note that $16 = 2^4$. Thus we can write each side as a power of the same number:

$$2^{3x-5} = 2^4.$$

Since the base is the same, 2, the exponents must be the same. Thus,

$$3x - 5 = 4$$
$$3x = 9$$
$$x = 3.$$

Check:
$$\begin{array}{c|c} 2^{3x-5} = 16 \\ \hline 2^{3 \cdot 3 - 5} \; ? \; 16 \\ 2^{9-5} \\ 2^4 \\ 16 & \text{TRUE} \end{array}$$

The solution is 3.

Do Exercises 1 and 2.

ALGEBRAIC–GRAPHICAL CONNECTION

The solution, 3, of the equation $2^{3x-5} = 16$ in Example 1 is the x-coordinate of the point of intersection of the graphs of $y = 2^{3x-5}$ and $y = 16$, as we see in the graph on the left below.

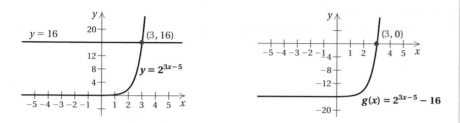

If we subtract 16 on both sides of $2^{3x-5} = 16$, we get $2^{3x-5} - 16 = 0$. The solution, 3, is then the x-coordinate of the x-intercept of the function $g(x) = 2^{3x-5} - 16$, as we see in the graph on the right above.

When it does not seem possible to write both sides of an equation as powers of the same base, we can use the following principle along with the properties developed in Section 8.4.

THE PRINCIPLE OF LOGARITHMIC EQUALITY

For any logarithm base a, and for x, $y > 0$,

$$\log_a x = \log_a y \longrightarrow x = y.$$

(If the logarithms, base a, of two expressions are the same, then the expressions are the same.)

Because calculators can generally find only common or natural logarithms (without resorting to the change-of-base formula), we usually take the common or natural logarithm on both sides of the equation.

The principle of logarithmic equality is useful anytime a variable appears as an exponent.

EXAMPLE 2   Solve: $5^x = 12$.

$$5^x = 12$$
$$\log 5^x = \log 12 \qquad \text{Taking the common logarithm on both sides}$$
$$x \log 5 = \log 12 \qquad \text{Property 2}$$
$$x = \frac{\log 12}{\log 5} \longleftarrow \boxed{\text{Caution!}}$$

This is not $\log \frac{12}{5}$!

This is an exact answer. We cannot simplify further, but we can approximate using a calculator:

$$x = \frac{\log 12}{\log 5} \approx 1.5440.$$

We can also partially check this answer by finding $5^{1.5440}$ using a calculator:

$$5^{1.5440} \approx 12.00078587.$$

We get an answer close to 12, due to the rounding. This checks.

---

3. Solve: $7^x = 20$.

Answer on page A-49

725

4. Solve: $e^{0.3t} = 80$.

5. Solve: $\log_5 x = 2$.

Answers on page A-49

Do Exercise 3 on the preceding page.

If the base is e, we can make our work easier by taking the logarithm, base e, on both sides.

 EXAMPLE 3   Solve: $e^{0.06t} = 1500$.

We take the natural logarithm on both sides:

$$e^{0.06t} = 1500$$
$$\ln e^{0.06t} = \ln 1500 \qquad \text{Taking ln on both sides}$$
$$\log_e e^{0.06t} = \ln 1500 \qquad \text{Definition of natural logarithms}$$
$$0.06t = \ln 1500 \qquad \text{Here we use Property 4: } \log_a a^k = k.$$
$$t = \frac{\ln 1500}{0.06}.$$

We can approximate using a calculator:

$$t = \frac{\ln 1500}{0.06} \approx \frac{7.3132}{0.06} \approx 121.89.$$

We can also partially check this answer using a calculator.

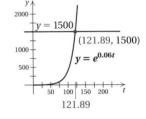

Check: $\dfrac{e^{0.06t} = 1500}{\underset{e^{7.3134}}{e^{0.06(121.89)}} \; ? \; 1500}$
$\qquad\qquad\quad 1500.269444 \; | \qquad$ TRUE

The solution is about 121.89.

Do Exercise 4.

b   Solving Logarithmic Equations

Equations containing logarithmic expressions are called **logarithmic equations.** We solved some logarithmic equations in Section 8.3 by converting to equivalent exponential equations.

EXAMPLE 4   Solve: $\log_2 x = 3$.

We obtain an equivalent exponential equation:

$$x = 2^3$$
$$x = 8.$$

The solution is 8.

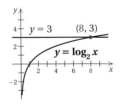

Do Exercise 5.

To solve a logarithmic equation, first try to obtain a single logarithmic expression on one side and then write an equivalent exponential equation.

CALCULATOR CORNER

Solving Exponential Equations   Use the INTERSECT feature to solve the equations in Examples 1–3 and Margin Exercises 1–4. (See the Calculator Corner on p. 246 for the procedure.)

EXAMPLE 5 Solve: $\log_4 (8x - 6) = 3$.

We already have a single logarithmic expression, so we write an equivalent exponential equation:

$$8x - 6 = 4^3 \qquad \text{Writing an equivalent exponential equation}$$
$$8x - 6 = 64$$
$$8x = 70$$
$$x = \tfrac{70}{8}, \text{ or } \tfrac{35}{4}$$

Check:
$$\frac{\log_4 (8x - 6) = 3}{\log_4 \left(8 \cdot \tfrac{35}{4} - 6\right) \overset{?}{\;} 3}$$
$$\log_4 (70 - 6)$$
$$\log_4 64$$
$$3 \quad \Big| \quad \text{TRUE}$$

The solution is $\tfrac{35}{4}$.

Do Exercise 6.

6. Solve: $\log_3 (5x + 7) = 2$.

EXAMPLE 6 Solve: $\log x + \log (x - 3) = 1$.

Here we have common logarithms. It will help us follow the solution to first write in the 10's before we obtain a single logarithmic expression on the left.

$$\log_{10} x + \log_{10} (x - 3) = 1$$
$$\log_{10} [x(x - 3)] = 1 \qquad \text{Using Property 1 to obtain a single logarithm}$$
$$x(x - 3) = 10^1 \qquad \text{Writing an equivalent exponential expression}$$
$$x^2 - 3x = 10$$
$$x^2 - 3x - 10 = 0$$
$$(x + 2)(x - 5) = 0 \qquad \text{Factoring}$$
$$x + 2 = 0 \quad or \quad x - 5 = 0 \qquad \text{Using the principle of zero products}$$
$$x = -2 \quad or \quad x = 5$$

7. Solve: $\log x + \log (x + 3) = 1$.

Check: For -2:
$$\frac{\log x + \log (x - 3) = 1}{\log (-2) + \log (-2 - 3) \overset{?}{\;} 1}$$

The number -2 does *not* check because negative numbers do not have logarithms.

For 5:
$$\frac{\log x + \log (x - 3) = 1}{\log 5 + \log (5 - 3) \overset{?}{\;} 1}$$
$$\log 5 + \log 2$$
$$\log (5 \cdot 2)$$
$$\log 10$$
$$1 \quad \Big| \quad \text{TRUE}$$

The solution is 5.

Do Exercise 7.

Answers on page A-49

8. Solve:

$$\log_3 (2x - 1) - \log_3 (x - 4) = 2.$$

EXAMPLE 7 Solve: $\log_2 (x + 7) - \log_2 (x - 7) = 3$.

$$\log_2 (x + 7) - \log_2 (x - 7) = 3$$

$$\log_2 \frac{x + 7}{x - 7} = 3 \qquad \text{Using Property 3 to obtain a single logarithm}$$

$$\frac{x + 7}{x - 7} = 2^3 \qquad \text{Writing an equivalent exponential expression}$$

$$\frac{x + 7}{x - 7} = 8$$

$$x + 7 = 8(x - 7) \qquad \text{Multiplying by the LCM, } x - 7$$

$$x + 7 = 8x - 56 \qquad \text{Using a distributive law}$$

$$63 = 7x$$

$$\frac{63}{7} = x$$

$$9 = x$$

Check:

$$\begin{array}{c|c} \log_2 (x + 7) - \log_2 (x - 7) = 3 \\ \hline \log_2 (9 + 7) - \log_2 (9 - 7) \overset{?}{} 3 \\ \log_2 16 - \log_2 2 \\ \log_2 \frac{16}{2} \\ \log_2 8 \\ 3 & \text{TRUE} \end{array}$$

The solution is 9.

Do Exercise 8.

Answer on page A-49

Study Tips

READ FOR SUCCESS

In his article "The Daily Dozen Disciplines for Massive Success in 2001 & Beyond," Jerry Clark comments, "Research has shown that 58% of high school graduates never read another book from cover to cover the rest of their adult lives, that 78% of the population have not been in a bookstore in the last 5 years, and that 97% of the population of the U.S. do not have library cards." Clark then suggests spending at least 15 minutes each day reading an empowering and uplifting book or article. The following books are some suggestions from your author. Their motivating words may empower you in your study of mathematics.

1. *Fish*, by Stephen C. Lundin, Harry Paul, and John Christensen (Hyperion). This was a Wall Street Journal Business Bestseller. Though it has a strange title, it discusses a remarkable way to boost morale and improve results.

2. *True Success: A New Philosophy of Excellence*, by Tom Morris (Grosset/Putnam). Morris was a well-loved philosophy professor at Notre Dame. Students, especially athletes, flocked to his classes.

3. *The Road Less Traveled, Abounding Grace*, by M. Scott Peck. Noted psychiatrist and author of many excellent books, Peck has amazing insights and wisdom about life.

"Far worse than not reading books is not realizing that it matters!"

Jim Rohn, motivational speaker

a Solve.

1. $2^x = 8$

2. $3^x = 81$

3. $4^x = 256$

4. $5^x = 125$

5. $2^{2x} = 32$

6. $4^{3x} = 64$

7. $3^{5x} = 27$

8. $5^{7x} = 625$

9. $2^x = 11$

10. $2^x = 20$

11. $2^x = 43$

12. $2^x = 55$

13. $5^{4x-7} = 125$

14. $4^{3x+5} = 16$

15. $3^{x^2} \cdot 3^{4x} = \dfrac{1}{27}$

16. $3^{5x} \cdot 9^{x^2} = 27$

17. $4^x = 8$

18. $6^x = 10$

19. $e^t = 100$

20. $e^t = 1000$

21. $e^{-t} = 0.1$

22. $e^{-t} = 0.01$

23. $e^{-0.02t} = 0.06$

24. $e^{0.07t} = 2$

25. $2^x = 3^{x-1}$

26. $3^{x+2} = 5^{x-1}$

27. $(3.6)^x = 62$

28. $(5.2)^x = 70$

b Solve.

29. $\log_4 x = 4$

30. $\log_7 x = 3$

31. $\log_2 x = -5$

32. $\log_9 x = \dfrac{1}{2}$

33. $\log x = 1$

34. $\log x = 3$

35. $\log x = -2$

36. $\log x = -3$

37. $\ln x = 2$

38. $\ln x = 1$

39. $\ln x = -1$

40. $\ln x = -3$

41. $\log_3 (2x + 1) = 5$

42. $\log_2 (8 - 2x) = 6$

43. $\log x + \log (x - 9) = 1$

44. $\log x + \log (x + 9) = 1$

45. $\log x - \log (x + 3) = -1$

46. $\log (x + 9) - \log x = 1$

47. $\log_2 (x + 1) + \log_2 (x - 1) = 3$

48. $\log_2 x + \log_2 (x - 2) = 3$

49. $\log_4 (x + 6) - \log_4 x = 2$

50. $\log_4 (x + 3) - \log_4 (x - 5) = 2$

51. $\log_4 (x + 3) + \log_4 (x - 3) = 2$

52. $\log_5 (x + 4) + \log_5 (x - 4) = 2$

53. $\log_3 (2x - 6) - \log_3 (x + 4) = 2$

54. $\log_4 (2 + x) - \log_4 (3 - 5x) = 3$

55. $\mathbf{D_W}$ Explain how Exercises 37–40 could be solved using only the graph of $f(x) = \ln x$.

56. $\mathbf{D_W}$ Christina first determines that the solution of $\log_3 (x + 4) = 1$ is -1, but then rejects it. What mistake do you think she might be making?

SKILL MAINTENANCE

Solve. [7.4c]

57. $x^4 + 400 = 104x^2$

58. $x^{2/3} + 2x^{1/3} = 8$

59. $(x^2 + 5x)^2 + 2(x^2 + 5x) = 24$

60. $10 = x^{-2} + 9x^{-1}$

61. Simplify: $(125x^3 y^{-2} z^6)^{-2/3}$. [6.2c]

62. Simplify: i^{79}. [6.8d]

SYNTHESIS

63. Find the value of x for which the natural logarithm is the same as the common logarithm.

64. Use a graphing calculator to check your answers to Exercises 4, 20, 36, and 54.

65. Use a graphing calculator to solve each of the following equations.

    a) $e^{7x} = 14$ **b)** $8e^{0.5x} = 3$
    c) $xe^{3x-1} = 5$ **d)** $4 \ln (x + 3.4) = 2.5$

66. Use the INTERSECT feature of a graphing calculator to find the points of intersection of the graphs of each pair of functions.

    a) $f(x) = e^{0.5x-7}, \quad g(x) = 2x + 6$
    b) $f(x) = \ln 3x, \quad g(x) = 3x - 8$
    c) $f(x) = \ln x^2, \quad g(x) = -x^2$

Solve.

67. $2^{2x} + 128 = 24 \cdot 2^x$

68. $27^x = 81^{2x-3}$

69. $8^x = 16^{3x+9}$

70. $\log_x (\log_3 27) = 3$

71. $\log_6 (\log_2 x) = 0$

72. $x \log \frac{1}{8} = \log 8$

73. $\log_5 \sqrt{x^2 - 9} = 1$

74. $2^{x^2+4x} = \frac{1}{8}$

75. $\log (\log x) = 5$

76. $\log_5 |x| = 4$

77. $\log x^2 = (\log x)^2$

78. $\log_3 |5x - 7| = 2$

79. $\log_a a^{x^2+4x} = 21$

80. $\sqrt{x} \cdot \sqrt[3]{x} \cdot \sqrt[4]{x} \cdot \sqrt[5]{x} = 146$

81. $3^{2x} - 8 \cdot 3^x + 15 = 0$

82. If $x = (\log_{125} 5)^{\log_5 125}$, what is the value of $\log_3 x$?

730

CHAPTER 8: Exponential and
Logarithmic Functions

copyrightCopyright © 2007 Pearson Education, Inc.

8.7 MATHEMATICAL MODELING WITH EXPONENTIAL AND LOGARITHMIC FUNCTIONS

Objectives

a Solve applied problems involving logarithmic functions.

b Solve applied problems involving exponential functions.

Exponential and logarithmic functions can now be added to our library of functions that can serve as models for many kinds of applications. Let's review some of their graphs.

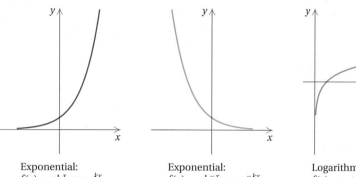

Exponential:
$f(x) = ab^x$, or ae^{kx}
$a, b > 0, k > 0$

Exponential:
$f(x) = ab^{-x}$, or ae^{-kx}
$a, b > 0, k > 0$

Logarithmic:
$f(x) = a + b \log_c x, c > 1$

a Applications of Logarithmic Functions

EXAMPLE 1 *Sound Levels.* To measure the "loudness" of any particular sound, the decibel scale is used. The loudness L, in decibels (dB), of a sound is given by

$$L = 10 \cdot \log \frac{I}{I_0},$$

where I is the intensity of the sound, in watts per square meter (W/m²), and $I_0 = 10^{-12}$ W/m². (I_0 is approximately the intensity of the softest sound that can be heard.)

a) It is common for the intensity of sound at live performances of rock music to reach 10^{-1} W/m² (even higher close to the stage). How loud, in decibels, is this sound level?

b) Audiologists and physicians recommend that earplugs be worn when one is exposed to sounds in excess of 90 dB. What is the intensity of such sounds?

a) To find the loudness, in decibels, we use the above formula:

$$L = 10 \cdot \log \frac{I}{I_0}$$

$$= 10 \cdot \log \frac{10^{-1}}{10^{-12}} \quad \text{Substituting}$$

$$= 10 \cdot \log 10^{11} \quad \text{Subtracting exponents}$$

$$= 10 \cdot 11 \quad \log 10^a = a$$

$$= 110.$$

The volume of the music is 110 decibels.

1. Acoustics. The intensity of sound in normal conversation is about 3.2×10^{-6} W/m². How high is this sound level in decibels?

b) We substitute and solve for I:

$$L = 10 \cdot \log \frac{I}{I_0}$$

$$90 = 10 \cdot \log \frac{I}{10^{-12}} \qquad \text{Substituting}$$

$$9 = \log \frac{I}{10^{-12}} \qquad \text{Dividing by 10}$$

$$9 = \log I - \log 10^{-12} \qquad \text{Using Property 3}$$

$$9 = \log I - (-12) \qquad \log 10^a = a$$

$$-3 = \log I \qquad \text{Adding } -12$$

$$10^{-3} = I. \qquad \text{Converting to an exponential equation}$$

Earplugs are recommended for sounds with intensities that exceed 10^{-3} W/m².

Do Exercises 1 and 2.

EXAMPLE 2 *Chemistry: pH of Liquids.* In chemistry the pH of a liquid is defined as:

$$pH = -\log [H^+],$$

where $[H^+]$ is the hydrogen ion concentration in moles per liter.

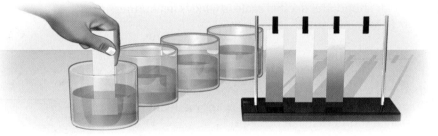

2. Audiology. Overexposure to excessive sound levels can diminish one's hearing to the point where the softest sound that is audible is 28 dB. What is the intensity of such a sound?

a) The hydrogen ion concentration of human blood is normally about 3.98×10^{-8} moles per liter. Find the pH.

b) The pH of seawater is about 8.3. Find the hydrogen ion concentration.

a) To find the pH of human blood, we use the above formula:

$$pH = -\log [H^+] = -\log [3.98 \times 10^{-8}]$$

$$\approx -(-7.400117) \qquad \text{Using a calculator}$$

$$\approx 7.4.$$

The pH of human blood is normally about 7.4.

b) We substitute and solve for $[H^+]$:

$$8.3 = -\log [H^+] \qquad \text{Using } pH = -\log [H^+]$$

$$-8.3 = \log [H^+] \qquad \text{Dividing by } -1$$

$$10^{-8.3} = [H^+] \qquad \text{Converting to an exponential equation}$$

$$5.01 \times 10^{-9} \approx [H^+]. \qquad \text{Using a calculator; writing scientific notation}$$

The hydrogen ion concentration of seawater is about 5.01×10^{-9} moles per liter.

Answers on page A-49

Do Exercises 3 and 4.

Do Exercises 3 and 4.

b   Applications of Exponential Functions

EXAMPLE 3   *Interest Compounded Annually.*   Suppose that $30,000 is invested at 4% interest, compounded annually. In t years, it will grow to the amount A given by the function

$$A(t) = 30{,}000(1.04)^t.$$

(See Example 6 in Section 8.1.)

a) How long will it take to accumulate $150,000 in the account?

b) Let $T =$ the amount of time it takes for the $30,000 to double itself; T is called the **doubling time.** Find the doubling time.

a) We set $A(t) = 150{,}000$ and solve for t:

$$150{,}000 = 30{,}000(1.04)^t$$

$$\frac{150{,}000}{30{,}000} = (1.04)^t \qquad \text{Dividing by 30,000}$$

$$5 = (1.04)^t$$

$$\log 5 = \log (1.04)^t \qquad \text{Taking the common logarithm on both sides}$$

$$\log 5 = t \log 1.04 \qquad \text{Using Property 2}$$

$$\frac{\log 5}{\log 1.04} = t \qquad \text{Dividing by log 1.04}$$

$$41.04 \approx t. \qquad \text{Using a calculator}$$

It will take about 41 yr for the $30,000 to grow to $150,000.

b) To find the *doubling time T*, we replace $A(t)$ with 60,000 and t with T and solve for T:

$$60{,}000 = 30{,}000(1.04)^T$$

$$2 = (1.04)^T \qquad \text{Dividing by 30,000}$$

$$\log 2 = \log (1.04)^T \qquad \text{Taking the common logarithm on both sides}$$

$$\log 2 = T \log 1.04 \qquad \text{Using Property 2}$$

$$T = \frac{\log 2}{\log 1.04} \approx 17.7. \qquad \text{Using a calculator}$$

The doubling time is about 17.7 yr.

Do Exercise 5.

Do Exercise 5.

The function in Example 3 illustrates exponential growth. Populations often grow exponentially according to the following model.

3. Coffee.   The hydrogen ion concentration of freshly brewed coffee is about 1.3×10^{-5} moles per liter. Find the pH.

4. Acidosis.   When the pH of a patient's blood drops below 7.4, a condition called *acidosis* sets in. Acidosis can be fatal at a pH level of 7.0. What would the hydrogen ion concentration of the patient's blood be at that point?

5. Interest Compounded Annually.   Suppose that $40,000 is invested at 4.3% interest, compounded annually.

a) After what amount of time will there be $250,000 in the account?

b) Find the doubling time.

Answers on page A-49

Answers on page A-49

733

6. Population Growth of the United States. What will the population of the United States be in 2008? in 2025?

Answer on page A-49

EXPONENTIAL GROWTH MODEL

An **exponential growth model** is a function of the form

$$P(t) = P_0 e^{kt}, \quad k > 0,$$

where P_0 is the population at time 0, $P(t)$ is the population at time t, and k is the **exponential growth rate** for the situation. The **doubling time** is the amount of time necessary for the population to double in size.

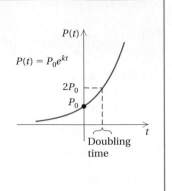

The exponential growth rate is the rate of growth of a population at any *instant* in time. Since the population is continually growing, the percent of total growth after one year will exceed the exponential growth rate.

EXAMPLE 4 *Population Growth of the United States.* In 2005, the population of the United States was 293 million, and the exponential growth rate was 0.8% per year.
Source: U.S. Bureau of the Census

a) Find the exponential growth function.

b) What will the population be in 2012?

a) We are trying to find a model. The given information allows us to create one. At $t = 0$ (2005), the population was 293 million. We substitute 293 for P_0 and 0.8%, or 0.008, for k to obtain the exponential growth function:

$$P(t) = P_0 e^{kt}$$
$$P(t) = 293 e^{0.008t}.$$

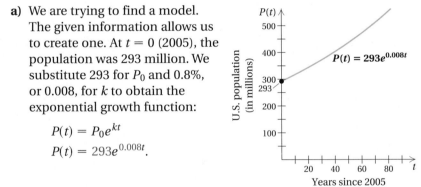

b) In 2012, we have $t = 7$. That is, 7 yr have passed since 2005. To find the population in 2012, we substitute 7 for t:

$$P(7) = 293 e^{0.008(7)} \qquad \text{Substituting 7 for } t$$
$$\approx 310 \text{ million.} \qquad \text{Using a calculator}$$

The population of the United States will be about 310 million in 2012.

Do Exercise 6.

EXAMPLE 5 *Interest Compounded Continuously.* Suppose that an amount of money P_0 is invested in a savings account at interest rate k, compounded continuously. That is, suppose that interest is computed every "instant" and added to the amount in the account. The balance $P(t)$, after t years, is given by the exponential growth model

$$P(t) = P_0 e^{kt}.$$

a) Suppose that \$30,000 is invested and grows to \$44,754.75 in 5 yr. Find the interest rate and then the exponential growth function.

b) What is the balance after 10 yr?

c) What is the doubling time?

a) We have $P(0) = 30{,}000$. Thus the exponential growth function is

$$P(t) = 30{,}000e^{kt}, \quad \text{where } k \text{ must still be determined.}$$

We know that $P(5) = 44{,}754.75$. We substitute and solve for k:

$$44{,}754.75 = 30{,}000e^{k(5)} = 30{,}000e^{5k}$$

$$\frac{44{,}754.75}{30{,}000} = e^{5k} \qquad \text{Dividing by 30,000}$$

$$1.491825 = e^{5k}$$

$$\ln 1.491825 = \ln e^{5k} \qquad \text{Taking the natural logarithm on both sides}$$

$$0.4 \approx 5k \qquad \begin{array}{l}\text{Finding } \ln 1.491825 \text{ on a calculator} \\ \text{and simplifying } \ln e^{5k}\end{array}$$

$$\frac{0.4}{5} = 0.08 \approx k.$$

The interest rate is about 0.08, or 8%, compounded continuously. Note that since interest is being compounded continuously, the interest earned each year is more than 8%. The exponential growth function is

$$P(t) = 30{,}000e^{0.08t}.$$

b) We substitute 10 for t:

$$P(10) = 30{,}000e^{0.08(10)} = 66{,}766.23.$$

The balance in the account after 10 yr will be $66,766.23.

c) To find the doubling time T, we replace $P(t)$ with 60,000 and solve for T:

$$60{,}000 = 30{,}000e^{0.08T}$$

$$2 = e^{0.08T} \qquad \text{Dividing by 30,000}$$

$$\ln 2 = \ln e^{0.08T} \qquad \text{Taking the natural logarithm on both sides}$$

$$\ln 2 = 0.08T$$

$$\frac{\ln 2}{0.08} = T \qquad \text{Dividing}$$

$$8.7 \approx T.$$

Thus the original investment of $30,000 will double in about 8.7 yr, as shown in the following graph of the growth function.

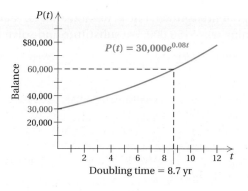

Doubling time = 8.7 yr

Do Exercise 7.

7. Interest Compounded Continuously.

a) Suppose that $5000 is invested and grows to $6356.25 in 4 yr. Find the interest rate and then the exponential growth function.

b) What is the balance after 1 yr? 2 yr? 10 yr?

c) What is the doubling time?

Answers on page A-49

Take a look at the graphs in Example 6 and in Margin Exercise 8. Note that the data tend to be rising in a manner that might fit an exponential function. To fit an exponential function to the data, we choose a pair of data points and use them to determine P_0 and k.

EXAMPLE 6 *1929 Pierce-Arrow.* The restoration of antique automobiles is a popular hobby among many people. The value of a 1929 Pierce-Arrow Roadster was $4800 when it was sold brand-new. Fully restored, the value of this automobile in 2001 was $150,000.

Source: Dr. Rex B. Gosnell; Gilmore Museum, Kalamazoo, MI

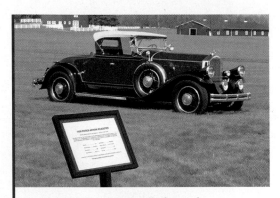

Dr. and Mrs. Rex B. Gosnell (the uncle and aunt of your author) donated this 1929 Pierce-Arrow Roadster to the Gilmore-Classic Car Club of America Museum in Hickory Corners, Michigan.

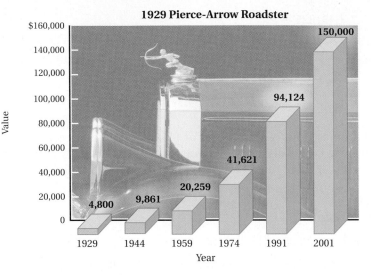

1929 Pierce-Arrow Roadster

a) If we let $t = 0$ correspond to 1929 and $t = 72$ to 2001, then t is the number of years since 1929. Use the data points $(0, 4800)$ and $(72, 150,000)$ to find the exponential growth rate and fit an exponential growth function to the data.

b) Use the function from part (a) to predict the value of the automobile in 2019.

c) Assuming exponential growth continues, predict the year in which the value of such an automobile will be $200,000.

a) We use the equation $P(t) = P_0 e^{kt}$, where $P(t)$ is the value of the automobile, in dollars, t years after 1929. In 1929, at $t = 0$, the value was $4800. Thus we substitute 4800 for P_0:

$$P(t) = 4800e^{kt}.$$

To find the exponential growth rate k, we note that 72 years later, at the end of 2001, the value was $150,000. We substitute and solve for k:

$$\left.\begin{array}{l} P(72) = 4800e^{k(72)} \\ 150,000 = 4800e^{k(72)} \end{array}\right\} \quad \text{Substituting}$$

$$31.25 = e^{72k} \qquad \text{Dividing by 4800}$$

$$\ln 31.25 = \ln e^{72k} \qquad \text{Taking the natural logarithm on both sides}$$

$$3.4420 \approx 72k \qquad \ln e^a = a$$

$$0.048 \approx k. \qquad \text{Dividing by 72}$$

The exponential growth rate was 0.048, or 4.8%, and the exponential growth function is $P(t) = 4800e^{0.048t}$.

b) Since 2019 is 90 yr from 1929, we substitute 90 for t:

$$P(90) = 4800e^{0.048(90)} \approx \$360,905.$$

The value of the automobile will be about $360,905 in 2019.

c) To predict when the value of the automobile will be $200,000, we substitute 200,000 for $P(t)$ and solve for t:

$$200{,}000 = 4800e^{0.048t}$$

$$41.6667 \approx e^{0.048t} \qquad \text{Dividing by 4800}$$

$$\ln 41.6667 \approx \ln e^{0.048t} \qquad \text{Taking the natural logarithm on both sides}$$

$$3.7297 \approx 0.048t \qquad \ln e^a = a$$

$$77.7 \approx t. \qquad \text{Dividing by 0.048}$$

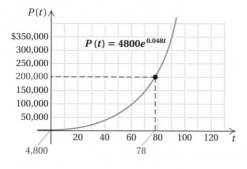

Rounding to 78 yr, we see that, according to this model, by the end of 1929 + 78, or 2007, the value of a 1929 Pierce-Arrow Roadster will be $200,000.

Do Exercise 8.

In some real-life situations, a quantity or population is *decreasing* or *decaying* exponentially.

EXPONENTIAL DECAY MODEL

An **exponential decay model** is a function of the form

$$P(t) = P_0 e^{-kt}, \quad k > 0,$$

where P_0 is the quantity present at time 0, $P(t)$ is the amount present at time t, and k is the **decay rate.** The **half-life** is the amount of time necessary for half of the quantity to decay.

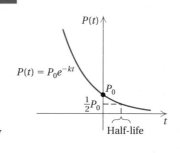

EXAMPLE 7 *Carbon Dating.* The radioactive element carbon-14 has a half-life of 5750 yr. The percentage of carbon-14 present in the remains of organic matter can be used to determine the age of that organic matter. Recently, while digging in Chaco Canyon, New Mexico, archaeologists found corn pollen that had lost 38.1% of its carbon-14. The age of this corn pollen was evidence that Indians had been cultivating crops in the Southwest centuries earlier than scientists had thought. What was the age of the pollen?

Source: *American Anthropologist*

---

8. Online Service Fees. The fees paid by small businesses for online services are projected to increase exponentially over the next few years, as shown in the graph below.

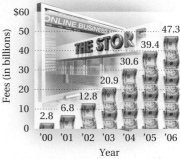

Online Fees for Small Businesses

Source: Keenan Vision

a) Let $P(t) = P_0 e^{kt}$, where $P(t)$ is the online fees, in billions of dollars, t years after 2000. Then $t = 0$ corresponds to 2000 and $t = 6$ corresponds to 2006. Use the data points $(0, 2.8)$ and $(6, 47.3)$ to find the exponential growth rate and fit an exponential growth function to the data.

b) Use the function found in part (a) to predict the online fees that will be paid in 2020.

c) Assuming exponential growth continues, predict the year in which the online fees will be $100 billion.

Answers on page A-49

9. Carbon Dating. How old is an animal bone that has lost 30% of its carbon-14?

We first find k. To do so, we use the concept of half-life. When $t = 5750$ (the half-life), $P(t)$ will be half of P_0. Then

$$0.5P_0 = P_0e^{-k(5750)}$$

$$0.5 = e^{-5750k} \qquad \text{Dividing by } P_0$$

$$\ln 0.5 = \ln e^{-5750k} \qquad \text{Taking the natural logarithm on both sides}$$

$$\ln 0.5 = -5750k$$

$$\frac{\ln 0.5}{-5750} = k$$

$$0.00012 \approx k.$$

Now we have a function for the decay of carbon-14:

$$P(t) = P_0e^{-0.00012t}. \qquad \text{This completes the first part of our solution.}$$

(*Note*: This equation can be used for any subsequent carbon-dating problem.)

If the corn pollen has lost 38.1% of its carbon-14 from an initial amount P_0, then 100% − 38.1%, or 61.9%, of P_0 is still present. To find the age t of the pollen, we solve the following equation for t:

$$61.9\%P_0 = P_0e^{-0.00012t} \qquad \text{We want to find } t \text{ for which } P(t) = 0.619P_0.$$

$$0.619 = e^{-0.00012t} \qquad \text{Dividing by } P_0$$

$$\ln 0.619 = \ln e^{-0.00012t} \qquad \text{Taking the natural logarithm on both sides}$$

$$\ln 0.619 = -0.00012t \qquad \ln e^a = a$$

$$\frac{\ln 0.619}{-0.00012} \approx t \qquad \text{Dividing by } -0.00012$$

$$4000 \approx t. \qquad \text{Rounding}$$

The pollen is about 4000 yr old.

Answer on page A-49   *Do Exercise 9.*

CALCULATOR CORNER

Mathematical Modeling: Fitting an Exponential Function to Data   We can use the exponential regression feature on a graphing calculator to fit an exponential function of the form $y = a \cdot b^x$ to a set of data. When we use the regression feature to find a mathematical model, we can use all the data points rather than only two points, as we did in Example 6.

The table at right lists the data from the graph in Margin Exercise 8 regarding fees paid by small businesses for online services.

YEARS SINCE 2000	FEES (in billions)
0	$ 2.8
1	6.8
2	12.8
3	20.9
4	30.6
5	39.4
6	47.3

Source: Keenan Vision

(continued)

a) Use the exponential regression feature on a graphing calculator to fit an exponential function to the data.

b) Make a scatterplot of the data. Then graph the function found in part (a) with the scatterplot.

c) Use the function found in part (a) to predict the fees that small businesses will pay for online services in 2020. Compare your answer with the one found in part (b) of Margin Exercise 8.

a) We enter the data on the STAT list editor screen as described in the Calculator Corner on p. 226. Then, to fit an exponential function to the data, we press **STAT** ▷ ⓪ **VARS** ▷ ① ① **ENTER** . The first three keystrokes display the STAT CALC menu and select EXPREG from that menu. The keystrokes **VARS** ▷ ① ① copy the regression equation to the equation-editor screen as y_1. We see that the regression equation is $y = 4.077027434 \cdot 1.583277276^x$.

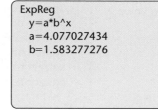

L1	L2	L3	3
0	2.8		
1	6.8		
2	12.8		
3	20.9		
4	30.6		
5	39.4		
6	47.3		
L3(1)=			

ExpReg
y=a*b^x
a=4.077027434
b=1.583277276

b) To make a scatterplot of the data and graph the regression equation along with the scatterplot, we first turn on Plot 1 as described in the Calculator Corner on p. 226. Then we press **ZOOM** ⑨ to see the graph.

c) We can use any of the methods described in the Calculator Corner on p. 177 to predict the fees that will be paid by small businesses for online services in 2020. We will use a table set in ASK mode. The year 2020 is 20 years after the year 2000, so we enter 20 for x and find that $y \approx 39{,}947$. Thus we predict that in 2020 small businesses will pay about $39,947 billion, or $39.947 trillion, for online services. The prediction found with exponential regression is approximately $5400 billion higher than the prediction found algebraically in Margin Exercise 8(b).

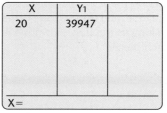

X	Y₁	
20	39947	
X=		

Exercises: As the following table shows, the number of Hispanic-owned businesses in the United States has soared in recent years.

YEARS SINCE 1982	NUMBER OF BUSINESSES
0	284,011
5	489,973
10	862,605
15	1,121,387
19	1,500,000

Source: U.S. Small Business Administration

1. Use the exponential regression feature on a graphing calculator to fit an exponential function to the data.

2. Make a scatterplot of the data. Then graph the function found in part (a) with the scatterplot.

3. Use the function found in part (a) to predict the number of Hispanic-owned businesses in 2010.

Translating for Success

1. *Grain Flow.* Grain flows through spout A four times faster than through spout B. When grain flows through both spouts, a grain bin is filled in 8 hr. How many hours would it take to fill the bin if grain flows through spout B alone?

2. *Rectangle Dimensions.* The perimeter of a rectangle is 50 ft. The width of the rectangle is 10 ft shorter than the length. Find the length and the width.

3. *Wire Cutting.* A 1086-in. wire is cut into three pieces. The second piece is 8 in. longer than the first. The third is four-fifths as long as the first. How long is each piece?

4. *iPod Sales.* Global sales of iPod music players increased exponentially from 0.1 million in 2002 to 4.6 million in 2005. Find the exponential growth rate k, and write an exponential growth function I for which $I(t)$ approximates the global sales of iPod music players t years after 2002.

5. *Child Spending.* In 2000, children age 12 and younger influenced $500 million in spending on family purchases. This was a 166% increase over the spending in 1997. How much was spent in 1997?

The goal of these matching questions is to practice step (2), *Translate*, of the five-step problem-solving process. Translate each word problem to an equation or a system of equations and select a correct translation from equations A–O.

A. $I(t) = 0.1e^{1.28t}$

B. $40x = 50(x - 3)$

C. $x^2 + (x - 10)^2 = 50^2$

D. $\dfrac{8}{x} + \dfrac{8}{4x} = 1$

E. $x + 166\%x = 500$

F. $\dfrac{500}{x} + \dfrac{500}{x - 2} = 8$

G. $x + y = 90,$
$0.1x + 0.25y = 16.50$

H. $x + (x + 1) + (x + 2) = 39$

I. $x + (x + 8) + \dfrac{4}{5}x = 1086$

J. $x + (x + 2) + (x + 4) = 39$

K. $I(t) = 1.28e^{0.1t}$

L. $x^2 + (x + 8)^2 = 1086$

M. $2x + 2(x - 10) = 50$

N. $\dfrac{500}{x} = \dfrac{500}{x + 2} + 8$

O. $x + y = 90,$
$0.1x + 0.25y = 1650$

Answers on page A-49

6. *Uniform Numbers.* The numbers on three baseball uniforms are consecutive integers whose sum is 39. Find the integers.

7. *Triangle Dimensions.* The hypotenuse of a right triangle is 50 ft. The length of one leg is 10 ft shorter than the other. Find the lengths of the legs.

8. *Coin Mixture.* A collection of dimes and quarters is worth $16.50. There are 90 coins in all. How many of each coin are there?

9. *Car Travel.* Emma drove her car 500 mi to see her friend. The return trip was 2 hr faster at a speed that was 8 mph more. Find her return speed.

10. *Train Travel.* An Amtrak train leaves a station and travels east at 40 mph. Three hours later, a second train leaves on a parallel track traveling east at 50 mph. After what amount of time will the second train overtake the first?

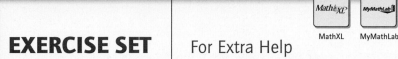
a Solve.

Sound Levels. Use the decibel formula from Example 1 for Exercises 1–4.

1. *Acoustics.* The intensity of sound of a riveter at work is about 3.2×10^{-3} W/m². How high is this sound level in decibels?

2. *Acoustics.* The intensity of sound of a dishwasher is about 2.5×10^{-6} W/m². How high is this sound level in decibels?

3. *Music.* At a performance of the rock group Phish, sound measurements of 105 dB were recorded. What is the intensity of such sounds?

4. *Jet Plane at Takeoff.* A jet plane at takeoff can generate sound measurements of 130 dB. What is the intensity of such sounds?

pH. Use the pH formula from Example 2 for Exercises 5–8.

5. *Milk.* The hydrogen ion concentration of milk is about 1.6×10^{-7} moles per liter. Find the pH.

6. *Mouthwash.* The hydrogen ion concentration of mouthwash is about 6.3×10^{-7} moles per liter. Find the pH.

7. *Alkalosis.* When the pH of a patient's blood rises above 7.4, a condition called *alkalosis* sets in. Alkalosis can be fatal at a pH level above 7.8. What would the hydrogen ion concentration of the patient's blood be at that point?

8. *Orange Juice.* The pH of orange juice is 3.2. What is its hydrogen ion concentration?

Walking Speed. In a study by psychologists Bornstein and Bornstein, it was found that the average walking speed w, in feet per second, of a person living in a city of population P, in thousands, is given by the function

$$w(P) = 0.37 \ln P + 0.05.$$

In Exercises 9–12, various cities and their populations are given. Find the walking speed of people in each city in 2000.

Source: *International Journal of Psychology*

9. Phoenix, Arizona: 3,251,876
(Metropolitan area)

10. Las Vegas, Nevada: 1,563,282
(Metropolitan area)

11. Fayetteville–Springdale–Rogers, Arkansas: 311,121
Source: U.S. Bureau of the Census, Census 2000

12. Naples, Florida: 251,377

13. *Pay It Forward.* The movie *Pay It Forward*, starring Kevin Spacey, Helen Hunt, and Haley Joel Osment, suggests an application of exponential functions. In the film, Osment's character does an act of kindness to each of three people. In return, each of these people must agree to "pay forward" three acts of kindness to other people, and so on. Let's assume that each person performs these acts of kindness within 1 month. Then the number N of people who receive the acts of kindness in month t is given by the function

$$N(t) = 3^t.$$

a) How many people would have received an act of kindness in month 5?
b) The population of the world is about 6.4 billion. After what amount of time will the acts of kindness reach the entire world population?
c) What is the doubling time for the number of people who receive an act of kindness?

14. *Spread of Rumor.* The number of people who hear a rumor increases exponentially. If 20 people start a rumor and if each person who hears the rumor repeats it to two people a day, the number of people N who have heard the rumor after t days is given by the function

$$N(t) = 20(3)^t.$$

a) How many people have heard the rumor after 5 days?
b) After what amount of time will 1000 people have heard the rumor?
c) What is the doubling time for the number of people who have heard the rumor?

15. *Projected College Costs.* The cost of tuition, books, room, and board at a state university is projected to follow the exponential function

$$C(t) = 11{,}054(1.06)^t,$$

where C is the cost, in dollars, and t is the number of years after 2000.

Source: College Board, Senate Labor Committee

a) Find the college costs in 2010.
b) In what year will the cost be $21,000?
c) What is the doubling time of the costs?

16. *Salvage Value.* A color photocopier is purchased for $5200. Its value each year is about 70% of its value in the preceding year. Its value in dollars after t years is given by the exponential function

$$V(t) = 5200(0.7)^t.$$

a) Find the salvage value of the copier after 3 yr.
b) After what amount of time will the salvage value be $1200?
c) After what amount of time will the salvage value be half the original value?

Growth. Use the exponential growth model $P(t) = P_0 e^{kt}$ for Exercises 17–22.

17. *World Population Growth.* In 2004, the population of the world reached 6.4 billion, and the exponential growth rate was 1.14% per year.

a) Find the exponential growth function.
b) What will the population be in 2010?
c) In what year will the population be 10 billion?
d) What is the doubling time?

18. *Population Growth of Laredo, Texas.* In 2000, with a population of 193,117, Laredo, Texas, was the ninth fastest-growing metropolitan area in the United States, with an exponential growth rate of 3.7% per year.

Source: U.S. Bureau of the Census

a) Find the exponential growth function.
b) What will the population be in 2010?
c) In what year will the population be 600,000?
d) What is the doubling time?

19. *Interest Compounded Continuously.* Suppose that P_0 is invested in a savings account in which interest is compounded continuously at 6% per year.

a) Express $P(t)$ in terms of P_0 and 0.06.
b) Suppose that $5000 is invested. What is the balance after 1 yr? 2 yr? 10 yr?
c) When will the investment of $5000 double itself?

20. *Interest Compounded Continuously.* Suppose that P_0 is invested in a savings account in which interest is compounded continuously at 5.4% per year.

a) Express $P(t)$ in terms of P_0 and 0.054.
b) Suppose that $10,000 is invested. What is the balance after 1 yr? 2 yr? 10 yr?
c) When will the investment of $10,000 double itself?

21. *Population Growth of Las Vegas, Nevada.* In 2000, the population of Las Vegas, Nevada, was 1,563,282. It had grown from a population of 852,737 in 1990, and was the fastest growing metropolitan area in the United States. Assume that the population increases according to an exponential growth function.

Source: U.S. Bureau of the Census

a) Let $t = 0$ correspond to 1990 and $t = 10$ correspond to 2000. Then t is the number of years since 1990. Use the data points $(0, 852,737)$ and $(10, 1,563,282)$ to find the exponential growth rate and fit an exponential growth function $P(t) = P_0 e^{kt}$ to the data, where $P(t)$ is the population of Las Vegas t years after 1990.
b) Use the function found in part (a) to predict the population of Las Vegas in 2010.
c) When will the population reach 8 million?

22. *First-Class Postage.* First-class postage (for the first ounce) was 20¢ in 1981 and 37¢ in 2002. Assume the cost increases according to an exponential growth function.

Source: U.S. Postal Service

a) Let $t = 0$ correspond to 1981 and $t = 21$ correspond to 2002. Then t is the number of years since 1981. Use the data points $(0, 20)$ and $(21, 37)$ to find the exponential growth rate and fit an exponential growth function $P(t) = P_0 e^{kt}$ to the data, where $P(t)$ is the cost of first-class postage, in cents, t years after 1981.
b) Use the function found in part (a) to predict the cost of first-class postage in 2008.
c) When will the cost of first-class postage be $1.00 or 100¢?

Carbon Dating. Use the carbon-14 decay function $P(t) = P_0 e^{-0.00012t}$ for Exercises 23 and 24.

23. *Carbon Dating.* When archaeologists found the Dead Sea scrolls, they determined that the linen wrapping had lost 22.3% of its carbon-14. How old was the linen wrapping?

24. *Carbon Dating.* In 1998, researchers found an ivory tusk that had lost 18% of its carbon-14. How old was the tusk?

Decay. Use the exponential decay function $P(t) = P_0 e^{-kt}$ for Exercises 25 and 26.

25. *Chemistry.* The decay rate of iodine-131 is 9.6% per day. What is the half-life?

26. *Chemistry.* The decay rate of krypton-85 is 6.3% per day. What is the half-life?

27. *Home Construction.* The chemical urea formaldehyde was found in some insulation used in houses built during the mid to late 1960s. Unknown at the time was the fact that urea formaldehyde emitted toxic fumes as it decayed. The half-life of urea formaldehyde is 1 yr. What is its decay rate?

28. *Plumbing.* Lead pipes and solder are often found in older buildings. Unfortunately, as lead decays, toxic chemicals can get in the water resting in the pipes. The half-life of lead is 22 yr. What is its decay rate?

29. *Decline of Discarded Yard Waste.* The amount of discarded yard waste has declined considerably in recent years because of increased recycling and composting. In 1996, 17.5 million tons were discarded, but by 1998 the figure dropped to 14.5 million tons. Assume the amount of discarded yard waste is decreasing according to the exponential decay model.

Source: *Statistical Abstract of the United States, 2000*

a) Find the value k, and write an exponential function that describes the amount of yard waste discarded t years after 1996.

b) Estimate the amount of discarded yard waste in 2010.

c) In what year (theoretically) will only 1 ton of yard waste be discarded?

30. *Decline in Cases of Mumps.* The number of cases of mumps has dropped exponentially from 5300 in 1990 to 300 in 2000.

Source: *Statistical Abstract of the United States, 2003*

a) Find the value k, and write an exponential function that can be used to estimate the number of cases t years after 1990.

b) Estimate the number of cases of mumps in 2008.

c) In what year (theoretically) will there be only 1 case of mumps?

31. *Value of a Sports Card.* Because he objected to smoking, and because his first baseball card was issued in cigarette packs, the great shortstop Honus Wagner halted production of his card before many were produced. One of these cards was sold in 1991 for $451,000 and again in 1996 for $640,500. For the following questions, assume that the card's value increases exponentially.

WAGNER, PITTSBURG

a) Find an exponential function $V(t)$, where t is the number of years after 1991, if $V_0 = 451,000$.

b) Predict the card's value in 2015.

c) What is the doubling time of the value of the card?

d) In what year will the value of the card first exceed $1,000,000?

e) In 2001, this card drew a bid of $1.1 million on eBay. How does this correlate with what can be predicted from this function? Would you buy the card?

32. *Art Masterpieces.* As of August 2004, the most ever paid for a painting is $104,168,000, paid in 2004 for Pablo Picasso's "Garçon à la Pipe." The same painting sold for $30,000 in 1950.

Source: BBC News, 5/6/04

a) Find the exponential growth rate k, and determine the exponential growth function V, for which $V(t)$ is the painting's value, in millions of dollars, t years after 1950.

b) Estimate the value of the painting in 2009.

c) What is the doubling time for the value of the painting?

d) How long after 1950 will the value of the painting be $1 billion?

33. *Population Decay of Pittsburgh.* The population of the metropolitan area of Pittsburgh declined from 2,394,811 in 1990 to 2,242,798 in 2000. Assume the population decreases according to an exponential decay function.

Source: U.S. Bureau of the Census

a) Let $t = 0$ correspond to 1990 and $t = 10$ correspond to 2000. Then t is the number of years since 1990. Use the data points $(0, 2{,}394{,}811)$ and $(10, 2{,}242{,}798)$ to find the exponential decay rate and fit an exponential decay function $P(t) = P_0 e^{-kt}$ to the data, where $P(t)$ is the population of Pittsburgh t years after 1990.

b) Use the function found in part (a) to predict the population of Pittsburgh in 2010.

c) When will the population decline to 1 million?

Pittsburgh   PA

34. *Skateboarding.* The number of skateboarders of age x (with $x \geq 16$), in thousands, can be approximated by
$$N(x) = 600(0.873)^{x-16}.$$

Sources: Based on figures from the U.S. Bureau of the Census and *Statistical Abstract of the United States*, 2000

a) Estimate the number of 41-yr-old skateboarders.
b) At what age are there only 2000 skateboarders? 200 skateboarders?

35. $\mathbf{D_W}$ Write a problem for a classmate to solve in which data that seem to fit an exponential growth function are provided. Try to find data in a newspaper to make the problem as realistic as possible.

36. $\mathbf{D_W}$ Do some research or consult with a chemist to determine how carbon dating is carried out. Write a report.

SKILL MAINTENANCE

Compute and simplify. Express answers in the form $a + bi$, where $i^2 = -1$.   [6.8c, d, e]

37. i^{46} **38.** i^{48} **39.** i^{53} **40.** i^{97} **41.** $i^{14} + i^{15}$

42. $i^{18} - i^{16}$ **43.** $\dfrac{8 - i}{8 + i}$ **44.** $\dfrac{2 + 3i}{5 - 4i}$ **45.** $(5 - 4i)(5 + 4i)$ **46.** $(-10 - 3i)^2$

SYNTHESIS

Use a graphing calculator to solve each of the following equations.

47. $2^x = x^{10}$ **48.** $(\ln 2)x = 10 \ln x$ **49.** $x^2 = 2^x$ **50.** $x^3 = e^x$

51. *Sports Salaries.* In 2001, Derek Jeter of the New York Yankees signed a $189 million 10-yr contract that will pay him $21 million in 2010. How much would Yankee owner George Steinbrenner need to invest in 2001 at 5% interest compounded continuously, in order to have the $21 million for Jeter in 2010?

52. *Nuclear Energy.* Plutonium-239 (Pu-239) is used in nuclear energy plants. The half-life of Pu-239 is 24,360 yr. How long will it take for a fuel rod of Pu-239 to lose 90% of its radioactivity?

Source: *Microsoft Encarta 97 Encyclopedia*

The review that follows is meant to prepare you for a chapter exam. It consists of three parts. The first part, Concept Reinforcement, is designed to increase understanding of the concepts through true/false exercises. The second part is a list of important properties and formulas. The third part is the Review Exercises. These provide practice exercises for the exam, together with references to section objectives so you can go back and review. Before beginning, stop and look back over the skills you have obtained. What skills in mathematics do you have now that you did not have before studying this chapter?

↶ CONCEPT REINFORCEMENT

Determine whether the statement is true or false. Answers are given at the back of the book.

_____ **1.** $\ln(ab) = \ln a - \ln b$

_____ **2.** $\log_a \dfrac{m}{n} = \log_a m - \log_a n$

_____ **3.** If we find that $\log(78) \approx 1.8921$ on a scientific calculator, we also know that $10^{1.8921} \approx 78$.

_____ **4.** $\log_a 0 = 1$

_____ **5.** $\log_a 1 = 0, \quad a > 0$

_____ **6.** $\ln(35) = \ln 7 + \ln 5$

_____ **7.** $\ln(64) = 6 \ln 2$

_____ **8.** The functions $f(x) = e^x$ and $g(x) = \ln x$ are inverses of each other.

IMPORTANT PROPERTIES AND FORMULAS

Exponential Functions: $\quad f(x) = a^x, \quad f(x) = e^x$

Composition of Functions: $\quad f \circ g(x) = f(g(x))$

Definition of Logarithms: $\quad y = \log_a x$ is that number y such that $x = a^y$,
where $x > 0$ and a is a positive constant other than 1.

Properties of Logarithms:

$\log M = \log_{10} M, \qquad \log_a 1 = 0, \qquad \ln M = \log_e M, \qquad \log_a a = 1,$

$\log_a MN = \log_a M + \log_a N, \qquad \log_a a^k = k, \qquad \log_a M^k = k \cdot \log_a M, \qquad \log_b M = \dfrac{\log_a M}{\log_a b},$

$\log_a \dfrac{M}{N} = \log_a M - \log_a N, \qquad e \approx 2.7182818284\ldots$

Growth: $\quad P(t) = P_0 e^{kt}$

Decay: $\quad P(t) = P_0 e^{-kt}$

Carbon Dating: $\quad P(t) = P_0 e^{-0.00012t}$

Interest Compounded Annually: $\qquad A = P(1 + r)^t$

Interest Compounded n Times per Year: $\quad A = P\left(1 + \dfrac{r}{n}\right)^{nt}$

Interest Compounded Continuously: $\quad P(t) = P_0 e^{kt}, \quad$ where P_0 dollars are invested for t years at interest rate k

Review Exercises

1. Find the inverse of the relation   [8.2a]

$$\{(-4, 2), (5, -7), (-1, -2), (10, 11)\}.$$

Determine whether the function is one-to-one. If it is, find a formula for its inverse.   [8.2b, c]

2. $f(x) = 4 - x^2$

3. $g(x) = \dfrac{2x - 3}{7}$

4. $f(x) = 8x^3$

5. $f(x) = \dfrac{4}{3 - 2x}$

6. Graph the function $f(x) = x^3 + 1$ and its inverse using the same set of axes.   [8.2c]

Graph.

7. $f(x) = 3^{x-1}$ [8.1a]

x	f(x)
0	
1	
2	
3	
−1	
−2	
−3	

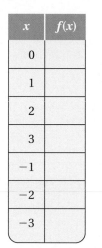

8. $f(x) = \log_3 x$, or $y = \log_3 x$ [8.3a]

$y = \log_3 x \rightarrow x = $ _____

x, or 3^y	y
	0
	1
	2
	3
	−1
	−2
	−3

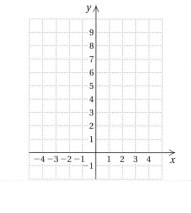

9. $f(x) = e^{x+1}$ [8.5c]

x	f(x)
0	
1	
2	
3	
−1	
−2	
−3	

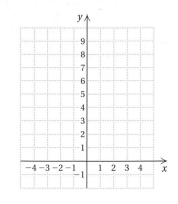

10. $f(x) = \ln (x - 1)$ [8.5c]

x	f(x)

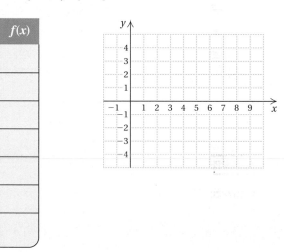

11. Find $f \circ g(x)$ and $g \circ f(x)$ if $f(x) = x^2$ and $g(x) = 3x - 5$.   [8.2d]

12. If $h(x) = \sqrt{4 - 7x}$, find $f(x)$ and $g(x)$ such that $h(x) = f \circ g(x)$.   [8.2d]

Convert to a logarithmic equation.   [8.3b]

13. $10^4 = 10,000$

14. $25^{1/2} = 5$

Convert to an exponential equation.   [8.3b]

15. $\log_4 16 = x$

16. $\log_{1/2} 8 = -3$

Find each of the following.   [8.3c]

17. $\log_3 9$

18. $\log_{10} \frac{1}{10}$

19. $\log_m m$

20. $\log_m 1$

Find the common logarithm, to four decimal places, using a calculator.   [8.3d]

21. $\log\left(\dfrac{78}{43,112}\right)$

22. $\log(-4)$

Express in terms of logarithms of x, y, and z.   [8.4d]

23. $\log_a x^4 y^2 z^3$

24. $\log \sqrt[4]{\dfrac{z^2}{x^3 y}}$

Express as a single logarithm.   [8.4d]

25. $\log_a 8 + \log_a 15$

26. $\frac{1}{2}\log a - \log b - 2\log c$

Simplify.   [8.4e]

27. $\log_m m^{17}$

28. $\log_m m^{-7}$

Given $\log_a 2 = 1.8301$ and $\log_a 7 = 5.0999$, find each of the following.   [8.4d]

29. $\log_a 28$

30. $\log_a 3.5$

31. $\log_a \sqrt{7}$

32. $\log_a \frac{1}{4}$

Find each of the following, to four decimal places, using a calculator.   [8.5a]

33. $\ln 0.06774$

34. $e^{-0.98}$

35. $e^{2.91}$

36. $\ln 1$

37. $\ln 0$

38. $\ln e$

Find the logarithm using the change-of-base formula. [8.5b]

39. $\log_5 2$

40. $\log_{12} 70$

Solve. Where appropriate, give approximations to four decimal places.   [8.6a, b]

41. $\log_3 x = -2$

42. $\log_x 32 = 5$

43. $\log x = -4$

44. $3\ln x = -6$

45. $4^{2x-5} = 16$

46. $2^{x^2} \cdot 2^{4x} = 32$

47. $4^x = 8.3$

48. $e^{-0.1t} = 0.03$

49. $\log_4 16 = x$

50. $\log_4 x + \log_4 (x - 6) = 2$

51. $\log_2 (x + 3) - \log_2 (x - 3) = 4$

52. $\log_3 (x - 4) = 3 - \log_3 (x + 4)$

Solve.   [8.7a, b]

53. *Acoustics.*   The intensity of the sound at the foot of Niagara Falls is about 10^{-3} W/m². How high is this sound level in decibels? [Use $L = 10 \cdot \log(I/I_0)$.]

54. *TiVo Subscriptions.* A TiVo is a digital video recorder usually connected to a satellite dish for which a subscriber pays a monthly fee. The number S, in millions, of TiVo subscriptions t years after 2003 can be approximated by the exponential function

$$S(t) = 0.15(1.4)^t.$$

Source: TiVo Company data

a) Predict the number of TiVo subscriptions in 2008, 2010, and 2020.
b) In what year will the number of TiVo subscriptions be 20 million?
c) What is the doubling time of the number of subscriptions?
d) Graph the function.

55. *Real-Estate Appraisal.* In 2005, Lucille invested in a building lot that cost $40,000. By 2008, fair-market value for the lot was $53,000. Assume that the value of the lot is increasing exponentially.

a) Find the value k, and write an exponential function that describes the value of Lucille's lot t years after 2005.
b) Predict the value of the lot in 2015.
c) In what year will the value of the lot first reach $85,000?

56. The value of Jose's stock market portfolio doubled in 3 yr. What was the exponential growth rate?

57. How long will it take $7600 to double itself if it is invested at 8.4%, compounded continuously?

58. How old is a skeleton that has lost 34% of its carbon-14? (Use $P(t) = P_0 e^{-0.00012t}$.)

59. $\mathbf{D_W}$ Explain why you cannot take the logarithm of a negative number. [8.3a]

60. $\mathbf{D_W}$ Explain why $\log_a 1 = 0$. [8.3a]

SYNTHESIS

Solve. [8.6a, b]

61. $\ln (\ln x) = 3$

62. $5^{x+y} = 25$, $2^{2x-y} = 64$

Graph.

1. $f(x) = 2^{x+1}$

x	$f(x)$

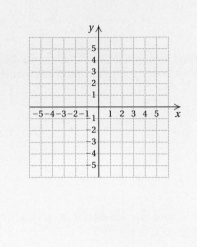

2. $y = \log_2 x$

x	y

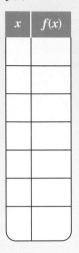

3. $f(x) = e^{x-2}$

x	$f(x)$

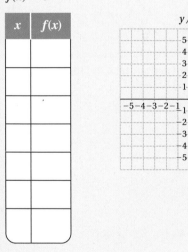

4. $f(x) = \ln(x - 4)$

x	$f(x)$

5. Find the inverse of the relation $\{(-4, 3), (5, -8), (-1, -3), (10, 12)\}$.

Determine whether the function is one-to-one. If it is, find a formula for its inverse.

6. $f(x) = 4x - 3$

7. $f(x) = (x + 1)^3$

8. $f(x) = 2 - |x|$

9. Find $f \circ g(x)$ and $g \circ f(x)$ if $f(x) = x + x^2$ and $g(x) = 5x - 2$.

10. Convert to a logarithmic equation:
$$256^{1/2} = 16.$$

11. Convert to an exponential equation:
$$m = \log_7 49.$$

Find each of the following.

12. $\log_5 125$

13. $\log_t t^{23}$

14. $\log_p 1$

Find the common logarithm, to four decimal places, using a calculator.

15. $\log 0.0123$

16. $\log(-5)$

17. Express in terms of logarithms of a, b, and c:
$$\log \frac{a^3 b^{1/2}}{c^2}.$$

18. Express as a single logarithm:
$$\tfrac{1}{3} \log_a x - 3 \log_a y + 2 \log_a z.$$

Given $\log_a 2 = 0.301$, $\log_a 6 = 0.778$, and $\log_a 7 = 0.845$, find each of the following.

19. $\log_a \frac{2}{7}$

20. $\log_a 12$

Find each of the following, to four decimal places, using a calculator.

21. $\ln 807.39$

22. $e^{4.68}$

23. $\ln 1$

24. Find $\log_{18} 31$ using the change-of-base formula.

Solve. Where appropriate, give approximations to four decimal places.

25. $\log_x 25 = 2$

26. $\log_4 x = \frac{1}{2}$

27. $\log x = 4$

28. $\ln x = \frac{1}{4}$

29. $7^x = 1.2$

30. $\log(x^2 - 1) - \log(x - 1) = 1$

31. $\log_5 x + \log_5 (x + 4) = 1$

32. *Tomatoes.* What is the pH of tomatoes if the hydrogen ion concentration is 6.3×10^{-5} moles per liter? (Use $\text{pH} = -\log[\text{H}^+]$.)

33. *Projected College Costs.* The cost of tuition, books, room, and board at a private university (four-year institutions) is projected to follow the exponential function

$$C(t) = 21{,}856(1.045)^t,$$

where C is the cost, in dollars, and t is the number of years after 2000.

Sources: U.S. Department of Education; National Center for Education Statistics

a) Find the college costs in 2008.
b) In what year will the cost be \$33,942?
c) What is the doubling time of the costs?

34. *Population Growth of Canada.* The population of Canada was 31.902 million in 2002 and 32.508 million in 2004. Assume the number N is growing exponentially according to the equation

$$N(t) = N_0 e^{kt}.$$

Source: Census Bureau of Canada

a) Find k and write the exponential growth function.
b) Estimate the population in 2008 and in 2015.
c) When will the population be 50 million?
d) What is the doubling time?

35. An investment with interest compounded continuously doubled in 15 yr. What is the interest rate?

36. How old is an animal bone that has lost 43% of its carbon-14? (Use $P(t) = P_0 e^{-0.00012t}$.)

SYNTHESIS

37. Solve: $\log_3 |2x - 7| = 4$.

38. If $\log_a x = 2$, $\log_a y = 3$, and $\log_a z = 4$, find

$$\log_a \frac{\sqrt[3]{x^2 z}}{\sqrt[3]{y^2 z^{-1}}}.$$

CHAPTER 8: Exponential and
Logarithmic Functions

1. *Time Interval of Appointments.* In an effort to minimize waiting time for patients in a doctor's office without increasing the physician's idle time, the formula

$$I = 1.08\frac{T}{N},$$

was developed to compute the time interval of appointments, where I is the interval time, in minutes, between scheduled appointments, T is the total number of minutes that a physician spends with patients per day, and N is the total number of scheduled appointments.

a) A doctor determines that she has a total of 8 hr each day to see patients and wants to keep 35 appointments. What is the interval time I of each appointment?

b) Suppose the doctor decides she wants 17.28 min of interval time for appointments. How many appointments can she keep per day?

c) Solve the interval-time formula for T.

d) Solve the interval-time formula for N.

e) Explain why you might want to use the new formulas found in parts (c) and (d).

Simplify.

2. $\left| -\frac{5}{2} + \left(-\frac{7}{2} \right) \right|$

3. $(-5x^4 y^{-3} z^2)(-4x^2 y^2)$

4. $2x - 3 - 2[5 - 3(2 - x)]$

5. $3^3 + 2^2 - (32 \div 4 - 16 \div 8)$

Solve.

6. $8(2x - 3) = 6 - 4(2 - 3x)$

7. $x(x - 3) = 10$

8. $\begin{aligned} x + y - 3z &= -1, \\ 2x - y + z &= 4, \\ -x - y + z &= 1 \end{aligned}$

9. $\begin{aligned} 4x - 3y &= 15, \\ 3x + 5y &= 4 \end{aligned}$

10. $\dfrac{7}{x^2 - 5x} - \dfrac{2}{x - 5} = \dfrac{4}{x}$

11. $\sqrt{x - 1} = \sqrt{x + 4} - 1$

12. $x - 8\sqrt{x} + 15 = 0$

13. $x^4 - 13x^2 + 36 = 0$

14. $\log_8 x = 1$

15. $3^{5x} = 7$

16. $\log x - \log(x - 8) = 1$

17. $x^2 + 4x > 5$

18. $|2x - 3| \ge 9$

19. If $f(x) = x^2 + 6x$, find a such that $f(a) = 11$.

20. Solve $D = \dfrac{ab}{b + a}$ for a.

21. Solve $\dfrac{1}{p} + \dfrac{1}{q} = \dfrac{1}{f}$ for q.

22. Find the domain of the function f given by

$$f(x) = \dfrac{-4}{3x^2 - 5x - 2}.$$

23. Evaluate $\dfrac{x^0 + y}{-z}$ when $x = 6$, $y = 9$, and $z = -5$.

24. *Chocolate Making.* Good's Candies makes all their chocolates by hand. It takes Anne 10 min to coat a tray of candies in chocolate. It takes Clay 12 min to coat a tray of candies. How long would it take Anne and Clay, working together, to coat the candies?

25. *Forgetting.* Students in a biology class took a final examination. A forgetting formula for determining what the average exam score would be on a retest t months later is

$$S(t) = 78 - 15 \log (t + 1).$$

a) The average score when the students first took the test occurs when $t = 0$. Find the students' average score on the final exam.

b) What would the average score be on a retest after 4 months?

26. *Marine Travel.* A fishing boat with a trolling motor can move at a speed of 5 km/h in still water. The boat travels 42 km downstream in the same time that it takes to travel 12 km upstream. What is the speed of the stream?

27. *Acid Mixtures.* Swim Clean is 30% muriatic acid. Pure Swim is 80% muriatic acid. How many liters of each should be mixed together in order to get 100 L of a solution that is 50% muriatic acid?

28. *Population Growth of Mozambique.* The population of Mozambique was approximately 19 million in 2004, and the exponential growth rate was 1.2% per year.

a) Write an exponential function describing the growth of the population of Mozambique.
b) Estimate the population in 2009 and in 2015.
c) What is the doubling time of the population?

29. *Landscaping.* A rectangular lawn measures 60 ft by 80 ft. Part of the lawn is torn up to install a sidewalk of uniform width around it. The area of the new lawn is 2400 ft^2. How wide is the sidewalk?

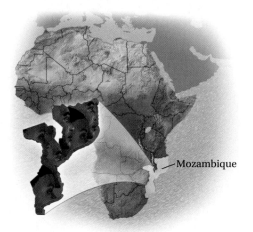
Mozambique

Graph.

30. $5x = 15 + 3y$

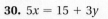

31. $f(x) = 2x^2 - 4x - 1$

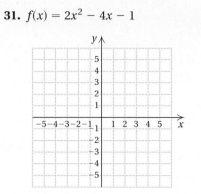

32. $f(x) = \log_3 x$

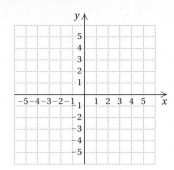

33. $f(x) = 3^x$

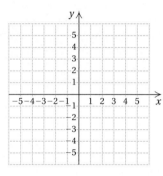

34. $-2x - 3y \le 6$

35. Given that y varies directly as the square of x and inversely as z, and $y = 2$ when $x = 5$ and $z = 100$, what is y when $x = 3$ and $z = 4$?

Perform the indicated operations and simplify.

36. $(11x^2 - 6x - 3) - (3x^2 + 5x - 2)$

37. $(3x^2 - 2y)^2$

38. $(5a + 3b)(2a - 3b)$

39. $\dfrac{x^2 + 8x + 16}{2x + 6} \div \dfrac{x^2 + 3x - 4}{x^2 - 9}$

40. $\dfrac{1 + \dfrac{3}{x}}{x - 1 - \dfrac{12}{x}}$

41. $\dfrac{3}{x + 6} - \dfrac{2}{x^2 - 36} + \dfrac{4}{x - 6}$

Factor.

42. $1 - 125x^3$

43. $6x^2 + 8xy - 8y^2$

44. $x^4 - 4x^3 + 7x - 28$

45. $2m^2 + 12mn + 18n^2$

46. $x^4 - 16y^4$

47. For the function described by
$$h(x) = -3x^2 + 4x + 8,$$
find $h(-2)$.

48. Divide: $(x^4 - 5x^3 + 2x^2 - 6) \div (x - 3)$.

49. Multiply $(5.2 \times 10^4)(3.5 \times 10^{-6})$. Write scientific notation for the answer.

For the radical expressions that follow, assume that all variables represent positive numbers.

50. Divide and simplify: $\dfrac{\sqrt[3]{40xy^8}}{\sqrt[3]{5xy}}$.

51. Multiply and simplify: $\sqrt{7xy^3} \cdot \sqrt{28x^2y}$.

52. Rationalize the denominator: $\dfrac{3 - \sqrt{y}}{2 - \sqrt{y}}$.

53. Multiply these complex numbers:
$$\left(1 + i\sqrt{3}\right)\left(6 - 2i\sqrt{3}\right).$$

54. Find the inverse of f if $f(x) = 7 - 2x$.

55. Find an equation of the line containing the point $(-3, 5)$ and perpendicular to the line whose equation is $2x + y = 6$.

56. Express as a single logarithm:
$$3 \log x - \tfrac{1}{2} \log y - 2 \log z.$$

57. Convert to an exponential equation:
$$\log_a 5 = x.$$

Find each of the following using a calculator. Round answers to four decimal places.

58. $\log 0.05566$ **59.** $10^{2.89}$ **60.** $\ln 12.78$ **61.** $e^{-1.4}$

In each of Questions 62–65, choose the correct answer from the selections given.

62. What is the inverse of the function $f(x) = 5^x$, if it exists?

a) $f^{-1}(x) = x^5$ **b)** $f^{-1}(x) = \log_x 5$

c) $f^{-1}(x) = \dfrac{5}{x}$ **d)** $f^{-1}(x) = \log_5 x$

e) None of these

63. Complete the square: $f(x) = -2x^2 + 28x - 9$.

a) $f(x) = 2(x - 7)^2 - 89$
b) $f(x) = -2(x - 7)^2 - 58$
c) $f(x) = -2(x - 7)^2 - 107$
d) $f(x) = -2(x - 7)^2 + 89$
e) None of these

64. Solve: $\log(x^2 - 9) - \log(x + 3) = 1$.

a) 12   **b)** 7
c) 27   **d)** 4
e) None of these

65. Solve $B = 2a(b^2 - c^2)$ for c.

a) $c = \sqrt{\dfrac{c}{2b} - 3}$ **b)** $c = -\sqrt{\dfrac{B}{2a} - b^2}$

c) $c = 2a(b^2 - B^2)$ **d)** $c = \sqrt{\dfrac{2ab^2 - B}{2a}}$

e) None of these

SYNTHESIS

Solve.

66. $\dfrac{5}{3x - 3} + \dfrac{10}{3x + 6} = \dfrac{5x}{x^2 + x - 2}$

67. $\log \sqrt{3x} = \sqrt{\log 3x}$

68. *Train Travel.* A train travels 280 mi at a certain speed. If the speed had been increased by 5 mph, the trip could have been made in 1 hr less time. Find the actual speed.

Conic Sections

Real-World Application

The spotlight on a musician casts an ellipse of light on the floor below him that is 6 ft wide and 10 ft long. Find an equation of that ellipse if the performer is in its center, x is the distance from the performer to the side of the ellipse, and y is the distance from the performer to the top of the ellipse.

This problem appears as Exercise 40 in Section 9.2.

Objectives

a Graph parabolas.

b Use the distance formula to find the distance between two points whose coordinates are known.

c Use the midpoint formula to find the midpoint of a segment when the coordinates of its endpoints are known.

d Given an equation of a circle, find its center and radius and graph it; and given the center and radius of a circle, write an equation of the circle and graph the circle.

This section and the next two examine curves formed by cross sections of cones. These curves are graphs of second-degree equations in two variables. Some are shown below.

CONIC SECTIONS IN THREE DIMENSIONS

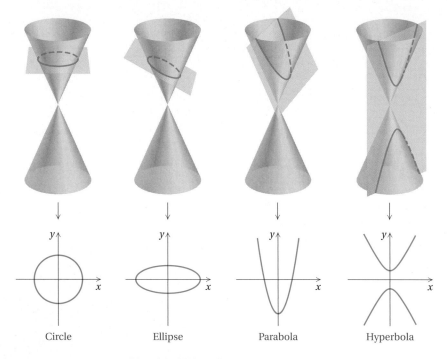

| Circle | Ellipse | Parabola | Hyperbola |

CONIC SECTIONS GRAPHED IN A PLANE

a Parabolas

When a cone is cut by a plane parallel to a side of the cone, as shown in the third figure above, the conic section formed is a **parabola.** General equations of parabolas are quadratic. Parabolas have many applications in electricity, mechanics, and optics. A cross section of a satellite dish is a parabola, and arches that support certain bridges are parabolas. (Free-hanging cables have a different shape, called a *catenary.*) An arc of a spray can have part of the shape of a parabola.

EQUATIONS OF PARABOLAS

$y = ax^2 + bx + c$ (Line of symmetry is parallel to the y-axis.)

$x = ay^2 + by + c$ (Line of symmetry is parallel to the x-axis.)

Recall from Chapter 7 that the graph of $f(x) = ax^2 + bx + c$ (with $a \neq 0$) is a parabola.

■ **EXAMPLE 1** Graph: $y = x^2 - 4x + 9$.

First, we must locate the vertex. To do so, we can use either of two approaches. One way is to complete the square:

$$
\begin{aligned}
y &= (x^2 - 4x) + 9 \\
&= (x^2 - 4x + 0) + 9 && \text{Adding 0} \\
&= (x^2 - 4x + 4 - 4) + 9 && \tfrac{1}{2}(-4) = -2; (-2)^2 = 4; \\
& && \text{substituting } 4 - 4 \text{ for } 0 \\
&= (x^2 - 4x + 4) + (-4 + 9) && \text{Regrouping} \\
&= (x - 2)^2 + 5. && \text{Factoring and simplifying}
\end{aligned}
$$

The vertex is $(2, 5)$.

A second way to find the vertex is to recall that the x-coordinate of the vertex of the parabola given by $y = ax^2 + bx + c$ is $-b/(2a)$:

$$
x = -\frac{b}{2a} = -\frac{-4}{2(1)} = 2.
$$

To find the y-coordinate of the vertex, we substitute 2 for x:

$$
y = x^2 - 4x + 9 = 2^2 - 4(2) + 9 = 5.
$$

Either way, the vertex is $(2, 5)$. Next, we calculate and plot some points on each side of the vertex. Since the x^2-coefficient, 1, is positive, the graph opens up.

x	y	
2	5	← Vertex
0	9	← y-intercept
1	6	
3	6	
4	9	

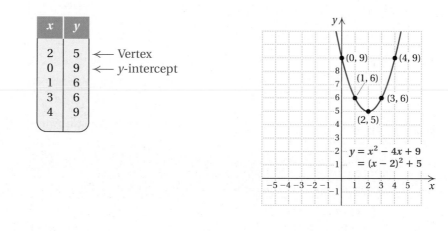

$$y = x^2 - 4x + 9$$
$$= (x - 2)^2 + 5$$

GRAPHING $y = ax^2 + bx + c$

To graph an equation of the type $y = ax^2 + bx + c$ (see Section 7.6):

1. Find the vertex (h, k) either by completing the square to find an equivalent equation

$$
y = a(x - h)^2 + k,
$$

or by using $x = -b/(2a)$ for the x-coordinate and substituting to find the y-coordinate.
2. Choose other values for x on each side of the vertex, and compute the corresponding y-values.
3. The graph opens up for $a > 0$ and down for $a < 0$.

Do Exercise 1.

1. Graph: $y = x^2 + 4x + 7$.

Answer on page A-51

2. Graph: $x = y^2 + 4y + 7$.

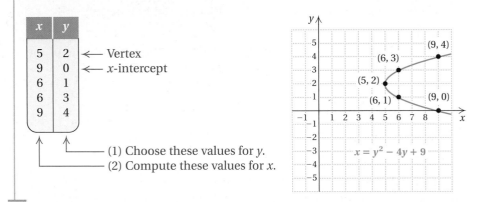

Equations of the form $x = ay^2 + by + c$ represent horizontal parabolas. These parabolas open to the right for $a > 0$ and open to the left for $a < 0$ and have lines of symmetry parallel to the x-axis.

EXAMPLE 2 Graph: $x = y^2 - 4y + 9$.

This equation is like that in Example 1 except that x and y are interchanged. That is, the equations are inverses of each other (see Section 8.2). The vertex is $(5, 2)$ instead of $(2, 5)$. To find ordered pairs, we choose values for y on each side of the vertex. Then we compute values for x. Note that the x- and y-values of the table in Example 1 are interchanged. The graph in Example 2 is the reflection of the graph in Example 1 across the line $y = x$. You should confirm that, by completing the square, we get $x = (y - 2)^2 + 5$.

x	y
5	2	← Vertex
9	0	← x-intercept
6	1
6	3
9	4

↑ ↑
(1) Choose these values for y.
(2) Compute these values for x.

GRAPHING $x = ay^2 + by + c$

To graph an equation of the type $x = ay^2 + by + c$:

1. Find the vertex (h, k) either by completing the square to find an equivalent equation

   $$x = a(y - k)^2 + h,$$

   or by using $y = -b/(2a)$ for the y-coordinate and substituting to find the x-coordinate.
2. Choose other values for y that are above and below the vertex, and compute the corresponding x-values.
3. The graph opens to the right if $a > 0$ and to the left if $a < 0$.

Do Exercise 2.

EXAMPLE 3 Graph: $x = -2y^2 + 10y - 7$.

We use the method of completing the square to find the vertex:

$$
\begin{aligned}
x &= -2y^2 + 10y - 7 \\
&= -2(y^2 - 5y) - 7 \\
&= -2(y^2 - 5y + 0) - 7 & \text{Adding 0} \\
&= -2\left(y^2 - 5y + \tfrac{25}{4} - \tfrac{25}{4}\right) - 7 & \tfrac{1}{2}(-5) = -\tfrac{5}{2}; \left(-\tfrac{5}{2}\right)^2 = \tfrac{25}{4}; \\
& & \text{substituting } \tfrac{25}{4} - \tfrac{25}{4} \\
&= -2\left(y^2 - 5y + \tfrac{25}{4}\right) + (-2)\left(-\tfrac{25}{4}\right) - 7 & \text{Using the distributive law} \\
&= -2\left(y^2 - 5y + \tfrac{25}{4}\right) + \tfrac{25}{2} - 7 \\
&= -2\left(y - \tfrac{5}{2}\right)^2 + \tfrac{11}{2}. & \text{Factoring and simplifying}
\end{aligned}
$$

The vertex is $\left(\tfrac{11}{2}, \tfrac{5}{2}\right)$.

Answer on page A-51

For practice, we also find the vertex by first computing its y-coordinate, $-b/(2a)$, and then substituting to find the x-coordinate:

$$y = -\frac{b}{2a} = -\frac{10}{2(-2)} = \frac{5}{2}$$

$$x = -2y^2 + 10y - 7 = -2\left(\tfrac{5}{2}\right)^2 + 10\left(\tfrac{5}{2}\right) - 7$$
$$= \tfrac{11}{2}.$$

To find ordered pairs, we first choose values for y and then compute values for x. A table is shown below, together with the graph. The graph opens to the left because the y^2-coefficient, -2, is negative.

x	y	
$\frac{11}{2}$	$\frac{5}{2}$	← Vertex
-7	0	← x-intercept
5	2	
5	3	
1	1	
1	4	
-7	5	

(1) Choose these values for y.

(2) Compute these values for x.

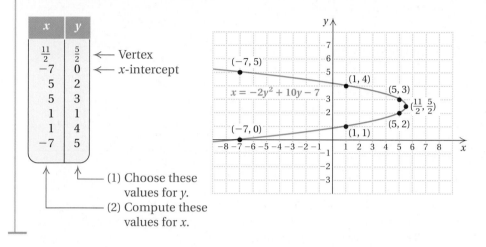

Do Exercise 3.

3. Graph: $x = 4y^2 - 12y + 5$.

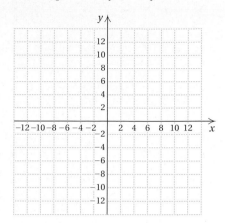

Answer on page A-52

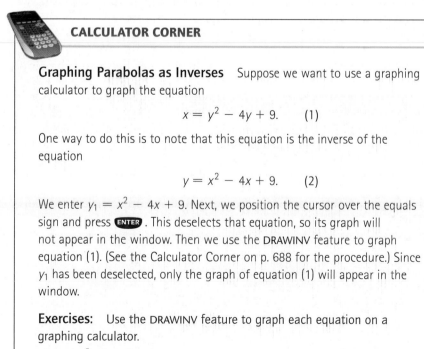

CALCULATOR CORNER

Graphing Parabolas as Inverses Suppose we want to use a graphing calculator to graph the equation

$$x = y^2 - 4y + 9. \qquad (1)$$

One way to do this is to note that this equation is the inverse of the equation

$$y = x^2 - 4x + 9. \qquad (2)$$

We enter $y_1 = x^2 - 4x + 9$. Next, we position the cursor over the equals sign and press **ENTER**. This deselects that equation, so its graph will not appear in the window. Then we use the DRAWINV feature to graph equation (1). (See the Calculator Corner on p. 688 for the procedure.) Since y_1 has been deselected, only the graph of equation (1) will appear in the window.

Exercises: Use the DRAWINV feature to graph each equation on a graphing calculator.

1. $x = y^2 + 4y + 7$
2. $x = -2y^2 + 10y - 7$
3. $x = 4y^2 - 12y + 5$

b   The Distance Formula

Suppose that two points are on a horizontal line, and thus have the same second coordinate. We can find the distance between them by subtracting their first coordinates. This difference may be negative, depending on the order in which we subtract. So, to make sure we get a positive number, we take the absolute value of this difference. The distance between two points on a horizontal line (x_1, y_1) and (x_2, y_1) is thus $|x_2 - x_1|$. Similarly, the distance between two points on a vertical line (x_2, y_1) and (x_2, y_2) is $|y_2 - y_1|$.

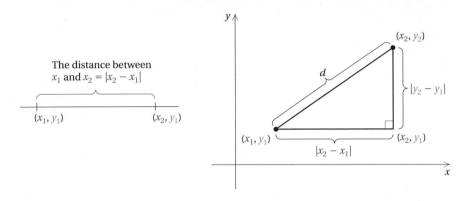

Now consider *any* two points (x_1, y_1) and (x_2, y_2). If $x_1 \neq x_2$ and $y_1 \neq y_2$, these points are vertices of a right triangle, as shown. The other vertex is then (x_2, y_1). The lengths of the legs are $|x_2 - x_1|$ and $|y_2 - y_1|$. We find d, the length of the hypotenuse, by using the Pythagorean theorem:

$$d^2 = |x_2 - x_1|^2 + |y_2 - y_1|^2.$$

Since the square of a number is the same as the square of its opposite, we don't need these absolute-value signs. Thus,

$$d^2 = (x_2 - x_1)^2 + (y_2 - y_1)^2.$$

Taking the principal square root, we obtain the distance between two points.

THE DISTANCE FORMULA

The distance between any two points (x_1, y_1) and (x_2, y_2) is given by
$$d = \sqrt{(x_2 - x_1)^2 + (y_2 - y_1)^2}.$$

This formula holds even when the two points *are* on a vertical or a horizontal line.

EXAMPLE 4 Find the distance between $(4, -3)$ and $(-5, 4)$. Give an exact answer and an approximation to three decimal places.

We substitute into the distance formula:

$$d = \sqrt{(-5 - 4)^2 + [4 - (-3)]^2} \quad \text{Substituting}$$
$$= \sqrt{(-9)^2 + 7^2}$$
$$= \sqrt{130} \approx 11.402. \quad \text{Using a calculator}$$

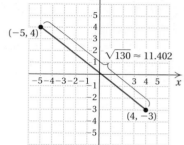

Do Exercises 4 and 5.

Find the distance between the pair of points. Where appropriate, give an approximation to three decimal places.

4. $(2, 6)$ and $(-4, -2)$

5. $(-2, 1)$ and $(4, 2)$

C | Midpoints of Segments

The distance formula can be used to derive a formula for finding the midpoint of a segment when the coordinates of the endpoints are known.

THE MIDPOINT FORMULA

If the endpoints of a segment are (x_1, y_1) and (x_2, y_2), then the coordinates of the midpoints are

$$\left(\frac{x_1 + x_2}{2}, \frac{y_1 + y_2}{2} \right).$$

(To locate the midpoint, determine the average of the x-coordinates and the average of the y-coordinates.)

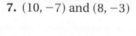

EXAMPLE 5 Find the midpoint of the segment with endpoints $(-2, 3)$ and $(4, -6)$.

Using the midpoint formula, we obtain

$$\left(\frac{-2 + 4}{2}, \frac{3 + (-6)}{2} \right), \quad \text{or} \quad \left(\frac{2}{2}, \frac{-3}{2} \right), \quad \text{or} \quad \left(1, -\frac{3}{2} \right).$$

Do Exercises 6 and 7.

Find the midpoint of the segment with the given endpoints.

6. $(-3, 1)$ and $(6, -7)$

7. $(10, -7)$ and $(8, -3)$

Answers on page A-52

8. Find the center and the radius of the circle

$$(x - 5)^2 + \left(y + \tfrac{1}{2}\right)^2 = 9.$$

Then graph the circle.

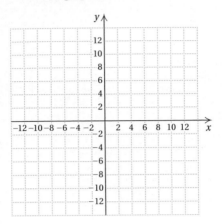

d Circles

Another conic section, or curve, shown in the figure at the beginning of this section is a *circle*. A **circle** is defined as the set of all points in a plane that are a fixed distance from a point in that plane.

Let's find an equation for a circle. We call the center (h, k) and let the radius have length r. Suppose that (x, y) is any point on the circle. By the distance formula, we have

$$\sqrt{(x - h)^2 + (y - k)^2} = r.$$

Squaring both sides gives an equation of the circle in standard form: $(x - h)^2 + (y - k)^2 = r^2$. When $h = 0$ and $k = 0$, the circle is centered at the origin. Otherwise, we can think of that circle being translated $|h|$ units horizontally and $|k|$ units vertically from the origin.

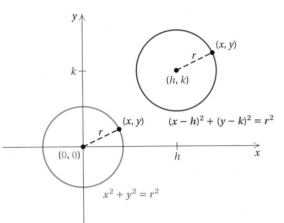

EQUATIONS OF CIRCLES

A circle centered at the origin with radius r has equation

$$x^2 + y^2 = r^2.$$

A circle with center (h, k) and radius r has equation

$$(x - h)^2 + (y - k)^2 = r^2. \qquad \text{(Standard form)}$$

---

9. Find the center and the radius of the circle $x^2 + y^2 = 64$.

EXAMPLE 6 Find the center and the radius and graph this circle:

$$(x + 2)^2 + (y - 3)^2 = 16.$$

First, we find an equivalent equation in standard form:

$$[x - (-2)]^2 + (y - 3)^2 = 4^2.$$

Thus the center is $(-2, 3)$ and the radius is 4. We draw the graph, shown at right, by locating the center and then using a compass, setting its radius at 4, to draw the circle.

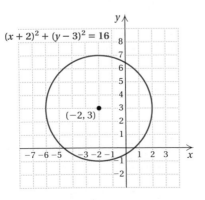

Answers on page A-52

Do Exercises 8 and 9 on the preceding page.

EXAMPLE 7 Write an equation of a circle with center $(9, -5)$ and radius $\sqrt{2}$.

We use standard form $(x - h)^2 + (y - k)^2 = r^2$ and substitute:

$(x - 9)^2 + [y - (-5)]^2 = (\sqrt{2})^2$ Substituting

$(x - 9)^2 + (y + 5)^2 = 2.$ Simplifying

Do Exercise 10.

With certain equations not in standard form, we can complete the square to show that the equations are equations of circles. We proceed in much the same manner as we did in Section 7.6.

EXAMPLE 8 Find the center and the radius and graph this circle:

$$x^2 + y^2 + 8x - 2y + 15 = 0.$$

First, we regroup the terms and then complete the square twice, once with $x^2 + 8x$ and once with $y^2 - 2y$:

$x^2 + y^2 + 8x - 2y + 15 = 0$

$(x^2 + 8x) + (y^2 - 2y) + 15 = 0$ Regrouping

$(x^2 + 8x + 0) + (y^2 - 2y + 0) + 15 = 0$ Adding 0

$(x^2 + 8x + 16 - 16) + (y^2 - 2y + 1 - 1) + 15 = 0$ $\left(\frac{8}{2}\right)^2 = 4^2 = 16;$ $\left(\frac{-2}{2}\right)^2 = 1;$ substituting $16 - 16$ and $1 - 1$

$(x^2 + 8x + 16) + (y^2 - 2y + 1) - 16 - 1 + 15 = 0$ Regrouping

$(x + 4)^2 + (y - 1)^2 - 2 = 0$ Factoring and simplifying

$(x + 4)^2 + (y - 1)^2 = 2$ Adding 2

$[x - (-4)]^2 + (y - 1)^2 = (\sqrt{2})^2.$ Writing standard form

The center is $(-4, 1)$ and the radius is $\sqrt{2}$.

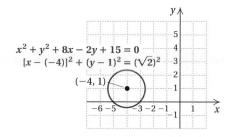

$x^2 + y^2 + 8x - 2y + 15 = 0$
$[x - (-4)]^2 + (y - 1)^2 = (\sqrt{2})^2$

$(-4, 1)$

Do Exercise 11.

10. Find an equation of a circle with center $(-3, 1)$ and radius 6.

11. Find the center and the radius of the circle

$$x^2 + 2x + y^2 - 4y + 2 = 0.$$

Then graph the circle.

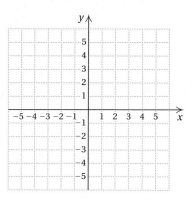

Answers on page A-52

Study Tips

FINAL STUDY TIP

You are arriving at the end of your course in Intermediate Algebra. If you have not begun to prepare for the final examination, be sure to read the comments in the Study Tips on pp. 535, 593, 692, and 725.

"The will to prepare to win is much more important than the will to win."

Bud Wilkinson,
former college football coach

Graphing Circles Equations of circles are not functions, so they cannot be entered directly in "$y =$" form on a graphing calculator. Nevertheless, there are two methods for graphing circles.

Suppose we want to graph the circle $(x - 3)^2 + (y + 1)^2 = 16$. One way to graph this circle is to use the CIRCLE feature from the DRAW menu. The center of the circle is $(3, -1)$ and its radius is 4. To graph it using the CIRCLE feature from the DRAW menu, we first press ⟨Y=⟩ and clear all previously entered equations. Then we select a square window. (See p. 212 for a discussion on squaring the viewing window.) We will use $[-3, 9, -5, 3]$. We press ⟨2ND⟩ ⟨QUIT⟩ to go to the home screen and then ⟨2ND⟩ ⟨DRAW⟩ ⟨9⟩ to display "Circle(." We enter the coordinates of the center and the radius, separating the entries by commas, and close the parentheses: ⟨3⟩ ⟨,⟩ ⟨(-)⟩ ⟨1⟩ ⟨,⟩ ⟨4⟩ ⟨)⟩ ⟨ENTER⟩ .

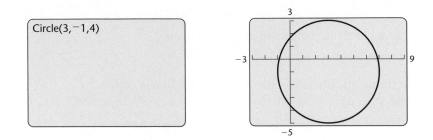

When the Graph screen is displayed, we can use the CLRDRAW operation from the DRAW menu to clear this circle before we graph another circle. To do this, we press ⟨2ND⟩ ⟨DRAW⟩ ⟨1⟩ .

Another way to graph a circle is to solve the equation for y first. Consider the equation above:

$$(x - 3)^2 + (y + 1)^2 = 16$$
$$(y + 1)^2 = 16 - (x - 3)^2$$
$$y + 1 = \pm\sqrt{16 - (x - 3)^2}$$
$$y = -1 \pm \sqrt{16 - (x - 3)^2}.$$

Now we can write two functions, $y_1 = -1 + \sqrt{16 - (x - 3)^2}$ and $y_2 = -1 - \sqrt{16 - (x - 3)^2}$. When we graph these functions in the same window, we have the graph of the circle. The first equation produces the top half of the circle and the second produces the lower half. The graphing calculator does not connect the two parts of the graph because of approximations made near the endpoints of each graph.

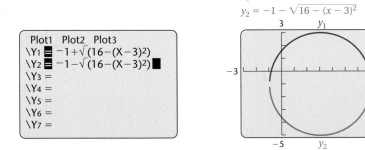

Exercises: Graph the circle using both methods above.

1. $(x - 1)^2 + (y + 2)^2 = 4$

2. $(x + 2)^2 + (y - 2)^2 = 25$

3. $x^2 + y^2 - 16 = 0$

4. $4x^2 + 4y^2 = 100$

5. $x^2 + y^2 - 10x - 11 = 0$

a   Graph the equation.

1. $y = x^2$

2. $x = y^2$

3. $x = y^2 + 4y + 1$

4. $y = x^2 - 2x + 3$

5. $y = -x^2 + 4x - 5$

6. $x = 4 - 3y - y^2$

7. $x = -3y^2 - 6y - 1$

8. $y = -5 - 8x - 2x^2$

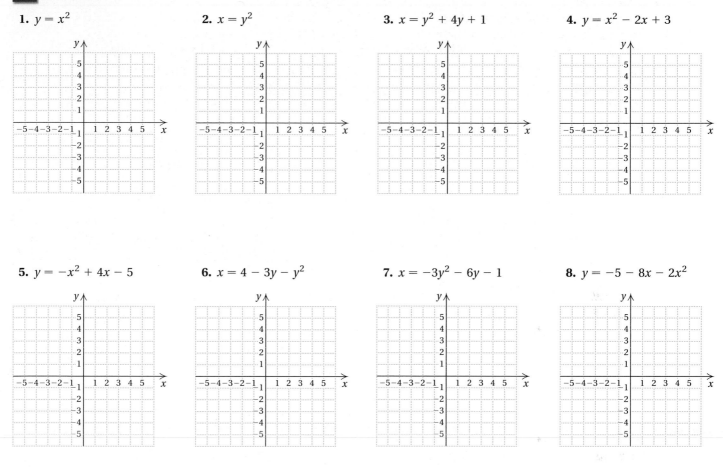

b   Find the distance between the pair of points. Where appropriate, give an approximation to three decimal places.

9. $(6, -4)$ and $(2, -7)$

10. $(1, 2)$ and $(-4, 14)$

11. $(0, -4)$ and $(5, -6)$

12. $(8, 3)$ and $(8, -3)$

13. $(9, 9)$ and $(-9, -9)$

14. $(2, 22)$ and $(-8, 1)$

15. $(2.8, -3.5)$ and $(-4.3, -3.5)$

16. $(6.1, 2)$ and $(5.6, -4.4)$

17. $\left(\dfrac{5}{7}, \dfrac{1}{14}\right)$ and $\left(\dfrac{1}{7}, \dfrac{11}{14}\right)$

18. $\left(0, \sqrt{7}\right)$ and $\left(\sqrt{6}, 0\right)$

19. $(-23, 10)$ and $(56, -17)$

20. $(34, -18)$ and $(-46, -38)$

21. (a, b) and $(0, 0)$

22. $(0, 0)$ and (p, q)

23. $\left(\sqrt{2}, -\sqrt{3}\right)$ and $\left(-\sqrt{7}, \sqrt{5}\right)$

24. $\left(\sqrt{8}, \sqrt{3}\right)$ and $\left(-\sqrt{5}, -\sqrt{6}\right)$

25. $(1000, -240)$ and $(-2000, 580)$

26. $(-3000, 560)$ and $(-430, -640)$

C   Find the midpoint of the segment with the given endpoints.

27. $(-1, 9)$ and $(4, -2)$

28. $(5, 10)$ and $(2, -4)$

29. $(3, 5)$ and $(-3, 6)$

30. $(7, -3)$ and $(4, 11)$

31. $(-10, -13)$ and $(8, -4)$

32. $(6, -2)$ and $(-5, 12)$

33. $(-3.4, 8.1)$ and $(2.9, -8.7)$

34. $(4.1, 6.9)$ and $(5.2, -6.9)$

35. $\left(\dfrac{1}{6}, -\dfrac{3}{4}\right)$ and $\left(-\dfrac{1}{3}, \dfrac{5}{6}\right)$

36. $\left(-\dfrac{4}{5}, -\dfrac{2}{3}\right)$ and $\left(\dfrac{1}{8}, \dfrac{3}{4}\right)$

37. $\left(\sqrt{2}, -1\right)$ and $\left(\sqrt{3}, 4\right)$

38. $\left(9, 2\sqrt{3}\right)$ and $\left(-4, 5\sqrt{3}\right)$

d Find the center and the radius of the circle. Then graph the circle.

39. $(x + 1)^2 + (y + 3)^2 = 4$

40. $(x - 2)^2 + (y + 3)^2 = 1$

41. $(x - 3)^2 + y^2 = 2$

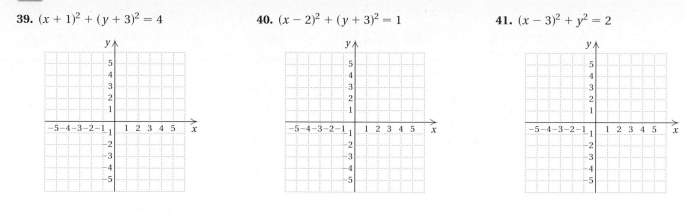

42. $x^2 + (y - 1)^2 = 3$

43. $x^2 + y^2 = 25$

44. $x^2 + y^2 = 9$

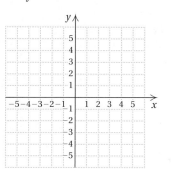
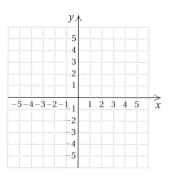

Find an equation of the circle having the given center and radius.

45. Center $(0, 0)$, radius 7

46. Center $(0, 0)$, radius 4

47. Center $(-5, 3)$, radius $\sqrt{7}$

48. Center $(4, 1)$, radius $3\sqrt{2}$

Find the center and the radius of the circle.

49. $x^2 + y^2 + 8x - 6y - 15 = 0$

50. $x^2 + y^2 + 6x - 4y - 15 = 0$

51. $x^2 + y^2 - 8x + 2y + 13 = 0$

52. $x^2 + y^2 + 6x + 4y + 12 = 0$

53. $x^2 + y^2 - 4x = 0$

54. $x^2 + y^2 + 10y - 75 = 0$

55. D_W Is the center of a circle part of the circle? Why or why not?

56. D_W 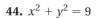 How could a graphing calculator be used to graph an equation of the form $x = ay^2 + by + c$?

Solve.   [3.2a], [3.3a]

57. $x - y = 7$,
$x + y = 11$

58. $x + y = 8$,
$x - y = -24$

59. $y = 3x - 2$,
$2x - 4y = 50$

60. $2x + 3y = 8$,
$x - 2y = -3$

61. $-4x + 12y = -9$,
$x - 3y = 2$

Factor.   [4.6b]

62. $4a^2 - b^2$

63. $x^2 - 16$

64. $a^2 - 9b^2$

65. $64p^2 - 81q^2$

66. $400c^2d^2 - 225$

Find an equation of a circle satisfying the given conditions.

67. Center $(0, 0)$, passing through $\left(1/4, \sqrt{31}/4\right)$

68. Center $(-4, 1)$, passing through $(2, -5)$

69. Center $(-3, -2)$, and tangent to the y-axis

70. The endpoints of a diameter are $(7, 3)$ and $(-1, -3)$.

Find the distance between the given points.

71. $(-1, 3k)$ and $(6, 2k)$

72. (a, b) and $(-a, -b)$

73. $(6m, -7n)$ and $(-2m, n)$

74. $\left(-3\sqrt{3}, 1 - \sqrt{6}\right)$ and $\left(\sqrt{3}, 1 + \sqrt{6}\right)$

If the sides of a triangle have lengths a, b, and c and $a^2 + b^2 = c^2$, then the triangle is a right triangle. Determine whether the given points are vertices of a right triangle.

75. $(-8, -5)$, $(6, 1)$, and $(-4, 5)$

76. $(9, 6)$, $(-1, 2)$, and $(1, -3)$

77. Find the midpoint of the segment with the endpoints $\left(2 - \sqrt{3}, 5\sqrt{2}\right)$ and $\left(2 + \sqrt{3}, 3\sqrt{2}\right)$.

78. Find the point on the y-axis that is equidistant from $(2, 10)$ and $(6, 2)$.

79. *Snowboarding.*   Each side edge of a Burton® Twin snowboard is an arc of a circle. The Twin 53 has a "running length" of 1150 mm and a "sidecut depth" of 19.5 mm (see the figure below).

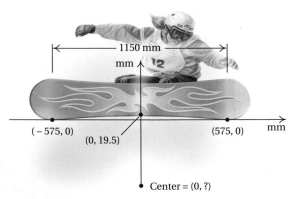

a) Using the coordinates shown, locate the center of the circle. (*Hint*: Equate distances.)
b) What radius is used for the edge of the board?

80. *Doorway Construction.*   Ace Carpentry is to cut an arch for the top of an entranceway. The arch needs to be 8 ft wide and 2 ft high. To draw the arch, the carpenters will use a stretched string with chalk attached at an end as a compass.

a) Using a coordinate system, locate the center of the circle.
b) What radius should the carpenters use to draw the arch?

ELLIPSES

a   Ellipses

When a cone is cut at an angle, as shown below, the conic section formed is an *ellipse*.

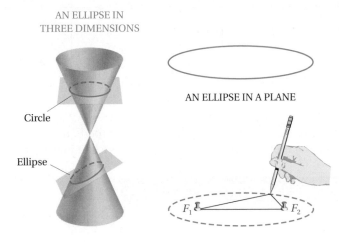

AN ELLIPSE IN THREE DIMENSIONS

Circle

Ellipse

AN ELLIPSE IN A PLANE

You can draw an ellipse by securing two tacks in a piece of cardboard, tying a string around them, placing a pencil as shown, and drawing with the string kept taut. The formal mathematical definition is related to this method of drawing.

An **ellipse** is defined as the set of all points in a plane such that the *sum* of the distances from two fixed points F_1 and F_2 (called the **foci;** singular, **focus**) is constant. In the preceding drawing, the tacks are at the foci. Ellipses have equations as follows.

EQUATION OF AN ELLIPSE

An ellipse with its center at the origin has equation

$$\frac{x^2}{a^2} + \frac{y^2}{b^2} = 1, \quad a, b > 0, \quad a \neq b. \qquad \text{(Standard form)}$$

We can think of a circle as a special kind of ellipse. A circle is formed when $a = b$ and the cutting plane is perpendicular to the axis of the cone. It is also formed when the foci, F_1 and F_2, are the same point. An ellipse with its foci close together is very nearly a circle.

When graphing ellipses, it helps to first find the intercepts. If we replace x with 0 in the standard form of the equation, we can find the y-intercepts:

$$\frac{0^2}{a^2} + \frac{y^2}{b^2} = 1$$

$$\frac{y^2}{b^2} = 1$$

$$y^2 = b^2$$

$$y = \pm b.$$

Graph the ellipse.

1. $\dfrac{x^2}{9} + \dfrac{y^2}{4} = 1$

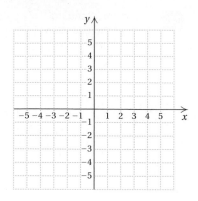

2. $\dfrac{x^2}{9} + \dfrac{y^2}{25} = 1$

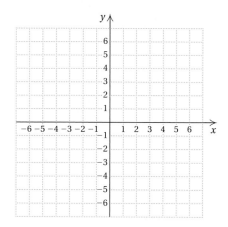

Answers on page A-53

Thus the y-intercepts are $(0, b)$ and $(0, -b)$. Similarly, the x-intercepts are $(a, 0)$ and $(-a, 0)$. If $a > b$, the ellipse is horizontal and $(-a, 0)$ and $(a, 0)$ are **vertices** (singular, **vertex**). If $a < b$, the ellipse is vertical and $(0, -b)$ and $(0, b)$ are the vertices.

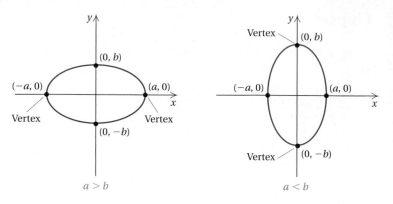

Plotting these points and filling in an oval-shaped curve, we get a graph of the ellipse. If a more precise graph is desired, we can plot more points.

INTERCEPTS AND VERTICES OF AN ELLIPSE

For the ellipse

$$\frac{x^2}{a^2} + \frac{y^2}{b^2} = 1,$$

the **x-intercepts** are $(-a, 0)$ and $(a, 0)$, and the **y-intercepts** are $(0, -b)$ and $(0, b)$. If $a > b$, then $(-a, 0)$ and $(a, 0)$ are the vertices. If $a < b$, then $(0, -b)$ and $(0, b)$ are the vertices.

EXAMPLE 1 Graph: $\dfrac{x^2}{4} + \dfrac{y^2}{9} = 1$.

Note that

$$\frac{x^2}{4} + \frac{y^2}{9} = \frac{x^2}{2^2} + \frac{y^2}{3^2}. \qquad a = 2, b = 3$$

Thus the x-intercepts are $(-2, 0)$ and $(2, 0)$, and the y-intercepts are $(0, -3)$ and $(0, 3)$. The vertices are $(0, -3)$ and $(0, 3)$. We plot these points and connect them with an oval-shaped curve. To be accurate, we might find some other points on the curve. We let $x = 1$ and solve for y:

$$\frac{1^2}{4} + \frac{y^2}{9} = 1$$

$$36\left(\frac{1}{4} + \frac{y^2}{9}\right) = 36 \cdot 1$$

$$36 \cdot \frac{1}{4} + 36 \cdot \frac{y^2}{9} = 36$$

$$9 + 4y^2 = 36$$

$$4y^2 = 27$$

$$y^2 = \frac{27}{4}$$

$$y = \pm\sqrt{\frac{27}{4}} \approx \pm 2.6.$$

Thus, $(1, 2.6)$ and $(1, -2.6)$ can also be plotted and used to draw the graph. Similarly, the points $(-1, -2.6)$ and $(-1, 2.6)$ can also be computed and plotted.

3. Graph: $16x^2 + 9y^2 = 144$.

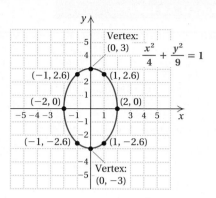

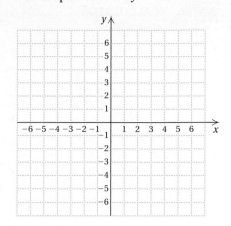

Do Exercises 1 and 2 on the preceding page.

EXAMPLE 2 Graph: $4x^2 + 25y^2 = 100$.

To write the equation in standard form, we multiply both sides by $\frac{1}{100}$:

$$\frac{1}{100}(4x^2 + 25y^2) = \frac{1}{100}(100) \qquad \text{Multiplying by } \frac{1}{100} \text{ to get 1 on the right side}$$

$$\left.\begin{array}{c} \dfrac{1}{100}(4x^2) + \dfrac{1}{100}(25y^2) = 1 \\[2mm] \dfrac{x^2}{25} + \dfrac{y^2}{4} = 1 \end{array}\right\} \qquad \text{Simplifying}$$

$$\frac{x^2}{5^2} + \frac{y^2}{2^2} = 1. \qquad a = 5, b = 2$$

4. Graph:

$$\frac{(x + 2)^2}{16} + \frac{(y - 3)^2}{9} = 1.$$

The x-intercepts are $(-5, 0)$ and $(5, 0)$, and the y-intercepts are $(0, -2)$ and $(0, 2)$. The vertices are $(-5, 0)$ and $(5, 0)$. We plot the intercepts and connect them with an oval-shaped curve. Other points can also be computed and plotted.

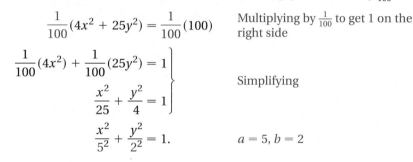

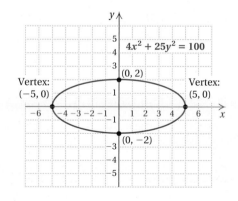

Do Exercise 3.

Answer on page A-53

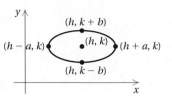

CALCULATOR CORNER

Graphing Ellipses In a Calculator Corner on p. 766, we graphed a circle by first solving the equation of the circle for y. We can graph an ellipse in the same way. Consider the ellipse in Example 2:

$$4x^2 + 25y^2 = 100$$
$$25y^2 = 100 - 4x^2$$
$$y^2 = \frac{100 - 4x^2}{25}$$
$$y = \pm\sqrt{\frac{100 - 4x^2}{25}}.$$

Now we enter

$$y_1 = \sqrt{\frac{100 - 4x^2}{25}} \quad \text{and}$$
$$y_2 = -\sqrt{\frac{100 - 4x^2}{25}}$$

and graph these equations in a square window.

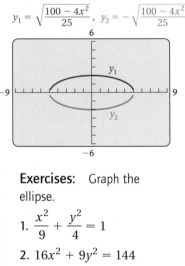

Exercises: Graph the ellipse.

1. $\dfrac{x^2}{9} + \dfrac{y^2}{4} = 1$

2. $16x^2 + 9y^2 = 144$

3. $\dfrac{(x-1)^2}{4} + \dfrac{(y+2)^2}{9} = 1$

4. $\dfrac{(x+2)^2}{16} + \dfrac{(y-3)^2}{9} = 1$

Horizontal and vertical translations, similar to those used in Chapter 7, can be used to graph ellipses that are not centered at the origin.

STANDARD FORM OF AN ELLIPSE

The standard form of a horizontal or vertical ellipse centered at (h, k) is

$$\frac{(x-h)^2}{a^2} + \frac{(y-k)^2}{b^2} = 1.$$

The vertices are $(h + a, k)$ and $(h - a, k)$ if horizontal; $(h, k + b)$ and $(h, k - b)$ if vertical.

EXAMPLE 3 Graph: $\dfrac{(x-1)^2}{4} + \dfrac{(y+5)^2}{9} = 1$.

Note that

$$\frac{(x-1)^2}{4} + \frac{(y+5)^2}{9} = \frac{(x-1)^2}{2^2} + \frac{(y+5)^2}{3^2}.$$

Thus, $a = 2$ and $b = 3$. To determine the center of the ellipse, (h, k), note that

$$\frac{(x-1)^2}{2^2} + \frac{(y+5)^2}{3^2} = \frac{(x-1)^2}{2^2} + \frac{(y-(-5))^2}{3^2}.$$

Thus the center is $(1, -5)$. We locate $(1, -5)$ and then plot $(1 + 2, -5)$, $(1 - 2, -5)$, $(1, -5 + 3)$, and $(1, -5 - 3)$. These are the points $(3, -5)$, $(-1, -5)$, $(1, -2)$, and $(1, -8)$. The vertices are $(1, -8)$ and $(1, -2)$.

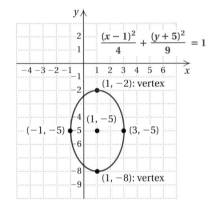

Note that this ellipse is the same as the ellipse in Example 1 but translated 1 unit to the right and 5 units down.

Do Exercise 4 on the preceding page.

Ellipses have many applications. The orbits of planets and some comets around the sun are ellipses. The sun is located at one focus. Whispering galleries are also ellipses. A person standing at one focus will be able to hear the whisper of a person standing at the other focus. One example of a whispering gallery is found in the rotunda of the capital building in Washington, D.C.

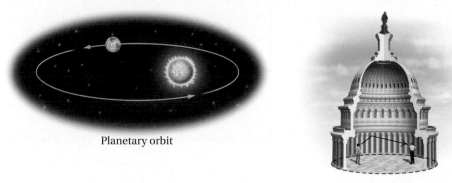

Planetary orbit

Whispering gallery

A medical instrument, the lithotripter, uses shock waves originating at one focus to crush a kidney stone located at the other focus.

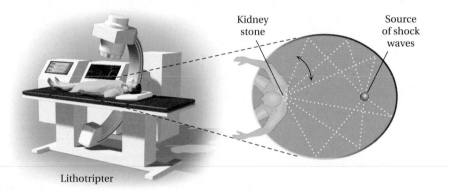

Kidney stone

Source of shock waves

Lithotripter

Wind-driven forest fires can be roughly approximated as the union of "half"-ellipses. Shown below is an illustration of such a forest fire. The smaller half-ellipse on the left moves into the wind, and the elongated half-ellipse on the right moves out in the direction of the wind. The wind tends to spread the fire to the right, but part of the fire will still spread into the wind.

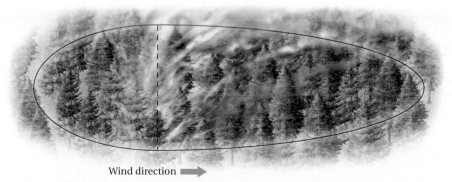

Wind direction ➡

Source: "Predicting Wind-Driven Wild Land Fire Size and Shape," Hal Anderson, Research Paper INT-305, U.S. Department of Agriculture, Forest Service, February 1983.

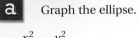
a   Graph the ellipse.

1. $\dfrac{x^2}{9} + \dfrac{y^2}{36} = 1$

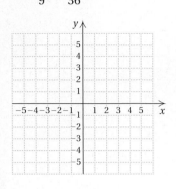

2. $\dfrac{x^2}{16} + \dfrac{y^2}{25} = 1$

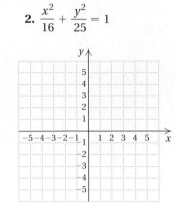

3. $\dfrac{x^2}{1} + \dfrac{y^2}{4} = 1$

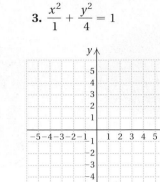

4. $\dfrac{x^2}{4} + \dfrac{y^2}{1} = 1$

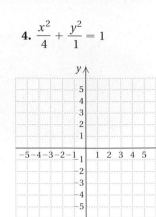

5. $4x^2 + 9y^2 = 36$
(*Hint*: Divide by 36.)

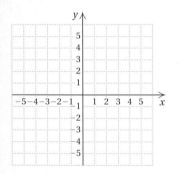

6. $9x^2 + 4y^2 = 36$

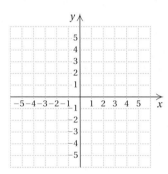

7. $x^2 + 4y^2 = 4$

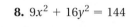

8. $9x^2 + 16y^2 = 144$

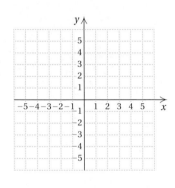

9. $2x^2 + 3y^2 = 6$

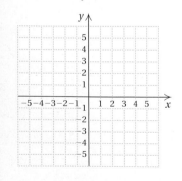

10. $5x^2 + 7y^2 = 35$

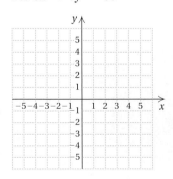

11. $12x^2 + 5y^2 - 120 = 0$

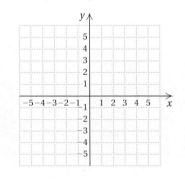

12. $3x^2 + 7y^2 - 63 = 0$

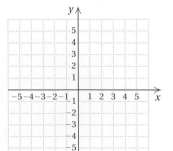

13. $\dfrac{(x-2)^2}{9} + \dfrac{(y-1)^2}{25} = 1$

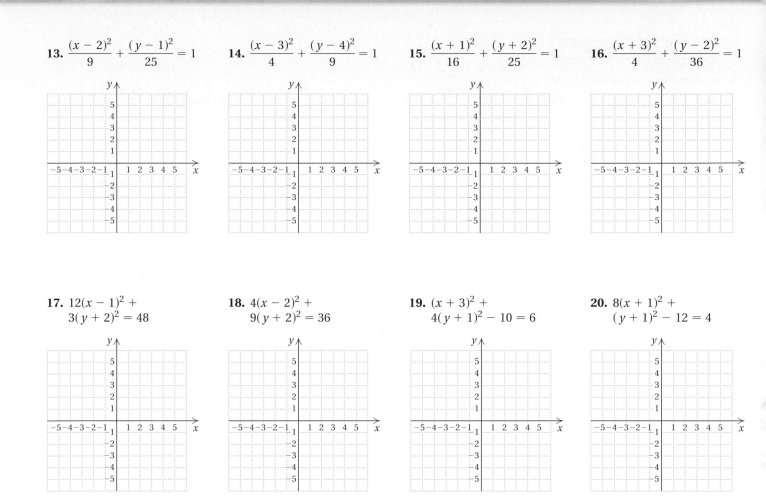

14. $\dfrac{(x-3)^2}{4} + \dfrac{(y-4)^2}{9} = 1$

15. $\dfrac{(x+1)^2}{16} + \dfrac{(y+2)^2}{25} = 1$

16. $\dfrac{(x+3)^2}{4} + \dfrac{(y-2)^2}{36} = 1$

17. $12(x-1)^2 + 3(y+2)^2 = 48$

18. $4(x-2)^2 + 9(y+2)^2 = 36$

19. $(x+3)^2 + 4(y+1)^2 - 10 = 6$

20. $8(x+1)^2 + (y+1)^2 - 12 = 4$

21. DW *Wind-Driven Forest Fires.*

a) Graph the wind-driven fire formed as the union of the following two curves:

$$\dfrac{x^2}{10.3^2} + \dfrac{y^2}{4.8^2} = 1, \quad x \ge 0; \qquad \dfrac{x^2}{3.6^2} + \dfrac{y^2}{4.8^2} = 1, \quad x \le 0.$$

b) What other factors do you think affect the shape of forest fires?

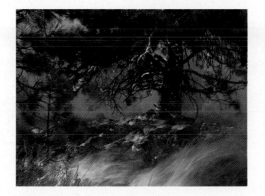

22. DW An eccentric person builds a pool table in the shape of an ellipse with a hole at one focus and a tiny dot at the other. Guests are amazed at how many bank shots the owner of the pool table makes. Explain how this can happen.

Solve. Give exact solutions.   [7.2a]

23. $3x^2 - 2x + 7 = 0$ **24.** $3x^2 - 12x + 7 = 0$ **25.** $x^2 + x + 2 = 0$ **26.** $x^2 + 2x = 10$

Solve. Give both exact and approximate solutions to the nearest tenth.   [7.2a]

27. $x^2 + 2x - 17 = 0$ **28.** $x^2 - 2x = 10$ **29.** $3x^2 - 12x + 7 = 10 - x^2 + 5x$ **30.** $2x^2 + 3x - 4 = 0$

Convert to a logarithmic equation.   [8.3b]

31. $a^{-t} = b$ **32.** $8^a = 17$

Convert to an exponential equation.   [8.3b]

33. $\ln 24 = 3.1781$ **34.** $p = \log_e W$

Find an equation of an ellipse that contains the following points.

35. $(-9, 0)$, $(9, 0)$, $(0, -11)$, and $(0, 11)$ **36.** $(-7, 0)$, $(7, 0)$, $(0, -5)$, and $(0, 5)$

37. $(-2, -1)$, $(6, -1)$, $(2, -4)$, and $(2, 2)$

38. *Planetary Motion.*   The maximum distance of the planet Mars from the sun is 2.48×10^8 mi. The minimum distance is 3.46×10^7 mi. The sun is at one focus of the elliptical orbit. Find the distance from the sun to the other focus.

39. Complete the square as needed and find an equivalent equation in standard form:
$$x^2 - 4x + 4y^2 + 8y - 8 = 0.$$

40. *Theatrical Lighting.*   The spotlight on a musician casts an ellipse of light on the floor below him that is 6 ft wide and 10 ft long. Find an equation of that ellipse if the performer is in its center, x is the distance from the performer to the side of the ellipse, and y is the distance from the performer to the top of the ellipse.

41.  Use a graphing calculator to check your answers to Exercises 7, 12, and 20.

9.3 HYPERBOLAS

Objectives

  Graph the standard form of the equation of a hyperbola.

b Graph equations (nonstandard form) of hyperbolas.

a Hyperbolas

A **hyperbola** looks like a pair of parabolas, but the actual shapes are different. A hyperbola has two **vertices** and the line through the vertices is known as an **axis.** The point halfway between the vertices is called the **center.**

Parabola

Hyperbola in three dimensions

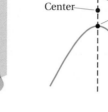

Hyperbola in a plane

EQUATIONS OF HYPERBOLAS

Hyperbolas with their centers at the origin have equations as follows:

$$\frac{x^2}{a^2} - \frac{y^2}{b^2} = 1; \qquad \text{(Axis horizontal)}$$

$$\frac{y^2}{b^2} - \frac{x^2}{a^2} = 1. \qquad \text{(Axis vertical)}$$

To graph a hyperbola, it helps to begin by graphing two lines called **asymptotes.** Although the asymptotes themselves are not part of the graph, they serve as guidelines for an accurate sketch.

ASYMPTOTES OF A HYPERBOLA

For hyperbolas with equations as given above, the **asymptotes** are the lines

$$y = \frac{b}{a}x \quad \text{and} \quad y = -\frac{b}{a}x.$$

---

1. Graph: $\dfrac{x^2}{16} - \dfrac{y^2}{25} = 1$.

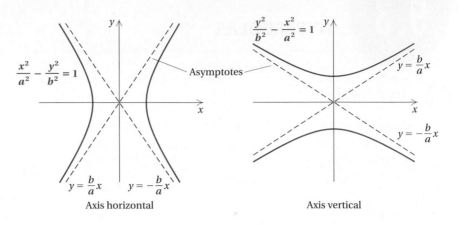

Axis horizontal   Axis vertical

As a hyperbola gets further away from the origin, it gets closer and closer to its asymptotes. The larger $|x|$ gets, the closer the graph gets to an asymptote. The asymptotes act to "constrain" the graph of a hyperbola. On the other hand, parabolas are *not* constrained by any asymptotes.

The next thing to do after sketching asymptotes is to plot vertices. Then it is easy to sketch the curve.

EXAMPLE 1   Graph: $\dfrac{x^2}{4} - \dfrac{y^2}{9} = 1$.

Note that

$$\frac{x^2}{4} - \frac{y^2}{9} = \frac{x^2}{2^2} - \frac{y^2}{3^2}, \qquad \text{Identifying } a \text{ and } b$$

so $a = 2$ and $b = 3$. The asymptotes are thus

$$y = \frac{3}{2}x \quad \text{and} \quad y = -\frac{3}{2}x.$$

We sketch them, as shown in the graph on the left below.

For horizontal or vertical hyperbolas centered at the origin, the vertices also serve as intercepts. Since this hyperbola is horizontal, we replace y with 0 and solve for x. We see that $x^2/2^2 = 1$ when $x = \pm 2$. The intercepts are $(2, 0)$ and $(-2, 0)$. You can check that no y-intercepts exist. The vertices are $(-2, 0)$ and $(2, 0)$.

Finally, we plot the intercepts and sketch the graph. Through each intercept, we draw a smooth curve that approaches the asymptotes closely, as shown.

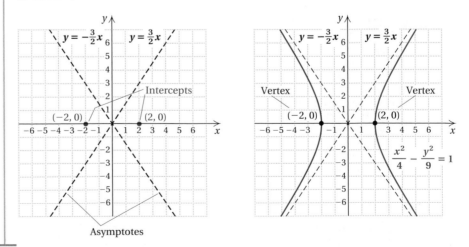

Do Exercise 1.

EXAMPLE 2   Graph: $\dfrac{y^2}{36} - \dfrac{x^2}{4} = 1$.

Note that

$$\dfrac{y^2}{36} - \dfrac{x^2}{4} = \dfrac{y^2}{6^2} - \dfrac{x^2}{2^2} = 1.$$

> The intercept distance is found in the term without the minus sign. Here there is a y in this term, so the intercepts are on the y-axis.

The asymptotes are thus $y = \frac{6}{2}x$ and $y = -\frac{6}{2}x$, or $y = 3x$ and $y = -3x$.

The numbers 6 and 2 can be used to sketch a rectangle that helps with graphing. Using ± 2 as x-coordinates and ± 6 as y-coordinates, we form all possible ordered pairs: $(2, 6)$, $(2, -6)$, $(-2, 6)$, and $(-2, -6)$. We plot these pairs and lightly sketch a rectangle through them. The asymptotes pass through the corners (see the figure on the left below). Since the hyperbola is vertical, we plot its y-intercepts, $(0, 6)$ and $(0, -6)$. The vertices are $(0, -6)$ and $(0, 6)$. Finally, we draw curves through the intercepts toward the asymptotes, as shown below.

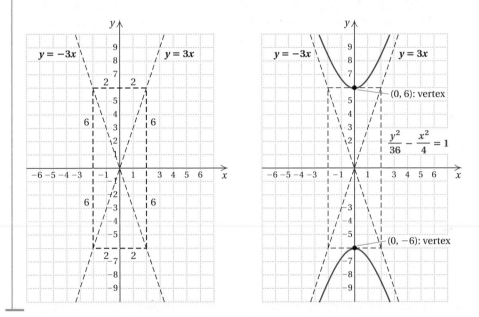

Although we will not consider these equations here, hyperbolas with center at (h, k) are given by

$$\dfrac{(x - h)^2}{a^2} - \dfrac{(y - k)^2}{b^2} = 1 \quad \text{or} \quad \dfrac{(y - k)^2}{b^2} - \dfrac{(x - h)^2}{a^2} = 1.$$

Do Exercise 2.

2. Graph.

a) $\dfrac{y^2}{9} - \dfrac{x^2}{49} = 1$

b) $\dfrac{x^2}{49} - \dfrac{y^2}{9} = 1$

3. Graph: $xy = 8$.

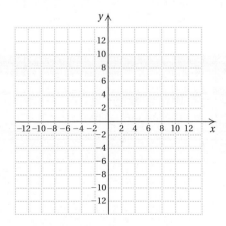

Answers on page A-54

Graphing Hyperbolas

Graphing hyperbolas is similar to graphing circles and ellipses. First, we solve the equation of the hyperbola for y and then graph the two resulting functions. Consider the hyperbola

$$\frac{x^2}{25} - \frac{y^2}{49} = 1.$$

Solving for y, we get

$$y_1 = \frac{7}{5}\sqrt{x^2 - 25} \quad \text{and}$$

$$y_2 = -\frac{7}{5}\sqrt{x^2 - 25}.$$

Now, we graph these equations in a square viewing window.

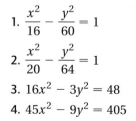

$$y_1 = \frac{7}{5}\sqrt{x^2 - 25}, \; y_2 = -\frac{7}{5}\sqrt{x^2 - 25}$$

Exercises: Graph the hyperbola.

1. $\dfrac{x^2}{16} - \dfrac{y^2}{60} = 1$

2. $\dfrac{x^2}{20} - \dfrac{y^2}{64} = 1$

3. $16x^2 - 3y^2 = 48$

4. $45x^2 - 9y^2 = 405$

b Hyperbolas (Nonstandard Form)

The equations for hyperbolas just examined are the standard ones, but there are other hyperbolas. We consider some of them.

> Hyperbolas having the x- and y-axes as asymptotes have equations as follows:
>
> $$xy = c, \quad \text{where } c \text{ is a nonzero constant.}$$

EXAMPLE 3 Graph: $xy = -8$.

We first solve for y:

$$y = -\frac{8}{x}. \qquad \text{Dividing by } x \text{ on both sides. Note that } x \neq 0.$$

Next, we find some solutions, keeping the results in a table. Note that x cannot be 0 and that for large values of $|x|$, y will be close to 0. Thus the x- and y-axes serve as asymptotes. We plot the points and draw the hyperbola.

x	y
2	-4
-2	4
4	-2
-4	2
1	-8
-1	8
8	-1
-8	1

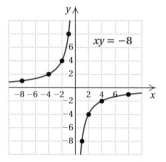

Do Exercise 3 on the preceding page.

Hyperbolas have many applications. A jet breaking the sound barrier creates a sonic boom with a wave front the shape of a cone. The intersection of the cone with the ground is one branch of a hyperbola. Some comets travel in hyperbolic orbits, and a cross section of certain lenses may be hyperbolic in shape.

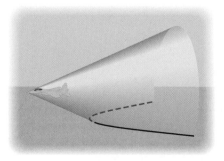

Visualizing for Success

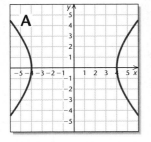

A

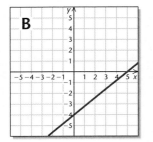

B

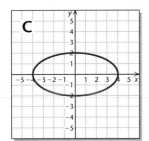

C

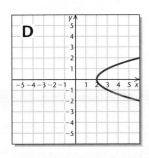

D

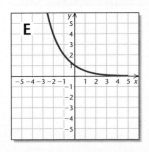

E

Match each equation with its graph.

1. $4y^2 + x^2 = 16$

2. $y = \left(\dfrac{1}{2}\right)^x$

3. $y - x^2 = 2$

4. $y^2 - 4x^2 = 16$

5. $4x - 5y = 20$

6. $x^2 + y^2 + 2x - 6y + 6 = 0$

7. $x^2 - y^2 = 16$

8. $4x^2 + y^2 = 16$

9. $y = \log_2 x$

10. $x - y^2 = 2$

Answers on page A-54

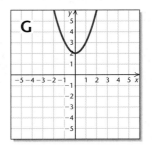

F

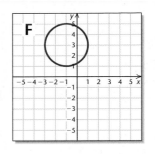

G

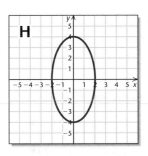

H

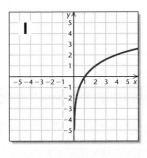

I

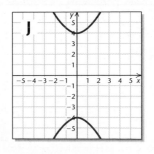

J

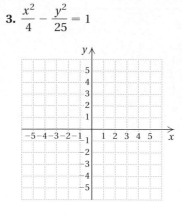

a Graph the hyperbola.

1. $\dfrac{y^2}{9} - \dfrac{x^2}{9} = 1$

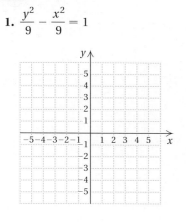

2. $\dfrac{x^2}{16} - \dfrac{y^2}{16} = 1$

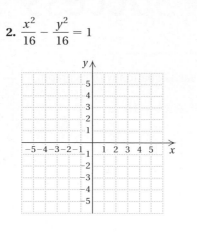

3. $\dfrac{x^2}{4} - \dfrac{y^2}{25} = 1$

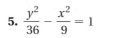

4. $\dfrac{y^2}{16} - \dfrac{x^2}{9} = 1$

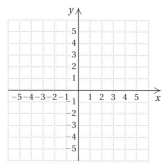

5. $\dfrac{y^2}{36} - \dfrac{x^2}{9} = 1$

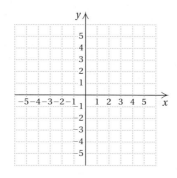

6. $\dfrac{x^2}{25} - \dfrac{y^2}{36} = 1$

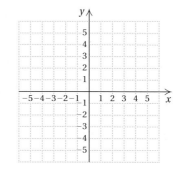

7. $y^2 - x^2 = 25$

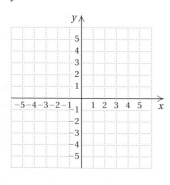

8. $x^2 - y^2 = 4$

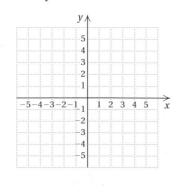

9. $x^2 = 1 + y^2$

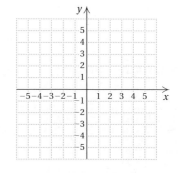

10. $9y^2 = 36 + 4x^2$

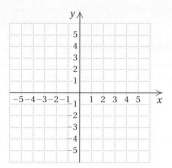

11. $25x^2 - 16y^2 = 400$

12. $4y^2 - 9x^2 = 36$

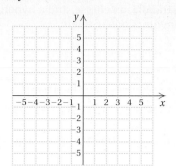

 Graph the hyperbola.

13. $xy = -4$

14. $xy = 6$

15. $xy = 3$

16. $xy = -9$

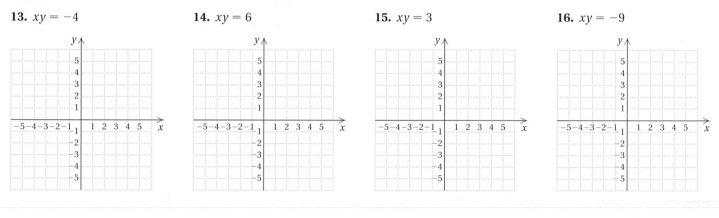

17. $xy = -2$

18. $xy = -1$

19. $xy = \dfrac{1}{2}$

20. $xy = \dfrac{3}{4}$

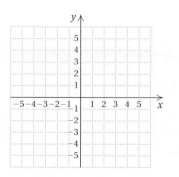

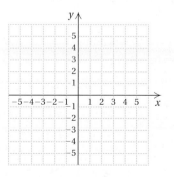

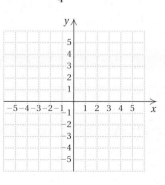

21. D_W Consider the standard equations of a circle, a parabola, an ellipse, and a hyperbola. Which, if any, are functions? Explain.

22. D_W If, in

$$\frac{x^2}{a^2} - \frac{y^2}{b^2} = 1,$$

$a = b$, what are the asymptotes of the graph? Explain.

SKILL MAINTENANCE

↪ VOCABULARY REINFORCEMENT

In each of Exercises 23–30, fill in the blank with the correct term from the given list. Some of the choices may not be used.

23. The expression $b^2 - 4ac$ in the quadratic formula is called the _____ . [7.4a]

24. The x-coordinate of the _____ of a parabola, $y = ax^2 + bx + c$, is $-b/(2a)$. [7.6a]

25. A graph represents a function if it is impossible to draw a(n) _____ that intersects the graph more than once. [2.2d]

26. Base-10 logarithms are called _____ logarithms. [8.3d]

27. The logarithm of a number is a(n) _____ . [8.3a]

28. If it is possible for a(n) _____ to intersect the graph of a function more than once, then the function is not one-to-one and therefore its inverse is not a function. [8.2b]

29. A(n) _____ is a correspondence between a first set, called the domain, and a second set, called the range, such that each member of the domain corresponds to exactly one member of its range. [2.2a]

30. The _____ is the amount of time necessary for half of a quantity to decay. [8.7b]

horizontal line

vertical line

natural

common

half-life

doubling time

base

exponent

vertex

discriminant

relation

function

SYNTHESIS

31. 📉 Use a graphing calculator to check your answers to Exercises 1, 8, 12, and 20.

32. Graph: $\dfrac{(x-2)^2}{16} - \dfrac{(y-2)^2}{9} = 1.$

Classify the graph of each of the following equations as a circle, an ellipse, a parabola, or a hyperbola.

33. $x^2 + y^2 - 10x + 8y - 40 = 0$

34. $y + 1 = 2x^2$

35. $1 - 3y = 2y^2 - x$

36. $9x^2 - 4y^2 - 36x + 24y - 36 = 0$

37. $4x^2 + 25y^2 - 8x - 100y + 4 = 0$

38. $\dfrac{x^2}{7} + \dfrac{y^2}{7} = 1$

39. $x^2 + y^2 = 8$

40. $y = \dfrac{2}{x}$

41. $x - \dfrac{3}{y} = 0$

42. $y + 6x = x^2 + 5$

43. $3x^2 + 5y^2 + x^2 = y^2 + 49$

44. $56x^2 - 17y^2 = 234 - 13x^2 - 38y^2$

9.4 NONLINEAR SYSTEMS OF EQUATIONS

 a Solve systems of equations where at least one equation is nonlinear.

b Solve applied problems involving nonlinear systems.

All the systems of equations we studied in Chapter 3 were linear. We now consider systems of two equations in two variables in which at least one equation is nonlinear.

a Algebraic Solutions

We first consider systems of one first-degree and one second-degree equation. For example, the graphs may be a circle and a line. If so, there are three possibilities for solutions.

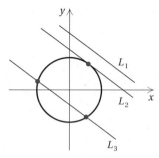

For L_1 there is no point of intersection of the line and the circle, hence no solution of the system in the set of real numbers. For L_2 there is one point of intersection, hence one real-number solution. For L_3 there are two points of intersection, hence two real-number solutions.

These systems can be solved graphically by finding the points of intersection. In solving algebraically, we use the substitution method.

EXAMPLE 1 Solve this system:

$$x^2 + y^2 = 25, \quad \textbf{(1)} \quad \text{(The graph is a circle.)}$$
$$3x - 4y = 0. \quad \textbf{(2)} \quad \text{(The graph is a line.)}$$

We first solve the linear equation (2) for x:

$$x = \tfrac{4}{3}y. \quad \textbf{(3)}$$

We then substitute $\tfrac{4}{3}y$ for x in equation (1) and solve for y:

$$\left(\tfrac{4}{3}y\right)^2 + y^2 = 25$$
$$\tfrac{16}{9}y^2 + y^2 = 25$$
$$\tfrac{25}{9}y^2 = 25$$
$$y^2 = 9$$
$$y = \pm 3.$$

Now we substitute these numbers for y in equation (3) and solve for x:

$$x = \tfrac{4}{3}(3) = 4; \qquad x = \tfrac{4}{3}(-3) = -4.$$

787

Solve. Sketch the graphs to confirm the solutions.

1. $x^2 + y^2 = 25,$
$y - x = -1$

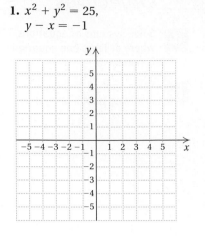

2. $y = x^2 - 2x - 1,$
$y = x + 3$

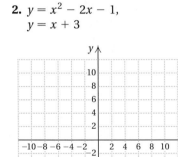

3. Solve:

$$y + 3x = 1,$$
$$x^2 - 2xy = 5.$$

Check: For (4, 3):

$$\begin{array}{c|c} x^2 + y^2 = 25 & 3x - 4y = 0 \\ \hline 4^2 + 3^2 \ \overset{?}{\,} \ 25 & 3(4) - 4(3) \ \overset{?}{\,} \ 0 \\ 16 + 9 & 12 - 12 \\ 25 \quad \text{TRUE} & 0 \quad \text{TRUE} \end{array}$$

For (−4, −3):

$$\begin{array}{c|c} x^2 + y^2 = 25 & 3x - 4y = 0 \\ \hline (-4)^2 + (-3)^2 \ \overset{?}{\,} \ 25 & 3(-4) - 4(-3) \ \overset{?}{\,} \ 0 \\ 16 + 9 & -12 + 12 \\ 25 \quad \text{TRUE} & 0 \quad \text{TRUE} \end{array}$$

The pairs $(4, 3)$ and $(-4, -3)$ check, so they are solutions. We can see the solutions in the graph. The graph of equation (1) is a circle, and the graph of equation (2) is a line. The graphs intersect at the points $(4, 3)$ and $(-4, -3)$.

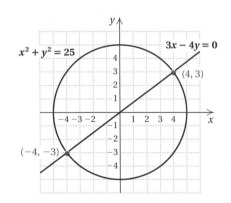

Do Exercises 1 and 2.

EXAMPLE 2 Solve this system:

$$y + 3 = 2x, \qquad \textbf{(1)}$$
$$x^2 + 2xy = -1. \qquad \textbf{(2)}$$

We first solve the linear equation (1) for y:

$$y = 2x - 3. \qquad \textbf{(3)}$$

We then substitute $2x - 3$ for y in equation (2) and solve for x:

$$x^2 + 2x(2x - 3) = -1$$
$$x^2 + 4x^2 - 6x = -1$$
$$5x^2 - 6x + 1 = 0$$
$$(5x - 1)(x - 1) = 0 \qquad \text{Factoring}$$
$$5x - 1 = 0 \quad or \quad x - 1 = 0 \qquad \text{Using the principle of}$$
$$\text{zero products}$$
$$x = \tfrac{1}{5} \quad or \qquad x = 1.$$

Now we substitute these numbers for x in equation (3) and solve for y:

$$y = 2\left(\tfrac{1}{5}\right) - 3 = -\tfrac{13}{5}; \qquad y = 2(1) - 3 = -1.$$

The check is left to the student. The pairs $\left(\tfrac{1}{5}, -\tfrac{13}{5}\right)$ and $(1, -1)$ are solutions.

Do Exercise 3.

EXAMPLE 3 Solve this system:

$$x + y = 5, \quad \text{(The graph is a line.)}$$
$$y = 3 - x^2. \quad \text{(The graph is a parabola.)}$$

We substitute $3 - x^2$ for y in the first equation:

$$x + 3 - x^2 = 5$$
$$-x^2 + x - 2 = 0$$
$$x^2 - x + 2 = 0. \quad \text{Multiplying by } -1$$

To solve this equation, we need the quadratic formula:

$$x = \frac{-b \pm \sqrt{b^2 - 4ac}}{2a} = \frac{-(-1) \pm \sqrt{(-1)^2 - 4(1)(2)}}{2(1)}$$

$$= \frac{1 \pm \sqrt{1 - 8}}{2} = \frac{1 \pm \sqrt{-7}}{2} = \frac{1}{2} \pm \frac{\sqrt{7}}{2}i.$$

Then solving the first equation for y, we obtain $y = 5 - x$. Substituting values for x gives us

$$y = 5 - \left(\frac{1}{2} + \frac{\sqrt{7}}{2}i\right) = \frac{9}{2} - \frac{\sqrt{7}}{2}i$$

and

$$y = 5 - \left(\frac{1}{2} - \frac{\sqrt{7}}{2}i\right) = \frac{9}{2} + \frac{\sqrt{7}}{2}i.$$

The solutions are

$$\left(\frac{1}{2} + \frac{\sqrt{7}}{2}i, \frac{9}{2} - \frac{\sqrt{7}}{2}i\right)$$

and

$$\left(\frac{1}{2} - \frac{\sqrt{7}}{2}i, \frac{9}{2} + \frac{\sqrt{7}}{2}i\right).$$

There are no real-number solutions. Note in the figure at right that the graphs do not intersect. Getting only nonreal complex-number solutions tells us that the graphs do not intersect.

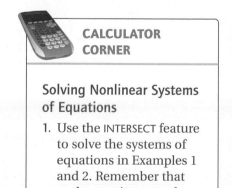

Do Exercise 4.

Two second-degree equations can have common solutions in various ways. If the graphs happen to be a circle and a hyperbola, for example, there are six possibilities, as shown below.

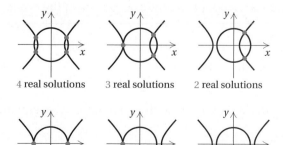

4 real solutions   3 real solutions   2 real solutions

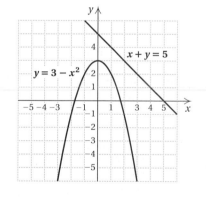

2 real solutions   1 real solution   0 real solutions

4. Solve:

$$9x^2 - 4y^2 = 36,$$
$$5x + 2y = 0.$$

Answer on page A-55

CALCULATOR CORNER

Solving Nonlinear Systems of Equations

1. Use the INTERSECT feature to solve the systems of equations in Examples 1 and 2. Remember that each equation must be solved for y before it is entered in the calculator.

2. Use the INTERSECT feature to solve the systems of equations in Margin Exercises 2 and 3.

5. Solve:

$$2y^2 - 3x^2 = 6,$$
$$5y^2 + 2x^2 = 53.$$

To solve systems of two second-degree equations, we can use either the substitution method or the elimination method. The elimination method is generally used when each equation is of the form $Ax^2 + By^2 = C$. Then we can eliminate an x^2- or a y^2-term in a manner similar to the procedure that we used for systems of linear equations in Chapter 3.

EXAMPLE 4   Solve:

$$2x^2 + 5y^2 = 22,   \textbf{(1)}$$
$$3x^2 - y^2 = -1.   \textbf{(2)}$$

In this case, we use the elimination method:

$$
\begin{array}{ll}
2x^2 + 5y^2 = 22 & \\
\underline{15x^2 - 5y^2 = -5} & \text{Multiplying by 5 on both sides of equation (2)} \\
17x^2 \qquad\; = 17 & \text{Adding} \\
\end{array}
$$
$$x^2 = 1$$
$$x = \pm 1.$$

If $x = 1$, $x^2 = 1$, and if $x = -1$, $x^2 = 1$, so substituting either 1 or -1 for x in equation (2) gives us

$$
\begin{array}{ll}
3x^2 - y^2 = -1 & \\
3 \cdot 1 - y^2 = -1 & \text{Substituting 1 for } x^2 \\
3 - y^2 = -1 & \\
-y^2 = -4 & \\
y^2 = 4 & \\
y = \pm 2. &
\end{array}
$$

Thus if $x = 1$, $y = 2$ or $y = -2$, yielding the pairs $(1, 2)$ and $(1, -2)$. If $x = -1$, $y = 2$ or $y = -2$, yielding the pairs $(-1, 2)$ and $(-1, -2)$.

Check: Since $(2)^2 = 4$, $(-2)^2 = 4$, $(1)^2 = 1$, and $(-1)^2 = 1$, we can check all four pairs at one time.

$$
\begin{array}{c}
\underline{2x^2 + 5y^2 = 22} \\
2(\pm 1)^2 + 5(\pm 2)^2 \;?\; 22 \\
2 + 20 \\
22 \quad \text{TRUE}
\end{array}
\qquad
\begin{array}{c}
\underline{3x^2 - y^2 = -1} \\
3(\pm 1)^2 - (\pm 2)^2 \;?\; -1 \\
3 - 4 \\
-1 \quad \text{TRUE}
\end{array}
$$

The solutions are $(1, 2)$, $(1, -2)$, $(-1, 2)$, and $(-1, -2)$.

Do Exercise 5.

When one equation contains a product of variables and the other equation is of the form $Ax^2 + By^2 = C$, we often solve for one of the variables in the equation with the product and then substitute in the other.

EXAMPLE 5   Solve:

$$x^2 + 4y^2 = 20,   \textbf{(1)}$$
$$xy = 4.   \textbf{(2)}$$

Here we use the substitution method. First, we solve equation (2) for y:

$$y = \frac{4}{x}.$$

Answer on page A-55

Then we substitute $4/x$ for y in equation (1) and solve for x:

$$x^2 + 4\left(\frac{4}{x}\right)^2 = 20$$

$$x^2 + \frac{64}{x^2} = 20$$

$$x^4 + 64 = 20x^2 \qquad \text{Multiplying by } x^2$$

$$x^4 - 20x^2 + 64 = 0 \qquad \begin{array}{l}\text{Obtaining standard form. This}\\ \text{equation is quadratic in form.}\end{array}$$

$$u^2 - 20u + 64 = 0 \qquad \text{Letting } u = x^2$$

$$(u - 16)(u - 4) = 0 \qquad \text{Factoring}$$

$$u = 16 \quad or \quad u = 4. \qquad \text{Using the principle of zero products}$$

Next, we substitute x^2 for u and solve these equations:

$$x^2 = 16 \quad or \quad x^2 = 4$$

$$x = \pm 4 \quad or \quad x = \pm 2.$$

Then $x = 4$ or $x = -4$ or $x = 2$ or $x = -2$. Since $y = 4/x$, if $x = 4$, $y = 1$; if $x = -4$, $y = -1$; if $x = 2$, $y = 2$; and if $x = -2$, $y = -2$. The ordered pairs $(4, 1)$, $(-4, -1)$, $(2, 2)$, and $(-2, -2)$ check. They are the solutions.

Do Exercise 6.

b   Solving Applied Problems

We now consider applications in which the translation is to a system of equations in which at least one equation is nonlinear.

EXAMPLE 6 *Architecture.* For a building at a community college, an architect wants to lay out a rectangular piece of ground that has a perimeter of 204 m and an area of 2565 m². Find the dimensions of the piece of ground.

1. **Familiarize.** We make a drawing of the area, labeling it using l for the length and w for the width.

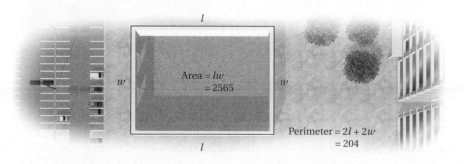

2. **Translate.** We then have the following translation:

    Perimeter: $2l + 2w = 204$;

        Area: $lw = 2565$.

3. **Solve.** We solve the system

    $2l + 2w = 204$,   (The graph is a line.)

    $lw = 2565$.   (The graph is a hyperbola.)

6. Solve:

$$x^2 + xy + y^2 = 19,$$

$$xy = 6.$$

Answer on page A-55

7. The perimeter of a rectangular field is 34 m, and the length of a diagonal is 13 m. Find the dimensions of the field.

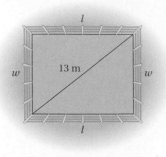

We solve the second equation for l and get $l = 2565/w$. Then we substitute $2565/w$ for l in the first equation and solve for w:

$$2\left(\frac{2565}{w}\right) + 2w = 204$$

$$2(2565) + 2w^2 = 204w \qquad \text{Multiplying by } w$$

$$2w^2 - 204w + 2(2565) = 0 \qquad \text{Standard form}$$

$$w^2 - 102w + 2565 = 0 \qquad \text{Dividing by 2}$$

$$w = \frac{-(-102) \pm \sqrt{(-102)^2 - 4 \cdot 1 \cdot 2565}}{2 \cdot 1}$$

Quadratic formula. Factoring could also be used, but the numbers are quite large.

$$w = \frac{102 \pm \sqrt{144}}{2} = \frac{102 \pm 12}{2}$$

$$w = 57 \quad or \quad w = 45.$$

If $w = 57$, then $l = 2565/w = 2565/57 = 45$. If $w = 45$, then $l = 2565/w = 2565/45 = 57$. Since length is generally considered to be greater than width, we have the solution $l = 57$ and $w = 45$, or $(57, 45)$.

4. Check. If $l = 57$ and $w = 45$, the perimeter is $2 \cdot 57 + 2 \cdot 45$, or 204. The area is $57 \cdot 45$, or 2565. The numbers check.

5. State. The length is 57 m and the width is 45 m.

Do Exercise 7.

EXAMPLE 7 *HDTV Dimensions.* The ratio of the width to the height of the screen of an HDTV (high-definition television) is 16 to 9. Suppose a large-screen HDTV has a 70-in. diagonal screen. Find the height and the width of the screen.

8. HDTV Dimensions. The ratio of the width to the height of the screen of an HDTV is 16 to 9. Suppose an HDTV screen has a 46-in. diagonal screen. Find the width and the height of the screen.

1. Familiarize. We first make a drawing and label it. Note that there is a right triangle in the figure. We let $h = $ the height and $w = $ the width.

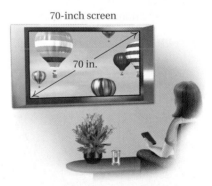

70-inch screen

70 in.

16-to-9 ratio, width to height HDTV

2. Translate. Next, we translate to a system of equations:

$$h^2 + w^2 = 70^2, \text{ or } 4900, \qquad \textbf{(1)}$$

$$\frac{w}{h} = \frac{16}{9}. \qquad \textbf{(2)}$$

3. Solve. We solve the system and get $(h, w) \approx (34, 61)$ and $(-34, -61)$.

4. Check. Widths cannot be negative, so we need check only $(34, 61)$. In the right triangle, $34^2 + 61^2 = 1156 + 3721 = 4877 \approx 4900 = 70^2$. Also, $\frac{61}{34} \approx \frac{16}{9}$.

5. State. The height is about 34 in., and the width is about 61 in.

Do Exercise 8.

Answers on page A-55

9.4

EXERCISE SET

For Extra Help

MathXL   MyMathLab   InterAct   Math Tutor   Digital Video   Student's
                     Math   Center   Tutor CD 4   Solutions
                                             Videotape 10   Manual

a   Solve.

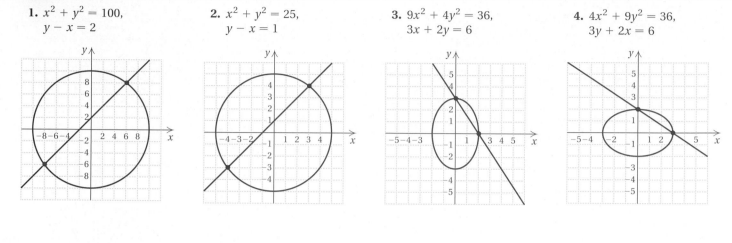

1. $x^2 + y^2 = 100,$
$y - x = 2$

2. $x^2 + y^2 = 25,$
$y - x = 1$

3. $9x^2 + 4y^2 = 36,$
$3x + 2y = 6$

4. $4x^2 + 9y^2 = 36,$
$3y + 2x = 6$

5. $y^2 = x + 3,$
$2y = x + 4$

6. $y = x^2,$
$3x = y + 2$

7. $x^2 - xy + 3y^2 = 27,$
$x - y = 2$

8. $2y^2 + xy + x^2 = 7,$
$x - 2y = 5$

9. $x^2 - xy + 3y^2 = 5,$
$x - y = 2$

10. $a^2 + 3b^2 = 10,$
$a - b = 2$

11. $a + b = -6,$
$ab = -7$

12. $2y^2 + xy = 5,$
$4y + x = 7$

13. $2a + b = 1$,
      $b = 4 - a^2$

14. $4x^2 + 9y^2 = 36$,
      $x + 3y = 3$

15. $x^2 + y^2 = 5$,
      $x - y = 8$

16. $4x^2 + 9y^2 = 36$,
      $y - x = 8$

17. $x^2 + y^2 = 25$,
      $y^2 = x + 5$

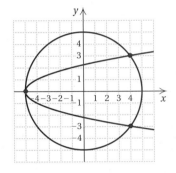

18. $y = x^2$,
      $x = y^2$

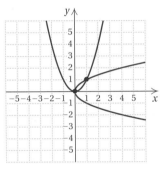

19. $x^2 + y^2 = 9$,
      $x^2 - y^2 = 9$

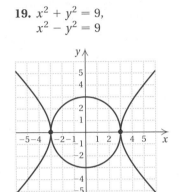

20. $y^2 - 4x^2 = 4$,
      $4x^2 + y^2 = 4$

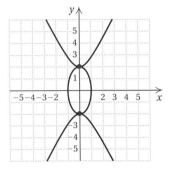

21. $x^2 + y^2 = 20$,
      $xy = 8$

22. $x^2 + y^2 = 5$,
      $xy = 2$

23. $x^2 + y^2 = 13$,
      $xy = 6$

24. $x^2 + y^2 + 6y + 5 = 0$,
      $x^2 + y^2 - 2x - 8 = 0$

25. $2xy + 3y^2 = 7,$
$3xy - 2y^2 = 4$

26. $xy - y^2 = 2,$
$2xy - 3y^2 = 0$

27. $4a^2 - 25b^2 = 0,$
$2a^2 - 10b^2 = 3b + 4$

28. $m^2 - 3mn + n^2 + 1 = 0,$
$3m^2 - mn + 3n^2 = 13$

29. $ab - b^2 = -4,$
$ab - 2b^2 = -6$

30. $a^2 + b^2 = 14,$
$ab = 3\sqrt{5}$

31. $x^2 + y^2 = 25,$
$9x^2 + 4y^2 = 36$

32. $x^2 + y^2 = 1,$
$9x^2 - 16y^2 = 144$

b Solve.

33. *Design of a Van.* The cargo area of a delivery van must be 60 ft², and the length of a diagonal of the van must accommodate a 13-ft board. Find the dimensions of the cargo area.

34. *Computer Parts.* Dataport Electronics needs a rectangular memory board that has a perimeter of 28 cm and a diagonal of length 10 cm. What should the dimensions of the board be?

35. A rectangle has an area of 14 in² and a perimeter of 18 in. Find its dimensions.

36. A rectangle has an area of 40 yd² and a perimeter of 26 yd. Find its dimensions.

37. The diagonal of a rectangle is 1 ft longer than the length of the rectangle and 3 ft longer than twice the width. Find the dimensions of the rectangle.

38. It will take 210 yd of fencing to enclose a rectangular field. The area of the field is 2250 yd². What are the dimensions of the field?

39. *Computer Screens.* The ratio of the length to the height of the screen on a computer monitor is 4 to 3. An IBM Thinkpad laptop has a 31-cm diagonal screen. Find the dimensions of the screen.

40. *HDTV Screens.* The ratio of the length to the height of an HDTV screen (see Example 7) is 16 to 9. The Remton Lounge has an HDTV screen with a 42-in. diagonal screen. Find the dimensions of the screen.

41. *Garden Design.* A garden contains two square peanut beds. Find the length of each bed if the sum of their areas is 832 ft^2 and the difference of their areas is 320 ft^2.

42. *Investments.* An amount of money invested for 1 yr at a certain interest rate yielded $225 in interest. If $750 more had been invested and the rate had been 1% less, the interest would have been the same. Find the principal and the rate.

43. The area of a rectangle is $\sqrt{2}$ m^2, and the length of a diagonal is $\sqrt{3}$ m. Find the dimensions.

44. The area of a rectangle is $\sqrt{3}$ m^2, and the length of a diagonal is 2 m. Find the dimensions.

45. $\mathbf{D_W}$ Write a problem for a classmate to solve that translates to a system of two equations in which at least one equation is nonlinear.

46. $\mathbf{D_W}$ Write a problem for a classmate to solve that translates to a system of two equations in which both equations are nonlinear.

SKILL MAINTENANCE

Find a formula for the inverse of the function, if it exists. [8.2c]

47. $f(x) = 2x - 5$

48. $f(x) = \dfrac{3}{2x - 7}$

49. $f(x) = \dfrac{x - 2}{x + 3}$

50. $f(x) = \dfrac{3x + 8}{5x - 4}$

51. $f(x) = |x|$

52. $f(x) = 4 - x^2$

53. $f(x) = 10^x$

54. $f(x) = e^x$

55. $f(x) = x^3 - 4$

56. $f(x) = \sqrt[3]{x + 2}$

57. $f(x) = \ln x$

58. $f(x) = \log x$

SYNTHESIS

59. Use a graphing calculator to check your answers to Exercises 1, 8, and 12.

60. Find the equation of an ellipse centered at the origin that passes through the points $(2, -3)$ and $\left(1, \sqrt{13}\right)$.

61. A piece of wire 100 cm long is to be cut into two pieces and those pieces are each to be bent to make a square. The area of one square is to be 144 cm^2 greater than that of the other. How should the wire be cut?

62. Find the equation of a circle that passes through $(-2, 3)$ and $(-4, 1)$ and whose center is on the line $5x + 8y = -2$.

63. *Railing Sales.* Fireside Castings finds that the total revenue R from the sale of x units of railing is given by the function

$$R(x) = 100x + x^2.$$

Fireside also finds that the total cost C of producing x units of the same product is given by the function

$$C(x) = 80x + 1500.$$

A break-even point is a value of x for which total revenue is the same as total cost; that is, $R(x) = C(x)$. How many units must be sold to break even? (See Section 3.8a.)

Solve.

64. $p^2 + q^2 = 13,$
$\dfrac{1}{pq} = -\dfrac{1}{6}$

65. $a + b = \dfrac{5}{6},$
$\dfrac{a}{b} + \dfrac{b}{a} = \dfrac{13}{6}$

9 Summary and Review

The review that follows is meant to prepare you for a chapter exam. It consists of three parts. The first part, Concept Reinforcement, is designed to increase understanding of the concepts through true/false exercises. The second part is a list of important properties and formulas. The third part is the Review Exercises. These provide practice exercises for the exam, together with references to section objectives so you can go back and review. Before beginning, stop and look back over the skills you have obtained. What skills in mathematics do you have now that you did not have before studying this chapter?

✎ CONCEPT REINFORCEMENT

Determine whether the statement is true or false. Answers are given at the back of the book.

_____ **1.** The graph of $y - x^2 = 5$ is a parabola opening upward.

_____ **2.** The graph of $\dfrac{x^2}{10} + \dfrac{y^2}{12} = 1$ is an ellipse with its center at the origin.

_____ **3.** The graph of $\dfrac{x^2}{10} - \dfrac{y^2}{12} = 1$ is a hyperbola with a vertical axis.

_____ **4.** The graph of $\dfrac{(x-1)^2}{10} + \dfrac{(y-4)^2}{8} = 1$ is an ellipse with its center not at the origin

_____ **5.** The graph of $(x-1)^2 + (y-4)^2 = 16$ is a circle with its center at $(-1, -4)$.

_____ **6.** The graph of $x - 2y^2 = 3$ is a parabola opening to the left.

_____ **7.** A system of equations that represent a parabola and a circle can have up to four real solutions.

_____ **8.** A system of equations representing a hyperbola and a circle can have no fewer than two real solutions.

IMPORTANT PROPERTIES AND FORMULAS

Distance Formula: $\quad d = \sqrt{(x_2 - x_1)^2 + (y_2 - y_1)^2}$ $\qquad$ _Midpoint Formula:_ $\left(\dfrac{x_1 + x_2}{2}, \dfrac{y_1 + y_2}{2}\right)$

Circle: $\quad (x - h)^2 + (y - k)^2 = r^2$

Ellipse: $\quad \dfrac{(x - h)^2}{a^2} + \dfrac{(y - k)^2}{b^2} = 1$, centered at (h, k), $a, b > 0$, $a \neq b$

$\qquad \dfrac{x^2}{a^2} + \dfrac{y^2}{b^2} = 1$, centered at $(0, 0)$, $a, b > 0$, $a \neq b$

Parabola

$y = ax^2 + bx + c, a > 0$ $\qquad\qquad$ $y = ax^2 + bx + c, a < 0$
$\quad = a(x - h)^2 + k$, opens up; $\qquad\quad = a(x - h)^2 + k$, opens down;
$x = ay^2 + by + c, a > 0$ $\qquad\qquad$ $x = ay^2 + by + c, a < 0$
$\quad = a(y - k)^2 + h$, opens to the right; $\qquad = a(y - k)^2 + h$, opens to the left

Hyperbola

$\dfrac{x^2}{a^2} - \dfrac{y^2}{b^2} = 1$, axis is horizontal; $\qquad \dfrac{y^2}{b^2} - \dfrac{x^2}{a^2} = 1$, axis is vertical

$xy = c, c \neq 0$, x- and y-axes as asymptotes

Review Exercises

Find the distance between each pair of points. Where appropriate, give an approximation to three decimal places. [9.1b]

1. $(2, 6)$ and $(6, 6)$

2. $(-1, 1)$ and $(-5, 4)$

3. $(1.4, 3.6)$ and $(4.7, -5.3)$

4. $(2, 3a)$ and $(-1, a)$

Find the midpoint of the segment with the given endpoints. [9.1c]

5. $(1, 6)$ and $(7, 6)$

6. $(-1, 1)$ and $(-5, 4)$

7. $\left(1, \sqrt{3}\right)$ and $\left(\frac{1}{2}, -\sqrt{2}\right)$

8. $(2, 3a)$ and $(-1, a)$

Find the center and the radius of the circle. [9.1d]

9. $(x + 2)^2 + (y - 3)^2 = 2$

10. $(x - 5)^2 + y^2 = 49$

11. $x^2 + y^2 - 6x - 2y + 1 = 0$

12. $x^2 + y^2 + 8x - 6y - 10 = 0$

13. Find an equation of the circle with center $(-4, 3)$ and radius $4\sqrt{3}$. [9.1d]

14. Find an equation of the circle with center $(7, -2)$ and radius $2\sqrt{5}$. [9.1d]

Graph.

15. $\dfrac{x^2}{16} + \dfrac{y^2}{4} = 1$ [9.2a]

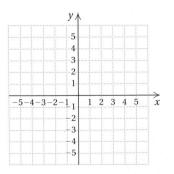

16. $\dfrac{y^2}{9} - \dfrac{x^2}{4} = 1$ [9.3a]

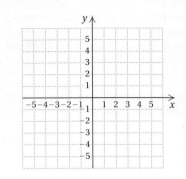

17. $x^2 + y^2 = 16$ [9.1d]

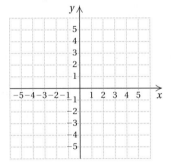

18. $x = y^2 + 2y - 2$ [9.1a]

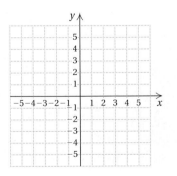

19. $y = -2x^2 - 2x + 3$ [9.1a]

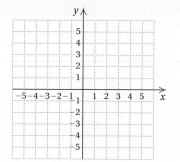

20. $x^2 + y^2 + 2x - 4y - 4 = 0$ [9.1d]

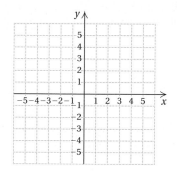

21. $\dfrac{(x-3)^2}{9} + \dfrac{(y+4)^2}{4} = 1$ [9.2a]

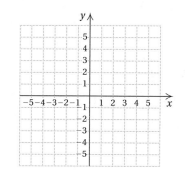

22. $xy = 9$ [9.3b]

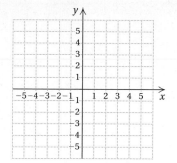

23. $x + y^2 = 2y + 1$ [9.1a]

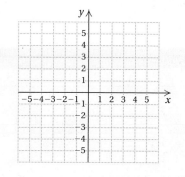

24. $\dfrac{x^2}{4} - \dfrac{y^2}{4} = 1$ [9.3a]

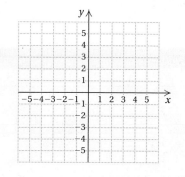

Solve. [9.4a]

25. $x^2 - y^2 = 33$,
$\quad x + y = 11$

26. $x^2 - 2x + 2y^2 = 8$,
$\quad 2x + y = 6$

27. $x^2 - y = 3$,
$\quad 2x - y = 3$

28. $x^2 + y^2 = 25$,
$\quad x^2 - y^2 = 7$

29. $x^2 - y^2 = 3$,
$\quad y = x^2 - 3$

30. $x^2 + y^2 = 18$,
$\quad 2x + y = 3$

31. $x^2 + y^2 = 100$,
$\quad 2x^2 - 3y^2 = -120$

32. $x^2 + 2y^2 = 12$,
$\quad xy = 4$

Solve. [9.4b]

33. *Vegetable Garden.* A rectangular vegetable garden has a perimeter of 38 m and an area of 84 m². What are the dimensions of the field?

34. Find two positive integers whose sum is 12 and the sum of whose reciprocals is $\frac{3}{8}$.

35. *Carton Dimensions.* One type of carton used by a manufacturer of table products exactly fits both a folded napkin of area 108 in² and a candle of length 15 in., laid diagonally on the bottom of the carton. What are the dimensions of the carton?

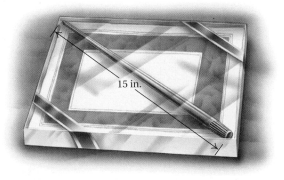

15 in.

36. *Flower Beds.* The sum of the areas of two circular flower beds is 130π ft². The difference of the circumferences is 16π ft. Find the radius of each flower bed.

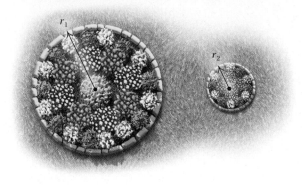

r_1

r_2

37. D_W We have studied techniques for solving systems of equations in this chapter. How do the equations differ from those systems that we studied earlier in the text? [9.4a]

38. D_W How does the graph of a hyperbola differ from the graph of a parabola? [9.1a], [9.3a]

SYNTHESIS

39. Solve: [9.4a]
$$4x^2 - x - 3y^2 = 9,$$
$$-x^2 + x + y^2 = 2.$$

40. Find an equation of the circle that passes through $(-2, -4)$, $(5, -5)$, and $(6, 2)$. [9.1d]

41. Find an equation of the ellipse with the intercepts $(-7, 0)$, $(7, 0)$, $(0, -3)$, and $(0, 3)$. [9.2a]

42. Find the point on the x-axis that is equidistant from $(-3, 4)$ and $(5, 6)$. [9.1b]

Classify the graph as a circle, an ellipse, a parabola, or a hyperbola.

43. $-y + 4x^2 = 5 - 2x$ [9.1a]

44. $\dfrac{x^2}{23} + \dfrac{y^2}{23} = 1$ [9.1d]

45. $43 - 12x^2 + y^2 = 21x^2 + 2y^2$ [9.2a]

46. $3x^2 + 3y^2 = 170$ [9.3a]

Chapter Test

For Extra Help

Work It Out!
Chapter Test Video
on CD

Find the distance between each pair of points. Where appropriate, give an approximation to three decimal places.

1. $(-6, 2)$ and $(6, 8)$

2. $(3, -a)$ and $(-3, a)$

Find the midpoint of the segment with the given endpoints.

3. $(-6, 2)$ and $(6, 8)$

4. $(3, -a)$ and $(-3, a)$

Find the center and the radius of the circle.

5. $(x + 2)^2 + (y - 3)^2 = 64$

6. $x^2 + y^2 + 4x - 6y + 4 = 0$

7. Find an equation of the circle with center $(-2, -5)$ and radius $3\sqrt{2}$.

Graph.

8. $y = x^2 - 4x - 1$

9. $x^2 + y^2 = 36$

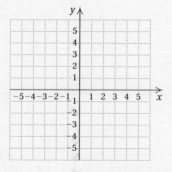

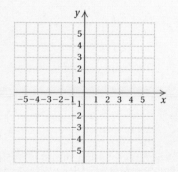

10. $\dfrac{x^2}{9} - \dfrac{y^2}{4} = 1$

11. $\dfrac{(x + 2)^2}{16} + \dfrac{(y - 3)^2}{9} = 1$

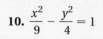

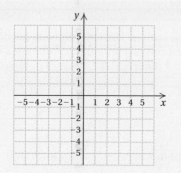

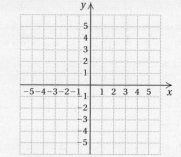

12. $x^2 + y^2 - 4x + 6y + 4 = 0$

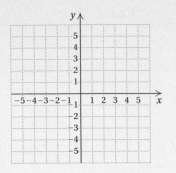

13. $9x^2 + y^2 = 36$

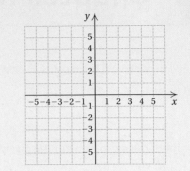

14. $xy = 4$

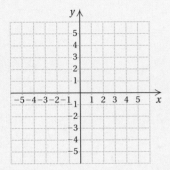

15. $x = -y^2 + 4y$

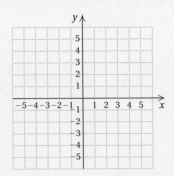

Solve.

16. $\dfrac{x^2}{16} + \dfrac{y^2}{9} = 1,$

$3x + 4y = 12$

17. $x^2 + y^2 = 16,$

$\dfrac{x^2}{16} - \dfrac{y^2}{9} = 1$

18. *Home Office.* A rectangular home office has a diagonal of 20 ft and a perimeter of 56 ft. What are the dimensions of the office?

19. *Investments.* Nikki invested a certain amount of money for 1 yr and earned $72 in interest. Erin invested $240 more than Nikki at an interest rate that was $\frac{5}{6}$ of the rate given to Nikki, but she earned the same amount of interest. Find the principal and the interest rate of Nikki's investment.

20. A rectangle with a diagonal of length $5\sqrt{5}$ yd has an area of 22 yd^2. Find the dimensions of the rectangle.

21. *Water Fountains.* The sum of the areas of two square water fountains is 8 m^2 and the difference of their areas is 2 m^2. Find the length of a side of each square.

SYNTHESIS

22. Find an equation of the ellipse passing through $(6, 0)$ and $(6, 6)$ with vertices at $(1, 3)$ and $(11, 3)$.

23. Find the points whose distance from $(8, 0)$ is 10.

24. The sum of two numbers is 36, and the product is 4. Find the sum of the reciprocals of the numbers.

25. Find the point on the y-axis that is equidistant from $(-3, -5)$ and $(4, -7)$.

CHAPTER 9: Conic Sections

Simplify.

1. $\left| \dfrac{2}{3} - \dfrac{4}{5} \right|$

2. $\dfrac{63x^2y^3}{-7x^{-4}y}$

3. $1000 \div 10^2 \cdot 25 \div 4$

4. $5x - 3[4(x - 2) - 2(x + 1)]$

Solve.

5. $\dfrac{1}{3}x - \dfrac{1}{5} \geq \dfrac{1}{5}x - \dfrac{1}{3}$

6. $|x| > 6.4$

7. $3 \leq 4x + 7 < 31$

8. $\begin{aligned} 3x + y &= 4, \\ -6x - y &= -3 \end{aligned}$

9. $\begin{aligned} x - y + 2z &= 3, \\ -x \quad + z &= 4, \\ 2x + y - z &= -3 \end{aligned}$

10. $2x^2 = x + 3$

11. $3x - \dfrac{6}{x} = 7$

12. $\sqrt{x + 5} = x - 1$

13. $x(x + 10) = -21$

14. $2x^2 + x + 1 = 0$

15. $x^4 - 13x^2 + 36 = 0$

16. $\dfrac{3}{x - 3} - \dfrac{x + 2}{x^2 + 2x - 15} = \dfrac{1}{x + 5}$

17. $\begin{aligned} -x^2 + 2y^2 &= 7, \\ x^2 + y^2 &= 5 \end{aligned}$

18. $\log_2 x + \log_2 (x + 7) = 3$

19. $7^x = 30$

20. $\log_3 x = 2$

21. $x^2 - 1 \geq 0$

22. $\dfrac{x + 1}{x - 2} > 0$

23. $P = \dfrac{3}{4}(M + 2N)$, for N

24. $\dfrac{1}{p} + \dfrac{1}{q} = \dfrac{1}{f}$, for p

Solve.

25. *World Demand for Lumber.* The world is experiencing an exponential demand for lumber. The amount of timber N, in billions of cubic feet, consumed t years after 2000, can be approximated by

$$N(t) = 65(1.018)^t,$$

where $t = 0$ corresponds to 2000.

Source: U. N. Food and Agricultural Organization; American Forest and Paper Association

a) How much timber is projected to be consumed in 2008? in 2015?
b) What is the doubling time?
c) Graph the function.

26. *Interest Compounded Annually.* Suppose that $50,000 is invested at 4% interest, compounded annually.

a) Find a function A for the amount in the account after t years.
b) Find the amount of money in the account at $t = 0$, $t = 4$, $t = 8$, and $t = 10$.
c) Graph the function.

Simplify.

27. $(2x + 3)(x^2 - 2x - 1)$

28. $(3x^2 + x^3 - 1) - (2x^3 + x + 5)$

29. $\dfrac{2m^2 + 11m - 6}{m^3 + 1} \cdot \dfrac{m^2 - m + 1}{m + 6}$

30. $\dfrac{x}{x - 1} + \dfrac{2}{x + 1} - \dfrac{2x}{x^2 - 1}$

31. $\dfrac{1 - \dfrac{5}{x}}{x - 4 - \dfrac{5}{x}}$

32. $(x^4 + 3x^3 - x + 4) \div (x + 1)$

33. $\dfrac{\sqrt{75x^5y^2}}{\sqrt{3xy}}$

34. $4\sqrt{50} - 3\sqrt{18}$

35. $(16^{3/2})^{1/2}$

36. $\left(2 - i\sqrt{2}\right)\left(5 + 3i\sqrt{2}\right)$

37. $\dfrac{5 + i}{2 - 4i}$

38. *Food and Drink Sales.* Total sales of food and beverages in U.S. restaurants S, in billions of dollars, can be modeled by the function

$$S(t) = 18t + 344.7,$$

where t is the number of years since 2000.

a) Find the total sales of food and drink in U.S. restaurants in 2005, in 2008, and in 2010.
b) Graph the function.
c) Find the y-intercept.
d) Find the slope.
e) Find the rate of change.

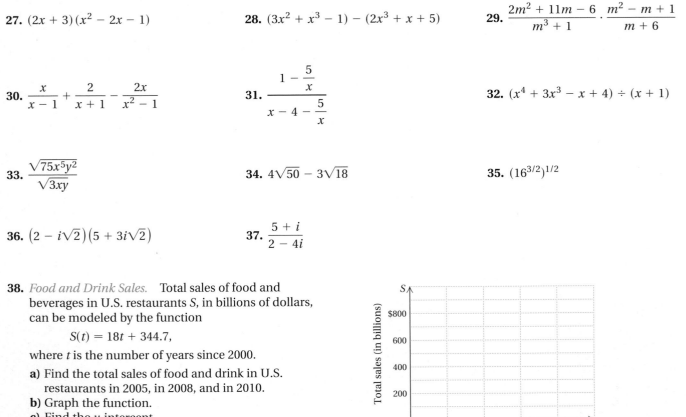

Graph.

39. $4y - 3x = 12$

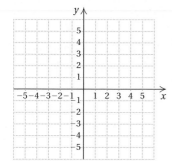

40. $y < -2$

41. $x + y \leq 0,$
$\quad x \geq -4,$
$\quad y \geq -1$

42. $f(x) = 2x^2 - 8x + 9$

43. $(x - 1)^2 + (y + 1)^2 = 9$

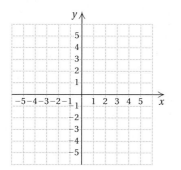

44. $x = y^2 + 1$

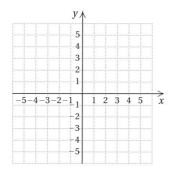

45. $f(x) = e^{-x}$

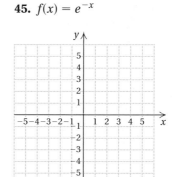

46. $f(x) = \log_2 x$

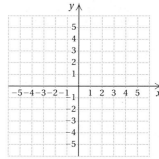

Factor.

47. $2x^4 - 12x^3 + x - 6$

48. $3a^2 - 12ab - 135b^2$

49. $x^2 - 17x + 72$

50. $81m^4 - n^4$

51. $16x^2 - 16x + 4$

52. $81a^3 - 24$

53. $10x^2 + 66x - 28$

54. $6x^3 + 27x^2 - 15x$

55. Find an equation of the line containing the points $(1, 4)$ and $(-1, 0)$.

56. Find an equation of the line containing the point $(1, 2)$ and perpendicular to the line whose equation is $2x - y = 3$.

57. Find the center and the radius of the circle
$$x^2 - 16x + y^2 + 6y + 68 = 0.$$

58. Find $f^{-1}(x)$ when $f(x) = 2x - 3$.

59. z varies directly as x and inversely as the cube of y, and $z = 5$ when $x = 4$ and $y = 2$. What is z when $x = 10$ and $y = 5$?

60. Given the function f described by $f(x) = x^3 - 2$, find $f(-2)$.

61. Find the distance between the points $(2, 1)$ and $(8, 9)$.

62. Find the midpoint of the segment with endpoints $(-1, -3)$ and $(3, 0)$.

63. Rationalize the denominator: $\dfrac{5 + \sqrt{a}}{3 - \sqrt{a}}$.

64. Find the domain: $f(x) = \dfrac{4x - 3}{3x^2 + x}$.

65. Given that $f(x) = 3x^2 + x$, find a such that $f(a) = 2$.

Solve.

66. *Music Club.* A music club offers two types of membership. Limited members pay a fee of $10 a year and can buy CDs for $10 each. Preferred members pay $20 a year and can buy CDs for $7.50 each. For what numbers of annual CD purchases would it be less expensive to be a preferred member?

67. *Train Travel.* A passenger train travels at twice the speed of a freight train. The freight train leaves a station at 2 A.M. and travels north at 34 mph. The passenger train leaves the station at 11 A.M. traveling north on a parallel track. How far from the station will the passenger train overtake the freight train?

68. *Perimeters of Polygons.* A pentagon with all five sides the same size has a perimeter equal to that of an octagon in which all eight sides are the same size. One side of the pentagon is 2 less than three times one side of the octagon. What is the perimeter of each figure?

69. *Ammonia Solutions.* A chemist has two solutions of ammonia and water. Solution A is 6% ammonia and solution B is 2% ammonia. How many liters of each solution are needed in order to obtain 80 L of a solution that is 3.2% ammonia?

70. *Air Travel.* An airplane can fly 190 mi with the wind in the same time that it takes to fly 160 mi against the wind. The speed of the wind is 30 mph. How fast can the plane fly in still air?

71. *Work.* Bianca can do a certain job in 21 min. Dahlia can do the same job in 14 min. How long would it take to do the job if the two worked together?

72. *Centripetal Force.* The centripetal force F of an object moving in a circle varies directly as the square of the velocity v and inversely as the radius r of the circle. If $F = 8$ when $v = 1$ and $r = 10$, what is F when $v = 2$ and $r = 16$?

73. *Rectangle Dimensions.* The perimeter of a rectangle is 34 ft. The length of a diagonal is 13 ft. Find the dimensions of the rectangle.

74. *Dimensions of a Rug.* The diagonal of a Persian rug is 25 ft. The area of the rug is 300 ft^2. Find the length and the width of the rug.

75. *Maximizing Area.* A farmer wants to fence in a rectangular area next to a river. (Note that no fence will be needed along the river.) What is the area of the largest region that can be fenced in with 100 ft of fencing?

76. *Carbon Dating.* Use the function $P(t) = P_0 e^{-0.00012t}$ to find the age of a bone that has lost 25% of its carbon-14.

77. *Beam Load.* The weight W that a horizontal beam can support varies inversely as the length L of the beam. If a 14-m beam can support 1440 kg, what weight can a 6-m beam support?

78. Fit a linear function to the data points $(2, -3)$ and $(5, -4)$.

79. Fit a quadratic function to the data points $(-2, 4)$, $(-5, -6)$, and $(1, -3)$.

80. Convert to a logarithmic equation: $10^6 = r$.

81. Convert to an exponential equation: $\log_3 Q = x$.

82. Express as a single logarithm:
$$\tfrac{1}{5}(7 \log_b x - \log_b y - 8 \log_b z).$$

83. Express in terms of logarithms of x, y, and z:
$$\log_b \left(\frac{xy^5}{z} \right)^{-6}.$$

84. What is the maximum product of two numbers whose sum is 26?

85. Determine whether the function $f(x) = 4 - x^2$ is one-to-one.

86. For the graph of function f shown here, determine **(a)** $f(2)$; **(b)** the domain; **(c)** all x-values such that $f(x) = -5$; and **(d)** the range.

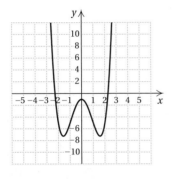

87. *Population Growth of Naples, Florida.* In 2000, the population of Naples, Florida, was 251,377. It had grown from a population of 152,099 in 1990. Naples was the second fastest growing metropolitan area in the United States. Assume the population growth increases according to an exponential growth function.

Source: U.S. Bureau of the Census

a) Let $t = 0$ correspond to 1990 and $t = 10$ correspond to 2000. Then t is the number of years since 1990. Use the data points $(0, 152{,}099)$ and $(10, 251{,}377)$ to find the exponential growth rate and fit an exponential growth function $P(t) = P_0 e^{kt}$ to the data, where $P(t)$ is the population of Naples t years after 1990.
b) Use the function found in part (a) to predict the population of Naples in 2015.
c) When will the population reach 2 million?

---

88. Solve: $\dfrac{9}{x} - \dfrac{9}{x + 12} = \dfrac{108}{x^2 + 12x}$.

89. Solve: $\log_2 (\log_3 x) = 2$.

90. Describe the graph of
$$\frac{x^2}{a^2} + \frac{y^2}{b^2} = 1$$
when $a^2 = b^2$.

91. Diaphantos, a famous mathematician, spent $\frac{1}{6}$ of his life as a child, $\frac{1}{12}$ as a young man, and $\frac{1}{7}$ as a bachelor. Five years after he was married, he had a son who died 4 yr before his father at half his father's final age. How long did Diaphantos live?

92. Solve:
$$x^2 + y^2 = 208,$$
$$xy = 96.$$

Appendixes

 HANDLING DIMENSION SYMBOLS

 DETERMINANTS AND CRAMER'S RULE

 ELIMINATION USING MATRICES

 THE ALGEBRA OF FUNCTIONS

Objectives

a Perform calculations with dimension symbols.

b Make unit changes.

1. A truck travels 210 mi in 3 hr. What is its average speed?

Perform these calculations.

2. $\dfrac{100 \text{ m}}{4 \text{ sec}}$

3. 7 yd + 9 yd

4. 24 in. · 3 in.

5. Calculate: 6 men · 11 hours.

Answers on page A-58

810

a Calculating with Dimension Symbols

In many applications, we add, subtract, multiply, and divide quantities having units, or dimensions, such as ft, km, sec, and hr. For example, to find average speed, we divide total distance by total time. What results is notation very much like a rational expression.

EXAMPLE 1 A car travels 150 km in 2 hr. What is its average speed?

$$\text{Speed} = \frac{150 \text{ km}}{2 \text{ hr}}, \text{ or } 75 \frac{\text{km}}{\text{hr}}$$

(The standard abbreviation for km/hr is km/h, but it does not suit our present discussion well.)

The symbol km/hr makes it look as though we are dividing kilometers by hours. It can be argued that we can divide only numbers. Nevertheless, we treat dimension symbols, such as km, ft, and hr, as if they were numerals or variables, obtaining correct results mechanically.

Do Exercise 1.

EXAMPLES Compare the following.

2. $\dfrac{150x}{2y} = \dfrac{150}{2} \cdot \dfrac{x}{y} = 75\dfrac{x}{y}$ with $\dfrac{150 \text{ km}}{2 \text{ hr}} = \dfrac{150}{2}\dfrac{\text{km}}{\text{hr}} = 75\dfrac{\text{km}}{\text{hr}}$

3. $3x + 2x = (3 + 2)x = 5x$ with $3 \text{ ft} + 2 \text{ ft} = (3 + 2) \text{ ft} = 5 \text{ ft}$

4. $5x \cdot 3x = 15x^2$ with $5 \text{ ft} \cdot 3 \text{ ft} = 15 \text{ ft}^2$ (square feet)

Do Exercises 2–4.

If 5 men work 8 hours, the total amount of labor is 40 man-hours.

EXAMPLE 5 Compare

$$5x \cdot 8y = 40xy \quad \text{with} \quad 5 \text{ men} \cdot 8 \text{ hours} = 40 \text{ man-hours}.$$

Do Exercise 5.

EXAMPLE 6 Compare

$$\frac{300x \cdot 240y}{15t} = 4800\frac{xy}{t} \quad \text{with} \quad \frac{300 \text{ kW} \cdot 240 \text{ hr}}{15 \text{ da}} = 4800\frac{\text{kW-hr}}{\text{da}}.$$

If an electrical device uses 300 kW (kilowatts) for 240 hr over a period of 15 days, its rate of usage of energy is 4800 kilowatt-hours per day. The standard abbreviation for kilowatt-hours is kWh.

Do Exercise 6.

b | Making Unit Changes

We can treat dimension symbols much like numerals or variables, because we obtain correct results that way. We can change units by substituting or by multiplying by 1, as shown below.

EXAMPLE 7 Convert 3 ft to inches.

METHOD 1.   We have 3 ft. We know that 1 ft = 12 in., so we substitute 12 in. for ft:

$$3 \text{ ft} = 3 \cdot 12 \text{ in.} = 36 \text{ in.}$$

METHOD 2.   We want to convert from "ft" to "in." We multiply by 1 using a symbol for 1 with "ft" on the bottom since we are converting from "ft," and with "in." on the top since we are converting to "in."

$$3 \text{ ft} = 3 \text{ ft} \cdot \frac{12 \text{ in.}}{1 \text{ ft}}$$

$$= \frac{3 \cdot 12}{1} \cdot \frac{\text{ft}}{\text{ft}} \cdot \text{in.} = 36 \text{ in.}$$

Do Exercise 7.

We can multiply by 1 several times to make successive conversions. In the following example, we convert mi/hr to ft/sec by converting successively from mi/hr to ft/hr to ft/min to ft/sec.

EXAMPLE 8 Convert 60 mi/hr to ft/sec.

$$60\frac{\text{mi}}{\text{hr}} = 60\frac{\text{mi}}{\text{hr}} \cdot \frac{5280 \text{ ft}}{1 \text{ mi}} \cdot \frac{1 \text{ hr}}{60 \text{ min}} \cdot \frac{1 \text{ min}}{60 \text{ sec}}$$

$$= \frac{60 \cdot 5280}{60 \cdot 60} \cdot \frac{\text{mi}}{\text{mi}} \cdot \frac{\text{hr}}{\text{hr}} \cdot \frac{\text{min}}{\text{min}} \cdot \frac{\text{ft}}{\text{sec}} = 88\frac{\text{ft}}{\text{sec}}.$$

Do Exercise 8.

---

6. Calculate:

$$\frac{200 \text{ kW} \cdot 140 \text{ hr}}{35 \text{ da}}.$$

7. Convert 7 ft to inches.

8. Convert 90 mi/hr to ft/sec.

Answers on page A-58

811

a Add the measures.

1. 45 ft + 23 ft

2. 55 km/hr + 27 km/hr

3. 17 g + 28 g

4. 3.4 lb + 5.2 lb

Find the average speeds, given total distance and total time.

5. 90 mi, 6 hr

6. 640 km, 20 hr

7. 9.9 m, 3 sec

8. 76 ft, 4 min

Perform the calculations.

9. $\dfrac{3 \text{ in.} \cdot 8 \text{ lb}}{6 \text{ sec}}$

10. $\dfrac{60 \text{ men} \cdot 8 \text{ hr}}{20 \text{ da}}$

11. $36 \text{ ft} \cdot \dfrac{1 \text{ yd}}{3 \text{ ft}}$

12. $55\dfrac{\text{mi}}{\text{hr}} \cdot 4 \text{ hr}$

13. $5 \text{ ft}^3 + 11 \text{ ft}^3$

14. $\dfrac{3 \text{ lb}}{14 \text{ ft}} \cdot \dfrac{7 \text{ lb}}{6 \text{ ft}}$

15. Divide $4850 by 5 days.

16. Divide $25.60 by 8 hr.

b Make the unit changes.

17. Change 3.2 lb to oz (16 oz = 1 lb).

18. Change 6.2 km to m.

19. Change 35 mi/hr to ft/min.

20. Change $375 per day to dollars per minute.

21. Change 8 ft to in.

22. Change 25 yd to ft.

23. How many years ago is 1 million sec ago? Let 365 days = 1 yr.

24. How many years ago is 1 billion sec ago?

25. How many years ago is 1 trillion sec ago?

26. Change 20 lb to oz.

27. Change $60 \dfrac{\text{lb}}{\text{ft}}$ to $\dfrac{\text{oz}}{\text{in.}}$.

28. Change $44 \dfrac{\text{ft}}{\text{sec}}$ to $\dfrac{\text{mi}}{\text{hr}}$.

29. Change 2 days to seconds.

30. Change 128 hr to days.

31. Change 216 in^2 to ft^2.

32. Change 1440 man-hours to man-days.

33. Change $80 \dfrac{\text{lb}}{\text{ft}^3}$ to $\dfrac{\text{ton}}{\text{yd}^3}$.

34. Change the speed of light, 186,000 mi/sec, to mi/yr.

B

DETERMINANTS AND CRAMER'S RULE

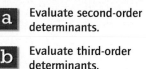

Objectives

a   Evaluate second-order determinants.

b   Evaluate third-order determinants.

c   Solve systems of equations using Cramer's rule.

In Chapter 3, you probably noticed that the elimination method concerns itself primarily with the coefficients and constants of the equations. Here we learn a method for solving a system of equations using just the coefficients and constants. This method involves *determinants*.

a   Evaluating Determinants

The following symbolism represents a **second-order determinant:**

$$\begin{vmatrix} a_1 & b_1 \\ a_2 & b_2 \end{vmatrix}.$$

To evaluate a determinant, we do two multiplications and subtract.

EXAMPLE 1   Evaluate:

$$\begin{vmatrix} 2 & -5 \\ 6 & 7 \end{vmatrix}.$$

We multiply and subtract as follows:

$$\begin{vmatrix} 2 & -5 \\ 6 & 7 \end{vmatrix} = 2 \cdot 7 - 6 \cdot (-5) = 14 + 30 = 44.$$

Determinants are defined according to the pattern shown in Example 1.

Evaluate.

1. $\begin{vmatrix} 3 & 2 \\ 4 & 1 \end{vmatrix}$

SECOND-ORDER DETERMINANT

The determinant $\begin{vmatrix} a_1 & b_1 \\ a_2 & b_2 \end{vmatrix}$ is defined to mean $a_1 b_2 - a_2 b_1$.

The value of a determinant is a *number*. In Example 1, the value is 44.

Do Exercises 1 and 2.

2. $\begin{vmatrix} 5 & -2 \\ -1 & -1 \end{vmatrix}$

b   Third-Order Determinants

A **third-order determinant** is defined as follows.

Note the minus sign here.

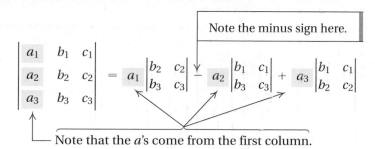

$$\begin{vmatrix} a_1 & b_1 & c_1 \\ a_2 & b_2 & c_2 \\ a_3 & b_3 & c_3 \end{vmatrix} = a_1 \begin{vmatrix} b_2 & c_2 \\ b_3 & c_3 \end{vmatrix} - a_2 \begin{vmatrix} b_1 & c_1 \\ b_3 & c_3 \end{vmatrix} + a_3 \begin{vmatrix} b_1 & c_1 \\ b_2 & c_2 \end{vmatrix}$$

Note that the *a*'s come from the first column.

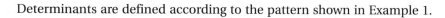

Answers on page A-58

813

Evaluate.

3. $\begin{vmatrix} 2 & -1 & 1 \\ 1 & 2 & -1 \\ 3 & 4 & -3 \end{vmatrix}$

Note that the second-order determinants on the right can be obtained by crossing out the row and the column in which each a occurs.

For a_1: $\begin{vmatrix} a_1 & b_1 & c_1 \\ a_2 & b_2 & c_2 \\ a_3 & b_3 & c_3 \end{vmatrix}$ For a_2: $\begin{vmatrix} a_1 & b_1 & c_1 \\ a_2 & b_2 & c_2 \\ a_3 & b_3 & c_3 \end{vmatrix}$

For a_3: $\begin{vmatrix} a_1 & b_1 & c_1 \\ a_2 & b_2 & c_2 \\ a_3 & b_3 & c_3 \end{vmatrix}$

EXAMPLE 2   Evaluate this third-order determinant:

$$\begin{vmatrix} -1 & 0 & 1 \\ -5 & 1 & -1 \\ 4 & 8 & 1 \end{vmatrix} = -1 \begin{vmatrix} 1 & -1 \\ 8 & 1 \end{vmatrix} - (-5) \begin{vmatrix} 0 & 1 \\ 8 & 1 \end{vmatrix} + 4 \begin{vmatrix} 0 & 1 \\ 1 & -1 \end{vmatrix}.$$

We calculate as follows:

$$-1 \begin{vmatrix} 1 & -1 \\ 8 & 1 \end{vmatrix} - (-5) \begin{vmatrix} 0 & 1 \\ 8 & 1 \end{vmatrix} + 4 \begin{vmatrix} 0 & 1 \\ 1 & -1 \end{vmatrix}$$
$$= -1[1 \cdot 1 - 8(-1)] + 5(0 \cdot 1 - 8 \cdot 1) + 4[0 \cdot (-1) - 1 \cdot 1]$$
$$= -1(9) + 5(-8) + 4(-1)$$
$$= -9 - 40 - 4$$
$$= -53.$$

4. $\begin{vmatrix} 3 & 2 & 2 \\ -2 & 1 & 4 \\ 4 & -3 & 3 \end{vmatrix}$

Do Exercises 3 and 4.

C   Solving Systems Using Determinants

Here is a system of two equations in two variables:

$$a_1 x + b_1 y = c_1,$$
$$a_2 x + b_2 y = c_2.$$

We form three determinants, which we call D, D_x, and D_y.

$$D = \begin{vmatrix} a_1 & b_1 \\ a_2 & b_2 \end{vmatrix} \qquad \text{In } D, \text{ we have the coefficients of } x \text{ and } y.$$

$$D_x = \begin{vmatrix} c_1 & b_1 \\ c_2 & b_2 \end{vmatrix} \qquad \text{To form } D_x, \text{ we replace the } x\text{-coefficients in } D \text{ with the constants on the right side of the equations.}$$

$$D_y = \begin{vmatrix} a_1 & c_1 \\ a_2 & c_2 \end{vmatrix} \qquad \text{To form } D_y, \text{ we replace the } y\text{-coefficients in } D \text{ with the constants on the right.}$$

It is important that the replacement be done *without changing the order of the columns.* Then the solution of the system can be found as follows. This is known as **Cramer's rule.**

Answers on page A-58

APPENDIX B: Determinants and
Cramer's Rule

$$x = \frac{D_x}{D}, \qquad y = \frac{D_y}{D}$$

5. Solve using Cramer's rule:

$$4x - 3y = 15,$$
$$x + 3y = 0.$$

EXAMPLE 3 Solve using Cramer's rule:

$$3x - 2y = 7,$$
$$3x + 2y = 9.$$

We compute D, D_x, and D_y:

$$D = \begin{vmatrix} 3 & -2 \\ 3 & 2 \end{vmatrix} = 3 \cdot 2 - 3 \cdot (-2) = 6 + 6 = 12;$$

$$D_x = \begin{vmatrix} 7 & -2 \\ 9 & 2 \end{vmatrix} = 7 \cdot 2 - 9(-2) = 14 + 18 = 32;$$

$$D_y = \begin{vmatrix} 3 & 7 \\ 3 & 9 \end{vmatrix} = 3 \cdot 9 - 3 \cdot 7 = 27 - 21 = 6.$$

Then

$$x = \frac{D_x}{D} = \frac{32}{12}, \text{ or } \frac{8}{3} \quad \text{and} \quad y = \frac{D_y}{D} = \frac{6}{12} = \frac{1}{2}.$$

The solution is $\left(\frac{8}{3}, \frac{1}{2}\right)$.

Do Exercise 5.

Cramer's rule for three equations is very similar to that for two.

$$a_1 x + b_1 y + c_1 z = d_1$$
$$a_2 x + b_2 y + c_2 z = d_2$$
$$a_3 x + b_3 y + c_3 z = d_3$$

$$D = \begin{vmatrix} a_1 & b_1 & c_1 \\ a_2 & b_2 & c_2 \\ a_3 & b_3 & c_3 \end{vmatrix} \qquad D_x = \begin{vmatrix} d_1 & b_1 & c_1 \\ d_2 & b_2 & c_2 \\ d_3 & b_3 & c_3 \end{vmatrix}$$

$$D_y = \begin{vmatrix} a_1 & d_1 & c_1 \\ a_2 & d_2 & c_2 \\ a_3 & d_3 & c_3 \end{vmatrix}$$

D is again the determinant of the coefficients of x, y, and z. This time we have one more determinant, D_z. We get it by replacing the z-coefficients in D with the constants on the right:

$$D_z = \begin{vmatrix} a_1 & b_1 & d_1 \\ a_2 & b_2 & d_2 \\ a_3 & b_3 & d_3 \end{vmatrix}.$$

Answer on page A-58

APPENDIX B: Determinants and
Cramer's Rule

6. Solve using Cramer's rule:

$$x - 3y - 7z = 6,$$
$$2x + 3y + z = 9,$$
$$4x + y = 7.$$

The solution of the system is given by

EXAMPLE 4   Solve using Cramer's rule:

$$x - 3y + 7z = 13,$$
$$x + y + z = 1,$$
$$x - 2y + 3z = 4.$$

We compute D, D_x, D_y, and D_z:

$$D = \begin{vmatrix} 1 & -3 & 7 \\ 1 & 1 & 1 \\ 1 & -2 & 3 \end{vmatrix} = -10; \qquad D_x = \begin{vmatrix} 13 & -3 & 7 \\ 1 & 1 & 1 \\ 4 & -2 & 3 \end{vmatrix} = 20;$$

$$D_y = \begin{vmatrix} 1 & 13 & 7 \\ 1 & 1 & 1 \\ 1 & 4 & 3 \end{vmatrix} = -6; \qquad D_z = \begin{vmatrix} 1 & -3 & 13 \\ 1 & 1 & 1 \\ 1 & -2 & 4 \end{vmatrix} = -24.$$

Then

$$x = \frac{D_x}{D} = \frac{20}{-10} = -2;$$

$$y = \frac{D_y}{D} = \frac{-6}{-10} = \frac{3}{5};$$

$$z = \frac{D_z}{D} = \frac{-24}{-10} = \frac{12}{5}.$$

The solution is $\left(-2, \frac{3}{5}, \frac{12}{5}\right)$.

In Example 4, we would not have needed to evaluate D_z. Once we found x and y, we could have substituted them into one of the equations to find z. In practice, it is faster to use determinants to find only two of the numbers; then we find the third by substitution into an equation.

Do Exercise 6.

In using Cramer's rule, we divide by D. If D were 0, we could not do so.

INCONSISTENT SYSTEMS; DEPENDENT EQUATIONS

If $D = 0$ and at least one of the other determinants is not 0, then the system does not have a solution, and we say that it is *inconsistent*.

If $D = 0$ and all the other determinants are also 0, then there is an infinite set of solutions. In that case, we say that the equations in the system are *dependent*.

Answer on page A-58

APPENDIX B: Determinants and Cramer's Rule

a Evaluate.

1. $\begin{vmatrix} 3 & 7 \\ 2 & 8 \end{vmatrix}$

2. $\begin{vmatrix} 5 & 4 \\ 4 & -5 \end{vmatrix}$

3. $\begin{vmatrix} -3 & -6 \\ -5 & -10 \end{vmatrix}$

4. $\begin{vmatrix} 4 & 5 \\ -7 & 9 \end{vmatrix}$

5. $\begin{vmatrix} 8 & 2 \\ 12 & -3 \end{vmatrix}$

6. $\begin{vmatrix} 1 & 1 \\ 9 & 8 \end{vmatrix}$

7. $\begin{vmatrix} 2 & -7 \\ 0 & 0 \end{vmatrix}$

8. $\begin{vmatrix} 0 & -4 \\ 0 & -6 \end{vmatrix}$

b Evaluate.

9. $\begin{vmatrix} 0 & 2 & 0 \\ 3 & -1 & 1 \\ 1 & -2 & 2 \end{vmatrix}$

10. $\begin{vmatrix} 3 & 0 & -2 \\ 5 & 1 & 2 \\ 2 & 0 & -1 \end{vmatrix}$

11. $\begin{vmatrix} -1 & -2 & -3 \\ 3 & 4 & 2 \\ 0 & 1 & 2 \end{vmatrix}$

12. $\begin{vmatrix} 1 & 2 & 2 \\ 2 & 1 & 0 \\ 3 & 3 & 1 \end{vmatrix}$

13. $\begin{vmatrix} 3 & 2 & -2 \\ -2 & 1 & 4 \\ -4 & -3 & 3 \end{vmatrix}$

14. $\begin{vmatrix} 2 & -1 & 1 \\ 1 & 2 & -1 \\ 3 & 4 & -3 \end{vmatrix}$

15. $\begin{vmatrix} 3 & 2 & 4 \\ 1 & 1 & 1 \\ 1 & 1 & 1 \end{vmatrix}$

16. $\begin{vmatrix} -1 & 6 & -5 \\ 2 & 4 & 4 \\ 5 & 3 & 10 \end{vmatrix}$

c Solve using Cramer's rule.

17. $3x - 4y = 6,$
    $5x + 9y = 10$

18. $5x + 8y = 1,$
    $3x + 7y = 5$

19. $-2x + 4y = 3,$
    $3x - 7y = 1$

20. $5x - 4y = -3,$
    $7x + 2y = 6$

21. $4x + 2y = 11,$
    $3x - y = 2$

22. $3x - 3y = 11,$
    $9x - 2y = 5$

23. $x + 4y = 8,$
    $3x + 5y = 3$

24. $x + 4y = 5,$
    $-3x + 2y = 13$

25. $2x - 3y + 5z = 27,$
    $x + 2y - z = -4,$
    $5x - y + 4z = 27$

26. $x - y + 2z = -3,$
    $x + 2y + 3z = 4,$
    $2x + y + z = -3$

27. $r - 2s + 3t = 6,$
    $2r - s - t = -3,$
    $r + s + t = 6$

28. $a - 3c = 6,$
    $b + 2c = 2,$
    $7a - 3b - 5c = 14$

29. $4x - y - 3z = 1,$
    $8x + y - z = 5,$
    $2x + y + 2z = 5$

30. $3x + 2y + 2z = 3,$
    $x + 2y - z = 5,$
    $2x - 4y + z = 0$

31. $p + q + r = 1,$
    $p - 2q - 3r = 3,$
    $4p + 5q + 6r = 4$

32. $x + 2y - 3z = 9,$
    $2x - y + 2z = -8,$
    $3x - y - 4z = 3$

Objective

a Solve systems of two or three equations using matrices.

C ELIMINATION USING MATRICES

The elimination method concerns itself primarily with the coefficients and constants of the equations. In what follows, we learn a method for solving systems using just the coefficients and the constants. This procedure involves what are called *matrices*.

a In solving systems of equations, we perform computations with the constants. The variables play no important role until the end. Thus we can simplify writing a system by omitting the variables. For example, the system

$$\begin{matrix} 3x + 4y = 5, \\ x - 2y = 1 \end{matrix} \quad \text{simplifies to} \quad \begin{matrix} 3 & 4 & 5 \\ 1 & -2 & 1 \end{matrix}$$

if we omit the variables, the operation of addition, and the equals signs. The result is a rectangular array of numbers. Such an array is called a **matrix** (plural, **matrices**). We ordinarily write brackets around matrices. The following are matrices.

$$\begin{bmatrix} 4 & 1 & 3 & 5 \\ 1 & 0 & 1 & 2 \\ 6 & 3 & -2 & 0 \end{bmatrix}, \quad \begin{bmatrix} 6 & 2 & 1 & 4 & 7 \\ 1 & 2 & 1 & 3 & 1 \\ 4 & 0 & -2 & 0 & -3 \end{bmatrix}, \quad \begin{bmatrix} 1 & 2 \\ 145 & 0 \\ -7 & 9 \\ 8 & 1 \\ 0 & 0 \end{bmatrix}.$$

The **rows** of a matrix are horizontal, and the **columns** are vertical.

$$\begin{bmatrix} 5 & -2 & 2 \\ 1 & 0 & 1 \\ 0 & 1 & 2 \end{bmatrix} \begin{matrix} \longleftarrow \text{row 1} \\ \longleftarrow \text{row 2} \\ \longleftarrow \text{row 3} \end{matrix}$$
$$\begin{matrix} \uparrow & \uparrow & \uparrow \\ \text{column 1} & \text{column 2} & \text{column 3} \end{matrix}$$

Let's now use matrices to solve systems of linear equations.

EXAMPLE 1 Solve the system

$$\begin{matrix} 5x - 4y = -1, \\ -2x + 3y = 2. \end{matrix}$$

We write a matrix using only the coefficients and the constants, keeping in mind that x corresponds to the first column and y to the second. A dashed line separates the coefficients from the constants at the end of each equation:

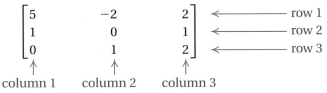

The individual numbers are called *elements* or *entries*.

Our goal is to transform this matrix into one of the form

$$\begin{bmatrix} a & b & \vdots & c \\ 0 & d & \vdots & e \end{bmatrix}.$$

The variables can then be reinserted to form equations from which we can complete the solution.

We do calculations that are similar to those that we would do if we wrote the entire equations. The first step, if possible, is to multiply and/or interchange the rows so that each number in the first column below the first number is a multiple of that number. In this case, we do so by multiplying Row 2 by 5. This corresponds to multiplying the second equation by 5.

$$\begin{bmatrix} 5 & -4 & \vdots & -1 \\ -10 & 15 & \vdots & 10 \end{bmatrix}$$ New Row 2 = 5(Row 2)

Next, we multiply the first row by 2 and add the result to the second row. This corresponds to multiplying the first equation by 2 and adding the result to the second equation. Although we write the calculations out here, we generally try to do them mentally:

$$2 \cdot 5 + (-10) = 0; \qquad 2(-4) + 15 = 7; \qquad 2(-1) + 10 = 8.$$

$$\begin{bmatrix} 5 & -4 & \vdots & -1 \\ 0 & 7 & \vdots & 8 \end{bmatrix}$$ New Row 2 = 2(Row 1) + (Row 2)

If we now reinsert the variables, we have

$$5x - 4y = -1, \qquad \textbf{(1)}$$
$$7y = 8. \qquad \textbf{(2)}$$

We can now proceed as before, solving equation (2) for y:

$$7y = 8 \qquad \textbf{(2)}$$
$$y = \tfrac{8}{7}.$$

Next, we substitute $\tfrac{8}{7}$ for y back in equation (1). This procedure is called *back-substitution*.

$$5x - 4y = -1 \qquad \textbf{(1)}$$
$$5x - 4 \cdot \tfrac{8}{7} = -1 \qquad \text{Substituting } \tfrac{8}{7} \text{ for } y \text{ in equation (1)}$$
$$x = \tfrac{5}{7} \qquad \text{Solving for } x$$

The solution is $\left(\tfrac{5}{7}, \tfrac{8}{7}\right)$.

Do Exercise 1.

EXAMPLE 2   Solve the system

$$2x - y + 4z = -3,$$
$$x \qquad - 4z = 5,$$
$$6x - y + 2z = 10.$$

We first write a matrix, using only the coefficients and the constants. Where there are missing terms, we must write 0's:

$$\begin{bmatrix} 2 & -1 & 4 & \vdots & -3 \\ 1 & 0 & -4 & \vdots & 5 \\ 6 & -1 & 2 & \vdots & 10 \end{bmatrix}. \quad \begin{array}{l} \textbf{(P1)} \\ \textbf{(P2)} \\ \textbf{(P3)} \end{array}$$ (P1), (P2), and (P3) designate the equations that are in the first, second, and third position, respectively.

Our goal is to find an equivalent matrix of the form

$$\begin{bmatrix} a & b & c & \vdots & d \\ 0 & e & f & \vdots & g \\ 0 & 0 & h & \vdots & i \end{bmatrix}.$$

A matrix of this form can be rewritten as a system of equations from which a solution can be found easily.

1. Solve using matrices:
$$5x - 2y = -44,$$
$$2x + 5y = -6.$$

Answer on page A-58

819

2. Solve using matrices:

$$x - 2y + 3z = 4,$$
$$2x - y + z = -1,$$
$$4x + y + z = 1.$$

The first step, if possible, is to interchange the rows so that each number in the first column below the first number is a multiple of that number. In this case, we do so by interchanging Rows 1 and 2:

$$\begin{bmatrix} 1 & 0 & -4 & \vdots & 5 \\ 2 & -1 & 4 & \vdots & -3 \\ 6 & -1 & 2 & \vdots & 10 \end{bmatrix}.$$

This corresponds to interchanging the first two equations.

Next, we multiply the first row by -2 and add it to the second row:

$$\begin{bmatrix} 1 & 0 & -4 & \vdots & 5 \\ 0 & -1 & 12 & \vdots & -13 \\ 6 & -1 & 2 & \vdots & 10 \end{bmatrix}.$$

This corresponds to multiplying new equation (P1) by -2 and adding it to new equation (P2). The result replaces the former (P2). We perform the calculations mentally.

Now we multiply the first row by -6 and add it to the third row:

$$\begin{bmatrix} 1 & 0 & -4 & \vdots & 5 \\ 0 & -1 & 12 & \vdots & -13 \\ 0 & -1 & 26 & \vdots & -20 \end{bmatrix}.$$

This corresponds to multiplying equation (P1) by -6 and adding it to equation (P3).

Next, we multiply Row 2 by -1 and add it to the third row:

$$\begin{bmatrix} 1 & 0 & -4 & \vdots & 5 \\ 0 & -1 & 12 & \vdots & -13 \\ 0 & 0 & 14 & \vdots & -7 \end{bmatrix}.$$

This corresponds to multiplying equation (P2) by -1 and adding it to equation (P3).

Reinserting the variables gives us

$$x \quad\quad - 4z = 5, \quad\quad \textbf{(P1)}$$
$$-y + 12z = -13, \quad\quad \textbf{(P2)}$$
$$14z = -7. \quad\quad \textbf{(P3)}$$

We now solve (P3) for z:

$$14z = -7 \quad\quad \textbf{(P3)}$$
$$z = -\tfrac{7}{14} \quad\quad \text{Solving for } z$$
$$z = -\tfrac{1}{2}.$$

Next, we back-substitute $-\tfrac{1}{2}$ for z in (P2) and solve for y:

$$-y + 12z = -13 \quad\quad \textbf{(P2)}$$
$$-y + 12\left(-\tfrac{1}{2}\right) = -13 \quad\quad \text{Substituting } -\tfrac{1}{2} \text{ for } z \text{ in equation (P2)}$$
$$-y - 6 = -13$$
$$-y = -7$$
$$y = 7. \quad\quad \text{Solving for } y$$

Since there is no y-term in (P1), we need only substitute $-\tfrac{1}{2}$ for z in (P1) and solve for x:

$$x - 4z = 5 \quad\quad \textbf{(P1)}$$
$$x - 4\left(-\tfrac{1}{2}\right) = 5 \quad\quad \text{Substituting } -\tfrac{1}{2} \text{ for } z \text{ in equation (P1)}$$
$$x + 2 = 5$$
$$x = 3. \quad\quad \text{Solving for } x$$

The solution is $\left(3, 7, -\tfrac{1}{2}\right)$.

Do Exercise 2.

Answer on page A-58

All the operations used in the preceding example correspond to operations with the equations and produce equivalent systems of equations. We call the matrices **row-equivalent** and the operations that produce them **row-equivalent operations.**

ROW-EQUIVALENT OPERATIONS

Each of the following row-equivalent operations produces an equivalent matrix:

a) Interchanging any two rows.
b) Multiplying each element of a row by the same nonzero number.
c) Multiplying each element of a row by a nonzero number and adding the result to another row.

The best overall method of solving systems of equations is by row-equivalent matrices; graphing calculators and computers are programmed to use them. Matrices are part of a branch of mathematics known as linear algebra. They are also studied in more detail in many courses in finite mathematics.

C EXERCISE SET

a Solve using matrices.

1. $4x + 2y = 11,$
$3x - y = 2$

2. $3x - 3y = 11,$
$9x - 2y = 5$

3. $x + 4y = 8,$
$3x + 5y = 3$

4. $x + 4y = 5,$
$-3x + 2y = 13$

5. $5x - 3y = -2,$
$4x + 2y = 5$

6. $3x + 4y = 7,$
$-5x + 2y = 10$

7. $2x - 3y = 50,$
$5x + y = 40$

8. $4x + 5y = -8,$
$7x + 9y = 11$

9. $4x - y - 3z = 1,$
$8x + y - z = 5,$
$2x + y + 2z = 5$

10. $3x + 2y + 2z = 3,$
$x + 2y - z = 5,$
$2x - 4y + z = 0$

11. $p + q + r = 1,$
$p - 2q - 3r = 3,$
$4p + 5q + 6r = 4$

12. $x + 2y - 3z = 9,$
$2x - y + 2z = -8,$
$3x - y - 4z = 3$

13. $x - y + 2z = 0,$
$x - 2y + 3z = -1,$
$2x - 2y + z = -3$

14. $4a + 9b = 8,$
$8a + 6c = -1,$
$6b + 6c = -1$

15. $3p + 2r = 11,$
$q - 7r = 4,$
$p - 6q = 1$

16. $m + n + t = 6,$
$m - n - t = 0,$
$m + 2n + t = 5$

17. $2x + 2y - 2z - 2w = -10,$
$x + y + z + w = -5,$
$x - y + 4z + 3w = -2,$
$3x - 2y + 2z + w = -6$

18. $2x - 3y + z - w = -8,$
$x + y - z - w = -4,$
$x + y + z + w = 22,$
$x - y - z - w = -14$

APPENDIX C: Elimination Using Matrices

THE ALGEBRA OF FUNCTIONS

a The Sum, Difference, Product, and Quotient of Functions

Suppose that a is in the domain of two functions, f and g. The input a is paired with $f(a)$ by f and with $g(a)$ by g. The outputs can then be added to get $f(a) + g(a)$.

EXAMPLE 1 Let $f(x) = x + 4$ and $g(x) = x^2 + 1$. Find $f(2) + g(2)$.

We visualize two function machines. Because 2 is in the domain of each function, we can compute $f(2)$ and $g(2)$.

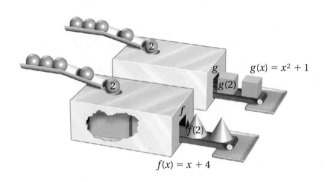

Since

$$f(2) = 2 + 4 = 6 \quad \text{and} \quad g(2) = 2^2 + 1 = 5,$$

we have

$$f(2) + g(2) = 6 + 5 = 11.$$

In Example 1, suppose that we were to write $f(x) + g(x)$ as $(x + 4) + (x^2 + 1)$, or $f(x) + g(x) = x^2 + x + 5$. This could then be regarded as a "new" function: $(f + g)(x) = x^2 + x + 5$. We can alternatively find $f(2) + g(2)$ with $(f + g)(x)$:

$$(f + g)(x) = x^2 + x + 5$$
$$(f + g)(2) = 2^2 + 2 + 5 \qquad \text{Substituting 2 for } x$$
$$= 4 + 2 + 5$$
$$= 11.$$

Similar notations exist for subtraction, multiplication, and division of functions.

THE SUM, DIFFERENCE, PRODUCT, AND QUOTIENT OF FUNCTIONS

For any functions f and g, we can form new functions defined as:

1. The **sum $f + g$:** $\quad (f + g)(x) = f(x) + g(x);$
2. The **difference $f - g$:** $\quad (f - g)(x) = f(x) - g(x);$
3. The **product fg:** $\quad (f \cdot g)(x) = f(x) \cdot g(x);$
4. The **quotient f/g:** $\quad (f/g)(x) = f(x)/g(x)$, where $g(x) \neq 0$.

---

1. Given $f(x) = x^2 + 3$ and $g(x) = x^2 - 3$, find each of the following.

a) $(f + g)(x)$

b) $(f - g)(x)$

c) $(f \cdot g)(x)$

d) $(f/g)(x)$

e) $(f \cdot f)(x)$

2. Given $f(x) = x^2 + x$ and $g(x) = 2x - 3$, find each of the following.

a) $(f + g)(-2)$

b) $(f - g)(4)$

c) $(f \cdot g)(-3)$

d) $(f/g)(2)$

EXAMPLE 2   Given f and g described by $f(x) = x^2 - 5$ and $g(x) = x + 7$, find $(f + g)(x)$, $(f - g)(x)$, $(f \cdot g)(x)$, $(f/g)(x)$, and $(g \cdot g)(x)$.

$$(f + g)(x) = f(x) + g(x) = (x^2 - 5) + (x + 7) = x^2 + x + 2;$$
$$(f - g)(x) = f(x) - g(x) = (x^2 - 5) - (x + 7) = x^2 - x - 12;$$
$$(f \cdot g)(x) = f(x) \cdot g(x) = (x^2 - 5)(x + 7) = x^3 + 7x^2 - 5x - 35;$$
$$(f/g)(x) = f(x)/g(x) = \frac{x^2 - 5}{x + 7};$$
$$(g \cdot g)(x) = g(x) \cdot g(x) = (x + 7)(x + 7) = x^2 + 14x + 49$$

Note that the sum, difference, and product of polynomials are also polynomial functions, but the quotient may not be.

Do Exercise 1.

EXAMPLE 3   For $f(x) = x^2 - x$ and $g(x) = x + 2$, find $(f + g)(3)$, $(f - g)(-1)$, $(f \cdot g)(5)$, and $(f/g)(-4)$.

We first find $(f + g)(x)$, $(f - g)(x)$, $(f \cdot g)(x)$, and $(f/g)(x)$.

$$(f + g)(x) = f(x) + g(x) = x^2 - x + x + 2$$
$$= x^2 + 2;$$
$$(f - g)(x) = f(x) - g(x) = x^2 - x - (x + 2)$$
$$= x^2 - x - x - 2$$
$$= x^2 - 2x - 2;$$
$$(f \cdot g)(x) = f(x) \cdot g(x) = (x^2 - x)(x + 2)$$
$$= x^3 + 2x^2 - x^2 - 2x$$
$$= x^3 + x^2 - 2x;$$
$$(f/g)(x) = \frac{f(x)}{g(x)} = \frac{x^2 - x}{x + 2}.$$

Then we substitute.

$$(f + g)(3) = 3^2 + 2 \quad \text{Using } (f + g)(x) = x^2 + 2$$
$$= 9 + 2 = 11;$$
$$(f - g)(-1) = (-1)^2 - 2(-1) - 2 \quad \text{Using } (f - g)(x) = x^2 - 2x - 2$$
$$= 1 + 2 - 2 = 1;$$
$$(f \cdot g)(5) = 5^3 + 5^2 - 2 \cdot 5 \quad \text{Using } (f \cdot g)(x) = x^3 + x^2 - 2x$$
$$= 125 + 25 - 10 = 140;$$
$$(f/g)(-4) = \frac{(-4)^2 - (-4)}{-4 + 2} \quad \text{Using } (f/g)(x) = (x^2 - x)/(x + 2)$$
$$= \frac{16 + 4}{-2} = \frac{20}{-2} = -10$$

Do Exercise 2.

Answers on page A-58

APPENDIX D: The Algebra of Functions

a

Let $f(x) = -3x + 1$ and $g(x) = x^2 + 2$. Find the following.

1. $f(2) + g(2)$

2. $f(-1) + g(-1)$

3. $f(5) - g(5)$

4. $f(4) - g(4)$

5. $f(-1) \cdot g(-1)$

6. $f(-2) \cdot g(-2)$

7. $f(-4)/g(-4)$

8. $f(3)/g(3)$

9. $g(1) - f(1)$

10. $g(2)/f(2)$

11. $g(0)/f(0)$

12. $g(6) - f(6)$

Let $f(x) = x^2 - 3$ and $g(x) = 4 - x$. Find the following.

13. $(f + g)(x)$

14. $(f - g)(x)$

15. $(f + g)(-4)$

16. $(f + g)(-5)$

17. $(f - g)(3)$

18. $(f - g)(2)$

19. $(f \cdot g)(x)$

20. $(f/g)(x)$

21. $(f \cdot g)(-3)$

22. $(f \cdot g)(-4)$

23. $(f/g)(0)$

24. $(f/g)(1)$

25. $(f/g)(-2)$

26. $(f/g)(-1)$

For each pair of functions f and g, find $(f + g)(x)$, $(f - g)(x)$, $(f \cdot g)(x)$, and $(f/g)(x)$.

27. $f(x) = x^2$,
$g(x) = 3x - 4$

28. $f(x) = 5x - 1$,
$g(x) = 2x^2$

29. $f(x) = \dfrac{1}{x - 2}$,
$g(x) = 4x^3$

30. $f(x) = 3x^2$,
$g(x) = \dfrac{1}{x - 4}$

31. $f(x) = \dfrac{3}{x - 2}$,
$g(x) = \dfrac{5}{4 - x}$

32. $f(x) = \dfrac{5}{x - 3}$,
$g(x) = \dfrac{1}{x - 2}$

Photo Credits

xxxi, Michael St.-Andre   **62 (margin),** Corbis   **62 (left),** © Underwood &
Underwood/CORBIS   **62 (right),** © Jim Craigmyle/CORBIS   **69,** © Arvind Garg/
CORBIS   **72,** Franco Vogt/Corbis   **90,** © Rufus F. Folkks/CORBIS; © Lucy Nicholson/
Reuters/Corbis; © Mitchell Gerber/CORBIS; Michael St.-Andre   **93,** Duncan Smith,
PhotoDisc   **97,** Corbis   **98,** © Randy Faris/CORBIS   **100,** PhotoDisc   **107,**
© Bettmann/CORBIS   **111,** Delta Queen   **120,** © David Butow/CORBIS SABA   **121,**
PhotoDisc   **129,** Keith Brofsky/PhotoDisc   **139,** © Charles O'Rear/CORBIS   **154,**
© Duomo/Corbis   **181,** © Royalty-Free/Corbis   **184,** Phil Schermeister, Corbis
201, Karen Bittinger   **217, 224, 225,** © Royalty-Free/Corbis   **230,** BrandXPictures/
Getty   **240,** © LE SEGRETAIN PASCAL/CORBIS SYGMA   **244,** EyeWire Collection
255, Angelo Hornak/Corbis   **269,** Susan Dawson   **280,** © Tom Stewart/CORBIS
293, © Viviane Moos/Corbis   **294,** Stephanie Maze/Corbis   **295,** © Grace/zefa/
Corbis   **296,** Karen Bittinger   **313,** Warren Morgan/Corbis   **322,** Dave G. Houser/
Corbis   **326,** Karen Bittinger   **329,** Zigy Kaluzny/Tony Stone Images   **353,** Karen
Bittinger   **400,** © Danny Lehman/CORBIS   **410,** Jose Luis Pelaez, Inc./Corbis
460, © Royalty-Free/Corbis   **464,** Brian Spurlock   **468,** Delta Queen   **471 (left),**
Cory Sorensen/Corbis   **471 (right),** Mark Douet/Tony Stone/Getty   **472 (top),** AP
Wide World Photos   **472 (bottom),** PhotoDisc   **473 (top),** Keith Wood/Corbis
473 (bottom), Marv Bittinger   **474,** © Franco Vogt/Corbis   **481,** © Royalty-Free/
Corbis   **482,** PhotoDisc   **485,** EyeWire Collection   **486,** Digital Vision/Getty   **487,**
Stockbyte/Getty   **504,** © Royalty-Free/Corbis   **541,** © 1999 Bill Ellzey   **545,** © Galen
Rowell/CORBIS   **546,** EyeWire Collection   **570,** © Royalty-Free/Corbis   **572,**
EyeWire Collection   **583,** AP/Photo   **586 (left),** Corbis   **586 (right),** PhotoDisc
596, PhotoDisc   **678 (top),** © Royalty-Free/Corbis   **678 (bottom),** Digital Vision
689, © Michael S. Yamashita/CORBIS   **731,** Sarah Lawless/Stone/Getty Images
733, © Royalty-Free/Corbis   **734,** Alan Schein Photography/Corbis   **736,** John Lacko
737, 741, © Royalty-Free/Corbis   **742,** © B. Kohlhas/zefa/Corbis   **744 (left),** From
Classic Baseball Cards, by Bert Randolph Sugar, copyright © 1977 by Dover
Publishing, Inc.   **744 (right),** © Francis G. Mayer/Corbis   **745,** © Royalty-Free/
Corbis   **748,** Stockbyte/Getty Images   **749,** © Franco Vogt/CORBIS   **753,** Chronis
Jons/Tony Stone/Getty Images   **758,** Michael S. Yamashita/Corbis   **777,** Mark
Gibson, Digital Vision/Getty   **804,** PhotoDisc

Answers

CHAPTER R

Margin Exercises, Section R.1, pp. 2–8

1. -9 **2.** 6   **3.** 0   **4.** $\{1, 3, 5, 7, 9, 11, 13\}$; $\{x \mid x$ is one of the first seven odd whole numbers$\}$
5. 0.6875; terminating   **6.** $1.\overline{076923}$; repeating
7. (a) 20; (b) 20, 0; (c) 20, -10, 0; (d) $\sqrt{7}$, $-\sqrt{2}$, $9.34334333433334\ldots$; (e) 20, -10, -5.34, 18.999, $\frac{11}{45}$, 0, $-\frac{2}{3}$; (f) 20, -10, -5.34, 18.999, $\frac{11}{45}$, $\sqrt{7}$, $-\sqrt{2}$, 0, $-\frac{2}{3}$, $9.34334333433334\ldots$ **8.** $<$ **9.** $>$ **10.** $>$
11. $<$ **12.** $>$ **13.** $>$ **14.** $<$ **15.** $>$ **16.** $>$
17. $6 < x$ **18.** $7 > -4$ **19.** True   **20.** True
21. False   **22.** True   **23.**
24.   **25.** (a)   **26.** (c)   **27.** (g)
28. (d)   **29.** $\frac{1}{4}$ **30.** 2   **31.** $\frac{3}{2}$ **32.** 2.3

Calculator Corner, p. 9

1. 3   **2.** 14   **3.** 0   **4.** 5.67   **5.** 18.4   **6.** 0.8
7. $\frac{2}{3}$ **8.** $\frac{7}{9}$

Exercise Set R.1, p. 10

1. 1, 12, $\sqrt{25}$ **3.** -6, 0, 1, $-\frac{1}{2}$, -4, $\frac{7}{9}$, 12, $-\frac{6}{5}$, 3.45, $5\frac{1}{2}$, $\sqrt{25}$, $-\frac{12}{3}$ **5.** -6, 0, 1, $-\frac{1}{2}$, -4, $\frac{7}{9}$, 12, $-\frac{6}{5}$, 3.45, $5\frac{1}{2}$, $\sqrt{3}$, $\sqrt{25}$, $-\frac{12}{3}$, $0.131331333133331\ldots$ **7.** 12, 0
9. -11, 12, 0   **11.** $-\sqrt{5}$, π, $-3.565665666566665\ldots$
13. $\{m, a, t, h\}$ **15.** $\{1, 2, 3, 4, 5, 6, 7, 8, 9, 10, 11, 12\}$
17. $\{2, 4, 6, 8, \ldots\}$ **19.** $\{x \mid x$ is a whole number less than or equal to 5$\}$ **21.** $\left\{\dfrac{a}{b} \,\middle|\, a \text{ and } b \text{ are integers and } b \neq 0\right\}$
23. $\{x \mid x > -3\}$ **25.** $>$ **27.** $<$ **29.** $<$ **31.** $<$
33. $>$ **35.** $<$ **37.** $>$ **39.** $<$ **41.** $x < -8$
43. $y \geq -12.7$ **45.** False   **47.** True
49.
51.
53.
55.   **57.** 6   **59.** 28   **61.** 35

63. $\frac{2}{3}$ **65.** 0   **67.** D_W **69.** $\leq$ **71.** $\leq$ **73.** $\frac{1}{8}\%$, 0.3%, 0.009, 1%, 1.1%, $\frac{9}{100}$, $\frac{1}{11}$, $\frac{99}{1000}$, 0.11, $\frac{1}{8}$, $\frac{2}{7}$, 0.286

Margin Exercises, Section R.2, pp. 13–19

1. 4   **2.** 2   **3.** -5 **4.** 0   **5.** -18 **6.** -18.6
7. $-\frac{29}{5}$ **8.** $-\frac{7}{10}$ **9.** 0   **10.** -7.4 **11.** -3 **12.** -3.3
13. $-\frac{1}{4}$ **14.** $\frac{1}{10}$ **15.** 14   **16.** $-\frac{2}{3}$ **17.** 0   **18.** -9
19. $\frac{3}{5}$ **20.** -5.9 **21.** $\frac{2}{3}$ **22.** (a) -11; (b) 17; (c) 0; (d) $-x$; (e) x **23.** 17   **24.** -16 **25.** -3 **26.** -29.6
27. 3   **28.** 1   **29.** $\frac{3}{2}$ **30.** (a) -6; (b) -40; (c) 6
31. 10, 5, 0, -5; -10, -15, -20, -25, -30 **32.** -24
33. -28.35 **34.** -8 **35.** -10, -5, 0; 5, 10, 15, 20, 25
36. 72   **37.** $\frac{8}{15}$ **38.** 42.77   **39.** 2   **40.** -25 **41.** -3
42. 0.2   **43.** Not defined   **44.** 0   **45.** Not defined
46. Not defined   **47.** $\frac{8}{3}$ **48.** $-\frac{5}{4}$ **49.** $\frac{1}{18}$
50. $-\frac{1}{4.3}$, or $-\frac{10}{43}$ **51.** $\frac{1}{0.5}$, or 2   **52.** $-\frac{4}{5}, \frac{5}{4}; \frac{3}{4}, -\frac{4}{3}$; $-0.25, 4$; $-8, \frac{1}{8}$; 5, $-\frac{1}{5}$; 0, does not exist   **53.** $-\frac{6}{7}$
54. $\frac{36}{7}$ **55.** $\frac{3}{40}$ **56.** -8

Calculator Corner, p. 20

1. -6 **2.** -0.8 **3.** -12 **4.** -10.4 **5.** -12 **6.** 2.4
7. -4 **8.** -1.2 **9.** -54 **10.** -1.28 **11.** 32
12. 26.68   **13.** -3 **14.** -0.5 **15.** 2   **16.** About 1.261

Exercise Set R.2, p. 21

1. -28 **3.** 5   **5.** -16 **7.** -4 **9.** -10 **11.** -26
13. 1.2   **15.** -8.86 **17.** $-\frac{1}{3}$ **19.** $-\frac{4}{3}$ **21.** $\frac{1}{10}$
23. $\frac{7}{20}$ **25.** 4   **27.** -3.7 **29.** -10 **31.** 0   **33.** -4
35. -14 **37.** 0   **39.** -46 **41.** 5   **43.** 15
45. -11.6 **47.** -29.25 **49.** $-\frac{7}{2}$ **51.** $-\frac{1}{4}$ **53.** $-\frac{19}{12}$
55. $-\frac{7}{15}$ **57.** -21 **59.** -8 **61.** 24   **63.** -112
65. 34.2   **67.** $-\frac{12}{35}$ **69.** 2   **71.** 60   **73.** 26.46   **75.** 1
77. $-\frac{8}{27}$ **79.** -2 **81.** -7 **83.** 7   **85.** 0.3
87. Not defined   **89.** 0   **91.** Not defined   **93.** $\frac{4}{3}$
95. $-\frac{8}{7}$ **97.** $\frac{1}{25}$ **99.** 5   **101.** $-\dfrac{b}{a}$ **103.** $-\frac{6}{77}$
105. 25   **107.** -6 **109.** 5   **111.** -120 **113.** $-\frac{9}{8}$

115. $\frac{5}{3}$ **117.** $\frac{3}{2}$ **119.** $\frac{9}{64}$ **121.** -2 **123.** $0.\overline{923076}$
125. -1.62 **127.** Not defined   **129.** $-\frac{2}{3}, \frac{3}{2}; \frac{5}{4}, -\frac{4}{5};$
0, does not exist; $-1, 1; 4.5, -\frac{1}{4.5}; -x, \frac{1}{x}$ **131.** $\mathbf{D_W}$
133. $26, 0$ **134.** 26 **135.** $-13, 26, 0$ **136.** $\sqrt{3}$, π,
$4.57557555755557\ldots$ **137.** $-12.47, -13, 26, 0, -\frac{23}{32}, \frac{7}{11}$
138. $\sqrt{3}, -12.47, -13, 26, \pi, 0, -\frac{23}{32}, \frac{7}{11}$,
$4.57557555755557\ldots$ **139.** $<$ **140.** $>$ **141.** $<$
142. $>$ **143.** $\frac{1}{4}$ **145.** $31{,}250$

Margin Exercises, Section R.3, pp. 25–30

1. 8^4 **2.** m^6 **3.** $\left(\frac{7}{8}\right)^3$ **4.** 81 **5.** $\frac{1}{16}$ **6.** $1{,}000{,}000$
7. 0.008 **8.** 1131.6496 **9.** -256 **10.** 81 **11.** 8
12. -31 **13.** 1 **14.** 1 **15.** 1 **16.** $\frac{1}{m^4}$ **17.** $-\frac{1}{64}$
18. x^3 **19.** 125 **20.** $\frac{16}{9}$ **21.** a^{-3} **22.** $(-5)^{-4}$
23. -4117 **24.** 27 **25.** 551 **26.** $9; 33$ **27.** -25
28. 6 **29.** $\frac{54}{241}$ **30.** 34

Calculator Corner, p. 26

1. 625 **2.** $4{,}782{,}969$ **3.** 576 **4.** 41.126569
5. -10.48576 **6.** 75.686967 **7.** $\frac{729}{4096}$ **8.** $-\frac{3125}{243}$

Calculator Corner, p. 30

1. 56 **2.** 61 **3.** 96 **4.** 665 **5.** -15.876
6. -395.357 **7.** 262.5 **8.** $-14{,}284{,}863.06$ **9.** 4
10. $-2.\overline{4}$, or $-\frac{22}{9}$ **11.** 782.5, or $\frac{1565}{2}$ **12.** 5176.097
13. 76; the calculator first finds $144 \div 3$ and then adds 36 and subtracts 8; -36

Exercise Set R.3, p. 31

1. 4^5 **3.** 5^6 **5.** m^3 **7.** $\left(\frac{7}{12}\right)^4$ **9.** $(123.7)^2$ **11.** 128
13. -32 **15.** $\frac{1}{81}$ **17.** -64 **19.** 31.36 **21.** 5 **23.** 1
25. 1 **27.** $\frac{7}{8}$ **29.** 16 **31.** $\frac{27}{8}$ **33.** $\frac{1}{y^5}$ **35.** a^2
37. $-\frac{1}{11}$ **39.** 3^{-4} **41.** b^{-3} **43.** $(-16)^{-2}$ **45.** -4
47. -117 **49.** 2 **51.** 8 **53.** -358 **55.** $144; 74$
57. -576 **59.** 2599 **61.** 36 **63.** 5619.712
65. $-200{,}167{,}769$ **67.** 3 **69.** 3 **71.** 16 **73.** -310
75. 2 **77.** 1875 **79.** 7804.48 **81.** 12 **83.** 8
85. 16 **87.** -86 **89.** 37 **91.** -1 **93.** 22 **95.** -39
97. 12 **99.** -549 **101.** -144 **103.** 2 **105.** $-\frac{31}{76}$
107. $\frac{61}{13}$ **109.** $\mathbf{D_W}$ **111.** $\frac{9}{7}$ **112.** 2.3 **113.** 0
114. 900 **115.** -33 **116.** -79 **117.** 33 **118.** -79
119. -23 **120.** 23 **121.** -23 **122.** $\frac{5}{8}$ **123.** $25\frac{1}{4}$
125. $9 \cdot 5 + 2 - (8 \cdot 3 + 1) = 22$ **127.** 3125
129. $(2 + 3)^{-1} = (5)^{-1} = \frac{1}{5}$; $2^{-1} + 3^{-1} = \frac{1}{2} + \frac{1}{3} = \frac{3}{6} + \frac{2}{6} = \frac{5}{6}$,
so $(2 + 3)^{-1} \neq 2^{-1} + 3^{-1}$.

Margin Exercises, Section R.4, pp. 35–39

1. (b) and (d)   **2.** $9.6 + x = 16.1$; 6.5 million   **3.** $x - 16$
4. $47 + y$, or $y + 47$ **5.** $16 - x$ **6.** $\frac{1}{4}t$
7. $8x + 6$, or $6 + 8x$ **8.** $99\% \cdot \frac{a}{b} - 8$, or $(0.99) \cdot \frac{a}{b} - 8$
9. 84 **10.** -126 **11.** -50 **12.** 120 **13.** 42
14. -146 **15.** $96\ \text{ft}^2$ **16.** 64 **17.** 64 **18.** 16

Calculator Corner, p. 39

1. 12.2 **2.** -7.6 **3.** -20 **4.** $-16{,}576.04$
5. 21.3 **6.** 0.1

Exercise Set R.4, p. 40

1. $b + 8$, or $8 + b$ **3.** $c - 13.4$ **5.** $5 + q$, or $q + 5$
7. $a + b$, or $b + a$ **9.** $x \div y$, or $\frac{x}{y}$ **11.** $x + w$, or $w + x$
13. $n - m$ **15.** $p + q$, or $q + p$ **17.** $3q$ **19.** $-18m$
21. $17\%s$, or $0.17s$ **23.** $75t$ **25.** $\$40 - x$ **27.** -92
29. 3 **31.** 4 **33.** $\frac{45}{2}$, or 22.5 **35.** 16 **37.** 19
39. 57 **41.** $\$440.70$ **43.** $A = 113.04\ \text{cm}^2$; $C = 37.68\ \text{cm}$
45. $\mathbf{D_W}$ **47.** 243 **48.** -243 **49.** 100 **50.** $10{,}000$
51. 28.09 **52.** $\frac{9}{25}$ **53.** 1 **54.** 4.5 **55.** $3x$ **56.** 1
57. $d = r \cdot t$ **59.** 9

Margin Exercises, Section R.5, pp. 42–47

1. $-10, -10, -18; 40, 40, 72; 0, 0, 0; 6x - x$ and $5x$ are
equivalent.   **2.** $1, 13; 64, 34; 60.84, 32.04$; the expressions
are not equivalent.   **3.** $\frac{2y}{7y}$ **4.** $\frac{8x}{44x}$ **5.** $\frac{2}{3}$ **6.** $-\frac{5}{3}$
7. $2; 2$ **8.** $-14; -14$ **9.** $21; 21$ **10.** $440; 440$
11. $5 + y$; ba; $mn + 8$, or $nm + 8$, or $8 + nm$
12. $2 \cdot (x \cdot y)$; $(2 \cdot y) \cdot x$; $(y \cdot 2) \cdot x$; answers may vary
13. $180; 180$ **14.** $27; 27; 27$ **15.** $10; 10$ **16.** $24; 24$
17. $8y - 80$ **18.** $ax + ay - az$ **19.** $40x - 60y + 5z$
20. $-5x, -7y, 67t, -\frac{4}{5}$ **21.** $9(x + y)$ **22.** $a(c - y)$
23. $6(x - 2)$ **24.** $5(7x - 5y + 3w + 1)$
25. $b(s + t - w)$

Exercise Set R.5, p. 48

1. $-10, -10, 2; 25, 25, -5; 0, 0, 0; 2x + 3x$ and $5x$ are
equivalent.   **3.** $-12, -16, -12; 38.4, 51.2, 38.4; 0, 0, 0;$
$4x + 8x$ and $4(x + 2x)$ are equivalent.   **5.** $\frac{7x}{8x}$ **7.** $\frac{6a}{8a}$
9. $\frac{5}{3}$ **11.** -4 **13.** $3 + w$ **15.** tr **17.** $cd + 4$, $dc + 4$,
or $4 + dc$ **19.** $x + yz$, $x + zy$, or $zy + x$
21. $(m + n) + 2$ **23.** $7 \cdot (x \cdot y)$ **25.** $a + (8 + b)$,
$(a + 8) + b$, $b + (a + 8)$; others are possible
27. $(7 \cdot b) \cdot a$, $b \cdot (a \cdot 7)$, $(b \cdot a) \cdot 7$; others are possible
29. $4a + 4$ **31.** $8x - 8y$ **33.** $-10a - 15b$
35. $2ab - 2ac + 2ad$ **37.** $2\pi rh + 2\pi r$ **39.** $\frac{1}{2}ha + \frac{1}{2}hb$
41. $4a, -5b, 6$ **43.** $2x, -3y, -2z$ **45.** $24(x + y)$
47. $7(p - 1)$ **49.** $7(x - 3)$ **51.** $x(y + 1)$
53. $2(x - y + z)$ **55.** $3(x + 2y - 1)$ **57.** $a(b + c - d)$

59. $\frac{1}{4}\pi r(r + s)$ **61.** **D$_W$** **63.** $(x + y)^2$ **64.** $x^2 + y^2$
65. -102 **66.** $-\frac{1}{2}$ **67.** $\frac{1}{8}$ **68.** -46 **69.** No
71. Yes

Margin Exercises, Section R.6, pp. 50–53

1. $20x$ **2.** $-7x$ **3.** $6x$ **4.** $-6x$ **5.** $23.4x + 3.9y$
6. $\frac{22}{15}x - \frac{19}{12}y + 23$ **7.** -24 **8.** 0 **9.** 10 **10.** $-9x$
11. $24t$ **12.** $y - 7$ **13.** $y - x$ **14.** $-9x - 6y - 11$
15. $-23x + 7y + 2$ **16.** $3x + 2y + 1$ **17.** $2x + 5z - 24$
18. $-3x + 2y$ **19.** $-\frac{1}{4}t - 41w + 5d + 23$ **20.** $3x - 8$
21. $4y + 1$ **22.** $-2x - 9y - 6$ **23.** $15x + 9y - 6$
24. $-x - 2y$ **25.** $23x - 10y$ **26.** $3a - 3b + \frac{23}{4}$
27. $-x - 3$ **28.** $2x - 51$ **29.** $23x + 52$ **30.** $12a + 12$

Exercise Set R.6, p. 54

1. $12x$ **3.** $-3b$ **5.** $15y$ **7.** $11a$ **9.** $-8t$ **11.** $10x$
13. $11x - 5y$ **15.** $-4c + 12d$ **17.** $22x + 18$
19. $1.19x + 0.93y$ **21.** $-\frac{2}{15}a - \frac{1}{3}b - 27$
23. $P = 2(l + w)$ **25.** $2c$ **27.** $-b - 4$
29. $-b + 3$, or $3 - b$ **31.** $-t + y$, or $y - t$
33. $-x - y - z$ **35.** $-8x + 6y - 13$
37. $2c - 5d + 3e - 4f$ **39.** $1.2x - 56.7y + 34z + \frac{1}{4}$
41. $3a + 5$ **43.** $m + 1$ **45.** $9d - 16$ **47.** $-7x + 14$
49. $-9x + 17$ **51.** $17x + 3y - 18$ **53.** $10x - 19$
55. $22a - 15$ **57.** -190 **59.** $12x + 30$ **61.** $3x + 30$
63. $9x - 18$ **65.** $-4x + 808$ **67.** $-14y - 186$
69. **D$_W$** **71.** -37 **72.** -71 **73.** -23.1 **74.** $\frac{5}{24}$
75. -16 **76.** 16 **77.** -16 **78.** $-\frac{1}{6}$ **79.** $8a - 8b$
80. $-16a + 24b - 32$ **81.** $6ax - 6bx + 12cx$
82. $16x - 8y + 10$ **83.** $24(a - 1)$ **84.** $8(3a - 2b)$
85. $a(b - c + 1)$ **86.** $5(3p + 9q - 2)$
87. $(3 - 8)^2 + 9 = 34$ **89.** $5 \cdot 2^3 \div (3 - 4)^4 = 40$
91. $23a - 18b + 184$ **93.** $-9z + 5x$ **95.** $-x + 19$

Margin Exercises, Section R.7, pp. 58–64

1. 8^4 **2.** y^5 **3.** $-18x^3$ **4.** $-\dfrac{75}{x^{14}}$ **5.** $-42x^{8n}$
6. $-\dfrac{10y^2}{x^{12}}$ **7.** $60y$ **8.** 4^3 **9.** 5^6 **10.** $\dfrac{1}{10^6}$ **11.** $-5x^{2n}$
12. $-\dfrac{2y^{10}}{x^4}$ **13.** $\frac{3}{2}a^3b^2$ **14.** 3^{42} **15.** z^{20} **16.** $\dfrac{1}{t^{14m}}$
17. $8x^3y^3$ **18.** $\dfrac{16y^{14}}{x^4}$ **19.** $-32x^{20}y^{10}$ **20.** $\dfrac{1000y^{21}}{x^{12}z^6}$
21. x^9y^{12} **22.** $\dfrac{9x^4}{y^{16}}$ **23.** $-\dfrac{8a^9}{27b^{21}}$ **24.** 5.88×10^{12} mi
25. 2.57×10^{-10} **26.** 0.0000000000004567
27. $93,000,000$ mi **28.** 7.462×10^{-13} **29.** 5.6×10^{-15}
30. 2.0×10^3 **31.** 5.5×10^2 **32.** 5×10^2 sec
33. 1.90164×10^{27} kg

Calculator Corner, p. 65

1. 1.2312×10^{-4} **2.** 2.4495×10^3 **3.** 2.8×10^5
4. 5.4×10^{14} **5.** 3×10^{-6} **6.** 6×10^5 **7.** 1.2×10^{-14}
8. 3×10^{12}

1. 3^9 **3.** $\dfrac{1}{6^4}$ **5.** $\dfrac{1}{8^6}$ **7.** $\dfrac{1}{b^3}$ **9.** a^3 **11.** $72x^5$
13. $-28m^5n^5$ **15.** $-\dfrac{14}{x^{11}}$ **17.** $\dfrac{105}{x^{2t}}$ **19.** $-\dfrac{8}{y^{6m}}$
21. 8^7 **23.** 6^5 **25.** $\dfrac{1}{10^9}$ **27.** 9^2 **29.** $\dfrac{1}{x^{10n}}$ **31.** $\dfrac{1}{w^{5q}}$
33. a^5 **35.** $-3x^5z^4$ **37.** $-\dfrac{4x^9}{3y^2}$ **39.** $\dfrac{3x^3}{2y^2}$ **41.** 4^6
43. $\dfrac{1}{8^{12}}$ **45.** 6^{12} **47.** $125a^6b^6$ **49.** $\dfrac{y^{12}}{9x^6}$ **51.** $\dfrac{a^4}{36b^6c^2}$
53. $\dfrac{1}{4^9 \cdot 3^{12}}$ **55.** $\dfrac{8x^9y^3}{27}$ **57.** $\dfrac{a^{10}b^5}{5^{10}}$ **59.** $\dfrac{6^{30}2^{12}y^{36}}{z^{48}}$
61. $\dfrac{64}{x^{24}y^{12}}$ **63.** $\dfrac{5^7b^{28}}{3^7a^{35}}$ **65.** 10^a **67.** $3a^{-x-4}$
69. $\dfrac{-5x^{a+1}}{y}$ **71.** 8^{4xy} **73.** 12^{6b-2ab}
75. $5^{2c}x^{2ac-2c}y^{2bc+2c}$, or $25^c x^{2ac-2c}y^{2bc+2c}$ **77.** $2x^{a+2}y^{b-2}$
79. 4.7×10^{10} **81.** 1.6×10^{-8} **83.** 7.585×10^9
85. 2×10^{-7} **87.** $673,000,000$ **89.** 0.000066 cm
91. $11,000,000$ **93.** 9.66×10^{-5} **95.** 1.3338×10^{-11}
97. 2.5×10^3 **99.** 5.0×10^{-4} **101.** About $\$4.09 \times 10^7$
103. 6.3072×10^{10} sec **105.** About 4.08 light-years
107. 3.33×10^{-2} **109.** About 2.2×10^{-3} lb **111.** **D$_W$**
113. $19x + 4y - 20$ **114.** $-23t + 21$ **115.** -11
116. -231 **117.** -8 **118.** 8 **119.** 2^{21} **121.** $\dfrac{1}{a^{14}b^{27}}$
123. $4x^{2a}y^{2b}$

Concept Reinforcement, p. 70

1. True **2.** False **3.** True **4.** True **5.** False
6. False **7.** True **8.** True

Summary and Review: Chapter R, p. 71

1. $2, -\frac{2}{3}, 0.45\overline{45}, -23.788$
2. $\{x \mid x$ is a real number less than or equal to 46$\}$ **3.** $<$
4. $x < 19$ **5.** False **6.** True
7.
8. **9.** 7.23 **10.** 0 **11.** -2
12. -7.9 **13.** $-\frac{31}{28}$ **14.** -5 **15.** -26.7 **16.** $\frac{19}{4}$
17. 10.26 **18.** $-\frac{3}{7}$ **19.** 168 **20.** -4 **21.** 21
22. -7 **23.** $-\frac{7}{12}$ **24.** $\frac{8}{3}$ **25.** Not defined **26.** -24
27. 7 **28.** -2.3 **29.** 0 **30.** a^5 **31.** $\left(-\frac{7}{8}\right)^3$
32. $\dfrac{1}{a^4}$ **33.** x^{-8} **34.** 59 **35.** -116 **36.** $5x$
37. $28\%y$, or $0.28y$ **38.** $t - 9$ **39.** $\dfrac{a}{b} - 8$ **40.** -17
41. -8 **42.** 84 ft^2 **43.** $-4, 16, 36, 6; 95, 225, 25, 105;$
$-5, 25, 25, 5;$ none is equivalent **44.** $-16, -9, -16, 12;$
$6, 13, 6, 34; -14, -7, -14, 14; 2x - 14$ and $2(x - 7)$ are
equivalent **45.** $\dfrac{21x}{9x}$ **46.** -12 **47.** $a + 11$ **48.** $y \cdot 8$
49. $9 + (a + b)$ **50.** $(8x)y$ **51.** $-6x + 3y$
52. $8abc + 4ab$ **53.** $5(x + 2y - z)$ **54.** $pt(r + s)$

55. $-3x + 5y$ **56.** $12c - 4$ **57.** $9c - 4d + 3$
58. $x + 3$ **59.** $6x + 15$ **60.** $22x - 14$
61. $-17m - 12$ **62.** $-\dfrac{10x^7}{y^5}$ **63.** $-\dfrac{3y^3}{2x^4}$ **64.** $\dfrac{a^8}{9b^2c^6}$
65. $\dfrac{81y^{40}}{16x^{24}}$ **66.** 6.875×10^9 **67.** 1.312×10^{-1}
68. 1.422×10^{-2} m³   **69.** $\$6.7 \times 10^4$ **70. D_W** The commutative laws deal with order and tell us that we can change the order when adding or when multiplying without changing the result. The associative laws deal with grouping and tell us that numbers can be grouped in any manner when adding or multiplying without affecting the result. The distributive laws deal with multiplication together with addition or subtraction. They tell us that adding or subtracting two numbers b and c and then multiplying the sum or difference by a number a is equivalent to first multiplying b and c by a and then adding or subtracting those products. Each of the commutative and associative laws involves only one operation. Each of the distributive laws involves two operations.   **71. D_W** The expressions $2x + 3x$ and $5x$ are equivalent. (We can show this using the distributive law.) The expressions $(x + 1)^2$ and $x^2 + 1$ are not equivalent, because when $x = 3$, for example, we have $(x + 1)^2 = (3 + 1)^2 = 16$ but $x^2 + 1 = 3^2 + 1 = 10$.
72. x^{12y} **73.** 32   **74.** (a), (i); (d), (f); (h), (j)

Test: Chapter R, p. 73

1. [R.1a] $\sqrt{7}, \pi$
2. [R.1a] $\{x \mid x$ is a real number greater than 20$\}$
3. [R.1b] $>$ **4.** [R.1b] $5 \geq a$ **5.** [R.1b] True
6. [R.1b] True   **7.** [R.1c]
8. [R.1d] 0   **9.** [R.1d] $\frac{7}{8}$ **10.** [R.2a] -2
11. [R.2a] -13.1 **12.** [R.2a] -6 **13.** [R.2c] -1
14. [R.2c] -29.7 **15.** [R.2c] $\frac{25}{4}$ **16.** [R.2d] -33.62
17. [R.2d] $\frac{3}{4}$ **18.** [R.2d] -528 **19.** [R.2e] 15
20. [R.2e] -5 **21.** [R.2e] $\frac{8}{3}$ **22.** [R.2e] -82
23. [R.2e] Not defined   **24.** [R.2b] 13   **25.** [R.2b] 0
26. [R.3a] q^4 **27.** [R.3b] a^{-9} **28.** [R.3c] 0
29. [R.3c] $-\frac{16}{7}$ **30.** [R.4a] $t + 9$, or $9 + t$
31. [R.4a] $\dfrac{x}{y} - 12$ **32.** [R.4b] 18   **33.** [R.4b] 3.75 cm²
34. [R.5a] Yes   **35.** [R.5a] No   **36.** [R.5b] $\dfrac{27x}{36x}$
37. [R.5b] $\frac{3}{2}$ **38.** [R.5c] qp **39.** [R.5c] $4 + t$
40. [R.5c] $(3 + t) + w$ **41.** [R.5c] $4(ab)$
42. [R.5d] $-6a + 8b$ **43.** [R.5d] $3\pi rs + 3\pi r$
44. [R.5d] $a(b - c + 2d)$ **45.** [R.5d] $h(2a + 1)$
46. [R.6a] $10y - 5x$ **47.** [R.6a] $21a + 14$
48. [R.6b] $9x - 7y + 22$ **49.** [R.6b] $-7x + 14$
50. [R.6b] $10x - 21$ **51.** [R.6b] $-a + 38$
52. [R.7a] $-\dfrac{3y^2}{2x^4}$ **53.** [R.7a] $-\dfrac{6a^9}{b^5}$ **54.** [R.7a] $-50a^{9n}$
55. [R.7a] $-\dfrac{5}{x^{4t}}$ **56.** [R.7b] $\dfrac{a^{12}}{81b^8c^4}$ **57.** [R.7b] $\dfrac{16a^{48}}{b^{48}}$
58. [R.7c] 4.37×10^{-5} **59.** [R.7c] 3.741×10^7
60. [R.7c] 1.875×10^{-6} **61.** [R.7c] 1.196×10^{22} kg
62. [R.7c] 4.8×10^9 **63.** (b), (e); (d), (f), (h); (i), (j)

CHAPTER 1

Margin Exercises, Section 1.1, pp. 76–84

1. (a), (f)   **2.** (b), (c)   **3.** (d), (e)   **4.** Yes   **5.** No
6. Yes   **7.** Yes   **8.** No   **9.** -7 **10.** $-\frac{17}{20}$ **11.** 38
12. 140.3   **13.** $\frac{5}{4}$ **14.** -49 **15.** $\frac{3}{14}$ **16.** -3
17. -136 **18.** 5   **19.** $\frac{4}{3}$ **20.** $4 - 5y = 2; \frac{2}{5}$
21. $63x - 91 = 30x; \frac{91}{33}$ **22.** 3   **23.** $-\frac{29}{5}$ **24.** $\frac{17}{2}$
25. No solution   **26.** All real numbers are solutions.
27. All real numbers are solutions.   **28.** No solution
29. $-\frac{5}{4}$ **30.** -2 **31.** $-\frac{19}{8}$

Calculator Corner, p. 85

1. Left to the student   **2.** Left to the student

Exercise Set 1.1, p. 86

1. Yes   **3.** No   **5.** No   **7.** No   **9.** Yes   **11.** No
13. 7   **15.** -8 **17.** 27   **19.** -39 **21.** 86.86
23. $\frac{1}{6}$ **25.** 6   **27.** -4 **29.** -147 **31.** 32   **33.** -6
35. $-\frac{1}{6}$ **37.** 10   **39.** 11   **41.** -12 **43.** 8   **45.** 2
47. 21   **49.** -12 **51.** No solution   **53.** -1 **55.** $\frac{18}{5}$
57. 0   **59.** 1   **61.** All real numbers   **63.** No solution
65. 7   **67.** 2   **69.** 7   **71.** 5   **73.** $-\frac{3}{2}$ **75.** All real
numbers   **77.** 5   **79.** $\frac{23}{66}$ **81.** $\frac{5}{32}$ **83.** $\frac{79}{32}$ **85. D_W**
87. a^{14} **88.** $\dfrac{1}{a^{32}}$ **89.** $-\dfrac{18x^2}{y^{11}}$ **90.** $-2x^8y^3$
91. $12 - 20x$ **92.** $-5 + 6x$ **93.** $-12x + 8y - 4z$
94. $-10x + 35y - 20$ **95.** $2(x - 3y)$ **96.** $-4(x + 6y)$
97. $2(2x - 5y + 1)$ **98.** $-5(2x - 7y + 4)$
99. $\{1, 2, 3, 4, 5, 6, 7, 8, 9\}; \{x \mid x$ is a positive integer less than 10$\}$ **100.** $\{-8, -7, -6, -5, -4, -3, -2, -1\}; \{x \mid x$ is a negative integer greater than $-9\}$
101. Approximately -4.176 **103.** $\frac{3}{2}$ **105.** 8

Margin Exercises, Section 1.2, pp. 91–94

1. (a) 17.8; (b) 192 lb; (c) left to the student   **2.** $m = \dfrac{rF}{v^2}$
3. $m = \dfrac{H - 2r}{3}$ **4.** $c = \dfrac{5}{3}P - 10$, or $\dfrac{5P - 30}{3}$
5. $b = \dfrac{2A - ha}{h}$, or $\dfrac{2A}{h} - a$ **6.** $Q = \dfrac{T}{1 + iy}$
7. (a) About 1251; (b) $W = \dfrac{NR - Nr + 400L}{400}$, or
$W = L + \dfrac{NR - Nr}{400}$

Exercise Set 1.2, p. 95

1. $r = \dfrac{d}{t}$ **3.** $h = \dfrac{A}{b}$ **5.** $w = \dfrac{P - 2l}{2}$, or $\dfrac{P}{2} - l$
7. $b = \dfrac{2A}{h}$ **9.** $a = 2A - b$ **11.** $m = \dfrac{F}{a}$ **13.** $t = \dfrac{I}{Pr}$
15. $c^2 = \dfrac{E}{m}$ **17.** $p = 2Q + q$ **19.** $y = \dfrac{c - Ax}{B}$

21. $N = \dfrac{1.08T}{I}$ **23.** $m = \dfrac{4}{3}C - 5$, or $\dfrac{4C - 15}{3}$

25. $b = 3n - a + c$ **27.** $R = \dfrac{d}{1 - st}$ **29.** $B = \dfrac{T}{1 + qt}$

31. (a) 2485 calories;
(b) $a = \dfrac{19.18w + 7h + 92.4 - K}{9.52}$;

$h = \dfrac{K - 19.18w + 9.52a - 92.4}{7}$;

$w = \dfrac{K - 7h + 9.52a - 92.4}{19.18}$ **33. (a)** 1614 g;

(b) $a = \dfrac{P + 299}{9.337d}$ **35. (a)** 50 mg; **(b)** $d = \dfrac{c(a + 12)}{a}$, or

$c + \dfrac{12c}{a}$ **37.** $\mathbf{D_W}$ **39.** -5 **40.** 250 **41.** -2

42. -25 **43.** $\frac{4}{5}$ **44.** 6 **45.** -6 **46.** -5

47. $s = \dfrac{A - \pi r^2}{\pi r}$ **49.** $V_1 = \dfrac{P_2 V_2 T_1}{P_1 T_2}$; $P_2 = \dfrac{P_1 V_1 T_2}{T_1 V_2}$

51. 0.8 yr **53. (a)** Approximately 120.5 horsepower;
(b) approximately 94.1 horsepower

Margin Exercises, Section 1.3, pp. 100–106

1. 22 in., 33 in., 51 in. **2.** $124 **3.** $228,000
4. (a) 1.43 billion digital prints, 6.53 billion digital prints;
(b) 2006 **5.** 48 in. and 72 in. **6.** 31°, 93°, 56°
7. 421, 422, 423 **8.** $2\frac{7}{9}$ hr; $1\frac{2}{3}$ hr

Exercise Set 1.3, p. 107

1. 6.55 mi **3.** $6.8 billion **5.** 26°, 130°, 24° **7.** $1120
9. Length: 94 ft; width: 50 ft **11.** $66\frac{2}{3}$ cm and $33\frac{1}{3}$ cm
13. $265,000 **15.** 9, 11, 13 **17.** 348 and 349
19. 843 sq ft **21.** $38,950 **23.** 30 million pounds of
coffee beans **25. (a)** About 17,439 fatalities, about
17,497 fatalities; **(b)** 2004 **27.** 6 min **29.** Downstream:
1.25 hr; upstream: 1.875 hr **31.** $\frac{1}{2}$ hr **33.** $\mathbf{D_W}$
35. -2442 **36.** 208 **37.** 49 **38.** -119 **39.** 49
40. -119 **41.** $\frac{78}{1649}$ **42.** $\frac{17}{10}$ **43.** -1 **44.** -256
45. 32 **46.** 2,097,152 **47.** $110,000 **49.** 130 novels
51. 25% increase **53.** 62,208,000 sec **55.** Answers may
vary. A piece of material 75 cm long is to be cut into 2
pieces, one of them $\frac{2}{3}$ as long as the other. How should the
material be cut? **57.** $m\angle 2 = 120°$; $m\angle 1 = 60°$

Margin Exercises, Section 1.4, pp. 113–121

1. Yes **2.** No **3.** Yes **4.** $[-4, 5)$ **5.** $(-\infty, -2]$
6. $(2, 6]$ **7.** $[10, \infty)$ **8.** $[-30, 30]$
9. $\{x\,|\,x > 3\}$, or $(3, \infty)$; **10.** $\{x\,|\,x \le 3\}$, or $(-\infty, 3]$;

11. $\{x\,|\,x \le -2\}$, or $(-\infty, -2]$; **12.** $\left\{y\,\middle|\,y \le \frac{3}{10}\right\}$, or $\left(-\infty, \frac{3}{10}\right]$;

13. $\{y\,|\,y < -5\}$, or $(-\infty, -5)$ **14.** $\{x\,|\,x \ge 12\}$, or $[12, \infty)$

15. $\left\{y\,\middle|\,y \le -\frac{1}{5}\right\}$, or $\left(-\infty, -\frac{1}{5}\right]$ **16.** $\left\{x\,\middle|\,x < \frac{1}{2}\right\}$, or $\left(-\infty, \frac{1}{2}\right)$
17. $\left\{y\,\middle|\,y \ge \frac{17}{9}\right\}$, or $\left[\frac{17}{9}, \infty\right)$ **18.** $s \ge 88$ **19.** $d \ge \$4000$
20. $p \le \$21,000$ **21.** 50 min $< t <$ 60 min
22. $d > 25$ mi **23.** $w < 110$ lb **24.** $n > -8$
25. $c \le \$135,000$ **26.** $d \le 11.6\%$ **27.** $p \ge 37$
28. $\{t\,|\,t > 9.1\}$, or years after 2009
29. For $\{n\,|\,n < 100$ hr$\}$, plan A is better.

Translating for Success, p. 122

1. F **2.** I **3.** C **4.** E **5.** D **6.** J **7.** O **8.** M
9. B **10.** L

Exercise Set 1.4, p. 123

1. No, no, no, yes **3.** No, yes, yes, no, no **5.** $(-\infty, 5)$
7. $[-3, 3]$ **9.** $(-8, -4)$ **11.** $(-2, 5)$ **13.** $\left(-\sqrt{2}, \infty\right)$
15. $\{x\,|\,x > -1\}$, or $(-1, \infty)$ **17.** $\{y\,|\,y < 6\}$, or $(-\infty, 6)$

19. $\{a\,|\,a \le -22\}$, or $(-\infty, -22]$

21. $\{t\,|\,t \ge -4\}$, or $[-4, \infty)$

23. $\{y\,|\,y > -6\}$, or $(-6, \infty)$ **25.** $\{x\,|\,x \le 9\}$, or $(-\infty, 9]$

27. $\{x\,|\,x \ge 3\}$, or $[3, \infty)$

29. $\{x\,|\,x < -60\}$, or $(-\infty, -60)$ **31.** $\{x\,|\,x > 3\}$, or $(3, \infty)$

33. $\{x\,|\,x \le 0.9\}$, or $(-\infty, 0.9]$
35. $\left\{x\,\middle|\,x \le \frac{5}{6}\right\}$, or $\left(-\infty, \frac{5}{6}\right]$ **37.** $\{x\,|\,x < 6\}$, or $(-\infty, 6)$
39. $\{y\,|\,y \le -3\}$, or $(-\infty, -3]$ **41.** $\left\{y\,\middle|\,y > \frac{2}{3}\right\}$, or $\left(\frac{2}{3}, \infty\right)$
43. $\{x\,|\,x \ge 11.25\}$, or $[11.25, \infty)$
45. $\left\{x\,\middle|\,x \le \frac{1}{2}\right\}$, or $\left(-\infty, \frac{1}{2}\right]$ **47.** $\left\{y\,\middle|\,y \le -\frac{75}{2}\right\}$, or $\left(-\infty, -\frac{75}{2}\right]$
49. $\left\{x\,\middle|\,x > -\frac{2}{17}\right\}$, or $\left(-\frac{2}{17}, \infty\right)$ **51.** $\left\{m\,\middle|\,m > \frac{7}{3}\right\}$, or $\left(\frac{7}{3}, \infty\right)$
53. $\{r\,|\,r < -3\}$, or $(-\infty, -3)$ **55.** $\{x\,|\,x \ge 2\}$, or $[2, \infty)$
57. $\{y\,|\,y < 5\}$, or $(-\infty, 5)$ **59.** $\left\{x\,\middle|\,x \le \frac{4}{7}\right\}$, or $\left(-\infty, \frac{4}{7}\right]$
61. $\{x\,|\,x < 8\}$, or $(-\infty, 8)$ **63.** $\left\{x\,\middle|\,x \ge \frac{13}{2}\right\}$, or $\left[\frac{13}{2}, \infty\right)$
65. $\left\{x\,\middle|\,x < \frac{11}{18}\right\}$, or $\left(-\infty, \frac{11}{18}\right)$ **67.** $\left\{x\,\middle|\,x \ge -\frac{51}{31}\right\}$, or $\left[-\frac{51}{31}, \infty\right)$
69. $\{a\,|\,a \le 2\}$, or $(-\infty, 2]$
71. $\{W\,|\,W < (\text{approximately}) \, 189.1$ lb$\}$ **73.** $\{S\,|\,S \ge 84\}$
75. $\{B\,|\,B \ge \$11,500\}$ **77.** $\{S\,|\,S > \$7000\}$
79. $\{n\,|\,n > 25\}$ **81.** $\{p\,|\,p > 80\}$ **83.** $\{s\,|\,s > 8\}$
85. (a) 8.398 gal, 12.063 gal, 15.728 gal; **(b)** years after 1999
87. $\mathbf{D_W}$ **89.** $-9a + 30b$ **90.** $32x - 72y$
91. $-8a + 22b$ **92.** $-6a + 17b$ **93.** $10(3x - 7y - 4)$
94. $-6a(2 - 5b)$ **95.** $-4(2x - 6y + 1)$

96. $5(2n - 9mn + 20m)$ **97.** -11.2 **98.** 6.6
99. -11.2 **100.** 6.6 **101.** (a) $\{p \mid p > 10\}$;
(b) $\{p \mid p < 10\}$ **103.** True   **105.** All real numbers
107. All real numbers

Margin Exercises, Section 1.5, pp. 130–136

1. $\{0, 3\}$ **2.**

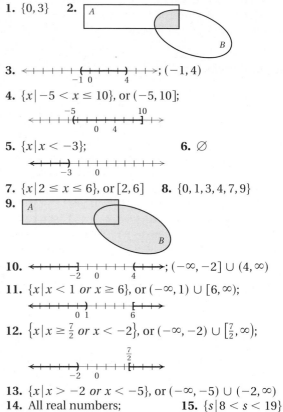

3. <++++++(++++++)+++++>; $(-1, 4)$
    $-1\,0$ 4

4. $\{x \mid -5 < x \le 10\}$, or $(-5, 10]$;
    -5 10
    <++++(++++++++++]+++>
        $0\quad 4$

5. $\{x \mid x < -3\}$;   **6.** $\varnothing$
    <++++)+++++++++++++>
    $-3\quad 0$

7. $\{x \mid 2 \le x \le 6\}$, or $[2, 6]$ **8.** $\{0, 1, 3, 4, 7, 9\}$
9.

10. <++++]++++++++(+>; $(-\infty, -2] \cup (4, \infty)$
    $-2\quad 0\qquad 4$

11. $\{x \mid x < 1 \ or \ x \ge 6\}$, or $(-\infty, 1) \cup [6, \infty)$;
    <+++++)+++++[+++>
        $0\,1\qquad\quad 6$

12. $\{x \mid x \ge \frac{7}{2} \ or \ x < -2\}$, or $(-\infty, -2) \cup [\frac{7}{2}, \infty)$;
                $\frac{7}{2}$
    <++++)+++++++[++>
    $-2\quad 0$

13. $\{x \mid x > -2 \ or \ x < -5\}$, or $(-\infty, -5) \cup (-2, \infty)$
14. All real numbers;   **15.** $\{s \mid 8 < s < 19\}$
    <+++++++++++++++>
            0

Exercise Set 1.5, p. 137

1. $\{9, 11\}$ **3.** $\{b\}$ **5.** $\{9, 10, 11, 13\}$
7. $\{a, b, c, d, f, g\}$ **9.** $\varnothing$ **11.** $\{3, 5, 7\}$
13. <+++(++++++]+++++>; $(-4, 1]$
    $-4\qquad 0\,1$

15. <+++++++++(++++++)>; $(1, 6)$
        $0\,1\qquad\qquad 6$

17. $\{x \mid -4 \le x < 5\}$, or $[-4, 5)$;
    <+++[++++++++)++>
        $-4\qquad 0\qquad 5$

19. $\{x \mid x \ge 2\}$, or $[2, \infty)$;   **21.** $\varnothing$
    <++++++++++[+++++>
                $0\quad 2$

23. $\{x \mid -8 < x < 6\}$, or $(-8, 6)$
25. $\{x \mid -6 < x \le 2\}$, or $(-6, 2]$
27. $\{y \mid -1 < y \le 5\}$, or $(-1, 5]$
29. $\{x \mid -\frac{5}{3} \le x \le \frac{4}{3}\}$, or $[-\frac{5}{3}, \frac{4}{3}]$
31. $\{x \mid -1 < x \le 6\}$, or $(-1, 6]$
33. $\{x \mid -\frac{3}{2} \le x < \frac{9}{2}\}$, or $[-\frac{3}{2}, \frac{9}{2})$
35. $\{x \mid 10 < x \le 14\}$, or $(10, 14]$
37. $\{x \mid -\frac{7}{2} < x < \frac{37}{2}\}$, or $(-\frac{7}{2}, \frac{37}{2})$

39. <++++++)++(++++++>; $(-\infty, -2) \cup (1, \infty)$
    $-2\quad 0\,1$

41. <++++]+++(+++++++>; $(-\infty, -3] \cup (1, \infty)$
    $-3\qquad 0\,1$

43. $\{x \mid x < -5 \ or \ x > -1\}$, or $(-\infty, -5) \cup (-1, \infty)$;
    <++++)+++(+++++++>
    $-5\qquad -1\,0$

45. $\{x \mid x \le \frac{5}{2} \ or \ x \ge 4\}$, or $(-\infty, \frac{5}{2}] \cup [4, \infty)$;
                $\frac{5}{2}$
    <++++++++]+[++++>
        $0\qquad 4$

47. $\{x \mid x \ge -3\}$, or $[-3, \infty)$;
    <+++[+++++++++++>
    $-3\qquad 0$

49. $\{x \mid x \le -\frac{5}{4} \ or \ x > -\frac{1}{2}\}$, or $(-\infty, -\frac{5}{4}] \cup (-\frac{1}{2}, \infty)$
51. All real numbers, or $(-\infty, \infty)$
53. $\{x \mid x < -4 \ or \ x > 2\}$, or $(-\infty, -4) \cup (2, \infty)$
55. $\{x \mid x < \frac{79}{4} \ or \ x > \frac{89}{4}\}$, or $(-\infty, \frac{79}{4}) \cup (\frac{89}{4}, \infty)$
57. $\{x \mid x \le -\frac{13}{2} \ or \ x \ge \frac{29}{2}\}$, or $(-\infty, -\frac{13}{2}] \cup [\frac{29}{2}, \infty)$
59. $\{d \mid 0 \text{ ft} \le d \le 198 \text{ ft}\}$ **61.** Between 23 and 27 beats
63. $\{W \mid 139.9 \text{ lb} < W < 189.1 \text{ lb}\}$
65. $\{d \mid 250 \text{ mg} < d < 500 \text{ mg}\}$ **67.** $\mathbf{D_W}$ **69.** 3.2
70. 12 **71.** 2 **72.** 0 **73.** $-\dfrac{32y^{30}}{x^{20}}$ **74.** $-\dfrac{20b}{a^7}$

75. $-\dfrac{4a^{17}}{5b^{15}}$ **76.** $25p^{12}q^{22}$ **77.** $\dfrac{a^6}{8b^6}$ **78.** $25p^{10}q^8$

79. $\{x \mid -4 < x \le 1\}$, or $(-4, 1]$
81. $\{x \mid \frac{2}{5} \le x \le \frac{4}{5}\}$, or $[\frac{2}{5}, \frac{4}{5}]$ **83.** $\{x \mid -\frac{1}{8} < x < \frac{1}{2}\}$, or $(-\frac{1}{8}, \frac{1}{2})$
85. $\{x \mid 10 < x \le 18\}$, or $(10, 18]$ **87.** True   **89.** False
91. All real numbers; $\varnothing$

Margin Exercises, Section 1.6, pp. 141–147

1. $7|x|$ **2.** x^8 **3.** $5a^2|b|$ **4.** $\dfrac{7|a|}{b^2}$ **5.** $9|x|$ **6.** 29

7. 5 **8.** $|p|$ **9.** $\{6, -6\}$;   **10.** $\varnothing$
    <+++++++++++++++>
    $-6\qquad 0\qquad 6$

11. $\{0\}$ **12.** $\{2, -2\}$ **13.** $\{\frac{17}{4}, -\frac{17}{4}\}$ **14.** $\{4, -4\}$
15. $\{3, 5\}$ **16.** $\{-\frac{13}{3}, 7\}$ **17.** $\varnothing$ **18.** $\{\frac{7}{4}, -\frac{1}{6}\}$
19. $\{-\frac{7}{2}\}$ **20.** $\{5, -5\}$;
    <++++++++++++++>
    $-5\qquad 0\qquad 5$

21. $\{x \mid -5 < x < 5\}$, or $(-5, 5)$;
    <++(++++++++++)++>
    $-5\qquad 0\qquad 5$

22. $\{x \mid x \le -5 \ or \ x \ge 5\}$, or $(-\infty, -5] \cup [5, \infty)$;
    <++]+++++++++[++>
    $-5\qquad 0\qquad 5$

23. $\{x \mid -2 < x < 5\}$, or $(-2, 5)$;
    <++++(+++++++)++>
    $-2\quad 0\qquad 5$

24. $\{x \mid 1 \le x \le \frac{11}{3}\}$, or $[1, \frac{11}{3}]$
25. $\{x \mid x \le -\frac{7}{3} \ or \ x \ge 1\}$, or $(-\infty, -\frac{7}{3}] \cup [1, \infty)$;
    <+++++]+++[+++++>
    $-\frac{7}{3}\quad 0\,1$

Exercise Set 1.6, p. 148

1. $9|x|$ **3.** $2x^2$ **5.** $2x^2$ **7.** $6|y|$ **9.** $\dfrac{2}{|x|}$ **11.** $\dfrac{x^2}{|y|}$

13. $4|x|$ **15.** $\dfrac{y^2}{3}$ **17.** 38 **19.** 19 **21.** 6.3 **23.** 5

25. $\{3, -3\}$ **27.** $\varnothing$ **29.** $\{0\}$ **31.** $\{15, -9\}$

33. $\left\{\frac{7}{2}, -\frac{1}{2}\right\}$ **35.** $\left\{\frac{23}{4}, -\frac{5}{4}\right\}$ **37.** $\{11, -11\}$

39. $\{291, -291\}$ **41.** $\{8, -8\}$ **43.** $\{7, -7\}$ **45.** $\{2, -2\}$

47. $\{8, -7\}$ **49.** $\{2, -12\}$ **51.** $\left\{\frac{7}{2}, -\frac{5}{2}\right\}$ **53.** $\varnothing$

55. $\left\{-\frac{13}{54}, -\frac{7}{54}\right\}$ **57.** $\left\{\frac{3}{4}, -\frac{11}{2}\right\}$ **59.** $\left\{\frac{3}{2}\right\}$ **61.** $\left\{5, -\frac{3}{5}\right\}$

63. All real numbers **65.** $\left\{-\frac{3}{2}\right\}$ **67.** $\left\{\frac{24}{23}, 0\right\}$ **69.** $\left\{32, \frac{8}{3}\right\}$

71. $\{x \mid -3 < x < 3\}$, or $(-3, 3)$

73. $\{x \mid x \le -2 \ or \ x \ge 2\}$, or $(-\infty, -2] \cup [2, \infty)$

75. $\{x \mid 0 < x < 2\}$, or $(0, 2)$

77. $\{x \mid -6 \le x \le -2\}$, or $[-6, -2]$

79. $\left\{x \mid -\frac{1}{2} \le x \le \frac{7}{2}\right\}$, or $\left[-\frac{1}{2}, \frac{7}{2}\right]$

81. $\left\{y \mid y < -\frac{3}{2} \ or \ y > \frac{17}{2}\right\}$, or $\left(-\infty, -\frac{3}{2}\right) \cup \left(\frac{17}{2}, \infty\right)$

83. $\left\{x \mid x \le -\frac{5}{4} \ or \ x \ge \frac{23}{4}\right\}$, or $\left(-\infty, -\frac{5}{4}\right] \cup \left[\frac{23}{4}, \infty\right)$

85. $\{y \mid -9 < y < 15\}$, or $(-9, 15)$

87. $\left\{x \mid -\frac{7}{2} \le x \le \frac{1}{2}\right\}$, or $\left[-\frac{7}{2}, \frac{1}{2}\right]$

89. $\left\{y \mid y < -\frac{4}{3} \ or \ y > 4\right\}$, or $\left(-\infty, -\frac{4}{3}\right) \cup (4, \infty)$

91. $\left\{x \mid x \le -\frac{5}{4} \ or \ x \ge \frac{23}{4}\right\}$, or $\left(-\infty, -\frac{5}{4}\right] \cup \left[\frac{23}{4}, \infty\right)$

93. $\left\{x \mid -\frac{9}{2} < x < 6\right\}$, or $\left(-\frac{9}{2}, 6\right)$

95. $\left\{x \mid x \le -\frac{25}{6} \ or \ x \ge \frac{23}{6}\right\}$, or $\left(-\infty, -\frac{25}{6}\right] \cup \left[\frac{23}{6}, \infty\right)$

97. $\{x \mid -5 < x < 19\}$, or $(-5, 19)$

99. $\left\{x \mid x \le -\frac{2}{15} \ or \ x \ge \frac{14}{15}\right\}$, or $\left(-\infty, -\frac{2}{15}\right] \cup \left[\frac{14}{15}, \infty\right)$

101. $\{m \mid -12 \le m \le 2\}$, or $[-12, 2]$

103. $\left\{x \mid \frac{1}{2} \le x \le \frac{5}{2}\right\}$, or $\left[\frac{1}{2}, \frac{5}{2}\right]$

105. $\{x \mid 0.49705 \le x \le 0.50295\}$, or $[0.49705, 0.50295]$

107. $\mathbf{D_W}$ **109.** Union **110.** Disjoint **111.** At least

112. $[a, b]$ **113.** Absolute value **114.** Equation

115. Equivalent **116.** Inequality

117. $\left\{d \mid 5\frac{1}{2} \ \text{ft} \le d \le 6\frac{1}{2} \ \text{ft}\right\}$ **119.** $\{x \mid x \ge -5\}$, or $[-5, \infty)$

121. $\left\{1, -\frac{1}{4}\right\}$ **123.** $\varnothing$ **125.** All real numbers

127. $|x| < 3$ **129.** $|x| \ge 6$ **131.** $|x + 3| > 5$

Concept Reinforcement, p. 152

1. True **2.** True **3.** False **4.** True **5.** False

Summary and Review: Chapter 1, p. 153

1. 8 **2.** $\frac{3}{7}$ **3.** $\frac{22}{5}$ **4.** $-\frac{1}{13}$ **5.** -0.2 **6.** 5

7. $d = \frac{11}{4}(C - 3)$ **8.** $b = \dfrac{A - 2a}{-3}$, or $\dfrac{2a - A}{3}$

9. 185 and 186 **10.** 15 m, 12 m **11.** 160,000

12. 40 sec **13.** $[-8, 9)$ **14.** $(-\infty, 40]$

15. ⟵++++]++++++++++⟶; $(-\infty, -2]$
      $\quad\quad\quad -2\ \ 0$

16. ⟵++++++++(++++++⟶; $(1, \infty)$
      $\quad\quad\quad\quad\quad 0\ 1$

17. $\{a \mid a \le -21\}$, or $(-\infty, -21]$

18. $\{y \mid y \ge -7\}$, or $[-7, \infty)$ **19.** $\{y \mid y > -4\}$, or $(-4, \infty)$

20. $\{y \mid y > -30\}$, or $(-30, \infty)$ **21.** $\{x \mid x > -3\}$, or $(-3, \infty)$

22. $\left\{y \mid y \le -\frac{6}{5}\right\}$, or $\left(-\infty, -\frac{6}{5}\right]$ **23.** $\{x \mid x < -3\}$, or $(-\infty, -3)$

24. $\{y \mid y > -10\}$, or $(-10, \infty)$

25. $\left\{x \mid x \le -\frac{5}{2}\right\}$, or $\left(-\infty, -\frac{5}{2}\right]$ **26.** $\left\{t \mid t > 4\frac{1}{4} \ \text{hr}\right\}$

27. \$10,000 **28.** ⟵++++[++++++++⟶; $[-2, 5)$
      $\quad\quad\quad\quad\quad -2\ \ 0 \quad\quad 5$

29. ⟵++++]++++++++(+⟶; $(-\infty, -2] \cup (5, \infty)$ **30.** $\varnothing$
      $\quad\quad\quad -2\ \ 0 \quad\quad 5$

31. $\{x \mid -7 < x \le 2\}$, or $(-7, 2]$

32. $\left\{x \mid -\frac{5}{4} < x < \frac{5}{2}\right\}$, or $\left(-\frac{5}{4}, \frac{5}{2}\right)$

33. $\{x \mid x < -3 \ or \ x > 1\}$, or $(-\infty, -3) \cup (1, \infty)$

34. $\{x \mid x < -11 \ or \ x \ge -6\}$, or $(-\infty, -11) \cup [-6, \infty)$

35. $\{x \mid x \le -6 \ or \ x \ge 8\}$, or $(-\infty, -6] \cup [8, \infty)$ **36.** $\dfrac{3}{|x|}$

37. $\dfrac{2|x|}{y^2}$ **38.** $\dfrac{4}{|y|}$ **39.** 62 **40.** $6, -6$ **41.** $\{9, -5\}$

42. $\left\{-14, \frac{4}{3}\right\}$ **43.** $\varnothing$ **44.** $\left\{x \mid -\frac{17}{2} < x < \frac{7}{2}\right\}$, or $\left(-\frac{17}{2}, \frac{7}{2}\right)$

45. $\{x \mid x \le -3.5 \ or \ x \ge 3.5\}$, or $(-\infty, -3.5] \cup [3.5, \infty)$

46. $\left\{x \mid x \le -\frac{11}{3} \ or \ x \ge \frac{19}{3}\right\}$, or $\left(-\infty, -\frac{11}{3}\right] \cup \left[\frac{19}{3}, \infty\right)$ **47.** $\varnothing$

48. $\{1, 5, 9\}$ **49.** $\{1, 2, 3, 5, 6, 9\}$ **50. (a)** About 9.62 sec;
(b) 2010; **(c)** $\{t \mid 3.23 < t < 7.85\}$, or approximately between
1991 and 1996 **51.** $\mathbf{D_W}$ **(1)** $-9(x + 2) = -9x - 18$, not
$-9x + 2$. **(2)** This would be correct if (1) were correct except
that the inequality symbol should not have been reversed.
(3) If (2) were correct, the right-hand side would be -5, not
8. **(4)** The inequality symbol should be reversed. The
correct solution is

$$7 - 9x + 6x < -9(x + 2) + 10x$$
$$7 - 9x + 6x < -9x - 18 + 10x$$
$$7 - 3x < x - 18$$
$$-4x < -25$$
$$x > \frac{25}{4}.$$

52. $\mathbf{D_W}$ "Solve" can mean to find all the replacements that
make an equation or inequality true. It can also mean to
express a formula as an equivalent equation with a given
variable alone on one side.
53. $\left\{x \mid -\frac{8}{3} \le x \le -2\right\}$, or $\left[-\frac{8}{3}, -2\right]$

Test: Chapter 1, p. 155

1. [1.1b] -2 **2.** [1.1c] $\frac{2}{3}$ **3.** [1.1b] $\frac{19}{15}$ **4.** [1.1d] 4

5. [1.1d] 1.1 **6.** [1.1d] -2 **7.** [1.2a] $B = \dfrac{A + C}{3}$

8. [1.2a] $n = \dfrac{m}{1 - t}$ **9.** [1.3a] Length: $14\frac{2}{5}$ ft; width: $9\frac{3}{5}$ ft

10. [1.3a] 43,333 copies **11.** [1.3a] 180,000

12. [1.3a] $59°, 60°, 61°$ **13.** [1.3b] $2\frac{2}{5}$ hr; 4 hr

14. [1.4b] $(-3, 2]$ **15.** [1.4b] $(-4, \infty)$

16. [1.4c] ⟵++++++++++++]⟶; $(-\infty, 6]$
      $\quad\quad\quad\quad 0 \quad\quad\quad 6$

17. [1.4c] ⟵++++]++++++++⟶; $(-\infty, -2]$
      $\quad\quad\quad -2\ \ 0$

18. [1.4c] $\{x \mid x \ge 10\}$, or $[10, \infty)$

19. [1.4c] $\{y \mid y > -50\}$, or $(-50, \infty)$

20. [1.4c] $\left\{a \mid a \le \frac{11}{5}\right\}$, or $\left(-\infty, \frac{11}{5}\right]$

21. [1.4c] $\{y \mid y > 1\}$, or $(1, \infty)$

22. [1.4c] $\left\{x \mid x > \frac{5}{2}\right\}$, or $\left(\frac{5}{2}, \infty\right)$

23. [1.4c] $\left\{x \mid x \le \frac{7}{4}\right\}$, or $\left(-\infty, \frac{7}{4}\right]$ **24.** [1.4d] $\left\{h \mid h > 2\frac{1}{10} \ \text{hr}\right\}$

25. [1.4d] $\{d \mid 33 \ \text{ft} \le d \le 231 \ \text{ft}\}$

26. [1.5a] ⟵++++[++++++]++++⟶; $[-3, 4]$
      $\quad\quad\quad\quad -3 \quad 0 \quad 4$

27. [1.5b] $\longleftrightarrow$; $(-\infty, -3) \cup (4, \infty)$

28. [1.5a] $\{x \mid x \geq 4\}$, or $[4, \infty)$

29. [1.5a] $\{x \mid -1 < x < 6\}$, or $(-1, 6)$

30. [1.5a] $\{x \mid -\frac{2}{5} < x \leq \frac{9}{5}\}$, or $\left(-\frac{2}{5}, \frac{9}{5}\right]$

31. [1.5b] $\{x \mid x < -4 \text{ or } x > -\frac{5}{2}\}$, or $(-\infty, -4) \cup \left(-\frac{5}{2}, \infty\right)$

32. [1.5b] All real numbers, or $(-\infty, \infty)$

33. [1.5b] $\{x \mid x < 3 \text{ or } x > 6\}$, or $(-\infty, 3) \cup (6, \infty)$

34. [1.6a] $\dfrac{7}{|x|}$ **35.** [1.6a] $2|x|$ **36.** [1.6b] 8.4

37. [1.5a] $\{3, 5\}$ **38.** [1.5b] $\{1, 3, 5, 7, 9, 11, 13\}$

39. [1.6c] $\{-9, 9\}$ **40.** [1.6c] $\{-6, 12\}$ **41.** [1.6d] $\{1\}$

42. [1.6c] $\varnothing$ **43.** [1.6e] $\{x \mid -\frac{7}{8} < x < \frac{11}{8}\}$, or $\left(-\frac{7}{8}, \frac{11}{8}\right)$

44. [1.6e] $\{x \mid x < -3 \text{ or } x > 3\}$, or $(-\infty, -3) \cup (3, \infty)$

45. [1.6e] $\{x \mid -99 \leq x \leq 111\}$, or $[-99, 111]$

46. [1.6e] $\{x \mid x \leq -\frac{13}{5} \text{ or } x \geq \frac{7}{5}\}$, or $\left(-\infty, -\frac{13}{5}\right] \cup \left[\frac{7}{5}, \infty\right)$

47. [1.6e] $\varnothing$ **48.** [1.5a] $\{x \mid \frac{1}{5} < x < \frac{4}{5}\}$, or $\left(\frac{1}{5}, \frac{4}{5}\right)$

CHAPTER 2

Margin Exercises, Section 2.1, pp. 159–165

1.–10.

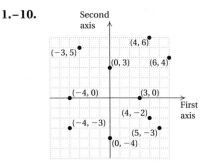

11. Both negative

12. First positive, second negative   **13.** No   **14.** Yes

15. $(-6, 4), (-2, 2)$; answers may vary

16.

x	y	(x, y)
-3	6	$(-3, 6)$
-1	2	$(-1, 2)$
0	0	$(0, 0)$
1	-2	$(1, -2)$
3	-6	$(3, -6)$

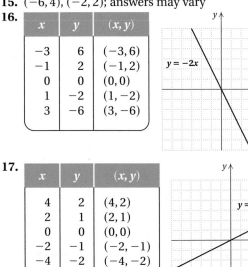

17.

x	y	(x, y)
4	2	$(4, 2)$
2	1	$(2, 1)$
0	0	$(0, 0)$
-2	-1	$(-2, -1)$
-4	-2	$(-4, -2)$
-1	$-\frac{1}{2}$	$(-1, -\frac{1}{2})$

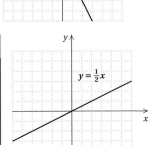

18.

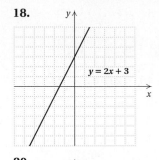

19.

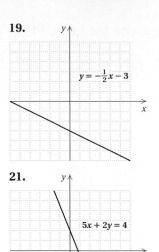

20.

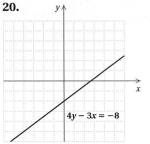

21.

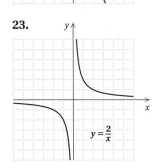

22.

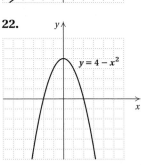

23.

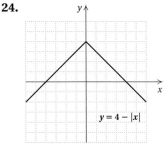

24.

Calculator Corner, p. 162

1.

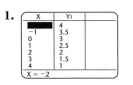

2.

Calculator Corner, p. 166

1. $y = 2x - 1$

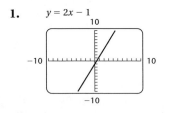

2. $y = -3x + 2$

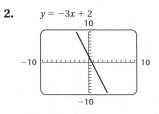

3. $y = 5x - 3$

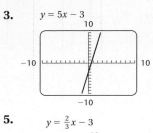

4. $y = -4x + 5$

5. $y = \frac{2}{3}x - 3$

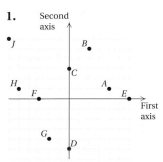

6. $y = -\frac{3}{4}x + 4$

7. $y = 3.104x - 6.21$

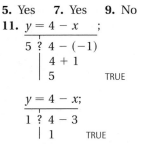

8. $y = -2.98x - 1.75$

15.

$$6x - 3y = 3;$$
$$\overline{6 \cdot 1 - 3 \cdot 1 \; ? \; 3}$$
$$6 - 3 \;\Big|$$
$$3 \;\Big|\quad \text{TRUE}$$

$$6x - 3y = 3;$$
$$\overline{6(-1) - 3(-3) \; ? \; 3}$$
$$-6 + 9 \;\Big|$$
$$3 \;\Big|\quad \text{TRUE}$$

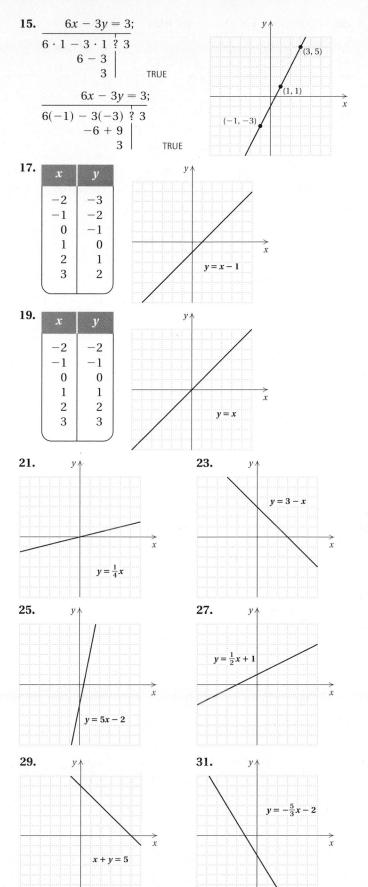

17.

x	y
-2	-3
-1	-2
0	-1
1	0
2	1
3	2

$y = x - 1$

19.

x	y
-2	-2
-1	-1
0	0
1	1
2	2
3	3

$y = x$

21. $y = \frac{1}{4}x$

23. $y = 3 - x$

25. $y = 5x - 2$

27. $y = \frac{1}{2}x + 1$

29. $x + y = 5$

31. $y = -\frac{5}{3}x - 2$

Exercise Set 2.1, p. 167

1.

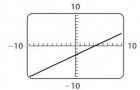

3.

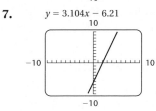

Triangle Area = 21 square units

5. Yes   **7.** Yes   **9.** No

11. $\dfrac{y = 4 - x}{5 \; ? \; 4 - (-1)}$
$$\Big|\; 4 + 1$$
$$\Big|\; 5 \qquad \text{TRUE}$$

$$\dfrac{y = 4 - x;}{1 \; ? \; 4 - 3}$$
$$\Big|\; 1 \qquad \text{TRUE}$$

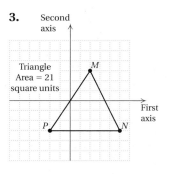

$(-1, 5)$
$(1, 3)$
$(3, 1)$

13. $\dfrac{3x + y = 7;}{3 \cdot 2 + 1 \; ? \; 7}$
$$6 + 1 \;\Big|$$
$$7 \;\Big|\quad \text{TRUE}$$

$$\dfrac{3x + y = 7;}{3 \cdot 4 + (-5) \; ? \; 7}$$
$$12 - 5 \;\Big|$$
$$7 \;\Big|\quad \text{TRUE}$$

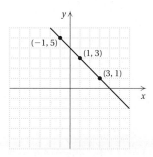

$(2, 1)$
$(3, -2)$
$(4, -5)$

33.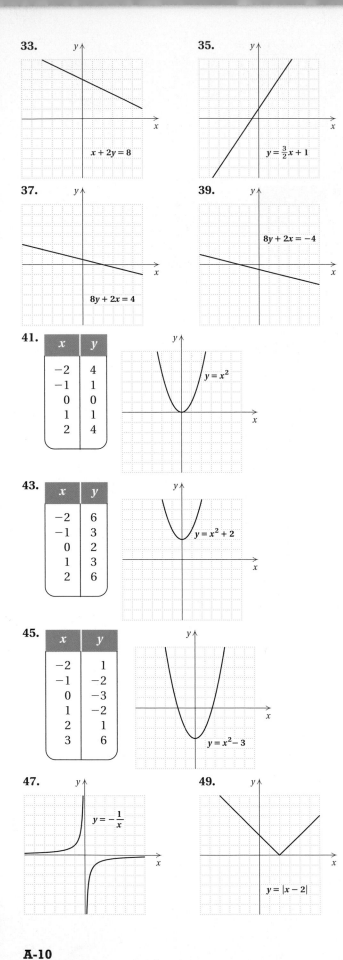

$x + 2y = 8$

35.

$y = \frac{3}{2}x + 1$

37.

$8y + 2x = 4$

39.

$8y + 2x = -4$

41.

x	y
−2	4
−1	1
0	0
1	1
2	4

$y = x^2$

43.

x	y
−2	6
−1	3
0	2
1	3
2	6

$y = x^2 + 2$

45.

x	y
−2	1
−1	−2
0	−3
1	−2
2	1
3	6

$y = x^2 - 3$

47.

$y = -\dfrac{1}{x}$

49.

$y = |x - 2|$

51.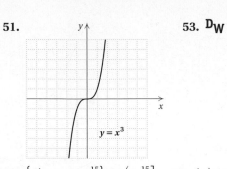

$y = x^3$

53. D_W

55. $\{x \mid 1 < x \le \frac{15}{2}\}$, or $\left(1, \frac{15}{2}\right]$ **56.** $\{x \mid x > -3\}$, or $(-3, \infty)$
57. $\{x \mid x \le -\frac{7}{3} \,or\, x \ge \frac{17}{3}\}$, or $\left(-\infty, -\frac{7}{3}\right] \cup \left[\frac{17}{3}, \infty\right)$
58. $\{x \mid -6 < x < 6\}$, or $(-6, 6)$ **59.** 25 ft
60. (a) \$2720; \$2040; \$680; \$0; **(b)** 2.8 yr   **61.** 4 mi
62. \$330,000   **63.** Left to the student   **65.** Left to the
student   **67.** $y = -x + 4$ **69.** $y = |x| - 3$

Margin Exercises, Section 2.2, pp. 175–181

1. Yes   **2.** No   **3.** Yes   **4.** No   **5.** Yes   **6.** Yes
7. No   **8. (a)** -4; **(b)** 148; **(c)** 31; **(d)** $8a^2 + 6a - 4$
9. (a) -33; **(b)** -3; **(c)** $-5a - 2$; **(d)** $6a - 6h - 7$
10.

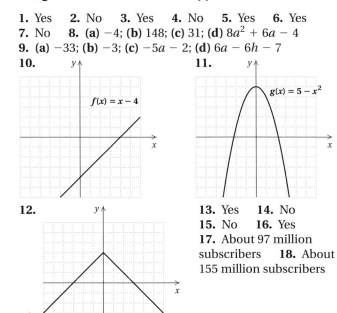

$f(x) = x - 4$

11.

$g(x) = 5 - x^2$

12.

$t(x) = 3 - |x|$

13. Yes   **14.** No
15. No   **16.** Yes
17. About 97 million
subscribers   **18.** About
155 million subscribers

Calculator Corner, p. 177

1. -13.3 **2.** -14.4 **3.** 14   **4.** 34

Calculator Corner, p. 180

1. 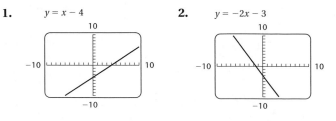 $y = x - 4$

2. $y = -2x - 3$

3. $y = -3x + 4$

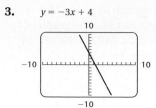

4. $y = x^2 + 2$

5. $y = 1 - x^2$

6. $y = 3x^2 - 4x + 1$

7. $y = x^3$

8. $y = |x + 3|$

9. $y = x^4 + 2$

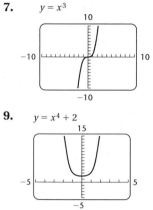

37.

$g(x) = -2x + 3$

39.

$f(x) = \frac{1}{2}x + 1$

41.

x	$f(x)$
-3	-1
-2	0
-1	1
0	2
1	1
2	0
3	-1

$f(x) = 2 - |x|$

43.

$g(x) = |x - 1|$

45.

x	$f(x)$
-3	9
-2	4
-1	1
0	0
1	1
2	4
3	9

$f(x) = x^2$

47.

x	$f(x)$
-3	10
-2	4
-1	0
0	-2
1	-2
2	0
3	4

$f(x) = x^2 - x - 2$

49.

$f(x) = 2 - x^2$

Exercise Set 2.2, p. 182

1. Yes   **3.** Yes   **5.** Yes   **7.** No   **9.** Yes   **11.** No
13. Yes   **15. (a)** 9; **(b)** 12; **(c)** 2; **(d)** 5; **(e)** 7.4; **(f)** $5\frac{2}{3}$
17. (a) -21; **(b)** 15; **(c)** 2; **(d)** 0; **(e)** $18a$; **(f)** $3a + 3$
19. (a) 7; **(b)** -17; **(c)** 6; **(d)** 4; **(e)** $3a - 2$; **(f)** $3a + 3h + 4$
21. (a) 0; **(b)** 5; **(c)** 2; **(d)** 170; **(e)** 65; **(f)** $32a^2 - 12a$
23. (a) 1; **(b)** 3; **(c)** 3; **(d)** 11; **(e)** $|a - 1| + 1$; **(f)** $|a + h| + 1$
25. (a) 0; **(b)** -1; **(c)** 8; **(d)** 1000; **(e)** -125; **(f)** $-27a^3$
27. (a) 159.48 cm; **(b)** 167.73 cm   **29.** $1\frac{20}{33}$ atm; $1\frac{10}{11}$ atm;
$4\frac{1}{33}$ atm   **31.** 1.792 cm; 2.8 cm; 11.2 cm

33.

x	$f(x)$
-2	4
-1	2
0	0
2	-4

$f(x) = -2x$

35.

x	$f(x)$
-1	-4
0	-1
1	2
2	5

$f(x) = 3x - 1$

51.

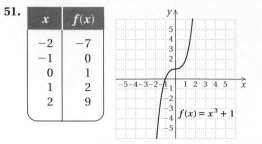

x	$f(x)$
-2	-7
-1	0
0	1
1	2
2	9

53. Yes **55.** Yes **57.** No **59.** No **61.** About 1150 stations **63.** About 1.9 billion images **65.** $\mathbf{D_W}$ **67.** Quadrants **68.** Relation **69.** Function; domain; range; domain; range **70.** Graph **71.** Inputs **72.** Solutions **73.** Addition principle **74.** Vertical-line test **75.** 26; 99 **77.** $g(x) = \frac{15}{4}x - \frac{13}{4}$

Margin Exercises, Section 2.3, pp. 189–191

1. Domain $= \{-3, -2, 0, 2, 5\}$; range $= \{-3, -2, 2, 3\}$ **2.** (a) -4; (b) $\{x \mid -3 \le x \le 3\}$, or $[-3, 3]$; (c) $-3, 3$; (d) $\{y \mid -5 \le y \le 4\}$, or $[-5, 4]$ **3.** Domain: all real numbers; range: all real numbers **4.** All real numbers **5.** $\{x \mid x$ is a real number *and* $x \ne -\frac{2}{3}\}$, or $\left(-\infty, -\frac{2}{3}\right) \cup \left(-\frac{2}{3}, \infty\right)$

Exercise Set 2.3, p. 192

1. (a) 3; (b) $\{-4, -3, -2, -1, 0, 1, 2\}$; (c) $-2, 0$; (d) $\{1, 2, 3, 4\}$ **3.** (a) $2\frac{1}{2}$; (b) $[-3, 5]$; (c) $2\frac{1}{4}$; (d) $[1, 4]$ **5.** (a) $2\frac{1}{4}$; (b) $[-4, 3]$; (c) 0; (d) $[-5, 4]$ **7.** (a) 1; (b) all real numbers; (c) 3; (d) all real numbers **9.** (a) 1; (b) all real numbers; (c) $-2, 2$; (d) $[0, \infty)$ **11.** (a) -1; (b) $[-6, 5]$; (c) $-4, 0, 3$; (d) $[-2, 2]$ **13.** $\{x \mid x$ is a real number *and* $x \ne -3\}$, or $(-\infty, -3) \cup (-3, \infty)$ **15.** All real numbers **17.** All real numbers **19.** $\{x \mid x$ is a real number *and* $x \ne \frac{14}{5}\}$, or $\left(-\infty, \frac{14}{5}\right) \cup \left(\frac{14}{5}, \infty\right)$ **21.** All real numbers **23.** $\{x \mid x$ is a real number *and* $x \ne \frac{3}{2}\}$, or $\left(-\infty, \frac{3}{2}\right) \cup \left(\frac{3}{2}, \infty\right)$ **25.** $\{x \mid x$ is a real number *and* $x \ne 1\}$, or $(-\infty, 1) \cup (1, \infty)$ **27.** All real numbers **29.** All real numbers **31.** $\{x \mid x$ is a real number *and* $x \ne \frac{5}{2}\}$, or $\left(-\infty, \frac{5}{2}\right) \cup \left(\frac{5}{2}, \infty\right)$ **33.** All real numbers **35.** $\{x \mid x$ is a real number *and* $x \ne -\frac{5}{4}\}$, or $\left(-\infty, -\frac{5}{4}\right) \cup \left(-\frac{5}{4}, \infty\right)$ **37.** $-8; 0; -2$ **39.** $\mathbf{D_W}$ **41.** $\{S \mid S > \$42,500\}$ **42.** $\{x \mid x \ge 90\}$ **43.** $\{-8, 8\}$ **44.** $\{\ \}$, or $\varnothing$ **45.** $\{-4, 18\}$ **46.** $\{-8, 5\}$ **47.** $\left\{\frac{1}{2}, 3\right\}$ **48.** $\left\{-1, \frac{9}{13}\right\}$ **49.** $\{\ \}$, or $\varnothing$ **50.** $\left\{\frac{8}{3}\right\}$ **51.** $(-\infty, 0) \cup (0, \infty)$; $[2, \infty)$; $[-4, \infty)$; $[0, \infty)$

Margin Exercises, Section 2.4, pp. 195–202

1.

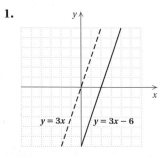

The graph of $y = 3x - 6$ looks just like the graph of $y = 3x$, but it is moved down 6 units.

2.

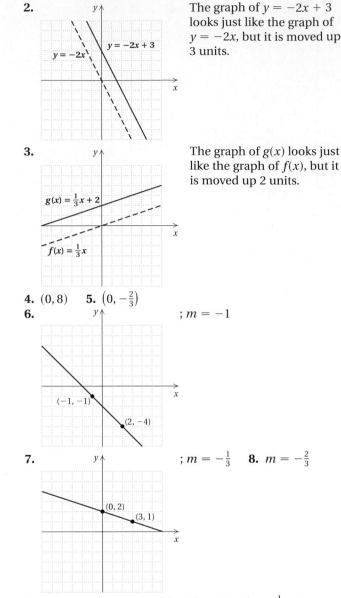

The graph of $y = -2x + 3$ looks just like the graph of $y = -2x$, but it is moved up 3 units.

3. The graph of $g(x)$ looks just like the graph of $f(x)$, but it is moved up 2 units.

4. $(0, 8)$ **5.** $\left(0, -\frac{2}{3}\right)$

6. ; $m = -1$

7. ; $m = -\frac{1}{3}$ **8.** $m = -\frac{2}{3}$

9. Slope: -8; y-intercept: $(0, 23)$ **10.** Slope: $\frac{1}{2}$; y-intercept: $\left(0, -\frac{5}{2}\right)$ **11.** The rate of change is 3 haircuts per hour. **12.** -0.141 reports per 1000 passengers per year

Calculator Corner, p. 194

1. The graph of $y_2 = x + 4$ is the same as the graph of $y_1 = x$, but it is moved up 4 units. **2.** The graph of $y_3 = x - 3$ is the same as the graph of $y_1 = x$, but it is moved down 3 units. **3.** The graph of $y = x + 8$ will be the same as the graph of $y_1 = x$, but it will be moved up 8 units. **4.** The graph of $y = x - 5$ will be the same as the graph of $y_1 = x$, but it will be moved down 5 units.

Calculator Corner, p. 198

1. The graph of $y = 10x$ will slant up from left to right. It will be steeper than the other graphs. **2.** The graph of $y = 0.005x$ will slant up from left to right. It will be less steep than the other graphs. **3.** The graph of $y = -10x$

will slant down from left to right. It will be steeper than the other graphs.   **4.** The graph of $y = -0.005x$ will slant down from left to right. It will be less steep than the other graphs.

Exercise Set 2.4, p. 203

1. $m = 4$; y-intercept: $(0, 5)$ **3.** $m = -2$; y-intercept: $(0, -6)$ **5.** $m = -\frac{3}{8}$; y-intercept: $\left(0, -\frac{1}{5}\right)$ **7.** $m = 0.5$; y-intercept: $(0, -9)$ **9.** $m = \frac{2}{3}$; y-intercept: $\left(0, -\frac{8}{3}\right)$
11. $m = 3$; y-intercept: $(0, -2)$ **13.** $m = -8$; y-intercept: $(0, 12)$ **15.** $m = 0$; y-intercept: $\left(0, \frac{4}{17}\right)$ **17.** $m = -\frac{1}{2}$
19. $m = \frac{1}{3}$ **21.** $m = 2$ **23.** $m = \frac{2}{3}$ **25.** $m = -\frac{1}{3}$
27. $\frac{2}{25}$, or 8%   **29.** 3.5%   **31.** The rate of change is 5.06 billion messages daily per year.   **33.** The rate of change is $-\$900$ per year.   **35.** The rate of change is one point per \$1000 of family income.   **37.** $\mathbf{D_W}$ **39.** -1323
40. $45x + 54$ **41.** $350x - 60y + 120$ **42.** 25
43. Square: 15 yd; triangle: 20 yd
44. $\left\{x \mid x \le -\frac{24}{5} \, or \, x \ge 8\right\}$, or $\left(-\infty, -\frac{24}{5}\right] \cup [8, \infty)$
45. $\left\{x \mid -\frac{24}{5} < x < 8\right\}$, or $\left(-\frac{24}{5}, 8\right)$ **46.** $\left\{-\frac{24}{5}, 8\right\}$
47. $\{\ \}$, or $\varnothing$

7. $m = 0$

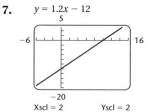

8.

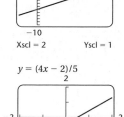

9.

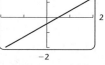

10. **(a)** Not defined; **(b)** $m = 0$; **(c)** $m = 0$; **(d)** not defined; **(e)** $m = 0$; **(f)** not defined   **11.** Yes   **12.** No   **13.** No   **14.** Yes   **15.** No

Margin Exercises, Section 2.5, pp. 206–211

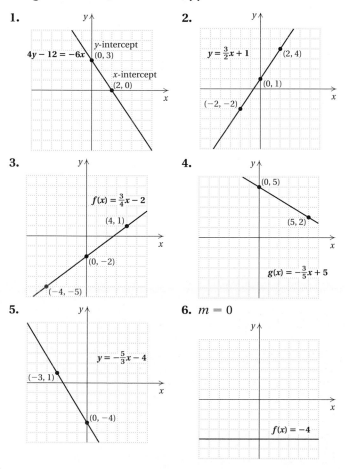

Calculator Corner, p. 206

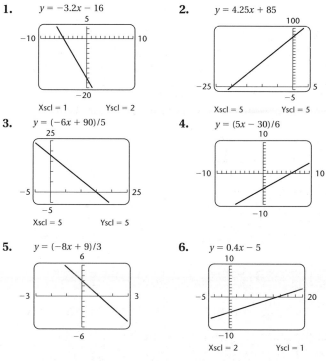

Calculator Corner, p. 212

1. Left to the student

Visualizing for Success, p. 213

1. D **2.** I **3.** H **4.** C **5.** F **6.** A **7.** G **8.** B
9. E **10.** J

Exercise Set 2.5, p. 214

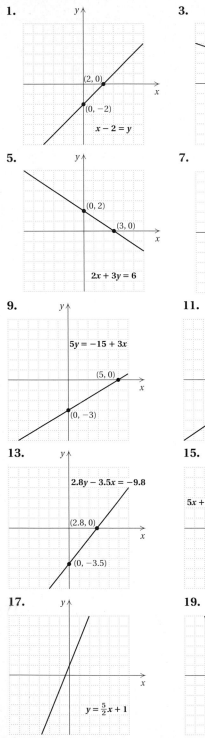

1. (graph: $x - 2 = y$, points $(2, 0)$, $(0, -2)$)

3. (graph: $x + 3y = 6$, points $(0, 2)$, $(6, 0)$)

5. (graph: $2x + 3y = 6$, points $(0, 2)$, $(3, 0)$)

7. (graph: $f(x) = -2 - 2x$, points $(-1, 0)$, $(0, -2)$)

9. (graph: $5y = -15 + 3x$, points $(5, 0)$, $(0, -3)$)

11. (graph: $2x - 3y = 6$, points $(3, 0)$, $(0, -2)$)

13. (graph: $2.8y - 3.5x = -9.8$, points $(2.8, 0)$, $(0, -3.5)$)

15. (graph: $5x + 2y = 7$, points $(0, \frac{7}{2})$, $(\frac{7}{5}, 0)$)

17. (graph: $y = \frac{5}{2}x + 1$)

19. (graph: $f(x) = -\frac{5}{2}x - 4$)

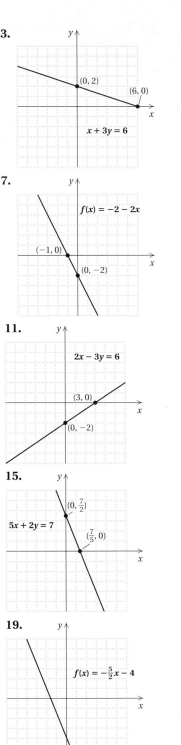

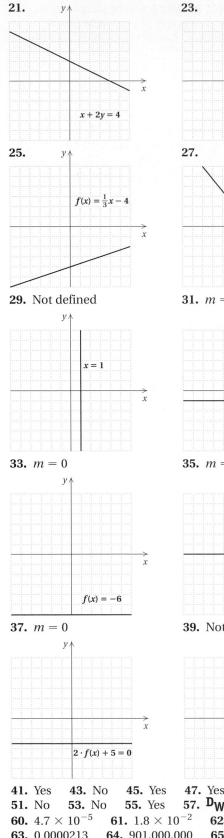

21. (graph: $x + 2y = 4$)

23. (graph: $4x - 3y = 12$)

25. (graph: $f(x) = \frac{1}{3}x - 4$)

27. (graph: $5x + 4 \cdot f(x) = 4$)

29. Not defined **31.** $m = 0$

(graph: $x = 1$) (graph: $y = -1$)

33. $m = 0$ **35.** $m = 0$

(graph: $f(x) = -6$) (graph: $y = 0$)

37. $m = 0$ **39.** Not defined

(graph: $2 \cdot f(x) + 5 = 0$) (graph: $7 - 3x = 4 + 2x$)

41. Yes   **43.** No   **45.** Yes   **47.** Yes   **49.** Yes
51. No   **53.** No   **55.** Yes   **57.** $\mathbf{D_W}$ **59.** 5.3×10^{10}
60. 4.7×10^{-5} **61.** 1.8×10^{-2} **62.** 9.9902×10^{7}
63. 0.0000213   **64.** 901,000,000   **65.** 20,000
66. 0.085677   **67.** $3(3x - 5y)$ **68.** $3a(4 + 7b)$
69. $7p(3 - q + 2)$ **70.** $64(x - 2y + 4)$ **71.** $y = 3$
73. $a = 2$ **75.** $y = \frac{2}{15}x + \frac{2}{5}$ **77.** $y = 0$; yes
79. $m = -\frac{3}{4}$ **81.** (a) II; (b) IV; (c) I; (d) III

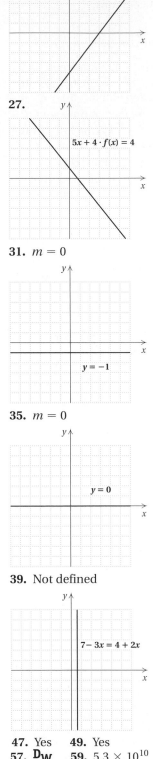

Margin Exercises, Section 2.6, pp. 219–225

1. $y = 3.4x - 8$ **2.** $y = -5x - 18$ **3.** $y = 3x - 5$
4. $y = 8x - 19$ **5.** $y = -\frac{2}{3}x + \frac{14}{3}$ **6.** $y = -\frac{5}{3}x + \frac{11}{3}$
7. $y = -17x - 56$ **8.** $y = 3x - 7$ **9.** $y = -2x + 14$
10. (a) $C(t) = 40t + 50$; **(b)**

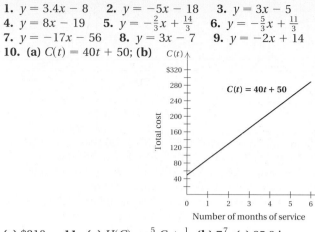

(c) $210   **11. (a)** $H(C) = \frac{5}{16}C + \frac{1}{8}$; **(b)** $7\frac{7}{8}$; **(c)** 25.2 in.

Calculator Corner, pp. 226–228

1. 5   **2. (a)** $y = 2.747311828x + 62.32258065$; **(b)** 90%

Exercise Set 2.6, p. 229

1. $y = -8x + 4$ **3.** $y = 2.3x - 1$ **5.** $f(x) = -\frac{7}{3}x - 5$
7. $f(x) = \frac{2}{3}x + \frac{5}{8}$ **9.** $y = 5x - 17$ **11.** $y = -3x + 33$
13. $y = x - 6$ **15.** $y = -2x + 16$ **17.** $y = -7$
19. $y = \frac{2}{3}x - \frac{8}{3}$ **21.** $y = \frac{1}{2}x + \frac{7}{2}$ **23.** $y = x$
25. $y = \frac{7}{4}x + 7$ **27.** $y = \frac{3}{2}x$ **29.** $y = \frac{1}{6}x$
31. $y = 13x - \frac{15}{4}$ **33.** $y = -\frac{1}{2}x + \frac{17}{2}$ **35.** $y = \frac{5}{7}x - \frac{17}{7}$
37. $y = \frac{1}{3}x + 4$ **39.** $y = \frac{1}{2}x + 4$ **41.** $y = \frac{4}{3}x - 6$
43. $y = \frac{5}{2}x + 9$ **45. (a)** $C(t) = 40t + 85$;
(b) ; **(c)** $345

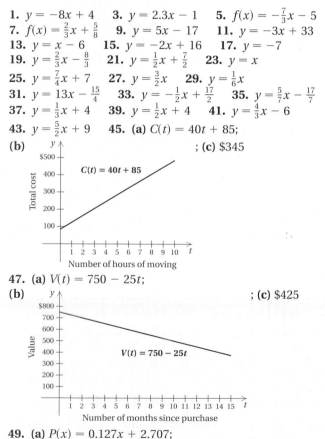

47. (a) $V(t) = 750 - 25t$;
(b) ; **(c)** $425

49. (a) $P(x) = 0.127x + 2.707$;
(b) 3.088 trillion prescriptions; 3.85 trillion prescriptions
51. (a) $R(t) = -0.075t + 46.8$; **(b)** 41.325 sec; 41.1 sec;
(c) 2021   **53. (a)** $M(t) = 0.236t + 71.8$; **(b)** about 75.8 yr
55. $\mathbf{D_W}$ **57.** $\{x \mid x > 24\}$, or $(24, \infty)$ **58.** $\{-27, 24\}$
59. $\{x \mid x \leq 24\}$, or $(-\infty, 24]$ **60.** $\{x \mid x \geq \frac{7}{3}\}$, or $\left[\frac{7}{3}, \infty\right)$

61. $\{x \mid -8 \leq x \leq 5\}$, or $[-8, 5]$ **62.** $\{-7, \frac{1}{3}\}$
63. $\{\ \}$, or $\varnothing$ **64.** $\{x \mid -\frac{15}{2} \leq x < 24\}$, or $\left[-\frac{15}{2}, 24\right)$
65. $1350   **66.** $30

Concept Reinforcement, p. 232

1. True   **2.** False   **3.** True   **4.** True   **5.** False
6. True   **7.** False   **8.** False

Summary and Review: Chapter 2, p. 233

1.

$$\frac{3x - y = 2}{3 \cdot 0 - (-2)\ ?\ 2}$$
$$0 + 2$$
$$2 \quad | \quad \text{TRUE}$$

$$\frac{3x - y = 2}{3(-1) - (-5)\ ?\ 2}$$
$$-3 + 5$$
$$2 \quad | \quad \text{TRUE}$$

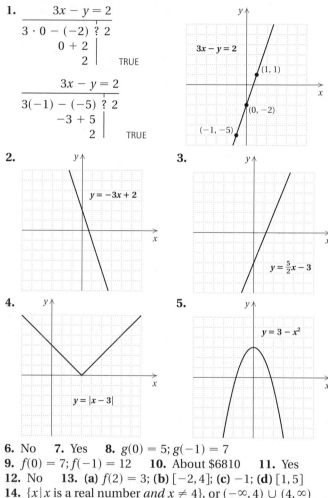

2.

3.

4.

5.

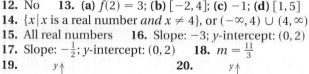

6. No   **7.** Yes   **8.** $g(0) = 5; g(-1) = 7$
9. $f(0) = 7; f(-1) = 12$ **10.** About $6810   **11.** Yes
12. No   **13. (a)** $f(2) = 3$; **(b)** $[-2, 4]$; **(c)** -1; **(d)** $[1, 5]$
14. $\{x \mid x$ is a real number *and* $x \neq 4\}$, or $(-\infty, 4) \cup (4, \infty)$
15. All real numbers   **16.** Slope: -3; y-intercept: $(0, 2)$
17. Slope: $-\frac{1}{2}$; y-intercept: $(0, 2)$ **18.** $m = \frac{11}{3}$
19.   **20.**

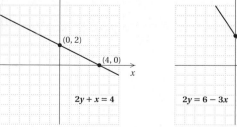

21.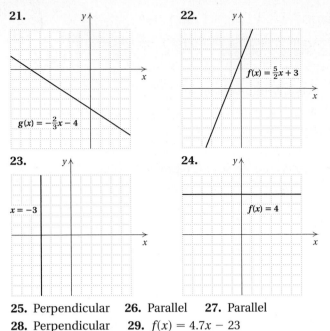

$g(x) = -\frac{2}{3}x - 4$

22. $f(x) = \frac{5}{2}x + 3$

23. $x = -3$

24. $f(x) = 4$

25. Perpendicular   **26.** Parallel   **27.** Parallel

28. Perpendicular   **29.** $f(x) = 4.7x - 23$

30. $y = -3x + 4$ **31.** $y = -\frac{3}{2}x$ **32.** $y = -\frac{5}{7}x + 9$

33. $y = \frac{1}{3}x + \frac{1}{3}$ **34. (a)** $R(x) = -0.064x + 46.8$;
(b) about 44.37 sec; 44.24 sec   **35. DW** The concept of slope is useful in describing how a line slants. A line with positive slope slants up from left to right. A line with negative slope slants down from left to right. The larger the absolute value of the slope, the steeper the slant.
36. DW The notation $f(x)$ can be read "f of x" or "f at x" or "the value of f at x." It represents the output of the function f for the input x. The notation $f(a) = b$ provides a concise way to indicate that for the input a, the output of the function f is b.   **37.** $f(x) = 3.09x + 3.75$

Test: Chapter 2, p. 236

1. [2.1b] Yes   **2.** [2.1b] No
3. [2.1c]

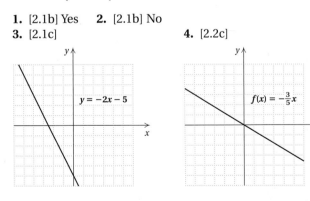

$y = -2x - 5$

4. [2.2c]

$f(x) = -\frac{3}{5}x$

5. [2.2c]

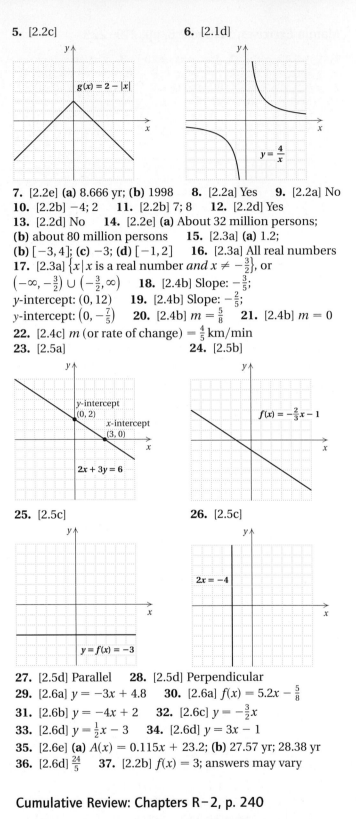

$g(x) = 2 - |x|$

6. [2.1d]

$y = \frac{4}{x}$

7. [2.2e] **(a)** 8.666 yr; **(b)** 1998   **8.** [2.2a] Yes   **9.** [2.2a] No
10. [2.2b] -4; 2   **11.** [2.2b] 7; 8   **12.** [2.2d] Yes
13. [2.2d] No   **14.** [2.2e] **(a)** About 32 million persons;
(b) about 80 million persons   **15.** [2.3a] **(a)** 1.2;
(b) $[-3, 4]$; **(c)** -3; **(d)** $[-1, 2]$ **16.** [2.3a] All real numbers
17. [2.3a] $\{x \mid x \text{ is a real number } and \ x \neq -\frac{3}{2}\}$, or
$\left(-\infty, -\frac{3}{2}\right) \cup \left(-\frac{3}{2}, \infty\right)$ **18.** [2.4b] Slope: $-\frac{3}{5}$;
y-intercept: $(0, 12)$ **19.** [2.4b] Slope: $-\frac{2}{5}$;
y-intercept: $\left(0, -\frac{7}{5}\right)$ **20.** [2.4b] $m = \frac{5}{8}$ **21.** [2.4b] $m = 0$
22. [2.4c] m (or rate of change) $= \frac{4}{5}$ km/min
23. [2.5a]   **24.** [2.5b]

y-intercept $(0, 2)$; x-intercept $(3, 0)$; $2x + 3y = 6$

$f(x) = -\frac{2}{3}x - 1$

25. [2.5c]   **26.** [2.5c]

$y = f(x) = -3$

$2x = -4$

27. [2.5d] Parallel   **28.** [2.5d] Perpendicular
29. [2.6a] $y = -3x + 4.8$ **30.** [2.6a] $f(x) = 5.2x - \frac{5}{8}$
31. [2.6b] $y = -4x + 2$ **32.** [2.6c] $y = -\frac{3}{2}x$
33. [2.6d] $y = \frac{1}{2}x - 3$ **34.** [2.6d] $y = 3x - 1$
35. [2.6e] **(a)** $A(x) = 0.115x + 23.2$; **(b)** 27.57 yr; 28.38 yr
36. [2.6d] $\frac{24}{5}$ **37.** [2.2b] $f(x) = 3$; answers may vary

Cumulative Review: Chapters R–2, p. 240

1. [2.6e] **(a)** $R(x) = -0.006x + 3.85$; **(b)** 3.50 min; 3.49 min
2. [2.3a] **(a)** 6; **(b)** $[0, 30]$; **(c)** 25; **(d)** $[0, 15]$ **3.** [R.3a] 31
4. [R.1d] 2.2   **5.** [R.1d] $\frac{2}{3}$ **6.** [R.2c] $\frac{1}{2}$ **7.** [R.2d] 36.48
8. [R.6b] $-10x + 32$ **9.** [R.7a] $\dfrac{32x^8}{y}$ **10.** [R.7a] $-\dfrac{9}{x^2 y^2}$
11. [R.6b] $23x + 31$ **12.** [R.3c] -328 **13.** [R.3c] -1

14. [1.1b] -22 **15.** [1.1d] $\frac{15}{88}$ **16.** [1.1c] 20

17. [1.1d] $-\frac{21}{4}$ **18.** [1.1d] -5 **19.** [1.2a] $x = \dfrac{W - By}{A}$

20. [1.2a] $A = \dfrac{M}{1 + 4B}$ **21.** [1.4c] $\{y \mid y \le 7\}$, or $(-\infty, 7]$

22. [1.4c] $\left\{x \mid x < -\frac{3}{2}\right\}$, or $\left(-\infty, -\frac{3}{2}\right)$

23. [1.4c] $\left\{x \mid x > -\frac{1}{11}\right\}$, or $\left(-\frac{1}{11}, \infty\right)$ **24.** [1.5b] All real numbers **25.** [1.5a] $\{x \mid -7 < x \le 4\}$, or $(-7, 4]$

26. [1.5a] $\left\{x \mid -2 \le x \le \frac{3}{2}\right\}$, or $\left[-2, \frac{3}{2}\right]$ **27.** [1.6c] $\{-8, 8\}$

28. [1.6e] $\{y \mid y < -4 \text{ or } y > 4\}$, or $(-\infty, -4) \cup (4, \infty)$

29. [1.6e] $\left\{x \mid -\frac{3}{2} \le x \le 2\right\}$, or $\left[-\frac{3}{2}, 2\right]$

30. [2.6d] $y = -4x - 22$ **31.** [2.6d] $y = \frac{1}{4}x - 5$

32. [2.1c] **33.** [2.5a]

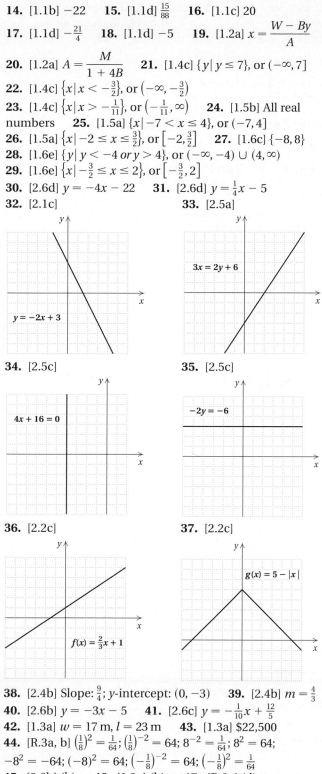

34. [2.5c] **35.** [2.5c]

36. [2.2c] **37.** [2.2c]

38. [2.4b] Slope: $\frac{9}{4}$; y-intercept: $(0, -3)$ **39.** [2.4b] $m = \frac{4}{3}$
40. [2.6b] $y = -3x - 5$ **41.** [2.6c] $y = -\frac{1}{10}x + \frac{12}{5}$
42. [1.3a] $w = 17\,\text{m},\, l = 23\,\text{m}$ **43.** [1.3a] \$22,500
44. [R.3a, b] $\left(\frac{1}{8}\right)^2 = \frac{1}{64}$; $\left(\frac{1}{8}\right)^{-2} = 64$; $8^{-2} = \frac{1}{64}$; $8^2 = 64$;
$-8^2 = -64$; $(-8)^2 = 64$; $\left(-\frac{1}{8}\right)^{-2} = 64$; $\left(-\frac{1}{8}\right)^2 = \frac{1}{64}$
45. [2.6b] (b) **46.** [1.3a] (b) **47.** [R.3a] (d)
48. [R.6b] (e) **49.** [2.6e] \$151,000 **50.** [2.5d] (1), (4)
51. [1.5a] $\{x \mid 6 < x \le 10\}$, or $(6, 10]$

CHAPTER 3

Margin Exercises, Section 3.1, pp. 245–249

1. $(0, 1)$ **2.** $(2, 1)$ **3.** No solution **4.** Consistent: 1, 2; inconsistent: 3 **5.** Infinitely many solutions
6. Independent: 1, 2, 3; dependent: 5
7. (a) -3; (b) , -3; (c) the same, -3

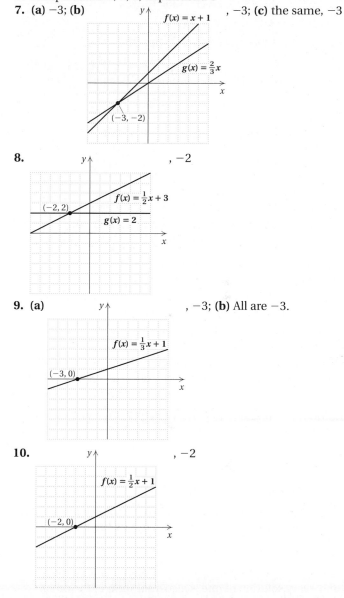

8. , -2

9. (a) , -3; (b) All are -3.

10. , -2

Calculator Corner, p. 246

1. $(2, 3)$ **2.** $(-4, -1)$ **3.** $(-1, 5)$ **4.** $(3, -1)$

Exercise Set 3.1, p. 250

1. $(3, 1)$; consistent; independent **3.** $(1, -2)$; consistent; independent **5.** $(4, -2)$; consistent; independent
7. $(2, 1)$; consistent; independent **9.** $\left(\frac{5}{2}, -2\right)$; consistent; independent **11.** $(3, -2)$; consistent; independent

13. No solution; inconsistent; independent   **15.** Infinitely many solutions; consistent; dependent   **17.** $(4, -5)$; consistent; independent   **19.** $(2, -3)$; consistent; independent   **21.** Consistent; independent; F   **23.** Consistent; dependent; B   **25.** Inconsistent; independent; D   **27.** $\mathbf{D_W}$ **29.** -3 **30.** -20 **31.** $\frac{9}{20}$ **32.** -38 **33.** $(2.23, 1.14)$ **35.** $(3, 3), (-5, 5)$

Margin Exercises, Section 3.2, pp. 254–256

1. $(2, 4)$ **2.** $(5, 2)$ **3.** $(-3, 2)$ **4.** $(13, 16)$ **5. (a)** No solution; **(b)** the same, no solution   **6.** Length: 160 ft; width: 100 ft

Calculator Corner, p. 254

Left to the student

Exercise Set 3.2, p. 257

1. $(2, -3)$ **3.** $\left(\frac{21}{5}, \frac{12}{5}\right)$ **5.** $(2, -2)$ **7.** $(-2, -6)$ **9.** $(-2, 1)$ **11.** $\left(\frac{1}{2}, \frac{1}{2}\right)$ **13.** $\left(\frac{19}{8}, \frac{1}{8}\right)$ **15.** No solution   **17.** Length: 40 ft; width: 20 ft   **19.** 48° and 132°   **21.** Wins: 23; ties: 14   **23.** $\mathbf{D_W}$ **25.** 1.3   **26.** $-15y - 39$ **27.** $p = \dfrac{7A}{q}$ **28.** $\frac{7}{3}$ **29.** -23 **30.** $\frac{29}{22}$ **31.** $m = -\frac{1}{2}$; $b = \frac{5}{2}$ **33.** Length: 57.6 in.; width: 20.4 in.

Margin Exercises, Section 3.3, pp. 260–265

1. $(1, 4)$ **2.** $\left(\frac{1}{3}, \frac{1}{2}\right)$ **3.** $(2, 3)$ **4.** $(-127, 100)$ **5.** $(-138, -118)$ **6.** No solution   **7.** Infinitely many solutions   **8.** $2x + 3y = 1, 3x - y = 7$; $(2, -1)$ **9.** $9x + 10y = 5, 9x - 4y = 3$; $\left(\frac{25}{63}, \frac{1}{7}\right)$ **10. (a)** Length: 160 ft; width: 100 ft; **(b)** the solutions are the same.

Calculator Corner, p. 263

1. We get $y = -3x + 5$ and $y = -3x - 2$. Since the graphs are parallel lines, the system of equations has no solution.

2. Each equation is equivalent to $y = \dfrac{2x + 6}{3}$, or $y = \frac{2}{3}x + 2$. Since the equations are equivalent, the graphs are the same and every solution of one is also a solution of the other. Thus we have an infinite number of solutions.

Exercise Set 3.3, p. 266

1. $(1, 2)$ **3.** $(-1, 3)$ **5.** $\left(\frac{128}{31}, -\frac{17}{31}\right)$ **7.** $(6, 2)$ **9.** $\left(\frac{140}{13}, -\frac{50}{13}\right)$ **11.** $(4, 6)$ **13.** $\left(\frac{110}{19}, -\frac{12}{19}\right)$ **15.** Infinitely many solutions   **17.** No solution   **19.** $\left(\frac{1}{2}, -\frac{1}{2}\right)$ **21.** $\left(-\frac{4}{3}, -\frac{19}{3}\right)$ **23.** $\left(\frac{1000}{11}, -\frac{1000}{11}\right)$ **25.** $(140, 60)$ **27.** Length: 110 m; width: 60 m   **29.** 14° and 76°   **31.** Coach-class: 131; first-class: 21   **33.** $\mathbf{D_W}$ **35.** 1   **36.** 5   **37.** 3   **38.** 291   **39.** 15   **40.** $12a^2 - 2a + 1$

41. 53   **42.** 8.92   **43.** $\{x \mid x$ is a real number *and* $x \neq -7\}$, or $(-\infty, -7) \cup (-7, \infty)$ **44.** Domain: all real numbers; range: $\{y \mid y \leq 5\}$, or $(-\infty, 5]$ **45.** $y = -\frac{3}{5}x - 7$ **46.** $\dfrac{a^3}{b}$ **47.** $(23.12, -12.04)$ **49.** $A = 2, B = 4$ **51.** $p = 2, q = -\frac{1}{3}$

Margin Exercises, Section 3.4, pp. 270–277

1. White: 22; red: 8

WHITE	RED	TOTAL
w	r	30	→ $w + r = 30$
\$18.95	\$19.50	
$18.95w$	$19.50r$	572.90	→ $18.95w + 19.50r = 572.90$

2. Kenyan: 8 lb; Sumatran: 12 lb
3. \$1800 at 7%; \$1900 at 9%

FIRST INVESTMENT	SECOND INVESTMENT	TOTAL
x	y	\$3700	→ $x + y = 3700$
7%	9%	
1 yr	1 yr	
$0.07x$	$0.09y$	\$297	→ $0.07x + 0.09y = 297$

4. Attack: 15 qt; Blast: 45 qt

ATTACK	BLAST	MIXTURE
a	b	60	→ $a + b = 60$
2%	6%	5%
$0.02a$	$0.06b$	0.05 × 60, or 3	→ $0.02a + 0.06b = 3$

5. 280 km

DISTANCE	RATE	TIME
d	35 km/h	t	→ $d = 35t$
d	40 km/h	$t - 1$	→ $d = 40(t - 1)$

6. 180 mph

DISTANCE	RATE	TIME
d	$r + 20$	4 hr	→ $d = 4(r + 20)$
d	$r - 20$	5 hr	→ $d = 5(r - 20)$

Translating for Success, p. 278

1. G **2.** E **3.** D **4.** A **5.** J **6.** B **7.** C **8.** I
9. F **10.** H

Exercise Set 3.4, p. 279

1. 32 brushes at $8.50; 13 brushes at $9.75
3. Humulin: 21 vials; Novolin Velosulin: 29 vials
5. 30-sec: 4; 60-sec: 8 **7.** 5 lb of each **9.** 25% acid: 4 L;
50% acid: 6 L **11.** 10 silk neckties **13.** $7500 at 6%;
$4500 at 9% **15.** Whole milk: $169\frac{3}{13}$ lb; cream: $30\frac{10}{13}$ lb
17. $5 bills: 7; $1 bills: 15 **19.** $7400 at 5.5%; $10,600
at 4% **21.** 375 mi **23.** 14 km/h **25.** 144 mi
27. 2 hr **29.** $1\frac{1}{3}$ hr **31.** About 1489 mi **33.** $\mathbf{D_W}$
35. -7 **36.** -11 **37.** -3 **38.** 33 **39.** -15
40. $8a - 7$ **41.** -23 **42.** 0.2 **43.** -4 **44.** -17
45. $-12h - 7$ **46.** 3993 **47.** $4\frac{4}{7}$ L **49.** City: 261 mi;
highway: 204 mi **51.** Brown: 0.8 gal; neutral: 0.2 gal

Margin Exercises, Section 3.5, pp. 284–287

1. $(2, 1, -1)$ **2.** $\left(2, -2, \frac{1}{2}\right)$ **3.** $(20, 30, 50)$

Exercise Set 3.5, p. 288

1. $(1, 2, -1)$ **3.** $(2, 0, 1)$ **5.** $(3, 1, 2)$ **7.** $(-3, -4, 2)$
9. $(2, 4, 1)$ **11.** $(-3, 0, 4)$ **13.** $(2, 2, 4)$ **15.** $\left(\frac{1}{2}, 4, -6\right)$
17. $\left(\frac{1}{2}, \frac{1}{3}, \frac{1}{6}\right)$ **19.** $\left(\frac{1}{2}, \frac{2}{3}, -\frac{5}{6}\right)$ **21.** $(15, 33, 9)$
23. $(4, 1, -2)$ **25.** $\mathbf{D_W}$ **27.** $a = \dfrac{F}{3b}$ **28.** $a = \dfrac{Q - 4b}{4}$,
or $\dfrac{Q}{4} - b$ **29.** $c = \dfrac{2F + td}{t}$, or $\dfrac{2F}{t} + d$ **30.** $d = \dfrac{tc - 2F}{t}$,
or $c - \dfrac{2F}{t}$ **31.** $y = \dfrac{Ax - c}{B}$ **32.** $y = \dfrac{c - Ax}{B}$ **33.** Slope:
$-\frac{2}{3}$; y-intercept: $\left(0, -\frac{5}{4}\right)$ **34.** Slope: -4; y-intercept: $(0, 5)$
35. Slope: $\frac{2}{5}$; y-intercept: $(0, -2)$ **36.** Slope: 1.09375;
y-intercept: $(0, -3.125)$ **37.** $(1, -2, 4, -1)$

Margin Exercises, Section 3.6, pp. 290–292

1. 64°, 32°, 84° **2.** $4000 at 5%; $5000 at 6%; $16,000
at 7%

Exercise Set 3.6, p. 293

1. 20-oz: 11; 32-oz: 15; 40-oz: 8 **3.** 32°, 96°, 52°
5. $-7, 20, 42$ **7.** Automatic transmission: $865; power
door locks: $520; air conditioning: $375 **9.** A: 1500;
B: 1900; C: 2300 **11.** First fund: $45,000; second fund:
$10,000; third fund: $25,000 **13.** Asian–American: 385;
African–American: 200; Caucasian: 154 **15.** Roast beef: 2;
baked potato: 1; broccoli: 2 **17.** Par-3: 6; par-4: 8; par-5: 4
19. Two-pointers: 32; three-pointers: 5; foul shots: 13
21. $\mathbf{D_W}$ **23.** $\mathbf{D_W}$ **25.** At most **26.** Empty set
27. Linear **28.** Negative **29.** Consistent
30. Perpendicular **31.** y-intercept **32.** Horizontal
33. 180° **35.** 464

Margin Exercises, Section 3.7, pp. 298–306

1. No **2.** Yes **3.**

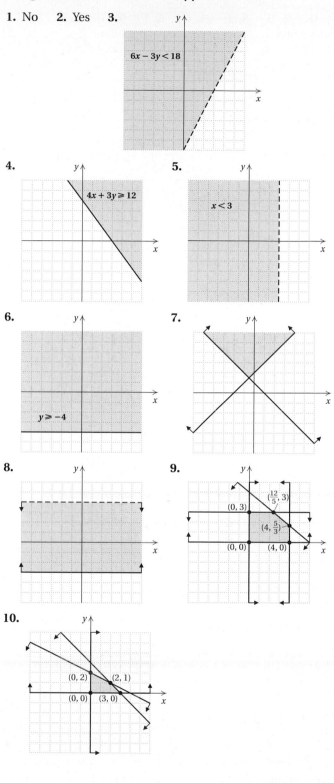

Calculator Corner, p. 307

1. Left to the student **2.** Left to the student

A-19

Chapter 3

Visualizing for Success, p. 308

1. D   **2.** B   **3.** E   **4.** C   **5.** I   **6.** G   **7.** F   **8.** H
9. A   **10.** J

Exercise Set 3.7, p. 309

1. Yes   **3.** Yes   **5.**
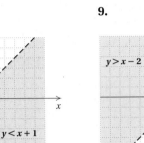

7.
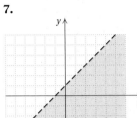

9.
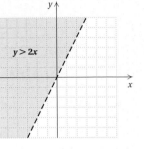

11.

13.
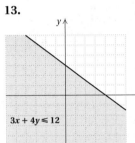

15.
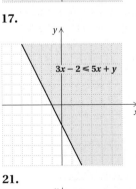

17.
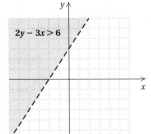

19.

21.
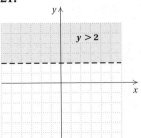

23.
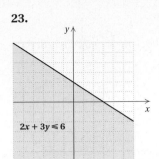

25. F
27. B
29. C

31.
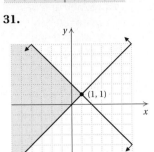

33.
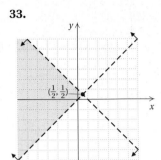

35.
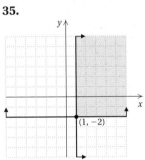

37.
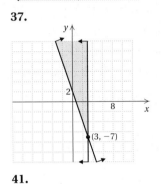

39.
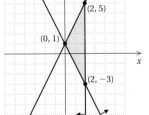

41.

43.
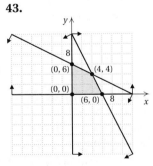

45.
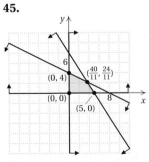

47. DW   **49.** $\frac{10}{17}$ **50.** $-\frac{14}{13}$ **51.** -2 **52.** $\frac{29}{11}$ **53.** -12
54. $\frac{333}{245}$ **55.** 2   **56.** 3   **57.** 1   **58.** 8   **59.** 4
60. $|2 - 2a|$, or $2|1 - a|$ **61.** 6   **62.** 0.2

A-20

Answers

63. $w > 0,$
$h > 0,$
$w + h + 30 \le 62,$ or
$w + h \le 32,$
$2w + 2h + 30 \le 130,$ or
$w + h \le 50$

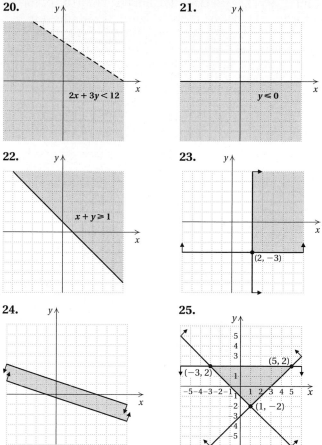

65. Left to the student   **67.** Left to the student
69. Left to the student

Margin Exercises, Section 3.8, pp. 315–317

1. (a) $C(x) = 80,000 + 20x;$ **(b)** $R(x) = 36x;$
(c) $P(x) = 16x - 80,000;$ **(d)** a loss of $16,000; a profit of
$176,000;
(e)

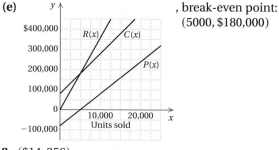

, break-even point:
(5000, $180,000)

2. ($14, 356)

Exercise Set 3.8, p. 318

1. (a) $P(x) = 45x - 270,000;$ **(b)** (6000, $420,000)
3. (a) $P(x) = 50x - 120,000;$ **(b)** (2400, $144,000)
5. (a) $P(x) = 80x - 10,000;$ **(b)** (125, $12,500)
7. (a) $P(x) = 18x - 16,000;$ **(b)** (889, $35,560)
9. (a) $P(x) = 75x - 195,000;$ **(b)** (2600, $325,000)
11. (a) $C(x) = 22,500 + 40x;$ **(b)** $R(x) = 85x;$
(c) $P(x) = 45x - 22,500;$ **(d)** $112,500 profit; $4500 loss;
(e) (500, $42,500)   **13. (a)** $C(x) = 16,404 + 6x;$
(b) $R(x) = 18x;$ **(c)** $P(x) = 12x - 16,404;$ **(d)** $19,596 profit;
$4404 loss; **(e)** (1367, $24,606)   **15.** ($70, 300)
17. ($22, 474)   **19.** ($50, 6250)   **21.** ($10, 1070)
23. ^{D}W   **25.** Slope: $\frac{3}{5}$; y-intercept: $\left(0, \frac{8}{5}\right)$ **26.** Slope: $-\frac{6}{7}$;
y-intercept: $\left(0, \frac{13}{7}\right)$ **27.** Slope: 1.7; y-intercept: (0, 49)
28. Slope: $-\frac{4}{3}$; y-intercept: (0, 4)

Concept Reinforcement, p. 320

1. True   **2.** False   **3.** False   **4.** True   **5.** True

Summary and Review: Chapter 3, p. 320

1. $(-2, 1)$; consistent; independent   **2.** Infinitely many
solutions; consistent; dependent   **3.** No solution;
inconsistent; independent   **4.** $\left(\frac{2}{5}, -\frac{4}{5}\right)$ **5.** No solution
6. $\left(-\frac{11}{15}, -\frac{43}{30}\right)$ **7.** $\left(\frac{37}{19}, \frac{53}{19}\right)$ **8.** $\left(\frac{76}{17}, -\frac{2}{119}\right)$ **9.** $(2, 2)$
10. Infinitely many solutions   **11.** CD: $14; cassette: $9
12. 5 L of each   **13.** $5\frac{1}{2}$ hr   **14.** $(10, 4, -8)$

15. $\left(-\frac{7}{3}, \frac{125}{27}, \frac{20}{27}\right)$ **16.** $(2, 0, 4)$ **17.** $\left(2, \frac{1}{3}, -\frac{2}{3}\right)$
18. $90°, 67\frac{1}{2}°, 22\frac{1}{2}°$ **19.** $20 bills: 5; $5 bills: 15; $1 bills: 19
20.

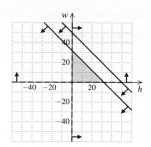

$2x + 3y < 12$

21.

$y \le 0$

22.

$x + y \ge 1$

23.

$(2, -3)$

24.

25.

$(-3, 2)$ $(5, 2)$
$(1, -2)$

26. (a) $C(x) = 35,000 + 175x;$ **(b)** $R(x) = 300x;$
(c) $P(x) = 125x - 35,000;$ **(d)** $115,000 profit; $10,000 loss;
(e) (280, $84,000)   **27.** ($3, 81)   **28.** ^{D}W The comparison
is summarized in the table in Section 3.3.   **29.** ^{D}W Many
problems that deal with more than one unknown quantity
are easier to translate to a system of equations than to a
single equation. Problems involving complementary or
supplementary angles, the dimensions of a geometric
figure, mixtures, and the measures of the angles of a triangle
are examples.
30. $d + q = 20,$
$25d + 10q = 90 + 10d + 25q;$
or $d + q = 20$
$d - q = 6;$
dimes: 13; quarters: 7
31. $(0, 2)$ and $(1, 3)$
32. (a) 165.91 cm; **(b)** 160.60 cm; **(c)** (120.857, 308.521);
(d) 120.857 cm

Test: Chapter 3, p. 323

1. [3.1a] $(-2, 1)$; consistent; independent   **2.** [3.1a] No
solution; inconsistent; independent   **3.** [3.1a] Infinitely
many solutions; consistent; dependent   **4.** [3.2a] $\left(3, -\frac{11}{3}\right)$

5. [3.2a] $\left(\frac{15}{7}, -\frac{18}{7}\right)$ **6.** [3.3a] $\left(-\frac{3}{2}, -\frac{3}{2}\right)$ **7.** [3.3a] No
solution   **8.** [3.4a] 34% solution: $48\frac{8}{9}$ mL; 61% solution:
$71\frac{1}{9}$ mL   **9.** [3.4a] Buckets: 17; dinners: 11   **10.** [3.4b]
120 km/h   **11.** [3.2b], [3.3b] Length: 93 ft; width: 51 ft
12. [3.5a] $\left(2, -\frac{1}{2}, -1\right)$ **13.** [3.8b] ($3, 55)   **14.** [3.6a]
3.5 hr   **15.** [3.8a] **(a)** $C(x) = 40,000 + 30x$; **(b)** $R(x) = 80x$;
(c) $P(x) = 50x - 40,000$; **(d)** $20,000 profit; $30,000 loss;
(e) (800, $64,000)   **16.** [3.7b]

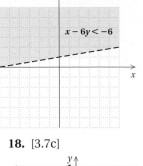

$x - 6y < -6$

17. [3.7c]   **18.** [3.7c]

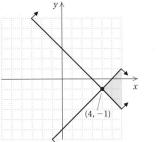

$(4, -1)$

$(-2, 0)$

$\left(-\frac{1}{2}, -\frac{9}{4}\right)$

19. [3.2a], [3.3a] $m = 7$; $b = 10$

CHAPTER 4

Margin Exercises, Section 4.1, pp. 328–332

1. $-92x^5, -8x^4, x^2, 5; -92x^5$ **2.** $5, -4, -2, 1, -1, -5; 5$
3. (a) $6x^2, -5x^3, 2x, -7; 2, 3, 1, 0; 3; -5x^3; -5; -7$;
(b) $2y, -4, -5x, 7x^2y^3z^2, 5xy^2; 1, 0, 1, 7, 3; 7; 7x^2y^3z^2; 7; -4$
4. Monomials: (c), (d), (e); binomials: (a), (f), (h);
trinomials: (b), (g)   **5. (a)** $5 + 10x - 6x^2 + 7x^3 - x^4$;
(b) $-x^4 + 7x^3 - 6x^2 + 10x + 5$
6. (a) $-5 + x^3 + 5x^4y - 3y^2 + 3x^2y^3$;
(b) $3x^2y^3 - 3y^2 + 5x^4y + x^3 - 5$ **7.** $5; 13; 13$
8. (a) $C(3) = 4.326$ mcg/mL; **(b)** $C(9) \approx 2$ mcg/mL
9. $3y - 4x + 4xy^2$ **10.** $8xy^3 + 2x^3y + 3x + 7 - 9x^2y$
11. $-4x^3 + 2x^2 - 4x + 2$ **12.** $10y^5 - 4y^2 + 5$
13. $5p^2q^4 - 8p^2q^2 + 5$ **14.** $-\left(4x^3 - 5x^2 + \frac{1}{4}x - 10\right)$;
$-4x^3 + 5x^2 - \frac{1}{4}x + 10$ **15.** $-\left(8xy^2 - 4x^3y^2 - 9x - \frac{1}{5}\right)$;
$-8xy^2 + 4x^3y^2 + 9x + \frac{1}{5}$
16. $-\left(-9y^5 - 8y^4 + \frac{1}{2}y^3 - y^2 + y - 1\right)$;
$9y^5 + 8y^4 - \frac{1}{2}y^3 + y^2 - y + 1$ **17.** $3x^2 + 5$
18. $14y^3 - 2y + 4$ **19.** $p^2 - 6p - 2$
20. $3y^5 - 3y^4 + 5y^3 - 2y^2 - 3$
21. $9p^4q - 10p^3q^2 + 4p^2q^3 + 9q^4$
22. $y^3 - y^2 + \frac{4}{3}y + 0.1$

1. Correct   **2.** Incorrect   **3.** Correct   **4.** Correct
5. Incorrect   **6.** Incorrect

Exercise Set 4.1, p. 333

1. $-9x^4, -x^3, 7x^2, 6x, -8; 4, 3, 2, 1, 0; 4; -9x^4; -9; -8$
3. $t^3, 4t^7, s^2t^4, -2; 3, 7, 6, 0; 7; 4t^7; 4; -2$
5. $u^7, 8u^2v^6, 3uv, 4u, -1; 7, 8, 2, 1, 0; 8; 8u^2v^6; 8; -1$
7. $-4y^3 - 6y^2 + 7y + 23$ **9.** $-xy^3 + x^2y^2 + x^3y + 1$
11. $-9b^5y^5 - 8b^2y^3 + 2by$ **13.** $5 + 12x - 4x^3 + 8x^5$
15. $3xy^3 + x^2y^2 - 9x^3y + 2x^4$
17. $-7ab + 4ax - 7ax^2 + 4x^6$ **19.** $45; 21; 5$
21. $-168; -9; 4; -7\frac{7}{8}$ **23.** $204; 140; 91; 55$
25. (a) About 340 mg; **(b)** about 190 mg; **(c)** $M(5) \approx 65$;
(d) $M(3) \approx 300$ **27. (a)** $18,750; **(b)** $24,000
29. $P(x) = -x^2 + 280x - 7000$ **31.** 17   **33.** 8
35. $2x^2$ **37.** $3x + 4y$ **39.** $7a + 14$ **41.** $-6a^2b - 2b^2$
43. $9x^2 + 2xy + 15y^2$ **45.** $-x^2y + 4y + 9xy^2$
47. $5x^2 + 2y^2 + 5$ **49.** $6a + b + c$
51. $-4a^2 - b^2 + 3c^2$ **53.** $-3x^2 + 2x + xy - 1$
55. $5x^2y - 4xy^2 + 5xy$ **57.** $9r^2 + 9r - 9$
59. $-\frac{2}{15}xy + \frac{19}{12}xy^2 + 1.7x^2y$ **61.** $-(5x^3 - 7x^2 + 3x - 6)$;
$-5x^3 + 7x^2 - 3x + 6$ **63.** $-(-13y^2 + 6ay^4 - 5by^2)$;
$13y^2 - 6ay^4 + 5by^2$ **65.** $11x - 7$ **67.** $-4x^2 - 3x + 13$
69. $2a - 4b + 3c$ **71.** $-2x^2 + 6x$
73. $-4a^2 + 8ab - 5b^2$ **75.** $16ab + 8a^2b + 3ab^2$
77. $0.06y^4 + 0.032y^3 - 0.94y^2 + 0.93$ **79.** $x^4 - x^2 - 1$
81. **D$_W$**
83.   **84.**

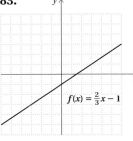

$f(x) = \frac{2}{3}x - 1$

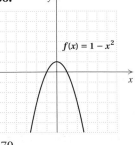

$g(x) = |x| - 1$

85.   **86.**

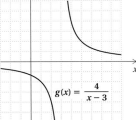

$g(x) = \dfrac{4}{x - 3}$

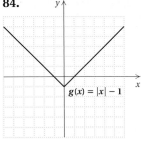

$f(x) = 1 - x^2$

87. $3y - 6$ **88.** $-10x - 20y + 70$
89. $-42p + 28q + 140$ **90.** $8w - 6t + 20$

91.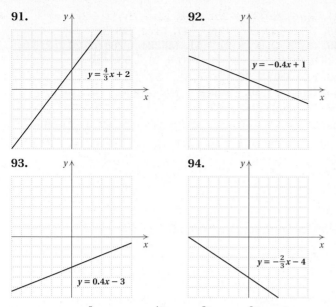
$y = \frac{4}{3}x + 2$

92.
$y = -0.4x + 1$

93.

94.
$y = -\frac{2}{3}x - 4$

$y = 0.4x - 3$

95. 494.55 cm^3 **97.** $47x^{4a} + 40x^{3a} + 30x^{2a} + x^a + 4$
99. Left to the student

Margin Exercises, Section 4.2, pp. 338–344

1. $-18y^3$ **2.** $24x^8y^3$ **3.** $-90x^4y^9z^{12}$ **4.** $-6y^2 - 18y$
5. $8xy^3 - 10xy$ **6.** $5x^3 + 15x^2 - 4x - 12$
7. $6y^2 + y - 12$ **8.** $p^4 + p^3 - 12p^2 - 5p + 15$
9. $2x^4 - 8x^3 + 4x^2 - 21x + 20$
10. $8x^5 + 2x^4 - 35x^3 + 34x^2 - 15x + 6$
11. $8x^5 + 12x^4 - 20x^3 + 4x^2 - 15x + 6$
12. $a^5 - 2a^4b + 3a^3b + a^3b^2 - 7a^2b^2 + 5ab^3 - b^4$
13. $y^2 + 6y - 40$ **14.** $2p^2 + 7pq - 15q^2$
15. $x^3y^3 + x^2y^3 + 2x^2y^2 + 2xy^2$ **16.** $a^2 - 2ab + b^2$
17. $x^2 + 16x + 64$ **18.** $9x^2 - 42x + 49$
19. $m^6 + \frac{1}{2}m^3n + \frac{1}{16}n^2$ **20.** $x^2 - 64$ **21.** $16y^2 - 49$
22. $7.84a^2 - 16.81b^2$ **23.** $9w^2 - \frac{9}{25}q^4$
24. $49x^4y^2 - 4y^2$ **25.** $4x^2 + 12x + 9 - 25y^2$
26. $81x^4 - 16y^4$ **27.** $f(a + 1) = a^2 + 4a - 4;$
$f(a + h) - f(a) = 2ah + h^2 + 2h$

Calculator Corner, p. 341

1. Correct   **2.** Incorrect   **3.** Correct   **4.** Incorrect
5. Incorrect   **6.** Correct

Exercise Set 4.2, p. 345

1. $24y^3$ **3.** $-20x^3y$ **5.** $-10x^6y^7$ **7.** $14z - 2zx$
9. $6a^2b + 6ab^2$ **11.** $15c^3d^2 - 25c^2d^3$ **13.** $15x^2 + x - 2$
15. $s^2 - 9t^2$ **17.** $x^2 - 2xy + y^2$ **19.** $x^6 + 3x^3 - 40$
21. $a^4 - 5a^2b^2 + 6b^4$ **23.** $x^3 - 64$ **25.** $x^3 + y^3$
27. $a^4 + 5a^3 - 2a^2 - 9a + 5$
29. $4a^3b^2 + 4a^3b - 10a^2b^2 - 2a^2b + 3ab^3 + 7ab^2 - 6b^3$
31. $x^2 + \frac{1}{2}x + \frac{1}{16}$ **33.** $\frac{1}{8}x^2 - \frac{2}{9}$
35. $3.25x^2 - 0.9xy - 28y^2$ **37.** $a^2 + 13a + 40$
39. $y^2 + 3y - 28$ **41.** $9a^2 + 3a + \frac{1}{4}$
43. $x^2 - 4xy + 4y^2$ **45.** $b^2 - \frac{5}{6}b + \frac{1}{6}$
47. $2x^2 + 13x + 18$ **49.** $400a^2 - 6.4ab + 0.0256b^2$

51. $4x^2 - 4xy - 3y^2$ **53.** $x^6 + 4x^3 + 4$
55. $4x^4 - 12x^2y^2 + 9y^4$ **57.** $a^6b^4 + 2a^3b^2 + 1$
59. $0.01a^4 - a^2b + 25b^2$ **61.** $A = P + 2Pi + Pi^2$
63. $d^2 - 64$ **65.** $4c^2 - 9$ **67.** $36m^2 - 25n^2$
69. $x^4 - y^2z^2$ **71.** $m^4 - m^2n^2$ **73.** $16p^4 - 9p^2q^2$
75. $\frac{1}{4}p^2 - \frac{4}{9}q^2$ **77.** $x^4 - 1$ **79.** $a^4 - 2a^2b^2 + b^4$
81. $a^2 + 2ab + b^2 - 1$ **83.** $4x^2 + 12xy + 9y^2 - 16$
85. $t^2 + 3t - 4, p^2 + 7p + 6, h^2 + 2ah + 5h,$
$t^2 + t + c - 6, a^2 + 5a + 5$ **87.** $3t^2 - 13t + 18,$
$3p^2 - p + 4, 3h^2 + 6ah - 7h, 3t^2 - 19t + 34 + c,$
$3a^2 - 7a + 13$ **89.** $-t^2 + 7t - 6, -p^2 + 3p + 4,$
$-h^2 - 2ah + 5h, -t^2 + 9t - 14 + c, -a^2 + 5a + 5$
91. $-t^2 + 5t, -p^2 + p + 6, -h^2 - 2ah + 3h,$
$-t^2 + 7t - 6 + c, -a^2 + 3a + 9$ **93.** $\mathbf{D_W}$
95. 5.5 hr **96.** 180 mph **97.** $\left(\frac{4}{3}, -\frac{14}{27}\right)$ **98.** $(1, 3)$
99. Infinitely many solutions   **100.** $\left(\frac{10}{21}, \frac{11}{14}\right)$
101. Left to the student   **103.** z^{5n5}
105. $r^8 - 2r^4s^4 + s^8$ **107.** $9x^{10} - \frac{30}{11}x^5 + \frac{25}{121}$
109. $x^{4a} - y^{4b}$ **111.** $x^6 - 1$

Margin Exercises, Section 4.3, pp. 349–352

1. (a) $x - 5$ and $x + 1$; (b) $x^2, -4x,$ and -5 **2.** $3(x^2 - 2)$
3. $4x^3(x^2 - 2)$ **4.** $3y^2(3y^2 - 5y + 1)$
5. $3x^2y(2 - 7xy + y^2)$ **6.** $-8(x - 4)$
7. $-3(x^2 + 5x - 3)$ **8.** (a) $h(t) = -16t(t - 6);$
(b) $h(2) = 128$ in each   **9.** $(p + q)(2x + y + 2)$
10. $(y + 3)(2y - 11)$ **11.** $(y^2 - 2)(5y + 2)$
12. Cannot be factored by grouping

Calculator Corner, p. 352

1. Correct   **2.** Correct   **3.** Incorrect   **4.** Incorrect
5. Incorrect   **6.** Correct   **7.** Incorrect   **8.** Correct

Exercise Set 4.3, p. 353

1. $3a(2a + 1)$ **3.** $x^2(x + 9)$ **5.** $4x^2(2 - x^2)$
7. $4xy(x - 3y)$ **9.** $3(y^2 - y - 3)$ **11.** $2a(2b - 3c + 6d)$
13. $5(2a^4 + 3a^2 - 5a - 6)$ **15.** $3x^2y^4z^3(5y - 4x^2z^4)$
17. $7a^3b^3c^3(2ac^2 + 3b^2c - 5ab)$ **19.** $-5(x + 9)$
21. $-6(a + 14)$ **23.** $-2(x^2 - x + 12)$ **25.** $-3y(y - 8)$
27. $-(a^4 - 2a^3 + 13a^2 + 1)$ **29.** $-3(y^3 - 4y^2 + 5y - 8)$
31. $N(x) = x\left(\frac{1}{3}x^2 + \frac{1}{2}x + \frac{1}{6}\right),$ or $\frac{1}{6}x(2x^2 + 3x + 1)$
33. (a) $h(t) = -8t(2t - 9);$ (b) $h(2) = 80$ in each
35. $R(x) = 0.4x(700 + x)$ **37.** $(a + c)(b - 2)$
39. $(x - 2)(2x + 13)$ **41.** $2a^2(x - y)$
43. $(a + b)(c + d)$ **45.** $(b^2 + 2)(b - 1)$
47. $(y^2 + 1)(y - 8)$ **49.** $12(x^2 + 3)(2x - 3)$
51. $a(a^3 - a^2 + a + 1)$ **53.** $(2y^2 + 5)(y^2 + 3)$ **55.** $\mathbf{D_W}$
57. Slope–intercept   **58.** Equivalent   **59.** Down
60. Inconsistent   **61.** Point–slope   **62.** Supplementary
63. Ascending   **64.** Constant
65. $x^5y^4 + x^4y^6 = x^3y^3(x^2y^3 + xy^5)$
67. $(x^2 - x + 5)(r + s)$ **69.** $(x^4 + x^2 + 5)(a^4 + a^2 + 5)$
71. $x^{1/3}(1 - 7x)$ **73.** $x^{1/3}(1 - 5x^{1/6} + 3x^{5/12})$
75. $3a^n(a + 2 - 5a^2)$ **77.** $y^{a+b}(7y^a - 5 + 3y^b)$

Margin Exercises, Section 4.4, pp. 356–359

1. $(x + 2)(x + 3)$ **2.** $(y + 2)(y + 5)$
3. $(m - 2)(m - 6)$ **4.** $(t - 3)(t - 8)$, or $(3 - t)(8 - t)$
5. (a) $(x - 5)(x + 4)$; (b) The product of each pair is positive. **6.** $x(x - 9)(x + 6)$ **7.** $2x(x - 7)(x + 6)$
8. $x(x + 6)(x - 2)$ **9.** $(y - 6)(y + 2)$
10. $(x + 10)(x - 11)$ **11.** Not factorable
12. $(x - 2y)(x - 3y)$ **13.** $(p - 8q)(p + 2q)$
14. $(x^2 - 2)(x^2 - 7)$ **15.** $(p^3 + 3)(p^3 - 2)$
16. $-(x + 5)(x - 2)$, or $(-x - 5)(x - 2)$, or $(x + 5)(-x + 2)$
17. $-(x - 4)(x - 4)$, or $(-x + 4)(x - 4)$

Exercise Set 4.4, p. 360

1. $(x + 4)(x + 9)$ **3.** $(t - 5)(t - 3)$ **5.** $(x - 11)(x + 3)$
7. $2(y - 4)(y - 4)$ **9.** $(p + 9)(p - 6)$
11. $(x + 3)(x + 9)$ **13.** $\left(y - \frac{1}{3}\right)\left(y - \frac{1}{3}\right)$
15. $(t - 3)(t - 1)$ **17.** $(x + 7)(x - 2)$
19. $(x + 2)(x + 3)$ **21.** $-1(x - 8)(x + 7)$, or $(-x + 8)(x + 7)$, or $(x - 8)(-x - 7)$
23. $-y(y - 8)(y + 4)$, or $y(-y + 8)(y + 4)$, or $y(y - 8)(-y - 4)$ **25.** $(x^2 + 16)(x^2 - 5)$
27. Not factorable **29.** $(x + 9y)(x + 3y)$
31. $-1(x - 9)(x + 5)$, or $(-x + 9)(x + 5)$, or $(x - 9)(-x - 5)$ **33.** $-1(z + 12)(z - 3)$, or $(-z - 12)(z - 3)$, or $(z + 12)(-z + 3)$
35. $(x^2 + 49)(x^2 + 1)$ **37.** $(x^3 + 9)(x^3 + 2)$
39. $(x^4 - 3)(x^4 - 8)$ **41.** $\mathbf{D_W}$ **43.** Countryside: $9\frac{3}{8}$ lb; Mystic: $15\frac{5}{8}$ lb **44.** 8 weekdays **45.** Yes **46.** No
47. No **48.** Yes **49.** All real numbers
50. All real numbers **51.** $\left\{x \mid x \text{ is a real number } and \ x \neq \frac{7}{4}\right\}$, or $\left(-\infty, \frac{7}{4}\right) \cup \left(\frac{7}{4}, \infty\right)$ **52.** All real numbers
53. $76, -76, 28, -28, 20, -20$ **55.** $x - 365$

Margin Exercises, Section 4.5, pp. 362–367

1. $(x - 7)(3x + 8)$ **2.** $(3x + 2)(x + 1)$
3. $2(4y - 1)(3y - 5)$ **4.** $2x^3(2x - 3)(5x - 4)$
5. $(3x + 4)(x + 5)$ **6.** $4(2x - 1)(2x + 3)$
7. $(7x - 4y)(3x + y)$ **8.** $3(4a + 9b)(5a - b)$
9. $(2x + 3)(2x - 1)$ **10.** $(4x + 1)(x + 9)$
11. $y^2(5y + 4)(2y - 3)$ **12.** $a(36a^2 + 21a + 1)$

Exercise Set 4.5, p. 369

1. $(3x + 1)(x - 5)$ **3.** $y(5y - 7)(2y + 3)$
5. $(3c - 8)(c - 4)$ **7.** $(5y + 2)(7y + 4)$
9. $2(5t - 3)(t + 1)$ **11.** $4(2x + 1)(x - 4)$
13. $3(3a - 1)(2a - 5)$ **15.** $5(3t + 1)(2t + 5)$
17. $x(3x - 4)(4x - 5)$ **19.** $x^2(7x + 1)(2x - 3)$
21. $(3a - 4)(a + 1)$ **23.** $(3x + 1)(3x + 4)$
25. $-1(z - 3)(12z + 1)$, or $(-z + 3)(12z + 1)$, or $(z - 3)(-12z - 1)$ **27.** $-1(2t - 3)(2t + 5)$, or $(-2t + 3)(2t + 5)$, or $(2t - 3)(-2t - 5)$
29. $x(3x + 1)(x - 2)$ **31.** $(24x + 1)(x - 2)$
33. $-2t(2t + 5)(2t - 3)$ **35.** $-x(24x + 1)(x - 2)$
37. $(7x + 3)(3x + 4)$ **39.** $4(10x^4 + 4x^2 - 3)$

Margin Exercises, Section 4.6, pp. 371–377

1. (a), (b), (d) **2.** $(x + 7)^2$ **3.** $(3y - 5)^2$ **4.** $(4x + 9y)^2$
5. $(4x^2 - 5y^3)^2$ **6.** $-2(2a - 3b)^2$ **7.** $3(a - 5b)^2$
8. $(y + 2)(y - 2)$ **9.** $(7x^2 + 5y^5)(7x^2 - 5y^5)$
10. $\left(m + \frac{1}{3}\right)\left(m - \frac{1}{3}\right)$ **11.** $(5xy + 2a)(5xy - 2a)$
12. $(3x + 4y)(3x - 4y)$ **13.** $5(2x + y)(2x - y)$
14. $y^2(9x^2 + 4)(3x + 2)(3x - 2)$
15. $(a + 4)(a - 4)(a + 1)$ **16.** $(x + 1 + p)(x + 1 - p)$
17. $(y - 4 + 3m)(y - 4 - 3m)$
18. $(x + 4 + 10t)(x + 4 - 10t)$
19. $[8p + (x + 4)][8p - (x + 4)]$, or $(8p + x + 4)(8p - x - 4)$ **20.** $(x - 2)(x^2 + 2x + 4)$
21. $(4 - y)(16 + 4y + y^2)$ **22.** $(3x + y)(9x^2 - 3xy + y^2)$
23. $(2y + z)(4y^2 - 2yz + z^2)$
24. $(m + n)(m^2 - mn + n^2)(m - n)(m^2 + mn + n^2)$
25. $2xy(2x^2 + 3y^2)(4x^4 - 6x^2y^2 + 9y^4)$
26. $(3x + 2y)(9x^2 - 6xy + 4y^2)(3x - 2y)(9x^2 + 6xy + 4y^2)$
27. $(x - 0.3)(x^2 + 0.3x + 0.09)$

Visualizing for Success, p. 378

1. A, E **2.** F, J **3.** G, K **4.** L, S **5.** P, Q **6.** C, I
7. D, H **8.** M, O **9.** N, T **10.** B, R

Exercise Set 4.6, p. 379

1. $(x - 2)^2$ **3.** $(y + 9)^2$ **5.** $(x + 1)^2$ **7.** $(3y + 2)^2$
9. $y(y - 9)^2$ **11.** $3(2a + 3)^2$ **13.** $2(x - 10)^2$
15. $(1 - 4d)^2$, or $(4d - 1)^2$ **17.** $(y + 2)^2(y - 2)^2$
19. $(0.5x + 0.3)^2$ **21.** $(p - q)^2$ **23.** $(a + 2b)^2$
25. $(5a - 3b)^2$ **27.** $(y^3 + 13)^2$ **29.** $(4x^5 - 1)^2$
31. $(x^2 + y^2)^2$ **33.** $(x + 4)(x - 4)$ **35.** $(p + 7)(p - 7)$
37. $(pq + 5)(pq - 5)$ **39.** $6(x + y)(x - y)$
41. $4x(y^2 + z^2)(y + z)(y - z)$ **43.** $a(2a + 7)(2a - 7)$
45. $3(x^4 + y^4)(x^2 + y^2)(x + y)(x - y)$
47. $a^2(3a + 5b^2)(3a - 5b^2)$ **49.** $\left(\frac{1}{6} + z\right)\left(\frac{1}{6} - z\right)$
51. $(0.2x + 0.3y)(0.2x - 0.3y)$
53. $(m + 2)(m - 2)(m - 7)$ **55.** $(a + b)(a - b)(a - 2)$
57. $(a + b + 10)(a + b - 10)$ **59.** $(20 - p)(4 + p)$
61. $(a + b + 3)(a + b - 3)$ **63.** $(r - 1 + 2s)(r - 1 - 2s)$
65. $2(m + n + 5b)(m + n - 5b)$
67. $[3 + (a + b)][3 - (a + b)]$, or $(3 + a + b)(3 - a - b)$
69. $(z + 3)(z^2 - 3z + 9)$ **71.** $(x - 1)(x^2 + x + 1)$
73. $(y + 5)(y^2 - 5y + 25)$ **75.** $(2a + 1)(4a^2 - 2a + 1)$

Margin Exercises, Section 4.4, pp. 356–359 (continued, right column)

41. $(4a - 3b)(3a - 2b)$ **43.** $(2x - 3y)(x + 2y)$
45. $2(3x - 4y)(2x - 7y)$ **47.** $(3x - 5y)(3x - 5y)$
49. $(3x^3 - 2)(x^3 + 2)$ **51.** (a) 224 ft; 288 ft; 320 ft; 288 ft; 128 ft; (b) $h(t) = -16(t - 7)(t + 2)$ **53.** $\mathbf{D_W}$
55. $(2, -1, 0)$ **56.** $\left(\frac{3}{2}, -4, 3\right)$ **57.** $(1, -1, 2)$
58. $(2, 4, 1)$ **59.** Parallel **60.** Parallel **61.** Neither
62. Perpendicular **63.** $y = -\frac{1}{7}x - \frac{23}{7}$ **64.** $y = -\frac{1}{3}x - \frac{7}{3}$
65. $y = -\frac{7}{17}x - \frac{19}{17}$ **66.** $y = -\frac{5}{2}x - \frac{2}{3}$ **67.** Left to the student **69.** $(pq + 4)(pq + 3)$ **71.** $\left(x + \frac{4}{5}\right)\left(x - \frac{1}{5}\right)$
73. $(y + 0.5)(y - 0.1)$ **75.** $(7ab + 6)(ab + 1)$
77. $3(x + 15)(x - 11)$ **79.** $6x(x + 9)(x + 4)$
81. $(x^a + 8)(x^a - 3)$

77. $(y - 2)(y^2 + 2y + 4)$ **79.** $(2 - 3b)(4 + 6b + 9b^2)$
81. $(4y + 1)(16y^2 - 4y + 1)$ **83.** $(2x + 3)(4x^2 - 6x + 9)$
85. $(a - b)(a^2 + ab + b^2)$ **87.** $\left(a + \frac{1}{2}\right)\left(a^2 - \frac{1}{2}a + \frac{1}{4}\right)$
89. $2(y - 4)(y^2 + 4y + 16)$
91. $3(2a + 1)(4a^2 - 2a + 1)$ **93.** $r(s + 4)(s^2 - 4s + 16)$
95. $5(x - 2z)(x^2 + 2xz + 4z^2)$
97. $(x + 0.1)(x^2 - 0.1x + 0.01)$
99. $8(2x^2 - t^2)(4x^4 + 2x^2t^2 + t^4)$
101. $2y(y - 4)(y^2 + 4y + 16)$
103. $(z - 1)(z^2 + z + 1)(z + 1)(z^2 - z + 1)$
105. $(t^2 + 4y^2)(t^4 - 4t^2y^2 + 16y^4)$ **107.** $\mathbf{D_W}$
109. $\left(-\frac{41}{53}, \frac{148}{53}\right)$ **110.** $\left(-\frac{26}{7}, -\frac{134}{7}\right)$ **111.** $(1, 13)$
112. No solution
113.

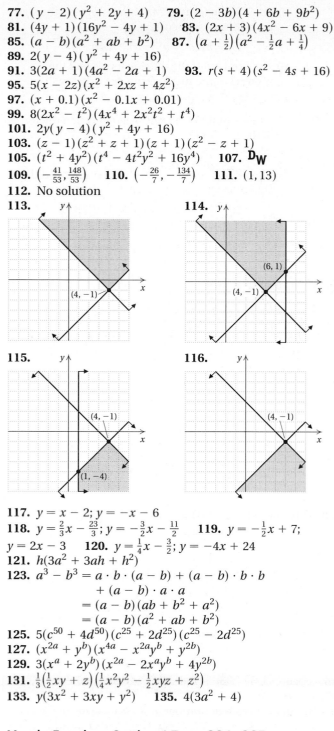

117. $y = x - 2$; $y = -x - 6$
118. $y = \frac{2}{3}x - \frac{23}{3}$; $y = -\frac{3}{2}x - \frac{11}{2}$ **119.** $y = -\frac{1}{2}x + 7$;
$y = 2x - 3$ **120.** $y = \frac{1}{4}x - \frac{3}{2}$; $y = -4x + 24$
121. $h(3a^2 + 3ah + h^2)$
123. $a^3 - b^3 = a \cdot b \cdot (a - b) + (a - b) \cdot b \cdot b$
$\qquad\qquad + (a - b) \cdot a \cdot a$
$\qquad\quad = (a - b)(ab + b^2 + a^2)$
$\qquad\quad = (a - b)(a^2 + ab + b^2)$
125. $5(c^{50} + 4d^{50})(c^{25} + 2d^{25})(c^{25} - 2d^{25})$
127. $(x^{2a} + y^b)(x^{4a} - x^{2a}y^b + y^{2b})$
129. $3(x^a + 2y^b)(x^{2a} - 2x^ay^b + 4y^{2b})$
131. $\frac{1}{3}\left(\frac{1}{2}xy + z\right)\left(\frac{1}{4}x^2y^2 - \frac{1}{2}xyz + z^2\right)$
133. $y(3x^2 + 3xy + y^2)$ **135.** $4(3a^2 + 4)$

Margin Exercises, Section 4.7, pp. 384–385

1. $3y(y + 2x)(y - 2x)$ **2.** $7(a - 1)(a^2 + a + 1)$
3. $(2x + 3y)(4x^2 - 6xy + 9y^2)(2x - 3y)(4x^2 + 6xy + 9y^2)$
4. $(x - 2)(3 - bx)$ **5.** $5(y^4 + 4x^6)$ **6.** $3(x - 2)(2x + 3)$
7. $(a - b)^2(a + b)$ **8.** $3(x + 3a)^2$
9. $2(x - 5 + 3b)(x - 5 - 3b)$

Exercise Set 4.7, p. 386

1. $(y + 15)(y - 15)$ **3.** $(2x + 3)(x + 4)$
5. $5(x^2 + 2)(x^2 - 2)$ **7.** $(p + 6)^2$ **9.** $2(x - 11)(x + 6)$

11. $(3x + 5y)(3x - 5y)$ **13.** $4(m^2 + 5)(m^2 - 5)$
15. $3(x + 12)(x - 7)$ **17.** $2x(y + 5)(y - 5)$
19. $(18 - a)(12 + a)$
21. $(m + 1)(m^2 - m + 1)(m - 1)(m^2 + m + 1)$
23. $(x + 3 + y)(x + 3 - y)$
25. $2(5x - 4y)(25x^2 + 20xy + 16y^2)$
27. $(m^3 + 10)(m^3 - 2)$ **29.** $(c - b)(a + d)$
31. $(5b - a)(10b + a)$ **33.** $(x^2 + 2)(2x - 7)$
35. $2(x + 2)(x - 2)(x + 3)$
37. $2(2x + 3y)(4x^2 - 6xy + 9y^2)$ **39.** $(6y - 5)(6y + 7)$
41. $(a^4 + b^4)(a^2 + b^2)(a + b)(a - b)$
43. $ab(a + 4b)(a - 4b)$ **45.** $\left(\frac{1}{4}x - \frac{1}{3}y^2\right)^2$
47. $5(x - y)^2(x + y)$ **49.** $(9ab + 2)(3ab + 4)$
51. $y(2y - 5)(4y^2 + 10y + 25)$
53. $(a - b - 3)(a + b + 3)$ **55.** $\mathbf{D_W}$ **57.** Correct: 55;
incorrect: 20   **58.** $\frac{80}{7}$ **59.** $(6y^2 - 5x)(5y^2 - 12x)$
61. $5\left(x - \frac{1}{3}\right)\left(x^2 + \frac{1}{3}x + \frac{1}{9}\right)$ **63.** $x(x - 2p)$
65. $y(y - 1)^2(y - 2)$
67. $(2x + y - r + 3s)(2x + y + r - 3s)$ **69.** $c(c^w + 1)^2$
71. $3x(x + 5)$ **73.** $(x - 1)^3(x^2 + 1)(x + 1)$
75. $y(y^4 + 1)(y^2 + 1)(y + 1)(y - 1)$

Margin Exercises, Section 4.8, pp. 389–395

1. $4, 2$ **2.** $\frac{1}{2}, -3$ **3.** $0, 2$ **4.** -5 **5.** $0, 2, -3$
6. $-\frac{3}{2}, \frac{1}{5}$ **7.** $\{x \mid x$ is a real number $and \ x \neq -4 \ and \ x \neq 7\}$
8. **(a)** $(2, 0)$ and $(4, 0)$; **(b)** $2, 4$; **(c)** The solutions of
$x^2 - 6x + 8 = 0$, 2 and 4, are the first coordinates of the
x-intercepts, $(2, 0)$ and $(4, 0)$, of the graph of
$f(x) = x^2 - 6x + 8$.   **9.** 11 sec   **10.** 12 cm and 13 cm

Calculator Corner, p. 392

1. Left to the student

Translating for Success, p. 396

1. Q   **2.** F   **3.** B   **4.** A   **5.** P   **6.** D   **7.** O   **8.** H
9. I   **10.** J

Exercise Set 4.8, p. 397

1. $-7, 4$ **3.** 3   **5.** -10 **7.** $-5, -4$ **9.** $0, -8$
11. $-5, 5$ **13.** $-12, 12$ **15.** $7, -9$ **17.** $-4, 8$
19. $-2, -\frac{2}{3}$ **21.** $\frac{1}{2}, \frac{3}{4}$ **23.** $0, 6$ **25.** $\frac{2}{3}, -\frac{3}{4}$ **27.** $-1, 1$
29. $\frac{2}{3}, -\frac{5}{7}$ **31.** $0, \frac{1}{5}$ **33.** $7, -2$ **35.** $0, -2, 3$
37. $0, -8, 8$ **39.** $5, -5, 1, -1$ **41.** $-6, 6$ **43.** $-\frac{7}{4}, \frac{4}{3}$
45. $-8, -4$ **47.** $-4, \frac{3}{2}$ **49.** $-9, -3$
51. $\{x \mid x$ is a real number $and \ x \neq -1 \ and \ x \neq 5\}$
53. $\{x \mid x$ is a real number $and \ x \neq -3 \ and \ x \neq 3\}$
55. $\{x \mid x$ is a real number $and \ x \neq \frac{1}{5}\}$
57. $\{x \mid x$ is a real number $and \ x \neq 0 \ and \ x \neq 2 \ and \ x \neq 5\}$
59. x-intercepts: $(-5, 0)$ and $(9, 0)$; solutions: $-5, 9$
61. x-intercepts: $(-4, 0)$ and $(8, 0)$; solutions: $-4, 8$
63. Length: 12 cm; width: 8 cm   **65.** Height: 6 ft; base: 4 ft
67. Length: 12 m; width: 9 m   **69.** $d = 12$ ft; $h = 16$ ft
71. 16, 18, 20   **73.** 6 cm   **75.** 3 cm   **77.** 150 ft by 200 ft

79. 40 m, 41 m   **81.** 24 ft, 25 ft   **83.** 7 sec   **85.** $\mathbf{D_W}$
87. 7   **88.** 1   **89.** 7   **90.** 2.5   **91.** 1.3   **92.** $\frac{19}{15}$
93. 691   **94.** 1023   **95.** $y = \frac{11}{6}x + \frac{32}{3}$
96. $y = -\frac{11}{10}x + \frac{24}{5}$ **97.** $y = -\frac{3}{10}x + \frac{32}{5}$
98. $y = \frac{26}{31}x + \frac{934}{31}$ **99.** $\{-3, 1\}$; $\{x \mid -4 \le x \le 2\}$, or $[-4, 2]$
101. Left to the student   **103. (a)** 1.2522305, 3.1578935;
(b) $-0.3027756, 0, 3.3027756$; **(c)** 2.1387475, 2.7238657;
(d) $-0.7462555, 3.3276509$

Concept Reinforcement, p. 402

1. True   **2.** True   **3.** False   **4.** False   **5.** False
6. True   **7.** False   **8.** False

Summary and Review: Chapter 4, p. 402

1. (a) 7, 11, 3, 2; 11; **(b)** $-7x^8y^3$; -7;
(c) $-3x^2 + 2x^3 + 3x^6y - 7x^8y^3$;
(d) $-7x^8y^3 + 3x^6y + 2x^3 - 3x^2$ **2.** $0; -6$ **3.** $4; -31$
4. (a) About 230,000; **(b)** about 389,000   **5.** $-x^2y - 2xy^2$
6. $ab + 12ab^2 + 4$ **7.** $-x^3 + 2x^2 + 5x + 2$
8. $x^3 + 6x^2 - x - 4$ **9.** $13x^2y - 8xy^2 + 4xy$
10. $9x - 7$ **11.** $-2a + 6b + 7c$ **12.** $16p^2 - 8p$
13. $6x^2 - 7xy + y^2$ **14.** $-18x^3y^4$
15. $x^8 - x^6 + 5x^2 - 3$ **16.** $8a^2b^2 + 2abc - 3c^2$
17. $4x^2 - 25y^2$ **18.** $4x^2 - 20xy + 25y^2$
19. $20x^4 - 18x^3 - 47x^2 + 69x - 27$
20. $x^4 + 8x^2y^3 + 16y^6$ **21.** $x^3 - 125$ **22.** $x^2 - \frac{1}{2}x + \frac{1}{18}$
23. $a^2 - 4a - 4$; $2ah + h^2 - 2h$ **24.** $3y^2(3y^2 - 1)$
25. $3x(5x^3 - 6x^2 + 7x - 3)$ **26.** $(a - 9)(a - 3)$
27. $(3m + 2)(m + 4)$ **28.** $(5x + 2)^2$
29. $4(y + 2)(y - 2)$ **30.** $(x - y)(a + 2b)$
31. $4(x^4 + x^2 + 5)$ **32.** $(3x - 2)(9x^2 + 6x + 4)$
33. $(0.4b - 0.5c)(0.16b^2 + 0.2bc + 0.25c^2)$
34. $y(y^2 + 1)(y + 1)(y - 1)$ **35.** $2z^6(z^2 - 8)$
36. $2y(3x^2 - 1)(9x^4 + 3x^2 + 1)$
37. $(1 + a)(1 - a + a^2)$ **38.** $4(3x - 5)^2$
39. $(3t + p)(2t + 5p)$ **40.** $(x + 3)(x - 3)(x + 2)$
41. $(a - b + 2t)(a - b - 2t)$ **42.** 10   **43.** $\frac{2}{3}, \frac{3}{2}$
44. $0, \frac{7}{4}$ **45.** $-4, 4$ **46.** $-4, 11$
47. $\{x \mid x \text{ is a real number } and\ x \ne \frac{2}{3} \text{ and } x \ne -7\}$
48. Length: 8 in.; width: 5 in.
49. $-7, -5, -3; 3, 5, 7$ **50.** 7
51. $\mathbf{D_W}$ The discussion could include the following points:
  a) We can now solve certain polynomial equations.
  b) Whereas most linear equations have exactly one
    solution, polynomial equations can have more than
    one solution.
  c) We used factoring and the principle of zero products
    to solve polynomial equations.
52. $\mathbf{D_W}$ **a)** The middle term, $2 \cdot a \cdot 3$, is missing.
$$(a + 3)^2 = a^2 + 6a + 9$$
  b) The middle term of the trinomial factor should
    be $-ab$.
$$a^3 + b^3 = (a + b)(a^2 - ab + b^2)$$

c) The middle term, $-2ab$, is missing and the sign
  preceding b^2 is incorrect.
$$(a - b)(a - b) = a^2 - 2ab + b^2$$
d) The product of the outside terms and the
  product of the inside terms are missing.
$$(x + 3)(x - 4) = x^2 - x - 12$$
e) There should be a minus sign between the terms
  of the product.
$$(p + 7)(p - 7) = p^2 - 49$$
f) The middle term, $-2 \cdot t \cdot 3$, is missing and the
  sign preceding 9 is incorrect.
$$(t - 3)^2 = t^2 - 6t + 9$$
53. $2(2x + y)(4x^2 - 2xy + y^2)(2x - y)(4x^2 + 2xy + y^2)$
54. $2(3x^2 + 1)$ **55.** $a^3 - (b - 1)^3$ **56.** $0, \frac{1}{8}, -\frac{1}{8}$

Test: Chapter 4, p. 405

1. [4.1a] **(a)** 4, 3, 9, 5; 9; **(b)** $5x^5y^4$; 5;
(c) $3xy^3 - 4x^2y - 2x^4y + 5x^5y^4$;
(d) $5x^5y^4 + 3xy^3 - 4x^2y - 2x^4y$ **2.** [4.1b] 4; 2
3. [4.1b] **(a)** About \$1.66 billion; **(b)** about \$2.8 billion
4. [4.1c] $3xy + 3xy^2$ **5.** [4.1c] $-3x^3 + 3x^2 - 6y - 7y^2$
6. [4.1c] $7a^3 - 6a^2 + 3a - 3$
7. [4.1c] $7m^3 + 2m^2n + 3mn^2 - 7n^3$ **8.** [4.1d] $6a - 8b$
9. [4.1d] $7x^2 - 7x + 13$ **10.** [4.1d] $2y^2 + 5y + y^3$
11. [4.2a] $64x^3y^3$ **12.** [4.2b] $12a^2 - 4ab - 5b^2$
13. [4.2a] $x^3 - 2x^2y + y^3$
14. [4.2a] $-3m^4 - 13m^3 + 5m^2 + 26m - 10$
15. [4.2c] $16y^2 - 72y + 81$ **16.** [4.2d] $x^2 - 4y^2$
17. [4.2e] $a^2 + 15a + 50$; $2ah + h^2 - 5h$
18. [4.3a] $x(9x + 7)$ **19.** [4.3a] $8y^2(3y + 2)$
20. [4.6c] $(y + 2)(y - 2)(y + 5)$
21. [4.4a] $(p - 14)(p + 2)$
22. [4.5a, b] $(6m + 1)(2m + 3)$
23. [4.6b] $(3y + 5)(3y - 5)$
24. [4.6d] $3(r - 1)(r^2 + r + 1)$ **25.** [4.6a] $(3x - 5)^2$
26. [4.6b] $(z + 1 + b)(z + 1 - b)$
27. [4.6b] $(x^4 + y^4)(x^2 + y^2)(x + y)(x - y)$
28. [4.6c] $(y + 4 + 10t)(y + 4 - 10t)$
29. [4.6b] $5(2a + b)(2a - b)$
30. [4.5a, b] $2(4x - 1)(3x - 5)$
31. [4.6d] $2ab(2a^2 + 3b^2)(4a^4 - 6a^2b^2 + 9b^4)$
32. [4.8a] $-3, 6$ **33.** [4.8a] $-5, 5$ **34.** [4.8a] $-\frac{3}{2}, -7$
35. [4.8a] 0, 5
36. [4.8a] $\{x \mid x \text{ is a real number } and\ x \ne -1\}$, or
$(-\infty, -1) \cup (-1, \infty)$ **37.** [4.8b] Length: 8 cm; width: 5 cm
38. [4.8b] 24 ft   **39.** [4.8b] 5
40. [4.3a] $f(n) = \frac{1}{2}n(n - 1)$
41. [4.7a] $(3x^n + 4)(2x^n - 5)$ **42.** [4.2c] 19

Cumulative Review: Chapters R–4, p. 407

1. [R.7c] 4000; 4×10^4; 300,000; 5.77×10^5 **2.** [R.4b] 1
3. [R.1d] 0   **4.** [R.2c] -4 **5.** [R.6b] $-a - 5$
6. [R.6b] $5x + 20$ **7.** [R.7b] $\dfrac{a^{60}}{(-3)^{10}b^{20}}$

8. [4.1c] $-2x^2 + x - xy - 1$ **9.** [4.1d] $-2x^2 + 6x$
10. [4.2a] $a^4 + a^3 - 8a^2 - 3a + 9$
11. [4.2b] $x^2 + 13x + 36$ **12.** [1.1d] 2   **13.** [1.1d] 13
14. [1.2a] $b = \dfrac{2A - ha}{h}$, or $\dfrac{2A}{h} - a$
15. [1.4c] $\left\{x \mid x \geq -\frac{7}{9}\right\}$, or $\left[-\frac{7}{9}, \infty\right)$
16. [1.5b] $\left\{x \mid x < \frac{5}{4} \text{ or } x > 4\right\}$, or $\left(-\infty, \frac{5}{4}\right) \cup (4, \infty)$
17. [1.6e] $\{x \mid -2 < x < 5\}$, or $(-2, 5)$ **18.** [3.5a] $(1, 3, -9)$
19. [3.3a] $(4, -2)$ **20.** [3.3a] $\left(\frac{19}{8}, \frac{1}{8}\right)$
21. [3.5a] $(-1, 0, -1)$ **22.** [3.3a] $\left(\frac{3}{2}, -\frac{1}{3}\right)$ **23.** [3.3a] $\left(\frac{1}{3}, \frac{1}{2}\right)$
24. [3.5a] $\left(\frac{1}{2}, \frac{1}{2}, \frac{1}{4}\right)$ **25.** [4.8a] $-3, -8$ **26.** [4.8a] $\frac{1}{2}, 7$
27. [4.8a] $\frac{2}{3}, -2$
28. [4.8a] $\{x \mid x \text{ is a real number } and \ x \neq 5 \text{ and } x \neq -3\}$
29. [4.3a] $3x^2(x - 4)$
30. [4.3b], [4.6d] $(2x + 1)(x + 1)(x^2 - x + 1)$
31. [4.4a] $(x - 2)(x + 7)$ **32.** [4.5a, b] $(4a - 3)(5a - 2)$
33. [4.6b] $(2x + 5)(2x - 5)$ **34.** [4.6a] $2(x - 7)^2$
35. [4.6d] $(a + 4)(a^2 - 4a + 16)$
36. [4.6d] $(2x - 1)(4x^2 + 2x + 1)$
37. [4.4a] $(a^3 + 6)(a^3 - 2)$
38. [4.6b] $x^2 y^2 (2x + y)(2x - y)$
39. [1.5b]

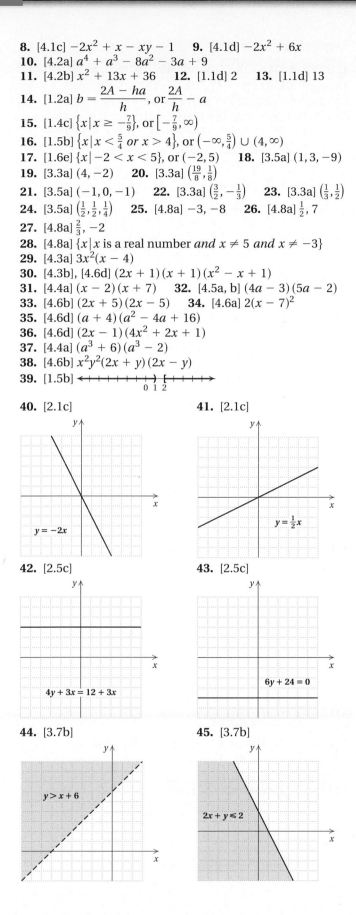

40. [2.1c]   **41.** [2.1c]

$y = -2x$ $y = \frac{1}{2}x$

42. [2.5c]   **43.** [2.5c]

$4y + 3x = 12 + 3x$ $6y + 24 = 0$

44. [3.7b]   **45.** [3.7b]

$y > x + 6$ $2x + y \leq 2$

46. [2.2c]   **47.** [2.2c]

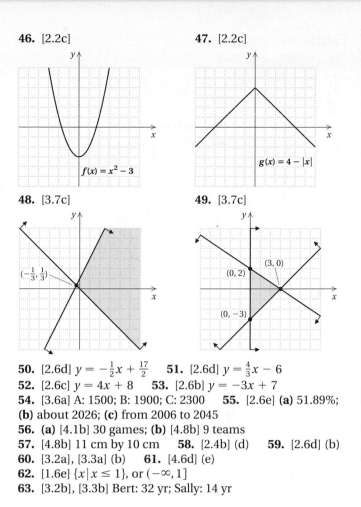

$f(x) = x^2 - 3$ $g(x) = 4 - |x|$

48. [3.7c]   **49.** [3.7c]

$\left(-\frac{1}{3}, \frac{1}{3}\right)$ $(0, 2)$ $(3, 0)$ $(0, -3)$

50. [2.6d] $y = -\frac{1}{2}x + \frac{17}{2}$ **51.** [2.6d] $y = \frac{4}{3}x - 6$
52. [2.6c] $y = 4x + 8$ **53.** [2.6b] $y = -3x + 7$
54. [3.6a] A: 1500; B: 1900; C: 2300   **55.** [2.6e] **(a)** 51.89%;
(b) about 2026; **(c)** from 2006 to 2045
56. **(a)** [4.1b] 30 games; **(b)** [4.8b] 9 teams
57. [4.8b] 11 cm by 10 cm   **58.** [2.4b] (d)   **59.** [2.6d] (b)
60. [3.2a], [3.3a] (b)   **61.** [4.6d] (e)
62. [1.6e] $\{x \mid x \leq 1\}$, or $(-\infty, 1]$
63. [3.2b], [3.3b] Bert: 32 yr; Sally: 14 yr

CHAPTER 5

Margin Exercises, Section 5.1, pp. 412–420

1. $-\frac{5}{2}$ **2.** $\left\{x \mid x \text{ is a real number } and \ x \neq -\frac{5}{2}\right\}$, or
$\left(-\infty, -\frac{5}{2}\right) \cup \left(-\frac{5}{2}, \infty\right)$ **3.** 2, 5   **4.** $\{t \mid t \text{ is a real number } and$
$t \neq 2 \text{ and } t \neq 5\}$, or $(-\infty, 2) \cup (2, 5) \cup (5, \infty)$ **5.** $\dfrac{(3x + 2y)x}{(5x + 4y)x}$
6. $\dfrac{(2x^2 - y)(3x + 2)}{(3x + 4)(3x + 2)}$ **7.** $\dfrac{-2a + 5}{-a + b}$, or $\dfrac{5 - 2a}{b - a}$
8. $\frac{4}{5}$ **9.** $7x$ **10.** $2a + 3$ **11.** $\dfrac{3x + 2}{2(x + 2)}$
12. $\dfrac{2(y + 2)}{y - 1}$ **13.** $\dfrac{5(a + 2b)}{4(a - 2b)}$ **14.** -1 **15.** -1
16. -1 **17.** $\dfrac{3(x - y)(x - y)}{x + y}$ **18.** $a - b$ **19.** $\dfrac{x - 5}{x + 3}$
20. $\dfrac{1}{x + 7}$ **21.** $y^3 - 9$ **22.** $\dfrac{x + 5}{2(x - 5)}$ **23.** $\dfrac{2ab(a + b)}{a - b}$
24. $a^2 - 4$

Calculator Corner, p. 418

1. Correct   **2.** Correct   **3.** Incorrect   **4.** Incorrect
5. Incorrect   **6.** Correct   **7.** Correct   **8.** Correct

Calculator Corner, p. 419

Left to the student

Exercise Set 5.1, p. 421

1. $-\frac{17}{3}$ **3.** $-7, -5$ **5.** $\{x \mid x \text{ is a real number } and\ x \neq -\frac{17}{3}\}$, or $\left(-\infty, -\frac{17}{3}\right) \cup \left(-\frac{17}{3}, \infty\right)$ **7.** $\{x \mid x \text{ is a real number } and\ x \neq -7\ and\ x \neq -5\}$, or $(-\infty, -7) \cup (-7, -5) \cup (-5, \infty)$
9. $\dfrac{7x(x+2)}{7x(x+8)}$ **11.** $\dfrac{(q-5)(q+5)}{(q+3)(q+5)}$ **13.** $3y$ **15.** $\dfrac{2}{3p^4}$
17. $a - 3$ **19.** $\dfrac{4x - 5}{7}$ **21.** $\dfrac{y - 3}{y + 3}$ **23.** $\dfrac{t + 4}{t - 4}$
25. $\dfrac{x - 8}{x + 4}$ **27.** $\dfrac{w^2 + wz + z^2}{w + z}$ **29.** $\dfrac{1}{3x^3}$
31. $\dfrac{(x-4)(x+4)}{x(x+3)}$ **33.** $\dfrac{y+4}{2}$ **35.** $\dfrac{(2x+3)(x+5)}{7x}$
37. $c - 2$ **39.** $\dfrac{1}{x+y}$ **41.** $\dfrac{3x^5}{2y^3}$ **43.** 3
45. $\dfrac{(y-3)(y+2)}{y}$ **47.** $\dfrac{2a+1}{a+2}$ **49.** $\dfrac{(x+4)(x+2)}{3(x-5)}$
51. $\dfrac{y(y^2+3)}{(y+3)(y-2)}$ **53.** $\dfrac{x^2+4x+16}{(x+4)(x+4)}$
55. $\dfrac{4y^2-6y+9}{(4y-1)(2y-3)}$ **57.** $\dfrac{2s}{r+2s}$ **59.** $\mathbf{D_W}$
61. Domain $= \{-4, -2, 0, 2, 4, 6\}$; range $= \{-3, -2, 0, 1, 3, 4\}$
62. Domain $= [-4, 5]$; range $= [-3, 2]$
63. Domain $= [-5, 5]$; range $= [-4, 4]$
64. Domain $= [-4, 5]$; range $= [0, 2]$
65. $(3a - 5b)(2a + 5b)$ **66.** $(3a - 5b)^2$
67. $10(x - 7)(x - 1)$ **68.** $(5x - 4)(2x - 1)$
69. $(7p + 5)(3p - 2)$ **70.** $2(3m + 1)(2m - 5)$
71. $2x(x - 11)(x + 3)$ **72.** $10(y + 13)(y - 5)$
73. $y = -\frac{2}{3}x - 5$ **74.** $y = -\frac{2}{7}x + \frac{48}{7}$
75. $\dfrac{x-3}{(x+1)(x+3)}$ **77.** $\dfrac{m-t}{m+t+1}$ **79.** $\frac{13}{19}$; -3; not defined; $\dfrac{2a+2h+3}{4a+4h-1}$

Margin Exercises, Section 5.2, pp. 425–431

1. 90 **2.** 72 **3.** $\frac{47}{60}$ **4.** $\frac{97}{72}$ **5.** $5a^3b^2$
6. $(y + 3)(y + 4)(y + 4)$ **7.** $2x^2(x - 3)(x + 3)(x + 2)$
8. $2(a + b)(a - b)$, or $2(a + b)(b - a)$ **9.** $\dfrac{12 + y}{y}$
10. $3x + 1$ **11.** $\dfrac{a - b}{b + 2}$ **12.** $\dfrac{y + 12}{x^2 + y^2}$ **13.** $\dfrac{9x^2 + 28y}{21x}$
14. $\dfrac{3}{x + y}$ **15.** $\dfrac{a + 12}{a(a + 3)}$ **16.** $\dfrac{3y^2 + 12y + 3}{(y - 4)(y - 3)(y + 5)}$
17. $\dfrac{1 - b^2}{3}$ **18.** $\dfrac{2x^2 + 11}{x - 5}$ **19.** $\dfrac{3 + 7x}{4y}$ **20.** $\dfrac{11x^2}{2x - y}$
21. $\dfrac{2}{x - 1}$

Calculator Corner, p. 430

Left to the student

Exercise Set 5.2, p. 432

1. 120 **3.** 144 **5.** 210 **7.** 45 **9.** $\frac{11}{10}$ **11.** $\frac{17}{72}$
13. $\frac{251}{240}$ **15.** $21x^2y$ **17.** $10(y - 10)(y + 10)$
19. $30a^3b^2$ **21.** $5(y - 3)^2$ **23.** $(y + 5)(y - 5)$, or $(y + 5)(5 - y)$ **25.** $(2r + 3)(r - 4)(3r - 1)(r + 4)$
27. $x^3(x - 2)^2(x^2 + 4)$ **29.** $10x^3(x - 1)^2(x + 1)(x^2 + 1)$
31. $\dfrac{2x + 7y}{x + y}$ **33.** $\dfrac{3y + 5}{y - 2}$ **35.** $a + b$ **37.** $\dfrac{13}{y}$
39. $\dfrac{1}{a + 7}$ **41.** $a^2 + ab + b^2$ **43.** $\dfrac{2y^2 + 22}{y^2 - y - 20}$
45. $\dfrac{x + y}{x - y}$ **47.** $\dfrac{3x - 4}{x^2 - 3x + 2}$ **49.** $\dfrac{8x + 1}{x^2 - 1}$
51. $\dfrac{2x - 14}{15(x + 5)}$ **53.** $\dfrac{-a^2 + 7ab - b^2}{a^2 - b^2}$ **55.** $\dfrac{y}{y^2 - 5y + 6}$
57. $\dfrac{3y - 10}{y^2 - y - 20}$ **59.** $\dfrac{3y^2 - 3y - 29}{(y + 8)(y - 3)(y - 4)}$
61. $\dfrac{2x^2 - 13x + 7}{(x + 3)(x - 1)(x - 3)}$ **63.** 0 **65.** $\dfrac{3}{x + 2}$
67. $\dfrac{-2y - 3}{(y + 4)(y - 4)}$ **69.** $\dfrac{-3x^2 - 3x - 4}{x^2 - 1}$
71. $\dfrac{-2}{x - y}$, or $\dfrac{2}{y - x}$ **73.** $\mathbf{D_W}$

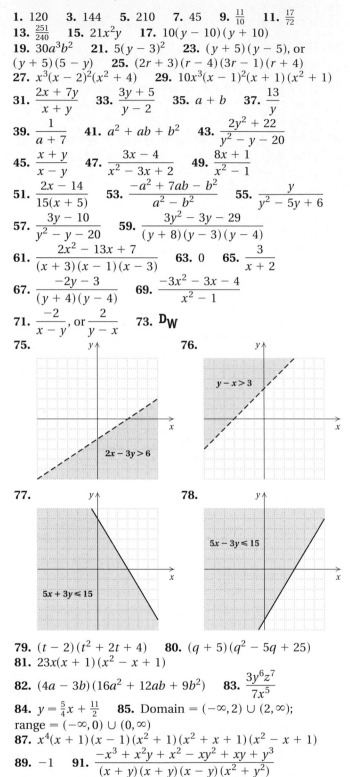

75. $y - x > 3$; $2x - 3y > 6$
76.
77. $5x + 3y \leqslant 15$
78. $5x - 3y \leqslant 15$

79. $(t - 2)(t^2 + 2t + 4)$ **80.** $(q + 5)(q^2 - 5q + 25)$
81. $23x(x + 1)(x^2 - x + 1)$
82. $(4a - 3b)(16a^2 + 12ab + 9b^2)$ **83.** $\dfrac{3y^6z^7}{7x^5}$
84. $y = \frac{5}{4}x + \frac{11}{2}$ **85.** Domain $= (-\infty, 2) \cup (2, \infty)$; range $= (-\infty, 0) \cup (0, \infty)$
87. $x^4(x + 1)(x - 1)(x^2 + 1)(x^2 + x + 1)(x^2 - x + 1)$
89. -1 **91.** $\dfrac{-x^3 + x^2y + x^2 - xy^2 + xy + y^3}{(x + y)(x + y)(x - y)(x^2 + y^2)}$

Margin Exercises, Section 5.3, pp. 436–441

1. $\dfrac{x^2}{2} + 8x + 3$ **2.** $5y^3 - 2y^2 + 6y$ **3.** $\dfrac{x^2y^2}{2} + 5xy + 8$

4. $x + 5$ **5.** $2x^3 - 5x^2 + 19x - 83$, R 341; or
$2x^3 - 5x^2 + 19x - 83 + \dfrac{341}{x + 4}$ **6.** $3y^3 - 2y^2 + 6y - 4$

7. $y^2 - 8y - 24$, R -66; or $y^2 - 8y - 24 + \dfrac{-66}{y - 3}$

8. $x^2 + 10x + 10$, R 5; or $x^2 + 10x + 10 + \dfrac{5}{x - 1}$

9. $y - 11$, R $3y - 27$; or $y - 11 + \dfrac{3y - 27}{y^2 - 3}$

10. $2x^2 + 2x + 14$, R 34; or $2x^2 + 2x + 14 + \dfrac{34}{x - 3}$

11. $x^2 - 4x + 13$, R -30; or $x^2 - 4x + 13 + \dfrac{-30}{x + 2}$

12. $y^2 - y + 1$

Exercise Set 5.3, p. 442

1. $4x^4 + 3x^3 - 6$ **3.** $9y^5 - 4y^2 + 3$
5. $16a^2b^2 + 7ab - 11$ **7.** $x + 7$

9. $a - 12$, R 32; or $a - 12 + \dfrac{32}{a + 4}$

11. $x + 2$, R 4; or $x + 2 + \dfrac{4}{x + 5}$

13. $2y^2 - y + 2$, R 6; or $2y^2 - y + 2 + \dfrac{6}{2y + 4}$

15. $2y^2 + 2y - 1$, R 8; or $2y^2 + 2y - 1 + \dfrac{8}{5y - 2}$

17. $2x^2 - x - 9$, R $(3x + 12)$; or $2x^2 - x - 9 + \dfrac{3x + 12}{x^2 + 2}$

19. $2x^3 + 5x^2 + 17x + 51$, R $152x$; or
$2x^3 + 5x^2 + 17x + 51 + \dfrac{152x}{x^2 - 3x}$

21. $x^2 - x + 1$, R -4; or $x^2 - x + 1 + \dfrac{-4}{x - 1}$

23. $a + 7$, R -47; or $a + 7 + \dfrac{-47}{a + 4}$

25. $x^2 - 5x - 23$, R -43; or $x^2 - 5x - 23 + \dfrac{-43}{x - 2}$

27. $3x^2 - 2x + 2$, R -3; or $3x^2 - 2x + 2 + \dfrac{-3}{x + 3}$

29. $y^2 + 2y + 1$, R 12; or $y^2 + 2y + 1 + \dfrac{12}{y - 2}$

31. $3x^3 + 9x^2 + 2x + 6$ **33.** $x^2 + 2x + 4$
35. $y^3 + 2y^2 + 4y + 8$ **37.** $\mathbf{D_W}$
39.   **40.**

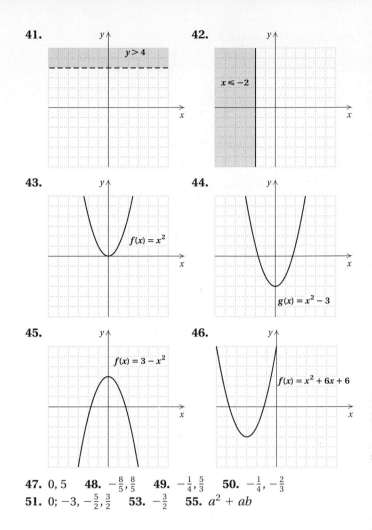

41.   **42.**

43.   **44.**

45.   **46.**

47. $0, 5$ **48.** $-\frac{8}{5}, \frac{8}{5}$ **49.** $-\frac{1}{4}, \frac{5}{3}$ **50.** $-\frac{1}{4}, -\frac{2}{3}$
51. $0; -3, -\frac{5}{2}, \frac{3}{2}$ **53.** $-\frac{3}{2}$ **55.** $a^2 + ab$

Margin Exercises, Section 5.4, pp. 445–448

1. $\dfrac{14y + 7}{14y - 2}$ **2.** $\dfrac{x}{x + 1}$ **3.** $\dfrac{b + a}{b - a}$ **4.** $\dfrac{a^2b^2}{b^2 + ab + a^2}$
5. $\dfrac{14y + 7}{14y - 2}$ **6.** $\dfrac{x}{x + 1}$ **7.** $\dfrac{b + a}{b - a}$ **8.** $\dfrac{a^2b^2}{b^2 + ab + a^2}$

Calculator Corner, p. 445

Left to the student

Exercise Set 5.4, p. 449

1. $\frac{26}{35}$ **3.** $\frac{88}{15}$ **5.** $\dfrac{x^3}{y^5}$ **7.** $\dfrac{3x + y}{x}$ **9.** $\dfrac{1 + 2a}{1 - a}$
11. $\dfrac{x^2 - 1}{x^2 + 1}$ **13.** $\dfrac{3y + 4x}{4y - 3x}$ **15.** $\dfrac{a^2(b - 3)}{b^2(a - 1)}$ **17.** $\dfrac{1}{a - b}$
19. $\dfrac{-1}{x(x + h)}$ **21.** $\dfrac{(x - 4)(x - 7)}{(x - 5)(x + 6)}$ **23.** $\dfrac{x + 1}{5 - x}$
25. $\dfrac{5x - 16}{4x + 1}$ **27.** $\dfrac{zw(w - z)}{w^2 - wz + z^2}$ **29.** $\dfrac{2x^2 - 11x - 27}{2x^2 + 21x + 13}$
31. $\mathbf{D_W}$ **33.** $2x(2x^2 + 10x + 3)$
34. $(y + 2)(y^2 - 2y + 4)$ **35.** $(y - 2)(y^2 + 2y + 4)$
36. $2x(x - 9)(x - 7)$ **37.** $(10x + 1)(100x^2 - 10x + 1)$
38. $(1 - 10a)(1 + 10a + 100a^2)$

39. $(y - 4x)(y^2 + 4xy + 16x^2)$
40. $\left(\frac{1}{2}a - 7\right)\left(\frac{1}{4}a^2 + \frac{7}{2}a + 49\right)$ **41.** $s = 3T - r$
42.

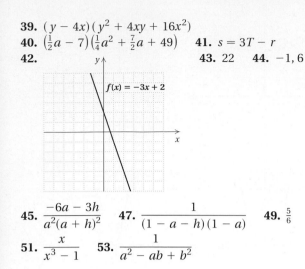

43. 22   **44.** $-1, 6$

45. $\dfrac{-6a - 3h}{a^2(a + h)^2}$ **47.** $\dfrac{1}{(1 - a - h)(1 - a)}$ **49.** $\frac{5}{6}$
51. $\dfrac{x}{x^3 - 1}$ **53.** $\dfrac{1}{a^2 - ab + b^2}$

Margin Exercises, Section 5.5, pp. 452–456

1. $\frac{2}{3}$ **2.** -31 **3.** No solution   **4.** -3 **5.** 7   **6.** -13
7. $4, -3$

Calculator Corner, p. 454

1. Left to the student   **2.** Left to the student

Study Tips, p. 457

1. Rational expression   **2.** Solutions
3. Rational expression   **4.** Rational expression
5. Rational expression   **6.** Solutions
7. Rational expression   **8.** Solutions   **9.** Solutions
10. Solutions   **11.** Rational expression   **12.** Solutions
13. Rational expression

Exercise Set 5.5, p. 458

1. $\frac{31}{4}$ **3.** $-\frac{12}{7}$ **5.** 144   **7.** $-1, -8$ **9.** 2   **11.** 11
13. 11   **15.** No solution   **17.** 2   **19.** 5   **21.** -145
23. $-\frac{10}{3}$ **25.** -3 **27.** $\frac{31}{5}$ **29.** $\frac{85}{12}$ **31.** $-6, 5$
33. No solution   **35.** $\frac{17}{4}$ **37.** No solution   **39.** $\frac{67}{35}, 4$
41. $-\frac{3}{2}, 2$ **43.** $\frac{17}{4}$ **45.** $\frac{3}{5}$ **47.** $\mathbf{D_W}$
48. $4(t + 5)(t^2 - 5t + 25)$
49. $(1 - t)(1 + t + t^2)(1 + t)(1 - t + t^2)$
50. $(a + 2b)(a^2 - 2ab + 4b^2)$
51. $(a - 2b)(a^2 + 2ab + 4b^2)$ **52.** 3   **53.** $-4, 3$
54. $-7, 7$ **55.** $\frac{1}{4}, \frac{2}{3}$
57. (a) $(-3.5, 1.\overline{3})$; **(b), (c)** Left to the student

Margin Exercises, Section 5.6, pp. 462–468

1. $2\frac{2}{5}$ hr   **2.** Pipe A: 32 hr; pipe B: 96 hr   **3.** 525 mg
4. 2440 wild horses   **5.** Jaime: 23 km/h; Mara: 15 km/h
6. 35.5 mph

Translating for Success, p. 469

1. N   **2.** B   **3.** A   **4.** C   **5.** E   **6.** G   **7.** I
8. K   **9.** M   **10.** O

Exercise Set 5.6, p. 470

1. $3\frac{3}{14}$ hr   **3.** $8\frac{4}{7}$ hr   **5.** $2\frac{19}{40}$ hr   **7.** Juan: 10 days; Ariel:
40 days   **9.** Skyler: 12 hr; Jake: 6 hr   **11.** 34 in. or less
13. About 50 home runs   **15.** 1160 trees   **17.** 954 deer
19. (a) 4.8 T; **(b)** 48 lb   **21.** 1   **23.** 12 mph   **25.** 2 mph
27. 5.2 ft/sec   **29.** Simone: $5\frac{1}{3}$ mph; Rosanna: $3\frac{1}{3}$ mph
31. $\mathbf{D_W}$ **32.** Domain: $\{-4, -2, 0, 1, 2, 4\}$;
range: $\{-2, 0, 2, 4, 5\}$ **33.** Domain: $[-5, 5]$; range: $[-4, 3]$
34. Domain: $[-5, 5]$; range: $[-5, 0]$
35. Domain: $[-5, 5]$; range: $[-5, 3]$
36.   **37.**

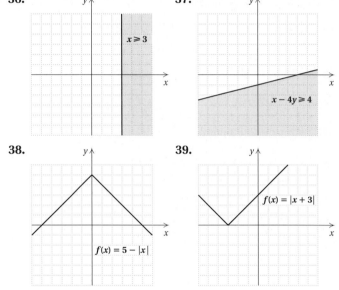

38.   **39.**

41. 30 mi   **43.** City: 261 mi; highway: 204 mi
45. $21\frac{9}{11}$ min after 4:00

Margin Exercises, Section 5.7, pp. 475–476

1. $T = \dfrac{PV}{k}$ **2.** $n = \dfrac{pT - IM}{Ip}$ **3.** $s = \dfrac{Fv}{g - F}$
4. $t = \dfrac{ab}{b + a}$

Exercise Set 5.7, p. 477

1. $W_2 = \dfrac{d_2 W_1}{d_1}$ **3.** $r_2 = \dfrac{Rr_1}{r_1 - R}$ **5.** $t = \dfrac{2s}{v_1 + v_2}$
7. $s = \dfrac{Rg}{g - R}$ **9.** $p = \dfrac{qf}{q - f}$ **11.** $a = \dfrac{bt}{b - t}$
13. $E = \dfrac{Inr}{n - I}$ **15.** $H^2 = \dfrac{704.5W}{I}$ **17.** $r = \dfrac{eR}{E - e}$
19. $R = \dfrac{3V + \pi h^3}{3\pi h^2}$ **21.** $r = \dfrac{A}{P} - 1,$ or $\dfrac{A - P}{P}$
23. $h = \dfrac{2gR^2}{V^2} - R,$ or $\dfrac{2gR^2 - RV^2}{V^2}$ **25.** $Q = \dfrac{2Tt - 2AT}{A - q}$
27. $\mathbf{D_W}$ **28.** Dimes: 2; nickels: 5; quarters: 5

29. 30-min: 4; 60-min: 8 **30.** -6 **31.** 6 **32.** 0
33. $8a^3 - 2a$ **34.** $-\frac{4}{5}$ **35.** $y = -\frac{4}{5}x + \frac{17}{5}$

Margin Exercises, Section 5.8, pp. 480–485

1. $\frac{2}{5}$; $y = \frac{2}{5}x$ **2.** 0.7; $y = 0.7x$ **3.** 50 volts
4. 1,575,000 tons **5.** 0.6; $y = \dfrac{0.6}{x}$ **6.** 16 hr
7. $y = 7x^2$ **8.** $y = \dfrac{9}{x^2}$ **9.** $y = \frac{1}{2}xz$ **10.** $y = \dfrac{5xz^2}{w}$
11. 490 m **12.** 1 ohm

Exercise Set 5.8, p. 486

1. 5; $y = 5x$ **3.** $\frac{2}{15}$; $y = \frac{2}{15}x$ **5.** $\frac{9}{4}$; $y = \frac{9}{4}x$
7. 285,360,000 cans **9.** $66\frac{2}{3}$ cm **11.** 90 g **13.** 40 kg
15. 98; $y = \dfrac{98}{x}$ **17.** 36; $y = \dfrac{36}{x}$ **19.** 0.05; $y = \dfrac{0.05}{x}$
21. 3.5 hr **23.** $\frac{2}{9}$ ampere **25.** 1.92 ft **27.** 160 cm³
29. $y = 15x^2$ **31.** $y = \dfrac{0.0015}{x^2}$ **33.** $y = xz$
35. $y = \frac{3}{10}xz^2$ **37.** $y = \dfrac{xz}{5wp}$ **39.** 36 mph **41.** 2.5 m
43. 50 earned runs **45.** 729 gal **47.** $\mathbf{D_W}$ **49.** Like
50. Complementary **51.** Opposites, additive
52. Vertical **53.** Intersection **54.** Linear
55. Multiplication principle **56.** $(0, a)$ **57.** $\dfrac{\pi}{4}$
59. Q varies directly as the square of p and inversely as the cube of q. **61.** $7.20

Concept Reinforcement, p. 490

1. False **2.** True **3.** True **4.** True **5.** False
6. False

Summary and Review: Chapter 5, p. 491

1. $-3, 3$ **2.** $\{x \mid x$ is a real number $and\ x \neq -3\ and\ x \neq 3\}$, or $(-\infty, -3) \cup (-3, 3) \cup (3, \infty)$ **3.** $\dfrac{x - 2}{3x + 2}$ **4.** $\dfrac{1}{a - 2}$
5. $48x^3$ **6.** $(x - 7)(x + 7)(3x + 1)$
7. $(x + 5)(x - 4)(x - 2)$ **8.** $\dfrac{y - 8}{2}$ **9.** $\dfrac{(x - 2)(x + 5)}{x - 5}$
10. $\dfrac{3a - 1}{a - 3}$ **11.** $\dfrac{(x^2 + 4x + 16)(x - 6)}{(x + 4)(x + 2)}$
12. $\dfrac{x - 3}{(x + 1)(x + 3)}$ **13.** $\dfrac{2x^3 + 2x^2y + 2xy^2 - 2y^3}{(x - y)(x + y)}$
14. $\dfrac{-y}{(y + 4)(y - 1)}$ **15.** $4b^2c - \frac{5}{2}bc^2 + 3abc$
16. $y - 14$, R -20; or $y - 14 + \dfrac{-20}{y - 6}$
17. $6x^2 - 9$, R $(5x + 22)$; or $6x^2 - 9 + \dfrac{5x + 22}{x^2 + 2}$
18. $x^2 + 9x + 40$, R 153; or $x^2 + 9x + 40 + \dfrac{153}{x - 4}$
19. $3x^3 - 8x^2 + 8x - 6$, R -1; or
$3x^3 - 8x^2 + 8x - 6 + \dfrac{-1}{x + 1}$ **20.** $\frac{3}{4}$

21. $\dfrac{a^2b^2}{2(a^2 - ab + b^2)}$ **22.** $\dfrac{(x - 9)(x - 6)}{(x - 3)(x + 6)}$
23. $\dfrac{4x^2 - 14x + 2}{3x^2 + 7x - 11}$ **24.** $\frac{28}{11}$ **25.** 6 **26.** No solution
27. 3 **28.** $-\frac{11}{3}$ **29.** 2 **30.** $5\frac{1}{7}$ hr
31.

	DISTANCE	SPEED	TIME
Downstream	50 mi	$x + 6$	t
Upstream	30 mi	$x - 6$	t

24 mph
32. 4000 mi **33.** $d = \dfrac{Wc}{c - W}$; $c = \dfrac{Wd}{d - W}$
34. $b = \dfrac{ta}{Sa - p}$; $t = \dfrac{Sab - pb}{a}$ **35.** $y = 4x$
36. $y = \dfrac{2500}{x}$ **37.** 20 min **38.** About 77.7
39. 500 watts **40.** $\mathbf{D_W}$ When adding or subtracting rational expressions, we use the LCM of the denominators (the LCD). When solving a rational equation or when solving a formula for a given letter, we multiply by the LCM of all the denominators to clear fractions. When simplifying a complex rational expression, we can use the LCM in either of two ways. We can multiply by a/a, where a is the LCM of all the denominators occurring in the expression. Or we can use the LCM to add or subtract as necessary in the numerator and in the denominator. **41.** $\mathbf{D_W}$ Rational equations differ from those previously studied because they contain variables in denominators. Because of this, possible solutions must be checked in the original equation to avoid division by 0. **42.** $a^2 + ab + b^2$ **43.** All real numbers except 0 and 13

Test: Chapter 5, p. 493

1. [5.1a] 1, 2 **2.** [5.1a] $\{x \mid x$ is a real number $and\ x \neq 1$ $and\ x \neq 2\}$, or $(-\infty, 1) \cup (1, 2) \cup (2, \infty)$
3. [5.1c] $\dfrac{3x + 2}{x - 2}$ **4.** [5.1c] $\dfrac{p^2 - p + 1}{p - 2}$
5. [5.2a] $(x + 3)(x - 2)(x + 5)$ **6.** [5.1d] $\dfrac{2(x + 5)}{x - 2}$
7. [5.2b] $\dfrac{x - 6}{(x + 4)(x + 6)}$ **8.** [5.1e] $\dfrac{y + 4}{2}$
9. [5.2b] $x + y$ **10.** [5.2c] $\dfrac{3x}{(x - 1)(x + 1)}$
11. [5.2c] $\dfrac{a^3 + a^2b + ab^2 + ab - b^2 - 2}{a^3 - b^3}$
12. [5.3a] $4s^2 + 3s - 2rs^2$ **13.** [5.3b] $y^2 - 5y + 25$
14. [5.3b] $4x^2 + 3x - 4$, R $(-8x + 2)$; or
$4x^2 + 3x - 4 + \dfrac{-8x + 2}{x^2 + 1}$ **15.** [5.3c] $x^2 + 6x + 20$, R 54;
or $x^2 + 6x + 20 + \dfrac{54}{x - 3}$ **16.** [5.3c] $4x^2 - 26x + 130$,
R -659; or $4x^2 - 26x + 130 + \dfrac{-659}{x + 5}$ **17.** [5.4a] $\dfrac{x + 1}{x}$

18. [5.4a] $\dfrac{b^2 - ab + a^2}{a^2b^2}$ **19.** [5.5a] $-1, 4$ **20.** [5.5a] 9

21. [5.5a] No solution   **22.** [5.5a] $-\dfrac{7}{2}, 5$

23. [5.5a] $\dfrac{17}{8}$ **24.** [5.6a] 2 hr   **25.** [5.6c] $3\dfrac{3}{11}$ mph

26. [5.6b] $14\dfrac{2}{17}$ gal   **27.** [5.7a] $a = \dfrac{Tb}{T - b}$; $b = \dfrac{Ta}{a + T}$

28. [5.7a] $a = \dfrac{2b}{Qb + t}$ **29.** [5.8e] $Q = \dfrac{5}{2}xy$

30. [5.8c] $y = \dfrac{250}{x}$ **31.** [5.8b] \$495   **32.** [5.8d] $7\dfrac{1}{2}$ hr

33. [5.8f] 615.44 cm^2 **34.** [5.5a] All real numbers except 0 and 15   **35.** [5.2a] $(1 - t^6)(1 + t^6)$

36. [5.4a], [5.5a] x-intercept: $(11, 0)$; y-intercept: $\left(0, -\dfrac{33}{5}\right)$

CHAPTER 6

Margin Exercises, Section 6.1, pp. 496–502

1. $3, -3$ **2.** $6, -6$ **3.** $11, -11$ **4.** 1 **5.** 6 **6.** $\dfrac{9}{10}$
7. 0.08 **8.** (a) 4; (b) -4; (c) does not exist as a real number   **9.** (a) 7; (b) -7; (c) does not exist as a real number   **10.** (a) 12; (b) -12; (c) does not exist as a real number   **11.** $\dfrac{5}{8}$ **12.** -0.9 **13.** 1.2 **14.** 4.123
15. 6.325 **16.** 33.734 **17.** -29.455 **18.** 0.793

19. -5.569 **20.** $28 + x$ **21.** $\dfrac{y}{y + 3}$ **22.** 2;

$\sqrt{22} \approx 4.690$; does not exist as a real number
23. -2; $-\sqrt{7} \approx -2.646$; does not exist as a real number
24. Domain $= \{x \mid x \ge 5\} = [5, \infty)$
25. Domain $= \left\{x \mid x \ge -\dfrac{3}{2}\right\} = \left[-\dfrac{3}{2}, \infty\right)$
26.   **27.**

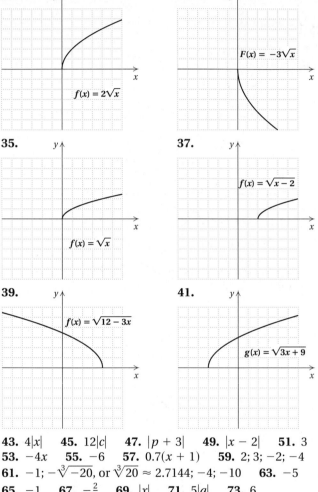

28. $|y|$ **29.** 24 **30.** $5|y|$ **31.** $4|y|$ **32.** $|x + 7|$
33. $2|x - 2|$ **34.** $7|y + 5|$ **35.** $|x - 3|$ **36.** -4
37. $3y$ **38.** $2(x + 2)$ **39.** $-\dfrac{7}{4}$
40. $-3; 0$; $\sqrt[3]{-5} \approx -1.710$; $\sqrt[3]{7} \approx 1.913$ **41.** 3 **42.** -3
43. x **44.** y **45.** 0 **46.** $-2x$ **47.** $3x + 2$ **48.** 3
49. -3 **50.** Does not exist as a real number   **51.** 0
52. $2|x - 2|$ **53.** $|x|$ **54.** $|x + 3|$ **55.** $x + 3$ **56.** $3x$

Calculator Corner, p. 503

1. 6.557 **2.** 102.308 **3.** -96.985 **4.** -0.804
5. 7.469 **6.** 9.283 **7.** 2.012 **8.** -2.812 **9.** 0.775
10. 0.775 **11.** 3.162 **12.** 4.378

Exercise Set 6.1, p. 504

1. $4, -4$ **3.** $12, -12$ **5.** $20, -20$ **7.** $-\dfrac{7}{6}$ **9.** 14
11. 0.06 **13.** Does not exist as a real number

15. 18.628 **17.** 1.962 **19.** $y^2 + 16$ **21.** $\dfrac{x}{y - 1}$

23. $\sqrt{20} \approx 4.472$; 0; does not exist as a real number; does not exist as a real number
25. $\sqrt{11} \approx 3.317$; does not exist as a real number; $\sqrt{11} \approx 3.317$; 12 **27.** Domain $= \{x \mid x \ge 2\} = [2, \infty)$
29. About 24.5 mph; about 54.8 mph
31.   **33.**

35.   **37.**

39.   **41.**

43. $4|x|$ **45.** $12|c|$ **47.** $|p + 3|$ **49.** $|x - 2|$ **51.** 3
53. $-4x$ **55.** -6 **57.** $0.7(x + 1)$ **59.** $2; 3; -2; -4$
61. -1; $-\sqrt[3]{-20}$, or $\sqrt[3]{20} \approx 2.7144$; -4; -10 **63.** -5
65. -1 **67.** $-\dfrac{2}{3}$ **69.** $|x|$ **71.** $5|a|$ **73.** 6
75. $|a + b|$ **77.** y **79.** $x - 2$ **81.** $\mathbf{D_W}$ **83.** $-2, 1$
84. $-1, 0$ **85.** $-\dfrac{7}{2}, \dfrac{7}{2}$ **86.** $4, 9$ **87.** $-2, \dfrac{5}{3}$ **88.** $\dfrac{5}{2}$
89. $0, \dfrac{5}{2}$ **90.** $0, 1$ **91.** $a^9b^6c^{15}$ **92.** $10a^{10}b^9$
93. Domain $= \{x \mid -3 \le x < 2\} = [-3, 2)$ **95.** $1.7; 2.2; 3.2$
97. (a) Domain: $(-\infty, \infty)$; range: $(-\infty, \infty)$; (b) domain: $(-\infty, \infty)$; range: $(-\infty, \infty)$; (c) domain: $[-3, \infty)$; range: $(-\infty, 2]$; (d) domain: $[0, \infty)$; range: $[0, \infty)$; (e) domain: $[3, \infty)$; range: $[0, \infty)$

Margin Exercises, Section 6.2, pp. 508–512

1. $\sqrt[4]{y}$ **2.** $\sqrt{3a}$ **3.** 2 **4.** 5 **5.** $\sqrt[5]{a^3b^2c}$
6. $(19ab)^{1/3}$ **7.** $19(ab)^{1/3}$ **8.** $\left(\dfrac{x^2y}{16}\right)^{1/5}$ **9.** $7(2ab)^{1/4}$

10. $\sqrt[5]{x^3}$ **11.** 4 **12.** 32 **13.** $(7abc)^{4/3}$ **14.** $6^{7/5}$
15. $\frac{1}{2}$ **16.** $\frac{1}{(3xy)^{7/8}}$ **17.** $\frac{1}{27}$ **18.** $\frac{7p^{3/4}}{q^{6/5}}$ **19.** $\left(\frac{7n}{11m}\right)^{2/3}$
20. $7^{14/15}$ **21.** $5^{1/3}$ **22.** $9^{2/5}$ **23.** $\frac{q^{1/8}}{p^{1/3}}$ **24.** $\sqrt{a}$
25. x **26.** $\sqrt{2}$ **27.** $\sqrt[4]{xy^2}$ **28.** $a^2\sqrt{b}$ **29.** ab^2
30. $\sqrt[4]{63}$ **31.** $\sqrt[6]{x^4y^3z^5}$ **32.** $\sqrt[4]{ab}$ **33.** $\sqrt[7]{5m}$
34. $\sqrt[6]{m}$ **35.** $a^{20}b^{12}c^4$ **36.** $\sqrt[10]{x}$

Calculator Corner, p. 510

1. 3.344 **2.** 3.281 **3.** 0.283 **4.** 11.053
5. 5.527×10^{-5} **6.** 2

Exercise Set 6.2, p. 513

1. $\sqrt[7]{y}$ **3.** 2 **5.** $\sqrt[5]{a^3b^3}$ **7.** 8 **9.** 343 **11.** $17^{1/2}$
13. $18^{1/3}$ **15.** $(xy^2z)^{1/5}$ **17.** $(3mn)^{3/2}$ **19.** $(8x^2y)^{5/7}$
21. $\frac{1}{3}$ **23.** $\frac{1}{1000}$ **25.** $\frac{3}{x^{1/4}}$ **27.** $\frac{1}{(2rs)^{3/4}}$ **29.** $\frac{2a^{3/4}c^{2/3}}{b^{1/2}}$
31. $\left(\frac{8yz}{7x}\right)^{3/5}$ **33.** $x^{2/3}$ **35.** $\frac{x^4}{2^{1/3}y^{2/7}}$ **37.** $\frac{7x}{z^{1/3}}$
39. $\frac{5ac^{1/2}}{3}$ **41.** $5^{7/8}$ **43.** $7^{1/4}$ **45.** $4.9^{1/2}$
47. $6^{3/28}$ **49.** $a^{23/12}$ **51.** $a^{8/3}b^{5/2}$ **53.** $\frac{1}{x^{2/7}}$
55. $\sqrt[3]{a}$ **57.** x^5 **59.** $\frac{1}{x^3}$ **61.** a^5b^5 **63.** $\sqrt{2}$
65. $\sqrt[3]{2x}$ **67.** x^2y^3 **69.** $2c^2d^3$ **71.** $\sqrt[12]{7^4 \cdot 5^3}$
73. $\sqrt[20]{5^5 \cdot 7^4}$ **75.** $\sqrt[6]{4x^5}$ **77.** a^6b^{12} **79.** $\sqrt[18]{m}$
81. $\sqrt[12]{x^4y^3z^2}$ **83.** $\sqrt[30]{\frac{d^{35}}{c^{99}}}$ **85.** $\mathbf{D_W}$ **87.** $a = \frac{Ab}{b-A}$
88. $s = \frac{Qt}{Q-t}$ **89.** $t = \frac{Qs}{s+Q}$ **90.** $b = \frac{ta}{t-a}$
91. Left to the student

Margin Exercises, Section 6.3, pp. 515–519

1. $\sqrt{133}$ **2.** $\sqrt{21pq}$ **3.** $\sqrt[4]{2821}$ **4.** $\sqrt[3]{\frac{10}{pq}}$
5. $\sqrt[6]{500}$ **6.** $\sqrt[6]{25x^3y^2}$ **7.** $4\sqrt{2}$ **8.** $2\sqrt[3]{10}$
9. $10\sqrt{3}$ **10.** $6y$ **11.** $2a\sqrt{3b}$ **12.** $2bc\sqrt{3ab}$
13. $2\sqrt[3]{2}$ **14.** $3xy^2\sqrt[3]{3xy^2}$ **15.** $3\sqrt{2}$ **16.** $6y\sqrt{7}$
17. $3x\sqrt[3]{4y}$ **18.** $7\sqrt{3ab}$ **19.** 5 **20.** $56\sqrt{xy}$ **21.** $5a$
22. $\frac{20}{7}$ **23.** $\frac{5}{6}$ **24.** $\frac{x}{10}$ **25.** $\frac{3x\sqrt[3]{2x^2}}{5}$ **26.** $\sqrt[12]{xy^2}$

Exercise Set 6.3, p. 520

1. $2\sqrt{6}$ **3.** $3\sqrt{10}$ **5.** $5\sqrt[3]{2}$ **7.** $6x^2\sqrt{5}$
9. $3x^2\sqrt[3]{2x^2}$ **11.** $2t^2\sqrt[3]{10t^2}$ **13.** $2\sqrt[4]{5}$ **15.** $4a\sqrt{2b}$
17. $3x^2y^2\sqrt[4]{3y^2}$ **19.** $2xy^3\sqrt[5]{3x^2}$ **21.** $5\sqrt{2}$ **23.** $3\sqrt{10}$
25. 2 **27.** $30\sqrt{3}$ **29.** $3x^4\sqrt{2}$ **31.** $5bc^2\sqrt{2b}$
33. $a\sqrt[3]{10}$ **35.** $2y^3\sqrt[3]{2}$ **37.** $4\sqrt[4]{4}$ **39.** $4a^3b\sqrt{6ab}$
41. $\sqrt[6]{200}$ **43.** $\sqrt[4]{12}$ **45.** $a\sqrt[4]{a}$ **47.** $b\sqrt[10]{b^9}$
49. $xy\sqrt[6]{xy^5}$ **51.** $3\sqrt{2}$ **53.** $\sqrt{5}$ **55.** 3 **57.** $y\sqrt{7y}$

59. $2\sqrt[3]{a^2b}$ **61.** $4\sqrt{xy}$ **63.** $2x^2y^2$ **65.** $\frac{1}{\sqrt[6]{a}}$
67. $\sqrt[12]{a^5}$ **69.** $\sqrt[12]{x^2y^5}$ **71.** $\frac{5}{6}$ **73.** $\frac{4}{7}$ **75.** $\frac{5}{3}$ **77.** $\frac{7}{y}$
79. $\frac{5y\sqrt{y}}{x^2}$ **81.** $\frac{3a\sqrt[3]{a}}{2b}$ **83.** $\frac{3x}{2}$ **85.** $\frac{2x\sqrt[5]{x^3}}{y^2}$
87. $\frac{x^2\sqrt[6]{x}}{yz^2}$ **89.** $\mathbf{D_W}$ **91.** $2\frac{2}{3}$ hr; 8 hr **92.** Height: 4 in.;
base: 6 in. **93.** 8 **94.** $\frac{15}{2}$ **95.** No solution
96. No solution **97.** (a) 1.62 sec; (b) 1.99 sec; (c) 2.20 sec
99. $2yz\sqrt{2z}$

Margin Exercises, Section 6.4, pp. 524–525

1. $13\sqrt{2}$ **2.** $10\sqrt[4]{5x} - \sqrt{7}$ **3.** $19\sqrt{5}$
4. $(3y+4)\sqrt[3]{y^2} + 2y^2$ **5.** $2\sqrt{x-1}$ **6.** $5\sqrt{6} + 3\sqrt{14}$
7. $a\sqrt[3]{3} - \sqrt[3]{2a^2}$ **8.** $-4 - 9\sqrt{6}$
9. $3\sqrt{ab} - 4\sqrt{3a} + 6\sqrt{3b} - 24$ **10.** -3 **11.** $p - q$
12. $20 - 4y\sqrt{5} + y^2$ **13.** $58 + 12\sqrt{6}$

Exercise Set 6.4, p. 526

1. $11\sqrt{5}$ **3.** $\sqrt[3]{7}$ **5.** $13\sqrt[3]{y}$ **7.** $-8\sqrt{6}$ **9.** $6\sqrt[3]{3}$
11. $21\sqrt{3}$ **13.** $38\sqrt{5}$ **15.** $122\sqrt{2}$ **17.** $9\sqrt[3]{2}$
19. $29\sqrt{2}$ **21.** $(1 + 6a)\sqrt{5a}$ **23.** $(2 - x)\sqrt[3]{3x}$
25. $(21x + 1)\sqrt{3x}$ **27.** $2 + 3\sqrt{2}$ **29.** $15\sqrt[3]{4}$
31. $(x + 1)\sqrt[3]{6x}$ **33.** $3\sqrt{a-1}$ **35.** $(x + 3)\sqrt{x-1}$
37. $4\sqrt{5} - 10$ **39.** $\sqrt{6} - \sqrt{21}$ **41.** $-12 + 6\sqrt{3}$
43. $2\sqrt{15} - 6\sqrt{3}$ **45.** -6 **47.** $3a\sqrt[3]{2}$ **49.** 1
51. -12 **53.** 44 **55.** 1 **57.** 3 **59.** -19
61. $a - b$ **63.** $1 + \sqrt{5}$ **65.** $7 + 3\sqrt{3}$ **67.** -6
69. $a + \sqrt{3a} + \sqrt{2a} + \sqrt{6}$ **71.** $2\sqrt[3]{9} - 3\sqrt[3]{6} - 2\sqrt[3]{4}$
73. $7 + 4\sqrt{3}$ **75.** $\sqrt[5]{72} + 3 - \sqrt[5]{24} - \sqrt[5]{81}$ **77.** $\mathbf{D_W}$
79. $\frac{x(x^2 + 4)}{(x + 4)(x + 3)}$ **80.** $\frac{(a + 2)(a + 4)}{a}$ **81.** $a - 2$
82. $\frac{(y - 3)(y - 3)}{y + 3}$ **83.** $\frac{4(3x - 1)}{3(4x + 1)}$ **84.** $\frac{x}{x + 1}$
85. $\frac{pq}{q + p}$ **86.** $\frac{a^2b^2}{b^2 - ab + a^2}$ **87.** $-\frac{29}{3}, 5$
88. $\left\{x \mid -\frac{29}{3} < x < 5\right\}$, or $\left(-\frac{29}{3}, 5\right)$
89. $\left\{x \mid x \le -\frac{29}{3} \text{ or } x \ge 5\right\}$, or $\left(-\infty, -\frac{29}{3}\right] \cup [5, \infty)$
90. $-12, -\frac{2}{5}$ **91.** Domain $= (-\infty, \infty)$ **93.** 6
95. $14 + 2\sqrt{15} - 6\sqrt{2} - 2\sqrt{30}$ **97.** $3\sqrt[3]{3} + 2\sqrt[3]{9} - 8$

Margin Exercises, Section 6.5, pp. 530–532

1. $\frac{\sqrt{10}}{5}$ **2.** $\frac{\sqrt[3]{10}}{2}$ **3.** $\frac{2\sqrt{3ab}}{3b}$ **4.** $\frac{\sqrt[4]{56}}{2}$ **5.** $\frac{x\sqrt[3]{12x^2y^2}}{2y}$
6. $\frac{7\sqrt[3]{2x^2y}}{2y^2}$ **7.** $c^2 - b$ **8.** $a - b$ **9.** $\frac{\sqrt{3} + y}{\sqrt{3} + y}$
10. $\frac{\sqrt{2} - \sqrt{3}}{\sqrt{2} - \sqrt{3}}$ **11.** $6 - 2\sqrt{2}$ **12.** $-7 - 6\sqrt{2}$

Exercise Set 6.5, p. 533

1. $\dfrac{\sqrt{15}}{3}$ **3.** $\dfrac{\sqrt{22}}{2}$ **5.** $\dfrac{2\sqrt{15}}{35}$ **7.** $\dfrac{2\sqrt[3]{6}}{3}$ **9.** $\dfrac{\sqrt[3]{75ac^2}}{5c}$

11. $\dfrac{y\sqrt[3]{9yx^2}}{3x^2}$ **13.** $\dfrac{\sqrt[4]{s^3t^3}}{st}$ **15.** $\dfrac{\sqrt{15x}}{10}$ **17.** $\dfrac{\sqrt[3]{100xy}}{5x^2y}$

19. $\dfrac{\sqrt[4]{2xy}}{2x^2y}$ **21.** $\dfrac{54 + 9\sqrt{10}}{26}$ **23.** $-2\sqrt{35} - 2\sqrt{21}$

25. $\dfrac{\sqrt{15} + 20 - 6\sqrt{2} - 8\sqrt{30}}{-77}$ **27.** $\dfrac{6 - 5\sqrt{a} + a}{9 - a}$

29. $\dfrac{3\sqrt{6} + 4}{2}$ **31.** $\dfrac{x - 2\sqrt{xy} + y}{x - y}$ **33.** $\mathbf{D_W}$ **35.** 30

36. $-\dfrac{19}{5}$ **37.** 1 **38.** $\dfrac{x - 2}{x + 3}$ **39.** Left to the student

41. $-\dfrac{3\sqrt{a^2 - 3}}{a^2 - 3}$

Margin Exercises, Section 6.6, pp. 536–541

1. 100 **2.** No solution **3.** 1 **4.** 2, 5 **5.** 4 **6.** 17
7. 9 **8.** 27 **9.** 5 **10.** About 1149.9 ft/sec
11. About 88°F

Calculator Corner, p. 537

1. Left to the student **2.** Left to the student

Exercise Set 6.6, p. 542

1. $\dfrac{19}{2}$ **3.** $\dfrac{49}{6}$ **5.** 57 **7.** $\dfrac{92}{5}$ **9.** -1 **11.** No solution
13. 3 **15.** 19 **17.** -6 **19.** $\dfrac{1}{64}$ **21.** 9 **23.** 15
25. 2, 5 **27.** 6 **29.** 5 **31.** 9 **33.** 7 **35.** $\dfrac{80}{9}$
37. 2, 6 **39.** -1 **41.** No solution **43.** 3 **45.** About
44.1 mi **47.** About 680 ft **49.** About 117 ft **51.** About
4.3 mi **53.** 151.25 ft; 281.25 ft **55.** About 25°F
57. About 0.81 ft **59.** $\mathbf{D_W}$ **61.** $4\frac{4}{9}$ hr **62.** Jeff: $1\frac{1}{3}$ hr;
Grace: 4 hr **63.** 2808 mi **64.** 84 hr **65.** $0, -2.8$
66. $0, \frac{5}{3}$ **67.** $-8, 8$ **68.** $-3, \frac{7}{2}$ **69.** $2ah + h^2$
70. $2ah + h^2 - h$ **71.** $4ah + 2h^2 - 3h$
72. $4ah + 2h^2 + 3h$ **73.** Left to the student
75. 6912 **77.** 0 **79.** $-6, -3$ **81.** 2 **83.** $0, \frac{125}{4}$
85. 2 **87.** $\frac{1}{2}$ **89.** 3

Margin Exercises, Section 6.7, pp. 547–548

1. $\sqrt{41}$; 6.403 **2.** $\sqrt{6}$; 2.449 **3.** $\sqrt{200}$; 14.142
4. $\sqrt{8200}$ ft; 90.554 ft **5.** 7.4 ft

Translating for Success, p. 549

1. J **2.** B **3.** O **4.** M **5.** K **6.** I **7.** G **8.** E
9. F **10.** A

Exercise Set 6.7, p. 550

1. $\sqrt{34}$; 5.831 **3.** $\sqrt{450}$; 21.213 **5.** 5 **7.** $\sqrt{43}$; 6.557
9. $\sqrt{12}$; 3.464 **11.** $\sqrt{n - 1}$ **13.** $\sqrt{116}$ ft; 10.770 ft

15. 7.1 ft **17.** $\sqrt{10{,}561}$ ft; 102.767 ft **19.** $s + s\sqrt{2}$
21. $\sqrt{181}$ cm; 13.454 cm **23.** 12 in. **25.** $(3, 0), (-3, 0)$
27. $\sqrt{340} + 8$ ft; 26.439 ft **29.** $\sqrt{420.125}$ in. ≈ 20.497 in.
31. $\mathbf{D_W}$ **33.** Flash: $67\frac{2}{3}$ mph; Crawler: $53\frac{2}{3}$ mph
34. $3\frac{3}{4}$ mph **35.** $-7, \frac{3}{2}$ **36.** 3, 8 **37.** 1 **38.** $-2, 2$
39. 13 **40.** 7 **41.** 26 packets **43.** $\sqrt{75}$ cm

Margin Exercises, Section 6.8, pp. 554–560

1. $i\sqrt{5}$, or $\sqrt{5}i$ **2.** $5i$ **3.** $-i\sqrt{11}$, or $-\sqrt{11}i$ **4.** $-6i$
5. $3i\sqrt{6}$, or $3\sqrt{6}i$ **6.** $15 - 3i$ **7.** $-12 + 6i$ **8.** $3 - 5i$
9. $12 - 7i$ **10.** -10 **11.** $-\sqrt{34}$ **12.** 42
13. $-9 - 12i$ **14.** $-35 - 25i$ **15.** $-14 + 8i$
16. $11 + 10i$ **17.** $5 + 12i$ **18.** $-i$ **19.** 1 **20.** i
21. -1 **22.** $8 - i$ **23.** 3 **24.** $-7 - 6i$ **25.** $-1 + i$
26. $6 - 3i$ **27.** $-9 + 5i$ **28.** $\pi + \frac{1}{4}i$ **29.** 53 **30.** 10
31. $2i$ **32.** $\frac{10}{17} - \frac{11}{17}i$
33.

$$x^2 + 1 = 0$$

$$\begin{array}{c|c} (-i)^2 + 1 & ?\ 0 \\ i^2 + 1 & \\ -1 + 1 & \\ 0 & \text{TRUE} \end{array}$$

Yes

34.

$$x^2 - 2x + 2 = 0$$

$$\begin{array}{c|c} (1 - i)^2 - 2(1 - i) + 2 & ?\ 0 \\ 1 - 2i + i^2 - 2 + 2i + 2 & \\ 1 - 2i - 1 - 2 + 2i + 2 & \\ 0 & \text{TRUE} \end{array}$$

Yes

Calculator Corner, p. 559

1. $-2 - 9i$ **2.** $20 + 17i$ **3.** $-47 - 161i$
4. $-\frac{151}{290} + \frac{73}{290}i$ **5.** -20 **6.** -28.373 **7.** $-\frac{16}{25} - \frac{1}{50}i$
8. 81 **9.** $117 + 118i$ **10.** $-\frac{14}{169} + \frac{34}{169}i$

Exercise Set 6.8, p. 561

1. $i\sqrt{35}$, or $\sqrt{35}i$ **3.** $4i$ **5.** $-2i\sqrt{3}$, or $-2\sqrt{3}i$
7. $i\sqrt{3}$, or $\sqrt{3}i$ **9.** $9i$ **11.** $7i\sqrt{2}$, or $7\sqrt{2}i$ **13.** $-7i$
15. $4 - 2\sqrt{15}i$, or $4 - 2i\sqrt{15}$ **17.** $(2 + 2\sqrt{3})i$
19. $12 - 4i$ **21.** $9 - 5i$ **23.** $7 + 4i$ **25.** $-4 - 4i$
27. $-1 + i$ **29.** $11 + 6i$ **31.** -18 **33.** $-\sqrt{14}$
35. 21 **37.** $-6 + 24i$ **39.** $1 + 5i$ **41.** $18 + 14i$
43. $38 + 9i$ **45.** $2 - 46i$ **47.** $5 - 12i$ **49.** $-24 + 10i$
51. $-5 - 12i$ **53.** $-i$ **55.** 1 **57.** -1 **59.** i
61. -1 **63.** $-125i$ **65.** 8 **67.** $1 - 23i$ **69.** 0
71. 0 **73.** 1 **75.** $5 - 8i$ **77.** $2 - \dfrac{\sqrt{6}}{2}i$ **79.** $\frac{9}{10} + \frac{13}{10}i$
81. $-i$ **83.** $-\frac{3}{7} - \frac{8}{7}i$ **85.** $\frac{6}{5} - \frac{2}{5}i$ **87.** $-\frac{8}{41} + \frac{10}{41}i$
89. $-\frac{4}{3}i$ **91.** $-\frac{1}{2} - \frac{1}{4}i$ **93.** $\frac{3}{5} + \frac{4}{5}i$
95.

$$x^2 - 2x + 5 = 0$$

$$\begin{array}{c|c} (1 - 2i)^2 - 2(1 - 2i) + 5 & ?\ 0 \\ 1 - 4i + 4i^2 - 2 + 4i + 5 & \\ 1 - 4i - 4 - 2 + 4i + 5 & \\ 0 & \text{TRUE} \end{array}$$

Yes

97.

$$x^2 - 4x - 5 = 0$$

$$\frac{\overline{(2+i)^2 - 4(2+i) - 5 \; \overset{?}{\vert} \; 0}}{\begin{array}{l} 4 + 4i + i^2 - 8 - 4i - 5 \\ 4 + 4i - 1 - 8 - 4i - 5 \end{array} \Big\vert}$$

$$ -10 \; \Big\vert \quad \text{FALSE}$$

No

99. D_W **101.** Rational **102.** Difference of squares
103. Coordinates **104.** Positive **105.** Proportion
106. Trinomial square **107.** Negative **108.** Zero
products **109.** $-4 - 8i$; $-2 + 4i$; $8 - 6i$ **111.** $-3 - 4i$
113. $-88i$ **115.** 8 **117.** $\frac{3}{5} + \frac{9}{5}i$ **119.** 1

Concept Reinforcement, p. 565

1. True **2.** False **3.** False **4.** True **5.** False
6. True

Summary and Review: Chapter 6, p. 566

1. 27.893 **2.** 6.378 **3.** $f(0)$, $f(-1)$, and $f(1)$ do not exist
as real numbers; $f\left(\frac{41}{3}\right) = 5$ **4.** Domain $= \left\{x \mid x \geq \frac{16}{3}\right\}$, or
$\left[\frac{16}{3}, \infty\right)$ **5.** $9|a|$ **6.** $7|z|$ **7.** $|c - 3|$ **8.** $|x - 3|$
9. -10 **10.** $-\frac{1}{3}$ **11.** 2; -2; 3 **12.** $|x|$ **13.** 3
14. $\sqrt[5]{a}$ **15.** 512 **16.** $31^{1/2}$ **17.** $(a^2 b^3)^{1/5}$ **18.** $\frac{1}{7}$
19. $\frac{1}{4x^{2/3}y^{2/3}}$ **20.** $\frac{5b^{1/2}}{a^{3/4}c^{2/3}}$ **21.** $\frac{3a}{t^{1/4}}$ **22.** $\frac{1}{x^{2/5}}$
23. $7^{1/6}$ **24.** x^7 **25.** $3x^2$ **26.** $\sqrt[12]{x^4 y^3}$ **27.** $\sqrt[12]{x^7}$
28. $7\sqrt{5}$ **29.** $-3\sqrt[3]{4}$ **30.** $5b^2\sqrt[3]{2a^2}$ **31.** $\frac{7}{6}$
32. $\frac{4}{3}x^2$ **33.** $\frac{2x^2}{3y^3}$ **34.** $\sqrt{15xy}$ **35.** $3a\sqrt[3]{a^2 b^2}$
36. $\sqrt[15]{a^5 b^9}$ **37.** $y\sqrt[3]{6}$ **38.** $\frac{5}{2}\sqrt{x}$ **39.** $\sqrt[12]{x^5}$ **40.** $7\sqrt[3]{x}$
41. $3\sqrt{3}$ **42.** $(2x + y^2)\sqrt[3]{x}$ **43.** $15\sqrt{2}$
44. $-43 - 2\sqrt{10}$ **45.** $8 - 2\sqrt{7}$ **46.** $9 - \sqrt[3]{4}$
47. $\frac{2\sqrt{6}}{3}$ **48.** $\frac{2\sqrt{a} - 2\sqrt{b}}{a - b}$ **49.** 13 **50.** 4 **51.** 4
52. 9 cm **53.** $\sqrt{24}$ ft; 4.899 ft **54.** About 4166 rpm
55. 4480 rpm **56.** 25 **57.** $\sqrt{46}$; 6.782
58. $\left(5 + 2\sqrt{2}\right)i$ **59.** $-2 - 9i$ **60.** $1 + i$ **61.** 29
62. i **63.** $9 - 12i$ **64.** $\frac{2}{5} + \frac{3}{5}i$ **65.** 3 **66.** No
67.

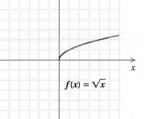

$f(x) = \sqrt{x}$

68. D_W The procedure for solving radical equations is to
isolate one of the radical terms, use the principle of powers,
repeat these steps if necessary until all radicals are
eliminated, solve, and then check the possible solutions.
A check is necessary since the principle of powers does not
always yield equivalent equations. **69.** -1 **70.** 3

Test: Chapter 6, p. 568

1. [6.1a] 12.166 **2.** [6.1a] 2; does not exist as a real
number **3.** [6.1a] Domain $= \{x \mid x \leq 2\}$, or $(-\infty, 2]$
4. [6.1b] $3|q|$ **5.** [6.1b] $|x + 5|$ **6.** [6.1c] $-\frac{1}{10}$
7. [6.1d] x **8.** [6.1d] 4 **9.** [6.2a] $\sqrt[3]{a^2}$ **10.** [6.2a] 8
11. [6.2a] $37^{1/2}$ **12.** [6.2a] $(5xy^2)^{5/2}$ **13.** [6.2b] $\frac{1}{10}$
14. [6.2b] $\frac{8a^{3/4}}{b^{3/2}c^{2/5}}$ **15.** [6.2c] $\frac{x^{8/5}}{y^{9/5}}$ **16.** [6.2c] $\frac{1}{2.9^{31/24}}$
17. [6.2d] $\sqrt[4]{x}$ **18.** [6.2d] $2x\sqrt{x}$ **19.** [6.2d] $\sqrt[15]{a^6 b^5}$
20. [6.2d] $\sqrt[12]{8y^7}$ **21.** [6.3a] $2\sqrt{37}$ **22.** [6.3a] $2\sqrt[4]{5}$
23. [6.3a] $2a^3 b^4\sqrt[3]{3a^2 b}$ **24.** [6.3b] $\frac{2x\sqrt[3]{2x^2 y^2}}{y^3}$
25. [6.3b] $\frac{5x}{6y^2}$ **26.** [6.3a] $\sqrt[3]{10xy^2}$ **27.** [6.3a] $xy\sqrt[4]{x}$
28. [6.3b] $\sqrt[5]{x^2 y^2}$ **29.** [6.3b] $2\sqrt{a}$ **30.** [6.4a] $38\sqrt{2}$
31. [6.4b] -20 **32.** [6.4b] $9 + 6\sqrt{x} + x$
33. [6.5b] $\frac{13 + 8\sqrt{2}}{-41}$ **34.** [6.6a] 35 **35.** [6.6b] 7
36. [6.6a] 5 **37.** [6.7a] 7 ft **38.** [6.6c] 3600 ft
39. [6.7a] $\sqrt{98}$; 9.899 **40.** [6.7a] 2 **41.** [6.8a] $11i$
42. [6.8b] $7 + 5i$ **43.** [6.8c] $37 + 9i$ **44.** [6.8d] $-i$
45. [6.8e] $-\frac{77}{50} + \frac{7}{25}i$ **46.** [6.8f] No **47.** [6.8c, e] $-\frac{17}{4}i$
48. [6.6b] 3

Cumulative Review: Chapters R–6, p. 570

1. [2.2b] 30 games **2.** [4.8b] 9 teams **3.** [R.2d] $\frac{21}{50}$
4. [R.2c] -37.2 **5.** [R.6b] $25c - 31$ **6.** [R.3c] 174
7. [R.7b] $\frac{64x^{21}}{27y^{15}}$ **8.** [4.1c] $-3x^3 + 9x^2 + 3x - 3$
9. [4.2c] $4x^4 - 4x^2 y + y^2$
10. [4.2a] $15x^4 - x^3 - 9x^2 + 5x - 2$
11. [5.1d] $\frac{(x + 4)(x - 7)}{x + 7}$ **12.** [5.2c] $\frac{-2x + 4}{(x + 2)(x - 3)}$
13. [5.4a] $\frac{y - 6}{y - 9}$ **14.** [5.3b, c] $y^2 + y - 2 + \frac{-1}{y + 2}$
15. [6.1c] $-2x$ **16.** [6.3a] $4(x - 1)$ **17.** [6.4a] $57\sqrt{3}$
18. [6.3a] $4xy^2\sqrt{y}$ **19.** [6.5b] $\sqrt{30} + \sqrt{15}$
20. [6.1d] $\frac{m^2 n^4}{2}$ **21.** [6.2c] $6^{8/9}$ **22.** [6.8b] $3 + 5i$
23. [6.8e] $\frac{7}{61} - \frac{16}{61}i$ **24.** [1.1d] 2 **25.** [1.2a] $c = 8M + 3$
26. [1.4c] $\{a \mid a > -7\}$, or $(-7, \infty)$
27. [1.5a] $\{x \mid -10 < x < 13\}$, or $(-10, 13)$ **28.** [1.6c] $\frac{4}{3}, \frac{8}{3}$
29. [3.3a] $(5, 3)$ **30.** [3.5a] $(-1, 0, 4)$ **31.** [4.8a] $\frac{25}{7}, -\frac{25}{7}$
32. [5.5a] -5 **33.** [5.5a] $\frac{1}{3}$ **34.** [5.7a] $R = \dfrac{nE - nrI}{I}$
35. [6.6a] 6 **36.** [6.6b] $-\frac{1}{4}$ **37.** [6.6a] 5

38. [2.2c]

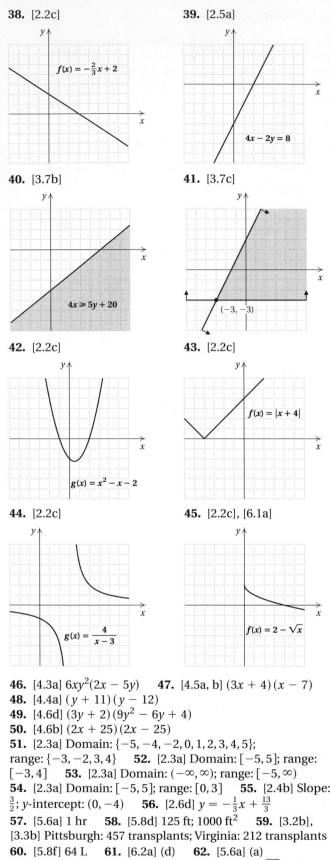

$f(x) = -\frac{2}{3}x + 2$

39. [2.5a]

$4x - 2y = 8$

40. [3.7b]

41. [3.7c]

$4x \geqslant 5y + 20$

$(-3, -3)$

42. [2.2c]

43. [2.2c]

$g(x) = x^2 - x - 2$

$f(x) = |x + 4|$

44. [2.2c]

45. [2.2c], [6.1a]

$g(x) = \dfrac{4}{x - 3}$

$f(x) = 2 - \sqrt{x}$

46. [4.3a] $6xy^2(2x - 5y)$ **47.** [4.5a, b] $(3x + 4)(x - 7)$
48. [4.4a] $(y + 11)(y - 12)$
49. [4.6d] $(3y + 2)(9y^2 - 6y + 4)$
50. [4.6b] $(2x + 25)(2x - 25)$
51. [2.3a] Domain: $\{-5, -4, -2, 0, 1, 2, 3, 4, 5\}$;
range: $\{-3, -2, 3, 4\}$ **52.** [2.3a] Domain: $[-5, 5]$; range:
$[-3, 4]$ **53.** [2.3a] Domain: $(-\infty, \infty)$; range: $[-5, \infty)$
54. [2.3a] Domain: $[-5, 5]$; range: $[0, 3]$ **55.** [2.4b] Slope:
$\frac{3}{2}$; y-intercept: $(0, -4)$ **56.** [2.6d] $y = -\frac{1}{3}x + \frac{13}{3}$
57. [5.6a] 1 hr **58.** [5.8d] 125 ft; 1000 ft^2 **59.** [3.2b],
[3.3b] Pittsburgh: 457 transplants; Virginia: 212 transplants
60. [5.8f] 64 L **61.** [6.2a] (d) **62.** [5.6a] (a)
63. [5.3b, c] (e) **64.** [6.6a] (b) **65.** [6.7a] $\sqrt{45}$ ft; 6.708 ft
66. [6.6b] $-\frac{8}{9}$

CHAPTER 7

Margin Exercises, Section 7.1, pp. 574–583

1. (a) $(2, 0), (4, 0)$; **(b)** 2, 4; **(c)** The solutions of
$x^2 - 6x + 8 = 0$, 2 and 4, are the first coordinates of the
x-intercepts, $(2, 0)$ and $(4, 0)$, of the graph of
$f(x) = x^2 - 6x + 8$. **2. (a)** 4 and -4, or ± 4;
(b) $(-4, 0), (4, 0)$ **3. (a)** $0, -\frac{7}{2}$; **(b)** $\left(-\frac{7}{2}, 0\right), (0, 0)$
4. (a) $\frac{3}{5}, 1$; **(b)** $\left(\frac{3}{5}, 0\right), (1, 0)$ **5.** $\sqrt{3}$ and $-\sqrt{3}$, or $\pm\sqrt{3}$;
1.732 and -1.732, or ± 1.732 **6.** $\dfrac{2\sqrt{6}}{3}$ and $-\dfrac{2\sqrt{6}}{3}$, or
$\pm\dfrac{2\sqrt{6}}{3}$; 1.633 and -1.633, or ± 1.633 **7.** $\dfrac{\sqrt{2}}{2}i$ and $-\dfrac{\sqrt{2}}{2}i$,
or $\pm\dfrac{\sqrt{2}}{2}i$ **8. (a)** $1 \pm \sqrt{5}$; **(b)** $\left(1 - \sqrt{5}, 0\right), \left(1 + \sqrt{5}, 0\right)$
9. $-8 \pm \sqrt{11}$ **10.** $-2, -4$ **11.** $10, -2$
12. $-3 \pm \sqrt{10}$ **13.** $-\dfrac{3}{2} \pm \dfrac{\sqrt{19}}{2}$ **14.** $\dfrac{1}{3} \pm \dfrac{\sqrt{22}}{3}$
15. $\dfrac{1}{3} \pm i\dfrac{\sqrt{2}}{3}$ **16.** About 33.5 in. **17.** 1 sec

Calculator Corner, p. 575

Left to the student

Calculator Corner, p. 578

The calculator returns an ERROR message because the graph
of $y = 4x^2 + 9$ has no x-intercepts. This indicates that the
equation $4x^2 + 9 = 0$ has no real-number solutions.

Exercise Set 7.1, p. 584

1. (a) $\sqrt{5}, -\sqrt{5}$, or $\pm\sqrt{5}$; **(b)** $\left(-\sqrt{5}, 0\right), \left(\sqrt{5}, 0\right)$ **3. (a)** $\frac{5}{3}i$,
$-\frac{5}{3}i$, or $\pm\frac{5}{3}i$; **(b)** no x-intercepts **5.** $\pm\dfrac{\sqrt{6}}{2}$; ± 1.225
7. $5, -9$ **9.** $8, 0$ **11.** $11 \pm \sqrt{7}$; 13.646, 8.354
13. $7 \pm 2i$ **15.** $18, 0$ **17.** $\dfrac{3}{2} \pm \dfrac{\sqrt{14}}{2}$; 3.371, -0.371
19. $5, -11$ **21.** $9, 5$ **23.** $-2 \pm \sqrt{6}$ **25.** $11 \pm 2\sqrt{33}$
27. $-\dfrac{1}{2} \pm \dfrac{\sqrt{5}}{2}$ **29.** $\dfrac{5}{2} \pm \dfrac{\sqrt{53}}{2}$ **31.** $-\dfrac{3}{4} \pm \dfrac{\sqrt{57}}{4}$
33. $\dfrac{9}{4} \pm \dfrac{\sqrt{105}}{4}$ **35.** $2, -8$ **37.** $-11 \pm \sqrt{19}$
39. $5 \pm \sqrt{29}$ **41. (a)** $-\dfrac{7}{2} \pm \dfrac{\sqrt{57}}{2}$; **(b)** $\left(-\dfrac{7}{2} - \dfrac{\sqrt{57}}{2}, 0\right)$,
$\left(-\dfrac{7}{2} \pm \dfrac{\sqrt{57}}{2}, 0\right)$ **43. (a)** $\dfrac{5}{4} \pm i\dfrac{\sqrt{39}}{4}$; **(b)** no x-intercepts
45. $\dfrac{3}{4} \pm \dfrac{\sqrt{17}}{4}$ **47.** $\dfrac{3}{4} \pm \dfrac{\sqrt{145}}{4}$ **49.** $\dfrac{2}{3} \pm \dfrac{\sqrt{7}}{3}$
51. $-\dfrac{1}{2} \pm i\dfrac{\sqrt{7}}{2}$ **53.** $2 \pm 3i$ **55.** 0.866 sec
57. About 8.1 sec **59.** About 6.8 sec **61.** $\mathbf{D_W}$
63. (a) $T(t) = 33.75t + 241.25$, where t is the number of
years since 1995; **(b)** 680,000 people; **(c)** 2020

64.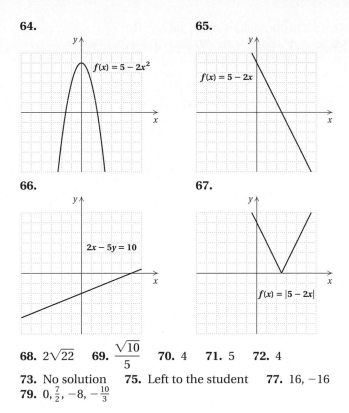

graph labeled $f(x) = 5 - 2x^2$

65.

graph labeled $f(x) = 5 - 2x$

66.

graph labeled $2x - 5y = 10$

67.

graph labeled $f(x) = |5 - 2x|$

68. $2\sqrt{22}$ **69.** $\dfrac{\sqrt{10}}{5}$ **70.** 4   **71.** 5   **72.** 4

73. No solution   **75.** Left to the student   **77.** $16, -16$

79. $0, \dfrac{7}{2}, -8, -\dfrac{10}{3}$

Margin Exercises, Section 7.2, pp. 590–592

1. (a) $-\dfrac{1}{2}, 4$; **(b)** $-\dfrac{1}{2}, 4$ **2.** $\dfrac{-1 \pm \sqrt{22}}{3}$; $1.230, -1.897$

3. $\dfrac{1 \pm i\sqrt{7}}{2}$, or $\dfrac{1}{2} \pm i\dfrac{\sqrt{7}}{2}$ **4.** $\dfrac{5 \pm \sqrt{73}}{6}$; $2.257, -0.591$

Calculator Corner, p. 590

1.–3. Left to the student

Calculator Corner, p. 592

1. The graph has no x-intercepts.   **2.** Yes; 2   **3.** Yes; 2
4. Yes; 1

Calculator Corner, p. 593

1. $-3, 0.8$ **2.** $-1.5, 5$ **3.** $3, 8$ **4.** $2, 4$

Exercise Set 7.2, p. 594

1. $-4 \pm \sqrt{14}$ **3.** $\dfrac{-4 \pm \sqrt{13}}{3}$ **5.** $\dfrac{1}{2} \pm i\dfrac{\sqrt{3}}{2}$ **7.** $2 \pm 3i$

9. $\dfrac{-3 \pm \sqrt{41}}{2}$ **11.** $-1 \pm 2i$ **13. (a)** $0, -1$;

(b) $(0,0), (-1,0)$ **15. (a)** $\dfrac{3 \pm \sqrt{229}}{22}$; **(b)** $\left(\dfrac{3 + \sqrt{229}}{22}, 0\right)$,

$\left(\dfrac{3 - \sqrt{229}}{22}, 0\right)$ **17. (a)** $\dfrac{2}{5}$; **(b)** $\left(\dfrac{2}{5}, 0\right)$ **19.** $-1, -2$

21. $5, 10$ **23.** $\dfrac{17 \pm \sqrt{249}}{10}$ **25.** $2 \pm i$ **27.** $\dfrac{2}{3}, \dfrac{3}{2}$

29. $2 \pm \sqrt{10}$ **31.** $\dfrac{3}{4}, -2$ **33.** $\dfrac{1}{2} \pm \dfrac{3}{2}i$

35. $1, -\dfrac{1}{2} \pm i\dfrac{\sqrt{3}}{2}$ **37.** $-3 \pm \sqrt{5}$; $-0.764, -5.236$

39. $3 \pm \sqrt{5}$; $5.236, 0.764$ **41.** $\dfrac{3 \pm \sqrt{65}}{4}$; $2.766, -1.266$

43. $\dfrac{4 \pm \sqrt{31}}{5}$; $1.914, -0.314$ **45.** D_W **47.** 2   **48.** 3

49. 10   **50.** 8   **51.** No solution   **52.** No solution
53. $\dfrac{15}{4}$ **54.** $\dfrac{17}{3}$ **55.** Left to the student; $-0.797, 0.570$

57. $\dfrac{1 \pm \sqrt{1 + 8\sqrt{5}}}{4}$ **59.** $\dfrac{-i \pm i\sqrt{1 + 4i}}{2}$

61. $\dfrac{-1 \pm 3\sqrt{5}}{6}$ **63.** $3 \pm \sqrt{13}$

Margin Exercises, Section 7.3, pp. 596–601

1. 10 ft   **2.** Distance d is 12 ft; distance to top of ladder is
16 ft.   **3.** Distance d is about 4.782 ft; distance to top of
ladder is about 8.782 ft.

4.

	DISTANCE	SPEED	TIME
Faster Ship	3000	$r + 10$	$t - 50$
Slower Ship	3000	r	t

20 knots, 30 knots

5. $w_2 = \dfrac{w_1}{A^2}$ **6.** $r = \sqrt{\dfrac{V}{\pi h}}$ **7.** $t = \dfrac{-g + \sqrt{g^2 + 64s}}{32}$

8. $b = \dfrac{ta}{\sqrt{1 + t^2}}$

Translating for Success, p. 602

1. B   **2.** G   **3.** F   **4.** L   **5.** N   **6.** C   **7.** J   **8.** E
9. K   **10.** A

Exercise Set 7.3, p. 603

1. Length: 9 ft; width: 2 ft   **3.** Length: 18 yd; width: 9 yd
5. Height: 16 m; base: 7 m   **7.** Height: $4 + 8\sqrt{2}$ ft; base:
$8\sqrt{2} - 4$ ft   **9.** 3 cm   **11.** 6 ft, 8 ft   **13.** 28 and 29
15. Length: $2 + \sqrt{14}$ ft ≈ 5.742 ft;
width: $\sqrt{14} - 2$ ft ≈ 1.742 ft

17. $\dfrac{17 - \sqrt{109}}{2}$ in. ≈ 3.280 in.

19. $7 + \sqrt{239}$ ft ≈ 22.460 ft; $\sqrt{239} - 7$ ft ≈ 8.460 ft
21. First part: 60 mph; second part: 50 mph
23. 40 mph   **25.** Cessna: 150 mph; Beechcraft: 200 mph;
or Cessna: 200 mph; Beechcraft: 250 mph   **27.** To
Hillsboro: 10 mph; return trip: 4 mph   **29.** About 11 mph

31. $s = \sqrt{\dfrac{A}{6}}$ **33.** $r = \sqrt{\dfrac{Gm_1 m_2}{F}}$ **35.** $c = \sqrt{\dfrac{E}{m}}$

37. $b = \sqrt{c^2 - a^2}$ **39.** $k = \dfrac{3 + \sqrt{9 + 8N}}{2}$

41. $r = \dfrac{-\pi h + \sqrt{\pi^2 h^2 + 2\pi A}}{2\pi}$ **43.** $g = \dfrac{4\pi^2 L}{T^2}$

45. $H = \sqrt{\dfrac{704.5W}{I}}$ **47.** $v = \dfrac{c\sqrt{m^2 - m_0^2}}{m}$ **49.** $\mathbf{D_W}$

51. $\dfrac{1}{x-2}$ **52.** $\dfrac{x^3 + x^2 + 2x + 2}{(x-1)(x^2+x+1)}$ **53.** $\dfrac{-x}{(x+3)(x-1)}$

54. $3x^2\sqrt{x}$ **55.** $2i\sqrt{5}$ **56.** $\dfrac{3x+3}{3x+1}$ **57.** $\dfrac{4b}{a(3b^2-4a)}$

59. $\pm\sqrt{2}$ **61.** $A(S) = \dfrac{\pi S}{6}$ **63.** $l = \dfrac{w + w\sqrt{5}}{2}$

Margin Exercises, Section 7.4, pp. 609–612

1. Two real **2.** One real **3.** Two nonreal
4. $x^2 - 5x - 14 = 0$ **5.** $3x^2 + 7x - 20 = 0$
6. $x^2 + 25 = 0$ **7.** $x^2 + \sqrt{2}x - 4 = 0$ **8.** $\pm 3, \pm 1$
9. 4 **10.** $-\frac{1}{3}, \frac{1}{2}$ **11.** $(-3, 0), (-1, 0), (2, 0), (4, 0)$

Exercise Set 7.4, p. 613

1. One real **3.** Two nonreal **5.** Two real **7.** One real
9. Two nonreal **11.** Two real **13.** Two real **15.** One
real **17.** $x^2 - 16 = 0$ **19.** $x^2 + 9x + 14 = 0$
21. $x^2 - 16x + 64 = 0$ **23.** $25x^2 - 20x - 12 = 0$
25. $12x^2 - (4k + 3m)x + km = 0$
27. $x^2 - \sqrt{3}x - 6 = 0$ **29.** $\pm\sqrt{3}$ **31.** 1, 81
33. $-1, 1, 5, 7$ **35.** $-\frac{1}{4}, \frac{1}{9}$ **37.** 1 **39.** $-1, 1, 4, 6$
41. $\pm 1, \pm 3$ **43.** $-1, 2$ **45.** $\pm\dfrac{\sqrt{15}}{3}, \pm\dfrac{\sqrt{6}}{2}$
47. $-1, 125$ **49.** $-\frac{11}{6}, -\frac{1}{6}$ **51.** $-\frac{3}{2}$
53. $\dfrac{9 \pm \sqrt{89}}{2}, -1 \pm \sqrt{3}$ **55.** $\left(\frac{4}{25}, 0\right)$ **57.** $(4, 0), (-1, 0),$
$\left(\dfrac{3 + \sqrt{33}}{2}, 0\right), \left(\dfrac{3 - \sqrt{33}}{2}, 0\right)$ **59.** $\mathbf{D_W}$ **61.** Kenyan:
30 lb; Peruvian: 20 lb **62.** Solution A: 4 L; solution B: 8 L
63. $4x$ **64.** $3x^2$ **65.** $3a\sqrt[4]{2a}$ **66.** 4
67. **68.**

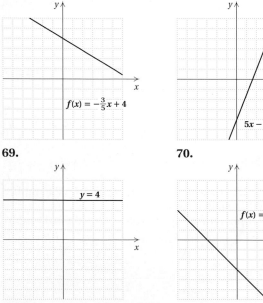

$f(x) = -\frac{3}{5}x + 4$

$5x - 2y = 8$

69. **70.**

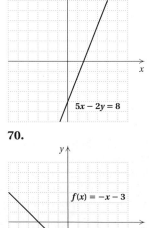

$y = 4$

$f(x) = -x - 3$

71. Left to the student **73.** (a) $-\frac{3}{5}$; (b) $-\frac{1}{3}$
75. $x^2 - \sqrt{3}x + 8 = 0$ **77.** $a = 1, b = 2, c = -3$
79. $\frac{100}{99}$ **81.** 259 **83.** 1, 3

Margin Exercises, Section 7.5, pp. 618–622

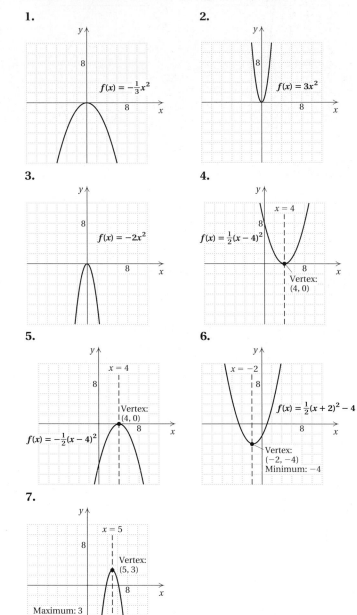

1.

$f(x) = -\frac{1}{3}x^2$

2.

$f(x) = 3x^2$

3.

$f(x) = -2x^2$

4.

$x = 4$

$f(x) = \frac{1}{2}(x - 4)^2$

Vertex:
$(4, 0)$

5.

$x = 4$

Vertex:
$(4, 0)$

$f(x) = -\frac{1}{2}(x - 4)^2$

6.

$x = -2$

$f(x) = \frac{1}{2}(x + 2)^2 - 4$

Vertex:
$(-2, -4)$
Minimum: -4

7.

$x = 5$

Vertex:
$(5, 3)$

Maximum: 3

$f(x) = -2(x - 5)^2 + 3$

Calculator Corner, p. 617

Left to the student

Calculator Corner, p. 618

1. For $a > 0$, the graph opens up. For $a > 1$, the graph of
$y = ax^2$ is narrower than the graph of $y = x^2$. For $0 < a < 1$,
the graph of $y = ax^2$ is wider than the graph of $y = x^2$.

2. For $a < 0$, the graph opens down. For $a < -1$, the graph of $y = ax^2$ is narrower than the graph of $y = x^2$. For $-1 < a < 0$, the graph of $y = ax^2$ is wider than the graph of $y = x^2$.

Calculator Corner, p. 619

Left to the student

Calculator Corner, p. 619

If h is positive, the graph of $y = a(x - h)^2$ is the graph of $y = ax^2$ shifted h units to the right. If h is negative, the graph of $y = a(x - h)^2$ is the graph of $y = ax^2$ shifted $|h|$ units to the left.

Calculator Corner, p. 621

If k is positive, the graph of $y = a(x - h)^2 + k$ is the graph of $y = a(x - h)^2$ shifted up k units. If k is negative, the graph of $y = a(x - h)^2 + k$ is the graph of $y = a(x - h)^2$ shifted down $|k|$ units.

Exercise Set 7.5, p. 624

1.

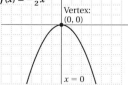

x	f(x)
0	0
1	4
2	16
−1	4
−2	16

3.

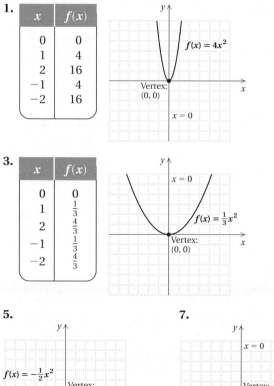

x	f(x)
0	0
1	$\frac{1}{3}$
2	$\frac{4}{3}$
−1	$\frac{1}{3}$
−2	$\frac{4}{3}$

5.

7.

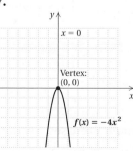

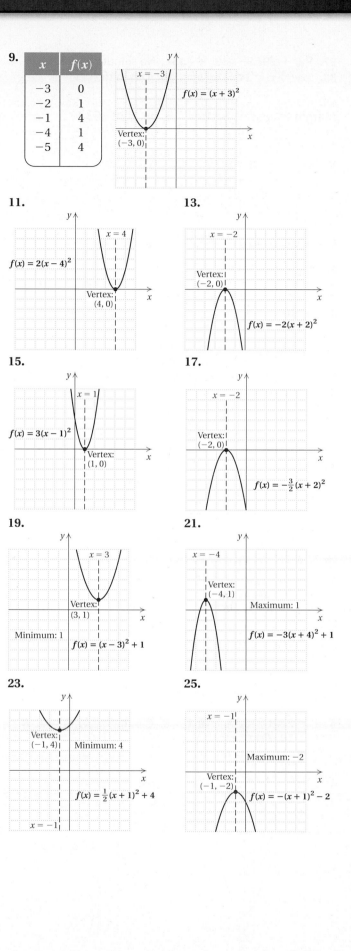

9.

x	f(x)
−3	0
−2	1
−1	4
−4	1
−5	4

11.

13.

15.

17.

19.

21.

23.

25.

27. DW **29.** 9 **30.** $\frac{11}{3}$ **31.** No solution **32.** 15
33. $5xy^2\sqrt[4]{x}$ **34.** $12a^2b^2$ **35.** Left to the student

Margin Exercises, Section 7.6, pp. 628–632

1.

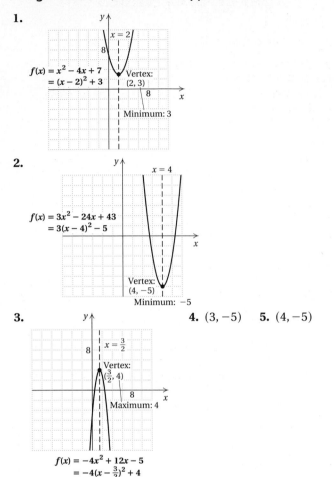

$f(x) = x^2 - 4x + 7$
$= (x - 2)^2 + 3$

$x = 2$

Vertex: (2, 3)

Minimum: 3

2.

$x = 4$

$f(x) = 3x^2 - 24x + 43$
$= 3(x - 4)^2 - 5$

Vertex: (4, −5)

Minimum: −5

3.

$x = \frac{3}{2}$

Vertex: $\left(\frac{3}{2}, 4\right)$

Maximum: 4

$f(x) = -4x^2 + 12x - 5$
$= -4\left(x - \frac{3}{2}\right)^2 + 4$

4. (3, −5) **5.** (4, −5)

6. $\left(\frac{3}{2}, 4\right)$ **7.** y-intercept: (0, −3); x-intercepts: (−3, 0), (1, 0)
8. y-intercept: (0, 16); x-intercept: (−4, 0) **9.** y-intercept:
(0, 1); x-intercepts: $\left(2 + \sqrt{3}, 0\right)$, $\left(2 - \sqrt{3}, 0\right)$, or (3.732, 0),
(0.268, 0)

Visualizing for Success, p. 633

1. F **2.** H **3.** A **4.** I **5.** C **6.** J **7.** G **8.** B
9. E **10.** D

Exercise Set 7.6, p. 634

1.

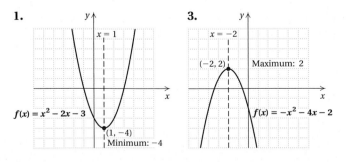

$x = 1$

$f(x) = x^2 - 2x - 3$

(1, −4)

Minimum: −4

3.

$x = -2$

(−2, 2) Maximum: 2

$f(x) = -x^2 - 4x - 2$

5.

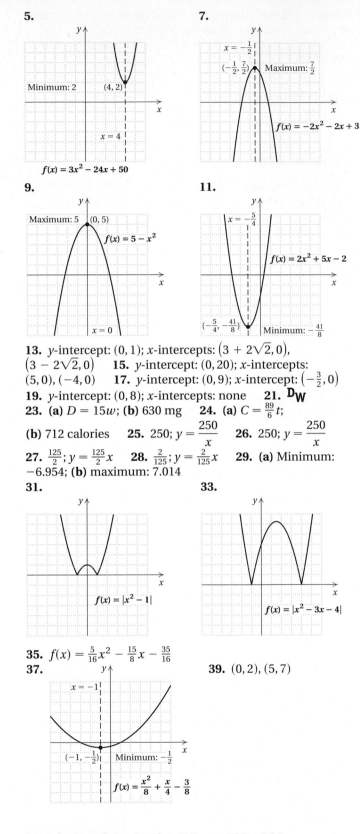

Minimum: 2 (4, 2)

$x = 4$

$f(x) = 3x^2 - 24x + 50$

7.

$x = -\frac{1}{2}$

$\left(-\frac{1}{2}, \frac{7}{2}\right)$ Maximum: $\frac{7}{2}$

$f(x) = -2x^2 - 2x + 3$

9.

Maximum: 5 (0, 5)

$f(x) = 5 - x^2$

$x = 0$

11.

$x = -\frac{5}{4}$

$f(x) = 2x^2 + 5x - 2$

$\left(-\frac{5}{4}, -\frac{41}{8}\right)$ Minimum: $-\frac{41}{8}$

13. y-intercept: (0, 1); x-intercepts: $\left(3 + 2\sqrt{2}, 0\right)$,
$\left(3 - 2\sqrt{2}, 0\right)$ **15.** y-intercept: (0, 20); x-intercepts:
(5, 0), (−4, 0) **17.** y-intercept: (0, 9); x-intercept: $\left(-\frac{3}{2}, 0\right)$
19. y-intercept: (0, 8); x-intercepts: none **21. DW**
23. (a) $D = 15w$; **(b)** 630 mg **24. (a)** $C = \frac{89}{6}t$;
(b) 712 calories **25.** 250; $y = \frac{250}{x}$ **26.** 250; $y = \frac{250}{x}$
27. $\frac{125}{2}$; $y = \frac{125}{2}x$ **28.** $\frac{2}{125}$; $y = \frac{2}{125}x$ **29. (a)** Minimum:
−6.954; **(b)** maximum: 7.014

31.

$f(x) = |x^2 - 1|$

33.

$f(x) = |x^2 - 3x - 4|$

35. $f(x) = \frac{5}{16}x^2 - \frac{15}{8}x - \frac{35}{16}$
37.

$x = -1$

$\left(-1, -\frac{1}{2}\right)$ Minimum: $-\frac{1}{2}$

$f(x) = \frac{x^2}{8} + \frac{x}{4} - \frac{3}{8}$

39. (0, 2), (5, 7)

Margin Exercises, Section 7.7, pp. 639–644

1. 25 yd by 25 yd **2.** $f(x) = ax^2 + bx + c, a > 0$
3. $f(x) = mx + b$ **4.** Polynomial, neither quadratic nor
linear **5.** Polynomial, neither quadratic nor linear

6. (a)

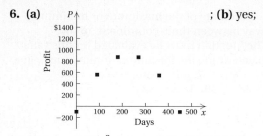

; (b) yes;

(c) $f(x) = -0.02x^2 + 9x - 100$; **(d)** \$912.50

Calculator Corner, p. 640

1. Minimum: 1   **2.** Minimum: 4.875   **3.** Maximum: 6
4. Maximum: 0.5625

Calculator Corner, pp. 645–646

1. (a)

(b) $y = -0.0082833093x^2 + 0.8242996891x + 0.2121786608$;
(c) **(d)** 15.4 ft; the estimates of 15.4 ft and 15 ft are relatively close.

Exercise Set 7.7, p. 647

1. 180 ft by 180 ft   **3.** 3.5 in.   **5.** 3.5 hundred, or 350
7. 200 ft^2; 10 ft by 20 ft   **9.** 11 days after the concert was announced; about 62   **11.** $P(x) = -x^2 + 980x - 3000$; \$237,100 at $x = 490$ **13.** 121; 11 and 11   **15.** -4; 2 and -2 **17.** 36; -6 and -6 **19.** $f(x) = mx + b$
21. $f(x) = ax^2 + bx + c, a > 0$ **23.** Polynomial, neither quadratic nor linear   **25.** $f(x) = ax^2 + bx + c, a < 0$
27. $f(x) = 2x^2 + 3x - 1$ **29.** $f(x) = -\frac{1}{4}x^2 + 3x - 5$
31. (a) $A(s) = \frac{3}{16}s^2 - \frac{135}{4}s + 1750$; **(b)** about 531 per 200,000,000 kilometers driven
33. $h(d) = -\frac{1}{147}d^2 + \frac{6}{7}d \approx -0.0068d^2 + 0.8571d$
35. $\mathbf{D_W}$ **37.** Radical; radicand   **38.** Dependent
39. Sum   **40.** At least one   **41.** Inverse
42. Independent   **43.** Descending   **44.** x-intercept
45. $y = -0.290x^4 + 2.699x^3 - 8.306x^2 + 9.190x + 12.235$, where x is the number of years after 1997

Margin Exercises, Section 7.8, pp. 652–657

1. $\{x \mid x < -3 \ or \ x > 1\}$, or $(-\infty, -3) \cup (1, \infty)$
2. $\{x \mid -3 < x < 1\}$, or $(-3, 1)$ **3.** $\{x \mid -3 \le x \le 1\}$, or $[-3, 1]$ **4.** $\{x \mid x < -4 \ or \ x > 1\}$, or $(-\infty, -4) \cup (1, \infty)$
5. $\{x \mid -4 \le x \le 1\}$, or $[-4, 1]$ **6.** $\{x \mid x < -1 \ or \ 0 < x < 1\}$, or $(-\infty, -1) \cup (0, 1)$ **7.** $\{x \mid 2 < x \le \frac{7}{2}\}$, or $\left(2, \frac{7}{2}\right]$
8. $\{x \mid x < 5 \ or \ x > 10\}$, or $(-\infty, 5) \cup (10, \infty)$

Calculator Corner, p. 653

1. $\{x \mid x < -4 \ or \ x > 1\}$, or $(-\infty, -4) \cup (1, \infty)$
2. $\{x \mid -2 < x < 3\}$, or $(-2, 3)$
3. $\{x \mid x \le -2 \ or \ 0 \le x \le 0.5\}$, or $(-\infty, -2] \cup [0, 0.5]$
4. $\{x \mid -4 \le x \le 0 \ or \ x \ge 4\}$, or $[-4, 0] \cup [4, \infty)$

Exercise Set 7.8, p. 658

1. $\{x \mid x < -2 \ or \ x > 6\}$, or $(-\infty, -2) \cup (6, \infty)$
3. $\{x \mid -2 \le x \le 2\}$, or $[-2, 2]$ **5.** $\{x \mid -1 \le x \le 4\}$, or $[-1, 4]$ **7.** $\{x \mid -1 < x < 2\}$, or $(-1, 2)$ **9.** All real numbers, or $(-\infty, \infty)$ **11.** $\{x \mid 2 < x < 4\}$, or $(2, 4)$
13. $\{x \mid x < -2 \ or \ 0 < x < 2\}$, or $(-\infty, -2) \cup (0, 2)$
15. $\{x \mid -9 < x < -1 \ or \ x > 4\}$, or $(-9, -1) \cup (4, \infty)$
17. $\{x \mid x < -3 \ or \ -2 < x < 1\}$, or $(-\infty, -3) \cup (-2, 1)$
19. $\{x \mid x < 6\}$, or $(-\infty, 6)$ **21.** $\{x \mid x < -1 \ or \ x > 3\}$, or $(-\infty, -1) \cup (3, \infty)$ **23.** $\{x \mid -\frac{2}{3} \le x < 3\}$, or $\left[-\frac{2}{3}, 3\right)$
25. $\{x \mid 2 < x < \frac{5}{2}\}$, or $\left(2, \frac{5}{2}\right)$ **27.** $\{x \mid x < -1 \ or \ 2 < x < 5\}$, or $(-\infty, -1) \cup (2, 5)$ **29.** $\{x \mid -3 \le x < 0\}$, or $[-3, 0)$
31. $\{x \mid 1 < x < 2\}$, or $(1, 2)$ **33.** $\{x \mid x < -4 \ or \ 1 < x < 3\}$, or $(-\infty, -4) \cup (1, 3)$ **35.** $\{x \mid 0 < x < \frac{1}{3}\}$, or $\left(0, \frac{1}{3}\right)$
37. $\{x \mid x < -3 \ or \ -2 < x < 1 \ or \ x > 4\}$, or
$(-\infty, -3) \cup (-2, 1) \cup (4, \infty)$ **39.** $\mathbf{D_W}$ **41.** $\frac{5}{3}$ **42.** $\frac{5}{2a}$
43. $\frac{4a}{b^2}\sqrt{a}$ **44.** $\frac{3c}{7d}\sqrt[3]{c^2}$ **45.** $\sqrt{2}$ **46.** $17\sqrt{5}$
47. $(10a + 7)\sqrt[3]{2a}$ **48.** $3\sqrt{10} - 4\sqrt{5}$ **49.** Left to the student   **51.** $\{x \mid 1 - \sqrt{3} \le x \le 1 + \sqrt{3}\}$, or $\left[1 - \sqrt{3}, 1 + \sqrt{3}\right]$ **53.** All real numbers except 0, or $(-\infty, 0) \cup (0, \infty)$ **55.** $\{x \mid x < \frac{1}{4} \ or \ x > \frac{5}{2}\}$, or $\left(-\infty, \frac{1}{4}\right) \cup \left(\frac{5}{2}, \infty\right)$ **57. (a)** $\{t \mid 0 < t < 2\}$, or $(0, 2)$;
(b) $\{t \mid t > 10\}$, or $(10, \infty)$

Concept Reinforcement, p. 660

1. False   **2.** True   **3.** False   **4.** True   **5.** False
6. True

Summary and Review: Chapter 7, p. 660

1. (a) $\pm\frac{\sqrt{14}}{2}$; **(b)** $\left(-\frac{\sqrt{14}}{2}, 0\right), \left(\frac{\sqrt{14}}{2}, 0\right)$ **2.** $0, -\frac{5}{14}$
3. 3, 9   **4.** $-\frac{3}{8} \pm i\frac{\sqrt{7}}{8}$ **5.** $\frac{7}{2} \pm i\frac{\sqrt{3}}{2}$ **6.** 3, 5
7. $-2 \pm \sqrt{3}$; $-0.268, -3.732$ **8.** 4, -2 **9.** $4 \pm 4\sqrt{2}$
10. $\frac{1 \pm \sqrt{481}}{15}$ **11.** $-2 \pm \sqrt{3}$ **12.** 0.901 sec
13. Length: 14 cm; width: 9 cm   **14.** 1 in.   **15.** First part: 50 mph; second part: 40 mph   **16.** Two real   **17.** Two nonreal   **18.** $25x^2 + 10x - 3 = 0$ **19.** $x^2 + 8x + 16 = 0$
20. $p = \frac{9\pi^2}{N^2}$ **21.** $T = \sqrt{\frac{3B}{2A}}$ **22.** 2, -2, 3, -3

23. 3, −5 **24.** ±√7, ±√2 **25.** 81, 16 **26. (a)** (1, 3);
(b) $x = 1$; **(c)** maximum: 3;
(d)

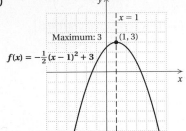

27. (a) $\left(\frac{1}{2}, \frac{23}{4}\right)$; **(b)** $x = \frac{1}{2}$; **(c)** minimum: $\frac{23}{4}$;
(d)

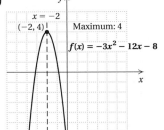

28. (a) (−2, 4); **(b)** $x = -2$; **(c)** maximum: 4;
(d)

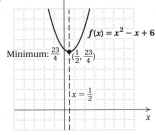

29. y-intercept: (0, 14); x-intercepts: (2, 0), (7, 0)
30. −121; 11 and −11 **31.** $f(x) = -x^2 + 6x - 2$
32. (a) $N(x) = -0.720x^2 + 38.211x - 393.127$; **(b)** about
105 **33.** $\{x \mid -2 < x < 1 \text{ or } x > 2\}$, or $(-2, 1) \cup (2, \infty)$
34. $\{x \mid x < -4 \text{ or } -2 < x < 1\}$, or $(-\infty, -4) \cup (-2, 1)$
35. $^{\mathbf{D}}\mathbf{W}$ The graph of $f(x) = ax^2 + bx + c$ is a parabola with
vertex $\left(-\frac{b}{2a}, \frac{4ac - b^2}{4a}\right)$. The line of symmetry is $x = -\frac{b}{2a}$.

If $a < 0$, then the parabola opens down and $\frac{4ac - b^2}{4a}$ is the
maximum function value. If $a > 0$, the parabola opens up
and $\frac{4ac - b^2}{4a}$ is the minimum function value. The
x-intercepts are $\left(\frac{-b - \sqrt{b^2 - 4ac}}{2a}, 0\right)$ and
$\left(\frac{-b + \sqrt{b^2 - 4ac}}{2a}, 0\right)$, for $b^2 - 4ac > 0$. When

$b^2 - 4ac = 0$, there is just one x-intercept, $\left(-\frac{b}{2a}, 0\right)$.
When $b^2 - 4ac < 0$, there are no x-intercepts.

36. $^{\mathbf{D}}\mathbf{W}$ The x-coordinate of the maximum or minimum
point lies halfway between the x-coordinates of the
x-intercepts. The function must be evaluated for this value
of x in order to determine the maximum or minimum value.
37. $f(x) = \frac{7}{15}x^2 - \frac{14}{15}x - 7$; minimum: $-\frac{112}{15}$
38. $h = 60$, $k = 60$ **39.** 18 and 324

Test: Chapter 7, p. 663

1. [7.1a] **(a)** $\pm\frac{2\sqrt{3}}{3}$; **(b)** $\left(\frac{2\sqrt{3}}{3}, 0\right), \left(-\frac{2\sqrt{3}}{3}, 0\right)$

2. [7.2a] $-\frac{1}{2} \pm i\frac{\sqrt{3}}{2}$ **3.** [7.4c] 49, 1 **4.** [7.2a] 9, 2

5. [7.4c] $\pm\sqrt{\frac{5 + \sqrt{5}}{2}}, \pm\sqrt{\frac{5 - \sqrt{5}}{2}}$ **6.** [7.2a] $-2 \pm \sqrt{6}$;

0.449, −4.449 **7.** [7.2a] 0, 2 **8.** [7.1b] $2 \pm \sqrt{3}$
9. [7.1c] About 6.7 sec **10.** [7.3a] About 2.89 mph
11. [7.3a] 7 cm by 7 cm **12.** [7.1c] About 0.866 sec
13. [7.4a] Two nonreal **14.** [7.4b] $x^2 - 4\sqrt{3}x + 9 = 0$
15. [7.3b] $T = \sqrt{\frac{V}{48}}$, or $\frac{\sqrt{3V}}{12}$ **16.** [7.6a] **(a)** (−1, 1);
(b) $x = -1$; **(c)** maximum: 1; **(d)**

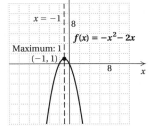

17. [7.6a] **(a)** (3, 5); **(b)** $x = 3$; **(c)** minimum: 5;
(d)

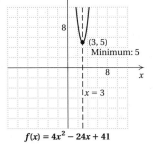

18. [7.6b] y-intercept: (0, −1); x-intercepts:
$\left(2 - \sqrt{3}, 0\right), \left(2 + \sqrt{3}, 0\right)$ **19.** [7.7a] −16; 4 and −4
20. [7.7b] $f(x) = \frac{1}{5}x^2 - \frac{3}{5}x$
21. [7.7b] **(a)** $C(t) = 2.405t^2 + 19.05t + 349$; **(b)** about
\$2617 billion; about \$3085 billion
22. [7.8a] $\{x \mid -1 < x < 7\}$, or $(-1, 7)$
23. [7.8b] $\{x \mid -3 < x < 5\}$, or $(-3, 5)$
24. [7.8b] $\{x \mid -3 < x < 1 \text{ or } x \geq 2\}$, or $(-3, 1) \cup [2, \infty)$
25. [7.6a, b], [7.7b] $f(x) = -\frac{4}{7}x^2 + \frac{20}{7}x + 8$; maximum: $\frac{81}{7}$
26. [7.2a] $\frac{1}{2}$
27. [7.4c] ± 2, $\pm\sqrt{2}$, $\pm 2i$, $\pm\sqrt{2}i$

CHAPTER 8

Margin Exercises, Section 8.1, pp. 667–673

1. (a)

x	$f(x)$
0	1
1	3
2	9
3	27
−1	$\frac{1}{3}$
−2	$\frac{1}{9}$
−3	$\frac{1}{27}$

(b)

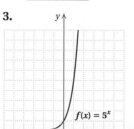
$f(x) = 3^x$

2. (a)

x	$f(x)$
0	1
1	$\frac{1}{3}$
2	$\frac{1}{9}$
3	$\frac{1}{27}$
−1	3
−2	9
−3	27

(b)

$f(x) = \left(\frac{1}{3}\right)^x$

3.

$f(x) = 4^x$

4.

$f(x) = \left(\frac{1}{4}\right)^x$

5.

$f(x) = 2^{x+2}$

6.

$f(x) = 2^x - 4$

7.

$x = 3^y$

8. \$42,000; \$44,100   **9. (a)** $A(t) = \$40,000(1.05)^t$;
(b) \$40,000, \$48,620.25, \$59,098.22, \$65,155.79;
(c)   **10.** \$9925.67

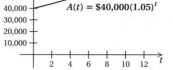
$A(t) = \$40,000(1.05)^t$

Calculator Corner, p. 670

1. Left to the student   **2.** Left to the student

Calculator Corner, p. 674

1. \$1125.51   **2.** \$1127.16   **3.** \$30,372.65
4. (a) \$10,540; **(b)** \$10,547.29; **(c)** \$10,551.03; **(d)** \$10,554.80;
(e) \$10,554.84

Exercise Set 8.1, p. 675

1.

x	$f(x)$
0	1
1	2
2	4
3	8
−1	$\frac{1}{2}$
−2	$\frac{1}{4}$
−3	$\frac{1}{8}$

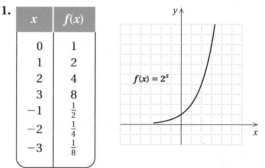
$f(x) = 2^x$

3.

$f(x) = 5^x$

5.

$f(x) = 2^{x+1}$

7.

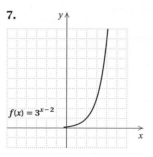
$f(x) = 3^{x-2}$

9.

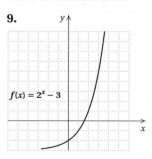
$f(x) = 2^x - 3$

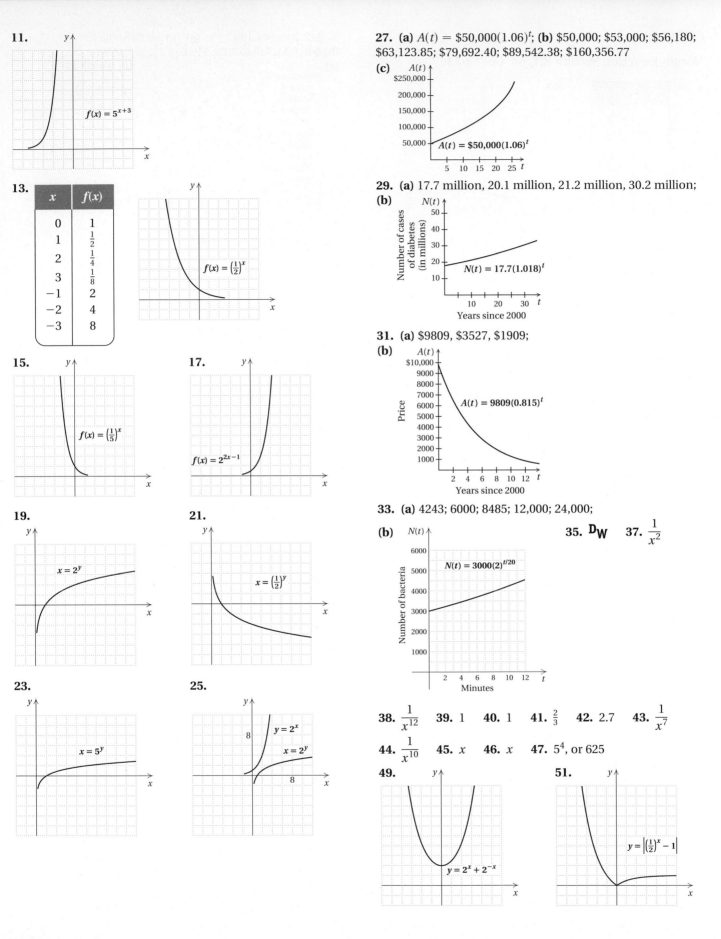

11.

$f(x) = 5^{x+3}$

13.

x	f(x)
0	1
1	$\frac{1}{2}$
2	$\frac{1}{4}$
3	$\frac{1}{8}$
−1	2
−2	4
−3	8

$f(x) = \left(\frac{1}{2}\right)^x$

15. $f(x) = \left(\frac{1}{5}\right)^x$

17. $f(x) = 2^{2x-1}$

19. $x = 2^y$

21. $x = \left(\frac{1}{2}\right)^y$

23. $x = 5^y$

25. $y = 2^x$ $x = 2^y$

27. (a) $A(t) = \$50{,}000(1.06)^t$; **(b)** \$50,000; \$53,000; \$56,180; \$63,123.85; \$79,692.40; \$89,542.38; \$160,356.77

(c) $A(t) = \$50{,}000(1.06)^t$

29. (a) 17.7 million, 20.1 million, 21.2 million, 30.2 million;

(b) $N(t) = 17.7(1.018)^t$

Number of cases of diabetes (in millions)

Years since 2000

31. (a) \$9809, \$3527, \$1909;

(b) $A(t) = 9809(0.815)^t$

Price

Years since 2000

33. (a) 4243; 6000; 8485; 12,000; 24,000;

(b) $N(t) = 3000(2)^{t/20}$

Number of bacteria

Minutes

35. $\mathbf{D_W}$ **37.** $\dfrac{1}{x^2}$

38. $\dfrac{1}{x^{12}}$ **39.** 1 **40.** 1 **41.** $\dfrac{2}{3}$ **42.** 2.7 **43.** $\dfrac{1}{x^7}$

44. $\dfrac{1}{x^{10}}$ **45.** x **46.** x **47.** 5^4, or 625

49. $y = 2^x + 2^{-x}$

51. $y = \left|\left(\frac{1}{2}\right)^x - 1\right|$

53.

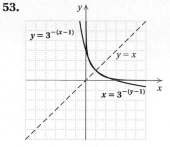

$y = 3^{-(x-1)}$
$y = x$
$x = 3^{-(y-1)}$

55. Left to the student

Margin Exercises, Section 8.2, pp. 680–691

1. Inverse of $g = \{(5, 2), (4, -1), (1, -2)\}$

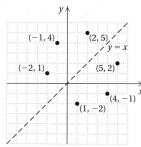

$(-1, 4)$ $(2, 5)$ $y = x$
$(-2, 1)$ $(5, 2)$
$(4, -1)$
$(1, -2)$

2. Inverse: $x = 6 - 2y$;

x	y
6	0
2	2
0	3
−4	5

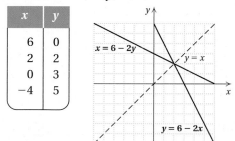

$x = 6 - 2y$
$y = x$
$y = 6 - 2x$

3. Inverse: $x = y^2 - 4y + 7$;

x	y
7	0
4	1
3	2
4	3
7	4

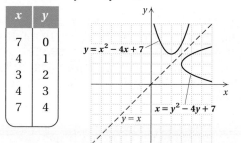

$y = x^2 - 4x + 7$
$x = y^2 - 4y + 7$
$y = x$

4. Yes **5.** No **6.** Yes **7.** No **8.** (a) Yes;
(b) $f^{-1}(x) = 3 - x$; **(c)**

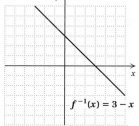

$f^{-1}(x) = 3 - x$

9. (a) Yes; (b) $g^{-1}(x) = \dfrac{x + 2}{3}$;

(c)

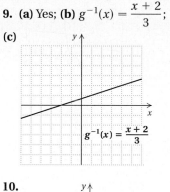

$g^{-1}(x) = \dfrac{x + 2}{3}$

10.

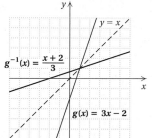

$y = x$
$g^{-1}(x) = \dfrac{x + 2}{3}$
$g(x) = 3x - 2$

11. (a) Yes; (b) $f^{-1}(x) = \sqrt[3]{x - 1}$;
(c)   **12.** $x^2 + 4; x^2 + 10x + 24$

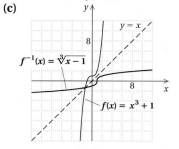

$f^{-1}(x) = \sqrt[3]{x - 1}$
$y = x$
$f(x) = x^3 + 1$

13. $4\sqrt[3]{x} + 5; \sqrt[3]{4x + 5}$ **14.** (a) $f(x) = \sqrt[3]{x}, g(x) = x^2 + 1$;
(b) $f(x) = \dfrac{1}{x^4}, g(x) = x + 5$

15. $f^{-1} \circ f(x) = f^{-1}(f(x)) = f^{-1}\left(\frac{2}{3}x - 4\right)$

$= \dfrac{3\left(\frac{2}{3}x - 4\right) + 12}{2} = \dfrac{2x - 12 + 12}{2}$

$= \dfrac{2x}{2} = x;$

$f \circ f^{-1}(x) = f(f^{-1}(x)) = f\left(\dfrac{3x + 12}{2}\right)$

$= \dfrac{2}{3}\left(\dfrac{3x + 12}{2}\right) - 4 = \dfrac{6x + 24}{6} - 4$

$= x + 4 - 4 = x$

Calculator Corner, p. 688

1.   **2.**

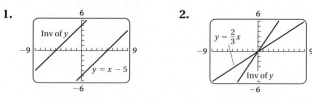

6
Inv of y
−9   9
$y = x - 5$
−6

6
$y = \frac{2}{3}x$
−9   9
Inv of y
−6

3.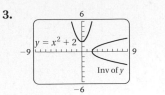

4.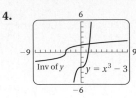

Calculator Corner, p. 692

1. Left to the student   **2.** 4; 7   **3.** 9; 9   **4.** 1; 25

Exercise Set 8.2, p. 693

1. Inverse: $\{(2, 1), (-3, 6), (-5, -3)\}$

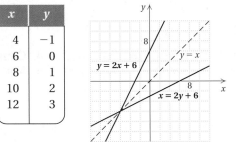

3. Inverse: $x = 2y + 6$

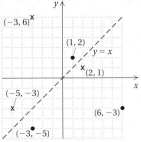

x	y
4	-1
6	0
8	1
10	2
12	3

5. Yes   **7.** No   **9.** No   **11.** Yes   **13.** $f^{-1}(x) = \dfrac{x + 2}{5}$

15. $f^{-1}(x) = \dfrac{-2}{x}$ **17.** $f^{-1}(x) = \frac{3}{4}(x - 7)$

19. $f^{-1}(x) = \dfrac{2}{x} - 5$ **21.** Not one-to-one

23. $f^{-1}(x) = \dfrac{1 - 3x}{5x - 2}$ **25.** $f^{-1}(x) = \sqrt[3]{x + 1}$

27. $f^{-1}(x) = x^3$
29. $f^{-1}(x) = 2x + 6$

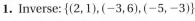

31. $f^{-1}(x) = \sqrt[3]{x}$

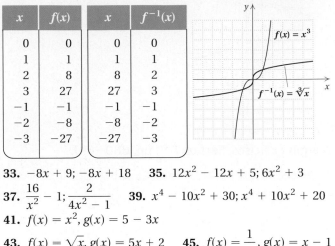

x	$f(x)$	x	$f^{-1}(x)$
0	0	0	0
1	1	1	1
2	8	8	2
3	27	27	3
-1	-1	-1	-1
-2	-8	-8	-2
-3	-27	-27	-3

33. $-8x + 9$; $-8x + 18$ **35.** $12x^2 - 12x + 5$; $6x^2 + 3$

37. $\dfrac{16}{x^2} - 1$; $\dfrac{2}{4x^2 - 1}$ **39.** $x^4 - 10x^2 + 30$; $x^4 + 10x^2 + 20$

41. $f(x) = x^2$, $g(x) = 5 - 3x$

43. $f(x) = \sqrt{x}$, $g(x) = 5x + 2$ **45.** $f(x) = \dfrac{1}{x}$, $g(x) = x - 1$

47. $f(x) = \dfrac{1}{\sqrt{x}}$, $g(x) = 7x + 2$

49. $f(x) = x^4$, $g(x) = \sqrt{x} + 5$

51. $f^{-1} \circ f(x) = f^{-1}(f(x)) = f^{-1}\left(\frac{4}{5}x\right) = \frac{5}{4}\left(\frac{4}{5}x\right) = x$;
$f \circ f^{-1}(x) = f(f^{-1}(x)) = f\left(\frac{5}{4}x\right) = \frac{4}{5}\left(\frac{5}{4}x\right) = x$

53. $f^{-1} \circ f(x) = f^{-1}(f(x)) = f^{-1}\left(\dfrac{1 - x}{x}\right)$

$= \dfrac{1}{\dfrac{1 - x}{x} + 1} = \dfrac{1}{\dfrac{1}{x}} = x$;

$f \circ f^{-1}(x) = f(f^{-1}(x)) = f\left(\dfrac{1}{x + 1}\right)$

$= \dfrac{1 - \dfrac{1}{x + 1}}{\dfrac{1}{x + 1}} = \dfrac{\dfrac{x}{x + 1}}{\dfrac{1}{x + 1}} = x$

55. $f^{-1}(x) = \frac{1}{3}x$ **57.** $f^{-1}(x) = -x$ **59.** $f^{-1}(x) = x^3 + 5$
61. (a) 40, 42, 46, 50; (b) $f^{-1}(x) = x - 32$; (c) 8, 10, 14, 18
63. $\mathbf{D_W}$ **65.** $\sqrt[3]{a}$ **66.** $\sqrt[3]{x^2}$ **67.** a^2b^3 **68.** $2t^2$
69. $\sqrt{3}$ **70.** $2\sqrt[4]{2}$ **71.** $\sqrt{2xy}$ **72.** $\sqrt[4]{p^2t}$ **73.** $2a^3b^8$
74. $10x^3y^6$ **75.** $3a^2b^2$ **76.** $3pq^3$ **77.** No   **79.** Yes
81. (1) C; (2) A; (3) B; (4) D   **83.**

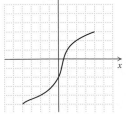

85. $f(x) = \frac{1}{2}x + 3$; $g(x) = 2x - 6$; yes

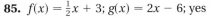

Margin Exercises, Section 8.3, pp. 698–704

1. $\log_2 64$ is the power to which we raise 2 to get 64; 6

2.

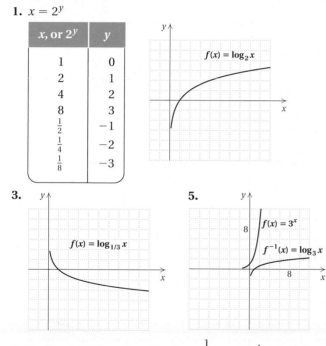

3. $0 = \log_6 1$
4. $-3 = \log_{10} 0.001$
5. $0.25 = \log_{16} 2$
6. $T = \log_m P$

7. $2^5 = 32$ **8.** $10^3 = 1000$ **9.** $a^7 = Q$ **10.** $t^x = M$
11. 10,000 **12.** 3 **13.** $\frac{1}{4}$ **14.** 4 **15.** -4 **16.** 0
17. 0 **18.** 1 **19.** 1 **20.** 0 **21.** 4.8934
22. -4.5100 **23.** Does not exist as a real number
24. $\log 1000 = 3$, $\log 10,000 = 4$; between 3 and 4; 3.9945
25. $8 = 10^{0.9031}$; $947 = 10^{2.9763}$; $634,567 = 10^{5.8025}$;
$0.00708 = 10^{-2.1500}$; $0.000778899 = 10^{-3.1085}$;
$18,600,000 = 10^{7.2695}$; $1860 = 10^{3.2695}$ **26.** 78,234.8042

Exercise Set 8.3, p. 705

1. $x = 2^y$

x, or 2^y	y
1	0
2	1
4	2
8	3
$\frac{1}{2}$	-1
$\frac{1}{4}$	-2
$\frac{1}{8}$	-3

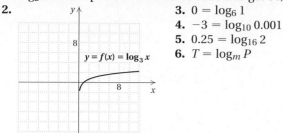

$f(x) = \log_2 x$

3.

$f(x) = \log_{1/3} x$

5.

$f(x) = 3^x$

$f^{-1}(x) = \log_3 x$

7. $3 = \log_{10} 1000$ **9.** $-3 = \log_5 \frac{1}{125}$ **11.** $\frac{1}{3} = \log_8 2$

13. $0.3010 = \log_{10} 2$ **15.** $2 = \log_e t$ **17.** $t = \log_Q x$
19. $2 = \log_e 7.3891$ **21.** $-2 = \log_e 0.1353$ **23.** $4^w = 10$
25. $6^2 = 36$ **27.** $10^{-2} = 0.01$ **29.** $10^{0.9031} = 8$
31. $e^{4.6052} = 100$ **33.** $t^k = Q$ **35.** 9 **37.** 4 **39.** 4
41. 3 **43.** 25 **45.** 1 **47.** $\frac{1}{2}$ **49.** 2 **51.** 2
53. -1 **55.** 0 **57.** 4 **59.** 2 **61.** 3 **63.** -2
65. 0 **67.** 1 **69.** $\frac{2}{3}$ **71.** 4.8970 **73.** -0.1739
75. Does not exist as a real number **77.** 0.9464
79. $6 = 10^{0.7782}$; $84 = 10^{1.9243}$; $987,606 = 10^{5.9946}$;
$0.00987606 = 10^{-2.0054}$; $98,760.6 = 10^{4.9946}$;
$70,000,000 = 10^{7.8451}$; $7000 = 10^{3.8451}$ **81.** $\mathbf{D_W}$
83. Conjugate **84.** Direct **85.** Leading term

---

86. Quadratic; discriminant **87.** Inconsistent
88. Parabolas **89.** Line of symmetry **90.** $a + bi$
91. **93.** 25 **95.** 32

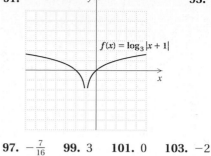

$f(x) = \log_3 |x + 1|$

97. $-\frac{7}{16}$ **99.** 3 **101.** 0 **103.** -2

Margin Exercises, Section 8.4, pp. 709–712

1. $\log_5 25 + \log_5 5$ **2.** $\log_b P + \log_b Q$ **3.** $\log_3 35$
4. $\log_a (JAM)$ **5.** $5 \log_7 4$ **6.** $\frac{1}{2} \log_a 5$
7. $\log_b P - \log_b Q$ **8.** $\log_2 5$
9. $\frac{3}{2} \log_a z - \frac{1}{2} \log_a x - \frac{1}{2} \log_a y$
10. $2 \log_a x - 3 \log_a y - \log_a z$
11. $3 \log_a x + 4 \log_a y - 5 \log_a z - 9 \log_a w$
12. $\log_a \dfrac{x^5 z^{1/4}}{y}$ **13.** $\log_a \dfrac{1}{b\sqrt{b}}$, or $\log_a b^{-3/2}$ **14.** 0.602
15. 1 **16.** -0.398 **17.** 0.398 **18.** -0.699 **19.** $\frac{3}{2}$
20. 1.699 **21.** 1.204 **22.** 6 **23.** 3.2 **24.** 12

Calculator Corner, p. 710

1. Not correct **2.** Correct **3.** Not correct **4.** Correct
5. Not correct **6.** Correct **7.** Not correct **8.** Not correct

Exercise Set 8.4, p. 713

1. $\log_2 32 + \log_2 8$ **3.** $\log_4 64 + \log_4 16$
5. $\log_a Q + \log_a x$ **7.** $\log_b 252$ **9.** $\log_c Ky$
11. $4 \log_c y$ **13.** $6 \log_b t$ **15.** $-3 \log_b C$
17. $\log_a 67 - \log_a 5$ **19.** $\log_b 2 - \log_b 5$ **21.** $\log_c \frac{22}{3}$
23. $2 \log_a x + 3 \log_a y + \log_a z$
25. $\log_b x + 2 \log_b y - 3 \log_b z$
27. $\frac{4}{3} \log_c x - \log_c y - \frac{2}{3} \log_c z$
29. $2 \log_a m + 3 \log_a n - \frac{3}{4} - \frac{5}{4} \log_a b$
31. $\log_a \dfrac{x^{2/3}}{y^{1/2}}$, or $\log_a \dfrac{\sqrt[3]{x^2}}{\sqrt{y}}$ **33.** $\log_a \dfrac{2x^4}{y^3}$ **35.** $\log_a \dfrac{\sqrt{a}}{x}$
37. 2.708 **39.** 0.51 **41.** -1.609 **43.** $\frac{3}{2}$ **45.** 2.609
47. t **49.** 5 **51.** 7 **53.** -7 **55.** $\mathbf{D_W}$ **57.** i
58. -1 **59.** 5 **60.** $\frac{3}{5} + \frac{4}{5}i$ **61.** $23 - 18i$ **62.** $10i$
63. $-34 - 31i$ **64.** $3 - 4i$ **65.** Left to the student
67. $\log_a (x^6 - x^4 y^2 + x^2 y^4 - y^6)$
69. $\frac{1}{2} \log_a (1 - s) + \frac{1}{2} \log_a (1 + s)$ **71.** False **73.** True
75. False

Margin Exercises, Section 8.5, pp. 716–718

1. 11.2675 **2.** -10.3848 **3.** Does not exist as a real
number **4.** Does not exist **5.** 2.3026 **6.** 78,237.1596

7. 0.1353   **8. (a)** 1.0860; **(b)** 1.0860   **9.** 5.5236

10.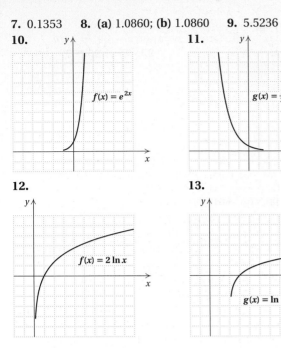

$f(x) = e^{2x}$

11.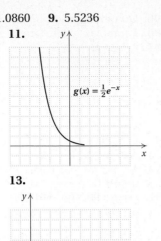

$g(x) = \frac{1}{2}e^{-x}$

12.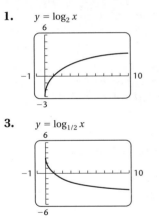

$f(x) = 2\ln x$

13.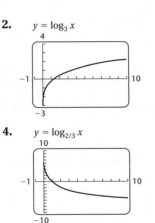

$g(x) = \ln(x-2)$

Calculator Corner, p. 719

1. $y = \log_2 x$

6

−1 ⎯⎯⎯⎯ 10

−3

2. $y = \log_3 x$

4

−1 ⎯⎯⎯⎯ 10

−3

3. $y = \log_{1/2} x$

6

−1 ⎯⎯⎯⎯ 10

−6

4. $y = \log_{2/3} x$

10

−1 ⎯⎯⎯⎯ 10

−10

Visualizing for Success, p. 720

1. J   **2.** B   **3.** O   **4.** G   **5.** N   **6.** F   **7.** A   **8.** H
9. I   **10.** K

Exercise Set 8.5, p. 721

1. 0.6931   **3.** 4.1271   **5.** 8.3814   **7.** −5.0832
9. −1.6094   **11.** Does not exist   **13.** −1.7455   **15.** 1
17. 15.0293   **19.** 0.0305   **21.** 109.9472   **23.** 5
25. 2.5702   **27.** 6.6439   **29.** 2.1452   **31.** −2.3219
33. −2.3219   **35.** 4.6284

37.

x	$f(x)$
0	1
1	2.7
2	7.4
3	20.1
−1	0.4
−2	0.1
−3	0.05

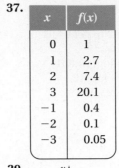

$f(x) = e^x$

39.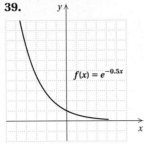

$f(x) = e^{-0.5x}$

41.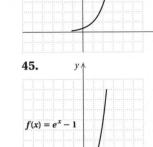

$f(x) = e^{x-1}$

43.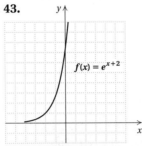

$f(x) = e^{x+2}$

45.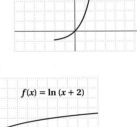

$f(x) = e^x - 1$

47.

x	$f(x)$
0	0.7
1	1.1
2	1.4
3	1.6
−0.5	0.4
−1	0
−1.5	−0.7

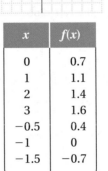

$f(x) = \ln(x+2)$

49.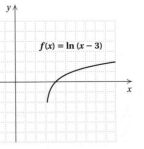

$f(x) = \ln(x-3)$

51.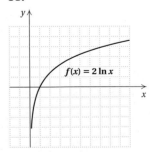

$f(x) = 2\ln x$

53.

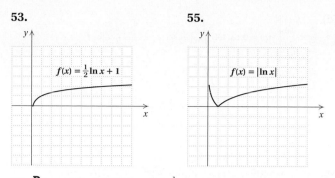

$f(x) = \frac{1}{2}\ln x + 1$

55.

$f(x) = |\ln x|$

57. $\mathbf{D_W}$ **59.** 16, 256   **60.** $\frac{1}{4}$, 9   **61.** 49, 121
62. $\pm 3, \pm 4$ **63.** Domain: $(-\infty, \infty)$; range: $[0, \infty)$
65. Domain: $(-\infty, \infty)$; range: $(-\infty, 100)$ **67.** $\left(\frac{5}{2}, \infty\right)$

Margin Exercises, Section 8.6, pp. 724–728

1. 1   **2.** 3   **3.** 1.5395   **4.** 14.6068   **5.** 25   **6.** $\frac{2}{5}$
7. 2   **8.** 5

Calculator Corner, p. 726

Left to the student

Exercise Set 8.6, p. 729

1. 3   **3.** 4   **5.** $\frac{5}{2}$ **7.** $\frac{3}{5}$ **9.** 3.4594   **11.** 5.4263
13. $\frac{5}{2}$ **15.** $-3, -1$ **17.** $\frac{3}{2}$ **19.** 4.6052   **21.** 2.3026
23. 140.6705   **25.** 2.7095   **27.** 3.2220   **29.** 256
31. $\frac{1}{32}$ **33.** 10   **35.** $\frac{1}{100}$ **37.** $e^2 \approx 7.3891$
39. $\dfrac{1}{e} \approx 0.3679$ **41.** 121   **43.** 10   **45.** $\frac{1}{3}$ **47.** 3
49. $\frac{2}{5}$ **51.** 5   **53.** No solution   **55.** $\mathbf{D_W}$ **57.** $\pm 10, \pm 2$
58. $-64, 8$ **59.** $-2, -3, \dfrac{-5 \pm \sqrt{41}}{2}$ **60.** $-\frac{1}{10}, 1$
61. $\dfrac{y^{4/3}}{25x^2z^4}$ **62.** $-i$ **63.** 1   **65.** **(a)** 0.3770;
(b) -1.9617; **(c)** 0.9036; **(d)** -1.5318 **67.** 3, 4   **69.** -4
71. 2   **73.** $\pm\sqrt{34}$ **75.** $10^{100,000}$ **77.** 1, 100
79. 3, -7 **81.** 1, 1.465

Margin Exercises, Section 8.7, pp. 732–738

1. About 65 decibels   **2.** $10^{-9.2}$ W/m²   **3.** About 4.9
4. 10^{-7} moles per liter   **5.** **(a)** 43.5 yr; **(b)** 16.5 yr
6. About 300 million; about 344 million   **7.** **(a)** $k \approx 6\%$,
$P(t) = 5000e^{0.06t}$; **(b)** \$5309.18; \$5637.48; \$9110.59; **(c)** about
11.6 yr   **8.** **(a)** $k \approx 0.471$, $P(t) = 2.8e^{0.471t}$, where $P(t)$ is in
billions of dollars and t is the number of years after 2000;
(b) about \$34,531 billion, or \$34.531 trillion; **(c)** 2008
9. 2972 yr

Calculator Corner, pp. 738–739

1. $y = 309{,}870.3567 \cdot 1.090787419^x$
2. 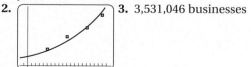   **3.** 3,531,046 businesses

Translating for Success, p. 740

1. D   **2.** M   **3.** I   **4.** A   **5.** E   **6.** H   **7.** C   **8.** G
9. N   **10.** B

Exercise Set 8.7, p. 741

1. About 95 dB   **3.** $10^{-1.5}$ W/m², or about
3.2×10^{-2} W/m²   **5.** About 6.8   **7.** 1.58×10^{-8} moles
per liter   **9.** 3.04 ft/sec   **11.** 2.17 ft/sec
13. **(a)** 243 people; **(b)** about 20.6 months; **(c)** about
0.6 month   **15.** **(a)** \$19,796; **(b)** 2011; **(c)** 11.9 yr
17. **(a)** $P(t) = 6.4e^{0.0114t}$; **(b)** 6.9 billion; **(c)** 2043; **(d)** 60.8 yr
19. **(a)** $P(t) = P_0e^{0.06t}$; **(b)** \$5309.18; \$5637.48; \$9110.59;
(c) in 11.6 yr   **21.** **(a)** $k \approx 0.061$, $P(t) = 852{,}737e^{0.061t}$;
(b) 2,888,380; **(c)** 2027   **23.** About 2103 yr   **25.** About
7.2 days   **27.** 69.3% per year   **29.** **(a)** $k \approx 0.094$,
$W(t) = 17.5e^{-0.094t}$; **(b)** 4.7 million tons; **(c)** 2173
31. **(a)** $V(t) = 451{,}000e^{0.07t}$; **(b)** \$2,419,866; **(c)** 9.9 yr;
(d) 2002; **(e)** The function predicts that the card's value will
be about \$908,202 in 2001. According to this, the card would
not be a good buy at \$1.1 million.   **33.** **(a)** $k \approx 0.007$,
$P(t) = 2{,}394{,}811e^{-0.007t}$; **(b)** about 2,081,949; **(c)** 2115
35. $\mathbf{D_W}$ **37.** -1 **38.** 1   **39.** i **40.** i **41.** $-1 - i$
42. -2 **43.** $\frac{63}{65} - \frac{16}{65}i$ **44.** $-\frac{2}{41} + \frac{23}{41}i$ **45.** 41
46. $91 + 60i$ **47.** $-0.937, 1.078, 58.770$
49. $-0.767, 2, 4$ **51.** \$13.4 million

Concept Reinforcement, p. 746

1. False   **2.** True   **3.** True   **4.** False   **5.** True
6. True   **7.** True   **8.** True

Summary and Review: Chapter 8, p. 746

1. $\{(2, -4), (-7, 5), (-2, -1), (11, 10)\}$ **2.** Not one-to-one
3. $g^{-1}(x) = \dfrac{7x + 3}{2}$ **4.** $f^{-1}(x) = \frac{1}{2}\sqrt[3]{x}$
5. $f^{-1}(x) = \dfrac{3x - 4}{2x}$ **6.**

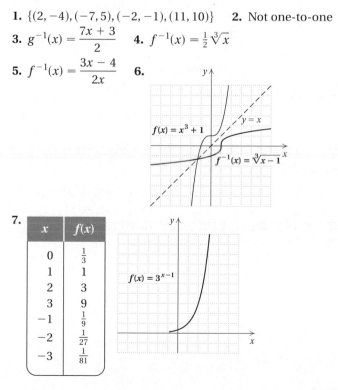

$f(x) = x^3 + 1$

$y = x$

$f^{-1}(x) = \sqrt[3]{x - 1}$

7.

x	$f(x)$
0	$\frac{1}{3}$
1	1
2	3
3	9
-1	$\frac{1}{9}$
-2	$\frac{1}{27}$
-3	$\frac{1}{81}$

$f(x) = 3^{x-1}$

8. $x = 3^y$

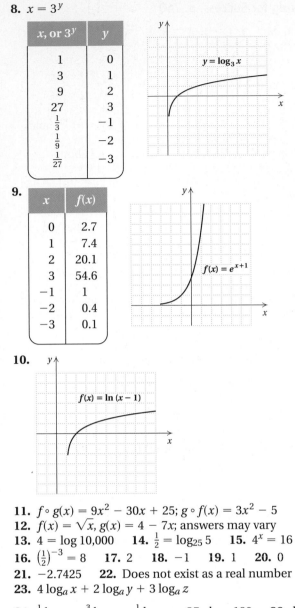

x, or 3^y	y
1	0
3	1
9	2
27	3
$\frac{1}{3}$	-1
$\frac{1}{9}$	-2
$\frac{1}{27}$	-3

$y = \log_3 x$

9.

x	$f(x)$
0	2.7
1	7.4
2	20.1
3	54.6
-1	1
-2	0.4
-3	0.1

$f(x) = e^{x+1}$

10.

$f(x) = \ln(x - 1)$

11. $f \circ g(x) = 9x^2 - 30x + 25;\ g \circ f(x) = 3x^2 - 5$
12. $f(x) = \sqrt{x},\ g(x) = 4 - 7x$; answers may vary
13. $4 = \log 10{,}000$ **14.** $\frac{1}{2} = \log_{25} 5$ **15.** $4^x = 16$
16. $\left(\frac{1}{2}\right)^{-3} = 8$ **17.** 2 **18.** -1 **19.** 1 **20.** 0
21. -2.7425 **22.** Does not exist as a real number
23. $4 \log_a x + 2 \log_a y + 3 \log_a z$

24. $\frac{1}{2} \log z - \frac{3}{4} \log x - \frac{1}{4} \log y$ **25.** $\log_a 120$ **26.** $\log \dfrac{a^{1/2}}{bc^2}$

27. 17 **28.** -7 **29.** 8.7601 **30.** 3.2698 **31.** 2.54995
32. -3.6602 **33.** -2.6921 **34.** 0.3753 **35.** 18.3568
36. 0 **37.** Does not exist **38.** 1 **39.** 0.4307

40. 1.7097 **41.** $\frac{1}{9}$ **42.** 2 **43.** $\dfrac{1}{10{,}000}$

44. $e^{-2} \approx 0.1353$ **45.** $\frac{7}{2}$ **46.** 1, -5 **47.** 1.5266
48. 35.0656 **49.** 2 **50.** 8 **51.** $\frac{17}{5}$ **52.** $\sqrt{43}$
53. 90 dB **54.** (a) 0.807 million; 1.581 million;
45.737 million; (b) in 14.5 yr; (c) 2 yr;

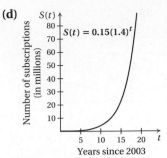
$S(t) = 0.15(1.4)^t$
Years since 2003

55. (a) $k \approx 0.094,\ V(t) = 40{,}000e^{0.094t}$; (b) \$102,399;
(c) after 8 yr, or in 2013 **56.** $k \approx 0.231$
57. About 8.25 yr **58.** About 3463 yr **59.** $\mathbf{D_W}$ You
cannot take the logarithm of a negative number because
logarithm bases are positive and there is no real-number
power to which a positive number can be raised to yield a
negative number. **60.** $\mathbf{D_W}$ $\log_a 1 = 0$ because $a^0 = 1$.
61. e^{e^3} **62.** $\left(\frac{8}{3}, -\frac{2}{3}\right)$

Test: Chapter 8, p. 750

1. [8.1a]

$f(x) = 2^{x+1}$

2. [8.3a]

$y = \log_2 x$

3. [8.5c]

$f(x) = e^{x-2}$

4. [8.5c]

$f(x) = \ln(x - 4)$

5. [8.2a] $\{(3, -4), (-8, 5), (-3, -1), (12, 10)\}$
6. [8.2b, c] $f^{-1}(x) = \dfrac{x + 3}{4}$ **7.** [8.2b, c] $f^{-1}(x) = \sqrt[3]{x} - 1$
8. [8.2b] Not one-to-one
9. [8.2d] $f \circ g(x) = 25x^2 - 15x + 2,\ g \circ f(x) = 5x^2 + 5x - 2$
10. [8.3b] $\log_{256} 16 = \frac{1}{2}$ **11.** [8.3b] $7^m = 49$ **12.** [8.3c] 3
13. [8.3c] 23 **14.** [8.3c] 0 **15.** [8.3d] -1.9101
16. [8.3d] Does not exist as a real number **17.** [8.4d]

$3 \log a + \frac{1}{2} \log b - 2 \log c$ **18.** [8.4d] $\log_a \dfrac{x^{1/3} z^2}{y^3}$

19. [8.4d] -0.544 **20.** [8.4d] 1.079 **21.** [8.5a] 6.6938
22. [8.5a] 107.7701 **23.** [8.5a] 0 **24.** [8.5b] 1.1881
25. [8.6b] 5 **26.** [8.6b] 2 **27.** [8.6b] 10,000
28. [8.6b] $e^{1/4} \approx 1.2840$ **29.** [8.6a] 0.0937 **30.** [8.6b] 9

31. [8.6b] 1   **32.** [8.7a] 4.2   **33.** [8.7b] **(a)** $31,081;
(b) 2010; **(c)** 15.7 yr   **34.** [8.7b] **(a)** $k \approx 0.009$,
$N(t) = 31.902e^{0.009t}$, where t is the number of years since
2000 and N is in millions; **(b)** 33.672 million, 35.862 million;
(c) 2052; **(d)** 77.0 yr   **35.** [8.7b] About 4.6%
36. [8.7b] About 4684 yr   **37.** [8.6b] 44, -37
38. [8.4d] 2

Cumulative Review: Chapters R–8, p. 753

1. [1.2a], [5.7a] **(a)** About 14.8 min; **(b)** 30 appointments;
(c) $T = \dfrac{IN}{1.08}$; **(d)** $N = 1.08\,\dfrac{T}{I}$; **(e)** The formula found in
part (c) would be useful if we knew the interval time and the
number of appointments desired and wanted to find the
total number of minutes the doctor would have to spend
with patients per day in order to achieve this. The formula
found in part (d) would allow us to find the total number of
appointments that could be scheduled given a particular
interval time and the total number of minutes the doctor
can spend with patients per day.   **2.** [R.1d], [R.2a] 6
3. [R.7a] $\dfrac{20x^6z^2}{y}$ **4.** [R.6b] $-4x - 1$ **5.** [R.3c] 25
6. [1.1d] $\frac{11}{2}$ **7.** [4.8a] $-2, 5$ **8.** [3.5a] $(1, -2, 0)$
9. [3.3a] $(3, -1)$ **10.** [5.5a] $\frac{9}{2}$ **11.** [6.6b] 5
12. [7.4c] 9, 25   **13.** [7.4c] $\pm 3, \pm 2$ **14.** [8.6b] 8
15. [8.6a] 0.3542   **16.** [8.6b] $\frac{80}{9}$
17. [7.8a] $\{x \mid x < -5 \text{ or } x > 1\}$, or $(-\infty, -5) \cup (1, \infty)$
18. [1.6e] $\{x \mid x \le -3 \text{ or } x \ge 6\}$, or $(-\infty, -3] \cup [6, \infty)$
19. [7.2a] $-3 \pm 2\sqrt{5}$ **20.** [5.7a] $a = \dfrac{Db}{b - D}$
21. [5.7a] $q = \dfrac{pf}{p - f}$
22. [5.1a] $\left(-\infty, -\frac{1}{3}\right) \cup \left(-\frac{1}{3}, 2\right) \cup (2, \infty)$ **23.** [R.4b] 2
24. [5.6a] $\frac{60}{11}$ min, or $5\frac{5}{11}$ min   **25.** [8.7a] **(a)** 78; **(b)** 67.5
26. [5.6c] $2\frac{7}{9}$ km/h   **27.** [3.4a] Swim Clean: 60 L; Pure
Swim: 40 L   **28.** [8.7b] **(a)** $P(t) = 19e^{0.012t}$, where $P(t)$ is in
millions and t is the number of years after 2004; **(b)** about
20.2 million, about 21.7 million; **(c)** about 57.8 yr
29. [4.8b] 10 ft   **30.** [2.5a]

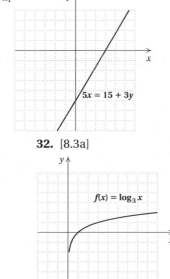

31. [7.6a]
32. [8.3a]

33. [8.1a]

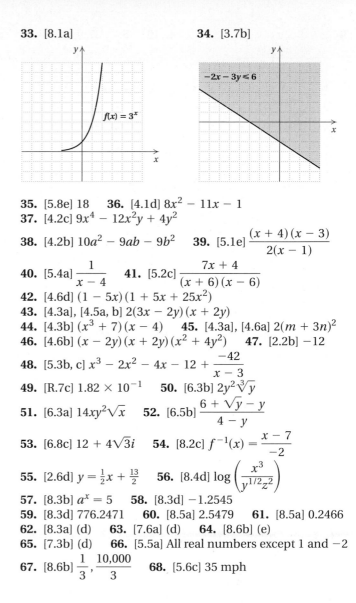

34. [3.7b]

35. [5.8e] 18   **36.** [4.1d] $8x^2 - 11x - 1$
37. [4.2c] $9x^4 - 12x^2y + 4y^2$
38. [4.2b] $10a^2 - 9ab - 9b^2$ **39.** [5.1e] $\dfrac{(x + 4)(x - 3)}{2(x - 1)}$
40. [5.4a] $\dfrac{1}{x - 4}$ **41.** [5.2c] $\dfrac{7x + 4}{(x + 6)(x - 6)}$
42. [4.6d] $(1 - 5x)(1 + 5x + 25x^2)$
43. [4.3a], [4.5a, b] $2(3x - 2y)(x + 2y)$
44. [4.3b] $(x^3 + 7)(x - 4)$ **45.** [4.3a], [4.6a] $2(m + 3n)^2$
46. [4.6b] $(x - 2y)(x + 2y)(x^2 + 4y^2)$ **47.** [2.2b] -12
48. [5.3b, c] $x^3 - 2x^2 - 4x - 12 + \dfrac{-42}{x - 3}$
49. [R.7c] 1.82×10^{-1} **50.** [6.3b] $2y^2 \sqrt[3]{y}$
51. [6.3a] $14xy^2 \sqrt{x}$ **52.** [6.5b] $\dfrac{6 + \sqrt{y} - y}{4 - y}$
53. [6.8c] $12 + 4\sqrt{3}i$ **54.** [8.2c] $f^{-1}(x) = \dfrac{x - 7}{-2}$
55. [2.6d] $y = \frac{1}{2}x + \frac{13}{2}$ **56.** [8.4d] $\log\left(\dfrac{x^3}{y^{1/2}z^2}\right)$
57. [8.3b] $a^x = 5$ **58.** [8.3d] -1.2545
59. [8.3d] 776.2471   **60.** [8.5a] 2.5479   **61.** [8.5a] 0.2466
62. [8.3a] (d)   **63.** [7.6a] (d)   **64.** [8.6b] (e)
65. [7.3b] (d)   **66.** [5.5a] All real numbers except 1 and -2
67. [8.6b] $\dfrac{1}{3}, \dfrac{10,000}{3}$ **68.** [5.6c] 35 mph

CHAPTER 9

Margin Exercises, Section 9.1, pp. 759–765

1.
2.

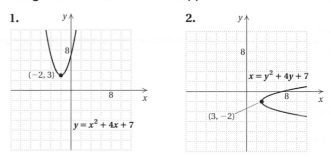

3.

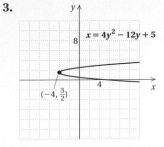
$x = 4y^2 - 12y + 5$

$(-4, \frac{3}{2})$

4. 10   **5.** $\sqrt{37} \approx 6.083$

6. $\left(\frac{3}{2}, -3\right)$ **7.** $(9, -5)$ **8.**

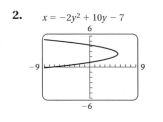

Center: $(5, -\frac{1}{2})$
Radius: 3

9. $(0,0)$; $r = 8$ **10.** $(x + 3)^2 + (y - 1)^2 = 36$
11. Center: $(-1, 2)$; radius: $\sqrt{3}$;

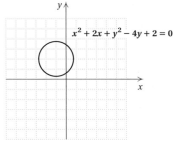
$x^2 + 2x + y^2 - 4y + 2 = 0$

Calculator Corner, p. 761

1. $x = y^2 + 4y + 7$

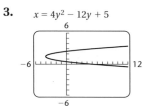

2. $x = -2y^2 + 10y - 7$

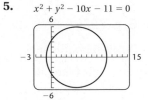

3. $x = 4y^2 - 12y + 5$

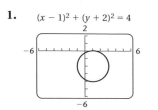

Calculator Corner, p. 766

1. $(x - 1)^2 + (y + 2)^2 = 4$

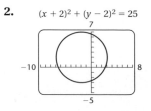

2. $(x + 2)^2 + (y - 2)^2 = 25$

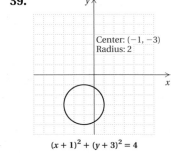

3. $x^2 + y^2 - 16 = 0$

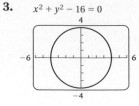

4. $4x^2 + 4y^2 = 100$

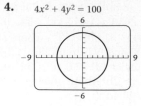

5. $x^2 + y^2 - 10x - 11 = 0$

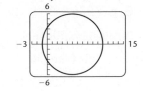

Exercise Set 9.1, p. 767

1.

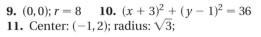
$y = x^2$

3.

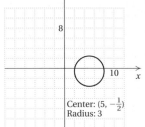
$x = y^2 + 4y + 1$

5.

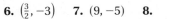
$y = -x^2 + 4x - 5$

7.

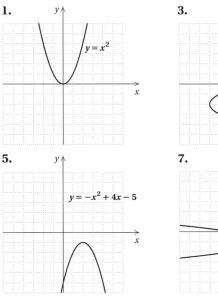
$x = -3y^2 - 6y - 1$

9. 5   **11.** $\sqrt{29} \approx 5.385$ **13.** $\sqrt{648} \approx 25.456$ **15.** 7.1
17. $\frac{\sqrt{41}}{7} \approx 0.915$ **19.** $\sqrt{6970} \approx 83.487$ **21.** $\sqrt{a^2 + b^2}$
23. $\sqrt{17 + 2\sqrt{14} + 2\sqrt{15}} \approx 5.677$
25. $\sqrt{9{,}672{,}400} \approx 3110.048$ **27.** $\left(\frac{3}{2}, \frac{7}{2}\right)$ **29.** $\left(0, \frac{11}{2}\right)$
31. $\left(-1, -\frac{17}{2}\right)$ **33.** $(-0.25, -0.3)$ **35.** $\left(-\frac{1}{12}, \frac{1}{24}\right)$
37. $\left(\frac{\sqrt{2} + \sqrt{3}}{2}, \frac{3}{2}\right)$ **39.**

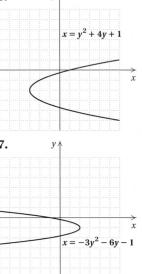

Center: $(-1, -3)$
Radius: 2

$(x + 1)^2 + (y + 3)^2 = 4$

41.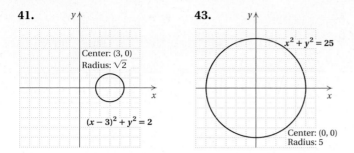

Center: (3, 0)
Radius: $\sqrt{2}$

$(x - 3)^2 + y^2 = 2$

43.

$x^2 + y^2 = 25$

Center: (0, 0)
Radius: 5

3.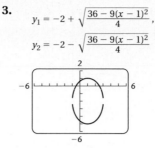

$y_1 = -2 + \sqrt{\dfrac{36 - 9(x - 1)^2}{4}}$,

$y_2 = -2 - \sqrt{\dfrac{36 - 9(x - 1)^2}{4}}$

4.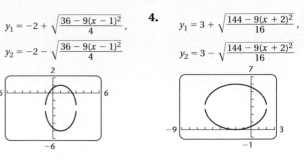

$y_1 = 3 + \sqrt{\dfrac{144 - 9(x + 2)^2}{16}}$,

$y_2 = 3 - \sqrt{\dfrac{144 - 9(x + 2)^2}{16}}$

45. $x^2 + y^2 = 49$ **47.** $(x + 5)^2 + (y - 3)^2 = 7$
49. $(-4, 3), r = 2\sqrt{10}$ **51.** $(4, -1), r = 2$
53. $(2, 0), r = 2$ **55. D_W**   **57.** $(9, 2)$ **58.** $(-8, 16)$
59. $\left(-\frac{21}{5}, -\frac{73}{5}\right)$ **60.** $(1, 2)$ **61.** No solution
62. $(2a + b)(2a - b)$ **63.** $(x - 4)(x + 4)$
64. $(a - 3b)(a + 3b)$ **65.** $(8p - 9q)(8p + 9q)$
66. $25(4cd - 3)(4cd + 3)$ **67.** $x^2 + y^2 = 2$
69. $(x + 3)^2 + (y + 2)^2 = 9$ **71.** $\sqrt{49 + k^2}$
73. $8\sqrt{m^2 + n^2}$ **75.** Yes   **77.** $\left(2, 4\sqrt{2}\right)$
79. (a) $(0, -8467.8)$; **(b)** 8487.3 mm

Margin Exercises, Section 9.2, pp. 772–773

1.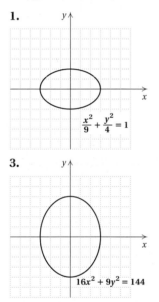

$\dfrac{x^2}{9} + \dfrac{y^2}{4} = 1$

2.

$\dfrac{x^2}{9} + \dfrac{y^2}{25} = 1$

3.

$16x^2 + 9y^2 = 144$

4.

$\dfrac{(x + 2)^2}{16} + \dfrac{(y - 3)^2}{9} = 1$

Calculator Corner, p. 774

1.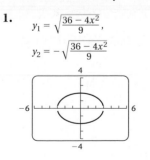

$y_1 = \sqrt{\dfrac{36 - 4x^2}{9}}$,

$y_2 = -\sqrt{\dfrac{36 - 4x^2}{9}}$

2.

$y_1 = \sqrt{\dfrac{144 - 16x^2}{9}}$,

$y_2 = -\sqrt{\dfrac{144 - 16x^2}{9}}$

Exercise Set 9.2, p. 776

1.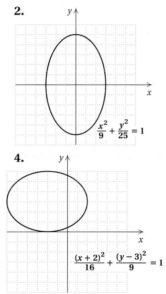

$\dfrac{x^2}{9} + \dfrac{y^2}{36} = 1$

3.

$\dfrac{x^2}{1} + \dfrac{y^2}{4} = 1$

5.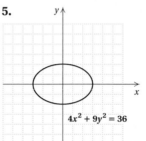

$4x^2 + 9y^2 = 36$

7.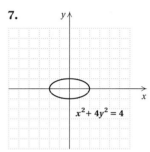

$x^2 + 4y^2 = 4$

9.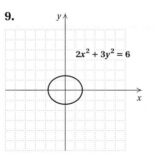

$2x^2 + 3y^2 = 6$

11.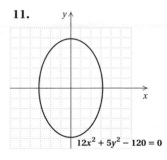

$12x^2 + 5y^2 - 120 = 0$

13.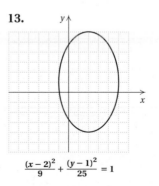

$\dfrac{(x - 2)^2}{9} + \dfrac{(y - 1)^2}{25} = 1$

15.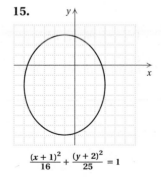

$\dfrac{(x + 1)^2}{16} + \dfrac{(y + 2)^2}{25} = 1$

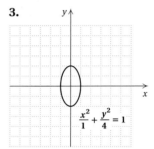
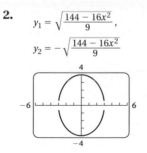
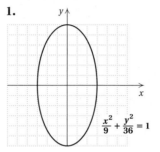

17.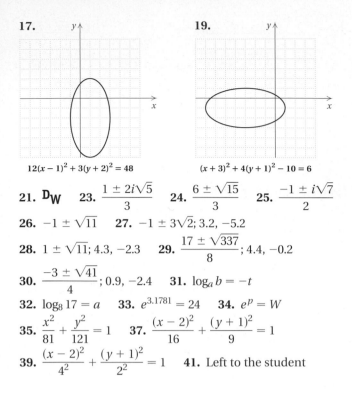

$12(x-1)^2 + 3(y+2)^2 = 48$

19.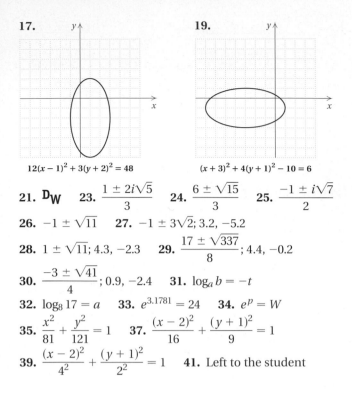

$(x+3)^2 + 4(y+1)^2 - 10 = 6$

21. $\mathbf{D_W}$ **23.** $\dfrac{1 \pm 2i\sqrt{5}}{3}$ **24.** $\dfrac{6 \pm \sqrt{15}}{3}$ **25.** $\dfrac{-1 \pm i\sqrt{7}}{2}$

26. $-1 \pm \sqrt{11}$ **27.** $-1 \pm 3\sqrt{2}$; 3.2, −5.2

28. $1 \pm \sqrt{11}$; 4.3, −2.3 **29.** $\dfrac{17 \pm \sqrt{337}}{8}$; 4.4, −0.2

30. $\dfrac{-3 \pm \sqrt{41}}{4}$; 0.9, −2.4 **31.** $\log_a b = -t$

32. $\log_8 17 = a$ **33.** $e^{3.1781} = 24$ **34.** $e^p = W$

35. $\dfrac{x^2}{81} + \dfrac{y^2}{121} = 1$ **37.** $\dfrac{(x-2)^2}{16} + \dfrac{(y+1)^2}{9} = 1$

39. $\dfrac{(x-2)^2}{4^2} + \dfrac{(y+1)^2}{2^2} = 1$ **41.** Left to the student

Margin Exercises, Section 9.3, pp. 780–781

1.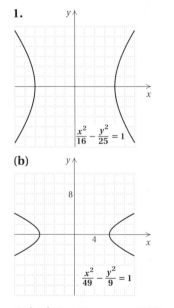

$\dfrac{x^2}{16} - \dfrac{y^2}{25} = 1$

2. (a)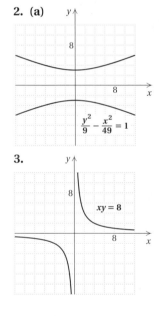

$\dfrac{y^2}{9} - \dfrac{x^2}{49} = 1$

(b)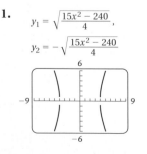

$\dfrac{x^2}{49} - \dfrac{y^2}{9} = 1$

3.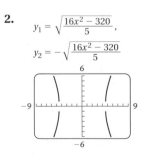

$xy = 8$

Calculator Corner, p. 782

1. $y_1 = \sqrt{\dfrac{15x^2 - 240}{4}}$,

$y_2 = -\sqrt{\dfrac{15x^2 - 240}{4}}$

2. $y_1 = \sqrt{\dfrac{16x^2 - 320}{5}}$,

$y_2 = -\sqrt{\dfrac{16x^2 - 320}{5}}$

3. $y_1 = \sqrt{\dfrac{16x^2 - 48}{3}}$,

$y_2 = -\sqrt{\dfrac{16x^2 - 48}{3}}$

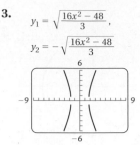

4. $y_1 = \sqrt{\dfrac{45x^2 - 405}{9}}$,

$y_2 = -\sqrt{\dfrac{45x^2 - 405}{9}}$

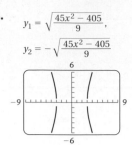

Visualizing for Success, p. 783

1. C **2.** E **3.** G **4.** J **5.** B **6.** F **7.** A **8.** H
9. I **10.** D

Exercise Set 9.3, p. 784

1.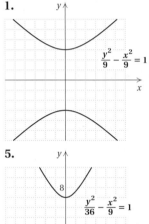

$\dfrac{y^2}{9} - \dfrac{x^2}{9} = 1$

3.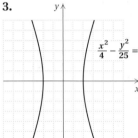

$\dfrac{x^2}{4} - \dfrac{y^2}{25} = 1$

5.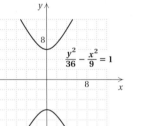

$\dfrac{y^2}{36} - \dfrac{x^2}{9} = 1$

7.

$y^2 - x^2 = 25$

9.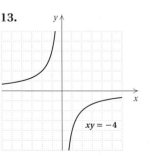

$x^2 = 1 + y^2$

11.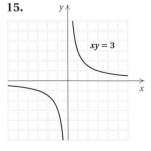

$25x^2 - 16y^2 = 400$

13.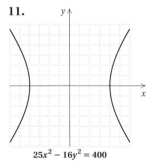

$xy = -4$

15.

$xy = 3$

17.

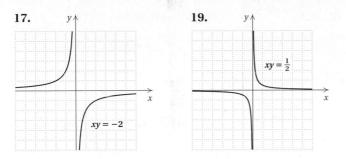

19.

$$xy = \tfrac{1}{2}$$
$$xy = -2$$

21. **D**_W   **23.** Discriminant   **24.** Vertex
25. Vertical line   **26.** Common   **27.** Exponent
28. Horizontal line   **29.** Function   **30.** Half-life
31. Left to the student   **33.** Circle   **35.** Parabola
37. Ellipse   **39.** Circle   **41.** Hyperbola   **43.** Circle

Margin Exercises, Section 9.4, pp. 788–792

1.

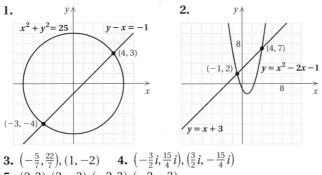

$$x^2 + y^2 = 25 \qquad y - x = -1$$
$(4, 3)$
$(-3, -4)$

2.

$(4, 7)$
$(-1, 2)$
$$y = x^2 - 2x - 1$$
$$y = x + 3$$

3. $\left(-\tfrac{5}{7}, \tfrac{22}{7}\right), (1, -2)$ **4.** $\left(-\tfrac{3}{2}i, \tfrac{15}{4}i\right), \left(\tfrac{3}{2}i, -\tfrac{15}{4}i\right)$
5. $(2, 3), (2, -3), (-2, 3), (-2, -3)$
6. $(3, 2), (-3, -2), (2, 3), (-2, -3)$ **7.** 12 m by 5 m
8. Width: about 40 in.; height: about 22.5 in.

Calculator Corner, p. 789

1. Left to the student   **2.** Left to the student

Exercise Set 9.4, p. 793

1. $(-8, -6), (6, 8)$ **3.** $(2, 0), (0, 3)$ **5.** $(-2, 1)$
7. $\left(\dfrac{5 + \sqrt{70}}{3}, \dfrac{-1 + \sqrt{70}}{3}\right), \left(\dfrac{5 - \sqrt{70}}{3}, \dfrac{-1 - \sqrt{70}}{3}\right)$
9. $\left(\tfrac{7}{3}, \tfrac{1}{3}\right), (1, -1)$ **11.** $(-7, 1), (1, -7)$ **13.** $(3, -5), (-1, 3)$
15. $\left(\dfrac{8 + 3i\sqrt{6}}{2}, \dfrac{-8 + 3i\sqrt{6}}{2}\right), \left(\dfrac{8 - 3i\sqrt{6}}{2}, \dfrac{-8 - 3i\sqrt{6}}{2}\right)$
17. $(-5, 0), (4, 3), (4, -3)$ **19.** $(3, 0), (-3, 0)$
21. $(2, 4), (-2, -4), (4, 2), (-4, -2)$ **23.** $(2, 3), (-2, -3),$
$(3, 2), (-3, -2)$ **25.** $(2, 1), (-2, -1)$ **27.** $(5, 2), (-5, 2),$
$\left(2, -\tfrac{4}{5}\right), \left(-2, -\tfrac{4}{5}\right)$ **29.** $\left(\sqrt{2}, -\sqrt{2}\right), \left(-\sqrt{2}, \sqrt{2}\right)$
31. $\left(\dfrac{8i\sqrt{5}}{5}, \dfrac{3\sqrt{105}}{5}\right), \left(-\dfrac{8i\sqrt{5}}{5}, \dfrac{3\sqrt{105}}{5}\right),$
$\left(\dfrac{8i\sqrt{5}}{5}, -\dfrac{3\sqrt{105}}{5}\right), \left(-\dfrac{8i\sqrt{5}}{5}, -\dfrac{3\sqrt{105}}{5}\right)$
33. Length: 12 ft; width: 5 ft   **35.** Length: 7 in.; width: 2 in.
37. Length: 12 ft; width: 5 ft   **39.** Length: 24.8 cm;
height: 18.6 cm   **41.** 24 ft, 16 ft   **43.** Length: $\sqrt{2}$ m;
width: 1 m   **45.** **D**_W   **47.** $f^{-1}(x) = \dfrac{x + 5}{2}$

48. $f^{-1}(x) = \dfrac{7x + 3}{2x}$ **49.** $f^{-1}(x) = \dfrac{3x + 2}{1 - x}$
50. $f^{-1}(x) = \dfrac{4x + 8}{5x - 3}$ **51.** Does not exist   **52.** Does not
exist   **53.** $f^{-1}(x) = \log x$ **54.** $f^{-1}(x) = \ln x$
55. $f^{-1}(x) = \sqrt[3]{x + 4}$ **56.** $f^{-1}(x) = x^3 - 2$
57. $f^{-1}(x) = e^x$ **58.** $f^{-1}(x) = 10^x$ **59.** Left to the
student   **61.** One piece: $38\tfrac{12}{25}$ cm; other piece: $61\tfrac{13}{25}$ cm
63. 30   **65.** $\left(\tfrac{1}{2}, \tfrac{1}{3}\right), \left(\tfrac{1}{3}, \tfrac{1}{2}\right)$

Concept Reinforcement, p. 797

1. True   **2.** True   **3.** False   **4.** True   **5.** False
6. False   **7.** True   **8.** False

Summary and Review: Chapter 9, p. 798

1. 4   **2.** 5   **3.** $\sqrt{90.1} \approx 9.492$ **4.** $\sqrt{9 + 4a^2}$ **5.** $(4, 6)$
6. $\left(-3, \tfrac{5}{2}\right)$ **7.** $\left(\dfrac{3}{4}, \dfrac{\sqrt{3} - \sqrt{2}}{2}\right)$ **8.** $\left(\tfrac{1}{2}, 2a\right)$
9. $(-2, 3), \sqrt{2}$ **10.** $(5, 0), 7$ **11.** $(3, 1), 3$
12. $(-4, 3), \sqrt{35}$ **13.** $(x + 4)^2 + (y - 3)^2 = 48$
14. $(x - 7)^2 + (y + 2)^2 = 20$
15.

$$\dfrac{x^2}{16} + \dfrac{y^2}{4} = 1$$

16.

$$\dfrac{y^2}{9} - \dfrac{x^2}{4} = 1$$

17.

$$x^2 + y^2 = 16$$

18.

$$x = y^2 + 2y - 2$$

19.

$$y = -2x^2 - 2x + 3$$

20.

$$x^2 + y^2 + 2x - 4y - 4 = 0$$

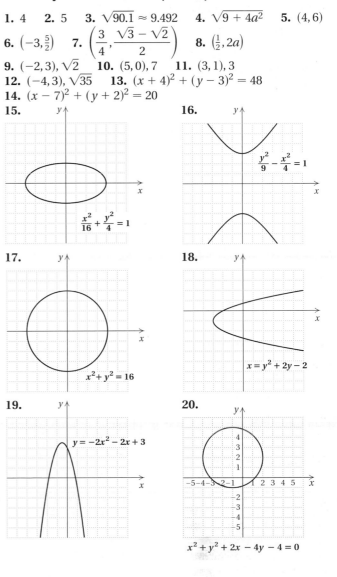

21.

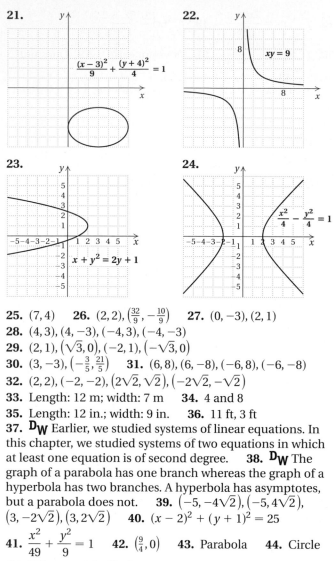

$\dfrac{(x-3)^2}{9} + \dfrac{(y+4)^2}{4} = 1$

22.

$xy = 9$

23.

$x + y^2 = 2y + 1$

24.

$\dfrac{x^2}{4} - \dfrac{y^2}{4} = 1$

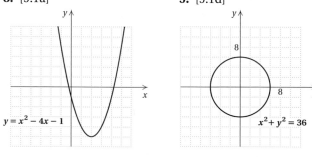

25. $(7,4)$ **26.** $(2,2), \left(\frac{32}{9}, -\frac{10}{9}\right)$ **27.** $(0,-3), (2,1)$
28. $(4,3), (4,-3), (-4,3), (-4,-3)$
29. $(2,1), \left(\sqrt{3},0\right), (-2,1), \left(-\sqrt{3},0\right)$
30. $(3,-3), \left(-\frac{3}{5}, \frac{21}{5}\right)$ **31.** $(6,8), (6,-8), (-6,8), (-6,-8)$
32. $(2,2), (-2,-2), \left(2\sqrt{2}, \sqrt{2}\right), \left(-2\sqrt{2}, -\sqrt{2}\right)$
33. Length: 12 m; width: 7 m **34.** 4 and 8
35. Length: 12 in.; width: 9 in. **36.** 11 ft, 3 ft
37. $\mathbf{D_W}$ Earlier, we studied systems of linear equations. In this chapter, we studied systems of two equations in which at least one equation is of second degree. **38.** $\mathbf{D_W}$ The graph of a parabola has one branch whereas the graph of a hyperbola has two branches. A hyperbola has asymptotes, but a parabola does not. **39.** $\left(-5, -4\sqrt{2}\right), \left(-5, 4\sqrt{2}\right),$
$\left(3, -2\sqrt{2}\right), \left(3, 2\sqrt{2}\right)$ **40.** $(x-2)^2 + (y+1)^2 = 25$

41. $\dfrac{x^2}{49} + \dfrac{y^2}{9} = 1$ **42.** $\left(\frac{9}{4}, 0\right)$ **43.** Parabola **44.** Circle

45. Ellipse **46.** Circle

Test: Chapter 9, p. 801

1. [9.1b] $\sqrt{180} \approx 13.416$ **2.** [9.1b] $\sqrt{36 + 4a^2}$, or $2\sqrt{9 + a^2}$
3. [9.1c] $(0,5)$ **4.** [9.1c] $(0,0)$ **5.** [9.1d] Center: $(-2,3)$;
radius: 8 **6.** [9.1d] Center: $(-2,3)$; radius: 3
7. [9.1d] $(x+2)^2 + (y+5)^2 = 18$
8. [9.1a] **9.** [9.1d]

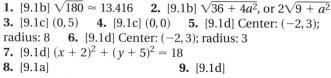

$y = x^2 - 4x - 1$ $x^2 + y^2 = 36$

10. [9.3a]

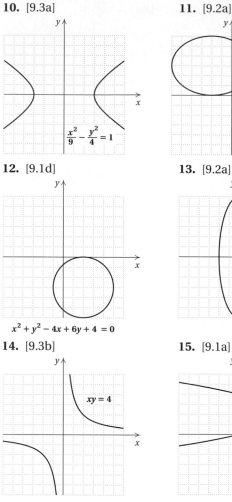

$\dfrac{x^2}{9} - \dfrac{y^2}{4} = 1$

11. [9.2a]

$\dfrac{(x+2)^2}{16} + \dfrac{(y-3)^2}{9} = 1$

12. [9.1d]

$x^2 + y^2 - 4x + 6y + 4 = 0$

13. [9.2a]

$9x^2 + y^2 = 36$

14. [9.3b]

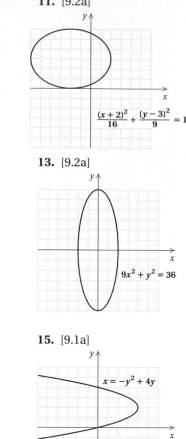

$xy = 4$

15. [9.1a]

$x = -y^2 + 4y$

16. [9.4a] $(0,3), (4,0)$ **17.** [9.4a] $(4,0), (-4,0)$
18. [9.4b] 16 ft by 12 ft **19.** [9.4b] $1200, 6\%$
20. [9.4b] 11 yd by 2 yd **21.** [9.4b] $\sqrt{5}$ m, $\sqrt{3}$ m

22. [9.2a] $\dfrac{(x-6)^2}{25} + \dfrac{(y-3)^2}{9} = 1$
23. [9.1d] $\{(x,y) \mid (x-8)^2 + y^2 = 100\}$
24. [9.4b] 9 **25.** [9.1b] $\left(0, -\frac{31}{4}\right)$

Cumulative Review/Final Examination, p. 803

1. [R.1d] $\frac{2}{15}$ **2.** [R.7a] $-9x^6y^2$ **3.** [R.3c] 62.5
4. [R.6b] $-x + 30$ **5.** [1.4c] $\{x \mid x \geq -1\}$, or $[-1, \infty)$
6. [1.6e] $\{x \mid x < -6.4 \text{ or } x > 6.4\}$, or $(-\infty, -6.4) \cup (6.4, \infty)$
7. [1.5a] $\{x \mid -1 \leq x < 6\}$, or $[-1, 6)$
8. [3.2a], [3.3a] $\left(-\frac{1}{3}, 5\right)$ **9.** [3.5a] $(-1, 2, 3)$
10. [4.8a] $-1, \frac{3}{2}$ **11.** [5.5a] $-\frac{2}{3}, 3$ **12.** [6.6a] 4

13. [4.8a] $-7, -3$ **14.** [7.2a] $-\dfrac{1}{4} \pm i\dfrac{\sqrt{7}}{4}$

15. [7.4c] $-3, -2, 2, 3$ **16.** [5.5a] -16
17. [9.4a] $(1,2), (-1,2), (1,-2), (-1,-2)$
18. [8.6b] 1 **19.** [8.6a] 1.748 **20.** [8.6b] 9
21. [7.8a] $\{x \mid x \leq -1 \text{ or } x \geq 1\}$, or $(-\infty, -1] \cup [1, \infty)$
22. [7.8b] $\{x \mid x < -1 \text{ or } x > 2\}$, or $(-\infty, -1) \cup (2, \infty)$

23. [1.2a] $N = \dfrac{4P - 3M}{6}$ **24.** [5.7a] $p = \dfrac{qf}{q - f}$

25. [8.1c], [8.7b] **(a)** About 74.97 billion ft^3, about 84.94 billion ft^3; **(b)** about 39 yr;
(c)

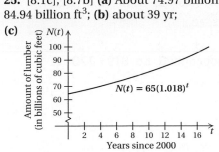

26. [8.7b] **(a)** $A(t) = \$50,000(1.04)^t$; **(b)** $50,000; $58,492.93; $68,428.45; $74,012.21;
(c)

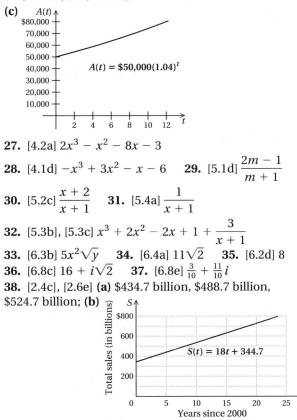

27. [4.2a] $2x^3 - x^2 - 8x - 3$

28. [4.1d] $-x^3 + 3x^2 - x - 6$ **29.** [5.1d] $\dfrac{2m-1}{m+1}$

30. [5.2c] $\dfrac{x+2}{x+1}$ **31.** [5.4a] $\dfrac{1}{x+1}$

32. [5.3b], [5.3c] $x^3 + 2x^2 - 2x + 1 + \dfrac{3}{x+1}$

33. [6.3b] $5x^2\sqrt{y}$ **34.** [6.4a] $11\sqrt{2}$ **35.** [6.2d] 8
36. [6.8c] $16 + i\sqrt{2}$ **37.** [6.8e] $\frac{3}{10} + \frac{11}{10}i$
38. [2.4c], [2.6e] **(a)** $434.7 billion, $488.7 billion, $524.7 billion; **(b)**

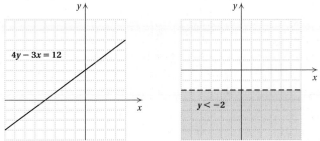

(c) $(0, 344.7)$; **(d)** 18; **(e)** an increase of $18 billion per year
39. [2.5a]   **40.** [3.7b]

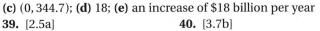

41. [3.7c]   **42.** [7.6a]
43. [9.1d]   **44.** [9.1a]
45. [8.5c]   **46.** [8.3a]

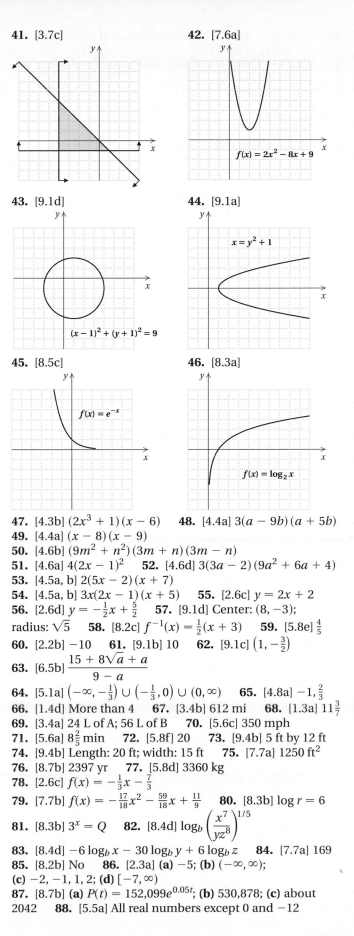

47. [4.3b] $(2x^3 + 1)(x - 6)$ **48.** [4.4a] $3(a - 9b)(a + 5b)$
49. [4.4a] $(x - 8)(x - 9)$
50. [4.6b] $(9m^2 + n^2)(3m + n)(3m - n)$
51. [4.6a] $4(2x - 1)^2$ **52.** [4.6d] $3(3a - 2)(9a^2 + 6a + 4)$
53. [4.5a, b] $2(5x - 2)(x + 7)$
54. [4.5a, b] $3x(2x - 1)(x + 5)$ **55.** [2.6c] $y = 2x + 2$
56. [2.6d] $y = -\frac{1}{2}x + \frac{5}{2}$ **57.** [9.1d] Center: $(8, -3)$;
radius: $\sqrt{5}$ **58.** [8.2c] $f^{-1}(x) = \frac{1}{2}(x + 3)$ **59.** [5.8e] $\frac{4}{5}$
60. [2.2b] -10 **61.** [9.1b] 10   **62.** [9.1c] $\left(1, -\frac{3}{2}\right)$
63. [6.5b] $\dfrac{15 + 8\sqrt{a} + a}{9 - a}$
64. [5.1a] $\left(-\infty, -\frac{1}{3}\right) \cup \left(-\frac{1}{3}, 0\right) \cup (0, \infty)$ **65.** [4.8a] $-1, \frac{2}{3}$
66. [1.4d] More than 4   **67.** [3.4b] 612 mi   **68.** [1.3a] $11\frac{3}{7}$
69. [3.4a] 24 L of A; 56 L of B   **70.** [5.6c] 350 mph
71. [5.6a] $8\frac{2}{5}$ min   **72.** [5.8f] 20   **73.** [9.4b] 5 ft by 12 ft
74. [9.4b] Length: 20 ft; width: 15 ft   **75.** [7.7a] 1250 ft^2
76. [8.7b] 2397 yr   **77.** [5.8d] 3360 kg
78. [2.6c] $f(x) = -\frac{1}{3}x - \frac{7}{3}$
79. [7.7b] $f(x) = -\frac{17}{18}x^2 - \frac{59}{18}x + \frac{11}{9}$ **80.** [8.3b] $\log r = 6$
81. [8.3b] $3^x = Q$ **82.** [8.4d] $\log_b \left(\dfrac{x^7}{yz^8}\right)^{1/5}$
83. [8.4d] $-6 \log_b x - 30 \log_b y + 6 \log_b z$ **84.** [7.7a] 169
85. [8.2b] No   **86.** [2.3a] **(a)** -5; **(b)** $(-\infty, \infty)$;
(c) $-2, -1, 1, 2$; **(d)** $[-7, \infty)$
87. [8.7b] **(a)** $P(t) = 152,099e^{0.05t}$; **(b)** 530,878; **(c)** about 2042   **88.** [5.5a] All real numbers except 0 and -12

A-57

89. [8.6b] 81 **90.** [9.1d] Circle centered at the origin with radius a **91.** [1.3a] 84 yr
92. [9.4a] $(-12, -8), (-8, -12), (8, 12), (12, 8)$

APPENDIXES

Margin Exercises, Appendix A, pp. 810–811

1. $70\dfrac{\text{mi}}{\text{hr}}$ **2.** $25\dfrac{\text{m}}{\text{sec}}$ **3.** 16 yd **4.** 72 in^2

5. 66 man-hours **6.** $800\dfrac{\text{kW-hr}}{\text{da}}$ **7.** 84 in. **8.** $132\dfrac{\text{ft}}{\text{sec}}$

Exercise Set A, p. 812

1. 68 ft **3.** 45 g **5.** $15\dfrac{\text{mi}}{\text{hr}}$ **7.** $3.3\dfrac{\text{m}}{\text{sec}}$ **9.** $4\dfrac{\text{in.-lb}}{\text{sec}}$

11. 12 yd **13.** 16 ft^3 **15.** $\dfrac{\$970}{\text{day}}$ **17.** 51.2 oz

19. 3080 ft/min **21.** 96 in. **23.** Approximately 0.03 yr

25. Approximately 31,710 yr **27.** $80\dfrac{\text{oz}}{\text{in.}}$ **29.** 172,800 sec

31. $\frac{3}{2}$ ft^2 **33.** $1.08\dfrac{\text{ton}}{\text{yd}^3}$

Margin Exercises, Appendix B, pp. 813–816

1. -5 **2.** -7 **3.** -6 **4.** 93 **5.** $(3, -1)$
6. $(1, 3, -2)$

Exercise Set B, p. 817

1. 10 **3.** 0 **5.** -48 **7.** 0 **9.** -10 **11.** -3
13. 5 **15.** 0 **17.** $(2, 0)$ **19.** $\left(-\frac{25}{2}, -\frac{11}{2}\right)$

21. $\left(\frac{3}{2}, \frac{5}{2}\right)$ **23.** $(-4, 3)$ **25.** $(2, -1, 4)$ **27.** $(1, 2, 3)$
29. $\left(\frac{3}{2}, -4, 3\right)$ **31.** $(2, -2, 1)$

Margin Exercises, Appendix C, pp. 819–820

1. $(-8, 2)$ **2.** $(-1, 2, 3)$

Exercise Set C, p. 822

1. $\left(\frac{3}{2}, \frac{5}{2}\right)$ **3.** $(-4, 3)$ **5.** $\left(\frac{1}{2}, \frac{3}{2}\right)$ **7.** $(10, -10)$
9. $\left(\frac{3}{2}, -4, 3\right)$ **11.** $(2, -2, 1)$ **13.** $(0, 2, 1)$
15. $\left(4, \frac{1}{2}, -\frac{1}{2}\right)$ **17.** $(w, x, y, z) = (1, -3, -2, -1)$

Margin Exercises, Appendix D, p. 824

1. (a) $2x^2$; (b) 6; (c) $x^4 - 9$; (d) $\dfrac{x^2 + 3}{x^2 - 3}$; (e) $x^4 + 6x^2 + 9$
2. (a) -5; (b) 15; (c) -54; (d) 6

Exercise Set D, p. 825

1. 1 **3.** -41 **5.** 12 **7.** $\frac{13}{18}$ **9.** 5 **11.** 2
13. $x^2 - x + 1$ **15.** 21 **17.** 5
19. $-x^3 + 4x^2 + 3x - 12$ **21.** 42 **23.** $-\frac{3}{4}$ **25.** $\frac{1}{6}$
27. $x^2 + 3x - 4$; $x^2 - 3x + 4$; $3x^3 - 4x^2$; $\dfrac{x^2}{3x - 4}$
29. $\dfrac{1}{x - 2} + 4x^3$; $\dfrac{1}{x - 2} - 4x^3$; $\dfrac{4x^3}{x - 2}$; $\dfrac{1}{4x^3(x - 2)}$
31. $\dfrac{3}{x - 2} + \dfrac{5}{4 - x}$; $\dfrac{3}{x - 2} - \dfrac{5}{4 - x}$; $\dfrac{15}{(x - 2)(4 - x)}$;
$\dfrac{3(4 - x)}{5(x - 2)}$

Glossary

A

Abscissa The first coordinate in an ordered pair of numbers

Absolute value The distance that a number is from 0 on the number line

ac-method A method for factoring trinomials of the type $ax^2 + bx + c$, $a \neq 1$, involving the product, ac, of the leading coefficient a and the last term c

Additive identity The number 0

Additive inverse A number's opposite; two numbers are additive inverses of each other if their sum is 0

Algebraic expression An expression consisting of variables, constants, numerals, and operation signs

Ascending order When a polynomial is written with the terms arranged according to degree from least to greatest, it is said to be in ascending order.

Associative law of addition The statement that when three numbers are added, regrouping the addends gives the same sum

Associative law of multiplication The statement that when three numbers are multiplied, regrouping the factors gives the same product

Asymptote A line that a graph approaches more and more closely as x increases or as x decreases

Axes Two perpendicular number lines used to identify points in a plane

Axis of symmetry A line that can be drawn through a graph such that the part of the graph on one side of the line is an exact reflection of the part on the opposite side

B

Base In exponential notation, the number being raised to a power

Binomial A polynomial composed of two terms

Break-even point In business, the point of intersection of the revenue function and the cost function

C

Circle A set of points in a plane that are a fixed distance r, called the radius, from a fixed point (h, k), called the center

Circumference The distance around a circle

Coefficient The numerical multiplier of a variable

Common logarithm A logarithm with base 10

Commutative law of addition The statement that when two numbers are added, changing the order in which the numbers are added does not affect the sum

Commutative law of multiplication The statement that when two numbers are multiplied, changing the order in which the numbers are multiplied does not affect the product

Completing the square Adding a particular constant to an expression so that the resulting sum is a perfect square

Complex number Any number that can be written as $a + bi$, where a and b are real numbers

Complex rational expression A rational expression that has one or more rational expressions within its numerator and/or denominator

Complex-number system A number system that contains the real-number system and is designed so that negative numbers have square roots

Composite function A function in which a quantity depends on a variable that, in turn, depends on another variable

Compound inequality A statement in which two or more inequalities are combined using the word *and* or the word *or*

Compound interest Interest computed on the sum of an original principal and the interest previously accrued by that principal

Conic section A curve formed by the intersection of a plane and a cone

Conjugates Pairs of radical terms, like $\sqrt{a} + \sqrt{b}$ and $\sqrt{a} - \sqrt{b}$ or $c + \sqrt{d}$ and $c - \sqrt{d}$, for which the product does not have a radical term

Conjunction A sentence in which two statements are joined by the word *and*

Consecutive even integers Even integers that are two units apart

Consecutive integers Integers that are one unit apart

Consecutive odd integers Odd integers that are two units apart

Consistent system of equations A system of equations that has at least one solution

Constant A known number

Constant function   A function given by an equation of the form $f(x) = b$, where b is a real number

Constant of proportionality   The constant in an equation of direct or inverse variation

Coordinates   The numbers in an ordered pair

Cube root   The number c is called a cube root of a if $c^3 = a$.

D

Degree of a polynomial   The degree of the term of highest degree in a polynomial

Degree of a term   The sum of the exponents of the variables

Demand function   A function modeling the relationship between the price of a good and the quantity of that good demanded

Denominator   The number below the fraction bar in a fraction

Dependent equations   The equations in a system are dependent if one equation can be removed without changing the solution set.

Descending order   When a polynomial is written with the terms arranged according to degree from greatest to least, it is said to be in descending order.

Determinant   The determinant of a two-by-two matrix $\begin{bmatrix} a & c \\ b & d \end{bmatrix}$ is denoted $\begin{vmatrix} a & c \\ b & d \end{vmatrix}$ and represents $ad - bc$.

Diameter   A segment that passes through the center of a circle and has its endpoints on the circle

Difference of cubes   Any expression that can be written in the form $a^3 - b^3$

Difference of squares   Any expression that can be written in the form $a^2 - b^2$

Direct variation   A situation that translates to an equation of the form $y = kx$, with k a positive constant

Discriminant   The radicand, $b^2 - 4ac$, from the quadratic formula

Disjoint sets   Two sets with an empty intersection

Disjunction   A sentence in which two statements are joined by the word *or*

Distributive law of multiplication over addition   The statement that multiplying a factor by the sum of two numbers gives the same result as multiplying the factor by each of the two numbers and then adding

Distributive law of multiplication over subtraction   The statement that multiplying a factor by the difference of two numbers gives the same result as multiplying the factor by each of the two numbers and then subtracting

Domain   The set of all first coordinates of the ordered pairs in a function

Doubling time   The time necessary for a population to double in size

E

Elimination method   An algebraic method that uses the addition principle to solve a system of equations

Ellipse   The set of all points in a plane for which the sum of the distances from two fixed points F_1 and F_2 is constant

Empty set   The set without members

Equation   A number sentence that says that the expressions on either side of the equals sign, $=$, represent the same number

Equation of direct variation   An equation, described by $y = kx$ with k a positive constant, used to represent direct variation

Equation of inverse variation   An equation, described by $y = k/x$ with k a positive constant, used to represent inverse variation

Equilibrium point   The point of intersection between the demand function and the supply function

Equivalent equations   Equations with the same solution set

Equivalent expressions   Expressions that have the same value for all allowable replacements

Equivalent inequalities   Inequalities that have the same solution set

Evaluate   To substitute a value for each occurrence of a variable in an expression

Exponent   In expressions of the form a^n, the number n is an exponent. For n a natural number, a^n represents n factors of a.

Exponential decay   A decrease in quantity over time that can be modeled by an exponential equation of the form $P(t) = P_0 e^{-kt}, k > 0$

Exponential equation   An equation in which a variable appears as an exponent

Exponential function   A function that can be described by an exponential equation

Exponential growth   An increase in quantity over time that can be modeled by an exponential function of the form $P(t) = P_0 e^{kt}, k > 0$

Exponential notation   A representation of a number using a base raised to a power

F

Factor   *Verb*: To write an equivalent expression that is a product. *Noun*: A multiplier

Factorization of a polynomial   An expression that names the polynomial as a product

Fixed costs   In business, costs that are incurred whether or not a product is produced

Focus   One of two fixed points that determine the points of an ellipse

FOIL   To multiply two binomials by multiplying the First terms, the Outside terms, the Inside terms, and then the Last terms

Formula   An equation that uses numbers or letters to represent a relationship between two or more quantities

Fraction equation   An equation containing one or more rational expressions; also called a *rational equation*

Fraction expression   A quotient, or ratio, of polynomials; also called a *rational expression*

Fraction notation   A number written using a numerator and a denominator

Function   A correspondence that assigns to each member of a set called the domain exactly one member of a set called the range

G

Grade   The measure of a road's steepness

Graph   A picture or diagram of the data in a table; a line, curve, or collection of points that represents all the solutions of an equation

Greatest common factor (GCF)   The common factor of a polynomial with the largest possible coefficient and the largest possible exponent(s)

H

Half-life   The amount of time necessary for half of a quantity to decay

Hypotenuse   In a right triangle, the side opposite the right angle

I

Identity Property of 1   The statement that the product of a number and 1 is always the original number

Identity Property of 0   The statement that the sum of a number and 0 is always the original number

Imaginary number   A number that can be named bi, where b is some real number and $b \neq 0$

Imaginary number i   The square root of -1; that is, $i = \sqrt{-1}$ and $i^2 = -1$

Inconsistent system of equations   A system of equations for which there is no solution

Independent equations   Equations that are not dependent

Index   In the radical $\sqrt[n]{a}$, the number n is called the index.

Inequality   A mathematical sentence using $<, >, \leq, \geq,$ or $\neq$

Input   A member of the domain of a function

Integers   The whole numbers and their opposites

Intercept   The point at which a graph intersects the x- or y-axis

Intersection of two sets   The set of all elements that are common to both sets

Interval notation   The use of a pair of numbers inside parentheses and brackets to represent the set of all numbers between those two numbers

Inverse relation   The relation formed by interchanging the members of the domain and the range of a relation

Inverse variation   A situation that translates to an equation of the form $y = k/x$, with k a positive constant

Irrational number   A real number that cannot be named as a ratio of two integers

J

Joint variation   A situation that translates to an equation of the form $y = kxz$, with k a constant

L

Leading coefficient   The coefficient of the term of highest degree in a polynomial

Leading term   The term of highest degree in a polynomial

Least common denominator (LCD)   The least common multiple of the denominators

Legs

Legs   In a right triangle, the two sides that form the right angle

Like terms   Terms that have exactly the same variable factors

Line of symmetry   A line that can be drawn through a graph such that the part of the graph on one side of the line is an exact reflection of the part on the opposite side

Linear equation   Any equation that can be written in the form $y = mx + b$ or $Ax + Bx = C$, where x and y are variables

Linear function   A function that can be described by an equation of the form $y = mx + b$, where x and y are variables

Linear inequality   An inequality whose related equation is a linear equation

Linear programming   A branch of mathematics involving graphs of inequalities and their constraints

Logarithmic equation   An equation containing a logarithmic expression

Logarithmic function, base a   The inverse of an exponential function with base a

M

Maximum value   The largest function value (output) achieved by a function

Minimum value   The smallest function value (output) achieved by a function

Monomial   A constant, a variable, or a product of a constant and one or more variables

Motion problem   A problem that deals with distance, speed, and time

Multiplication property of 0   The statement that the product of 0 and any real number is 0

Multiplicative identity   The number 1

Multiplicative inverses   Reciprocals; two numbers whose product is 1

N

Natural logarithm   A logarithm with base e

Natural numbers   The counting numbers: 1, 2, 3, 4, 5, …

Nonlinear function   A function whose graph is not a straight line

Numerator   The number above the fraction bar in a fraction

O

One-to-one function   A function for which different inputs have different outputs

Opposite   The opposite, or additive inverse, of a number a is written $-a$. Opposites are the same distance from 0 on the number line but on different sides of 0.

Opposite of a polynomial   To find the opposite of a polynomial, replace each term with its opposite—that is, change the sign of every term.

Ordered pair   A pair of numbers of the form (h, k) for which the order in which the numbers are listed is important

Ordinate   The second coordinate in an ordered pair of numbers

Origin   The point on a graph where the two axes intersect

Output   A member of the range of a function

P

Parabola   A graph of a quadratic equation

Parallel lines   Lines in the same plane that never intersect; two lines are parallel if they have the same slope.

Perfect square   A rational number p for which there exists a number a for which $a^2 = p$

Perfect-square trinomial   A trinomial that is the square of a binomial

Perimeter   The sum of the lengths of the sides of a polygon

Perpendicular lines   Lines that form a right angle

Pi (π)   The number that results when the circumference of a circle is divided by its diameter; $\pi \approx 3.14$, or 22/7

Point–slope equation   An equation of the type $y - y_1 = m(x - x_1)$, where x and y are variables

Polynomial   A monomial or sum of monomials

Polynomial equation   An equation in which two polynomials are set equal to each other

Polynomial inequality   An inequality that is equivalent to an inequality with a polynomial as one side and 0 as the other

Prime polynomial   A polynomial that cannot be factored using only integer coefficients

Principal square root   The nonnegative square root of a number

Principle of zero products   The statement that an equation $ab = 0$ is true if and only if $a = 0$ is true or $b = 0$ is true, or both are true

Proportion   An equation stating that two ratios are equal

Proportional numbers   Two pairs of numbers having the same ratio

Pythagorean theorem   In any right triangle, if a and b are the lengths of the legs and c is the length of the hypotenuse, then $a^2 + b^2 = c^2$.

Q

Quadrants   The four regions into which the axes divide a plane

Quadratic equation   An equation of the form $ax^2 + bx + c = 0$, where $a \neq 0$

Quadratic formula   The solutions of $ax^2 + bx + c = 0$, $a \neq 0$, are given by the equation $x = \dfrac{-b \pm \sqrt{b^2 - 4ac}}{2a}$.

Quadratic function   A second-degree polynomial function in one variable

Quadratic inequality   A second-degree polynomial inequality in one variable

R

Radical equation   An equation in which a variable appears in a radicand

Radical expression   An algebraic expression in which a radical symbol appears

Radical symbol   The symbol $\sqrt{}$

Radicand   The expression under the radical symbol

Radius   A segment with one endpoint on the center of a circle and the other endpoint on the circle

Range   The set of all second coordinates of the ordered pairs in a function

Ratio   The quotient of two quantities

Rational equation   An equation containing one or more rational expressions

Rational expression   A quotient of two polynomials

Rational inequality   An inequality containing a rational expression

Rational number   A number that can be written in the form a/b, where a and b are integers and $b \neq 0$

Rationalizing the denominator   A procedure for finding an equivalent expression without a radical in the denominator

Real numbers   All rational and irrational numbers

Reciprocal   A multiplicative inverse; two numbers are reciprocals if their product is 1.

Rectangle   A four-sided polygon with four right angles

Reflection   The mirror image of a graph

Relation   A correspondence between a first set, the domain, and a second set, the range, such that each member of the domain corresponds to at least one member of the range

Repeating decimal   A decimal in which a number pattern repeats indefinitely

Right triangle   A triangle that includes a right angle

Rise   The change in the second coordinate between two points on a line

Roster notation   A way of naming sets by listing all the elements in the set

Run   The change in the first coordinate between two points on a line

S

Scientific notation   A representation of a number of the form $M \times 10^n$, where n is an integer, $1 \leq M < 10$, and M is expressed in decimal notation

Set   A collection of objects

Set-builder notation   The naming of a set by describing basic characteristics of the elements in the set

Simplify   To rewrite an expression in an equivalent, abbreviated, form

Slope   The ratio of the rise to the run for any two points on a line

Slope–intercept equation   An equation of the form $y = mx + b$, where x and y are variables

Solution   A replacement or substitution that makes an equation or inequality true

Solution set   The set of all solutions of an equation, an inequality, or a system of equations or inequalities

Solve   To find all solutions of an equation, an inequality, or a system of equations or inequalities; to find the solution(s) of a problem

Speed   The ratio of distance traveled to the time required to travel that distance

Square   A four-sided polygon with four right angles and all sides of equal length

Square of a number   A number multiplied by itself

Square root   The number c is a square root of a if $c^2 = a$.

Substitute   To replace a variable with a number

Substitution method   An algebraic method for solving systems of equations

Sum of cubes   An expression that can be written in the form $a^3 + b^3$

Sum of squares   An expression that can be written in the form $a^2 + b^2$

Supply function   A function modeling the relationship between the price of a good and the quantity of that good supplied

System of equations   A set of two or more equations that are to be solved simultaneously

T

Term   A number, a variable, or a product or a quotient of numbers and/or variables

Terminating decimal   A decimal that can be written using a finite number of decimal places

Total cost   The amount spent to produce a product

Total profit   The amount taken in less the amount spent, or total revenue minus total cost

Total revenue   The amount taken in from the sale of a product

Trinomial   A polynomial that is composed of three terms

Trinomial square   The square of a binomial expressed as three terms

U

Union of sets A and B   The set of all elements belonging to either A or B

V

Value   The numerical result after a number has been substituted into an expression

Variable   A letter that represents an unknown number

Variable expression   An expression containing a variable

Variation constant   The constant in an equation of direct or inverse variation

Vertex   The point at which the graph of a quadratic equation crosses its axis of symmetry

Vertical-line test   The statement that a graph represents a function if it is impossible to draw a vertical line that intersects the graph more than once

W

Whole numbers   The natural numbers and 0: 0, 1, 2, 3, …

X

***x*-intercept**   The point at which a graph crosses the x-axis

Y

***y*-intercept**   The point at which a graph crosses the y-axis

Index

Geometric Formulas

PLANE GEOMETRY

Rectangle

Area: $A = l \cdot w$
Perimeter: $P = 2 \cdot l + 2 \cdot w$

Square

Area: $A = s^2$
Perimeter: $P = 4 \cdot s$

Triangle

Area: $A = \frac{1}{2} \cdot b \cdot h$

Sum of Angle Measures

$A + B + C = 180°$

Right Triangle

Pythagorean Theorem:
$a^2 + b^2 = c^2$

Parallelogram

Area: $A = b \cdot h$

Trapezoid

Area: $A = \frac{1}{2} \cdot h \cdot (a + b)$

Circle

Area: $A = \pi \cdot r^2$
Circumference:
$C = \pi \cdot d = 2 \cdot \pi \cdot r$
$\left(\frac{22}{7}\right.$ and 3.14 are different
approximations for $\left.\pi\right)$

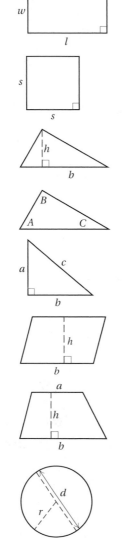

SOLID GEOMETRY

Rectangular Solid

Volume: $V = l \cdot w \cdot h$

Cube

Volume: $V = s^3$

Right Circular Cylinder

Volume: $V = \pi \cdot r^2 \cdot h$
Surface Area:
$S = 2 \cdot \pi \cdot r \cdot h + 2 \cdot \pi \cdot r^2$

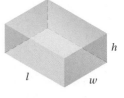

Right Circular Cone

Volume: $V = \frac{1}{3} \cdot \pi \cdot r^2 \cdot h$
Surface Area: $S = \pi \cdot r^2 + \pi \cdot r \cdot s$

Sphere

Volume: $V = \frac{4}{3} \cdot \pi \cdot r^3$
Surface Area: $S = 4 \cdot \pi \cdot r^2$